# Advances in the Identification & Analysis of Organic Pollutants in Water Volume 1

Edited by
Lawrence H. Keith

230 Collingwood, P.O. Box 1425, Ann Arbor, Michigan 48106

Library of Congress Card Catalog Number 81-68031
ISBN 0-250-40397-8

Manufactured in the United States of America


Butterworths, Ltd., Borough Green, Sevenoaks, Kent TN15 8PH England

# PREFACE

Five years ago the first book on identification and analysis of organic pollutants in water was published. The genesis of that book was primarily a symposium of the same title held during the First Chemical Congress of the North American Continent in Mexico City. In the ensuing years, advances in the identification and analysis of organic pollutants in water have continued at an accelerated pace. Thus, it was only natural that the second symposium on this subject be held during the Second Chemical Congress of the North American Continent in Las Vegas, Nevada.

As in the initial publication, all but a few of the chapters in this book were contributed by the many distinguished authors who participated in that symposium. These scientists represent the leading analytical and environmental chemists on the North American continent and abroad. The advances described in these volumes represent the current state-of-the-art methodology for analysis of organic compounds, often at trace levels, in all types of water samples ranging from the very pure (drinking water) to the extremely "dirty" (untreated industrial wastewaters).

Knowledge of specific organic pollutants in water has increased exponentially over the past five years. This has been largely due to the methodology that has enabled these analyses to be conducted efficiently, scientific and governmental interest in the subject, and the fact that knowledge of the subject area five years ago was minimal. Despite intense work over the past five years, little "brand new" methodology has been introduced for analyzing trace levels of organic pollutants in water. However, refinements of the techniques then in use have made these methods even better and more efficient and have contributed significantly to our current methodology. Solvent extractions, resins or gas purging techniques are still the primary methods used to concentrate organics from water, and gas or high-pressure liquid chromatography with various detectors (including mass spectrometers) are still used to identify and quantify them. One new method, using tandem mass spectrometers (MS/MS), offers promise of analyzing for specific compounds in a complex matrix without the tedious and expensive concentration and separation requirements of conventional instrumental analyses. At present, however, MS/MS is quite expensive and still in its initial stage of exploration and exploitation.

One significant difference in the direction of environmental analyses today, as compared to five years ago, is the emphasis on improving both accuracy and reliability of quantitative data. This is being accomplished through more stringent quality assurance/quality control programs and the use of numerous and more sophisticated quantitative methods such as the use of multiple internal standards, often isotopically labled. Five to ten years ago, compounds identified in water often were not even quantified. Two driving forces appear to be the primary reason for this change of emphasis: (1) legal and regulatory ramifications and (2) the realization that adverse effects of all organic pollutants are related to their concentration in water.

Predictably, environmental analyses seem to be moving increasingly into the lawyer's domain. This makes life more complicated for analytical environmental chemists, who have the sometimes difficult problem of explaining that 18 and 24 parts per billion are really the same number (within experimental error), or that it is not a worthwhile environmental goal to try to reduce the pH of water to near zero. Nevertheless, the overall effect of these legal and regulatory pressures on scientists is a positive one. It requires an emphasis on careful work and on improving the accuracy and precision of measurements that are often pushed to their limits. But, past history has shown that growth and improvement in most disciplines increase more rapidly when pressure is applied.

Another significant difference in the direction of environmental analyses today is an emphasis on automation. Analyses of organic pollutants in water are becoming more prevalent as a result of new regulations and modifications of old ones (e.g., the National Pollutant Discharge Elimination System industrial wastewater discharge permits). The result is an increasing competitiveness and pressure to produce large numbers of analyses for decreased costs. This, of course, is a natural and healthy trend as long as the pressure for cheap analyses does not compromise accuracy and precision, and it is realized that not all objectives can be met using cheap analyses for a "laundry list" of pollutants.

There still remains much to be learned about the kinds of pollutants in water and the chemistry involved. Until we have the correct methodology for analyzing organic compounds, we will continue to miss some even though they may be literally right under our noses. An example is the recent discovery that dihaloacetonitriles are apparently a new class of anthropoaqueous pollutants in drinking water. Despite the fact that countless chemists have been analyzing for trihalomethanes in drinking water these past five to six years, the fact was missed that dihaloacetonitriles are also probably present in many of those same samples. Why? Because the methodology being used was not conducive for the analysis of dihaloacetonitriles.

I think we have reached the stage in the evolution of analysis of organic pollutants in water where the easy methods and the easy compounds have been exploited. Now we will have to work harder and be more clever to produce newer and better methodology with which to uncover the more difficult organic pollutants. Is this necessary? I think so. How can one make an intelligent decision until the pertinent facts are known?

Five years ago, I said, "much still remains to be done. We have only begun to learn and to apply what we have learned to provide a better and cleaner environment." I think we have applied what we have learned very well over the past five years. We know a great deal more about the prevalence, distribution and concentrations of organic pollutants in water now than we did then. However, much still remains to be done. Now we must refine our techniques, verify our quantitative accuracy and turn our attention to developing new methodology for analysis of the more elusive and difficult organic pollutants.

L. H. Keith

**Lawrence H. Keith's** current technical interests continue to center around analyses of organic pollutants in the environment, with emphasis on developing new methods or improving on old ones. Techniques for the safe handling of carcinogenic and/or extremely toxic materials are also an important aspect of Dr. Keith's current research efforts.

Dr. Keith was formerly involved with the selection of many of the initial U.S. Environmental Protection Agency's 129 Priority Pollutants, and he also helped to formulate some of the initial methodology for analyzing for these pollutants. He is presently involved in the selection of representative compounds and methodologies for the Appendix C Priority Pollutants and for the synfuel industry.

A member of the American Chemical Society's Division of Environmental Chemistry Executive Committee, Dr. Keith has served as secretary, alternate councilor, program chairman and chairman of the division. He is also a past chairman of the Central Texas Section of the American Chemical Society and past secretary and councilor of the Northeast Georgia Section of the American Chemical Society.

In other professional activities, Dr. Keith served as Vice-Chairman of the Gordon Research Conference on Environmental Sciences: Water, and is currently a delegate of the U.S. National Committee to the International Association of Water Pollution Research. He is also a member of the National Research Council Committee on Military Environmental Research.

Dr. Keith and his wife, Virginia, reside in Austin, Texas.

**Ronald G. Webb** was born in Weatherford, Oklahoma, and began his college studies at the Central Christian College in Oklahoma City (1957–1959). He completed his BA at David Lipscomb College in Nashville, Tennessee (1961), and then went on to earn his PhD at the University of Georgia in Athens in 1968. After graduation, Ron stayed in Athens and began his professional career at the Southeast Water Laboratory which was then a part of the U.S. Department of the Interior. This laboratory changed names several times and finally became one of the regional research laboratories when the U.S. Environmental Protection Agency was created in 1970.

Ron is probably best known for his work with the chromatographic separation and identification of many of the individual isomers comprising Arclor (polychlorinated biphenyl) mixtures. He was also involved in developing methodology for extracting and concentrating organic pollutants from water in the latter years of his short career. Thus, it was natural that Ron, early on, became involved with helping to develop methodology for extracting, concentrating and chromatographing the moderately volatile Priority Pollutants after they had been chosen. For his part Ron received the EPA Silver Medal for Superior Service "In recognition of outstanding accomplishment in developing a new and technically sound approach for identifying and measuring priority toxic pollutants in industrial wastes to establish Best Available Technology guidelines."

However, chemistry only occupied one segment of Ron's busy life. With his wife, Gretchen, and his two sons, they maintained an active participation in the membership and affairs of the Church of Christ. And, as long as he was able, Ron also was an active member of the Society for the Preservation and Encouragement of Barbershop Quartet Singing in America.

Ron loved to sing, and he and Gretchen kept a positive outlook on life even though the cancer that eventually took his life was discovered back when he was still in graduate school. Ron and I shared an office for many years while we were together at the Athens Environmental Research Laboratory. I am proud to have known and worked with him there; he was one of the bravest men I have ever known.

It is with a deep sense of loss for a friend and a colleague that this volume is dedicated to the memory of Ron Webb and to his optimism and everlasting courage.

# ACKNOWLEDGMENTS

As in the previous book, most, but not all, of the chapters were derived from a symposium sponsored by the American Chemical Society (ACS) Division of Environmental Chemistry. The editor is grateful for the support of the ACS and the division for providing the forum from which this book is primarily derived. The many authors who have diligently written these chapters are leading scientists in their fields and, without their work, ideas, and contributions, these books could never have been produced.

The editor is also very grateful for the help and support from his wife, Virginia, who did most of the work compiling the extensive index. This index is the heart of these volumes when they are used as reference books, and thus, helps make these books unique in their field. No other works of this kind meld analytical methodologies with specific organic pollutants that have been identified in water. It is, in fact, this dual purpose of providing a single source for both the identification and the methods for analysis of organic pollutants in water that has proven to be so useful to workers in the field of environmental chemistry.

## CONTENTS

## Section 2: High-Resolution Gas Chromatography



## Section 3: Stable Labeling



## Section 4: Microextraction

**Section 5: Resin Adsorption**



**Section 6: High-Performance Liquid Chromatography**

# CHAPTER 1

# VALIDATION AND PRIORITY POLLUTANT ANALYSIS

**R. O. Kagel**
Environmental Quality
Dow Chemical USA
Midland, Michigan

The modern analytical chemist has become so proficient in detecting compounds at such trace concentration levels that, under certain circumstances and with particular compounds, he might expect to find everything in anything. This was, in fact, pointed out several years ago by Donaldson [1], for wastewater analysis:

> the number of compounds detected in a sample of water is related to the detection level. As the detection level decreases an order of magnitude, the number of compounds detected increases an order of magnitude. Based on the number of compounds detected by current methods, one would expect to find every known compound at concentration $10^{-12}$ g/l or higher.

More recently, quantitative results for 2,3,7,8-tetrachlorodibenzo-*p*-dioxin (TCDD) in a variety of matrices were reported at the subpart-per-trillion level [2], which approaches Donaldson's $10^{-12}$ g/L. The finding that TCDD is ubiquitous would reinforce Donaldson's observation. Priority pollutant analyses in wastewaters have routinely been reported in the $10^{-8}$- to $10^{-9}$-g/l range and below. Donaldson notes that for drinking water, the ability of an analyst to detect compounds is marginal at $10^{-8}$ g/L, and a higher detection limit would probably be selected for wastewater effluents where the constituents are generally more concentrated and matrix effects more severe.

The first question we should ask ourselves involves the validity of an analytical measurement performed near the limit of detection, or for that matter, any analytical measurement involving a sample complicated by gross matrix effects. These situations are typically encountered in priority pollutant analyses. It behooves us as analytical chemists to know what we are doing and how we are doing it, for as other scientific disciplines learn more about the effects of chemicals in the environment, there is an ever-increasing demand for lower detection limits and for technology transfer from one sample matrix to another.

In this chain of events, Wessel [3] notes: "Analytical methodology tends to be one of the sciences at the very forefront of many environmental disputes." Wessel, a well-known environmental attorney, who is concerned with translating science to a language easily understood by the general public, goes on to say:

> Understandably, therefore, there are differences of opinion among chemists as to just what is involved [in data interpretation]. One, using so-called "thin layer chromatography," will see a signal which he considers to be that of a particular compound for which he is searching; a second, examining precisely the same chromatogram, will be unable to totally distinguish what he sees from a whole host of other signals which are "noise," or the routine, ordinary background emanations of compounds which are omnipresent in the environment. Each will describe what he sees, and let others examine the evidence for themselves. Where there is no scientific consensus, the issue as to what is in fact present must be resolved by whoever the arbiter may be and ultimately, as in all socio-scientific disputes, by the public. The public certainly will not be able to understand all of what each of these analytical chemists has done, but it can be made to appreciate just how difficult it is to find trace concentration quantities at these extremely low levels.

In Wessel's case, the arbiter in all socioscientific disputes is ultimately the general public. However, this is not necessarily true in the case of priority pollutant analyses. A more immediate arbiter is almost certainly the courts. If the general public has a problem understanding analytical data, imagine what the magnitude of the problem must be within the legal profession! At least one other attorney has come to grips with the problem of data validity, and argues, on constitutional grounds, a case for methods validation [4].

So, once we have answered our first question regarding what constitutes methods validation, we should concern ourselves with the legal aspects of this subject. It is almost a foregone conclusion that methods validity will be a central issue in future litigation when effluent guidelines for the priority pollutants are issued.

The scientific and legal aspects of methods validation are the two subjects to which my remarks will be addressed.

## SCIENTIFIC VALIDATION

Validation of an analytical procedure, of course, involves the statistical treatment of data to determine precision, accuracy, sensitivity and reproducibility of the procedure from laboratory to laboratory or even from analyst to analyst within a laboratory. In the jargon of the field, a validated analytical procedure is an analytical method. Validation provides a common denominator for agreement on just what an analytical result really means. We expect the civil and criminal sanctions for noncompliance with priority pollutant effluent limitations to be severe; therefore, there must be unequivocal methodology for analyzing effluents.

The question of unequivocal measurement methodology is by no means a new one. In fact, it is an age-old issue with implications much more serious than mere civil and criminal sanctions. The issue, in truth, is a moral issue, and the first known set of guidelines to deal with it are found in Proverbs 11:1: "A false balance is abomination to the Lord: But a just weight is his delight." These are words for environmental analytical chemists to live by.

Over the years, other criteria have been developed that are generally recognized as either necessary or highly desirable to assure that results will be accurate and, it is to be hoped, precise. Rogers [5] has detailed several criteria necessary for methods validation:

1. Sampling: The sampling step is crucial. Unless precautions are taken to assure that a sample is representative, even the most careful analysis will be misleading. Sample handling, including storage and transporation, is critical to minimizing errors from contamination or loss. A well-designed sampling program must be part of the validation protocol.
2. Independent Analysis: Different portions of the same sample should be analyzed using at least two procedures that are as nearly independent as possible. This criterion cannot always be followed because of sample size limitations, and at times there is no other equivalent procedure available, especially when state-of-the-art technology is used. In these cases, one must rely on two or more physical properties, such as gas chromatographic retention time or use of two or more different column packings.
3. Material Balance: This criterion is useful for major component analysis, but is not relevant when dealing with trace analysis.

4. Recovery: There are a number of ways to determine recovery. In trace analysis, the method of "standard addition" is commonly used. Here, a known amount of the material of interest is added to a real sample. The sample is analyzed before and after the addition. The difference of the two results provides an estimate of the amount originally present. The "internal-standard" method involves addition of an isotopically labeled species. Isotopic dilution is a third method.
5. Interferences: The number of interfering species increases as the concentration of the material of interest decreases. To minimize interferences, it is desirable to use highly selective steps to remove interfering species before measurement. It is also desirable to use a highly discriminating measurement technique.
6. Replicate Analysis: Replicate analysis lends confidence to a result when the analysis is performed by an expert who is using an already proven method. The number of replicates is an important factor in estimating the confidence level of the results.
7. Interlaboratory Reproducibility: Two or more laboratories applying the same procedures should obtain precise results with confidence limits that overlap to a large degree.
8. Limits of Detection (LOD)/Limits of Determination (LD). Rogers did not include LOD or LD in his original list of criteria. LOD and LD are tied to precision, accuracy and reproducibility, and define the working limits of an analytical procedure.

Methods validation for priority pollutant analyses has been debated since 1975. In 1977 the Chemical Manufacturers Association's Environmental Monitoring Task Group (CMA/EMTG) developed a validation protocol for the U.S. Environmental Protection Agency (EPA) Effluent Guidelines Division (EGD). The protocol was based on the EPA Environmental Monitoring and Support Laboratory (EMSL) procedure [6].

The EGD also recognized the need for validation in its protocol [7]:

> None of the methods are known to work in all of the effluent types presently being studied. Therefore, all procedures that are utilized will have to be validated.

Along with the Section 304h Compliance Methods [8], EMSL also published quality control/quality assurance procedures which involved validation protocol for gas chromatography/mass spectrometry (GC/MS).

All of these efforts, however, lacked uniformity. National and even international agreement is needed on what constitutes a valid environmental measurement. If the scientific community does not resolve these issues, the legal community certainly will.

The American Chemical Society (ACS) has taken the first positive step toward resolving these issues. The ACS Committee on Environmental Improvement directed its Subcommittee on Environmental Analytical Chemistry to develop guidelines to improve the overall process of environmental analytical methods, with emphasis on trace organic analysis.

The subcommittee has produced a draft document [9] designed to be (1) compatible with U.S. regulatory requirements, (2) good analytical laboratory practices and (3) consistent with those already proposed by the International Union of Pure and Applied Chemistry. These guidelines should be followed for priority pollutant analysis. This document represents the most authoritative review of the subject to date.

Historically, there are several ways to determine the number of samples needed to produce a given precision and accuracy. The pure statistician says that each experiment should be designed statistically. The more measurements in replicate, the better the statistics. Unfortunately, statistical design of each experiment is a luxury that cannot always be afforded in the real world. Pertinent data are not always available. The answer has been compromise with what the purist would call "government statistics."

Government statistics have their beginnings in residue analysis. Validation protocols were developed by the government in connection with pesticide or herbicide registration procedures. Patterson and Lehman [10] described the 10-10-10 validation principle, later advocated by Harris and Cummings [11], EPA Pesticide Registration, as the minimal data requirement necessary to support the registration of a pesticide. The mechanics of the 10-10-10 principle involve making 10 determinations on a control sample to determine interference, 10 determinations on fortified samples to determine recoveries and 10 determinations on different aliquots of the same sample to determine precision. The statistics of the analytical procedure are usually verified by independent laboratories. This procedure has been applied widely to residue studies involving crops, soil, plants and animals.

In the 1970s EPA/EMSL adopted a 7-7-7 scheme for validating wastewater [6]. Wastewater analysis is in fact a form of residue analysis. The only difference is that the host matrix is an effluent, a complex mixture of organic and inorganic matter in water, rather than plant or animal tissue.

The ACS draft document recommends the mathematical approach of Walpole and Meyers [12], to determine the minimum number of samples (n) required for certain confidence limits when data are available.

$$n = \left( \frac{Z \bullet \sigma}{E} \right)^2$$

| Z | Confidence |
|---|---|
| 3 | 99.7% |
| 1.96 | 99.5% |
| 1.64 | 90 % |
| 1.0 | 68 % |

where Z = value from Z tables
$\sigma$ = standard deviation
E = error

When data are not available to calculate the minimum number of samples, use of the n-n-n rule is recommended.

The ACS document recommends that methods be selected on the basis of high recovery, minimum contamination and the use of controls and calibrations to prevent random and systematic errors. It also recommends the Youden and Steiner [13] ruggedness test to establish the influence of possible interferences. The validity and quality assurance of the measurement process is established by arranging six key samples into a protocol:

1. calibration standards;
2. field blanks;
3. spiked field blanks (blind to laboratory);
4. spiked laboratory blanks (known to laboratory);
5. working standards (known to project leader but blind to analysts); and
6. field samples.

The frequency and order for running a sequence of blanks, controls and field samples are decided on the merits of each protocol or measurement situation [14].

Working standards are used for performance tests. The values of working standards are well known and traceable to round-robin studies or, ideally, to standard reference materials. Spiked samples are used for recovery calculations. Field blanks are for the purpose of determining interferences.

Relative standard deviations or coefficients of variability are calculated in the normal way from these data. As the concentration of analyte decreases, the relative standard deviation increases. There are three distinct regions of variability described by the heteroscadastic curve (Figure 1) which defines the behavior of a system as the limit of detection is approached.

According to the ACS draft document,

> The limit of detection (LOD) is the lowest concentration of an analyte in a given matrix that the analytical process is capable of detecting.

Here, the analytical process means the entire analytical method from start to finish. LOD is usually defined in terms of the gross analyte signal $S_o$, being significantly larger than the blank signal $S_b$. A minimum requirement for a detectable signal is:

$$S_o > S_b + 3\sigma_b$$

where $\sigma_b$ is the variability in the field blank. Results in the region of:

$$3_{\sigma b} < S_x < 10_{\sigma x}$$

imply the presence of an analyte in accordance with the method used.

The ACS draft document further stipulates that when a suitable blank sample cannot be obtained, the LOD is based on the signal-to-noise ratio S/N. Noise is measured peak-to-peak, and the LOD is 3 times the noise in the signal.

$$\text{LOD} = 3\text{N}$$

Reagent blanks should be monitored, regardless of whether a sample blank is available.

The LOD is a qualitative parameter that only defines the presence or absence of an analyte. The analytical significance of the apparent concentration near the LOD is very uncertain as shown by the heteroscadastic curve (Figure 1). Quantitative numbers, those with analytical significance, are obtained only after a clear safety margin has been exceeded. The safety margin is the LD. A minimum criterion recommended by the ACS draft guidelines is that a quantitative determination be based on:

$$S_o > S_b + 10_{\sigma b}$$

In the case where a blank is not available, LD is taken as ten times the noise:

$$\text{LD} = 10\text{N}$$

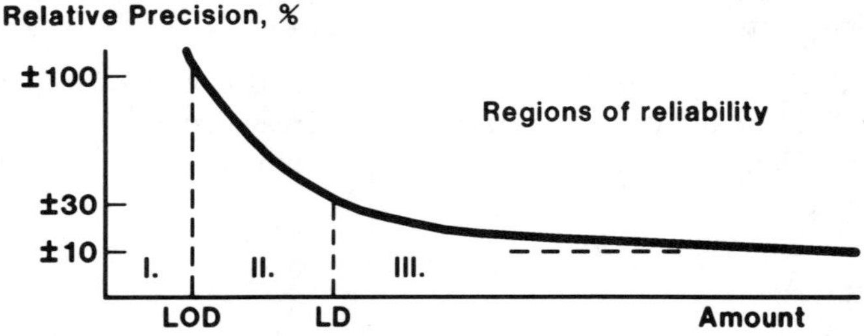


**Figure 1.** The relative variability of analytical measurements characteristically increases as the analyte concentration diminishes. I. Region of nondetection. II. Region of detection (extends upward). III. Region of determination.

The ACS draft guidelines also recommend use of the notation:

Detected but unconfirmed: $S_b + 3\sigma_b < S_o < S_b + 10\sigma_b$

and

None detected: $S_o < S_b + 3\sigma_b$

The precision and accuracy of results in the region $S_b < S_o < S_b + 3\sigma_b$ are usually extremely poor, should not be considered as numerically meaningful or reproducible, and are difficult or impossible to confirm by another analytical method.

The region, detected but unconfirmed, is merely a mathematical expression of what Rogers [15] calls Feigl's [16] "region of uncertain reaction." Feigl, very prominent in spot tests in the 1930s and 1940s, attributes the "region of uncertain reaction" to Emich [17].

The process is shown in Table I. Take two portions of sample and 10 aliquots of each portion. Dilute the original sample and repeat the process, finding a positive result for eight of the replicates but not for the other two. For the other portion, six of the ten replicates were positive. Three more

**Table I. Region of Uncertain Reaction[a]**

| | No. Positives/10 Reps. | | |
|---|---|---|---|
| | Series 1 | Series 2 | Decision |
| Original Sample | 10 | 10 | "Always" found |
| One Dilution | 8 | 6 | Sometimes found |
| Two Successive Dilutions | 5 | 7 | |
| Three Successive Dilutions | 2 | 1 | |
| Four Successive Dilutions | 0 | 1 | |
| Five Successive Dilutions | 0 | 0 | Never found |

[a]Applied by Feigl to qualitative detection.

dilutions were performed before a concentration was reached where no positive result was found in either series. Out of a million replicates, one might find a positive result. The number of replicates and the concentration of the sample affect the location of that detection limit.

The detection limit and regulatory limit that everyone uses is not clear-cut. The amount represented by the upper limit of the uncertain region represents the maximum amount that can be added to or subtracted from the amount actually present in a sample. If, for example, the region of uncertain reaction represents 10 ppb, and there are 11 ppb in the sample, one can find 21 ppb if the error is positive or 1 ppb if the error is negative. This is a moving region of uncertainty so that individual results can go back and forth between the two extremes when replicates are analyzed.

Rogers also uses an interesting analogy to describe how he sees these principles applying to real-world situations.

> In the lawyer's world, below the regulatory limit, one is in compliance; above the limit, one is in violation. The limit itself is infinitesimal in thickness. Conformance or non-conformance is limited only by the number of figures your computer prints out. In other words, by stating a regulatory limit as it is now done, the lawyers have repealed the laws of probability insofar as recognizing that legitimate scientific factors can introduce uncertainties into measurements. Instead, all uncertainty is assigned to human error; this makes laws easy to write and comprehend but scientifically untenable to enforce.

The physical scientist's ideal world is much like the lawyer's world. The limit of detection replaces the regulatory limit. To the left of the line, it is not detected and to the right, it is always detected.

## LEGAL VALIDATION

Lawyers have their own special way of viewing the technical community, and the technical community, for its part, often frowns on those views. Duncan, the Health and Safety Executive for England, noted at a recent U.S. conference on the toxic substance regulatory process [18]: "You have lawyers the way other people have mice."

Nevertheless, with the possibility of future litigation over methods validation, the technical community should be aware of such terms as "arbitrary and capricious," and, lest there be any doubt, "invalid," "void," unenforceable" and "of no legal effect." Judicial rulings in these terms can negate years of technical research.

Lawyers, indeed, have been paying attention to analytical matters. At a recent American Society for Testing and Materials (ASTM) symposium [19], Davis presented an interesting legal analysis of the implications of validation for the inductively coupled plasma method proposed by EPA [20]. His arguments have broader implications within the context of the Clean Water Act (CWA) for all priority pollutant analysis.

Davis argues that nowhere in the CWA or its legislative history are legal requirements set forth for test procedures developed under Section 304h of the act. The statute delegates the task of defining the "guidelines" for establishing test procedures to EPA.

The Section 304h regulations promulgated or proposed thus far by EPA do not give detailed descriptions of criteria used by EPA to evaluate and approve test procedures. These regulations, however, do identify acceptable test procedures as those which generate accurate and reliable data. Compliance use of these test procedures necessitates that they be reliably accurate; that is, each person must have a reliable way to confirm that the discharges of his facility are in compliance with the terms of his National Pollutant Discharge Elimination System (NPDES) permit. Each person has a right to know where he stands.

> This is the self-evident requirement of simple fairness which is embodied in the most fundamental principles of law. Courts have repeatedly held that constitutional due process requires that persons be given adequate notice as to whether or not their conduct is proscribed by some provision of the law. Statutes or regulations not clearly enough defined or sufficiently explicit to give such notice have been held to be unconstitutionally void for vagueness.

Davis then cites numerous cases in which the courts have applied the vagueness doctrine to environmental law on a number of occasions when regulations were remanded because they were "too vague to warn the industry of the scope of prohibited conduct." Davis goes on to say that:

> When the administration and enforcement of a statute or regulation depends on the evaluation of data, the clear and explicit definition of responsibilities demanded by the constitution requires that accurate test methods be employed to collect and assess the pertinent data.

In other words,

> The law demands that regulatory authorities validate the adequacy of the test procedures and other scientific methods they use to collect and evaluate data.

To support this argument, Davis cites court cases involving data gathering by police using radar to provide evidence that a motor vehicle had been operated in excess of the speed limit. American radar devices are accurate to ∓2 mph, but this error does not prevent authorities from using radar. The law accepts the underlying scientific principle of radar and the general capability of such devices to measure speed. However, the court requires verification of the accuracy of a particular instrument to ensure that the measurement taken was not beyond its capabilities. Without proof of this accuracy, evidence of speed as registered by radar is inadmissable.

Similarly, in another case, the court found certain automobile air test bag procedures inadequate because they could not "objectively" be used to determine compliance. "Tests to determine compliance must be capable of producing identical results when test conditions are exactly duplicated."

All of these legal arguments, of course, eventually come back to methods validation. The arguments point out the course of future legal challenges in this regulatory arena. They emphasize the importance of sample handling; precision and accuracy, replication and reproducibility; and inter- and intralaboratory variability and verification. In other words, there is legal justification for all of the scientific factors that go into validation.

## CONCLUSIONS

There are sound technical arguments for the validation of priority pollutant methodology. These arguments are intricately entwined with legal issues that will ultimately be argued in the courts. The technical issues should be resolved by the technical community. The merits of these issues are what lawyers should be left to resolve, rather than the issues themselves.

Roisman [21], former environmentalist lawyer who heads the Justice Department's toxic waste section, recently gave some guiding philosophy behind the government's legal-environmental movement in that area when he said: "Government is perfectly prepared to punish the innocent for the sins of the guilty."

An analytical result may very well be the infinitesimally fine line that separates the guilty from the innocent. Unless the analytical methodology is properly validated and the result statistically meaningful, that fine dividing line might as well be determined by chance, a flip of the coin or a roll of the dice.

## REFERENCES

1. Donaldson, W. T. "Identification and Measurement of Trace Organics in Water: An Overview," *Environ. Sci. Technol.* 11:348 (1977).
2. Bumb, R. R., W. B. Crummett, S. S. Cutie, J. R. Gledhill, R. A. Hummel, R. O. Kagel, L. L. Lamparski, E. V. Luoma, D. L. Miller, T. J. Nestrick, L. A. Shadoff, R. H. Stehl and J. S. Woods. *Science* (in press).
3. Wessel, M. R. *Science and CONscience* (New York: Columbia University Press, in press).
4. Hunton and Williams, Inc. "Comments on EPA's Proposed ICP Test Method for the Analysis of Pollutants," Personal communication (April 1980).
5. Cowgill, U. M., H. Freiser, M. L. Gross, L. B. Rogers and D. Schuetzle. "Evaluation of the Analytical Chemical Procedures of the Dow Chemical Company for Determination of Polychlorinated Dibenzo-p-dioxins," Personl communication (June 1979).
6. "Handbook for Analytical Quality Control in Water and Wastewater Laboratories," U.S. EPA, Environmental Monitoring and Support Laboratory, Cincinnati, OH (1979).
7. "Analytical Methods for the Verification Phase of the BAT Review," U.S. EPA Effluent Guidelines Division (1977).
8. *Federal Register* (December 3, 1979).
9. "Guidelines for Data Acquisition and Data Quality Evaluation in Environmental Chemistry," American Chemical Society, Committee on Environmental Improvement, Subcommittee on Environmental Analytical Chemistry (in preparation).
10. Patterson, W. I., and A. J. Lehman. *Assoc. Food Drug Off. U.S.* XVII (1) (1953).
11. Harris, T. H., and J. G. Cummings. *Residue Rev.* 6:104 (1964).
12. Walpole, R., and R. Meyers, Probability and Statistics for Engineers *and Scientists* (New York: Macmillan Publishing Co., 1972), p. 190.
13. Youden, W. J., and E. H. Steiner. *Statistical Manual of the Association of Official Analytical Chemists* (1975).

14. "Sixth Materials Research Symposium, 1973," National Bureau of Standards, Spl. Publ. 408 (1975), p. 806.
15. Rogers, L. B. "Validation of Analytical Data: Problems Encountered at Trace-Level Concentrations," Chemical Manufacturers Association, Seminar on Priority Pollutants, Norfolk, VA, January 16, 1980.
16. Feigle, F. *Chemistry of Specific Selective and Sensitive Reactions* (New York: Academic Press, 1949).
17. Emich, F. *Ber.* 43:10 (1919).
18. Smith, R. J. *Science* 203:32 (1978).
19. Davis, J. K. "Legal Issues in Compliance Monitoring," ASTM Symposium on Legal Implications of Environmental ASTM Standards for Forensic Purposes, Milwaukee, WI, June 13, 1980.
20. *Federal Register* (December 3, 1979).
21. Alexander, T. "The Hazardous-Waste Nightmare," *Fortune* (April 21, 1980).

# SECTION 1

# PROTOCOLS

# CHAPTER 2

# THE MASTER ANALYTICAL SCHEME: AN OVERVIEW OF INTERIM PROCEDURES

**A. W. Garrison, A. L. Alford, J. S. Craig, J. J. Ellington, A. F. Haeberer, J. M. McGuire, J. D. Pope and W. M. Shackelford**

Analytical Chemistry Branch
Environmental Research Laboratory
U.S. Environmental Protection Agency
Athens, Georgia

**E. D. Pellizzari**

Analytical Sciences Division
Research Triangle Institute
Research Triangle Park, North Carolina

**J. E. Gebhart**

Department of Analytical Chemistry
Gulf South Research Institute
New Orleans, Louisiana

Comprehensive analytical methodology for organic compounds in water has long been needed by chemists working in governmental, industrial and academic laboratories. Special techniques have been available, and continue to emerge, to analyze a sample for a specific compound or groups of similar compounds. In addition, "survey" methods have evolved over the past 5-10 years based on identification by gas chromatography/mass spectrometry (GC/MS) of a fairly broad spectrum of extractable and purgeable organic components. The Master Analytical Scheme (MAS) discussed in this and related chapters in this book [1-4], however, is the first effort to develop a comprehensive qualitative-quantitative scheme to include organics of all volatility classes (that will, or can be derivitized to, pass through a gas chromatograph), of almost all functional group types, in almost any water sample.

In developing the MAS, existing techniques were evaluated and modified, and new techniques were developed to produce a comprehensive qualitative-quantitative protocol. Some of the new techniques developed include scouting measurements to determine sample water quality before analysis, a new way to package and dose purgeable internal standards, a flow-under extractor for continuous liquid-liquid extraction (LLE) that avoids emulsion formation and several new isolation procedures for ionic intractable compounds. Other chapters in this book describe the experimental development of the scheme [1,2] and give some preliminary results [3,4].

This chapter presents an overview of the current interim MAS, for which all analytical operations and techniques have not been tested thoroughly. A refined version of the MAS will be published after tests on environmental samples [3,4] are completed and necessary improvements are incorporated.

## PROCEDURES

Figure 1 shows procedures for implementation of the MAS. Each step is summarized below.

### Sample Handling

Three subsamples are required for a comprehensive sample analysis: one each for purgeable, extractable and intractable organics. Procedures are prescribed for sample collection, storage and preservation. Extractable samples are collected in bottles that can be used later for LLE; septum-capped bottles are used for purgeables. Preservation is primarily by storage at 4°C. Chlorine determination (e.g., using a Hach Chlorine Test Kit) indicates the level of sodium thiosulfate necessary to reduce stoichiometrically any residual chlorine left from water treatment.

Various water quality scouting measurements [4] help in the selection of appropriate analytical procedures, which are optimized according to water quality rather than sample "type" (e.g., drinking water or municipal effluent). Headspace gas analysis by GC of a separate small sample is employed to determine the dilution necessary for purge-and-trap analysis, and indicates levels at which internal standards are to be added. A trial shake-out with methylene chloride of a small aliquot of the extractable sample shows whether emulsion formation is a problem, and thus whether the flow-under extractor [1] must

# SECTION 1

# PROTOCOLS

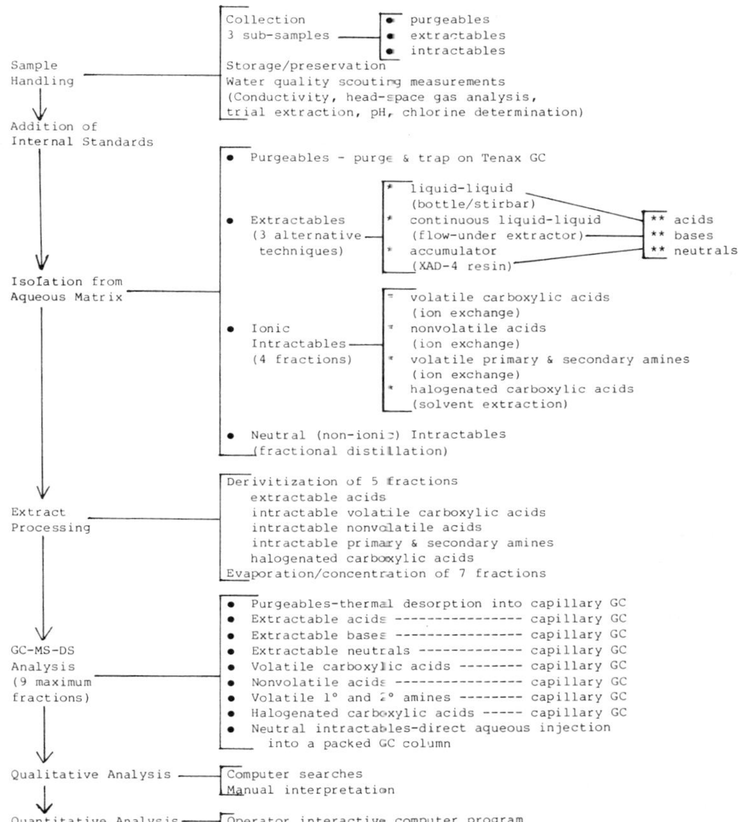


**Figure 1.** MAS flow diagram.

be used. Conductivity measurements indicate maximum sample volume allowable for isolation of ionic intractables by ion exchange resins.

## Internal Standards

Deuterated internal standards (Table I) are used to monitor recovery, quantify sample components and measure relative GC retention times. The initial sets of MAS standards were prepared by the National Bureau of Stan-

Table 1. Internal Standards for the MAS

| Standards, by Sample Aliquot | $\mu g/L$ in $H_2O$[a] Low | High |
|---|---|---|
| Purgeables | | |
| Bromoethane-$d_5$ | 0.3 | 2.9 |
| Diethyl Ether-$d_{10}$ | 2.7 | 27.0 |
| Anisole-2,4,6-$d_3$ | 2.5 | 24.9 |
| Chlorobenzene-$d_5$ | 0.3 | 3.4 |
| *n*-Decane-$d_{22}$[b] | 2.8 | 27.8 |
| Naphthalene-$d_8$ | 2.5 | 24.7 |
| Extractables | | |
| *o*-Xylene-$d_{10}$ | 11.3 | 551 |
| Nitrobenzene-$d_5$ | 519 | 2463 |
| Naphthalene-$d_8$ | 95.0 | 383 |
| Anthracene-$d_{10}$ | 41.1 | 40.0 |
| Di-*n*-octyl Ether-$d_{34}$ | 590 | 2561 |
| Acridine-$d_9$ | 98.2 | 499 |
| Phenylethylamine-$d_4$[c] | 10.0 | 517 |
| Hexamethylenetetramine-$d_{12}$ | 98.8 | 500 |
| Phenol-$d_5$ | 460 | 2134 |
| Benzoic Acid-$d_5$ | 11.4 | 499 |
| Bisphenol A-$d_{16}$[b] | 499 | 2502 |
| Intractables | | |
| *n*-Butylamine-$d_9$[d] | 570 | 2058 |
| *n*-Butyric Acid-$d_7$[e] | 51 | 2342 |
| 2-Naphthalenesulfonic Acid·$H_2O$-$d_7$[f] | 99 | 501 |
| *n*-Dodecyl Phosphate-$d_{25}$[f] | 484 | 2047 |
| Ethanol-$d_5$[g] | 120 | 512 |
| Acetonitrile-$d_3$[g] | 544 | 2709 |
| *t*-Butanol-$d_9$[g] | 112 | 519 |

[a]Concentration after dosing in the appropriate volume of sample.
[b]*n*-Decane-$d_{22}$ and bisphenol A-$d_{16}$ have since been determined unsuitable as internal standards.
[c]Phenylethylamine-$d_4$ will be transferred to the intractable primary and secondary amine fraction in the future.
[d]For the primary and secondary amine fraction.
[e]For the volatile acid fraction.
[f]For the nonvolatile acid fraction.
[g]For the neutral intractable fraction.

dards (NBS). After assuring chemical and isotopic purity of each standard compound, NBS determined its stability and compatibility with the other internal standards. Methanol (or water for the intractables) solutions were prepared and packaged for sample dosing.

For samples to be analyzed for purgeable components, internal standards are added at a concentration indicated by the results of scouting analysis of a separately collected sample aliquot. Purgeable internal standards are packaged in glass capillary ampules that are added to the sample and crushed with a magnetic stirbar. For sample aliquots to be analyzed for extractable and intractable organics, internal standards are packaged in vials such that emptying the entire contents of the vial into the prescribed sample volume produces the concentrations of standards shown in Table I.

Internal standards were selected and are pacakged so that two or three standards appear in each extract ready for GC/MS analysis; retention times are such that the standards span the chromatographic window.

## Isolation of Organics [1]

After addition of internal standards, the three subsamples are processed as follows.

### *Purgeables*

Highly volatile (purgeable) organics are analyzed by a modification of the Bellar-Lichtenberg [U.S. Environmental Protection Agency (EPA) volatile organic analysis (VOA)] method [1,3,4]. Sodium sulfate is used to "salt out" the organics in a 200-mL sample which is purged at 30°C. The sample is diluted, if necessary, in accordance with the total concentration of purgeable organics as indicated by GC scouting of the headspace gas. Dilution prevents saturation of the GC/MS data system (DS) and decreases foaming potential. (A 200-mL volume is purged in any case.) Organic vapors are collected on a Tenax GC sorbent trap, from which they are thermally desorbed into a liquid nitrogen cold trap, then flash-evaporated into a glass capillary GC/MS/DS. An "external" standard, perfluorotoluene, is added in the vapor state to the Tenax GC sorbent, using a special cartridge-loading device, before collection of the purged sample components. Comparison of MS signals for the external standard with those for the internal standards purged from the sample allows calculation of recoveries of the internal standards, thus monitoring performance of the entire analytical operation.

The refined version of the MAS will include an optional closed system for purgeables using a currently available purge-and-trap apparatus (such as the Tekmar LSC-2®), modified to interface with a capillary GC column.

*Extractables*

Water-insoluble compounds of low to intermediate volatility have traditionally been isolated by LLE in a separatory funnel. Batch LLE of a one-liter sample using a magnetic stirbar and solvent (methylene chloride) in the same vessel used for sample collection is the basic MAS method for the extractables [1]. A special valve head for solvent removal from the vessel has been developed.

For some samples, however, batch LLE is not suitable. An initial trial solvent extraction in a stoppered graduated cylinder indicates whether emulsion formation is likely to be a problem. If it is, continuous LLE with methylene chloride in a flow-under extractor [1] should be used. For samples whose extractable organic concentration is expected to be low, as with drinking water and some surface waters, XAD–4 resin sorbent columns are designated [1] for sorption/concentration from larger volumes of water. For all extractable samples, pH adjustment is recommended for extraction/fractionation of acids, bases and neutrals.

Adjustment of pH to 2 or less allows extraction of weak to moderately strong acids along with neutral compounds, with an option to back-extract with aqueous base to separate acids from neutrals. Some acids, e.g., sulfonic acids, are too strong or water-soluble to be extracted at this pH and are included in the nonvolatile acid fraction of the intractables (see Ionic Intractables). Other acids, mostly the lower-molecular-weight carboxylic type, are too volatile to be recovered efficiently during LLE and subsequent extract processing; these are included in the volatile acid fraction of the intractables. Certain α-halogenated carboxylic acids (e.g., trichloroacetic acid) are either too strong to be extracted at pH 1 or 2, or they decarboxylate during basic back-extraction; these also comprise a separate fraction in the interim MAS (see Ionic Intractables). Future modification of the scheme may allow these compounds to be extracted with the "extractable" acids. Finally, some bases appear in both the extractable basic fraction and the ionic intractable fraction called "volatile primary and secondary amines." The two fractions, however, are generated from two separate sample aliquots. Primary and secondary amines are measured in the ionic intractable fraction, and other amines are measured in the extractable basic fraction.

Acids that are solvent-extracted at this stage of the protocol are derivatized with pentafluorobenzyl bromide to form the corresponding pentafluorobenzyl esters.

External standards (4-fluoro-2-iodotoluene and 2-fluorobiphenyl) are added to all final extracts just before GC/MS analysis to confirm the recovery of the deuterated internal standards extracted from the water.

*Neutral (Nonionic) Intractables*

Low-molecular-weight, water-soluble, nonextractable and nonpurgeable compounds (e.g., butanol, acetonitrile and acetone) are not removed from the water but are concentrated by fractional or azeotropic distillation of the sample [1]. Sodium chloride (100 g) is added to 500 mL of sample, which is distilled using a Claisen distillation head. The first 25 mL of distillate, enriched in organics, is collected for analysis by direct aqueous injection into the GC/MS. Further concentration can be achieved by redistilling the collected distillate. Because the concentration factor for this technique is low, the detection limit is high, about 100 μg/L. Improved distillation techniques to increase sensitivity are being developed for the refined version of the MAS.

*Ionic Intractables*

Compounds that are easily dissociated in water have not been included in analytical schemes because of difficulties with extraction and chromatography. (An exception is the "acid fraction" obtained on LLE from an acidified water sample.) New techniques, however, were developed to allow inclusion of these compounds into the MAS. Ion exchange resins are used to separate three classes of ionic intractables from the sample matrix using three separate aliquots of the sample. Procedures for elution of these compounds from the resin column, cleanup, concentration and derivatization vary. A fourth class of ionic intractables, α-halogenated carboxylic acids, is separated from another sample aliquot by LLE.

*Volatile Carboxylic Acids [1,3,4].* Butyric, methacrylic and other volatile carboxylic acids are separated from the water on Biorad AG 1-X8 anion exchange resin, then eluted with sodium bisulfate in acetonitrile:water solution. The volatile acids are distilled from the eluate, converted to their nonvolatile salts, then derivatized with pentafluorobenzyl bromide to form the pentafluorobenzyl esters.

*Nonvolatile Acids [1].* These compounds, e.g., napthalene sulfonic acid, are also separated from the water on Biorad AG 1-X8 resin. They are eluted with HCl in methanol, the solvent is evaporated and the acids are methylated with diazomethane.

*Volatile Primary and Secondary Amines [1].* Compounds such as hexylamine and diphenylamine are isolated from the water sample on Biorad AG 50W-X8 cation exchange resin, then eluted with sodium hydroxide in acetonitrile:water solution. The eluent is acidified, the solution is evaporated to dryness, and the amine hydrochloride salts are dissolved in base and extracted with methyl-*t*-butyl ether. The extract is split; half is derivatized with pentafluorobenzyl bromide to make pentafluorobenzyl tertiary amines from the secondary amines, and half is derivatized with pentafluorobenzaldehyde to make Schiff bases of the primary amines.

*α-Halogenated Carboxylic Acids [1,3].* These compounds (e.g., trichloroacetic acid and 2-chloropropionic acid) are extracted from an acidified sample aliquot with methyl-*t*-butyl ether, then back-extracted into a small volume of aqueous buffer solution. This solution is evaporated to dryness and the acids are methylated with HCl:methanol reagent.

External standards (4-fluoro-2-iodotoluene and 2-fluorobiphenyl) are added to each final ionic intractable extract just before GC/MS analysis to confirm recovery of the deuterated internal standards that were added to the original water samples.

## Extract Processing

Extractable and ionic intractable fractions require further processing before GC/MS analysis. The necessary derivatization steps are described above and summarized in Figure 1. Derivatization procedures may be necessary for extractable bases, but have not yet been developed. If appropriate capillary GC columns become available, both extractable amines and volatile primary and secondary amines may be chromatographed in their free forms.

Cleanup techniques may be necessary for certain fractions, especially the neutral extractables. Such techniques will be developed for the refined version of the MAS.

Concentration of as many as seven of the nine final extract solutions—all except purgeables and neutral intractables—is necessary before GC/MS analysis. Kuderna-Danish (KD) evaporation is used to concentrate the extracts to 4 mL, followed by nitrogen blowdown to 0.5 mL using a modified Snyder column [1].

## Gas Chromatography

As shown in Figure 1, as many as nine extracts or fractions can be obtained from one sample set if the entire MAS protocol is applied. The neutral intractable fraction is analyzed by direct aqueous injection into a packed or micropacked GC column. All other columns, including the one for analysis of purgeables, are glass capillaries. Fused-silica columns are recommended; the newly available wide-bore columns are preferred to avoid compound overloading with complex extracts. Performance standards rather than specific columns are specified (see Quality Assurance), but some commercially available columns that meet these standards are designated. A typical capillary column found suitable for most of these fractions is a SE-30 or SP-2100, 25-M or 50-M-wide bore (0.3 mm). fused-silica column. A typical column for the neutral intractables is a 180 cm x 2 mm, 0.2% Carbowax 1500 M on 80/100 mesh Carbopack C. No more than three different GC columns should be necessary for the entire MAS. It is expected that extracts of some fractions can be combined for GC/MS. The MAS protocol for each fraction prescribes optimum GC conditions and column performance standards and suggests appropriate columns.

### Qualitative Analysis

Sample components are identified by established GC/MS/DS techniques; no research was conducted on MAS identification procedures. GC/MS data are stored on tape or disk; internal standards in each extract are used as reference points for retention time measurements as well as for quantification. Compounds are identified by computer searching of mass spectra databanks or by manual interpretation. Preliminary MAS analysis at the Athens Environmental Research Laboratory used an automated peak-finding routine and a probability-based reverse-search spectral matching computer program developed for identification of EPA priority pollutants [5], while similar work at Research Triangle Institute was done using a combination of the INCOS data system on a Finnigan 4000-series GC/MS and manual verification of data system output.

### Quantitative Analysis [2,3]

Extensive recovery studies were conducted during development of the MAS. Approximately 375 model compounds from a wide variety of chemical classes (Table II) were dosed into representative samples of all major types of water (drinking water, surface water, and treated municipal, industrial and energy-related effluents); recoveries were determined, and recovery factors will be stored in a computer databank now under development. These factors will be made available to MAS users. Relative molar response (RMR) factors (relative to the deuterated internal standards) were also determined and will be stored in the databank. The MAS user can apply the associated computer

**Table II. Types of Model Compounds Chosen for Recovery and Relative Response Studies**

| Physical/ Chemical Property Class | Chemical Function Class | Number of Compounds Studied |
|---|---|---|
| Strong Acids | Phosphoric/phosphonic | 4 |
| | Sulfonic | 5 |
| | Group I phenols | 8 |
| | Carboxylic acids | 35 |
| Water–Soluble Alcohols | Carbohydrates | 1 |
| | Group I alcohols | 15 |
| Organometallics | | 8 |
| Basic Nitrogen Compounds | Alkyl/aromatic amines (1°, 2°, 3°) | 50 |
| Weak Acids | Group II phenols | 12 |
| | Amides | 1 |
| Water–Insoluble Alcohols | | 4 |
| Polar Neutrals | Aldehydes | 19 |
| | Ketones | 14 |
| | Phosphates | 2 |
| | Phosphonates | 1 |
| | Sulfonates | 3 |
| | Sulfones | 1 |
| | Sulfoxides | 1 |
| Nonpolar Neutrals | Esters/ethers | 48 |
| | Sulfides | 4 |
| | Nonbasic nitrogen compounds | 16 |
| | Aliphatic/aromatic compounds | 93 |
| | Polynuclear aromatics | 19 |
| Pesticides and Polychlorinated Biphenyls | | 7 |
| Carbamates | | 2 |
| Total | | 373 |

program to calculate the concentration in the original water sample of these model compounds as they are identified. For compounds that are not in the databank, concentration can be estimated by using RMR and recovery factors for structurally similar compounds in the databank.

During development of the MAS, it was determined that GC/MS data should be collected in the full scan mode rather than by selected ion monitoring. It was also shown, however, that quantification should be based on the peak areas (or heights) of selected ions of a compound rather than on the total ion current for that compound [2,3]. Thus, RMR factors involve ratios of ion peak areas and moles of the model compounds (those in the databank) to ion areas and moles of the internal standards selected for quantification (usually the closest eluting standard). These ion areas depend strongly on the GC/MS instrumentation and MS tuning. Thus, it is necessary for each MAS user either to develop his or her own RMR values, or, alternatively, to measure a few RMR at the MS tune specified by the MAS and use a linear regression plot of these against the tabulated RMR to develop a correction factor (slope) for the tabulated RMR. Tabulated RMR values will be provided as part of the MAS. RMR values developed by this linear regression correction procedure will not be as accurate as those measured directly by the user. In any case, it is necessary for the user to check the tune of the mass spectrometer daily by chromatographing a performance standard prescribed by the MAS. If the tune is significantly different from that used when the RMR values were measured, the same linear regression procedure should be used to correct the measured or tabulated RMR for the current tune. (The linear regression and correction can be done by the quantification computer program to be provided with the refined MAS.)

RMR factors are determined using solutions of model compounds within the same extract class (e.g., purgeables, neutral intractables and nonvolatile acids) containing all the appropriate internal standards for that class. For the first version of the MAS, 19 solutions containing up to 35 components each were analyzed in replicate by GC/MS (Finnigan 4000, Finnigan 3300 and Varian MAT 44) to determine RMR for the initial computer databank. Precision of RMR measurement must meet certain statistical criteria, expressed as coefficients of variation of replicates. This precision depends on GC/MS stability, operator error and other factors.

To use the computer quantification program, the user must provide sample component identities, some information as to the specific MAS analytical steps used, selected ion peak areas for each sample component and the appropriate (nearest) internal standard, and the concentration of the internal standard in the original water sample. If the identified sample component is one of the model compounds, the computation program will select the appropriate RMR and recovery factor for that component and calculate its concentration in the original sample. (At present, the raw data must be provided to the computer. In the future, however, a computer program may be available to automatically transfer raw data from the GC/MS data files.)

Errors for each step of the scheme have been estimated and summed to give total error propagation for each extract class of the MAS, e.g., ±11% for volatile acids [4]. No estimates have yet been developed for the error involved in quantifying compounds not in the databank, in which case RMR and recovery factors can only be estimated based on those for structurally related compounds. An additional error is involved in using recovery factors from the databank; because sample matrices used for recovery studies were only representative of the various water types, errors will occur in applying these factors to other samples dependent on the matrix differences between the sample being analyzed and the representative recovery sample.

## Quality Assurance [4]

Emphasis on quality assurance during development, testing and improvement of the MAS produced an analytical method having known limits, and, within those limits, statistical estimates of error. During development (phase II) of the MAS, statistical analysis techniques were applied to all experiments [1]; this allowed determination of an error propagation factor for the entire scheme [4]. In the testing phase (phase III) of the MAS contract, three separate laboratories will analyze nine pairs (unspiked and spiked with model compounds) of environmental samples by the MAS. This will provide statistics on precision and accuracy. Each laboratory will also analyze replicates of some of the samples and extracts and calculate intralaboratory statistics. Some results of these inter- and intralaboratory analyses are now available [3].

Of most importance are the quality assurance steps prescribed as part of the MAS procedure. Some of these steps are outlined below.

### *Internal and External Standards*

Comparison of the recovered quantity of deuterated internal standard to the quantity of external standard added to the extract just before GC/MS analysis reveals recovery deficiencies, which indicate malfunction of the MAS procedure. The primary use of deuterated internal standards, however, is for quantification; reference to internal standards is generally accepted as the most accurate quantification technique available for GC/MS analysis of organics in water. These standards are also useful as retention time indices for an aid in compound identification.

### *System Performance Standards*

The use of internal vs external standards as described above is one system performance standard. More obvious is the use of standard performance solutions to check the performance of the GC/MS/DS. The MAS prescribes such standard solutions and corresponding criteria of acceptance for each sample fraction. Criteria include those for GC peak asymmetry, separation number, resolution and column acidity and basicity; inertness of the GC-to-MS transfer line; and tune of the MS. These standard solutions also contain the deuterated internal standards, appropriate to each fraction, for RMR verification and periodic determination of the RMR correction factor by linear regression, if necessary (see Quantitative Analysis).

*Sample Scouting [4]*

As described above (see Sample Handling), several sample scouting measurements are prescribed to characterize water quality and, in turn, allow selection of the appropriate and optimal analytical techniques for a particular water sample. Scouting also indicates the correct level (relative to levels of endogenous sample components) of internal standards to add to the sample, thus increasing quantitative accuracy.

*Propagation of Error [4]*

Summation of the estimated error for each step of the scheme gives the total propagated error for the method.

*Blanks*

Procedures for field and procedural blanks and for quality control of each analytical operation are prescribed by the MAS.

*Standard Samples*

For some procedures of the MAS, instructions are provided for preparing test mixtures for the user to dose into distilled water. A certified standard water sample may eventually be available for test analysis by MAS users.

## CONCLUSIONS

The MAS provides protocols for the GC/MS analysis of purgeable, extractable, neutral intractable and ionic intractable organics in surface and drinking water and in industrial, energy-related and municipal effluents. Nominal lower quantifiable limits are 0.1 $\mu$g/L for drinking water, 1 $\mu$g/L for surface water and 10 $\mu$g/L for the effluents. One unique feature of the MAS is its comprehensiveness. Another is its qualitative-quantitative aspect—an extensive database of response and recovery factors allows computer estimation of concentration without recourse to standards for each analyte. This interim

MAS will be refined to correct deficiencies observed during analysis of environmental samples.

## ACKNOWLEDGMENTS

Development of the MAS would not have been possible without the efforts and contributions of R. D. Coney and T. A. Scott of the Analytical Chemistry Branch, Athens Environmental Research Laboratory; K. B. Tomer, L. S. Sheldon, L. C. Michael and J. T. Bursey of Research Triangle Institute; J. F. Ryan, R. D. Cox, D. L. Perry and L. C. Rando of Gulf South Research Institute; and H. S. Hertz and E. White of the National Bureau of Standards.

## REFERENCES

1. Gebhart, J. E. et al. "The Master Analytical Scheme: Development of Effective Techniques for Isolation and Concentration of Organics in Water," Chapter 3, this volume.
2. Ryan, J. F. et al. "The Master Analytical Scheme: An Assessment of Factors Influencing Precision and Accuracy of Gas Chromatography/Mass Spectrometry Data," Chapter 4, this volume.
3. Tomer, K. B. et al. "Quantitative Aspects of the Master Analytical Scheme for Organics in Water," Chapter 5, this volume.
4. Michael, L. C. et al. "Quality of Master Analytical Scheme Data: Purgeables and Volatile Organic Acids," Chapter 6, this volume.
5. Shackelford, W. M. et al. "A Computer Survey of GC/MS Data Acquired in the EPA Priority Pollutant Screening Analysis: Systems and Results," in *Advances in the Identification and Analysis of Organic Pollutants in Water, Vol. 2*, L. H. Keith, Ed. (Ann Arbor, MI: Ann Arbor Science Publishers, Inc., 1981).

# CHAPTER 3

# THE MASTER ANALYTICAL SCHEME: DEVELOPMENT OF EFFECTIVE TECHNIQUES FOR ISOLATION AND CONCENTRATION OF ORGANICS IN WATER

**J. E. Gebhart, J. F. Ryan and R. D. Cox**

Analytical Chemistry Department
Gulf South Research Institute
New Orleans, Louisiana

**E. D. Pellizzari, L. C. Michael and L. S. Sheldon**

Analytical Sciences Division
Chemistry and Life Sciences Group
Research Triangle Institute
Research Triangle Park, North Carolina

The major objective of the Master Analytical Scheme (MAS) has been to identify or develop, evaluate, and validate methodologies which can be combined into a comprehensive approach for qualitative and quantitative analysis of volatile organic compounds in water using gas chromatography/mass spectrometry/computer (GC/MS/COMP) as the major analytical tool. To achieve this objective, the program was divided into the following major phases:

- Phase I: literature survey and planning of experiments
- Phase II: experiments to develop the MAS
- Phase III: application and improvement of the MAS

## PHASE I

A schematic of phase I activities is presented in Figure 1. During this period of the program, the current literature was reviewed, government documents were searched and more than 250 researchers in the field of trace organic analysis of water were contacted. The survey was intended to be a comprehensive and critical review of research applicable to the analysis of organics in water and was designed to identify deficiencies in current methodologies and instrumentation. A state-of-the-art literature report was prepared [1] that discussed the operating principles, amenability of chemical classes and aqueous matrices to analysis, and inherent advantages and disadvantages of each technique. As part of the phase I effort, a matrix of analytical operations report was also prepared [2] that addressed the expected applicability of each technique to each major volatility and functional group class of organic compounds.

These reports were presented and discussed at a public meeting held in Atlanta, GA, in January 1979. This meeting was attended by approximately 100 researchers from the government, academic and private sectors who are experienced in the analysis of aqueous samples for organics. The matrix report and the suggestions of researchers served to guide the planning of experiments necessary to arrive at the final analytical scheme.

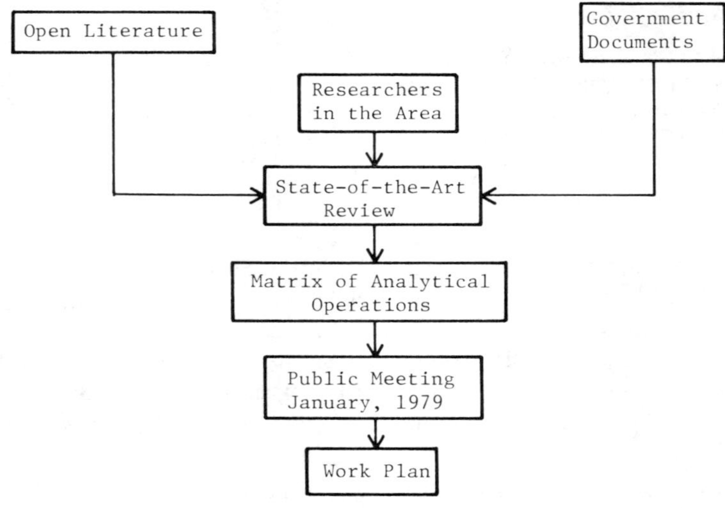


**Figure 1.** Schematic of phase I activities.

## PHASE II

During phase II, experiments were conducted to compare existing analytical techniques, investigate the applicability of new techniques, optimize promising techniques for recovery of each class of compounds from water and identify those techniques most compatible for inclusion in a comprehensive scheme.

The approach used in phase II experimental work is presented in Table I. Level I experiments were conducted with radiolabeled ($^{14}C$ or $^{3}H$) compounds spiked at known levels in distilled water. Only a few compounds were used in the evaluation of each technique; the goal was to optimize the operating parameters for each technique. These evaluations were conducted in triplicate using test solutions fortified at two concentrations.

As a result of level I work, several techniques were eliminated from further consideration. The criteria used in selecting techniques for further evaluation were:

1. recovery and reproducibility;
2. time required;
3. difficulty;
4. cost; and
5. compatibility with a comprehensive scheme.

**Table I. Design of Phase II Experiments**

| Development of Scheme |
|---|
| Level I |
| Distilled water |
| Few isotopes |
| Determine optimum analytical parameters |
| Recovery at two concentrations: LOD and 100 x LOD |
| Triplicate analysis |
| Level II |
| Distilled water, municipal effluent and energy effluent |
| Complete range of isotopes |
| Recovery of each step of the procedure |
| Recovery at two concentrations: LOD and 10 x LOD |
| Triplicate analysis |
| Level III |
| Distilled water, municipal effluent and energy effluent |
| Analyze unfortified samples |
| Analyze samples fortified with unlabeled compounds |
| Recovery for entire procedure |
| Recovery at one concentration: 1 to 2 times levels of endogenous compounds |

The normal areas of concern are addressed by the first criterion: accuracy and precision. The value of each method was judged primarily on these considerations. In addition, attempts were made to identify for further evaluation techniques that (1) are not labor-intensive, (2) are amenable to processing a large number of samples, (3) do not require a highly trained individual to produce reliable data, and (4) are not expensive to set up or run. The final criterion for further evaluation was that the technique be complementary to other techniques that would be included in a comprehensive scheme. Examples of the application of these criteria are given below.

## Example 1

Methods for preparation of aqueous samples for analysis of organics include, as a final step, a procedure to remove excess solvent to concentrate compounds of interest prior to instrumental analysis. The most common methods for solvent removal are Kuderna-Danish (KD) evaporation and rotary evaporation. Level I experiments were conducted to compare these two techniques and to choose one for inclusion in the final scheme.

The $^{14}C$-labeled compounds used in these studies include 2,4,5-trichlorophenol, malathion, benzidine, ($\alpha$)naphthol, naphthalene, benze[a]pyrene-(B[a]P) and bromobenzene. A known amount of each of these compounds was added to methylene chloride, and the volume of methylene chloride was reduced from 150 to 4 mL by using KD techniques or rotary evaporation. For each compound the experiment was conducted in triplicate by each technique. The results of this investigation are presented in Table II.

These results indicate that the accuracy and precision of the two techniques are essentially equivalent and certainly acceptable for the compounds

**Table II. Results of Solvent Evaporation Experiment: 150 to 4 mL**

| Compound | Concentration (ppb) | Percent Recovery with Rotary Evaporation | Percent Recovery with KD Evaporation |
|---|---|---|---|
| 2,4,5-Trichlorophenol | 50 | 98 ± 4 | 98 ± 2 |
| Malathion | 50 | 84 ± 3 | 86 ± 5 |
| Benzidine | 7 | 105 ± 3 | 98 ± 4 |
| ($\alpha$)-Naphthol | 14 | 80 ± 8 | 97 ± 2 |
| Naphthalene | 37 | 93 ± 4 | 111 ± 4 |
| Benzo[a]pyrene | 15 | 100 ± 3 | 97 ± 9 |
| Bromobenzene | 6 | 94 ± 4 | 97 ± 3 |

used in the evaluation. For a single sample, the time requirements are approximately the same, and both techniques require about the same degree of skill to perform. The cost of the rotary evaporation apparatus, however, is significantly greater than the cost of a KD setup. For most laboratories, the cost factor would mitigate against purchase of multiple units to facilitate batch processing of a large number of samples. Therefore, further evaluation of rotary evaporation as a method for removing excess solvent was not conducted and the KD evaporation technique was incorporated into the scheme.

### Example 2

Among the methods which have been reported for the isolation of organics from water is the closed-loop gas-stripping, or Grob, technique [3-7]. This procedure involves continuous gas-stripping of water samples followed by trapping on a small activated carbon filter. The sensitivity of this method is excellent (part-per-trillion), and the applicability of the procedure extends to compounds in the $C_{24}$ molecular-weight range.

This technique, however, has not been found to be effective in the isolation of highly volatile compounds (e.g., chloroform) [4], and therefore cannot replace the purge-and-trap method in a comprehensive scheme. Similarly, the applicability of the Grob technique does not extend to intermediately volatile compounds that can be isolated by liquid extraction procedures (e.g., phenol). Therefore, for the Grob technique to be incorporated into the comprehensive scheme, applicability must be demonstrated for some group of compounds that are not recovered by either purge-and-trap or extraction techniques.

The Grob technique was evaluated for highly water-soluble compounds such as acetone and methanol. Again, $^{14}C$-labeled compounds were used in these investigations. However, the purge efficiency for these compounds was quite low; in most cases less than 5% of the material was removed from the water. The trapping efficiency of the activated carbon was also unacceptably low for acetone (less than 50%) and methanol (less than 7%).

Therefore, the Grob technique did not eliminate the need for a procedure for highly volatile compounds, some extractable compounds or water-soluble compounds, and was not evaluated further.

Procedures which were selected for level II studies were evaluated by using a much wider range of labeled compounds added to distilled water, treated municipal effluent and energy-related effluent. Again, evaluations were conducted in triplicate and at two concentrations for each compound/matrix combination. The goal of this work was to determine the recovery of the organic compounds at each step of the sample preparation procedure and to optimize any step that might compromise the overall recovery of the method.

Level III experiments were designed to permit assessment of the overall recovery for each procedure. Aliquots of distilled water, treated municipal effluent and energy-related effluent were analyzed with and without fortification with mixtures of unlabeled model compounds. Evaluations were conducted in triplicate at one concentration; gas chromatographic analyses were performed using conditions that had been determined to be optimum with standard mixtures.

## PHASE III

Phase III research, which is now in progress, involves application of the methods developed and optimized in phase II to real-world samples. The sample matrices used in these validation studies include:

- distilled water (1),
- finished drinking waters (4),
- surface waters (2),
- biologically treated municipal effluents (2),
- industrial effluents (5), and
- energy-related effluents (5)

In these studies, aliquots of unfortified samples are analyzed by GC/MS/COMP. The organic compounds in each sample are identified and quantified. Another aliquot of the sample that has been fortified with a number of model compounds, including about two-thirds of the U.S. Environmental Protection Agency (EPA) priority pollutants, is then qualitatively and quantitatively analyzed by GC/MS/COMP. The results of these analyses will be used to determine what improvements are needed for each method and to finalize the scheme.

This chapter presents the details of isolation and concentration procedures incorporated into the MAS in its phase II form and the results of phase II experiments. Research is in progress to apply the scheme and to improve these methods when necessary; therefore, the methods presented here are preliminary. Each method will be modified somewhat as a result of phase III work.

## EXPERIMENTAL

For each water sample, a minimum of three aliquots is necessary for qualitative and quantitative analysis of organics. One aliquot will be analyzed for purgeable or highly volatile compounds; one for intractable or water-soluble, nonextractable, nonpurgeable compounds; and one for extractable organics. The details of each of these methodologies are presented below.

## Purgeables

The procedure for the analysis of purgeable organics in water utilizes gas-stripping to remove semisoluble and insoluble compounds followed by trapping of these materials on a Tenax GC cartridge. Exposed cartridges are thermally desorbed; released organics are trapped in a liquid nitrogen cold trap, from which they are flash-evaporated and analyzed by GC/MS/COMP using capillary columns.

The apparatus used to purge highly volatile organic compounds from the aqueous matrix is shown in Figure 2. A sample volume of 200 mL is used in the analysis of clean samples such as finished drinking water. For more contaminated waters, dilution is necessary, and a 20-mL sample is diluted 10:1 for analysis. To the sample are added 60 g of anhydrous sodium sulfate, and the entire apparatus is maintained at 30°C with a water bath during the purging procedure. The sample is purged with 500 mL of helium, and a flowrate between 10 and 100 mL/min is used. Once the organics have been trapped on the cartridge, the cartridge is either analyzed immediately or stored in a culture vial with a Teflon®-faced screw cap at 20°C until analysis.

Injection of adsorbed materials from the Tenax GC cartridge onto a gas chromatographic column is accomplished by using the thermal desorption system (Nutech Corporation, Durham, NC) illustrated in Figure 3. This system consists of four main components: a desorption chamber; a six-port, two-position, high-temperature, low-volume valve (Valco Instruments, Inc.); a nickel capillary cryogenic trap; and a temperature controller. In a typical desorption/injection cycle, an exposed cartridge is placed in the preheated (250°C) desorption chamber with a flow of helium gas (15 mL/min) through the cartridge to transfer the organics to the cryogenic trap. During this step, the valve is in position A (Figure 3). After eight minutes of thermal desorption, the valve is rotated to position B, the temperature of the capillary trap is increased rapidly (greater than 100°C/min), and the carrier gas sweeps the organics onto the GC column. When the maximum trap temperature is reached, the trap heater is turned off; however, the valve is retained in position B with the cartridge in the desorption chamber. After approximately 30 min, the cartridge is removed, and the valve is returned to position A. These operations are performed as closely together as possible.

Analysis of purgeable organics is performed on a glass capillary column such as a 50-m x 0.4-mm i.d. SE-30 ($BaCO_3$) WCOT. The temperature program is 30 to 250°C at 4°C/min.

## Intractables

Intractable compounds are water-soluble, nonextractable and nonpurgeable. Inclusion of this group of compounds is a significant challenge to a

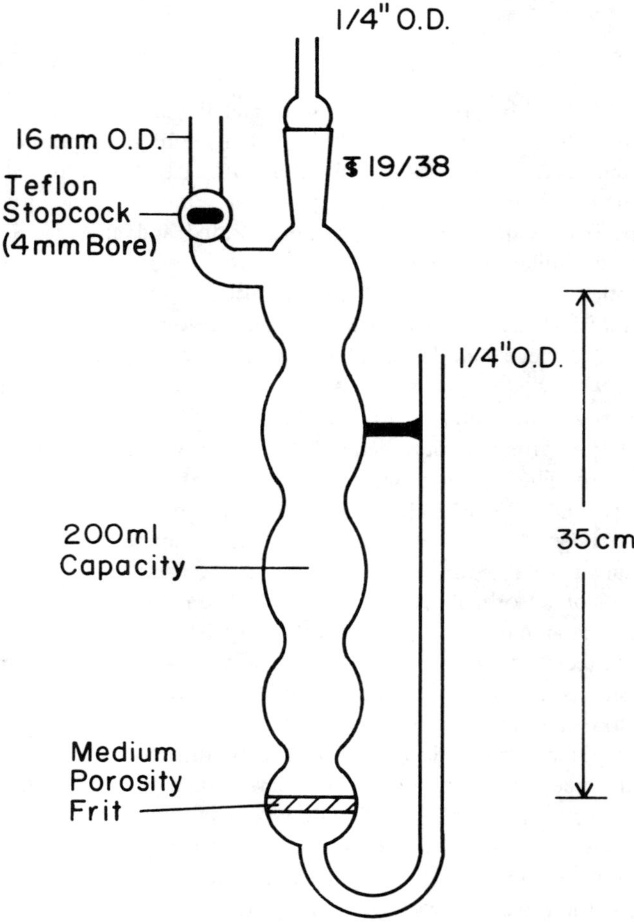


**Figure 2.** Purge flask, 200-ml capacity.

comprehensive scheme. In this research program, this group of compounds was first divided according to characteristics in an aqueous matrix, i.e., into neutral (nonionic) compounds and ionic compounds. The ionic intractables were further divided according to chemical and physical characteristics. The categories of ionic intractables include volatile primary and secondary amines, volatile carboxylic acids, nonvolatile organic acids, and halogenated carboxylic acids.

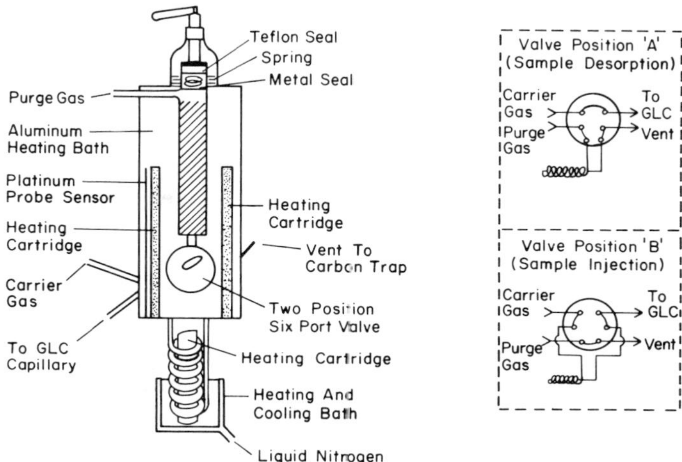


**Figure 3.** Thermal desorption system for Tenax GC cartridges.

*Nonionic Intractables*

This group of compounds, which includes compounds such as methanol, ethanol, acetone, acetonitrile and butanone, are isolated from an aqueous matrix by fractional distillation. The apparatus used in this procedure is illustrated in Figure 4.

To a 500-mL sample, 100 g of sodium chloride is added. This mixture is distilled at a rate of about 1 mL/min. Collection of 25 mL of distillate in a chilled graduated cylinder results in good recovery and a concentration factor of 20. For more dilute samples, the distillate may be collected in a small round-bottom flask and subjected to a second distillation procedure.

Nonionic intractables are analyzed by direct aqueous injection techniques isothermally at 100°C on a Carbowax 1500 packed or micropacked column.

*Ionic Intractables*

*Volatile Primary and Secondary Amines.* This group of compounds includes such primary amines as *n*-butylamine, allylamine, hexylamine and hexadecylamine, and secondary amines such as dibutylamine, diallylamine, di-*n*-propylamine and piperidine. The procedure for the analysis of these compounds involves adsorption on the cation exchange resin Biorad AG

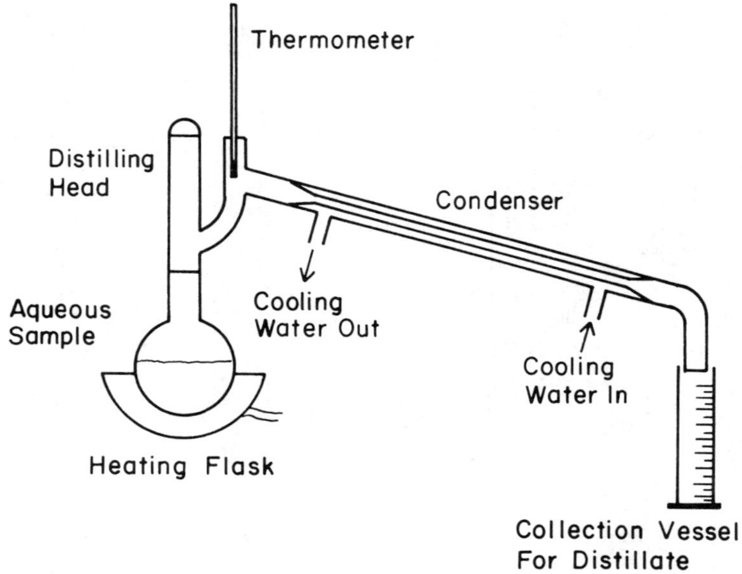


**Figure 4.** Fractional distillation apparatus.

W-X8 in the $H^+$ form followed by elution with potassium hydroxide in acetonitrile:water. The eluted amines are converted to their HCl salts, and the solution is evaporated to dryness. The residue is resuspended in water, and sufficient KOH is added to give a 0.1 *N* solution. The amines are extracted into methyl-*t*-butyl ether. Primary amines are derivatized with pentafluorobenzaldehyde. Secondary amines are derivatized with pentafluorobenzyl bromide. Because the volatile amines are converted to less volatile derivatives, the derivatized extract may be concentrated by using nitrogen blowdown prior to analysis. Because the primary and secondary amines are not derivatized under similar conditions, the extract must be split prior to derivatization.

Derivatized sample concentrates are analyzed by GC/MS/COMP by using, for example, a 20-m x 0.5-mm i.d. SP-2100 fused-silica capillary column. The temperature program is 70 to 250°C at 4°C/min.

*Volatile Carboxylic Acids.* This group of compounds is analyzed by a procedure involving adsorption from the basified sample on the anion exchange resin Biorad AG 1-X8 in the $Cl^-$ form. The acids are eluted from the resin using sodium bisulfate in an acetonitrile:water solution. Volatile acids are separated from the sodium bisulfate and nonvolatile sample components by distillation. The distilled acids are converted to their nonvolatile salts and the solution is evaporated to dryness. Acids are then derivatized using pentafluorobenzyl bromide prior to GC/MS analysis.

The derivatized extract concentrates are analyzed by using, for example, a 20-m x 0.5-mm i.d. SP-2100 fused-silica capillary column temperature-programmed from 70 to 250°C at 4°C/min.

*Nonvolatile Organic Acids.* This group of compounds includes such chemicals as benzene sulfonic, malonic, azealic and benzene phosphonic acid. The procedure for the analysis of these compounds in water utilizes adsorption on the anion exchange resin Biorad AG 1-X8 in the Cl form followed by elution with hydrochloric acid in methanol. The solvent is removed by evaporation, and the acids are converted to their methyl esters by using diazomethane.

The derivatized sample extracts are analyzed by GC/MS/COMP by using, for example, a 25-m x 0.5 mm i.d. SP-2100 fused-silica capillary column temperature-programmed from 30 to 250°C at 4°C/min.

*Halogenated Carboxylic Acids.* The procedure for the analysis of this group of compounds involves extraction of the acidified aqueous sample with methyl-*t*-butyl ether. The volume of the extract is reduced, and the acids are back-extracted into a small volume of an aqueous buffer solution. The solution is evaporated to dryness, and the acids are converted to their methyl ester derivatives using HCl:methanol.

The derivatized sample extracts are analyzed by GC/MS/COMP, using a column such as a 20-m x 0.5-mm i.d. fused-silica capillary column temperature-programmed from 70 to 250°C at 4°C/min.

## Extractables

Three procedures have been evaluated for the analysis of extractable organic compounds in water. The choice of the appropriate extraction procedure is based on the degree of contamination of the sample. Clean samples are analyzed by using resin columns because this procedure permits convenient processing of a large sample volume. More contaminated samples are prepared for analysis using a bottle-stirbar technique for batch liquid-liquid extraction (LLE). Finally, samples that may form emulsions are analyzed using a flow-under continuous LLE method.

*Resin Column*

Resin sorbent columns are used to prepare clean samples for analysis for extractable organics. For the isolation of basic and neutral compounds, the sample is adjusted to pH 7 and allowed to pass through a 10-mL bed of Amberlite XAD-4 (Rohm and Haas) at about 10 mL/min. Adsorbed organics are then eluted with four 10-mL portions of diethyl ether. A 10-min equilibration time is allowed for each aliquot of the eluting solvent. The eluates are pooled and dried with anhydrous sodium sulfate. The excess solvent is removed using Kuderna-Danish (KD) techniques followed by nitrogen blowdown.

The aqueous sample that passed through the resin is then adjusted to pH 2 and passed through a fresh column of XAD-4 at 10 mL/min. The adsorbed acids are eluted with four 10-mL portions of ethyl acetate. Again, each aliquot of solvent is allowed to equilibrate for 10 min before it is drawn off. The eluants are pooled, dried, derivitized with pentafluorobenzyl bromide and concentrated as described above.

*Batch LLE*

Batch LLE is the most convenient method for isolation of extractable organics. This technique involves the use of a stirbar and magnetic stirrer to bring the sample and solvent into contact. This permits extraction of the sample in the same vessel used for collection and minimizes possible losses due to transfer. For basic and neutral compounds, the sample is adjusted to pH 11, and extraction is performed using methylene chloride in a volume ratio of 1:5 with the sample. The mixture is stirred for two hours, the phases are allowed to separate, and the organic layer is drawn off by attaching the valve device pictured in Figure 5 and inverting the bottle. The extract is dried and concentrated as described above.

The isolation of acids is performed in a similar manner. The same sample is adjusted to pH 2, and sufficient sodium chloride is added to make the sample 30% (w/v) in this salt. Methylene chloride is added, and the mixture is stirred for two hours. Again the phases are allowed to separate, the organic layer is drawn off, dried and concentrated. The acid fraction is derivatized with pentafluorobenzyl bromide and concentrated for analysis.

*Flow-Under Continuous LLE*

Samples which have a tendency to form emulsions are prepared for analysis using the flow-under extractor illustrated in Figure 6. This apparatus extracts organics by bringing the sample and solvent into contact without

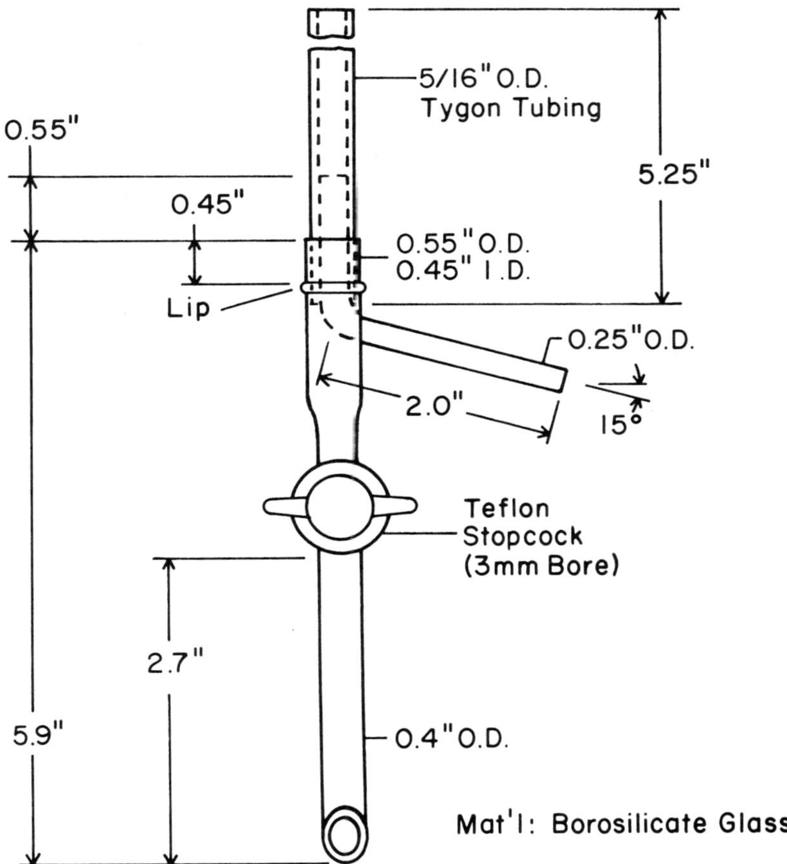


**Figure 5.** Valve device for removing solvent after batch LLE.

breaking the interface between the two phases. Stirring the aqueous layer prevents the process from being diffusion-limited, and the distillation feature cycles fresh solvent into contact with the sample.

Isolation of basic and neutral compounds is accomplished with the sample adjusted to pH 11, and acids are extracted after the addition of sodium chloride and with the sample adjusted to pH 2. Methylene chloride is the extracting solvent. When the extraction is complete (about 2 hr for each fraction), the phases are allowed to separate, and the methylene chloride is drawn off, dried and concentrated. The acid fraction is derivatized with pentafluorobenzyl bromide.

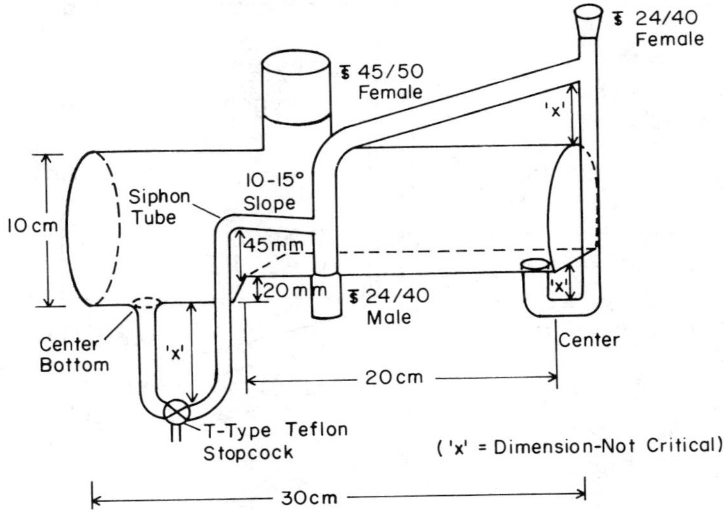


**Figure 6.** Flow-under continuous LLE.

Sample extracts prepared by any of these extraction procedures are analyzed by GC/MS/COMP using a fused silica capillary column, 25-m x 0.5-mm i.d., SP-2100, programmed from 30 to 250°C at 4°C/min.

## RESULTS

Data presented in this section were obtained in phase II experiments. Distilled water, treated municipal effluent and energy-related effluents were fortified with known amounts of a large number of model compounds. Samples were then prepared and analyzed by using the methods described above. Results of the analysis of fortified samples were compared with those of unfortified samples to determine percent recoveries for each compound. Data presented here represent the average of triplicate analysis.

### Purgeables

Recovery data for select model compounds are presented in Table III. These results demonstrate the accuracy and precision of the purgeable procedure. The lower recovery for ethyl acetate is attributed to the high water solubility of this compound. However, this material can be identified and quantified by this procedure.

Table III. Recovery Data for Select Model Compounds Using Purge-and-Trap

| Compound | Percent Recovery ± SD | | |
|---|---|---|---|
| | Distilled Water | Municipal Effluent | Energy Effluent |
| Ethyl Acetate | 23 ± 2 | 29 ± 4 | 30 ± 9 |
| Ethyl Hexanoate | 108 ± 13 | 95 ± 3 | 100 ± 3 |
| Phenyl Ether | 97 ± 14 | 70 ± 4 | 85 ± 4 |
| Methylene Chloride | 99 ± 3 | 105 ± 15 | 106 ± 5 |
| Chloroform | 79 ± 11 | 105 ± 18 | 95 ± 6 |
| Benzene | 76 ± 7 | 104 ± 5 | 111 ± 8 |
| Diphenyl Methane | 63 ± 6 | 66 ± 3 | 64 ± 14 |

Table IV. Recovery Data for Nonionic Intractable Organics Using Fractional Distillation

| Compound | Percent Recovery ± SD | | |
|---|---|---|---|
| | Distilled Water | Municipal Effluent | Energy Effluent |
| Ethanol | 88 ± 6 | 90 ± 3 | 100 ± 10 |
| 1-Butanol | 81 ± 3 | 89 ± 10 | 89 ± 9 |
| Acetone | 83 ± 14 | 69 ± 17 | 60 ± 22 |
| 2-Butanone | 79 ± 9 | 67 ± 14 | 83 ± 8 |
| Acetonitrile | 50 ± 5 | 59 ± 7 | 74 ± 1 |

## Intractables

### *Nonionic Intractables*

The results of the analysis of fortified distilled water, treated municipal effluent and energy-related effluent for select nonionic intractable compounds are presented in Table IV. In general, the accuracy and precision of these analyses are quite good. The higher standard deviation for acetone is probably due to inefficient collection of this highly volatile compound.

### *Ionic Intractables*

*Volatile Primary and Secondary Amines.* The results of the application of this procedure to distilled water fortified with select model compounds

Table V. Recovery Data for Select Volatile Primary and Secondary Amines from Deionized Water

| Compound | Percent Recovery ± SD |
|---|---|
| *t*-Butylamine | 88[a] |
| Allylamine | 109[a] |
| Hexylamine | 77[a] |
| Cyclohexylamine | 91[a] |
| Benzylamine | 89[a] |
| Diallylamine | 110 ± 21 |
| Dipropylamine | 87 ± 20 |
| Piperdine | 14 ± 2 |
| Morpholine | 12 ± 5 |
| 2-Methylpiperdine | 52 ± 5 |
| Dibutylamine | 70 ± 15 |
| Dicyclohexylamine | 41 ± 8 |

[a] Single determination.

Table VI. Recovery Data for Select Volatile Carboxylic Acids

| | Percent Recovery ± SD | | |
|---|---|---|---|
| Acid | Distilled Water | Municipal Effluent | Energy Effluent |
| Butyric | 90 ± 15 | 95 ± 17 | 93 ± 12 |
| Cratonic | 74 ± 16 | 88 ± 16 | 85 ± 12 |
| Hexanoic | 61 ± 6 | 56 ± 11 | 60 ± 9 |
| Isobutyric | 79 ± 12 | 87 ± 13 | 81 ± 9 |
| Trimethylacetic | 66 ± 5 | 66 ± 11 | 58 ± 5 |
| Methacrylic | 78 ± 12 | 84 ± 6 | 76 ± 6 |
| Ethylbutyric | 59 ± 4 | 56 ± 8 | 49 ± 6 |

are presented in Table V. Low recoveries were obtained for piperdine and morpholine. However, these compounds can be identified and quantified in the base/neutral extractable fraction.

*Volatile Carboxylic Acids.* The results of the analysis of fortified samples for volatile carboxylic acids are presented in Table VI. The recoveries of these compounds are acceptably high and do not exhibit significant matrix effects.

*Nonvolatile Organic Acids.* The applicability of this procedure for model compounds is illustrated in Table VII. While the recovery by this method is acceptable, the overall precision is not high. Evaluation of the procedure using $^{14}C$-labeled compounds indicates that the lack of reproducibility is probably attributable to the derivatization step.

*Halogenated Carboxylic Acids.* The recovery of chloroacetic acid from distilled water is 78%. Recovery of other halogenated acids is approximately equivalent.

## Extractables

Recovery data for extractable model compounds from municipal effluent using each of the three procedures are presented in Table VIII. One group of compounds not well recovered by any of these methods was alkanes. Recovery for heptadecane was 17% using sorbent columns or batch LLE.

**Table VII. Recovery Data for Select Nonvolatile Acids**

| Acid | Percent Recovery ± SD |
|---|---|
| Phosphanous | 80 ± 42 |
| Benzene Sulfonic | 90 ± 33 |
| Phosphanic | 117 ± 15 |
| Toluene Sulfonic | 95 ± 33 |
| Azealic | 36 ± 9 |

**Table VIII. Recovery Data for Extractable Organics in Municipal Effluent**

| Compound | Percent Recovery ± SD | | |
|---|---|---|---|
| | Sorbent Column | BLLE | CLLE |
| Phenol | 71 ± 2 | 75 ± 3 | 71 ± 8 |
| 2,4,5-Trichlorophenol | 73 ± 9 | 99 ± 8 | 64 ± 11 |
| Aniline | 61 ± 7 | 58 ± 2 | 54 ± 15 |
| *p*-Nitroaniline | 94 ± 13 | 75 ± 1 | 67 ± 9 |
| Phenanthrene | 70 ± 4 | 59 ± 3 | 48 ± 26 |
| Benzofuran | 76 ± 3 | 84 ± 9 | 54 ± 7 |
| Heptadecane | 17 ± 8 | 17 ± 9 | 2 ± 1 |
| Dichloroanisole | 68 ± 8 | 89 ± 9 | 45 ± 9 |
| Benzonitrile | 50 ± 6 | 52 ± 3 | 48 ± 3 |
| Butyl Benzyl Phthalate | 48 ± 7 | 56 ± 1 | 20 ± 3 |
| Nitrobenzene | 90 ± 2 | 90 ± 8 | 100 ± 13 |

When the flow-under system was used, recovery of this compound was only 2%. However, as illustrated by these data, the accuracy and precision of these procedures for other compounds are quite good.

## SUMMARY

The methodologies which comprise a comprehensive scheme for the analysis of organics in water have been described. Data resulting from the application of these procedures to fortified aqueous samples have been presented. This program has been designed to develop a protocol applicable to the screening of any aqueous sample for volatile (gas chromatographable) organic compounds. The data presented here demonstrate that this protocol is applicable to the identification and quantification of a wide variety of organics in water.

## ACKNOWLEDGMENT

This research was supported by EPA Contract No. 68-03-2704. Mention of trade names or commercial products does not constitute endorsement or recommendation for use.

## REFERENCES

1. Bursey, J. T. et al. "Master Scheme for the Analysis of Organic Compounds in Water, Preliminary Draft Report. Part I: State-of-the-Art Review of Analytical Operations," U.S. EPA, Contract No. 68-03-2704, Environmental Research Laboratory, Office of Research and Development, Athens, GA, (1979).
2. Pellizzari, E. D. et al. "Master Scheme for the Analysis of Organic Compounds in Water, Preliminary Draft Report. Part II: Matrix of Analytical Operations."
3. Grob, K., *J. Chromatog.* 84:255 (1973).
4. Grob, K., K. Grob, Jr., and G. Grob. *J. Chromatog.* 106:299 (1975).
5. Grob, K., and G. Grob. *J. Chromatog.* 90:303 (1974).
6. Grob, K., and F. Zurcher. *J. Chromatog.* 117:285 (1976).
7. Grob, K., and G. Grob. *Fifth Int. TOB. Sci. Congr.*, Hamburg (1970).

# CHAPTER 4

# THE MASTER ANALYTICAL SCHEME: AN ASSESSMENT OF FACTORS INFLUENCING PRECISION AND ACCURACY OF GAS CHROMATOGRAPHY/ MASS SPECTROMETRY DATA

**J. F. Ryan, J. E. Gebhart, L. C. Rando and D. L. Perry**

Department of Analytical Chemistry
Gulf South Research Institute
New Orleans, Louisiana

**K. E. Tomer, E. D. Pellizzari and J. T. Bursey**

Analytical Sciences Division
Research Triangle Institute
Research Triangle Park, North Carolina

For as long as mass spectrometrists have been generating gas chromatography/mass spectrometry (GC/MS) data, they have performed mathematical manipulations designed to quantify levels of those chemicals qualitatively identified. A number of methodologies have been used for quantification, among which are those using internal and external standards, regression analysis, and single point calibration. Some quantification methods have been used to assess levels only in a final solvent extract, while other methods have been designed to assess overall recovery and levels from the original sample. The choice of quantification methodology available to the analyst is wide, but according to initial literature surveys, the variability in precision and accuracy is unknown [1].

The use of GC/MS in regulatory activities makes it imperative to examine the reliability and reproducibility of GC/MS data. Because of this importance, reliability of quantification became one of the major tasks in the development of the U.S. Environmental Protection Agency (EPA) Master Analytical Scheme (MAS). Internal standard methods were chosen to enhance reliability and accuracy, and comprehensive experiments were performed to assess the precision and accuracy of quantified GC/MS data.

To quantify the amount of a compound of interest based on GC/MS techniques, response factors must be determined. This factor is the mathematical ratio of the integrated peak areas of known amounts of the compound of interest and an internal standard. Subsequent quantification of unknown amounts of the compounds of interest depends only on knowing the amount of internal standard and the previously determined response ratio. In calculating the response factor, either weight or molar values can be used as is shown in the following equations:

$$\text{Relative molar response (RMR) factor:}\quad \text{RMR} = \frac{\text{Area}_{\text{cmpd}}}{\text{Area}_{\text{IS}}} \times \frac{\text{moles}_{\text{IS}}}{\text{moles}_{\text{cmpd}}} \tag{1}$$

$$\text{Relative response ratio (RRR):}\quad \text{RRR} = \frac{\text{Area}_{\text{cmpd}}}{\text{Area}_{\text{IS}}} \times \frac{\text{grams}_{\text{IS}}}{\text{grams}_{\text{cmpd}}} \tag{2}$$

A number of parameters were examined to determine the extent of their impact on the quantification precision of the internal standard method. Among the factors examined were:

1. chemical class of internal standard;
2. retention time separation between an internal standard and the compound of interest;
3. chromatographic peak area integration using both total ion current and selected ion current; and
4. daily and long-term stability and reproducibility of quantification response factors.

A number of unanticipated results were obtained from these experiments:

1. The quantification precision of chemicals identified in GC/MS data files is not significantly improved by using internal standards chemically similar to the unknown compounds.

2. The quantification precision for compounds identified in GC/MS data files *is* improved by using an internal standard that elutes close to the compound, rather than an internal standard that elutes farther away.
3. The precision of GC/MS data is not improved by the use of total ion current peak area integrations, rather than selected ion current integrations; also, the use of selected ion current integrations minimized peak heterogeneity.
4. Response factors used for quantitative calculations should be checked daily and corrected if necessary.

The following sections of this chapter present data leading to these conclusions.

## RESULTS AND DISCUSSION

### Influence of Chemical Class on Precision of Quantification

To assess the influence of chemical class on quantitative precision, a standard mixture which included 23 compounds in 4 distinct chemical classes was prepared. The compounds used for this evaluation are shown in Table I.

This mixture was analyzed with GC/MS 9 times each day on 3 separate days; a total of 27 individual chromatograms was produced. A typical reconstructed chromatogram is shown in Figure 1 along with compound identification. The conditions that produced this chromatogram are shown in Table II.

**Table I. Standard Chemical Mixture**

| Hydrocarbon Class | Phenol Class | Polynuclear Aromatic Hydrocarbon Class | Amine Class |
|---|---|---|---|
| *n*-Nonane | Phenol | Naphthalene | Aniline |
| *n*-Decane | *m*-Cresol | Methylnaphthalene | *p*-Toluidine |
| *n*-Dodecane | 3,4-Dimethylphenol | Fluorene | Hexamethylenetetraamine |
| *n*-Hexadecane | *p*-tert-Butylphenol | Phenanthrene | Quinoline |
| *n*-Octadecane | *p*-Phenylphenol | Pyrene | Carbazole |
| *n*-Eicosane | | Triphenylene | Tribenzylamine |

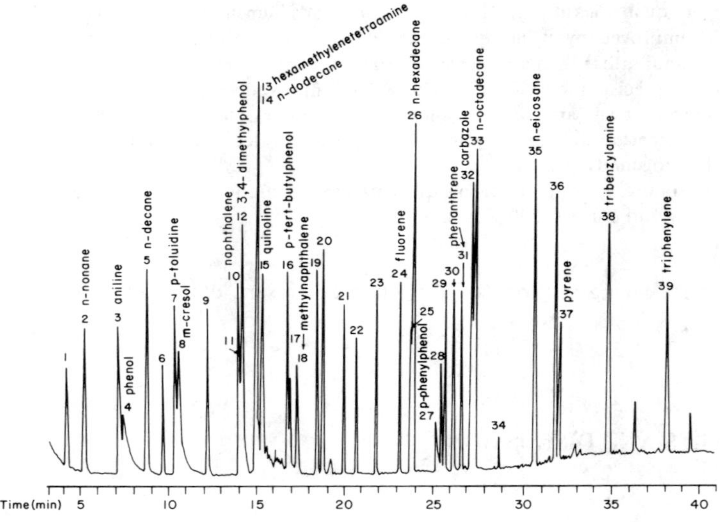


**Figure 1.** Reconstructed ion chromatogram of standard mixture.

**Table II. Conditions of GC/MS Analysis**

| | |
|---|---|
| GC/MS Model | Hewlett-Packard 5993 (splitless) |
| Column | 50-meter OV-101 support coated open tubular glass column: 0.5-mm i.d. |
| Carrier Gas | UHP helium |
| Column Flow at 70°C | 4 mL/min |
| Backflush Flow at 70°C | 33 mL/min |
| Temperature Program | 70°C (2 min) to 250°C at 6°C/min |
| Injection Port Temperature | 250°C |
| Scan Delay | 3 min |
| Scan Range | 35–450 amu |
| A/D Samples/0.1 amu | 3 |
| Scan Speed | 241.7 amu/sec |
| Ion Source Temperature | 150°C |
| Ionization Energy | 70 eV |
| Sample Volume | 1 $\mu$L |

With this vast amount of data, any one chemical can be picked as an internal standard, and RRR values for all other chromatographed compounds in the four classes can be calculated. Tables III to V present information collected on three separate days and show the RRR values with a corresponding coefficient of variation (CV) for each amine as measured against a member of each of the four chemical classes. (CV is simply the standard deviation expressed as a percentage of the arithmetic mean.) The four specific chemical class members chosen as internal standards were quinoline, fluorene, *p*-tert-butylphenol and *n*-hexadecane. These compounds were selected since they all eluted midway in the chromatogram and enabled a chemical class comparison between compounds with approximately the same retention time. A summary of the geometric mean CV data for amines is shown in Table VI. The geometric mean G of a set of n numbers $x_1, x_2, x_3, ...x_n$ is the nth root of the product of the numbers:

$$G = (x_1 x_2 x_3 ... x_n)^{1/n}$$

G is used since these data yield an asymmetrical frequency curve.

These data indicate that the geometric mean CV of the relative response ratio is not discernibly lower when an amine standard is used than when a different chemical class internal standard is used. This trend holds for the other three classes of compounds, as demonstrated in Tables VII to IX, which present summary data analogous to those of Table VI.

As these tables show, criteria for selection of an internal standard in a GC/MS quantitative analysis need not include chemical class.

## Assessment of the Effect of Retention Time Separation on Quantitative Precision

Because the chemical class of the internal standard had no significance on RRR precision, the effect of retention time separation was determined using the same 27-chromatogram data set described above.

An internal standard was chosen to represent each chemical class (amines, phenols, polynuclear aromatics and hydrocarbons). The four standards selected were the same ones used in the chemical class study: quinoline, *n*-hexadecane, *p*-tert-butylphenol and fluorene. For each daily set of nine GC/MS runs, the RRR coefficients of variation for each compound in each chemical class were calculated and compared to the retention time separation (expressed as ΔRRT) between the standard and the individual compound. ΔRRT is the absolute time difference between the internal standard and the compound divided by the internal standard retention time.

$$\Delta RRT = \frac{RT_{IS} - RT_{chemical}}{RT_{IS}} \tag{3}$$

**Table III. Data Set I: Subset Amines. RRR Factors as a Function of Chemical Class Measured on Day 1**

| | Internal Standard | | | | | | | |
|---|---|---|---|---|---|---|---|---|
| | Quinoline | | Fluorene | | *p*-tert-Butylphenol | | *n*-Hexadecane | |
| Compound | Mean RRR[a] | CV | Mean RRR[a] | CV | Mean RRR[a] | CV | Mean RRR[a] | CV |
| Aniline | 0.98 | 4 | 1.08 | 5 | 1.63 | 5 | 0.78 | 5 |
| *p*-Toluidine | 1.07 | 4 | 1.19 | 5 | 1.79 | 4 | 0.85 | 4 |
| Hexamethylenetetraamine | 0.36 | 6 | 0.40 | 5 | 0.61 | 7 | 0.29 | 7 |
| Quinoline | | | 1.11 | 3 | 1.67 | 1 | 0.80 | 1 |
| Carbazole | 1.03 | 5 | 1.14 | 3 | 1.72 | 5 | 0.82 | 5 |
| Tribenzylamine | 1.38 | 4 | 1.53 | 4 | 2.31 | 4 | 1.10 | 5 |
| Geometric Mean CV[b] | | 5 | | 4 | | 4 | | 4 |

[a] $n = 9$.

[b] Example calculation: for the column of numbers, 4, 4, 6, 5, 4, the geometric mean is $(4 \times 4 \times 6 \times 5 \times 4)^{1/5} = 5$.

Table IV. Data Set II: Subset Amines. RRR Factors as a Function of Chemical Class Measured on Day 29

| Compound | Internal Standard | | | | | | | |
|---|---|---|---|---|---|---|---|---|
| | Quinoline | | Fluorene | | *p*-tert-Butylphenol | | *n*-Hexadecane | |
| | Mean RRR[a] | CV | Mean RRR[a] | CV | Mean RRR[a] | CV | Mean RRR[a] | CV |
| Aniline | 0.95 | 5 | 0.75 | 7 | 1.08 | 6 | 0.97 | 9 |
| *p*-Toluidine | 1.11 | 4 | 0.88 | 6 | 1.27 | 4 | 1.14 | 8 |
| Hexamethylenetetraamine | 0.44 | 2 | 0.34 | 3 | 0.50 | 2 | 0.45 | 4 |
| Quinoline | | | 0.79 | 3 | 1.14 | 1 | 1.03 | 5 |
| Carbazole | 1.08 | 15 | 0.86 | 14 | 1.24 | 15 | 1.11 | 15 |
| Tribenzylamine | 0.83 | 16 | 0.65 | 14 | 0.95 | 16 | 0.85 | 12 |
| Geometric Mean CV | | 6 | | 6 | | 5 | | 8 |

[a]n = 9.

Table V. Data Set II: Subset Amines. RRR Factors as a Function of Chemical Class Measured on Day 40

| | Internal Standard | | | | | | | |
|---|---|---|---|---|---|---|---|---|
| | Quinoline | | Fluorene | | *p*-tert-Butylphenol | | *n*-Hexadecane | |
| Compound | Mean RRR[a] | CV | Mean RRR[a] | CV | Mean RRR[a] | CV | Mean RRR[a] | CV |
| Aniline | 0.96 | 3 | 0.86 | 6 | 1.14 | 3 | 0.84 | 4 |
| *p*-Toluidine | 1.13 | 3 | 1.01 | 4 | 1.34 | 1 | 0.99 | 2 |
| Hexamethylenetetraamine | 0.44 | 2 | 0.40 | 5 | 0.53 | 2 | 0.39 | 3 |
| Quinoline | | | 0.90 | 4 | 1.19 | 3 | 0.87 | 2 |
| Carbazole | 1.31 | 4 | 1.18 | 4 | 1.57 | 3 | 1.15 | 3 |
| Tribenzylamine | 1.45 | 12 | 1.30 | 14 | 1.72 | 12 | 1.26 | 11 |
| Geometric Mean CV | | 4 | | 5 | | 3 | | 3 |

[a] $n = 9$.

Table VI. Geometric Mean of the CV for RRR Factors of Amines as a Function of Chemical Class and Time

| Internal Standard | Day 1 | Day 29 | Day 40 |
|---|---|---|---|
| Quinoline | 5[a] | 6 | 4 |
| Fluorene | 4 | 6 | 5 |
| *p*-tert-Butylphenol | 4 | 5 | 3 |
| *n*-Hexadecane | 4 | 8 | 3 |

[a]Refer to Table III for sample calculation.

Table VII. Geometric Mean of the CV for RRR Response Factors of Polynuclear Aromatic Hydrocarbons as a Function of Chemical Class and Time

| Internal Standard | Day 1 | Day 29 | Day 40 |
|---|---|---|---|
| Quinoline | 5 | 6 | 4 |
| Fluorene | 5 | 9 | 4 |
| *p*-tert-Butylphenol | 5 | 6 | 3 |
| *n*-Hexadecane | 5 | 8 | 3 |

Table VIII. Geometric Mean of the CV for RRR Factors of Hydrocarbons as a Function of Chemical Class and Time

| Internal Standard | Day 1 | Day 29 | Day 40 |
|---|---|---|---|
| Quinoline | 4 | 6 | 4 |
| Fluorene | 4 | 6 | 5 |
| *p*-tert-Butylphenol | 4 | 5 | 3 |
| *n*-Hexadecane | 4 | 5 | 4 |

Table IX. Geometric Mean of the CV for RRR Factors of Phenols as a Function of Chemical Class and Time

| Internal Standard | Day 1 | Day 29 | Day 40 |
|---|---|---|---|
| Quinoline | 3 | 3 | 3 |
| Fluorene | 5 | 5 | 5 |
| *p*-tert-Butylphenol | 5 | 5 | 3 |
| *n*-Hexadecane | 4 | 7 | 2 |

Table X presents these data for the amine category. Although these data look complex, they are readily understood when graphically presented (Figure 2), where in each case, the CV is a minimum at ΔRRT at zero.

In the other three chemical classes the trend continues. These data are presented in Figures 3 to 5.

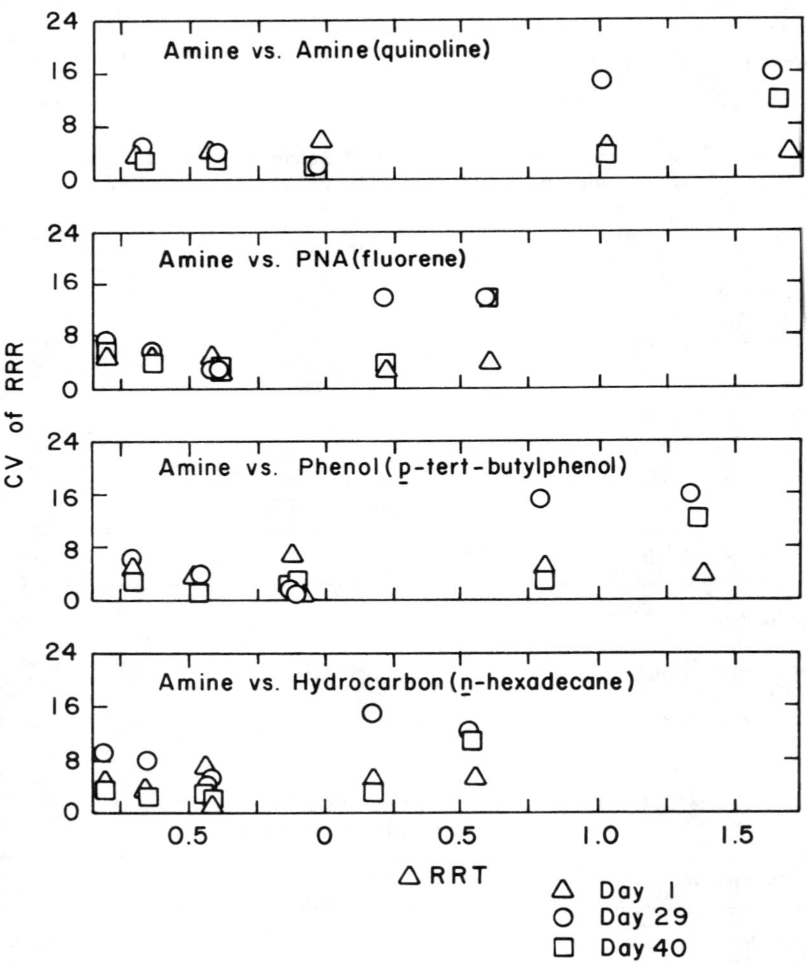


Figure 2. Subset amines: precision of RRR as a function of retention time (symbols represent maximum percent positive deviation).

**Table X. Subset Amines: Comparison of RRR CV with Retention Time Separation**

| Compound | Internal Standard | | | | | | | |
|---|---|---|---|---|---|---|---|---|
| | Quinoline[a] | | Fluorene[a] | | *p*-tert-Butylphenol[a] | | *n*-Hexadecane[a] | |
| | $CV_{RRR}$ | $\Delta RRT$ | $CV_{RRR}$ | $\Delta RRT$ | $CV_{RRR}$ | $\Delta RRT$ | $CV_{RRR}$ | $\Delta RRT$ |
| Aniline | 5 | 0.67 | 7 | 0.80 | 6 | 0.71 | 9 | 0.81 |
| *p*-Toluidine | 4 | 0.40 | 6 | 0.64 | 4 | 0.46 | 8 | 0.65 |
| Hexamethylenetetraamine | 2 | 0.033 | 3 | 0.42 | 2 | 0.14 | 4 | 0.44 |
| Quinoline | | | 3 | 0.40 | 1 | 0.11 | 5 | 0.42 |
| Carbazole | 15 | 1.01 | 14 | 0.21 | 15 | 0.79 | 15 | 0.17 |
| Tribenzylamine | 16 | 1.64 | 14 | 0.59 | 16 | 1.34 | 12 | 0.53 |

[a] n = 9, day 29.

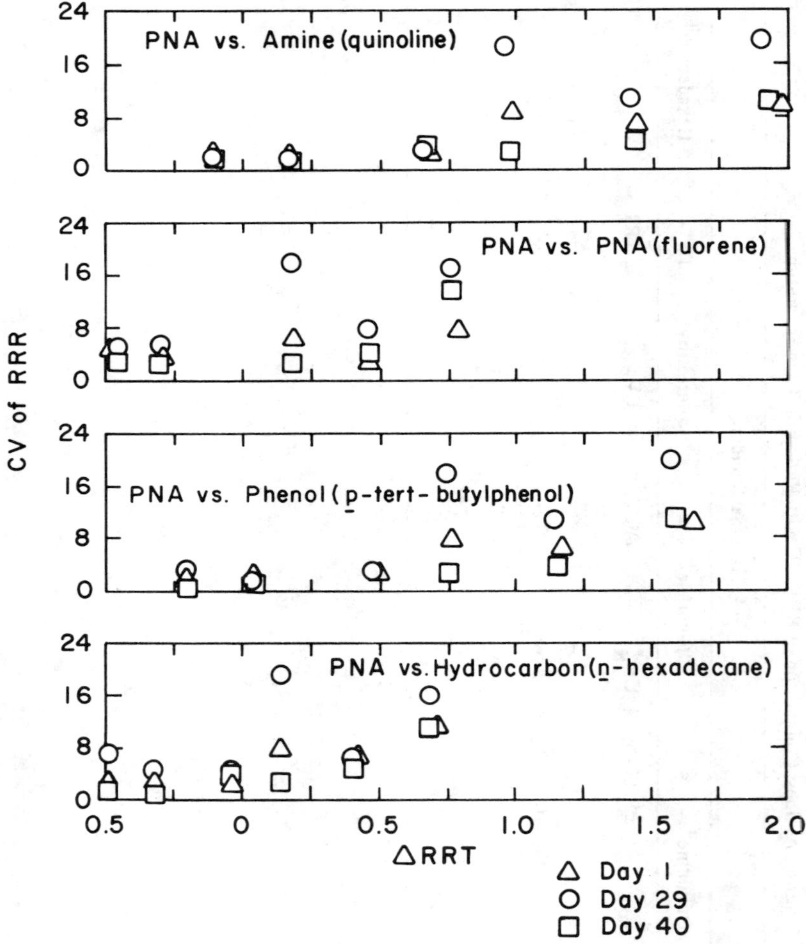


Figure 3. Subset polynuclear aromatic hydrocarbons: RRR as a function of retention time (symbols represent maximum percent positive deviation).

## Time Stability of Relative Response Ratios

If a RRR is used for quantification of GC/MS data files, the quantified data resulting from such calculations will only be as accurate as the chromatographic area measurements and the response ratio. A key factor in assessing this accuracy is time stability of RRR.

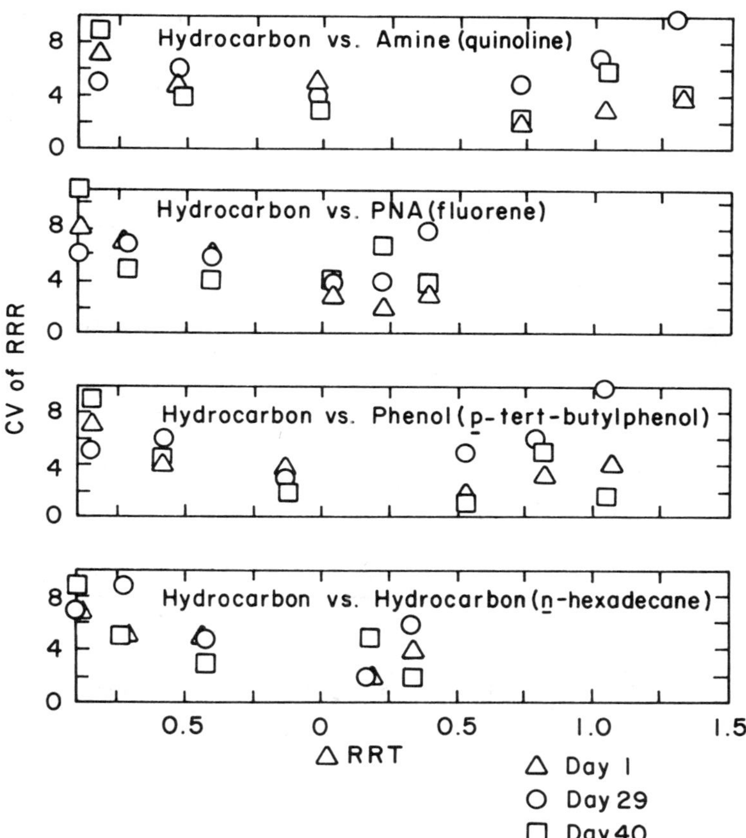


**Figure 4.** Subset hydrocarbons: precision of RRR as a function of retention time (symbols represent maximum percent positive deviation).

A study of the precision of the relative response ratios with time was performed using the data set of 27 GC/MS analyses described in the earlier studies. After a typical internal standard (*n*-hexadecane) was chosen, RRR were calculated for each of the other individual compounds in the four classes relative to the chosen standard: a mean RRR for the nine chromatograms collected on day 1, a mean RRR for the nine collected on day 29 and a mean RRR for the nine collected on day 40. These data, showing both the daily mean RRR and the CV for each of the three days along with a composite CV, are presented in Table XI. The composite CV is a representation of the relative standard deviation of the RRR from days 1, 29 and 40.

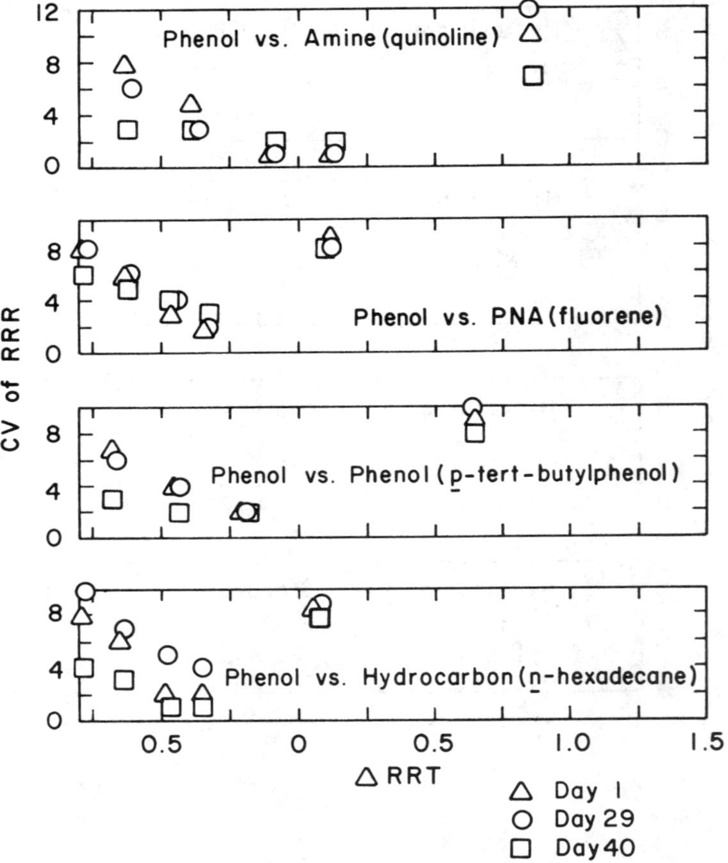


**Figure 5.** Subset phenols: precision of RRR as a function of retention time (symbols represent maximum percent positive deviation).

The data show that precision within any given day is quite good. It ranged from the best cases of 1% for quinoline (day 1), methylnaphthalene (day 40), 3,4-dimethylphenol (day 40), and *p*-tert-butylphenol (day 40), to the worst case of 19% for phenanthrene on day 29. The composite CV, however, for each compound determined over all for the 27 GC/MS analyses is not as precise. For example, the best case is *n*-nonane, which had a composite CV of 4%; the response ratio varied from 0.79 to 0.85 on the 3 days. In the worst cases, fluorene and *p*-phenylphenol had composite CV of 29%, where the response ratios varied from 0.72 to 1.30 and from 0.34 to 0.61, respectively.

Table XI. Stability of RRR[a] and CV for Individual Chemicals over 8-hr and 40-day Time Spans

| Chemical | Day 1 | | Day 29 | | Day 40 | | Composite CV |
|---|---|---|---|---|---|---|---|
| | Mean RRR | CV | Mean RRR | CV | Mean RRR | CV | |
| Amines | | | | | | | |
| Aniline | 0.78 | 5 | 0.97 | 9 | 0.84 | 4 | 12[b] |
| *p*-Toluidine | 0.85 | 4 | 1.14 | 8 | 0.99 | 2 | 15 |
| Hexamethylenetetraamine | 0.29 | 7 | 0.45 | 4 | 0.39 | 3 | 21 |
| Quinoline | 0.80 | 1 | 1.03 | 5 | 0.87 | 2 | 13 |
| Carbazole | 0.82 | 5 | 1.11 | 15 | 1.15 | 3 | 17 |
| Tribenzylamine | 1.10 | 5 | 0.85 | 12 | 1.26 | 11 | 20 |
| Polynuclear Aromatic Hydrocarbons | | | | | | | |
| Naphthalene | 1.16 | 3 | 1.70 | 7 | 1.38 | 2 | 19 |
| Methylnaphthalene | 0.73 | 3 | 1.13 | 5 | 0.89 | 1 | 22 |
| Fluorene | 0.72 | 3 | 1.30 | 4 | 0.97 | 4 | 29 |
| Phenanthrene | 0.93 | 8 | 1.67 | 19 | 1.43 | 3 | 28 |
| Pyrene | 0.83 | 7 | 1.43 | 7 | 1.31 | 5 | 27 |
| Triphenylene | 0.76 | 11 | 0.76 | 16 | 0.64 | 11 | 10 |
| Hydrocarbons | | | | | | | |
| *n*-Nonane | 0.83 | 7 | 0.85 | 7 | 0.79 | 9 | 4 |
| *n*-Decane | 0.99 | 5 | 1.11 | 9 | 1.02 | 5 | 6 |
| *n*-Dodecane | 1.04 | 5 | 1.15 | 5 | 1.06 | 3 | 6 |
| *n*-Hexadecane | | | | | | | |
| *n*-Octadecane | 0.90 | 2 | 0.87 | 2 | 0.94 | 5 | 4 |
| *n*-Eicosane | 0.89 | 4 | 0.81 | 6 | 0.96 | 2 | 9 |

Table XI, continued

| Chemical | Day 1 | | Day 29 | | Day 40 | | |
|---|---|---|---|---|---|---|---|
| | Mean RRR | CV | Mean RRR | CV | Mean RRR | CV | Composite CV |
| Phenols | | | | | | | |
| Phenol | 0.64 | 8 | 0.83 | 10 | 0.76 | 4 | 14 |
| *m*-Cresol | 0.50 | 6 | 0.68 | 7 | 0.62 | 3 | 15 |
| 3,4-Dimethylphenol | 0.63 | 2 | 0.83 | 5 | 0.72 | 1 | 14 |
| *p*-tert-Butylphenol | 0.61 | 2 | 0.90 | 4 | 0.73 | 1 | 20 |
| *p*-Phenylphenol | 0.34 | 9 | 0.53 | 9 | 0.61 | 8 | 29 |

[a]Internal standard = *n*-hexadecane.

[b]Example calculation: for the numbers, 0.78, 0.97, 0.84, the mean ± SD is 0.86 ± 0.10; therefore,the composite CV is 0.10/0.86 x 100% = 12.

Thus, good precision can be obtained for a given 8- or 10-hr period, but over a series of days, the response ratios can vary considerably, with a corresponding loss in precision. The variation of the RRR with time is mainly a function of the variation of the tune of the ion source with time. For a discussion of a RRR correction factor to compensate for these tuning variations, the reader is referred to a paper in this series [2].

## SUMMARY

The data in this paper present the most significant aspects of the work on precision and accuracy in quantification. Results of a number of additional experiments that produced additional conclusions are:

1. Total ion current (TIC) RRR values are no more precise than selected ion current (SIC) RRR values; additionally, TIC quantification measurements are susceptible to error because of chromatographic peak heterogeneity. For these reasons, the MAS is based on SIC RRR values.
2. Mass difference between the selected quantifying ion of an internal standard and the selected quantifying ion of the unknown compound had no effect. No reduction was found in the percent CV when the Dalton value separation was small rather than large.
3. No concentration dependence of relative response ratio was observed over a two-order-of-magnitude range.

As shown in this chapter, RRR precision is not enhanced by a chemical similarity between internal standard and compound of interest, but is enhanced by the retention time proximity between the two. Finally, RRR can vary considerably over different days (and be corrected when necessary) but be precise on a single day.

## ACKNOWLEDGMENTS

Support by the U.S. Environmental Protection Agency under Contract Number 68-03-2704 is gratefully acknowledged.

## DISCLAIMER

Mention of commercial products, trade names and companies is for informational purposes only and does not imply endorsement by the U.S. Environmental Protection Agency or the Gulf South Research Institute Laboratories.

## REFERENCES

1. "Master Scheme for the Analysis of Organic Compounds in Water. Part I: State-of-the-Art Review of Analytical Operations," EPA Contract No. 68-03-2704 (1979).
2. Tomer, K. B. et al. "Quantitative Aspects of the Master Analytical Scheme for Organics in Water," Chapter 5, this volume.

## CHAPTER 5

# QUANTITATIVE ASPECTS OF THE MASTER ANALYTICAL SCHEME FOR ORGANICS IN WATER

**K. B. Tomer, J. T. Bursey, L. C. Michael, L. S. Sheldon and E. D. Pellizzari***

Analytical Sciences Division
Chemistry and Life Sciences Group
Research Triangle Institute
Research Triangle Park, North Carolina

**A. L. Alford, J. D. Pope and A. W. Garrison**

Analytical Chemistry Branch
Environmental Research Laboratory
U.S. Environmental Protection Agency
Athens, Georgia

The objectives of the Master Scheme for the Analysis of Organics in Water and the general approach employed in defining the methodology to be used in attaining these goals have been reported [1-4]. The application of the Master Analytical Scheme (MAS), as it is now defined, to spiked and unspiked water samples has also been reported [2].

In the MAS the gas chromatography/mass spectrometry/computer (GC/MS/COMP) laboratory is charged with providing quantitative data for all the organic compounds which have been identified in the various fractions. Quantification is to be performed using molar response factors relative to specified internal standards; these response factors for several instruments will be provided to the user in a computerized database.

*Author to whom correspondence should be addressed.

We have used relative molar responses (RMR) as our response factors:

$$\text{RMR} = \frac{(\text{response of unknown/moles of unknown})}{(\text{response of standard/moles of standard})}$$

Although selected ion minitoring, in which a limited number (approx. 18) of ions are monitored, provides the most precise quantitative GC/MS data and the most instrumental sensitivity, it also necessitates foreknowledge of exactly what is to be quantified. For a comprehensive scheme such as the MAS, selected ion monitoring is too restrictive to use for quantification because the MAS is designed to identify and quantify all chromatographable components. Thus, all quantification in the MAS is based on mass spectral data acquired in the full scan mode. During the initial phases of the development of this program, experiments were directed toward optimization and standardization of GC/MS performance and toward an understanding of (and therefore control of) factors affecting the precision of quantitative data acquired by GC/MS/COMP systems operating in the full scan mode. Standardization and verification of instrumental performance is of prime importance to the analytical methodology of this scheme because: (1) changes in column performance, especially adsorptivity and tailing, can adversely affect quantification, especially near the detection limits; (2) changes in instrumental tune will change actual RMR significantly from those contained in the database, again adversely affecting quantification; and (3) gross changes in tune can also affect the identification of unknowns.

Our results during this phase of the program can be summarized as follows [3]:

1. System performance mixtures have been designated for testing column performance, acidity/basicity, polarity, resolution, adsorptivity of transfer lines and high mass/low mass balance of the mass spectrometer [5].
2. Because relative abundances of ions in the mass spectrum of a given compound may vary significantly while the instrument still exhibits a suitable high mass/low mass balance, experiments directed toward minimizing quantification errors due to mass spectral variance have been carried out. This has been done by investigating the changes in RMR over a period of time while trying to keep the tune constant and also the changes in RMR caused by introducing major changes in the tune. Even with software control of the instrumental tune, response factors varied as much as a factor of two over a six-week period. A linear relationship between RMR determined under different tuning conditions was observed, however, and this relationship can be used to reduce quantification errors by generating appropriate corrections for RMR to compensate for changing instrumental conditions. This linear relationship is illustrated by plots of RMR at one tune vs the same

RMR at a second tune. Figures 1 and 2 present representative plots for a Finnigan 3300 and a Hewlett-Packard 5993A, respectively (an application of this relationship is given later). It should be emphasized here that this is a means of reducing, not eliminating, differences in response caused by tuning changes.

3. Because quantification uses system response, peak heights or peak areas could be used. Quantification via full scan data, however, has been found to be more precise based on peak areas rather than peak heights.
4. Precision of quantitative GC/MS data is unaffected by the chemical class of the internal standard used, e.g., in quantifying phenols, a hydrocarbon used as a standard will give measurements as precise as those obtained using a phenol standard.
5. Precision of quantitative GC/MS data is enhanced by using an internal standard that elutes in close proximity to the unknown.
6. Precision of quantitative GC/MS data is unaffected by the m/z value of the ions used in quantification.
7. The use of reconstructed total ion current for quantification is no more precise than the use of selected ion current.
8. RMR determined at one concentration are valid over a wide concentration range (from the limits of detection to multiplier saturation); i.e., relative concentration of compound and standard within the dynamic range of the instrument does not affect RMR significantly.
9. At present, estimation of RMR based on chemical class or spectral similarity can introduce errors greater than 100%. For example, the RMR of the base peak of *o*-toluidine is half that of the base peak of *p*-toluidine. Within a chemical class the RMR can vary greatly; e.g., the RMR of a group of five polynuclear aromatic hydrocarbons (PAH) relative to fluorene ranged from 0.66 to 1.47.

During the third phase of this program, the MAS is to be applied to a variety of water types and chemical compounds. Spiked and unspiked samples of distilled, surface and drinking water, and industrial and municipal effluents will be analyzed in experiments in which the analyst does not know the number, identity or quantity of the spiked components. Two sets of experiments are being performed. For the first set, only one laboratory is performing the complete set of analyses. For the second set, however, three laboratories are analyzing identical samples. In this manner, the precision and accuracy of the quantitative methodologies developed for the MAS can be tested in laboratories that have not been involved in the development of all the techniques. Some early qualitative results are reported in another chapter in this volume [2]. Also during this phase, improvements in the scheme are to be made where necessary. This chapter reports on the initial quantitative results of the third phase.

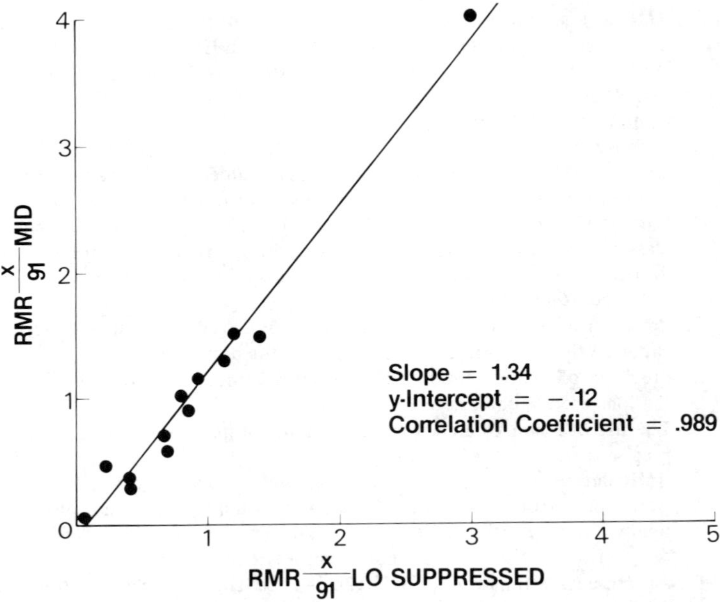


**Figure 1.** Plot of selected molar responses relative to m/z 91 at one tune vs the same response at a second tune (Finnigan 3300).

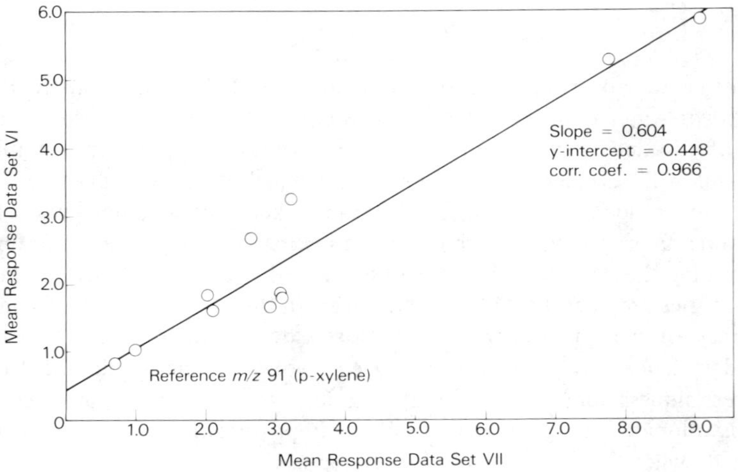


**Figure 2.** Plot of selected responses relative to m/z 91 at one tune vs the same responses at a second tune (Hewlett–Packard 5993).

## EXPERIMENTAL PROCEDURES

The procedures used for the isolation and concentration of the organic constituents of the water samples have been detailed elsewhere [4]. The internal standards used in the MAS were chosen to cover the elution range expected for the fraction in which they appear and also to represent typical recoveries of compounds found in that fraction. Additionally, the different standards are added at two concentration levels, one ten times the other, so that a large concentration range can be covered.

As part of the analytical procedures a series of quality control/quality assurance samples are analyzed prior to environmental sample analysis. These include a system performance solution for each final fraction type to test high mass/low mass balance, the transfer line and the condition of the GC column. This system performance solution can also contain the internal and external standards for a given fraction, which can be used to help correct RMR for changes in tune (see below). The reconstructed ion chromatogram of a typical system performance solution is shown in Figure 3. A mixture of known amounts of typical compounds in a class and the standard for the class, if the standards are not included in the system performance solution, and a procedural blank(s) are also analyzed as part of the experimental protocol.

The following procedure was used for quantification of identified compounds. The mass spectrum of the compound in the system performance solution chosen as a check of the high mass/low mass balance of the mass spectrometer was checked against the spectrum of that compound observed when the RMR in the database were obtained. If the relative abundances are within ±20%, no corrections are made. If, however, the relative abundances varied by a greater amount, the performance solution was used to help minimize the quantification error introduced by the spectral variation. The RMR of the ions in the internal and external standards contained in the system performance solution relative to one of the standard ions are determined (Table I). A linear regression analysis of the RMR relative to this specific standard ion, as y, is carried out against the same RMR determined when the data bank RMR were determined, as x (Table II). The slope and intercept are then used to modify the databank RMR. Using the data in Tables I and II for m/z 236 gives a correlation coefficient of 0.995 with a slope of 1.31 and an intercept of -0.13. Thus, a RMR of 0.81 from the databank tables would be corrected to 0.93, which would be the corrected RMR to be used in the quantification procedure.

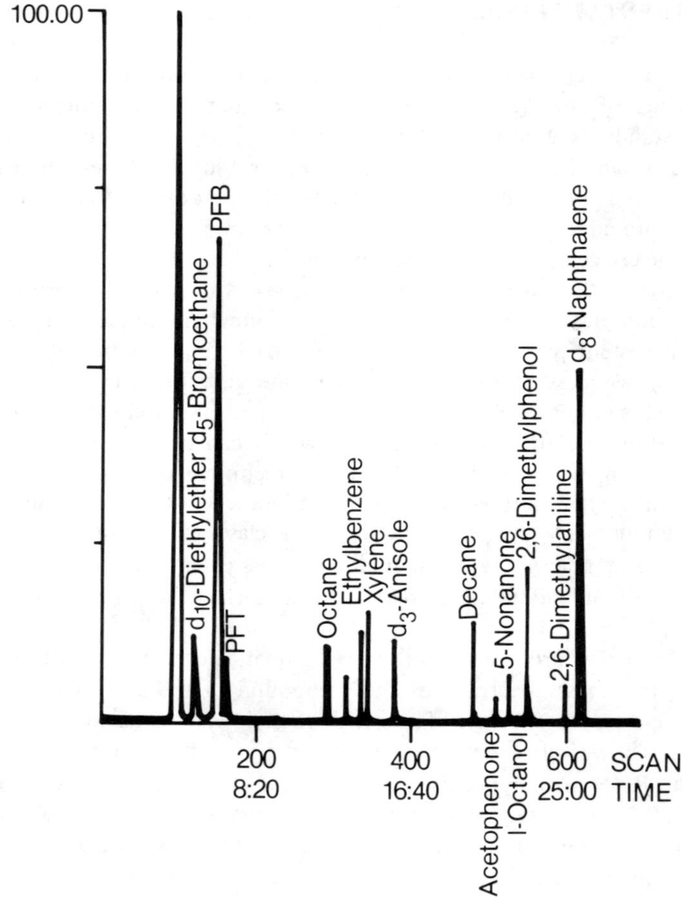


**Figure 3.** Column performance mixture for purgeable compounds.

## RESULTS AND DISCUSSION

In sample analyses, the concentration of the internal standard spiked into the water on collection is determined using an external standard spiked into the extract at a known concentration just prior to analysis. The observed internal standard concentration is then checked against the expected concentration. The expected concentration is based on the amount added to the water sample, volume analyzed and historical recovery. If the internal stan-

Table I. RMR of Standard Ions from Purgeable Performance Solution Relative to m/z 236 of Perfluorotoluene

| Compound | Ion | RMR |
| --- | --- | --- |
| Perfluorotoluene | 186 | 0.92 |
| $d_{10}$-Diethyl Ether | 50 | 0.11 |
| | 66 | 0.275 |
| $d_3$–Anisole | 68 | 0.45 |
| | 81 | 0.45 |
| | 111 | 0.68 |
| $d_8$–Naphthalene | 136 | 3.05 |

dard is 20% lower than that expected, the results are flagged to indicate poor recovery for the analysis. The concentration of each identified compound is then calculated based on the responses of an ion of the compound and an ion of the nearest eluting standard and the databank RMR corresponding to these ions. (The internal standards have been chosen to cover the elution range.) Each calculated concentration is then corrected for the historical recovery by using the databank recovery factor previously determined for that compound.

The error in accuracy associated with the isolation procedures utilized in the MAS has been reported [2], but the error in accuracy associated with the mass spectral quantification also must be propagated to determine the overall error in accuracy. For example, with purgeable compounds, although replicates in one day gave coefficients of variation (CV) for the precision of the RMR determination of less than ±10%, replicate injections of a solution of RMR standards over a five-day period while keeping the tune as reproducible as possible still gave coefficients of variation for the precision of the RMR of approximately ±25%. The exact error in accuracy introduced by the application of the linear regression analysis for RMR correction is not known, but is predicted to be ±20%. A mean deviation in precision of ±19% was determined on the basis of the duplicate determinations made on the purgeable fraction of the intralaboratory study. These deviations in precision for mass spectral quantification and isolation procedures gave an estimated overall error in accuracy of ±44% as determined from the square root of the sum of the squares of the CV of the precisions.

Table III presents some RMR determined under varying conditions of mass spectrometer tune. Previous data indicated that the RMR of the m/z 91 ions of toluene and of the $C_2$-alkylbenzene isomers with respect to any given standard were quite similar, as were the RMR of the m/z 120 ions of the $C_3$-alkylbenzene isomers. In this data set, differences of up to a factor of 20 are observed for one standard, perfluorotoluene, while a reasonable distribution

Table II. RMR of Standard Ions for Purgeables (Finnigan 3300)

| | | Standard | | | | | | | | |
|---|---|---|---|---|---|---|---|---|---|---|
| | | Perfluoro-toluene | | $d_3$-Anisole | | $d_8$-Naphtha-lene | $d_{10}$-Diethyl ether | | $d_5$-Chloro-benzene | |
| **Compound** | **Ion** | **236** | **186** | **111** | **81** | **136** | **50** | **66** | **82** | **117** |
| Perfluorotoluene | 236 | | 1.04 | 1.60 | 2.35 | 0.42 | 6.82 | 3.11 | 1.44 | 0.90 |
| | 186 | 0.96 | | 1.53 | 2.25 | 0.40 | 6.54 | 2.99 | 1.38 | 0.87 |
| $d_3$-Anisole | 111 | 0.63 | 0.68 | | 1.48 | 0.27 | 4.28 | 1.95 | 0.92 | 0.58 |
| | 81 | 0.42 | 0.44 | 0.68 | | 0.18 | 2.84 | 1.30 | 0.59 | 0.37 |
| | 68 | 0.39 | 0.40 | 0.61 | 0.90 | 0.16 | 2.62 | 1.19 | 0.56 | 0.34 |
| $d_8$-Naphthalene | 136 | 2.38 | 2.44 | 3.77 | 5.58 | | 15.99 | 7.28 | 3.31 | 2.09 |
| $d_{10}$-Diethyl ether | 50 | 0.15 | 0.15 | 0.24 | 0.36 | 0.063 | | 0.46 | 0.21 | 0.13 |
| | 66 | 0.32 | 0.34 | 0.52 | 0.78 | 0.139 | 2.19 | | 0.46 | 0.29 |
| $d_5$-Chloro-benzene | 82 | 0.69 | 0.72 | 1.093 | 1.72 | 0.31 | 4.83 | 2.20 | | 0.63 |
| | 117 | 1.11 | 1.10 | 1.73 | 2.71 | 0.48 | 7.65 | 3.49 | 1.59 | |
| | 119 | 0.43 | 0.45 | 0.69 | 1.05 | 0.19 | 2.95 | 1.35 | 0.61 | 0.39 |

Table III. Representative RMR[a] Determined on a Finnigan 4000

| Compound | Ion | Standard: m/z 236 Perfluoro-toluene | Standard: m/z 111 2,4,6,-$d_3$-Anisole |
|---|---|---|---|
| Toluene | 91 | 1.95 | 1.03 |
| Ethylbenzene | 91 | 5.71 | 1.44 |
| *m*-Xylene | 91 | 2.02 | 1.08 |
| *o*-Xylene | 91 | 0.15 | 1.37 |
| 1,3,5-Trimethylbenzene | 120 | 1.81 | 1.39 |
| 1,2,4-Trimethylbenzene | 120 | 3.51 | 0.88 |
| 1,2,3-Trimethylbenzene | 120 | 0.69 | 1.01 |

[a]Mean of nine replicate determinations.

Table IV. Corrected RMR Determined on a Finnigan 4000

| Compound | Ion | Standard: m/z 236 Perfluoro-toluene | Standard: m/z 111 2,4,6-$d_3$-Anisole |
|---|---|---|---|
| Toluene | 91 | 1.79 | 1.36 |
| Ethylbenzene | 91 | 2.36 | 1.86 |
| *m*-Xylene | 91 | 1.86 | 1.43 |
| *o*-Xylene | 91 | 1.61 | 1.61 |
| 1,3,5-Trimethylbenzene | 120 | 1.81 | 1.39 |
| 1,2,4-Trimethylbenzene | 120 | 1.47 | 1.16 |
| 1,2,3-Trimethylbenzene | 120 | 1.02 | 0.78 |

was observed for $d_3$-anisole as a standard. The relative abundances of the major ions of perfluorotoluene were observed to vary significantly during these determinations. To correct the RMR for the changes in tune, linear regression analyses were performed, as indicated in the experimental section, using one of the data sets as x in which the RMR best approached the mean observed for all data sets. The slopes and intercepts thus determined were used to recalculate the observed RMR. These results are given in Table IV. The CV for the perfluorotoluene data have been reduced from 95 to 17 for m/z 91 and from 71 to 27 for m/z 120, but the CV of the $d_3$-anisole data have not been altered significantly. Thus, the RMR corrections based on the linear regression analyses have significantly reduced large variations without increasing the variation for data sets with already acceptable variations.

An alternative procedure for some specific chemical classes and fractions is to include in the system performance solution control compounds whose RMR reflect the mean RMR of the base peaks of the compounds in that chemical class. For example, the mean RMR of the base peaks of the purgeable aromatic compounds relative to the external standard was 1.22 with a CV of ±21% for the entire class. This CV was approximately the same as that observed when RMR were determined over a period of several days. The RMR of *p*-xylene was 1.29,very close to the mean. *p*-Xylene is a component of the purgeable system performance standard and can be used to determine the mean RMR for that day for aromatic compounds.

At this time we are in the third phase of MAS development, in which spiked and unspiked waters are to be analyzed. Purgeable components, intractable halogenated acids and volatile intractable acids in ten water samples have been analyzed in the intralaboratory study. Analyses of the purgeable fractions of the interlaboratory study samples are in progress. As pointed out earlier, the ten intralaboratory waters analyzed include deionized water, drinking water, two municipal effluents, one municipal influent, three manufacturing effluents and two energy-related effluents. The interlaboratory samples include distilled, drinking and surface waters, and municipal, energy and industrial effluents.

Tables V and VI present some of the results obtained for the intralaboratory purge and trap analyses. For a drinking water sample (Table V), the observed accuracy of the analysis for only trichloroethylene, tetrachloroethylene, dichlorobenzene, dichloropropene and diphenyl ether was beyond the estimated accuracy error limit of ±44% for the purgeable method. Dichlorobenzene consistently gave low results. For a municipal waste effluent (Table VI), trichloroethane, tetrachloroethylene, tetrachlorobenzene and dichlorobenzene gave results outside these limits. The data in Tables V and VI are for some of the first samples analyzed. The overall precision of replicate analyses of the ten intralaboratory samples was ±19%, showing improvement in the precision of the analysis over the course of the analyses. This is well within our estimated reproducibility. It needs to be stressed at this point that these results were calculated using RMR that had been determined six months previous to sample analysis.

Table VII presents the mean accuracies [(observed/spiked) x 100] of spiked compounds in all water types. Recoveries of carbon tetrachloride, dichloropropene, *n*-nonane, bromobenzene, dichlorobenzene, dichloropropane, trichlorobenzene, 1,2-dichloroethane, tetrachlorobenzene and diphenyl ether are outside our predicted error in accuracy of ±44%. The remainder were not.

Table V. Purgeable Components in Drinking Water

| Compound | ng Observed[a] (CV) | ng Spiked | $100 \times \frac{\text{ng Observed}}{\text{ng Spiked}}$ |
|---|---|---|---|
| 1,2,–Dichloroethylene | 318 (44) | 387 | 82 |
| 1,2–Dichloroethane | 311 (25) | 553 | 56 |
| Benzene | 42 (37) | 44 | 95 |
| 1,1,1–Trichloroethane | 711 (19) | 540 | 130 |
| Trichloroethylene | 23 (17) | 12 | 192 |
| Dichloropropene Isomer | 219 (39) | 493 | 44 |
| Tetrachloroethylene | 1170 (42) | 659 | 177 |
| Bromobenzene | 420 (34) | 676 | 62 |
| Iodobenzene | 659 (33) | 802 | 82 |
| Dichlorobenzene Isomer | 16 (43) | 88 | 18 |
| 1,2,4,5–Tetrachlorobenzene | 1086 (53) | 1055 | 103 |
| 1,4–Dibromobutane | 95 (126) | 72 | 131 |
| Toluene | 43 (24) | 45 | 96 |
| Ethylbenzene | 50 (64) | 52 | 96 |
| *n*–Nonane | 315 (39) | 442 | 71 |
| Diphenyl Ether | 16 | 55 | 29 |

[a]Mean of three determinations corrected for unspiked samples and historical recovery.

Table VI. Purgeable Components in Municipal Effluent

| Compound | ng Observed[a] | ng Spiked | $100 \times \frac{\text{ng Observed}}{\text{ng Spiked}}$ |
|---|---|---|---|
| 1,2-Dichloroethylene | 381 ± 169 | 387 | 98 |
| 1,1,1-Trichloroethane | 1272 ± 514 | 540 | 235 |
| Carbon Tetrachloride | 174 ± 18 | 320 | 54 |
| 1,2-Dichloropropane | 144 ± 8 | 117 | 123 |
| Dichloropropene Isomer | 306 ± 48 | 493 | 62 |
| Tetrachloroethylene | 2337 ± 1023 | 659 | 355 |
| 1,4-Dibromobutane | 271 ± 86 | 360 | 75 |
| Ethylbenzene | 285 ± 158 | 260 | 109 |
| Benzene | 228 ± 48 | 220 | 103 |
| Bromobenzene | 704 ± 142 | 676 | 104 |
| Iodobenzene | 795 ± 197 | 802 | 99 |
| Dichlorobenzene Isomer | 201 ± 59 | 440 | 46 |
| 1,2,4,5-Tetrachlorobenzene | 302 ± 77 | 1055 | 29 |
| Diphenyl Ether | 200 ± 55 | 275 | 73 |
| 1,3,5-Trichlorobenzene | T[b] | 130 | |

[a]Mean of duplicate determinations; corrected for unspiked samples and historical recovery.
[b]Trace.

**Table VII. Mean Accuracy [as 100 x (ng Observed/ng Spiked)] – Purgeables**

| Compound | Mean Accuracy[a] |
|---|---|
| 1,2-Dichloroethylene | 102 |
| 1,1-Dichloroethylene | 94 |
| 1,1,1-Trichloroethane | 80 |
| Benzene | 74 |
| Carbon Tetrachloride | 43 |
| 1,2-Dichloroethane | 156 |
| Dichloropropane | 50 |
| Dichloropropene | 22 |
| 1,1,2-Trichloroethane | 71 |
| Tetrachloroethylene | 96 |
| Ethylbenzene | 97 |
| *n*-Nonane | 43 |
| Bromobenzene | 42 |
| Dichlorobenzene Isomer | 44 |
| Dichlorobenzene Isomer | 27 |
| Iodobenzene | 64 |
| 1,2,4,5-Tetrachlorobenzene | 47 |
| Trichloroethylene | 129 |
| Tetrachloroethane | 63 |
| Toluene | 79 |
| Diphenyl Ether | 38 |
| 1,2,4-Trichlorobenzene | 38 |
| 1,4-Dibromobutane | 61 |

[a]Mean of 10 samples, 20 analyses; corrected for unspiked samples and historical recoveries.

One aspect of Table VII needs to be pointed out in relationship to quality assurance/quality control programs associated with method development. If one uses only spiked vs unspiked water samples to determine recoveries for a developed procedure, serious problems can arise. For example, in the case of *n*-nonane, the overall recovery appears to be quite low. Because the individual steps have been shown to give good recoveries by use of radioisotopes, however, the low recovery may be associated with volatility/solubility problems in the addition of the spike and/or changes in instrument response with time.

The intractable halogenated acids represent another fraction whose analyses have also been completed in the intralaboratory study. These compounds represent a class that is isolatable with difficulty. Tables VIII and IX present results with this fraction for several water types. The observed quantities have not been corrected for historical recoveries because they have not been determined at this time. Except for those cases (e.g., drinking water) in which the spiked compounds were observed in the unspiked samples, the reproducibility was good.

**Table VIII. Volatile Halogenated Acids as Methyl Esters–Distilled Water**

| Compound | μg Observed[a] | μg Spiked | % Accuracy |
|---|---|---|---|
| Chloroacetate | 16 ± 2 | 29 | 55 |
| Dichloroacetate | 49 ± 6 | 44 | 111 |
| Trichloroacetate | 42 ± 16 | 28 | 150 |
| 2-Chloropropionate | 21 ± 2 | 35 | 60 |
| 3-Chloropropionate | 42 ± 9 | 56 | 75 |

[a]Mean of two determinations; corrected for unspiked samples, recoveries undetermined.

**Table IX. Volatile Halogenated Acids as Methyl Esters–Municipal Waste Effluent**

| Compound | μg Observed[a] (CV) | μg Spiked | % Accuracy |
|---|---|---|---|
| Chloroacetate | 171 (92) | 147 | 115 |
| Dichloroacetate | 386 (40) | 234 | 165 |
| Trichloroacetate | 71 (48) | 140 | 51 |
| 2-Chloropropionate | 160 (100) | 177 | 90 |
| 3-Chloropropionate | 115 ± 4[b] | 280 | 41 |

[a]Mean of three determinations; corrected for unspiked samples, recoveries undetermined.
[b]Found in only two analyses.

The analyses of volatile intractable acid fractions have also been completed for the intralaboratory samples. These compounds were analyzed as their pentafluorobenzyl esters. Tables X and XI give the results for a drinking water and an industrial effluent, respectively. Observed and spiked amounts compare very favorably, with the observed amounts of only three compounds lying outside a ±39% deviation from the spiked amounts. Propagation of error predicts an error in accuracy of ±39%.

At present, the purge-and-trap fraction from three sample types has been analyzed in the interlaboratory studies: distilled, drinking and surface water. Two of the laboratories determined their response factors close to the time of analysis, the third laboratory's responses were determined several weeks prior to the analyses. At this point, few comparative data are available. In each laboratory, however, replicate analyses of a given sample were performed, and coefficients of variation for each set of replicates were calculated. If all of these coefficients of variation are summed and averaged, an overall mean precision of ±12% is observed.

Tables XII to XIV give the results of the Research Triangle Institute (RTI) analysis of the purgeable fractions of the interlaboratory samples obtained at

**Table X. Volatile Intractable Acids as PFB[a] Esters–Drinking Water**

| Ester | μg Observed[b] | μg Spiked | % Accuracy |
|---|---|---|---|
| Isobutyrate | 29 ± 3 | 25 | 116 |
| Methacrylate | 19 ± 2 | 25 | 76 |
| *n*-Butyrate | 23 ± 2 | 25 | 92 |
| Crotonate | 31 ± 3 | 32.5 | 95 |
| Ethylbutyrate | 18 ± 2 | 25 | 72 |
| Hexanoate | 21 ± 1 | 27.5 | 76 |
| Ethylhexanoate | 15 ± 2 | 27.5 | 55 |
| Cyclohexanecarboxylate | 25 ± 5 | 27.5 | 91 |
| Cyclohexylacetate | 21 ± 4 | 30 | 71 |
| Octanoate | 13 ± 4 | 27.5 | 47 |

[a]PFB = perfluorobenzyl moiety.
[b]Mean of two determinations; corrected for unspiked samples and historical recovery.

**Table XI. Volatile Intractable Acids as PFB Esters–Manufacturing Effluent (Soap)**

| Ester | μg Observed[c] | μg Spiked | % Accuracy |
|---|---|---|---|
| Isobutyrate | 31 ± 1 | 50 | 62 |
| Methacrylate | 13 ± 9 | 50 | 126 |
| Butyrate[a,b] | 142 ± 36 | 50 | 284 |
| Crotonate | 43 ± 19 | 65 | 66 |
| Ethylbutyrate | 46 ± 19 | 50 | 92 |
| Hexanoate[a] | 50 ± 19 | 55 | 91 |
| Ethylhexanoate | 70 ± 40 | 55 | 127 |
| Cyclohexanecarboxylate | 48 ± 2 | 55 | 87 |
| Octanoate[a] | 57 ± 2 | 55 | 104 |
| Cyclohexylacetate | 61 ± 2 | 60 | 102 |

[a]Found in unspiked.
[b]Unspiked > spiked quantity.
[c]Corrected for unspiked samples and historical recovery; mean of two determinations.

this point. The observed amounts and spiked amounts are within the expected limits of accuracy with 45 out of 64 analytes observed to be within ±44% of the spiked amount. Of the 19 observations outside the error in accuracy, 2 are toluene and methylene chloride, which are often present as background, and 7 are hydrocarbons. The hydrocarbons are easily lost in transfer/dilution steps. This can be seen from Table XIV, in which the observed concentrations of the hydrocarbons significantly higher when the undiluted sample is included.

Table XII. Purgeable Components in Interlaboratory Deionized Water

| Compound | μg/L Observed[a] | μg/L Spiked | % Accuracy |
|---|---|---|---|
| 1,1-Dichloroethylene | 6.2 | 5.62 | 110 |
| 1,1-Dichloroethane | 6.3 | 6.52 | 97 |
| Chloroform | 4.0 | 8.16 | 49 |
| Carbon Disulfide | 4.2 | 5.80 | 72 |
| *n*-Hexane | 4.1 | 3.64 | 112 |
| Benzene | 3.6 | 4.84 | 74 |
| Thiophene | 4.5 | 4.88 | 92 |
| Toluene | 3.5 | 6.40 | 55 |
| 1,2-Dibromoethane | 22.9 | 12.04 | 190 |
| 1,3-Dichloroethane | 5.8 | 8.98 | 65 |
| *n*-Octane | 3.7 | 3.86 | 97 |
| 1,1,1,2-Tetrachloroethane | 14.1 | 11.34 | 124 |
| Chlorobenzene | 9.4 | 8.16 | 115 |
| Ethylbenzene | 3.2 | 4.00 | 80 |
| 1,4-Dichlorobutane | 2.0 | 8.40 | 24 |
| *n*-Nonane | 0.9 | 3.32 | 27 |
| Benzonitrile | 5.1 | 7.44 | 69 |
| 1,2,4-Trimethylbenzene | 3.4 | 4.86 | 70 |
| Benzyl Chloride | 3.8 | 6.08 | 63 |
| *n*-Decane | 2.6 | 3.32 | 79 |
| 1,2,4-Trichlorobenzene | 6.3 | 10.68 | 59 |

[a]Corrected for unspiked samples and historical recovery.

## SUMMARY AND CONCLUSIONS

The first-generation MAS has been applied to a series of water samples in one laboratory and to replicate water samples analyzed independently by three laboratories. The accuracy found was, for the most part, within the limits expected on the basis of propagation of error. The MAS, therefore, promises to be an extremely useful tool for the qualitative-quantitative analysis of organics in a broad range of water samples.

**Table XIII. Purgeable Components in Interlaboratory Drinking Water**

| Compound | μg/L Observed[a] | μg/L Spiked | % Accuracy |
|---|---|---|---|
| 1,1-Dichloroethylene | 4.3 ± 1.3 | 5.62 | 77 |
| 1,1-Dichloroethane | 4.4 ± 0.7 | 6.52 | 67 |
| Carbon Disulfide | 4.4 ± 1.8 | 5.80 | 76 |
| Benzene | 3.8 ± 0.5 | 4.84 | 79 |
| Thiophene | 4.4 ± 1.2 | 4.88 | 90 |
| Toluene | 11.6 ± 3.2 | 6.40 | 180 |
| 1,2-Dibromoethane | 15.4 ± 4.2 | 12.04 | 127 |
| 1,3-Dichloroethane | 4.8 ± 1.2 | 8.98 | 54 |
| *n*-Octane | 1.2 ± 0.4 | 3.86 | 31 |
| 1,1,1,2-Tetrachloroethane | 18.9 ± 2.6 | 11.34 | 166 |
| Chlorobenzene | 10.1 ± 1.2 | 8.6 | 123 |
| Ethylbenzene | 2.5 ± 0.1 | 4.00 | 63 |
| 1,4-Dichlorobutane | 5.2 ± 1.4 | 8.40 | 62 |
| *n*-Nonane | 0.6 ± 0.5 | 3.32 | 18 |
| Benzonitrile | 28.4 ± 6.8[b] | 7.44 | 381 |
| 1,2,4-Trimethylbenzene | 6.7 ± 1.3 | 4.86 | 137 |
| Benzyl Chloride | 8.0 ± 1.1 | 6.08 | 131 |
| *n*-Decane | 2.3 ± 0.5 | 3.32 | 70 |
| 1,2,4-Trichlorobenzene | 7.5 ± 0.7 | 10.68 | 70 |

[a]Two determinations; corrected for unspiked samples and historical recovery.
[b]Recovers poorly by purge-and-trap.

## ACKNOWLEDGMENTS

We wish to acknowledge support for this work under EPA Contract No. 68-03-2704. The excellent technical assistance provided by Mr. R. Porch, Ms. B. Bickford, Ms. D. DiStefano, Mr. P. Dodd, Mr. M. Parker, Mr. R. Wiseman, Ms. J. Storm and Mr. J. Turlington of the Research Triangle Institute; Mr. J. Craig, Dr. J. McGuire and Dr. W. Shackelford of the Athens Environmental Research Laboratory; and Dr. J. Gebhart and Dr. J. Ryan of Gulf South Research Institute is gratefully acknowledged.

**Table XIV. Purgeable Components in Interlaboratory Surface Water**

| | μg/L Observed | | μg/L | Percent Accuracy | |
|---|---|---|---|---|---|
| | Two Determinations[a] | Three Determinations[b] | Spiked | Two Determinations | Three Determinations |
| *n*–Pentane | 10.0 (4)[c] | 12.7 (34)[d] | 76 | 13 | 17 |
| Dichloromethane | 110.0 (12) | 82.5 (55) | 160 | 69 | 52 |
| 1,1–Dichloroethylene | 111.3 (14) | 82.3 (21) | 146 | 76 | 56 |
| 1,1–Dichloroethane | 72.7 (10) | 129.4 (26) | 142 | 51 | 91 |
| Chloroform | 78.7 (6) | 72.1 (17) | 178 | 44 | 41 |
| Carbon Disulfide | 138.3 (9) | 153.3 (13) | 151 | 92 | 105 |
| *n*–Hexane | 35.6 (11) | 43.6 (32) | 79 | 45 | 55 |
| Benzene | 53.8 (13) | 65.4 (31) | 106 | 51 | 62 |
| Thiophene | 84.6 (9) | 122.2 (14) | 127 | 67 | 96 |
| Dichloropropene Isomer | 75.1 (2) | 100.2 (43) | 146 | 51 | 69 |
| Toluene | 55.4 (2) | 54.4 (3) | 104 | 53 | 52 |
| 1,2–Dibromoethane | 215.4 (2) | 250 (11) | 261 | 83 | 96 |
| *n*–Octane | 21.4 (69) | 55.5 (107) | 84 | 25 | 66 |
| 1,1,1,2–Tetrachloroethane | 155.6 (23) | 182.6 (18) | 185 | 84 | 99 |
| Chlorobenzene | 154.4 (1) | 195 (36) | 160 | 97 | 122 |
| Ethylbenzene | 71.0 (1) | 94.0 (42) | 104 | 68 | 90 |
| 1,4–Dichlorobutane | 52.1 (11) | 54.8 (14) | 137 | 38 | 40 |
| *n*–Nonane | 9.1 (10) | 48.7 (126) | 86 | 11 | 57 |
| 1,2,4–Trimethylbenzene | 111.7 (3) | 107.4 (12) | 106 | 105 | 101 |
| *o*–Chlorotoluene | 76.4 (27) | 123 (61) | 132 | 58 | 93 |
| *n*–Decane | 4.8 (24) | 52.6 (145) | 88 | 5.5 | 60 |
| 1,2,4–Trichlorobenzene | 145.7 (6) | 154.7 (5) | 174 | 84 | 89 |
| *n*–Dodecane | 3.3 (10) | 29.8 (144) | 72 | 4.6 | 41 |
| *n*–Tetradecane | 2.9 (12) | 30.3 (126) | 91 | 3.2 | 33 |

[a]Mean of two determinations (diluted samples); corrected for unspiked samples and historical recoveries.
[b]Mean of three determinations (two diluted samples and one undiluted sample); corrected for unspiked samples and historical recoveries.
[c]Percent deviation (±) of the two determinations.
[d]CV.

## REFERENCES

1. Garrison, A. W. "The Master Analytical Scheme: An Overview of Interim Procedures," Chapter 2, this volume.
2. Michael, L. et al. "Quality of Master Analytical Scheme Data: Purgeables and Volatile Organic Acids," Chapter 6, this volume.
3. Ryan, J. F. et al. "The Master Analytical Scheme: An Assessment of Factors Influencing Precision and Accuracy of Gas Chromatography/Mass Spectrometry Data," Chapter 4, this volume.
4. Gebhart, J. E. et al. "The Master Analytical Scheme: Development of Effective Techniques for Isolation and Concentration of Organics in Water," Chapter 3, this volume.
5. Eichelberger, J. W., L. E. Harris and W. L. Budde. "Reference Compound to Calibrate Ion Abundance Measurements in Gas Chromatography–Mass Spectrometry Systems," *Anal. Chem.* 47:995 (1975).

# CHAPTER 6

# QUALITY OF MASTER ANALYTICAL SCHEME DATA: PURGEABLES AND VOLATILE ORGANIC ACIDS

**L. C. Michael, R. Wiseman, L. S. Sheldon, J. T. Bursey, K. B. Tomer and E. D. Pellizzari***

Analytical Sciences Division
Chemistry and Life Sciences Group
Research Triangle Institute
Research Triangle Park, North Carolina

**T. A. Scott,** R. Coney and A. W. Garrison**

Analytical Chemistry Branch
Environmental Research Laboratory
U.S. Environmental Protection Agency
Athens, Georgia

**J. E. Gebhart and J. F. Ryan**

Analytical Chemistry Department
Gulf South Research Institute
New Orleans, Louisiana

The overall objective of the Master Analytical Scheme (MAS) is to develop and validate analytical techniques that can be combined into a comprehensive protocol. This protocol provides for qualitative and quantitative analysis of volatile organic compounds in water using gas chromatography/mass spectrometry/computer (GC/MS/COMP). Capillary columns are used for optimum resolution and sensitivity. This scheme includes isolation techniques for three major component categories: purgeables, extractables and intractables. The development of analytical protocols for these classifications has been discussed elsewhere [1,2].

---

*Author to whom correspondence should be addressed.
**Present address: Radian Corp., Sacramento, CA.

Throughout method development, validation and application phases of this program, a high degree of importance has been attached to quality assurance (QA). QA is defined here as those efforts that combine to provide the foundation for the integrity of the analytical data. These include assessment of the data in terms of precision and accuracy. Precision in the context of this program is the agreement between replicate measurements made on a sample either by one laboratory or among several laboratories. Accuracy describes the agreement of a measurement with an accepted or true value.

During method development and validation phases of the MAS, extensive effort was expended on generation of recovery and precision data for each step in the analytical method. The purpose of this approach was to identify recovery-limiting factors, to determine method applicability to different compound classes and sample matrices, and to predict the reliability of the ultimate analytical result. Results of this QA program are discussed for two of the analytical isolation categories: purgeables and volatile organic acids. Purgeables were isolated by purge and trap onto Tenax® GC (Enka N.V., The Netherlands). Volatile acids were isolated by ion exchange. These two very different analytical methods were selected to demonstrate the breadth of the applicability of the MAS.

Another important feature of the MAS is the use of preanalysis (scouting) techniques to determine optimum sample volume and amount of internal standards needed. Two scouting techniques were used. Headspace analysis was used to determine total purgeable component concentration in samples to be analyzed by the purge-and-trap technique. For samples in which volatile organic acids are to be analyzed, conductivity measurements permit selection of an optimum sample volume that will not exceed the capacity of the ion exchange resin. These scouting techniques permit more accurate selection of optimum sample volumes than predictions based on sample type (e.g., drinking water or industrial effluent). Optimum volume selection is especially important for samples to be analyzed with the purge-and-trap technique because no extract exists to permit dilution or concentration for subsequent GC/MS/COMP analysis. This chapter discusses the application of QA techniques to the analysis of environmental samples for purgeables and volatile acids using the MAS.

## EXPERIMENTAL

### Quality Assurance

#### *Method Development*

Method development experiments for purgeables and volatile acids were conducted using $^{14}$C-labeled compounds (methyl bromide, chloroform, thiophene, toluene, bromobenzene, nitrobenzene, acetic acid and hexanoic acid). Recovery was assessed by liquid scintillation counting of 1-mL aliquots of the sample at various stages of analysis. A mass balance was constructed for the entire isolation procedure using distilled water, 10% municipal effluent (in distilled water), 10% energy effluent (in distilled water), 500 ppm activated sludge (in distilled water) and 500 ppm river-bottom particulate (in distilled water). Radiolabeled purgeable compounds were used at approximately 0.1 and 1 ppb in distilled water and 10 ppb in the effluent and particulate matrices to evaluate storage stability, purge efficiency and recovery (the amount desorbed from the Tenax GC cartridge). Recovery of $^{14}C$-labeled compounds from Tenax GC was accomplished by thermal desorption of the cartridges at 260°C directly into scintillation counting cocktail (Figure 1).

Experiments with $^{14}C$-labeled volatile acids at 50 and 250 ppb revealed recoveries for isolation from water, elution from the ion exchange resin and solvent reduction. No attempt was made to determine radioactivity remaining on the resin. All radiolabeled experiments were performed in triplicate, and the data were analyzed statistically.

### *Method Validation*

On completion of method development studies, the purge-and-trap and volatile acid procedures were validated in distilled water, municipal effluent and energy effluent using an extensive list (presented in Results section) of

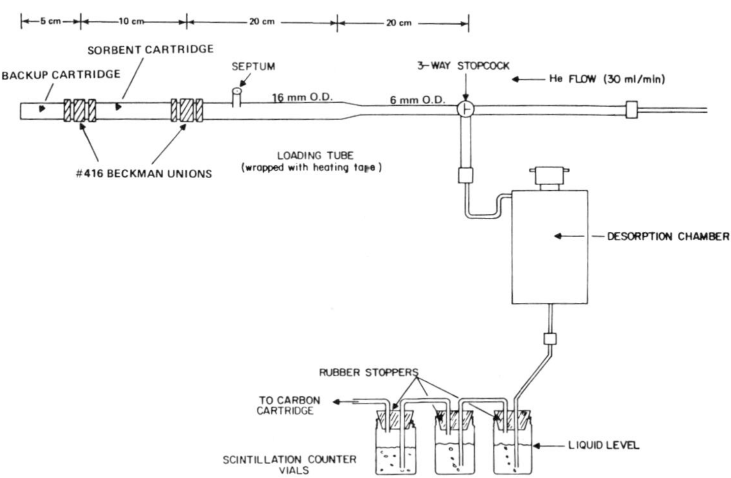


**Figure 1.** Loading apparatus for external standards on Tenax GC cartridge.

model compounds encompassing the range of applicability of the specific technique. The purge-and-trap procedure involved purging 200 mL of distilled water plus 60 g of sodium sulfate at 30°C with 500 mL helium onto a Tenax GC cartridge. Distilled water was spiked at 1 ppb; municipal and energy effluents were spiked at 10 ppb and subsequently diluted to 10% with distilled water. The volatile acid procedure was validated using the same media as for purge-and-trap spiked at 50 and 250 ppb. Volatile acid analysis involved adsorption of spiked compounds in 200 mL of sample on 10 mL Biorad AG 1-X8 resin, elution with 60 mL of 0.67 *N* $NaHSO_4$ in acetonitrile:water (1:2) and distillation of the eluent to concentrate the components. Sufficient 1 *N* $K_2CO_3$ is added to the distillate to raise the pH to 8 or greater. This distillate is subsequently evaporated to dryness using rotary evaporation and nitrogen blow-down and derivatized with pentafluorobenzyl bromide. A detailed description of the purgeable and volatile acid techniques is presented elsewhere [2,3].

These validation experiments provided recovery data for compounds to be used ultimately for correction of GC/MS/COMP quantitative data and provided a measure of precision that could be assimilated into an overall MAS error propagation assessment.

*Method Application*

The application phase involved analysis of ten environmental samples (Table I) by the Research Triangle Institute (RTI). These samples were specifically selected to represent a broad spectrum of water types. Each sample was split into two aliquots, one spiked with model compounds and one unspiked. An interlaboratory comparison study was also conducted in which nine environmental samples (Table II) were analyzed, spiked and unspiked, by three separate laboratories. These laboratories were Research Triangle Institute, Gulf South Research Institute (GSRI) and the U.S. Environmental Protection Agency (EPA) Environmental Research Laboratory in Athens, GA. In both cases, replicate analyses were performed on selected samples.

Environmental water samples were spiked with representative model compounds (Tables III and IV). Samples were spiked at the time of collection or as soon thereafter as possible to most closely mimic actual sampling conditions. The analyst was unaware of the number, identity or quantity of spiked compounds.

Internal standards (Table V), prepared at several concentrations in methanol by the National Bureau of Standards (NBS) in Washington, DC, consisted of representative deuterated compounds. GC was used by NBS to verify the concentration of these standards. Storage experiments were also

Table I. Environmental Samples Analyzed by RTI Using the MAS

| Sample Type | Source | Replicate Analyses |
|---|---|---|
| Distilled Water | Milli-Q system | 2 |
| Drinking Water | RTI cafeteria | 3 |
| Energy Effluent | Oil refinery lagoon | 2 |
| Industrial Effluent | Steel manufacturer | 2 |
| Municipal Effluent | Durham, NC (city) | 2 |
| Municipal Effluent | Durham, NC (county) | 2 |
| Industrial Effluent | Burlington, NC (WWTP influent[a]) | 2 |
| Energy Effluent | In situ coal gasification | 2 |
| Energy Effluent | Low-Btu coal gasification | 2 |
| Industrial Effluent | Soap and detergent manufacturer | 2 |

[a] 85% industrial; 15% municipal.

Table II. Environmental Samples Analyzed by RTI, GSRI and the EPA Athens Enviornmental Research Laboratory

| Sample Type | Source | Replicate Analyses |
|---|---|---|
| Distilled Water | GSRI | 1 |
| Drinking Water | GSRI | 2 |
| Municipal Effluent | Jefferson Parish, LA | 1 |
| Energy Effluent | Oil refinery (location A) | 1 |
| Energy Effluent | Oil refinery (location B) | 1 |
| Industrial Effluent | Aluminum manufacturer (location A) | 2 |
| Industrial Effluent | Aluminum manufacturer (location B) | 1 |
| Surface Water | Bayou St. John | 1 |
| Surface Water | Mississippi River (New Orleans) | 3 |

performed to detect decomposition losses with time. Volatile acid internal standard solutions were contained in sealed ampules and were delivered to the sample (1-4 $\mu$L) by pipet or syringe. Purgeable internal standard solutoins were contained in 1-$\mu$L micropipets sealed in glass tubes (approx. 1.5 x 30 mm). These were delivered to the sample (500 mL) by addition of the entire capsule, which was subsequently smashed by vigorous magnetic stirring.

Table III. Model Compounds Spiked into Purge-and-Trap Samples Analyzed by RTI

| Compound | bp (°C) |
|---|---|
| 1,1-Dichloroethene | 37 |
| Methylene Chloride | 40 |
| *trans*-1,2-Dichloroethene | 48 |
| Bromochloromethane | 49 |
| Acrolein | 53 |
| 1,1-Dichloroethane | 57 |
| Chloroform | 61 |
| Hexane | 68 |
| 1,1,1-Trichloroethane | 74 |
| Carbon Tetrachloride | 77 |
| Benzene | 80 |
| 1,2-Dichloroethane | 83 |
| Trichloroethene | 87 |
| 1,2-Dichloropropane | 96 |
| 1,2-Dichloropropene | 104 |
| Toluene | 111 |
| 1,1,2-Trichloroethane | 113 |
| Tetrachloroethene | 121 |
| Ethylbenzene | 136 |
| 1,1,2,2-Tetrachloroethane | 147 |
| Nonane | 151 |
| Bromobenzene | 156 |
| 1,4-Dichlorobutane | 161 |
| Ethyl Hexanoate | 168 |
| 1,3-Dichlorobenzene | 173 |
| 1,4-Dichlorobenzene | 174 |
| 1,2-Dichlorobenzene | 181 |
| Iodobenzene | 189 |
| 1,4-Dibromobutane | 197 |
| 1,2,4-Trichlorobenzene | 214 |
| 1,2,4,5-Tetrachlorobenzene | 244 |

The external standards, perfluorotoluene (for purgeables) and 4-fluoro-2-iodotoluene (for volatile acids), were added to the Tenax GC cartridge and to the concentrated sample extract, respectively, to assess recovery of the internal standards. Perfluorotoluene (200 ng) was loaded onto the cartridge prior to sample analysis in a helium stream at 30 mL/min using the heated chamber illustrated in Figure 1. 4-Fluoro-2-iodotoluene (200 ng) was introduced to the concentrated volatile acid extract immediately prior to GC/MS/COMP analysis. Tenax GC cartridges were thermally desorbed [2] and the desorbate chromatographed on a 50-m x 0.32-mm i.d. SP-2100 fused-

**Table IV. Model Compounds Spiked into Samples Analyzed for Volatile Acids by RTI**

**Table IV. Model Compounds Spiked into Samples Analyzed for Volatile Acids by RTI**

| | |
|---|---|
| 2-Methylpropanoic Acid | Hexanoic Acid |
| 2-Methylpropenoic Acid | Ethylhexanoic Acid (Isomer) |
| Methacrylic Acid | Cyclohexanoic Acid |
| Butanoic Acid | Octanoic Acid |
| 2-Butenoic Acid | Ethylcyclohexanoic Acid (Isomer) |
| Ethylbutanoic Acid (Isomer) | |

**Table V. Deuterated Internal Standards Used in the MAS**

Purgeables
- Bromoethane-$d_5$
- Diethyl Ether-$d_{10}$
- Anisole-2,4,6,-$d_3$
- $n$-Decane-$d_{22}$
- Naphthalene-$d_8$
- Chlorobenzene-$d_5$

Volatile Acids
- $n$-Butyric Acid-$d_7$

silica capillary column temperature-programmed from 30°C for 2 min to 250°C at 4°C/min. Helium was used as the carrier gas at a flowrate of 2.6 mL/min. Volatile acid extracts were chromatographed on 25-m x 0.32-mm i.d. SP-2100 fused-silica capillary column (Hewlett-Packard) temperature-programmed from 70 to 250°C at 4°C/min, with a helium flow of 2.6 mL/min.

## Sample Scouting

Samples intended for purge-and-trap analysis were subjected to a head-space scouting technique to assess the approximate total purgeable component concentration. In the procedure, a 10-mL aliquot of each sample collected for scouting was transferred to a 3-dram (12-mL) septum-capped vial containing 3 g of anhydrous sodium sulfate. After dissolving the sodium sulfate and equilibrating at 50°C for 1 hr, 100 $\mu$L of headspace was withdrawn and analyzed by GC with flame ionization detection (FID). The column (180 cm x 0.2 cm i.d., 2% OV-17 on 80/100 Gas Chrom Q) was temperature-programmed from 30 to 200°C at 20°C/min after a 5-min initial hold. A helium carrier gas flow of 30 mL/min was used. The total chromatographic peak area was compared to that resulting from headspace analysis of standard solutions with individual volatile component concentrations of 0.01, 0.1, 1 and 10 ppm. Standard solution components were selected to include a range of compound classes and volatilities (Table VI). A calibration curve, presented in Figure 2, was prepared concurrently with sample scouting by summing the individual peak areas from a chromatogram obtained at a given concentration using a Varian CDS 111 chromatography system operated in the $10^{-11}$-A full-scale range. This calibration curve, as well as the degree of sample dilution required to provide successful GC/MS/COMP analysis of the actual sample, must be determined by each laboratory performing the sample analysis. Sample dilution was determined relative to the dynamic concentration range of the mass spectrometer being employed. In our case this range is approximately 5–1000 ng. For a 200-mL sample, this represents a purgeable concentration range of 0.025–5 ppb per component or 2.5–500 ppb total for 100 components. Samples requiring dilutions were diluted to provide a total purgeable concentration of 50–100 ppb in the final 200-mL volume purged.

Conductivity measurements were performed on samples to be analyzed for volatile acids using a conductivity cell and bridge (Model RC 16B2, Industrial Instruments, Inc., Cedar Grove, NJ). The maximum sample size,

**Table VI. Standard Mixture for Calibrating Headspace Analysis**

| Compound | bp (°C) | Class |
|---|---|---|
| Diethyl Ether | 35 | Aliphatic ether |
| Chloroform | 61 | Aliphatic halogenated hydrocarbon |
| Methyl Ethyl Ketone | 80 | Ketone |
| Thiophene | 84 | Sulfur-containing |
| Toluene | 111 | Aromatic hydrocarbon |
| Ethyl Butyrate | 120 | Aliphatic ester |
| Octane | 125 | Aliphatic hydrocarbon |
| Chlorobenzene | 132 | Aromatic halogenated hydrocarbon |
| Anisole | 156 | Aromatic ether |
| Decane | 174 | Aliphatic hydrocarbon |
| 1,4-Bromobutane | 197 | Aliphatic halogenated hydrocarbon |
| Naphthalene | 218 | Aromatic hydrocarbon |

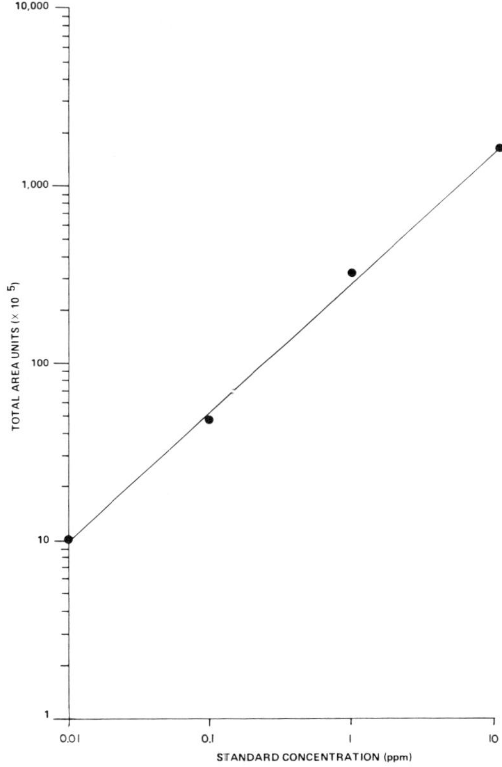


**Figure 2.** Calibration curve for headspace scouting.

and hence the detection limit (Table VII) was dictated by the sample conductivity because the equivalents of ion exchange resin were constant for all water types.

## Sample Analysis

Samples were analyzed for purgeables and volatile acids using a detailed analytical protocol presented elsewhere [2].

Table VII. Acceptable Sample Size and Corresponding Detection Limits for Sample Conductivity Ranges in Volatile Acid Analysis

| Conductivity Range (ohm/cm) | Sample Size[a] (mL) | Detection Limit (ppb) |
|---|---|---|
| 150-300 | 1,000 | 3 |
| 300-600 | 500 | 6 |
| 600-1,200 | 250 | 12 |
| 1,200-3,000 | 100 | 30 |
| 3,000-6,000 | 50 | 60 |
| 6,000-12,000 | 25 | 120 |
| 12,000-30,000 | 10 | 300 |
| 30,000-60,000 | 5 | 600 |

[a] Based on a 10-mL volume of ion exchange resin.

## RESULTS AND DISCUSSION

### Quality Assurance

Recoveries of $^{14}$C-labeled compounds from distilled water, municipal effluent and two particulate types by purge-and-trap are shown in Figure 3. Distilled water was examined at two concentrations, 1 and 0.1 ppb (limit of detection). For all compounds except bromobenzene and in all media tested, recoveries are comparable. The suppressed recovery of bromobenzene at the limit of detection reflects its lower volatility and increased tendency for adsorption. This probability for adsorption is further indicated by the low recovery of bromobenzene from 500-ppm activated sludge and to a lesser extent from 500-ppm river-bottom particulate. Activated sludge severely affects the recovery of methyl bromide, a compound known to be unstable in water. Total radioactivity, accounted for on the Tenax GC and in the purged water, was approximately 90–100%. The percentage that was unaccounted for was presumed to be lost as a result of leaks in the apparatus, nonquantitative desorption from the Tenax GC cartridge and isotopic impurity of the radiolabeled compound.

Figure 4 shows recoveries for $^{14}$C-labeled acetic and hexanoic acids in the same matrices as the purgeable compounds (except for distilled water at the limit of detection). Recoveries for acetic acid show significant variation with water type and are particularly low in the municipal effluent and activated sludge media. Furthermore, it appears that a significant portion of

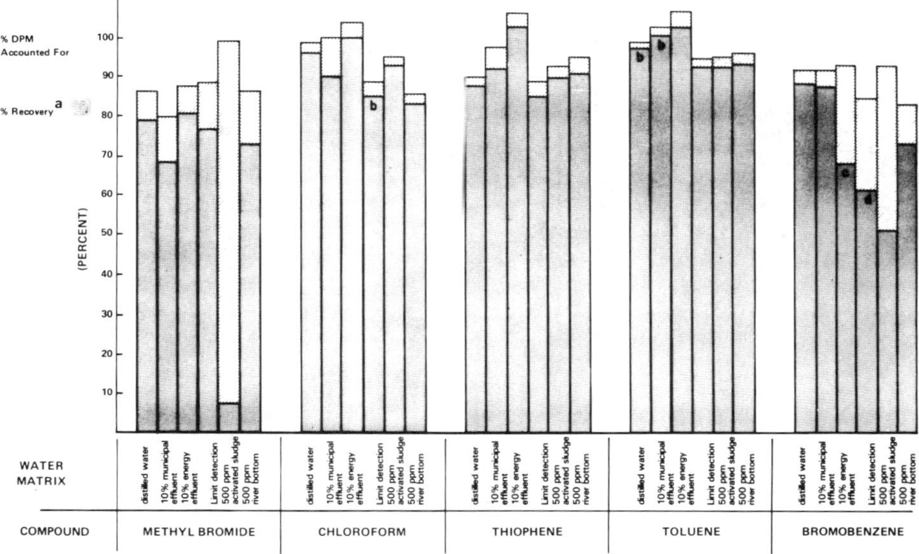


**Figure 3.** Summary of recoveries for radiolabeled purgeable compunds in various water matrices. Spiking levels: distilled water, 1 ppb; 10% municipal effluent, 10 ppb; 10% energy effluent, 10 ppb; limit of detection, 0.2 ppb; 500 ppm activated sludge, 10 ppb; 500 ppm river-bottom, 10 ppb. (a) Mean of triplicate determinations; (b) mean of duplicate determinations; (c) mean of six determinations; (d) single determination.

the acetic acid was retained by the ion exchange resin in these two media since approximately 25% of the radioactivity is unaccounted for. This is also observed for hexanoic acid in all media studied.

Recovery data representing a compilation of recoveries for nonradiolabeled purgeables and volatile acids in distilled water, municipal effluent and energy effluent are shown in Tables VIII and IX, respectively. Although individual compound recoveries vary from undetected to more than 100%, the precision associated with the individual recovery values is ±18% or less. Recoveries of aldehydes and ketones (Table VIII) were not determined in municipal and energy effluent because corresponding values in distilled water were very low. Recoveries of purgeable compounds, in general, increase with increasing volatility and decreasing water solubility: aldehydes and ketones are recovered poorly, esters and ethers show intermediate recovery (20–70%), and the hydrocarbons and halogenated hydrocarbons are recovered at greater than 80%. With only a few exceptions, volatile acids (Table IX) were recovered at 50% or greater, with a mean recovery of 83% for all compounds in all media.

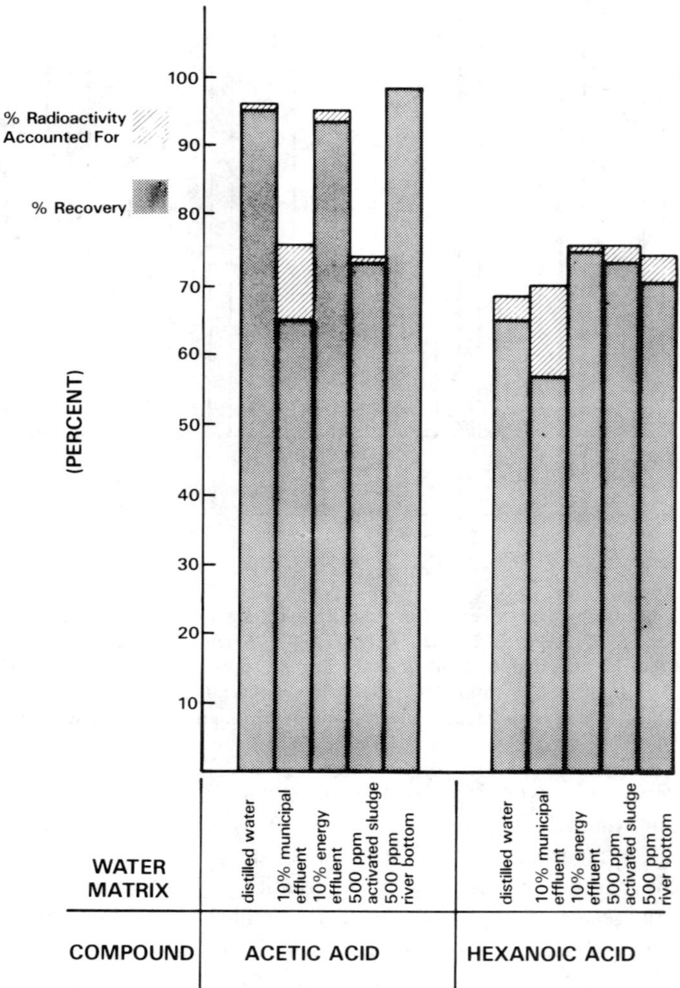


**Figure 4.** Summary of recoveries for radiolabeled volatile acids (2.5 ppb) in various water matrices.

Storage of solutions of radiolabeled compounds in water showed that extended storage can significantly affect recovery of purgeable compounds (Table X). Consequently, samples to be analyzed for purgeables were stored for a minimum period, generally less than five days. This decision was not based on storage data for a 6-day period but was made by assuming a constant loss rate from 0 to 3 weeks and a maximum acceptable storage loss of 15%.

Table VIII. Summary of Recovery Data for Nonradiolabeled Purgeable Compounds in Various Water Matrices[a,b]

| Class | Compound | bp (°C) | Mol Wt | Density (g/mL) | Water Solubility[a] | % Recovery ± SD (CV)[b,c] Distilled Water | Municipal Effluent (10%) | Energy Effluent (10%) |
|---|---|---|---|---|---|---|---|---|
| Aldehydes | Propionaldehyde | 40 | 58 | 0.8071 | 1/5 | ND[d] | | |
| | *n*-Butyraldehyde | 75 | 72 | 0.8016 | 7% | 13 ± 2 (14) | | |
| | Crotonaldehyde | 104 | 70 | 0.853 | 18% | ND[d] | | |
| | Heptaldehyde | 153 | 114 | 0.809 | S | 34 ± 6 (16) | | |
| | Furfural | 162 | 96 | 1.5261 | 1/11 | ND[e] | | |
| | Benzaldehyde | 179 | 106 | 1.043 | 1/350 | ND[e] | | |
| | Salicylaldehyde | 196 | 122 | 1.167 | S | ND[e] | | |
| | *p*-Tolualdehyde | 200 | 120 | 1.5430 | | ND[e] | | |
| | Anisaldehyde | 248 | 136 | 1.119 | S | ND[e] | | |
| Ketones | Acetone | 57 | 58 | 1.3591 | MIS | ND[d] | ND[d] | |
| | Methyl ethyl ketone | 80 | 72 | 0.805 | 12.5% | ND[d] | | |
| | Cyclopentanone | 131 | 84 | 0.9509 | I | ND[e] | | |
| | Acetylacetone | 141 | 100 | 0.976 | 1/8 | ND[e] | | |
| | 2-Heptanone | 151 | 114 | 0.8111 | VS | 53 ± 5 (10)[f] | | |
| | Cyclohexanone | 157 | 98 | 0.9978 | SS | 53 ± 5 (10)[f] | | |
| | 2-Octanone | 173 | 128 | 0.8185 | SS | ND[e] | | |
| | Fenchone | 194 | 152 | 0.948 | I | ND[e] | | |
| | Acetophenone | 202 | 120 | 1.5339 | SS | ND[d] | | |
| | Phenyl acetone | 217 | 134 | 1.0157 | I | ND[e] | | |
| | *trans*-4-Phenyl-3-butene | | | | | ND[e] | | |
| Esters | Methyl formate | 32 | 60 | 0.9742 | VS | 36 ± 11 (31)[g] | 95 ± 24 | 110 ± 25 (23) |
| | Methyl acetate | 57 | 74 | 0.9723 | VS | 23 ± 5 (24) | ND[d] | ND[d] |

**Table VIII, continued**

| Class | Compound | bp (°C) | Mol Wt | Density (g/mL) | Water Solubility[a] | % Recovery ± SD (CV)[b, c] Distilled Water | Municipal Effluent (10%) | Energy Effluent (10%) |
|---|---|---|---|---|---|---|---|---|
| | Ethyl acetate | 77 | 88 | 0.9005 | S | 23 ± 2 (9) | 29 ± 4 | 30 ± 9 (31) |
| | *t*-Butyl acetate | 98 | 116 | 0.8620 | VSS | 34 ± 2 (3)[f] | 98 ± 12 (13) | 74 ± 19 |
| | Propyl acetate | 102 | 102 | 0.8884 | SS | 34 ± 2 (7)[f] | 64 ± 13 | 58 ± 14 (23)[h] |
| | Allyl acetate | 104 | 100 | 0.9276 | SS | 34 ± 2 (7) | 22 ± 3 | 26 ± 8 (31) |
| | Ethyl butyrate | 120 | 116 | 0.879 | 0.7% | 90 ± 4 | 82 ± 3 | 85 ± 8 (9) |
| | Butyl propionate | 146 | 130 | 0.8818 | SS | 126 ± 2 | 94 ± 3 | 94 ± (4) |
| | Ethoxyethyl acetate | 156 | 132 | 0.9749 | US | ND[e] | ND[e] | ND[e] |
| | Ethyl hexanoate | 168 | 144 | 0.8710 | I | 108 ± 13 | 95 ± 3 | 100 ± 3 (3) |
| | Furfuryl acetate | 177 | 140 | 1.1175 | I | ND[e] | ND[e] | ND[e] |
| | Diethyl oxalate | 186 | 146 | 1.0785 | SS | ND[e] | ND[e] | ND[e] |
| | Phenyl acetate | 196 | 136 | 1.0927 | S | ND[e] | ND[e] | ND[e] |
| | Dimethyl adipate | ≫200 | 174 | 1.0600 | I | 20 ± 5 (26) | ND[e] | ND[e] |
| | Benzyl acetate | 216 | 150 | 1.0563 | S | 94 ± 1 (10) | 4 ± 1 [f,g,h] | 5 ± 1 (20) |
| Ethers | Diethyl ether | 34.6 | 74 | 0.7077 | 6% | 100 ± 14 (14) | 100 ± 54 (49) | ND[d] |
| | Propylene oxide | 35 | 58 | 0.859 | 1% | 12 ± 2 (12) | | |
| | 2-Methylfuran | 63 | 82 | 0.9132 | I | ND[e] | 28 ± 26 (90) | 87 ± 106 (121) |
| | Tetrahydrofuran | 66 | 72 | 1.4070 | MIS | 24 ± 10 (39) | 24 ± 5 (22) | 11 ± 1 (10) |
| | Isopropyl ether | 69 | 102 | 0.7241 | SS | ND[d] | 109 ± 12 (11) | 100 ± 4 (4) |
| | Allyl ether | 94 | 98 | 0.805 | I | ND[d] | 100 ± 8 (8) | 98 ± 10 (11) |
| | 1,4-Dioxane | 101 | 88 | 1.0329 | S | ND[d] | 6 ± 3 (40) | 3 ± 0.4 (12) |
| | Epichlorohydrin | 118 | 92 | 1.1750 | I | 26 ± 2 (9) | 3 ± 1 (33) | 4 ± 0.6 (16) |
| | Butyl ether | 142 | 130 | 0.769 | I | ND[d] | ND[d] | ND[d] |
| | Anisole | 156 | 108 | 0.98 | I | 76 ± 5 (7) | 83 ± 6 (7) | 65 ± 6 (10) |
| | Hexyl ether | 223 | 186 | 0.7936 | I | 66 ± 15 (23) | 40 ± 5 (12) | 45 ± 5 (11) |

| | | | | | | | | | | | |
|---|---|---|---|---|---|---|---|---|---|---|---|
| | Phenyl ether | 259 | 170 | 1.075 | I | 97 ± 14 | (14) | 70 ± 4 | (6) | 85 ± 4 | (4) |
| | Benzyl ether | 298 | 198 | 1.0428 | I | 78 ± 92 | (116) | 5 ± 3 | (50) | 10 ± 6 | (60) |
| Aliphatic | Pentane | 36 | 72 | 0.6262 | VS | 185 ± 108 | (58) | ND[d,i] | | ND[d] | |
| Hydrocarbons | Cyclopentane | 49 | 70 | 0.7510 | I | 60 ± 29 | (48)[i] | ND[c] | | | |
| | | | | | | 115 ± 12 | (10) | | | 109 ± 15 | (14) |
| | Hexane | 68 | 86 | 0.6594 | I | 47 ± 17 | (36) | ND[d,i] | | | |
| | | | | | | 97 ± 7 | (7) | | | 124 ± 32 | (26) |
| | Cyclohexane | 83 | 82 | 0.8110 | I | 76 ± 26 | (35) | | | | |
| | | | | | | 81 ± 9 | (11) | 124 ± 5[i] | | 115 ± 116 | (14) |
| | Heptane | 98 | 100 | 0.6838 | I | 50 ± 15 | (31) | | | | |
| | | | | | | 110 ± 11 | (10) | 121 ± 33[i] | | 130 ± 13 | (10) |
| | 1-Octane | ≅122 | 112 | 0.715 | I | 104 ± 7 | (7)[f] | 110 ± 23[i] | | 143 ± 29 | (20) |
| | Octane | 125 | 114 | 0.7025 | I | 104 ± 7 | (7)[f] | 98 ± 23 | (23)[i] | 102 ± 23 | (22) |
| | Nonane | 151 | 143 | 0.7176 | I | 40 ± 12 | (29) | | | | |
| | | | | | | 57 ± 7 | (11) | 70 ± 3 | (5)[i] | 70 ± 14 | (20) |
| | Dipentene | ≅170 | 138 | 0.75 | SS | 61 ± 18 | (29)[g] | | | | |
| | | | | | | 87 ± 9 | (11) | 92 ± 5 | (5)[i] | 79 ± 9 | (12) |
| | Decane | 174 | 142 | 0.951 | SS | 23 ± 8 | (34) | | | | |
| | | | | | | 38 ± 8 | (21) | 66 ± 4 | (7)[i] | 66 ± 12 | (17) |
| | Dodecane | ≅216 | 170 | 0.7847 | I | 53 ± 21 | (40) | | | | |
| | | | | | | 48 ± 15 | (30) | 64 ± 2 | (4)[i] | 65 ± 6 | (9) |
| | Tetradecane | 254 | 198 | 0.7627 | I | 12 ± 2 | (12) | | | | |
| | | | | | | 27 ± 11 | (40) | 22 ± 5 | (22)[i] | 36 ± 4 | (10) |
| | Hexadecane | 287 | 266 | 0.7733 | I | ND | | 5 ± 1 | (25)[i] | 10 ± 3 | (26) |
| Halogenated | Methylene chloride | 40 | 84 | 1.335 | SS | 99 ± 33[g,k] | | 105 ± 15 | (14)[g] | | |
| Aliphatic | Allyl chloride | 45 | 76 | 0.9397 | I | 113 ± 12 | | ND[d] | | ND[d] | |
| | *trans*-1,2-Dichloroethylene | 48 | 96 | 1.2565 | SS | 116 ± 15.0 | | 103 ± 10 | (10) | 106 ± 5 | (5) |
| | Chloroform | 61 | 119 | 1.484 | 0.5% | 79 ± 11[f] | | 105 ± 18 | (18) | 95 ± 6 | (7) |
| | Bromochloromethane | 68 | 129 | 1.991 | I | 79 ± 11[f] | | | | 110 ± 20 | (18) |
| | 1,2-Dichloroethane | 84 | | | | 80 ± 7 | | 82 ± 5 | (6) | | |

Table VIII, continued

| Class | Compound | bp (°C) | Mol Wt | Density (g/mL) | Water Solubility[a] | % Recovery ± SD (CV)[b,c] Distilled Water | | Municipal Effluent (10%) | | Energy Effluent (10%) | |
|---|---|---|---|---|---|---|---|---|---|---|---|
| | Trichloroethylene | 87 | 131 | 1.462 | SS | 87 ± 2 | | 84 ± 10 | (12) | 96 ± 7 | (7) |
| | 1,2-Dichloropropane | 96 | 113 | 1.1558 | SS | 84 ± 11 | | 111± 12 | (11) | 97 ± 5 | (5) |
| | 1,1,2-Trichloroethane | 113 | 133 | 1.4405 | SS | 86 ± 6 | | 87 ± 3 | | 85 ± 2 | (2) |
| | 1-Bromo-3-methylbutane | 122 | 151 | 1.2609 | I | 100 ± 12 | | 101 ± 4 | (4) | 113 ± 4 | (12) |
| | 1,2-Dibromoethane | 131 | 188 | 2.1792 | SS | 90 ± 5 | | 80 ± 5 | (6) | 96 ± 6 | (7) |
| | 1-Chlorohexane | 132 | 120 | 0.8784 | I | 100 ± 7 | | | | 103 ± 10 | (10) |
| | 1,2-Dibromopropane | 141 | 202 | 1.9366 | I | | | 99 ± 3 | (3) | 98 ± 5 | (5) |
| | 3-Bromo-1-chloropropane | 142.5 | 157 | 1.537 | I | | | 101 ± 2 | (2) | 102 ± 4 | (4) |
| | 1,4-Dichlorobutane | 161.3 | 127 | 1.1408 | I | 86 ± 2 | | 91 ± 2 | (2) | 91 ± 5 | (6) |
| | 1,4-Dibromobutane | 197 | 216 | 1.808 | I | 92 ± 10 | | 86 ± 4 | (4) | | |
| | 1-Bromodecane | 238 | 221 | 1.066 | I | 27 ± 4 | | 45 ± 5 | (12) | 38 ± 11 | (30) |
| Aromatics | Benzene | 80 | 78 | 0.8787 | 1/1430 | 76 ± 7 | (9) | 104 ± 5 | | 111 ± 8 | (7) |
| | Toluene | 111 | 92 | 0.866 | VSS | 74 ± 11 | (15) | 103 ± 4 | | 107 ± 6 | (5) |
| | Ethylbenzene | 136 | 106 | 0.866 | I | 66 ± 10 | (15)[f] | 99 ± 4 | | 101 ± 4 | (4) |
| | Xylene | 139 | 106 | 0.8684 | I | 66 ± 10 | (15)[f] | 100 ± 1 | | 100 ± 5 | (5) |
| | Cumene | 152 | 120 | 0.864 | I | 59 ± 12 | (20) | 99 ± 1 | | 96 ± 9 | (10) |
| | *t*-Butylbenzene | 134 | | 0.8671 | I | ND[h] | | ND[h] | | 97 ± 10 | (11)[f] |
| | 1,2,4-Trimethylbenzene | 170 | 120 | 0.89 | I | 84 ± 8 | | 99 ± 2 | (2) | 97 ± 10 | (11)[f] |
| | Diethylbenzene | ≅182 | 134 | 0.87 | I | 60 ± 10 | (16) | 86 ± 1 | | 81 ± 12 | (15) |
| | Triethylbenzene | 217 | 162 | 0.87 | I | 42 ± 10 | (24) | 68 ± 2 | | 66 ± 14 | (22) |
| | Naphthalene | 218 | 128 | 1.162 | I | 83 ± 7 | | 85 ± 2 | (2) | | |
| | Diphenylmethane | 266 | 168 | 1.0008 | I | 63 ± 6 | (9) | 66 ± 3 | | 64 ± 14 | (22) |
| Halogenated | Fluorobenzene | 85 | 96 | 1.0244 | I | 91 ± 9[k] | | 107 ± 5 | (5) | 101 ± 9 | |

| | | | | | | | | | | |
|---|---|---|---|---|---|---|---|---|---|---|
| Aromatics | α,α,α-Trifluorotoluene | 103 | 146 | 1.1886 | I | 83 ± 10 | | 108 ± 8 | (8) | 101 ± 9 |
| | Chlorobenzene | 132 | 112 | 1.1064 | I | 93 ± 7 | | 101 ± 4 | (4) | 95 ± 3 |
| | Bromobenzene | 156 | 157 | 1.5219 | I | 92 ± 4 | | 95 ± 2 | (2) | 91 ± 6 |
| | *p*-Bromotoluene | 184 | 171 | 1.3898 | I | 92 ± 2 | | 91 ± 3 | (4) | 86 ± 12 |
| | Iodobenzene | 189 | 204 | 1.8230 | I | 92 ± 4 | | 91 ± 3 | (3) | 87 ± 8 |
| | 1,2,4-Trichlorobenzene | 214 | 181 | 1.4542 | I | 103 ± 7 | | 84 ± 7 | (8)[n] | 93 ± 1 |
| | α,α,α-Trichlorotoluene | 221 | 195 | 1.3723 | I | | | | | |
| | 1,2,4,5-Tetrachlorobenzene | 244 | 216 | 1.858 | I | 79 ± 7 | | 56 ± 7 | (12) | |
| Sulfur (Misc.) | Carbon disulfide | 47 | 76 | 1.2632 | I | 93 ± 27 | | 93 ± 14 | (15)[g] | |
| | Thiophene | 84 | 81 | 1.0573 | I | 64 ± 8 | | | | |
| | *t*-Butyl disulfide | ≅200 | 178 | 0.909 | | ND[e] | | | | |
| | Benzyl sulfide | 296 | 186 | 1.118 | I | | | | | |
| Nitrogen (Misc.) | Propionitrile | 97 | 55 | 0.7818 | 12% | 25 ± 8 | (34) | | | |
| | Nitroethane | 114 | 75 | 1.041 | 4.5 | 6 ± 1 | (7) | | | |
| | 1-Nitropropane | 132 | 89 | 0.9934 | 1.59 | 9 ± 0.3 | (4) | | | |
| | Benzonitrile | 191 | 103 | 1.010 | SS | ND[e] | | | | |
| | Nitrobenzene | 210 | 123 | 1.204 | 0.2% | 30 ± 2 | (7)[g] | | | |

[a] Abbreviations: bp = boiling point; mol wt = molecular weight; SD = standard deviation; CV = coefficient of variation; S = soluble; MIS = miscible; I = insoluble; VS = very soluble; SS = slightly soluble; VSS = very slightly soluble.

[b] Based on triplicate determinations; omission of coefficient of variation indicates duplicate data. Spiking levels: distilled water, 1 ppb; 10% municipal effluent, 10 ppb; 10% energy effluent, 10 ppb.

[c] Chromatographic conditions: 30°C for 5 min; 30–230°C at 4°C/min.

[d] Recovery not determined–severe background interference.

[e] Recovery not determined–no peak detected in purge.

[f] Composite recovery–coeluting peaks.

[g] Slight background interference.

[h] Poor peak resolution.

[i] Exposed cartridges stores 10 days at 0°C; analysis on 30-m x 0.25 mm OV-101 WCOT column.

[j] Top value includes transfer–bottom value does not.

[k] Includes transfer (pouring) to purge flask.

[l] Compound impure–several peaks observed.

**Table IX. Summary of Recovery Data for Nonradiolabeled Volatile Acids in Various Water Matrices**

| Acid | Matrix | % Recovery ± SD (CV) 50 ppb | % Recovery ± SD (CV) 250 ppb |
|---|---|---|---|
| Butyric | Distilled water | 125 ± 8 (6) | 90 ± 15 (17) |
| | Municipal effluent | 65 ± 14 (22) | 95 ± 17 (18) |
| | Energy effluent | 103 ± 11 (11) | 93 ± 12 (13) |
| Crotonic | Distilled water | 83 ± 10 (12) | 74 ± 16 (22) |
| | Municipal effluent | 65 ± 17 (26) | 88 ± 16 (18) |
| | Energy effluent | 95 ± 17 (18) | 85 ± 12 (14) |
| Hexanoic | Distilled water | 78 ± 22 (28) | 61 ± 6 (10) |
| | Municipal effluent | 43 ± 20 (47) | 56 ± 11 (20) |
| | Energy effluent | 75 ± 24 (32) | 60 ± 9 (15) |
| Isobutyric | Distilled water | 105 ± 9 (9) | 79 ± 12 (15) |
| | Municipal effluent | 93 ± 12 (13) | 87 ± 13 (15) |
| | Energy effluent | 98 ± 13 (13) | 81 ± 9 (11) |
| Trimethylacetic | Distilled water | 97 ± 6 (6) | 66 ± 5 (8) |
| | Municipal effluent | 77 ± 8 (10) | 66 ± 11 (17) |
| | Energy effluent | 78 ± 8 (10) | 58 ± 5 (9) |
| Methacrylic | Distilled water | 92 ± 7 (8) | 78 ± 12 (15) |
| | Municipal effluent | 81 ± 8 (10) | 84 ± 6 (7) |
| | Energy effluent | 86 ± 11 (13) | 76 ± 6 (8) |
| Ethyl butyric | Distilled water | 64 ± 3 (5) | 59 ± 4 (7) |
| | Municipal effluent | 63 ± 6 (10) | 56 ± 8 (14) |
| | Energy effluent | 72 ± 11 (15) | 49 ± 6 (12) |
| Ethyl hexanoic | Distilled water | 71 ± 6 (8) | 47 ± 4 (9) |
| | Municipal effluent | 48 ± 7 (15) | 40 ± 6 (15) |
| | Energy effluent | 72 ± 5 (7) | 40 ± 6 (15) |

**Table X. Storage Study[a] with Radiolabeled Compounds**

| Compound | Storage Period (weeks) | Mean % Recovery ± SD (CV) |
|---|---|---|
| Acetone | 2 | 98 ± 2 (2) |
| Naphthalene | 2 | 109 ± 6 (5) |
| Acetonitrile | 3 | 109 ± 2 (2) |
| Chloroform | 3 | 57 ± 20 (35) |
| Toluene | 3 | 55 ± 19 (35) |
| Bromobenzene | 3 | 94 ± 3 (3) |

[a] Determined by removal of sample aliquots followed by liquid scintillation counting.

Sample storage and transfer steps in purge-and-trap have only negative deviations because they are susceptible only to losses, not to gains. The overall error of +23 to -28% for purge-and-trap (Table XI) reflects errors associated with addition of model compounds and internal standards, and the determination of recoveries from storage, transfer and the purge-and-trap step itself. Volatile acid studies (Table XII) show an overall error of ±11%. Storage studies were not performed with this group of compounds, but associated losses are not considered to be as significant as with the purgeables. Sample volume measurement is less precise with purgeables because of the nature of the delivery process [2].

Chromatograms of replicate purge-and-trap analyses of spiked drinking water and unspiked municipal influent (Figures 5 and 6) qualitatively illustrate analytical precision through similarity of chromatographic profiles. Each of these chromatograms represents analysis of samples from separate containers. Ion current intensities are normalized with the most intense peak at 100%. Peak intensities of multiple chromatograms displayed in a single figure are normalized to the top chromatogram.

**Table XI. Estimation of Error Propagated in Purge-and-Trap Analysis (Excluding GC/MS/COMP)**

| Operation | Relative SD (%) |
|---|---|
| Sample Collection Volume | ±1 |
| Internal Standard Addition (Concentration) | ±5 |
| (Delivery) | ±5 |
| Model Compound Addition (Concentration) | ±5 |
| (Delivery) | ±1 |
| Sample Storage | -10 |
| Analysis Volume Measurement | ±2 |
| Transfer to Purge Flask | -13 |
| Recovery[a] | ±18 |
| External Standard Addition (Concentration) | ±2 |
| (Delivery) | ±10 |
| Overall[b] | +23; -28 |

[a] Includes purge efficiency, trapping efficiency and recovery from trap.

[b] Calculated according to the formula: $RSD_T = \left( \sum_{i=1}^{n} (RSD_i)^2 \right)^{1/2}$.

**Table XII. Estimation of Error Propagated in Volatile Acid Analysis (Excluding GC/MS/COMP)**

| Operation | Relative SD (%) |
|---|---|
| Sample Collection Volume | ±1 |
| Internal Standard Addition | ±3 |
| Model Compound Addition (Concentration) | ±1 |
| (Delivery) | ±1 |
| Sample Storage | ±2[a] |
| Analysis Volume Measurement | ±1 |
| Recovery[b] | ±10 |
| External Standard Addition (Concentration) | ±1 |
| (Delivery) | ±1 |
| Overall | ±11 |

[a] Estimated.
[b] Steps including extraction through derivatization.

## Sample Scouting

Headspace scouting analysis showed that a municipal effluent sample required no dilution, but an energy (coal gasification) effluent required dilution by a factor of 50 prior to purge-and-trap analysis (Figure 7). Total peak area expressed in peak integrator units for samples to be analyzed was determined by headspace scouting analysis (Table XIII). Each of these samples was spiked with internal standards prior to headspace analysis. Approximate concentrations of sample components were estimated from the calibration curve (Figure 2) and appropriate dilutions were made to give approximately 50–100 ppb total purgeable sample component concentration before purging. The high area count for distilled water is accounted for in four peaks which elute early in the chromatogram. This reflects background from the syringe used for the headspace sample as well as spiked internal standards in the distilled water itself.

Conductivity measurements (Table XIV) on samples to be analyzed for volatile acids were used to determine maximum acceptable sample size according to previously established criteria (Table VII). No measurements were made on distilled and drinking water as these media were not expected to contain significant ion content.

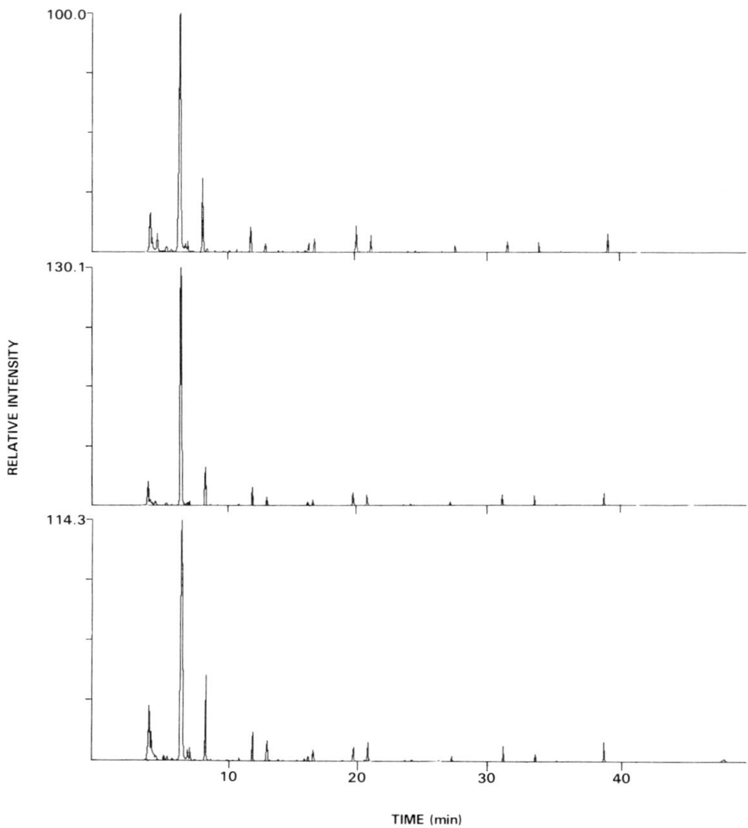


**Figure 5.** Replicate analyses of drinking water (spiked).

## Sample Analysis

A primary objective of the MAS application phase was to accurately detect compounds spiked into various water types. Identification of sample components, both spiked and endogenous, was achieved by a combination of computerized mass spectral search systems and manual interpretation. Chromatograms of municipal effluent, municipal influent, industrial effluent and energy effluent (Figures 8 to 11) illustrate the utility of both the purge-and-trap and volatile acid isolation techniques of the MAS for diverse and complex water types.

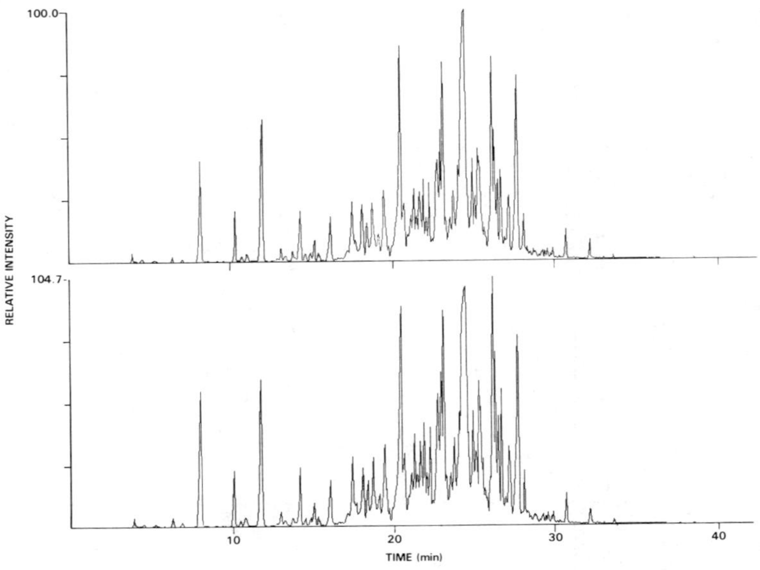


**Figure 6.** Replicate analyses of municipal wastewater influent (unspiked).

## SUMMARY

Quality assurance procedures were applied to method development, validation and application phases for purgeable and volatile acid segments of a MAS for the analysis of organic compounds in water. Recovery data demonstrate the applicability of the various techniques to diverse water types. Error propagation estimates for purge-and-trap indicate a relative standard deviation of +23 to –28%, excluding GC/MS/COMP. A similar estimation for volatile acid analysis reveals a relative standard deviation of ±11%.

For each sample to be analyzed, preanalysis sample scouting (headspace analysis for purgeables and conductivity measurements for volatile acids) was used to determine the optimum sample volume and level of internal standard to be added.

GC/MS/COMP analysis was performed to demonstrate the applicability of the purge-and-trap and volatile acid MAS protocols to complex and diverse water types. A total of 19 environmental water samples, unspiked and spiked with model compounds and deuterated internal standards, were analyzed. The results of some of these analyses are presented elsewhere [5].

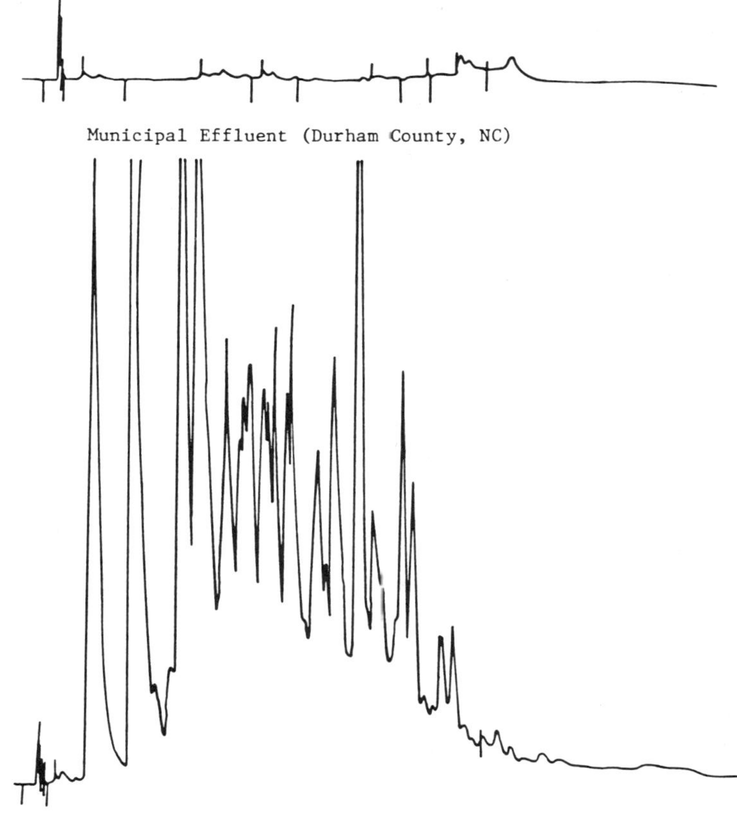


**Figure 7.** Headspace analysis of municipal and energy effluent.

## ACKNOWLEDGMENT

The authors acknowledge support for this work under EPA Contract No. 68-03-2704. Technical assistance provided by Ms. J. Storm, Ms. S. Lee, Mr. J. Turlington, Mr. R. Porch, Ms. B. Bickford, Ms. D. DiStefano, Mr. P. Dodd, Mr. F. McKinney and Mr. M. Parker of the Research Triangle Institute;

Table XIII. Headspace Scouting of Environmental Water Samples

| Sample | Source | Integrator Units |
|---|---|---|
| Distilled Water | GSRI | 275,000 |
| Drinking Water | GSRI | 3,260,000 |
| Municipal Effluent | Jefferson Parish, LA | 22,000,000 |
| Energy Effluent | Oil refinery (location A) | 21,300,000 |
| Energy Effluent | Oil refinery (location B) | 25,500,000 |
| Industrial Effluent | Aluminum manufacturer (location A) | 347,000 |
| Industrial Effluent | Aluminum manufacturer (location B) | 173,000 |
| Surface Water | Bayou St. John | 124,000 |
| Surface Water | Mississippi River | 11,200,000 |
| Municipal Effluent | Durham, NC (county) | 134,000 |
| Industrial Effluent | Burlington, NC (county) (treatment plant influent) | 10,700,000 |
| Energy Effluent | Low-Btu coal gasification | 32,600,000 |
| Energy Effluent | In situ coal gasification | 74,000,000 |
| Standard (0.01 ppm) | | 1,014,000 ± 525,000 (52) |
| Standard (0.1 ppm) | | 4,850,000 ± 1,010,000 (2) |
| Standard (1 ppm) | | 33,400,000 ± 991,000 (3.0) |
| Standard (10 ppm) | | 169,000,000 ± 34,600,000 (2.0) |

Table XIV. Conductivity Measurements Made on Environmental Samples Analyzed by RTI Using the MAS

| Sample Type | Source | Conductivity (ohm/cm) | Maximum Sample Volume (mL) |
|---|---|---|---|
| Distilled Water | Milli-Q system | <150 | 2,500 |
| Drinking Water | RTI cafeteria | <150 | 2,500 |
| Energy Effluent | Oil refinery lagoon | 6,400 | 25 |
| Industrial Effluent | Steel manufacturer | 5,500 | 50 |
| Municipal Effluent | Durham, NC (city) | 330 | 500 |
| Municipal Effluent | Durham, NC (county) | 500 | 500 |
| Industrial Effluent | Burlington, NC (WWTP[a] influent) | 600 | 500 |
| Energy Effluent | In situ coal gasification | 34,000 | 5 |
| Energy Effluent | Low-Btu coal gasification | 600 | 500 |
| Industrial Effluent | Soap and detergent industry | 490 | 500 |

[a] Wastewater treatment plant.

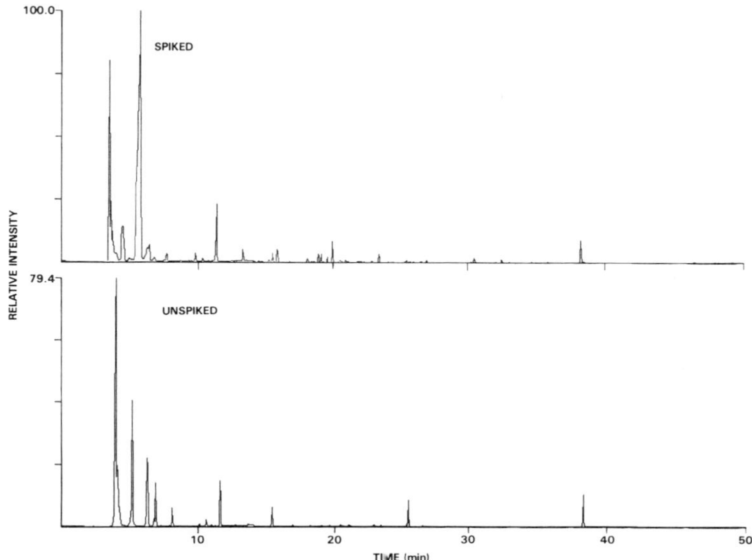


**Figure 8.** Reconstructed ion chromatograms of compounds purged from municipal effluent (spiked and unspiked).

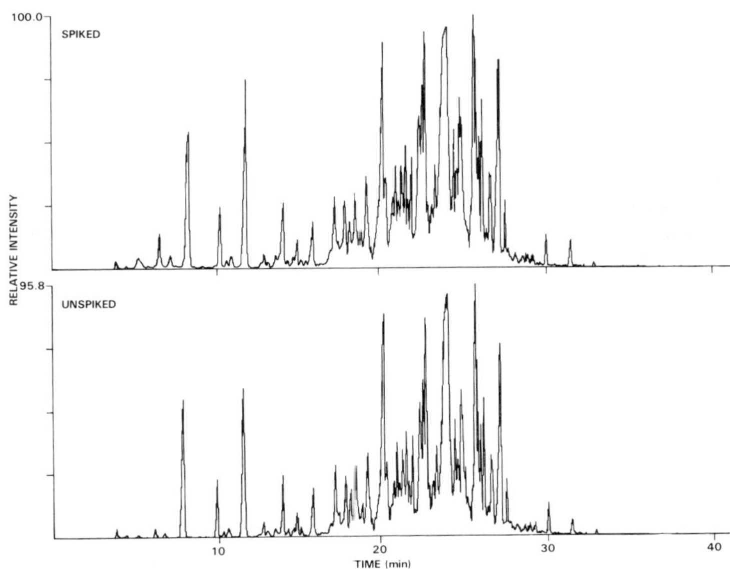


**Figure 9.** Reconstructed ion chromatograms of compounds purged from municipal influent (spiked and unspiked).

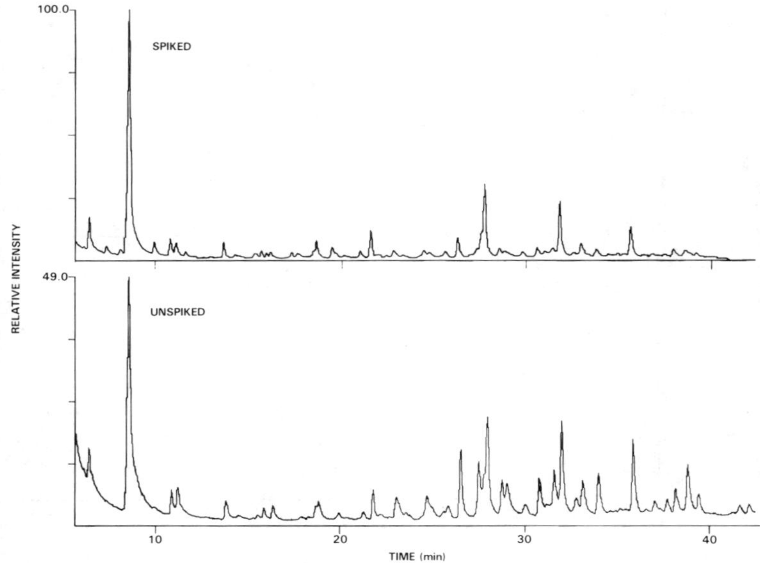


**Figure 10.** GC/MS/COMP analysis of perfluorobenzene derivatives of volatile acids in a soap and detergent sample (spiked and unspiked).

and Mr. J. Craig, Dr. J. McGuire, Mr. J. Pope and Dr. W. Shackelford of the Athens Environmental Research Laboratory is gratefully acknowledged. Appreciation is also extended to Ms. A. Alford, Mr. B. Loy and Mr. R. Ryan for their assistance in editing the manuscript.

Mention of trade names or commercial products does not constitute endorsement or recommendation for use by the U.S. Environmental Protection Agency.

## REFERENCES

1. Pellizzari, E. D. "Master Scheme for the Analysis of Organic Compounds in Water, Preliminary Draft Report. Part III: Experimental Development and Results," U.S. Environmental Protection Agency, Contract No. 68-03-2704, Environmental Research Laboratory, Office of Research and Development, Athens, GA (1980).

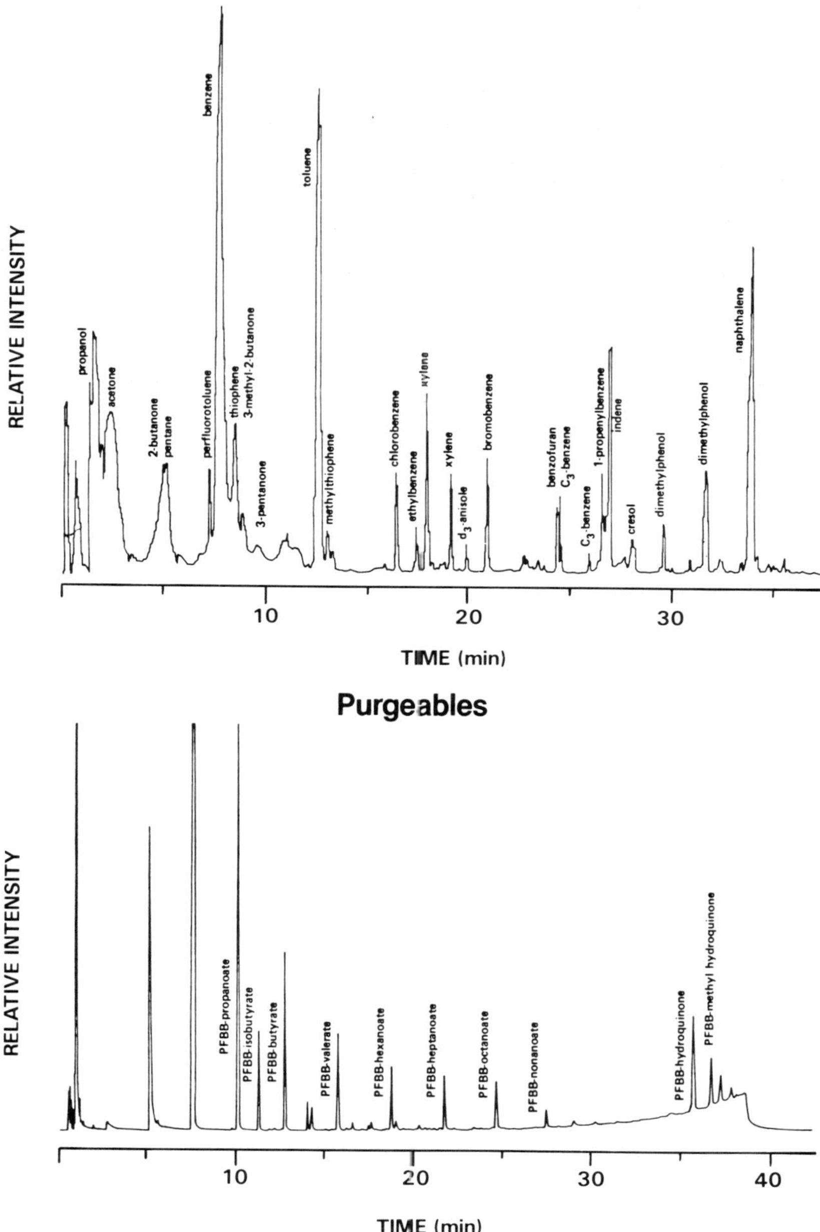


**Figure 11.** Reconstructed ion chromatogram of purged compounds and gas chromatogram of intractable acid compounds (PFBB derivatives) in an energy effluent sample.

2. Gebhart, J. E. et al. "The Master Analytical Scheme: Development of Effective Techniques for the Isolation and Concentration of Organics in Water," Chapter 3, this volume.
3. Garrison, A. W. "The Master Analytical Scheme: An Overview of Interim Procedures," Chapter 2, this volume.
4. Michael, L. C., M. D. Erickson, S. P. Parks and E. D. Pellizzari. "Volatile Environmental Pollutants in Biological Matrices with a Headspace Purge Technique," *Anal. Chem.* 52:1836-1841 (1980).
5. Tomer, K. B. et al. "Quantitative Aspects of the Master Analytical Scheme for Organics in Water," Chapter 5, this volume.

## CHAPTER 7

# DEVELOPMENT OF METHODS FOR PESTICIDES IN WASTEWATER: APPLICABILITY OF A GENERAL APPROACH

**Hope Miller, Paul Cramer, Arbor Drinkwine, Alice Shan, Glenn Trischan and John Going**

Midwest Research Institute
Kansas City, Missouri

Section 304h of the Clean Water Act (CWA) requires the U.S. Environmental Protection Agency (EPA) to "promulgate guidelines establishing test procedures for the analysis of pollutants." These test procedures then must be used for compliance monitoring under the CWA, as well as for filing for National Pollutant Discharge Elimination System (NPDES) and state certificates. When EPA entered into a settlement agreement requiring it to study and if necessary regulate 65 priority pollutants and classes of pollutants, necessary Section 304h methods were not available. In many cases, this was because the compounds were relatively unknown outside the scientific community and had only rarely been monitored by industry or regulated by EPA. To implement the CWA, 129 target pollutants were selected, and literature searches and programs were begun to provide the required methods for these compounds. In December 1979 a series of 12 new test procedures were proposed for quantitative measurement of specific organic materials. These methods, numbered 601 to 612, were developed through in-house and contracted research by the EPA Environmental Monitoring and Support Laboratory (EMSL), Cincinnati, OH. This chapter details the continuation of this research effort involving the development of methods for the determination of 53 pesticides in manufacturers' wastewater. The discussion includes the philosophy of and approach to the development of these procedures, a guide to their applicability and limitations, and a case study on the development of a test procedure for triazine pesticides. Also included are results of chromatography, clean water extraction, stability and cleanup determinations as well as the verification of the suitability of the methods for analysis of each compound in a relevant wastewater.

## EXPERIMENTAL

The objectives of this method development program were (1) to keep to a minimum the number of unique multiresidue procedures required to analyze for the 53 pesticides in untreated and treated production wastewaters; (2) to employ common techniques and equipment with gas chromatography (GC) or high-pressure liquid chromatography (HPLC) as the determination step; and (3) to develop test procedures that would have detection limits of 10 $\mu$g/L for HPLC determinations and 1 $\mu$g/L for GC determinations and an overall 85% recovery efficiency.

The approach taken to accomplish these objectives was first to categorize the compounds to be studied according to similarities in structure and then to test each class member against a standard method. This was done in the expectation that if any modifications of the standard method were required for the analysis of a given pesticide, these modifications would also be applicable to the other compounds in its class. A single procedure could thereby be developed which would be suitable for the analysis of the entire class. The compounds studied are listed by class in Table I.

In this chapter, a general discussion of the protocol for developing test procedures is followed by a detailed account of the applications of this protocol to the pesticide class, triazines. This case study is presented to illustrate some of the problems encountered and the measures taken to overcome them.

The protocol is briefly outlined in the flow diagram given in Figure 1. The standard method, patterned after existing 304h methods, utilizes 1-L sample volumes extracted with three 60-mL aliquots of methylene chloride followed by Kuderna-Danish (KD) concentration and standardized Florisil cleanup technique [1] if required.

**Table I. Pesticides Selected for Study**

| | |
|---|---|
| Phenols | Ethoprop |
| Dinoseb | Fensulfothion |
| Organochlorine and Polychlorinated Biphenyl (PCB) | Fenthion |
| Chlorobenzilate | Mevinphos |
| Chloroneb | Naled |
| Chloropropylate | Methyl Parathion |
| Dibromochloropropane | Phorate |
| Etridiazole | Ronnel |
| PCNB | Stirofos |
| Triazines | Trichloronate |
| Ametryn | Carbamate and Urea (Labile Compounds–HPLC Determinations) |
| Atrazine | Benomyl |
| Prometon | Carbendazim |
| Prometryn | Carbofuran |
| Propazine | Diuron |
| Simetryn | Fluometuron |
| Simazine | Linuron |
| Terbuthylazine | Methomyl |
| Terbutryn | Oxamyl |
| Cyanazine | Propachlor |
| Organophosphorus | Propoxur |
| Azinphosmethyl | Organonitrogen |
| Bolstar | Bromacil |
| Chlorpyrifos | Deet |
| Coumaphos | Hexazinone |
| Demeton-O | Metribuzin |
| Demeton-S | Terbacil |
| Diazinon | Triadimefon |
| Dichlorvos | Tricyclazole |
| Disulfoton | |

## Chromatography

Determination of a chromatographic system included selection of a column packing and detector which met the sensitivity (1–10 $\mu$g/L) and selectivity requirements for those compounds to be studied as a group. Grouping within a compound class was based on co-occurrence in the industrial wastewater which was to be used for method verification. Initially, operating parameters (carrier flow, oven temperature, column length) were chosen for optimum resolution of this group of compounds. The separation

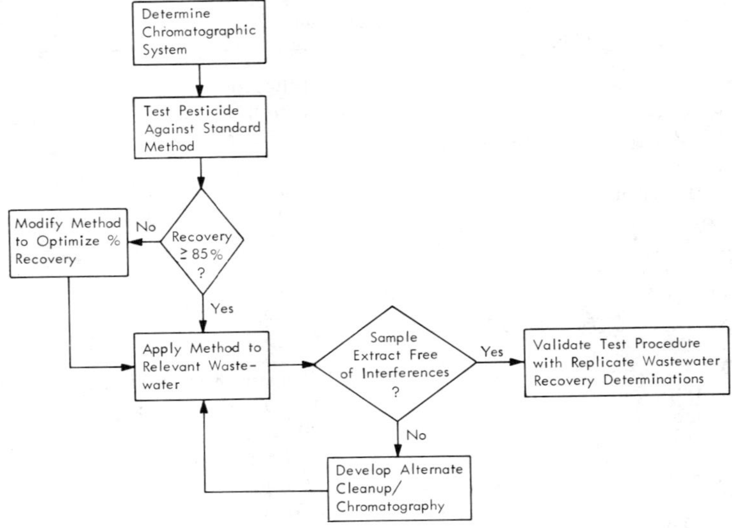


**Figure 1.** Flow diagram of protocol for development of test procedure.

had to be sufficient for quantification while keeping the analysis time under 1 hr. The systems, which were initially developed using standard solutions, were used to evaluate the efficiency of model procedure. They often required some modification when applied to a wastewater to allow the resolution of pesticides from structurally similar starting materials and degradation products.

## Verification or Modification of Standard Method

### *Extraction and Concentration*

The percent recovery of each compound from deionized water at neutral pH was determined, and adjustments in the pH of the water or the volume of the extraction solvent were made to maximize the percent recovery value of this step.

### *Stability*

Deionized water fortified with each compound was stored in the light at room temperature and neutral pH for at least seven days, and then the compounds were extracted and concentrated according to the verified procedure. Losses during storage were documented, but no experiments were run to isolate the causes of losses or to determine satisfactory storage conditions.

### *Cleanup*

An attempt was made to develop a cleanup system (solid sorbent and elution mixtures) for each of the compounds studied. The problem was approached in two stages. During the first stage, hexane fortified with each compound was applied to a standardized fully activated Florisil column, and the column was eluted with 200 mL each of 6, 15, 50 and 100% diethyl ether/petroleum ether mixtures. The elution pattern and percent recovery value for this procedural step were determined for each compound. Variations from the cleanup step of the standard method included changes in composition of the elution mixture, water deactivation of the Florisil and use of alumina as the adsorbent. In the second stage, an extract of a relevant wastewater was used to verify the selected system. Modifications were made when necessary and possible. Since manufacturers' wastewater generally contains structurally similar compounds (i.e., starting materials, by-products and degradation products), the development of a cleanup for extracts of this matrix is most important but is also most difficult. Successful recovery of the analyte from the adsorbent is a requirement that can be satisfied during the evaluation and modification of a model using interference-free reagents. However, caution must be used when attempting to apply to one matrix a cleanup developed for another matrix. Often, further method modifications are required to separate the compounds of interest from interferences found in the relevant wastewater.

### *Application to Relevant Wastewater*

After extraction and cleanup procedures were developed using fortified water and reagents, industrial wastewaters which reflected the matrices in which the studied compounds would likely be present were analyzed. Further modifications such as those noted above were made as necessary. Selected compound groups were then fortified at detectable levels into a relevant wastewater sample, and percent recovery values from this matrix were determined.

## CASE STUDY–TRIAZINE PESTICIDES

The 10 triazines studied are listed in Table II according to the similarity of functionalities in the "6" position.

### Chromatography

Matisova et al. [2] and others have reported satisfactory results using Carbowax 20M and an alkali flame detector for the analysis of five of these triazines. Therefore, the first chromatographic system used was a 5% Carbowax 20M-TPA on a Supelcoport packed column coupled with a thermionic nitrogen-specific detector (TSD). Figure 2 (left) shows a GC/TSD chromatogram of the standard mixture of triazine pesticides (1.1 ng) on the 5% Carbowax 20M-TPA run at 200°C (isothermal). Cyanazine, not shown in this chromatogram, was retained 85 min with this system. Since the 1-hr retention time limit was exceeded for cyanazine and it would be unlikely that a single stationary phase could be selected which would meet the criteria for all 10 triazines, a second system (3% SP-2250, 230°C) was used to analyze cyanazine. Figure 2 (right) shows the chromatogram of the 10-pesticide mixture on the 3% SP-2250 column.

### Extraction Efficiency

Deionized water (1 L) fortified with 1 μg of each triazine was extracted with three 60-mL portions of methylene chloride. Table III shows that the percent recovery values for duplicate determinations were all 88% or greater.

**Table II. Studied Triazines Classified According to "6" Position Function**

| 6-(Methylthio) | 6-Chloro[a] | 6-Methoxy |
|---|---|---|
| Ametryn | Atrazine | Prometon |
| Prometryn | Propazine | |
| Simetryn | Simazine | |
| Terbutryn | Terbuthylazine | |
| | Cyanazine | |

[a]The chlorine is *meta* to the amine substituents.

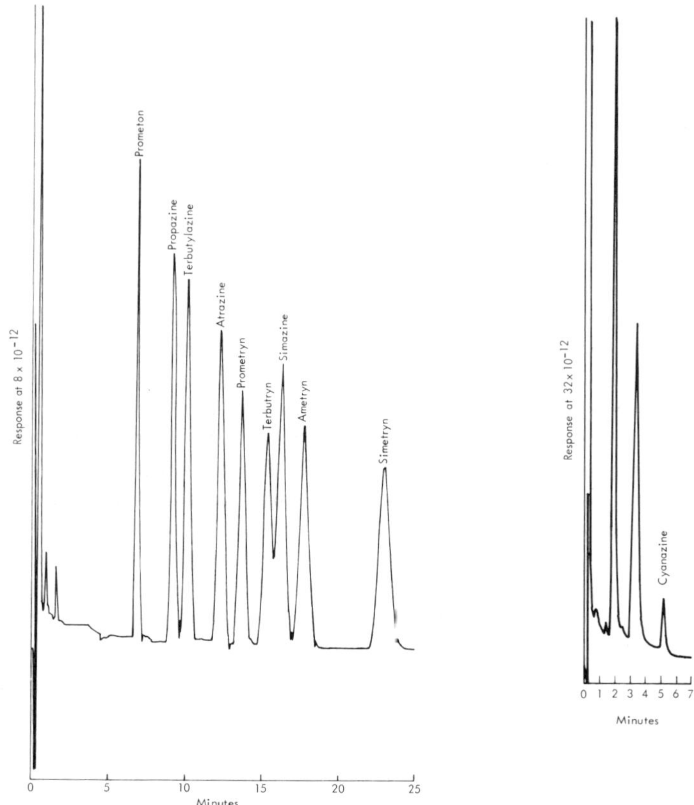


**Figure 2.** GC/DSD chromatogram of mixed triazine standard (1.1 ng) on a Carbowax-TPA column (left); GC/TSD chromatogram of mixed triazine standard (1.1 ng) on the SP-2250 column (right).

**Table III. Extraction Efficiency of Selected Triazine Compounds for 1 μg/L Fortification–Duplicate Determinations**

| Compound | % Recovery | |
|---|---|---|
| Prometon | 96 | 99 |
| Propazine | 97 | 100 |
| Terbuthylazine | 95 | 102 |
| Atrazine | 101 | 101 |
| Prometryn | 96 | 101 |
| Terbutryn | 93 | 99 |
| Simazine | 94 | 103 |
| Ametryn | 103 | 100 |
| Simetryn | 102 | 102 |
| Cyanazine | 88 | 101 |

## Stability

Deionized water (1 L) was fortified with 1 μg of each pesticide. The solution was stored at room temperature in the light for seven days prior to extraction. All the compounds were recovered at 90% or better.

## Column Cleanup

A 10-mL sample of hexane fortified with 1 μg of each triazine was loaded onto a fully activated Florisil column, and the column was eluted with 6, 15, 50 and 100% mixtures of diethyl ether/petroleum ether. The results in Table IV show that those triazines with a 6-chloro group exhibited recoveries of 83% or greater. Prometon, with a 6-methoxy group, gave 66% recovery, and no triazine with a 6-(methylthio) group gave better than 25% recovery.

A series of more polar acetone/hexane eluants was tried to improve recoveries for the 6-(methylthio) compounds, but with no success.

Based on a report [3] of the use of alumina for cleanup of extracts containing triazines, a 10% deactivated alumina column was investigated in an effort to develop a single cleanup for all 10 triazines. Successive elutions were made with 200 mL of 6, 15, 50 and 100% diethyl ether/hexane. Using this system, 9 of the 10 triazines were recovered at 84% or greater. As shown in Table V, cyanazine was not successfully recovered from the deactivated column.

**Table IV. Recovery of Triazines from Florisil with Diethyl Ether/Petroleum Ether**

| Compound | % Recovery | Fraction(s) % Diethyl Ether |
|---|---|---|
| 6-Chloro | | |
| Atrazine | 96 | 15, 50 |
| Propazine | 91 | 15, 50 |
| Simazine | 92 | 50 |
| Terbuthylazine | 91 | 15, 50 |
| Cyanazine | 83 | 50 |
| 6-Methoxy | | |
| Prometon | 66 | 100 |
| 6-(Methylthio) | | |
| Ametryn | 7 | 50 |
| Prometryn | 10 | 15, 50 |
| Simetryn | 23 | 100 |
| Terbutryn | Not detected | |

**Table V. Recovery of Triazines from 10% Water Deactivated Alumina with Diethyl Ether/Hexane**

| Compound | % Recovery | Fraction(s) % Diethyl Ether |
|---|---|---|
| 6-Chloro | | |
| Atrazine | 91 | 15, 50 |
| Propazine | 94 | 15, 50 |
| Simazine | 94 | 15 |
| Terbuthylazine | 92 | 15, 50 |
| Cyanazine | 29 | 15 |
| 6-Methoxy | | |
| Prometon | 84 | 15 |
| 6-(Methylthio) | | |
| Ametryn | 91 | 15, 50 |
| Prometryn | 111 | 15, 50 |
| Simetryn | 89 | 15 |
| Terbutryn | 85 | 50 |

Therefore, if after prescreening the raw extract by GC/TSD, it is determined that cleanup is needed, the extract has be to split to accomodate the analyses of both cyanazine and the 6-(methylthio) compounds.

## Wastewater Analysis

Untreated and treated waste streams were extracted, concentrated and analyzed by GC/TSD. Since cleanup was required only for the analysis of cyanazine, recovery determinations for ametryn, prometryn, simetryn, terbutryn, atrazine, propazine, simazine, terbuthylazine and prometon were completed prior to application of the extract to the Florisil column. The recovery values given in Table VI show the average of four determinations in relevant wastewaters. However, as shown in Figure 3, even after Florisil cleanup, an interference remained which prevented quantification of cyanazine. At this point the protocol called for development of alternative cleanup or chromatographic systems. Since GC coupled with solid adsorbent cleanup had already been investigated extensively, it was decided to attempt the separation and analysis with a HPLC/ultraviolet (UV) system.

The HPLC system consisted of a $\mu$-Bondapak $C_{18}$ (10 $\mu$m, 4 x 30 cm i.d.) column and a 50:50 methanol:water mobile phase. The retention volume was 10.2 mL with a flowrate of 1 mL/min. Figure 4 shows the HPLC chromato-

**Table VI. Recovery of Selected Triazine Pesticides from Wastewater**

| Compound | % Recovery[a] | Compound | % Recovery[a] |
|---|---|---|---|
| Prometon | 119 | Terbutryn | 102 |
| Propazine | 103 | Simazine | 106 |
| Terbuthylazine | 106 | Ametryn | 111 |
| Atrazine | 121 | Simetryn | 182[b] |
| Prometryn | 93 | | |

[a]Average of four determinations of triazines fortified at background levels.

[b]Poor peak shape of this late eluting compound may have led to a quantification error.

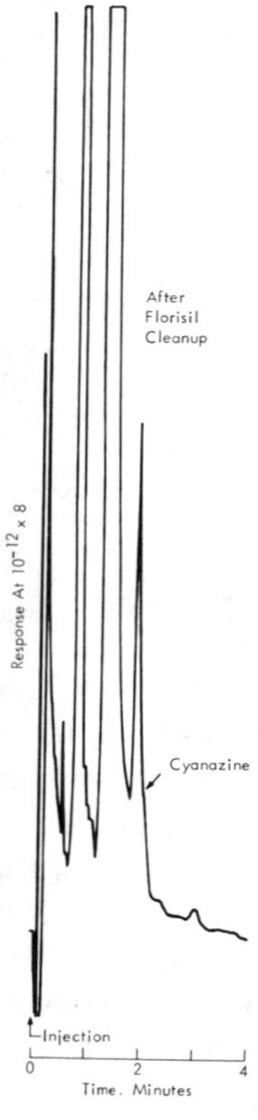


**Figure 3.** GC/TSD chromatogram of Wastewater extract (after Florisil cleanup) on SP-2250 column.

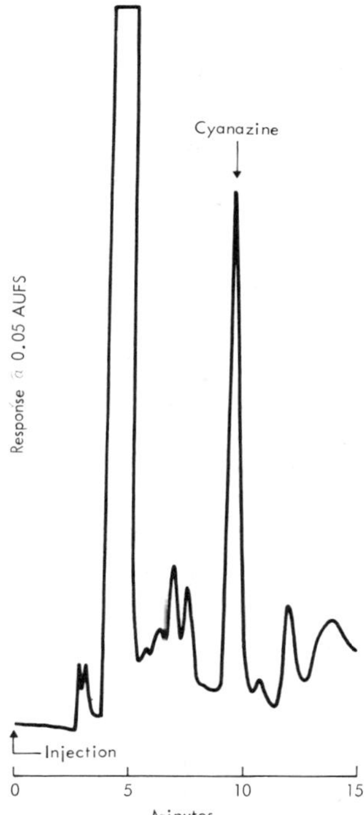


**Figure 4.** HPLC chromatogram of wastewater extract fortified with cyanazine.

gram of an untreated wastewater extract with added cyanazine which demonstrated good resolution of the triazine from any interference. Unfortunately, the wastewater supply had at this point been depleted, and verification of this method could not be carried out.

## RESULTS

### Chromatography

The chromatographic columns, detectors and operating parameters were selected to separate and detect specified groups of compounds in the presence of one another and other interferences associated with the wastewater analyzed.

The chromatographic conditions, detection limits and retention times for each compound are summarized in Tables VII and VIII. The dimensions of all GC columns were 1.83 mm x 2 mm i.d., except the 5% SP-2401 column, which was 0.92 m x 2 mm i.d. Although 16 GC systems were developed for the separation and detection of the 54 pesticides, it should be noted that only 6 different stationary phases were used for these systems. The modifications in loading, column length and oven temperatures were made to separate co-occurring pesticides and their interferences in the minimum time. The number and type of systems would be likely to vary with a change in compound groupings. For example, the 1-m column and rapid program rate employed for 10 organophosphorus compounds were used in this case to reduce on-column residence time for azinphosmethyl, which apparently decomposed on the longer column. A single HPLC column ($\mu$-Bondapak $C_{18}$) with either water/methanol or water/acetonitrile as the mobile phase was used for the analysis of all 10 labile pesticides. This was a Waters Associates $\mu$-Bondapak $C_{18}$ (10 $\mu$m, 30 cm x 4 mm i.d.) preceded by a guard column of Whatman Co:PELL ODS (30–38 $\mu$m, 7 cm x 4 mm i.d.).

Two sets of operating parameters are given for cyanazine, dinoseb and Deet. The second cyanazine method using HPLC was required to separate a wastewater interference which could not be removed by solid sorbent cleanup techniques. Method 604, which was developed prior to this work, was specified for the analysis of dinoseb. In the absence of interfering phenols, analysis time was reduced by operating at 160°C isothermal. A thermionic nitrogen-specific detector, which might also improve sensitivity, was not evaluated for this analysis. Wastewater interferences required modification of the isothermal conditions initially developed for the analysis of DEET. Temperature programming was required to effect the necessary separation. The detection limit goals of 1 $\mu$g/L GC and 10 $\mu$g/L HPLC were met for 48 of the 53 compounds.

## Extraction

All the studied compounds were successfully extracted from water at pH 7 with three 60-mL portions of methylene chloride, with the exceptions of carbendazim, benomyl and dinoseb. An increase in solvent volume from 60 to 350 mL resulted in an increase in the recovery of carbendazim from 15 to 83%. Since benomyl slowly hydrolyzes to carbendazim [4], it was decided to develop the method for benomyl around the total conversion to carbendazim. Neutral wastewater (1 L) and HCl (10 mL) were stirred for

Table VII. Gas Chromatographic Systems for Analysis of 44 Selected Pesticides

| Column | Conditions | Detector | Compound | Detection Limit (µg/L) | Retention Time |
|---|---|---|---|---|---|
| 5% Carbowax 20M TPA | 200°C | TSD | Terbutryn | 0.05 | 15.4 |
| | | | Simazine | 0.05 | 16.3 |
| | | | Terbuthylazine | 0.03 | 10.2 |
| | | | Simetryn | 0.07 | 23.0 |
| | | | Propazine | 0.03 | 9.2 |
| | | | Prometryn | 0.03 | 13.8 |
| | | | Prometon | 0.03 | 6.9 |
| | | | Atrazine | 0.03 | 12.4 |
| | | | Ametryn | 0.06 | 17.7 |
| | | | Cyanazine | 0.04 | 85.0 |
| 1% SP-1240 DA | From 80 to 180°C at 10°C/min, hold at 180°C for 10 min | FID[a] | Dinoseb | 20 | 15.8 |
| | 160°C | FID | | | 13.5 |
| 1% SP-2250 | 150°C | EC[b] | Chloroneb | 0.001 | 10 |
| 3% SP-2250 | 230°C | TSD | Cyanazine | 0.04 | 5.2 |
| 3% SP-2250 DB | 230°C for 3 min, 10°C/min to 250°C | TSD | Bromacil | 0.02 | 3.7 |
| | | | Terbacil | 0.5 | 2.1 |
| | | | Hexazinone | 0.5 | 7.6 |
| | 240°C | TSD | Tricyclazole | 0.1 | 3.5 |
| 1.5% SP-2250/1.95% SP-2401 | 160°C | EC | PCNB | 0.02 | 3.1 |
| | 140°C | | Etridiazole | 0.003 | 1.3 |
| | 100°C | | Dibromochloropropane | 0.001 | 3.1 |
| | 215°C | | Chloropropylate | 0.001 | 3.7 |
| | 215°C | | Chlorobenzilate | 0.001 | 3.8 |

**Table VII, continued**

| Column | Conditions | Detector | Compound | Detection Limit (μg/L) | Retention |
|---|---|---|---|---|---|
| 3% SP-2401 | 180°C | TSD | Deet | 0.1 | 1.6 |
| | 130°C for 1 min, 12°C/min to 200°C | | Deet | | 4.6 |
| | 240°C | | Triadimefon | 0.7 | 4.1 |
| | | | Metribuzin | 0.7 | 2.4 |
| 5% SP-2401 (1 m) | 150°C for 1 min, 25°C/min to 220°C for 9 min | FPD[c] | Azinphosmethyl | 1.5 | 6.8 |
| | | | Bolstar | 0.15 | 4.2 |
| | | | Coumaphos | 1.5 | 11.6 |
| | | | Demeton-O | 0.25 | 2.5 |
| | | | Demeton-S | 0.25 | 1.2 |
| | | | Disulfoton | 0.2 | 2.1 |
| | | | Fensulfothion | 1.5 | 6.4 |
| | | | Fenthion | 0.1 | 3.1 |
| | | | Phorate | 0.15 | 1.4 |
| | | | Trichloronate | 0.15 | 2.9 |
| 5% SP-2401 (2 m) | 160°C to 220°C at 10°C/min | FPD | Chlorpyrifos | 0.3 | 5.8 |
| | | | Ronnel | 0.3 | 4.9 |
| | 170°C for 2 min, 20°C/min to 220°C, hold for 10 min | FPD | Stirofos | 0.5 | 8.5 |
| | | | Naled | 0.5 | 3.9 |
| | | | Mevinphos | 0.3 | 2.4 |
| | | | Dichlorvos | 0.1 | 0.8 |
| | 190°C | FPD | Diazinon | 0.8 | 2.0 |
| | | | Ethoprop | 0.03 | 1.2 |
| | | | Methyl parathion | 0.75 | 9.0 |

[a]FID = flame ionization detection.

[b]EC = electron capture detection.

[c]FPD = flame photometric detection.

**Table VIII. High-Pressure Liquid Chromatographic System ($\mu$-Bondapak $C_{18}$) for Analysis of Ten Selected Pesticides**

| Mobile Phase | Detector | Compound | Detection Limit ($\mu$g/L) | Retention Volume (mL) | Flowrate (mL/min) |
|---|---|---|---|---|---|
| Methanol/water 25/75% | UV-254 nm | Oxamyl | 1.5 | 8 | 1 |
| Methanol/water 50/50% | UV-254 nm | Cyanazine | 6 | 10.2 | 1 |
| | UV-280 nm | Carbendazim (Benomyl) | 3 | 7.8 | 2 |
| Acetonitrile/water 50/50% | UV-254 nm | Fluometuron | 0.5 | 7.2 | 2 |
| | | Propachlor | 16 | 9.4 | 2 |
| | UV-280 nm | Propoxur | 16 | 6.8 | 2 |
| | | Carbofuran | 5 | 7 | 2 |
| Acetonitrile/water 10/90% to acetonitrile 100% in 30 min with linear gradient | UV-254 nm | Methomyl | 4 | 13 | 2 |
| | | Linuron | 0.3 | 35.8 | 2 |
| | | Diuron | 0.3 | | |

24 hr to assure the complete hydrolysis of benomyl to carbendazim. The pH was then raised to 7 for sample extraction with three 350-mL portions of methylene chloride.

The previously developed test procedure for phenols (No. 604) was evaluated for the analysis of dinoseb. Method 604 requires an initial extraction (three 60-mL portions of $CH_2Cl_2$) at pH 11 to remove basic interferences and a final extraction at pH 2 for the partitioning of dinoseb and other acidic phenols. The goal of 85% extraction efficiency was met for 95% of the studied compounds.

## Stability

Deionized water fortified with each compound was extracted on day 0 and day 7 after storage at ambient conditions and neutral pH. The percent recovery values are included in Table IX. Comparison of the two values would indicate a need for some means of preservation for Bolstar, demeton-S, disulfoton, fenthion, phorate, trichloronate, ronnel and dichlorvos. It should be noted, however, that no effort was made to determine the cause of analyte losses during storage.

**Table IX. Extraction Efficiency and Seven-day Stability Results**

| Compound | Concentration (μg/L) | % Recovery | |
|---|---|---|---|
| | | Day 0 | Day 7 |
| Dinoseb | 100 | 94; 95 | 92 |
| Chloroneb | 5 | 87; 73 | 70 |
| Chlorobenzilate | 5 | 96 | 96 |
| Chloropropylate | 5 | 91 | 94 |
| Dibromochloropropane | 5 | 86; 84 | 84 |
| Etridiazole | 1 | 99 | 100 |
| PCNB | 1 | 68 | 65 |
| Ametryn | 1 | 103; 100 | 90 |
| Atrazine | 1 | 101; 101 | 92 |
| Prometon | 1 | 96; 99 | 94 |
| Prometryn | 1 | 96; 101 | 93 |
| Propazine | 1 | 97; 100 | 94 |
| Simetryn | 1 | 102; 102 | 93 |
| Simazine | 1 | 94; 103 | 91 |
| Terbutylazine | 1 | 95; 102 | 96 |
| Terbutryn | 1 | 93; 93 | 93 |
| Azinphosmethyl | 1 | 96 | 87 |

Table IX, continued

| Compound | Concentration ($\mu g/L$) | % Recovery | |
|---|---|---|---|
| | | Day 0 | Day 7 |
| Bolstar | 1 | 100 | 79 |
| Coumaphos | 1 | 99 | 97 |
| Demeton-O | 1 | 91 | 75 |
| Demeton-S | 1 | 97 | 0 |
| Disulfoton | 1 | 111 | 71 |
| Fensulfothion | 1 | 102 | 91 |
| Fenthion | 1 | 87 | 61 |
| Phorate | 1 | 89 | 29 |
| Trichloronate | 1 | 107 | 55 |
| Chlorpyrifos | 1 | 78 | 70 |
| Ronnel | 1 | 91 | 58 |
| Stirofos | 1 | 83 | 80 |
| Naled | 1 | 95 | 91 |
| Mevinphos | 1 | 92 | 88 |
| Dichlorvos | 1 | 110 | 70 |
| Diazinon | 1 | 91 | 89 |
| Ethoprop | 1 | 101 | 100 |
| Methyl Parathion | 1 | 99 | 94 |
| Carbofuran | 10 | 101; 101 | 102 |
| Cyanazine | 1 | 88; 101 | 94 |
| Carbendazim | 10 | 83 | ND[a] |
| Benomyl | 150 | 81; 72 | ND |
| Diuron | 10 | 95; 89 | 93 |
| Linuron | 10 | 91; 92 | 96 |
| Methomyl | 100 | 78; 77 | 95 |
| Propachlor | 100 | 104; 109 | 104 |
| Propoxur | 100 | 91; 89 | 91 |
| Fluometuron | 10 | 98; 87 | 99 |
| Oxamyl | 10 | 98 | 82 |
| Metribuzin | 1 | 95; 100 | 102; 88 |
| Triadimefon | 1 | 93; 93 | 88 |
| Deet | 1 | 97 | 96 |
| Tricyclazole | 1 | 100; 100 | 81 |
| Bromacil | 1 | 91 | 100 |
| Hexazinone | 1 | 102 | 83 |
| Terbacil | 1 | 97 | 99 |

[a]ND = value not determined.

## Cleanup

The greatest deviation from the standard method and the major source of reduced recovery were in the area of cleanup. Only about 60% of the pesticides could be recovered from Florisil by the standard elution system. However, because of the use of selective detector systems, it was only necessary to clean up the extracts of half of the analyzed wastewaters representing only 12 of the 54 studied compounds. Column materials and elution mixtures were evaluated using the predetermined order given in Table X.

It was observed that the adsorptivity varied greatly between determinations made with fortified hexane and actual wastewater extracts. It is advisable to retain all fractions and to determine the elution pattern of the compound in the presence of any new set of matrix interferences.

Since the existing phenol Method 604 evaluated for dinoseb does not include a cleanup step, no method was developed.

Triazines containing the function group S-$CH_3$ could be recovered successfully only from 10% deactivated alumina. Although nonsulfonated triazines were recovered from Florisil with ether/petroleum ether, the alumina column was used because both types of triazine were present in the wastewater.

Organophosphorus pesticides as a class exhibited extremely poor recovery from Florisil. Because of the specificity of the flame photometric detector, solid sorbent cleanup was not required in the analysis of wastewater, and only system 1 was evaluated for those compounds because of time considerations.

Satisfactory recovery of labile compounds such as carbamates and ureas generally required substitution of the more polar acetone/hexane mixture, as would be expected. No solid sorbent technique was successful both in recovering cyanazine and removing the interferences which appeared in a relevant wastewater. An HPLC method was ultimately developed for this purpose.

**Table X. Column Cleanup Systems**

| System No. | Solid Sorbent | Elution Mixtures and Percentages | |
|---|---|---|---|
| 1 | Florisil | Ethyl ether/petroleum ether | 6, 15, 50, 100% |
| 2 | Florisil | Acetone/hexane | 6, 15, 50, 100% |
| 3 | 2% water deactivated Florisil | Acetone/hexane | 6, 15, 50, 100% |
| 4 | 6% water deactivated Florisil | Acetone/hexane | 6, 15, 50, 100% |
| 5 | 10% water deactivated alumina | Ethyl ether/petroleum ether | 6, 15, 50, 100% |

Table XI summarizes the recovery results from solid sorbents and indicates which compounds were determined in wastewater without the need for cleanup.

### Wastewater Analyses

When possible, wastewater was obtained from industrial sites that manufactured one or more of the compounds studied. These water samples were utilized to allow for needed method modifications due to matrix effects and to verify the efficiency of final procedures.

Percent recovery values for the final procedure were determined on wastewater that had been fortified at levels relevant to those observed in the background. Table XII provides these recovery data. Background and fortification levels which are on file at EMSL are not available for release at this time.

## SUMMARY

The degree of success that was achieved in applying a standard approach was fairly good. Table XIII lists the average recoveries for the five general classes of pesticides that were achieved for the various steps in the procedure. Generally speaking, recoveries for clean water extraction, seven-day stability and spiked wastewater were good. The greatest deviation from a single method and the major source of reduced recovery was in the area of cleanup. The degree of success in meeting the sensitivity goal is given in the data shown in Table XIV. For 80% of the studied pesticides, a detection limit of 1 $\mu$g/L or better was achieved by the basic protocol.

## ACKNOWLEDGMENTS

The authors wish to acknowledge the contributions of Dr. Mark Marcus, Dr. David Hall, Ms. Carol Mellom, Mr. David Mills, Ms. Janine Neils, Mrs. Gail Rehagen, Mr. Russ Shaffenberg and Dr. James Spigarelli.

This research was sponsored by Environmental Monitoring and Support Laboratory, Cincinnati, OH, of the U.S. Environmental Protection Agency.

Table XI. Percent Recovery of Studied Compounds from Selected Solid Sorbent Cleanup Systems

| Compound | Added Amount ($\mu g$) | Sorbent | Elution Mixture | Fraction(s) | % Recovery | Required for Wastewater |
|---|---|---|---|---|---|---|
| Chloroneb | 10 | Florisil | Ether/petroleum | 6 | 93 | Yes |
| Chlorobenzilate | 10 | | ether, 6, 15 | 15, 50 | 85 | Yes |
| Chloropropylate | 10 | | 50, 100a, 100b % | 15, 50 | 93 | Yes |
| Dibromochloropropane | 10 | | | 6, 15, 50 | 71 | Yes |
| Etridiazole | 1 | | | 6 | 100 | Yes |
| PCNB | 1 | | | 6 | 75 | |
| Azinphosmethyl | 1 | | | 100[a] | 33 | No |
| Bolstar | 1 | | | 6 | 35 | No |
| Coumaphos | 1 | | | 100[a] | 44 | No |
| Demeton-O | 1 | | | | 0 | No |
| Demeton-S | 4 | | | | 0 | No |
| Disulfoton | 1 | | | 6, 15, 50 | 52 | No |
| Fensulfothion | 1 | | | | 0 | No |
| Fenthion | 1 | | | 6, 15 | 25 | No |
| Phorate | 1 | | | 6 | 34 | No |
| Trichloronate | 1 | | | 6 | 67 | No |
| Chlorpyrifos | 1 | | | 6 | 100 | No |
| Ronnel | 1 | | | 6 | 82 | No |
| Stirofos | 1 | | | 15 | 63 | No |
| Naled | 1 | | | | 0 | No |
| Mevinphos | 1 | | | 100[a] | 14 | No |
| Dichlorvos | 1 | | | 100[a] | 24 | No |
| Diazinon | 1 | | | 15, 50, 100[a] | 76 | No |
| Ethoprop | 1 | | | 100[a] | 70 | No |
| Methyl Parathion | 1 | | | 50 | 90 | No |
| Cyanazine | 1 | | | 50 | 83 | Yes[a] |

| | | | | | | |
|---|---|---|---|---|---|---|
| Carbofuran | 10 | | | 50, 100[a] | 104 | No |
| Propoxur | 100 | | | 100[a] | 89 | Yes |
| Fluometuron | 10 | | | 100[a], 100[b] | 95 | Yes |
| Carbendazim/Benomyl | 100 | Florisil | Acetone/hexane | 50 | 46 | No |
| Diuron | 10 | | 6, 15, 50, 100% | 15, 50 | 82 | No |
| Linuron | 10 | | | 6, 15 | 96 | |
| Methomyl | 100 | | | 50 | 84 | No |
| Oxamyl | 100 | | | 50 | 92 | No |
| Metribuzin | 1 | 2% water | Acetone/hexane | 6, 15 | 67 | Yes |
| Triadimefon | 1 | deactivated | 6, 15, 50, 100% | 15 | 100 | Yes |
| Deet | 1 | Florisil | | 6, 15 | 81 | No |
| Tricyclazole | 1 | | | 6, 15, 50, 60 | 101 | No |
| Hexazinone | 1 | | | 50 | 82 | No |
| Terbacil | 1 | 6% water | Acetone/hexane | 15 | 62 | No |
| Bromacil | 1 | deactivated Florisil | 6, 15, 50, 100% | 15, 50 | 48 | No |
| Ametryn | 1 | 10% water | Ether/petroleum | 15, 50 | 91 | No |
| Atrazine | 1 | deactivated alumina | 6, 5, 50, 100% | 15 | 99 | No |
| Prometon | 1 | | | 15 | 84 | No |
| Prometryn | 1 | | | 15, 50 | 111 | No |
| Propazine | 1 | | | 15, 50 | 94 | No |
| Simetryn | 1 | | | 15 | 89 | No |
| Simazine | 1 | | | 15 | 94 | No |
| Terbuthylazine | 1 | | | 15, 50 | 92 | No |
| Terbutryn | 1 | | | 50 | 85 | No |
| Propachlor | | Florisil | 20% ether/hexane then 6, 15, 50 100% acetone/hexane | 6 | 94 | Yes |
| Dinoseb[b] | | | | | | |

[a]No solid sorbent evaluated was effective in removing a wastewater interference, and the analysis was completed by HPLC.

[b]Previously developed Method 604, evaluated for dinoseb, had no sorbent cleanup.

Table XII. Percent Recovery of Studied Compounds from Relevant Wastewater

| Compound | Influent % Recovery | Effluent % Recovery |
|---|---|---|
| Dinoseb | 86; 98 | 69; 74 |
| Chlorobenzilate | 26; 48 | 118; 172 |
| Chloropropylate | | 129; 130 |
| Chloroneb | 48 | 63 |
| Dibromochloropropane | 74; 87 | 73; 32 |
| Etridiazole | 81; 113 | 92; 100 |
| PCNB | 20; 36 | |
| Ametryn | 111; 96 | 125; 111 |
| Atrazine | 115; 100 | 142; 129 |
| Prometon | 126; 98 | 135; 117 |
| Prometryn | 80; 72 | 123; 97 |
| Propazine | 105; 77 | 122; 109 |
| Simetryn | 198; 168 | 194; 169 |
| Simazine | 123; 103 | 104; 83 |
| Terbuthylazine | 126; 102 | 105; 94 |
| Terbutryn | 131; 103 | 104; 83 |
| Azinphosmethyl | 101 | 69; 101 |
| Bolstar | 77, 111 | 85; 94 |
| Coumaphos | 131; 256 | 213; 255 |
| Demeton-O | 95; 78 | 72; 80 |
| Demeton-S | 17 | 0; 0 |
| Disulfoton | 129; 141 | 93; 93 |
| Fensulfothion | 89; 139 | 72; 78 |
| Fenthion | 43; 41 | 60; 42 |
| Phorate | 47 | 42; 69 |
| Trichloronate | 51; 44 | 34; 40 |
| Chlorpyrifos | 97; 81 | 104; 102 |
| Ronnel | 93; 80 | 68; 60 |
| Diazinon | 110; 137 | 108; 108 |
| Dichlorvos | 120; 110 | 89; 78 |
| Mevinphos | 86; 92 | 120; 78 |
| Naled | 60; 80 | 95; 90 |
| Stirofos | 110; 120 | 120; 98 |
| Methyl Parathion | 60; 85 | 80; 90 |
| Carbofuran | | 120; 108 |
| Cyanazine[a] | | |
| Carbendazim/Benomyl | | 138; 93 |
| Diuron[a] | | |
| Linuron | 108; 212 | 91; 101 |
| Methomyl | 66; 130 | 56; 42 |
| Fluometuron | 148; 170 | 59; 55 |
| Oxamyl | 90; 94 | 80; 76 |
| Propachlor | | 120; 97 |
| Propoxur | 56; 68 | 98; 89 |
| | 63; 69 | 71 |

Table XII, continued

| Compound | Influent % Recovery | Effluent % Recovery |
|---|---|---|
| Bromacil | 102; 88 | 109; 126 |
| Hexazinone | 137; 83 | 95; 108 |
| Terbacil | | 104; 119 |
| DEET | 110; 110 | 106; 103 |
| Metribuzin | 79; 41 | 41; 50 |
| Triadmefon | 74; 63 | 100; 94 |
| Tricyclazole | 82; 74 | 94; 97 |
| Ethoprop[b] | | |

[a]Insufficient sample for quantification.
[b]Relevant wastewater sample was unavailable.

Table XIII. Summary of Results by General Class

| Class | % Recovery[a,b] | | | |
|---|---|---|---|---|
| | Clean Water Extraction | Seven-day Stability (25°C plus light) | Cleanup[c] | Spiked Wastewater |
| Organochlorine | 88 ± 12 | 84 ± 14 | 88 ± 12 (1) | 92 ± 21 |
| Organonitrogen | 96 ± 4 | 95 ± 8 | 76 ± 15 (4) | 93 ± 22 |
| Organophosphorus | 95 ± 9 | 77 ± 17 | 45 ± 32 (1) | 83 ± 20 |
| Triazines | 99 ± 2 | 93 ± 2 | 93 ± 8 (1) | 104 ± 14 |
| Carbamates/Ureas | 91 ± 9 | 94 ± 7 | 90 ± 5 (2) | 84 ± 19 |

[a]Recovery experiments made at 1 $\mu$g/L when possible.

[b]Percent recovery ± standard deviation.

[c]Number of cleanup systems given in parentheses.

Table XIV. Summary of Detection Limits

| Detection Limit Ranges | | Percent of Pesticides in Detection Limit Range |
|---|---|---|
| Mass (ng)[a] | Concentration (μg/L)[b] | |
| 0.001 - 0.01 | 0.001 - 0.01 | 5 |
| 0.01 - 0.1 | 0.01 - 0.1 | 35 |
| 0.1 - 1.0 | 0.1 - 1.0 | 80 |
| 1.0 - 10.0 | 1.0 - 10.0 | 94 |
| 10.0 - 100.0 | 10.0 - 100.0 | 100 |

[a]Quantity required for a 10:1 signal-to-noise ratio.

[b]Based on quantity required for 10:1 signal-to-noise ratio, and assuming a 1-L sample, 5 mL final extract volume and 5 μL injection.

## REFERENCES

1. "Manual for Analytical Quality Control for Pesticides and Related Compounds in Human and Environmental Samples," U.S. EPA, EPA-600/1-76-017 (1976).
2. Matisova, E., J. Krupcik and O. Liska. *J. Chromatog.* 173:136 (1979).
3. Erickson, M. D., C. W. Frank and D. P. Morgan. *J. Agric. Food Chem.* 27:740 (1979).
4. Austin, D. J., A. Lord and I. H. Williams. *Pestic. Sci.* 7:211 (1976).

# SECTION 2

# HIGH-RESOLUTION GAS CHROMATOGRAPHY

CHAPTER 8

# DETERMINATION OF PHENOLIC WATER POLLUTANTS BY GLASS CAPILLARY GAS CHROMATOGRAPHY

**Walter Giger and Christian Schaffner**

Swiss Federal Institute for Water Resources
and Water Pollution Control (EAWAG)
Dübendorf, Switzerland

Phenolic chemicals are of growing concern as water pollutants. Some phenols have very low taste and odor thresholds (e.g., 2 $\mu$g/L for 2-chlorophenol; others are highly persistent and toxic (e.g., pentachlorophenol [PCP]. Buikema et al. [1] reviewed the occurrence and biological effects of phenolic compounds in aquatic ecosystems. More than 90% of phenols are produced by industrial synthesis, but some phenolic compounds are natural products. Phenols are used, for example, for wood preserving or as chemical intermediates for dye manufacturing. Natural phenols are formed in aquatic and terrestrial vegetation and can be released as pollutants by the pulp and paper industry.

Phenolic chemicals have been studied in various types of industrial wastewaters [1,2]. Hexachlorophene and PCP have been investigated in sewage and water by Buhler et al. [3]. Dietz and Traud [4] have determined a large number of phenols, particularly chlorinated derivatives in domestic sewage effluents, various wastewaters and surface waters. Chlorinated phenols in surface waters of the Netherlands have been studied by Wegman and Hofstee [5].

The priority pollutant list issued by the U.S. Environmental Protection Agency (EPA) includes 11 phenols [6]. A panel of the American Chemical Society, Division of Environmental Chemistry has suggested a list of chemicals as consensus voluntary reference compounds (CVRC) for organic water pollutants [7]. These CVRC should be used widely to compare improvements in new or modified techniques with existing methodology. This list contains 13 phenols, four of which are U.S. Environmental Protection Agency (EPA) priority pollutants.

Many analytical methods for the determination of phenolic compounds in waters and wastewaters have been published [1,2]. The procedure used to isolate phenolic compounds from environmental samples is of critical importance. Most methods apply multistep extractions at varying pH levels (e.g., 2, 3, 4 and 5). Some phenols are quite hydrophilic and are therefore not efficiently extracted from aqueous solutions by organic solvents. Anion exchange has been used successfully to concentrate phenols from wastewaters [8]. Many phenolic compounds are amenable to gas chromatographic (GC) analyses either as free phenols or as phenol derivatives such as ethers and esters. In most studies packed-column GC has been applied, although the composition of many phenolic mixtures is usually extremely complex. There is no doubt that the high separation power of capillary GC has to be applied to resolve the compositional complexity of phenolic mixtures, particularly for the analysis of wastewaters. Krijgsman and van de Kemp [9] and Wegman and Hofstee [5] have demonstrated that the determination of chlorophenols in surface waters is feasible by capillary GC. They have formed phenol acetates to facilitate GC determinations.

In this chapter we describe a new, reliable method for the quantitative determination of lipophilic phenols, including priority pollutants and CVRC. An efficient isolation scheme to concentrate phenols from trace levels (as low as 0.1 $\mu$g/L) in different types of waters has been developed. Furthermore, the phenols are determined as free phenols by glass capillary GC using a new type of column. Quantitative aspects, advantages over other procedures and limitations will be discussed. Examples of analyses of some sewage effluents, wastewaters and riverwaters are given.

## EXPERIMENTAL

### Materials

Solvents were "nanograde" quality (Mallinckrodt, St. Louis, MO) and were used without further purification. In part of this investigation "all glass distilled" solvent from Burdick & Jackson (distributed by Fluka, Buchs, Switzerland) was used. This solvent contained a number of low-boiling impurities (mainly xylenes). Sulfuric acid, sodium sulfate, sodium chloride and potassium hydroxide were "analytical grade" (Merck, Darmstadt, Germany). Sodium chloride and cotton were preextracted with methylene chloride. Sodium sulfate was dried and purified at 800°C over night. Silica (Merck, Kieselgel-40, 70-230 mesh) was activated at 250°C overnight.

The phenolic reference compounds were obtained from Fluka, Buchs, Switzerland and Chem Service, West Chester, PA. These chemicals were used without further purification.

## Extraction and Isolation

Figure 1 outlines the procedure for the isolation of phenolic compounds from water samples. The samples were acidified to about pH 2 with 50% sulfuric acid. Distilled water was added to a volume of 1.5 L if a smaller sample volume was analyzed (e.g., in the case of industrial wastewaters). The sample was distilled following the procedure described in *Standard Methods* [10]. A 500-ml distillate was collected and made basic (pH = 11.5) with solid potassium hydroxide pellets. Approximately 5 g of sodium chloride were added, and the alkaline distillate was then placed in a 1-L separatory funnel and extracted twice with 40-mL portions of methylene chloride.

The methylene chloride layers were separated and discarded. The remaining aqueous solution was acidified with 50% sulfuric acid to pH 2 and extracted twice with 40-mL portions of methylene chloride. The combined organic extracts of the acidic solutions were dried with sodium sulfate and filtered through a plug of cotton into a 100-mL pear-shaped round flask. At this point an internal standard, 2,4,6-tribromophenol (TBP), was added. For this purpose a solution in the range of 5 mg TBP in 5 mL of acetone had been prepared. A 1- to 4-$\mu$L aliquot of this solution was added to the extract with a 5$\mu$L syringe. The extract was then concentrated to 1-2 mL using a micro-Snyder column. The concentrated extract was applied to a glass column (20 x 1.0 cm i.d.) containing 5 mL of silica gel (deactivated with 15% w/w of water). The phenolic fraction was eluted with 20 mL of methylene chloride. The eluate was finally concentrated to 0.5-1 mL using a micro-Snyder column.

## Gas Chromatography

Gas chromatography was performed on a Carlo Erba apparatus (Model 4150) equipped with a flame ionization detector (FID). An injection port

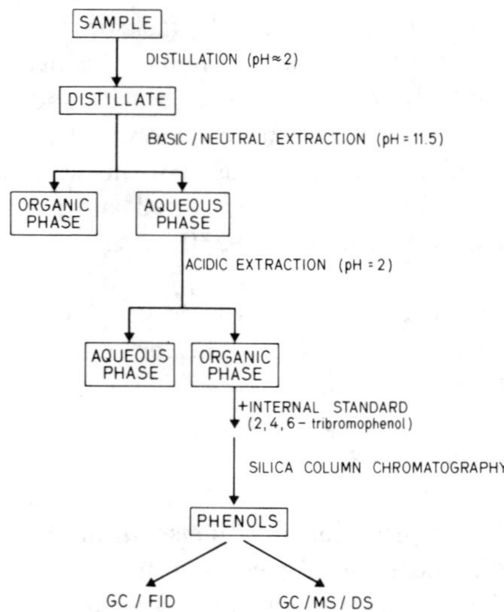


**Figure 1.** Analytical scheme for the determination of phenolic water pollutants.

with built-in septum flushing was used (Model 76, purchased from Brechbühler, Urdorf, Switzerland).

The glass capillary columns (26-32 m x 0.3 mm) were kindly supplied by K. and G. Grob. The columns had been deactivated by persilylation and coated with OV-73 [11,12].

Aliquots of 1-2 $\mu$L were injected without stream splitting onto the column at ambient temperature. The injector temperature was 275°C. After 30 sec, the split valve was opened, allowing the septum and injection port to be purged at a flowrate of approximately 30 mL/min. After the elution of the solvent the oven temperature was raised at 2.5°C/min to 180°C. Hydrogen was used as the carrier gas with a back-pressure of 0.6 atm.

Electronic integration of the GC peak areas was performed by a digital integrator (Minigrator, Spectra Physics).

## Gas Chromatography/Mass Spectrometry (GC/MS)

For mass spectrometric identifications, a Finnigan GC/MS (Model 1015 D) interfaced with a dedicated data system (Model 6000) was used.

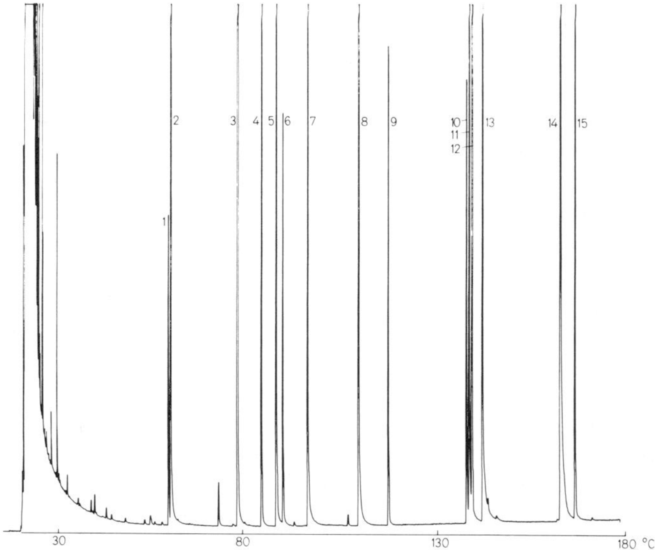


**Figure 2.** Glass capillary gas chromatogram of 15 phenols contained in the lists of priority pollutants and consensus voluntary reference compounds. (1) 2-Chlorophenol, (2) phenol, (3) 2-methoxyphenol (guaiacol), (4) 2-nitrophenol, (5) 2,4-dimethylphenol, (6) 2,4-dichlorophenol, (7) 2-hydroxyphenol (catechol), (8) 4-chloro-3-methylphenol, (9) 2,4,6-trichlorophenol, (10) 2,4-dinitrophenol, (11) 1-naphthol, (12) 2,6-di-tert-butyl-4-methylphenol, (13) 4-nitrophenol, (14) 4-phenylphenol and (15) pentachlorophenol.

The glass capillary column was directly coupled with the mass spectrometer by a fused-silica capillary. Helium was used as the carrier gas. Ionization was performed by electron impact at an electron energy of 70 eV.

## RESULTS AND DISCUSSION

### Gas Chromatography

Figure 2 shows the gas chromatogram of a reference mixture of 15 phenols which were selected from the lists of priority pollutants and/or CVRC. The gas chromatogram clearly demonstrates that all of these compounds can be determined as free phenols on the persilylated glass capillary columns used in this investigation [11,12]. All compounds produce peaks with good shapes (symmetric, little tailing), a prerequisite for quantitative determinations. It should be pointed out that nitrophenols (compounds 4, 10 and 13) and 2-hydroxyphenol (compound 7) can be gas chromatographed without derivatization. This is a significant progress in the analysis of phenolic compounds by capillary GC. The "true" neutrality, or optimum inertness [11,12], of these glass capillaries makes derivatization of the phenolic hydroxyl groups unnecessary. Therefore problems associated with derivatization steps (introduction of impurities, incomplete reaction, etc.) can be avoided.

It should be noticed that some of these phenolic compounds are not easily separated. 2-Chlorophenol and phenol (peaks 1 and 2) or 2,4-dinitrophenol, 1-naphthol and 2,6-di-tert-butyl-4-methylphenol (peaks 10, 11 and 12) have minimum differences in retention times which are necessary for quantitative determinations. The use of less efficient separation columns (packed columns) would certainly hamper reliable determinations.

In Table I, GC response factors relative to TBP are listed for phenol and six chlorinated phenols. The large range (from 0.97 for PCP to 0.23 for phenol) is caused by the varying FID sensitivity for the compounds with different numbers of halogen substituents. The relative standard deviation of the response factors was 4-9%, which can be taken as a first evidence for the feasibility of quantitative determinations.

**Table I. Gas Chromatographic Response Factors and Recoveries for Phenol and Chlorinated Phenols**

| Compound | Response Factor[a] ± SD | Recovery[b] (% ± SD) |
|---|---|---|
| Phenol | 0.23 ± 0.02 | 29 ± 4 |
| 2-Chlorophenol | 0.32 ± 0.03 | 83 ± 4 |
| 2,4-Dichlorophenol | 0.48 ± 0.03 | 92 ± 7 |
| 4-Chloro-3-methylphenol | 0.36 ± 0.03 | 46 ± 3 |
| 2,4,6-Trichlorophenol | 0.54 ± 0.03 | 89 ± 3 |
| 2,3,4,6-Tetrachlorophenol | 0.74 ± 0.03 | 92 ± 2 |
| Pentachlorophenol | 0.97 ± 0.07 | 83 ± 1 |

[a]Relative to the internal standard (2,4,6-tribromophenol), three to six determinations, directly after silica column chromatography as described in Experimental.

[b]Three determinations of spiked groundwater samples (approximately 1.5 μg of one individual phenol per liter of water).

## Isolation and Recoveries

Table I contains recovery rates for phenol and chlorinated phenols. The phenolic chemicals were spiked at approximately 1.5 $\mu$g of one individual compound per liter of a groundwater containing no detectable amounts of these substances. The recoveries for all the chlorinated phenols shown in Table I are good (83-92%). 4-Chloro-3-methylphenol (46%) and phenol (29%) show significantly lower recoveries. The reproducibilities of the recovery determination (±7-1% relative standard deviation) are good to excellent.

The precision for the compounds with low recoveries is also good (less than ±4%). In addition, for 4-chlorophenol a recovery of only 26 ± 3% has been found. 1-Naphthol and 4-phenylphenols were recovered at less than 5%. 2,6-Di-tert-butyl-4-methylphenol was recovered at 60%.

In the development of this method it became clear that the applied liquid/liquid extraction (LLE) procedure can only achieve satisfactory recoveries for lipophilic phenols with low water solubilities. The low recovery of phenol is caused by its low extraction efficiency from water into methylene chloride. Other hydrophilic phenols, including nitrophenols, polyhydroxyphenols (like catechol) and 1-naphthol, are recovered even less efficiently. The different recoveries of 2- and 4-chlorophenol can be explained by the fact that these two isomers have very different water solubilities (less than 0.1 and 2.7 g/100 g, respectively) [1]. Replacing methylene chloride by diethyl ether did not help much, although the partition coefficient from water into ether is much higher for many phenols [4]. A more promising way to concentrate hydrophilic phenols is the application of an anion exchange resin for enrichment of these compounds as done by Richard and Fritz [8].

Distillation of the sample, as the initial step of the isolation procedure, allows one to overcome problems associated with emulsions forming, particularly with extractions performed at high pH. Because of this distillation, analyses of very dirty samples such as raw industrial wastewaters are feasible. On the other hand, distillation limits the group of compounds amenable to this determination procedure because only steam-distillable compounds can be determined. We attribute the low recoveries of 1-naphthol and 4-phenylphenol to the insufficient steam volatility of these chemicals. At least three other members of the CVRC list would probably behave similarly (bisphenol A, tetrabromobisphenol A, and 4-chloro-1-naphthol).

TBP proved to be well suited as an internal standard, particularly if PCP is a major target compound. This polybrominated phenol is commercially available, is chemically very similar and elutes close to PCP. TBP is probably not present in waters and wastewaters. For the use of an electron capture detector (ECD), the high ECD sensitivity of this compound is an additional advantage.

It should be noted that some operations significantly reduce the recoveries of phenols. Concentration in a rotary evaporator or column chromatography on fully activated silica gel are two examples worth mentioning. Therefore micro-Snyder columns and deactivated silica were applied.

Less acidic phenols are partially extracted in the extraction at high pH and are therefore insufficiently recovered by our procedure. We suppose that this is partially the cause for the low recovery of 2,6-di-tert-butyl-4-methylphenol.

The silica column chromatography removes acids from the acidic extracts. In many samples $C_{10}$- to $C_{16}$-carboxylic acids were present in the acidic extracts. These acids could also be gas chromatographed as free acids. Good peak shapes, however, were obtained only on very new columns. The removal of these acids should help prolong the life of the capillary column.

## Reproducibility

Table II lists the reproducibilities of the determinations of four phenolic compounds in a primary sewage effluent. An excellent relative standard deviation of ±6% has been obtained for PCP present at a concentration of 0.72 μg/L. Considering the low recoveries for phenol and presumably also the methylphenols, good precisions have been achieved for these compounds.

**Table II. Reproducibility of the Determination of Pentachlorophenol, Phenol and Methylphenols in Primary Sewage Effluent**

| Compound | Concentrations (μg/L) | | | | Relative Standard Deviation (±%) |
|---|---|---|---|---|---|
| | Triplicate Determinations | | | Average ± SD | |
| Pentachlorophenol[a] | 0.77 | 0.71 | 0.69 | 0.72 ± 0.04 | 6 |
| Phenol[a] | 1.7 | 2.3 | 2.1 | 2.0 ± 0.3 | 15 |
| 2-Methylphenol[b] | 2.6 | 2.8 | 2.8 | 2.7 ± 0.1 | 4 |
| 3- and 4-Methylphenol[b] | 2.1 | 2.3 | 2.3 | 2.2 ± 0.1 | 5 |

[a]Concentrations for pentachlorophenol and phenol are corrected for recovery rates of 83 and 29%, respectively.

[b]2-, 3- and 4-methylphenols are determined by using response factor (0.23) and recovery (29%) of phenol. 3- and 4-methylphenols are not separated by the GC column used in this work.

## Applications

Figure 3 shows the gas chromatograms of the phenolic fraction extracted from primary (Figure 3A) and secondary (Figure 3B) sewage effluents. These analyses were made on 24-hour composite samples from a sewage treatment plant in Zurich. The sampling of the secondary effluent was started six hours later than the primary effluent to allow for the residence time in the sludge aerator. Thus, it should be possible to correlate the analyses of the two effluents. In the sewage after the primary treatment (settling tank), PCP and 2,3,4,6-tetrachlorophenol have been found at 0.72 and 0.28 $\mu$g/L, respectively. In the effluent of the aerated sludge treatment plant (secondary effluent) the respective concentrations were 0.47 and 0.18 $\mu$g/L, reflecting 35 and 36% removal of these compounds. These results are similar to the data which have been reported by Buhler et al. [3]. These authors have found 4-28% removal for PCP which was present at 1-5 $\mu$g/L in the raw sewage. Dietz and Traud [4] have determined comparable levels of chlorophenols in sewage effluents.

Phenol was only reduced from 1.9 to 1.5 $\mu$g/L (21% elimination). The methylphenols, however, were efficiently removed at elimination rates of 95%.

In addition to these quantitative determinations, we have identified several phenolic compounds which allow some interesting comparisons between primary and secondary effluents. Dimethylphenols (peak 6) and 2-methoxyphenol (guaiacol, peak 4) were detectable only in the primary effluent. 2-Nitrophenol (peak 5) was a minor constituent in the primary effluent but became a major constituent in the secondary effluent. m/2-Nitromethylphenols (peak 9) and 2-nitrodimethylphenol (peak 10) were not evident from the FID chromatogram of the influent to the aerated sludge treatment, but they were noticeable in the FID trace of the extract of the effluent from the secondary treatment. However, by applying mass chromatography (single ion monitoring of the molecular ions 153 and 167 m/e) 2-nitromethyl- and 2-nitrodimethylphenols could be detected as minor constituents in the phenolic extracts of the primary effluent.

These concentration differences between primary and secondary sewage effluents must be caused by different removal efficiencies for these chemicals in the biological sludge treatment. From an environmental point of view the bioresistant chemicals are of special concern because they survive biological sewage treatment and are discharged into natural waters. It should be mentioned that the two composite samples had been taken during rainy weather. During dry weather significantly higher concentrations of phenolics were found (e.g., 2.1 $\mu$g/L of PCP in the primary effluent).

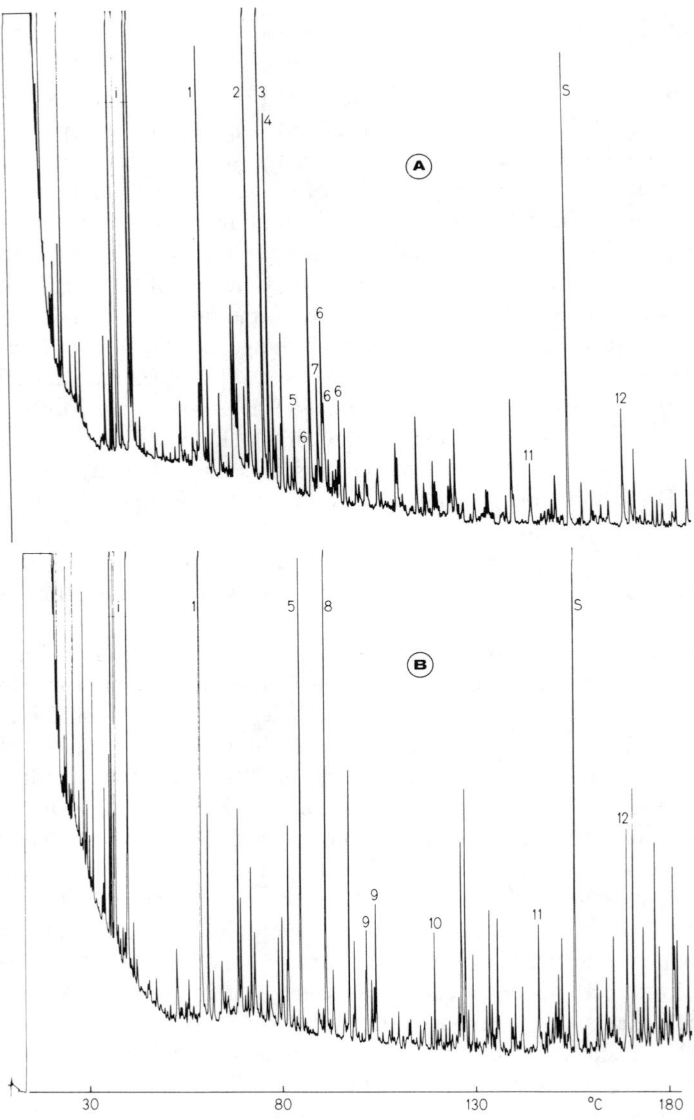


**Figure 3.** Gas chromatograms of phenols extracted from sewage effluents. (A) primary effluent; (B) secondary effluent. (1) Phenol, (2) 2-methylphenol, (3) 3/4-methylphenols, (4) 2-methoxyphenol, (5) 2-nitrophenol, (6) dimethylphenols, (7) 2,4-dichlorophenol, (8) unknown, (9) 2-nitro-methylphenols, (10) 2-nitro-dimethylphenol, (11) 2,3,4,6-tetrachlorophenol, (12) pentachlorophenol, (S) internal standard (2,4,6-di-bromophenol) and (i) impurities of the solvent.

Figure 4 shows the gas chromatogram of phenols extracted from the secondary effluent of the sewage treatment plant at Dübendorf. This chromatogram represents the analysis of a grab sample taken during dry weather. The following concentrations were determined: 1.0 μg/L PCP, 0.9 μg/L 2,3,4,6-tetrachlorophenol, 0.6 μg/L 1,4-dichlorophenol and 0.5 μg/L phenol. 2-Nitrophenol and nitromethylphenols were also identified in this sample.

In water samples from three different Swiss rivers we have determined between 0.14 and 0.16 μg of PCP per liter of riverwater. These values are close to the practical detection limits of our method (approximately 0.1 μg/L).

Figure 5 shows the GC analysis of the phenolic fraction of an industrial wastewater of a pharmaceutical company. The major phenolic constituent of this wastewater was phenol at an approximate concentration of 20 mg/L. This value is probably too low because the recovery of phenol at this concentration is expected to be less than recovery in the μg/L range (for which we have determined it) (Table I). In addition to the most abundant phenol,

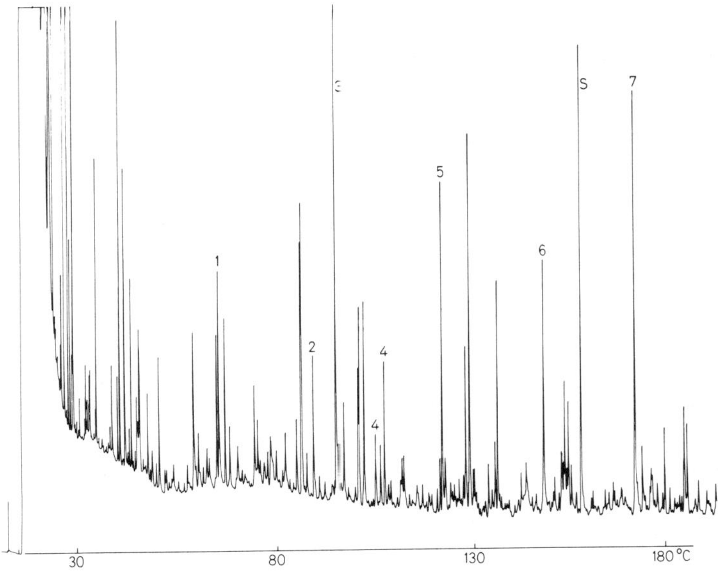


**Figure 4.** Gas chromatogram of phenols extracted from secondary sewage effluent. (1) Phenol, (2) 2-nitrophenol, (3) 2,4-dichlorophenol, (4) 2-nitro-methylphenol, (5) 2,4,6-trichlorophenol, (6) 2,3,4,6-tetrachlorophenol, (7) pentachlorophenol, and (S) internal standard (2,4,6-tribromophenol).

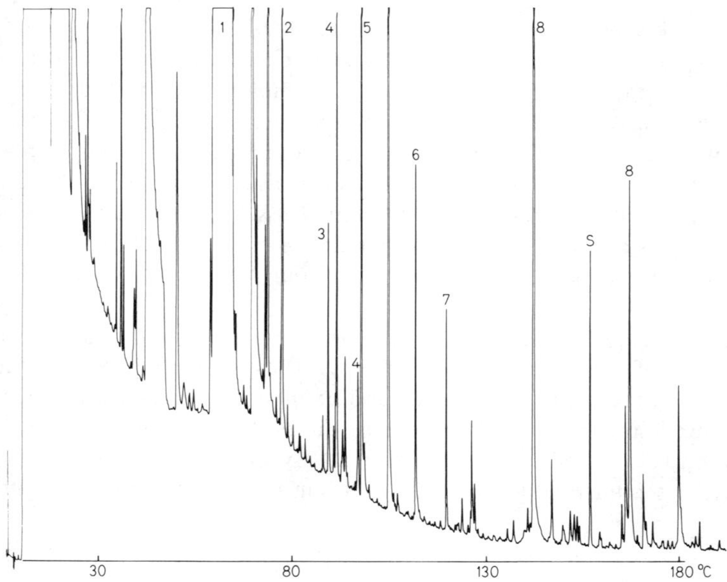


**Figure 5.** Gas chromatogram of phenols extracted from wastewater of a pharmaceutical plant. (1) Phenol, (2) methylphenol, (3) ethylphenol, (4) dichlorophenols, (5) 4-chlorophenol, (6) 4-chloro-3-methylphenol, (7) trichlorophenol, (8) hydroxybiphenyls, and (S) internal standard (2,4,6-tribromophenol).

we have identified a number of other phenols some of them with halogen substituents. The gas chromatogram of Figure 5 demonstrates that our method can be successfully applied to this type of wastewater. Due to the separation efficiency of the column, minor constituents could be determined even if they were present at concentrations more than $10^3$ times lower than the most abundant constituent. 4-Chloro-3-methylphenol (peak 6), for example, was determined at a concentration of 9.8 $\mu$g/L.

Figure 6 shows the total ion current chromatograms of phenolic compounds extracted from wastewater effluents of a sulfite pulp mill. Analyses of grab samples from the raw wastewater (mainly effluents from the chlorine bleachery) (Figure 6A) and from the effluent of a double-stage biological treatment plant (Figure 6B) are depicted. These are very preliminary results, and a detailed investigation of these wastewaters is projected. Although direct comparison of grab samples is difficult, it can be seen that chlorinated phenols (shaded peaks) are relatively more abundant in the effluent from the biological treatment than in the raw wastewater. This is caused by the higher

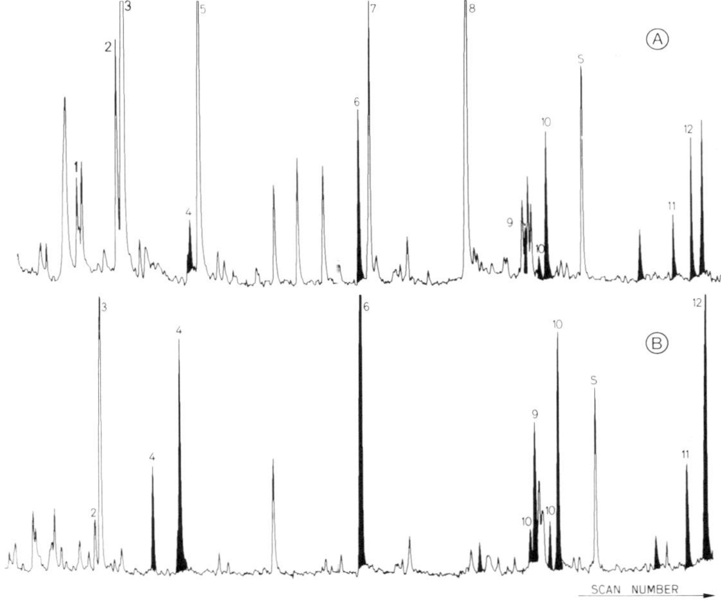


**Figure 6.** Total ion current chromatograms of phenols extracted from wastewaters of a sulfite pulp mill. (A) Raw wastewater; (B) biologically treated wastewater. (1) 2-Methylphenol, (2) 3/4-methylphenol, (3) 2-methoxyphenol (guaiacol), (4) dichlorophenols, (5) ethylphenol, (6) 2,4,6-trichlorophenol, (7) $C_5$-phenol, (8) $C_6$-phenol, (9) 2,3,4,6-tetrachlorophenol, (10) trichloroguaiacols, (11) pentachlorophenol, (12) tetrachloroguaiacol, (S) internal standard (2,4,6-tribromophenol) and (shaded peaks) chlorinated phenols.

biological resistance of the chlorinated phenols. The unchlorinated phenols are removed quite efficiently by the wastewater treatment process. Of special significance (bioaccumulation, toxicity) are the tri- and tetrachloroguaiacols which have been already described in the literature [13,14]. The two chromatograms of Figure 6 again demonstrate that our procedures are well suited for analyses of wastewaters which contain extremely complex mixtures of organic compounds. In the phenolic concentrates provided by our isolation scheme only a few nonphenolic interferences are observed, e.g., some early eluting aromatic hydrocarbons and some phthalates eluting after the last peaks shown in Figure 6.

## ACKNOWLEDGMENTS

The authors thank K. and G. Grob for supplying glass capillary columns and for very valuable advice. The composite samples of sewage effluents were collected by C. Liebi.

This work was supported in part by the Swiss Department of Commerce (Commission of the European Communities Project COST 64b).

## REFERENCES

1. Buikema, A. L., Jr., M. J. McGinniss and J. Cairns, Jr. *Marine Environ. Res.* 2:87-181 (1979).
2. Van Hall, C. E., Ed. *Measurement of Organic Pollutants in Water and Wastewater*, ASTM Spec. Pub. 686 (Philadelphia: American Society for Testing and Materials, 1979).
3. Buhler, D. R., M. E. Rasmusson and H. S. Nakane. *Environ. Sci. Technol.* 7:929-934 (1973).
4. Dietz, F., and J. Traud. *Vom Wasser* 51:235-257 (1978).
5. Wegmann, R. C. C., and A. W. M. Hofstee. *Water Res.* 13:651-657 (1979).
6. Keith, L. H., and W. A. Telliard. *Environ. Sci. Technol.* 13:416-423 (1979).
7. Keith, L. H. *Environ. Sci. Technol.* 13:1469-1471 (1979).
8. Richard, J. J., and J. S. Fritz. *J. Chromatog. Sci.* 18:35-38 (1980).
9. Krijgsman, W., and C. G. Van de Kemp. *J. Chromatog.* 131:412-416 (1977).
10. *Standard Methods for the Examination of Water and Wastewater*, 14th ed. (New York: American Public Health Association, 1976).
11. Grob, K., G. Grob and K. Grob, Jr. *J. High Resolution Chromatog. Chromatog. Commun.* 2:31-35 (1979).
12. Grob, K., G. Grob and K. Grob, Jr. *J. High Resolution Chromatog. Chromatog. Commun.* 2:677-678 (1979).
13. Rogers, I. H., and L. H. Keith. In: *Identification and Analysis of Organic Pollutants in Water*, L. H. Keith, Ed. (Ann Arbor, MI: Ann Arbor Science Publishers, Inc., 1976), pp. 625-639.
14. Lindstrom, K., and J. Nordin. *J. Chromatog.* 128:13-26 (1976).

# CHAPTER 9

# CHROMATOGRAPHIC CONSIDERATIONS FOR THE ANALYSIS OF ACID-EXTRACTABLE PRIORITY POLLUTANTS

**R. R. Freeman**
Hewlett-Packard Co.
Avondale, Pennsylvania

The U.S. Environmental Protection Agency (EPA) has proposed regulations establishing test procedures for the analysis of selected pollutants [1]. These pollutants have been specified as priority pollutants and include 14 metals and 114 organic compounds. The organic priority pollutants may be divided into two classes: volatiles and extractables. The extractables can be further divided into base/neutral and acid fractions. The acid extractables include 11 phenolic compounds, and Method 604 is applicable to the determination of these compounds in municipal and industrial discharges. The method involves a concentration step [liquid/liquid extraction (LLE)] and a derivatization step (pentafluorobenzyl derivatives) to remove interference and improve the overall sensitivity of the analysis. While the base/neutral pollutants can easily be chromatographed at sub-ppm levels on a high-resolution 30-m open tubular column, the acidic pollutants have proven to be difficult to chromatograph even at relatively high concentrations (approx. 20–50 ng) on either conventional packed or high-resolution open tubular columns [2].

A special phosphoric acid deactivated polyester stationary phase (SP-1240 DA) improved the quality of the separation over the original Tenax GC column; however, the quantitative, reproducible analysis of the 11 phenols continues to be a difficult analytical problem.

Recent developments in the area of high-resolution gas chromatography (GC) have provided new tools to the analyst which should facilitate the determination of phenols. The purpose of this investigation is to determine the chromatographic parameters which have a significant effect on the quantitative analysis of the acid extractable priority pollutants. Both column and extracolumn effects will be studied.

All columns used during the course of this investigation were either manufactured by Hewlett-Packard (Avondale, PA) or J&W Scientific (Orangevale, CA). Throughout the study a Hewlett-Packard 5880A Gas Chromatograph was used. All phenol standards were made gravimetrically using a Mettler ME-30 Laboratory Balance. Four primary standard solutions were used during the course of this investigation in an attempt to eliminate any bias associated with the preparation of the samples. The sampling with respect to the phenol concentrations was random to prevent column priming.

## RESULTS AND DISCUSSION

### Column Effects

Column effects include solute interactions with:

1. the column surface,
2. the surface deactivating agent, and
3. the stationary phase.

Columns made from conventional glasses such as borosilicate or soda-lime cannot be used. These glasses all contain alkali metal oxides which have a demonstrable adverse effect on the chromatography of the priority pollutant phenols [3]. Figure 1 shows the separation of six phenols on two different Se-54 columns. One column is made from fused-silica [4]; the other column is made from borosilicate glass. The underivatized phenols are chromatographed with excellent peak shape at the 20-ng level on the fused-silica column, whereas on the borosilicate column only 2-nitrophenol, 2,4,-dichlorophenol and 4-chloro-3-methyl phenol can be detected. There is no evidence of 2,4-dinitrophenol, 4-nitrophenol or pentachlorophenol (PCP). Clearly, the inertness of the fused-silica column is the determining factor in this analysis. A fused-silica column must be used if these phenols are to be analyzed at the ppm level.

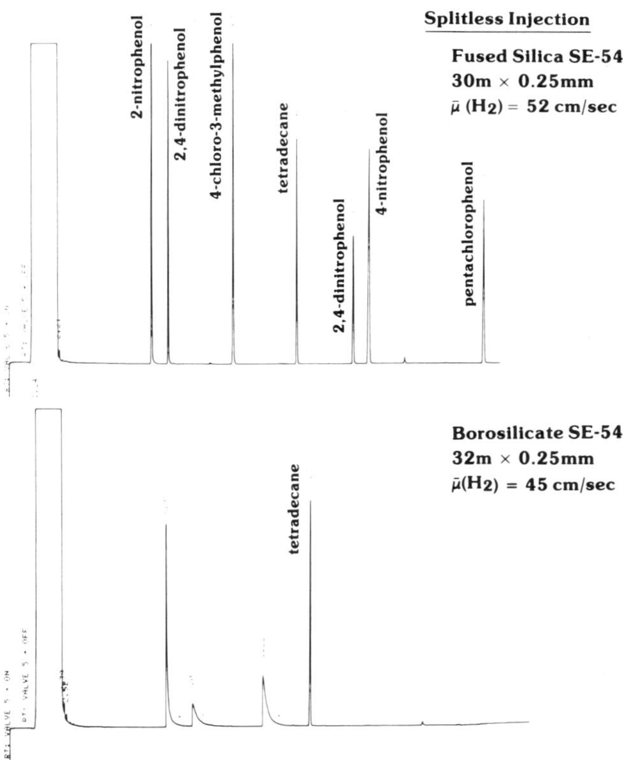


**Figure 1.** Free phenol analysis (20 ng/component) on borosilicate and fused-silica open tubular columns. In both cases a fused-silica injection port liner was used, and the column extended to within 5 mm of the flame. Both columns were manufactured by J&W Scientific.

The presence of silanol, siloxane and oxygen anions on the surface of the fused-silica column produce a surface which is acidic in character. Consequently, various deactivating techniques have evolved to generate a better defined surface. These include Carbowax 20M®* [5], silylations [6] and siloxane reactions [7]. Each of these techniques produce a modified surface which is chemically well defined. In practice, the Carbowax treatment produces a basic surface, and the remaining techniques appear to leave the surface acidic. Naturally, the phenols will experience less surface inter-

*Carbowax is a registered trademark of Union Carbide Corp.

action on the acidic surface than on a basic surface; this can easily be demonstrated by analyzing the phenols on columns prepared with the various deactivating agents. This set of experiments is best done by installing two columns (each prepared using a different deactivation procedure) simultaneously in the injection port. This removes any bias associated with the injection process. Two of these experiments are summarized in Figure 2. Two columns with an acidic surface (SP-2100: DMCS deactivated, and SE-54: no deactivation) are compared to a column with a basic surface (SP-2100: Carbowax deactivated). Both 2,4-dinitrophenol and PCP are easily chromatographed on the two acidic columns, whereas neither compound is eluted from the Carbowax deactivated column. At the 5 to 10-ng level these two fairly acidic compounds are completely adsorbed on the

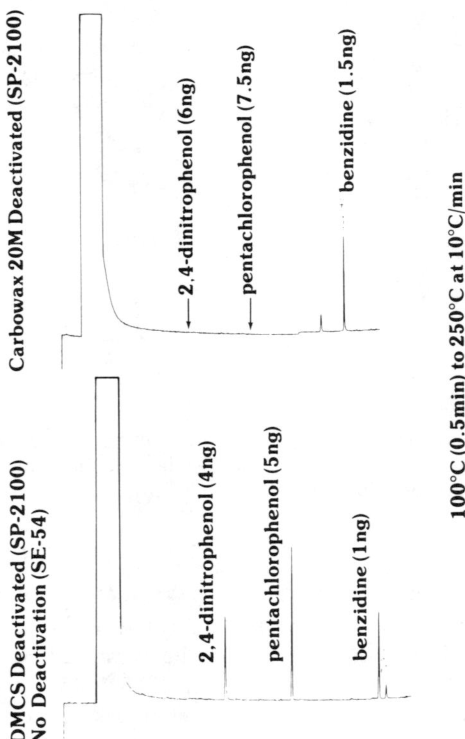


**Figure 2.** Effect of surface deactivation on the separation of acidic priority pollutants. With either an undeactivated or a silylated surface, the free phenols can be chromatographed at the nanogram level. A Carbowax 20M deactivated surface is basic, which appears to adsorb the acidic phenols at the nanogram level. All analyses were made using splitless injection and hydrogen carrier gas. The columns were 20- to 25-m x 0.2-mm fused-silica prepared by Hewlett-Packard.

Carbowax deactivated surface. It is not surprising that the second criteria for the analysis of the acid extractable priority pollutants is that the column surface be neutral-to-acidic.

The choice of the stationary phase is of secondary importance in this case. Seclected phenols were analyzed on SE-30 (a methyl silicone gum) and SE-54 (a 1% vinyl, 5% phenyl, methyl silicone gum). The degree of selectivity of the SE-54 for the phenols is illustrated by plotting the difference in the retention times on the two stationary phases as a function of oven temperature program rate (Figure 3). Two conclusions can be drawn from the data. First of all, the SE-54 shows a fair amount of selectivity for the nitrophenols as evidenced by the dramatic difference in the retention times at low program rates. Secondly, it is not surprising that at high program rates this degree of selectivity is lost, for at high program rates (i.e., higher elution temperatures) each solute spends less time in the liquid phase than at low program rates.

A stationary phase should be selected on the basis of stability. Gum phases such as SE-54 and SE-30 have demonstrated a resistance to water and oxygen along with exhibiting excellent thermal stability.

## Extracolumn Effects

Extracolumn effects can be described as any interaction between the solute(s) and the chromatographic system other than those occurring on the column. In practice these can be limited to those interactions which occur during the sampling process and those which occur at the detector interface.

Of the various sampling techniques used for trace analysis, perhaps splitless and on-column are recognized as the most universal. This is because the necessary hardware is readily available and the operational parameters are well understood [8,9]. Splitless injection requires a flash vaporization of the sample. This gaseous mixture is then transferred onto the column where the solute bandwidths are condensed using either a "solvent-effect" or a cold trapping mechanism. With on-column injection the sample is placed directly on the column without a volatilization step. The discrimination which may occur either because of:

1. prevolatilization of the solvent in the syringe,
2. thermal instability of the solute, or
3. catalytic effects of the injection port liner

can be reduced significantly by eliminating the flash vaporization of the sample.

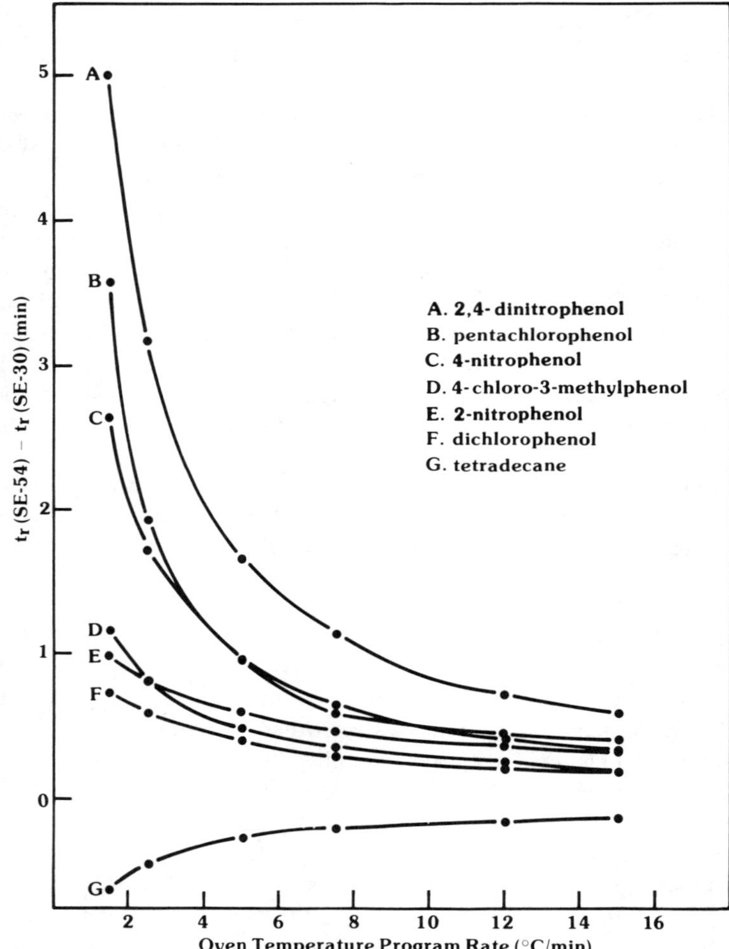


**Figure 3.** Stationary phase selectivity for free phenols. The greater the differences between the retention times of the phenol on the two stationary phases the greater the selectivity of the active phase (i.e., SE-54) for the particular phenol. Note that selectivity is greatest for the more acidic phenols. Columns: 30-m x 0.25-mm SE-30 and 30-m x 0.25-mm SE-54. $\bar{\mu}$ ($H_2$) = 50 cm/sec. Temp 1 (2 min) = 75°C. Columns manufactured by J&W Scientific.

The experimental conditions for the analysis of free phenols using splitless injection have been investigated. Kalman [10] has reported injection port temperature (180–350°C) effects on the analysis of phenols using splitless injection. Results indicate that for the 11 priority pollutant phenols the least amount of scatter in the data (expressed as a percent relative standard deviation) occurs at an injection port temperature between 200 and 225°C.

All splitless data generated during the course of this investigation were obtained with the injection port at 220°C.

A second study dealt with the effect the injection port liner configuration has on the analysis. First of all it was found that even a small amount (38 mg) of coated (2% OV-101) Chromosorb-WHP can strongly influence the analytical results. For example, at the 20-ng level 2,4-dinitrophenol and PCP could not be detected with such a "packed" injection port liner. Standard splitless injection port liners ("open" 2-mm i.d. liners) made from borosilicate and fused-silica glasses were compared. At the 20-ng level no differences in the peak shape or height of the individual phenols could be determined. At the 1-ng level the responses (expressed as pA/ng) for 2,4 dinitrophenol and PCP were 22 and 10% lower, respectively, with the borosilicate liner. This effect is reproducible; however, the magnitude of the effect varies with the condition of the splitless liner. It seems to be strongly dependent on the cleanliness, and the degree of deactivation of the splitless liner.

An indication of how the flash vaporization process affects the analysis of the free phenols can be obtained by comparing the response of the phenols relative to an internal standard using splitless and on-column injection. Phenol discrimination as a function of injection mode is shown in Figures 4 and 5. Note that in each intance the phenol/tetradecane area ratio is larger when on-column injection is used. This can be attributed to a loss occurring during the splitless injection because of the flash vaporization process. At low levels (i.e., less than 1 ng) the area ratio decreases dramatically when using splitless injection. The rate of decrease is inversely proportional to the amount of the various phenols injected. In every instance, the phenol/tetradecane ratio remains more constant (indicating a lower level of discrimination) with on-column rather than splitless injection. The ratio also provides an indication of the phenol adsorption occurring on the column surface. 2,4-Dinitrophenol provides an excellent example where the differentiation between discrimination (splitless) and adsorption (on-column) mechanisms can be observed.

Perhaps the most outstanding characteristics of on-column injection is the high precision realized. Table I shows the precision of on-column and splitless injection for two of the phenols as a function of concentration. These two phenols represent the extremes in analytical difficulty. The precision of the data with on-column sampling is observed to be 2–10 times better, depending on the phenol concentration, than data obtained using splitless injection.

A second area of concern lies in the detector interface. With fused-silica columns it is a simple matter to extend the column to the tip of the flame jet, in essence, eliminating the detector interface. This is essential, for it can easily be demonstrated (Figure 6) that hot metallic surfaces have an adverse effect on the peak shape of free phenols—especially 2,4-dinitrophenol.

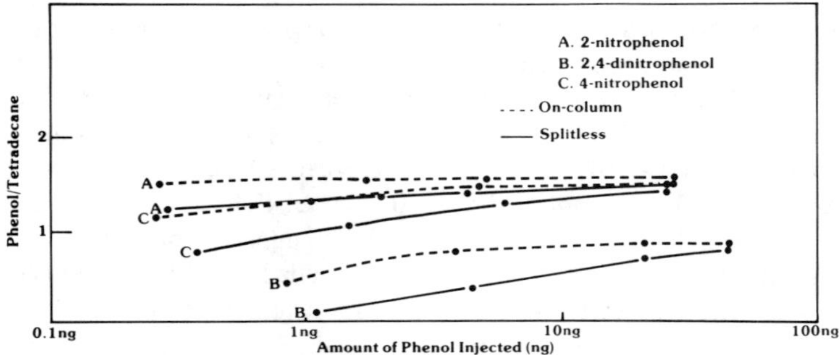


**Figure 4.** Phenol adsorption as a function of injection mode for selected nitrophenols. The area ratio (active compound/passive compound) is a sensitive indicator of column activity. Column: fused-silica SE-54 (18-m x 0.32-mm), $\bar{\mu}$ ($H_2$) = 50 cm/sec. Oven profile: 75 (1 min) to 220°C at 7.5°C/min.

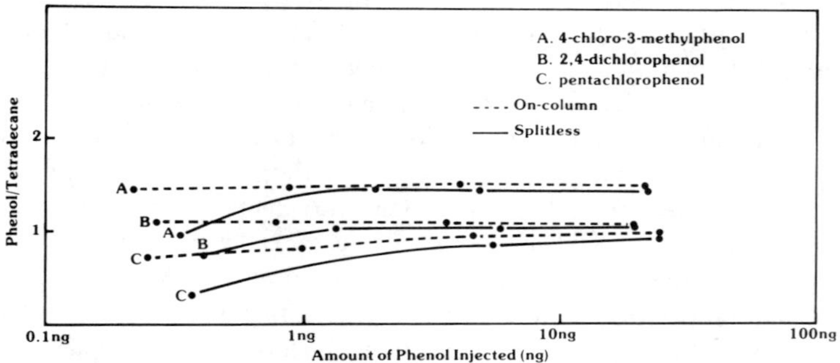


**Figure 5.** Phenol adsorption as a function of injection mode for selected chlorophenols. For conditions see Figure 4.

## SUMMARY

The analysis of the priority pollutant phenols can be described as a three-tiered problem. As the phenol concentration decreases the nature of the analytical problem becomes more involved. This situation is, of course, typical in the realm of trace analysis. The chromatographic conditions vary as the concentration decreases. Above 5 nanograms the analysis can be performed using either on-column (Figure 7) or splitless injection (with either a borosilicate or silica injection port liner) (Figure 8) and a fused-silica open

Table I. Precision of the Two Sampling Modes as a Function of Phenol Concentration: Phenol/Tetradecane Area Ratios for One of the Easier and the Most Difficult Phenols

| Concentration (ng) | 4-Chloro-3-methylphenol/ Tetradecane Ratio | | 2,4-Dinitrophenol/ Tetradecane Ratio | |
|---|---|---|---|---|
| | Splitless | On-column | Splitless | On-column |
| 25 | 1.454 ± 0.021 (1.4)[a] | 1.489 ± 0.006 (0.4) | 0.694 ± 0.129 (19) | 0.851 ± 0.012 (1.4) |
| 5 | 1.460 ± 0.062 (4.3) | 1.473 ± 0.006 (0.4) | 0.351 ± 0.043 (12) | 0.765 ± 0.014 (1.8) |
| 1 | 1.522 ± 0.063 (4.5) | 1.477 ± 0.028 (2.0) | 0.126 ± 0.021 (17) | 0.393 ± 0.048 (12) |
| 0.25 | 0.939 ± 0.135 (14) | 1.440 ± 0.004 (0.3) | | |

[a] Numbers in parentheses = percent relative standard deviation (n = 14).

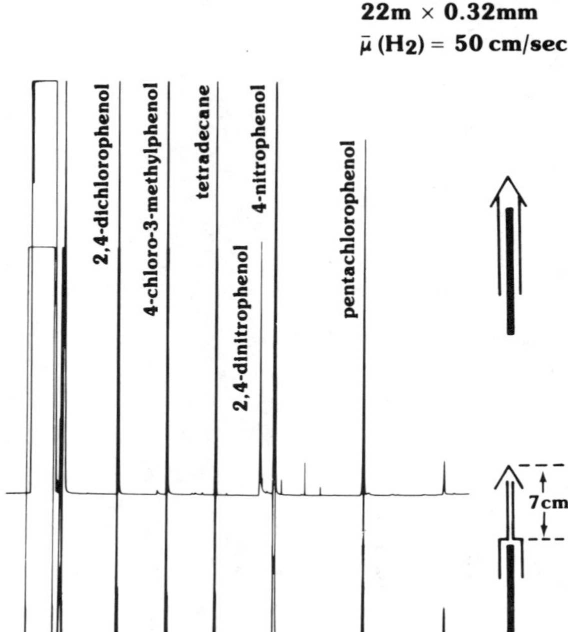


**Figure 6.** Free phenol adsorption at the detector interface. In the upper chromatogram the column is extended to the flame tip. In the lower chromatogram the column effluent passed through 7 cm of a 0.010-in. 316 stainless flame jet. Note that the peak tailing is more evident on the lower trace and that the 2,4-dinitrophenol is almost completely adsorbed.

tubular column characterized by an acidic column surface and a nonpolar liquid phase. At the 5-ng level, column adsorption starts to become significant for 2,4-dinitrophenol. The remaining phenols can be analyzed routinely at this level. At the nanogram level this analysis can be done with splitless (with a silica injection port liner) (Figure 9) or on-column (Figure 10) injection; however, better precision and higher signal/noise will be realized, especially for 2,4-dinitrophenol, with on-column injection. Obviously a fused-silica open tubular column (acid-to-neutral surface) coated with a gum nonpolar stationary phase must be used. The gum phase provides a stable column which is necessary for the analysis. At the nanogram level special care must be paid to the glassware and the storage of the samples [11]. The free chloro- and nitrophenols are chemically active, and experience indicates that the glassware should be acid-deactivated and the samples stored at low temperatures. At the sub-nanogram level this analysis is extremely difficult on a routine basis. Subtle differences in the column can have a strong effect on the phenol adsorption. Nevertheless, with on-column injection, a nonpolar (gum) fused-silica column and a silica detector interface, the analysis can be done with less than 2% relative standard deviation between analyses (Figure 11). 2,4-Dinitrophenol presents a real problem at this level and at least with the columns used during the course of this investigation it is extremely difficult to detect below 1 ng.

At the subnanogram level consideration should be given to the possibility of derivatizing the phenols either before [12] or after the extraction [13]. This will reduce the glassware and storage problems and provide the chemical selectivity inherent in a derivatization procedure. In addition, the phenol

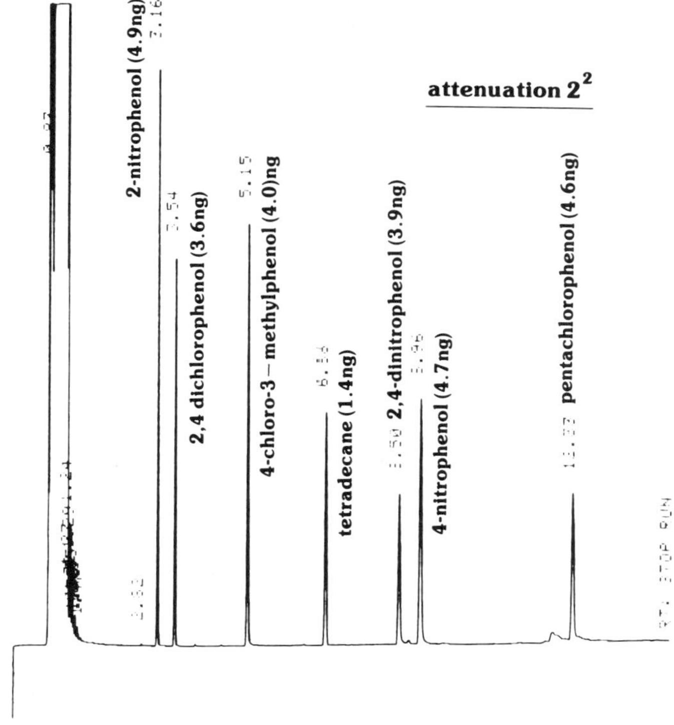


Figure 7. Free phenol analysis: on-column injection. Column: fused-silica SE-54 (18-m x 0.32-mm), $\bar{\mu}$ ($H_2$) = 50 cm/sec. Oven profile: 90 (1 min) to 220°C at 7.5°C/min.

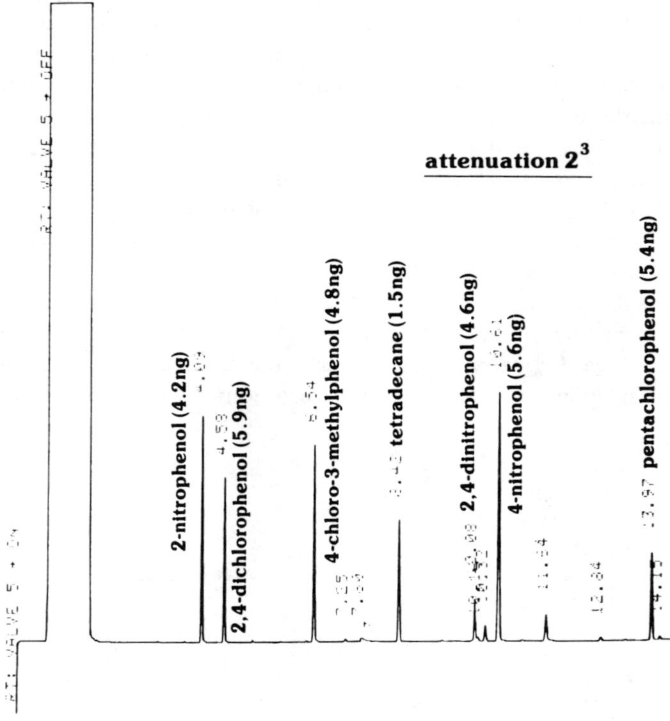


**Figure 8.** Free phenol analysis: splitless injection. Column: fused-silica SE-54 (18-m x 0.32-mm), $\bar{\mu}$ ($H_2$) = 50 cm/sec. Oven profile: 75 (1 min) to 220°C at 7.5°C/min.

derivative may enhance the chromatographic selectivity and sensitivity if atom-selective detectors can be used. This will greatly improve the reliability of the entire analytical method for phenols at the subnanogram levels.

## CONCLUSION

The advent of fused-silica open tubular columns and the development of a "cool" on-column sampling system have made possible the quantitative

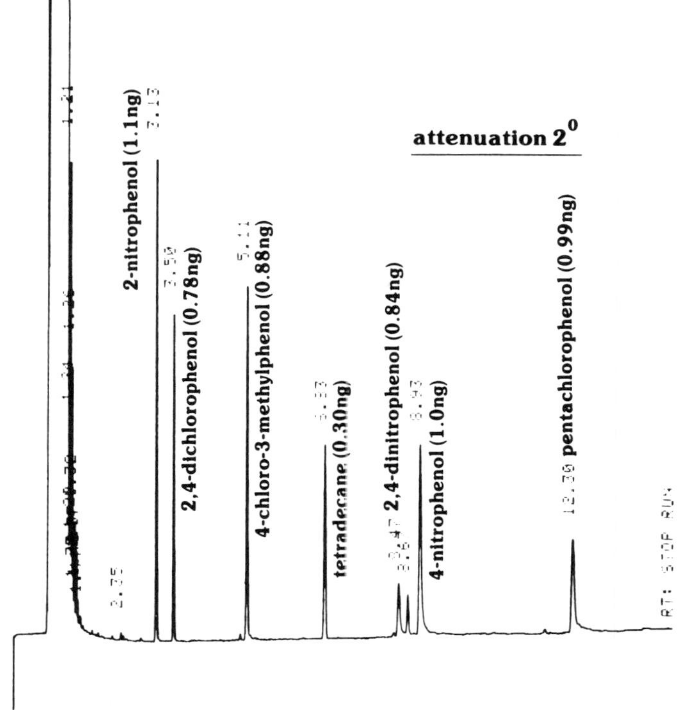


**Figure 9.** Free phenol analysis: splitless injection. See Figure 8 for conditions.

analysis of underivatized priority pollutant phenols at the $\mu$g/L level. The joining of these two techniques provides:

1. a high-resolution separation of often complex matrices,
2. the overall inertness necessary for the analysis of these polar acidic compounds, and
3. a quantitative, high-precision sampling technique which eliminates the discrimination historically associated with open tubular columns.

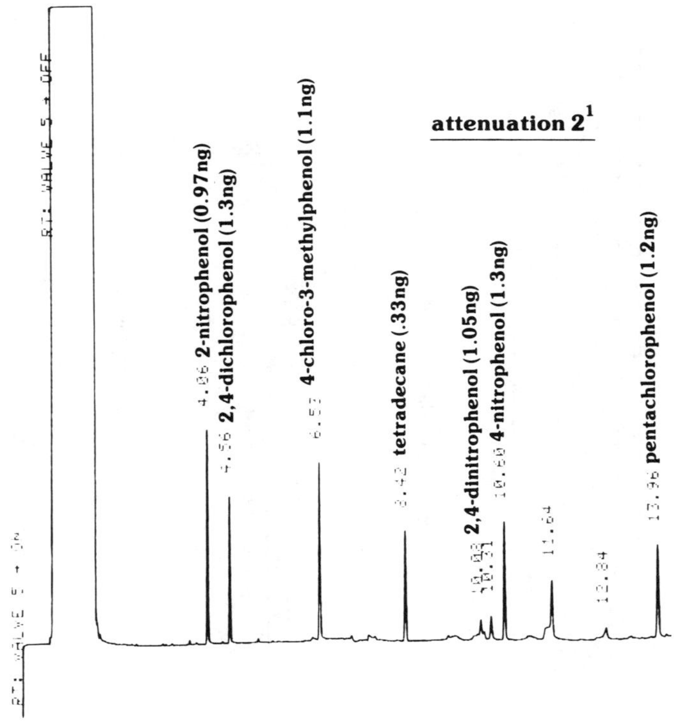


**Figure 10.** Free phenol analysis: on-column injection. See Figure 7 for conditions.

## REFERENCES


1. *Federal Register* (December 3, 1979), p. 69464.
2. Van Hall, C. E., Ed. *Measurement of Organic Pollutants in Water and Wastewater*, ASTM Spec. Tech. Pub. 686 (Philadelphia: American Society for Testing and Materials, 1979), p. 191.
3. Dandeneau, R., P. Bente, T. Rooney and R. Hiskes. *Am. Lab.* (June 1979).
4. Dandeneau, R., and E. H. Zerenner. *J. High Resolution Chromatog. Chromatog. Commun.* 3:351 (1979).
5. Cronin, D. A. *J. Chromatog.* 97:263 (1974).
6. Grob, K., and G. Grob. *J. High Resolution Chromatog. Commun.* 3:197 (1980).
7. Schomburg, G., H. Husmann and H. Behlau. *Chromatographia* 13:321 (1980).

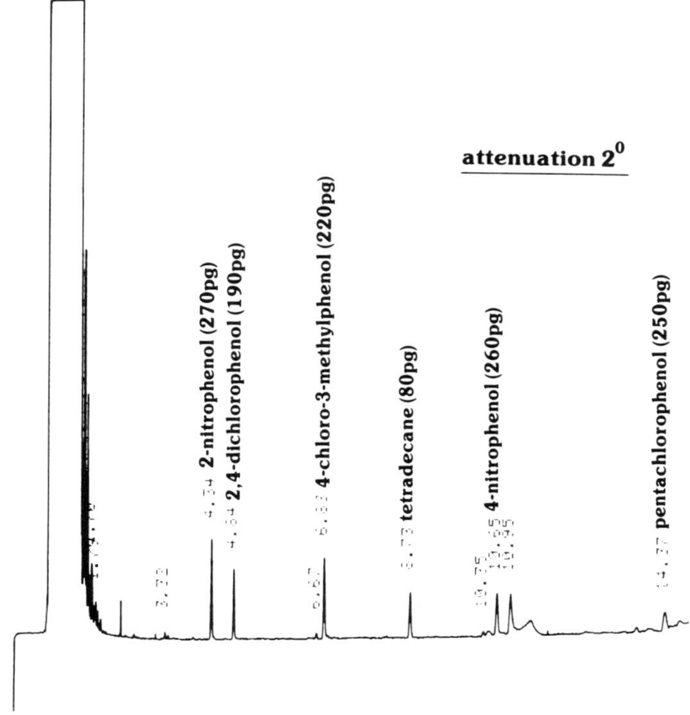


**Figure 11.** Free phenol analysis: on-column injection. Conditions similar to Figure 7, except the program rate is 5°C/min.


8. Freeman, R. R., Ed. *High Resolution Gas Chromatography*, No. 5950-3562 (Avondale, PA: Hewlett-Packard Co.).
9. Freeman, R. R., K. B. Augenblick and R. H. Phillips. Technical Paper No. 88, Hewlett-Packard Co., Avondale, PA.
10. Kalman, D., Department of Environmental Health, University of Washington. Personal communication (January 1980).
11. Hall, J. R., Hydroscience Associates, Inc. Personal communication (September 1979).
12. Lamparski, L. L., and T. J. Nestrick. *J. Chromatog.* 156-143 (1978).
13. Coutts, R. T., E. E. Hargesheimer and F. M. Pasutto. *J. Chromatog.* 179:291 (1979).

# CHAPTER 10

# PART-PER-TRILLION ANALYSIS OF VOLATILE, BASE/NEUTRAL AND ACIDIC WATER CONTAMINANTS ON A SINGLE FUSED-SILICA CAPILLARY COLUMN

**Albert R. Trussell and Fong-Yi Lieu**

Environmental Research Laboratory
James M. Montgomery, Consulting Engineers, Inc.
Pasadena, California

**James G. Moncur**

Lockheed Missiles and Space Co.
Palo Alto, California

Rapidly changing technology in the last 20 years has led to the development of a number of advances in analytical water chemistry. Notable among these advances are improvements in computers, electronic circuitry, instrumentation, capillary columns, and purity of solvents and standards. In summation, these events have taken trace organic chemists from the part-per-million era into the part-per-trillion era. Most recently, the advent of high-quality, fused-silica open tubular (FSOT) capillary columns has led to a number of new capabilities. The flexibility of the FSOT capillary given by its polyimide coating allows the column to be interfaced directly into the ion source of a msass spectrometer for added sensitivity. This same flexibility allows the column to be looped easily for focused cryogenic cooling during the desorption step of purgeable organic analysis. The ability of the FSOT column to withstand intensely low temperatures combined with its insensitivity to water permits efficient purgeable organic analysis. Furthermore, the inert nature of the FSOT column and its lack of polar active sites for adsorption of acidic and basic compounds allow the analysis of base/neutral and acid fractions [1]. The FSOT column also has low bleed at 275–300°C, which facilitates analysis of polynuclear aromatic (PAH) compounds. The summation of these capabilities lead to the analysis of purgeable, base/neutral and acid extractable organic water contaminants on a single column. Additional features of the same column allow the resolution of pesticides, and the column can resolve selected base/neutral compounds during the purge-and-trap technique. The experimental procedures, detection limits, and applications of these techniques are detailed below.

## VOLATILES

### Experimental

Volatile organics were analyzed on a Finnigan 4021 GC/MS (Finnigan Corporation, Sunnyvale, CA) interfaced (Figure 1) with a Tekmar LSC-2 liquid sample concentrator (Tekmar, Inc., Cincinnati, OH). Primary standards were made gravimetrically in methanol at the 1-$\mu$g/$\mu$L level, with the exception of the volatile gases, which were purchased (Supelco, Inc., Supelco Park, PA). Blank water was produced by purging deionized water at 60°C overnight with ultrapure nitrogen. The sample size utilized in this study was 25 mL. The sample was purged for eight minutes with a flow of 15 mL per minute into a Tenax/silica gel trap at 25°C [2]. The compounds were desorbed at 180°C for 2.5 minutes with a 15-mL/min flowrate. During the desorption step, the volatile organics were trapped on a loop at the front of the capillary column with the use of liquid nitrogen in a Dewar flask. The column utilized was a FSOT capillary column 0.25 mm x 30 m, coated with the liquid phase SE-54 (J&W Scientific, Rancho Cordova, CA). The column head pressure was held constant at 10 psi by setting the GC flow controller to 15 mL/min. A split ratio of 10:1 was used during the desorption step. The gas chromatograph was held isothermal at 30°C for 4.5 min. Then, the oven was programmed to 65°C at 4°C/min followed by 8°C/min to 150°C, where temperature was held for the duration of the run. Between runs, the Tenax trap was baked out at 225°C for 10 min. Injector, interface and transfer line temperatures were maintained at 250°C. The mass spectrometer was run in the electron impact, direct mode, with an emission current of 0.5 mA, an electron energy of 70 eV, the electron multiplier at 1100 V, the dynodes

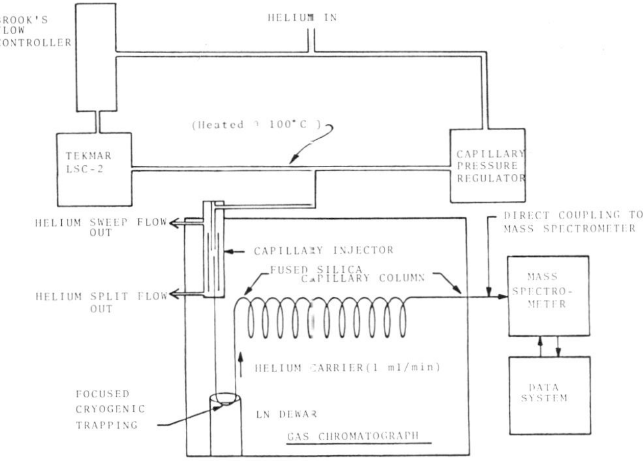


**Figure 1.** Interfacing schematic.

at 3000 V and the ion source at 250°C. The data system was programmed to scan from m/e 33 to 283 in 0.75 sec, with a 0.05-sec hold at the bottom.

## Results

A systematic study of this purge and trap capillary technique is shown in Table I. Aqueous standards containing 35 organic compounds were made at concentrations of 20, 2, 0.2 and 0.02 $\mu$g/L. These four water standards and the blank water in which they were made were analyzed by the technique discussed above. The results are graphically represented in Figure 2, which shows the chromatogram of the 200-ng/L run. This chromatogram demonstrates the excellent resolution and sensitivity of the technique. It also demonstrates the presence of several artifact peaks. Carbon dioxide, which is absorbed in the Tenax trap, elutes at 4 min, 20 sec. Dichlorodifluoromethane (4:25) and dichlorodifluoroethane (4:52 were trapped and concentrated from the carrier gas in all runs. Tetrahydrofuran was a system contaminant which appeared at 7 min, 50 sec. Hexamethylcyclotrisiloxane elutes at 11 min, 10 sec and is a column artifact. Dimethylbenzyl alcohol eluted at 24 min and was a system contaminant. None of these interferences coeluted with the 35 compounds of interest.

**Table I. Analysis of Volatile Organics Standards**

| Entry No. | Compound | Retention Time (min:sec) | Quantification Ion | Ion Areas | | | | | External Standard Quantification | | | | |
|---|---|---|---|---|---|---|---|---|---|---|---|---|---|
| | | | | Water Blank | 0.02 μg/L | 0.2 μg/L | 2.0 μg/L | 20 μg/L | Water Blank | 0.02 μg/L | 0.2 μg/L | 2.0 μg/L[a] | 20 μg/L |
| 1 | Chloromethane | 4:25 | 50 | ND[b] | 560 | 4,475 | 53,804 | 352,995 | ND | 0.021 | 0.166 | 2.00 | 13.1 |
| 2 | Vinyl chloride | 4:27 | 62 | ND | 295 | 3,904 | 42,843 | 531,061 | ND | 0.014 | 0.182 | 2.00 | 16.4 |
| 3 | Bromomethane | 4:32 | 94 | ND | 223 | 2,558 | 24,059 | 226,998 | ND | 0.018 | 0.212 | 2.00 | 18.8 |
| 4 | Chloroethane | 4:34 | 64 | ND | 197 | 2,796 | 29,417 | 286,956 | ND | 0.013 | 0.190 | 2.00 | 19.5 |
| 5 | Trichlorofluoromethane | 4:40 | 101 | ND | 225 | 1,443 | 17,720 | 183,055 | ND | 0.022 | 0.138 | 1.70 | 17.5 |
| 6 | 1,1-Dichloroethane | 4:49 | 61 | ND | 303 | 3,702 | 43,771 | 348,131 | ND | 0.013 | 0.161 | 1.90 | 15.1 |
| 7 | Dichloromethane | 4:54 | 84 | 5,817[c] | 6,571[c] | 9,652[c] | 40,374 | 330,750 | 0.326 | 0.374 | 0.549 | 2.30 | 18.8 |
| 8 | *trans*-1,2-Dichloroethene | 5:04 | 96 | ND | 367 | 2,900 | 31,032 | 285,091 | ND | 0.024 | 0.188 | 2.02 | 18.5 |
| 9 | 1,1-Dichloroethane | 5:10 | 63 | ND | 462 | 5.735 | 59,067 | 431,032 | ND | 0.013 | 0.165 | 1.70 | 12.4 |
| 10 | Bromochloromethane | 5:33 | 130 | ND | 98 | 1,503 | 15,496 | 151,344 | ND | 0.018 | 0.272 | 2.80 | 27.3 |
| 11 | Chloroform | 5:34 | 83 | 603[c] | 1,126[c] | 5,983 | 55,975 | 446,118 | 0.025 | 0.050 | 0.267 | 2.50 | 19.9 |
| 12 | 1,1,1-Trichloroethane | 5:54 | 97 | ND | 769 | 3,202 | 30.919 | 370,254 | ND | 0.045 | 0.186 | 1.80 | 21.6 |
| 13 | Carbon tetrachloride | 5:54 | 117 | ND | 63 | 403 | 4,768 | 61,297 | ND | 0.026 | 0.169 | 2.00 | 25.7 |
| 14 | 1,2-Dichloroethane | 5:58 | 62 | ND | 341 | 2,509 | 23,738 | 279,659 | ND | 0.026 | 0.190 | 1.80 | 21.2 |
| 15 | Benzene | 6:10 | 78 | 30,885[c] | 35,677[c] | 55,066[c] | 293,113 | 895,116[d] | 0.211 | 0.341 | 0.526 | 2.80 | 8.5 |
| 16 | 1,2-Dichloropropane | 6:50 | 63 | ND | 648 | 6,649 | 66,123 | 557,820 | ND | 0.023 | 0.241 | 2.40 | 20.2 |
| 17 | Trichloroethene | 6:50 | 130 | 152[c] | 582 | 4,349 | 45,401 | 431,205 | 0.007 | 0.029 | 0.220 | 2.30 | 21.8 |
| 18 | Bromodichloromethane | 7:00 | 83 | ND | 168 | 2,614 | 28,928 | 361,575 | ND | 0.012 | 0.184 | 2.00 | 25.4 |
| 19 | *trans*-1,3-Dichloropropene | 7:54 | 75 | ND | 285 | 3,274 | 33,666 | 394,363 | ND | 0.016 | 0.183 | 1.90 | 22.1 |
| 20 | Toluene | 8:42 | 92 | 6,541[c] | 11,167[c] | 19,784[c] | 99,849 | 783,919[d] | 0.224 | 0.190 | 0.337 | 1.70 | 13.3 |

| | | | | | | | | | | | | | |
|---|---|---|---|---|---|---|---|---|---|---|---|---|---|
| 21 | *cis*-1,3-Dichloropropene | 8:43 | 75 | ND | 177 | 2,305 | 27,367 | 350,349 | ND | 0.011 | 0.145 | 1.74 | 22.0 |
| 22 | 1,1,2-Trichloroethane | 8:53 | 83 | ND | 180 | 1,809 | 17,154 | 182,636 | ND | 0.020 | 0.200 | 1.90 | 20.2 |
| 23 | Dibromochloromethane | 9:44 | 129 | ND | 21 | 915 | 15,683 | 220,672 | ND | 0.004 | 0.169 | 2.90 | 40.8 |
| 24 | Tetrachloroethene | 10:19 | 164 | ND | 201 | 2,711 | 32,429 | 376,641 | ND | 0.017 | 0.226 | 2.70 | 31.4 |
| 25 | Chlorobenzene | 12:04 | 112 | ND | 936 | 10,552 | 108,454 | 982,910 | ND | 0.020 | 0.214 | 2.20 | 19.9 |
| 26 | Ethylbenzene | 12:55 | 91 | 577[c] | 2,429 | 18,869 | 193,870 | 1,432,130[d] | 0.006 | 0.026 | 0.204 | 2.10 | 15.5 |
| 27 | *m,p*-Xylene[e] | 13:22 | 91 | 1,322[c] | 4,163 | 27,235 | 272,079 | 1,903,580[d] | 0.010 | 0.063 | 0.410 | 4.10 | 28.7 |
| 28 | Bromoform | 14:10 | 173 | ND | ND | 120 | 4,593 | 84,753 | ND | ND | 0.079 | 3.04 | 56.1 |
| 29 | *o*-Xylene | 14:45 | 91 | 679[c] | 1,139[c] | 17,761 | 175,283 | 1,360,690[d] | 0.008 | 0.016 | 0.243 | 2.40 | 18.6 |
| 30 | 1,1,2,2-Tetrachloroethane | 16:05 | 83 | ND | 334 | 1,801 | 18,598 | 221,683 | ND | 0.052 | 0.281 | 2.90 | 34.3 |
| 31 | Propylbenzene | 18:38 | 91 | 817[c] | 2,235 | 20,538 | 223,884 | 1,688,590[d] | 0.007 | 0.021 | 0.193 | 2.1 | 15.8 |
| 32 | *p*-Chlorotoluene | 18:43 | 126 | 246[c] | 289[c] | 3,740 | 41,718 | 441,475 | 0.014 | 0.015 | 0.188 | 2.10 | 22.2 |
| 33 | *m*-Dichlorobenzene | 20:54 | 146 | ND | 563 | 5,446 | 55.805 | 571,372 | ND | 0.016 | 0.156 | 1.62 | 16.4 |
| 34 | *o*-Dichlorobenzene | 22:02 | 146 | ND | 781 | 6,890 | 67,081 | 640,135 | ND | 0.027 | 0.239 | 2.34 | 22.2 |
| 35 | 1,2-Dibromo-3-chloropropane | 23:46 | 157 | ND | 58 | 205 | 1,944 | 23,558 | ND | 0.077 | 0.274 | 2.60 | 31.5 |

[a] Initialized amount.
[b] ND = not detected.
[c] Carrier gas/trapping system/blank water contamination.
[d] Ion area saturated.
[e] Coeluting isomers.

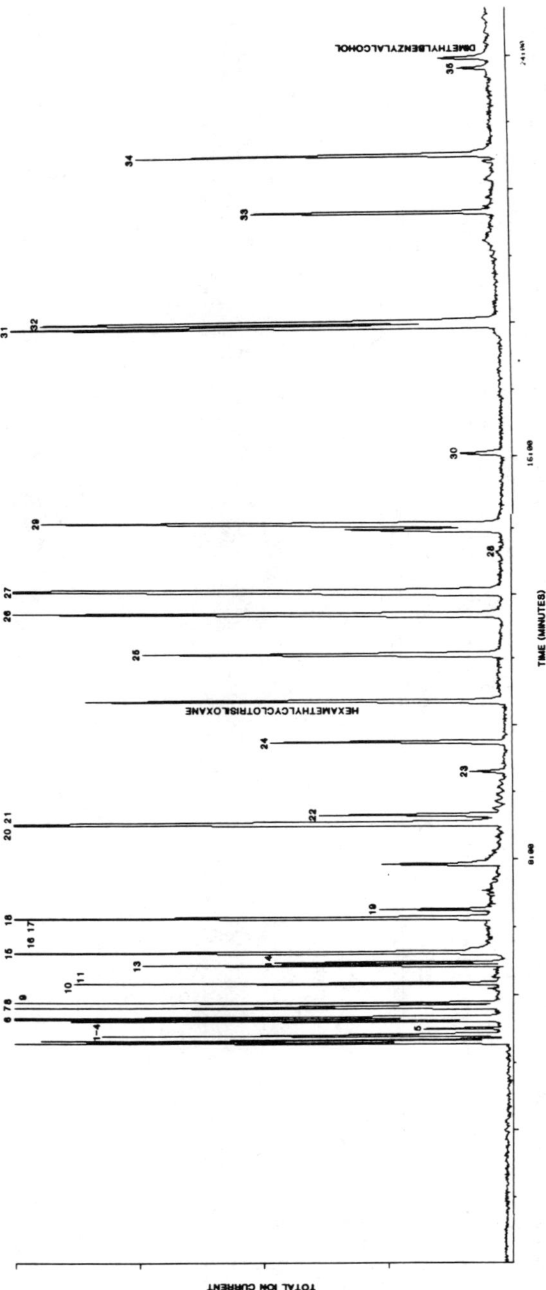


Figure 2. Volatile organics at approx. 200 ng/L.

Table I demonstrates that the sensitivity for all the compounds except bromoform was less than or equal to 20 ppt. However, precision is poorer at this level for some compounds, and the presence of nine of the compounds in the blank run precludes their accurate quantification. A review of the ion areas listed in the table shows that the linear range of this technique is approximately 100–1000. This linear range could be extended by the use of secondary quantification ions. The retention time for the compounds listed in the table are reproducible within ±3 sec.

Although only the analysis of standards is shown here, this method has been validated on environmental samples over a period of six months. Under heavy daily usage with this technique, the capillary column was found to last for approximately four months before there was a noticeable degradation in resolving power. No instrumentation difficulties have resulted from the use of this technique. In fact, the mass spectrometer ion source stayed clean throughout the six-month period. Work is currently underway on the definition of matrix effects on the purging efficiency of volatile organics and on the improvement of detection limits by Tenax cleanup of the carrier gas, use of internal standards and single ion monitoring by variable window.

## BASE/NEUTRALS/ACIDS

### Experimental

Organic extracts were analyzed on a Finnigan 4021 GC/MS (Finnigan Corporation, Sunnyvale, CA). Base/neutral and acid standards were purchased (Supelco, Inc., Supelco Park, PA). Two-liter water samples were manually extracted with four sequential aliquots of 50 mL each of methylene chloride (Baker, Resi-Analyzed, J. T. Baker Co., Phillipsburg, NJ). The extracts were concentrated to 0.5 mL in a Kuderna-Danish apparatus and stored in 1-mL silane-treated vials. One microliter each of both the acid and base/neutral fractions were coinjected with a small air pocket left between the solutions in the syringe. The injector, separator and transfer line temperatures were 290°C. The septum sweep was set to 4 ml/min, and the split flow was turned on at 1 min to 20 mL/min. The column utilized for the chromatographic separation was a 0.25-mm x 30-m FSOT column coated with SE-54 (J & W Scientific, Rancho Cordova, CA). The column head pressure was kept constant at 7.5 psi. The 2 $\mu$L are injected slowly between 12 and 36 sec into the run. During the first minute of the run, the column is maintained at 30°C with the oven door open. Then, the GC door is closed and the column is programmed from 65 to 95°C at 2°C/min and then from 95 to 300°C at 8°C/min where the temperature is held for the remainder of the run. The mass spectrometer is run in the electron impact direct mode, with an emission current of 0.5 mA, an electron energy of 70 eV, the electron multiplier at 1200 V, the dynodes at 3000 V and the ion source at 250°C. The data system is set to scan from m/e 34 to 345 in 0.95 sec with a 0.05-sec hold at the bottom.

### Results

An example of the base/neutral acid coinjection technique is shown in Figure 3 which displays the results of an extraction of a 5 μg/L water standard. This figure shows the ability of the SE-54 FSOT column to sharply resolve phenols, amines, hydrocarbons, and polynuclear aromatic compounds. It was necessary to install a new capillary injector designed according to the criteria developed by Grob and Grob [3] to resolve N-nitrosodimethylamine from methylene chloride (Figure 3). Table II shows a systematic study of the recovery of base/neutral and acidic organic compounds from aqueous standards at the concentration of 5, 0.5 and 0.05 μg/L. Of the extractable priority pollutants which were present in our concentrated standards, 51 were recovered from water at the 5-μg/L level. Three phenolic compounds, benzidine and dichlorobenzidine were present in the primary standards and were chromatographable by direct injection at an earlier date. Their absence here is attributed to their decomposition in the standards and/or poor extraction efficiency. Twelve compounds dropped out at the 0.5-μg/L level. Another 18 compounds were no longer visible after dropping down to the 0.05-μg/L level. However, this still leaves 23 compounds which were visible at the 50 ppt level. It should be noted that based on ion areas in the 5-μg/L runs, one would have expected to see more compounds at lower levels. It is likely that the phenomenon occurring here is nonlinear losses and adsorption of these compounds. Because of these problems and questionable reliability of standards, the present minimum detection limits for most of these compounds is 50–500 ppt. The precision of the base/neutral/acid analysis is apparently poorer than that of the purge-and-trap technique; however, the linear range is still $\geqslant$100 on most compounds. The retention time repeatability of this splitless injection technique is ±10 sec. The authors are pursuing three areas to improve the sensitivity and precision of the analysis of extractable water contaminants: (1) the use of improved extraction devices, (2) preparation of in-house standards, and (3) the use of multiple surrogate internal standards.

## PESTICIDES

### Experimental

Pesticide analysis was also conducted on the Finnigan 4021 GC/MS. Pesticide standards were prepared gravimetrically by adding 99% pure com-

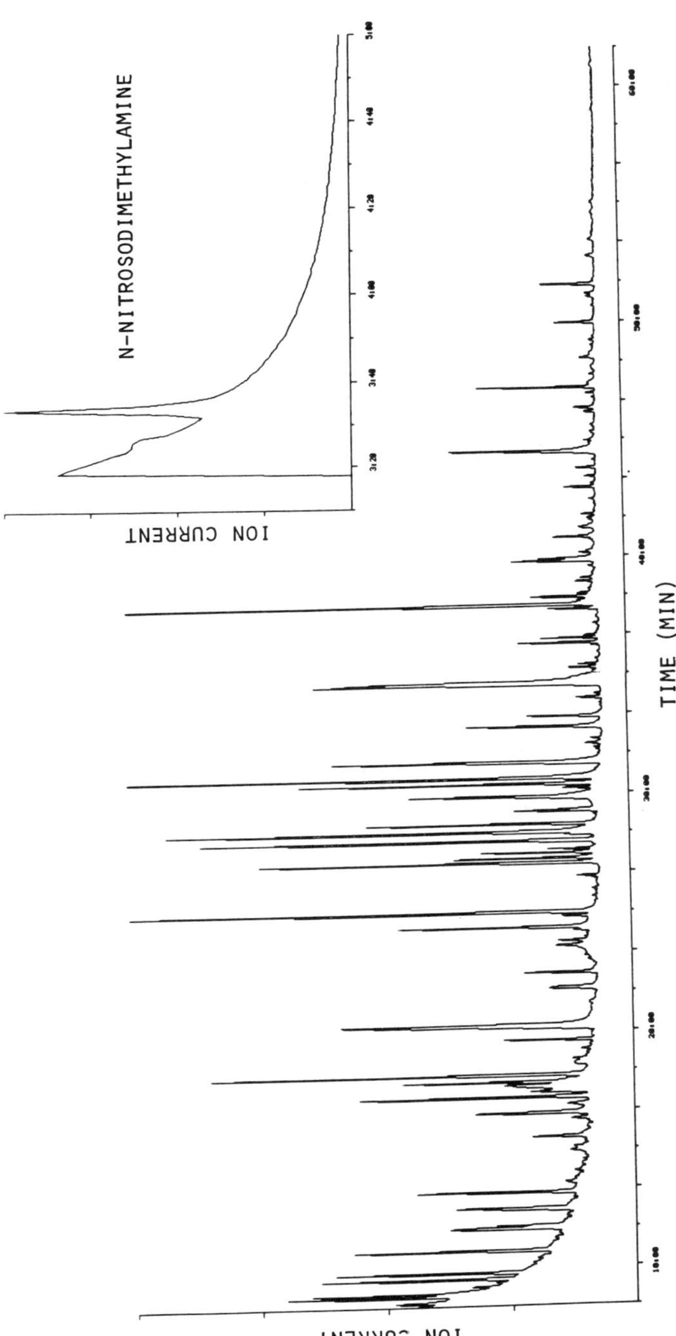


**Figure 3.** Base/neutral/acid coinjection of extract from 5-μg/L water standard.

Table II. Analysis of Base/Neutral/Acid Extracts

| Index No. | Compound | Retention Time | Quantification Ion | Ion Area | | | |
|---|---|---|---|---|---|---|---|
| | | | | Water Blank | 0.05 μg/l | 0.5 μg/l | 5 μg/l |
| 1 | N-nitrosodimethylamine | 3:34 | 74 | | | 995 | 9.790 |
| 2 | Phenol | 8:30 | 94 | | | 2269 | 15,979 |
| 3 | *bis*(2-Chloroethyl) ether | 8:47 | 93 | | 278 | 2689 | 45,206 |
| 4 | 2-Chlorophenol | 8:53 | 128 | | 121 | 3544 | 25,931 |
| 5 | 1,3-Dichlorobenzene[a] | 9:30 | 146 | | 117 | 2439 | 32,924 |
| 6 | 1,4-Dichlorobenzene[a] | 9:45 | 146 | | 120 | 2132 | 31,489 |
| 7 | 1,2-Dichlorobenzene[a] | 10:42 | 146 | | 175 | 2609 | 34,569 |
| 8 | *bis*(2-Chloroisopropyl) ether | 11:39 | 45 | | 172 | 2558 | 86,299 |
| 9 | Hexachloroethane | 12:26 | 117 | | | 514 | 7,398 |
| 10 | N-Nitrosodipropylamine | 12:30 | 70 | | | 745 | 19,985 |
| 11 | Nitrobenzene | 13:11 | 77 | | | 730 | 41,380 |
| 12 | Isophorone | 14:58 | 82 | | | 345 | 6,462 |
| 13 | 4-Nitrophenol | 15:32 | 139 | | | 481 | 6,034 |
| 14 | 2-Nitrophenol | b | 139 | | | | |
| 15 | 2,4-Dimethyl phenol | 16:33 | 122 | | 85 | 1647 | 22,216 |
| 16 | *bis*(2-Chloroethoxy) methane | 17:12 | 93 | | 200 | 3389 | 69,567 |
| 17 | 2,4-Dichlorophenol | 17:25 | 162 | | | 1190 | 13,302 |
| 18 | 1,2,4-Trichlorobenzene | 17:51 | 180 | | 89 | 1651 | 27,154 |
| 19 | Naphthalene | 18:08 | 128 | 328 | 787 | 9263 | 143,076 |
| 20 | Hexachlorobutadiene | 19:37 | 225 | | 49 | 398 | 8,238 |
| 21 | 4-Chloro-3-methyl phenol | 22:27 | 142 | | | 1346 | 9.181 |
| 22 | Hexachlorocyclopentadiene | 23:48 | 237 | | | 36 | 1,751 |
| 23 | 2,4,6-Trichlorophenol | 24:21 | 196 | | | 866 | 7,608 |
| 24 | 2-Chloronaphthalene | 25:00 | 162 | | 914 | 3838 | 67,664 |
| 25 | Acenaphthylene | 27:04 | 152 | | 375 | 6170 | 108,443 |

| | | | | | | | |
|---|---|---|---|---|---|---|---|
| 26 | Acenaphthene | 27:59 | 154 | 513 | 705 | 5349 | 72,688 |
| 27 | Dimethyl phthalate | 27:15 | 163 | 271 | 547 | 1694 | 39,476 |
| 28 | 2,4-Dinitrophenol | b | 184 | | | | |
| 29 | 2-Methyl-4,6-dinitrophenol | b | 198 | | | | |
| 30 | 2,6-Dinitrotoluene | 27:28 | 165 | | | 126 | 8,701 |
| 31 | 2,4-Dinitrotoluene | 29:14 | 165 | | | 153 | 9,786 |
| 32 | Fluorene | 30:22 | 166 | | 200 | 5444 | 64,926 |
| 33 | 4-Chlorophenylphenyl ether | 30:31 | 204 | | | 1957 | 25,518 |
| 34 | Diethyl phthalate | 30:38 | 149 | 2810 | 2751 | 9741 | 85,884 |
| 35 | N-Nitrosodiphenylamine | 31:14 | 169 | | 1874 | 3428 | 21,704 |
| 36 | 1,2-Diphenyl hydrazine | 31:17 | 77 | | 183 | 3489 | 88,102 |
| 37 | Hexachlorobenzene | 33:13 | 284 | 51 | 67 | 627 | 8,506 |
| 38 | Pentachlorophenol | 34:10 | 266 | | | | 603 |
| 39 | Phenanthrene | 34:38 | 178 | | | 5128 | 52,122 |
| 40 | Anthracene | 34:49 | 178 | | | 301 | 3,525 |
| 41 | Dibutyl phthalate | 37:58 | 149 | 6210 | 6337 | 5829 | 291,220 |
| 42 | Fluoranthene[c] | 39:55 | 202 | | 50 | 1038 | 14,554 |
| 43 | Pyrene[c] | 40:51 | 202 | | 100 | 859 | 11,655 |
| 44 | Benzidine | b | 184 | | | | |
| 45 | Benzyl butyl phthalate | 44:27 | 149 | | | | 8,670 |
| 46 | *bis*(2-Ethyl hexyl) phthalate | 47:00 | 149 | | | | 2,301 |
| 47 | Chrysene[c] | 46:12 | 228 | | | | 4,231 |
| 48 | Benzo[a]anthracene[c] | 46:22 | 228 | | | | 4,988 |
| 49 | 3,3-Dichlorobenzidine | b | 252 | | | | |
| 50 | Dioctyl phthalate | 47:14 | 149 | | | 188 | 16,546 |
| 51 | Benzo[b]fluroanthene[c] | 51:03 | 252 | | | | 4,436 |
| 52 | Benzo[k]fluoranthene[c] | 51:10 | 252 | | | | 4,384 |
| 53 | Benzo[a]pyrene[c] | 52:45 | 252 | | | | 3,852 |
| 54 | Indeno [1,2,3-c,d] pyrene[c] | 60:53 | 276 | | | | 2,068 |
| 55 | Dibenzo [a,h] anthracene[c] | 61:19 | 278 | | | | 1,552 |
| 56 | Benzo [g,h,i] perylene[c] | 62:17 | 276 | | | | 2,226 |

[a] Elution order of isomers not confirmed.
[b] Not present in primary standard solution/not extracted.
[c] Concentrations one-half other components (2.5, 0.25, 0.025 μg/l).

pounds to toluene to make a 1-$\mu$g/$\mu$L primary standard. The analysis of pesticides in this study was done on standards in hexane rather than on extracts from water. Injections (1 $\mu$L) were made in the split mode with a sweep flow of 4 mL/min and a split flow of 20 mL/min, giving a split ratio of approximately 20:1. The FSOT column was held isothermal at 250°C during the analysis. The injector, separator and transfer lines were maintained at 290°C. The mass spectrometer was run in the electron impact during mode with an emission current of 0.5 mA, electron energy of 70 eV, electron multiplier at 1000 V, dynodes at 3000 V and ion source at 250°C. The data system scanned from m/e 34 to 445 in 0.95 sec, with a hold at the bottom of 0.05 sec.

## Results

Figure 4 displays a chromatogram of a pesticide standard containing 10 identified components (5 ng each component) on the SE-54 FSOT column. The excellent resolution of these pesticides demonstrates this column's effectiveness for confirmation of pesticides found in high concentrations by electron capture gas chromatographic work. Therefore, when conducting volatile organic and base/neutral acid extractable work on the SE-54 column,

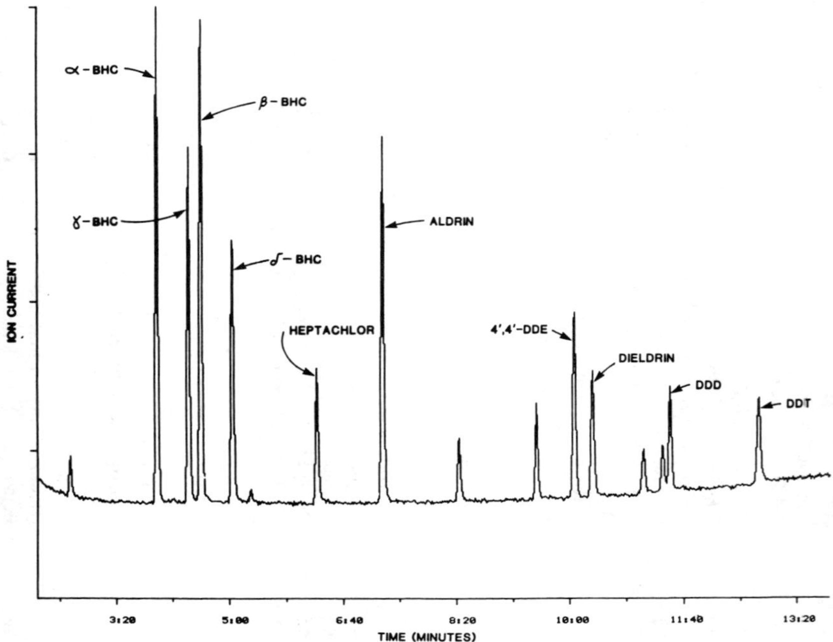


**Figure 4.** Pesticide standard on SE-54 column (5 ng of each component).

it would not be necessary to change columns to confirm pesticides found in the same sample.

## APPLICATIONS

One consequence of the use of this capillary column in the analysis of purgeable compounds in water is that a number of higher-molecular-weight, higher-boiling compounds can be analyzed within a reasonable period of time at the end of the chromatographic run. An example of this new feature is shown in Figure 5. The use of the capillary column allows more overlaps of the purge-and-trap technique with the extraction procedure. Examples of some of these compounds which can be anayzed at the end of the volatile run are: dichlorobenzenes, 4-chlorophenyl phenyl ether, *bis*(2-chloroisopropyl) ether, hexachloroethane, trichlorobenzene, naphthalene, hexachlorobutadiene, 2-chloronaphthalene, acenaphthylene and acenaphthalene. Another example of this extended capability is demonstrated in Figure 6, which shows the analysis of a wastewater by purge-and-trap for the presence of odorous compounds. Over 200 compounds were found in this sample and range from compounds as light as pentane to as heavy as N-phenylbenzamine. The notable odor-causing compounds which showed up in this run were dimethyldisulfide and 2-methylthiophene. The authors have also found this technique useful for the part-per-trillion analysis of $C_6$-$C_{12}$ aldehydes produced by ozonation of drinking water. The purge-and-trap capillary technique has been demonstrated to be useful in the analysis of water for compounds from the following categories:

- aliphatic hydrocarbons;
- aromatic hydrocarbons;
- alicyclic hydrocarbons;
- arenes;
- aldehydes;
- ketones; and
- certain heterocyclic compounds.

Further research is warranted to determine its utility for the analysis of other classes of compounds and other types of compounds within the classes mentioned above.

## CONCLUSIONS

The most commonly used chromatographic techniques for the analysis of purgeable and extractable organic water contaminants call for the use of four different packed columns. The techniques presented above represent a simplified chromatographic procedure which yields improved sensitivity and resolution. A single FSOT capillary column can be used for the analysis

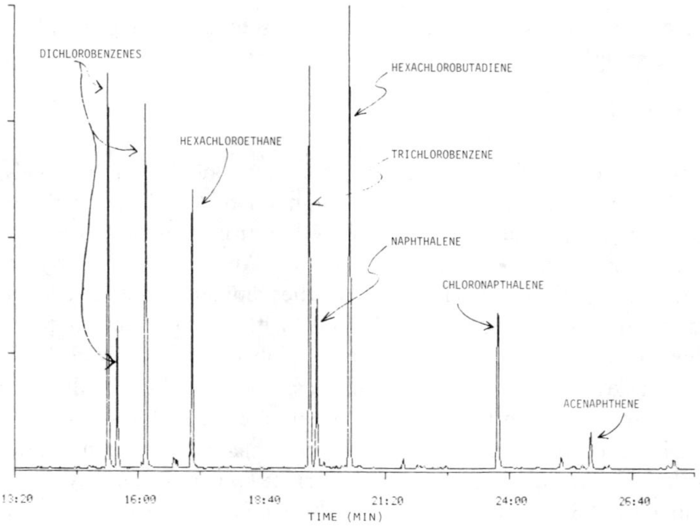


**Figure 5.** Selected base/neutrals by purge-and-trap.

of volatile, base/neutral acidic and pesticide fraction organic contaminants in water. Many of the compounds tested could be recovered from water at the low part-per-trillion level range. Considerable time-saving is afforded by the use of the single column for both volatile and extractable water contaminants. Coinjection of both base/neutral and acid fractions in the same chromatographic run allows further time-saving. The authors are pursuing analysis at the sub-part-per-trillion range by the development of purer blanks and standards and by the optimization of extraction methodology. Future work will be conducted to determine the effect of sample matrix on recovery of water contaminants at low levels and the extension of these techniques to other classes of organic compounds found in water.

## ACKNOWLEDGEMENTS

The authors would like to acknowledge the assistance of Ms. Carol Swart in the JMM Environmental Research Laboratory Extraction Lab and the support of Dr. Lawrence Leong, Laboratory Director. Additional thanks go to Mr. Paul E. Kelly of the Finnigan Instruments Applications Laboratory for his helpful suggestions.

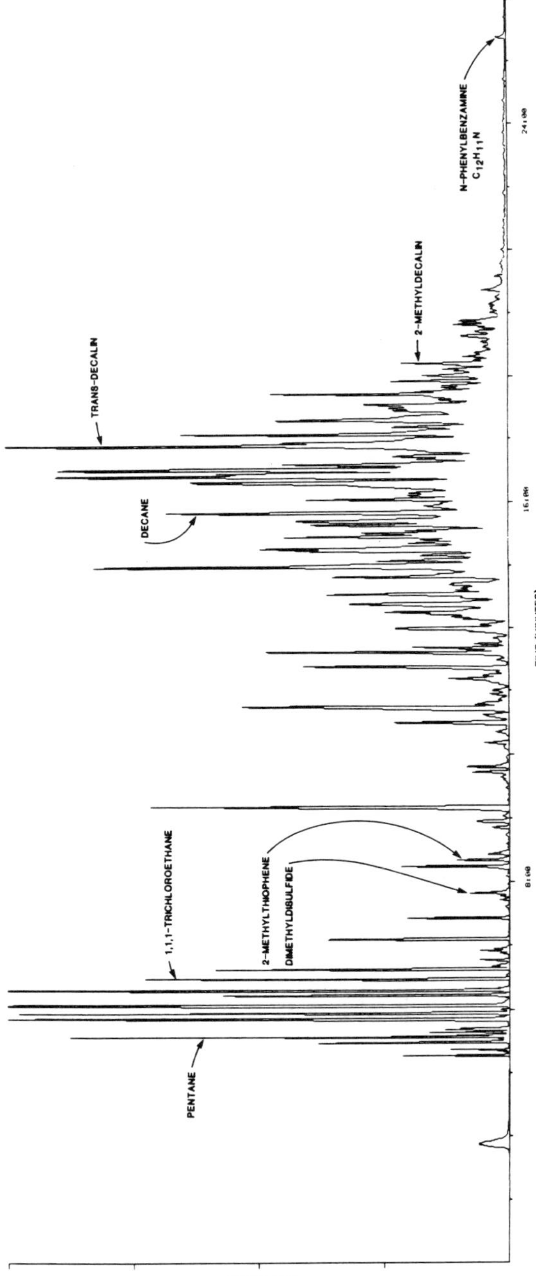


**Figure 6.** Odorous compounds in wastewater by purge-and-trap.

## REFERENCES

1. Sauter, A. D., L. D. Betwoski and E. P. Meier. "Quantitative Aspects of GC/MS Analysis," Proceedings of the 28th annual American Society of Mass Spectrometry meeeting on Mass Spectroscopy and Allied Topics, New York, NY, May 1980, p. 169.
2. Bellar, T. A., and J. J. Lichtenburg. "Semi-automated Headspace Analysis of Drinking Waters and Industrial Waters for Purgeable Volatile Organic Compounds," Proceedings of the ASTM Symposium on Measurement of Organic Pollutants in Water and Wastewater, June 1978.
3. Grob, K., and K. Grob, Jr. "Splitless Injection and the Solvent Effect," *J. High Resolution Chromatog. Chromatog. Commun.* (July 1978).

# CHAPTER 11

# ANALYSIS OF POLYCHLORINATED BIPHENYLS BY GLASS CAPILLARY AND PACKED-COLUMN CHROMATOGRAPHY

**Michael D. Mullin**

U.S. Environmental Protection Agency
Large Lakes Research Station
Grosse Ile, Michigan

**John C. Filkins**

Cranbrook Institute of Science
Bloomfield Hills, Michigan

Polychlorinated biphenyls (PCB) have received much attention in recent years in the area of environmental contamination and toxicology. This is due, in part, to the persistence of this class of compounds in the environment and their ubiquitous nature, i.e., they are detectable in most types of environmental samples [1-10]. Any attempt to study PCB is complicated by the fact that the routinely reported commercial mixtures are not single compounds but are complex mixtures of up to 60 or more individual compounds [11-13] with widely varying toxicological, environmental and analytical properties.

The most common way to describe PCB is to refer to them by Aroclor number, e.g., Aroclor 1016, Aroclor 1242 or Aroclor 1260. These identities refer to the commercial formulations previously manufactured in the United States by Monsanto Company (St. Louis, MO) and sold as the individual Aroclor or in combination with other formulations. Aroclor, therefore, is a general, nonspecific term referring to a commercial mixture of PCB. To refer to components of a commercial mixture or environmental sample, the terms "homolog" or "isomer" are more specific and should be used. Homolog refers to all the PCB compounds containing the same number of chlorine atoms per biphenyl molecule, regardless of ring position. For example, reference to the hexachlorobiphenyl homologs denotes all the PCB with six chlorines per molecule. Isomer refers to the individual, specific compounds. Examples are the 42 different hexachlorobiphenyl isomers theoretically possible, or it is possible to refer to the 2,2′,4,4′,5,5′-hexachlorobiphenyl isomer as compared to the 2,2′,3,4,4′,5′-hexachlorobiphenyl isomer. A detailed explanation of the proper rules of nomenclature concerning PCB is given in Hutzinger et al. [47].

The vast majority of North American laboratories analyzing environmental samples for PCB utilize some form of gas chromatography (GC) with conventional packed-column chromatography and an electron capture detector (ECD). A growing number of these facilities are also equipped with mass spectrometers for more detailed, but less sensitive, analyses. The alternative to these techniques is the use of high-resolution glass capillary GC. This method has been used extensively in Europe for more than a decade and is increasing in use in North America, but more for analysis of petroleum hydrocarbons, tobacco components and steroids than for analysis of persistent halogenated hydrocarbons such as PCB, DDE and lindane [15-21].

## EXPERIMENTAL

### Equipment

#### *Capillary Column*

A Varian 3760 gas chromatograph equipped with an ECD was used for the capillary analyses. The PCB were separated on a 0.25-mm i.d. x 56-m glass capillary column coated with C-87 [22] to a thickness of 0.2-0.3 μm (Quad-Rex Corp., New Haven, CT). The oven temperature was programmed at a rate of 1.3°C/min. from 140 to 210°C for the pressure optimization chromatograms (Figures 3-6), and from 140 to 230°C for the sample and analytical standard chromatograms (Figures 2 and 8). The injector was held at 300°C, and the ECD was held at 320°C. Sample volume injected was 2.5 μL with no sample splitting in the injector. The nitrogen and hydrogen carrier gases were held at constant pressure as indicated in the appropriate figures.

*Packed Column*

A Hewlett-Packard 5730 gas chromatograph equipped with an ECD was used for the packed-column analyses. The PCB were separated on a 2-mm i.d. x 1.83-m glass column packed wtih 3% SE-30 on 80-100 mesh Chromosorb W HP (Supelco, Bellefonte, PA). The oven temperature was held initially at 170°C for 8 min, then programmed from 170 to 200°C at a rate of 4°C/min. The injection port was held at 200°C, the ECD at 300°C and the 5% methane in argon carrier gas flow was maintained at 30 mL/min. Sample volume injected was 4.6 $\mu$L.

**Materials**

Isomeric PCB standards were obtained from RFR Corporation (Hope, RI), Analabs (North Haven, CT) or Safe [23] and were used as received. Organic solvents used to make the analytical standard solutions were obtained from Burdick & Jackson (Muskegon, MI) and were used as received.

## RESULTS AND DISCUSSION

Figures 1 and 2 compare the use of a packed column with a glass capillary column. The resolution of a mixture containing 80 and 79 ng/mL, respectively, of Aroclors 1242 and 1260 as chromatographed on a packed column and a glass capillary column is illustrated. Depending on the number of very small peaks utilized, there are only 25-30 resolvable peaks in the packed-column chromatogram, whereas more than 100 easily identifiable peaks are available in the glass capillary column chromatogram.

Optimization of a glass capillary system differs somewhat from packed-column optimization. Column temperature will have an effect similar to that observed in packed-column systems: increasing temperature shortens the analysis time while decreasing temperature lengthens the analysis time. The carrier gas in capillary systems is at constant pressure, whereas it is flowrate-controlled in packed-column systems. The choice of a carrier gas can also make a substantial difference in the quality of the resulting chromatogram in capillary column systems. There are three main carrier gases in use for capillary systems: nitrogen, helium and hydrogen. Helium is more expensive to use than either nitrogen or hydrogen, and the resultant chromatograms are intermediate in quality between them [24]; therefore, this discussion will focus on nitrogen and hydrogen.

Figure 3 shows the effect of varying column pressure on the resulting chromatograms of a 1:1 mixture of Aroclors 1242 and 1260 using nitrogen as

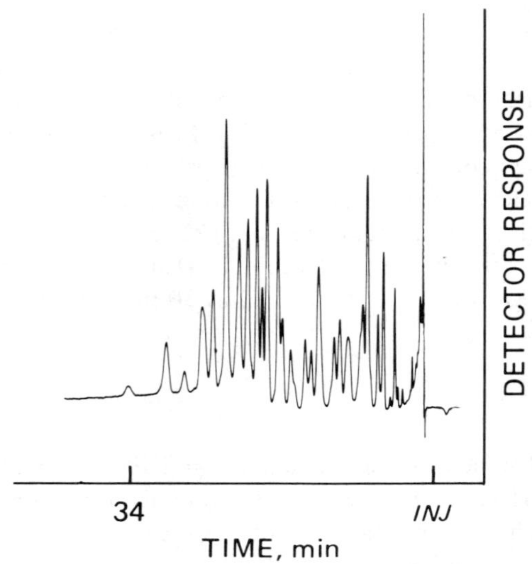


**Figure 1.** Packed–column chromatogram of Aroclors 1242 and 1260. See text for instrumental operating conditions.

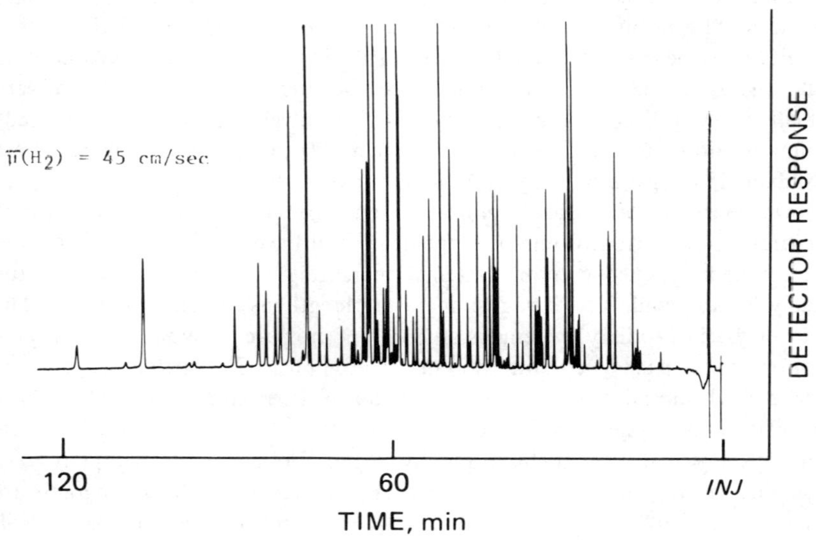


**Figure 2.** Glass capillary chromatograms of Aroclors 1242 and 1260. See text for instrumental operating conditions.

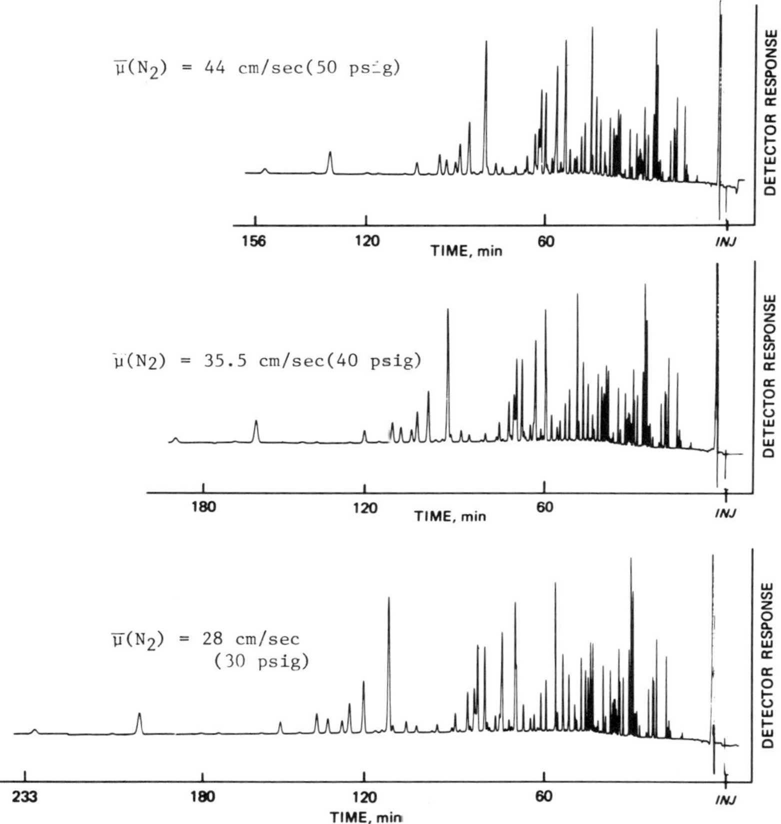


**Figure 3.** Effect of varying column pressure with nitrogen as the carrier gas. See text for instrumental operating conditions.

the carrier gas. The higher pressure results in a shorter analysis time, but also compresses the peaks, increasing the difficulty in making subsequent identification.

Figure 4 shows a similar effect of varying column pressure on the resulting chromatograms of the same sample using hydrogen as the carrier gas. Again, the higher pressure results in a decreased analysis time, but also causes some of the peaks to coalesce, hindering identification of these peaks. If time is a major consideration, the use of higher column pressures will shorten the analysis time, but at the expense of resolution quality. If, on the other hand, maximum resolution is the major criterion, the lower column pressure should be used to facilitate compound identification. Figures 5 and 6 compare the use of hydrogen and nitrogen as carrier gases with both systems optimized for maximum resolution. The use of hydrogen results in a much shorter analysis time and greater sensitivity.

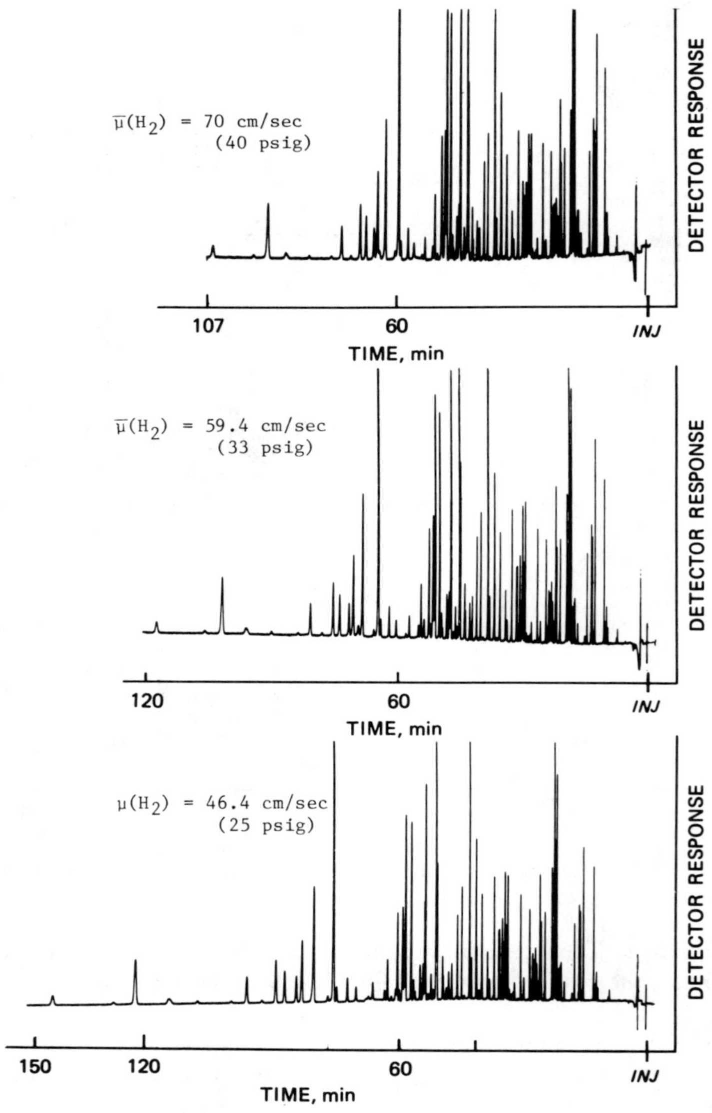


**Figure 4.** Effect of varying column pressure with hydrogen as the carrier gas. See text for instrumental operating conditions.

The advantages of glass capillary over packed-column GC are greater selectivity (resolution of adjacent peaks), increased resolution (more peaks observed per sample), greater sensitivity, faster analysis to yield equivalent resolution to packed-column gas chromatography (although for optimum resolution glass capillary GC may require longer analysis time [25]), and the

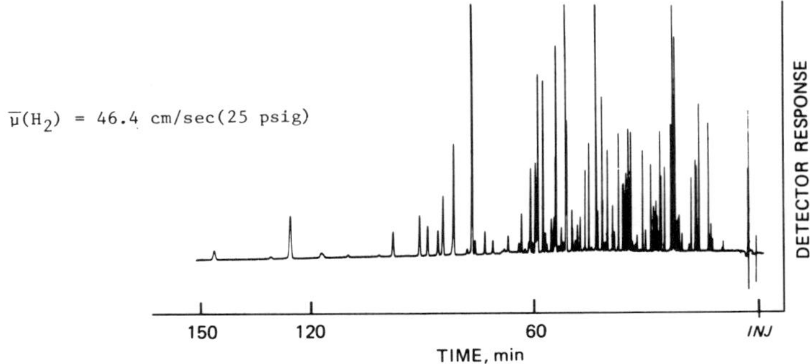


**Figure 5.** Glass capillary chromatogram optimized for maximum resolution with hydrogen as the carrier gas. See text for instrumental operating conditions.

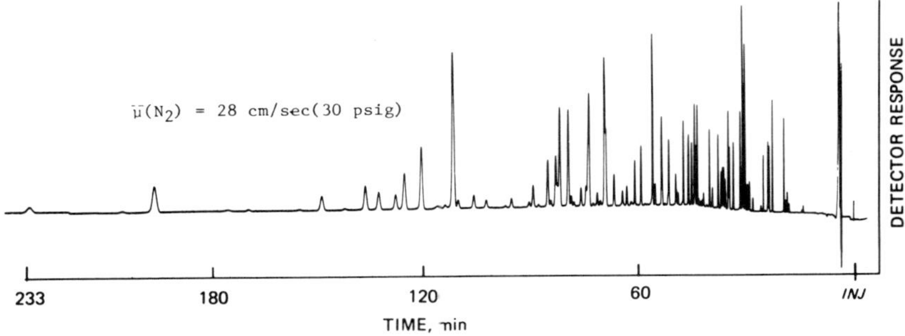


**Figure 6.** Glass capillary chromatogram optimized for maximum resolution with nitrogen as the carrier gas. See text for instrumental operating conditions.

use of fewer types of column materials to analyze a wide range of analytes of interest [24].

The disadvantages include the relative fragility of the columns (familiarity with the column makes this a minor disadvantage), greater cost per column, longer analysis times for optimum resolution and the larger initial cost to modify current instrumentation from packed to capillary column. In light of the additional gain in information these limitations are offset by the advantages of the capillary system.

The use of glass capillary GC for the analysis of PCB emphasizes the complexity of the samples involved and indicates the requirement for analytical standards of known composition with which to calibrate the instrumentation. Currently available Aroclor standards are largely of unknown composition. It is not sufficient to know that there are a given number of peaks in a specific standard or sample. Rather, isomeric identity is needed for two reasons: (1) the varying molecular responses elicited by different isomers from the ECD make it difficult to calibrate an instrument for accurate quantification; and (2) the widely varying toxicological properties of different

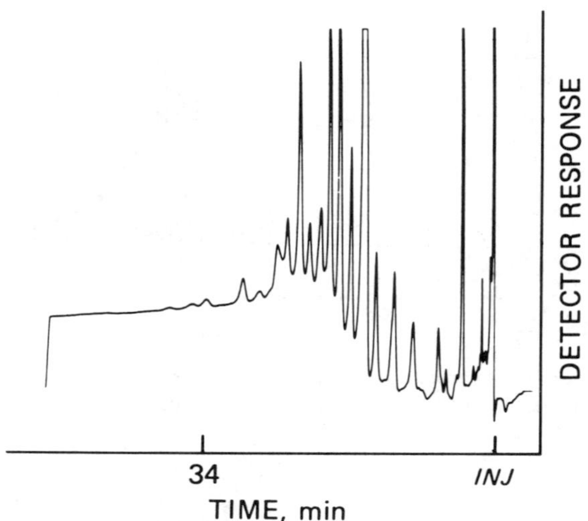


**Figure 7.** Packed-column chromatogram of composite breast milk sample. See text for instrumental conditions.

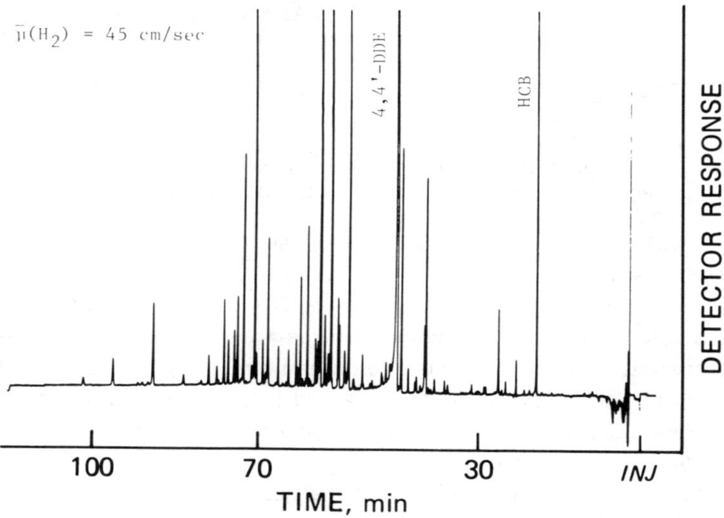


**Figure 8.** Glass capillary chromatogram of composite breast milk sample. See text for instrumental operating conditions.

isomers mandate the availability of information concerning the presence or absence of these isomers from the commercial mixtures and, therefore, the environmental samples.

Figures 7 and 8 compare chromatograms obtained from a packed column to a glass capillary column for a composite breast milk sample of unknown concentration (the dilutions performed to arrive at the final extract are not known). Several items of importance may be discerned. The two labeled peaks in Figure 8—4,4′-DDE (1,1-dichloro-2,2′-bis(4-chlorophenyl)ethene) and hexachlorobenzene (HCB)—account for more than 50% of the total area integrated in the sample. With a sensitivity about ten times that of PCB to an ECD, this suggests that the 4,4′-DDE accounts for approximately 5% of the chlorinated hydrocarbons in this chromatogram. The lower resolution obtained using the packed column shown in Figure 7 does not separate the 4,4′-DDE from the PCB, making qualitative identification and subsequent quantification difficult. Including the peaks which includes the 4,4′-DDE will greatly overestimate the quantity of PCB present, but it would be difficult to exclude this peak a priori as the chromatogram does not exhibit deviations from what would be expected from this sample matrix. The glass capillary column chromatogram presents a clearer, more unambiguous description of the sample composition than the packed-column chromatogram.

An interesting comparison can be made between the latter portions of the chromatograms, starting after 70 min from injection, in Figures 5 and 8; the number of peaks and their relative intensities are very similar in both examples. While not an absolute match, they are quite close, whereas the same observation cannot be made for the earlier portions of the chromatograms. This may be due to selective uptake, degradation of lower chlorinated isomers in the body, or the body may have been exposed to a source(s) already weathered with respect to the earlier-eluting isomers.

## CONCLUSION

The use of glass capillary GC enables the identification of individual components in complex environmental samples which will be invaluable to toxicologists in aiding them to evaluate compounds for potential toxicological importance that are known to be present in environmental samples. In addition, glass capillary GC will enable analysts to better quantify the compounds in their samples by minimizing interferences and more accurately quantifying the resulting chromatograms.

## ACKNOWLEDGMENTS

This research was partially supported by U.S. Environmental Protection Agency Grant No. R-806579. The technical assistance of Gail R. Sawka in the preparation of the analytical standard solutions was greatly appreciated. The breast milk sample was made available by J. Mes, Department of Health and Welfare, Ottawa, Canada.

## DISCLAIMER

The mention of brand or trade names does not constitute endorsement by the U.S. Environmental Protection Agency.

## REFERENCES


1. Bowes, G. W., and C. J. Jonkel. *J. Fish. Res. Board Can.* 32:2111 (1975).
2. Eisenreich, S. J., G. J. Hollod and T. C. Johnson. *Environ. Sci. Technol.* 13:569 (1979).
3. Greischus, U. et al. *Bull. Environ. Contam. Toxicol.* 11:113 (1974).
4. Humphrey, H. E. B. Final Report on FDA Contract 223-73-2209. Michigan Department of Public Health, Lansing, MI.
5. Jensen, S., and G. Sunderstrom. *Ambio* 3:70 (1974).
6. Mes, J. et al. *Bull Environ. Contam. Toxicol.* 17:196 (1977).
7. Mes, J., and D. Davies. *Bull. Environ. Contam. Toxicol.* 21:381 (1979).
8. Murphy, T. J. U.S. EPA Ecological Research Series, EPA-600/3-78-071 (1978).
9. Swain, W. R. *J. Great Lakes Res.* 4:398 (1978).
10. Veith, G. et al. *Pestic. Monit. J.* 13(1):1 (1979).
11. Neu, H., M. Zell and K. Ballschmiter. *Fres. Z. Anal. Chem.* 293:193 (1978).
12. Sissons, D., and D. Welti. *J. Chromatog.* 60:15 (1971).
13. Zell, M., H. New and K. Ballschmiter. *Chemosphere* 2/3:69 (1977).
14. Hutzinger, O., S. Safe and V. Zitko. *The Chemistry of PCB's* (Cleveland, OH: CRC Press, 1974), p. 3.
15. Destry, D. H., and A. Goldup. *Gas Chromatography*, R. D. W. Scott, Ed. (London: Butterworths, 1960).
16. Gouw, T. H., I. M. Wittemore and R. E. Jentoft. *Anal. Chem.* 42:1394 (1970).
17. Grob, K., and G. Grob. *Chromatographia* 5:3 (1972).
18. Koenig, W. A., and A. J. Nicholson. *Anal. Chem.* 47:951 (1975).
19. Lin, S.-N. and E. C. Horning. *J. Chromatog.* 112:465 (1975).
20. Novotny, M., R. Segura and A. Zlatkis. *Anal. Chem.* 44:9 (1972).
21. Novotny, M., M. P. Maskarinec, A. T. G. Steverink and R. Farlow. *Anal. Chem.* 48:468 (1976).
22. Kovats, E. S. et al. *J. Chromatog.* 126:3 (1976).
23. Safe, S. H. Prepared under Cooperative Agreement No. CR 806928 with the U.S. Environmental Protection Agency, Large Lakes Research Station, Grosse Ile, MI.
24. Novotny, M., and S. P. Cram. *Capillary Gas Chromatography* (Washington, DC: American Chemical Society, 1977), pp. 1-5.
25. Novotny, M., and A. Zlatkis. *Chromatog. Rev.* 14:1 (1971).

# CHAPTER 12

# HIGH-RESOLUTION GAS CHROMATOGRAPHY AND NEGATIVE ION CHEMICAL IONIZATION MASS SPECTROMETRY OF POLYCHLORINATED BIPHENYLS

**E. D. Pellizzari, K. B. Tomer and M. A. Moseley**

Analytical Sciences Division
Chemistry and Life Sciences Group
Research Triangle Institute
Research Triangle Park, North Carolina

A review on the comparison of packed and capillary gas chromatography (GC) of polychlorinated biphenyls (PCB) can be found elsewhere in this book [1]. The advantages of glass capillaries for PCB analysis becomes important when the individual isomers are to be measured. The differences in toxicological properties between PCB isomers lend impetus to their measurement. For many years the quantification of PCB followed a very conventional method which was based on general reference "profiling" techniques using selected Aroclor mixtures and packed GC columns.

Although glass capillaries coated with C-87 stationary phase provide excellent resolution of individual PCB isomers, this phase has a temperature limit of approximately 220°C. Thus, the C-87 phase coated columns are inadequate for the analysis of polychlorinated terphenyls (PCT) and polybrominated biphenyls (PBB). Temperatures of 280-290°C are required for complete elution of PCT and PBB. Nevertheless, a thorough investigation of stationary phases and glass capillary column preparation has not been conducted for developing the best capillary for PCB, PCT and PBB resolution and electron capture detection (ECD).

Another significant limitation of current techniques for PCB measurement is the potential interferences associated with ECD. A method which provides, in addition to sensitive detection, a verification of the structural entity being measured is highly desirable. This is typically why mass spectrometry (MS) is the method of choice. However, conventional MS suffers from inadequate sensitivity for PCB analysis in environmental and biological samples. The limits of detection between electron impact MS and ECD differ by as much as 2–3 orders of magnitude. For this reason, electron-impact MS has not been widely successful for verifying or quantifying PCB in environmental samples.

A relatively new and promising technique, negative ion chemical ionization (NICI)-MS, has been introduced into the analytical chemist's arsenal of tools. It possesses the inherent sensitivity of ECD. Because NICI-MS is a new technique, the operating parameters for analysis of PCB, as well as for other halogenated compounds, still remain to be optimized.

The research described here addresses two areas: high-resolution GC and NICI-MS of PCB. The objectives of this investigation were:

1. to investigate methods for preparing better capillary columns for PCB, PCT, PBB and pesticide analysis. This objective encompasses two specific aims. First, it involves a systematic study of the variables associated with capillary column preparation, i.e., choice of glass type, deactivation/pretreatment methods, stationary phase, stationary phase film thickness, and internal diameter and column length which affect maximum PCB resolution. Secondly, it involves evaluating prepared capillaries using PCB standards and extracts of environmental samples (e.g., fish, mother's milk and rainwater);
2. to study and optimize NICI-MS for PCB analysis; and
3. to combine the high-resolution GC development with NICI-MS for both qualitative and quantitative analysis of PCB in environmental samples.

## EXPERIMENTAL PROCEDURE

### Preparation of Capillary Columns

Pyrex glass capillaries were drawn from two tube sizes: 8.0 mm o.d. x 4.0 mm i.d. x 1.0 m and 7.5 mm o.d. x 1.6 mm i.d. x 1.0 m. Pyrex tubing was rinsed with 50 mL each of methylene chloride, acetone, 1% NaOH, methanol and acetone prior to drawing. The glass capillaries were prepared using either a Shimadzu GDM-1 (Kyoto, Japan) or a Hupe and Busch (Karlsruhe, Germany) glass drawing machine. Fused-silica capillaries in lengths of 25 and 50 m (0.23 mm i.d.) were obtained from Hewlett-Packard (Avondale, PA 19311).

After drawing, Pyrex glass capillaries were acid-leached by filling with 20% HCl. An 8% liquid volume was removed, both capillary ends were flame-sealed and the column was heated at ~ 175°C overnight.

Three different pretreatment methods were investigated. Silanization of Pyrex glass capillaries was accomplished using the method of Grob et al. [2–4]. A modification of this method was used for the fused-silica columns. After acid leaching, capillaries were dehydrated by placing in an oven such that both ends were accessible to the outside of the oven. The capillary was heated to 240°C for 20 min and then both ends were connected to an aspirator. The capillary was further dehydrated at 240°C for 10 min for 25-m, 20–30 min for 50-m. Before disconnecting the vacuum from the capillary ends, a small flame was used to warm the capillary ends external to the oven to remove any water vapor. A 20% column volume of hexamethyldisilazane and diphenyltetramethyldisilazane in pentane (1/1/2 v/v/v) was passed through the capillary at 2 cm/sec. After the entire solution was flushed from the capillary, both ends were connected to a vacuum source (5 min for 25-m, 20 min for 50-m). Subsequently, the capillary ends were flame-sealed under vacuum, then heated in an oven at 400°C for 10–14 hr. For fused-silica capillaries, the column was heated at 320°C for 60 hr. On cooling, one end of the column was opened under toluene until ~5% of the column volume was filled; the toluene was then replaced with methanol until again ~5% more of the column was filled; finally the methanol was replaced with diethyl ether. When liquid uptake ceased the percentage of coils remaining empty relative to total number of coils was calculated (proper silanization will give 30–50% of column volume as gaseous ammonia). If more than 50% of the coils are filled with ammonia it is indicative of insufficient dehydration, while less than 30% indicates partial dehydration or an insufficient vacuum. The solvent mixture is passed through the column with the diethyl ether eluting last.

Roughening of the Pyrex glass capillary wall by producing a $BaCO_3$ surface was accomplished by the method of Grob et al. [5–7]. After acid leaching, the capillary was rinsed with a 10% column volume of distilled water and again with methanol. The capillary was dehydrated by heating at 280°C for 5 hr under a low nitrogen flow. After cooling and removing from the oven the capillary was connected to a 15-m buffer column. The capillary was then rinsed with a 10% column volume of 0.05% HCl followed by a 15% column volume of a barium hydroxide solution (1.5 mg Marlazin L10, 1.5 mL saturated barium hydroxide, 45 mL $CO_2$-free distilled water). A 40-cm

$N_2$ gas plug was introduced behind the barium hydroxide solution, and the gas plug was followed by $CO_2$. The solution was passed through the capillary at 1 cm/sec, and the $CO_2$ flow was continued for an additional 30 min. The capillary was heated to 90°C in an oven for 30 min, cooled and dried and deactivated by passing first a 15% column volume of acetone, then 0.05% Carbowax 20M in methylene chloride. After flushing the solutions from the capillary, it was dried under a vacuum for 10 min, then heated in an oven at 280°C for 20 min under a low $N_2$ flow. After cooling, the Carbowax 20M deactivation procedure was repeated, omitting the acetone rinse. To enhance the wettability of the capillary surface, it was treated with Emulphor ON-870 by passing through it a 20% column volume of 0.1% Emulphor ON-870 in methylene chloride. Subsequently, the capillary was dried in vacuo for 20 min and then heated at 240°C under a low $N_2$ flow. The capillary was cooled, rinsed with methylene chloride and dried under $N_2$ for 20 min.

Superox-4 deactivation of Pyrex and fused-silica capillaries described by Arrendale et al. [8] was modified. After acid leaching, the column was rinsed with a 40% column volume of distilled water and then with a 40% column volume of methanol. After dehydrating the capillary at 280°C for 5 hr under a low $N_2$ flow and cooling the column, a 30% column volume of 0.2% Superox-4 in methylene chloride was passed through it at 1 cm/sec. The capillary was then dried under a $N_2$ flow for 30 min, the ends flame-sealed and heated at 320°C for 1 hr. After cooling, the capillary was treated with a 20% column volume of 1% Emulphor ON-870 in methylene chloride at 1 cm/sec. The capillary was dried with $N_2$ for 30 min, flame-sealed at both ends and heated at 240°C for 1 hr. Finally, the capillary was rinsed with a column volume of methylene chloride.

Capillary columns were coated using the static method [9, 10]. Stationary phases in pentane were used [10]. Using a positive pressure the capillaries were filled, and one end was sealed with a solution of sodium tetrasilicate, taking care that no air bubbles were introduced at the interface. After allowing the capillary to sit overnight, it was placed in a water bath at 25°C with both ends external. To the open end was attached a vacuum (aspirator) and the temperature of the water bath slowly increased to 30–32°C. After the pentane was completely evacuated the water glass end was broken and residual solvent vapor was evacuated.

Before straightening the capillary ends, 3–4 coils were rinsed with a mixture of methylene chloride and pentane (1/1, v/v). After straightening, the ends were deactivated with a 0.05% solution of Superox-4 in methylene chloride.

Capillaries were conditioned by programming from ambient temperature to within 10°C of the phase limits at 1°C/min.

### Gas Chromatography

Column evaluations were performed on a Carlo Erba 4160 (Milan, Italy) high-resolution gas chromatograph equipped with a flame ionization detector (FID) and a Grob injection port. Helium velocity was 28 cm/sec with an injection port split ratio of 20:1. Temperature program rate was 1 and 0.5°C/min for 25- and 50-m capillaries, respectively. Injector and detector temperatures were 290 and 300°C, respectively.

Analysis was also conducted using a Varian 3700 GC equipped with a $^{63}$Ni ECD. The optimum helium linear velocity was determined from Van Deemter plots for each column using a 1°C/min program rate. Nitrogen scavenger gas at 35 mL/min into the detector base and 5 mL/min into the detector-capillary interface was employed. Starting column temperature was 150°C with a 1°C/min program rate.

### Performance Criteria

For calculating height equivalent to an effective theoretical plate (HEETP), a pure PCB isomer, 2,2′, 4′,5-tetrachlorobiphenyl (TCB) was used. HEETP was calculated as follows:

$$N = 5.54 \left( \frac{RT'_A}{W_{1/2A}} \right)^2$$

$$HEETP = \frac{L}{N}$$

where L = column length (mm)
N = number of effective plates
$RT'_A$ = retention distance of 2,2′4′,5-tetrachlorobiphenyl minus retention of methane
$W_{1/2A}$ = peak width at half-height

The second test parameter used for evaluating capillaries was the "Trennzahl" (TZ) or separation number. The TZ was determined by chromatographing 2,2′,4′,5-tetrachlorobiphenyl and 2,2′,4,4′,6,6′-hexachlorobiphenyl. TZ was calculated from the following relationship:

$$TZ = \frac{RT_B - RT_A}{W_{1/2A} + W_{1/2B}} - 1$$

where $RT_A$ = retention of 2,2′4′,5-tetrachlorobiphenyl
$RT_B$ = retention of 2,2′,4,4′,6,6′-hexachlorobiphenyl
$W_{1/2A}$ = peak width at half-height
$W_{1/2B}$ = peak width at half-height

The third parameter, resolution (R), was determined between two isomers of tetrachlorobiphenyl according to the relationship:

$$R = \frac{2(RT_A - RT_B)}{1.699\,(W_{1/2A} + W_{1/2B})}$$

where $RT_A$ = retention of 2,2′,5,5′-tetrachlorobiphenyl
$RT_B$ = retention of 2,2′,4′,5-tetrachlorobiphenyl

The Grob et al. [11] mixture was also used for capillary column evaluations in our initial work.

## RESULTS AND DISCUSSION

### High-Resolution Chromatography

Several dozen capillaries were prepared and evaluated for resolution of PCB isomers. Table I lists performance characteristics for a selected number of columns. Six variables which affect HEETP, TZ and R were investigated. These were: stationary phase, glass type, pretreatment, film thickness, internal diameter and length. Inspection of these data reveals that C-87 and Apiezon L capillaries yield the best results. Attempts to coat the C-87 phase on a barium carbonate-pretreated Pyrex or Superox-4-pretreated fused-silica gave very poor results. A mixed phase (SE-54/C-87, 1:4 w/w) was successfully prepared on a Superox-4 column with good performance characteristics. Also, the mixed-phase coated capillary was slightly more thermally stable than C-87 coated alone. Mixed-phase preparations apparently exhibited better wetting characteristics on the glass surface than pure C-87 coated columns.

Even though the SP-2100 capillaries gave the highest TZ values, the HEETP and R values were not comparably as good as C-87 and Apiezon L coated columns. In general these data (Table I) indicate that thin stationary phase film thickness and small i.d. capillaries coated with C-87 or Apiezon L on silanized Pyrex glass give optimum separation of PCB isomers.

A comparison of phase selectivity (McReynold's constants), TZ number and R values was made and is given in Table II. Based on TZ and R values the best phases for PCB analysis are predicted to be Apiezons L and M. Both Apiezons L and M have reported temperature stabilities approaching 300°C, which make them good candidates for PCT and PBB analyses.

Figure 1 depicts a chromatogram of Aroclors 1242, 1260 and 5460, each introduced onto the column at a 400-pg level. An Apiezon L capillary 50 m in length was temperature-programmed from 150 to 290°C at 1°C/min with ECD. Several peaks have been identified as to the specific PCB isomer by comparison to the retention times of individual known isomers chromatographed under identical conditions. The latter half of the chromatogram represents PCT. Even though complete resolution of the PCT isomers was not achieved, the elution pattern can be an aid to their fingerprinting in environmental samples.

**Table I. Performance Characteristics for Several Capillary Columns**

| Stationary Phase | Glass Type | Pretreatment | Film Thickness (μm) | Length (m) | i.d. (mm) | TZ | HEETP | R |
|---|---|---|---|---|---|---|---|---|
| Se-54 | Pyrex | $BaCO_3$ | 0.12 | 20 | 0.3 | 23 | 0.200 | 1.3 |
| | Fused silica | Carbowax 20M | 0.13 | 25 | 0.2 | 23 | 0.280 | 1.6 |
| SE-54/C-87 | Pyrex | Superox-4 | 0.20 | 25 | 0.3 | ND | 0.120 | 5.0 |
| | Fused silica | | 0.12 | 18 | 0.2 | 27 | 0.096 | 3.1 |
| C-87 | Pyrex | Silanized | 0.20 | 25 | 0.4 | 20 | 0.230 | 3.2 |
| | | | 0.20 | 25 | 0.3 | 29 | 0.180 | 3.7 |
| | Soft glass | None | ND | 50 | 0.2 | ND | 0.080 | 7.0 |
| SP-2100 | Fused silica | Carbowax 20M | 0.15 | 25 | 0.2 | 33 | 0.230 | 2.6 |
| | | | 0.12 | 50 | 0.2 | 39 | 0.300 | 2.5 |
| | | | 0.26 | 50 | 0.3 | 36 | 0.230 | 2.8 |
| Apiezon L | Pyrex | Silanized | 0.20 | 25 | 0.4 | 32 | 0.160 | 4.2 |
| | | | 0.10 | 50 | 0.2 | 48 | 0.210 | 5.8 |
| | Fused silica | | 0.10 | 17 | 0.2 | 26 | 0.180 | 3.1 |

Table II. Comparison of Phase Selectivity, TZ and Resolution

| Phase[a] | ΔI Values | | | | | | | TZ | R |
|---|---|---|---|---|---|---|---|---|---|
| | 1 | 2 | 3 | 4 | 5 | 6 | 7 | | |
| Squalane | 0 | 0 | 0 | 0 | 0 | 0 | 0 | | |
| Paraffin Oil | 9 | 5 | 2 | 6 | 11 | 2 | 2 | | |
| C-87 | 21 | 10 | 3 | 12 | 25 | 0 | 0 | 29 | 3.7 |
| Apiezon M | 31 | 22 | 15 | 30 | 40 | 12 | 10 | | |
| Apiezon L | 32 | 22 | 15 | 32 | 42 | 13 | 11 | 32 | 4.2 |
| OV-101 | 17 | 57 | 45 | 67 | 43 | 33 | 23 | 33 | 2.6 |
| SE-54 | 33 | 72 | 66 | 99 | 67 | 46 | 36 | 23 | 1.6 |

[a]Phases coated on 25-m columns, all 0.2 mm i.d. except for Apiezon L, which was 0.4 mm.

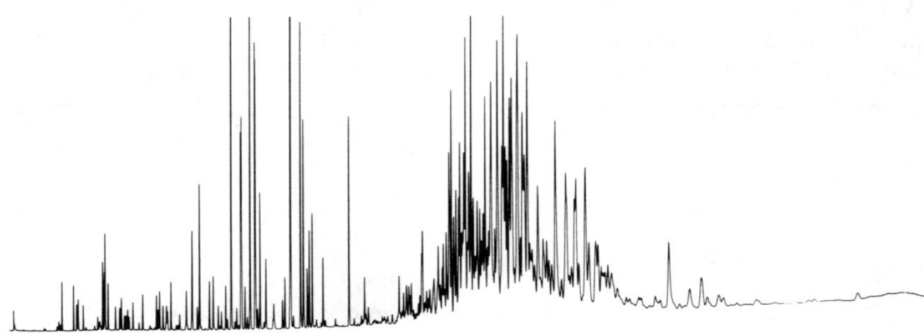

**Figure 1.** ECD of Aroclors 1242, 1260 and 5460 (400 pg each) chromatographed on an Apiezon L (WCOT) silanized Pyrex glass capillary, 0.20 mm i.d. x 50 m in length. The carrier was 53 cm/sec, capillary was temperature programmed from 150 to 390°C at 1°C/min. Atten. $1.6 \times 10^{-11}$ AUFS.

Figures 2 to 4 depict chromatograms of environmental samples analyzed on an Apiezon L capillary column. Figures 2, 3 and 4 represent extracts of yellow perch, rainwater and Canadian mother's breast milk, respectively. In addition to PCB, the rainwater sample possibly contains traces of PCT.

## Negative Ion Chemical Ionization Mass Spectrometry

The major process in NICI-MS is often dissociative electron capture [12, 13]. Compared to conventional electron-impact MS, less information about

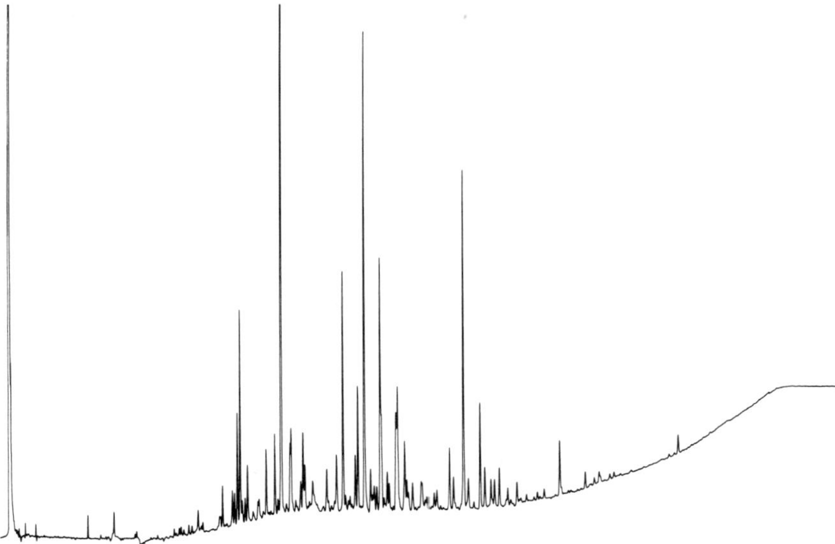

**Figure 2.** ECD of PCB in an extract of yellow perch. See Figure 1 for chromatographic conditions. Atten. 4 x $10^{-12}$ AUFS.

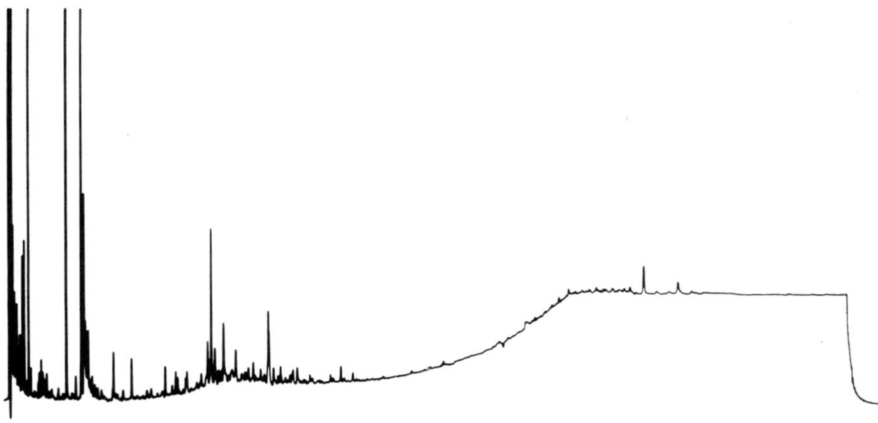

**Figure 3.** ECD in an extract of rainwater. See Figure 1 for chromatographic conditions. Attn; 4 x $10^{-12}$ AUFS.

the structure of the compound is obtained. The specific aim of this research has been to optimize NICI-MS. Several moderating and reagent gases were studied to enhance (1) the formation of molecular ions of the individual PCB; or (2) their dissociation to yield chloride 35 and 37 isotopic masses. In addition, a comparison of two ion source designs was made, i.e., an LKB 2091 magnetic instrument and a quadrupole Finnigan 4023 instrument.

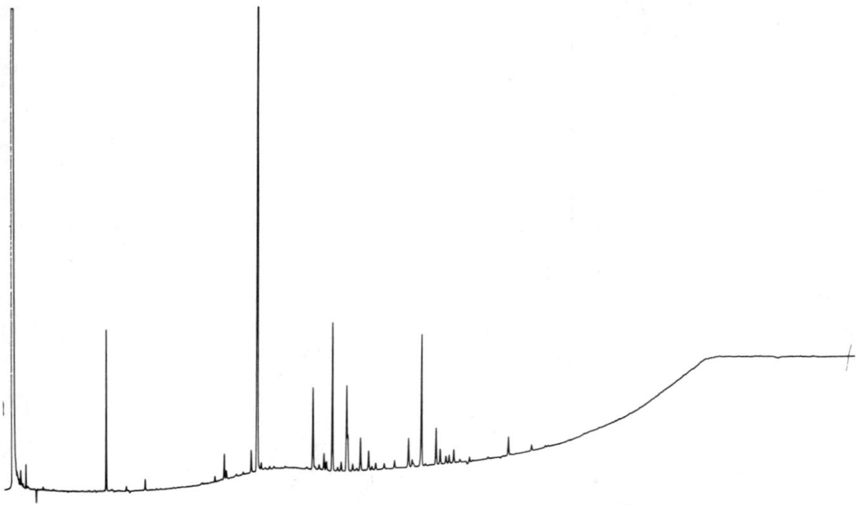

**Figure 4.** ECD of PCB in an extract of Canadian mother's milk. See Figure 1 for chromatographic conditions. Atten. 4 x $10^{-12}$ AUFS.

The LKB 2091 GC/MS does not have a pulsed positive-negative capability. Also, typical NICI conditions on this instrument utilize an open rather than a closed source design.

The composition of a solution of five PCB covering the molecular weight range of interest is given in Table III. Using this mixture a number of moderating gases were investigated. Two of which yielded no useful information were difluorodichloromethane and tetrafluoromethane. Tetrafluoromethane decreased the overall sensitivity, probably because it was competing too effectively with the PCB for available thermal electrons. No chemical reaction species were observed between tetrafluoromethane and the PCB (Table III). When the moderating gases were used with a closed source, similar results were obtained, except that the sensitivity was decreased still further.

**Table III. Test Solution for Optimization of NICI-MS**

| |
|---|
| 2-Chlorobiphenyl |
| 2,3,4-Trichlorobiphenyl |
| 2,2′,3,3′,4,4′-Hexachlorobiphenyl |
| 2,2′,3,3′4,4′5,5′-Octachlorobiphenyl |
| Decachlorobiphenyl |

The moderating gas typically used in NICI-MS is methane. In studies on the effects of source pressure on sensitivity increasing the pressure of methane also increased the sensitivity. As a compromise between optimum sensitivity and excessive pressure, a reagent gas pressure of 4 x $10^{-5}$ torr was normally employed. This pressure was measured at the throat of the diffusion pump and thus the actual source pressure may have been significantly higher.

Figure 5 depicts the mass spectrum of octachlorobiphenyl utilizing methane reagent gas and the LKB 2091 GC/MS system. The molecular ion (nominal m/z 428) of octachlorobiphenyl was relatively low in abundance, the predominant ion was mass 35 and 37.

Several investigators have demonstrated that methane-oxygen increases the intensity of the higher mass ions in the NICI spectra of halogenated compounds [12,13]. Table IV presents results of varying the concentrations of oxygen in nitrogen vs pure methane. Decachlorobiphenyl and octachlorobiphenyl were examined under various conditions to determine the percent relative abundance of the molecular ion relative to mass 35. Using a 40% $O_2$/60% $N_2$ mixture at two different partial pressures yielded only a slight increase in the relative molecular ion abundance for decachlorobiphenyl and

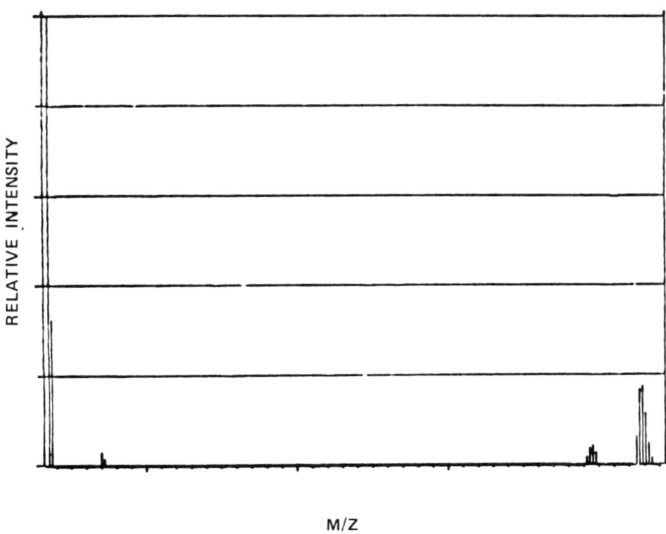


Figure 5. Mass spectrum of octachlorobiphenyl obtained for methane NICI-MS using the LKB 2091 magnetic sector instrument.

octachlorobiphenyl with an increase in pressure. On the other hand, increasing the $O_2$ concentration from 40 to 60% substantially increased the percent relative abundance of the molecular anion (Table IV). Oxygen in $N_2$ also significantly enhanced the relative molecular ion abundance as compared to methane alone. Thus, increasing the partial pressure of $O_2$ in the reagent gas mixture led to increased intensity of high-mass ions; however, no molecular ion species were detected for the lower chlorine substituted biphenyls.

Nitrous oxide at ~1000 ppm in $N_2$ has been reported to enhance electron capture sensitivity of compounds which normally do not exhibit a high degree of sensitivity toward capture [14]. This reagent gas was also investigated with the LKB 2091 GC/MS system. In general, the background level was enhanced as expected. PCB were also observed, but inspection of the mass spectra revealed that the predominant mechanism was still dissociative electron capture. Figure 6 represents a spectrum for trichlorobiphenyl obtained with $N_2O$ in $N_2$ under NICI conditions. The predominant masses were 35/37 for the chloride ion isotopes.

The best conditions for obtaining structural information with the LKB 2091 GC/MS system was with oxygen-enhanced methane. However, even these conditions hold little promise for yielding molecular ion species for the lower chlorinated PCB. On the other hand, the detection limits were very good for the chloride 35/37 ions.

**Table IV. Results of Oxygen-Enhanced Methane NICI (at $4 \times 10^{-5}$ torr)**

| Reagent | Partial Pressure (torr) | Decachlorobiphenyl | | Octachlorobiphenyl |
|---|---|---|---|---|
| | | Ion | %RA[a] | %RA[a] |
| 40% $O_2$/60% $N_2$ | $4.2 \times 10^{-6}$ | $M^+$ | 22.4 | 7.2 |
| | | M–Cl+O | 3.5 | 2.8 |
| | | 35 | 100 | 100 |
| | 1.2 | | | |
| | $1.2 \times 10^{-5}$ | $M^+$ | 24 | 7.6 |
| | | M–Cl+O | 5.6 | 2.2 |
| | | 35 | 100 | 100 |
| 60% $O_2$/40% $N_2$ | $4.2 \times 10^{-6}$ | $M^+$ | 33 | 28.8 |
| | | M–Cl+O | 8.7 | 9.2 |
| | | 35 | 100 | 100 |
| $CH_4$ | $4.0 \times 10^{-5}$ | $M^+$ | 11.7 | 7.5 |
| | | M–Cl | 4.8 | 1.1 |
| | | 35 | 100 | 100 |

[a]Maximim intensity of ion observed in chromatographic peak relative to the maximum intensity of the base peak.

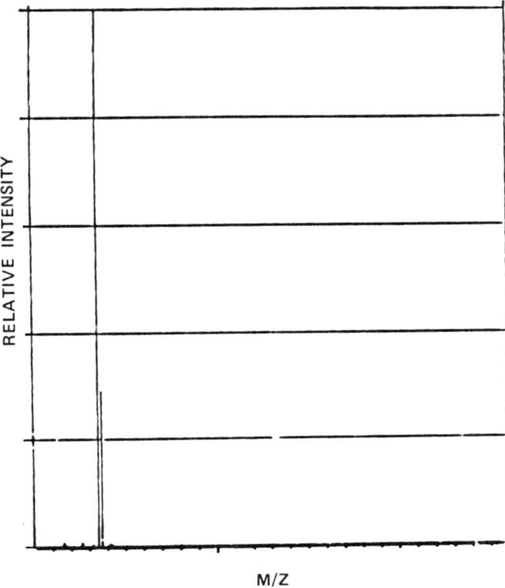


Figure 6. Mass spectrum of trichlorobiphenyl obtained with $N_2O/N_2$ NICI-MS using the LKB 2091.

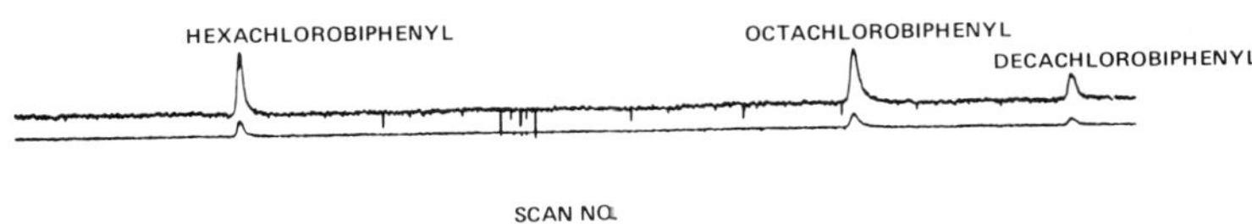


Figure 7. Selected ion monitoring trace of m/z 35 and 37 for a 150-pg injection of hexachlorobiphenyl, octachlorobiphenyl and decachlorobiphenyl. Methane was the reagent gas for NICI-MS. Injection split ratio was 15:1.

Using conditions which yield predominantly dissociative electron capture the limits of detection were investigated. The test mixture (Table III) was chromatographed on a fused-silica column (SP-2100, 25 m). Figure 7 depicts these results with approximately 10 pg of each component introduced onto the capillary.

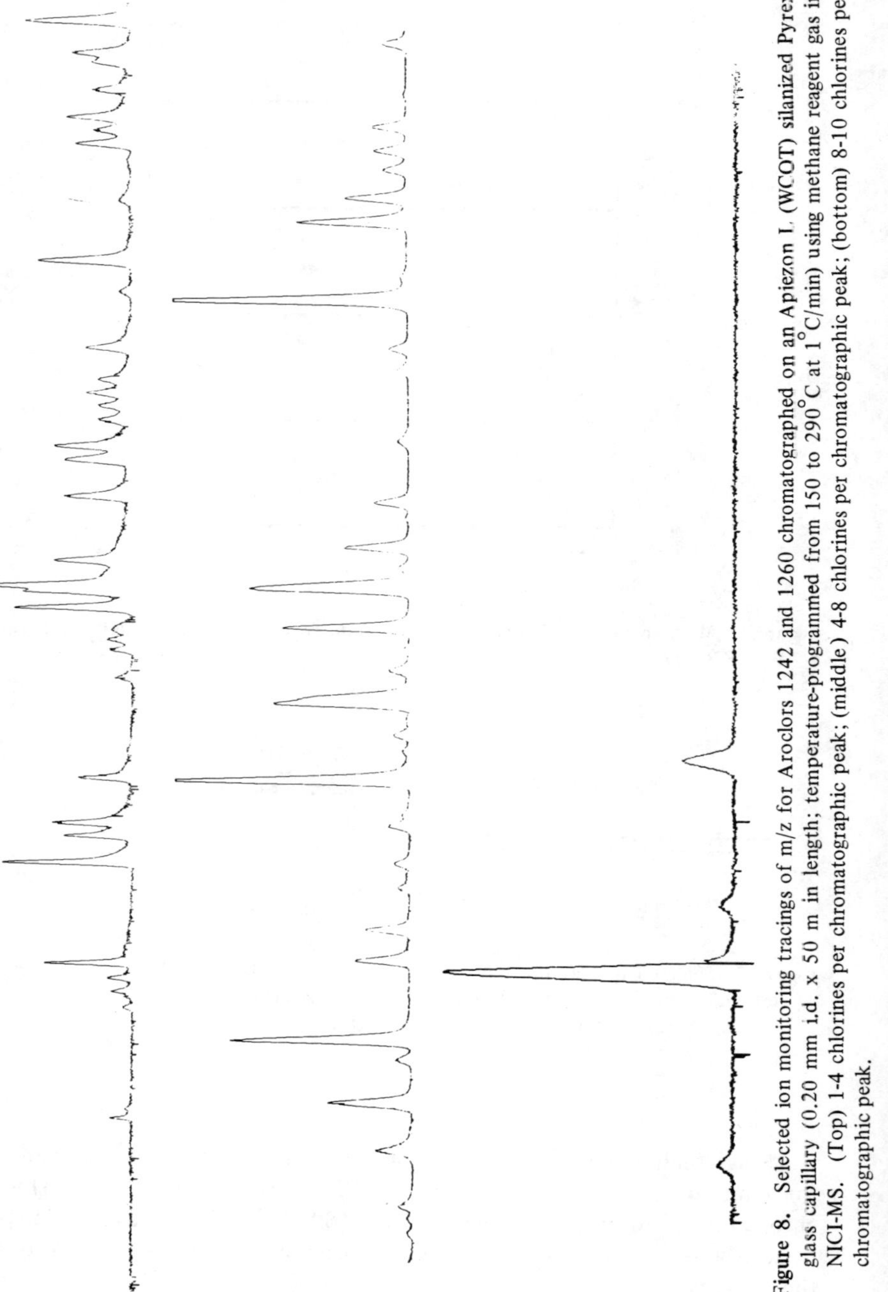

**Figure 8.** Selected ion monitoring tracings of m/z for Aroclors 1242 and 1260 chromatographed on an Apiezon L (WCOT) silanized Pyrex glass capillary (0.20 mm i.d. x 50 m in length; temperature-programmed from 150 to 290°C at 1°C/min) using methane reagent gas in NICI-MS. (Top) 1-4 chlorines per chromatographic peak; (middle) 4-8 chlorines per chromatographic peak; (bottom) 8-10 chlorines per chromatographic peak.

Using the previously described Apiezon L capillary, the Aroclor 1242/1260 mixture was examined under NICI conditions and selected ion monitoring. Figure 8 depicts the chloride 35 ion tracings for 400 pg of each on column.

The common variables in chemical ionization (CI) MS are source pressure, temperature and voltages. The chemical thermodynamics are also affected by the physical design of the GC/MS source where the CI takes place. Utilizing methane as a common reagent gas and maintaining comparable ionizer temperature and pressure a Finnigan 4023 GC/MS system was examined. The source is of a switchable EI/CI design and during CI operation the extractor lenses are given a positive potential of IV to increase the residence time of reagent gas ions and sample molecules. The instrument is capable of pulsed positive-negative ion chemical ionization (PPNICI). Thus, a more precise comparison of spectra and sensitivity between positive and negative chemical ionization (by eliminating the variables of differences in quantity injected, column and source temperatures, source pressure and background variation) could be made. The rapid switching between positive and negative CI (12 kHz) allowed simultaneous information to be acquired.

Figure 9 represents a PPNICI analysis of an Aroclor 1260 on an SP-2100 fused-silica capillary. Methane was the moderating gas. As might be expected,

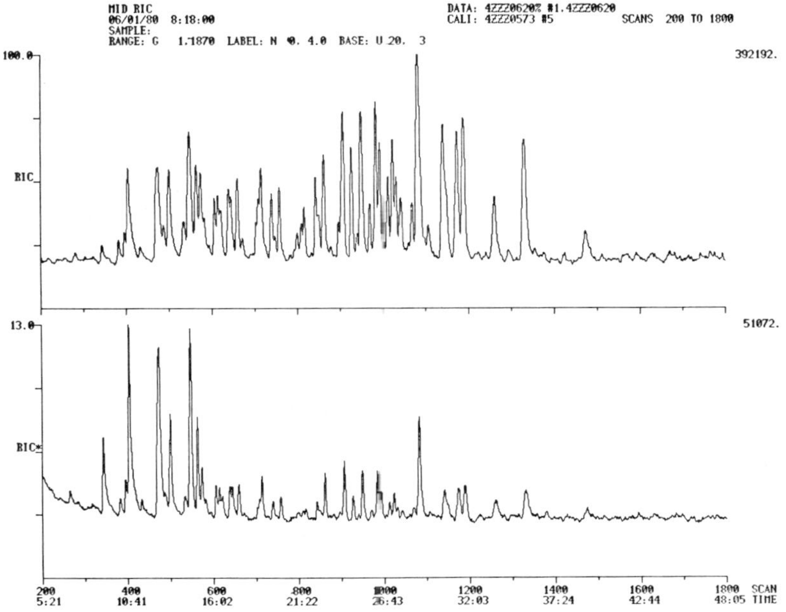


**Figure 9.** PPNICI analysis of Aroclor 1260 on an SP-2100 fused-silica capillary (0.23 mm i.d. x 50 m) using methane reagent gas. Upper tracing for pulsed negative bottom for pulsed positive CI/MS.

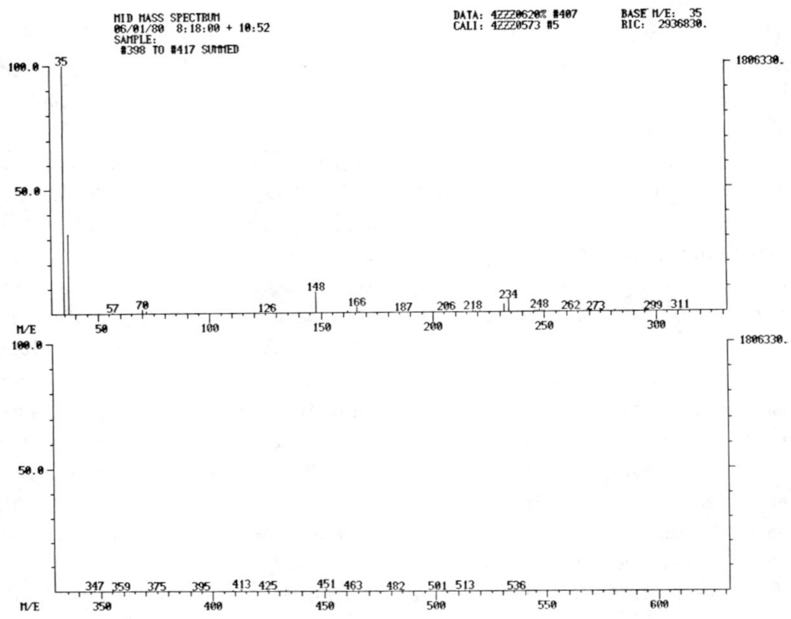


**Figure 10.** NICI mass spectra from Aroclor 1260 analyzed in Figure 9. One to five chlorines.

significant differences in sensitivity were exhibited between the pulsed positive (lower tracing) and negative (upper) modes. Mass spectra for compounds (i.e., individual chromatographic peaks) with one to five chlorines are shown in Figure 10. The predominant ion abundance is for m/z 35/37. Very little molecular ion abundance was observed. Beginning with PCB isomers containing six or more chlorines, the molecular ion becomes apparent, and for higher degrees of chlorination it becomes dominant (Figures 11 to 15).

The significant feature of a PPNICI analysis was obvious when ion tracings were developed for PCB containing one to ten chlorines. The regions of the chromatogram containing isomers with one to seven chlorines as observed in the pulsed positive CI mode are shown in Figure 16. Simultaneous information for the pulsed negative CI mode is depicted in Figures 17 and 18. The latter figure presents the region of the chromatogram for isomers with five to ten chlorines.

Oxygen-enhanced methane experiments were also performed with the PPNICI system. The enhancement of molecular ions for the PCB isomers was not as evident as with the magnetic sector system. Also, the intensity of the chloride 35/37 ions using only methane gas was not as great as with the sector system. However, only PCB isomers with 8, 9 and 10 chlorines exhibited molecular ions on the sector instrument.

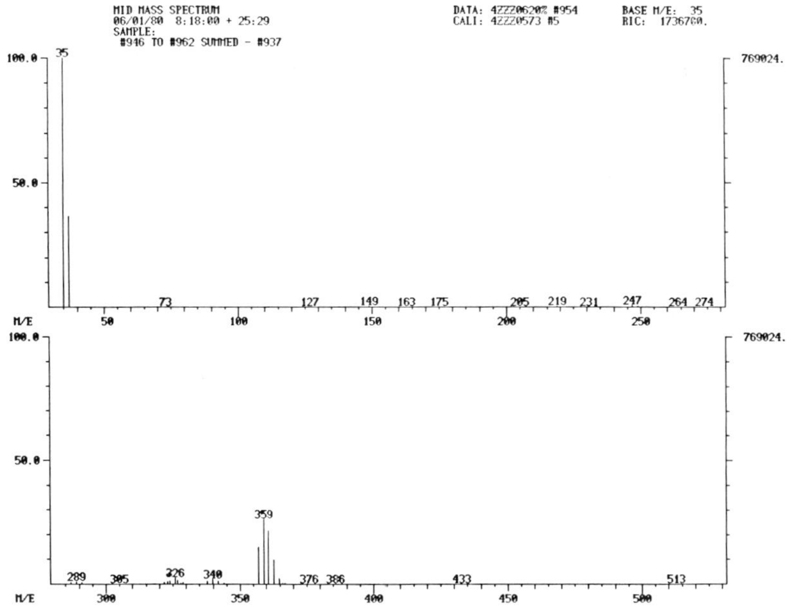


Figure 11. NICI mass spectrum for PCB with six chlorines (Figure 9).

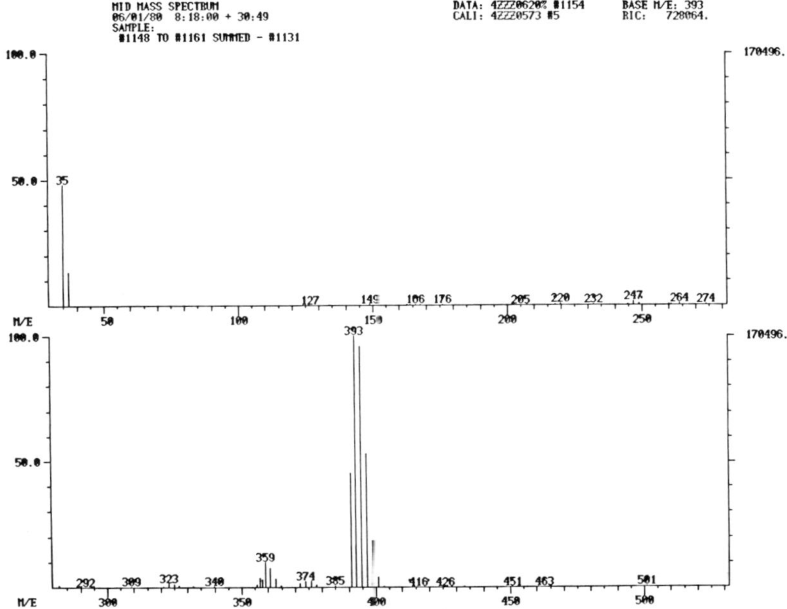


Figure 12. NICI mass spectrum for PCB with seven chlorines (Figure 9).

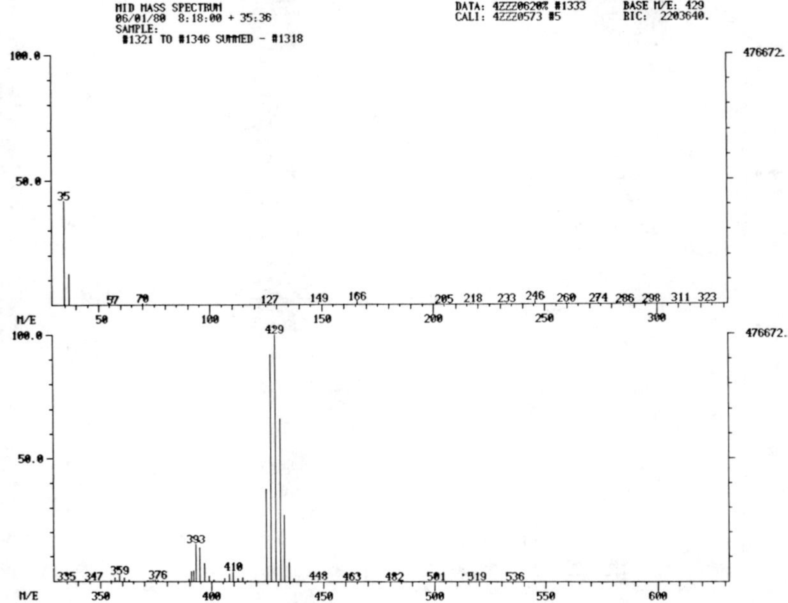


Figure 13. NICI mass spectrum for PCB with eight chlorines (Figure 9).

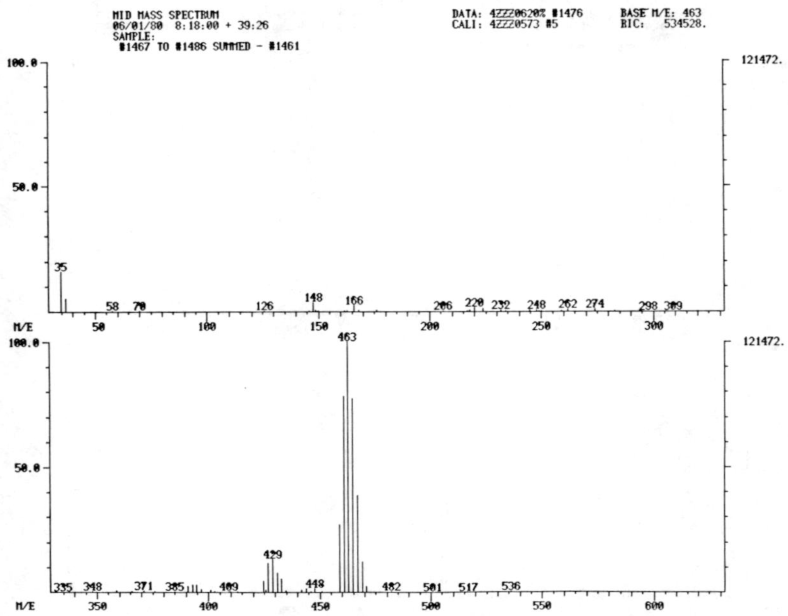


Figure 14. NICI mass spectrum for PCB with nine chlorines (Figure 9).

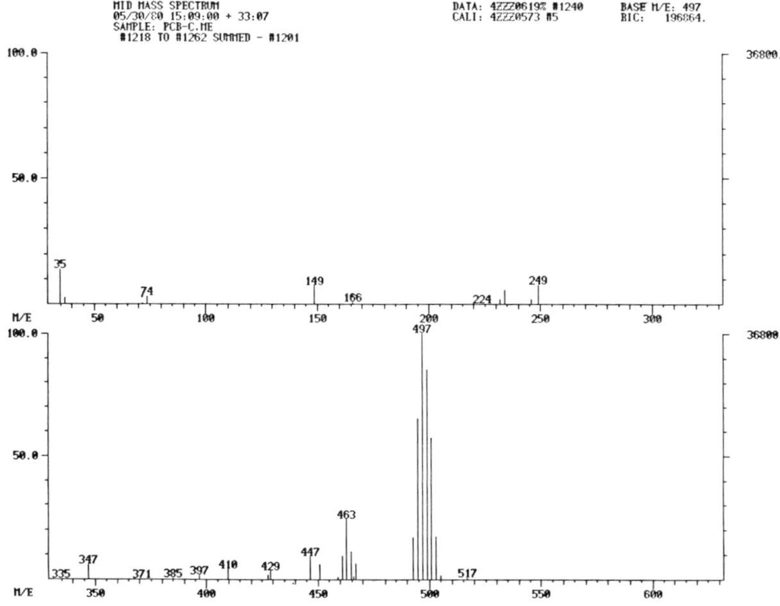


**Figure 15.** NICI mass spectrum for PCB with ten chlorines (Figure 9).

The use of pulsed positive-negative CI reveals several important aspects. First is the ability to verify the identity and/or degree of chlorination for each chromatographic peak in a complex mixture. By selecting the appropriate reagent gas, sensitive detection of the individual isomers is possible, so quantification should ultimately be possible.

## SUMMARY

An investigation into the preparation of better capillary columns for resolving PCB isomers was conducted. Excellent resolution and temperature stability were obtained for an Apiezon L phase (0.1 $\mu$m film thickness) coated on a silanized surface of a Pyrex capillary 0.20 mm i.d. x 50 m in length.

Studies were conducted on moderating and reagent gases used in NICI-MS for enhancing either m/z 35/37 or molecular ions of PCB using a magnetic sector and quadrupole GC/MS systems. Open and closed sources were compared. Oxygen-enhanced methane gave best conditions for optimum molecular ion intensities, while pure methane was most suitable with an open source design for enhanced m/z 35/37.

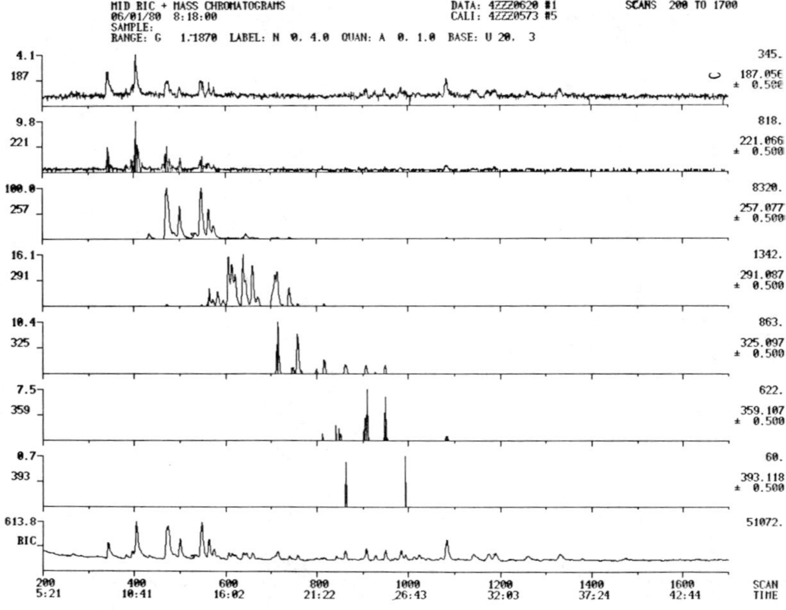


**Figure 16.** Ion tracings for Aroclor 1260 analyzed in Figure 9. Pulsed positive CI using methane reagent gas. One to seven chlorines.

The combination of high-resolution GC and NICI-MS provides an instrumental method for environmental analysis which can be used for:

1. identifying or verifying PCB isomers;
2. determining the number of chlorines in a chromatographic peak; and
3. quantifying each individual isomer in a sample.

## ACKNOWLEDGMENT

The authors express their deep appreciation to Dr. M. Mullin, whose encouragement and suggestions provided the necessary perspective to this research program. We also wish to thank Dr. Mullin for providing the pure individual PCB isomers and extracts of environmental sample used in this program. The excellent technical assistance provided by Ms. P. Gentry and B. Bickford for the NICI-MS analyses is gratefully appreciated. This research was supported by EPA Grant No. CR807167-01-1, Larger Lakes Research Station, Grosse Ile, MI.

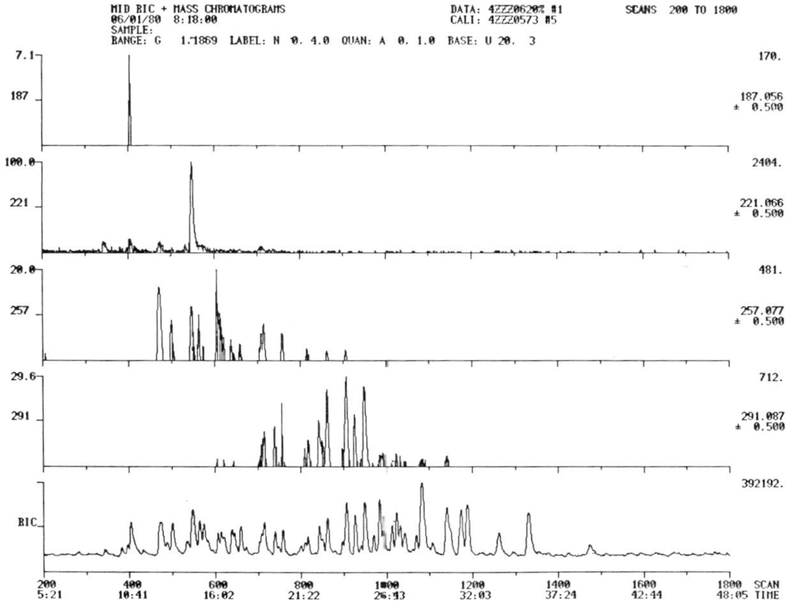


Figure 17. Ion tracings for Aroclor 1260 analyzed in Figure 9. Pulsed negative CI. One to four chlorines.

## REFERENCES


1. Mullin, M. D., and J. C. Filkins. "Analysis of Polychlorinated Biphenyls by Glass Capillary and Packed Column Gas Chromatography," Chapter 11, this volume.
2. Grob, K., G. Grob and K. Grob, Jr. "Deactivation of Glass Capillary Columns by Silylation," *J. High Resolution Chromatog. Chromatog. Commun.* 2:31 (1979).
3. Grob, K., G. Grob and K. Grob, Jr. "Deactivation of Glass Capillaries by Persilylation, Part 2," *J. High Resolution Chromatog. Chromatog. Commun.* 2:677 (1979).
4. Grob, K. G. Grob and K. Grob, Jr. "Deactivation of Glass Capillaries by Perslylation, Part 3," *J. High Resolution Chromatog. Chromatog. Commun.* 3:197 (1980).
5. Grob, K., G. Grob and K. Grob, Jr. "The Barium Carbonate Procedure for the Preparation of Glass Capillary Columns: Further Information," *Chromatographia* 10:191 (1977).
6. Grob, K., and G. Grob. "A New Generally Applicable Procedure for the Preparation of Glass Capillary Columns," *J. Chromatog.* 125:471 (1976).

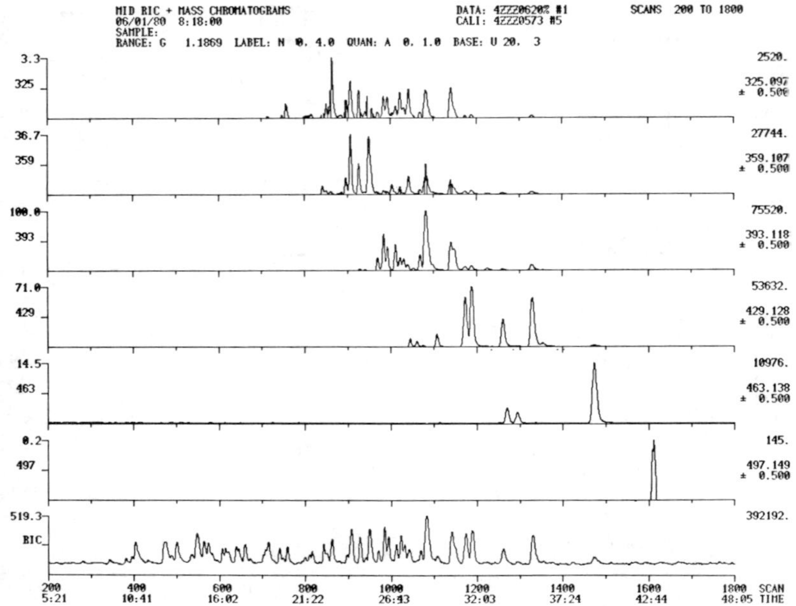


**Figure 18.** Ion tracings for Aroclor 1260 analyzed in Figure 9. Pulsed negative CI. Five to ten chlorines.


7. Grob, K., G. Grob and K. Grob, Jr. "Preparation of Apolar Glass Capillary Columns by the Barium Carbonate Procedure," *J. High Resolution Chromatog. Chromatog. Commun.* 3:149 (1978).
8. Arrendale, R. F., L. B. Smith and L. B. Rogers. "Comparison of Dynamically Coated SE-54, SP-2250, and Carbowax 20M WCOT Glass Capillary Columns, Prepared After Surface Pretreatment with Superox-4 or with $BaCo_3$," *J. High Resolution Chromatog. Chromatog. Commun.* 3:115 (1980).
9. Bouche, J., and M. Verzele. "Static Coating Procedure for Glass Capillary Columns," *J. Gas Chromatog.* 6:501 (1968).
10. Grob, K. "Static Coating of Glass Capillary Columns," *J. High Resolution Chromatog. Chromatog. Commun.* 2:93 (1978).
11. Grob, K., Jr., G. Grob and K. Grob. "A Comprehensive, Standardized Quality Test for Glass Capillaries," *J. Chromatog.* 156:1 (1978).
12. Hass, J. R., M. D. Frieser and M. K. Hoffman. "The Mass Spectrometry of Polychlorinated Dibenzo-*p*-Dioxin, OR6," *Mass Spectrom.* 14:9 (1979).
13. Busch, K. L., A. Norstrom, M. M. Bursey, J. R. Hass and C.-A. Nilsson. "Methane-Oxygen Enhanced Negative Ion Mass Spectra of Polychlorinated Diphenyl Ethers," *Biomed. Mass Spectrom.* 6:157 (1979).
14. Phillips, M. P., R. E. Sievers, P. D. Goldan, W. C. Kuster and F. C. Fehsenfeld. "Enhancement of Electron Capture Detection Sensitivity to Nonelectron Attaching Compounds by Addition of Nitrous Oxide to the Carrier Gas," *Anal. Chem.* 51:1819 (1979).

# SECTION 3

# STABLE LABELING

# CHAPTER 13

# DETERMINATION OF PRIORITY POLLUTANTS IN INDUSTRIAL WASTEWATERS BY STABLE-ISOTOPE DILUTION GAS CHROMATOGRAPHY/MASS SPECTROMETRY

**B. N. Colby and A. E. Rosecrance**

Chemistry and Chemical Engineering
Systems, Science and Software
La Jolla, California

As a result of a court decision in 1976, commonly known as the Consent Decree, a list of specific priority pollutants was established by the U.S. Environmental Protection Agency (EPA) to be monitored and regulated in aqueous industrial discharges. At present, gas chromatography (GC) and gas chromatography/mass spectroscopy (GC/MS) analysis methods are available for monitoring 114 organic priority pollutant compounds [1]. Two GC/MS methods, Methods 624 and 625, which EPA may specify for priority pollutant determinations, allow the analyst to select external or internal standard quantitative approaches. A third approach to GC/MS quantification (i.e., isotope dilution) exists, but is not currently specified as acceptable. A method for 2,3,7,8-tetrachlorodibenzo-*p*-dioxin (TCDD), i.e., Method 613, incorporates a $^{37}$Cl-labeled analog of TCDD as an internal standard as do isotope dilution methods, but, because the $^{37}Cl_4$-TCDD is not added to the sample prior to extraction, the method does not acutally use isotope dilution for quantification. Because three fundamental approaches to GC/MS quantification exist, a study was initiated by this laboratory to investigate the practical merits of each. That study was subsequently expanded to pursue a more complete investigation of isotope dilution as it applies to priority pollutant determinations in aqueous industrial discharges. The purpose of this chapter is to report on the results of this study.

The initial phase of this investigation involved the preparation of three sets of water solutions spiked with varying concentrations of benzene, toluene, chloroform and 1,2-dichloroethane. These were prepared from organic-free water. Two of the solution series were also spiked with a soap solution to induce foaming when the sample was purged according to the analysis protocol of Method 624. Foaming is a common problem known to induce changes in recovery and, consequently, the accuracy of the analysis. Foaming in one of the soap solutions was eliminated by adding a silicone surfactant and in the other by using a heat dispersion technique [2]. The third solution was used as a reference standard as specified in Method 624. All solutions were additionally spiked to 20 $\mu$g/L each of the chlorobromomethane, 1,4-dichlorobutane and stable isotopically labeled analogs of the four test compounds.

After the samples had been analyzed, the acquired data were reduced using the external standard, internal standard and isotope dilution quantification techniques. Figure 1 shows the calibration approaches. The data for each approach were subsequently reduced using four methods for calibration curve generation. These were:

- direct proportionality, i.e., data point and the origin;
- two data points to provide an experimental intercept;
- standard additions, to compensate for matrix effects; and
- linear regression.

When the results were tabulated, the analytical accuracy, expressed in terms of the error factor [2], exhibited some very striking trends. These are shown in Figure 2. With all data reduction approaches, the error was greatest when no internal standard was employed and least when isotope dilution was employed. The reason for this result is that isotope dilution internal standards are isotopically labeled analogs of the compounds determined. These exactly mimic the behavior of the determined compounds and consequently, concentrations are corrected on a per-sample basis for matrix and method induced variations in recovery. Use of nonstructurally related internal standards, such as the chlorobromomethane and 1,4-dichlorobutane incorporated here, had a positive influence on accuracy but not to the extent that the labeled analogs did. When no internal standard was incorporated, matrix

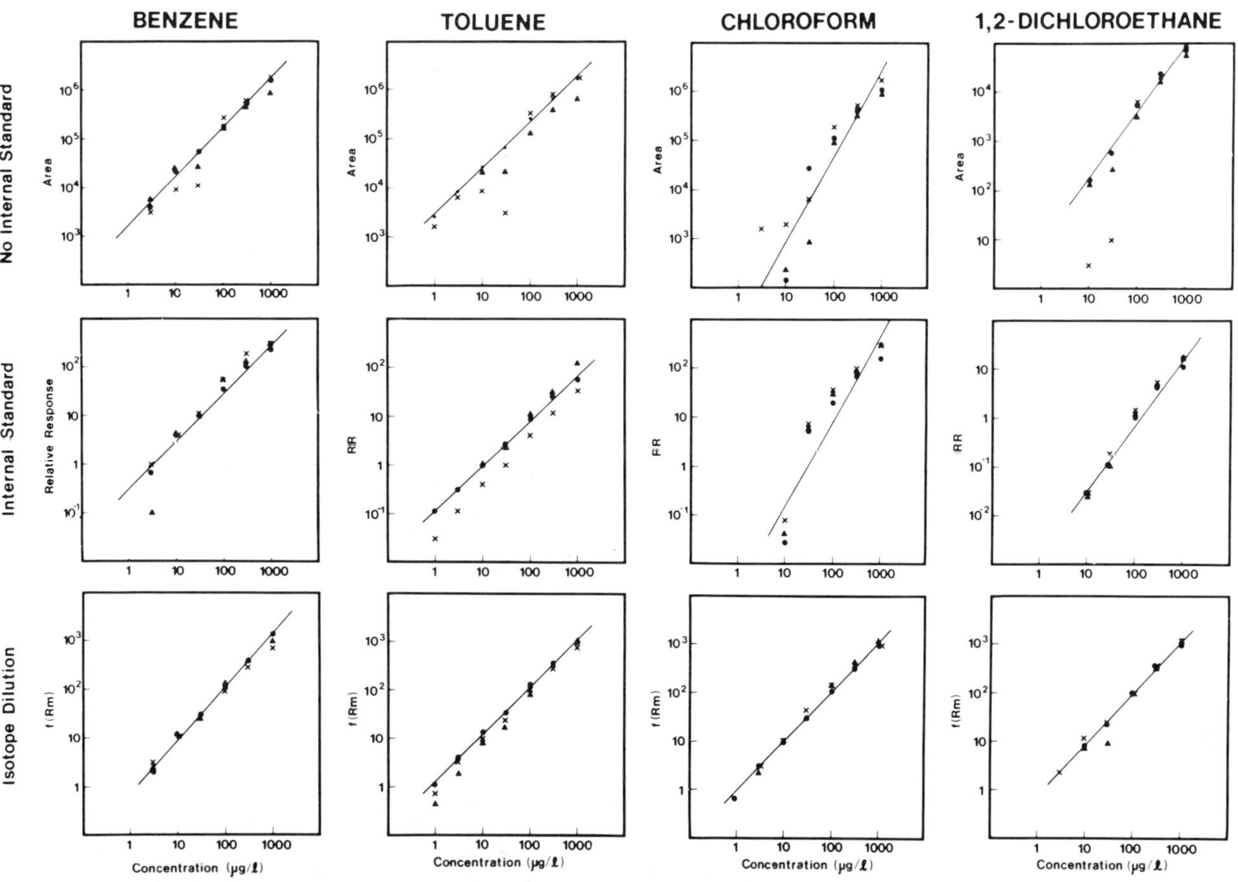


Figure 1. Analysis data for pure water solution (●) and soap solutions using silicone antifoaming agent (▲) and heat dispersion (x).

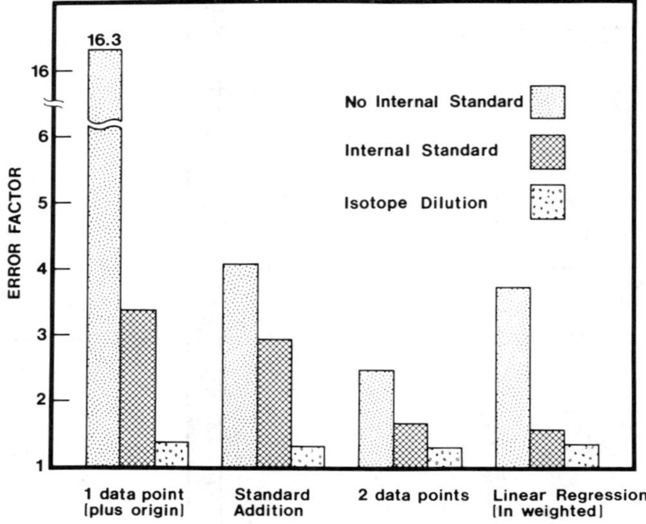


**Figure 2.** Analytical accuracy for soap solution analyses.

effects were not compensated and thus accuracy was poor. It was interesting to note that the method of standard additions did not yield accurate compensation for matrix effects. This was apparently a result of propagated errors.

Because isotope dilution was clearly demonstrated as the most accurate quantification approach, the study was expanded toward a more complete investigation of that method. Before describing those studies, however, it is appropriate to give in detail the steps of an isotope dilution GC/MS analysis. They are much like those of an internal standard method except that a stable isotopically labeled analog of each compound to be determined is added to the sample prior to preparation. For example, 100 $\mu$g of each base/neutral and acid fraction priority pollutant would be added to a 1-L water sample before extracting the sample. In this way, the recoveries of the labeled analogs can be used to compensate or correct the recoveries of the unlabeled (naturally abundant) compounds which were originally present in the sample. Sample preparation and GC/MS analysis procedures are unchanged from Methods 624 and 625 with the exception of quantification. Quantification is accomplished by creating calibration curves of the relative response (RR) of unlabeled (naturally abundant) compound-to-labeled analog vs concentration. The RR is calculated from extracted ion current profile (EICP) areas according to:

$$RR = \frac{(R_y - R_m)(R_x + 1)}{(R_m - R_x)(R_y + 1)} \qquad (1)$$

where $R = \dfrac{\text{EICP area of unlabeled compound quantification mass}}{\text{EICP area of labeled compound quantification mass}}$

m = ratio for the quantification standard (or unknown) analysis mixture

x = ratio for the pure unlabeled (naturally abundant) compound

y = ratio for the pure labeled compound

Equation 1 corrects the RR value for the potential isotopic overlap between the labeled and unlabeled species [3]. It should only be used when:

$$2R_y < R_m < \tfrac{1}{2}R_x \qquad (2)$$

or propagated errors will become large. $R_x$ and $R_y$ should be determined experimentally with each set of samples to provide up-to-date corrections based on instrument performance and istopic purity. An example of a calibration curve is given in Figure 3 for chloroform using $^{13}$C-chloroform as the labeled analog.

To demonstrate that a compound and its labeled analog exhibit identical recoveries when spiked into actual wastewater samples, 13 base/neutral compounds and their labeled analogs were spiked into the effluents of four aqueous industrial wastes. Two industries were selected which involved organic chemical materials, a pesticide manufacturer and a leather tanner; and two were selected which involved inorganic processing, a molybdenum mining and processing facility, and a uranium mine. The unlabeled base/neutral priority pollutants were added to give 20 μg/L and the labeled analogs, to give

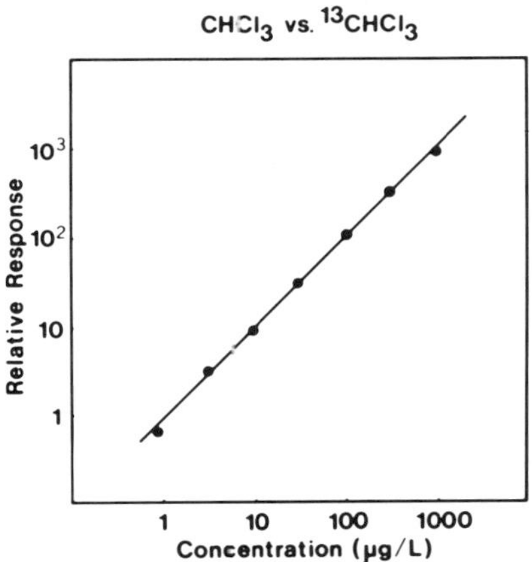


**Figure 3.** Example of an isotope dilution calibration curve for $CHCl_3$ (m/z = 85) using $^{13}CHCl_3$ (m/z = 86) as the internal standard.

Table I. Mean Recoveries for Four Spiked Industrial Effluents

| Compound | Recovery (%) | | Recovery Ratio (Unlabeled/Labeled) |
|---|---|---|---|
| | Unlabeled | Labeled | |
| Acenaphthene | 50.8 | 50.9 | 0.998 |
| 1,2-Dichlorobenzene | 39.3 | 39.5 | 0.995 |
| 1,4-Dichlorobenzene | 71.6 | 70.9 | 1.01 |
| Naphthalene | 32.1 | 30.6 | 1.05 |
| Benzo[a]anthracene plus Chrysene | 51.2 | 49.2 | 1.05 |
| Benzo[a]pyrene | 61.9 | 59.5 | 1.04 |
| Anthracene plus Phenanthrene | 51.9 | 48.1 | 1.08 |
| Fluorene | 40.2 | 40.0 | 1.00 |
| Pyrene | 80.1 | 71.5 | 1.12 |
| 4-Chlorophenyl Phenyl Ether | 48.4 | 44.4 | 1.09 |
| N-Nitrosodiphenylamine | 70.7 | 63.1 | 1.12 |
| Mean | 54.4 | 51.6 | 1.05 |
| Standard Deviation | 15.1 | 13.2 | 0.05 |

Table II. Mean Precisions for Four Spiked Industrial Effluents

| Compound | Precision (% SD) | | |
|---|---|---|---|
| | Unlabeled | Labeled | Ratio |
| Acenaphthene | 22 | 16 | 6.4 |
| 1,2-Dichlorobenzene | 36 | 35 | 4.7 |
| 1,4-Dichlorobenzene | 69 | 57 | 7.8 |
| Naphthalene | 30 | 52 | 9.1 |
| Benzo[a]anthracene plus Chrysene | 20 | 24 | 11.8 |
| Benzo[a]pyrene | 31 | 28 | 5.0 |
| Anthracene plus Phenanthrene | 18 | 17 | 3.5 |
| Fluorene | 15 | 17 | 3.2 |
| Pyrene | 27 | 25 | 3.0 |
| 4-Chlorophenyl Phenyl Ether | 18 | 17 | 1.4 |
| N-Nitrosodiphenylamine | 24 | 21 | 3.4 |
| Mean | 28 | 28 | 5.1 |

100 μg/L. One hour was allowed for the samples to equilibrate before extracting according to Method 625. After GC/MS analysis of the extracts, percent recoveries for the unlabeled and labeled species were calculated using the external standard technique. Also calculated was the recovery ratio of unlabeled-to-labeled compound. The results are shown in Table I. On a compound-for-compound basis in each sample, there was a nearly perfect one-to-one relation between recoveries calculated using the external standard approach. This was in complete agreement with the recovery ratios.

The precision of the data given in Table I is provided in Table II. It was interesting to note that although percent standard deviations were relatively high for the external standard determinations, approximately 30% SD, the precision for the recovery ratios was extremely good at approximately 5% SD. These precision data serve to further establish an anticipated high level of performance for isotope dilution GC/MS analyses of priority pollutants in industrial effluents.

To establish further the precision and accuracy of isotope dilution in applications to priority pollutant determinations, a series of 20 industrial effluents was analyzed. Each of these was divided into two aliquots; one was spiked at 20 μg/L with 36 unlabeled priority pollutants and the other was analyzed as received to establish background levels for the 36 compounds of interest. Both aliquots were spiked with labeled analogs of the 36 compounds of interest to act as isotope dilution internal standards. The semivolatile labeled compounds were spiked at 100 μgL and the volatile labeled compounds at 20 μg/L, except for acrylonitrile, which was 100 μg/L. The industries involved were: municipal wastewater treatment (two samples), water supply, coal mining (two samples), gasohol manufacture, leather tanning, hard rock mining (two samples), pesticide manufacture (two samples), blast furnace, steam electric, coal gasification, organic chemical manufacture (five samples) and a pulp/paper mill. When more than one sample from an industrial type was utilized, each was selected such that either the manufacturing processes or the treatment processes was different. Ten of the above samples were prepared and analyzed in the authors' laboratory; the other ten were prepared and analyzed by TRW. All samples were supplied courtesy of the Effluent Guidelines Division of EPA.

The accuracies and precisions for determining the 36 spiked priority pollutants are given in Table III and summarized in Table IV. Based on these results, it is clear that isotope dilution quantification results are unaffected by sample matrix. Also, although it is not obvious from Table III, there was no significant difference in either the precision or accuracy of the results obtained in the two laboratories.

Although the high levels of precision and accuracy generated by isotope dilution quantification are important, the potential for incorporating the

**Table III. Accuracy of Recovery Corrected Concentrations for 20 Spiked Industrial Effluents**

| Compound | Accuracy (%) | % SD |
|---|---|---|
| Acenaphthene | 98.9 | 8.2 |
| 1,2-Dichlorobenzene | 99.2 | 14.7 |
| 1,4-Dichlorobenzene | 102.9 | 14.2 |
| Naphthalene | 103.8 | 10.9 |
| Nitrobenzene | 99.2 | 25.2 |
| Benzo[a] anthracene and Chrysene | 98.3 | 16.9 |
| Benzo[a] pyrene | 98.4 | 13.3 |
| Anthracene and Phenanthrene | 101.0 | 10.1 |
| Fluorene | 97.5 | 7.3 |
| Pyrene | 100.2 | 9.4 |
| 4-Chlorophenyl Phenyl Ether | 100.0 | 7.7 |
| N-Nitrosodiphenylamine | 101.0 | 6.9 |
| 2-Nitrophenol | 108.3 | 13.5 |
| 2,4-Dichlorophenol | 102.9 | 16.7 |
| 2,4-Dinitrophenol | 102.3 | 30.5 |
| 2-Chlorophenol | 104.1 | 9.1 |
| 2,4-Dimethylphenol | 76.3 | 29.9 |
| Phenol | 95.9 | 20.3 |
| Pentachlorophenol | 99.1 | 21.6 |
| 2,4,6-Trichlorophenol | 103.8 | 10.0 |
| Acrylonitrile | 100.4 | 4.8 |
| Benzene | 99.9 | 6.6 |
| Carbon Tetrachloride | 97.6 | 11.9 |
| Chlorobenzene | 102.8 | 5.8 |
| 1,2-Dichloroethane | 100.0 | 6.2 |
| 1,1,1-Trichloroethane | 101.4 | 8.7 |
| 1,1,2-Trichloroethane | 99.0 | 6.9 |
| 1,1,2,2-Tetrachloroethane | 99.4 | 6.7 |
| 1,1-Dichloroethylene | 95.6 | 6.9 |
| Ethylbenzene | 101.0 | 14.9 |
| Methylene Chloride | 99.3 | 9.1 |
| Bromoform | 99.1 | 7.4 |
| Toluene | 90.7 | 15.6 |
| Chloroform | 92.1 | 17.9 |

isotopically labeled compounds into a quality assurance quality control (QA/QC) scheme is equally important. QA/QC data for each priority pollutant in the form of its labeled analog can be monitored for each sample analyzed. That is, by spiking the labeled analogs and monitoring their recovery via conventional internal (or external) standard techniques, it is possible to generate compound-specific recovery data for each sample analyzed without the necessity of preparing and analyzing additional aliquots of each sample

Table IV. Mean Accuracies and Precisions for Recovery Corrected Concentrations

| Fraction | Accuracy (%) | Precision (% SD) |
|---|---|---|
| Base/Neutral | 100.0 | 12.1 |
| Acid | 99.1 | 19.0 |
| Volatile | 98.5 | 9.2 |

Table V. Validation QA/QC Comparison (Four-day Episode)

| | Sample Data Sets | QA/QC Data Sets | |
|---|---|---|---|
| | | Existing QA | Isotope Dilution |
| Initial QA/QC | | | |
| Pretreatment | 0 | 1 | 0 |
| Validation (1 day) | 1 | 11 | 3 |
| Ongoing QA | | | |
| Full QA (1 day) | 1 | 2 | 1 |
| Surrogate Only (2 days) | 2 | 0 | 0 |
| Totals | 4 | 14 | 4 |

spiked with unlabeled (naturally abundant) priority pollutants. The feasibility of this approach is clearly illustrated by the data given in Table I. The result of incorporating priority pollutant recovery monitoring into each sample without having to increase the number of sample preparations and GC/MS analyses would reduce costs and provide a more complete QA/QC.

For a typical four-day priority pollutant validation episode, four sample data sets are produced (Table V). Using the existing QA/QC protocol, these four sample data sets would be accompanied by 14 QA/QC data sets associated with establishing analytical precision and accuracy or percent recovery. The majority of this QA/QC effort is associated with the initial sample data set. It involves a screening (pretreatment) data set to establish approximate priority pollutant background levels, six replicate data sets to establish precision and six spiked aliquot data sets to establish recovery. The spiked aliquot data sets consist of three spiking levels prepared and anlyzed in duplicate. Also, a series of surrogate compounds is spiked into each sample. The percent recoveries and precisions of these surrogates are later used to infer consistent behavior of priority pollutants in samples for which no priority pollutant spiking is employed, i.e., during "ongoing QA." During ongoing QA, analytical precision is determining on an intermittent basis through the analysis of replicates.

For isotope dilution analyses, QA/QC could be incorporated into samples analyses with the exception of precision evaluation for recovery corrected concentrations. For this precision information, the day 1 sample could be analyzed in quadruplicate, and a replicate could be analyzed with each subsequent third sample (Table V). For a four-day episode, this would generate an additional four data sets for QA/QC. This corresponds to a reduction of sample analyses by greater than 70% compared with the existing procedure.

In brief summary, isotope dilution quantification can provide three improvements in the determination of priority pollutants in industrial wastewaters compared with existing methods. These are:

1. increased analytical accuracy;
2. increased analytical precision; and
3. more extensive QA/QC at a low cost.

## ACKNOWLEDGMENTS

This work was supported in part by the Environmental Protection Agency under Contract Number 68-02-3153. The authors would like to thank William Telliard of EPA for his continued interest in this work and assistance in providing samples, and R. G. Beimer and staff at TRW for analyzing the ten wastewater samples (EPA Contract Number 68-02-2689).

## REFERENCES

1. *Federal Register*, 44(233):69464–69575 (1979).
2. Rose, M. E., and B. N. Colby. *Anal. Chem.* 51:2176 (1979).
3. Colby, B. N., and M. W. McCaman. *Biomed. Mass Spectrom.* 6:225 (1979).

## CHAPTER 14

# STABLE ACTIVABLE TRACERS FOR ENVIRONMENTALLY SIGNIFICANT ORGANIC MOLECULES

**L. M. Ghannam and W. D. Loveland**

Department of Chemistry
Oregon State University
Corvallis, Oregon

**D. J. Baumgartner**

U.S. Environmental Protection Agency
Corvallis Environmental Research Laboratory
Corvallis, Oregon

Current emphasis on the environmental impact of agricultural and industrial organic chemicals has resulted in a need to assess their physical and chemical behavior in the environment. Often, the need for such an assessment may pose problems because of the extreme toxicity of the chemicals in question, a fact which hampers widespread field experiments and model validation studies with these chemicals, and which necessitates elaborate and time-consuming extraction and analysis procedures for these chemicals.

In this chapter, we report the development, use, testing and possible applications of *stable activable tracers* (SAT) to mimic the physical and chemical behavior of toxic, biologically significant, organic molecules in the environment. Such tracers are stable, inexpensive and nontoxic at the concentrations used in field experiments. They are injected into the environment, allowed to disperse and their concentration is measured by post-sampling neutron activation analysis (NAA) [1,2]. SAT can be detected in amounts as small as 1 pg, allowing one to follow long-range, large-scale pollutant dispersal. Several different sources of pollutants can be marked with different SAT and their emissions tracked over many miles after mixing.

To mimic the behavior of various environmentally significant organic molecules, we have synthesized metal chelates of similar structure where central metal atoms can be detected easily by NAA. Our first candidate for such SAT were the dysprosium $\beta$-diketonates, (1) dysprosium(III) tribenzoylmethane monohydrate, $Dy(dbm)_3 \cdot H_2O$, (2) dysprosium(III) triacetylacetonate trihydrate, $Dy(acac)_3 \cdot 3H_2O$. Such molecules were expected to be relatively insoluble in water, soluble in lipids, highly stable due to chelation effects [3] and easily detectable (detection sensitivity 1 pg) by NAA.

To decide which organic molecules, if any, might behave like the abovementioned SAT in the environment, we have relied on the fact that the *n*-octanol/water partition coefficient (P) is highly correlated [4–8] with the biological activity for a number of organic molecules. Neely et al. [9] have shown that the bioconcentration factor (BF) of various organic chemicals in trout is correlated to log P as:

$$\log BF = 0.542 \log P + 0.124$$

It is also noted that a number of investigators have also established a relationship between log P and the solubility S in water, for various chemicals [10].

Our plan, therefore, was to synthesize the SAT, test their stability in the environment, measure their S and P values, and see if these values could be related to their bioaccumulation properties in a manner consistent with known data for environmentally significant organic molecules.

## EXPERIMENTAL

Dysprosium(III) trisdibenzolmethane monohydrate and dysprosium(III) triacetylacetonate trihydrate were synthesized using standard chemical procedures [11,12].

### Stability

To measure the stability of these tracers in the marine environment, stock solutions of $Dy(dbm)_3 \cdot H_2O$ at concentrations of 4 and 2 $\mu$g/L were prepared in a matrix of 27 °/oo seawater at 10 °/oo in acetone. At definite time intervals, 1-mL aliquots were withdrawn from the solution, evaporated and extracted with 2 mL of octanol of which 1 mL was prepared for NAA. The extraction step was necessary to minimize sodium chloride contamination from the seawater which, on activation, would give copious quantities of long-lived (t½ = 15 hr) $^{24}Na$, making it impossible to detect the small quantities of short-lived (t½ = 2.65 hr) $^{165}Dy$.

The stability of $Dy(acac)_3.3H_2O$ was assessed similarly at a concentration level of 10 μg/L. Two samples were taken, on the first and tenth days. They were extracted and analyzed similarly, and no significant change in concentration was observed.

## Partition Coefficients

The *n*-octanol/water partition coefficients for the Dy chelates were determined by measuring their solubility in octanol and water (Table I). Reagent grade *n*-octanol was purified by washing with (1) 0.1 *N* $H_2SO_4$ and 0.1 *N* NaOH, (2) ultrapure water until neutral, and (3) $CaCl_2$, and distilling twice through a Vigreux column. Distilled, deionized and organic-free water was used for the water-solubility measurements. Octanol and water in two separate volumetric flasks were saturated with the Dy chelates and shaken for a period of 24 hr. The solutions were centrifuged, and 1 mL of each solution was pipetted into activation vials and prepared along with Dy standards for NAA.

## Bioconcentration Experiments

The European oyster (*Ostrea edulis*) was chosen as a model system for bioconcentration experiments due to its ease of analysis and low metabolism rate. Previous experiments with oysters have shown that a period of 3-5 days

**Table I. Solubilities and Octanol/Water Partition at 25°C of Dy Chelates Determined by NAA**

| | Solubility | | | |
|---|---|---|---|---|
| | in Octanol | in Water | P | Log P |
| $Dy(dbm)_3 \cdot H_2O$ | 6.176 | 0.006218 | 994 | 2.997 |
| $Dy(acac)_3 \cdot 3H_2O$ | 34.441 | 1.742 | 19.77 | 1.296 |

is long enough to reach equilibrium of bulky organic molecules [13]. Oysters of comparable size (5 cm in diameter) were prepared for bioassay. The shell was cut following standard procedures [14] so as to expose the gills and cloaca.

The oysters were left to bioaccumulate the Dy chelates in a solution of 27 °/oo seawater at 10 °/oo in acetone. The Dy chelate concentrations were 60 and 6 μg/L. After an exposure time of 5 days without any noticeable toxic effect, the oysters were removed and placed separately in fresh seawater. They were left to equilibrate for 12 days and then sacrificed for analysis.

Analysis of the oysters was as follows: the shell was opened, and the whole body was dried and weighed accurately. The chelates were extracted by blending the whole-body tissues with $CaCl_2$ and 50 mL of acetone for about 20 min. The resulting extract was filtered and evaporated. Octanol (2 mL) was used to pick up the Dy chelates of which 1 mL was prepared for NAA. The seawater was extracted as above.

## RESULTS AND DISCUSSION

### Stability of the Tracers

The results of the tracer stability tests are shown in Figure 1. One can see that the tracers remain soluble and undissociated for a period of ten days. This property should enhance the attractiveness of the tracers for long-term tracer experiments.

### Tracer Partition Coefficients

The tracer partition coefficients are shown in Table I and fall within the general range of a number of pesticides. To relate these coefficients to the bioconcentration factors of environmentally significant organic molecules, we were forced to gather together the relatively meager data showing the relationship between the bioconcentration factor and the partition coefficient, P, for this species of oyster. Table II shows data gathered from several sources [15-18]. The bioconcentration factors were found to correlate (Figure 2) with the values of the partition coefficients in the following manner:

$$\log BF = 0.5978 \log P + 0.9906; R = 0.9280, S = 0.2771$$

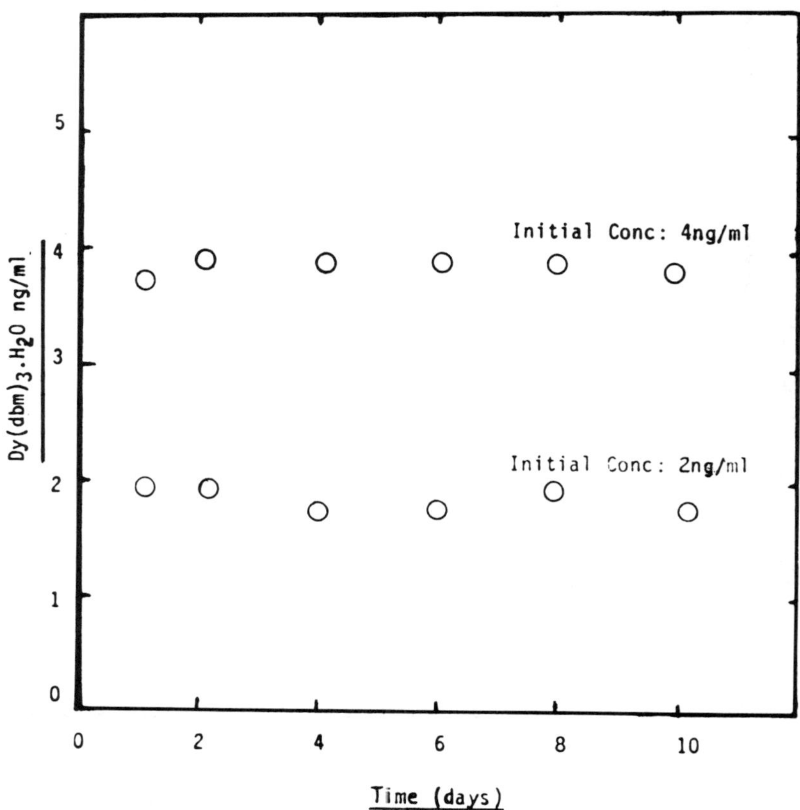


**Figure 1.** Stability of $Dy(dbm)_3 \cdot H_2O$ at two initial concentrations.

**Table II. Bioconcentration Factor (BF) of Selected Pesticides in the European Oyster,** ***Ostrea Edulis***

| Pesticide | BF | Log BF | Log P | Reference |
|---|---|---|---|---|
| Toxaphene | 15,200 | 4,181 | 5.3 | 15 |
| Dieldrin | 2,880 | 3,459 | 4.2 | 15 |
| Endrin | 2,780 | 3,44 | 4.10 | 15 |
| DDT | 70,000 | 4,845 | 6.19 | 16 |
| Hexachlorobiphenyl | 160,000 | 5,204 | 6.72 | 16 |
| Methoxychlor | 5,780 | 3,761 | 4.30 | 15 |
| Heptachlor | 8,500 | 3,929 | 5.8 | 15 |

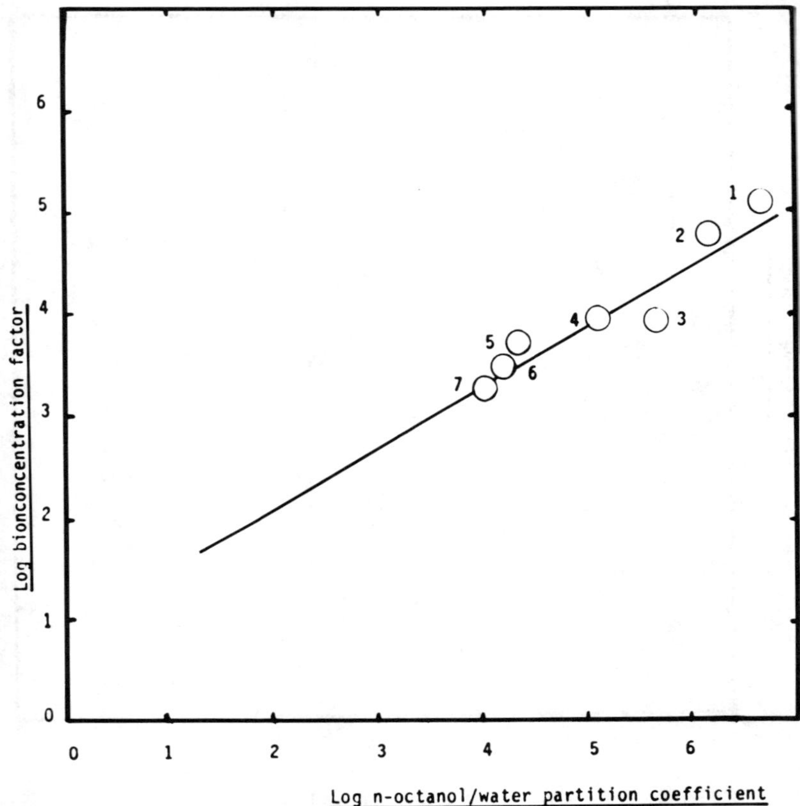


**Figure 2.** Linear regression between logarithms of partition coefficient (P) and bioconcentration factor (BF) of selected pesticides in oysters. Compounds plotted are (1) hexachlorobiphenyl, (2) DDT, (3) toxaphene, (4) heptachlor, (5) methoxychlor, (6) dieldrin and (7) endrin.

Based on this correlation and the values of P for the Dy chelates shown in Table I, one would predict the log BF to be 2.778 and 1.765 for the Dy $(dbm)_3 \cdot H_2O$ and $Dy(acac)_3 \cdot 3H_2O$ species, respectively.

## Bioconcentration Experiments

The experimental values of log BF for the Dy chelates are shown in Figure 3. As one can see, bioconcentration factors of the SAT appear to follow the same systematic dependence on log P as do a number of environ-

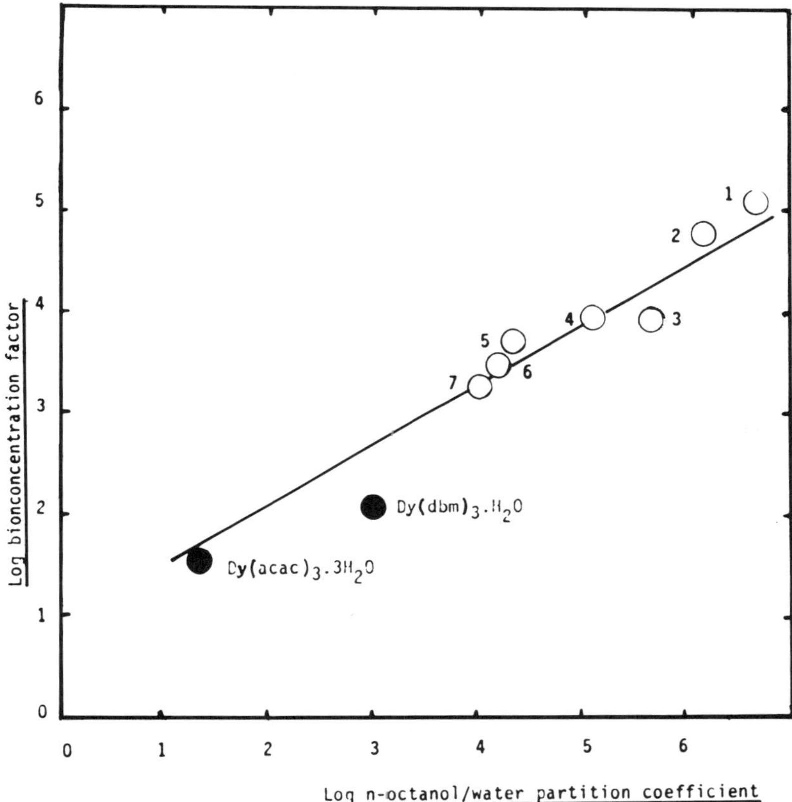


**Figure 3.** Bioconcentration factors of $Dy(dbm)_3 \cdot H_2O$ and $Dy(acac)_3 \cdot 3H_2O$.

mentally significant organic molecules. Thus, it might be expected that these tracers would show similar bioconcentration behavior as organic analogs with similar values of log P.

Further work will involve the simultaneous exposure of marine organisms, such as the oysters studied here, to the SAT and their organic analogs. The ultimate extension of this work, however, is to be able to design an organometallic molecule from first principles to mimic the behavior of any particular organic compound in the environment. An intermediate approach to this goal might involve use of the structure-activity relationships developed by Hansch and Leo [18] and adapted for computer use by Jurs et al. [19] and Chou and Jurs [20]. Research is now in progress to develop sets of fragment constants for metal atoms in chelates to allow this a priori prediction of which stable activable tracer to use to trace a given organic molecule.

## ACKNOWLEDGMENTS

We gratefully acknowledge the financial support of the U.S. Department of the Interior and the Water Resources Research Institute of Oregon State University for this work, and for the use of facilities at the U.S. Environmental Protection Agency, Corvallis Environmental Research Laboratory. Dr. Ron Riley was especially helpful in supplying oysters and analysis methods.

## REFERENCES

1. Shum, Y. S., W. D. Loveland and E. W. Hewson. *APCA J.* 25(11): 1123 (1975).
2. Chick, S. W., and W. D. Loveland. In: *Proceedings of the Third International Conference on Nuclear Methods in Environmental and Energy Research*. J. R. Vogt, Ed., p. 233 (1978).
3. Bjerrum, J., G. Schwarzenbach and L. G. Sillin. *Stability Constants of Metal Ion Complexes, Parts I and II* (London: Chemical Society, 1957).
4. Valtenberg, V., Ed. *Biological Correlations–The Hansch Approach* (Washington, DC: American Chemical Society, 1972).
5. Dunn, W. J., III, and C. Hansch. *Chem.-Biol. Interact.* 9:75 (1974).
6. Hansch, C., and J. M. Clayton. *J. Pharm. Sci.* Vol. 62 (1973).
7. Kenaga, E. E. *Res. Rev.* 44:73 (1972).
8. Lu, P. Y., and R. L. Metcalf. *Environ. Health Perspec.* 10:269 (1975).
9. Neely, W. B., D. R. Branson and G. E. Blau. *Environ. Sci. Technol.* 8: 1113 (1974).
10. Chiou, C. T., V. H. Freed, D. W. Schemedding and R. L. Kohnert. *Environ. Sci. Technol.* (June 1977), p. 475.
11. Charles, R. G., and A. Perroto *J. Inorg. Nucl. Chem.* 26:373 (1964).
12. Pope, G. W., J. F. Steinbach and W. F. Wagner. *J. Inorg. Nucl. Chem.* 20:304 (1961).
13. Riley, R. T., M. A. Shirazi and R. C. Swartz. U.S. EPA, Corvallis. Unpublished data.
14. Galtsoff, P. S. "The American Oyster," *Fish. Bull.* Vol 64 (1964).
15. Reisch, D. J. *J. Water Poll. Control Fed.* 50(6):1424 (1978).
16. Butler, P. A. *J. Appl. Ecol.* 3:253 (1966).
17. Parrish, P. R., U.S. EPA, Gulf Breeze, FL. Unpublished data.
18. Leo, A., C. Hansch and D. Elkins. *Chem. Rev.* 71(6):525 (1971).
19. Jurs, P. C., J. T. Chou and M. Yuan. *J. Med. Chem.* 22(5):476 (1979).
20. Chou, J. T., and P. C. Jurs. *J. Chem. Inf. Computer Sci.* 19(3):172 (1979).

# SECTION 4

# MICROEXTRACTION

## CHAPTER 15

# MICROEXTRACTION AS AN APPROACH TO ANALYSIS FOR PRIORITY POLLUTANTS IN INDUSTRIAL WASTEWATER

**J. W. Rhodes and C. P. Nulton**

Southwest Research Institute
Division of Chemistry and Chemical Engineering
San Antonio, Texas

The first step in the analysis of water and wastewater for organic priority pollutants is to isolate the analytes in a water-immiscible solvent. Since quantitative recovery is stressed [1], multiple extractions with relatively large volumes of solvent are used. However, this exhaustive extraction approach to wastewater analysis is often the source of difficulties (e.g., emulsions, interferences and false positives). This chapter describes a simple, one-step extraction method (henceforth referred to as microextraction), based on techniques previously described [2] for the analysis of distilled flavor volatiles, in which the solvent-to-water ratio ranges from 1:40 to 1:1000. This approach was developed for use in the U.S. Environmental Protection Agency (EPA) verification sampling and analysis program, which is currently being carried out for the purpose of quantifying by gas chromatography (GC) the priority pollutants in a wide variety of process waters and effluents from the organic chemicals and plastics industries. Because the microextraction technique was easy to perform, flexible, and required minimal glassware and sample handling, this approach not only fulfilled the practical requirements of this program but also gave reliable data.

The use of this technique for extracting priority pollutants is justified, and laboratory procedures and resulting data are presented. Also, a method based on simple extraction theory for determining recoveries without spiking is discussed, and preliminary data are presented.

## EXPERIMENTAL

To minimize contamination, the glassware, sodium chloride and sodium sulfate were heated to 400°C for 8 hr prior to use. Solvents used included pentane, hexane, dichloromethane (all Burdick and Jackson, distilled in glass, pesticide grade) and diisopropyl ether (Fisher certified).

GC analyses were performed on Hewlett-Packard 5700 and Tracor 560 models equipped with flame ionization (FID), Hall electrolytic conductivity, electron capture (ECD), alkali flame (AFID) and photoionization (PID) detectors (see Rhoades and Nulton [3] for more details).

Standard compounds were obtained from EPA (Washington, DC), Chem Service, Inc. (West Chester, PA), RFR, Inc. (Hope, RI) and Supelco, Inc. (Bellefonte, PA).

All samples were stored at 4°C in glass containers with Teflon® lined lids. Samples for the analysis of volatile compounds were collected in 40-mL serum vials, which were filled to overflowing and sealed with a Teflon-lined crimp cap.

Although most of the spiked samples were analyzed within several hours after being fortified, some were held up to 96 hours before analysis. Duplicate sample analyses were usually performed within one hour.

### Microextraction Procedures

Wastewater samples (10–100 mL) were extracted in volumetric flasks with 200–1000 $\mu$L of solvent. In most cases the sample was saturated with NaCl prior to extraction. An additional sample was prepared and extracted in exactly the same manner, except a known amount of each compound of interest was spiked into the aqueous phase prior to the addition of solvent.

After agitation of the flask by either mechanical rotation or hand shaking, the phases were allowed to separate so that the organic phase collected in the neck of the flask. A few microliters could be removed by a microsyringe and injected directly onto the column of a gas chromatograph, or a portion of the extract could be removed with a Pasteur pipette and stored in a sealed vial for later analysis. When emulsions formed, a small volume of the emulsion was passed through a tube packed with glass wool; this always provided enough extract for analytical purposes.

Prior to analyzing for phenols, the base-neutrals were removed from samples by adjusting 200 mL of wastewater to pH 10 and extracting three times with 20 mL of dichloromethane and once with 10 mL of hexane. A 90-mL portion of the aqueous layer was then adjusted to pH 2, added to a 100-mL volumetric flask containing 30 g of NaCl, and extracted with 1 mL of diisopropyl ether. Spiking was done prior to the dichloromethane extraction.

In some cases, an internal standard was added to the extraction solvent to compensate for injection errors and solvent evaporation.

When extracts from unspiked and spiked samples are injected in equal amounts, the difference in detector response between these two extracts is due to the known concentration of the added spike. Consequently, analyte concentrations already adjusted for extraction efficiency are calculated directly by:

$$[S]_{uw} = \frac{[S]_{sw}}{R_s - R_u} R_u$$

where $[S]_{uw}$ = adjusted concentration of analyte in the unspiked wastewater
$[S]_{sw}$ = concentration of spiked analyte in the wastewater
$R_u$ = detector response for unspiked sample extract
$R_s$ = detector response for spiked sample extract

Extraction efficiency (%E) can be obtained indirectly after determining the apparent concentration from comparison to an external standard.

Throughout this study, duplicate unspiked analyses were performed on samples to establish method variability.

## RESULTS

Tables I and II present the results obtained on effluents and process waters from the production of organic chemicals and plastics. Shown along with the microextraction data gathered by Southwest Research Institute (SRI) are the data generated by several laboratories (including SRI) which used exhaustive extraction methods. It is stressed, however, that these results do not provide an ideal basis for comparing methods since they were obtained on different wastewater samples and, in many cases, by different laboratories. Nevertheless, the microextraction approach gave rise to less variability in both spike

Table I. Summary of Spike Recovery Data

| Compound | Microextraction Method | | | Other Methods[a] | | | Method Code[c] |
|---|---|---|---|---|---|---|---|
| | Average % Recovery | Standard[b] Deviation | No. of Spiked Samples | Average % Recovery | Standard[b] Deviation | No. of Spiked Samples | |
| Benzene | 101 | 19 | 62 | 96 | 22 | 47 | 8 |
| Toluene | 95 | 17 | 86 | 109 | 27 | 47 | 8 |
| Ethylbenzene | 93 | 17 | 49 | 100 | 18 | 41 | 8 |
| Naphthalene | 90 | 18 | 24 | 36 | 17 | 12 | 17-4 |
| Acenaphthylene | 88 | 25 | 24 | 74 | 27 | 11 | 17-4 |
| Fluorene | 91 | 21 | 24 | 88 | 24 | 9 | 17-4 |
| Fluoranthene | 95 | 14 | 24 | 86 | 48 | 9 | 17-4 |
| Phenanthrene | 95 | 17 | 24 | 66 | 24 | 10 | 17-4 |
| Anthracene | 97 | 14 | 24 | 94 | 41 | 9 | 17-4 |
| Pyrene | 94 | 14 | 24 | 72 | 36 | 9 | 17-4 |
| Benzo[a]anthracene | 87 | 22 | 24 | | | | |
| 1,1-Dichloroethylene | 73 | 18 | 21 | | | | |
| 1,1-Dichloroethane | 65 | 13 | 21 | | | | |
| *trans*-1,2-Dichloroethylene | 65 | 15 | 21 | | | | |
| Chloroform | 79 | 21 | 21 | | | | |
| 1,2-Dichloroethane | 65 | 14 | 18 | | | | |
| Carbon Tetrachloride | 71 | 14 | 14 | | | | |
| Dibromodichloromethane | 62 | 29 | 12 | | | | |
| 1,2-Dichloropropane | 67 | 14 | 19 | | | | |
| Trichloroethylene | 69 | 22 | 21 | | | | |
| 1,1,2-Trichloroethane | 68 | 14 | 21 | | | | |
| 1,1,2,2-Tetrachloroethane | 97 | 25 | 21 | | | | |
| Phenol | 55 | 15 | 73 | 46 | 46 | 120 | 5 |
| *p*-Chloro-*m*-cresol | 82 | 16 | 10 | | | | |

| | | | | | | | |
|---|---|---|---|---|---|---|---|
| 2-Nitrophenol | 53 | 13 | 13 | 123 | 66 | 8 | 5 |
| 4-Nitrophenol | 38 | 17 | 11 | 68 | 72 | 10 | 5 |
| 2,4-Dinitrophenol | 24 | 12 | 12 | 24 | | 2 | 5 |
| Pentachlorophenol | 60 | 21 | 41 | 49 | 48 | 18 | 5 |
| bis(2-Ethylhexyl) Phthalate | 85 | 25 | 50 | 60 | 59 | 80 | 3-3 |
| bis(*n*-Butyl) Phthalate | 61 | 29 | 13 | 66 | 66 | 55 | 3-3 |
| bis(*n*-Octyl) Phthalate | 82 | 21 | 13 | 92 | | 2 | 3-3 |
| Nitrobenzene | 75 | | 3 | | | | |
| *o*-Dichlorobenzene | 86 | | 8 | | | | |
| N-Nitrosodiphenylamine | 96 | | 6 | | | | |
| Hexachloroethane | 71 | 17 | 22 | | | | |
| Hexachlorobutadiene | 74 | 25 | 22 | | | | |
| Hexachlorobenzene | 87 | 18 | 22 | | | | |

[a]These data were supplied by William F. Cowen, Catalytic, Inc., and include data obtained by Southwest Research Institute and several other laboratories taking part in the EPA Effluent Guidelines (Organic Chemicals Branch) verification program.

[b]Absolute standard deviation.

[c]EPA-Effluent Guidelines method code: 8 = purge-and-trap with FID detection (EPA Method 603 with FID); 17-4 = EPA Method 610, Florisil cleanup and FID detection; 5 = Amberlite A-26 resin and FID detection [5]; 3-3 = similar to EPA Method 606 except 15% dichloromethane in hexane (v:v) used instead of dichloromethane for extraction; Florisil cleanup and ECD detection.

## Table II. Summary of Duplicate Analyses

| | Microextraction Method | | | Other Methods[a] | | | |
|---|---|---|---|---|---|---|---|
| | Variability[b] Average | Variability[b] Range | No. of Duplicates | Variability[b] Average | Variability[b] Range | No. of Duplicates | Method Code[c] |
| Benzene | 9 | 0–27 | 33 | | | | |
| Toluene | 12 | 0–67 | 42 | | | | |
| Ethylbenzene | 17 | 3–48 | 14 | | | | |
| Naphthalene | 12 | 0–25 | 4 | 26 | 10–47 | 4 | 17-4 |
| Acenaphthylene | 10 | 2–18 | 3 | 29 | 6–73 | 4 | 17-4 |
| Fluorene | 12 | 0.3–23 | 3 | 38 | 4–112 | 4 | 17-4 |
| Fluoranthene | | 14–23 | 2 | 24 | 22–26 | 2 | |
| Phenanthrene | 12 | 4–21 | 3 | 42 | 11–110 | 4 | 17-4 |
| Anthracene | | 12–46 | 2 | 16 | 7–24 | 3 | 17-4 |
| Pyrene | 17 | 0–33 | 4 | 27 | 6–51 | 3 | 17-4 |
| Phenol | 16 | 0–66 | 31 | 84 | 17–200 | 21 | 5 |
| *p*-Chloro-*m*-cresol | 11 | | 1 | 43 | 5–123 | 8 | 15 |
| Pentachlorophenol | 8 | 0–22 | 4 | | | | |
| bis(2-Ethylhexyl) Phthalate | 40 | 0–143 | 27 | 110 | 5–200 | 31 | 3-3 |
| bis(*n*-Butyl) Phthalate | 32 | 0–67 | 11 | 86 | 4–200 | 14 | |
| bis(*n*-Octyl) Phthalate | 22 | 0–35 | 8 | | | | |
| Nitrobenzene | 6 | 0–14 | 4 | | | | |
| *o*-Dichlorobenzene | 18 | 0–67 | 6 | | | | |
| Hexachloroethane | 9 | 0–34 | 8 | | | | |
| Hexachlorobutadiene | 16 | 0–40 | 9 | | | | |
| Hexachlorobenzene | 29 | 0–84 | 8 | | | | |
| *trans*-1,2-Dichloroethylene | 19 | 0–76 | 10 | | | | |
| Chloroform | 12 | 0–66 | 20 | | | | |
| 1,2-Dichloroethane | 7 | 0–35 | 10 | | | | |
| Trichloroethylene | 18 | 0–57 | 11 | | | | |
| 1,1,2,2-Tetrachloroethylene | 14 | 0–61 | 12 | | | | |

[a]Data supplied by William F. Cowen, Catalytic, Inc., and include data obtained by Southwest Research Institute and several other laboratories taking part in an EPA Effluent Guidelines (Organic Chemicals Branch) verification program.

[b]All values in absolute difference between duplicates expressed as percent of mean value ($|X_1 - X_2| / \overline{X} \times 100$);

[c]EPA Effluent Guidelines Method Code: 17-4 = EPA Method 610, Florisil cleanup and FID detection; 5 = Amberlite A-26 resin and FID detection [5]; 15 = acid extraction (3 x 60 with dichloromethane) cleanup with Amberlite A-26 resin and FID detection; 3-3 = similar to EPA Method 606 except 15% dichloromethane in hexane (v:v) used instead of dichloromethane for extraction; Florisil cleanup and ECD detection.

recoveries and duplicate analyses than was seen for the exhaustive extraction methods.

Moreover, recoveries with the microtechnique are as good as (and often better than) the recoveries obtained by the exhaustive extraction procedures. This is due to elimination of the sample handling steps (solvent concentration, column cleanup, etc.) that usually accompany exhaustive extractions, and the use of NaCl and more polar extraction solvents (e.g., diisopropyl ether) in the microextraction procedure.

High recoveries (70% or greater) are not essential for reliable data. If an analytical approach leads to only a 30-60% recovery, but does so consistently, reliable data can still be obtained. Conversely, if an extraction method produces quantitative recoveries but leads to false positives and/or requires extensive cleanup procedures, the quality of the resulting data is often poor.

Insight into the encouraging results obtained by the microextraction technique is gained from considering the simplest form of Nernst's distribution law which states "a solute dissolved in one phase in equilibrium with another immiscible phase will distribute itself between the two phases so a ratio of the concentration in the two phases is constant at a given temperature" [4]. That is,

$$\frac{[S]_{ol}}{[S]_{wl}} = K_d \text{ or } [S]_{ol} = K_d [S]_{wl}$$

where $K_d$ is the distribution constant and $[S]_{ol}$ and $[S]_{wl}$ are the concentrations of the solute in the organic phase and water phase, respectively.

The amount of solute in each phase is the product of the concentration in that phase and the volume of that phase. The total amount of solute in the system can be expressed as the sum of the amounts of solute in each phase:

$$C_s = [S]_{ol} V_{ol} + [S]_{wl} V_w \quad (1)$$

where $C_s$ = total amount of solute
$V_{ol}$ = volume of organic solvent
$V_w$ = volume of water

Substituting $K_d$ $[S]_{wl}$ for $[S]_{ol}$ in Equation 1 gives:

$$C_s = K_d [S]_{wl} V_{ol} + [S]_{wl} V_w \quad (2)$$

The percent of the solute extracted into the solvent phase (%E) can be determined by dividing the amount in the solvent by the total amount of solute and multiplying by 100.

$$\%E = \frac{K_d[S]_{w1} V_{o1}}{K_d d[S]_{wl} V_{ol} + [S]_{wl} V_w} \times 100 \quad (3)$$

which simplifies to

$$\%E = \frac{100\, K_d}{K_d + \dfrac{V_w}{V_{o1}}} \quad (4)$$

If the distribution constant is known, the percent of the solute extracted can be calculated for a given water/solvent ratio.

Figure 1 shows a series of curves relating %E to the ratio of solvent to water for a number of different $K_d$ values. The relative concentration of the solute in the organic extract is also shown as a function of these same variables. It can be seen that for easily extracted compounds (i.e., $K_d$ greater than 200, which is the case for most priority pollutants under the extraction conditions described previously) increasing the volume of solvent results in small gains in the amount of analyte extracted and large decreases in its concentration. In contrast, the total amount of a poorly extracted compound increases in direct proportion to the volume of solvent used. Consequently, exhaustive extraction for easily extracted compounds will provide little improvement in recovery of the analyte while increasing the recovery of potential interferences. This accounts for the observation that, generally, microextracts have contained fewer interferences than extracts obtained by exhaustive extraction.

Microextraction has a number of practical advantages over the exhaustive techniques:

1. Extract preparation is easily and readily accomplished by the analyst; there is no need for a "sample processing crew." Errors and accidents (spills, etc.) do not pose serious problems because a new extract can be prepared immediately.
2. Minimal use of glassware and solvent reduces cleaning problems and minimizes sample contamination.
3. Emulsions are not a problem since only an analytical portion of extract (1–5 $\mu$L) is required.

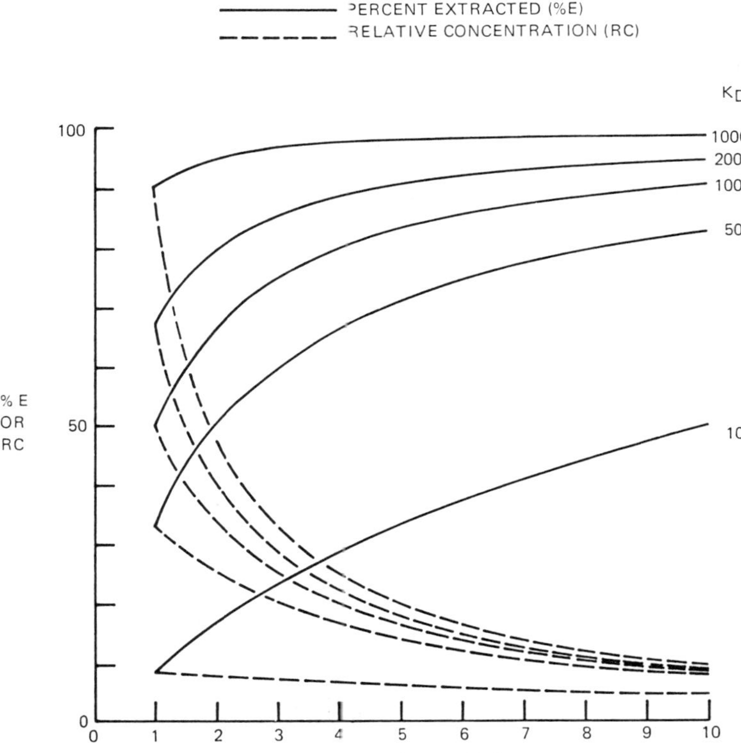


**Figure 1.** Effect of the solvent/water ratio on the percent extraction and relative concentration of analytes for different distribution coefficients ($K_d$) (reprinted from Rhoades and Nulton [3] courtesy of Marcel Dekker, Inc.)

4. Solvent concentration is not necessary; thus, volatile materials can be analyzed in the same extract as the semivolatiles.
5. Column cleanup of extracts is not usually required; all the microextraction data presented in Tables I and II were obtained without using a column cleanup of the solvent extracts.

There is one other feature of the microextraction technique which may have some practical significance. From Equation 2, we have defined the total amount of solute in the system as the sum of the amount in the solvent and the amount in the water. If additional solvent is added to the system, the total amount of solute, the volume of water, and the distribution constant will remain the same after equilibrium. If $V_{o2}$ is equal to the new total volume of solvent and $[S]_{w2}$ is equal to the concentration of solute in the water at $V_{o2}$ solvent volume, then the total amount of solute, $C_s$, can be expressed as:

$$C_s = K_d [S]_{w2} V_{o2} + [S]_{w2} V_w \tag{5}$$

Subtracting Equation 2 from Equation 5 gives

$$K_d [S]_{w2} V_{o2} + [S]_{w2} V_w - K_d [S]_{w1} V_{o1} - [S]_{w1} V_w = 0$$

therefore

$$K_d = \frac{V_w\left([S]_{w1} - [S]_{w2}\right)}{[S]_{w2} V_{o2} - [S]_{w1} V_{o1}} \tag{6}$$

If the above expression for $K_d$ is substituted in Equation 4, the percent extracted into $V_{o1}$ becomes

$$\%E_{V_{o1}} = \frac{100\, V_{o1}}{V_{o2} - V_{o1}}\left(\frac{[S]_{w1}}{[S]_{w2}} - 1\right) \tag{7}$$

Since

$$\frac{[S]_{w1}}{[S]_{w2}} = \frac{[S]_{V_{o1}}}{[S]_{V_{o2}}}$$

then

$$\%E_{V_{o1}} = \frac{100\, V_{o1}}{V_{o2} - V_{o1}}\left[\frac{[S]_{V_{o1}}}{[S]_{V_{o2}}} - 1\right] \tag{8}$$

Thus, the extraction efficiency of a solute from a water sample can be determined solely from the ratio of the concentration of the solute in the two known volumes of extracting solvent. Area or peak height measurements may be substituted for solute concentrations if the detector response is linear. Consequently, for extracting volumes of 1 and 5 mL, Equation 8 can be simplified to:

$$\%E_1 = 25\left[\frac{P_1}{P_5} - 1\right] \tag{9}$$

or

$$\%E_5 = 125 \left[ \frac{P_5}{P_1} - 1 \right] \tag{10}$$

where $\%E_1$ = percent extracted into 1 mL solvent
$\%E_5$ = percent extracted into 5 mL solvent
$P_1$ = peak height or area from 1 mL extract
$P_2$ = peak height or area from 5 mL extract

Preliminary studies were done to test the validity of the two solvent volume (TV) approach (Equation 9). Two samples of deionized water (one saturated with NaCl) spiked with the methyl, ethyl and butyl phthalate esters (500 μg/L) and a third water sample spiked with a series of polynuclear aromatic hydrocarbons (PAH) and 1,2,3,4-tetrachlorodibenzo-*p*-dioxin (100 μ/gL) were microextracted with 1 mL of hexane. After analyzing a portion of the resulting extracts by GC, an additional 4 mL of hexane was added to each sample. The 5-mL extraction and analyses were carried out in a manner identical to the 1-mL extraction. Table III shows that the recoveries calculated by both the external standard (ESTD) and the TV method generally were comparable with each other.

**Table III. Percent Extracted into 1 mL Hexane from 100 mL Spiked Deionized Water**

| Compound | Percent Extracted | |
|---|---|---|
| | TV Method | ESTD |
| Experiment 1 (No NaCl) | | |
| Dimethyl Phthalate | 8 | 6 |
| Diethyl Phthalate | 52 | 49 |
| Dibutyl Phthalate | 90 | 99 |
| Experiment 2 (Saturated with NaCl) | | |
| Dimethyl Phthalate | 66 | 63 |
| Diethyl Phthalate | 96 | 99 |
| Dibutyl Phthalate | 65 | 73 |
| Experiment 3 (Saturated with NaCl) | | |
| Phenanthrene | 82 | 93 |
| Anthracene | 101 | 92 |
| Fluoranthene | 95 | 94 |
| Pyrene | 96 | 93 |
| TCDD[a] | 89 | 92 |
| Chrysene | 97 | 94 |

[a]1,2,3,4-Tetrachlorodibenzo-*p*-dioxin.

Determining extraction efficiency by spiking experiments can be time-consuming and complicated, especially if recoveries for a number of compounds at different concentrations are sought (e.g., a priority pollutant screen). On the other hand, the TV method gives an extraction efficiency for each compound at the concentration at which it occurs. Since the preliminary data indicate that such an approach is feasible, further investigations with wastewaters should be initiated to determine problem areas and to make suitable modifications. Work in this area could lead to an efficient and cost-effective method for analysis of pollutants in water.

## ACKNOWLEDGMENTS

A portion of this work was done under U.S. Environmental Protection Agency Contract 68-01-5011. The authors wish to thank Edward McGovern and Mike Taylor of Southwest Research Institute for their suggestions; William F. Cowen of Catalytic, Inc., for his advice and his help in organizing the data; and Robert L. Grob, Villanova University, and Gerald Umbreit, Greenwood Laboratories, for their advice and support.

## REFERENCES

1. "Guidelines Establishing Test Procedures for the Analysis of Pollutants," *Federal Register* (December 3, 1979).
2. Rhoades, J. W., and J. D. Millar. *Agric. Food Chem.* 13:5 (1965).
3. Rhoades, J. W., and C. P. Nulton. *J. Environ. Sci. Health* 5:467 (1980).
4. Nernst, Z. *J. Physik* 8:110 (1891).
5. Henderson, J. E., G. R. Peyton and W. H. Glaze. In: *Identification and Analysis of Organic Pollutants in Water*, L. H. Keith, Ed. (Ann Arbor, MI: Ann Arbor Science Publishers, Inc., 1976).

# CHAPTER 16

# EVALUATION OF THE MICROEXTRACTION TECHNIQUE TO ANALYZE ORGANICS IN WATER

**K. E. Thrun and J. E. Oberholtzer**

Arthur D. Little, Inc.
Cambridge, Massachusetts

Solvent extraction is a widely employed cleanup/concentration step in the procedures for analysis of a variety of organic substances in water and wastewater. Extraction procedures may be varied in a number of ways, including the nature of the extracting solvent used, whether extraction is performed continuously or batch-wise, and ratio of the volume of organic solvent used to volume of aqueous sample taken. The sample-to-solvent ratio is one major procedural variation. Exhaustive extraction procedures can involve extracting 1 L of aqueous sample as many as three times with 200-250 mL of an organic solvent each time. Typically, the large volume of extract must be concentrated prior to analysis to achieve the desired sensitivity. On the other hand, microextraction procedures involve a single equilibration of 10-100 mL of aqueous sample using sample-to-solvent ratios on the order of 100:1 or even greater. Because a large sample-to-solvent ratio is used, concentration of the analytes in the organic phase accompanies the extraction and a separate concentration step can be avoided.

Murray [1] recently compared a microextraction procedure with two macro methods for the analysis of chlorinated pesticides, alkanes and phthalates in samples of tapwater. With a single extraction, at a sample:solvent (hexane) ratio of 2000:1, he reported an average recovery of 58% and relative standard deviations ranging from 2.9 to 10.2%. The macro methods involved a tenfold reduction in extract volume by using either a micro-Snyder column or a rotary evaporator. Substantial losses (40-90%) of compounds with boiling points less than 250°C occurred during the concentration step. Murray concluded that overall, the micro extract gave as good a recovery as did the macro methods and it was a much more rapid procedure.

The microextraction technique evaluated here is based on procedures developed by Rhoades and Nulton at Southwest Research Institute.

Various effects on extraction efficiencies when using a microextraction technique to extract benzene, toluene, ethyl benzene and *o*-xylene from water into pentane were studied. The effects of sample-to-solvent ratio (20:1 and 100:1), salting out with sodium sulfate and the presence of other organic substances in the matrix were all evaluated. The applicability of the microextraction technique to the analysis of benzene, toluene and ethyl benzene (BTE) in a sample of latex waste which contained high levels of suspended solids, was studied.

Further, the applicability of the microextraction technique for analyzing phenolic and phthalate priority pollutants in water was briefly examined.

## APPARATUS AND PROCEDURES

Extractions of BTE were performed on 90-mL aliquots of clean water standards and the latex waste samples. The extractions were performed in 100 mL-volumetric flasks. Throughout this work, clean water blanks and standards were prepared using deionized, distilled water which had been contacted with activated carbon to remove trace organics. The latex waste samples were analyzed as received and after making the standard additions of the analytes of interest. The extraction of *o*-xylene from clean water was also studied. All samples were extracted with pentane (5mL for 20:1 and 1 mL for 100:1) and shaken by hand or with a Burrell wrist-action shaker in the inverted position. In experiments where salt was added to clean water samples, sufficient $Na_2SO_4$ was added to saturate the 90-mL sample (approximately 14 g). All latex waste samples were saturated with approximately 30 g NaCl.

Extractions of phenolic compounds were performed on 90-mL aliquots of a clean water standard containing the 16 phenolic compounds of interest. The aqueous samples were adjusted to a pH less than 2 with 1:1 $H_2SO_4$:$H_2O$

and were then saturated with 30 g of NaCl. The samples were extracted with 1 mL distilled diisopropyl ether. Samples were shaken in the inverted position using a Burrell wrist-action shaker.

Extractions of phthalate compounds were performed on 90-mL aliquots of a standard solution containing phthalate compounds. The samples were saturated with 30 g NaCl and extracted with 1 mL hexane. Samples were shaken in the inverted position using a Burrell wrist-action shaker.

Extracted samples were analyzed for aromatic hydrocarbons using a Hewlett-Packard 5840A gas chromatograph equipped with a flame ionization detector (FID) and an 8-ft by 1/8-in. stainless steel column packed with 10% 1,2,3-tris-(2-cyano-ethoxy)-propane on 100/120 mesh Chromasorb PAW. (Supelco, Inc.) The column temperature was 95°C (isothermal). The helium carrier gas flowrate was 30 mL/min, the injector temperature was 190°C, and the FID temperature was 275°C.

Analyses of the phenolic compounds were performed using a Varian 2700 gas chromatograph equipped with a FID and a 6-ft x 2-mm glass column packed with 1% SP 1240 DA, 100/120 mesh Supelcoport. The column temperature was programmed from 85 (isothermal, 4 min) to 190°C (isothermal, 20 min) at a rate of 10°C/min. The helium carrier gas flowrate was 30 mL/min, the injector temperature was 200°C, and the FID temperature was 275°C.

Phthalate extracts were analyzed using a Hewlett-Packard 5840A gas chromatograph equipped with a $^{63}$Ni electron capture detector and a 6-ft x 4-mm i.d. glass column packed with 1.5% SP 2250/1.95% SP 2401, 100/120 mesh Supelcoport. The column temperature was programmed from 190 (isothermal, 3 min) to 235°C (isothermal, 20 min) at a rate of 15°C/min. The 95% argon/15% methane carrier gas flowrate was 75 mL/min. The electron capture detector (ECO) was maintained at 300°C, and the injector temperature was 250°C.

## RESULTS AND DISCUSSION

### Extraction of Benzene and Related Compounds from Water

The factors affecting the extracting efficiency of benzene, toluene, ethyl benzene and *o*-xylene (BTEX) from clean water studied were the sample-to-solvent ratio, and addition of salt and organics other than the analytes. All extractions discussed in this section were carried out on aliquots of a clean water standard containing BTEX at the following levels: benzene, 440 ppb; toluene, 430 ppb; ethyl benzene, 430 ppb; and *o*-xylene, 440 ppb. All extractions were carried out in duplicate, and the recoveries reported are the averages for the duplicate pairs. The range between individual measurements for a duplicate pair was, in all cases, less than 8% of the mean value.

The effect of sample:solvent ratio on the recovery of BTEX from water, to which no salt or additional organic substances had been added, can be seen from Figure 1. At a sample:solvent ratio of 20:1, benzene was about 90% extracted, and the other three analytes were extracted, essentially, completely. At a ratio of 100:1 only 66% of the benzene was extracted during a single equilibration.

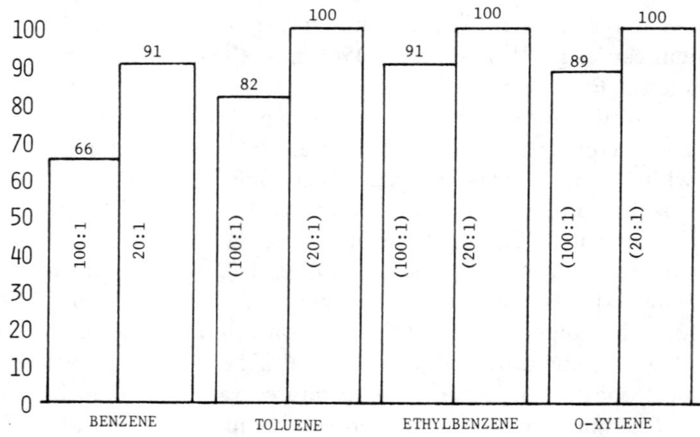


**Figure 1.** Effects of sample:solvent ratio on percent recovery of BTEX from water. No added salt or organic matrix.

**Table I. Single-Stage Extraction Efficiency**

| Substance | Distribution Coefficient | 100:1 | | 20:1 | |
|---|---|---|---|---|---|
| | | Calc. | Exp. | Calc. | Exp. |
| Benzene | 182 | 65 | 66 | 90 | 91 |
| Toluene | 708 | 88 | 82 | 97 | 100 |
| *o*-Xylene | 2818 | 96 | 89 | 99 | 100 |

One important question which is sometimes raised, concerning microextractions with high sample:solvent ratios and relatively short equilibrium times, is whether or not the system has, in fact, reached equilibrium. DeLingy et al. [2] reported equilibrium distribution coefficients for benzene, toluene and xylene for the system water/*n*-heptane. In Table I, their distribution coefficient values and calculated extraction efficiencies one would predict for sample:solvent ratios of 100:1 and 20:1 are tabulated. With the exceptions of toluene and xylene at 100:1, which were 6–7% lower than calculated, the agreement between the efficiencies obtained from microextraction and those predicted from the DeLigny et al. equilibrium studies is excellent.

Figures 2 to 6 show the effects of the presence of salt, three levels of acetonitrile (MeCN) and 1 ppm of carbon tetrachloride on the recoveries of BTEX from a single microextraction equilibration. For a sample:solvent ratio of 20:1, the recovery of benzene, shown in Figure 2, was not significantly affected by the presence of organics. Addition of salt did increase recoveries somewhat. The recoveries of the other analytes were always essentially complete after one equilibration whether or not salt or organics were present at the 20:1 ratio.

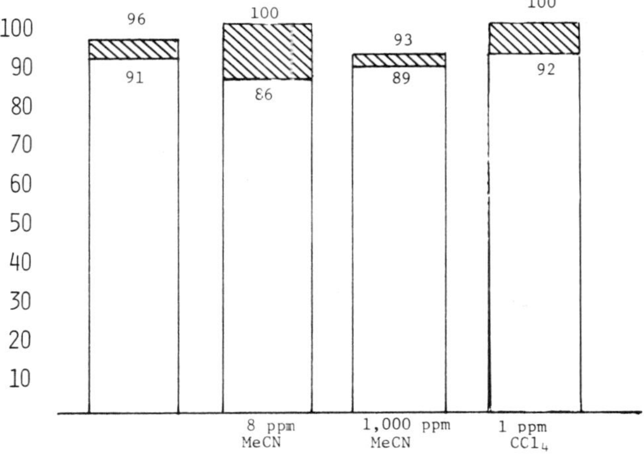


**Figure 2.** Effects of salt and organics on percent recovery of benzene. Sample:solvent ratio = 20:1.

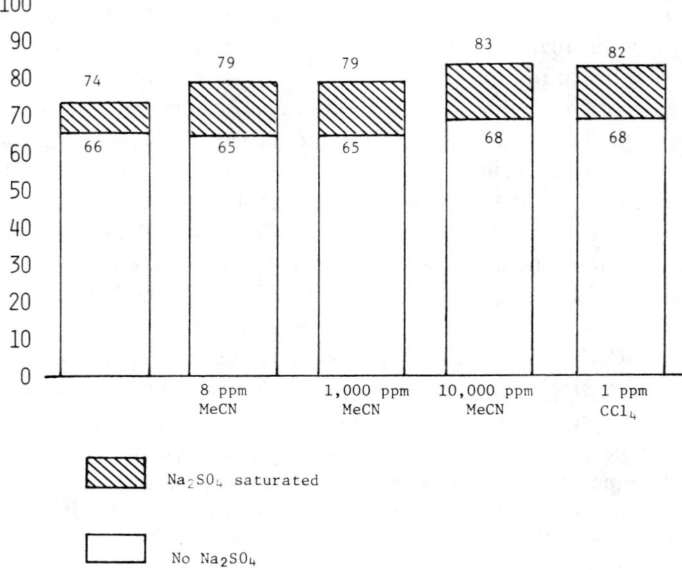


**Figure 3.** Effects of salt and organics on percent recovery of benzene. Sample:solvent ratio = 100:1.

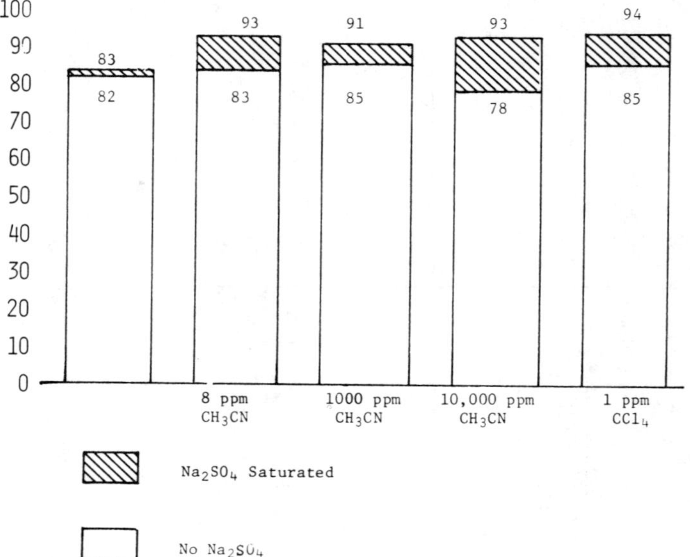


**Figure 4.** Effects of salt and organics on percent recovery of toluene. Sample:solvent ratio = 100:1.

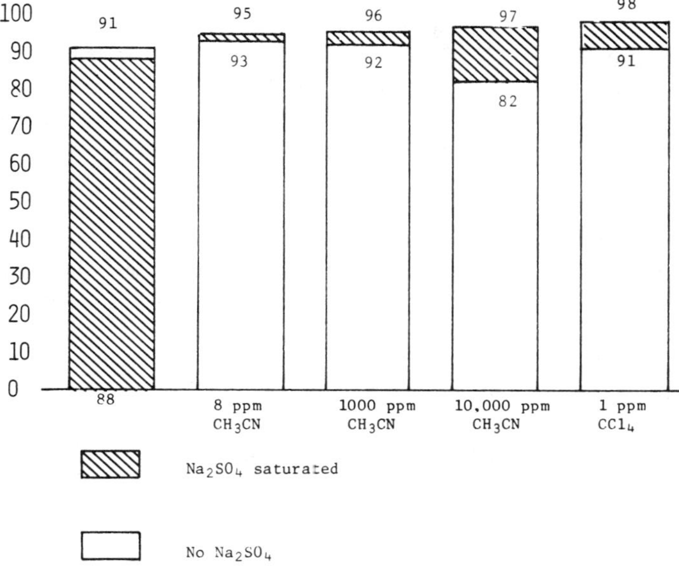


**Figure 5.** Effects of salt and organics on percent recovery of ethyl benzene. Sample: solvent ratio = 100:1.

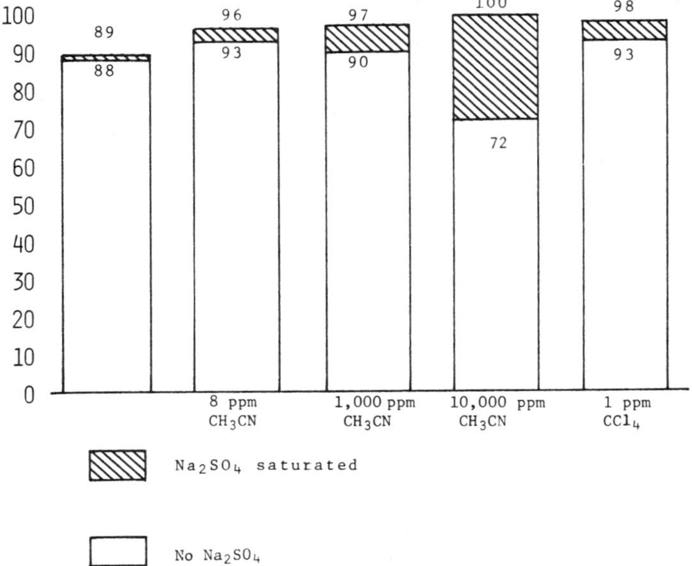


**Figure 6.** Effects of salt and organics cn percent recovery of *o*-xylene. Sample:solvent ratio = 100:1.

Figure 3 shows that at a 100:1 sample-to-solvent ratio, the presence of other organic matrix components did not affect the recovery of benzene significantly. However, addition of salt did raise recoveries substantially. Comparing Figures 4 to 6 for toluene, ethyl benzene and *o*-xylene, which are more easily extracted from water than is benzene, the addition of salt had relatively little effect when no other organics were present. In contrast to the behavior observed for benzene, however, the addition of acetonitrile, particularly at the 10,000-ppm level, seemed to reduce recoveries substantially, most noticeably for *o*-xylene. However, in those cases where substantial reductions in recovery due to the presence of acetonitrile were observed, saturating the solution with salt more than counteracted the decrease due to acetonitrile. Incidentally, small increases in the extractability of acetonitrile were observed when aqueous samples were saturated with salt.

## Extraction of Benzene and Related Compounds from a Latex Wastewater

To evaluate the performance of the microextraction procedure on samples more representative of those which might be encountered in wastewater monitoring, a sample containing a substantial amount of suspended latex solids was obtained from a plant which produces acrylonitrile/butadiene/styrene latices. The sample was found to contain 450 mg/L total suspended solids (TSS) and had a total organic carbon (TOC) content of 730 mg/L. The fact that TOC content significantly exceeded the TSS level suggested that there was a substantial amount of organic materials dissolved in the liquid phase. The latex sample, which supposedly contained significant quantities of benzene, toluene and ethyl benzene, was first analyzed by purge-and-trap gas chromatography/mass spectrometry (GC/MS) using a procedure similar to that described by Bellar and Lichtenburg [3]. The sample was purged for 4 min at about 43°C.

Of the three analytes of interest, only ethyl benzene was found at a substantial level (110 $\mu$g/L). Benzene was found at 1 $\mu$g/L; toluene was not detected. Although the concentrations were not quantified, GC/MS analysis indicated that, as one might suspect, large amounts of acrylonitrile, styrene and butadiene were present in the sample.

To study the effect of shaking time on extraction efficiency, samples of the latex waste were spiked with BTE at 241 $\mu$g/L and subjected to several shaking protocols. The results of this study are presented in Table II. Spike recoveries for BTE from the latex waste after a single 2-min period of shaking were quite low (44, 24 and 9%, respectively). Allowing the samples to stand quietly in contact with the extracting solvent between a series of 2- min shakes showed increased recovery with each period of standing/shaking. Con-

Table II. Percent Recoveries of Benzene, Toluene and Ethyl Benzene Spiked into Clean Water and a Latex Waste[a]

| | Analyte/Sample | | |
|---|---|---|---|
| | Benzene/Latex Waste No. 2 | Toluene/Latex Waste No. 2 | Ethyl Benzene/Latex Waste No. 2 |
| Single 2-min Shake | 44 (2) | 24 (4) | 108 (30) |
| Intermittent 2-min Shakes | | | |
| 15 min | 56 (12) | 40 (22) | 28 (39) |
| 1 hr | 80 (24) | 64 (17) | 52 (25) |
| 2 hr | 92 (15) | 80 (14) | 76 (36) |
| 4 hr | 98 (10) | 91 (11) | 97 (2) |
| 30 hr | 96 (16) | 98 (9) | 106 (10) |
| Continuous Shaking for: | | | |
| 15 min | 91 (22) | 78 (24) | 82 (6) |
| 1 hr | 91 (0) | 91 (6) | 82 (13) |
| 2 hr | 100 (2) | 98 (1) | 90 (23) |
| 4 hr | 100 (4) | 99 (8) | 90 (23) |
| 30 hr | 104 (12) | 106 (21) | 118 (2) |

[a]Each analyte spiked at the 241-μg/L level. Percent recoveries are the average of duplicate determinations. Numbers in parentheses are the ranges between the two measurements expressed as a percent of the mean.

tinuous shaking proved the most effective of all. A single 15-min period of shaking resulted in recoveries similar to those obtained after three 2-min shakes and 2 hr of standing. Extending the period of continuous shaking to one hour and longer resulted in only small increases in the recovery. Spike recoveries, most notably in the case of benzene, were greater than those found in the clean water studies discussed earlier. The reason for the increased recoveries is not known with certainty; however, the substantial amount of total dissolved organics in the latex sample may have played a role. In the clean water studies, the presence of other organics along with salt resulted in higher recoveries as compared to experiments carried out with only BTE and salt.

The unspiked latex waste sample was subsequently extracted with both intermittent and continuous shaking. Toluene was not detected in any of the extracts, and only traces of benzene were found in a few. The amounts of ethyl benzene extracted increased with extended periods of both intermittent and continuous extraction as shown in Figure 7. After 1 hr of continuous shaking or 30 hr of intermittent shaking, levels in the range of 140-170 μg/L were measured in the solution. Apparently, a 4-min purge was insufficient time to remove all of the ethyl benzene from the suspended solids, because only 110 μg/L was found from the GC/MS purge-and-trap measurement.

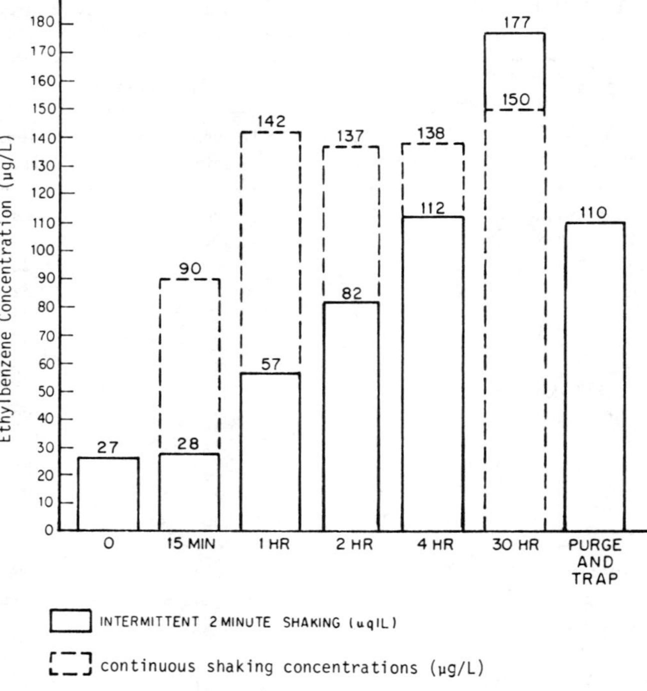


Figure 7. Release of ethyl benzene from latex waste.

## Extraction of Priority Pollutant Phenols

The extraction efficiencies measured for 11 priority pollutant phenols subjected to microextraction are given in Table III. Included in the table are the recoveries observed for five other phenols which were found to be useful GC internal standards for priority pollutant phenol analyses using the GC conditions employed in this study. Extractions were performed at a sample-to-solvent ratio of 100:1 after adjusting the pH of the aqueous sample to 2 or less. Extending the period of shaking beyond two minutes did not increase recoveries substantially.

The single equilibration extraction efficiencies observed in this work are compared in Table IV with efficiencies calculated from *n*-octanol/water partition coefficient data [4] using a sample:solvent ratio of 100:1. The recoveries agree qualitatively. Those measured in this work are generally higher,

Table III. Microextraction Recoveries of Priority Pollutant Phenols from Clean Water Standard Solutions[a]

| Compound | Percent Recovery[b] | | | |
|---|---|---|---|---|
| | 2 min-Shake | 15-min Shake | 60-min Shake | 120-min Shake |
| 2-Chlorophenol | 76 (6.5)[c] | 82 (19.5) | 87 (6.9) | 89 (20.9) |
| 2-Nitrophenol | 68 (9.4) | 69 (22.3) | 76 (12.6) | 75 (35.6) |
| *o*-Bromophenol (IS)[d] | 81 (4.9) | 83 (13.2) | 87 (4.1) | 84 (24.9) |
| Phenol | 54 (1.1) | 54 (7.4) | 55 (9.9) | 57 (18.4) |
| 2-Bromo-4-methyl-phenol (IS) | 84 (2.8) | 88 (2.6) | 88 (8.6) | 95 (17.3) |
| 2,4-Dimethylphenol | 83 (0) | 82 (0) | 83 (8.2) | 89 (12.8) |
| 2,4-Dichlorophenol | 89 (2.5) | 90 (1.1) | 88 (9.3) | 94 (17.9) |
| 2,4,6-Trichlorophenol | 89 (3.2) | 92 (3.3) | 91 (7.4) | 97 (14.8) |
| 2,4-Dibromophenol (IS) | 90 (0) | 92 (1.6) | 91 (7.6) | 98 (10.8) |
| 4-Chloro-*m*-cresol | 91 (0) | 92 (3.9) | 87 (2.3) | 93 (6.5) |
| 2,3,4,5-Tetrachlorophenol (IS) | 87 (3.2) | 91 (2.9) | 90 (12.5) | 94 (15.3) |
| 2,4-Dinitrophenol | 42 (0) | 40 (7.5) | 36 (4.7) | 34 (28.1) |
| 4,6-Dinitro-*o*-cresol | 77 (3.6) | 75 (0) | 77 (11.9) | 79 (15.4) |
| Pentachlorophenol | 68 (7.4) | 73 (5.8) | 74 (20.8) | 78 (18.7) |
| *p*-Benzylphenol (IS) | 87 (0) | 94 (6.9) | 90 (2.0) | 97 (6.4) |
| 4-Nitrophenol | 66 (6.1) | 54 (5.9) | 59 (6.4) | 66 (4.9) |

[a]A 90-mL sample aliquot was taken; all phenols were at a concentration of 50 μg/L, except the dinitrophenols, which were present at 160 μg/L.
[b]Average recovery based on duplicate samples.
[c]Difference between duplicate recoveries expressed as a percent of the mean.
[d]Internal standard.

Table IV. Comparisons of Recoveries Observed for Phenols with Those Calculated from Octanol/Water Partition Coefficients

| Compound | *n*-Octanol | | Diisopropyl Ether |
|---|---|---|---|
| | Log Partition Coefficient[a] | Calculated Recovery (%) | Experimental Recovery (%) |
| 2-Chlorophenol | 2.19 | 61 | 76 |
| 2-Nitrophenol | 1.79 | 38 | 68 |
| *o*-Bromophenol (IS) | 2.35 | 69 | 81 |
| Phenol | 1.46 | 22 | 54 |
| 2,4,6-Trichlorophenol | 3.06 | 92 | 89 |
| 2,4-Dibromophenol (IS) | 2.56 | 78 | 90 |
| 4-Chloro-*m*-cresol | 3.10 | 93 | 91 |
| 2,4-Dinitrophenol | 1.54 | 26 | 42 |

[a]Source: Hansch and Leo [4].
[b]Sample-to-solvent ratio, 100:1.

Table V. Phthalate Microextraction Recovery Data [a]

| Compound | 15-min Shaking | | 1-hr Shaking | |
|---|---|---|---|---|
| | Percent Recovery[b] | Range[c] | Percent Recovery[b] | Range[c] |
| Priority Pollutants | | | | |
| Dimethyl Phthalate | 43 | 14 | 43 | 0 |
| Diethyl Phthalate | 96 | 10 | 88 | 0 |
| Di-*n*-butyl Phthalate | 85 | 2 | 86 | 1 |
| Butyl-benzyl Phthalate | 90 | 23 | 100 | 26 |
| *bis*-(2-Ethylhexyl) Phthalate | 90 | 10 | 89 | 4 |
| Di-*n*-octyl Phthalate | 88 | 8 | 90 | 3 |
| **Internal Standards** | | | | |
| Diisobutyl Phthalate | 93 | 2 | 96 | 3.1 |
| Di(4-Methylpentyl) Phthalate | 78 | 3.8 | 93 | 0 |
| Diamyl Phthalate | 98 | 4.5 | 95 | 4.2 |

[a]Based on extraction of 90-mL aqueous samples (phthalate concentration range 44-833 μg/L.
[b]Based on the average of duplicate samples.
[c]Difference between duplicate recoveries expressed as a percent of the mean recovery.

particularly in the case of those phenols which are the more poorly extracted. The increased recoveries are due to the combination of pH adjustment, addition of salt and use of an ether solvent in this work.

### Extraction of Priority Pollutant Phthalates from Clean Water Standards

Extraction efficiencies determined for phthalate compounds analyzed by microextraction are given in Table V. These recoveries were calculated from data obtained when aqueous samples were shaken with hexane for 15 min and 1 hr. All the phthalates were recovered efficiently after shaking for 15 min. No significant increases were observed after shaking for one hour.

Phthalate partition coefficient data were available in the literature [4] for the system toluene/water. Recoveries calculated from those values, for three of the phthalates studied, are compared with the recoveries using *n*-hexane determined in this work (Table VI). The two sets of data are generally in good agreement.

## CONCLUSIONS

The microextraction technique can be used effectively as a cleanup/concentration step, in analytical procedures for a variety of compound types, in aqueous samples. Recoveries of most of the compounds studied were high if the solutions were saturated with salt. Even in cases where recoveries were lower, they were quite reproducible from run to run. By analyzing a spiked sample in parallel with the unknown, accurate results can be obtained.

**Table VI. A Comparison of Experimental Phthalate Extraction Efficiencies with Calculated Values**

| | Toluene | | *n*-Hexane |
|---|---|---|---|
| Compound | Log Partition Coefficient[a] | Calculated Recovery (%) | Experimental Recovery[b] (%) |
| Dimethyl Phthalate | 2.22 | 65 | 43 |
| Diethyl Phthalate | 3.22 | 95 | 96 |
| Dibutyl Phthalate | 3.70 | 98 | 85 |

[a]Source: Hansch and Leo [4].
[b]Sample-to-solvent ratio, 100:1; 15-min shaking.

Thus, if the actual recovery is determined by analyzing a spiked sample, experimental measurements can be corrected for incomplete recovery.

The microextraction technique is convenient. The sample size required is small, as is the extraction apparatus. As soon as the extraction is completed, an aliquot may be taken for analysis; no further concentration is necessary. The microextraction technique is especially suitable for use with reasonably nonvolatile substances that have relatively high partition coefficients (greater than 150 to obtain a 60% extraction efficiency at 100:1 sample-to-solvent ratio). However, when dealing with "unusual" samples, such as the latex waste studied in this work, microextraction (or any other extraction procedure, for that matter), should be tested to ensure that the contact period is sufficient to achieve equilibrium.

## ACKNOWLEDGMENT

This work was supported by the U.S. Environmental Protection Agency, Effluent Guidelines Division under Contract No. 68-01-5011 with Catalytic, Inc., Philadelphia, PA. The guidance of Dr. William F. Cowen of Catalytic, Inc., and suggestions from Dr. Carter Nulton of Southwest Research Institute are gratefully acknowledged.

## REFERENCES

1. Murray, D. A. J. *J. Chromatog.*, 177:135–140 (1979).
2. DeLingy, C. L., and H. J. H. Kreutzer and G. F. Visserman. *Recuil Travaux Chimiques Pays-BAS*, 85:5 (1966).
3. Bellar, T. A., and J. J. Lichtenberg. *J. Am. Water Works Assoc.*, 66:739 (1974).
4. Hansch, C., and J. J. Leo. *Substituent Constants for Correlation Analysis in Chemistry and Biology* (New York: John Wiley & Sons, 1979), pp. 171–319.

## CHAPTER 17

# FURTHER OPTIMIZATION OF THE PENTANE LIQUID-LIQUID EXTRACTION METHOD FOR THE ANALYSIS OF TRACE ORGANIC COMPOUNDS IN WATER

**W. H. Glaze,* R. Rawley, J. L. Burleson, D. Mapel and D. R. Scott****

Department of Chemistry and
Institute of Applied Sciences
North Texas State University
Denton, Texas

The analysis of trihalomethanes (THM) and other very volatile (purgeable) organic compounds in dilute aqueous solutions may be accomplished by at least five procedures, each of which utilizes gas chromatography (GC) as the analytical method:

1. purge-and-trap/desorption/GC [1,2]
2. liquid-liquid extraction (LLE)/GC [3-6]
3. headspace sampling/GC [7,8]
4. direct aqueous injection/GC [9,10]
5. resin adsorption/elution/GC [11]

---

*Present address: Graduate Program in Environmental Sciences, The University of Texas at Dallas, Richardson TX.

**Present address: U.S. Environmental Protection Agency, Research Triangle Park, NC.

At the First Chemical Congress in Mexico City, our group described a pentane LLE/electron capture detection (ECD) GC method which is more simple to utilize than the purge methods and which involves the use of less expensive, less elaborate equipment [3]. In brief, the method utilizes a 125-mL headspace-free sample taken according to EPA specifications [6]. From the sample bottle one removes with a syringe 5 mL of water and simultaneously injects with a second syringe 3 mL of *n*-pentane. The bottle is shaken on a mechanical shaker for several minutes. An important element of the procedure is the use of an internal standard (1,2-dibromoethane) in the pentane, which serves to improve the precision of the GC analysis. While this procedure does not quantitatively extract the THM from the aqueous phase, good reproducibility is obtained on a variety of water matrices, and the method compares favorably with the purge-and-trap procedure.

Subsequent to the original report, two other papers appeared which describe LLE methods for THM analysis [4,5]. The U.S. Environmental Protection Agency (EPA) has recently suggested the use of either the Bellar/ Lichtenberg procedure [1] or an LLE method adapted from Mieure [4]. Pentane is specified as the solvent of choice, although Mieure used methylcyclohexane. As noted in the EPA-recommended procedure, however, a solvent other than pentane may be required when laboratory temperatures are high and/or barometric pressures are low [6].

In the intervening months since the original report of the method, LLE procedures have been used by our laboratory and many others with good results [12,13]. The method is not without its limitations, however, the most significant of which are [6]:

1. a tendency toward high blank THM levels due to solvent, reagent water or laboratory atmosphere contamination;
2. inability to absolutely confirm components by GC/MS below about 50-$\mu$g/L levels; and
3. recommended applicability only for the THM, i.e., not for other volatile organic compounds in the sample.

This chapter describes a second-generation LLE method which is more convenient and rapid for routine THM analysis. Also described is a third-generation method which is capable of the analysis of other purgeable organics when coupled with capillary GC/ECD or GC/flame ionization detection (FID) analysis. Extensive data on matrix effects have been obtained which illustrate the relative insensitivity of the second-generation method to changes in pH and ionic strength.

## EXPERIMENTAL

### Sampling

The water sample should be obtained without headspace, thereby preventing loss of volatile compounds. The rigorously cleaned 25- or 120-mL bottle is filled just to overflowing with sample, and the septum (Supelco 3-3200) and cap are placed on the bottle with the Teflon® (shiny) side against the sample. Samples should be maintained in an ice chest until the extraction step, at which time they should be warmed to room temperature. Analysis should be conducted as soon after extraction as possible.

### Buffer-Quench Solution

A solution of phosphate buffer and sodium sulfite is added to each bottle just prior to sampling. The buffer assures that the sample will be analyzed at a pH of 6.5, while the sulfite quenches residual chlorine. The buffer-quench solution is prepared by mixing four volumes of 1.0 *M* $NaH_2PO_4$/0.1 *M* $Na_2SO_3$ and six volumes of 1.0 *M* $Na_2HPO_4$; 0.25 or 1 mL is added to each 25- or 120-mL bottle just prior to sampling. Bottles containing standards and blanks should also have this solution added prior to separation.

### Standards

Standards are spiked into sample bottles containing very pure water. Organic impurities may be removed by passage of distilled or deionized water through activated carbon, XAD-2 macroreticular resin or both. Commercial systems for preparation of very pure water are available from Millipore, Culligan and other suppliers. Stock solutions of the organohalides are prepared gravimetrically in nanograde methanol according to the EPA method [6]. With storage at 4°C, these stock solutions are stable for four weeks.

Two secondary dilution mixtures are prepared in methanol from the stock solutions. The first, the quality check standard mixture, should contain 5.0 ng/$\mu$L of each compound, so that a 10-$\mu$L injection into 25 mL of very pure water yields a solution containing 2.0 $\mu$g/L of each compound. The second, called the calibration standard mixture, should contain a quantity of each compound so that a 10-μL injection into 25 mL of very pure water will generate a calibration standard very similar (±10%) to the unknown. The aqueous calibration standards are not stable and should be made up fresh for each sample set.

### Internal Standard/*n*-Pentane

The internal satandard (IS) used in this procedure is 1,2-dibromoethane, although others may be preferred. A 200-μg/L solution of the internal standard in pentane should be prepared and stored in the dark in a volumetric flask with a ground-glass stopper. A sufficient quantity of this mixture for several runs may be prepared in advance, provided that the solution is chromatographed prior to each use as a check against contaminants. The pentane should be checked by GC/ECD prior to making up this solution. If contaminants are present, they may be removed by distillation from sodium or by a basic alumina column cleanup step. The column used should be 15-in. tall x 0.75-in. i.d., and is sufficient to purify several hundred milliliters of solvent. No preparation of the alumina is required. The pentane is poured through the column, with no pressure applied; flows of about 3 mL/min should be achieved with gravity feed.

### Extraction: Second Generation LLE Method (Figure 1)

The extraction procedure involves two steps: water removal and pentane addition. To allow a headspace for addition and mixing of the solvent, 5 mL of water are removed from each sample through the septum with a glass syringe (20 gauge). A second needle allows air to replace the water. The syringe should first be flushed with very pure water so that the needle contains no air. This enables the water to be removed slowly with accuracy, without the complication of bubbles in the syringe during this measurement.

Pentane is added to the sample with the bottle upright, and with the second needle still in the bottle; the needle point should be well above the water surface, so that air is expelled as the pentane/internal standard solution is added (Figure 1, right). After this addition, the syringe and needles are removed, and the bottle is shaken vigorously for 30 sec on a vortex mixer (Sybron/Thermolyne Model M16715). Alternatively, the bottle may be shaken by hand for 1 min. Following this, the bottle is inverted and allowed to stand for at least 1 min, enabling the phases to separate. The bottle should remain with the cap down, except when the pentane is sampled for chromatography.

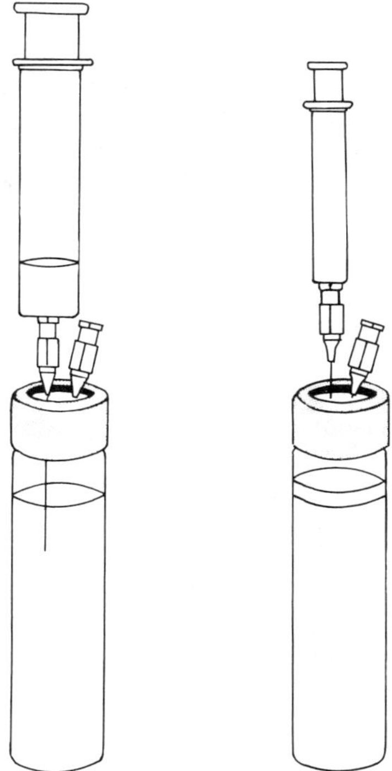

**Figure 1.** Left, removal of water (5 mL). Right, addition of hydrocarbon (1 mL).

## Extraction: Third-Generation LLE Method

The extraction procedure in the third-generation LLE method involves the use of a 100- or 120-mL bottle depending on availability. Bottles are spiked with quench as before and with 5 g of sodium chloride (analytical reagent grade). One milliliter of water is removed from the sample bottle as described in the previous section. Next, 0.5 mL of pentane is added slowly while the sample bottle is sitting in a small ultrasonic bath (Heat Systems Ultrasonics, Inc., Model D-50). The pentane disperses into fine droplets immediately on being injected into the water. The bottle is then mixed for one minute on a vortex shaker and allowed to sit until the two layers separate. The latter process is facilitated by the addition of salt as noted above.

## Packed-Column Gas Chromatography

Several packed columns have been found to resolve THM; some of these include Squalene, SP-2100, SP-1000 and porous polymers. The column used routinely in this laboratory is 10% Squalene on 80/100 mesh Chromasorb W AW. The column is glass, 9 ft x 2 mm i.d. Chromatographic conditions are as follows: oven, 68°C; injection port, 100°C; $^{63}$Ni detector, 350°C; argon-methane (90:10) carrier flow, 20 mL/min; amount of sample injected, 2 $\mu$L.

The recommended detector is a linearized $^{63}$Ni ECD. An electronic integrator is recommended for quantification of compounds, but other methods for peak height or area measurement will suffice.

A second chromatographic column may be used for confirmation purposes and to resolve other halogenated organics. This column is also 9 ft x 2 mm i.d., and contains a mixed phase 6% OV-11 and 4% SP-2100 on 100/120 mesh Supelcoport. The column is programmed from 45°C after 12 min, at 1°/min, to 75°C. With this program, the column will resolve THM and six other organohalides: carbon tetrachloride, 1,1,1-trichloroethane, 1,2-dichloroethane, trichloroethylene, tetrachloroethylene and dichloroiodomethane. The last compound to elute, bromoform, is seen at 39 min under these conditions.

## Capillary Gas Chromatography

The column used in this work is a 60-m x 0.25-mm i.d. glass column with SE-54 wall coating (Supelco). Conditons for GC/FID chromatography on the Hewlett-Packard 5840 are as follows: initial temperature: 0°C; purge delay for splitless injection: 0.5 min; septum purge 4 mL/min; inlet purge: 70 mL/min; temperature program: hold at 0°C 1 min, then to 120°C at 3°/min, stop at 40 min; linear helium velocity: 31 cm/sec at 0°C; helium makeup: 40 mL/min. Conditions for ECD are the same except that 50 mL/min of argon/methane makeup gas is used.

# RESULTS AND DISCUSSION

Figure 2 is a typical ECD chromatogram of THM analyzed by the second-generation THM method. The method is rapid and reproducible in our hands. Recently, 58 duplicate analyses were made for instantaneous THM levels. The average range for the duplicate values was 1.7% for $CHCl_3$ (mean 250 $\mu$g/L), and 2.3% for $CHClBr_2$ (mean 13.1 $\mu$g/L). The precision for bromo-

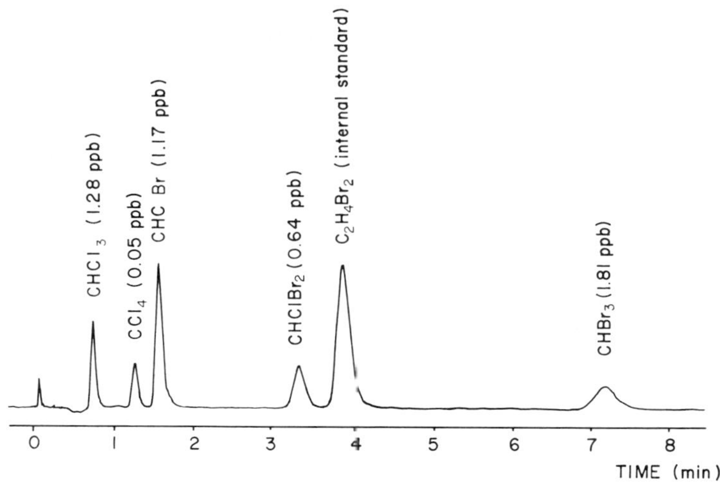


**Figure 2.** $^{63}Ni$ ECD gas chromatogram of mixture of THM, $CCl_4$ and 1,2-$C_2H_4Br_2$ (internal standard). 10% Squalene column and conditions described in text.

form was not as good because of its low level (mean of 0.70 $\mu$g/L; avg range for duplicates: 68%). These results illustrate the precision of which the method is capable for several samples in the same concentration range. For diverse samples run by student labor, we find median RSD (%) values in the 7–14% range in most cases.

Figure 3 shows the chromatogram of a nine-compound standard mixture on the alternative column used for confirmation purposes. Lower limits of detection of the nine compounds are shown in Table I using this column. For the chloroform the lower detection limit is largely determined by blank levels in reagent water and the extraction solvent. Table I distinguishes between the more realistic "Method DL" and the "Instrument DL," the latter of which only relates to the GC analysis step. The two are identical except for chloroform, 1,1,1-trichloroethane, carbon tetrachloride and tetrachloroethylene, which are found in blanks. As noted in the table, the 20:1 LLE method is not particularly useful for situations where low levels of 1,2-dichloroethane must be detected; an obvious liability of the method. The separation capability of the mixed phase column is illustrated in Figure 4 which shows a reconstructed GC/MS scan of combined Supelco purgeable standards.

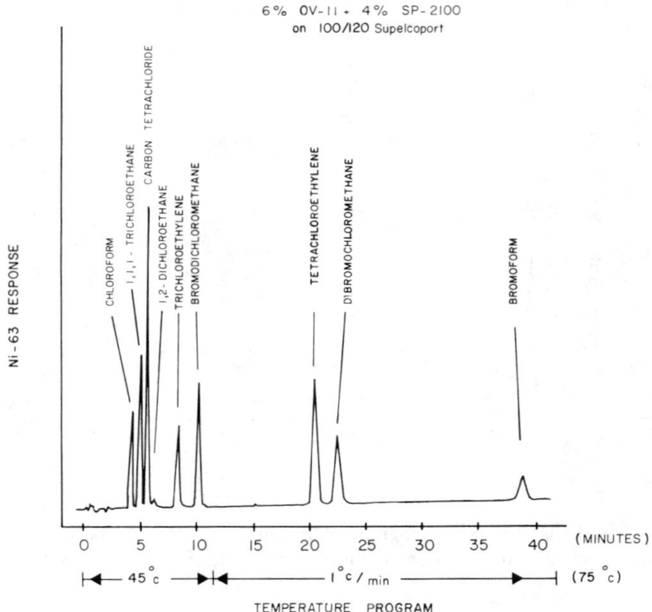


**Figure 3.** $^{63}$Ni ECD gas chromatogram of nine-compound mixture of halogenated organics. 4% SP-2100/6% OV-11 column and conditions described in text.

## pH and Other Matrix Effects

Tables II and III summarize some of many experiments conducted using the 20:1 second-generation LLE method with various pH and ionic strength values. The data show no significant effects of changing these matrix parameters. Other experiments have shown no statistically significant effects of added sodium sulfite quench over the range from 0 to 126 mg/L, or methanol over the range from 0 to 1.0% (v/v). (Methanol is usually added in the process of preparing standard solutions.)

## 120:0.5 LLE Method

The third-generation LLE method was developed for those cases where compounds other than THM are of interest. The water-to-solvent ration of 240 was chosen as the result of several experiments which tested reproduci-

Table I. 20:1 LLE Method Detection Limits[a]

| Compound | Detection Limit (μg/L) | |
|---|---|---|
| | Instrument DL[b] | Method DL[c] |
| Chloroform | 0.07 | 0.46 |
| 1,1,1-Trichloroethane | 0.03 | 0.07 |
| Carbon Tetrachloride | 0.01 | 0.014 |
| 1,2-Dichloroethane | 8.7 | 8.7 |
| Trichloroethylene | 0.09 | 0.09 |
| Bromodichloromethane | 0.04 | 0.04 |
| Tetrachloroethylene | 0.05 | 0.46 |
| Chlorodibromomethane | 0.07 | 0.07 |
| Bromoform | 0.2 | 0.2 |

[a]Second-generation LLE method.

[b]Noise determined with the mixed phase column on the HP 5840A at an attenuation of $2^2$. The baseline to peak high frequency noise was converted into a peak height equivalent S/N of two.

[c]Using mean blank +3 standard deviations.

Table II. 20:1 LLE Method: pH Effects[a]

| Compound | μg/L | pH 4 | pH 6 | pH 8 | pH 10 |
|---|---|---|---|---|---|
| $CHCl_3$ | 200 | 2.53 ± 0.10 | 2.44 ± 0.05 | 2.440 ± 0 | 2.53 ± 0.03 |
| | 40 | 0.353 ± 0.039 | 0.403 ± 0.016 | 0.396 ± 0.017 | 0.388 ± 0.008 |
| $CHCl_2Br$ | 200 | 9.40 ± 0.35 | 9.15 ± 0.16 | 9.160 ± 0 | 9.41 ± 0.09 |
| | 40 | 1.57 ± 0.02 | 1.77 ± 0.06 | 1.73 ± 0.07 | 1.70 ± 0.04 |
| $CHClBr_2$ | 200 | 8.70 ± 0.37 | 8.44 ± 0.13 | 8.47 ± 0.02 | 8.70 ± 0.10 |
| | 40 | 1.35 ± 0.05 | 1.54 ± 0.06 | 1.50 ± 0.06 | 1.48 ± 0.10 |
| | 8 | 0.263 ± 0.014 | 0.262 ± 0.011 | 0.255 ± 0.001 | 0.26 ± 0.014 |
| $CHBr_3$ | 40 | 0.537 ± 0.008 | 0.612 ± 0.023 | 0.593 ± 0.027 | 0.586 ± 0.012 |
| | 8 | 0.106 ± 0.054 | 0.105 ± 0.041 | 0.102 ± 0 | 0.104 ± 0.057 |

[a]Values reported as area counts from electronic integrator x $10^{-6}$; duplicate determinations at 0.01 ionic strength.

**Table III. 20:1 LLE Method: Ionic Strength Effects**[a]

| | Ionic Strength[b] | | | |
|---|---|---|---|---|
| **Compound** | **0.01 (D)** | **0.1 (D)** | **0.5 (T)** | **1.0 (T)** |
| $CHCl_3$ | 15.3 ± 3.0 | 12.2 ± 0.4 | 11.1 ± 0.40 | 11.2 ± 0.4 |
| $CHCl_2Br$ | 21.4 ± 4.0 | 18.5 ± 2.0 | 15.8 ± 0.50 | 16.0 ± 0.8 |
| $CHClBr_2$ | 23.8 ± 4.0 | 21.8 ± 2.0 | 18.0 ± 0.70 | 18.1 ± 0.7 |
| $CHBr_3$ | 4.8 ± 0.5 | 6.0 ± 0.5 | 3.77 ± 0.06 | 3.7 ± 0.3 |

[a]Concentrations in μg/L in Denton, TX, tapwater spiked with various quantities of NaCl to give desired ionic strength.

[b]Duplicate (D) or triplicate (T) determinations.

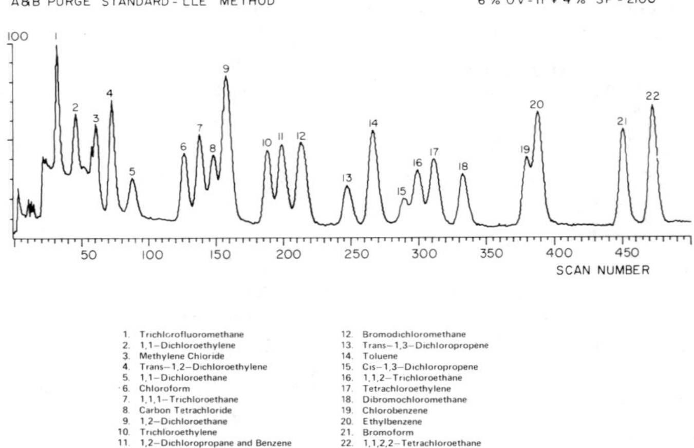


**Figure 4.** Reconstructed GC/MS chromatogram of purgeable standards A and B obtained from Supelco.

bility, ease of manipulation, availability of containers, etc. In recent work, 120-mL bottles with crimped tops are used [3], but other choices are possible. Three features of the method are notable: (1) the use of 5 g of NaCl to improve extraction efficiency [5] and separability of the phases; (2) the use of a sonicator to facilitate mixing of the solvent and water (see Experimental section); and (3) the use of capillary columns to facilitate the analysis step.

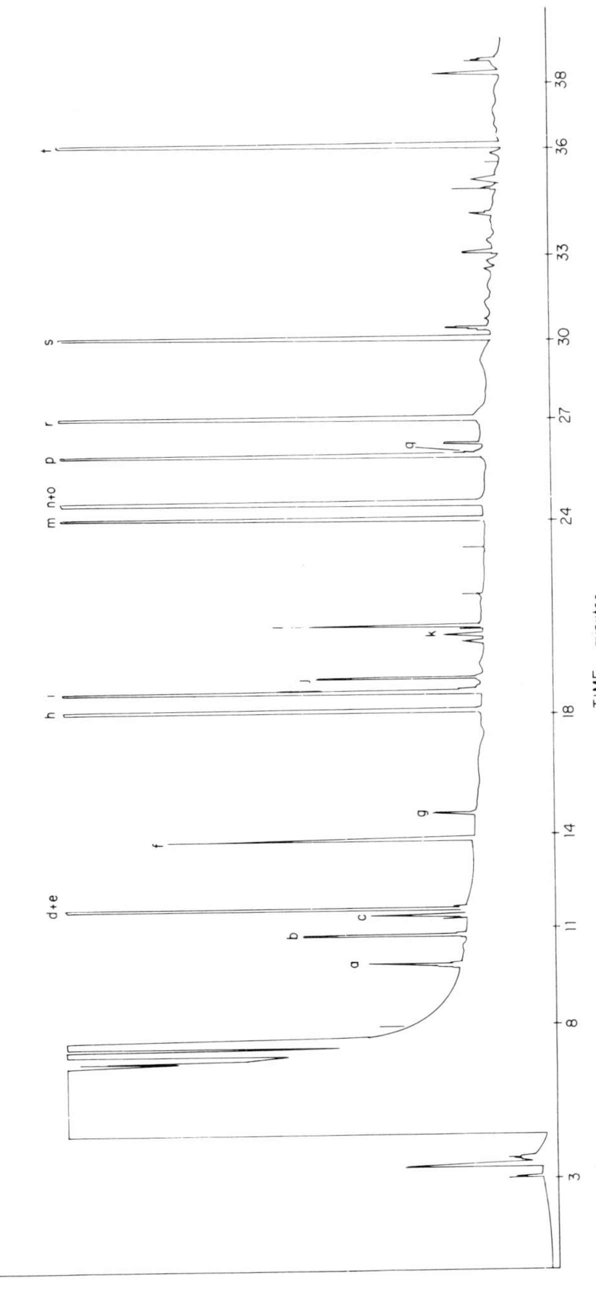


**Figure 5.** FID chromatogram of volatile organic compound standard SE-54 capillary column. See Table IV for compound codes.

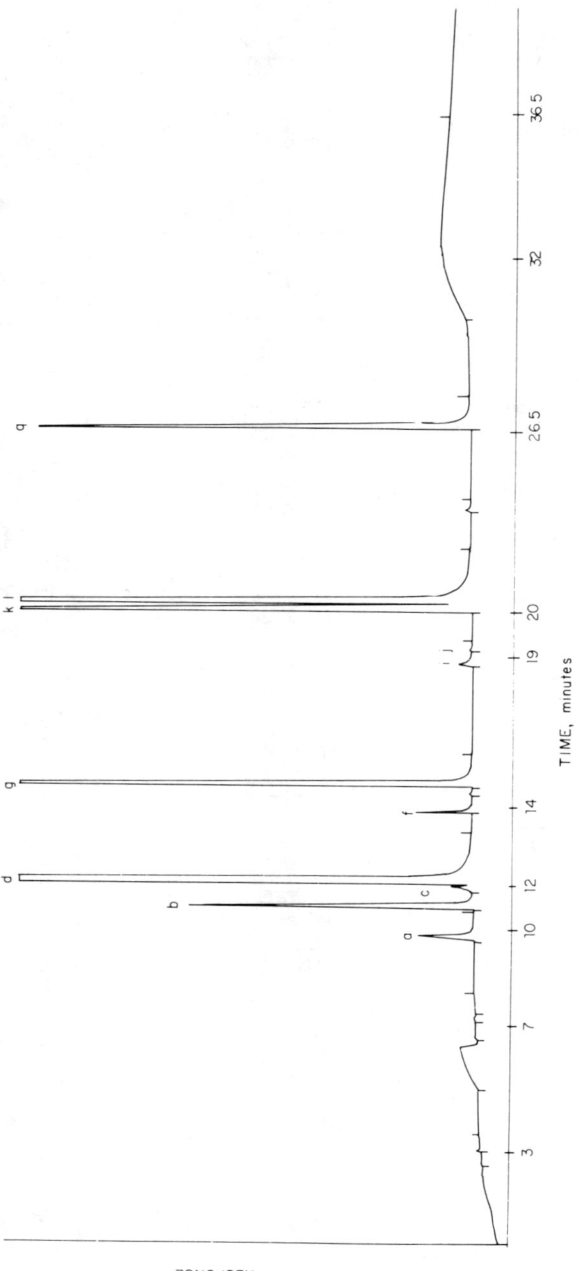


**Figure 6.** $^{63}$Ni chromatogram of volatile organic compound standard code listed in Table IV.

Figures 5 and 6 show the chromatograma of an actual water extract containing standard compounds. Table IV lists the concentrations of the compounds in the water before extraction, and estimates the lower detection limit for ECD and FID. Two retention-time redundancies occur, between the pairs carbon tetrachloride/benzene and bromoform/*o*-xylene. Fortunately, the relative response factors of the members of each pair on FID and ECD are convenient. Under most conditions, one can assume with confidence that FID and ECD will allow quantification of the hydrocarbon or the halide, respectively, without significant interference from the other member of the pair.

**Table IV. 120:0.5 LLE Method: Estimated Detection Limits**

| | | | Detection Limit (μg/L) | |
|---|---|---|---|---|
| **Code** | **Compound**[a] | **μg/L** | **FID** | **EC** |
| a | Chloroform | 20 | 14 | 0.16 |
| b | 1,1,1-Trichloroethane | 21 | 5 | 0.10 |
| c | 1,2-Dichloroethane | 23 | 8 | 2.2 |
| d | Carbon tetrachloride | 23 | b | 0.005 |
| e | Benzene | 24 | 1 | b |
| f | Trichloroethylene | 26 | 6 | 0.03 |
| g | Bromodichloromethane | 27 | 8 | 0.06 |
| h | Toluene | 23 | 0.9 | b |
| i | 1,Chloro-2-bromopropane | c | | |
| j | 1,1,2-Trichloroethane | d | d | d |
| k | Chlorodibromomethane | 25 | 9 | b |
| l | Tetrachloroethylene | 20 | 4 | 0.04 |
| m | Ethylbenzene | 22 | 0.6 | b |
| n | *p*-Xylene | 24 | 0.6 | b |
| o | *m*-Xylene | 28 | 0.6 | b |
| p | *o*-Xylene | 23 | 0.6 | b |
| q | Bromoform | 27 | 14 | 0.09 |
| r | 1,4-Dichlorobutane | c | | |
| s | *n*-Propylbenzene | 30 | 0.8 | b |
| t | *n*-Butylbenzene | 28 | 0.8 | b |

[a]Reference Figures 4 and 5.

[b]Not measured, but large.

[c]Internal standard.

[d]Values not known.

### Current Research

The LLE method in various modifications is currently being used for the study of ozone and ozone/ultraviolet reaction kinetics; for routine measurements of THM formation potential in a pilot water treatment facility; and for trace organic monitoring as the method is applicable. Both capillary and packed column applications continue, although the former is increasingly preferred. Recent studies have compared the second- and third-generation LLE methods for VOA analysis in a variety of matrices. Further extension and optimization of the method will be pursued in the coming year.

## ACKNOWLEDGMENT

This research was sponsored by the U.S. Environmental Protection Agency. Special thanks are given to Dr. Thomas Bellar, EPA/MERL, for assistance to this work as project officer and to Ellen Zuefeldt for laboratory assistance.

## REFERENCES

1. Bellar, T. A., and J. A. Lichtenberg. *J. Am. Water Works Assoc.* 66:739 (1974).
2. Grob, K., and F. Zurcher. *J. Chromatog.* 117:285 (1976).
3. Henderson, J. E., G. R. Peyton and W. H. Glaze. In: *Identification and Analysis of Organic Pollutants in Water*, L. H. Keith, Ed. (Ann Arbor, MI: Ann Arbor Science Publishers, Inc., 1976), p. 105.
4. Mieure, J. P. *J. Am. Water Works Assoc.* 69:60 (1977).
5. Richard, J. J., and G. A. Junk. *J. Am. Water Works Assoc.* 69:62 (1977).
6. U.S. Environmental Protection Agency. "Analysis of Trihalomethanes in Drinking Water by Liquid/Liquid Extraction," *Federal Register* 44(231):68683 (1979).
7. Rook, J. J. *Water Treat. Exam.* 23:234 (1974).
8. Kaiser, K. K., Jr., and B. G. Oliver. *Anal. Chem.* 48:2207 (1976).
9. Nicholson, A. A., and O. Meresz. *Bull. Environ. Contam. Toxicol.* 14:453 (1975).
10. Pfaender, F. K., R. B. Jonas, A. A. Stevens, L. Moore and J. R. Haas. *Environ. Sci. Technol.* 12:438 (1978).
11. Glaze, W. H., G. R. Peyton and R. Rawley. *Environ. Sci. Technol.* 11: 685 (1977).
12. Glaze, W. H., and R. Rawley. *J. Am. Water Works Assoc.* 71:509 (1979).

CHAPTER 18

# EXTRACTION OF ORGANIC COMPOUNDS FROM WATER USING SMALL AMOUNTS OF SOLVENT

**G. A. Junk, I. Ogawa and H. J. Svec**

Ames Laboratory
and Department of Chemistry
Iowa State University
Ames, Iowa

Organic solvents are used extensively to extract trace amounts of organic components from aqueous samples because the extraction procedure is simple, convenient and familiar to most analytical chemists. The factors normally considered in the choice of the solvent are availability, purity, cost, volatility, water solubility, safety and a favorable distribution coefficient for the components to be extracted. After the choice of an organic solvent has been made, chemists have traditionally used relatively large amounts of the solvent in single, sequential or continuous extractions. However, the use of large amounts of solvent has the following disadvantages:

1. a concentration step is required to remove most of the solvent to improve the detection sensitivity;
2. the blank level is greater because the impurities in the solvent are also concentrated in the concentration step;
3. the more volatile components are lost during the concentration step;
4. contamination and losses of sample are more probable because of the added sampling handling and transfers; and
5. the cost of analyses increases because more solvent is used and the concentration step adds to the time per analysis.

These advantages can be circumvented if a small amount of solvent is used to yield a high water-to-solvent-volume ratio. No concentration step, with its associated difficulties, is then necessary. This makes the choice of solvent much less complicated because the influence of availability, purity, cost and volatility are minimized.

The technique of using a high water-to-solvent-volume ratio was first reported by Rhoades and Millar [1]. Grob et al. [2] tested a similar procedure for extracting trace components from 1-L water samples. More recently, Murray [3] described a variation of Grob's approach. Other investigators [4-6] have used comparable techniques to extract a number of organic components from water.

An extraction technique similar to these others [1-6] has been used by us to test the recovery of many different organic components from water using volume ratios of 100 to 500. These recovery results are interpreted relative to theoretical and practical considerations and the the analytical results obtained by employing more traditional extraction techniques.

## EXPERIMENTAL

### Solvents and Chemicals

Pentane, chloroform and acetone were obtained from Burdick and Jackson (Muskegon, MI) and used without further purification. Analytical reagent grade methanol and anhydrous diethyl ether from Mallinckrodt (St. Louis, MO) were distilled prior to use. Acenaphthylene was obtained from Aldrich (Milwaukee, WI). All other organic compounds were obtained from Chem Service (West Chester, PA).

### Extraction Procedure

The extractions were performed in modified 100-mL or 1-L volumetric flasks having precalibrated necks with inside diameters (i.d.) of 2-10 mm. A 100-mL extraction vessel is shown in Figure 1. When this vessel was used, only 85 mL of water were extracted. This allows for about 15 mL of headspace, which is necessary to achieve sufficient contact of the water and the organic solvent phase during the short time of vigorous shaking. However, this headspace is detrimental to subsequent operations in the analysis, so backfilling with water, as described later, is essential to the success of the method.

Pentane was chosen as the extracting solvent for most of the analyses because its solubility of 0.04 g/L in water is low [2] and its distribution coef-

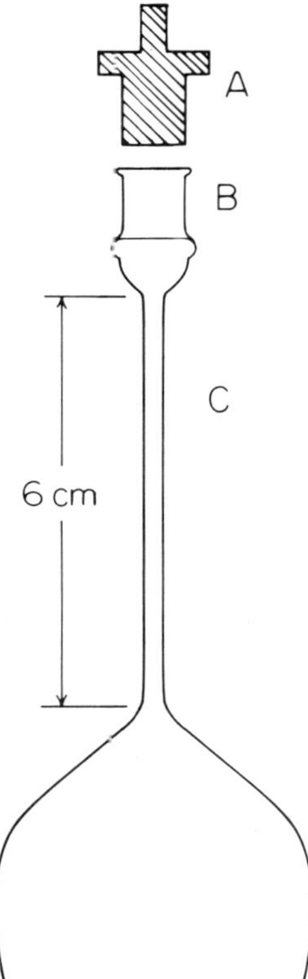


**Figure 1.** 100-mL extraction vessel equipped with a Teflon® stopper (A), 19/38 standard taper (B), and a precalibrated neck of 4 mm i.d. for measuring small volumes of solvent.

ficients for many organic solutes are favorable. Pentane is also available in excellent quality and is suitable for gas chromatographic (GC) analyses. Its high vapor pressure of 420 mm Hg at 20°C [2] is desirable for concentration and separation purposes, but this same volatility also causes solvent losses. The greatest loss occurs when the extraction vessel is opened to backfill with water. Under our experimental conditions the headspace temperature at the start of the backfilling operation was 6-12°C while the water temperature near the pentane-water interface was approximately 6°C. Under these conditions, the pentane loss was 17-54 μL when the vessel was opened for the 1- to 3-min required for backfilling. To minimize this loss, the backfilling operation was completed as rapidly as possible. The amount of water used to backfill was carefully measured to allow for the expansion which occurs when both the water and pentane equilibrate to room temperature. Theoretically, the pentane losses are reduced by reducing both the headspace volume and temperature. Practically, compromises are made to select a headspace volume sufficient to achieve efficient extraction and a temperature high enough to ensure no breakage of the vessel.

The narrow neck of the extraction vessel made possible the accurate measurement of the final volume of pentane. However, very narrow-necked vessels are difficult to fill and empty, so the ultimate dimensions of the vessels are dictated by the volume of water to be extracted, the desired accuracy of the pentane measurements and the desired convenience of the extraction operations.

For most of the recovery tests, concentrated solutions of model organic compounds in methanol or acetone were added to the 100-mL extraction vessel containing 85 mL of organic-free water. The test compounds were dissolved by shaking the vessel vigorously for 30 sec. After adding 0.15 mL of pentane, the vessel was capped tightly with the Teflon® stopper and shaken very vigorously for 2 min. The 100-mL vessel was cooled by plunging it into the liquid $N_2$ up to the stopper for approx. 0.5 min. The Teflon stopper was then removed and the vessel was backfilled rapidly, using 4°C water in a long-needled syringe, until the lower level of the pentane was in the lower part of the neck. The volume of pentane was determined by measuring the height in the precalibrated neck after equilibration to room temperature. Samples for GC analyses were taken directly from the pentane in the neck of the extraction vessel.

For some of the recovery tests at the lowest concentrations, the model compounds were added to 850 mL of water in modified 1-L volumetric flasks. A 10-mm i.d. neck was suitable for pentane volumes greater than 1 mL while 2- to 4-mm i.d. necks were used for smaller volumes of pentane. An ice bath was more practical for cooling the 1-L flasks than the liquid nitrogen used for the 100-mL flasks, but both cooling methods are satisfactory for all vessels.

For natural waters, the flask size was dictated by the volume to be extracted and the desired analytical sensitivity. The same extraction procedures as used for the recovery tests were employed but without the spiking step.

## Analysis

Perkin-Elmer 3920B and Varian 1200 gas chromatographs were used with OV-17 and Carbowax 20M packed columns for separation. Flame ionization (FID) and electron capture detection (ECD) were used to detect the organic solutes.

## RESULTS AND DISCUSSION

### Theoretical Considerations

An important consideration in analytical work concerns the fraction of organic solute which is extracted by the chosen procedure. This fraction is usually symbolized by percent extracted, %E, and is related to the distribution coefficient (D) and the volume ratio (R) by the equation,

$$\%E = \frac{100\ D}{D + R} \tag{1}$$

where D = the ratio of concentration in the organic phase over the concentration in the aqueous phase

R = the ratio of the water volume ($V_W$) over the organic solvent volume ($V_O$)

Part of this relationship is frequently illustrated in graphical form by plotting the logarithm of the distribution coefficient against percent recovery of solute for a specified volume ratio [7]. Eight of these plots for volume ratios frequently employed in analytical work and for distribution coefficients from 1 to $10^5$ are reproduced in Figure 2. These plots are useful for making general analytical decisions. For example, if the D is estimated to be 100 and an R of 5 is chosen, the first extraction yields more than 95% recovery. Little is gained by resorting to successive extractions. If the actual value of D is greater than 100, even less is gained. Even if D is as low as 5, where 50% recovery is obtained using an R of 5, the value of successive extractions for monitoring purposes is questionable provided the 50% recovery is reproducible. For those cases where the distribution coefficient is less than about 5, the use of successive extractions and large volume ratios is warranted.

Many organic components of interest have distribution coefficients between 100 and 100,000. Inspection of the R = 100 plot shows that all these components will be extracted with greater than 50% efficiency. This is quite acceptable for most analytical work. As D approaches 100,000, efficient extraction is obtained with even larger values of R. The practical range for R is 100–1000 for water volumes of 100 mL to 1 L. This allows for the direct use of the extract for separation and analysis of the organic solutes without resorting to a concentration step to enhance sensitivity.

The plots in Figure 2 can also be used as a convenient substitute for calculations using Equation 1. For example, if %E is determined at a particular R for a model compound added to water, the plots can be used to determine D. In those cases where D is known or can be estimated, the plots can be used

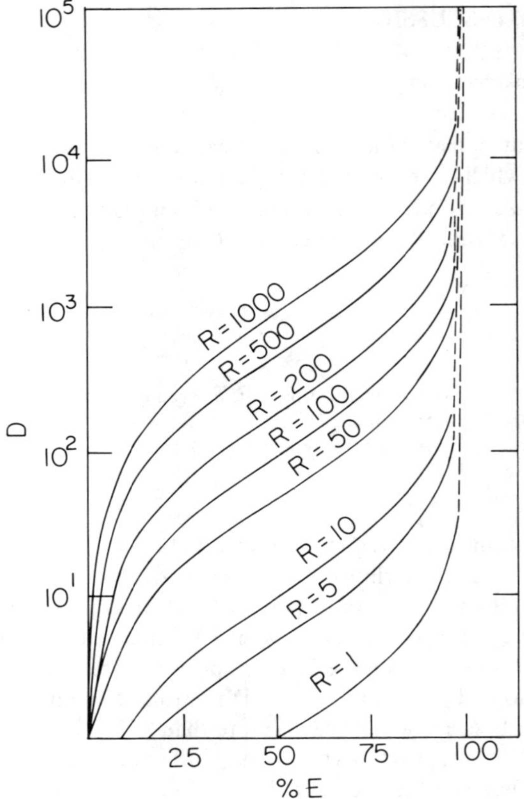


**Figure 2.** Plots of percent extracted (%E) vs distribution coefficient (D) for some volume ratios (R) between 1 and 1000 where R = volume of water over volume of organic solvent.

to determine the value of R required to achieve the desired %E. Conversely if R has been predetermined then %E can be estimated.

Some experimental data for five haloalkanes were accumulated to illustrate one of these uses of the plots in Figure 2. Known amounts of chloroform, bromodichloromethane, chlorodibromomethane, 1,3-dichloropropane and bromoform in 100 mL of water were extracted using an R of 100; the distribution coefficients were taken from the R = 100 plot. These coefficients were then used to predict the percent recoveries from the R = 10 and R = 200 plots. The same components in 100 mL of water were then extracted at these two volume ratios. The percent recoveries measured from the experimental data and predicted from the plots are given in Table I. The excellent agreement is more evidence of the validity of the extraction theory and of the value of the plots shown in Figure 2.

Table I. Measured and Predicted Recoveries of Haloalkanes at Volume Ratios (R) of 10 and 200

| Haloalkane | Percent Recovery (E) | | | |
|---|---|---|---|---|
| | R = 10 | | R = 200 | |
| | Measured | Predicted | Measured | Predicted |
| Chloroform | NM[a] | 88 | 24 | 27 |
| Bromodichloromethane | 92 | 92 | 31 | 35 |
| Chlorodibromomethane | 96 | 93 | 36 | 38 |
| 1,3-Dichloropropane | 96 | 96 | 54 | 60 |
| Bromoform | 98 | 96 | 43 | 46 |

[a]NM = not measured.

## Recovery Tests

Recoveries of model compounds added to water are listed in Tables II and III. Halocarbons, aromatic hydrocarbons, alcohols, alkanes, ketones and several pesticides were tested at different concentrations and for volume ratios of water to pentane of 100 and 500. In general, the observed recoveries are good, except for those components which are more soluble in water and thus are expected to have distribution coefficients of less than 10. Primarily due to the large uncertainty in the reported solubilities of organic compounds in water it is not possible to predict even approximate extraction efficiencies based on solubility data. In some cases where poor recoveries are obtained with pentane, as in the case with many of the ketones and some of the polar pesticides, it is possible to obtain an improvement in the extraction efficiency by changing to more polar solvents such as methylene chloride, diethyl ether and chloroform. However, the improvement is never impressive, so multiple extractions with larger volumes of solvent followed by concentration steps are a better approach to monitor such compounds.

The high recoveries shown in Table II for the alkanes at 1000 and 100 $\mu$g/L may be biased because these concentrations exceed the reported solubilities in water. At 10 and 1 $\mu$g/L the alkanes could be extracted efficiently only from water which had been filtered through a 0.45-$\mu$m silver membrane. Even then, prolonged and extremely vigorous shaking was necessary to achieve high extraction efficiency. Reduced recoveries of alkanes from water have been observed by others [9,10]. Our experimental evidence suggests that these low recoveries may be due to strong adsorption of the alkanes onto the microparticulate matter in the water. An alternative to prolonged and vigorous shaking to provide sufficient contact of pentane and the water

Table II. Recovery of Model Compounds Using Water to Pentane Volume Ratios of 100 and 500[a]

| | % Recovery | | | |
|---|---|---|---|---|
| Components | 1000 μg/L | 100 μg/L | 10 μg/L | 1 μg/L |
| Halocarbons | | | | |
| Chloroform | 43 | | | |
| Bromodichloromethane | 51 | 50 | | |
| Chlorodibromomethane | 56 | 58 | 60 | 54 |
| 1,3-Dichloropropane | 74 | 76 | | |
| Bromoform | 63 | 64 | 62 | 58 |
| Chlorobenzene | 86 | 78 | (63) | |
| 1,2-Dichlorobenzene | 90 | 92 | (95) | 91 |
| 1,2,4-Trichlorobenzene | 92 | 88 | 100 | 92 |
| Aromatic Hydrocarbons | | | | |
| Toluene | 69 | 68 | (44) | |
| Ethyl Benzene | 80 | 82 | (64) | |
| Cumene | 82 | 88 | (87) | |
| Naphthalene | 82 | 93 | (87) | |
| 2-Methylnaphthalene | 92 | 93 | (97) | |
| 1-Methylnaphthalene | 94 | | | |
| Acenaphthene | 94 | 94 | (94) | |
| Alcohols | | | | |
| 1-Octanol | 46 | 45 | | |
| 2-Octanol | 64 | | | |
| 1-Decanol | 95 | 98 | | |
| 1-Dodecanol | 98 | 94 | | |
| Alkanes | | | | |
| Decane | 94 | 92 | (88) | |
| Dodecane | 98 | 94 | (90) | (94) |
| Tetradecane | 100 | 98 | (90) | (91) |
| Hexadecane | 100 | 100 | (94) | (93) |
| Octadecane | | | (97) | (96) |
| Ketones | | | | |
| 4-Heptanone | 12 | | | |
| 3-Heptanone | 54 | 51 | (51) | |
| Cyclohexanone | 14 | | | |
| Fenchone | 24 | 30 | (33) | |
| Camphor | 16 | 19 | (24) | |
| Phenyl-2-propanone | 11 | 13 | | |
| Acetophenone | 16 | 13 | | |
| Benzophenone | 90 | 94 | 90 | 91 |
| Benzil | 91 | 94 | 96 | 95 |

[a]An R of 100 and a water volume of 85 mL was used for all extractions except those in parenthesis, where R was 500 and the water volume was 1 L. Reproducibility was ±10%.

Table III. Recovery of Pesticides Added to Water at 10 μg/L and Extracted with Pentane at R = 100

| Pesticide | % Recovery | Pesticide | % Recovery |
|---|---|---|---|
| Alachlor | 60 | DDT | 87 |
| Aldrin | 95 | Dichlobenil | 53 |
| Atrazine | 6 | Dieldrin | 82 |
| Azinphos Methyl | 45 | Endosulfan II | 87 |
| BHC | 79 | Endrin | 80 |
| Captan | 3 | Heptachlor | 85 |
| Chlordane A | 81 | Heptachlor Epoxide | 80 |
| Chlordane B | 95 | Kelthane | 94 |
| Chlordane C | 94 | Lindane | 86 |
| Chlordane D | 97 | Malathion | 52 |
| Chlordane E | 96 | Methoxychlor | 89 |
| Chlordecone | 46 | Mirex | 85 |
| Chlorpropham | 79 | Parathion Ethyl | 83 |
| 2,4-D Ester | 80 | Phosmet | 44 |
| DCPA | 98 | Propachlor | 13 |
| DDE | 91 | Simazine | 5 |
| | | Trifluralin | 96 |

is the use of larger volumes of pentane. For example 0.1 μg of octadecane can be efficiently extracted from 1 L of water if an R of 40 is used.

The influence of other organic constituents, present at high concentrations in water, on the extractability of organic solutes has been a concern expressed by both proponents and opponents of the solvent extraction method. Our initial concern was with the relatively large amount of methanol in the spiking solution added to the water. To ascertain whether methanol causes a measurable change in the extraction efficiency, several tests were made at successively higher levels of methanol in the test water. The test results are recorded in Table IV. The recoveries of five halocarbons at 1, 5 and 10% methanol in the water are the same, within experimental error, as those given in Table II. Surprisingly, a measurable difference in extraction efficiency is observed only at the very high concentration of 20% methanol. These data suggest unaltered extraction efficiency even when large quantities of some other water soluble components are present.

Table IV. Recovery of Haloalkanes at 1000 μg/L from Water Containing Different Amounts of Methanol[a]

| Haloalkane | Percent Recovery at 1% $CH_3OH$ | 5% $CH_3OH$ | 10% $CH_3OH$ | 20% $CH_3OH$ |
|---|---|---|---|---|
| Chloroform | 40 | 50 | NM[b] | NM |
| Bromodichloromethane | 47 | 47 | 48 | 41 |
| Chlorodibromomethane | 53 | 60 | 53 | 42 |
| 1,3-Dichloropropane | 67 | 71 | 67 | 58 |
| Bromoform | 56 | 57 | 57 | 47 |

[a]All extractions with pentane at R = 100.
[b]NM = not measured.

Table V. Analysis of Wellwater Using Large and Small Volumes of Pentane for the Extraction

| Compound | Concentration (μg/L) R = 10 | R = 500 |
|---|---|---|
| Indene | 94 | 92 |
| 3-Methylindene | 36 | 32 |
| 2-Methylindene | 46 | 43 |
| 2-Methylnaphthalene | 138 | 136 |
| Acenaphthene | 31 | 29 |
| Acenaphthylene | 172 | 168 |

## Applications

The extraction of selected organic components using small amounts of solvent is suited ideally to many monitoring aspects of our research programs. It has been used extensively for determining halocarbons, herbicides, insecticides and aromatic compounds in a variety of natural and wastewaters. As an example of its applicability, the data in Table V are cited. The wellwaters in Ames, IA, are analyzed for aromatic constituents on a regular basis. These analyses are accomplished using pentane or hexane and a volume ratio of 500. The data in the table show that the same results are achieved, but with much less effort and cost, as those obtained by more traditional extractions at volume ratios which require the use of a concentration step to achieve adequate sensitivity. All of these analyses are sufficiently reproducible for most monitoring programs. The results also compare favorably with values obtained by application of the independent resin sorption technique [11] for extraction of the organic components from the water.

## CONCLUSIONS

The experimental feasibility of using small amounts of organic solvent to extract trace levels of organic components from water has been well documented in this report and by other investigators [1-6]. This technique is applicable to many monitoring programs and is recommended for routine use because of the significant advantages highlighted in this report. However the potential user needs to be aware of the following disadvantages:

1. special vessels and manipulations not used in traditional solvent extractions are required;
2. more aggressive shaking is necessary to achieve the distribution equilibrium;
3. existing vessels accommodate only those solvents which are less dense than water; and
4. phase separation problems are apt to be more severe when only small amounts of solvent are used.

Both the advantages and the disadvantages should be considered in any decision regarding the routine use of small amounts of solvent for extraction purposes.

Although volume ratios as high as 7000 with 1-L water samples have been used here, an R of 100-500 for water samples of 100 mL to 1 L is recommended for more convenient routine work.

## ACKNOWLEDGMENT

We are grateful to Jim McDonald of the City of Ames for samples of well-water and to Ray Vick and Joyce Chow for technical assistance.

Ames Laboratory is operated for the U.S. Department of Energy by Iowa State University under Contract No. W-7405-Eng-82. This research was supported by the Assistant Secretary for Environment, Office of Health and Environmental Research, WPAS-HA-02-04-03 and in part by the National Science Foundation under grant CHE 75-71502-A03.

## REFERENCES


1. Rhoades, J. W., and J. D. Millar. *J. Agric. Food Chem.* 13:5 (1965).
2. Grob, K., K. Grob, Jr. and G. Grob. *J. Chromatog.* 106:299 (1975).
3. Murray, D. A. J. *J. Chromatog.* 177:135 (1979).
4. Rhoades, J. W., and C. P. Nulton. "Microextraction as an Approach to Analysis for Priority Pollutants in Industrial Wastewater," Chapter 15, this volume.
5. Thrun, K. E., and J. E. Oberholtzer. "Evaluation of the Microextraction Technique to Analyze Organics in Water," Chapter 16, this volume.
6. Glaze, W. H., R. Ralwey, J. L. Burleson, D. Mapel and D. R. Scott. "Further Optimization of the Pentane Liquid-Liquid Extraction Method for the Analysis of Trace Organic Compounds in Water," Chapter 17, this volume.
7. Morrison, G. H., and H. Freiser. *Solvent Extraction in Analytical Chemistry* (New York: John Wiley & Sons, Inc., 1957), p. 12.
8. Wilhoit, R. C., and B. J. Zwolinski. "Selected Values of Properties of Hydrocarbons and Related Compounds," in *Handbook of Vapor Pressures and Heats of Vaporization of Hydrocarbons and Related Compounds* (TX: Thermodynamics Research Center, 1971), p. 99.
9. Meyers, P. A., and T. G. Oas. *Environ. Sci. Technol.* 12:934 (1978).
10. Sutton, C., and J. A. Calder. *Environ. Sci. Technol.* 8:654 (1974).
11. Junk, G. A., J. J. Richard, M. D. Grieser, D. Witiak, J. L. Witiak, M. D. Arguello, R. Vick, H. Svec, J. S. Fritz and G. V. Calder. *J. Chromatog.* 99:745 (1974).

# SECTION 5

# RESIN ADSORPTION

# CHAPTER 19

# ANIONIC AND NEUTRAL ORGANIC COMPONENTS IN WATER BY ANION EXCHANGE

**Gregor A. Junk and John J. Richard**

Ames Laboratory, U.S. Department of Energy
Iowa State University
Ames, Iowa

The evolution of analytical schemes for the determination of trace levels of organic components in water has centered on the development of new techniques and the successful modification of traditional procedures for the isolation, concentration, separation, identification and quantification of organic compounds which are generally hydrophobic. In contrast to this remarkable evolution, the schemes for soluble, polar and ionic organic compounds, which are classified as hydrophilic, have not kept pace. The reasons for this developmental lag are many, two of which are the failure of the isolation methods and the almost exclusive use of gas chromatography (GC) for separation and measurement. The isolation methods are generally inefficient for trace levels of hydrophilic material, and much of this material is not amenable to separation by conventional GC techniques.

One approach to the solution of this problem with hydrophilic material is the adaptation of well-known ion exchange methods for isolating ionic material present in waters. These ion exchange methods should be useful for the determination of at least some of the many organic components which exist at different equilibrium levels as cations and anions in an aqueous medium.

The isolation of organic anions for analytical purposes from aqueous solution by ion exchange was first reported in 1944 [1]. Many other investigations [2-18] followed this pioneering effort. This emphasis on anion exchange, as opposed to cation exchange for isolation of organic cations, is probably due to the expected dominance of organic anions in most waters. This expectation was the primary motivation for the research reported in this chapter, where a prepared anion exchange resin with macroreticular characteristics is used in a unique modification of the resin sorption method [19] to separate and determine both the anionic (hydrophilic) and the neutral (hydrophobic) components in different waters.

The preparation of the anion exchange resin, the modification of the resin sorption method [19], some test results with anionic and neutral compounds, and applications to natural and wastewaters are reported and discussed.

## EXPERIMENTAL

The details of the procedure for the preparation of the anion exchange resin and parts of the procedures for the isolation of anions by anion exchange have already been published [15-17]. Only brief summaries of those procedures which are related to the test results and the applications in this paper are presented here.

### Preparation of Anion Exchange Resin

The starting material for the preparation of the anion exchange resin is the styrene-divinylbenzene copolymer XAD-4® (Rohm and Haas, Philadelphia, PA). This macroreticular resin is very similar to XAD-2®, which has been tested extensively for use in extracting hydrophobic organic components from water [19]. The XAD-4 as received from the supplier was cleaned according to procedures published previously [19], and the 80-100 mesh particles were subjected to chloromethylation using monochloromethyl ether followed by amination using trimethylamine as shown in Figure 1. The ammonium salt formed by the insertion mechanism was washed clean and settled into the small glass column where it was converted from the chloride to the hydroxide form by passing 75 mL of dilute base through the column. The column containing the anion exchange resin was used directly for test purposes and for the applications described in this report.

The resin prepared in this fashion from XAD-4 does not shrink, swell or dissolve when subjected to drastic changes in pH and different organic sol-

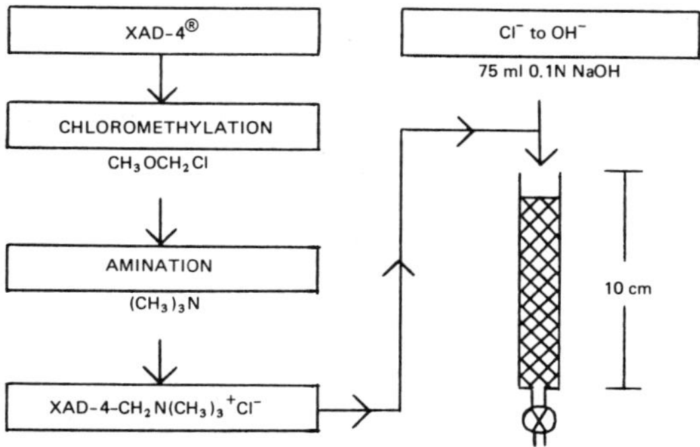


**Figure 1.** Flowchart for preparation of anion exchange resin and column.

vents. It also has the low blank level necessary for the measurement of trace amounts of organic components and retains a high efficiency for removing neutral organic components from water solutions. Commercial anion exchange resins familiar to the authors do not have these desirable characteristics.

## Isolation Procedure

The flowchart for isolation of neutral and anionic components from water is shown as part of Figure 2. The water to be tested or analyzed (1 L or less) is passed through the column containing 2 g of the anion exchange resin prepared from XAD-4.

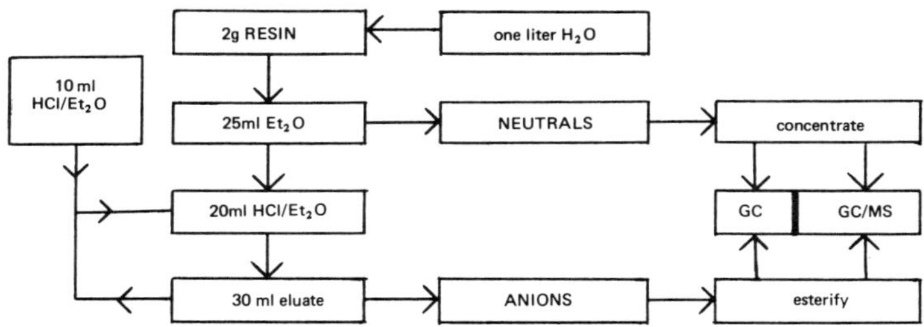


**Figure 2.** Analytical procedure for isolation, separation, identification and quantification of neutral and anionic organic components using the anion exchange resin.

The resin column is then eluted with 25 mL of diethyl ether to remove the neutral components which are separated, identified and quantified as described later on.

The anionic components are eluted from the column with 30 mL of HCl saturated ether. The first 20-mL portion is collected and passed through the column again. Then a fresh 10-mL portion of the HCl-saturated ether is passed through and combined with the original 20 mL. This eluate contains the anionic components which are derivatized prior to separation, identification and quantification.

The derivatization of the ionic components can best be accomplished if no trace of aqueous HCl is present in the ether eluate. The most convenient procedure for achieving an ether eluate free of aqueous HCl is to pass 20 mL of methanol through the column before the elution with HCl-saturated ether. This dries the column thoroughly, and the methanol eluate can be discarded since the neutral components have already been collected and the ionic components are still held on the column. The subsequent elution with HCl-saturated ether gives a dry anion extract which can be derivatized directly without resorting to other schemes for removing aqueous HCl. This on-column drying procedure with methanol is also necessary when HCl-saturated methanol is used to elute the anions from the column. This substitute eluent for the anions has been used successfully in some test and wastewater analyses.

## Neutral Analysis

The procedures used for the analysis of the neutral organic components in the ether eluate are shown as part of Figure 2. These are standard procedures involving concentration by microdistillation of most of the diethyl ether and then separation, identification and measurement of the neutral components by GC and combination gas chromatography/mass spectrometry (GC/MS).

## Anion Analysis

The organic material which was isolated as anions from the water is subsequently separated, identified and quantified as methylated derivatives. This is accomplished by taking the 30 mL of the diethyl ether eluate to dryness and methylating the residue with an excess of diazomethane. For most anions present in water this reaction is equivalent to esterification of acidic material and this term is used in the flowchart of Figure 2. When the methylation reaction has ceased as evidenced by the persistence of the yellow color of the reactant, the solution is diluted to a predetermined volume with diethyl ether. An aliquot of this solution is separated, identified and quantified in the same manner as that used for the neutral fraction.

This scheme discriminates against all components which are not gas chromatographable or which do not convert to chromatographable products by methylation using diazomethane.

## RESULTS AND DISCUSSIONS

The anion exchange procedure as described in the experimental section can be used for (1) the sole determination of organic compounds which form anions in water; (2) the sole determination of organic compounds which are neutral and hydrophobic; or (3) the combined determination of both anionic and neutral compounds. The procedure is almost as simple as the resin sorption method described previously [19,20]. Certain advantages occur automatically in this anion exchange procedure. For example, the removal of the anionic material from the neutral fraction results in better GC separations, particularly when capillary columns are used. Likewise, the separation of the neutral components from the ionic material results in much less complicated chromatograms, improving the chances for successful identification and quantification. These advantages are present when the procedure is used for anionic, neutral or both anionic and neutral components.

Test results described in this section suggest that these procedural advantages are supplemented by improved isolation of acidic anions over that which is achieved by solvent extraction or resin sorption of acidic organic components from water solutions adjusted to about pH 2 with inorganic acids.

### Recovery Efficiency for Anions and Neutrals

The anion exchange procedure as outlined in Figure 2 was tested for the recovery of trace amounts of both anionic and neutral components from different waters. The results of these efficiency experiments for different types of components in the various test waters are presented and discussed separately in the sections which follow.

#### *Anion Components in Distilled Water*

The efficiency of the anion exchange procedure of isolating organic components which form anions in water has been checked by passing standard

**Table I. Summary of the Anion Exchange Recovery of Acidic Components Added to Distilled Water at the 100-μg/L Level**

| Compound Type | Number Tested | Avg % Recovery[a] |
|---|---|---|
| Aliphatic Acids | 13 | 97 |
| Nitrophenols | 4 | 86 |
| Chlorophenols | 4 | 90 |
| Sub. Sulfonic Acids | 6 | 93 |
| Miscellaneous | 4 | 86 |
| Total | 31 | |
| Weighted Average | | 93 |

[a]Individual values are given in Richard and Fritz [16].

solutions of distilled water containing 100 μg of model compounds per liter of water through the anion exchange column. The exhaust water was discarded, and the elution, concentration, separation and quantification steps outlined in Figure 2 were followed. A summary of some test results published previously [16] for individual compounds is listed in Table I. The weighted average recovery of 93% for 31 compounds which are difficult to isolate at trace levels using conventional procedures was highly encouraging. In particular, the procedure was shown to be very efficient for aliphatic and aromatic acids and phenolic compounds.

The procedure has been optimized recently, and the extraction efficiencies of some pesticides and primary decomposition products were measured. These results are tabulated in Table II for nine acidic pesticides and eight decomposition products. All of these organic chemicals are of interest in environmental and wastewater samples. The high average recovery of 93% is also encouraging.

### *Anionic Components in Wastewaters*

The applicability of the anion isolation procedure to wastewater was tested by standard additions of 2,4-dichlorophenoxyacetic acid, tetrachloroterephthalic acid and N-1-naphthylphthalamic acid to a wastewater sample taken from a pesticide disposal pit. These three acids were initially determined by the anion exchange procedure to be present at 3, 5 and 40 mg/L. Known amounts of these acids were then added to separate aliquots of this same sample from the disposal pit, and the acids in these spiked samples were also determined by anion exchange. The results of these standard additions analyses are plotted in Figure 3. The linearity of these plots shows applica-

**Table II. Recovery of Selected Pesticides and Decomposition Products Added to Distilled Water at 200 μg/L and Extracted by the Anion Exchange Procedure**

| Pesticide or Decomposition Product | % Recovered |
|---|---|
| 2,4–Dichlorophenoxyacetic Acid (2,4–D) | 93 |
| 2,4,5–Trichlorophenoxyacetic Acid (2,4,5–T) | 102 |
| 2–(4–Chloro–2–methylphenoxy)–propionic Acid (MCPP) | 93 |
| Pentachlorophenol | 100 |
| 3–Amino–2,5–dichlorobenzoic Acid (Chloramben) | 86 |
| 2–Methoxy–3,6–dichlorobenzoic Acid (Dicamba) | 98 |
| 3–Isopropyl–1H–benzo–2,1,3–thiadiazin–4–one–2,2–dioxide (Bentazon) | 91 |
| N–(Phosphonomethyl)–glycine (Glyphosate) | 103 |
| Sodium salt of N–1–naphthylphthalimide (Naptalam) | 95 |
| 2,4–Dichlorophenol | 100 |
| 2,4,5–Trichlorophenol | 95 |
| 2,4–Dichlorobenzoic Acid | 91 |
| Diethylphosphoric Acid | 88 |
| Dimethylphosphoric Acid | 81 |
| Dimethylthiophosphoric Acid | 94 |
| Tetrachloroterephthalic Acid | 83 |
| 4–Cyclohexene–1,2–dicarboximide | 95 |
| Average % Recovery | 93 |

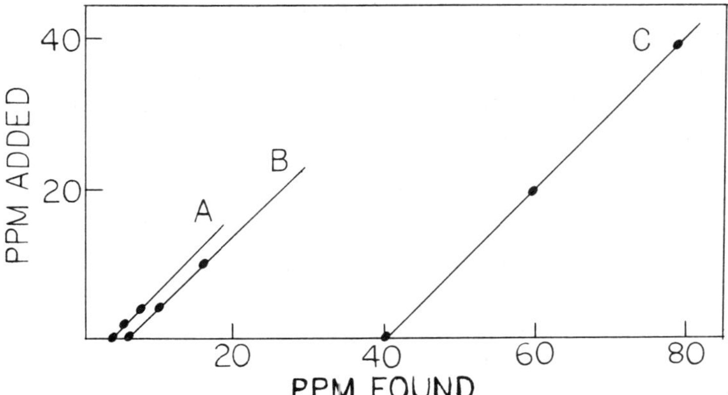


**Figure 3.** Plots of standard additions results for 2,4-dichlorophenoxyacetic (A), tetrachloroterephthalic (B) and N-1-naphthalphthalamic (C) acids in water taken from a pesticide disposal pit.

bility over the concentration range which was tested, and the slopes of one prove near 100% extraction efficiency of the spiked components from water containing all sorts of other ionic and neutral organic material. These standard addition data still do not prove absolutely that the extraction of the anionic components originally present in the wastewater was 100%. However, the data do show applicability to real waters much better than do data from the addition of model compounds to distilled water or simulated wastewater.

### *Anionic Pesticides in Saltwater*

The determination of acidic material in sea-, estuarine and brackish waters has been an environmental concern for many years. The high inorganic salt content of these waters tends to improve the extraction efficiency by conventional solvent extraction methods but may be detrimental to the anion exchange procedure due to competition for the exchange sites and overloading the column. Even though the affinity of the organic anions to the resin may be greater than the affinity of the common inorganic anions, the mass action effect of high concentrations of chloride in sea- and estuarine waters and sulfate in certain brackish waters may destroy the extraction efficiency. To test whether this would occur, three common pesticides which form anions in water and are recovered efficiently from distilled and wastewater were added to distilled water which also was spiked with 1 and 5% sodium chloride by weight. This concentration of inorganic salt covers the range normally found in seawater and is above that expected in most estuarine waters. Comparable tests were also run using sodium sulfate as the inorganic salt. This system simulates the inorganic anion content of some brackish waters. The recovery test results are tabulated in Table III. Based on these preliminary data for a very limited number of components, it appears as if the anion exchange procedure is also applicable to natural waters containing high inorganic salt levels.

**Table III. Anion Exchange Recovery of 200 $\mu$g of Organic Acids per Liter of Water Containing 1 and 5% NaCl and $Na_2SO_4$**

| | % Recovered From | | | |
|---|---|---|---|---|
| | NaCl | | $Na_2SO_4$ | |
| Organic Acid | 1% | 5% | 1% | 5% |
| 2,4-Dichlorophenoxyacetic | 88 | 84 | 96 | 94 |
| 2,4,5-Trichlorophenoxyacetic | 90 | 85 | 91 | 89 |
| Tetrachloroterephthalic | 51 | 20 | 93 | 88 |

### *Neutral Components–Comparative Results*

One of the reasons for the choice of XAD-4 as the starting material for the preparation of the anion exchange resin was the documented efficiency of styrene-divinyl benzene copolymers for the removal of neutral organic components from water [18]. If part or all of this efficiency could be retained after the addition of the ion exchange sites, a single resin could be used for the removal of both anionic and neutral material. The two kinds of organic components could then be separated and measured as shown in Figure 2 and as described in the Experimental section.

One means of testing this concept would be to prepare different concentrations of model compounds in water solution and to measure the extraction efficiency as was done for neutral components, using XAD-2 and the resin sorption procedure [19,20]. This approach was also used, although less exhaustively, for the measurement of the recovery efficiency for anionic components as described earlier in this chapter.

For the neutral components in water solutions a separate approach was chosen because of the documented validity of resin sorption [19,20] and solvent extraction procedures. Selected neutral components known to be present in real waters were analyzed by the anion exchange procedure, and these results were compared to those obtained from analysis of the same waters using either solvent extraction or resin sorption procedures.

Some analytical data for resin sorption and anion exchange procedures for pesticides in different kinds of water are listed in Table IV. The comparison of analytical results shown in column pairs is very favorable. The poorest comparisons are for the pesticides in the digester water, where filtration through a 0.45-$\mu$m filter did not produce a clear solution. The microparticulate matter in these samples probably influenced the adsorption and elution mechanisms in both procedures so that valid data for dissolved organic matter were not obtained.

The comparison of analytical results for solvent extraction and anion exchange procedures can be deduced from the two chromatograms shown in Figure 4. The wastewater sample here was from a pit used for the disposal of the pesticide, trifluralin, as a dilute aqueous solution of the emulsifiable concentrate sold as Treflan®. For the comparison a 500-mL sample of the wastewater was divided equally. The first 250 mL was passed through the anion exchange column, and the neutral eluate was collected and concentrated to 1 mL. The chromatogram from an aliquot of this extract is shown on the left of Figure 4. On the right, is the chromatogram obtained after conventional solvent extraction of the second 250-mL portion of the wastewater. All conditions were identical in the concentration step and chromatography except for the attenuation listed on the chromatograms. The very

Table IV. Selected Neutral Pesticides ($\mu g/L$) in Different Waters as Determined by Anion Exchange (AE) and Resin Sorption (RS) Procedures

| | Des Moines River[a] | | Des Moines River[b] | | Skunk River | | Saylorville Reservoir | | Red Rock Reservoir | | Urban Runoff[c] | | Digestor Wastewater | |
|---|---|---|---|---|---|---|---|---|---|---|---|---|---|---|
| Pesticide | AE | RS | AE | RS | AE | RS | AE | RS | AE | RS | AE | RS | AE | RS |
| Atrazine | 0.6 | 0.5 | 0.7 | 0.7 | 1.8 | 1.8 | 0.8 | 0.7 | 1.5 | 2.3 | 1.7 | 1.3 | | |
| Cyanazine | 0.8 | 0.4 | 1.0 | 0.9 | 2.3 | 2.0 | 1.2 | 1.0 | 2.3 | 2.4 | | | | |
| Alachlor | 0.8 | 0.6 | 1.4 | 1.3 | 4.7 | 4.0 | 1.4 | 1.3 | 0.7 | 0.9 | | | | |
| Propachlor | 0.3 | 0.2 | 0.3 | 0.3 | | | 0.3 | 0.3 | 0.7 | 1.0 | | | | |
| Dieldrin | | | | | | | | | | | 0.02 | 0.02 | 0.04 | 0.01 |
| DDT | | | | | | | | | | | | | 0.11 | 0.02 |
| DDD | | | | | | | | | | | | | 0.1 | 0.1 |
| Metribuzin | 0.2 | 0.1 | 0.2 | 0.2 | | | 0.2 | 0.2 | 0.1 | 0.1 | | | | |
| Dacthal | 0.1 | 0.01 | 0.01 | 0.01 | 0.2 | 0.1 | 0.01 | 0.01 | | | 2.7 | 2.6 | | |
| PCB[d] | | | | | | | | | | | | | 9 | 6 |

[a]Des Moines location.
[b]Boone location.
[c]Located in agricultural area.
[d]Polychlorinated biphenyls as Aroclor 1248.

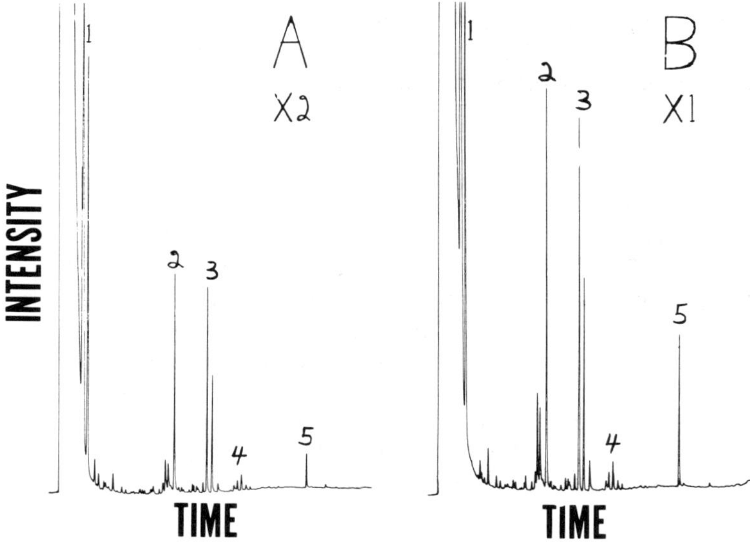


Figure 4. Capillary column gas chromatograms of the neutral components isolated by anion exchange (A) and solvent extraction (B) of a water sample taken from a pesticide disposal pit. Group 1 = xylenes + ethylbenzene; Group 2 = naphthalene; Group 3 = methylnaphthalenes; Group 4 = dimethyl- and ethylnaphthalenes; and Group 5 = trifluralin.

close similarity of these two chromatograms confirms the applicability of the anion exchange procedure for the measurement of the aromatic components listed in the figure caption.

Although a large number of different neutral components have not yet been tested, it appears that the efficiency of the styrene-divinylbenzene polymer for neutral components is not altered appreciably by the addition of the anion exchange sites. The predicted retention of this efficiency is confirmed by the limited number of comparative data presented here. Certainly, some change in surface interactions which affect adsorption do occur, but these may be minimal and of no serious consequence for many neutral components of environmental interest.

## Applications

The anion exchange procedure as described in this report has been used for the isolation of anionic organic material from a variety of surface and wastewaters. These investigations are still in progress, and results so far are incomplete. However, the procedure has been used with almost universal success and even the preliminary results as presented and discussed in the following sections may be of interest to scientists involved in the analyses of trace levels of organic components in different kinds of waters.

*Treflan Disposal Pit*

Experimental pits used for the disposal of pesticides have been described previously [17]. Containment of the waste pesticide is achieved in 110-L plastic garbage cans buried partially in the ground and filled with 60 L of water and 15 kg of soil. Trifluralin as the Treflan formulation (15 g) was added to one of these pits. After two years under ambient conditions, two 500-mL water samples were taken from the pit and extracted separately; one by anion exchange and the other by traditional solvent extraction of the acidified water. Both extracts were methylated and chromatographed under identical conditions. The chromatograms are shown in Figure 5. The anion

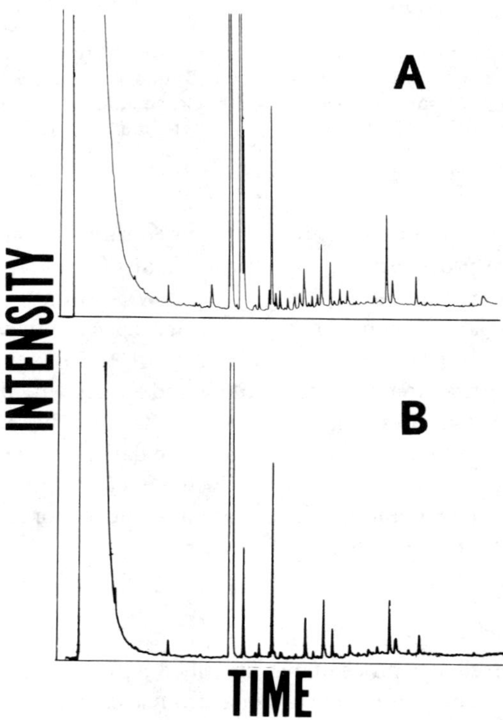


**Figure 5.** Gas chromatograms of methylated components from direct anion exchange (A) and acidified solvent extraction (B) of wastewater from a disposal pit.

exchange chromatogram has peaks which correspond to all those present in the solvent extraction chromatogram. In all cases the peak intensities are greater, which suggests better efficiency for many anions present in this wastewater. In a few cases, particularly in the first half of the chromatograms, the differences are dramatic, with the highest efficiency always observed for the anion exchange procedure. Identifications of the specific compounds responsible for the peaks in these chromatograms are to be published elsewhere [21].

### *Herbicide Disposal Pit*

A plastic lined earthen pit at the agronomy farm at Iowa State University had been used for two years for disposal of dilute aqueous solutions of pesticides, almost all of which were herbicides. A water sample was taken from this disposal pit, and the anionic and neutral components were isolated using the anion exchange scheme outlined in Figure 2.

The complex chromatogram obtained for the neutral components separated on a glass capillary column is reproduced in Figure 6. Obviously, a large number of neutral components are present in this wastewater. Complete and confirmed identification of these compounds are not yet available but the preliminary interpretations of GC/MS data show most are unreacted pesticides and extraneous organic components related to different formulating agents. The chromatogram is presented here not to discuss these identifications but rather to show applicability of the anion exchange method to neutral components in real wastewaters and to emphasize the incomplete nature of wastewater analyses which rely solely on the determination of neutral components.

The incomplete nature of the analysis for neutral components is illustrated by the chromatogram in Figure 7, which shows a large number of methylated products of the anionic components isolated from the same wastewater. Many of these components would not be identified and could not be quantified accurately without resorting to some means of efficient isolation and separation of the ionic from the neutral components prior to derivatization to products amenable to standard GC. The anion exchange method accomplishes the isolation efficiently and the separation conveniently.

Tentative identification of 48 compounds in this extract have been made from interpretation of GC/MS data. Ten individual components which represent a cross section of the total are listed in the figure caption. These and the other 38 tentative identifications are summarized as part of Table V. Discussion of this summary is combined with the discussion of the insecticide summary at the end of the next section.

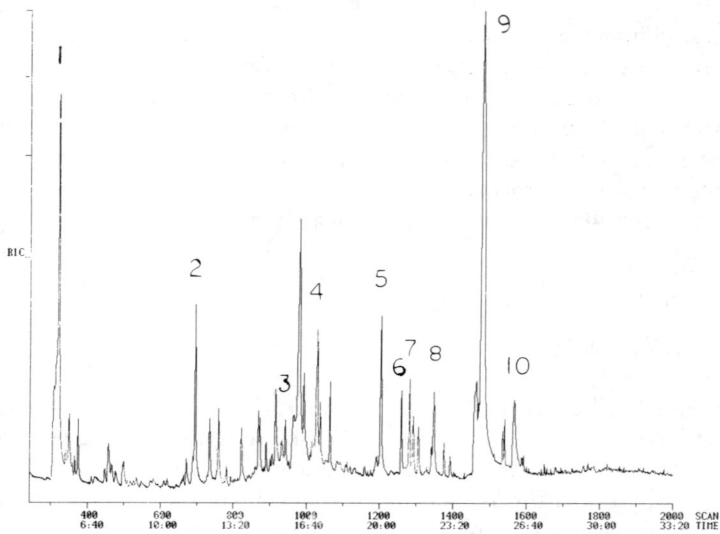


**Figure 6.** GC/MS reconstructed ion chromatogram (RIC) of the neutral components isolated by anion exchange of water taken from a disposal pit for herbicides (see Figure 7 for anionic components). (1) chlorobenzene; (2) 3-methylindene; (3) S-ethyl-N,N-dipropylthiocarbamate; (4) S-ethyl-N, N-diisobutylthiocarbamate; (5) 2-chloro-N-isopropyl-acetanilide; (6) isopropyl-N-(3-chlorophenyl) carbamate; (7) α,α,α-trifluoro-2,6-dinitro-N,N-dipropyl-P-toluidine; (8) 2-chloro-4-(ethylamino)-6-(isopropylamino)-S-triazine; (9) 2-chloro-2′,6′-diethyl-N-(methoxymethyl)acetanilide; (10) 2-[(4-chloro-6-(ethylamino)-S-trizain-2-yl)amino]-2-methylpropionitrile.

## *Insecticide Disposal Pit*

A concrete-lined disposal pit located at the horticulture station at Iowa State University had also been used for the disposal of dilute aqueous solutions of pesticides, most of which were insecticides. Water samples were taken from this disposal pit and treated in the same manner as that described above for the herbicide disposal pit. The chromatogram of the neutral components is not shown here because it is almost identical in complexity to the herbicide chromatogram in Figure 6, and the very preliminary identification data will neither be presented nor discussed in this chapter.

The chromatogram obtained for the methylated products of the anionic components is reproduced in Figure 8. As with the herbicide disposal pit, a large number of organic chemicals which form anions in water and which methylate to gas chromatographable components are present.

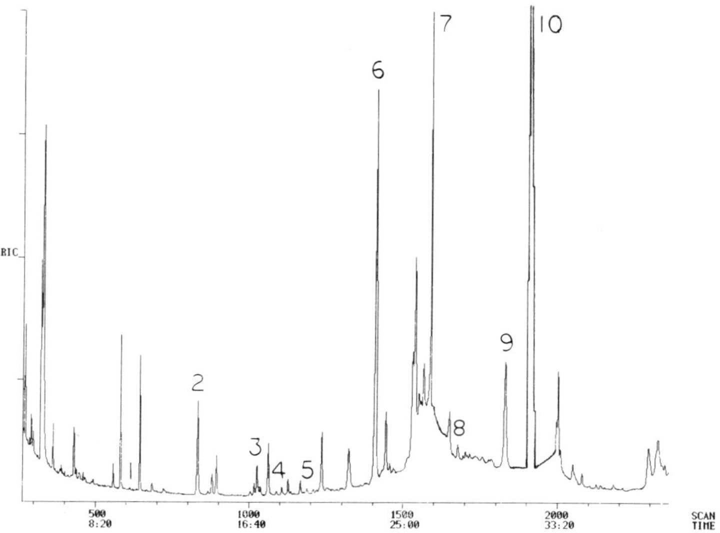


**Figure 7.** GC/MS reconstructed ion chromatogram (RIC) of the methylated products of the anionic components isolated by anion exchange of the same water described in Figure 6. (1) trichloropropionic acid; (2) 2-methylbenzoic acid; (3) 2,4-dichlorophenol; (4) 2,4,5-trichlorophenol; (5) 2,4-dichlorobenzoic acid; (6) 2-methoxy-3,6-dichlorobenzoic acid; (7) naphthoic acid; (8) 2,4,5-trichlorophenoxyacetic acid; (9) 3-amino-2,5-dichlorobenzoic acid; (10) 3-isopropyl-1H-benzo-2,1,3-thiadiazin-4-one-2,2-dioxide. (All identified as the methylated derivatives.)

Tentative identifications have also been made from interpretation of GC/MS data, and 10 of the more interesting components are listed in the figure caption. These 10 and 38 other identifications are combined with the identifications from the herbicide pit and are summarized in Table V.

The combined identification summary of Table V reflects information about pesticide usage, degradation of formulating agents, hydrolysis of nonpersistent pesticides and increased amounts of degradation with time. Examples of this information are:

1. The appreciable numbers of acidic pesticides reflect the increased use of this kind of pesticide and the parent esters which hydrolyze rapidly;
2. The aromatic solvent mixtures used to produce liquid formulations of both insecticides and herbicides are degraded microbiologically to substituted aromatic acids and phenols.
3. The substitution of the nonpersistent organophosphorus insecticides for the organochlorine insecticides used previously results in the forma-

**Table V. Summary of the Anionic Compounds Identified in the Herbicide and Insecticide Disposal Pits**

| Compound Type | Number Identified | |
|---|---|---|
| | Herb. Pit | Insect. Pit |
| Acidic Pesticides | 15 | 10 |
| Sub. Aromatic Acids | 15 | 9 |
| Phenolics | 9 | 5 |
| Organophosphorus | 2 | 12 |
| Miscellaneous | 7 | 12 |

tion of organophosphorus acids by chemical and biological hydrolysis mechanisms.

4. The eight-year-old insecticide pit has a greater number of identified compounds which are classified as miscellaneous because more time was available for decomposition to products only remotely related to the relatively persistent parent compounds.

### *Waste to Methane Digester Water*

This water sample was taken from an anaerobic digester used to produce methane from processed municipal refuse. The sample was filtered through a 0.45-$\mu$m filter, and the anion components were isolated, separated from the neutral components and methylated as described in the experimental section. An aliquot of the solution of methylated products was analyzed by GC/MS. The reconstructed ion chromatogram, which is very similar to that obtained by conventional GC with a flame ionization detector (FID) is reproduced as Figure 9. The anionic components identified as the methyl esters are listed as part of the figure caption. Fatty acids formed by the action of acid-forming anaerobic bacteria on the processed refuse, acidic herbicides formed by hydrolysis of the highly used esters and resin acids as indigenous components of woody refuse are the significant components identified so far. Other components have also been identified in the neutral fraction isolated from this digester water, the most significant of which are some chlorinated neutral pesticides and the polychlorinated biphenyls (PCB).

### *Urban Runoff Water*

Runoff water samples were collected after late spring rains in urban areas located near the center of an intensive agricultural region. These water sam-

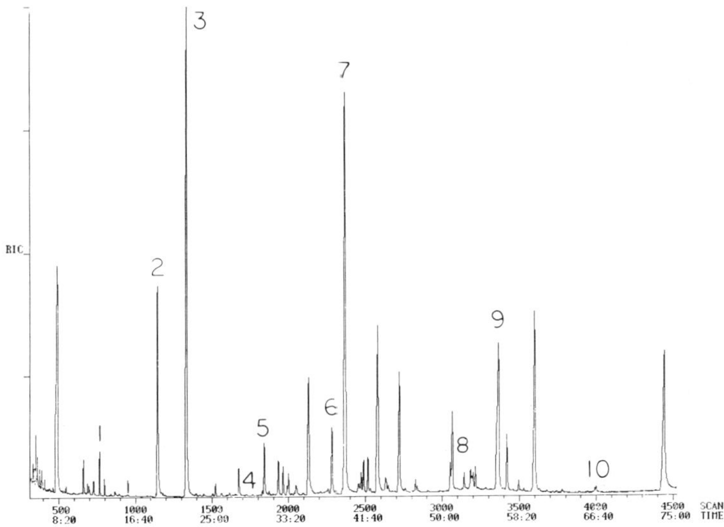


**Figure 8.** GC/MS RIC of the methylated products of the anionic components isolated by anion exchange of water taken from a disposal pit for insecticides. (1) 0,0,0–phosphorothioic acid; (2) 0,0,S–phosphorothioic acid; (3) 0,0,S–phosphorodithioic acid; (4) dimethylbenzoic acid; (5) 0,S,S–phosphorodithioic acid; (6) 2–methoxy–3,6–dichlorobenzoic acid; (7) 2–(4–chloro–2–methylphenoxy)–propionic acid; (8) trichloroterephthalic acid; (9) tetrachloroterephthalic acid; (10) dehydroabietic acid. (all identified as the methylated derivatives.)

ples were subjected to the anion exchange procedure shown in Figure 2. The results for the neutral components, atrazine, dieldrin and dacthal, in one of these water samples have already been discussed in the section on recovery efficiency, and the quantitative data are tabulated in the urban runoff column of Table IV.

Quantitative data by anion exchange were also obtained for two anionic components formed by the ambient hydrolysis of dacthal and the various esters of 2,4-dichlorophenoxyacetic acid (2,4-D). The two anionic acids, tetrachloroterephthalic acid and 2,4-D, were isolated by anion exchange, methylated, and separated and detected by electron capture gas chromatography. The analytical results are tabulated in Table VI.

## *River- and Impoundment Waters*

Water samples taken from two rivers and two surface water reservoirs were subjected to the anion exchange procedure. The quantitative results for six

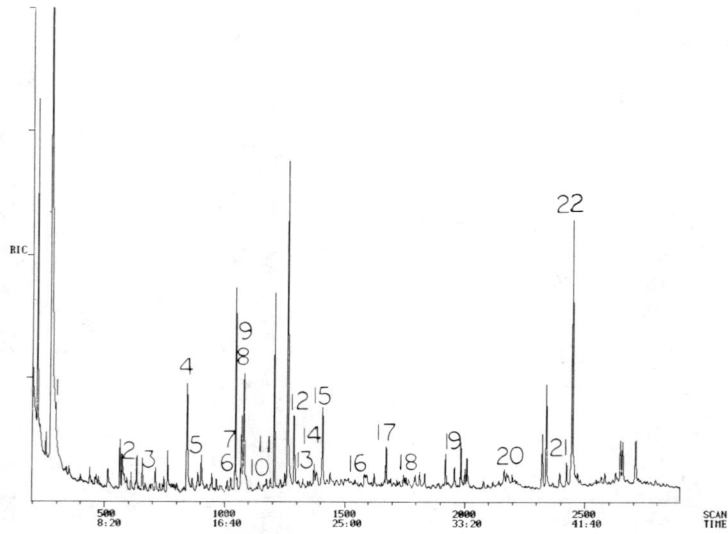


**Figure 9.** GC/MS RIC of the methylated products of the anonic components isolated by anion exchange of water from an anaerobic digester. Identified acids as methyl esters are: (1) caproic; (2) caprylic; (3) benzoic; (4) phenylacetic; (5) x-methylbenzoic; (6) phenylpropionic; (7) capric; (8) x,x-dimethylbenzoic; (9) β,β-dimethylphenylpropionic; (10) x,x,x-trimethylbenzoic; (11) phenylbutyric; (12) *o*-phthalic; (13) 2,4-dichlorophenoxyacetic; (14) 2-(4-chloro-2-methylphenoxy)propionic acid; (15) lauric; (16) 2,4,5-trichlorophenoxyacetic acid; (17) myristic; (18) tetrachloroterephthalic; (19) palmitic; (20) stearic; (21) dehydroabietic; (22) abietic.

**Table VI. Tetrachloroterephthalic (TCTP) and 2,4-Dichlorophenoxyacetic Acid (2,4-D) in Urban Runoff Waters**

| Urban Site Number | TCTP ($\mu$g/L) | 2,4-D ($\mu$g/L) |
|---|---|---|
| 1 | 4.8 | 3.1 |
| 2 | 6.9 | $<$0.01 |
| 3 | 8.4 | 4.8 |
| 4 | 2.8 | 0.02 |

neutral components (atrazine, cyanazine, alachlor, propachlor, metribuzin and dacthal) are reported in Table IV.

Quantitative data were also obtained for the two hydrolysis products of dachthal and 2,4-D esters. These results are given in Table VII. These same water samples were acidified and subjected to analyses for the same two acids using the XAD-2 resin sorption method [20]. In all five samples, these two acids could not be detected by this method.

## CONCLUSIONS

The conclusions based on interpretation of data presented in this report and the authors' experiences in measuring trace levels of organic components in different waters are:

1. The anion exchange procedure is more convenient and efficient than either the solvent extraction or resin sorption procedure for isolating the organic components which form anions in water.
2. The simultaneous isolation of both neutral and anionic components from water and the subsequent partition of these components in separate fractions is a very definite advantage in water analyses.
3. The anion exchange procedure is adaptable to the incorporation of a cation exchange resin so that a dual column could be used to isolate into separate fractions the neutral, anionic and cationic components.
4. Wide applicability is suggested by the documented success of the anion exchange procedure for a variety of environmental and wastewaters.
5. The organic components in the anion eluate can be separated, identified and quantified by liquid chromatography (LC) and LC/MS to avoid the discrimination of the derivatization in the GC and GC/MS approaches.

**Table VII. TCTP and 2,4-D in River and Impoundment Waters**

| River or Impoundment | TCTP ($\mu$g/L) | 2,4-D ($\mu$g/L) |
|---|---|---|
| Lower Des Moines River | 0.04 | 2.6 |
| Upper Des Moines River | 0.03 | 1.3 |
| Skunk River | 0.10 | 0.6 |
| Saylorville Reservoir | >0.01 | 0.7 |
| Red Rock Reservoir | 0.03 | 3.0 |

6. Although the techniques described in this report are useful for the isolation of many organic compounds which form anions in water, an indeterminate number of other anions and cations are not isolated, separated and measured by this or any other technique.
7. Unlike the resin sorption method for neutral organic components, the anion exchange procedure is not amenable to the isolation of large amounts of organic anions by passing large volumes of water through the anion exchange column.
8. The large number of organic components which form anions in water and are measured as methylated derivatives confirm the expected degradation of neutral to anionic components in environmental waters.
9. The use of pits for the disposal of pesticide wastes as dilute aqueous solutions is inexpensive and effective.

## ACKNOWLEDGMENTS

Ames Laboratory is operated for the U.S. Department of Energy by Iowa State University under Contract No. W-7405-Eng-82. Parts of this research were supported by the Assistant Secretary for the Environment, Office of Health and Environmental Safety, WPAS-GK-01-02-04-04. Other financial support was received from the U.S. Environmental Protection Agency under Contract R804533010. Special thanks go to Mike Avery for providing the GC/MS data and to C. Chriswell for collecting the digester water samples.

## REFERENCES

1. Cannan, K. "The Estimation of the Dicarboxylic Amino Acids in Protein Hydrolysates," *J. Biol. Chem.* 152:401 (1944).
2. Busch, H., R. B. Hurlbert and V. R. Potter. "Anion Exchange Chromatography of Acids of the Citric Acid Cycle," *J. Biol. Chem.* 196:717 (1952).
3. Reinbothe, H. "Anion Exchangers for the Separation and Quantitative Determination of Organic Acids," *Pharmazie* 12:732 (1957).
4. Plapp, F. W., and J. E. Casida. "Ion Exchange Chromatography for Hydrolysis Products of Organophospate Insecticides" *Anal. Chem.* 30:1622 (1958).
5. Burchfield, H. P., and E. E. Storrs. *Biochemical Applications of Gas Chromatography*, (New York: Academic Press, 1962), pp. 588–595.
6. Canvin, D. T. "Analysis of Some Organic Acids by Gas–Liquid Chromatography," *Can. J. Biochem.* 43:1281 (1965).
7. Kuksis, A., and P. Prioreschi. "Isolation of Krebs Cycle Acids from Tissues for Gas Chromatography," *Anal. Biochem.* 19:468 (1967).
8. Gee, M. "Methyl Esterification of Nonvolatile Plant Acids for Gas Chromatographic Analysis," *Anal. Chem.* 37:926 (1965).

9. Marinelli, L., M. F. Feil and A. Schait. "Isolation and Identification of Organic Acids in Beer," *Proc. Am. Soc. Brew. Chem.* Vol. 113 (1968).
10. Horning, M. G. In: *Biomedical Applications of Mass Spectrometry*, H. A. Szymanski, Ed. (New York: Plenum Press, 1968), p. 53.
11. Backer, D. W. "Analysis of Organic Acids in Fruit Products by Anion Exchange Isolation and Gas Chromatographic Determination," *J. Assoc. Off. Anal. Chem.* 56:1257 (1973).
12. Chriswell, C. D., R. C. Chang and J. S. Fritz. "Chromatographic Determination of Phenols in Water," *Anal. Chem.* 47:1325 (1975).
13. Verweij, A., H. L. Boter and C. E. A. M. Degenhardt. "Chemical Warfare Agents: Verification of Compounds Containing the Phosphorus-Methyl Linkage in Waste Water," *Science* 204:618 (1979).
14. Lores, E. M., and D. E. Bradway. "Extraction and Recovery of Organophosphorus Metabolites from Urine Using an Anion Exchange Resin," *J. Agric. Food Chem.* 25:75 (1977).
15. Gabriel, D. M., and V. J. Mulley. "Detection, Separation and Isolation of Anionic Surfactants," in *Anionic Surfactants–Chemical Analysis, Vol. 8,* J. Cross, Ed. (New York: Marcel Dekker, 1977), pp. 28-37.
16. Richard, J. J., and J. S. Fritz. "The Concentration, Isolation, and Determination of Acidic Material from Aqueous Solution," *J. Chromatog. Sci.* 18:35 (1980).
17. Junk, G. A., and J. J. Richard. "Anion Exchange Isolation and Identification of Degradation Products from Pesticide Disposal Pits," *Water Qual. Bull.* 6:40 (1981).
18. Richard, J. J., C. D. Chriswell and J. S. Fritz. "Concentration and Determination of Organic Acids in Complex Aqueous Samples," *J. Chromatog.* 199:143 (1980).
19. Junk, G. A., J. J. Richard, M. D. Grieser, D. Witiak, J. L. Witiak, M. D. Arguello, R. Vick, H. J. Svec, J. S. Fritz and G. V. Calder. "Use of Macroreticular Resins in the Analysis of Water for Trace Organic Contaminants," *J. Chromatog.* 99:745 (1974).
20. Junk, G. A., J. J. Richard, J. S. Fritz and H. J. Svec. "Resin Sorption Methods for Monitoring Selected Contaminants in Water," in *Identification and Analysis of Organic Pollutants in Water*, L. H. Keith, Ed. (Ann Arbor, MI: Ann Arbor Science Publishers, Inc., 1976), pp. 135-153.
21. Junk, G. A., and J. J. Richard. "Degradation of Herbicides in Disposal Pits Containing Soil and Water," (in preparation).

# CHAPTER 20

# THERMAL DESORPTION FROM A XAD-4 ADSORPTION SUBSTRATE FOR WATER ANALYSIS BY CAPILLARY GAS CHROMATOGRAPHY

**John Paul Ryan and James S. Fritz**

Ames Laboratory, U.S. Department of Energy
and Department of Chemistry
Iowa State University
Ames, Iowa

Organic compounds in water samples are commonly analyzed by gas chromatography (GC) following a preconcentration step. The preconcentration may be achieved by gas purging [1,2], solvent extraction [2,3] or sorption on a resin column with subsequent elution with a volatile solvent such as diethyl ether [4]. A method has been published in which organic impurities in water were concentrated on a tube containing XAD resin. The sorbed substances were then thermally desorbed into a gas chromatograph which was specially equipped for a preliminary separation of sample components from any residual water [5]. This method worked quite well for a broad range of model organic compounds, but a drawback was the limited resolution that could be obtained with the packed column of the gas chromatograph.

Now a system for thermal desorption of organic compounds from XAD resin is described that employs a capillary GC column. This retains the advantages of our earlier system [5], but the glass capillary GC column provides for much greater resolution of sample compounds. The new method has been studied primarily for the concentration and determination of aromatic hydrocarbons, but it should also be applicable to the analysis of other organic impurities in aqueous samples.

## EXPERIMENTAL

### Apparatus

*Adsorption Tubes (Minicolumns)*

Pyrex® glass tubing (0.25-in. o.d. x 2-mm i.d.) was cut into 8-cm lengths and fired at the ends. After rinsing with acetone, each tube was plugged at one end with silanized glass wool (silanizing the glass tubes is not required). A methanolic slurry of 100-120 mesh AIBN-initiated XAD–4 resin (Rohm and Haas, Philadelphia, PA) was drawn up to a depth of about 5 cm by a syringe connected to the plugged end. A glass wool plug was then inserted into the opposite end of the tube, holding the resin bed in place.

*Conditioning*

The minicolumns were conditioned in four heating cycles. These consisted of filling the minicolumns with distilled water and placing them in a heated zone. During conditioning, helium gas was forced through the minicolumns at about 20 mL/min. The conditioning cycles were 8–10 min long. The first was carried out at 240°C and the rest at a temperature of 5–10°C hotter than the highest temperature to be reached during the analysis run. The conditioning process was carried out in a manifold which could accommodate four minicolumns at a time.

*Thermal Desorption Apparatus and Gas Chromatograph Inlet System*

The thermal desorption apparatus was essentially the same as the one used earlier [5]. It consisted of an XAD-4 sample tube and a Tenax®-filled loop with appropriate heaters, two inlets for carrier gas, and 4- and 6-port valves. All this is arranged so that organics can be desorbed from the XAD tube onto the Tenax loop and the water vapor vented; the organics are then desorbed from the Tenax loop, in the opposite sense of the adsorption step, into the inlet system of a gas chromatograph (Figure 1). After this study, the transfer lines and Tenax loop were converted from stainless steel to deactivated nickel

[6]. Reactivity of the compounds studied with stainless steel was noted in the case of tritan, which has a very high boiling point and decomposed as much as 15%. Low recoveries of some chlorinated hydrocarbons were also observed with the stainless steel system.

The approximately tenfold difference in the diameters of the Tenax loop and the capillary chromatographic column dictated the use of either a splitter or a cold trap. The former alternative was chosen because an inlet splitter would be easy to construct and versatile to use. Also, a splitter would permit higher desorption flowrates from the Tenax loop, making desorption more rapid.

The most essential features of the inlet system are shown in Figure 1. The black, hourglass shape is the glass injection-port sleeve (Pyrex 6 mm o.d. x 1.8 mm i.d.). This is held firmly in place with a 0.25-in. brass Swagelok nut and graphite ferrule combination at the base of the injection part. A second nut and ferrule combination connect the Pyrex sleeve to a specially constructed T. The T consists of a 1/8-in. stainless steel tube silver-soldered into the side of a normal 1/4- to 1/16-in. reducing union. The 1/8-tube is connected to the split exit valve, restrictor and vent. The front of the glass capillary column is inserted into the sleeve through the T and is sealed at the 1/16-in. fitting with a Swagelok nut and graphite ferrule. A restriction in the injection port sleeve (shown in the figure) promotes the mixing and volatilization of samples which are injected directly into the chromatograph, and is not necessary for thermal desorption injection. For direct injections, the best linearity is achieved if the top half is packed with silanized glass wool. Splitter construction is reviewed by Schomburg et al. [7].

The injection-port block is solid above the Pyrex glass sleeve except for a 1-mm hole passing up to the septum. The thermal-desorption inlet is located just above the top of the glass sleeve, which is butted up against the 1-mm inlet hole. The trasnfer line between the thermal-desorption apparatus and the injection port was of the same tubing used in the construction of the Tenax precolumn (1/8-in. stainless steel for the work presented here but was later converted to 1/16-in. deactivated nickel).

### *Gas Chromatography*

A Tracor 560 gas chromatograph was modified for use with glass capillary columns. This included the construction of an interface connecting the thermal desorption apparatus to the gas chromatographic injection port. A flame ionization detector (FID) was used.

Separations were done on J and W glass capillary columns (Supelco, Inc., Bellefonte, PA). Columns were 30 m long and were either Carboxwax 20M or SE-30, depending on the application.

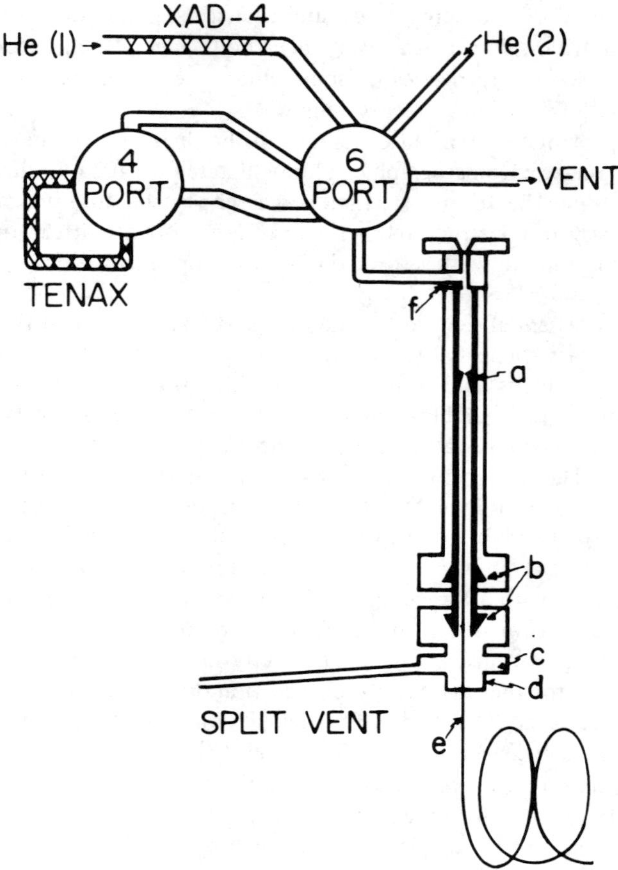


**Figure 1.** Schematic of the thermal desorption system and splitter interface to the glass capillary column. (a) Injection port sleeve, 6 mm o.d. x 2 mm i.d. Pyrex; (b) 1/4-in. brass nut and graphite ferrule combination; (c) 1/4- to 1/16-in. reducing union; (d) 1/16-in. nut and graphite ferrule; (e) WCOT column; (f) thermal desorption inlet.

Temperature programs were run at either 2 or 4°C/min. The initial temperature ranged from 45-75°C. The final temperature was 150-240°C, depending on the sample and desorption parameters.

## Procedure

### *Sampling*

Tapwater samples were taken by attaching a minicolumn to a faucet with a special adaptor, turning on the faucet and allowing the water to drip through the minicolumn at a rate of 6-12 mL/min. A collecting vessel was used to measure sample volume, and flowrates were checked periodically. Variations in flowrate over the 6- to 12-mL/min range had no discernible effect on the recovery of the hydrocarbons that were tested.

A standard determination consisted of injecting a standard acetone solution of the analyte 1.0 cm into the resin bed from the front of the minicolumn. The resin bed was then wetted by forcing 2–4 mL of distilled water through the minicolumn from front to back. The moist minicolumn was then analyzed by the normal thermal desorption procedure. This procedure was validated by comparison with standards injected directly into the gas chromatograph, and with water samples containing known, added impurities.

### *Desorption*

The XAD-4 sample tube is placed in the heating block and connected to the rest of the system with 0.25-in. Swagelok fittings and graphite ferrules. The organic sample components are desorbed thermally and carried in a stream of helium through the valves to the Tenax precolumn. The block heater temperature is typically 180-210°C, while the precolumn temperature is initially 45°C. The water vapor from the wet minicolumn passes through the warm Tenax precolumn in about 5 min while all but the most volatile and/or hydrophilic components are retained almost indefinitely by the Tenax.

When the XAD-4 desorption is deemed complete for the desired sample constituents, the Tenax precolumn is closed off (four-port valve), the temperature is raised to 275°C and the gas chromatographic carrier gas is diverted (six-port valve + four-port valve) to backflush the precolumn, injecting the organic components into the gas chromatograph. Closing off the Tenax precolumn makes it possible to inject moderately volatile compounds as a plug instead of allowing them to elute from the Tenax at a slower rate, which would be determined by the heat transfer characteristics of the system. The result of this injection process is that these compounds, which might otherwise spread out along the analytical column during the injection process, elute in nice sharp peaks.

## RESULTS AND DISCUSSION

### Effect of Desorption Conditions

To test the procedure on compounds of low volatility, a mixture of aromatic hydrocarbons in acetone was injected into the XAD-4 sample tube and carried through the desorption and chromatographic analysis procedure. Figure 2 shows excellent resolution, including good separation of anthracene and phenanthrene.

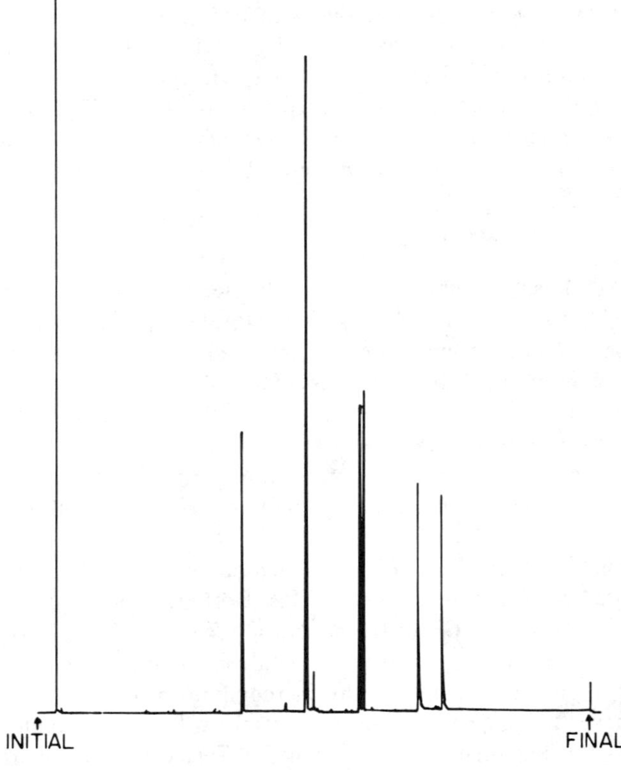


**Figure 2.** Thermal desorption chromatogram. The peaks are acetone (the diluent), biphenyl, fluorene, phenanthrene, anthracene, tritan and pyrene. The small peak next to fluorene is a degradation product of tritan. About 500 ng of each of the compounds above except flourene, of which 1000 ng were present, were desorbed through 1 cm of an XAD-4 minicolumn (wet). The desorption time was 15 min, temperature 200 ± 1°C and the flowrate was 42 mL/min of helium. The Tenax desorption temperature was 300°C. The injection split ratio was 45:1. Chromatography was done on a 30-m WCOT column (SE-30). Program from 50 to 240°C at 4°C/min.

A study was carried out to ascertain the effect of experimental parameters on the peak heights of these high-boiling compounds. The effect of flowrate for thermal desorption from the XAD-4 tube is illustrated in Table I. The final desorption temperature (210°C) was near the maximum safe temperature for XAD-4, and the desorption time was 20 min. The absolute peak heights varied with the sample volume, but compared to an internal standard (biphenyl) the first three compounds gave essentially constant peak heights over the range of flowrates in the table. Pyrene and tritan, which are less volatile than the others, showed progressively improved recovery as the flowrate was increased.

Table I also gives an isothermal desorption at 203°C. Except for the partial loss of tritan, presumably to thermal degradation, the conditions were sufficient for maximal recovery of the test compounds. The values are actually higher than those of the direct injection sample. This can be accounted for by splitter discrimination against high-molecular-weight compounds during direct injection, as a consequence of incomplete volatilization and mixing. Linear splitting is observed with thermal desorprtion injections, because all sample components must be volatilized before they can reach the splitter.

For many samples it is helpful to use even higher or lower flowrates. Compounds which are more volatile can often be desorbed efficiently at flowrates of a few mL/min. The use of high desorption flowrates appears to be limited by the more volatile constituents of the sample, which are retained less well by the Tenax precolumn and can be "blown through" at high flowrates. However, an isothermal desorption (200 ± 1°C) gave maximum recovery of fluorene through pyrene in just 15 min when desorption flow was 80 mL/min.

**Table I. Peak Height Ratios with Respect to Biphenyl for Thermal Desorption as a Function of Desorption Flowrate[a]**

| | Peak Height Ratio | | | | | |
|---|---|---|---|---|---|---|
| | Flowrate (mL/min) | | | | | |
| Compound | 13 | 29 | 42 | 58 | 42[b] | Direct Injection |
| Fluorene | 2.37 | 2.33 | 2.40 | 2.36 | 2.54 | 2.27 |
| Phenanthrene | 1.08 | 1.12 | 1.21 | 1.09 | 1.09 | 0.97 |
| Anthracene | 1.09 | 1.07 | 1.15 | 1.14 | 1.19 | 1.01 |
| Tritan | 0.18 | 0.42 | 0.66 | 0.80 | 1.26 | 1.32 |
| Pyrene | 0.28 | 0.56 | 0.68 | 0.76 | 1.17 | 0.85 |

[a]Average desorption temperature was 195°C (180°C initial, 210°C final); desorption time was 20 min; other parameters were as in Figure 2.
[b]Isothermal desorption (200 ± 3°C).

Table II shows the effect of varying the desorption time. A plot of peak height vs desorption time indicated incomplete recovery of biphenyl after 10 min, but a plateau region with apparently complete recovery extended from 15 to 30 min. Plots for fluorene, phenanthrene and anthracene showed similar behavior, indicating that a desorption of around 15 min is adequate. Pyrene and tritan require a combination of maximal temperature, fast flowrate and a somewhat extended period of time for efficient desorption.

## Application to Water Analysis

Figure 3 shows a chromatograph obtained for the analysis of a 3-L sample of Ames, IA tapwater. The major peaks were identified as follows: (86°) unknown, (101°) indene, (118°) alkyl indanes, (132°) naphthalene, (150°) 1-methylnaphthalene, (175°) acenapthene, and (180°) acenaphthylene. The chromatogram in Figure 3 has much sharper peaks and a better baseline than the chromatogram from an Ames, IA, tapwater sample reported earlier using a packed column [5].

The levels of three impurities in the present tapwater sample were estimated by comparison with standards desorbed from wet XAD-4 resin tubes: indene, 0.70 ppb; naphthalene, 0.28 ppb; and 1-methylnaphthalene, 0.35 ppb.

In testing the recovery of organic compounds from Ames, IA, tapwater, it was found that virtually all of the major sample components were recovered with uniform efficiency from samples ranging from 0.2 to 5.0 L in volume. However, some loss of efficiency for the recovery of specific compounds was occasionally observed when adsorption tubes (minicolumns) were reused.

**Table II. Peak Height Ratios with Respect to Biphenyl as a Function of Desorption Time[a]**

| Compound | Desorption Time (min) | | | | |
|---|---|---|---|---|---|
| | 10 | 15 | 20 | 25 | 30 |
| Fluorene | 2.73 | 2.29 | 2.32 | 2.26 | 2.28 |
| Phenanthrene | 1.19 | 1.03 | 1.09 | 1.10 | 1.13 |
| Anthracene | 1.09 | 1.08 | 1.12 | 1.14 | 1.12 |
| Tritan | 0.09 | 0.12 | 0.42 | 0.96 | 1.12 |
| Pyrene | 0.09 | 0.18 | 0.45 | 0.83 | 0.85 |

[a]Desorption flowrates were 26 ± 2 mL/min; other parameters were as in Table I.

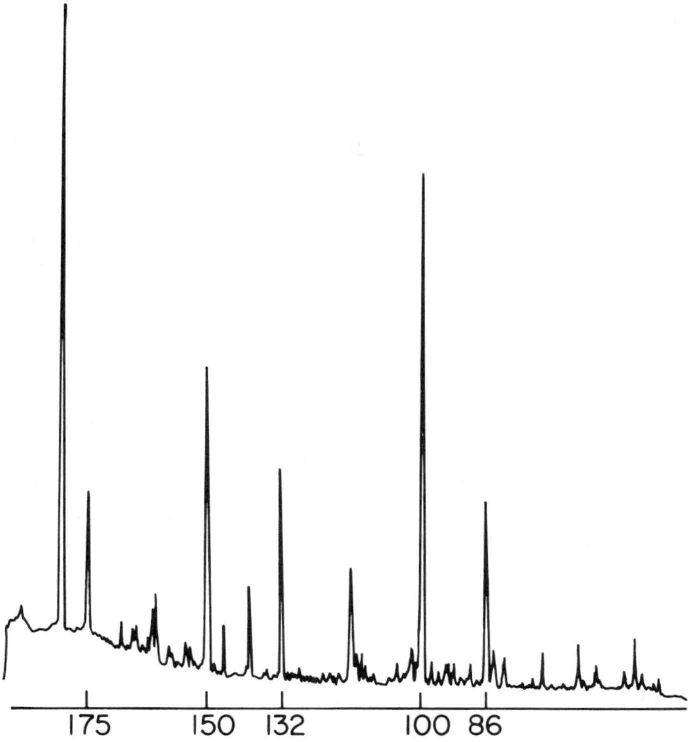


**Figure 3.** Chromatogram of a 3-L sample of Ames, IA, tapwater. The sampling flowrate was 12 mL/min, desorption temperature was 180–210°C, desorption time was 10 min, flowrate was 25 mL/min. The glass capillary column was 25 m x 0.25 mm i.d., coated with Carbowax 20M. The column temperature was 45°C for 2 min, then 2°C/min to 175°C, held for 10 min and recycled. The attenuation was 2 x 10, and the split ratio 25:1.

## Effect of Sample Tube Reuse

In analyses of tapwater it was observed that the resin in sample tubes often became progressively discolored with continued use. As the resin became more discolored, the peak heights of indene and alkyl indenes decreased with respect to the peak heights of the other sample components. In one case the recovery of indene dropped to nearly 50% when a particularly well-used and discolored sample tube was used for the analysis. It was also observed that indene standards that had been injected directly into used, discolored resin tubes could not be efficiently desorbed from them.

A series of analyses was then performed with sample tubes that had been used to analyze measured amounts of Ames, IA, tapwater. The results of these analyses were compared with each other and with those obtained using new tubes. The used resin tubes gave consistently low recovery values for indene and the alkyl indanes, and the magnitude of the loss in recovery efficiency was directly related to the amount of water that had passed through each minicolumn in its previous sampling cycles. This memory effect appeared to be related to a gradual discoloration (yellow to orange to brown) of the resin bed that occurred as the tapwater sample passed through the sample tube.

The colored material could not be eluted with organic solvents alone, but after acidification with 0.5 *M* HCl a few milliliters of diethylether was sufficient to elute almost all of the colored material from the minicolumn. We therefore concluded that the loss of indene and indanes was caused by their reacting with acidic humic material on the XAD-4 tubes. The most probable interfering mechanism would be the formation of charge-transfer complexes between the analytes and the phenolic moieties of the humic material [8].

Since each sample tube contains only a small amount of resin, we recommend that the XAD-4 resin in the sample tube be discarded after running a sample.

## ACKNOWLEDGMENT

This work was supported by the U.S. Department of Energy, under contract No. 7405-eng-82, Office of Basic Energy Sciences, Chemical Sciences Division. The authors express their appreciation to Rohm and Haas for providing the resin used in this work.

## REFERENCES

1. Bertsch, W., E. Anderson and G. Holzer. "Trace Analysis of Organic Volatiles in Water by Gas Chromatography-Mass Spectroscopy with Glass Capillary Columns," *J. Chromatog.* 112:701-718 (1975).
2. Grob, K., K. Grob, Jr. and G. Grob. "Organic Substances in Potable Water and in its Precursor, III. The Closed-Loop Stripping Procedure Compared with Rapid Liquid Extraction," *J. Chromatog.* 106:299-315 (1975).
3. Budde, W. L., and J. W. Eichelberger. "Report: Organics in the Environment," *Anal. Chem.* 51:567A-574A (1979).
4. Junk, G. A., J. J. Richard, M. D. Grieser, D. Witiak, J. L. Witiak, M. D. Arguello, R. Vick, H. J. Svec, J. S. Fritz and G. V. Calder. "Use of Macroreticular Resins in the Analysis of Water for Trace Organic Contaminants," *J. Chromatog.* 99:745-762 (1974).

5. Ryan, J. P., and J. S. Fritz. "Determination of Trace Organic Impurities in Water Using Thermal Desorption from XAD Resin," *J. Chromatog. Sci.* 16:488–492 (1978).
6. Fenimore, D. C., J. H. Whitford, C. M. Davis and A. Zlatkis. "Nickel Gas Chromatographic Columns: An Alternative to Glass for Biological Samples," *J. Chromatog.* 140:9-15 (1977).
7. Schomburg, G., H. Behlau, R. Dielmann, F. Weeke and H. Husman. "Sampling Techniques in Capillary Gas Chromatography," *J. Chromatog.* 142:87–102 (1977).
8. Aiken, G. R., E. M. Thurman, R. L. Malcolm and H. F. Walton. "Comparison of XAD Macroporous Resins for the Concentration of Fulvic Acid from Aqueous Solution," *Anal. Chem.* 51:1799-1803 (1979).

# CHAPTER 21

# COMPARISON OF SEVERAL METHODS FOR COLLECTION AND QUANTIFICATION OF POLYCYCLIC AROMATIC HYDROCARBONS FROM AQUEOUS MEDIA

**J. E. Caton, Z. K. Barnes, H. Kubota, W. H. Griest and M. P. Maskarinec**

Analytical Chemistry Division
Oak Ridge National Laboratory
Oak Ridge, Tennessee

The general analytical methodology for the analyses of aqueous samples for polycyclic aromatic hydrocarbons (PAH) may be divided into three areas: (1) isolation of the components of interest from the bulk of the aqueous media; (2) concentration of the isolated sample into the appropriate solvent; and (3) determination of the PAH of interest in the concentrated sample. Of these general areas more analytical problems arise from the isolation step. Thus, this present study has emphasized the comparison of several different isolation methods. Methods for isolation of PAH from water have been quite varied; but some of the generally employed techniques include solvent extraction [1-5] and collection on Tenax [6,7], XAD resins [8-13], polyurethane [14-17], carbon [18,19] or $C_{18}$ reverse-phase liquid chromatography (RPLC) media [20]. The comparative advantages and disadvantages of these techniques will not be reviewed here; however, it should be noted that the general attribute of a good PAH isolation procedure should be to deliver a maximal PAH recovery with a minimal amount of sample manipulation. With this criterion as one of the guidelines for evaluating the efficacy of the isolation step, methods were examined for the isolation of PAH from water involving liquid-liquid extraction (LLE), polystyrene sorbent resins and silica which had been chemically modified for reverse-phase liquid chromatography separations.

## EXPERIMENTAL

### Materials

$^{14}$C-labeled PAH used in this study were 1-$^{14}$C-naphthalene (5.10 mCi/mmol) and 7,10-$^{14}$C-benzo[a]pyrene (18.3 mCi/mmol) (Amersham/Searle, Arlington Heights, IL). Spiking solutions for the water samples were prepared by diluting the appropriate amount of labeled compound in acetone. PAH compounds used in the preparation of standards and standard mixtures were obtained from several different sources: Aldrich Chemical Co. (Milwaukee, WI), K and K Laboratories, Division of ICN Pharmaceuticals, Inc. (Plainview, NY), Pfaltz and Bauer (Flushing, NY), the Illinois Institute of Technology Research Institute Chemical Repository (Chicago, IL), and Dr. E. Sawicki of the U.S. Environmental Protection Agency (Research Triangle Park, NC). Stated purity of all compounds used was 97% or greater, and routine analysis by gas chromatography (GC) or liquid chromatography (LC) indicated that the purity indicated by the label was valid. All solvents were either glass-distilled l.c. grade or were redistilled in our laboratory from reagent-grade solvents.

### Collection and Preparation of Water Samples

The procedure for collection of streamwater samples has been described elsewhere [21]. Water samples containing known amounts of PAH were prepared by adding an acetone solution containing the appropriate PAH to an appropriate volume of distilled water which was being rapidly stirred either with a motor-driven glass stirring rod or a stirring bar.

### Preparation of Standard PAH Solution

Individual stock solutions (1 mg/mL) were prepared for each of the following PAH: naphthalene, acenaphthylene, acenaphthene, phenanthrene, anthracene, fluorene, fluoranthene, pyrene, 1,2-benzanthracene, chrysene, 1,2-benzopyrene, 1,2,5,6-dibenzanthracene and 1,12-benzoperylene. All were dissolved in acetonitrile except chrysene, 1,2,5,6-dibenzanthracene and 1,12-benzoperylene, which were dissolved in methylene chloride. A working solution was prepared by combining 1 mL of each of the above PAH solutions and adding enough acetone to bring the final volume to 100 mL. This acetone solution was divided into 1-mL aliquots, which were sealed in 1-dram vials and stored in the dark at 0°C. These aliquots were then used as a standard source of PAH throughout the study. Standard solutions of PAH were prepared by adding a 1-mL aliquot of the acetone solution to 1000 mL of distilled water to yield water samples containing 10 ng/mL of each of the 13 PAH.

## Tracer Addition and Recovery

An acetone solution containing approximately 1 $\mu$Ci of $^{14}C$ tracer was added to the appropriate volume of water with vigorous stirring. The volume of tracer solution never exceeded 1 mL/500 mL of water. The activity in the spiked water sample was then determined by counting 1-mL aliquots with 10 mL of Aquasol-2 (New England Nuclear, Boston, MA). After isolation of the PAH the residual water was again subjected to liquid scintillation counting with Aquasol-2. The isolated PAH samples which were in an organic solvent medium were counted with a toluene-based liquid scintillator containing 4 g of Omnifluor (New England Nuclear, Boston, MA) dissolved in 1 L of toluene. Recoveries were calculated as the fraction of activity initially present in the water that was found in the isolated PAH sample. Although such recovery determinations are in the absolute sense indicative only of the fate of the tracer involved, they permit the estimation of recovery at each step of the entire analytical procedure.

## Liquid-Liquid Extraction

### *Extraction Method 1*

A 1000-mL sample was placed in a 2-L pear-shaped separatory funnel equipped with a Teflon® stopcock. A motor-driven glass stirring rod was positioned as low as possible in the funnel. With the water being stirred very vigorously 50 mL of methylene chloride was added. The vigorous stirring was continued for 10 min, after which the methylene chloride was carefully drained through the Teflon stopcock. This extraction was repeated three times on a given water sample. The combined triple extract was concentrated under dry flowing nitrogen at reduced temperature and pressure.

Two other extraction procedures which have been more completely described elsewhere [21] were also employed:

*Extraction Method 2*

A 3.8-L water sample was double extracted by vigorously stirring with 100 mL of methylene chloride in a 4-L Pyrex Erlenmeyer flask for 24 hr.

*Extraction Method 3*

A 500-mL water sample at carefully adjusted pH and ionic strength was extracted five times with 100 mL of either methylene chloride or cyclohexane.

## Isolation on XAD-2 Resin

The nonpolar polystyrene adsorbent resin XAD-2 was used in two different approaches to isolate PAH from aqueous solution.

*Resin Method 1*

The first approach to using XAD-2 has been described previously [21]. In summary, this method involved 30 g of purified XAD-2 resin (acetone extracted; Polysciences, Inc., Warrington, PA) packed in a 30- x 1.7-cm i.d. glass column. Water was drained through the XAD-2 resin at approximately 30 mL/min. The collected PAH were eluted either with 100 mL of tetrahydrofuran followed by 125 mL of diethyl ether or with 50 mL of methanol, 100 mL of tetrahydrofuran, 150 mL of diethyl ether and 30 mL of methylene chloride.

*Resin Method 2*

A one-hole stopper was placed in the end of an Isolab Quick-Sep XAD-2 Resin Column (Isolab, Inc., Akron, OH) which had been prewashed with 10 mL of water. Teflon tubing connected this stopper and a second, two-holed stopper in the mouth of a 1-qt bottle. Air was allowed to enter the bottle via a Teflon tube inserted through the second stopper into the bottle. Inversion of the bottle allowed 900 mL of water to drain through the resin. The collected PAH were backwashed from the column with two 10-mL portions of methanol and three 10-mL portions of methylene chloride. Aliquots (1-mL) were removed from each of these solutions for liquid scintillation analysis before the remaining wash solution was concentrated almost to dryness under dry, flowing nitrogen and reduced pressure. The concentrated sample was diluted to a carefully measured volume with methylene chloride in preparation for further analysis.

**Isolation on Reverse-Phase (RP) Media**

Typical media for RPLC separations (silica containing chemically bonded octadecylsilane) were employed in three different forms to isolate PAH from water.

*RP Method 1*

A Waters Associates (Milford, MA) Sep-pak $C_{18}$ cartridge was prewashed with 50 mL methanol followed by 50 mL of water. The cartridge was connected to the mouth of a 1-L bottle that also had an air bleed tube. Inversion of the bottle allowed 600-900 mL of water to drain through the cartridge. Using a syringe with a Luer tip, the PAH were washed from the cartridge with two 10-mL portions of methanol and three 10-mL portions of methylene chloride. Recoveries were evaluated for each individual wash solution by liquid scintillation counting before all washes were combined and concentrated for further analysis.

*RP Method 2*

A Rheodyne (Berkeley, CA) $C_{18}$ cartridge was washed with 100 mL of methanol before 150-600 mL of water was pumped through it at 1.6 mL/min with a Milton Roy Model 396 minipump (Riviera Beach, FL). The cartridge was then connected to a 25-cm x 4.6-mm i.d. Zorbax ODS column (Alltech Associates, Arlington Heights, IL). The absorbed PAH were eluted onto this column with an acetonitrile:water solution (70:30 by volume) flowing at 1.6 mL/min. A Perkin Elmer Model 203 Fluorescence Spectrometer with an excitation wavelength of 294 nm and an emission wavelength of 400 nm was connected in series with a LDC ultraviolet (UV) analyzer (Model 1275) to monitor the eluate. Recovery was determined by counting a measured aliquot of the column effluent collected in the region where the compound containing the $^{14}C$ label was known to elute.

*RP Method 3*

A 5-cm x 2-mm i.d. stainless steel column was dry-packed with Whatman (Clifton, NJ) CO:Pell ODS and washed with 100 mL of methanol. Subsequently, 100-250 mL of water was pumped through this guard column at 1.6 mL/min. The guard column was then connected to the 25-cm x 4.6-nm i.d. Zorbax ODS column and the absorbed PAH were eluted as described above for RP Method 2.

### Liquid Chromatography Analyses

LC analyses of the PAH in the wash concentrate from the XAD-2 columns and $C_{18}$ Sep-paks were carried out on a 25-cm x 4.6-mm i.d. Zorbax BP-ODS column (Alltech Associates, Arlington Heights, IL). Samples (20 $\mu$L) were eluted with 70/30 acetonitrile/water at a flowrate of 1.6 mL/min. A Waters Fluorescence Detector, Model H20-E, with an excitation wavelength of 365 nm and an emission wavelength of 425 nm, was used in series with a Varian Varichrom detector (at 250 nm) to monitor the eluate.

### Liquid Scintillation Analysis of Solids

Any residual $^{14}C$ tracer remaining on the XAD-2 or RP media was determined by oxidation of a measured portion of these solids followed by liquid scintillation analysis of the liberated $CO_2$. This oxidation was carried out on a Packard Model 306 Sample Oxidizer (Packard Instrument Co., Inc., Downers Grove, IL) where 50 mg of the solid sample was combined with 200 mg of cellulose and 200 $\mu$L of Combustaid™ (Packard Instrument Co., Inc., Downers Grove, IL). This mixture was oxidized for 1.5 min, after which the $CO_2$ was retained in 6 mL of Carbasorb (Packard Instrument Co., Inc., Downers Grove, IL)and 12 mL of Permafluor V (Packard Instrument Co., Inc., Downers Grove, IL). Subsequently, the $^{14}C$ activity in this solution was determined on a Packard Model C 2425 (Packard Instrument Co., Inc., Downers Grove, IL) liquid scintillation spectrometer. (Counting time was 20 min; quenching correction was made from automatic external standard ratio.)

## RESULTS AND DISCUSSION

As indicated earlier, there are three general steps which one must successfully accomplish in most schemes designed to analyze an aqueous sample for PAH: (1) the PAH must be isolated from the bulk of the aqueous media; (2) the isolated PAH sample must be concentrated into some appropriate solvent; and (3) the concentrated sample must be analyzed by some appropriate technique [usually GC, gas chromatography/mass spectrometry (GC/MS), or high-performance liquid chromatography (HPLC)]. The emphasis in the present work has been to study the isolation step by several different approaches. The recovery of $^{14}C$-tracer compounds is summarized in Table I. All the isolation methods have merits and disadvantages; however, extraction method 1 presents the best combination of simplicity, consistent high recovery and low analysis time. Table II lists recoveries of several different

**Table I. Recovery of $^{14}C$ Tracers from Water Samples by Several Different Methods[a]**

| Method | Percentage Recovery | |
|---|---|---|
| | $^{14}C$-Benzo[a]pyrene | $^{14}C$-Naphthalene |
| Extraction Method 1 | 96 | 73 |
| Extraction Method 2 | 91 | 87 |
| Extraction Method 3 | 100 | 100 |
| Resin Method 1 | 83 | ND[b] |
| Resin Method 2 | 89 | 50 |
| RP Method 1 | 81 | ND |
| RP Method 2 | 43 | ND |
| RP Method 3 | 40 | ND |

[a] See Experimental section for description of each section.
[b] ND = no recovery determination was made.

**Table II. Percentage Recovery of Several PAH from Water Samples by Different LLE Methods**

| PAH | Percentage Recovery by Extraction Method | |
|---|---|---|
| | Method 1 | Method 2 |
| Naphthalene | 73 | 87 |
| Acenaphthylene | 67 | ND[a] |
| Acenaphthene | 70 | ND |
| Phenanthrene | 73 | 14 |
| Anthracene | 78 | ND |
| Fluorene | 81 | ND |
| Fluoranthene | 85 | 100 |
| Pyrene | 82 | 90 |
| 1,2-Benzanthracene | 84 | 92 |
| Chrysene | 89 | ND |
| Benzo[a]pyrene | 96 | 9 |
| 1,2,5,6-dibenzanthracene | 93 | ND |
| Benzo[g,h,i]perylene | 91 | ND |

[a] ND = recovery not determined.

PAH from water by extraction methods 1 and 2. Of the three LLE methods studied, extraction method 3 yields the highest and most consistent recovery. However, compared to the other two extraction methods, extraction method 3 is much more complex and more labor-intensive. Extraction method 2 is easily the least complex of the three extraction methods; however, it is also the least consistent with recoveries in different trials ranging from values lower than 10% for a given PAH to greater than 90% for a different PAH compound.

In general, the resin concentration methods are less labor-intensive than solvent extraction. Resin sorption tubes are easily packed and transported to and from field sampling sites. In addition, because of the low labor-intensity, many samples can be isolated at the same time. However, isolation on XAD-2 resins has two disadvantages [21]: (1) careful control and adjustment of the pH and ionic strength of the water sample are required to get highly reproducible results and, (2) if allowed to go dry, the resin may degrade and contaminate the sample [7]. In addition, the isolated sample may not readily be released from the resin when it is washed with an organic solvent. In the case of resin method 2 the disposable column was backwashed sequentially with two 10-mL portions of methanol followed by three 10 mL portions of methylene chloride. Each 10-mL wash solution was analyzed for $^{14}$C-benzo[a]pyrene activity, and $^{14}$C activity was also determined in a 50-mg portion of the washed resin after oxidation. As indicated in Table III, some activity was retained on the resin even after what was considered thorough backwashing. Although this tracer activity directly indicates the fate of only the $^{14}$C-benzo[a]pyrene added to the water sample, it is quite

**Table III. Recovery of $^{14}$C-Benzo[a]pyrene from XAD-2 by Successive Backwashing**

| Wash No. | Solvent | Activity[a] (%) |
|---|---|---|
| 1 | 10 mL methanol | 17 |
| 2 | 10 mL methanol | 5 |
| 3 | 10 mL methylene chloride | 39 |
| 4 | 10 mL methylene chloride | 9 |
| 5 | 10 mL methylene chloride | 5 |
| Total Eluted | | 75 |
| Residual[b] | | 11 |
| Total Recovered | | 86 |

[a] Activity is expressed as the percentage of $^{14}$C activity found in a specific wash relative to the total amount of $^{14}$C activity in the water drained over the XAD-2.
[b] $^{14}$C activity found on the XAD-2 after the five successive 10-mL washes.

**Table IV. Percentage Recovery of Several PAH from Water Samples Using XAD-2 Resin**

| PAH | Percentage Recovery | |
|---|---|---|
| | Resin Method 1 | Resin Method 2 |
| Naphthalene | 49 | 61 |
| Acenaphthylene | ND[a] | 58 |
| Acenaphthene | ND | 59 |
| Phenanthrene | 12 | 64 |
| Anthracene | ND | 69 |
| Fluorene | ND | 73 |
| Fluoranthene | 75 | 74 |
| Pyrene | 80 | 77 |
| 1,2-Benzanthracene | 77 | 81 |
| Chrysene | ND | 86 |
| Benzo[a]pyrene | 21 | 91 |
| 1,2,5,6-dibenzanthracene | ND | 92 |
| Benzo[g,h,i]perylene | ND | 90 |

[a] ND = no determination of recovery was made.

likely to be indicative of the fate of similar PAH. The recoveries of several different PAH listed in Table IV as estimated from LC data would seem to indicate that the other PAH did indeed have a distribution similar to that shown for the $^{14}$C-benzo[a]pyrene tracer.

The isolation of PAH from water using RP media allows the possibility of avoiding the concentration step by directly eluting the isolated sample from the RP media onto an analytical RP column as illustrated in Figure 1 for 100 mL of a standard water solution (10 ng/mL of 13 PAH) concentrated on a RP guard column (RP method 3). In addition to avoiding the concentration step, isolation of PAH from aqueous samples on RP media may have a selectivity advantage relative to extraction and resin sorption. The extraction methods are likely to remove some significant proportion of all but the very highly polar (ionic) organic compounds in the water sample. In like manner the XAD-2 resin shows significantly reduced affinity only for those organic compounds that tend to have some ionic character in a given water sample [7,21]. However, RP media are more selective for neutral, nonpolar organic constituents. In fact, the guard column packed with RP media (RP method 3) retained only 8% of the $^{14}$C-hexadecane in a water sample, indicating selectivity for aromatic hydrocarbons relative to aliphatic hydrocarbons. Thus isolation of PAH from a real water sample on the RP media may yield a PAH sample requiring little or no cleanup prior

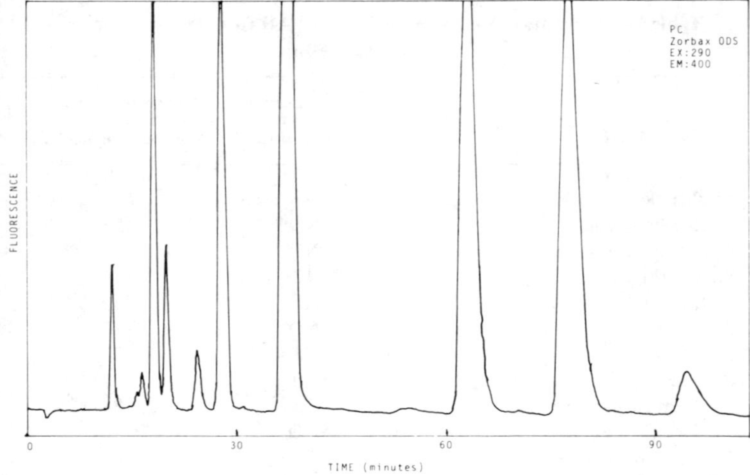


**Figure 1.** Chromatogram of PAH isolated from 100 mL of a standard water sample by RP Method 3. Isocratic elution from Zorbax ODS column with acetonitrile:water (70:30 v:v) at ambient temperature.

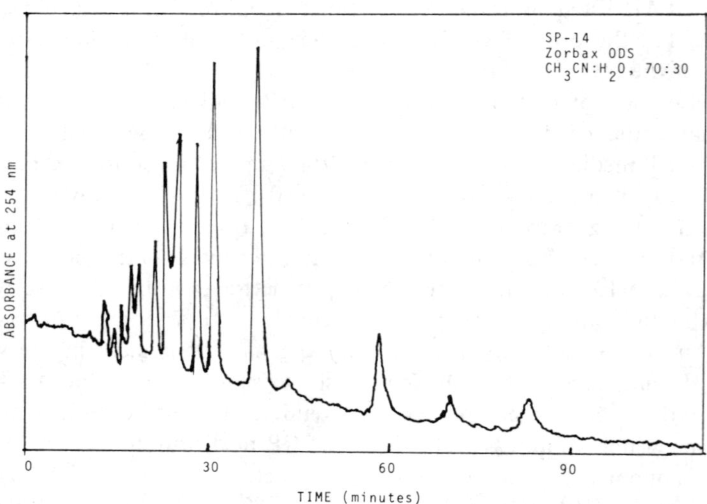


**Figure 2.** Liquid chromatogram of PAH isolated from an industrial wastewater by eluting sample directly onto a $C_{18}$ Sep-Pak (RP method 1). Isocratic elution with acetonitrile:water (70:30 v:v) at ambient temperature.

to analysis. This reduced necessity for sample cleanup is demonstrated in Figure 2 where the PAH in an industrial wastewater were isolated by RP method 1, concentrated into acetonitrile and analyzed by LC with no additional sample treatment. This chromatogram in Figure 2 compares favorably to a similar chromatogram prepared from standards shown in Figure 3. However, a sample isolated from the same industrial wastewater by extraction method 1 had to be subjected to a Florisil/alumina "cleanup" procedure [21] before enough non-PAH organic compounds were removed so that multicomponent analytical data could be obtained from the chromatogram (Figure 4). Although employing fluorescence detection or limiting analysis to the larger PAH (4 and 5 aromatic rings) might permit some samples to be analyzed without additional sample "cleanup," the major disadvantage of

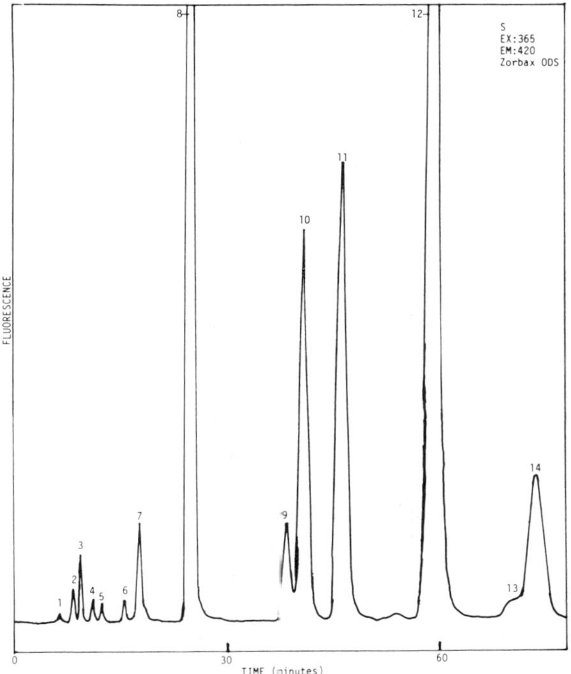


**Figure 3.** Chromatogram of standard PAH mixture. Isocratic elution from Zorbax ODS column with acetonitrile:water (70:30 v:v) at ambient temperature. (1) Naphthalene; (2) acenaphthylene; (3) acenaphthene; (4) phenanthrene; (5) anthracene + fluorene; (6) fluoranthene; (7) pyrene; (8) benzanthracene + chrysene; (9) benzo[b]fluoranthene; (10) benzo[k]fluoranthene; (11) benzo[a]pyrene; (12) dibenzanthracene; (13) indeno(1,2,3-c,d)pyrene; (14) benzo[g,h,i]perylene.

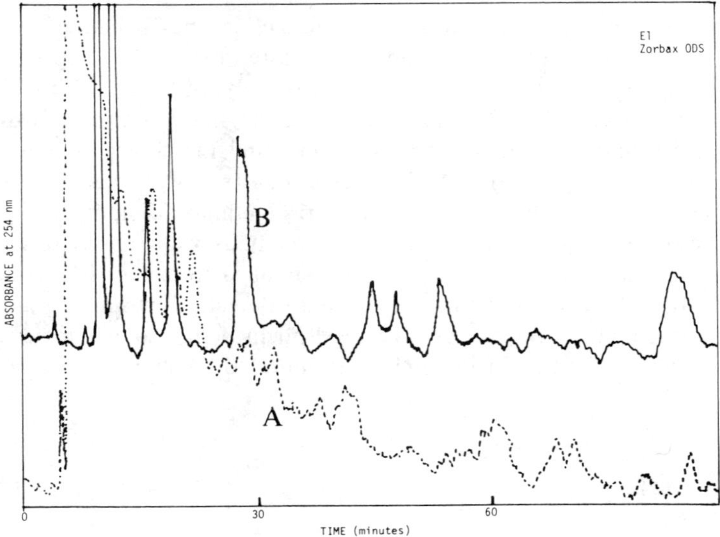


**Figure 4.** Liquid chromatogram of PAH isolated from an industrial wastewater by extraction method 1. Before (B) and after (A) additional sample cleanup. Isocratic elution from Zorbax ODS column with acetonitrile:water (70:30 v:v) at ambient temperature.

isolation on RP media in a guard column (RP methods 2 and 3) is that there is really no convenient way for additional sample "cleanup" prior to analyses by LC. Another disadvantage of the RP isolation methods is the lower recoveries listed in Table V. As shown in Table V the percentage of material recovered from the $C_{18}$ Sep-Pak (RP method 1) tends to increase with increasing molecular weight; however, the trend is just the opposite for the guard column methods (RP methods 2 and 3). The trend toward increasing recovery with higher molecular weight is typical for PAH samples concentrated at reduced pressure. In addition, in the case of the $C_{18}$ Sep-Pak, the capacity to remove PAH from water is likely to increase with the increasing molecular weight of the PAH. In the case of the guard columns (RP methods 2 and 3) the tendency for recovery to decrease with increasing molecular weight can be attributed to two factors: because there is no concentration step, there are no volatility losses; and there appears to be an increased tendency for the higher-molecular-weight PAH to be lost in the specific pumping system employed for this isolation step.

In contrast to the XAD-2 resin, sample oxidation and subsequent liquid scintillation analysis of the RP media after elution of the isolated sample indi-

**Table V. Percentage Recovery of Several PAH from Water Samples Using RP Media**

| PAH | Method RP Method 1 | RP Method 2 | RP Method 3 |
|---|---|---|---|
| Naphthalene | 37 | 34 | 43 |
| Acenaphthylene | 35 | 28 | 43 |
| Acenaphthene | 42 | 29 | 39 |
| Phenanthrene | 43 | 22 | 37 |
| Anthracene | 45 | 25 | 29 |
| Fluorene | 49 | 23 | 34 |
| Fluoranthene | 58 | 21 | 31 |
| Pyrene | 55 | 21 | 29 |
| 1,2-Benzanthracene | 58 | 18 | 19 |
| Chrysene | 65 | 19 | 21 |
| Benzo[a] pyrene | 63 | 15 | 24 |
| 1,2,5,6-Dibenzanthracene | 69 | 17 | 25 |
| Benzo[g,h,i] perylene | 67 | 17 | 23 |

**Table VI. Recovery of $^{14}C$-Benzo[a]pyrene from $C_{18}$ Sep-Pak by Successive Backwashing**

| Wash No. | Solvent | Activity[a] (%) |
|---|---|---|
| 1 | 10 mL methanol | 28 |
| 2 | 10 mL methanol | 1 |
| 3 | 10 mL methylene chloride | 22 |
| 4 | 10 mL methylene chloride | 3 |
| 5 | 10 mL methylene chloride | 1 |
| Total in Washes | | 55 |
| Residual[b] | | <1 |

[a] Activity is expressed as the percentage of $^{14}C$ activity found in a specific wash relative to the total amount of $^{14}C$ activity in the water drained over the $C_{18}$ Sep-Pak.

[b] $^{14}C$ activity found on the reversed phase media after the five successive 10-mL washes.

cated that less than 1% of the $^{14}C$-activity (naphthalenes or benzo[a]pyrene) was not removed for any of the RP isolation methods. However, as indicated in Table VI, the solvent sequence employed for removal of the PAH from $C_{18}$-Sep Pak is required to elute completely the isolated sample.

Some advantages and disadvantages of the eight isolation methods studied are summarized in Table VII. The "best" method for a given situation must be dictated by available equipment, skills and manpower. However, for simplicity and consistent high recovery some preference might be given to extraction method 1.

**Table VII. Comparison of Eight Different Methods for Isolating PAH from Water**

| Method[a] | Advantages | Disadvantages |
|---|---|---|
| Extraction Method 1 | Simple; good reovery | Labor-intensive |
| Extraction Method 2 | Simple | Variable recovery |
| Extraction Method 3 | High recovery | Very labor-intensive |
| Resin Method 1 | Not labor-intensive | Variable recovery; need pH and ionic strength control |
| Resin Method 2 | Not labor-intensive; widely used | Variable recovery; need pH and ionic strength control |
| Reverse-Phase Method 1 | Possibly isolates cleaner PAH sample; not labor intensive | Variable lower recovery |
| Reverse-Phase Methods 2 & 3 | Reusable; no concentration step | Recovery too low[b]; difficult to "clean up" isolated sample before analysis |

[a] Methods explained in Experimental section.
[b] Recovery may be peculiar to system employed rather than actual techniques; see text.

## ACKNOWLEDGMENT

G. M. Henderson and R. R. Reagan provided valuable technical assistance in carrying out the liquid-liquid extractions (RRR) and performing the liquid chromatographic analyses (GMH).

Research was sponsored by the Office of Health and Environmental Research, U.S. Department of Energy, under Contract W-7405-eng-26 with the Union Carbide Corporation.

## REFERENCES

1. Acheson, M. A., R. M. Harrison, R. Perry and R. A. Wellinsg. "Factors Affecting the Extraction and Analysis of Polynuclear Aromatic Hydrocarbons in Water," *Water Res.* 10:207 (1976).
2. Robbins, W. K. "Solvent Extraction of Polynuclear Aromatic Hydrocarbons," in *Polynuclear Aromatic Hydrocarbons: Chemistry and Biological Effects*, A. Bjorseth and A. J. Dennis, Eds. (Columbus, OH: Battelle Press, in press).
3. Wilkinson, J. E., P. E. Strup and P. W. Jones. "Quantitative Analysis of Selected PAH in Aqueous Effluent by High-Performance Liquid Chromatography," in *Polynuclear Aromatic Hydrocarbons*, P. W. Jones and P. Leber, Eds. (Ann Arbor, MI: Ann Arbor Science Publishers, Inc., 1979), pp. 217-229.
4. Ogan, K., E. Katz and W. Slavin. "Determination of Polycyclic Aromatic Hydrocarbons in Aqueous Samples by Reversed-Phase Liquid Chromatography," *Anal. Chem.* 51:1315 (1979).
5. Jungclaus, G. A., L. M. Games and R. A. Hites. "Identification of Trace Organic Compounds in Tire Manufacturing Plant Waste Waters," *Anal. Chem.* 48:1894 (1976).
6. Leoni, V., G. Puccetti and A. Grella. "Preliminary Results on the Use of Tenax for the Extraction of Pesticides and Polynuclear Aromatic Hydrocarbons from Surface and Drinking Waters for Analytical Purposes," *J. Chromatog.* 106:119 (1975).
7. Leoan, V., G. Pucetti, R. J. Columbo and A. M. D'Ovidio. "The Use of Tenax for the Extraction of Pesticides and PCBs from Water," *J. Chromatog.* 125:299 (1976).
8. Strup, P. E., J. E. Wilkinson and P. W. Jones. "Trace Analysis of Polynuclear Aromatic Hydrocarbons in Aqueous Systems Using XAD-2 Resin and Capillary Column Gas Chromatography-Mass Spectrometry Analysis," in *Carcinogenesis, Vol. 3: Polynuclear Aromatic Hydrocarbons*, P. W. Jones and R I. Freudenthal, Eds. (New York: Raven Press, 1978), pp. 131-138.
9. Burnham, A. K., G. V. Calder, J. S. Fritz, G. A. Junk, H. J. Svec and R. Willig. "Identification and Estimation of Neutral Organic Contaminants in Potable Water," *Anal. Chem.* 44:139 (1972).
10. Tateda, A., and J. S. Fritz. "Mini-column Procedure for Concentrating Organic Contaminants from Water," *J. Chromatog.* 152:329 (1978).
11. Van Rossum, P., and R. G. Webb. "Isolation of Organic Water Pollutants by XAD Resins and Carbon." *J. Chromatog.* 150:381 (1978).
12. Burnham, A. K., G. V. Calder, J. S. Fritz, G. A. Junk, H. J. Svec and R. Vick. "Trace Organics in Water: Their Isolation and Identification," *Water Technol./Qual.* (1973), p. 722.

13. Alben, K. "Coal Tar Coatings of Storage Tanks. A Source of Contamination of the Potable Water Supply," *Environ. Sci. Technol.* 14:468 (1980).
14. Basu, D. K., and J. Saxena. "Polynuclear Aromatic Hydrocarbons in Selected U.S. Drinking Waters and Their Raw Water Sources," *Environ. Sci. Technol.* 12:795 (1978).
15. Basu, D. K., and J. Saxena. "Monitoring of Polynuclear Aromatic Hydrocarbons in Water. II. Extraction and Recovery of Six Representative Compounds with Polyurethane Foams," *Environ. Sci. Technol.* 12:792 (1978).
16. Navratil, J. D., R. E. Sievers and H. F. Walton. "Open-Pore Polyurethane Columns for Collection and Preconcentrations of Polynuclear Aromatic Hydrocarbons from Water," *Anal. Chem.* 39:2260 (1977).
17. Saxena, J., J. Kozuchowski and D. K. Basu. "Monitoring of Polynuclear Hydrocarbons in Water. I. Extraction and Recovery of Benzo(a)pyrene with Porous Polyurethane Foam," *Environ. Sci. Technol.* 11:682 (1972).
18. Braus, H., F. M. Middleton and G. Walton. "Organic Chemical Compounds in Raw and Filtered Surface Waters," *Anal. Chem.* 23:1160 (1951).
19. Grob, K., K. Grob, Jr. and G. Grob. "Organic Substances in Potable Water and Its Precursor. III. The Closed-Loop Stripping Procedure Compared with Rapid Liquid Extraction," *J. Chromatog.* 106:299 (1975).
20. May, W. E., S. N. Chesler, S. P. Cram, B. H. Gump, H. S. Hertz, D. P. Enagonion and S. M. Dyszel. "Chromatographic Analysis of Hydrocarbons in Marine Sediments and Seawater," *J. Chromatog. Sci.* 13:535 (1975).
21. Griest, W. H., M. P. Maskarinec, S. E. Herbes and G. R. Southworth. "Multicomponent Methods for the Identification and Quantification of Polycyclic Aromatic Hydrocarbons in the Aqueous Environment," ASTM Committee Symposium on Analytical Methods Related to Waters Associated with Production of Fuels from Alternate Sources (in press).

# CHAPTER 22

# ANOMALOUS BREAKTHROUGH OF BENZO[A]PYRENE DURING CONCENTRATION WITH AMBERLITE® XAD-4 RESIN FROM AQUEOUS SOLUTIONS

**Peter F. Landrum* and John P. Giesy****

Savannah River Ecology Laboratory
Aiken, South Carolina

Because of the low environmental concentrations of polycyclic aromatic hydrocarbons (PAH) in surface waters, preconcentration is required before these compounds can be quantified. Liquid-liquid extractions (LLE) are difficult and generally not practical due to handling of large volumes and formation of emulsions [1-3] ; thus, preconcentrating media such as polyurethane foam have been employed [4-6] . Leoni et al. [7] describe the use of many matrices, such as activated carbon, polyethylene, Tenax and silicone adsorbents, for the concentration of PAH from water. Macroreticular polymeric resins such as XAD have been used by many investigators to concentrate trace organics from water [1,2,8-23] .

*Present address: Great Lakes Environmental Research Laboratory, National Oceanic and Atmospheric Administration, Ann Arbor, MI.

**Present address: Pesticide Research Center, Michigan State University, East Lansing, MI.

XAD-4 is a macroreticular polystyrene copolymer which is particularly suited to the concentration of hydrophobic PAH because it has an inert, hydrophobic surface with high porosity and surface area [16]. Also, like most other adsorbants XAD-4 can easily be regenerated. Thus, recoveries of adsorbed organics should be high and the resins can be reused repeatedly [14].

Grosser et al. [13] reported no breakthrough of nonpolar PAH on XAD-2 resins. Low recoveries from XAD resins for some PAH have been observed. This has been attributed to particulates [6,24], binding to glass walls of experimental vessels [2,22], reaction with residual chlorine [2] and incomplete extraction from the resin [18]. Also, large amounts of methanol reduced trapping efficiencies [23]. This study examined the effect of environmental and experimental conditions such as organic concentrations (humic and fulvic acids), iron concentrations, pH and ionic strength on trapping efficiencies of naphthalene, anthracene and benzo[a]pyrene (B[a]P) from water and recoveries from XAD-4 macroreticular resins.

## MATERIALS AND METHODS (EXPERIMENTAL)

The two natural waters investigated were from an artesian well and soft, black water stream. The wellwater contained less than 1 mg/l dissolved organic carbon (DOC), but the iron content was 0.2 mg/L. The soft, black water stream, Upper Three Runs Creek, had a DOC content of 7–10 mg/L [25]. All waters were centrifuged or filtered to remove particulates 0.45 $\mu$m or greater prior to the addition of $^{14}$C-PAH.

$^{14}$C-B[a]P (21.7 mCi/mmol, Amersham Searle), $^{14}$C-anthracene (3.3 mCi/mmol, California Bionuclear) and $^{14}$C-napthalene (5.0 mCi/mmol, California Bionuclear) were added in acetone to 1L of water and allowed to equilibrate for 1 hr. The water was dosed at or below the solubility of each component: B[a]P at 0.71 ± 0.16 $\mu$g/L; anthracene at 6.4 ± 1.7 $\mu$g/L and napthalene at 2.89 ± 0.13 $\mu$g/L. Prior to percolating the water through the resin columns, two 1-mL samples were taken for liquid scintillation counting.

The resin columns were 1.5-cm diam x 25-cm long equipped with a 250-mL reservoir and a stopcock. A plug of glass wool was placed in the column and 25 ± 1 mL of wet precleaned Amberlite XAD-4 (Rohm and Haas, Philadelphia, PA) resin was added to the column. The resin was precleaned by the method of Garnas [26] and stored under nanograde methanol. Prior to use the columns were drained and rinsed with 10 bed volumes of deionized water.

The water was percolated through the resin at a flowrate of 0.8–1.2 L/hr and the columns were drained. The water was collected and mixed, and two 1-mL samples were taken for liquid scintillation counting.

The columns were eluted sequentially with two bed volumes each of anhydrous diethyl ether, nanograde acetone and nanograde methanol. These were combined and mixed, two 1-mL samples were taken for liquid scintillation counting, and the volume of the combined organic eluates was measured.

Aldrich humic acid was purified by acid precipitation (2 times) and dialysis to remove salt [27]. Sodium chloride, calcium chloride, glacial acetic acid and hydrochloric acid were purchased from Mallinckrodt.

Thin layer chromatography (TLC) for purity of $^{14}C$ compounds was performed on (E. Merck) silica gel plates without fluorescent indicator using two solvent systems: hexane:benzene 9:1 and pentane:diethyl ether 9:1. Quantification was performed by scraping and liquid scintillation counting.

Research Products International 3a70B scintillation cocktail was used for all samples. Sample activities were determined by a Beckman model 8100 scintillation counter using preset windows for $^{14}C$ and 10-min counting times (maximum counting error $2\sigma \leqslant 10\%$). Samples were corrected for background, quenching and counting efficiency. The quenching and counting efficiency were corrected using a quench curve and samples channels ratio.

Iron concentrations were measured by an Instrumentation Laboratories model 351 atomic absorption spectrophotometer. DOC was determined using a Beckman 915 carbon analyzer.

Trapping efficiency was calculated by:

$$\text{Trapping efficiency} = 1 - \left(\frac{\text{DPOST}}{\text{DPRE}}\right) \quad 100\% \tag{1}$$

where DPOST = $^{14}C$ activity remaining in the water after percolation through the XAD-4

DPRE = $^{14}C$ activity in the water before percolation through the XAD-4

Recovery was calculated as the amount of $^{14}C$ material eluted from the column divided by the amount of $^{14}C$ present in the water prior to percolation through the resin column. Accountability was determined as the total amount of $^{14}C$ material found in the eluate and in the water after percolation through the resin column divided by the total $^{14}C$-activity in the water prior to the resin column.

## RESULTS AND DISCUSSION

The trapping efficiency of XAD-4 resin for $^{14}C$-B[a]P from distilled water ranged from 91 to 53% (73 ± 20%, $\bar{x} \pm 1$ SD). The range for deionized water was similar (55-94%, 75 ± 11%, $\bar{x} \pm 1$ SD). The same range of recoveries from

XAD-2 for PAH was observed by Benoit et al. [8,9]. The nearly quantitative results of several of the experiments indicate that the quantity of resin used has the capacity to adsorb the quantity of material used in the experiments. The large variability found with purified waters may be attributable to: (1) the presence of a soap film on the glassware, despite thorough rinsing (the soap could form micells or simply absorb B[a]P and carry it through the column); (2) the addition of dust particles in the glassware which accumulated during the time between washing and use; or (3) the failure to remove completely the methanol from the column.

The trapping efficiency of XAD-4 was less for B[a]P for the two natural waters than that from distilled and deionized waters (Table I). The lower trapping efficiency for B[a]P was postulated to result from the interaction of B[a]P with dissolved organics in the Upper Three Runs Creek water or with iron particles which formed after centrifugation or filtration of the wellwater. The interaction of B[a]P with humic or fulvic acids would result in a complex which carried a net negative charge at pH greater than 4 due to the carboxyl groups on the humic acid. Even though XAD resins will adsorb weakly acidic compounds [20], these resins have a low affinity for ionized organics [13].

**Table I. Trapping Efficiency and Recovery of Benzo[a]pyrene, Anthracene and Naphthalene from Two Natural Waters ($\bar{x} \pm 1$ SD)**

| Compound | Water | n | Trapping Efficiency (%) | Recovery (%) | Accountability (%) |
|---|---|---|---|---|---|
| Benzo[a]pyrene | Distilled | 6 | 73 ± 20 | 57 ± 14 | 85 ± 5 |
| | Deionized[a] | 11 | 75 ± 11[a] | 62 ± 11 | 88 ± 10 |
| | Brinkley Well | 6 | 68 ± 3 | 56 ± 10 | 88 ± 10 |
| | Upper Three Runs | 6 | 38 ± 5 | 29 ± 2 | 91 ± 6 |
| Anthracene | Deionized | 3 | 91 ± 4 | 83 ± 6 | 92 ± 7 |
| | Brinkley Well | 4 | 88 ± 6 | 93 ± 15 | 97 ± 11 |
| | Upper Three Runs | 3 | 89 ± 6 | 84 ± 3 | 94 ± 4 |
| Naphthalene | Deionized | 3 | 100[b] | 84 ± 3 | 84 ± 3 |
| | Brinkley Well | 3 | 100[b] | 81 ± 6 | 81 ± 6 |
| | Upper Three Runs | 3 | 100[b] | 82 ± 2 | 82 ± 2 |

[a] One data point eliminated as it exceeded 2 SD.
[b] No $^{14}C$ activity was detected in the water after the resin column.

The trapping efficiencies of XAD-4 for anthracene and naphthalene were greater than for B[a]P (Table I). The recoveries of anthracene from all three waters by XAD-4 were greater than those observed by van Rossum and Webb [2]. The recoveries of naphthalene observed in our study were the same as those observed for XAD-4 by other investigators [14,23]. Also, there was no effect of water quality on trapping efficiency for either anthracene or naphthalene. Snoeyink et al. [28] observed similar trapping efficiencies for anthracene in the presence of humic acids by activated charcoal. However, the trapping efficiencies for other organics, such as phenols, were affected.

Lowering the pH of the water increased trapping of B[a]P in the presence of humics (Table II). In Upper Three Runs water this may be as a B[a]P-humic complex. The lower pH causes protonation of the phenolic and carboxylic groups. This should increase the trapping efficiency of the humic and fulvic compounds by reducing their polarity [29]. Thus, any PAH associated with these compounds would be trapped as well. The trapping of the B[a]P as a humic acid complex is further substantiated by the lower elution efficiency as reflected in a lower accountability of $^{14}C$ for pH 2 Upper Three Runs water, despite the greatly improved trapping efficiency, because the solvent system used for these studies does not elute the humic acids.

The interactions of B[a]P with iron microparticulates in the wellwater would permit the B[a]P to be carried through the column. The improved trapping efficiency of B[a]P with lowered pH (Table II) is not well understood. However, it must result from formation of a neutral complex, dissociation of B[a]P from the iron microparticulates; or prevention of the formation of iron microparticulates in soluion.

**Table II. Effect of pH on Trapping Efficiency and Recovery of Benzo[a]pyrene from Two Natural Waters ($\bar{x} \pm 1$ SD)**

| Water | n | Trapping Efficiency (%) | Recovery (%) | Accountability (%) | pH |
|---|---|---|---|---|---|
| Brinkley Well | 6 | 68 ± 3 | 56 ± 10 | 88 ± 10 | 6 |
| | 3 | 81 ± 3 | 81 ± 12 | 99 ± 12 | 4 |
| | 3 | 76 ± 10 | 71 ± 10 | 96 ± 10 | 2 |
| Upper Three Runs | 3 | 36 ± 1 | 33 ± 9 | 96 ± 9 | 10 |
| | 6 | 38 ± 5 | 29 ± 2 | 91 ± 6 | 6 |
| | 3 | 100[a] | 62 ± 2 | 62 ± 2 | 2 |

[a] No $^{14}C$ activity was detected in the water after percolation through XAD-4.

Attempts to remove the iron from wellwater by aeration for 48 hours followed by centrifugation and filtration at 0.45 $\mu$m resulted in a decreased trapping efficiency with each step. The trapping efficiency decreased from 68 ± 3% for the wellwater to 46% (range 44–48, n = 2) for water which was preaerated and filtered, to 37 ± 2% for water which was preaerated, centrifuged and filtered. The iron concentrations for the above were determined to be 0.188, less than 0.1 and less than 0.1 mg/L, respectively. This decreased trapping efficiency with iron removal may result from the method of iron removal. This may create microparticulates which sorb B[a]P and allow it to pass through the XAD-4. More work is required on the effect of dissolved iron and other inorganics on the trapping efficiency of B[a]P by XAD-4.

Regeneration of XAD resin has been one of the prime factors supporting its use [10,14,15]. Decreased trapping efficiency for B[a]P was found with regenerated resin which had been used for trapping from the humic organic laden waters (Table III).

The presence of Aldrich humic acids in deionized water significantly decreases the trapping efficiency of XAD-4 for B[a]P (Figure 1). The decrease in trapping efficiency appears as a saturation-type curve, similar to that found in enzyme kinetics, with the addition of humic acid. Increased humic concentrations would yield a greater number of binding sites for B[a]P and reduce the fraction available to the resin. Changing the ionic strength of the solution containing 5 mg/L dissolved organics, with NaCl over a range of 3.45–86.2 meq/L had no effect on the trapping efficiency or recovery of B[a]P. This is in agreement with Junk et al. [14], who reported that increasing the ionic strength with NaCl up to 50 g/L did not improve recovery of organics using XAD-2. However, when 9–36 meq/L (0.5–2 g/L) of a divalent compound ($CaCl_2$) was added, the trapping efficiency (Figure 2) and recoveries (Table IV) were greater than from the humic-deionized water.

**Table III. Comparison of Trapping Efficiency for Benzo[a]pyrene from Deionized and Distilled Water on Regenerated and Fresh XAD-4 Resin ($\overline{x}$ ± 1 SD, n = 3)**

| Water | Resin | Trapping Efficiency | Recovery % | Accountability % |
|---|---|---|---|---|
| Distilled | New | 73 ± 20 | 57 ± 14 | 85 ± 5 |
| | Regenerated | 28 ± 5 | 22 ± 4 | 94 ± 5 |
| Deionized | New | 75 ± 11 | 62 ± 11 | 88 ± 10 |
| | Regenerated | 68 ± 12 | 54 ± 9 | 85 ± 3 |

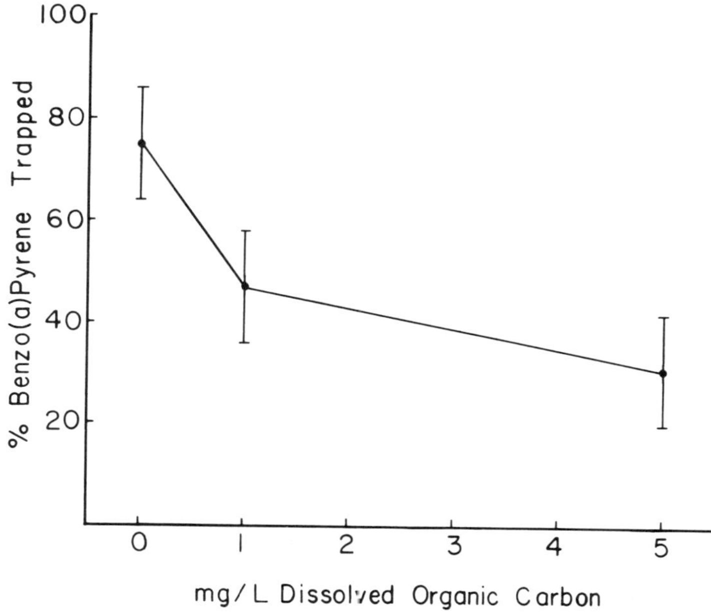


**Figure 1.** Effect of Aldrich humic acids on the trapping efficiency of XAD-4 for benzo-[a] pyrene ($\bar{x} \pm 1$ SD, n = 3).

The increased trapping efficiency appeared to result from the trapping of B[a]P as a humic acid complex. This trapping probably occurs due to the formation of an ion pair which can be trapped as a neutral species.

The brown humic material trapped with the use of $CaCl_2$ elutes from the resin column in two portions. The first fraction elutes with the ether solvent front. This material is apparently weakly associated with the resin. However, a large portion of the fraction preceding the ether solvent front adsorbs to the glass wool and is not eluted with any of the organic solvents. The second fraction which remained associated with the resin was eluted with strong base (1 *N* NaOH). This fraction contained 13.4 ± 2% if the recovered B[a]P (Table IV). This indicates that the mechanism of breakthrough of B[a] P under the influence of humic acids is the formation of a B[a]P-humic acid complex which passes through the resin column. Some fraction of the humics remain bound to the resin column, as noted by the dark color of the resin, and may account for the low accountability since the B[a]P would remain trapped with the humics (Table IV).

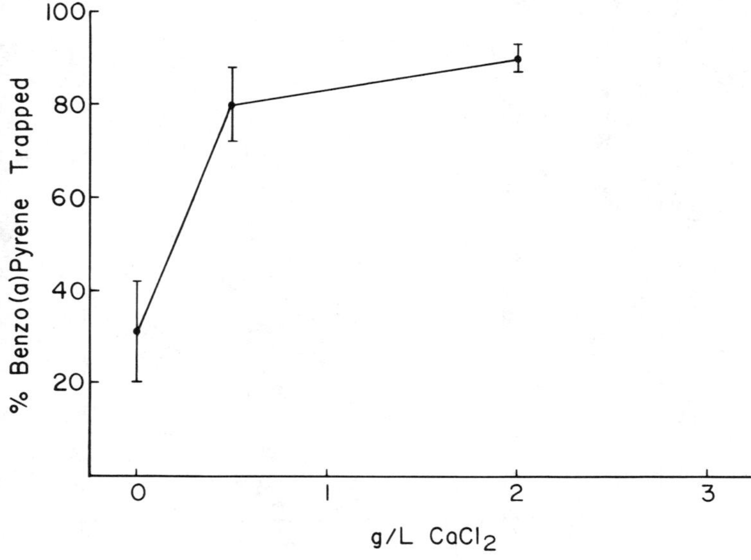


**Figure2.** Effects of $CaCl_2$ on the trapping efficiency of XAD-4 for benzo[a]pyrene in the presence of 5 mg/L humic acids (x ± 1 SD, n = 3).

**Table IV. Influence of $CaCl_2$ on the Trapping Efficiency of Benzo[a]pyrene by XAD-4 Resin in the Presence of 5.0 mg/L humic acids ($\bar{x}$ ± 1 SD, n = 3)**

| $CaCl_2$ (g/L) | Trapping Efficiency | Recovery (%) | Accountability (%) |
|---|---|---|---|
| 0 | 31 ± 11 | 26 ± 11 | 97 ± 5 |
| 0.2 | 26 ± 6 | 27 ± 3 | 100 ± 6 |
| 0.5 | 80 ± 8 | 55 ± 9 | 75 ± 16 |
| 2.0[a] | 90 ± 3 | 61 ± 4 | 70 ± 6 |
| 2.0[b] | 90 ± 3 | 74 ± 3 | 80 ± 5 |

[a] This eluate contains some particulate humic material.

[b] An additional elution with 50 ml of 1 *N* NaOH removed additional B[a]P from the resin column.

## SUMMARY

This study indicates that dissolved minerals and organics can drastically alter the trapping efficiency and recovery of B[a]P on Amberlite XAD-4 resin. This decreased trapping of a nonpolar compound is anomalous with the expected result. The influence of the abovementioned environmental factors appears to increase with the decreased water solubility of the compound. There seems to be a hydrophobicity limit below which no apparent effect occurs. The more insoluble compounds are sorbed to particulates and dissolved organics which alter their trapping and elution efficiency on Amberlite XAD-4 resin. Thus, monitoring studies evaluating the amount of B[a]P and perhaps PAH of equal or lower water solubility utilizing Amberlite XAD-4 resin for concentration must evaluate trapping and elution efficiency for the water to be studied. This conclusion is in agreement with that of Janardan et al. [3], who stated that current procedures are inadequate for broad-based monitoring for hydrophobic compounds in water.

## ACKNOWLEDGMENTS

This work was supported by an interagency agreement between the U.S. Environmental Protection Agency and U.S. Department of Energy, EPA-78-D-X0290, and contract DE-AC09-76SR00819 between the University of Georgia and U.S. Department of Energy. I wish to thank Ms. Marion Babcock for her technical assistance in the analysis of iron concentrations from well-water and Ms. Avis Newell for her technical assistance in determining the concentrations of dissolved organic carbon in experimental waters. J. Coleman drew the figures, and Dallis Mercer and Tonya Willingham typed the manuscript.

## REFERENCES

1. Osterroht, C. "Development of a Method for the Extraction and Determination of Non-polar Dissolved Organic Substances in Sea Water," *J. Chromatog.* 101:208-298 (1974).
2. Van Rossum, P., and R. G. Webb. "Isolation of Organic Water Pollutants by XAD Resins and Carbon," *J. Chromatog.* 150:381-392 (1978).
3. Janardan, K. G., D. J. Schaeffer and S. M. Somani. "Efficiencies of Liquid-Liquid Extraction, Carbon, and XAD-2 Adsorption in Isolating Organic Compounds from Environmental Sources," *Bull. Environ. Contam. Toxicol.* 24:145-151 (1980).

4. Basu, D. K., and J. Saxena. "Monitoring of Polynuclear Aromatic Hydrocarbons in Water. II. Extraction and Recovery of Six Representative Compounds with Polyurethane Foams," *Environ. Sci. Technol.* 12:791-794 (1978).
5. Basu, D. K., and J. Saxena. "Polynuclear Aromatic Hydrocarbons in Selected U.S. Drinking Waters and Their Raw Water Sources," *Environ. Sci. Technol.* 12:795-798 (1978).
6. Saxena, J., J. Kozuchowski, and D. K. Basu. "Monitoring Polynuclear Aromatic Hydrocarbons in Water. I. Extraction and Recovery of Benzo(a)pyrene with Proous Polyurethane Foam," *Environ. Sci. Technol.* 11:682-685 (1977).
7. Leoni, V., G. Puccetti and A. Grella. "Preliminary Results on the Use of Tenax® for Extraction of Pesticides and Polynuclear Aromatic Hydrocarbons from Surface and Drinking Waters for Analytical Purposes," *J. Chromatog.* 106:119-124 (1975).
8. Benoit, F. M., G. L. Lebel and D. T. Williams. "The Determination of Polycyclic Aromatic Hydrocarbons at the ng/L Level in Ottawa Tap Water," *Int. J. Environ. Anal. Chem.* 6:277-287 (1979).
9. Benoit, F. M., G. L. Lebel and D. T. Williams. "Polycyclic Aromatic Hydrocarbon Levels in Eastern Ontario Drinking Waters, 1978," *Bull. Environ. Contam. Toxicol.* 23:774-778 (1979).
10. Berkane, K., G. E. Caissie and V. N. Mallet. "The Use of Amberlite XAD-2 Resin for the Quantitative Recovery of Fenitrothion from Water–A Preservation Technique," *J. Chromatog.* 139:386-390 (1977).
11. Coburn, J. A., I. A. Valdmanis and S. Y. Chau. "Evaluation of XAD-2 for Multiresidue Extraction of Organochlorine Pesticides and Polychlorinated Biphenyls from Natural Waters," *J. Assoc. Off. Anal. Chem.* 60:224-228 (1977).
12. Dressler, M. "Extraction of Trace Amounts of Organic Compounds from Water with Porous Organic Polymers," *J. Chromatog.* 165:167-206 (1979).
13. Grosser, Z. A., J. C. Harris and P. L. Levins. "Quantitative Extraction of Polycyclic Aromatic Hydrocarbons and Other Hazardous Organic Species from Process Streams Using Macroreticular Resins," in *Polynuclear Aromatic Hydrocarbons*, P. W. Jones and P. Leber, Eds. (Ann Arbor, MI: Ann Arbor Science Publishers, Inc., 1979), pp. 67-79.
14. Junk, G. A., J. J. Richards, J. M. Grieser, D. Witrack, J. L. Witiak, M. D. Arguello, R. Vick, H. J. Svec, J. S. Fritz and G. V. Calder. "Use of Macroreticular Resins in the Analysis of Water for Trace Organic Contaminants," *J. Chromatog.* 99:745-762 (1974).
15. Kennedy, D. C. "Treatment of Effluent from Manufacture of Chlorinated Pesticides with a Synthetic, Polymeric Adsorbent, Amberlite XAD-4," *Environ. Sci. Technol.* 7:138-141 (1973).
16. Kunin, R. "The Use of Macroreticular Polymeric Adsorbents for the Treatment of Waste Effluents," *Pure Appl. Chem.* 46:205-211 (1976).
17. Lawrence, J., and H. M. Tosine. "Adsorption of Polychlorinated Biphenyls from Aqueous Solutions and Sewage," *Environ. Sci. Technol.* 10:381-383 (1976).
18. Olufsen, B., and G. E. Carlberg. "Polycyclic Aromatic Hydrocarbons in Water," Nordic PAH Project, Report No. 4 (1979).

19. Olsson, L., and O. Samuelson. "Sorption and Exclusion of Polar Solutes on Cross-Linked Polymers," *Chromatographia* 10:135-139 (1977).
20. Pietrzyk, D. J., E. P. Kroeff and T. D. Rotsch. "Effect of Solute Ionization on Chromatographic Retention on Porous Polystyrene Copolymers," *Anal. Chem.* 50:497-502 (1978).
21. Rees, G. A. V., and L. Au. "Use of XAD-2 Macroreticular Resin for the Recovery of Ambient Trace Levels of Pesticides and Industrial Organic Pollutants from Water," *Bull. Environ. Contam. Toxicol.* 22:561-566 (1979).
22. Strup, P. E., J. E. Wilkinson and P. W. Jones. "Trace Analysis of Polycyclic Aromatic Hydrocarbons in Aqueous Systems Using XAD-2 Resin and Capillary Column Gas Chromatography Mass Spectrometry Analysis," in *Carcinogenesis, Vol. 3: Polynuclear Aromatic Hydrocarbons*, P. W. Jones and R. I. Freudenthal, Eds. (New York: Raven Press, 1978), pp. 131-138.
23. Tateda, A., and J. S. Fritz. "Mini-column Procedure for Concentrating Organic Contaminants from Water," *J. Chromatog.* 152:329-340 (1978).
24. Mori, B. T., and K. J. Hall. "The Use of a Macroreticular Resin XAD-2 for the Recovery of Volatile Organic Compounds from Municipal Sewage," *J. Environ. Sci. Health* A12:341-351 (1977).
25. Giesy, J. P., and L. A. Briese. "Trace Metal Transport by Particulates and Organic Carbon in Two South Carolina Streams," *Verh. Int. Verein. Limnol.* 20:1401-1417 (1978).
26. Garnas, R. L. "Comparative Metabolism of Pesticides by Marine Invertebrates," PhD Dissertation, University of California, Davis (1975).
27. Alberts, J. J., Savannah River Ecology Laboratory, Aiken, SC. Personal communication (1980).
28. Snoeyink, V. L., J. J. McCreary and C. J. Murin. "Activated Carbon Adsorption of Trace Organic Compounds," NTIS Rep. 1977-PB-279253 (1977).
29. Aiken, G. R., E. M. Thurman and R. L. Malcom. "Comparison of XAD Macroporous Resins for the Concentration of Fulvic Acid from Aqueous Solution," *Anal. Chem.* 51:1799-1803 (1979).

# SECTION 6

# HIGH-PERFORMANCE LIQUID CHROMATOGRAPHY

## CHAPTER 23

# HIGH-PERFORMANCE LIQUID CHROMATOGRAPHIC ANALYSIS OF NITROPHENOLS

**A. F. Haeberer* and Trudy A. Scott***

U.S. Environmental Protection Agency
Environmental Research Laboratory
Analytical Chemistry Branch
Athens, Georgia

As a result of the now well known "EPA Consent Decree" settlement in the District Court of the District of Columbia, the U.S. Environmental Protection Agency (EPA) agreed to screen the wastewaters from 21 industrial categories for 129 toxic pollutants. These materials, generally referred to as "priority pollutants" include 11 phenols and nitrophenols, all of which are analyzed with difficulty on the gas chromatography (GC) column cited in the analytical protocol [1,2]. Because the GC behavior of the nitrophenols in question is particularly poor, the development of a liquid chromatographic (LC) procedure for nitrophenols in water was undertaken. These compounds are found primarily in the effluents from timber products, organic chemicals manufacturing, plastics and synthetic materials manufacturing, and pesticides manufacturing industries.

---

*Present address: U.S. Environmental Protection Agency, Washington, DC.
**Present address: Radian Corporation, Sacramento, CA.

## EXPERIMENTAL

### XAD Resin Accumulation

Mixed XAD-4 and XAD-8 neutral macroreticular resins have been shown to be efficient accumulators of a variety of organic compounds from dilute aqueous solution [3]. This mixed resin system was used in these experiments for the accumulation of nitrophenols from aqueous solutions ranging in concentration from 1 to 100 ppb. XAD-4 and XAD-8 resins were obtained from Chemical Dynamics Corp. (South Plainfield, NJ). The resins were extracted with methanol in a soxhlet extraction apparatus for 18 hr or until the concentrate (100:1) of the methanol extract was shown to contain no interferences when chromatographed using the same high-performance liquid chromatography (HPLC) conditions as the samples. Equal volumes of each resin were mixed and placed into the chromatographic columns under methanol until used. Two sizes of accumulator columns were used: one with bed dimensions of 1.6 x 12.6 cm, having a bed volume of 25 mL; the other, 0.7 x 6.4 cm, resulting in a 2.5-mL bed volume.

Aqueous samples, after adjustment to pH 2 with 3 *N* HCl, were extracted on these accumulator resin columns at a rate of about one-half bed volume per minute, or 10 and 1 mL/min, respectively. Accumulated organics were stripped from the resins by first eluting with one bed volume of acetone followed by four bed volumes of methylene chloride. The organic eluents were combined in a separatory funnel, the organic layer was transferred to a Kuderna-Danish (KD) evaporator and the supernatant water layer was discarded. One or two No. 12 Carborundum® boiling chips (Hengar granules, Arthur H. Thomas Co., Philadelphia, PA) were added to the KD flask. A Snyder column was fitted to the flask, and the solution was evaporated on a steam bath as rapidly as possible to a volume of about 2 mL.

This methylene chloride concentrate was now ready either to be converted into a methanol solution, compatible with the weak solvent system used in the reverse-phase HPLC of nitrophenols or, in the case of complex sample extracts, for cesium silicate cleanup of the acidic extract components. The former procedure required the addition of 20 mL of methanol at this point, along with another Carborundum boiling chip. The solution was thoroughly mixed with a vortex mixer. The methanol solution was again reduced via rapid KD evaporation and cooled to room temperature. This resulted in a final volume of approximately 1 mL. This final volume did not need to be precise since an internal standard, 20 $\mu$g of pyrene in 20 $\mu$l methanol, was introduced. The solution was then thoroughly mixed with a vortex mixer, transferred to a 2-mL vial and sealed with a septum closure, ready for HPLC analysis.

**Cesium Silicate Cleanup**

The application of cesium silicate to the isolation of carboxylic and phenolic compounds from complex matrices was introduced by Stalling and co-workers [4-6]. Cesium silicate was prepared by refluxing 15 g cesium hydroxide (Aldrich Chemical Co., Inc.) with 10 g of silicic acid (Bio-Sil® A 100-200 mesh, Bio Rad Laboratories) in 75 mL of methanol ("distilled in glass," Burdick and Jackson Laboratories, Inc.) for 3 hr. The material was subsequently cooled to room temperature, filtered and washed with clean, dry methanol. The air-dried cesium silicate-silicic acid was further dried for 18 hr at 110°C. Cleanup columns were prepared from disposable glass Pasteur pipettes, 150 x 5 mm i.d., dry packed for a length of about 65 mm and plugged at both ends with glass wool. The columns were washed with methylene chloride/cyclohexane (1/1, v/v) before use.

The methylene chloride concentrate from the primary KD evaporation was diluted with an equal volume (2 mL) of cyclohexane. The solution was mixed with a vortex mixer and applied to the cesium silicate cleanup column, using reduced pressure on the column effluent to speed elution. The KD evaporator and receiver were rinsed with an additional 2 mL of methylene chloride that was also applied to the cesium silicate column. The cleanup column was eluted with an additional 2 mL of methylene chloride, removing any remaining nonacidic interferences. The KD evaporator and receiver were then rinsed with 2 mL of methanol, which was also put through the cleanup column, and the eluent was collected. The internal standard, 20 μg of pyrene in 20 μL of methanol, was added, and the solution was mixed, transferred to a vial and sealed for HPLC.

**Liquid Chromatography**

The LC system consisted of an SP-8000 liquid chromatograph (Spectra-Physics, Santa Clara, CA), a Partisil-10 ODS-2 column (250 x 4.6 mm; Whatman Inc., Clifton, NJ), a SP 8310 UV/Visible detector (Spectra Physics) with the 254-nm band pass filter in place and a SP 8010 auto sampler (Spectra Physics). The injection volume was 25 μL in all instances (Autoinjector, 25-μL loop, Spectra Physics). The weak solvent for the elution gradient consisted of 0.386 g or 0.005 *M* ammonium acetate ("Baker Analyzed" HPLC Reagent, J. T. Baker Chemical Co., Phillipsburg, NJ) and 16 mL or 0.28 *M* glacial acetic acid (ACS Reagent, Fisher Scientific Co., Pittsburgh, PA) made up to 1 L with Milli-Q water (Millipore Corp., Bedford, MA). The strong solvent of the gradient was acetonitrile ("distilled in glass," ultraviolet (UV) grade, Burdick and Jackson Laboratories, Inc., Muskegon, MI).

### Reagents and Standard Solutions

The nitrophenols investigated were phenol, 2-nitrophenol (2-NP), 3-nitrophenol (3-NP), 4-nitrophenol (4-NP), 2,4-dinitrophenol (2,4-DNP), 2,4,6-trinitrophenol (2,4,6-TNP, picric acid) and 4,6-dinitro-*o*-cresol (4,6-DN-2C), all of which were supplied by Eastman Organic Chemicals, Rochester, NY. Individual concentrated standard solutions containing 1 mg nitrophenol/mL of solution were prepared from air-dried reagents and methanol. Dilute standards in the part-per-million and part-per-billion ranges were prepared by making the appropriate dilutions using polished water with a conductance of about 18 MΩ/cm from a Milli-Q water purification system.

The internal standard solution was prepared by dissolving 10.0 mg of pyrene (K & K Labs, ICN Pharmaceuticals, Inc., Plainview, NY), recrystallized from methanol, in 10.0 mL dry methanol. A 20-μL aliquot of this solution was added to each extract on completion of KD evaporative concentration.

## RESULTS AND DISCUSSION

### Development of Chromatographic Parameters

The chromatographic parameters for these separations were optimized by utilizing the microprocessor capabilities of the SP-8000 liquid chromatograph. To determine the most favorable acetic acid concentration for the weak solvent, a series of chromatographic runs were programmed for overnight execution. Each parameter set for these runs was carried out in triplicate to ensure column equilibration. The glacial acetic acid concentration was increased from 0 to 20 mL/L of 0.005 *M* ammonium acetate buffer in 2-mL increments. The chromatograms indicated that glacial acetic acid concentrations of 1 and 16 mL/L of buffer were best for the separation of these nitrophenols. From the manner in which the chromatographic peaks migrated on the chromatogram when the acetic acid concentrations were varied, it was concluded that other nitrophenols would, in all probability, require different acid concentrations for the optimization of their chromatographic separations. The results of these runs are presented graphically in Figure 1. Coelution is represented by crossing lines. This supports the work of Price et al. [7] on optimization of the chromatographic conditions for the HPLC separation of phenolic acids.

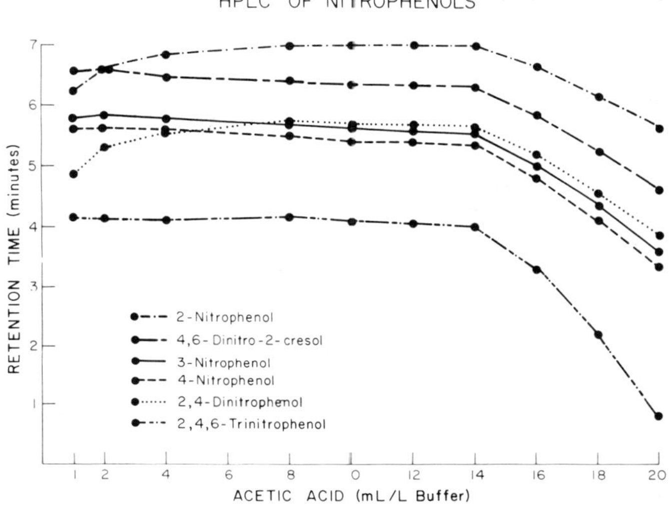


**Figure 1.** Dependence of nitrophenol retention on acetic acid concentration.

The best resolution was achieved with a mobile phase gradient running from 85% ammonium acetate-acetic acid buffer/15% acetonitrile linearly to 90% acetonitrile. The final strong solvent concentration, although not necessary for elution of the nitrophenols or the internal standard (pyrene), was necessary for column cleanup, especially after samples as complex as those obtained from the dye industry.

## Recoveries

Recovery studies using Athens, GA, tapwater were carried out at 100 and 10 ppb concentrations of each of the nitrophenols. The hydrogen ion activity (pH) was adjusted with 3 *N* HCl as indicated in Table I. The recovery levels for all six nitrophenols were 92% or better at pH 2. At higher pH values the recovery values were significantly lower, showing clearly that the nitrophenols must be in the undissociated form for efficient accumulation on the XAD 4/8 resins. Accumulation experiments at lower pH values were also conducted but did not result in greatly improved recoveries. The XAD 4/8 accumulations were also compared to the classical shake-out solvent-partition extraction at pH 2 with three 100-mL volumes of methylene chloride as

**Table I. Percent Recovery of Nitrophenols from Water on XAD 4/8 Resin (100 μg/L nitrophenols, 10 mL/min flowrate)**

| | Percent Recovery | | | | | | |
|---|---|---|---|---|---|---|---|
| pH | 2,4,6-TNP | 2,4-DNP | Phenol | 4-NP | 3-NP | 4,6-DN-2-C | 2-NP |
| 6–7 | 45 | 71 | | 60 | 59 | 77 | 68 |
| 3 | 86 | 91 | | 89 | 87 | 91 | 90 |
| 2 | 92 | 99 | 88 | 94 | 97 | 92 | 97 |
| 2[a] | 33 | 93 | 29 | 89 | 44 | 90 | 88 |

[a]Methylene chloride solvent-partition extraction.

extracting solvent. KD evaporation and chromatographic analysis of the combined extracts resulted in the overall yields reported on the last line of Table I.

Particular attention must be paid to both the rate of KD evaporation and to the kind of boiling stone used in the concentration of sample extract. Rapid KD evaporation results in much higher recoveries of solutes than slow, carefully controlled rates. Evidently the scrubbing action of the Snyder column chambers is optimized at higher throughput rates, when they remain partially filled with solvent.

Nitrophenols may be adsorbed on certain types of boiling stones as was shown by a variety of different experiments. This was particularly obvious with Boil Eazers®, a product of Fisher Scientific Company, that turned quite yellow, characteristic of adsorbed nitrophenols. The loss of nitrophenols due to adsorption on the boiling chip was also noted by reduced chromatographic peak intensities in the quantification of these samples. Boiling chips are not the only substrate that may adsorb the nitrophenols during the extract concentration step, however. The surface of the KD receiver itself may also adsorb these compounds particularly when solvents of low polarity (such as 50/50 methylene chloride/cyclohexane) are used in the primary concentration step prior to cesium silicate cleanup. It is necessary, therefore, to rinse the KD receiver with an additional 2 mL of methylene chloride to remove any adsorbed nitrophenols. This is particularly the case with picric acid, which is prone to low recoveries because of its high polarity or highly acidic nature ($pK_a = 0.38$).

## Cleanup

The application of a variety of cleanup techniques to complex samples and sample extracts was studied. These included the acid-base solvent partitioning of Klus and Kuhn [8], silica sample preparation cartridges and cesium silicate column chromatography.

Klus and Kuhn separated neutral and basic interferences from ethyl ether solutions of cigarette smoke condensate by solvent extraction of the acids with dilute NaOH solution. On acidification of the aqueous extract with dilute HCl and extraction with ethyl ether, the nitrophenol-containing "acidic fraction" was isolated in the organic layer. When this technique was applied to raw dye plant effluents spiked at 100-ppm levels of the six nitrophenols, recoveries averaged 60-70%. The resulting liquid chromatogram of this extract was still too complex to allow quantification of the nitrophenols at the desired low-ppb range. Further cleanup was mandatory.

Silica sample preparation cartridge (Sep-Pak, Waters, Inc., Millford, MA) cleanup of the $Na_2SO_4$-dried acidic fraction obtained from the Klus and Kuhn scheme further reduced nitrophenol recoveries to an unacceptable 40%. This technique was abandoned because of the low recoveries obtained from these silica cartridges.

Cesium silicate isolation of acidic components from the methylene chloride/acetone eluent (after KD concentration to 2 mL) from the XAD 4/8 accumulator resin columns resulted in the best recoveries of nitrophenols from finished dye plant effluents. At the 10-ppb level, recoveries of these compounds were 80% or greater. Because nitrophenols have rather low solubilities in nonpolar solvents and tend to be deposited on glass container walls, it is recommended practice to rinse the KD receiver with at least 2-mL portion of methylene chloride which is also applied to the cesium silicate cleanup column. The column is then eluted with 50/50 methylene chloride/cyclohexane to remove any neutral and basic interferences. Subsequent elution of the cesium silicate cleanup column with 2 mL of methanol strips off the carboxylic and phenolic compounds as the cesium salts. This eluent is used for HPLC quantification after addition of the internal standard (pyrene).

These techniques have been used with good success on samples as diverse as tapwater, finished dye plant effluents and finished municipal effluents. Chromatograms resulting from these analyses are presented in Figures 2 to 4. In the analysis of heavily loaded samples, one must use caution not to exceed the capacity of the XAD accumulator resin columns. This capacity is dependent on a variety of factors, such as the degree of dissociation of the solute molecule, and molecular structure, size and weight. The manufacturer cites

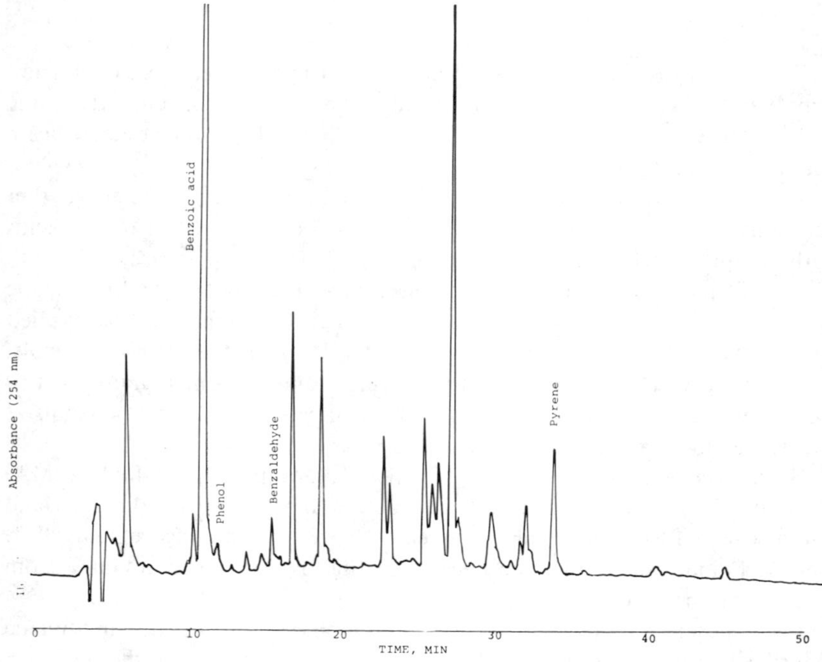


**Figure 2.** Chromatogram of unfortified dye-plant effluent extract after cesium silicate cleanup.

capacities of 2% phenol and 25% 2,4,6-trichlorophenol on XAD-4 [9]. The data presented in Table I clearly indicate that the solute molecule must be in the associated form to be accumulated efficiently by the XAD 4/8 resins. The recoveries obtained from a 1-L sample of water containing 100 $\mu$g of each of the nitrophenols and phenol extracted by the classical shake-out technique are also presented in Table I. Accumulation using XAD 4/8 resin has the obvious advantage of higher recoveries, and this technique may be automated much more easily than liquid-liquid extraction (LLE).

Figure 5 presents two chromatograms of a Consent Decree extract from the plastics and organic chemicals industry before and after cesium silicate cleanup. This illustrates the potential of this cleanup technique as well as the application of the HPLC method for the quantitative analysis of nitrophenols. A variety of these Consent Decree extracts have been analyzed by this HPLC method. Phenol, 2-nitrophenol, 4-nitrophenol, 2-nitro-*p*-cresol, 2,4,6-trichlorophenol and pentachlorophenol have been quantified in these

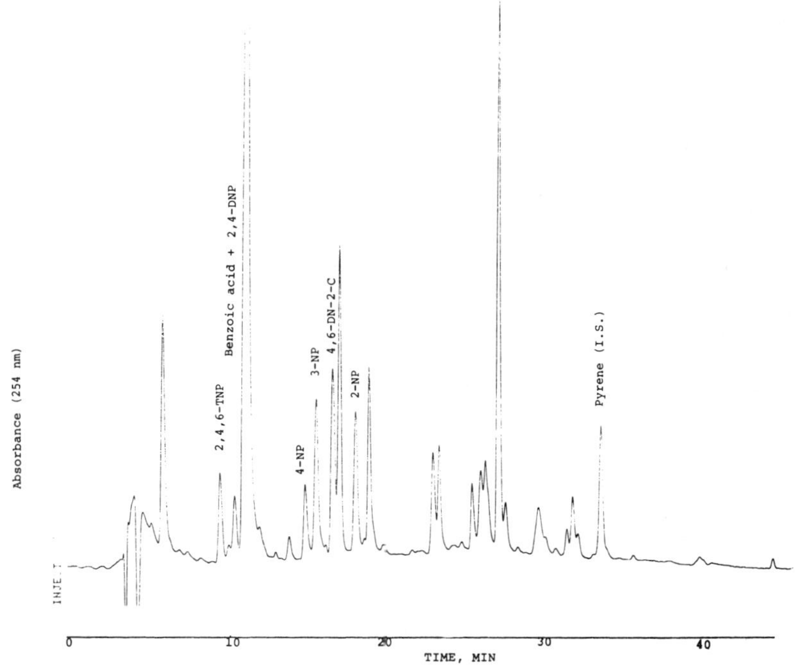


**Figure 3.** Chromatogram of fortified dye-plant effluent (10 ppm each nitrophenol) extract after cesium silicate cleanup.

samples. The concentrations range from 7 to 260 ppb of each phenol in the original effluent sample. These analyses support the analyses carried out with the Consent Decree protocol [1].

## CONCLUSION

A reverse-phase HPLC method has been developed for the analysis of nitrophenols. Accumulation using mixed XAD 4/8 macroreticular resins followed by KD evaporative concentration permits quantitative extraction of those compounds from aqueous matrices in the low ppb range. Cleanup procedures using cesium silicate chromatography facilitate the separation of phenolics from neutral and basic interferences.

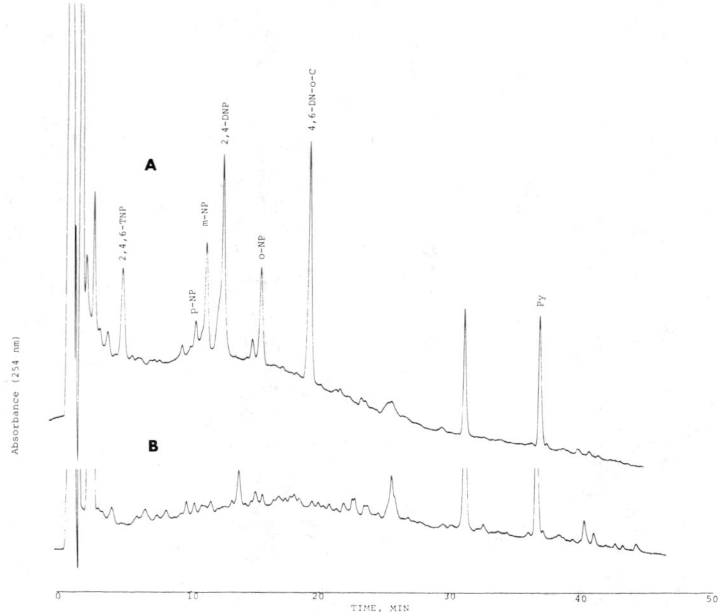


**Figure 4.** Chromatograms of municipal effluent extract. (A) Fortified at 10 ppm each nitrophenol; (B) Unfortified.

## DISCLAIMER

Mention of commercial products, sources and trade names is for informational purposes only and does not imply endorsement by the U.S. Environmental Protection Agency.

## REFERENCES

1. "Sampling and Analysis Procedures for Screening of Industrial Effluents for Priority Pollutants," U.S. EPA, Environmental Monitoring and Support Laboratory, Cincinnati, OH (1977).
2. Keith, L. H., and W. A. Telliard. *Environ. Sci. Technol.* 13:416 (1979).
3. Van Rossum, P., and R. G. Webb. *J. Chromatog.* 150:381-392 (1978).

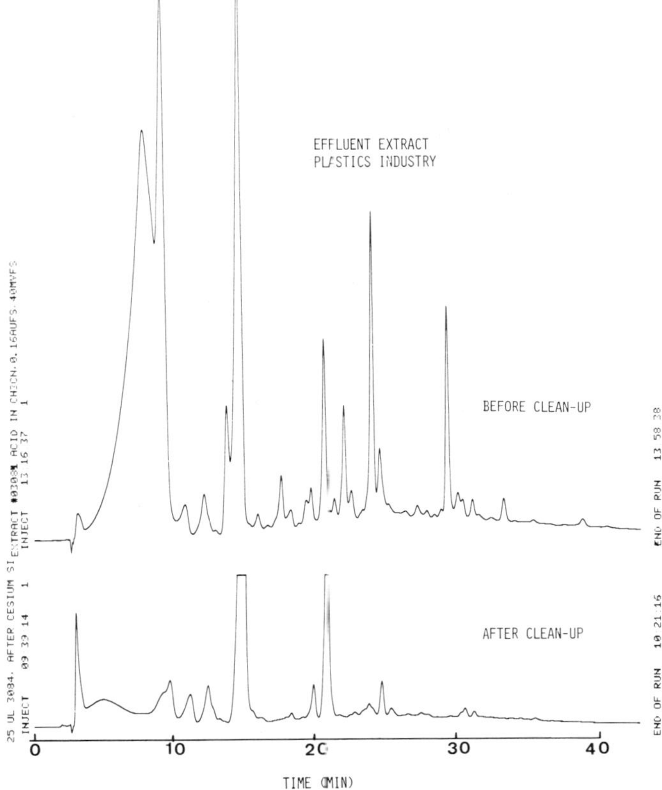


**Figure 5.** Chromatograms obtained from a Consent Decree extract of a plastics industry plant, before and after cesium silicate cleanup. Peak identification: at 14 minutes, 4-nitrophenol; at 21 minutes, 2-chloro-6-nitrophenol.

4. Stalling, D. L., L. M. Smith and J. D. Petty. "Approaches to Comprehensive Analysis of Persistent Halogenated Environmental Contaminants," paper presented before the ASTM Symposium on Organic Pollutants, Denver, CO, June 17-20, 1978.
5. Stalling, D. L., J. D. Petty, D. R. Dubay, L. M. Smith and D. M. Grev. "Carbon and Cs-Silicate Chromatography–Potential Separation Techniques for LC-MS," paper presented before the Ninth Annual Symposium on the Analytical Chemistry of Pollutants, Jekyll Island, GA, May 7-9, 1979.
6. Stalling, D. L., L. M. Smith and J. D. Petty. "GPC Enrichment and Carbon Chromatographic Fractionation of Hydrocarbons and Other Environmen-

tal Contaminants," paper presented before the 1979 Pittsburgh Conference on Analytical Chemistry and Applied Spectroscopy, Cleveland, OH March 5-9, 1979.
7. Price, W. P., Jr., R. Edens, D. L. Hendrix and S. N. Deming. *Anal. Biochem.* 93:233-237 (1979).
8. Klus, H., and H. Kuhn. *Mikrochim. Acta* (Vienna) I:405-412 (1975).
9. "Amberlite® XAD-4," technical bulletin, Industrial Chemicals, Rohm and Haas Co., Philadelphia, PA (1976).

## CHAPTER 24

# SIZE EXCLUSION, REVERSE-PHASE AND WEAK ANION EXCHANGE CHROMATOGRAPHY OF NATURAL ORGANICS IN WATER

**William H. Glaze,* P. C. Jones and Farida W. Saleh**

Department of Chemistry and
Institute of Applied Sciences
North Texas State University
Denton, Texas

The natural organic matrix in surface and groundwaters has been the subject of investigation for centuries, yet we are still largely ignorant of its molecular structure [1,2]. To say it is a complex matrix understates the problem. It is said to be seasonably and locationally variable, and its secondary and tertiary structure almost certainly changes with sample treatment [3]. Some feel there is no regular structure to the material as in other natural polymers such as proteins and carbohydrates.

Much of this opinion is based on information accumulated over a long period of time by several groups of workers, many of whom did not have at their disposal the most sophisticated analytical tools. Moreover, the approach often taken was to study unfractionated aquatic humus, or fractions obtained by the solubility scheme of Oden [4]. Kononova [5], among others, recognized that these fractions are molecularly heterogeneous.

*Present address: Graduate Program in Environmental Sciences, The University of Texas at Dallas, Richardson, TX.

The principal subject of this chapter is the soluble organic material in natural waters, often referred to as fulvic acid [4]. However, we shall attempt to refrain from using this nomenclature, since it implies some degree of uniformity which may be nonexistent. Rather, our approach will be to apply several analytical tools to the study of aquatic humus, hoping to see as we proceed a more complete picture of its molecular nature.

Central to this study is the use of modern high-performance separation methods. Indeed, it is our contention that we shall achieve our ultimate objective only by applying modern structure elucidation tools to aquatic humus *after* high-performance preseparation. To this end the present chapter describes the use of several high-performance liquid chromatography (HPLC) modes on a selected number of aquatic samples. The work is by no means complete, and should be taken only as a progress report of current work.

## EXPERIMENTAL

### HPLC Instrumentation

A Micromeretics Model 7000 B solvent delivery system was used in either the isocratic or gradient mode depending on the column and detector being used. The latter was either a Tracor scanning ultraviolet/visible (UV/vis) detector, Model 970A; a Shoeffel Model FS970 fluorescence detector; or a Bioanalytical Systems Inc. LC-4 Amperometric Controller with an Au/Hg amalgam electrode.

### HPLC Columns and Solvents

The following columns were used with the HPLC instruments and detectors described above:

1. a prepacked Whatman Partisil 100DS2 M9 preparative reverse-phase column 250 mm x 9.4 mm i.d.;
2. an analytical reverse-phase column, 25 cm x 4.6 mm i.d. slurry-packed in this laboratory with Whatman ODS2 10 μm microparticulate media;
3. two types of size exclusion columns (SEC), 25 cm x 4.6 mm i.d. and 25 cm x 9.4 mm i.d., packed with Partisil 10 (Whatman, 11 μ particle size 60-Å pore size), deactivated with bonded glycerylpropylsilane, coated and slurry-packed in this laboratory [6]; and
4. a weak anion exchange (WAX) column 1 m x 2 mm i.d. dry-packed with Whatman AL Pellionex WAX 40-μ pellicular media.

Solvents and special chemicals used were as follows:

- water: Milli-Q water from Millipore system with GAC, reverse osmosis (RO) and ion exchange purification;
- $Bu_4N^+ClO_4^-$: prepared in this laboratory from $LiClO_4$ and $Bu_4N^+OH^-$ (recrystallized);
- methanol: Baker HPLC grade;
- sodium polystyrene sulfonate MW standards: Pressure Chemical Co.; and
- Ovalbumin MW45,000: Pharmicia Fine Chemicals.

### Sample Pretreatment: Freeze-Drying

In some cases water samples were used without preconcentration; however, for some HPLC analyses a concentration step was necessary. This was accomplished by freeze-drying 50 mL of each water sample on a Labconco Freeze Dryer 3. The freeze-dried material was resolubilized in 5 mL Milli-Q water and filtered before injecting into the HPLC. Although this procedure resulted in a concentration factor of 100, some organics may be lost in the process of resolubilization.

### Sample Ozonation and Chlorination

Surface water samples from Ontario, Canada were ozonated at another location by personnel of International Environmental Consultants, Ltd. (Toronto) and shipped in ice by air freight to Denton, Texas. Two ozone doses were used, corresponding to $O_3$-to-TOC-ratios of 0.2 and 2.5. Other samples were chlorinated in Denton with a dose of 20 mg/L and stored at 26°C and at pH 6.5 (phosphate buffer) until $Na_2SO_3$ was added to quench the reaction.

## RESULTS AND DISCUSSION

### General Features of Size Exclusion Chromatograms of Raw-Water Samples

Size exclusion chromatograms of several unconcentrated water samples have been examined so far, using the glycerylpropylsilane CPG column described in earlier works [6–8]. It should be noted that this column should be used with discretion, as inaccurate data may be obtained due to ionic strength and column-solute effects [9]. Calibration should be repeated often, and we now recommend the use of solutions as dilute as possible and with tetrabutylammonium perchlorate eluent at 0.025 *M* in the best water available. Significant distortions are seen at high solute concentrations when one is studying polyelectrolytes [9].

Using these precautions we do not generally observe a polynodal molecular size distribution for aquatic humic material when the UV and fluorescence detectors and the 60-Å column are employed. This contrasts to other works using Sephadex gel filtration where polynodal patterns are obtained, though it is generally realized that Sephadex is subject to severe solute/column adsorption effects [1,10]. Recently Saito and Hayano [11] have observed some structure in the molecular size distribution curve using a new hydrophilic gel of undescribed composition. From a more general point of view, however, most works in this area suffer from the lack of a general carbon detector for HPLC application. The use of UV and fluorescence detectors as shown here and elsewhere [1,10,11] may not reveal the correct molecular weight distribution due to changes in molecular adsorptivity as the chromatogram develops. This factor is particularly important in studies of the effect of disinfection (oxidizing) agents such as ozone and chlorine. Figure 1 shows the size exclusion chromatogram with fluorescence detection ($\lambda_{ex} = 273$, $\lambda_{em} > 380$) of an unconcentrated Canadian surface-water sample. The molecular weight distribution curve shifts very slightly to lower apparent size values as the ozone dose is increased, but the effect is largely to attenuate the total response. Identical chromatograms are obtained with UV detection at 254 nm, but the samples must be preconcentrated. That the loss of response is not due to carbonation of the organics is shown by the fact that TOC is lowered only by about 5% during the ozonation process.

## Reverse-Phase Separations

Figures 2 to 4 show reverse-phase separation with fluorescence detection of a second surface-water source from Canada. The three figures show the separation of several components in the samples, none of which has been identified at this time. As in all samples, the ozonated samples show attenuated peak responses when the UV and fluoresence detectors are used. However, when the amperometric detector is used at -1.1 V with careful solvent degassing [12], two broad peaks are observed which are not attenuated by ozonation (Figure 5). In fact, the second peak at 4 mL retention volume appears to increase in area on ozonation. (The dead volume of column is 1 mL.) Unfortunately, the detector baseline responds drastically to solvent

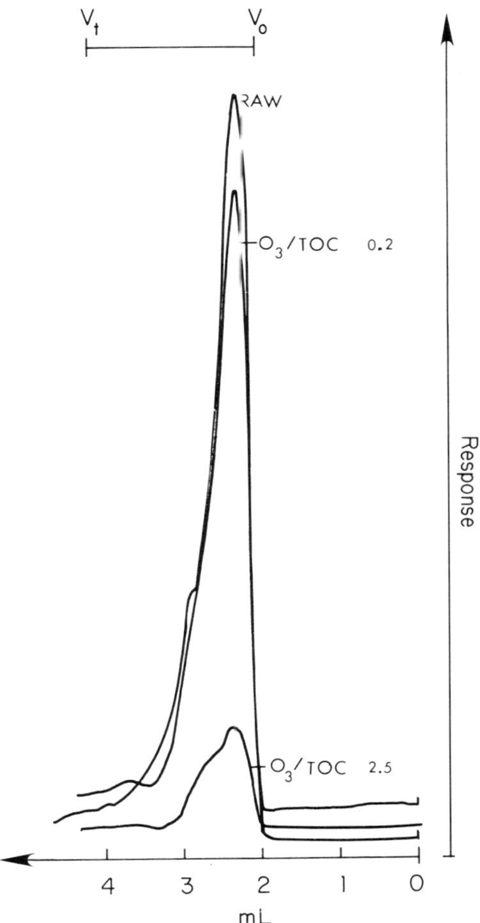


**Figure 1.** Size exclusion chromatograms (SEC) of unconcentrated Canadian surface water (0.25 *M* $Bu_4NClO_4$ eluent; fluorescence detector).

changes, so we have been forced so far to use HPLC conditions where much of resolution shown in Figure 2 is lost. Nonetheless, electrochemical detection appears to be a promising avenue for screening of disinfected natural waters for nonvolatile by-products. We are proceeding to optimize the method.

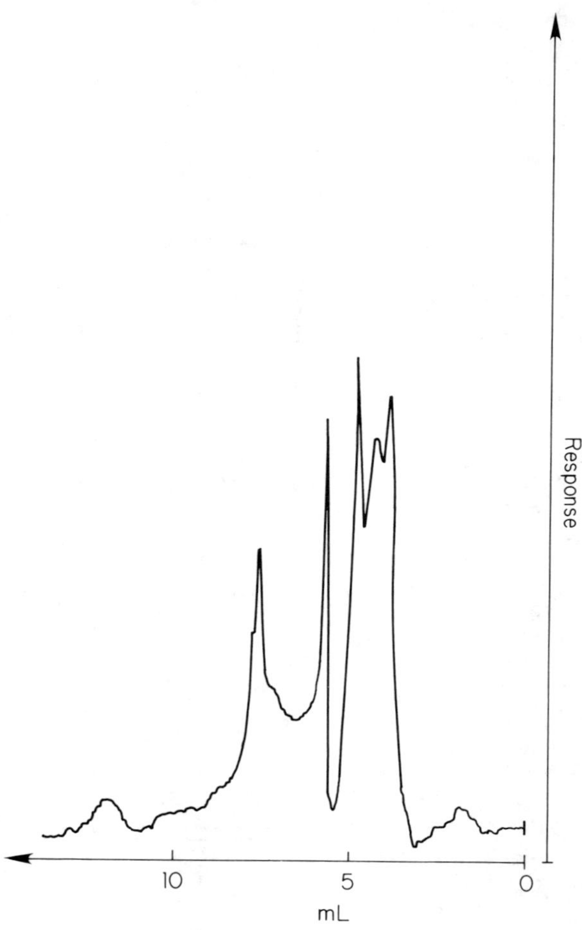


**Figure 2.** Reverse-phase HPLC chromatogram of unconcentrated Canadian surface water before ozonation (fluorescence detector: 0.01 μA range; flow 1 mL/min; isocratic pH 3 $H_2O$ 97%/MeOH 3%).

## Weak Anion Exchange Separations

Figure 6 shows the application of the weak anion exchange material AL Pellionex for the study of natural organics. The samples for these analyses were freeze-dried concentrates of the original water samples. A gradient

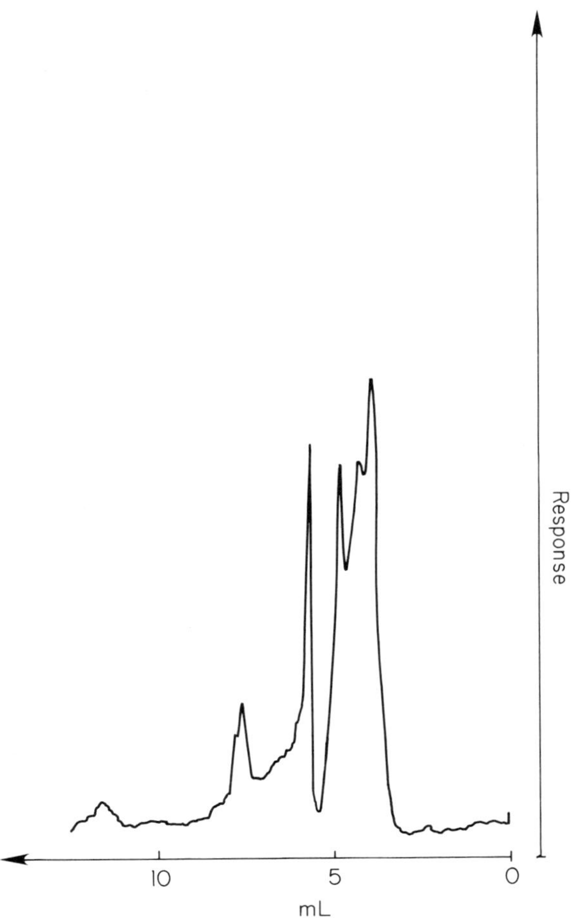


**Figure 3.** Same as Figure 2 after ozone dose of 0.2 x TOC value, unquenched.

system of 0-100% strong solvent in 10 min with curvature $N^3$ was used, with the 0% being pH 3 boric acid and the 100% being pH 6.5 borate buffer. The pH limitations of the WAX packing material are pH 2-7, so that a very basic solvent cannot be used. Three peaks have been seen in the limited number of cases we have studied, with one exception. The peak at longest retention volume was absent in water from one source.

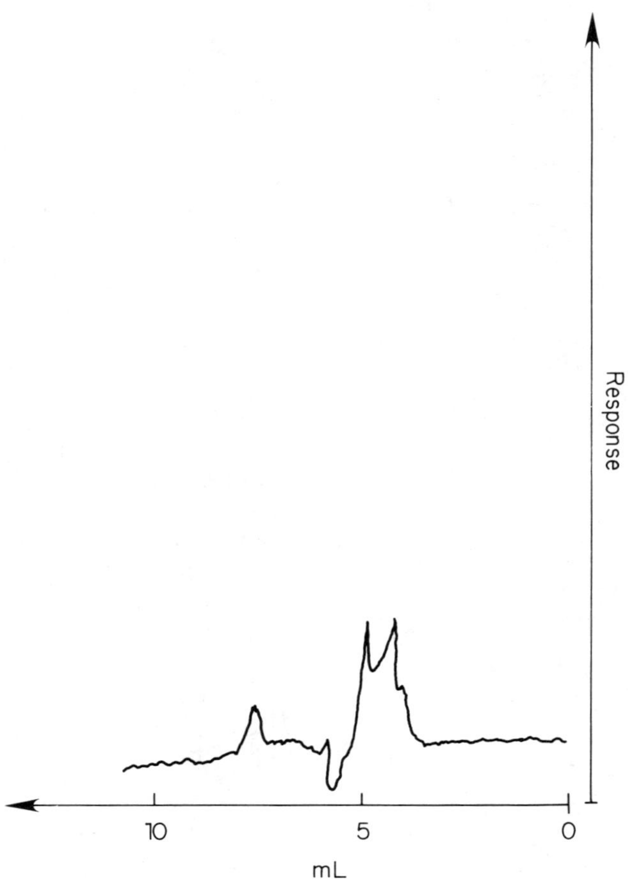


**Figure 4.** Same as Figure 2 after ozone dose of 2.5 x TOC value, unquenched.

Table I shows some characteristics of the three peaks collected by manual preparative HPLC. Trihalomethane formation potentials (THMFP) [13], total organic carbon and apparent molecular weights are listed for the three fractions. Several observations of these data are noteworthy. First, peaks I through III elute at increasing pH values indicating acidities in the order I < II < III. Figure 6 shows retention volumes of several model compounds which illustrate this trend. Of course the molecular nature of the three peaks

Table I. Characteristics of the Three Peaks Separated by Weak Anion Exchange HPLC

| | Mol Wt Range | Apparent Wt Average Mol Wt $M_w$ | Apparent No. Average Mol Wt $M_n$ | TOC (mg/L) | | THMFP (μg/L as Cl) | | THMFP/TOC (mg/mg) |
|---|---|---|---|---|---|---|---|---|
| | | | | $\overline{X}$ | S | $\overline{X}$ | S | |
| Peak I | $22.4 \times 10^3$–$0.5 \times 10^3$ | $4.99 \times 10^3$ | $2.2 \times 10^3$ | 1.32 | 0.02 | 15.6 | 2.1 | 0.012 |
| Peak II | $31.6 \times 10^3$–$0.5 \times 10^3$ | $8.1 \times 10^3$ | $3.3 \times 10^3$ | 1.31 | 0.01 | 23.5 | 0.04 | 0.018 |
| Peak III | $31.6 \times 10^3$–$0.6 \times 10^3$ | $9.3 \times 10^3$ | $4.4 \times 10^3$ | 1.71 | 0.01 | 43.4 | 2.6 | 0.025 |
| Before Fractionation | $31.6 \times 10^3$–$0.5 \times 10^3$ | $8.2 \times 10^3$ | $3.9 \times 10^3$ | 5.04 | 0.08 | 120.3 | 2.1 | 0.024 |

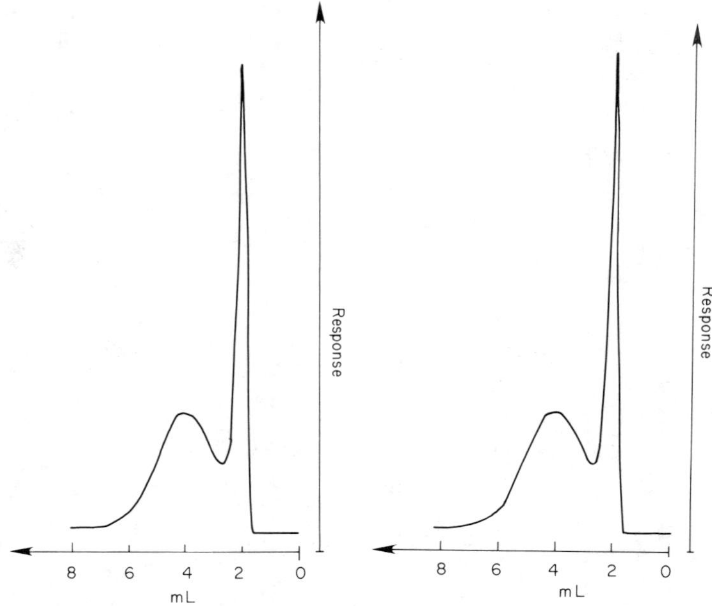


**Figure 5.** Reverse-phase HPLC chromatograms (with electrochemical detection) of unconcentrated Canadian surface water before and after ozonation (Bioanalytical Systems Detector; voltage setting-1.1 V; flow 1 mL/min); (right) raw water; (left) after ozone dose 2.5 x TOC value, unquenched.

is yet to be determined, but the separation process promises to provide fractions for such study. Second, the ratio of THMFP to TOC increases in the order I < II < III, and apparent size also increases in that order. Whether these observations are significant for water treatment considerations is not known at this time.

## ACKNOWLEDGMENTS

This work was supported in part by the grants from the U.S. Environmental Protection Agency (R-803007 and R-805822) and a subcontract from International Environmental Consultants, Ltd. (Toronto), to whom we are indebted for technical assistance in the ozonation studies. Funds for the IEC study were provided by the Ontario Ministry of the Environment.

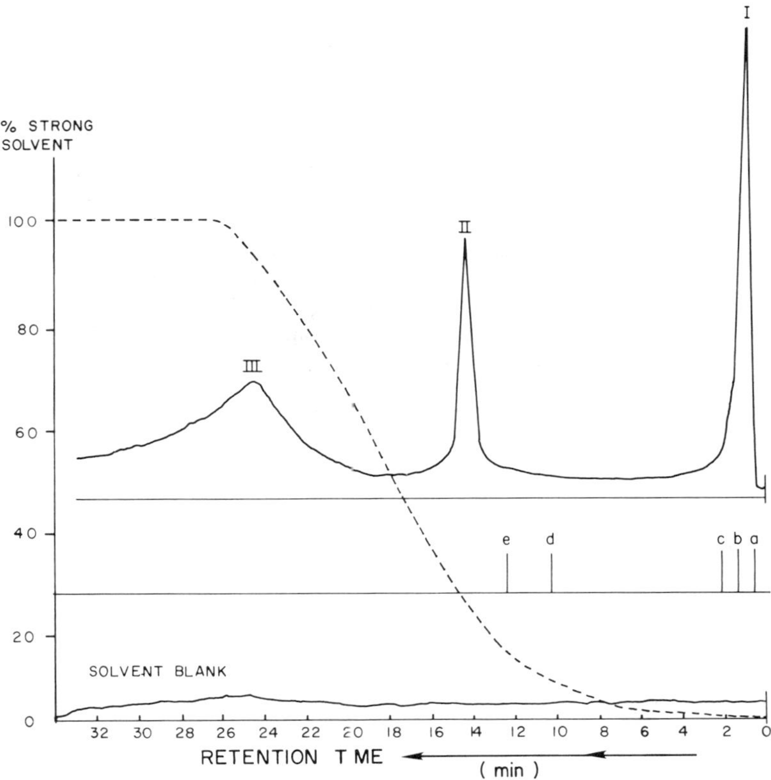


Figure 6. Weak anion exchange HPLC chromatogram of acid-soluble portion of freeze-dried concentrate, Cross Lake, LA, surface water. Borate buffer gradient - - - - - -; model compound retention volumes; (a) phenol (pKa 10); (b) 3-methylcatechol (pKa ~ 10); (c) vanillic acid (pKa 4.1); (d) 2,4-dihydroxybenzoic acid (pKa 3.0); 2,4,6-trihydroxybenzoic acid (pKa 1.7).

## REFERENCES

1. Gjessing, E. T. *Physical and Chemical Characteristics of Aquatic Humus* (Ann Arbor, MI: Ann Arbor Science Publishers, Inc., 1976).
2. Schnitzer, M., and S. U. Khan. *Humic Substances in the Environment* (New York: Marcel Dekker, 1972).
3. Worshaw, R. L., D. J. Pinckney and S. E. Booker. *J. Res. U.S. Geol. Survey* 5:565 (1977).

4. Oden, S. *Kollidchem. Beihefte* 11:74 (1919).
5. Kononova, M. M. *Soil Organic Matter*, 2nd (English) ed. (Oxford: Pergamon Press, 1966).
6. Regnier, F. E., and R. Noel. *J. Chromatog. Sci.* 14:316 (1976).
7. Glaze, W. H., G. R. Peyton, F. Y. Saleh and F. Y. Huang. *Int. J. Environ. Anal. Chem.* 7:143 (1979).
8. Glaze, W. H., F. Y. Saleh and W. Kinstley. In: *Water Chlorination: Environmental Impact and Health Effects, Vol. 3*, R. Jolley et al., Eds. (Ann Arbor, MI: Ann Arbor Science Publishers, Inc., 1980).
9. Barth, H., and F. E. Regnier. *J. Chromatog.* 192:275 (1980).
10. Chian, E. S., and F. B. DeWalle. *Environ. Sci. Technol.* 11:158 (1977).
11. Saito, Y., and S. Hayano. *J. Chromatog.* 177:390 (1979).
12. Funk, M. O., M. B. Keller and B. Levison. *Anal. Chem.* 52:771 (1980).
13. Stevens, A. A., and J. Symons. *J. Am. Water Works Assoc.* 69:546 (1977).

# CHAPTER 25

# DEVELOPMENT OF TECHNIQUES FOR THE ISOLATION AND IDENTIFICATION OF NONVOLATILE ORGANICS IN DRINKING WATER

**C. D. Watts, B. Crathorne,**
**R. I. Crane and M. Fielding**

Water Research Centre
Medmenham, Marlow
Buckinghamshire, England

In recent years a large amount of information has been published on the separation and identification of organic compounds extracted from drinking water [1]. These data have been obtained largely by the use of combined gas chromatography/mass spectrometry (GC/MS) and are therefore heavily biased toward volatile compounds. It is now accepted that only 10-20% of the organic compounds in raw and drinking water are amenable to GC/MS analysis. The characterization of the remainder presents a difficult and challenging problem for the analytical chemist, but the need for such work to be carried out is urgent and very real. Current research at this laboratory in important areas such as mutagenicity testing of water and characterization of chlorination and ozonation by-products is emphasizing the need for such techniques.

One approach to the problem of extraction, separation and identification of nonvolatile organics from drinking water has been developed at the Water Research Centre and reported in detail elsewhere [2]. In this method water samples were freeze-dried, and the solid residue was extracted with methanol. Separation of extracts was carried out by reverse-phase high-performance liquid chromatography (HPLC) both on an analytical and preparative scale. Selected fractions taken from preparative HPLC were then examined by electron impact (EI), field ionization (FI) and field desorption (FD) mass spectrometry to attempt identification of components.

Compounds identified in drinking water during the course of this work included 5-chlorouracil, 5-chlorouridine, 4-chlororesorcinol, 5-chlorosalicylic acid, 2-hydroxybenzothiazole, 2,4,6-trichlorophenol, dimethyl acridan, *p*-isopropyldiphenylamine and di-*n*-octylphthalate. Levels found ranged from 0.1 to 25 $\mu$g/L. While some of these compounds can be considered volatile in that they are amenable to GC/MS analysis without derivatization, the technique we have developed is capable of analyzing inherently nonvolatile organic compounds.

As a result of experiences gained with this method, refinements have been made, and these are presented in this chapter. Thus, the attempted use of an offline total organohalogen (TOX) monitor for the analysis of HPLC fractions is described, and an examination of methods for reducing the inorganic content of the extract is reported. Mass spectrometric analysis of various HPLC fractions using low-resolution EI and FD techniques is described.

## TECHNIQUES

### Sample Preparation

Drinking water samples (50 L or less) were freeze-dried to a solid residue using a centrifugal freeze-dryer (Edwards High Vacuum Ltd., Speedivac Model 30P2). About one week is required to dry 15 L of water to a solid residue. The solid residue is extracted with methanol, and the extract is concentrated to approximately 1 mL by rotary evaporation and finally under a stream of nitrogen. When required, water samples were spiked immediately before freeze-drying with some or all of the following compounds: uridine, benzoic acid, 2-hydroxybenzothiazole, di-*n*-propyl phthalate, 5-chlorouracil, 4-chlorobenzoic acid, 2,3,4-trichlorophenol, *p*-chlorophenylacetic acid and 3-hydroxy-2-methoxycinnamic acid.

Previously [2] it was found that freeze-drying provides good recoveries of total organic carbon (TOC). Subsequent work, however, has shown that freeze-drying also gives large amounts of salts in both the methanol extracts and HPLC fractions derived from them. These salts contain sodium and potassium ions, which interfere with field desorption mass spectrometry

(FD/MS) and halide ions, which interfere with organohalogen measurements. The removal of these interfering salts by ion exchange resins and a Sephadex LH-20 gel was evaluated.

Two distinct approaches to the use of ion exchange resins were used. The first involved removal of salts from the aqueous phase, prior to freeze-drying and extraction. This was tried using a column of weak anion exchange resin (Amberlite IR-45) followed by a column of weak cation exchange resin (Amberlite IR-40). Samples of drinking water (1 L) spiked with five selected compounds (uridine, 5-chlorouracil, 2-hydroxybenzothiazole, 2,3,4-trichlorophenol and di-*n*-propyl phthalate; 50 $\mu$g/L of each compound) were passed through these columns and monitored for inorganic halide before and after elution. The selected compounds in the column eluate were analyzed by freeze-drying, methanol extraction and HPLC.

The second approach was to remove the salts from the methanol extract of freeze-dried water samples. A mixed-bed column consisting of a strongly basic anion exchanger (Amberlyst A-26) and a strongly acid cation exchanger (Amberlyst 15) was used. Drinking-water samples (1 L) were freeze-dried and methanol-extracted, and the extract (50 mL) was spiked with the same five selected compounds. This solution was passed through the ion exchange column and monitored for inorganic halide before and after elution. The selected compounds in the eluate were analyzed by HPLC as described above.

The Sephadex LH-20 column was evaluated both for its ability to separate organic compounds from inorganic salts and as a possible fractionation step (based on molecular size) prior to HPLC. The gel column was calibrated for void volume (using Blue Dextran 2000) and salt elution volume (using sodium dichromate).

A 1-L sample of drinking water spiked with five selected organic compounds (uridine, 5-chlorouridine, 2-hydroxybenzimidazole, 3-hydroxy-2-methoxycinnamic acid and 2-hydroxybenzothiazole) was freeze-dried, and the methanol extract was passed through the gel column. Four fractions were collected corresponding to (1) high- and intermediate-molecular-weight (greater than 200 daltons); (2) low-molecular-weight (corresponding to the salt elution volume); (3) "interacting" compounds; and (4) strongly "interacting" compounds/column blank. The organic compounds in the fractions were monitored using HPLC.

A methanol extract of freeze-dried treated water (sample C) was then spiked with one marker compound (3-hydroxy-2-methoxycinnamic acid) and applied to the column. Elution was carried out using methanol/water (1/1 v/v) and fractions were collected as above. The fractions were monitored using HPLC.

## High-Performance Liquid Chromatography

The equipment used consisted of two solvent delivery systems (Waters Assoc., Model 6000A), a gradient former (Waters Assoc., Model 660) and two ultraviolet (UV) absorption detectors operated in series (Cecil Instruments, Model CE 212; LDC, Model 1203), at 280 and 254 nm. Syringe injections were made through a stop-flow septumless injection port for analytical separations (up to 10 $\mu$L injected) and via an injection valve (Rheodyne Inc., Model 7125) for preparative separations (up to 200 $\mu$L injected). Columns (20 cm x 7 mm i.d.) were packed in our laboratory [2] with 5-$\mu$m particle size Spherisorb-ODS (Phase Separations Ltd.) and generally had an efficiency of about 18,000 theoretical plates (HETP, 0.01 mm; reduced plate height, 2.2) as measured from a standard mixture of phenols [3].

A linear gradient was established from two solvent mixtures consisting of 1% methanol in 0.1% aq acetic acid (A) and 90% methanol in 0.1% aq acetic acid (B). The gradient was run over 20 min from 0 to 100% B. The flowrate was maintained at 2.0 mL/min.

## Organohalogen and Inorganic Halide Measurements

These measurements were made by titration against silver ion using a coulometer (KIWA NV, Rijswijk, Netherlands). To measure the amount of organically bound halogen present in samples, 10- to 90-$\mu$L portions were injected into a stream of oxygen and argon flowing through a quartz combustion tube heated to 875°C. The combustion process converts organohalogen to hydrogen halide. The effluent gases from the combustion tube are passed into the cell of the coulometer where the halide is titrated. On the instrument range used, the minimum detectable quantity of organohalogen is 0.2 nmol (7 ng as chlorine).

As far as we are aware, this is the first time that fractions from reverse-phase HPLC have been analyzed for organohalogen content using coulometric titration. Therefore it was established initially that methanol-water passed through an HPLC column did not pick up significant amounts of organohalogen. The results were:

| | Organohalogen | Relative Standard Deviation |
|---|---|---|
| Methanol-water (70/30 v/v) | 3.7 $\mu$mol/L | 50% |
| Methanol-water (70/30) after passage through HPLC column | 4.7 $\mu$mol/L | 34% |

These figures are acceptable since they lie close to the detection limit of the instrument. Efficiencies of combustion and titration of selected compounds dissolved in HPLC mobile phase were then determined:

- 5-Chlorouridine = 92%
- 2,3-Dichloronaphth-1-ol = 87–91%
- 4-Bromoanisole = 60–65%

The figures of around 90% for chlorinated compounds and around 60% for brominated compounds are normal for the combustion conditions used. The lower recovery of bromine is believed to be due to partial conversion of bromine to species more oxidized than bromide.

These experiments show that there is no basic incompatibility that invalidates the monitoring of organohalogen content of reverse-phase HPLC fractions by coulometry.

To monitor the inorganic halide content of water samples, HPLC fractions and the desalting performance of the ion exchange resins, 10-$\mu$L injections of solutions (aqueous or methanolic) were made directly into the coulometer cell. The minimum detectable concentration of inorganic halide in 10-$\mu$L injections was 20 $\mu$mol/L (0.7 mg/L as chloride).

### Mass Spectrometry

Preparative HPLC fractions were selected on the basis of a relatively high UV absorption of TOX level for mass spectrometric analysis. EI mass spectrometry was carried out on an AEI MS-30 instrument using a heated direct insertion probe. FD mass spectra were obtained on either an AEI MS-30 or a VG Micromass ZAB IF instrument. Samples were loaded by syringe onto FD emitters consisting of carbon microneedle activated tungsten wires (10-$\mu$m diameter). All mass spectra were acquired and processed on a VG Datasystems computer.

## RESULTS AND DISCUSSION

### Analysis of Freeze-Dried Water Extracts

#### *HPLC Separation*

Three drinking water samples (samples A–C) have been examined using the techniques outlined previously. Two of the samples have been fractionated by preparative HPLC, and certain fractions were examined by the TOX monitor and mass spectrometry.

Figure 1 shows a typical chromatogram obtained from an analytical HPLC separation of a drinking water extract (sample A). The initial analytical separation is carried out for several reasons. It provides a fingerprint of the sample for comparison with other samples analyzed and shows the position of the major UV-absorbing peaks in the chromatogram so that these can be collected as discrete fractions by preparative-scale HPLC. In addition, the identity of a compound found in an extract can be confirmed by co-injection of the sample and the authentic compound. Analytical HPLC also provides a method of quantitative analysis for any compound identified and enables spiking experiments to be monitored.

The large peak at the start of the chromatogram shown in Figure 1 is always observed when analyzing drinking water extracts under these conditions. It is essentially unretained and can therefore be regarded as an excluded peak. The composition of this peak is unknown, but it has been shown to contain several components by carrying out a simple form of exclusion chromatography using reverse-phase HPLC with water as eluent. It has also been established that the peak is not entirely made up of high-molecular-weight material (e.g., humic and fulvic acids) by fractionating a drinking water extract on Sephadex LH-20. The excluded peak was found to occur mainly in the low-molecular-weight fraction (less than 200 daltons), whereas humic and fulvic acids would be expected to elute in the high- and intermediate-molecular-weight fraction. In addition, injections of humic and fulvic acid concentrates (extracted from River Thames) did not produce

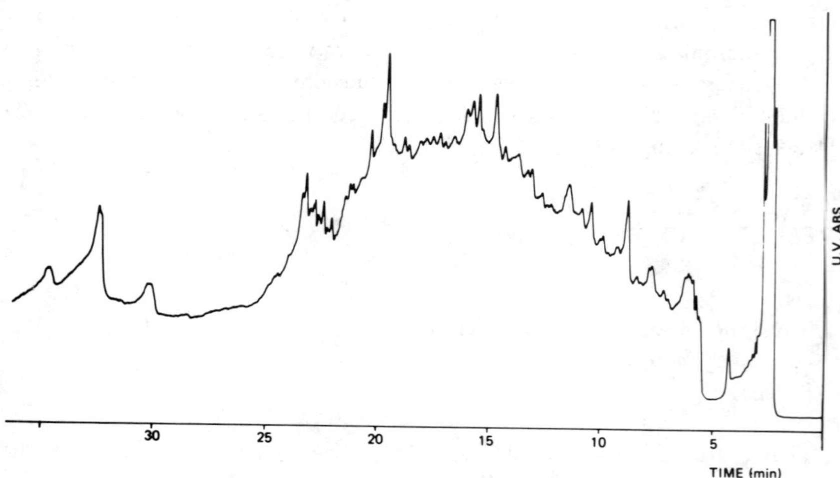


**Figure 1.** HPLC chromatogram of drinking water extract (sample A).

large excluded peaks on reverse-phase HPLC, but a series of broad and poorly resolved peaks spread over the chromatogram. Further work is in progress to try to characterize the excluded peak.

The HPLC system used in this work separates components essentially according to polarity. Thus, compounds of high polarity (e.g., purines and pyrimidines) are eluted at short retention times, whereas low polarity compounds (e.g., alkylphthalates or polycyclic aromatic hydrocarbons) are eluted at longer retention times. The presence of acetic acid suppresses the ionization of acidic solutes, giving better peak shapes for these components.

Preparative HPLC was carried out by injecting up to 200 $\mu$L of extract onto the column via a loop injector. Previous work [2] has shown that up to 200 $\mu$L of extract may be injected onto a 20-cm x 7-mm i.d. column with little loss of performance. Fractions were collected corresponding to the major peaks, as indicated by UV absorption, and also at roughly equal intervals throughout the rest of the analysis (see, for example, Figure 2).

*Organohalogen Analysis of HPLC Fractions*

Organohalogen measurements were made of HPLC fractions of the extract of sample B (see Figure 2). The largest values were found in fractions 1 and 2 (about 99% of the total found). Later fractions contained much lower levels, although certain fractions, for example 4 and 14, contained relatively high levels compared to neighboring fractions. Examination of such fractions by EI/MS indicated the presence of chlorine- and/or bromine-containing compounds. Conversely, EI/MS analysis of fractions which had relatively low

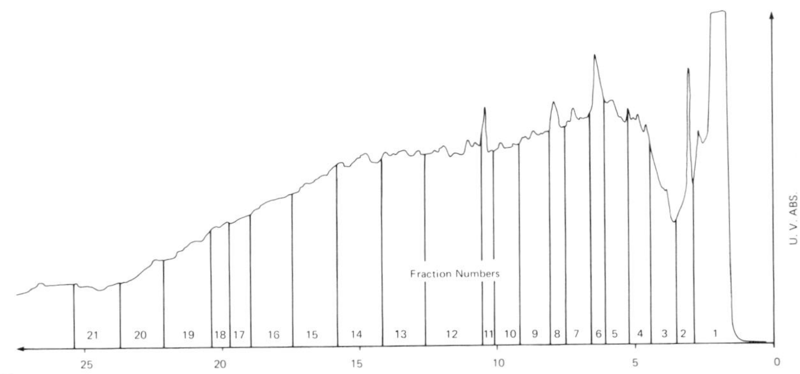


**Figure 2.** HPLC chromatogram from a drinking water extract (sample B) with indication of collected fractions.

levels of organohalogen, for example 7 and 11, showed no halogenated organics. It became apparent during this work that organohalogen levels found in fractions 1 and 2 were inflated by the presence of inorganic halide. An indication of this presence was provided by the abnormally slow (in comparison with organic halide) combustion-titration process observed for HPLC fractions, particularly the early eluting ones. This evidence was supported by FD/MS, where the presence of large amounts of sodium and potassium ions was found, particularly in early eluting HPLC fractions.

Results from organohalogen analysis of HPLC fractions taken from sample C are given in Table I. For these fractions an attempt was made to allow for the contributions of inorganic halides. Thus, a solution of sodium chloride was analyzed as for organohalogen samples. A recovery of 63% was obtained. The inorganic halide content of certain fractions was then measured by direct injection into the coulometer cell. Applying the factor of 63% to the results gave an estimate of the level of inorganic halide contribution to the organohalogen measurements. The corrected organohalogen results indicate that inorganic halide is more significant in the early eluting HPLC fractions. However the simple correction procedure applied here is clearly not satisfactory since the estimated interference in fraction 3 is larger than the original measured organohalogen value.

These experiments show that offline organohalogen measurements on HPLC fractions can give useful information on the presence of halogenated organics in drinking water.

Although the results are affected by the presence of inorganic halide, a crude method of correction can be applied. Further experiments may reveal that a less simple correction procedure will considerably improve the accuracy of the estimation. However it should be more satisfactory to desalt extracts before organohalogen measurement rather than attempting to correct for inorganic halide interference.

### *Mass Spectrometry*

Mass spectrometric analysis of preparative HPLC fractions generally indicated that these are composed of mixtures of several compounds. Although a few individual components have been identified [2] using EI/ and FI/MS, this is not generally possible with these techniques. Thus EI/MS of most HPLC fractions shows a large number of intense peaks over the mass range m/z 50–150, presumably resulting from pyrolysis of labile, higher-molecular-weight compounds but few, if any, higher mass ions. This is confirmed by FD/MS.

Table I. TOX and Inorganic Halide Levels in Fractions Taken from the HPLC Separation of a Drinking Water Extract (Sample C)

| Fraction | Measured Organohalogen (μg/fraction) | Estimated Interference Due to Inorganic Chloride[a] (μg/fraction) | Calculated Organohalogen (μg/fraction) |
|---|---|---|---|
| 1 | 66,200 | 47,000 | 19,200 |
| 2 | 9,700 | 5,500 | 4,200 |
| 3 | 44.6 | 67 | -23 |
| 4 | 37.1 | 12.3 | 24.8 |
| 5 | 33.0 | | |
| 6 | 40.1 | 16.2 | 23.9 |
| 7 | 17.9 | | |
| 8 | 23.1 | 5.2 | 17.9 |
| 9 | 22.3 | | |
| 10 | 20.8 | | |
| 11 | 20.0 | | |
| 12 | 22.9 | 2.0 | 20.9 |
| 13 | 25.2 | | |
| 14 | 25.3 | | |
| 15 | 17.5 | | |
| 16 | 15.9 | 0.9 | 15.0 |
| 17 | 16.8 | | |
| 18 | 16.7 | | |
| 19 | 14.7 | | |
| 20 | 22.3 | | |
| 21 | 19.1 | | |
| 22 | 21.5 | | |
| 23 | 19.6 | | |
| 24 | 14.9 | | |
| 25 | 10.7 | 0.8 | 9.9 |
| 26 | 9.8 | | |
| 27 | 7.0 | | |

[a]An aqueous solution of sodium chloride injected into the cell via the combustion tube gave a 63% recovery of chloride. This factor was applied to the inorganic halide content of certain fractions measured by direct injection into the cell, to estimate the amount of the measured organohalogen signal due to inorganic halide.

The most promising MS results have been obtained using FD/MS. Complex mass spectra consisting of many molecular and fragment ions have been obtained consistently on analysis of HPLC fractions. A typical example of the FD spectra obtained (Figure 3) shows a large number of intense low-mass ions, mostly resulting from fragmentation, and a spread of higher-mass ions, the majority of which are probably molecular ions. The complexity and large

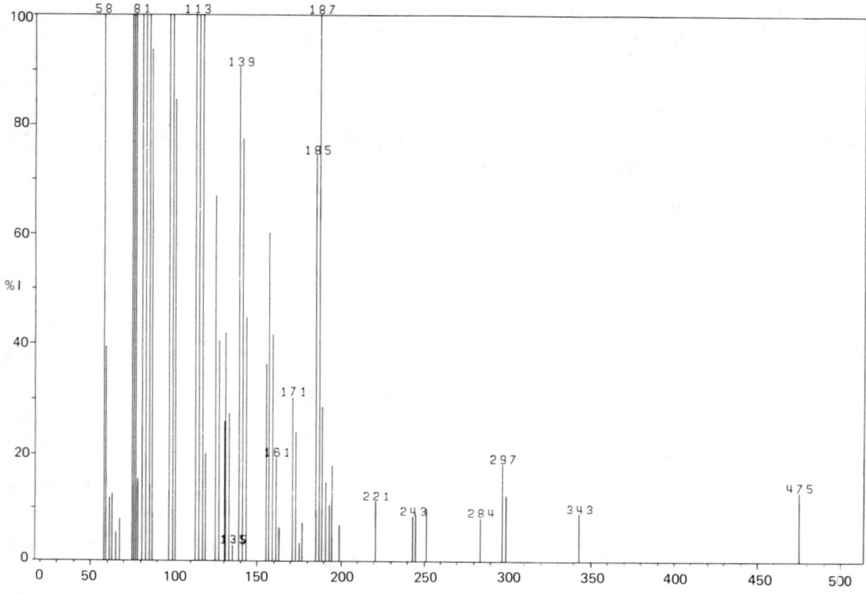


**Figure 3.** Low-molecular-weight scan of a treated water HPLC fraction by FD/MS.

molecular-weight range of these mixtures of compounds present in individual HPLC fractions is further confirmed when examining the mass range from m/z 500 to 1500. Several hundred ions are observed in this region, and there is an indication (from isotope patterns) that several of these contain bromine and/or chlorine. At present these compounds cannot be characterized further, but the techniques for doing this are available as is discussed later.

## Removal of Salts

### *Ion Exchange Chromatography*

The method using ion exchange resin columns on the water samples resulted in excellent removal of salts as indicated by a reduction in the inorganic halide of better than 90%. However, recoveries of the selected organics (uridine, 5-chlorouracil, benzoic acid and 2-hydroxybenzoic acid) from the column eluate were very poor, less than 2% with only two exceptions (2,3,4-trichlorophenol, 21% and di-*n*-propylphthalate, 78%). Similar results were obtained in repeat experiments, and this approach was therefore abandoned.

Removal of salts from a spiked methanol extract of a freeze-dried water sample was also very efficient, as indicated by monitoring the inorganic halide level. Once again the problem was that recovery of selected organics was also very poor, less than 10% with one exception (di-*n*-propylphthalate, 34%).

It was apparent from these experiments that while ion exchange columns were very efficient at desalting, they were almost equally efficient at removing the organic compounds added.

### *Gel Permeation Chromatography*

HPLC analysis of the gel fractions taken from the analysis of the spiked drinking water indicated that the selected organics eluted in the low-molecular-weight fraction. Recoveries were approximately 40% for uridine, 5-chlorouridine and 3-hydroxy-2-methoxycinnamic acid, while 2-hydroxybenzothiazole and 2-hydroxybenzimidazole were not recovered at all. However, since the recoveries of organics obtained were an improvement on those obtained with ion exchange resins, fractionation using gel permeation chromatography was carried out on a drinking water extract (from sample C). HPLC analysis of the four fractions collected produced the following results. The gel column was very efficient at separating the inorganic salts from the medium- to high-molecular-weight fraction. However, the marker compound added to the extract and the discrete components observed after HPLC separation and UV detection appear in the low-molecular-weight fraction. This does not necessarily mean that such components are all of low molecular weight, since some compounds will be retained on the column by "interaction" with the gel. The large peak appearing at the start of the HPLC chromatogram of a drinking water extract (see, for example, Figure 1) which has been termed the "excluded" peak, is also eluted from the gel column in the low-molecular-weight fraction. The two fractions collected after elution of salts/low-molecular-weight organics from the column contained a negligible portion of the extract.

These results indicate that Sephadex LH-20 does not separate inorganic salts from the bulk of organic material present in drinking water extracts. The reasons for this are that the low resolution of Sephadex LH-20 and the interaction of organic compounds with the gel result in the low-molecular-weight fraction (salt elution volume) containing much of the organic material. This can probably be overcome by changing to a high-resolution gel column which will provide both increased resolution to separate organic compounds from salts and give less "interaction" between the organics and the gel.

## Recovery of Selected Organic Compounds from Treated Water

Initial experiments [2] on the recovery of organic compounds from freeze-dried extracts involved the use of TOC measurements before and after freeze-drying a drinking water sample. These results indicated a high recovery of TOC. Similar TOC measurements on blanks consisting of double-distilled deionized water with added inorganic salts showed a negligible increase in TOC as a result of the freeze-drying process. However, it was essential to carry out some work on the recovery of individual compounds. In particular, this would test the efficiency of the methanol extraction of freeze-dried solids which cannot be evaluated using TOC measurements. Equivalent samples were also extracted with dichloromethane to determine recoveries using a less polar solvent.

The results (Table II) show a good recovery for most organics using methanol as the extraction solvent with the exception of *p*-chlorophenylacetic acid. The reason for the poor recovery of this particular compound is not apparent. Recoveries using dichloromethane as extraction solvent were generally poor. These experiments have only been done once, and further work is planned which will investigate a wider range of selected compounds and extraction solvents.

**Table II. Recovery of Selected Organic Compounds from Treated Water**

| | % Recovery | |
|---|---|---|
| | Methanol Extract | Dichloromethane Extract |
| Uridine | 75 | 0.8 |
| Benzoic Acid | 40 | 8 |
| 2-Hydroxybenzothiazole | 76 | 72 |
| Di-*n*-propylphthalate | 80 | 85 |
| 5-Chlorouracil | 76 | 0.9 |
| 4-Chlorobenzoic Acid | 69 | ND[a] |
| 2,3,4-Trichlorophenol | 80 | 10 |
| *p*-Chlorophenylacetic Acid | 18 | ND |

[a]Not detected.

## CONCLUSIONS AND FUTURE WORK

The isolation and concentration of nonvolatile organic compounds from drinking water by freeze-drying and methanol extraction of the residue has been shown to be an efficient method for a range of organic compound types. Processing of large samples is possible, but is a lengthy procedure.

A method of desalting the extracts, without also removing a large proportion of the organic material, is a desirable, though elusive, goal. Ion exchange resins were almost as efficient at removing organics as they were at removing inorganic salts. Sephadex gel gave the highest recoveries of organics, but did not separate either the selected organics or the majority of UV-detectable organics in a freeze-dried extract from salts, due to "interaction" and low resolution. High-resolution gel permeation chromatography appears to offer most hope for development of a suitable desalting method.

HPLC has been shown to be ideally suited to the separation of the extracted organic material. A reverse-phase system using a broad linear gradient provided good resolution. Packing of HPLC columns in our laboratory enables us to use highly efficient columns for this work. However, the methanol extract of a freeze-dried residue from drinking water contains a complex mixture, and even with the use of such columns mass spectrometric analysis shows that HPLC fractions rarely consist of a single major component. To overcome this problem an additional separation technique is required, and/or more efficient columns need to be used.

Initial work on the use of another separation technique has already been reported in this chapter, i.e., the use of gel permeation chromatography. Although the system used was only of low efficiency it was easily possible to separate a high/medium-molecular-weight fraction from the low-molecular-weight components of a drinking water extract. Further work will involve the use of high-efficiency aqueous compatible gels to separate the extract into narrow-range molecular-weight fractions. The further separation of each fraction by reverse-phase HPLC should produce much simpler fractions for analysis by mass spectrometry.

Preparation of more efficient columns will require the use of capillary liquid chromatography (LC). This is a recent development in LC, and there are still many practical problems to be solved before the technique could be used routinely. However, columns of extremely high efficiency have been reported [4] which indicates that capillary LC has great potential for the separation of complex mixtures.

The use of an offline TOX monitor for the analysis of HPLC fractions has been shown to provide an indication of the organohalogen content. In the continued absence of a suitable online chlorine detector for HPLC, this technique can provide useful data. However an efficient method of desalting the extract (to remove halide interference) would enable more accurate TOX analysis to be carried out.

The molecular weight of discrete compounds present in HPLC fractions has been shown by FD/MS to range up to at least m/z 1500. Identification of these trace nonvolatile compounds is a difficult problem, requiring the use of the most sophisticated chromatographic and mass spectrometric techniques available.

The promise shown by FD/MS for investigation of these nonvolatile compounds is being exploited, and will be used in combination with collisional activation and linked scanning techniques. With this approach it should be possible to obtain fragment ion mass spectra of the molecular ions observed in the FD mass spectrum. These data, together with high resolution FD/MS studies, which provide an empirical formula for each molecular ion, should yield valuable structural information on a large number of compounds.

Figure 4 shows the procedure which is emerging for characterization of nonvolatile organics in drinking water. While this procedure is slow, it is a

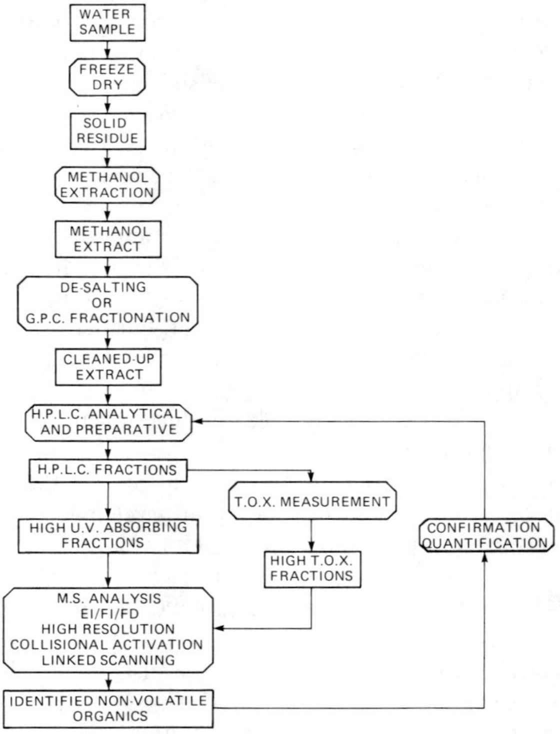


**Figure 4.** Scheme for characterization of nonvolatile organics in drinking water.

method with the required sensitivity and specificity for identification of a wide range of nonvolatile organics in drinking water.

## ACKNOWLEDGMENTS

This work was undertaken for and funded by the Department of the Environment whose permission to publish this chapter has been obtained. The authors would also like to thank the Director of the Water Research Centre, Medmenham Laboratory, for permission to publish.

## REFERENCES

1. Garrison, A. W., L. H. Keith and W. M. Shackleford. In: *Aquatic Pollutans: Transformations and Biological Effects* (Oxford: Pergamon Press, 1978), pp. 39-68.
2. Crathorne, B., C. D. Watts and M. Fielding. *J. Chromatog.* 185:671 (1979).
3. Bristow, P. A., and J. H. Knox. *Chromatographia* 10:279 (1977).
4. Scott, R. P. W., and P. Kucera. *J. Chromatog.* 169:51 (1979).

## CHAPTER 26

# EVALUATION OF TRACE ENRICHMENT METHODOLOGY IN CONJUNCTION WITH REVERSE-PHASE HIGH-PERFORMANCE LIQUID CHROMATOGRAPHY FOR PRECONCENTRATION, SEPARATION AND QUANTIFICATION OF TRACE NONVOLATILE ORGANIC COMPOUNDS IN AQUEOUS ENVIRONMENTAL SAMPLES

**Jeffrey A. Graham* and Arthur W. Garrison**

Analytical Chemistry Branch
Environmental Research Laboratory
U.S. Environmental Protection Agency
Athens, Georgia

Nonvolatile organic compounds, which may comprise up to 90% of dissolved organic compounds in water samples of environmental interest, have recently become increasingly important because of their possible adverse environmental health effects [1,2]. The analytical methodology employed for compounds readily separated by gas chromatography (GC) and subsequently identified and quantified by mass spectroscopy (MS) is difficult, and in many cases impossible to apply with any significant degree of confidence to determinations of nonvolatile compounds [2,3].

*National Research Council Research Associate, October 4, 1979 to September 1, 1980. Present address: Monsanto Research Corporation, Dayton Laboratory, Dayton, OH.

A separation of some type becomes mandatory when reliable analytical data are required on the identity and quantity of compounds present in a complex matrix, e.g., those encountered in various environmental water samples. The obvious approach to nonvolatile organic compounds, then, would be separation by liquid chromatography (LC). In 1972 Katz et al. [4] employed anion exchange chromatography for separation of nonvolatile organic compounds isolated from primary sewage, and ultraviolet (UV) absorption spectroscopy, GC and MS for quantification and identification. The authors commented, "Such treatments are tedious and become increasingly difficult as one progresses from more to less abundant constituents." Pitt and co-workers [5,6] and Jolley et al. [7] have also applied anion exchange chromatography in conjunction with other chromatographic, spectroscopic and radiochemical techniques for identification and quantification of nonvolatile organic compounds isolated from a number of environmental water samples.

Most recently, Crathorne et al. [3] separated a number of nonvolatile organic compounds isolated from drinking water by reverse-phase (RP) high-performance liquid chromatography (HPLC). Subsequent identification was accomplished via offline electron impact and field ionization mass spectroscopy. The authors stressed the need for "refinement of the extraction technique in order to obtain more of the organic matter in a suitable form for analysis."

The development of modern HPLC, especially in the RP mode, provided the analyst with a technique ideally suited for the separation of complex mixtures composed of a wide variety of compounds from many classes [8]. Presently available commercial RP columns (15 cm x 4.6 mm i.d., 5-$\mu$m diameter support) readily provide 8000–10,000 theoretical plates [9]. In addition, the columns provide increasingly reproducible relative retentions and have typical lifetimes of 6–12 months. Application of ion suppression techniques [10], or addition of an ion-pairing [11] or ion-interaction [12] reagent to the mobile phase readily allows utilization of RP columns for the separation of acidic, basic and zwitterionic organic species. Microcomputer-based liquid chromatographs provide highly reproducible mobile phase flowrates and gradients. Thus, modern HPLC methodology seems perfectly matched to the problem of separating the vast array of different nonvolatile organic compounds found in environmental water samples.

Kirkland [13] reviewed the preferred experimental conditions for trace analysis in HPLC and pointed out that very large sample volumes can be loaded onto LC columns if the solvent in which the sample is dissolved is of a low eluting power. If this results in large capacity factors, $k'$, for the components of interest, then they accumulate or are enriched on the front of the stationary support. After the desired volume of sample is loaded, the eluting strength of the mobile phase can be increased to the point at which the k′ values are small and the solute bands elute. Little and Fallick [14] first proposed an adaptation of this phenomenon for concentrating nonpolar organics from polar sources and coined the descriptive term "trace enrichment." They pumped 200 mL of riverwater through a RP column and subsequent applied binary solvent gradient to elute and separate the enriched components. Since these initial reports, several other workers have employed accumulation or trace enrichment of organic compounds from a variety of aqueous samples on octadecyl-derivatized silica (ODS) [15-26]. The majority of organic compounds that have been isolated with trace enrichment thus far, however, have been relatively nonpolar in nature and thus, unlike nonvolatile organic compounds, are easily separable by GC.

Recently, Saner et al. [27] evaluated a pellicular, porous ODS support for trace enrichment of oil-contaminated seawater. Additionally, Van Vliet et al. [28] addressed the optimization of the design of the enriching column to minimize extracolumn band broadening. Examining two phthalate esters in water at concentrations of 17–194 ppb, they demonstrated superiority of trace enrichment on small particle diameter ODS supports (i.e., 5 and 10 μm as opposed to 37–75 μm), independence of the efficiency of preconcentration on the volumetric enrichment rate and outstanding reproducibility for the method.

Trace enrichment on ODS, then, appears to be a promising first step in the approach to analysis of trace levels of nonvolatile organic compounds in water samples for several reasons. First, various operating parameters influence the lower limit of detection in HPLC. Taking into account the fact that the chromatographic process dilutes the injected sample, Karger et al. [29] derived an expression for the minimum detectable concentration as a function of the operating parameters in isocratic elution LC. Absorption spectroscopy is assumed to be employed as the detection method.

$$C_{min} = 5(2\pi)^{1/2} \times (H/L)^{-1/2} \times SNR^{-1} \times V_o(k' + 1) \times V_{inj}^{-1} \quad (1)$$

where $C_{min}$ = the minimal detectable concentration of solute, assuming a lower limit of detection equal to five times the RMS noise, a solute band of Gaussian profile and no interference from overlapping solute bands

$H$ = the height equivalent to a theoretical plate for a column of length $L$

$SNR$ = the signal-to-noise ratio

$V_o$ = the void volume of the column and associated connecting tubing

$k'$ = the capacity factor of the solute of interest

$V_{inj}$ = the volume of sample injected

Examination of Equation 1 reveals two readily controllable variables the analyst can adjust to a certain degree to achieve the detection limit necessary for a given analysis. Certainly the k′ value can be easily controlled by adjusting the eluting strength of the mobile phase. Additionally, by applying gradient elution, increasing the strong solvent component of the mobile phase as a function of time, the k′ value is effectively reduced for each solute as it elutes and thus, the detection limit is decreased while simultaneously decreasing analysis time [8,13,30-32]. The second variable, the volume of sample injected, can be increased substantially by trace enrichment.

Second, examination of the problems discussed by Crathorne et al. [3], "refinement of the extraction technique" and obtainment of a large amount of sample "in a suitable form for analysis," reveals that trace enrichment readily meets the demanding criteria. Finally, current sample collection, isolation and preconcentration techniques commonly applied to aqueous environmental samples can often be the limiting factor in obtaining accurate and reproducible quantitative data [33,34]. If trace enrichment were to take place at the sampling site, the enriching column could be appropriately sealed and shipped to the analytical laboratory. Once at the site of analysis, appropriate installation and application of a mobile phase gradient would provide selective elution of the enriched components directly onto the analytical column. The reduction in sample handling and pretreatment offered by this approach reduces a major source of error.

With these ideas in mind, the authors set out to evaluate the applicability of trace enrichment on ODS and subsequent elution and separation by gradient elution reverse-phase HPLC for the routine analytical determination of trace levels of nonvolatile organic compounds in aqueous environmental samples. The first step of the evaluation was to determine the efficiency of preconcentration or recovery, the fraction of the total amount of compound passed through the ODS accumulator column that was retained, for a group of model compounds under a given set of sampling conditions.

For a given compound, less than 100% recovery would be acceptable provided the value was reproducible. This was important, because it was anticipated that because of the highly polar and ionic nature of nonvolatile organic compounds, less than 100% recovery would be common. As pointed out earlier, most compounds isolated by trace enrichment thus far have been relatively nonpolar or hydrophobic in nature and, therefore, have been recovered with generally 100% efficiency. Enrichment of organic compounds from aqueous systems on ODS should be governed by the same phenomena as retention in RP LC. It is generally agreed that retention is caused by a hydrophobic [35] or a solvophobic effect [36-38]. Thus, organic compounds incapable of hydrogen bonding or interacting with the dipoles of the water molecule and that generally disturb the ordered structure of liquid water are strongly retained on ODS, i.e., recovered with almost 100% efficiency. As polar functional groups are attached, interaction with the surrounding water molecules becomes much more favorable, and the retention or recovery consequently decreases. The latter case is much more descriptive of the nature of nonvolatile organic compounds [3-7].

Many nonvolatile organic compounds are of an acidic, basic or zwitterionic nature and are sometimes exceedingly difficult to isolate efficiently from an aqueous matrix. The authors were interested in investigating the methodology of ion-pairing extractions [39] and ion-pairing chromatography [8,10,11,39] for increasing the recoveries of ionic organic compounds from aqueous matrices. Addition of so called "ion-pairing" reagents to mobile phases in reverse-phase HPLC may enhance retention by any of a number of mechanisms. In situ formation of an ion exchanger due to retention of the "ion-pairing" reagent on the RP support [40], ion-pair formation in the mobile phase between the ion-pairing reagent and the solute with subsequent sorption of the ion-solute pair onto the RP support [41], and several other mechanisms [12,42] have been proposed to explain empirically observed increased retention of ionic compounds in the presence of these so-called ion-pairing reagents. We were not so much interested in the mechanism as the end result.

Many aspects of the enrichment procedure could affect the reproducibility of fractional recoveries. Volumetric accumulation rate and total accumulation volume were the most obvious influences. Ideally, the concentration of compounds and the aqueous matrix should not affect the recovery. We anticipated, however, that variability in fractional recoveries of specific compounds from differing matrices would ultimately limit the accuracy with which quantitative measurements could be made. A thorough study of matrix effects would be a necessity.

The difficult task of qualitatively identifying unknown compounds isolated and separated by the techniques discussed thus far was not undertaken. This aspect still remains the ultimate limitation in identifying the vast array of nonvolatile organic compounds present in aqueous environmental samples. The trace enrichment methodology, however, was anticipated to be applicable to monitoring known nonvolatile organic compounds of interest in aqueous samples.

## EXPERIMENTAL

### Chemicals

Model compounds (Table I) were obtained from a wide variety of suppliers.

Acetonitrile and methanol, used as mobile phases and to dissolve various model compounds, were obtained from Burdick and Jackson Laboratories, Inc. (Muskegon, MI) and were used without further purification. Phosphoric

Table I. Model Compounds Examined in This Study

| Group | Compound |
|---|---|
| A | *m*-Nitroaniline |
| | 2-Chloro-ethylamino-6-isopropylamino-*s*-triazine (Atrazine) |
| | 2,6-Dichloroaniline |
| | N-Nitrosodiphenylamine |
| | Decafluorobiphenyl |
| | 2-Chloro-4-nitro-4′-[N-ethyl-N-(2-hydroxyethyl)] aminoazobenzene (Disperse Red Dye 13) |
| B | *p*-Hydroxyphenylacetic acid |
| | *o*-Chlorophenylacetic acid |
| | *p*-Chloromandelic acid |
| | *m*-Chlorosalicylic acid |
| | 1-Amino-8-naphthol-2,4-disulfonic acid (H-acid) |
| C | *m*-Hydroxyaniline |
| | Phenylurea |
| | Aniline |
| | *m*-Nitroaniline |
| | *p,p*′-Methylenedianiline |
| | 2,6-Dichloroaniline |
| | *p*-Phenylazoaniline (Solvent Yellow Dye 1) |

and perchloric acid, used to adjust the acidity of aqueous mobile phases and samples, were reagent grade, used without further purification and obtained from J. T. Baker Chemical Co. (Phillipsburg, NJ).

Ion-pairing reagents (tetrabutylammonium phosphate, sodium salts of 1-octanesulfonate and 1-dodecanesulfonate) were obtained from Regis Chemical Co. (Morton Grove, IL) and used without further purification. Sodium tetraborate, used to adjust the basicity of aqueous mobile phases and samples, was reagent grade, used without further purification and obtained from Curtin Matheson Scientific, Inc. (Atlanta, GA).

Water used in the preparation of aqueous mobile phases was purified from house reverse osmosis (RO) water. First, it was passed through a high-capacity mixed-bed ion exchange resin (Barnstead, D8901), then through an ultrapure mixed bed ion exchange resin (Barnstead 8902) and finally through an organic removal cartridge (Barnstead 8904). This configuration, which placed the organic removal cartridge after the ion exchange cartridges as opposed to before as the manufacturer recommended, produced water with somewhat higher conductivity but with significantly less UV-absorbing organic compounds (see Results and Discussion).

## Apparatus

The liquid chromatograph consisted of two Waters Assoc., Inc., (Milford, MA) Model M6000A pumps controlled by a Waters Assoc., Inc., Model 660 Solvent Programmer. Injections or column switching were accomplished by a Valco Instruments Co. (Houston, TX) Model CV-6-UHPa-N60 7000-psig six-port valve. A Waters Assoc., Inc., Model 440 fixed-wavelength absorption detector, operated at 254 nm, was used in series with a Waters Assoc., Inc., Model 450 variable-wavelength absorption detector, operated at 215 nm, for solute band detection. Chromatograms were recorded as a function of time on a Houston Instruments (Austin, TX) OmniScribe dual-pen recorder.

A Whatman Inc. (Clifton, NJ) Partisil-10 ODS-2, 250-mm x 4.5-mm i.d., reverse-phase HPLC column was employed as the analytical column throughout the study. A 40-mm x 2-mm i.d. precolumn, packed in-house with Partisil-10 ODS-2, was connected in series between the analytical column and the six-port injection valve. When mobile phases with a pH of 7.5 or greater were employed, a 40-mm x 4.6-mm i.d. ODS precolumn was connected between the outlet of the pumps and the solvent inlet of the six-port injection valve. This configuration provided a mobile phase saturated with silica and thereby retarded any dissolution of the precolumn or analytical column [43].

The trace enrichment or accumulator columns were constructed from 316 stainless steel tubing 70 x 2 mm i.d. and were terminated with 316 stainless steel ZDV fittings and 5.0-$\mu$m porosity, 316 stainless steel frits; all materials were obtained from Alltech Assoc., Inc. (Arlington Heights, IL). The accumulator columns were dry-packed with 0.155 g Waters Assoc., Inc. 80-100 mesh Porapak Q or 0.139 g of the ODS support obtained by cutting open a Waters Assoc., Inc., $C_{18}$ Sep-Pak®. No attempts to sieve the Sep-Pak ODS support into a narrow particle-diameter range were made.

The acidity of selected mobile phases was measured with a Markson Science Inc. (Atlanta, GA) Model 90 pH/temperature meter equipped with a Fisher Scientific Co. (Norcross, GA) Model 13-639-52 calomel reference electrode and with a Fisher Scientific Model 13-639-3 universal glass electrode.

## Procedures

All aqueous mobile phases were stirred and vigorously purged with high-purity helium for 5 min prior to addition of any ion-pairing reagents, buffers or organic modifiers. All mobile phases were filtered through a 0.45-$\mu$m filter prior to use.

Samples were prepared by dissolving an appropriate amount of compound in methanol. Each individual sample was chromatographed to ascertain its purity and to obtain molar absorbtivities at 254 and 215 nm. The ratio of the absorbtivities at these wavelengths aided in the identification of the individual compounds once spiked into real matrices, accumulated and separated.

Athens, GA, drinking water and treated municipal sewage from an Athens, GA, waste treatment facility were employed as two real sample matrices. Prior to spiking with model compounds and accumulation, each matrix was filtered through a 0.45-$\mu$m filter. Treated municipal sewage was not purged with helium, whereas drinking water was vigorously purged with high-purity helium prior to use to assure removal of residual chlorine and hypochlorite.

Prior to enriching a sample, each packed accumulator column was flushed with 100 void volumes (approximately 30 mL) of acetonitrile at 3.0 mL/min. Next, 10 mL of purified water was pumped through the column at 1.0 mL/min to eliminate the acetonitrile. The accumulator column, which was connected in place of the sample loop, was then switched out of series with the pump by the six-port valve. Next, the pump was primed and rinsed with the sample to be enriched, then the accumulator column was switched back in series with the pump and, at the desired flowrate, the sample was pumped through the accumulator column and exited to waste or to be collected for future study. Although the enriching was performed in the laboratory, it could have taken place equally well in the field with appropriate filtering and a portable pump. Once the components of the sample were enriched, the accumulator column was once again switched out of series with the pump. The analytical column was brought to the appropriate initial mobile phase composition, and the accumulator column was switched back in series with the pumping system and analytical column. The desired mobile phase gradient was then initiated, and the results were recorded. Each accumulator column could be used to enrich as many as 20 samples with no apparent degradation or compromise in performance. We elected, however, to repack each accumulator column after three samples had been enriched. As shown by Van Vliet et al. [28], the column-to-column reproducibility as measured by the peak heights of a given enriched sample component was on the order 3.9% relative standard deviation for five different columns.

### Calculations

Recoveries were calculated as the ratio of peak heights of each compound obtained from a direct injection of sample and that obtained from elution of the enriched compound from the accumulator column after the same amount as directly injected had been diluted and accumulated. All peak heights were measured manually to the closest 1/64 in.

## RESULTS AND DISCUSSION

### Sampling Dynamics

The procedure for further purifying the RO water was evaluated by enriching 100 mL at 5.0 mL/min on the ODS accumulator column. Figure 1 shows the chromatogram monitored at 254 and 215 nm resulting from gradient elution of the enriched components from the ODS accumulator column. The number and intensity of the background peaks from this purified water were less than those obtained from RO water further purified by a Milli-Q system or by connection of the ion exchange and organic removal cartridges as recommended by Barnstead. The sloping baseline, as shown by the 215-nm trace, was caused by refractive index changes between acetonitrile and water. We found this level of purity quite acceptable, and no further attempts at purification were made.

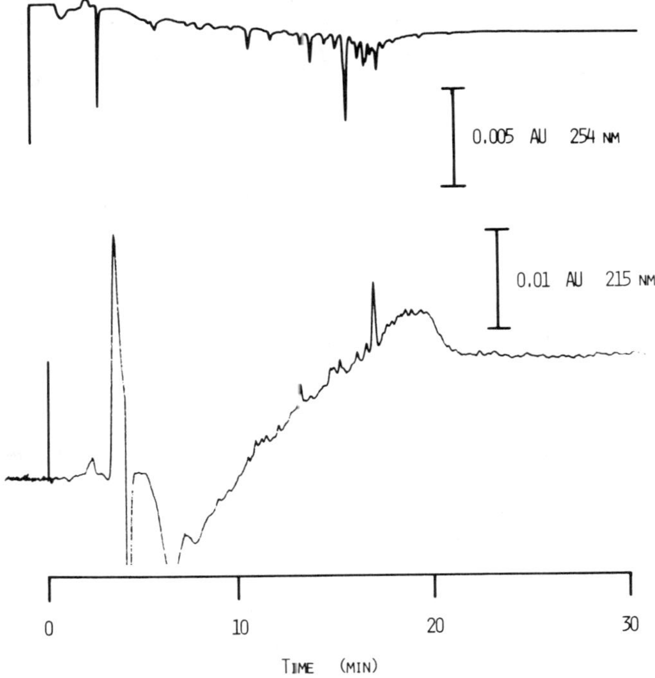


**Figure 1.** Chromatogram of components enriched from 100 mL of purified water on the ODS accumulator column and subsequently eluted and separated. Conditions: 10–90% acetonitrile in purified water at 5%/min and a flowrate of 1.0 mL/min. Column: Partisil-10, ODS-2, 250 x 4.6 mm i.d.

The effect of the volumetric accumulation rate was assessed by examination of the total sum of peak heights of the individual compounds, listed in Table I, Group A, that had been spiked in purified water and enriched at 1.0, 2.5, 5.0 and 7.5 mL/min. Figure 2 shows a chromatogram of a direct injection and that obtained from elution of an enriched sample that contained the same amount of each directly injected compound in 100 mL of purified water. Figure 3 shows the total recovery as a function of flowrate through the ODS accumulator column. Above volumetric accumulation rates of 1.0 mL/min, the amount of organic material recovered did not vary substantially with flowrate. This agreed quantitatively with the findings of Saner et al. [27] and Van Vliet et al. [28]. It was important to note, however, that the data were obtained with compounds that were substantially less hydrophobic than those examined by Saner et al. [27] and Van Vliet et al. [28]. This was important if the methodology was to be applied to routine analysis of nonvolatile organic compounds in aqueous samples. Additionally, this result meant that sampling could be done at rates of 5.0–10.0 mL/min, and that small variations in pumping rates should not adversely affect the recoveries of individual compounds.

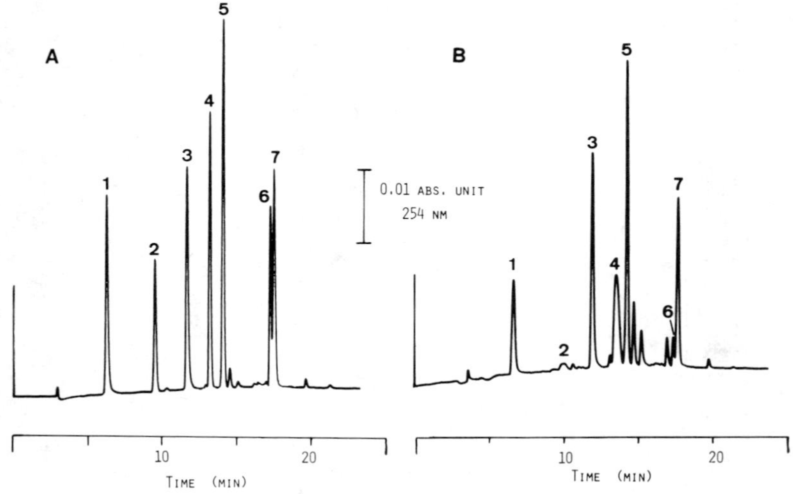


**Figure 2.** (A) Chromatogram of a 25-μL direct injection of (1) 0.26 μg caffeine; (2) 0.05 μg *m*-nitroaniline; (3) 0.44 μg atrazine; (4) 0.75 μg 2,6-dichloroaniline; (5) 0.43 μg N-nitrosodiphenylamine; (6) 0.85 μg decafluorobiphenyl; and (7) 0.41 μg disperse red dye 13. (B) Chromatogram of the same amount of each of the seven compounds spiked into 100 mL of purified water, enriched on the ODS accumulator column at 5.0 mL/min and subsequently eluted and separated. Conditions: (both A and B) 10–90% acetonitrile in purified water at 5%/min and a flowrate of 1.0 mL/min, Column: Partisil-10, ODS-2, 250 x 4.6 mm i.d.

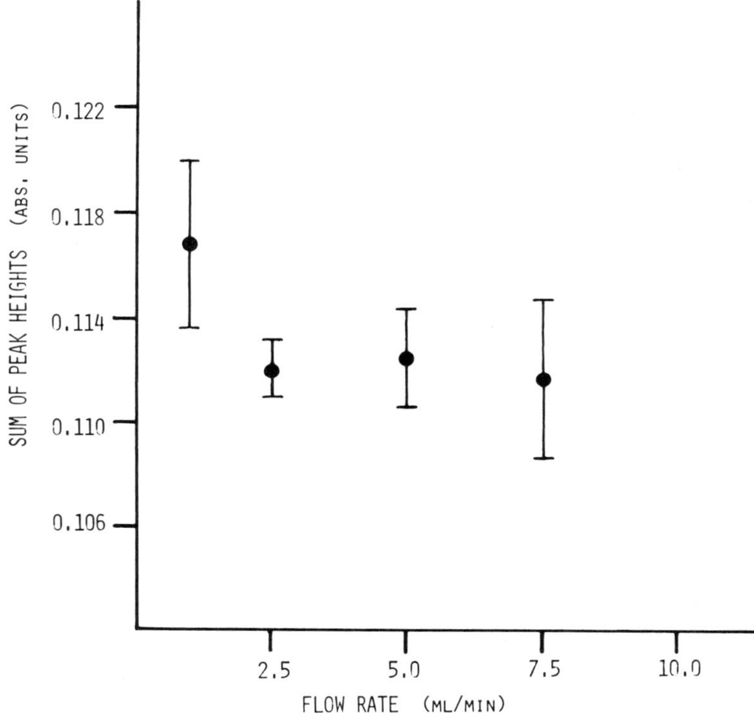


**Figure 3.** Sum of the peak heights of the seven compounds from Figure 2, spiked to the same concentration and enriched from 100 mL of purified water as a function of the flowrate through the ODS accumulation column.

The effect of different accumulation volumes on the recoveries of the seven compounds listed in Table I, Group A, were examined. Figures 4 and 5 show the recoveries of the individual compounds enriched on the ODS accumulator column at 5.0 mL/min from sample volumes of 25, 50, 100 and 200 mL. Note that atrazine and N-nitrosodiphenylamine were recovered at approximately the same efficiency from 50 to 200 mL. Disperse red dye 13 appeared to approach a minimum recovery at 100 mL, and at 200 mL was actually more efficiently recovered than at 25 mL. Decafluorobiphenyl displayed similar behavior to disperse red dye 13, except that the recovery at 200 mL was less than that at 25 mL. For all but disperse red dye 13, the maximum recovery was observed to occur at an accumulation volume of 25 mL. Caffeine, 2,6-dichloroaniline and *m*-nitroaniline all displayed decreasing recoveries with increasing accumulation volumes. Thus, it appeared that recoveries would have to be evaluated at constant accumulation volumes in order to be utilized for accurate quantification.

The recoveries of the same seven compounds were examined over three orders of magnitude of concentration. For these experiments 100 mL of sample was enriched at 5.0 mL/min. Figures 6 and 7 show that the recoveries for each of the seven compounds did not vary significantly with concentration. Van Vliet et al. [28] reported similar findings for somewhat more hydrophobic compounds.

The recovery for each compound at a given concentration was determined under the conditions in Figure 2 and shown in Table II. The averages were calculated from three determinations of 100 mL of sample enriched on the ODS accumulator column at 5.0 mL/min. The precision with which the recoveries were determined was outstanding considering the concentrations at which the compounds were present. The larger relative standard

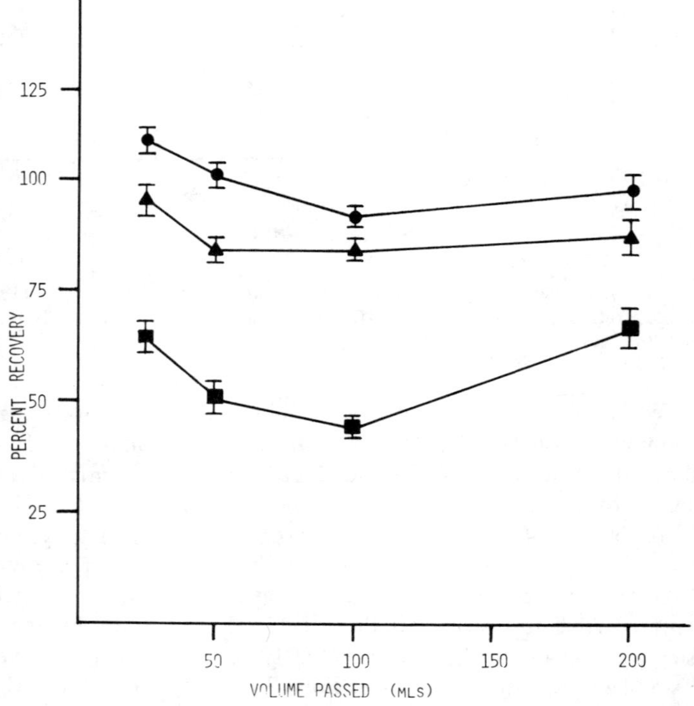


**Figure 4.** Percent recovery as a function of total sample volume passed through the ODS accumulator column at 5.0 mL/min with subsequent elution and separation as in Figure 2. ● = atrazine (4.4 ppb); ▲ = N-nitrosodiphenylamine (4.3 ppb); ■ = disperse red dye 13 (4.1 ppb).

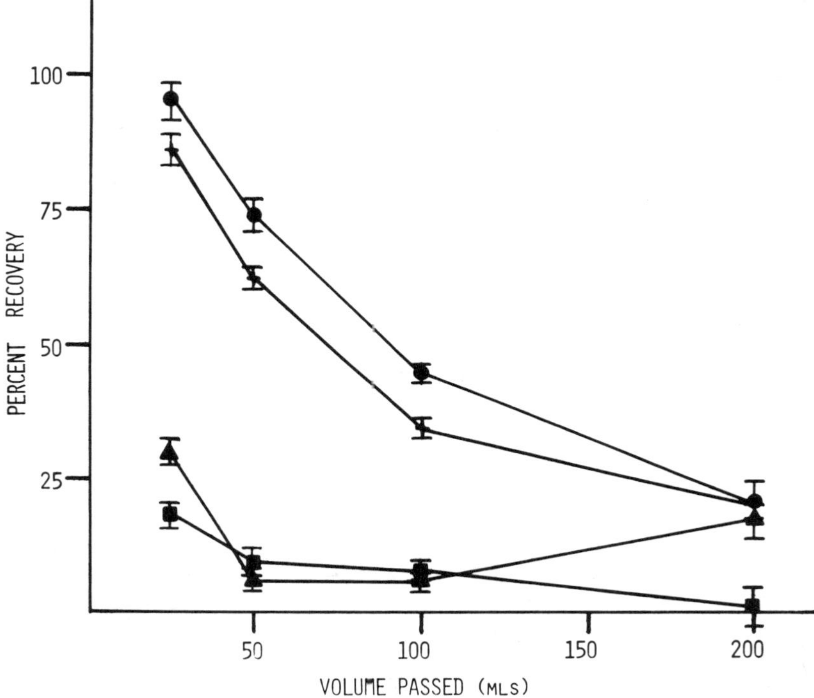


**Figure 5.** Percent recovery as a function of total sample volume passed through the ODS accumulator column at 5.0 mL/min with subsequent elution and separation as in Figure 2. ● = caffeine (2.6 ppb); + 2,6-dichloroaniline (7.5 ppb); ▲ = decafluorobiphenyl (8.5 ppb); ■ = *m*-nitroaniline (0.5 ppb).

deviation for caffeine could be explained in part by its elution volume. Because caffeine eluted early in relation to the other compounds, it may have not experienced the full effect of the mobile phase gradient. The peak height of a compound eluted under conditions of a mobile phase gradient is increased, relative to the corresponding height under isocratic conditions, because of the higher strength of the mobile phase at the time of elution. Thus, the k′ value is effectively smaller [30,31]. Additionally, when the rate of change of mobile phase strength is fast enough, the solute band can experience a compression effect due to the back of the band experiencing a stronger mobile phase composition than the front. Consequently, the back of the solute band moves at a somewhat higher velocity than the front and, thus, the solute band is effectively compressed [30,31].

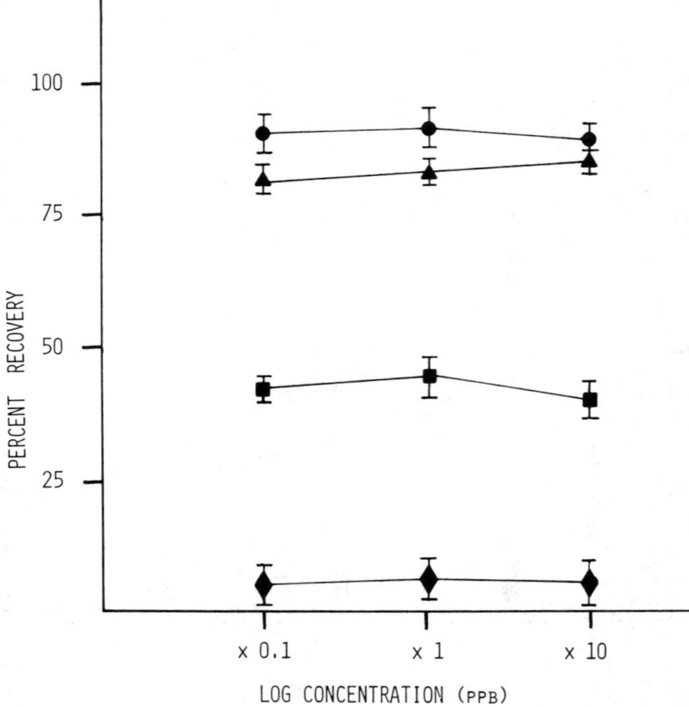


**Figure 6.** Percent recovery as a function of concentration in 100 mL of purified water prior to enrichment on the ODS accumulator column at 5.0 mL/min with subsequent elution and separation as in Figure 2. ● = atrazine (0.4–44 ppb); ▲ = N-nitrosodiphenyl amine (0.4–43 ppb); ■ = disperse red dye 13 (0.4–41 ppb); ◆ = decafluorobiphenyl (0.9–85 ppb).

The low value of and the large variation in the recovery of decafluorobiphenyl could not be justified on the same grounds. Therefore, a series of experiments was undertaken to ascertain whether this compound was not enriched or lost elsewhere prior to accumulation of the sample. It had been anticipated that, because of its high hydrophobicity, decafluorobiphenyl would be 100% recovered. This, in addition to its nonubiquitous nature, made it a candidate for possible use as an internal standard. Addition of a known amount of an internal standard to a grab sample at the sampling site could serve to monitor accurately the volume of sample enriched if the standard was 100% recovered. From the data in Figures 5 and 6 and Table II, however, it was evident that decafluorobiphenyl would be unsuitable. The data obtained thus far on decafluorobiphenyl had been from accumulations

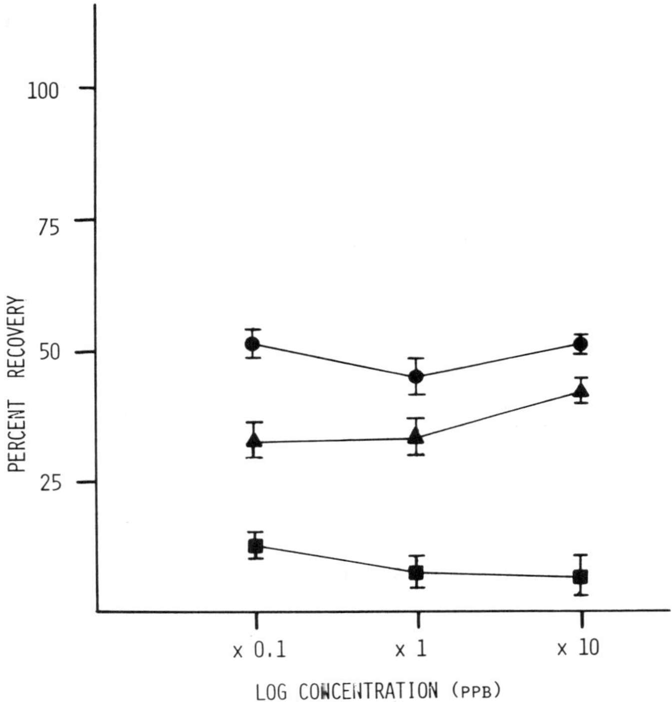


**Figure 7.** Percent recovery as a function of concentration in 100 mL of purified water prior to enrichment on the ODS accumulator column at 5.0 mL/min with subsequent elution and separation as in Figure 2. ● = caffeine (0.3–26 ppb); ▲ = 2,6-dichloroaniline (0.8–75 ppb); ■ = *m*-nitroaniline (0.05–5 ppb).

**Table II. Data on Model Compounds Enriched on ODS from Purified Water (Figure 2)**

| Solute | Concentration (μg/L) | Percent Recovery | % RSD[a] |
|---|---|---|---|
| Caffeine | 2.6 | 46 | 7.3 |
| *m*-Nitroaniline | 0.5 | 8.3 | 1.0 |
| Atrazine | 4.4 | 92 | 3.9 |
| 2,6-Dichloroaniline | 7.5 | 34 | 1.2 |
| *n*-Nitrosodiphenylamine | 4.3 | 84 | 2.8 |
| Decafluorobiphenyl | 8.5 | 7.4 | 11 |
| Disperse Red Dye 13 | 4.1 | 44 | 2.0 |

[a]n = 3.

of 100-mL samples in which the other six compounds in Table II had been coresident. When decafluorobiphenyl was accumulated from 100 mL where it alone was present, however, recoveries on the order of 100% were obtained. Further experimentation revealed that the mere addition of disperse red dye 13 at 4 ppb reduced the recovery of decafluorobiphenyl to around 10%; recoveries tended to vary widely. Adequate explanation of this phenomenon has evaded us thus far. It did, however, point out that if the trace enrichment method was to be successfully applied to quantitative determinations, an adequate study of the accumulation character of each compound under a wide range of conditions would be necessary.

## Enrichment of Acidic and Basic Compounds

Ion suppression [8,10] and ion-pairing [8,11,12,39-42], as utilized in reverse-phase HPLC, were considered as two possible approaches to increasing the recovery of organic acids and bases enriched on the ODS accumulator column. Six weakly acidic organic compounds, listed in Table I, Group B, were selected for preliminary experiments to assess the ion-pairing methodology. Tetrabutylammonium phosphate (TBAP) was selected as the basic counter-ion.

The contribution of TBAP to the background was examined by accumulating 100 mL of 0.5 m*M* TBAP at 5.0 mL/min on the ODS accumulator column. Figure 8 lists the conditions employed and shows the resulting chromatogram obtained by monitoring 215 and 254 nm. Next, 0.5 m*M* TBAP was spiked with the six model acidic compounds at levels listed in Table III. After enrichment of 100 mL of the spiked TBAP at 5.0 mL/min on the ODS accumulator column, the enriched model compounds were eluted with

**Table III. Data on Acidic Model Compounds Enriched on ODS from 0.5 m*M* TBAP (Figure 9)**

| Solute | Concentration ($\mu$g/L) | Percent Recovery | % RSD[a] |
|---|---|---|---|
| *p*-Hydroxyphenylacetic Acid | 2.8 | 40 | 21 |
| *o*-Chlorophenylacetic Acid | 3.4 | 78 | 5.4 |
| *p*-Chloromandelic Acid | 3.2 | 68 | 4.2 |
| *m*-Chlorobenzoic Acid | 2.6 | 98 | 5.1 |
| 5-Chlorosalicylic Acid | 3.9 | 43 | 1.7 |
| 1-Amino-8-naphthol-2,4-Disulfonic Acid | 5.1 | 94 | 3.8 |

[a]n = 3.

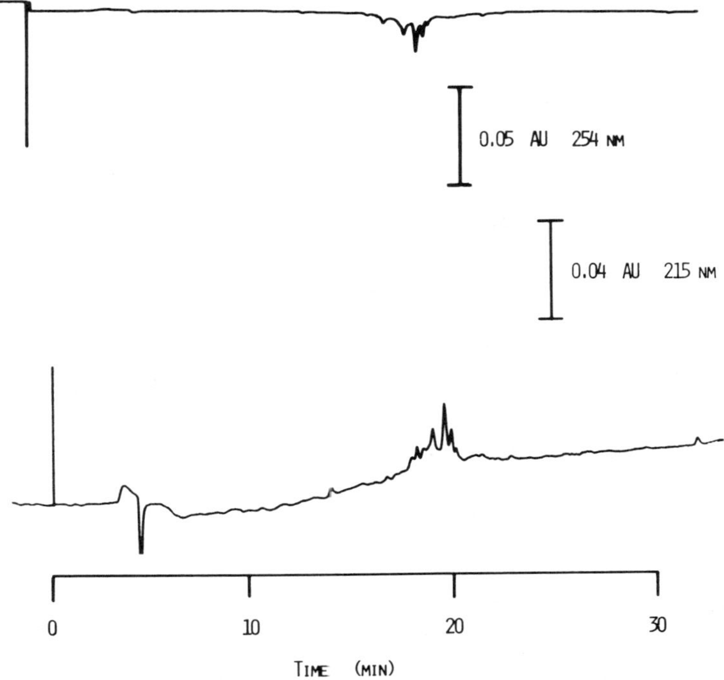


**Figure 8.** Chromatogram of components enriched from 100 mL of 0.5 m*M* TBAP in purified water and eluted from an ODS accumulator column. Conditions: 5–80% acetonitrile in 0.5 m*M* TBAP at 2.14%/min and a flowrate of 1.0 mL/min. Column: Partisil-10, ODS-2, 250 x 4.6 mm i.d.

the same mobile phase gradient as the enriched TBAP blank. Figure 9 shows the chromatogram of a direct injection of the six acidic model compounds and that obtained from elution of an enriched sample of 100 mL containing the same amount of each compound that had been directly injected. Each of the separations employed 0.5 m*M* TBAP, pH = 7.5, as the weak component of the mobile phase gradient. Table III lists the recoveries and percent relative standard deviations of each of the six acidic model compounds enriched on the ODS accumulator column from purified water containing 0.5 m*M* TBAP. Note that *p*-hydroxyphenylacetic acid, the most weakly retained of the six compounds, displayed the largest variation in recovery. This again could be explained in part by *p*-hydroxyphenylacetic acid not experiencing the full effect of the gradient. This was previously observed with caffeine, which was also weakly retained. The recoveries for the other five acidic compounds were quite acceptable and moderately precise.

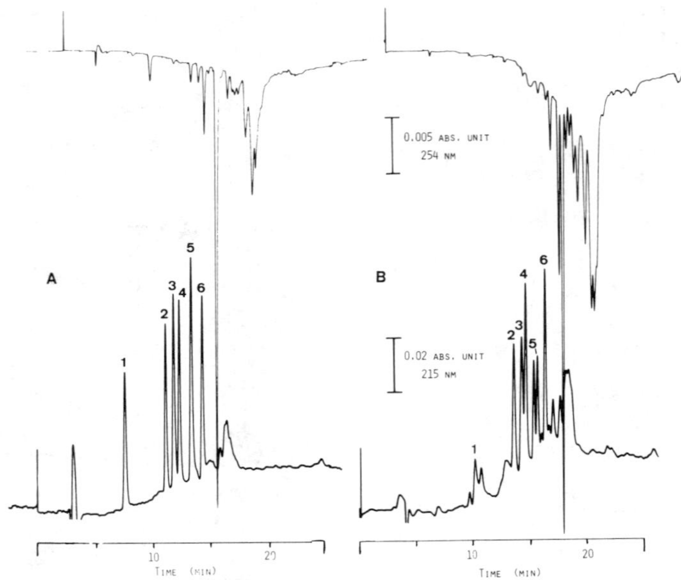


**Figure 9.** (A) Chromatogram of a 25-μL direct injection of (1) 0.28 μg *p*-hydroxyphenylacetic acid; (2) 0.34 μg *o*-chlorophenylacetic acid; (3) 0.32 μg *p*-chloromandelic acid; (4) 0.26 μg *m*-chlorobenzoic acid; (5) 0.39 μg 5-chlorosalicylic acid; and (6) 0.51 μg 1-amino-8-naphthol-2,4-disulfonic acid. (B) Chromatogram of the same amount of each of the six compounds spiked into 100 mL of 0.5 m*M* TBAP, enriched on the ODS accumulator column at 5.0 mL/min and subsequently eluted and separated. Conditions: (both A and B) 5–85% acetonitrile in 0.5 m*M* TBAP at 5%/min and a flowrate of 1.0 mL/min. Column: Partisil-10, ODS-2, 250 x 4.6 mm i.d.

An analogous approach to enrichment of the seven basic model compounds by addition of an alkyl-sulfonate to the sample, at an appropriate pH, prior to accumulation was attempted. Initially, 1 m*M* sodium octanesulfonate at pH 2.0 was examined for increased retention in LC. The seven basic model compounds examined are listed in Table I, Group C. No significant increase in retention was observed.

A more hydrophobic counter-ion, sodium dodecanesulfonate, was examined under comparable conditions for increased retention. Increased retention was observed for phenylurea, aniline and *m*-nitroaniline; however, an extremely large interfering peak eluted during the latter part of the separation when a 5–85% gradient of acetonitrile in 1 m*M* sodium dodecanesulfonate at pH 2.0 was run over 16 min. Rerun of the gradient without sample injection yielded a similar large peak. Thus, it appeared that the interference was due to the elution of the sodium dodecanesulfonate that was retained on the reverse-phase analytical column [40,44,45] and subsequently eluted as the mobile phase strength was increased. Addition of a comparable amount of sodium dodecanesulfonate to the acetonitrile was attempted in an effort to provide a relatively constant UV-absorbing background. This also failed, however, and thus it was concluded that the interference resulted from a sudden elution of retained sodium dodecanesulfonate.

An alternative approach to enhancing the retention of the seven weakly basic model compounds was to employ ion suppression. This would be accomplished by buffering the pH of the aqueous component of the mobile phase to a relatively high value. Silica will dissolve in aqueous solutions of pH greater than 7.6. Thus, it was necessary to saturate the incoming mobile phase with silica to prevent rapid deterioration of the analytical column. This was easily accomplished by installation of a column packed with a silica-based support after the pumping system and before the injection valve [43].

A 1 m*M* solution of sodium tetraborate (STB) was used to buffer the weak mobile phase to pH 9.2. The seven basic model compounds were spiked into purified water containing 1 m*M* STB, and a 100-mL sample was enriched on the ODS accumulator column at 5.0 mL/min. Figure 10 lists the conditions employed and shows the chromatogram obtained from a direct injection (A) and that obtained from the elution of the seven enriched basic model compounds from the ODS accumulator (B). Only *p*-phenylazoaniline and *p,p'*-methylenedianiline were recovered with any significant efficiency, whereas 2,6-dichloroaniline displayed the same broad peak at approximately the same efficiency as in Figure 2 and Table II. The remaining compounds (*m*-hydroxyaniline, phenylurea, aniline and *m*-nitroaniline) appeared to be poorly recovered. Because of the poor results, recoveries of the seven basic model compounds were not calculated. It is interesting to note, however, that the two basic model compounds that were recovered well also chromatographed well when purified water at pH 6.2 was used as the weak component of the mobile phase. Under these conditions the poorly recovered basic model compounds all severely tailed and at times disappeared entirely from the chromatogram. This seemed to indicate that the favorably recovered basic model compounds were hydrophobic enough, such that protonation of the amine functionality did not seriously reduce their partition coefficient.

## Macroreticular Resin Comparison

Since 1972 macroreticular resins have been used extensively for extraction of trace hydrophobic organic pollutants from aqueous matrices [46-48]. Most of the accumulations, however, have been associated with separation

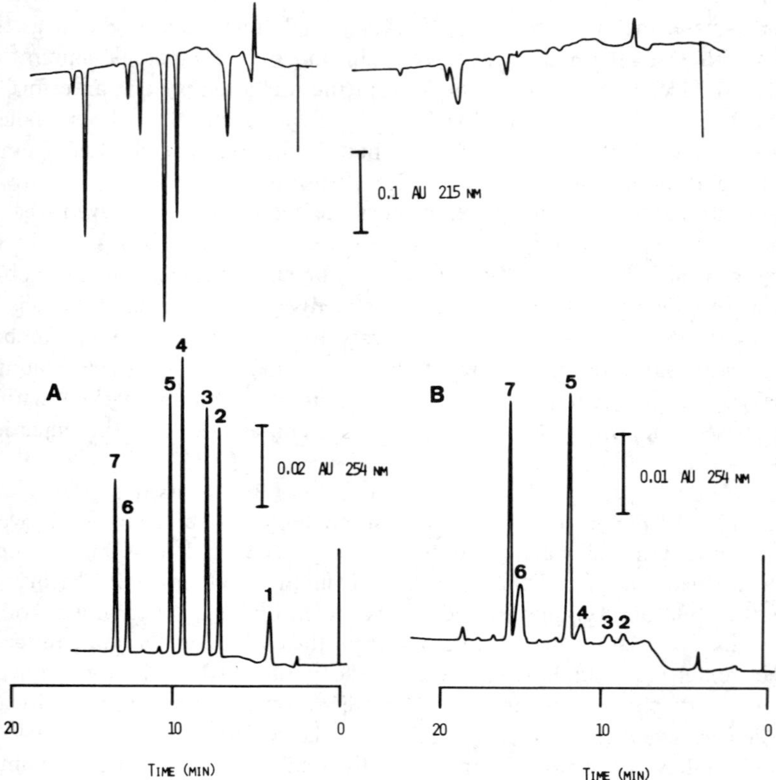


**Figure 10.** (A) Chromatogram of a 25-μL direct injection of (1) 0.53 μg *m*-hydroxyaniline; (2) 0.96 μg phenylurea; (3) 0.99 μg aniline; (4) 0.053 μg *m*-nitroaniline; (5) 0.20 μg *p,p'*-methylenedianiline; (6) 0.94 μg 2,6-dichloroaniline; and (7) 0.43 μg *p*-phenylazoaniline. (B) Chromatogram of the same amount of each of the seven compounds spiked into 100 mL of 1 m*M* STB, enriched on the ODS accumulator column at 5.0 mL/min and subsequently eluted and separated. Conditions: (both A and B) 5–85% acetonitrile in 1 m*M* STB at 5%/min and a flowrate of 1.0 mL/min. Column: Partisil-10, ODS-2, 250 x 4.6 mm i.d.

and analysis by GC/MS. Using the compounds for which we had accumulation data on the ODS accumulator column, an experimental comparison of ODS to a macroreticular resin or porous organic polymer revealed interesting results. Porapak Q, an ethylvinylbenzene-divinylbenzene copolymer, with a specific surface area of 630–840 $m^2/g$ and a mean pore diameter of 74.8 Å (46), in the 80–100 mesh range, was substituted for the ODS in identical accumulator columns. A 100-mL sample of purified water spiked with the compounds at the levels listed in Table II was enriched on the Porapak Q accumulator column at 5.0 mL/min. Subsequent elution and separation of

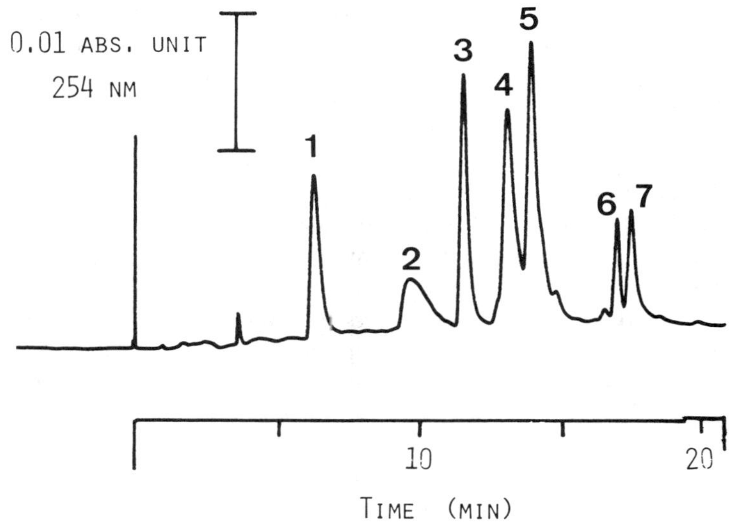


**Figure 11.** Chromatogram of a 100-mL spiked sample of purified water enriched on the Porapak Q accumulator column at 5.0 mL/min. (1) 2.6 ppb caffeine; (2) 0.5 ppb *m*-nitroaniline; (3) 4.4 ppb atrazine; (4) 7.5 ppb 2,6-dichloroaniline; (5) 4.3 ppb N-nitrosodiphenylamine; (6) 8.5 ppb decafluorobipenyl; and (7) 4.1 ppb disperse red dye 13. Conditions: 10–90% acetonitrile in purified water at 5%/min and a flowrate of 1.0 mL/min. Columns: Partisil-10, ODS-2, 250 x 4.6 mm i.d.

the enriched compounds utilizing the conditions in Figure 2 yielded the chromatogram shown in Figure 11. Examination readily revealed the broad character of the peaks when compared to Figure 2. Although the areas for which instrumentation was not immediately available to measure accurately seemed to indicate possibly higher recoveries, the increased extracolumn band broadening contributed by the macroreticular resin accumulator was deemed unacceptable. Presumably this broadening was caused by slow mass transfer from the stationary phase, Porapak Q, to the mobile phase as compared to that from the approximately monolayer bonded octadecyl moieties on the ODS accumulator.

When the Porapak Q accumulator column was used in conjunction with TBAP for enrichment of the six acidic model compounds an interesting result was obtained. Figure 12 shows the chromatogram obtained from direct injection and the chromatogram obtained from gradient elution of the same amount of each of the six acidic model compounds enriched on the Porapak Q accumulator column at 5.0 mL/min from 100 mL of purified water containing 0.5 m*M* TBAP. Note that the broad peaks observed previously (Figure 11) were absent. Consideration of the mechanism of retention in ion-pairing methodology provided a possible explanation. If the TBAP in the sample was retained by the Porapak Q, as indicated somewhat by the blank run of a 0.5 m*M* TBAP accumulation in Figure 8, the TBAP molecules could act as ion exchange sites. The retention of the six acidic model compounds would thus take place external to the polymer matrix, and mass transfer between the stationary phase and the mobile phase would no longer be the slow process observed with the neutral compounds in Figure 11. This in situ formation of an ion exchange support has been proposed previously [40,44] as a mechanism for increased retention when ion-pairing reagents were employed in the mobile phases with reverse-phase supports. Table IV lists the recoveries and precisions obtained from three determinations when the Porapak Q accumulator column was employed under the conditions listed in Figure 12. Except for 5-chlorosalicylic acid, all efficiencies were greater than those obtained on the ODS accumulator (Table III). The precision obtained with Porapak Q was comparable to that observed with the ODS support. Note also, that the precision for the early eluting *p*-hydroxyphenylacetic acid was again relatively poor.

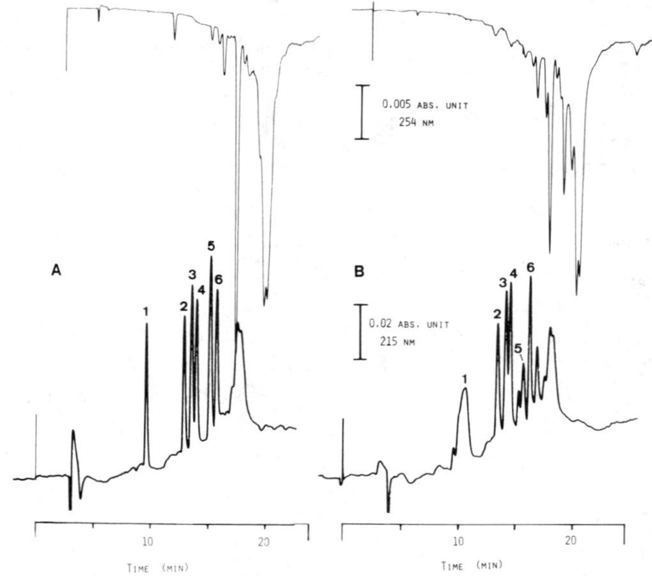


**Figure 12.** (A) Chromatogram of a 25-$\mu$L direct injection of (1) 0.28 $\mu$m *p*-hydroxyphenylacetic acid; (2) 0.34 $\mu$g *o*-chlorophenylacetic acid; (3) 0.32 $\mu$g *p*-chloromandelic acid; (4) 0.26 $\mu$g *m*-chlorobenzoic acid; (5) 0.39 $\mu$g 5-chlorosalicylic acid; and (6) 0.51 $\mu$g 1-amino-8-naphthol-2,4-disulfonic acid. (B) Chromatogram of the same amount of each of the six compounds spiked into 100 mL of 0.5 m*M* TBAP, enriched on the Porapak Q accumulator column at 5.0 mL/min and subsequently eluted and separated. Conditions: (both A and B) 5–85% acetonitrile in 0.5 m*M* TBAP at 5%/min and a flowrate of 1.0 mL/min. Column: Partisil-10, ODS-2, 250 x 4.6 mm i.d.

**Table IV. Data on Acidic Model Compounds Enriched on Porapak Q from 0.5 m*M* TBAP (Figure 12)**

| Solute | Concentration (μg/L) | Percent Recovery | % RSD[a] |
|---|---|---|---|
| *p*-Hydroxyphenylacetic Acid | 2.8 | 52 | 31 |
| *o*-Chlorophenylacetic Acid | 3.4 | 83 | 6.1 |
| *p*-Chloromandelic Acid | 3.2 | 87 | 4.7 |
| *m*-Chlorobenzoic Acid | 2.6 | 100 | 3.2 |
| 5-Chlorosalicylic Acid | 3.9 | 33 | 4.7 |
| 1-Amino-8-naphthol-2,4-disulfonic Acid | 5.1 | 104 | 5.1 |

[a]n = 3.

The basic group of model compounds was enriched on Porapak Q from a 100-mL solution of 1 m*M* STB, at pH 9.2 and 5.0 mL/min. Elution from the porous polymer accumulator with the same conditions as listed in Figure 10 yielded results similar to those in Figure 11: all peaks were relatively broad and thus unacceptable for online elution purposes. Under these conditions retention was presumably due to sorption by the polymer matrix and not due to an ion-pairing compound that was retained on the surface. Therefore, a limiting stationary phase to mobile phase mass transfer rate again explained the increased extracolumn band broadening observed with enrichment and online elution from Porapak-Q accumulator columns.

## Matrix Studies—Drinking Water

A 100-mL sample of Athens, GA, drinking (tap) water was enriched on the ODS accumulator column at 5.0 mL/min. A second 100-mL sample of the same was spiked with the compounds listed in Table II at the equivalent concentrations and enriched on a second ODS accumulator column at 5.0 mL/min. Chromatograms of the spiked enriched drinking water and the unspiked enriched drinking water (Figure 13), obtained from gradient elution, under the conditions listed, showed the spiked compounds clearly distinguishable from the components of the drinking water matrix. Table V lists the recoveries for the seven model compounds and the corresponding precisions for three determinations. Also shown are the data for purified water from Table II. Note that, except for *m*-nitroaniline and decafluorobiphenyl, the recoveries obtained for the model compounds from drinking

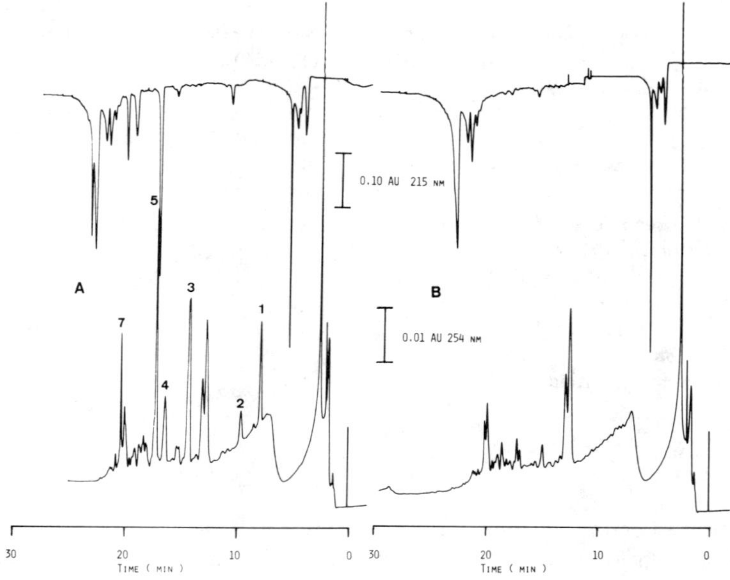


**Figure 13.** (A) Chromatogram from elution of a 100-mL enriched Athens, GA, drinking water sample spiked with (1) 2.6 ppb caffeine; (2) unidentified compound; (3) 4.4 ppb atrazine; (4) 7.5 ppb 2,6-dichloroaniline; (5) 4.3 ppb N-nitrosodiphenylamine; (6) 8.5 ppb decafluorobiphenyl (not detected); and (7) 4.1 ppb disperse red dye 13 (*m*-nitroaniline spiked at 0.5 ppb not detected). (B) Chromatogram from elution of a 100-mL sample of Athens, GA, drinking water enriched on the ODS accumulator column at 5.0 mL/min. Conditions: (both A and B) 10–90% acetonitrile in purified water at 5%/min and a flowrate of 1.0 mL/min. Column: Partisil-10, ODS-2, 250 x 4.6 mm i.d.

water and enriched on the ODS accumulator column were not statistically different from those enriched from highly purified water. Those compounds not detected (*m*-nitroaniline and decafluorobiphenyl) were poorly recovered from purified water to begin with. Based on the signal-to-noise and the intensity of the background due to the matrix, lower limits of detection in the 1- to 0.1-ppb range could easily be obtained for these compounds in drinking water.

To determine the recoveries of the six acidic model compounds from drinking water, a contribution from the matrix first had to be evaluated. When TBAP was added to drinking water to a concentration of 0.5 m*M*, and then enriched on the ODS accumulator column, the resulting chromatogram from the elution of enriched matrix components was extremely intense. As a result, only 10 mL of drinking water containing 0.5 m*M* TBAP could be enriched and subsequently eluted and chromatographed in the absorbance range that had been employed previously. Figure 14 shows the chromatogram

Table V. Data on Model Compounds Enriched on ODS from Various Matrices

| | | % Recovery (% RSD)[a] | | Treated Sewage, 10 mL | |
|---|---|---|---|---|---|
| Solute | Conc. (μg/L) | Purified Water, 100 mL | Drinking Water, 100 mL | Conc. (μg/L) | % Recovery (% RSD)[a] |
| Caffeine | 2.6 | 46 (7.3) | 45 (6.1) | 13 | 52 (7.7) |
| *m*-Nitroaniline | 0.5 | 8.3 (1.0) | ND[b] | 2.6 | ND |
| Atrazine | 4.4 | 92 (3.9) | 86 (3.7) | 22 | 47 (3.9) |
| 2,6 Dichloroaniline | 7.5 | 34 (1.2) | 32 (2.8) | 38 | 50 (1.9) |
| N-Nitrosodiphenylamine | 4.3 | 84 (2.8) | 87 (1.9) | 22 | 70 (2.1) |
| Decafluorobiphenyl | 8.5 | 7.4 (11) | ND | 42 | 39 (14) |
| Disperse Red Dye 13 | 4.1 | 44 (2.0) | 44 (2.0) | 21 | 22 (2.9) |

[a]n = 3.
[b]Not determined.

of the background of 10 mL of drinking water enriched on the ODS accumulator column at 2.0 mL/min (A), and 10 mL of the same drinking water to which TBAP had been added to a concentration of 0.5 m*M*, and enriched on the ODS accumulator column at 2.0 mL/min (B). The increased number and intensity of enriched components was readily evident.

Next, a drinking water sample containing 0.5 m*M* TBAP was spiked with the compounds previously enriched from purified water with this methodology. Table III lists those compounds; however, they were spiked into the drinking water at five times the concentration at which they had been examined in purified water. This was necessary to obtain discernible peak heights from the drinking water matrix because only 20% as much sample had been enriched compared to the purified water sample. Figure 15 shows the chromatogram obtained from direct injection of the acidic model compounds and the chromatogram obtained from elution of a sample of spiked drinking water containing 0.5 m*M* TBAP that had been enriched on

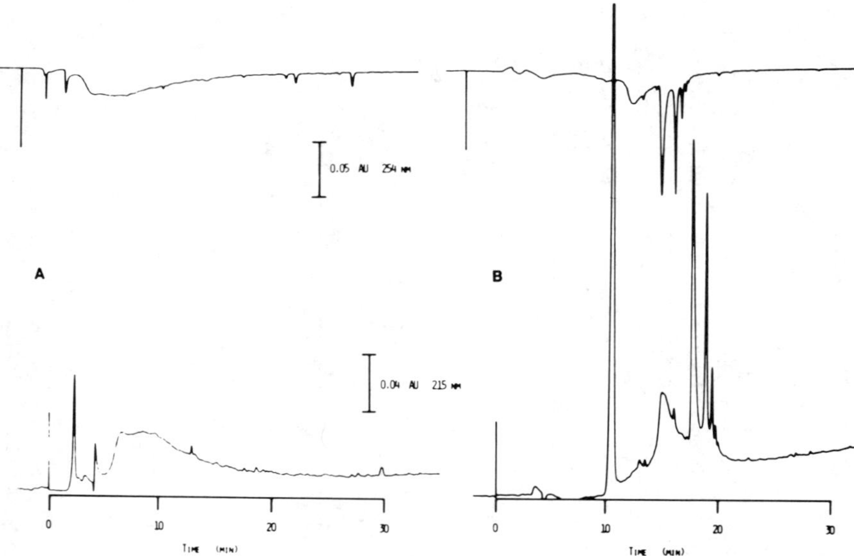


**Figure 14.** (A) Chromatogram from elution of a 10-mL sample of Athens, GA, drinking water enriched on the ODS accumulator column at 2.0 mL/min. Conditions: 5–80% acetonitrile in purified water at 2.14%/min and a flowrate of 1.0 mL/min. (B) Chromatogram from elution of a 10-mL sample of Athens, GA, drinking water to which TBAP had been added to a concentration of 0.5 m*M* and enriched on the ODS accumulator at 2.0 mL/min. Conditions: 5–80% acetonitrile in 0.5 m*M* TBAP at 2.14%/min and a flowrate of 1.0 mL/min. Column: (both A and B) Partisil-10, ODS-2, 250 x 4.6 mm i.d.

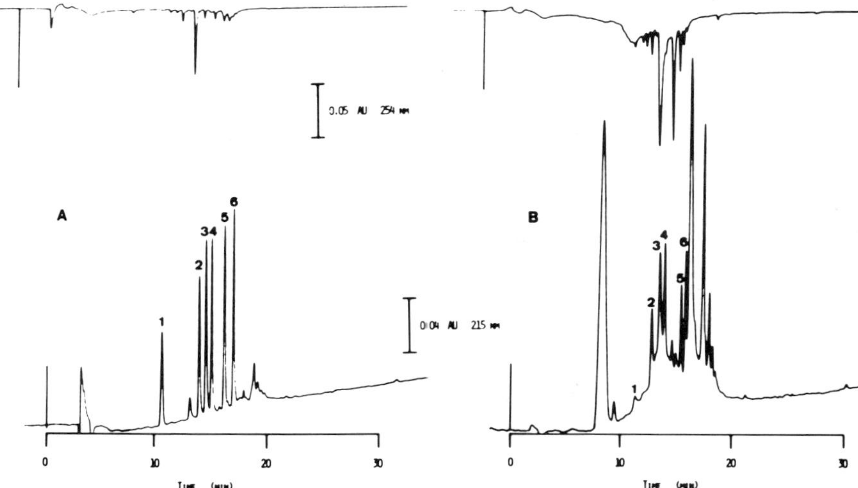


**Figure 15.** (A) Chromatogram of a 25-$\mu$L direct injection of (1) 0.28 $\mu$g *p*-hydroxyphenylacetic acid; (2) 0.34 $\mu$g *o*-chlorophenylacetic acid; (3) 0.32 $\mu$g *p*-chloromandelic acid; (4) 0.26 $\mu$g *m*-chlorobenzoic acid; (5) 0.39 $\mu$g 5-chlorosalicylic acid; and (6) 0.51 $\mu$g 1-amino-8-naphthol-2,4-disulfonic acid. (B) Chromatogram of the same six compounds spiked into 10 mL of Athens, GA, drinking water containing 0.5 m*M* TBAP at (1) 14, (2) 17, (3) 16, (4) 13, (5) 20 and (6) 26 ppb, and enriched on the ODS accumulator column at 2.0 mL/min. Conditions: (both A and B) 5–80% acetonitrile in 0.5 m*M* TBAP at 2.14%/min and a flowrate of 1.0 mL/min. Column: Partisil-10, ODS-2, 250 x 4.6 mm i.d.

the ODS accumulator column. Table VI lists the recoveries and precisions for three determinations of the acidic model compounds from drinking water. In all cases but one (5-chlorosalicylic acid), the recoveries were statistically less than those observed when enriched from purified water. Recall that only 10 mL of sample were enriched at a rate of 2.0 mL/min. Both of these operating conditions favored higher recoveries. Improved recoveries, however, were not observed.

The mechanism proposed earlier, by which retained TBAP acts as an in situ formed ion exchanger, could possibly explain these results. Presumably each retained TBAP molecule could act to retain only one solute ion. In a complex matrix, such as drinking water, the number of total solute ions available for interaction with TBAP was probably substantially increased. Additionally, many more acidic substances that could have displaced the acidic model compounds could have been present. Thus, at this fixed concentration of TBAP, competition could have occurred for the limited number of sites, reducing the overall recoveries. Addition of a higher concentration of TBAP to the sample may have well increased the recovery of the acidic model compounds enriched from drinking water. These experiments are planned for the future.

Table VI. Data on Acidic Model Compounds Enriched on ODS from Various Matrices Containing 0.5 m*M* TBAP

| Solute | Purified Water, 100 mL | | Drinking Water, 10 mL | |
|---|---|---|---|---|
| | Conc. ($\mu$g/L) | Percent Recovery (% RSD)[a] | Conc. ($\mu$g/L) | Percent Recovery (% RSD)[a] |
| *p*-Hydroxyphenylacetic Acid | 2.8 | 40 (21) | 14 | 22 (7.2) |
| *o*-Chlorophenylacetic Acid | 3.4 | 78 (5.4) | 17 | 46 (8.2) |
| *p*-Chloromandelic Acid | 3.2 | 68 (4.2) | 16 | 57 (3.9) |
| *m*-Chlorobenzoic Acid | 2.6 | 98 (5.1) | 13 | 64 (4.9) |
| 5-Chlorosalicylic Acid | 3.9 | 43 (1.7) | 20 | 49 (7.6) |
| 1-Amino-8-naphthol-2,4-disulfonic Acid | 5.1 | 94 (3.8) | 26 | 51 (7.4) |

[a] $n = 3$.

## Matrix Studies–Treated Municipal Sewage

Treated municipal sewage from an Athens, GA, municipal waste treatment facility was collected and vacuum-filtered through a 0.45-μm filter. The concentration of UV-absorbing compounds was so high that only 10 mL of the sample could be enriched. A spiked sample was prepared using the seven model compounds listed in Table V at five times the concentration that had been previously employed with purified and drinking water. Samples of the spiked and unspiked sewage (10 mL each) were enriched on the ODS accumulator column at 2.0 mL/min. Figure 16 shows the chromatogram obtained from the subsequent elution and separation of the enriched spiked sample and the enriched unspiked sample. Table V lists the recoveries and precisions for three determinations of the seven model compounds enriched from treated municipal sewage.

Specific trends in the variation of the recoveries of these compounds were not readily observable. Recall that smaller accumulation volumes tend to

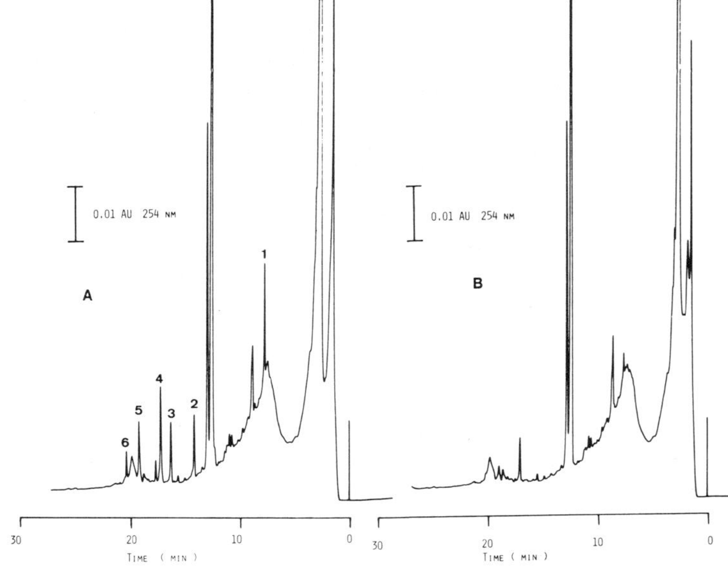


**Figure 16.** (A) Chromatogram of 10-mL sample of Athens, GA, treated municipal sewage spiked with (1) 13 ppb caffeine; (2) 22 ppb atrazine; (3) 38 ppb 2,6-dichloroaniline; (4) 22 ppb N-nitrosodiphenylamine; (5) 42 ppb decafluorobiphenyl; and (6) 21 ppb disperse red dye 13 (2.6 ppb *m*-nitroaniline not detected) and enriched on the ODS accumulator column at 2.0 mL/min. (B) Chromatogram of 10-mL sample of Athens, GA, sewage enriched on the ODS accumulator column at 2.0 mL/min. Conditions: (both A and B) 10–90% acetonitrile in purified water at 5%/min at a flowrate of 1.0 mL/min. Column: Partisil-10, ODS-2, 250 x 4.6 mm i.d.

increase the recoveries. This was the case for caffeine, 2,6-dichloroaniline and decafluorobiphenyl. However, *m*-nitroaniline was not detected, and atrazine, *N*-nitrosodiphenylamine and disperse red dye 13 all showed decreases in recovery relative to those in purified and drinking water.

Interestingly, though, the precision with which the recoveries were determined was comparable to those observed with purified and drinking water, except for the ever-anomalous decafluorobiphenyl. The variations in the trends of the recoveries were presumably due to unidentified matrix effects.

Evaluation of the enrichment of the six acidic model compounds from treated municipal sewage on the ODS accumulator was extremely difficult. Addition of TBAP to a concentration of 0.5 m*M* prior to enrichment produced the largest background observed with any sample. Figure 17 shows the chromatograms of 2.0 mL of enriched sewage samples with and without TBAP present. The large background obtained when TBAP was added indicated a high concentration of acidic compounds. Reliable compilation of data on the specific recoveries of individual acidic model compounds from the treated municipal sewage was impossible. Figure 17 also shows the resultant chromatogram after 2.0 mL of sewage spiked with the six acidic model compounds, at 50 times the concentration at which they were spiked in purified water, was enriched on the ODS accumulator column and subsequently eluted and separated. It appeared that the ODS accumulator column simply retained too much from the sewage sample in the presence of the TBAP. Thus, the inability to monitor accurately the model compounds at the desired ppb levels was in part due to insufficient resolution. Selective detection would have helped somewhat, but was unavailable at the time. Again the TBAP concentration was fixed, and competition for the limited number of ion exchange sites could have also affected the recoveries.

## CONCLUSIONS

The analytical capabilities of trace enrichment on ODS in conjunction with reverse-phase HPLC has been evaluated for routine analysis of trace nonvolatile organic compounds in aqueous samples. Model compounds could be determined in drinking water at 0.1- to 1-ppb levels easily with outstanding precision. Addition of an "ion-pairing" reagent to the sample increased the recoveries for acidic compounds. Variations in recoveries with adaptation of the "ion-pairing" methodologies occurred, however, when acidic model compounds were recovered from different matrices.

Studies of sampling dynamics showed that volumetric accumulation rates as high as 7.5 mL/min could be employed without a significant decrease in recovery. While large sample volumes, i.e., 500–2000 mL, could be enriched, examination of real matrices revealed that this was not generally necessary. These studies, however, also exposed two of the major problems with trace enrichment on ODS supports.

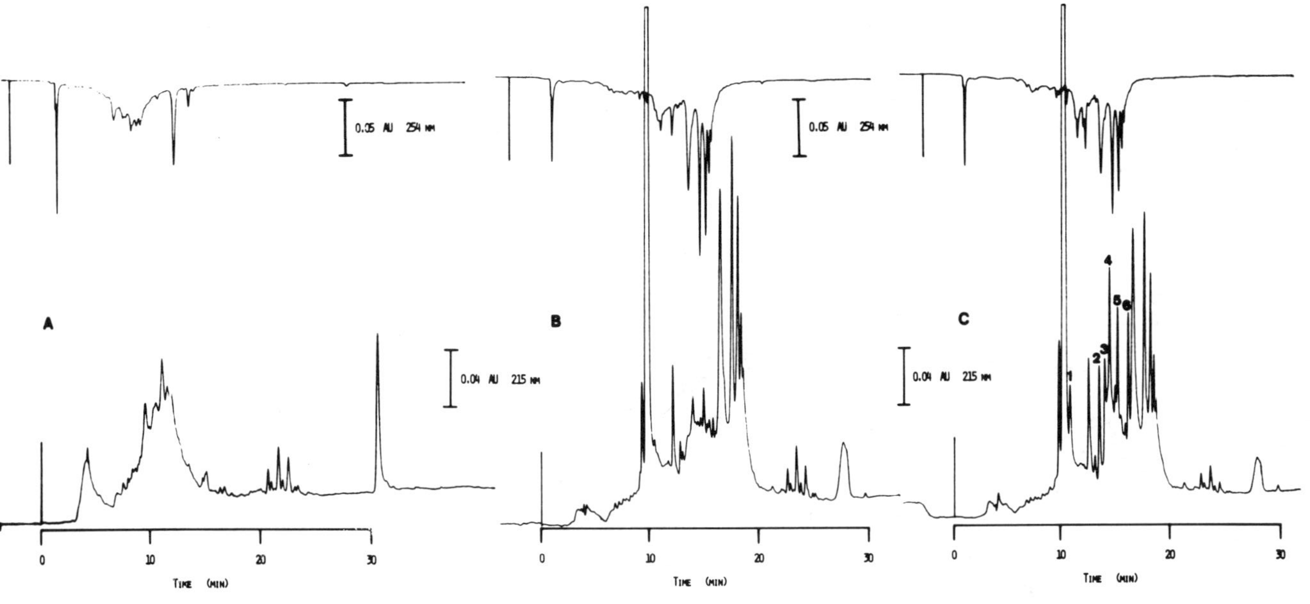


**Figure 17.** (A) Chromatogram of a 2-mL sample of Athens, GA, treated municipal sewage enriched on the ODS accumulator column at 1.0 mL/min. Conditions: 5–80% acetonitrile in purified water at 2.14%/min and a flowrate of 1.0 mL/min. (B) Chromatogram of a 2-mL sample of Athens, GA, sewage containing 0.5 m*M* TBAP, enriched on the ODS accumulator column at 1.0 mL/min. (C) Chromatogram of a 2-mL sample of Athens, GA, sewage containing 0.5 m*M* TBAP, spiked with (1) 140 ppb *p*-hydroxyphenylacetic acid; (2) 170 ppb *o*-chlorophenylacetic acid; (3) 160 ppb *p*-chloromandelic acid; (4) 130 ppb *m*-chlorobenzoic acid; (5) 220 ppb 5-chlorosalicylic acid; and (6) 260 ppb 1-amino-8-napthol-2,4-disulfonic acid and enriched on the ODS accumulator column at 1.0 mL/min. Conditions: (B and C) 5–80% acetonitrile in 0.5 m*M* TBAP at 2.14%/min and a flowrate of 1.0 mL/min. Column: Partisil-10, ODS-2, 250 x 4.6 mm i.d.

First, a large amount of material is retained by these ODS accumulator columns. The resolving capabilities of the liquid chromatographic system were found to be insufficient to handle the large amount of material recovered from treated municipal sewage. This problem was even greater when ion-pairing methodology was utilized to increase the retention of acidic organic compounds. Secondly, unexplainable variations in the recoveries of compounds enriched from different matrices limited the accuracy with which the methodology could be applied to quantitative determinations. The precision with which the recoveries could eventually be measured from a specific matrix, however, was found to be very good for the levels of compounds examined and, surprisingly, did not vary significantly from one matrix to another.

Comparison of the ODS accumulator column to one containing a hydrophobic macroreticular resin (Porapak Q) demonstrated the compromises necessary in applying an accumulation material to online elution with HPLC. Nonetheless, the methodology appeared quite promising. Further work would be necessary to reduce the variations of recoveries between matrices. Special attention should be paid to basic compounds because an acceptable method of increasing their recoveries was not successfully developed.

The data compiled thus far were obtained with a system where the aqueous sample was enriched in the laboratory. Adaptation of these procedures to field sampling, however, appears quite feasible. Solid support accumulation methods offer an added advantage of integration of the sample concentration over a certain time period. Further development of this technology will be important to address properly the problems of trace organic analysis that are now present and will continue to confront the analytical community.

## ACKNOWLEDGMENTS

J. Graham would like to thank the National Research Council and the U.S. Environmental Protection Agency for the opportunity to participate in the Resident Research Associate program. Both authors would like to thank Trudy A. Scott and Fred Haeberer for their many helpful suggestions, comments and discussions. Thanks also go to Mrs. E. McGarity for her help in preparing the final copy of the manuscript.

## DISCLAIMER

Mention of trade names or commercial products does not constitute endorsement or recommendation for use by the U.S. Environmental Protection Agency.

## REFERENCES

1. Cummings, R. B. In: *Water Chlorination: Environmental Impact and Health Effects, Vol. 1*, R. L. Jolley, Ed. (Ann Arbor, MI: Ann Arbor Science Publishers, Inc., 1976), p. 247.
2. Garrison, A. W. Technical Paper No. 9, WHO International Reference Centre for Community Water Supply, The Hague (1976).
3. Crathorne, B., C. D. Watts and M. Fielding. *J. Chromatog.* 185:671 (1979).
4. Katz, S., W. W. Pitt, Jr., C. D. Scott and A. A. Rosen. *Water Res.* 6:1029 (1972).
5. Pitt, W. W., Jr., R. L. Jolley and C. D. Scott. *Environ. Sci. Technol.* 9:1068 (1975).
6. Pitt, W. W., Jr., R. L. Jolley and S. Katz. In: *Identification and Analysis of Organic Pollutants in Water*, L. H. Keith, Ed. (Ann Arbor, MI: Ann Arbor Science Publishers, Inc., 1976), pp. 215-231.
7. Jolley, R. L., G. Jones, Jr., W. W. Pitt, Jr. and J. E. Thompson. In: *Identification and Analysis of Organic Pollutants in Water*, L. H. Keith, Ed. (Ann Arbor, MI: Ann Arbor Science Publishers, Inc., 1976), pp. 233-246.
8. Snyder, L. R., and J. J. Kirkland. *Introduction to Modern Liquid Chromatography*, 2nd ed. (New York: John Wiley & Sons, Inc., 1979).
9. Knox, J. H. *J. Chromatog. Sci.* 15:352 (1977).
10. Molnar, I., and C. Horvath. *J. Chromatog.* 143:391 (1977).
11. Tomlinson, E., T. M. Jeffries and C. J. Riley. *J. Chromatog.* 159:315 (1978).
12. Bidlingmeyer, B. A., S. N. Deming, W. P. Price, Jr., B. Sachok and M. Petrusek. *J. Chromatog.* 186:419 (1979).
13. Kirkland, J. J. *Analyst* 99:859 (1974).
14. Little, J. N., and G. J. Fallick. *J. Chromatog.* 112:389 (1975).
15. May, W. E., S. N. Chesler, S. P. Cram, B. H. Gump, H. S. Hertz, D. P. Enagonio and S. M. Dyszel. *J. Chromatog. Sci.* 13:535 (1975).
16. Erni, F., R. W. Frei and W. Linder. *J. Chromatog.* 125:265 (1976).
17. Frei, R. W. *Int. J. Environ. Anal. Chem.* 5:143 (1978).
18. Ogan, K., E. Katz and W. Slavin. *J. Chromatog. Sci.* 16:517 (1978).
19. Oyler, A. R., D. L. Bodenner, K. J. Welch, R. J. Luikkonen, R. M. Carlson, H. L. Koppermann and C. Caple. *Anal. Chem.* 50:837 (1978).
20. Kummert, R., E. Molnar-Kubica and W. Giger. *Anal. Chem.* 50:1637 (1978).
21. May, W. E., S. P. Wasik and D. H. Freeman. *Anal. Chem.* 50:175 (1978).
22. Lankelma, J., and H. Poppe. *J. Chromatog.* 149:587 (1978).
23. May, W. E., J. M. Brown, S. N. Chesler, F. Guenther, L. R. Hilpert, H. S. Hertz and S. A. Wise. NBS Special Publication 519, U.S. Dept. of Commerce (1979), pp. 219-224.
24. Walton, H. F., and G. A. Eiceman. NBS Special Publication 519, U.S. Dept. of Commerce (1979), pp. 185-190.
25. Edwards, R. W., K. A. Nonnemaker and R. L. Cotter. NBS Special Publication 519, U.S. Dept. of Commerce (1979), pp. 81-94.

26. Ishii, D., K. Hibi, K. Asai and M. Nagaya. *J. Chromatog.* 152:341 (1978).
27. Saner, W. A., J. R. Jadamec, R. W. Sager and T. J. Killeen. *Anal. Chem.* 51:2180 (1979).
28. Van Vliet, H. P. M., C. Bottsman, R. W. Frei and V. A. Brinkman. *J. Chromatog.* 185:483 (1979).
29. Karger, B. L., M. Martin and G. Guiochon. *Anal. Chem.* 46:1640 (1974).
30. Snyder, L. R., and D. L. Saunders. *J. Chromatog. Sci.* 7:195 (1969).
31. Snyder, L. R., J. W. Dolan and J. R. Grant. *J. Chromatog.* 165:3 (1979).
32. Dolan, J. W., J. R. Grant and L. R. Snyder. *J. chromatog.* 165:31 (1979).
33. Hertz, H. S., L. R. Hilpert, W. E. May, S. A. Wise, J. M. Brown, S. N. Chesler and F. R. Guenther. In: *Measurements of Organic Pollutants in Water and Wastewater*, ASTM STP 686, C. E. Van Hall, Ed. (Philadelphia, PA: American Society for Testing and Materials, 1979), pp. 291-301.
34. Albro, P. W. "Validation of Extraction and Cleanup Procedures for Environmental Analysis," paper presented at the Division of Environmental Chemistry, American Chemical Society, Washington, DC, September 9-14, 1979.
35. Karger, B. L., J. R. Gant, A. Hartkopf and P. H. Weiner. *J. Chromatog.* 128:65 (1976).
36. Horvath, C., W. Melander and I. Molnar. *J. Chromatog.* 125:129 (1976).
37. Horvath, C., W. Melander and I. Molnar. *Anal. Chem.* 49:142 (1977).
38. Horvath, C., and W. Melander. *J. Chromatog. Sci.* 15:393 (1977).
39. Schill, G., R. Modin, K. O. Borg and B. A. Persson. In: *Handbook of Derivatives for Chromatography*, K. Blan and G. King, Eds. (London: Heyden & Son, Ltd., 1977), pp. 500-529.
40. Scott, R. P. W., and P. Kucera. *J. Chromatog.* 175:51 (1979).
41. Horvath, C., W. Melander, I. Molnar and P. Molnar. *Anal. Chem.* 49:2295 (1977).
42. Knox, J. H., and J. Jurand. *J. Chromatog.* 149:297 (1978).
43. Atwood, J. G., G. J. Schmidt and W. Slavin. *J. Chromatog.* 171:109 (1977).
44. Kissinger, P. T. *Anal. Chem.* 49:883 (1977).
45. Graham, J. A., and L. B. Rogers. *J. Chromatog. Sci.* 18:614 (1980).
46. Burham, A. K., G. V. Calder, G. A. Junk, H. J. Svec and R. Willis. *Anal. Chem.* 44:139 (1972).
47. Van Rossum, P., and R. G. Webb. *J. Chromatog.* 150:381 (1978).
48. Dressler, M. *J. Chromatog.* 165:167 (1979).

# CHAPTER 27

# HIGH-PERFORMANCE LIQUID CHROMATOGRAPHIC DETERMINATION OF NITROSAMINES IN WATER AND WASTEWATER

**A. F. Haeberer and T. A. Scott***

U.S. Environmental Protection Agency
Analytical Chemistry Branch
Environmental Research Laboratory
Athens, Georgia

The carcinogenic nature of N-nitrosamines has been well publicized [1-3]. It is now thought that the exposure levels encountered via in vivo formation of N-nitrosamines may be more important in the development of adverse human health effects than the trace amounts ingested with food and drink, inhaled in tobacco smoke, or absorbed from cosmetics or pesticides [3,4]. Nevertheless, industry and municipal facilities may be required to establish the presence or absence of N-nitrosamines in their products and effluents. Therefore, researchers need to continue to assess the risk of environmental N-nitrosamine exposure. The need for the development of a rapid, economical and sensitive analytical method for nonvolatile and heat-sensitive N-nitrosamines in aqueous matrices became apparent when application of the now well-known Consent Decree protocol for the U.S.

*Present address: Radian Corp., Sacramento, CA.

Environmental Protection Agency (EPA) " priority pollutants" resulted in decomposition, with identical gas chromatographic (GC) retention times and mass fragmentation patterns for both N-nitrosodiphenylamine and its decomposition product, diphenylamine.

This work focused on the development of accumulation and chromatographic separation and detection methods for high-performance liquid chromatographic (HPLC) analysis of N-nitrosamines that would be more cost-effective than current methods utilizing GC separation and thermal energy analyzer (TEA) detection. Major emphasis was placed on the evaluation of HPLC detection methods that included the formation of a fluorescent derivative of the denitrosated amine and the utlization of the photoconductivity detector for the direct detection of the N-nitrosamines.

## EXPERIMENTAL

### Accumulation

Mixed XAD-4 and XAD-8 neutral macroreticular resins have been shown to be efficient accumulators of a variety of organic compounds from dilute aqueous solution [5]. This mixed resin system was used in these experiments for the accumulation of N-nitrosamines from aqueous solutions ranging in concentration from 0.2 to 100 ppb. XAD-4 and XAD-8 resins were obtained from Chemical Dynamics Corp. (South Plainfield, NJ). The resins were extracted with methanol in a soxhlet extraction apparatus for 18 hr or until the concentrate (100:1) of the methanol extract was shown to contain no interferences when chromatographed using the same HPLC conditions as the samples. Equal volumes of each resin were slurried in methanol and placed in the chromatographic columns under methanol until used. The accumulator column bed dimensions were 1.6 x 13 cm, resulting in a bed volume of approximately 25 mL. Aqueous samples were extracted on these accumulator resin columns at a rate of one-third bed volume per minute, or 8 mL/min. Accumulated organics were stripped from the resin by first eluting with one bed volume (25 mL) of acetone (removing any remaining water) followed by four bed volumes (100 mL) of methylene chloride. These eluents were combined in a separatory funnel, and the organic layer transferred to a 500-mL Kuderna-Danish (KD) evaporative concentrator which was then fitted with a three-section Snyder column (Kontes, Vineland, NJ). The supernatant water layer was discarded. One methylene chloride-preextracted No. 12 Carborundum® boiling chip (Hengar granules, Arthur H. Thomas Co., Philadelphia, PA) was added to the KD receiver. The solution was evaporated on a steam bath as rapidly as possible to a volume of about 2 mL.

## Extract Preparation

The methylene chloride concentrate was then ready to be derivitized or converted to a methanolic solution compatible with the photoconductivity detector. The latter procedure required the addition of 20 mL of methanol at this point, along with a fresh Carborundum boiling chip. The solution was thoroughly mixed with a vortex mixer. This solution was again reduced via rapid KD concentration and cooled to room temperature. This resulted in a final volume of approximately 1 mL. At this point the internal standard (2.0 μg of 1-bromo-2-chloroethane in 20 μL methanol) was introduced. The solution was mixed with a vortex mixer, transferred to a 2-mL vial and sealed with a septum closure, ready for HPLC analysis.

The derivitization of N-nitrosamines, using the method of Klimisch and Stadler [6], is illustrated in Figure 1. A 50-μL portion of the concentrated methylene chloride extract was placed into a 1-mL reaction vial (Reacti-Vial™, Pierce Chemical Co., Rockford, IL) to which 0.2 mL of 3% hydrobromic acid in glacial acetic acid was added. The vial was sealed and heated to 55°C for 30 min. The resulting mixture of denitrosated amine salts was dried with the aid of a stream of dry nitrogen and gentle warming (40°C). The residue was taken up in 50 μL of methanol and 50 μL of a 0.05% methanolic 7-chloro-4-nitrobenzo-2-oxa-1,3-diazole (NBD-chloride) solution, and 20 μL of a 0.1 *M* aqueous sodium bicaronate solution was added. The vial was again sealed and heated to 55°C for 90 min. The resulting solution, after cooling, was ready for quantification by reverse-phase HPLC.

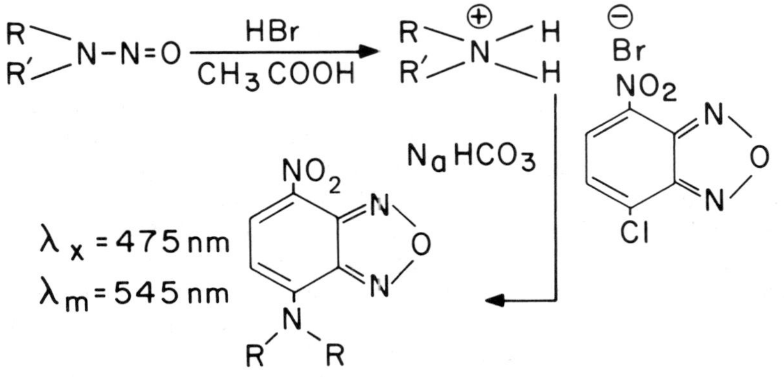

**Figure 1.** Denitrosation of N-nitrosamines and derivatization of their corresponding amine salts with NBD-chloride.

## Chromatography

A Spectra-Physics Model 8000 liquid chromatograph (Spectra-Physics, Santa Clara, CA) was used with a 4.6-mm x 250-mm Whatman Partisil-10 ODS-2 column (Whatman Inc., Clifton, NJ) for separation and quantification of the NBD derivatives. A linear mobile phase gradient was needed going from 60% 0.005 *M* aqueous ammonium acetate buffer/40% acetonitrile to 100% acetonitrile. The detector was a Farrand Model MK-1 spectrofluorometer with grating monochromators and a 10-μL flow cell. The excitation wavelength was set to 473 nm and the emission or fluroescence wavelength to 525 nm.

The underivitized N-nitrosamines were chromatographed isocratically using a Tracor Model 900 HPLC pump with a Rheodyne Model 725 injector. The reverse-phase column (4.6 x 150 mm) was packed in the authors' laboratory with 5 μm Spherisorb-ODS (Alltech Associates, Arlington Heights, IL). Detection was carried out with a Tracor Model 965 photoconductivity detector fitted with the mercury vapor lamp (254 nm). Two different isocratic mobile phases were needed to elute the various N-nitrosamines that have been examined to date. N-nitrosodimethylamine, N-nitrosomorpholine, N-nitrosopyrrolidine, N-nitrosodiethylamine and N-nitrosopiperidine all eluted within 17 min with a mobile phase consisting of 18% methanol in water. More hydrophobic compounds, such as N-nitroso-N-methylaniline, N-nitrosodipropylamine and N-n-butyl-N′-nitro-N-nitrosoguanidine, required a stronger mobile phase (40% methanol in water) for acceptable chromatographic elution.

## Reagents and Standard Solutions

NBD-chloride was purchased from Pierce Chemical Co., Rockford, IL, and was used without additional purification. Solutions containing 0.05% NBD-chloride in methanol were made fresh daily. A 3% solution of hydrobromic acid was prepared daily from 31% hydrobromic acid in acetic acid (Eastman Organic Chemicals, Rochester, NY) by 1:10 dilution with glacial acetic acid (J. T. Baker Chem. Co., Phillipsburg, NJ).

All N-nitrosamines were purchased from the Nanogens Co., Freedom, CA, as 100-ppm solutions either in cyclohexane for the NBD-derivatization study or in methanol for the photoconductivity detector work.

All solvents (methanol, acetone, acetonitrile and methylene chloride) were bought from Burdick and Jackson Laboratories, Inc. (Muskegon, MI), as the "distilled in glass" grade and used without additional purification.

The water that was used in all liquid chromatographic mobile phases as well as in accumulation and extraction experiments was purified by reverse-osmosis followed by additional polishing with a Barnstead hose-nipple cartridge system (Arthur H. Thomas Co., Philadelphia, PA). This system consisted of a "Standard High-Capacity" cartridge (D 8901) connected to an "Ultrapure" cartridge (D 8902) and finally an "Organic Removal" cartridge (D 8904). All connections were made with Teflon® tubing.

## RESULTS AND DISCUSSION

In these determinations of N-nitrosamines as their corresponding fluorescent amine derivatives, NBD-chloride was the reagent of choice because it reacts selectively with secondary as well as primary amines [5]. If any of these amines were in the water sample, however, they could also be accumulated on the XAD-4/8 resin column along with any uncharged organic compounds. This can be prevented by acidification of the sample solution to pH 2 with 3 *N* HCl, which will result in the formation of the amine-hydrochlorides. These compounds are not retained by the XAD-4/8 accumulator resin. NBD-chloride, unlike dansyl chloride, has the additional advantage of not fluorescing in its unreacted form. The detection linearity of these NBD-amine derivatives was more than adequate (linear over more than three orders of magnitude) and the lower detection limits for the N-nitrosamine derivatives tested was in the range of 50–100 pg. Results obtained with these derivatization experiments, however, proved to be quite variable. When results were compared from accumulated samples, standard N-nitrosamine solutions and amine hydrochloride solutions, it became clear that uncontrollable losses were incurred during the step in which the denitrosated amine salts were dried. This problem is particularly apparent with the very volatile N-nitrosamines such as N-nitrosodimethylamine, N-nitrosodiethylamine and N-nitrosodipropylamine that can vary from 80% recovery to not being detected at all if the drying procedure is carried out too long. Primarily for this reason and because this derivatization procedure requires considerably more manipulation than the determination of N-nitrosamines with the photoconductivity detector, this procedure was abandoned in favor of the latter method.

The photoconductivity detector response curves of three selected nitrosamines are presented in Figure 2. The response is linear over three orders of magnitude. Each data point represents the mean value of three determinations and the standard deviation is included within the symbol. Linearity cannot be assumed below 4 ng because the extrapolated lines do not pass through the origin.

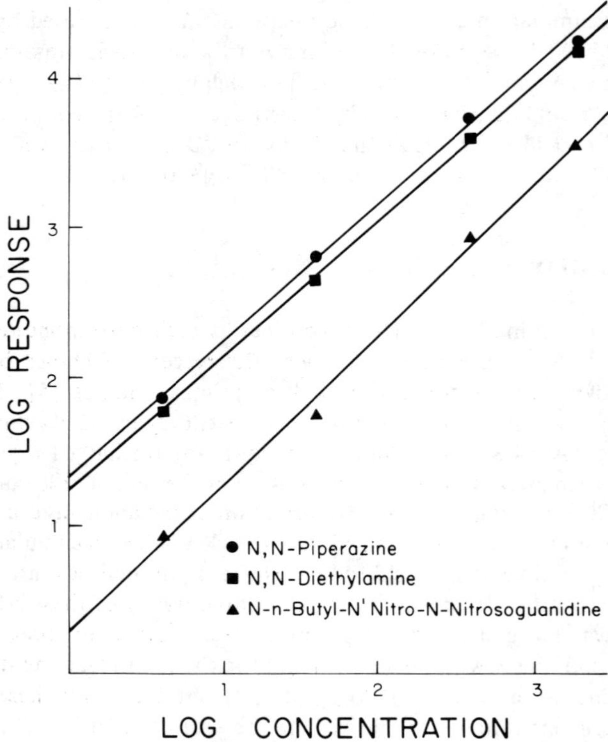


**Figure 2.** Response curves of three N-nitrosamines determined by photoconductivity detection.

The response of the photoconductivity detector to organohalogen compounds prompted the evaluation of a variety of these compounds as internal standards for the quantification of the N-nitrosamines. Bromotrichloromethane, bromochloromethane, ethylenebromide, bromoform, chloroacetone, iodomethane and bromodichloromethane were all found to be unsuitable either because of coelution with a N-nitrosamine, poor stability or because the compound had been found frequently in aqueous environmental samples. Although 1-bormo-2-chloroethane has been reported from a few environmental samples, it was chosen as the internal standard because it fitted all other criteria.

Because the difficulties associated with gradient elution encountered with the photoconductivity detector have not entirely been eliminated, all analyses were carried out isocratically. Nitrosamines of similar hydrophobicity were grouped together. Figure 3 shows the chromatogram of 40 ng of each compound in group 1 N-nitrosamines. The lower limit of detection (three times signal-to-noise ratio) for these compounds ranged from 0.4 to 2.5 ng.

Table I presents the recoveries of group 1 N-nitrosamines at 1 ppb in water, using the photoconductivity detector and comparing XAD 4/8 resin accumulation with classical liquid-liquid extraction (LLE) using methylene chloride as the extraction solvent. Resin accumulation was 25–30% higher than LLE for the more hydrophilic N-nitrosamines. The relative standard

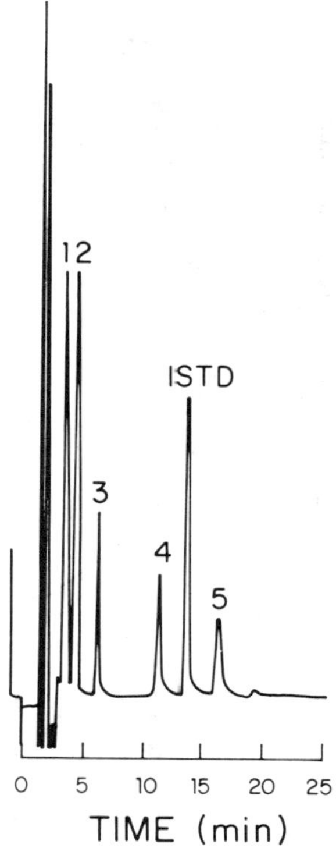


**Figure 3.** HPLC chromatogram of group 1 N-nitrosamines, 40 ng each. Column: 150 x 4.6 mm Spherisorb ODS-5 $\mu$. Mobile phase: 18% methanol/82% water. Peak identification: (1) N-nitrosodimethylamine; (2) N-nitrosomorpholine; (3) N-nitrosopyrrolidine; (4) N-nitrosodiethylamine; (ISTD) 1-bromo-2-chloroethane; (5) N-nitrosopiperidine.

Table I. Recovery of Group 1 Nitrosamines from 1-ppb $H_2O$ (Photoconductivity Detector)

| Nitrosamine | Resin Accumulation | | | Liquid Liquid Extraction | | |
|---|---|---|---|---|---|---|
| | No. of Determinations | Mean Recovery | RSD[a] | No. of Determinations | Mean Recovery | RSD |
| N-Nitrosodimethylamine | 7 | 46 | 20 | 5 | 22 | 19 |
| N-Nitrosomorpholine | 7 | 75 | 12 | 5 | 49 | 11 |
| N-Nitrosopyrrolidine | 3 | 80 | 10 | 3 | 48 | 14 |
| N-Nitrosodiethylamine | 7 | 78 | 9 | 5 | 74 | 9 |
| N-Nitrosopiperidine | 7 | 83 | 9 | 5 | 80 | 7 |

[a] RSD = relative standard deviation.

deviations obtained from the two methods are essentially identical. This indicates that the accumulation method, using XAD-4/8, permits the quantification of N-nitrosamines at lower levels than may be achieved by methylene chloride LLE.

A chromatogram of group 2 N-nitrosamines, eluting with a mobile phase of 40% methanol/60% water, is shown in Figure 4. Three compounds from group 1 were found to also elute conveniently with this group and were consequently included, demonstrating the range of compounds that can be determined with this mobile phase. The lowest level of detection ranged from 0.3 to 1.7 ng for the compounds of this group. Recovery studies for these N-nitrosamines, carried out at 0.4 ppb in water are presented in Table II.

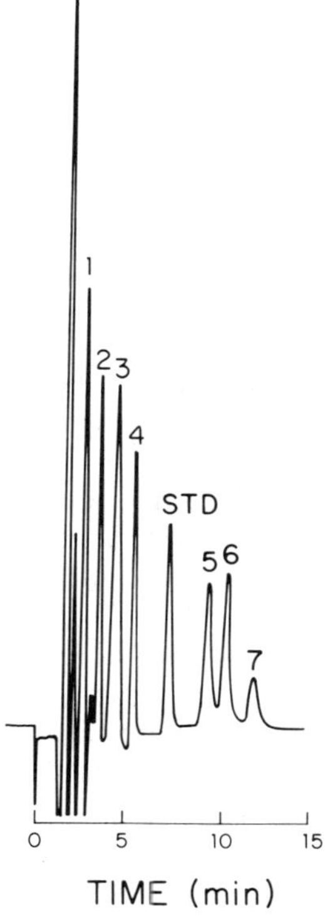


**Figure 4.** HPLC chromatogram of group 2 N-nitrosamines, 40 ng each. Column: 150 x 4.6 mm Spherisorb ODS-5 $\mu$. Mobile phase: 40% methanol/60% water. Peak identification: (1) N-nitrosopiperazine; (2) N-nitrosopyrrolidine; (3) N-nitrosodiethylamine; (4) N-nitrosopiperidine; (ISTD) 1-bromo-2-chloroethane; (5) N-nitroso-N-methylaniline; (6) N-nitrosodipropylamine; (7) N-*n*-butyl-N'-nitro-N-nitrosoguanidine.

**Table II. Recovery of Group 2 Nitrosamines from 0.4-ppb Water by XAD-4/8 Resin Accumulation (Photoconductivity Detector)**

| Nitrosamine | Mean Recovery[a] | RSD[b] |
|---|---|---|
| Nitrosopiperazine | 73 | 19 |
| N-Nitrosopyrrolidine | 70 | 15 |
| N-Nitrosodiethylamine | 73 | 12 |
| N-Nitrosopiperidene | 84 | 7 |
| N-Nitroso-N-methylaniline | 82 | 12 |
| N-Nitrosodipropylamine | 80 | 9 |
| N-*n*-Butyl-N′-nitro-N-nitrosoguanidine | 22 | 18 |

[a] Based on three replicate experiments.
[b] RSD = relative standard deviation.

The greater part of the N-nitrosamine losses was determined to occur during the KD evaporation and not in the accumulation step. The average recovery for the compounds of group 1 is 79% and group 2, 83% when they were subjected only to KD evaporation and the accumulation step was omitted. The average recoveries from accumulation on XAD 4/8 followed by KD evaporation was 72 and 77% for groups 1 and 2, respectively. This shows clearly that macroreticular resin accumulation is an efficient means of concentrating N-nitrosamines from very dilute aqueous solutions.

Figure 5 compares the HPLC chromatograms of group 1 N-nitrosamine standards, the resin accumulation extract from drinking water spiked at 0.2 ppb (0.2 $\mu$g/L) of each compound and unspiked drinking water. Figure 6 shows the chromatograms of the same compounds but with municipal effluent as matrix. The two major peaks are caused by the sample matrix, one peak coelutes with N-nitrosopyrrolidine, hindering the quantification of that compound. Figure 7 shows the chromatograms (standards, fortified at 0.4 ppb each compound, and unfortified) of group 2 N-nitrosamines with drinking water as the sample matrix. Loss of peak 7, N-*n*-butyl-N′-nitro-N-nitrosoguanidine, in the fortified sample, is due to decomposition and not to accumulation nor KD evaporation losses. Figure 8 is representative of the standards and recoveries of group 2 N-nitrosamines at the 10-ppb level from municipal effluent. The chromatograms in this figure resulted when a mobile phase consisting of 50% methyl alcohol and 50% water was used. Peak 4 is due to an impurity introduced with the internal standard solution.

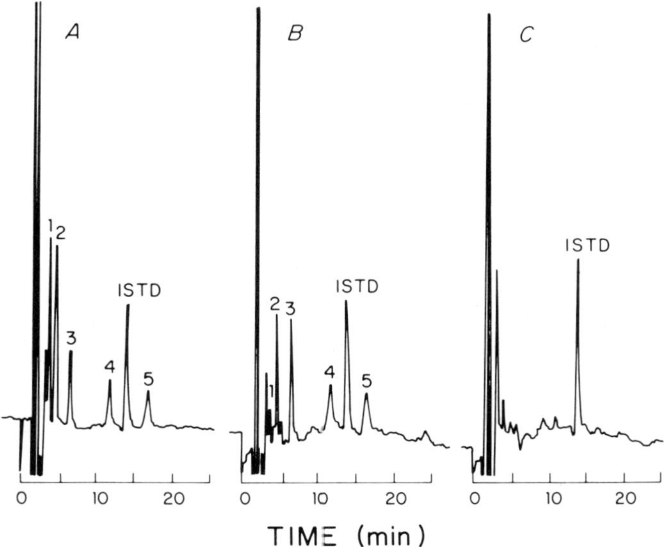


**Figure 5.** HPLC chromatograms of group 1 N-nitrosamines: (A) standards, 4 ng each; (B) extract from fortified drinking water; (C) extract from nonfortified drinking water. Peak identification: see Figure 3

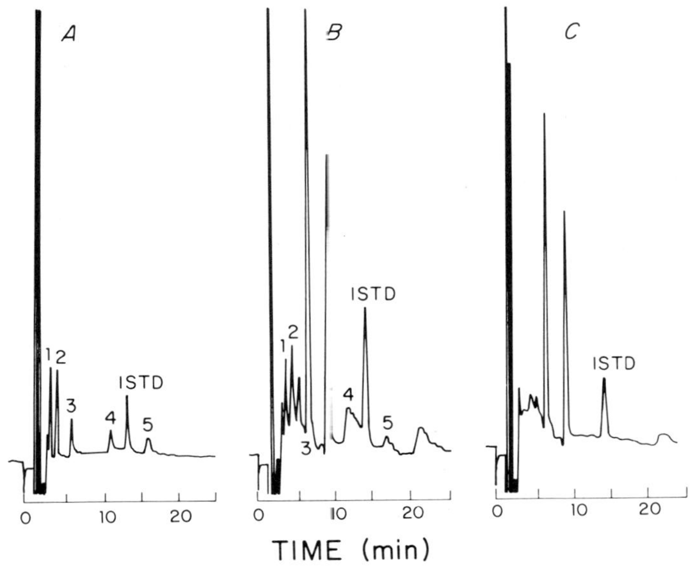


**Figure 6.** HPLC chromatograms of group 1 N-nitrosamines: (A) standards, 4 ng each; (B) extract from fortified municipal effluent; (C) extract from nonfortified municipal effluent. Peak identification: see Figure 3.

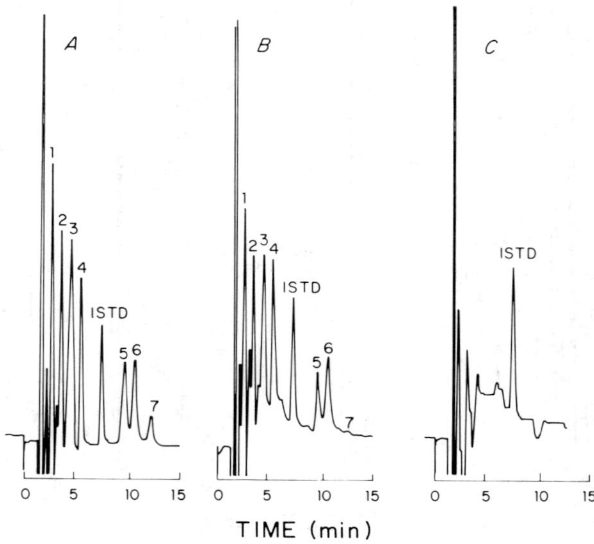


**Figure 7.** HPLC chromatograms of group 2 N-nitrosamines: (A) standards, 8 ng each; (B) extract from fortified drinking water; (C) extract from nonfortified drinking water. Peak identification: see Figure 4.

Future work will concentrate on improving the concentration step to reduce losses. Model compound test mixtures will be expanded to include even more hydrophobic N-nitrosamines such as N-nitrosodiphenylamine and N,N′-dinitrosoatrazine. Normal-phase HPLC will be investigated utilizing a cyano column with the aim of two-column as well as liquid chromatography-mass spectrometry confirmation of the N-nitrosamines. Other aqueous matrices, such as urine, will also be examined for N-nitrosamine recoveries.

## CONCLUSIONS

The determination of N-nitrosamines as the fluorescent NBD derivatives of the corresponding amines shows some promise if loss of the intermediate amine salt can be minimized during the drying procedure or if one is concerned only with nonvolatile N-nitrosamines. At present, there are no plans to further pursue this procedure.

The determination of N-nitrosamines by HPLC utlizing the photoconductivity detector offers considerable advantages and represents an effective means for analysis of these compounds.

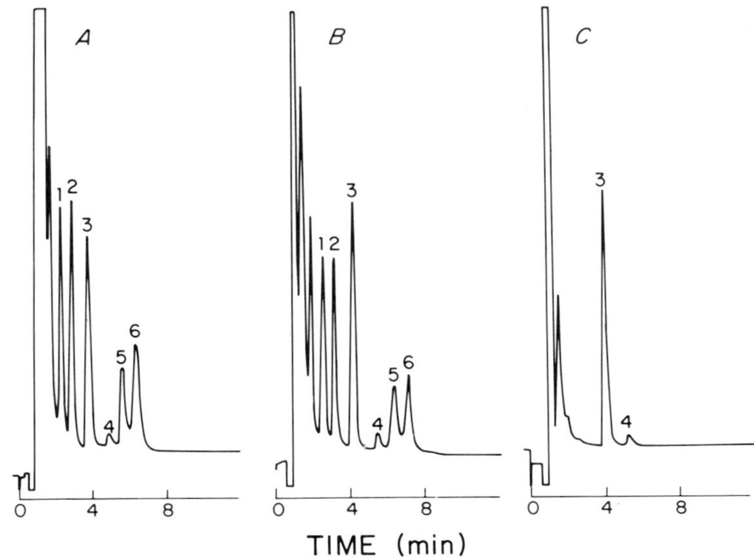


**Figure 8.** HPLC chromatograms of group 2 N-nitrosamines: (A) standard, 20 ng each; (B) extract from fortified municipal effluent; (C) extract from nonfortified municipal effluent. Peak identification: (1) N-nitrosodiethylamine; (2) N-nitrosopiperidene; (3,ISTD) 1-bromo-2-chloroethane; (4) unknown contaminant; (5) N-nitroso-N-methylaniline; (6) N-nitrosodipropylamine.

## DISCLAIMER

Mention of commercial products, sources and trade names is for informational purposes only and does not imply endorsement by the U.S. Environmental Protection Agency.

## REFERENCES

1. Magee, P. N., and J. M. Barnes. *Adv. Cancer Res.* 10:163 (1967).
2. Druckerey, H. R., R. Preussmann, E. Ivankovic and R. Schmahl. *L. Krebsforsch.* 69:103 (1967).
3. Ember, L. R. *Chem. Eng. News* (March 31, 1980), p. 20.
4. Lijinsky, W. *Chem. Eng. News* (July 3, 1978), p. 6
5. Van Rossum, P., and R. G. Webb. *J. Chromatog.* 150:381-392 (1978).
6. Klimish, H. J., and L. Stadler. *J. Chromatog.* 90:223 (1974).

## CHAPTER 28

# DEVELOPMENTS IN THE ANALYSIS OF PRIORITY POLLUTANTS BY LIQUID CHROMATOGRAPHY/MASS SPECTROMETRY AND IMMUNOASSAY PROCEDURES

**R. G. Christensen, H. S. Hertz,**
**D. J. Reeder and E. White V**

Center for Analytical Chemistry
National Bureau of Standards
Washington, DC

Several approaches to the rapid analysis of organic priority pollutants are being currently examined at the National Bureau of Standards (NBS). Two of these areas of research are presented in this chapter: combined liquid chromatography/mass spectrometry (LC/MS) and fluoroimmunoassay procedures.

## LIQUID CHROMATOGRAPHY/MASS SPECTROMETRY

The great potential of combined LC/MS as an analytical tool has fostered research by a considerable number of laboratories. In addition to the collection, evaporation and transfer of LC fractions to the mass spectrometer, a number of means have been used to combine LC and MS. Recent reviews by Arpino and Guiochon [1], Zerilli [2] and McFadden [3] (a review which contains examples of applications of LC/MS contributed by other authors) describe the operating principles, construction and performance of these devices. Two methods, direct liquid injection (DLI) and evaporation onto a moving wire or belt, are now offered commercially.

Although the DLI technique is intrinsically simple to implement, the moving belt interface has the advantages of concentrating the solute, thereby introducing a greater proportion of the sample into the ion source of the mass spectrometer and permitting both electron impact (EI) and chemical ionization (CI) modes of operation. A system which could preconcentrate the liquid stream and introduce the concentrate by DLI would combine some of the advantages of both ordinary DLI and the moving belt technique.

The interface implemented in this laboratory was designed and built with this aim in mind. The interface device (Figure 1) concentrates a liquid stream by allowing it to flow down a resistance-heated stationary wire. The residual liquid is drawn into the mass spectrometer through a capillary tube with a needle valve at the ion source end. The liquid sprays from the needle valve into the ion source of a conventional, differentially pumped, quadrupole mass spectrometer. Flowrates can be adjusted within the range of 2-20 $\mu$L/min, allowing spectra of either purely CI or mixed EI/CI character to be obtained.

As a test of the LC/MS system, a mixture of 40 ng each of phenanthrene, 9-methylanthracene and fluoranthene was injected onto an aminosilane column and eluted with a mixture of 10% methylene chloride in 2,2,4-trimethylpentane. The resultant chromatograms are shown in Figure 2. The upper trace is from the ultraviolet (UV) detector at 254 nm. The first peak contains the unresolved phenanthrene and 9-methylanthracene; the second peak represents the fluoranthene. The bottom trace shows the single ion record of masses (m/z) 192 (methylanthracene) and 202 (fluoranthene) without the concentrator wire in operation. The center trace shows the same single ion record with the concentrator in operation. There is about a 20-fold increase in sensitivity with no change in the noise level when the concentrator wire is used, indicating that about 95% of the solvent has been removed. Relative to the UV-detector recording, the peaks are delayed about 1 min and broadened by about 15 sec. Most of the delay and peak-broadening occurs in the DLI tube. It will be possible to reduce this broadening in the future by making the tube shorter. We also expect that it will be possible to increase the concentration enhancement from 20- to about 100-fold.

The capabilities of the system for performing quantitative analysis of individual compounds in complex mixtures was examined by measurement of phenol and *o*-cresol in the recently certified NBS Shale Oil Standard Reference Material (SRM 1580). Five determinations of each compound were made over a two-month period. Each determination consisted of a minimum

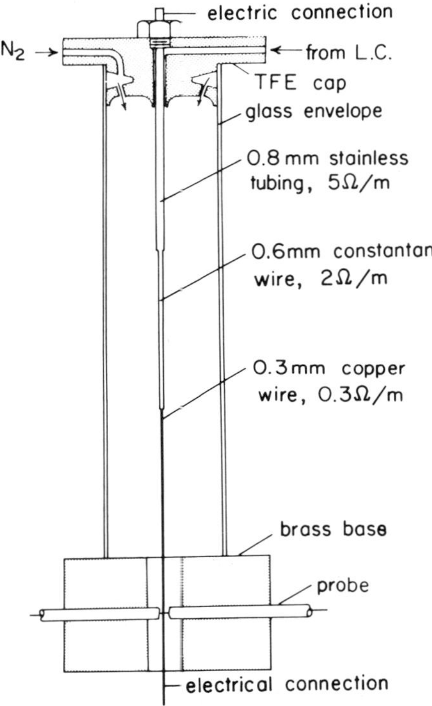


**Figure 1.** LC solvent concentrator wire for enrichment of the LC effluent (not drawn to scale).

of two LC analyses. The complexity of a shale oil sample is obvious, and the ability to quantify individual constituents without any prior sample handling is extremely encouraging. Peak height measurements of the appropriate ions gave the following concentrations and standard deviations: phenol, 419 ± 33 μg/g, and *o*-cresol, 450 ± 90 μg/g. These concentrations compare with certified values of 407 ± 50 μg/g for phenol and 385 ± 50 μg/g for *o*-cresol.

## FLUOROIMMUNOASSAY PROCEDURES

Immunoassays have been in use by research groups for nearly two decades. Their impact on the health care front has been significant; they are widely

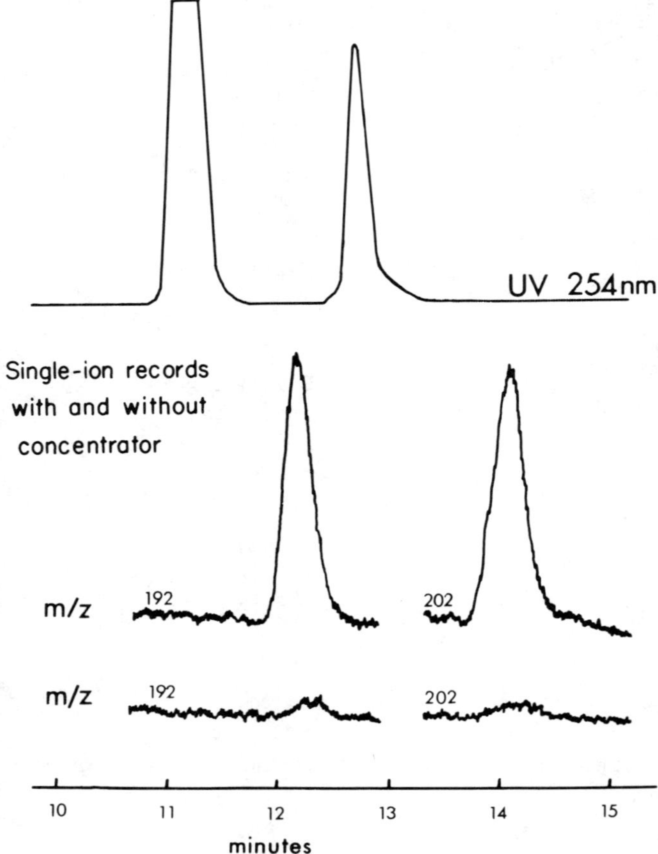


**Figure 2.** LC/MS analysis of a mixture of 40 ng each of phenanthrene, 9-methylanthracene and fluoranthene under EI conditions. UV detector recording at 254 nm (top); single ion records at m/z 192 (9-methylanthracene) and 202 (fluoranthene) with concentrator wire being heated (middle): single ion records obtained on the same sample without enrichment by the concentrator wire (bottom).

used as analytical tools by endocrinologists and clinical chemists. However, use of these assays is just beginning to affect pollutant analysis.

The appeal of immunoassays lies in their extreme sensitivity (detection of part-per-billion levels), ease of performance, rapidity, generally acceptable precision, potential for automation and relatively low cost per analysis.

As with any method, an immunoassay procedure must be evaluated thoroughly with respect to specificity, sensitivity and precision to establish and maintain valid conditions for analysis. In the establishment of an assay, many variables have to be controlled. Antisera production, antisera affinity, labeled antigens and separation techniques must be investigated.

In developing a model system for detection of dinitrophenols (DNP) in water, we prepared conjugates of DNP by reacting proteins, 2,4-dinitrobenzenesulfonic acid and $K_2CO_3$ in aqueous solution [4]. Extent of conjugation was calculated by optical density measurements at 360 nm, using 17,350 at pH 7.4 as the molar extinction coefficient for DNP-lysyl residues. Two proteins were used as protein bases: rabbit serum albumin (RSA) and bovine serum albumin (BSA).

Randomly bred New Zealand White rabbits were initially immunized via footpad and intramuscular route by injecting 2.0 mg of the specific DNP-BSA or DNP-RSA conjugates mixed in an emulsion with Freund's complete adjuvant. Subsequent injections were made subcutaneously and intramuscularly at monthly intervals with trial bleeds from ear veins taken just prior to the injections. Antibody response was measured vs homologous conjugates in a microdouble-gel diffusion slide system by observing the precipitating lines developed in the agar matrix. The rabbits were finally bled 25 weeks after initial immunization via cardiac puncture for optimal yield of serum.

Antibodies in the form of immunoglobulin G (IgG) were obtained from serum by use of a beaded, crosslinked agarose gel to which Cibicron Blue F3GA dye and diethylaminoethyl groups are attached. In this procedure, buffer-equilibrated serum was passed over a packed column (2.0 x 18 cm) of the gel. Protease-free IgG was obtained as the first peak eluted with a wash of the starting buffer. Other proteins, including albumin, were retained by the gel.

To obtain a specific solid-phase binding reagent for assay purposes, the purified IgG was covalently bound to hydrophilic polyacrylamide beads, 5–10 $\mu$m in diameter. Binding was through amide linkages between the carboxyl groups of the beads and the amino groups on the IgG, facilitated by a water-soluble carbodiimide coupling procedure. Immobilized antibody thus obtained was stable at 4°C for 4–6 months and could be freeze-dried to obtain a much longer shelf life.

Literature reports of affinity of 2,4-dinitrophenol for anti-DNP antibody suggest a wide diversity of binding capabilities [5]. A more homogeneous preparation of anti-DNP IgG may be prepared if the IgG is passed through an affinity column to which DNP is covalently bound. Subsequent elution of more avid and specific antibodies may be achieved by stripping the bound protein from the column with high concentrations (greater than 0.2 mol/L) of DNP or urea.

The technique of immunoassay is based on (1) the partial saturation of the antibody by the substance undergoing testing, and (2) the competition of the material (unlabeled ligand) with a chemically similar "indicator" material (labeled ligand) for the available binding sites on the antibody. In the model system we are developing the unlabeled ligand is 2,4-dinitrophenol. We have used as a labeled ligand a tritiated 2,4-dinitrophenol as well as 2,4-dinitrophenol that is labeled with a fluorescent moiety. The tritiated system (radioimmunoassay) has several disadvantages, including special handling and waste disposal, expensive counting instrumentation, and long periods of time required for accurate counting of the radioactivity. Shorter counting periods could be obtained by use of $^{125}I$ bound to the antibody, but the problem of relatively short reagent half-life that prevents long-term assay standardization is added to the abovementioned disadvantages.

The use of a fluorescently labeled ligand in a fluoroimmunoassay has the advantage of longer shelf life, less expensive instrumentation, ease of counting and less stringent disposal procedures. In fact, the use of fluorescence labels in the immunoassays has the potential to replace many existing radioimmunoassay procedures. A labeled ligand has been developed by reacting RSA with 2,4-dinitrobenzenesulfonic acid and fluorescein isothiocyanate (FITC) in a double labeling experiment.

Since the fluoroimmunoassay is based on competition of the unlabeled ligand with the labeled ligand for the antibody binding sites, we first determined the optimal amounts of immobilized antibody to be used in the assay. Increasing amounts of the immobilized antibody (0.05–1 mg) were added to constant amounts (about 100 μg) of labeled ligand (RSA-DNP-FITC). The final volume of each reaction tube was brought to 1.5 mL with phosphate-buffered saline (PBS). After one hour incubation at 37°C, the tubes were centrifuged at 1000 *g* for 10 min, and the supernatant fluid was decanted, leaving the solid-phase antibody (and attached, labeled ligand) in the bottom of the tubes. The pelleted antibody was washed with PBS, recentrifuged and separated from the wash fluid. Finally, the pellet was resuspended in 2.0 mL PBS.

The fluorescence of each reaction tube mixture was determined with a combination photon-counting fluorometer and microprocessor that automates the reading. Saturation of the beads with 100 μg of labeled ligand occurs with about 0.5 mg beads, adjusted to 100 μL.

Having established the optimal antibody concentration for use in the assay, we are now proceeding to qualify the assay in a competitive binding mode to optimize the sensitivity of the assay. Current detection limits are at the 100-μg/L level for DNP in water. To be useful in pollutant analysis, sensitivity must be at the sub-μg/L or even the pg/L level. While sensitivity of this order has been established for immunoassays in clinical use, work is still progressing to achieve these detection limits for pollutant analysis.

## ACKNOWLEDGMENTS

The authors gratefully acknowledge financial support from the Office of Energy, Minerals and Industry within the Office of Research and Development of the U.S. Environmental Protection Agency under the Interagency Energy/Environment Research and Development Program.

## DISCLAIMER

Identification of any commercial product does not imply recommendation or endorsement by the National Bureau of Standards, nor does it imply that the materials or equipment identified is necessarily the best available for the purpose.

## REFERENCES

1. Arpino, P., and G. Guiochon. *Anal. Chem.* 51:682A-701A (1979).
2. Zerilli, L. F. In: *Recent Development in Chromatography and Electrophoresis*, A. Frigerio and Z. Renoz, Eds. (Amsterdam: Elsvier Scientific Publishing Co., 1979), pp. 59-71.
3. McFadden, W. J. *J. Chromatog. Sci.* 50:97-115 (1980).
4. Williams, G. A., and M. Chase, Eds. *Methods in Immunology and Immunochemistry, Vol. 1* (New York: Academic Press, Inc., 1967), p. 1301.
5. Eisen, H. N. "Combining Sites of Anti 2,4-Dinitrophenyl Antibodies," in *Progress in Immunology*, B. Amos, Ed. (New York: Academic Press, Inc., 1971).

# SECTION 7

# DERIVATIZATION

# CHAPTER 29

# DETECTION AND QUANTIFICATION OF ELECTROPHILES IN ENVIRONMENTAL SAMPLES: LABELING OF POTENTIAL MUTAGENS IN DRINKING WATER BY 4-NITROTHIOPHENOL

**Albert M. Cheh* and Robert E. Carlson**

Gray Freshwater Biological Institute
University of Minnesota
Navarre, Minnesota

Direct acting mutagens have been found in many U.S. drinking water supplies [1-8]. Identification of the mutagens has proceeded slowly, however, because of their very low concentrations and because they comprise only a small part of a highly complex and diverse set of organics in drinking water. One approach to mutagen identification would be to use repeated fractionation together with detection of the mutagens by bioassay. Bioassay is a destructive detection method which rapidly consumes the samples containing mutagen. Another approach would be to identify any and all compounds present in the water. While useful for determining the general nature of the materials present, this method is not specific for the mutagens; most of the compounds identified will not be mutagens. Direct-acting mutagens are likely to be reactive electrophiles [9]; with both approaches they will be continually lost during analysis due to processes such as solvolysis.

---

*Present address: Department of Chemistry, American University, Washington, DC.

We have described a procedure for derivatizing reactive electrophiles by reaction with 4-nitrothiophenol (NTP) [10]. Any compounds which react with NTP are specifically labeled as electrophiles and are of interest as potential mutagens. The NTP thioether derivatives are stable and may be detected nondestructively by their absorbance at about 345 nm. In this chapter we describe the application of the NTP labeling procedure to concentrated organics from drinking water and show how the absorbance profiles for the high-performance liquid chromatography (HPLC)-separated NTP thioethers change with increasing NTP. By limiting the amount of NTP and the time of reaction, it is possible to distinguish between compounds which react rapidly and those which react slowly. With simultaneous monitoring of destruction of mutagenic activity, it should then be possible to suggest which changes in NTP thioether patterns are correlated with significant losses in mutagenic activity.

## METHODS

### Reagents

NTP was obtained from Aldrich. To remove the contaminating disulfide of NTP which formed slowly over time, it was purified by the method of Barnett and Jencks [11]. The purified material was stored at -20°C under argon. Other materials were available commercially and used without further purification.

### Drinking Water Concentrates

XAD adsorption methods were used to concentrate the organics from drinking water [4,8]. The residual nonvolatile organics recovered were dissolved in ethanol to give a 40,000-fold concentration.

### Reaction of NTP with Drinking Water Electrophiles

The reaction mixtures contained 130 $\mu$L of 0.05 $M$ potassium phosphate buffer (pH 7.4), 20 $\mu$L of ethanol containing varying amounts of NTP and 150 $\mu$L of 40,000X concentrated drinking water organics in ethanol. A series of reaction mixtures containing 0, $1.25 \times 10^{-4}$ $M$, $2.5 \times 10^{-4}$ $M$, $5 \times 10^{-4}$ $M$ and $2 \times 10^{-3}$ $M$ NTP were set up and allowed to react for 30 min at room temperature.

### HPLC Separation and Detection of NTP Thioether Products

A 100-$\mu$L sample of the reaction mixture was diluted with an equal volume of methyl ethyl ketone. The diluted material was subject to HPLC separation in a Hewlett Packard 1084B instrument containing a 0.46 x 25 cm RP-8 column. Gradient elution was performed as follows: Initial solvent, 20% methanol/80% water for 2 min, then the methanol concentration increased linearly to 30% methanol/70% water at 12 min. This was followed by the methanol concentration increasing linearly to 95% at 42 min. Elution at 95% continued another 8 min. The flowrate was 2 mL/min. Detection was by absorbance at 345 nm.

### Ames Assay

Aliquots (80-$\mu$L) of the reaction mixture were assayed in triplicate by the method of Ames [12]. Bacterial strain TA100 used in the absence of the liver post-mitochondrial supernatant activating system (S-9) has been found to give the highest response to drinking water mutagens [2,3,8,13]. S-9 was omitted in these experiments. However, NTP and NTP thioethers gave a positive response with TA100. Therefore, the nitroreductase deficient strain, TA100-FR1 [14], which did not respond to either NTP or NTP thioethers was substituted for TA100 and used throughout these experiments. The response of TA100 and TA100-FR1 to drinking water mutagens were found to be essentially identical.

## RESULTS AND DISCUSSION

The lowest line in Figure 1 shows the 345-nm absorbance profile of 100 $\mu$l of a drinking water concentrate (4-L volume prior to concentration) which was mixed with methyl ethyl ketone and separated by the standard HPLC gradient system (methods). The lines above it represent, in increasing order, the HPLC absorbance profiles after reaction with 1.25, 2.5, 5, 10 and 20 x $10^{-4}$ $M$ NTP for 30 min.

Figure 1 focuses on the region between 20 and 42 min, where the most significant changes occur. After 42 min all the chromatograms have returned to the background (zero NTP) level. In the region before 20 min increases in absorption with added NTP are observed close to the solvent front with added NTP; these are mainly due to unreacted NTP anion. With some water samples, but not the one shown, at the highest NTP levels occasional sizable peaks are observed early in the chromatogram, especially around 20 min. When these peaks are collected, concentrated and reinjected, they appear

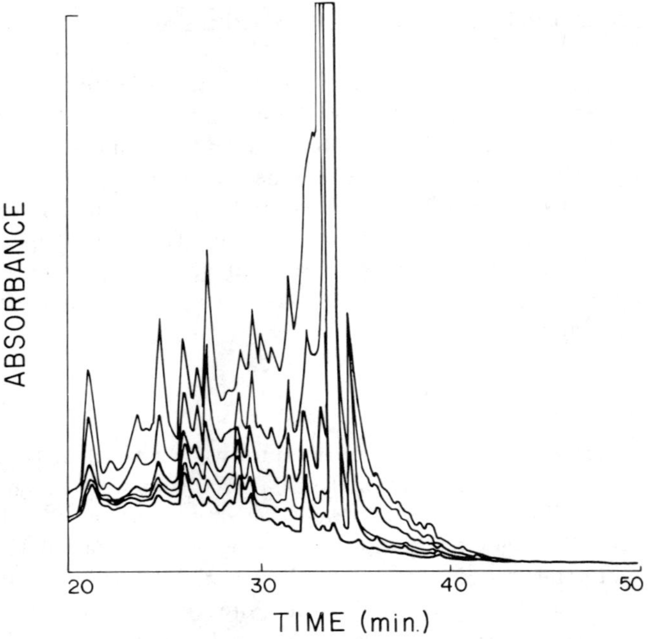


**Figure 1.** Changes in the HPLC chromatogram of a drinking water concentrate after reaction with NTP.

at 33-34 min and seem to be NTP disulfide. These peaks, therefore, appear to be matrix artifacts whereby some NTP disulfide cochromatographs with material that elutes around 20 min. Other than these phenomena, few increases in absorbance which might be attributed to thioether formation are seen before 20 min.

Two peaks in the HPLC profile between 20 and 42 min do not correspond to electrophile products. The peak at 33.9 min, which is present and off-scale in all samples treated with NTP, is NTP disulfide. The peak at 24, 8 min is un-ionized NTP. Despite a pKa of 4.57 for NTP [11], at pH 7.4 a small portion will remain un-ionized. The presence of organic solvent will further suppress, but not eliminate, NTP ionization.

All the other increases above the baseline (the lowest line in Figure 1, which corresponds to the inherent absorbance of the sample, without any added NTP) are presumed to be caused by NTP thioether products of electrophiles. In general, the entire region between 20 and 42 min increases with added NTP. This represents a mass of unresolved electrophile products. However, there are a number of distinct peaks.

Some NTP thioether peaks appear on reaction with the lowest doses of NTP, and do not appear to increase at higher doses. Other peaks continue to increase in size with increasing NTP reaction concentrations. A potential significance of these differences in electrophile reactivity may be seen in Figure 2, which illustrates the Ames assay for remaining mutagenic activity after the reactions with 1.25, 2.5, 5, 10, 20 and 100 x $10^{-4}$ *M* NTP. Because of the data scatter and error associated with a bioassay, it is not possible to establish a precise curve based on a single titration with the Ames assay (note the jog upward at $10^{-3}$ *M* NTP). Our experience with many different water samples [13], however, indicates that there is a general pattern of initial rapid decline in mutagenic activity on reaction with nucleophiles (such as what is seen from 0 to 5 x $10^{-4}$ *M* NTP in Figure 2) followed by a much slower decline with further reaction (from 5 x $10^{-4}$ *M* NTP to $10^{-2}$ *M* NTP in Figure 2)

Thus, it appears that those electrophiles which react first with low levels of NTP may be responsible for significant amounts of mutagenic activity, while those which react at much higher doses appear to be associated with much lower amounts of mutagenic activity.

Figure 3 is an expanded version of Figure 1 which allows more detailed inspection of the changes of the HPLC absorption profile after reaction with increasing NTP. Figure 3A (1.25 x $10^{-4}$ *M* NTP) shows a sizable peak at 34.9 min (right arrow) which is about the same area in Figure 3B (2.5 x $10^{-4}$ *M* NTP) and increases in Figure 3C (5 x $10^{-4}$ *M* NTP); no further increases in peak size occur at higher NTP concentrations (Figures 3D,E). A peak at 31.5 min is quite small at 1.25 x $10^{-4}$ *M* NTP (Figure 3A, left arrow), but also

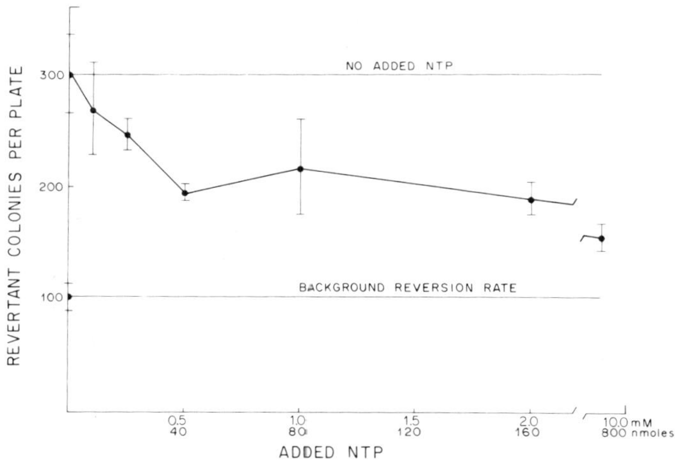


**Figure 2.** Titration of loss in Ames activity with added NTP.

increases to a maximum size at 5 x $10^{-4}$ $M$ NTP. A third at 26.8 min may increase up to $10^{-3}$ $M$ NTP (Figure 3D, left arrow) but ceases to rise thereafter. The peaks at 34.9 and 31.5 min (arrows in Figure 3A-D) are formed during the period of maximum reduction in Ames activity (Figure 2); thus they may be formed from significant mutagens.

Peaks at 21.1, 24.8, 26.1, 27.4 and 30.1 min (arrows, left to right in Figure 2E) increase substantially between 5 x $10^{-4}$ and 2 x $10^{-3}$ $M$ NTP, where there is little reduction in Ames activity (Figure 2). As noted above, the 24.8 peak is un-ionized NTP, the other peaks may be formed from electrophiles which are not highly mutagenic.

The overall increase in the absorption throughout the entire region from 20 to 42 min is readily observed in Figure 3. Unresolved thioethers in this mass of material could include some which, like the peaks at 34.9 and 31.5 min are fully formed by 5 x $10^{-4}$ $M$ NTP, and thus might have significant mutagenic activity. It is clear that there is such an abundance of organics in drinking water, that without preliminary fractionation, even a powerful technique such as capillary gas chromatography (GC) is only able to resolve a limited number of them [15]. We have encountered a similar problem in these studies. Even though NTP only labels electrophiles, and these are just a small fraction of the overall set of organics, the product thioethers are too numerous to be resolved well in a single HPLC run. It is suggested, following the comments of Coleman et al. [15] in their studies, that further fractionation before or after NTP reaction will be required.

It should be noted that the Ames assay only measures an aggregate change in mutagen level. Some of the thioethers whose formation is observed in the corresponding HPLC profile could be associated with far more loss in mutagenic activity than others formed simultaneously. Observation that a *specific* thioether is formed predominantly at low NTP concentration does not mean that the parent electrophile is highly mutagenic; it only suggests this. Electrophiles whose thioether formation occurs mainly at higher NTP levels, however, probably are much less likely to be highly mutagenic.

Once an HPLC injection is made, further reaction between NTP and the electrophiles stops because unreacted NTP anion immediately separates and comes out at the solvent front. Since comparisons are being made between the HPLC absorbance pattern and the Ames assay results, a question may be asked as to whether the NTP-electrophile reaction stops as well in the Ames assay.

On dilution of 80 $\mu$L of the reaction into a 2.6-mL volume consisting of 2 mL of top agar, 0.5 mL of 0.1 $M$ sodium phosphate (pH 7.4) and 0.1 mL of overnight culture of bacteria, both NTP and electrophiles are diluted 32.5-fold. Bimolecular reactions between NTP and electrophile will be slowed 1000-fold, although offsetting this will be a temperature rise from 25 to

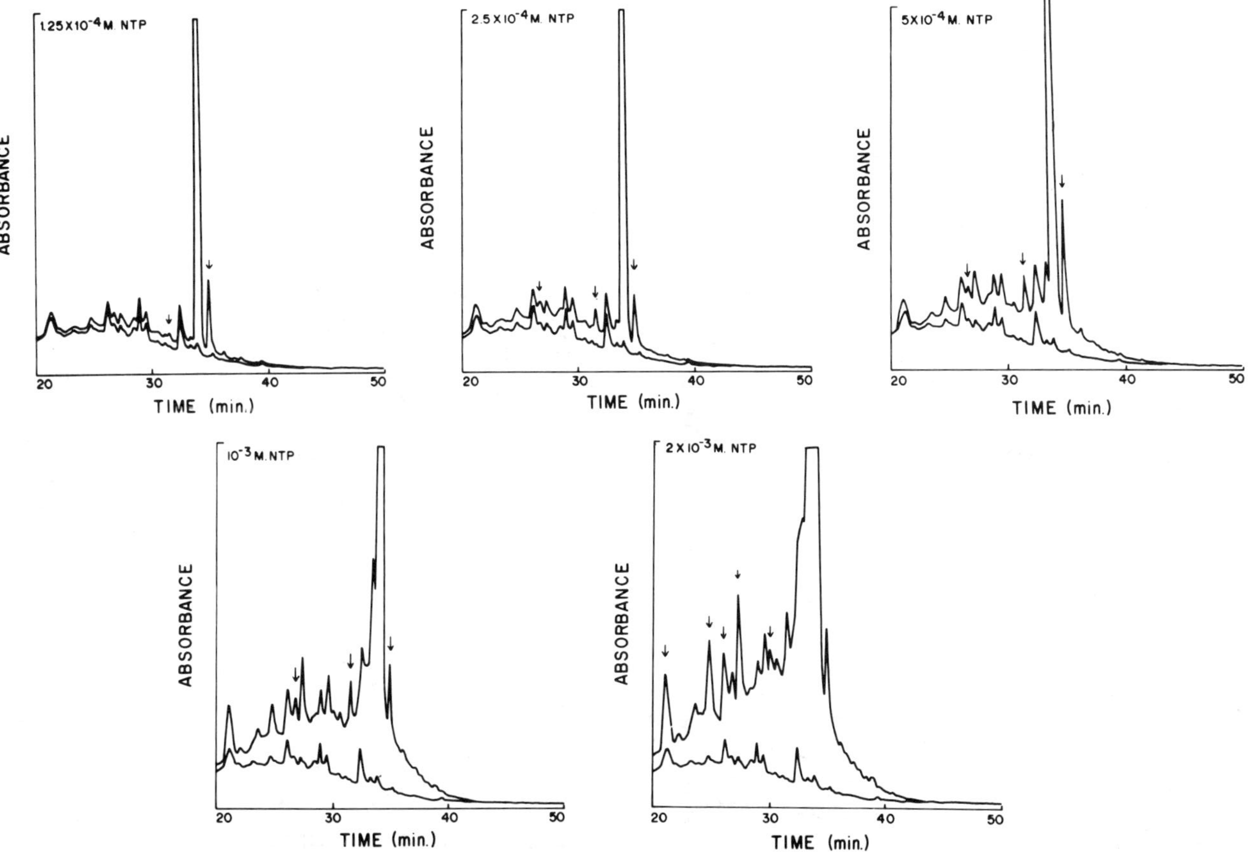


**Figure 3.** HPLC chromatograms after reaction with varying NTP.

37°C. This might result in an overall 500-fold reduction in rates. Mutagenesis may occur during the few bacterial cell divisions permitted by the limiting histidine present in the Ames assay [12]. If mutagenesis occurs in a time period of less than 8 hr, a 500-fold reduction in thioether formation rate will mean that the amount of further thioether formation might be roughly equal to 1/30 of the amount formed during the initial 30-min reaction.

Other factors also will suppress further NTP reaction with the electrophiles. As the electrophiles and NTP react, their levels drop. Reaction rates after the first 30 min are progressively slower than the initial ones. The addition of 0.1 mL of a full-grown culture of bacteria in nutrient broth means an abundance of added nucleophiles will be present to compete with further NTP reaction. Finally, plating of the 2.6 mL agar solution will result in diffusion of NTP and electrophile into the 25-mL VBE plate volume, resulting in further dilution of reactants.

In all, the dilution and competition factors should serve to sharply reduce further reaction between NTP and electrophiles. Therefore, the Ames titration and HPLC profiles should correspond fairly closely. If the NTP-electrophile reaction continues at a significant rate in the Ames assay, it would mean that the mutagen loss measured by the Ames assay would be greater than that which is observed in the HPLC profile.

As a final note we have observed that the $\epsilon_{345}$ for a large number of NTP thioethers is approximately the same [10]. By use of a cut-and-weigh procedure, we can measure the amount of thioether area above the background, e.g., in Figure 3C. Then, using areas for known thioethers for calibration, we can approximate the number of nanomoles of thioethers formed. In Figure 3C the increase in $\epsilon_{345}$ suggests the formation of about 5.4 nmol of NTP thioethers (disulfide peak omitted). If formation of this amount of thioether is responsible for the reduction in mutagenic activity of 131 revertants (in Figure 2 the reduction of 105 revertants from 0 to $5 \times 10^{-4}$ *M* NTP is divided by 0.8 to correct for the different volumes used in the HPLC separation and the Ames assays), the parent electrophiles have an aggregate mutagenic potency of $131/5.4 = 24$ revertant/nmol. Since mutagenic potencies of greater than 1.0 revertant/nmol may be considered high [13], these mutagens indeed are unusually potent. The amount of water sample in the HPLC profile corresponds to 4 L prior to concentration. From this we calculate an *aggregate* concentration of 5.4 nmol/4 L = $1.35 \times 10^{-9}$ *M*, or which when multiplied by molecular weights of 100–300 = 135–400 ng/L (parts per trillion); this indicates a very low concentration for the electrophiles in the original drinking water if there is complete recovery during concentration and complete reaction with NTP.

## SUMMARY

The work presented in this chapter shows that NTP may be used to label electrophiles in drinking water concentrates, and that the labeling may be used to distinguish between thioethers formed by rapid or slow electrophile reaction with NTP. Correlation of changes in the labeling pattern with changes in mutagenic activity measured by the Ames assay suggests that the more rapidly reacting electrophiles may possess unusual mutagenic activity and may be present at extremely low concentrations.

## REFERENCES

1. Pelon, W., B. F. Whitman and T. W. Beasley. *Environ. Sci. Technol.* 11:619-623 (1977).
2. Loper, J. C., D. R. Lang and C. C. Smith. In: *Water Chlorination: Environmental Impact and Health Effects, Vol. 2*, R. L. Jolley, H. Gorchev and D. H. Hamilton, Eds. (Ann Arbor, MI: Ann Arbor Science Publishers, Inc., 1978), pp. 433-450.
3. Loper, J. C., D. R. Lang, R. S. Schoeny and B. B. Richmond. *J. Toxicol. Environ. Health* 4:919-938 (1978).
4. Glatz, B. A., C. D. Chriswell, M. D. Arguello and H. J. Svec. *J. Am. Water Works Assoc.* 70:465-468 (1978).
5. Hooper, K., C. Gold and B. N. Ames. "Report to the Water Resources Control Board, State of California," (1978).
6. Bull, R. J., M. A. Pereira, and K. L. Blackburn. Paper presented at the Adsorption Technologies Conference, Washington, DC, April 30, 1979.
7. Schwartz, D. J., J. Saxena and F. C. Kopfler. *Environ. Sci. Technol.* 13:1138-1141 (1979).
8. Cheh, A. M., J. Skochdopole, P. Koski and L. Cole. *Science* 207:90-92 (1980).
9. Miller, E. C., and J. A. Miller. In: *Chemical Carcinogens*, C. E. Searle, Ed. (Washington, DC: American Chemical Society, 1976), pp. 737-762.
10. Cheh, A. M., and R. E. Carlson. (in preparation).
11. Barnett, R. E., and W. P. Jencks. *J. Am. Chem. Soc.* 91:6758-6765 (1969).
12. Ames, B. N., J. McCann and E. Yamasaki. *Mutat. Res.* 31:347-364 (1975).
13. Cheh, A. M., J. Skochdopole, C. Heilig, P. M. Koski and L. Cole. In: *Water Chlorination: Environmental Impact and Health Effects, Vol. 3*, R. L. Jolley, W. E. Brungs and D. H. Hamilton, Eds. (Ann Arbor, MI: Ann Arbor Science Publishers, Inc., 1980), pp. 803-815.
14. Rosenkranz, H. S., and W. T. Speck. *Biochem. Biophys. Res. Commun.* 66:520-525 (1975).
15. Coleman, W. E., R. G. Melton, F. C. Kopfler, K. A. Barone, T. A. Aurand and M. G. Jellison. *Environ. Sci. Technol.* 14:576-588 (1980).

## CHAPTER 30

# DETERMINATION OF LINEAR ALCOHOL ETHOXYLATES IN WASTE- AND SURFACE WATER

Victorio T. Wee

Procter and Gamble Company
Environmental Safety Department
Cincinnati, Ohio

Linear alcohol ethoxylates (AE) are a class of nonionic surfactants widely used industrially and commercially in the United States at the rate of 4 x $10^8$ lb/yr [1]. They have the general chemical formula $RO\text{-}(CH_2\text{-}CH_2\text{-}O)_nH$ where R represents an alkyl chain containing 6-16 carbon atoms and n represents the number of ethoxylate units, which can vary from 3 to 20 [2].

Several methods have been proposed for measuring nonionic surfactants in different environmental matrices. These methods include colorimetry [3], atomic absorption spectroscopy [4], nuclear magnetic resonance [5] and chromatographic techniques [5-8]. A common problem present in these methods is the interferences from other components of the environmental matrices. For instance, the reagents used in the cobaltothiocyanate-active substance (CTAS) method, a colorimetric method currently used for measuring linear AE, form complexes with other organic compounds. As a result, the CTAS measurement typically indicates a higher concentration of AE than is actually present.

This chapter describes a method for analyzing linear AE nonionic surfactants in natural environmental matrices. The method includes an effective cleanup procedure followed by gas chromatographic (GC) analysis of the hydrogen bromide (HBr) cleavage products of the surfactant. It is validated with one of the most widely used linear alcohol ethoxylates, Neodol 45-7, where R contains 14-15 carbon atoms and n has an average value of 7. The concentrations of linear AE obtained by the method described were then compared to concentrations of linear alcohol ethoxylates obtained by the CTAS procedure. This comparison demonstrated the nonspecificity of the CTAS procedure.

## MATERIALS AND METHODS

### Surfactant, Analytical Standards and Reagents

Neodol 45-7 was obtained from Shell Chemical Corp. Sample identity was confirmed by colorimetric technique (CTAS response) and by gas chromatography/mass spectrometry (GC/MS) analysis of its HBr-cleavage products, 1,2-dibromoethane, $C_{14}$ and $C_{15}$ alkyl bromides [9]. 1,2-dibromoethane, 1-bromooctane, 1-bromotetradecane, 1-bromopentadecane and 1-octanol were obtained from Eastman Organic Chemicals. Anion exchange resin (Amberlyst A-26) and cation exchange resin (Amberlyst 15) were obtained from Rohm & Haas. Silica gel, certified ACS Grade 12, 28-200 mesh was obtained from Fisher Scientific Co. Cobaltothiocyanate solution, 30 g $Co(NO_3)_2 \cdot 6H_2O$, 200 g $NH_4SCN$ dissolved and diluted to one liter with distilled water (stable for at least one month at 25°C) was also used.

### Sample Collection and Preparation

Wastewater samples were collected on different days and different times of the day from a local domestic sewage treatment plant. Sewage influent samples were collected after comminution, and effluent samples were collected before chlorination. The riverwater sample was obtained from a raw-water tap at a local water treatment plant. Samples were preserved with 1% formaldehyde. Samples containing suspended solids (SS) greater than 100 mg/L (e.g., activated sludge and sewage influent) were filtered through microfiber glass filters before extraction. This procedure minimizes the emulsion formed at the interface during extraction and improves the recovery of the nonionic surfactants. SS were measured by a standard procedure [10]. The solids were extracted with methanol for 16 hr, the methanol extract was evaporated, and the residue was redissolved in deionized water and combined with the initial filtrate.

Sample sizes varied from 400 to 600 mL for influent sewage and activated sludge to 1 L or more for surface water and sewage effluent. Sample volumes greater than 1 L were reduced to less than 1 L by rotary evaporation before further treatment.

## Isolation and Concentration

A cleanup procedure was developed for removing interfering material prior to analysis. The procedure consists of:

1. The sample is adjusted to 0.2 *M* $MgSO_4$ and extracted with four 60-mL portions of chloroform; the aqueous layer is discarded. The extract is taken to dryness under nitrogen on a steam bath. The residue is redissolved in 5 mL of methanol and placed on top of a 1-cm-diameter column packed (from bottom to top) with 10 cm of anion exchange resin, a plug of glass wool and 10 cm of cation exchange resin.
2. The sample is eluted through the column with 100 mL methanol. The eluate is then evaporated to dryness on a steam bath under nitrogen and the residue is redissolved in 10 mL methanol. Half of this solution is analyzed for total nonionic surfactants by the CTAS procedure according to Boyer et al. [3].
3. The other 5 mL of the methanol solution is prepared for GC analysis. It is evaporated to dryness; the residue is redissolved in 25 mL distilled water and transferred to a 250-mL separatory funnel. The flask is rinsed with 25 mL of distilled water and two 25-mL portions of 16% NaCl, and the washings are added to the separatory funnel.
4. The aqueous solution is extracted four times with 20 mL toluene. The toluene extracts are combined and dried over $Na_2SO_4$. The toluene is evaporated on a steam bath under nitrogen, and the residue is redissolved in 5 mL of $CHCl_3$ and transferred to a 1- x 15-cm silica gel column.
5. The column is rinsed with 100 mL of $CHCl_3$ and the eluent is discarded. The column is finally washed with 100 mL of methanol, and the eluate is collected for HBr-cleavage.

## HBr Cleavage and Gas Chromatographic Analysis (HBr-GC)

The eluate (step 5 above) is reduced to a volume of approximately 10 mL on a steam bath under nitrogen and transferred to a 12-mL centrifuge tube. A known quantity of internal standard (1-octanol) is added to the centrifuge tube. The mixture is evaporated to dryness on a steam bath and treated with 0.5 mL HBr as described by Luke [11]. The reaction products are taken up in 4 mL carbon disulfide ($CS_2$). The upper water layer is maintained in the tube to prevent evaporative loss of the internal standard. An aliquot of the $CS_2$ layer is then injected into the gas chromatograph.

An internal standard procedure (1-octanol derivatized to octyl bromide) was used to quantify the alkyl bromides derived from linear AE. An external reference mixture of 1,2-dibromethane, 1-bromooctane, 1-bromotetradecane and 1-bromopentadecane was used to establish response factors and occasionally to identify the expected reaction products of Neodol 45-7.

The efficiency of the HBr-cleavage reaction was checked using 10-μg to 4-mg samples of Neodol 45-7. Quantification of Neodol 45-7 based on the alkyl bromides formed (i.e., by GC) was then compared to the nominal weight initially added. The efficiency of the cleavage based on the analysis of four samples of Neodol 45-7 was 77% ± 1.5 of the theoretical. This factor (77%) was utilized in calculations where the initial weight of Neodol 45-7 by GC was required.

Gas chromatographic analyses were performed with a Hewlett-Packard 5840-A, microprocessor-based chromatographic system equipped with a flame ionization detector (FID). The GC column was a 183-cm x 3-mm i.d. stainless steel column, packed with 2% XE-60 on Chromosorb W-HP 80/100 mesh from Hewlett-Packard. Chromatographic runs were programmed from 60 to 160°C at 8–16°C/min. Helium was the carrier gas and the flowrate was 40 ml/min, at 60°C. Sample volume varied from 1 to 15 μL. A complete run usually lasted 25 min.

## Cobaltothiocyanate Active Substance

A 5-mL portion of the solution prepared in step 2 (above) is evaporated to dryness on a steam bath. The residue is redissolved in 5 mL $CH_2Cl_2$ and transferred to a 125-mL separatory funnel. Cobaltothiocyanate solution is then added and the measurement is done according to Boyer et al. [3].

## Calculations

In the HBr-GC method, AE are determined indirectly by quantification of their reaction products, the alkyl bromides and the application of appropriate gravimetric factors. For example, to quantify Neodol 45-7, the weights of $C_{14}Br$ and $C_{15}Br$ obtained from the GC data are substituted into the following equations to obtain the weight of the corresponding alkyl ethoxylate.

$$\text{weight of } C_{14}\,EO_n = \text{weight of } C_{14}\,Br \times \frac{MW\ C_{14}\,EO_n}{MW\ C_{14}\,Br} \qquad (1)$$

$$\text{weight of } C_{15}\ EO_n = \text{weight of } C_{15}\ Br \times \frac{MW\ C_{15}\ EO_n}{MW\ C_{15}\ Br} \qquad (2)$$

where $EO_n$ represents the ethoxylate portion of the AE molecule.

In this work, a value of 7 is assumed for n based on the widely used Neodol 45-7. A change from 7 to 14 in Equations 1 and 2 will increase the calculated AE concentration by only 35%. This slight increase is insignificant at the observed environmental levels.

## RESULTS AND DISCUSSION

### Precision and Sensitivity

Duplicate analyses of several wastewater samples showed the GC precision to be within 20%. For riverwater, the precision among unspiked samples containing less than 5 μg Neodol 45-7 was ±45%. At greater than 10 μg, the precision was ±20%. From these results it appears that at least 10 μg of Neodol 45-7 must be concentrated from the riverwater sample to obtain a precision of ±20% or better. The minimum detectable limit by GC is about 0.1 μg of each alkyl bromide peak.

For CTAS analysis, the precision was always greater than ±10%, provided 0.2 mg of AE was measured. The level of 0.2 mg is the detection limit of the CTAS procedure.

### Recovery

The recovery of added Neodol 45-7 in the environmental samples was followed through the entire cleanup and concentration procedure and was measured by both CTAS and HBr-GC with the exception of the riverwater sample. For the riverwater sample, the recovery was followed only by HBr-GC because the amount of CTAS material in a 5-L sample was far below the 0.2-mg detection limit of the CTAS method. No attempt was made to further concentrate larger volumes of riverwater to determine the CTAS level. The recovery efficiencies in different environmental matrices are shown in Table I along with their standard deviations.

Both CTAS and HBr-GC showed some variation in recovery efficiencies for samples containing different concentrations of solids. Recoveries were lower for samples containing "heavy solids" (i.e., influent and activated

Table I. Recovery of Neodol 45-7 by CTAS and GC

| Grab Sample | CTAS (ppm) | | | | GC (ppm) | | | |
|---|---|---|---|---|---|---|---|---|
| | Before Spike | Amt Spike | After Spike | % Recovery[a] | Before Spike | Amt Spike | After Spike | % Recovery |
| Morning Influent | 3.12 ± 0.07 (n = 2)[b] | 1.34 | 4.23 ± 0.19 (n = 2) | 83 | 0.19 (n = 1) | 1.02 | 1.0 ± 0.07 (n = 2) | 79 |
| Morning Effluent | 0.37 ± 0.01 (n = 2) | 0.2 | 0.56 ± 0.03 (n = 2) | 95 | 0.021 (n = 1) | 0.15 | 0.16 ± 0.03 (n = 2) | 94 |
| Activated Sludge | 5.32 ± 0.39 (n = 2) | 2.50 | 7.55 ± 0.14 (n = 2) | 88 | 0.22 ± 0.04 (n = 2) | 1.90 | 1.49 ± 0.4 (n = 2) | 67 |
| River Water | | | | | 0.0042 ± 0.002 (n = 2) | 0.008 | 0.016 ± 0.004 (n = 3) | 147 |

[a]Percent = (after spike - before spike)/amt spiked x 100.
[b]n = number of analyses.

sludge samples) than for samples containing "light solids" (i.e., effluent and riverwater). This is consistent with previous observations [3,12] that suggest a strong interaction between solids and nonionic surfactants. However, despite the lower recovery in "heavy solids" samples than in "light solids" samples, the range of recoveries obtained is acceptable for most environmental studies. The 147% recovery from the riverwater sample reflects the poor precision in the unspiked sample which had less than 5 $\mu$g total Neodol 45-7. Our experience with these samples has shown that the precision can be improved simply by concentrating a larger sample.

## Interferences

All previously proposed methods for analyzing nonionic surfactants in environmental samples are often overwhelmed by interfering substances. The present method overcomes this difficulty by incorporating an improved cleanup procedure. There are three key steps in the cleanup procedure; mixed-bed ion exchange, toluene extraction and silica gel adsorption. The mixed-bed ion exchange removes the cationic and anionic organic compounds that interfere with both colorimetric and GC analyses. The toluene extraction reduces the amount of polyethylene glycols (PEG) that can interfere with the GC analysis. Complete removal of the PEG is not necessary for obtaining the final chromatogram. The silica gel step, which is perhaps the most important of the three steps, further eliminates the bulk of the unidentified interferences.

Interference from one abundant nonlinear nonionic surfactant, nonylphenol ethoxylate, was also investigated. The results showed that the GC peaks resulting from the HBr-cleavage of nonylphenol ethoxylate eluted before the $C_{14}Br$ and $C_{15}Br$ peaks and did not interfere with the analysis of Neodol 45-7.

## GC Analysis of Environmental Samples

A typical chromatogram of the HBr cleavage products from Neodol 45-7 is shown in Figure 1. Peaks were identified by comparison of their retention times with reference standards and confirmed by combined GC/MS [9]. The other unlabeled peaks were impurities in the commercial sample of Neodol 45-7.

The environmental samples that were analyzed were grab samples and were prepared by using the isolation/concentration procedure. As mentioned earlier, all steps in this procedure were important to obtain an interference-limited chromatogram. Eliminating the silica gel step, for instance, leads to a chromatogram such as the one shown in Figure 2A. Note the presence of broad and unresolved peaks in the sample before the silica gel cleanup (Figure 2A) compared to the sample after the silica gel cleanup (Figure 2B).

Some typical chromatograms of samples from different environmental matrices are shown in Figure 3. These chromatograms illustrate the clean and complete separation of $C_{14}Br$ and $C_{15}Br$ peaks from other peaks. As would be expected, the sample obtained above the sewage outfall (Figure 3A) contained the least amount of AE, approx. 0.0005 ppm. In fact, only the $C_{14}Br$ peak was quantifiable. Although this analysis represents only one sample it suggests that the level of either manmade or naturally occurring organics resembling $C_{14}Br$ and $C_{15}Br$ peaks are very low.

Close examination of the chromatograms in Figure 3 indicated the presence of $C_{11}$, $C_{12}$, $C_{13}$ and $C_{16}$ alkyl bromide peaks. These peaks were identified by comparison with $C_{11}$, $C_{12}$, $C_{13}$ and $C_{16}$ alkyl bromide standards. The presence of these peaks suggests that linear AE with alkyl chain

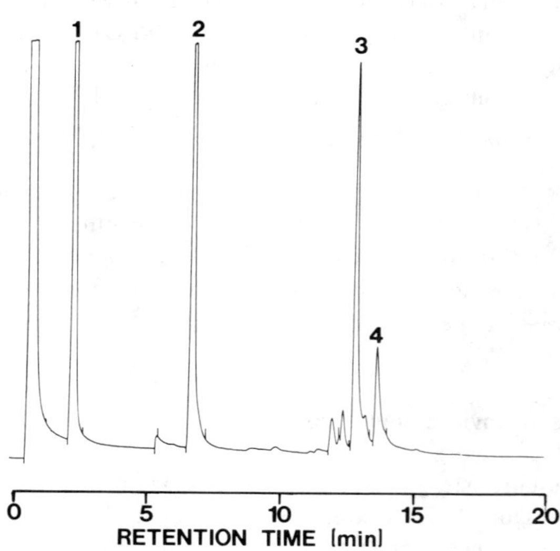


**Figure 1.** Chromatogram of the HBr cleavage products of Neodol 45-7. Peaks were identified by comparison of their retention time with those of alkyl bromide standards and by combined GC/MS. Peak assignment: (1) 1,2-dibromoethane; (2) 1-bromooctane (internal standard); (3) 1-bromotetradecane; (4) 1-bromopentadecane.

lengths ranging from $C_{11}$ to $C_{16}$ are present at very low levels and that they can be determined using this procedure.

The concentrations of linear AE ($C_{14}$-$C_{15}$, $EO_7$) measured in different environmental matrices are shown in Table II. Comparisons of the influent and effluent waste samples in Table II indicate excellent apparent removability of these surfactants within the treatment plant. Although these results are based on limited grab samples, they are consistent with literature reports that Neodol-type surfactants are biologically soft [12-16], and are readily removed by sewage treatment processes [1].

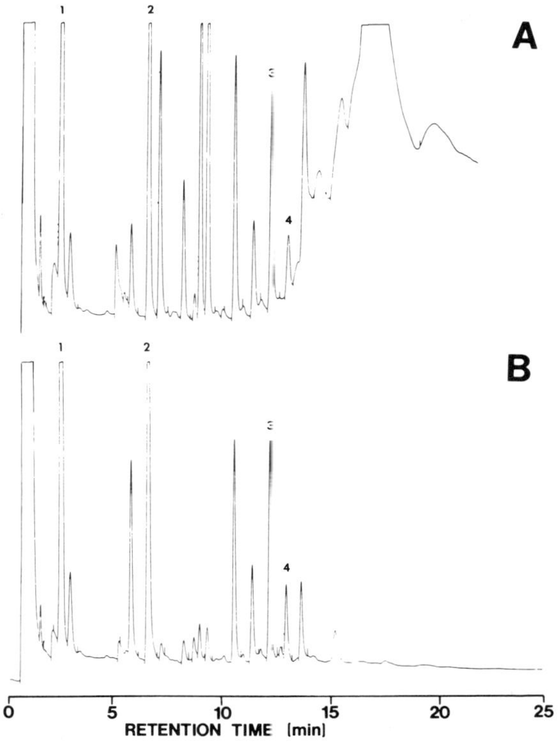


**Figure 2.** Chromatogram of sewage influent before and after the silica gel cleanup step. Peak identification: (1) 1,2-dibromoethane; (2) 1-bromooctane (internal standard); (3) 1-bromotetradecane and (4) 1-bromopentadecane. (A) Before and (B) after silica gel cleanup. Both samples were programmed at 13°C/min.

The results shown in Table II represent only a fraction of the CTAS value measured for the same samples (Table III). These results clearly demonstrate the nonspecificity of the CTAS method. Conclusions based on analyses by this colorimetric method in the past should now be reexamined in light of the present study.

## CONCLUSION

This chapter deals primarily with the development of a specific analytical method for linear AE ($C_{14}$-$C_{15}$, $EO_7$). The application of the method, although successful in a wide selection of environmental matrices, is done only on a limited number of samples.

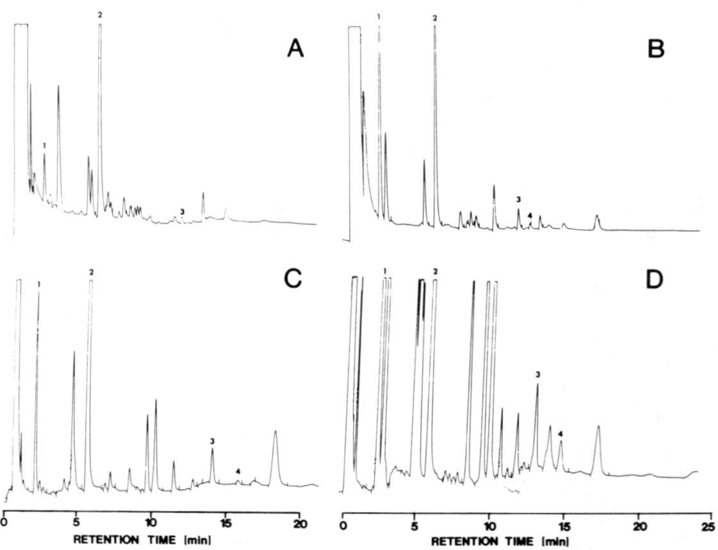


**Figure 3.** Chromatograms of samples from different environmental matrices. Peak identification: (1) 1,2-dibromoethane; (2) 1-bromooctane (internal standard); (3) 1-bromotetradecane and (4) 1-bromopentadecane. (A) Above sewage outfall, rate = 13°C/min; (B) sewage effluent, rate = 13°C/min; (C) riverwater, rate = 8°C/min; (D) Activated sludge, rate = 10°C/min.

**Table II. GC Analysis of $C_{14}$–$C_{15}$, $EO_7$ in Environmental Samples**

| Grab Samples | $C_{14}$–$C_{15}$, $EO_7$ (ppm) | No. of Analyses |
|---|---|---|
| Morning Influent | 0.37 ± 0.0028 | 2 |
| Morning Influent | 0.19 | 1 |
| Morning Effluent | 0.006 | 1 |
| Morning Effluent | 0.012 | 1 |
| Afternoon Influent | 0.47 ± 0.009 | 2 |
| Afternoon Effluent | 0.0085 ± 0.0007 | 2 |
| Activated Sludge | 0.22 ± 0.04 | 2 |
| Above Sewage Outfall | 0.0005 | 1 |
| Below Sewage Outfall | 0.0011 | 1 |
| River Water | 0.0042 ± 0.002 | 4 |

**Table III. Percent CTAS as $C_{14}$–$C_{15}$, $EO_7$**

| Grab Samples | CTAS (ppm) | No. of Analyses | % CTAS as $C_{14}$–$C_{15}$, $EO_7$ |
|---|---|---|---|
| Morning Influent | 5.4 ± 0 | 2 | 6.8 |
| Morning Influent | 3.1 ± 0.07 | 2 | 6.2 |
| Morning Effluent | 0.5 ± 0 | 2 | 1.2 |
| Morning Effluent | 0.4 ± 0.01 | 2 | 3.2 |
| Afternoon Influent | 4.3 ± 0.03 | 2 | 11.0 |
| Activated Sludge | 5.3 ± 0.4 | 2 | 4.0 |
| Above Sewage Outfall | 0.03 | 1 | 1.7 |
| Below Sewage Outfall | 0.24 | 1 | 0.5 |

The sensitivity of the method is at the low part-per-billion level. In addition, based on the samples investigated here, the method should also be applicable to the analysis of other linear AE with alkyl chain lengths ranging from $C_{11}$ to $C_{16}$ and EO ranging from 0 to 14. This study also demonstrates that the CTAS method overestimates the concentration of linear alcohol ethoxylates in the environment.

## ACKNOWLEDGMENT

The author thanks Michael A. Burchfield for laboratory assistance in performing several of the reported extractions and analyses.

## REFERENCES

1. Sykes, R. M., A. J. Rubin, A. J. Rath and M. C. Chang. "Treatability of a Nonionic Surfactant by Activated Sludge," *J. Water Poll. Control Fed.* 51:71 (1979).
2. A. D. Little, Inc. "Human Safety and Environmental Aspects of Major Surfactants," report to the Soap and Detergent Association (1977), p. 240.
3. Boyer, S. L., K. F. Guin, R. M. Kelley, M. L. Mausner, H. F. Robinson, T. M. Schmitt, C. R. Stahl and E. A. Setzkorn. "Analytical Method for Nonionic Surfactant in Laboratory Biodegradation and Environmental Studies," *Environ. Sci. Technol.* 11:1167-1171 (1977).
4. Crisp, P. T., J. M. Eckert and N. A. Gibson. "An Atomic Absorption Spectrometric Method for the Determination of Non-ionic Surfactants," *Anal. Chim. Acta* 104:93-98 (1979).
5. Jones, P., and G. Nickless. "Characterization of Non-ionic Detergents of the Polyethoxylated Type from Water Systems. II. Isolation and Examination of Polyethoxylated Material before and after Passage through a Sewage Plant," *J. Chromatog.* 156:99-110 (1978).
6. Favretto, L., B. Stancher and F. Tunis. "Extraction and Determination of Polyoxyethylene Alkyl Ether Non-ionic Surfactants in Water at Trace Levels," *Analyst* 103:955-962 (1978).
7. Stancher, B., F. Tunis and L. Favretto. "Chromatographic Studies on the Extraction of Polyoxy-ethylene Alkylphenyl Ether Non-ionic Surfactants from Water at Trace Levels," *J. Chromatog.* 131:309-316 (1977).
8. Cassidy, R. M., and C. M. Niro. "Application of High-Speed Liquid Chromatography to the Analysis of Polyoxyethylene Surfactants and Their Decomposition Products in Industrial Process Waters," *J. Chromatog.* 126:787-794 (1976).
9. Sammons, M. C., Procter and Gamble. Unpublished results.
10. *Standard Methods for the Examination of Water and Wastewater*, 14th ed. (New York: American Public Health Association, 1975), p. 94.
11. Luke, B. G. "The Analysis of Alkylene Oxide-Fatty Alcohol Condensates by Gas-Liquid Chromatography," *J. Chromatog.* 84:43-49 (1973).
12. Cook, K. A. "Degradation of the Non-ionic Surfactant Dobanol 45-7 by Activated Sludge," *Water Res.* 13:259-266 (1979).
13. Patterson, S. J., C. C. Scott and K. B. E. Tucker. "Nonionic Detergent Degradation. III. Initial Mechanism of the Degradation," *J. Am. Oil Chem. Soc.* 47:37-41 (1970).

14. Tobin, R. S., F. I. Onuska, B. G. Brownlee, D. H. J. Anthony and M. E. Comba. "The Application of an Ether Cleavage Technique to a Study of the Biodegradation of a Linear Alcohol Ethoxylate Nonionic Surfactant," *Water Res.* 10:529-535 (1976).
15. Larson, R. J. "The Role of Biodegradation Kinetics in Predicting Environmental Fate," in *Biotransformation and Fate of Chemicals in the Environment*, A. W. Maki, K. L. Dickson and J. Cairns, Eds. (Washington, DC: American Society for Microbiology, 1980).
16. Rudling, L., and P. Solyom. "The Investigation of Biodegradability of Branched Nonyl Phenol Ethoxylates," *Water Res.* 8:115-119 (1974).

# INDEX

# Advances in the Identification & Analysis of Organic Pollutants in Water Volume 2

Edited by
Lawrence H. Keith

230 Collingwood, P.O. Box 1425, Ann Arbor, Michigan 48106

Library of Congress Card Catalog Number 81-68031
ISBN 0-250-40398-6

Manufactured in the United States of America


Butterworths, Ltd., Borough Green, Sevenoaks, Kent TN15 8PH, England

# PREFACE

Five years ago the first book on identification and analysis of organic pollutants in water was published. The genesis of that book was primarily a symposium of the same title held during the First Chemical Congress of the North American Continent in Mexico City. In the ensuing years, advances in the identification and analysis of organic pollutants in water have continued at an accelerated pace. Thus, it was only natural that the second symposium on this subject be held during the Second Chemical Congress of the North American Continent in Las Vegas, Nevada.

As in the initial publication, all but a few of the chapters in this book were contributed by the many distinguished authors who participated in that symposium. These scientists represent the leading analytical and environmental chemists on the North American continent and abroad. The advances described in these volumes represent the current state-of-the-art methodology for analysis of organic compounds, often at trace levels, in all types of water samples ranging from the very pure (drinking water) to the extremely "dirty" (untreated industrial wastewaters).

Knowledge of specific organic pollutants in water has increased exponentially over the past five years. This has been largely due to the methodology that has enabled these analyses to be conducted efficiently, scientific and governmental interest in the subject, and the fact that knowledge of the subject area five years ago was minimal. Despite intense work over the past five years, little "brand new" methodology has been introduced for analyzing trace levels of organic pollutants in water. However, refinements of the techniques then in use have made these methods even better and more efficient and have contributed significantly to our current methodology. Solvent extractions, resins or gas purging techniques are still the primary methods used to concentrate organics from water, and gas or high-pressure liquid chromatography with various detectors (including mass spectrometers) are still used to identify and quantify them. One new method, using tandem mass spectrometers (MS/MS), offers promise of analyzing for specific compounds in a complex matrix without the tedious and expensive concentration and separation requirements of conventional instrumental analyses. At present, however, MS/MS is quite expensive and still in its initial stage of exploration and exploitation.

One significant difference in the direction of environmental analyses today, as compared to five years ago, is the emphasis on improving both accuracy and reliability of quantitative data. This is being accomplished through more stringent quality assurance/quality control programs and the use of numerous and more sophisticated quantitative methods such as the use of multiple internal standards, often isotopically labled. Five to ten years ago, compounds identified in water often were not even quantified. Two driving forces appear to be the primary reason for this change of emphasis: (1) legal and regulatory ramifications and (2) the realization that adverse effects of all organic pollutants are related to their concentration in water.

Predictably, environmental analyses seem to be moving increasingly into the lawyer's domain. This makes life more complicated for analytical environmental chemists, who have the sometimes difficult problem of explaining that 18 and 24 parts per billion are really the same number (within experimental error), or that it is not a worthwhile environmental goal to try to reduce the pH of water to near zero. Nevertheless, the overall effect of these legal and regulatory pressures on scientists is a positive one. It requires an emphasis on careful work and on improving the accuracy and precision of measurements that are often pushed to their limits. But, past history has shown that growth and improvement in most disciplines increase more rapidly when pressure is applied.

Another significant difference in the direction of environmental analyses today is an emphasis on automation. Analyses of organic pollutants in water are becoming more prevalent as a result of new regulations and modifications of old ones (e.g., the National Pollutant Discharge Elimination System industrial wastewater discharge permits). The result is an increasing competitiveness and pressure to produce large numbers of analyses for decreased costs. This, of course, is a natural and healthy trend as long as the pressure for cheap analyses does not compromise accuracy and precision, and it is realized that not all objectives can be met using cheap analyses for a "laundry list" of pollutants.

There still remains much to be learned about the kinds of pollutants in water and the chemistry involved. Until we have the correct methodology for analyzing organic compounds, we will continue to miss some even though they may be literally right under our noses. An example is the recent discovery that dihaloacetonitriles are apparently a new class of anthropoaqueous pollutants in drinking water. Despite the fact that countless chemists have been analyzing for trihalomethanes in drinking water these past five to six years, the fact was missed that dihaloacetonitriles are also probably present in many of those same samples. Why? Because the methodology being used was not conducive for the analysis of dihaloacetonitriles.

I think we have reached the stage in the evolution of analysis of organic pollutants in water where the easy methods and the easy compounds have been exploited. Now we will have to work harder and be more clever to produce newer and better methodology with which to uncover the more difficult organic pollutants. Is this necessary? I think so. How can one make an intelligent decision until the pertinent facts are known?

Five years ago, I said, "much still remains to be done. We have only begun to learn and to apply what we have learned to provide a better and cleaner environment." I think we have applied what we have learned very well over the past five years. We know a great deal more about the prevalence, distribution and concentrations of organic pollutants in water now than we did then. However, much still remains to be done. Now we must refine our techniques, verify our quantitative accuracy and turn our attention to developing new methodology for analysis of the more elusive and difficult organic pollutants.

L. H. Keith

**Lawrence H. Keith's** current technical interests continue to center around analyses of organic pollutants in the environment, with emphasis on developing new methods or improving on old ones. Techniques for the safe handling of carcinogenic and/or extremely toxic materials are also an important aspect of Dr. Keith's current research efforts.

Dr. Keith was formerly involved with the selection of many of the initial U.S. Environmental Protection Agency's 129 Priority Pollutants, and he also helped to formulate some of the initial methodology for analyzing for these pollutants. He is presently involved in the selection of representative compounds and methodologies for the Appendix C Priority Pollutants and for the synfuel industry.

A member of the American Chemical Society's Division of Environmental Chemistry Executive Committee, Dr. Keith has served as secretary, alternate councilor, program chairman and chairman of the division. He is also a past chairman of the Central Texas Section of the American Chemical Society and past secretary and councilor of the Northeast Georgia Section of the American Chemical Society.

In other professional activities, Dr. Keith served as Vice-Chairman of the Gordon Research Conference on Environmental Sciences: Water, and is currently a delegate of the U.S. National Committee to the International Association of Water Pollution Research. He is also a member of the National Research Council Committee on Military Environmental Research.

Dr. Keith and his wife, Virginia, reside in Austin, Texas.

**Aaron A. Rosen** was born in Cincinnati, Ohio, and began his college studies at the University of Cincinnati in 1931. After receiving his BA degree (1935) he completed his PhD there in 1938. In 1952, he joined the Federal Water Pollution Contol Administration in Cincinnati; this begame a part of the U.S. Environmental Protection Agency when it was created in 1970.

Aaron was one of the pioneers in developing methods for the analysis of low levels of organic pollutants in water. His work in this field is well known and he received two Superior Service Awards, in 1957 and 1965, from the U.S. Department of Health, Education and Welfare. The latter was for his work in identifying endrin as the cause of a huge fish kill in the Mississippi River. Other honors accorded to Aaron were the Bartlow Award (1959) from the American Chemical Society Water and Waste Division, the Gans Medal (1970) from the English Society for Water Treatment and Examination, Outstanding Cincinnati Chemist Award (1972) from the Cincinnati Section of the American Chemical Society, the Professional Accomplishment Award (1975) from the Engineering Society of Cincinnati and the Distinguished Service Award (1978) from the American Chemical Society Division of Environmental Chemistry.

I first met Aaron about 1968. Our interests in analysis of organic compounds in water overlapped, and Aaron was a great inspiration. Our paths continued to cross throughout the remaining years. Aaron had an endless capacity for living and great bravery in facing his illness. In their last years, he and Ron Webb became close friends in their common struggle to live a productive and active life in the shadow of cancer. Always an adventurer, Aaron left his doctors and the country to participate in the first Symposium on the Identification and Analysis of Organic Pollutants in Water in Mexico City. After that meeting he went snorkling in the clear waters off the Yucatan Peninsula with his wife, Gertrude. He later described it as a great experience and wondered why he had not done it before.

Aaron's scope of interests were limitless. His mind reached out to all knowledge, and he loved to discuss history, philosophy, law, economics, politics and literature, in addition to chemistry and other branches of science.

It is an honor to dedicate this volume to a friend, a colleague and a great pioneer of modern-day environmental chemistry.

## ACKNOWLEDGMENTS

As in the previous book, most, but not all, of the chapters were derived from a symposium sponsored by the American Chemical Society (ACS) Division of Environmental Chemistry. The editor is grateful for the support of the ACS and the division for providing the forum from which this book is primarily derived. The many authors who have diligently written these chapters are leading scientists in their fields and, without their work, ideas, and contributions, these books could never have been produced.

The editor is also very grateful for the help and support from his wife, Virginia, who did most of the work compiling the extensive index. This index is the heart of these volumes when they are used as reference books, and thus, helps make these books unique in their field. No other works of this kind meld analytical methodologies with specific organic pollutants that have been identified in water. It is, in fact, this dual purpose of providing a single source for both the identification and the methods for analysis of organic pollutants in water that has proven to be so useful to workers in the field of environmental chemistry.

## CONTENTS

**Section 9**
**Grob Closed-Loop Stripping**



**Section 10**
**Specialized Purging Techniques**

**Section 13**
**Industrial Wastewater Analysis**

# CHAPTER 31

# GLOBAL WATER QUALITY MONITORING

**Silvio Barabas**

National Water Research Institute
Canada Centre for Inland Waters
Burlington, Ontario
Canada

This chapter will attempt to underscore how limited our freshwater resources are, and how threatened they are by indiscriminate waste discharges from industrial, agricultural and urban sources. However, the world community within the United Nations (UN) system is trying to protect our environment in general, particularly our water resources, with an ambitious program known as Earthwatch.

The broad objective of Earthwatch is to monitor natural and manmade changes in the environment, and their impact on human health and well-being; to watch for and record the development and use of any potentially toxic chemicals; and to promote information exchange on pertinent environmental topics. The monitoring component of Earthwatch is known as the Global Environmental Monitoring System (GEMS).

## HYDROLOGIC CYCLE

There are three major factors that affect the availability (on a per-capita basis) and quality of freshwater on earth:

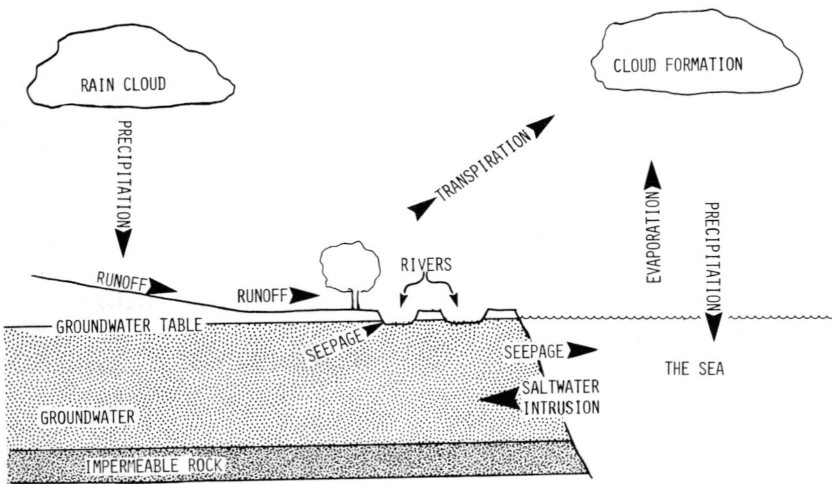


**Figure 1.** Global freshwater circulation.

1. the size of the earth's population;
2. its geographic distribution; and
3. the state of industrialization and urbanization of the peoples on earth.

The amount of freshwater at our disposal on a global and yearly basis has been the same throughout the millenia. It approximates 40,000 $km^3$/yr, and is maintained at this level by the hydrologic cycle (Figure 1).

All our freshwater on earth comes from the natural desalination process of the oceans where the needed energy is provided by the sun. We may visualize the sun, the oceans and the atmosphere as a huge water distillation plant and distribution system.

It is estimated that approximately 450,000 $km^3$ of ocean water is evaporated annually. However, more than 90% of this water, or about 410,000 $km^3$, is of no use to us as it precipitates back into the oceans. Only about 10% of the water evaporated from the oceans, or 40,000 $km^3$, is carried far enough to precipitate over the land. It is only this water that is available to us for our multiple domestic, agricultural and industrial uses. In reality, the total annual precipitation over the land is much more than that; it is of the order of 114,000 $km^3$; the balance of 74,000 $km^3$ comes from water permanently retained as moisture by the atmosphere, soil and vegetation in a never-ending cycle of evaporation from ground and water surfaces, transpiration from vegetation, and precipitation from the atmosphere. The excess of about 40,000 $km^3$ of water carried over from the evaporation of the oceans returns to the oceans as rivers and groundwater runoff. This closes the cycle.

## POPULATION AND GROSS NATIONAL PRODUCT (GNP)

The available freshwater is the same today as it was one billion years ago. However, the population on earth has dramatically increased over the last 150 years. The statistics are dumbfounding if we consder that it took the human species at least one million years to reach the first billion of population by about 1820, and then only about 110 years to double this figure by the year 1930. The third billion was added during the next 30 years, or by about 1960, and the fourth billion within the next 15 years, or by 1975. Even with the currently decreasing rate of population growth, at least 6 billion people will inhabit the earth by the year 2000.

However, the population growth alone is not fully indicative of the increased demand for water which is many times more than the increase in population. This is due to the fact that with the advent of the industrial era and the spectacular increase in GNP that this has brought about, the per-capita consumption of water has increased manyfold. However, our problems do not stop here. The industrial era has also brought about urbanization and indiscriminate industrial and domestic waste disposal. Thus today, and increasingly so in the near future, more people are and will be competing for the same water resources but of ever poorer quality—unless we do something today to stop and hopefully reverse the trend, toward the restoration of water quality to a degree at which it will no longer be a hazard to human health.

## PUBLIC AWARENESS

Public awareness of the great dangers which humankind was facing as a result of the wholesale attack that man himself was making against the environment has been a gradual process, which varied from country to country and within a country from urban to rural areas.

There has been, however, one particular event that has contributed greatly to the awakening of human conscience to the dangerous course man was pursuing that could not but lead to his own destruction. That event was the publication in 1962 of the environmental classic, *Silent Spring*, by Rachel Carson.

This book made great impact on the educated public. Of course, not all agreed with Carson's doomsday projections, and considerable controversy was aroused. However, because of the impact *Silent Spring* made, and the controversy it created, for the first time ever the state of our environment and ecology became a matter of public and international debate.

Of course, one did not have to belong to an international elite or to have read *Silent Spring* to recognize the unrelenting degradation of the environment. The spoilage of air, water, beaches and countryside as well as the poisoning and gradual disappearance of many kinds of fish, birds, and other aquatic and terrestrial species were obvious to all.

In the mid-1960s there was a growing outcry of the public at large across the continents, directed at their respective governments to do something to control the growing pollution and restore the quality of our environment.

## THE STOCKHOLM CONFERENCE

Since pollution does not recognize national borders, as there is nothing that can stop pollutants from "trespassing" from one country into another through the air and the rivers, aquifers and the sea, the need for coordinated international action became obvious.

The pressure for such action became strong and widespread enough to warrant extensive discussion of environmental problems and remedies in the forum of the UN General Assembly in the late 1960s.

The UN General Assembly recognized the seriousness and the complexity of the environmental problems and decided to convene a UN Conference on Human Environment. At that conference, held in Stockholm in June 1972, representatives of 113 nations agreed on a broad plan of action aimed at controlling environmental pollution of any kind.

The Stockholm Conference produced a Declaration of Principles and a series of Recommendations for Action. Of particular interest to us is Article 74 of the Recommendations for Action which deals explicitly with the need of developing a global environmental monitoring program. It calls for:

> . . . Drawing on the resources of the entire UN Systems. . . [and] . . . the active support of Governments and appropriate scientific and other international bodies to . . .
>
> Increase the capability of the UN System to provide awareness and advance warning of deleterious effects to human health and wellbeing from man-made pollutants . . .
>
> Provide this information in a form which is useful to policy makers at the national level . . .
>
> Assist those governments which desire to incorporate these and other environmental factors into national planning processes . . .
>
> Improve the international acceptability of procedures for testing pollutants and contaminants . . .

Thus, the main purpose of monitoring changes in environmental quality is to: (1) "Provide awareness and advance warning of any serious environmental deterioration" and (2) "Permit governments to initiate, individually and/or collectively corrective action for the protection and restoration of the environment".

The international recognition of the need to monitor on a continuing basis the long-term trends in the pollution of our environment has been a historic decision of the Stockholm Conference.

## UN ENVIRONMENT PROGRAM

To coordinate and financially support related international activities for the preservation and restoration of the environment, the UN General Assembly had created a new UN agency known as the UN Environment Programme (UNEP), and endowed it with a fund of approximately $100 million for the initial five-year period (1972–1977).

There are three functional activities within the UNEP: environmental assessment, environmental management and supporting measures.

Environmental assessment, also known as Earthwatch, comprises such priority subject areas as:

1. GEMS: In very general terms this activity comprises the monitoring of pollutants in air, water, soil, food and biota.
2. INFOTERRA (an International Information Referral System): The objective of this activity center is to collect, classify and code sources of environmental information and publish such information in a directory called International Directory of Sources for Environmental Information.
3. IRPTC (the International Register of Potentially Toxic Chemicals): The objective of this activity center should be obvious from its name.

Environmental management comprises the following five priority subject areas:

1. oceans;
2. terrestrial ecosystems;
3. energy;
4. natural disasters; and
5. environment and development.

Given the broadness of each subject area, it may easily be understood that each subject area may be subdivided into many, up to several hundred projects.

Supporting measures are the activities carried out in support of any specific environmental assessment or environmental program. Among them are:

1. education;
2. training;
3. public information; and
4. technical assistance.

To determine the priorities among the many possible environmental problems falling within each subject area, such problems undergo an analysis through what is called a "programmatic process." Essentially a problem is first identified, then it is assessed via a three-level scrutiny in terms of both needs and means, as shown below:

- State of the environment: identifies emerging problems (e.g., net loss of arable land);
- Level I: identifies gaps in knowledge and in actions (knowledge about rate of soil loss and methods of arresting soil loss);
- Level II: identifies activities and "actors" to fill those gaps (global assessment of soil loss; promotion of techniques to control soil loss; actors: FAO, UNESCO, etc.); and
- Level III: identifies actions within level II for fund support (global assessment: Africa north and south of Sahara).

The coordination and integration of the various activities is entrusted to specialized task forces. Their function is to monitor existing projects and to review new projects.

## GEMS GOALS

Even before the Stockholm Conference and in preparation for it, an Intergovernmental Working Group on Monitoring (IWGM) was convened in 1971 to define the objectives of monitoring, assess how these might be implemented and assign priorities for their implementation.

Subsequently, in 1974, an Intergovernmental Meeting on Monitoring (IMM) defined seven program goals for GEMS:

1. assessment of global atmospheric pollution and its impact on climate;
2. assessment of water pollution and environmental problems related to land use and agriculture;
3. assessment of the state of ocean pollution and its impact on the marine ecosystem;
4. assessment of the extent and distribution of contaminants in biological systems, particularly food chains;
5. assessment of the response of terrestrial ecosystems to pressures exerted on the environment.

6. monitoring factors necessary for forecasting environmental or ecological disasters; and
7. early warning systems for dangers to human health (environmentally induced diseases).

Some of these goals, like goal 1 (monitoring atmospheric pollutants) and goal 4 (monitoring the distribution of contaminants in biological systems and food chains) are often referred to as "pollution monitoring," since what is being monitored are the pollutants, their incidence and long-term trends. In other instances, as in goal 5 and partly in goal 3, what is being monitored are the ecological effects of the pressures exerted on the environment. Such monitoring is referred to as "ecological monitoring."

A breakdown of the two categories of monitoring is shown in Table I. The comprehensive GEMS program also comprises a third category of monitoring which is monitoring natural disasters (earthquakes, volcanic activities, etc.).

## GEMS/WATER

The specific responsibility entrusted to the National Water Research Institute in Canada is the worldwide coordination of the health-related pollution monitoring of fresh water known also as the GEMS/WATER project.

In summary, the objectives of the GEMS/WATER project are:

1. monitoring the incidence and trends of pollutants in water;
2. providing advance warning of any serious deterioration; and
3. prompting governments to initiate, individually and/or collectively, corrective action for the protection, restoration and improvement of the environment.

**Table I. Two Categories of Monitoring**

| Pollution Monitoring | Ecological Monitoring |
|---|---|
| Health-related<br>Air<br>Water<br>Food<br>Man (health effects of pollutants)<br>Climate-related<br>Effects of pollutants on climate<br>Ocean monitoring<br>Regional seas (incl. river mouth monitoring)<br>Open seas | Soil and vegetation cover monitoring (e.g., forests, desertification)<br>Water and resources monitoring (for living organisms, pollution indicator organisms, e.g., "mussel watch")<br>Biosphere (effect of UV; monitoring variables that can serve as indicators of biological productivity of, for example, forests and rangelands) |

## MONITORING

The term "monitoring" as used in this presentation implies the periodic quantitative measurement of preselected physical and chemical parameters in water samples taken at preselected sites in rivers, lakes and aquifers. Except for a limited number of physical and chemical parameters whose measurement is simple and straightforward and can be carried out on site (e.g., temperature, pH, conductivity, dissolved oxygen and chlorides) all other parameters are measured on samples brought to designated laboratories after ensuring proper preservation of the samples against subsequent change in the parameter concentration.

There are seven main components in the GEMS/WATER program:

1. establishment of a global network of monitoring stations at appropriate sites in rivers, lakes and aquifers;
2. adoption of uniform methodologies of water sampling and analysis;
3. application of a continuing data quality assurance program;
4. development and use of a comprehensive data storage and retrieval system;
5. organization of training courses for personnel engaged in all aspects of global water quality monitoring;
6. preparation of operators' manuals; and
7. provision of equipment in a limited number of cases.

The following is a brief description of the activities contemplated within each component of the GEMS/WATER program.

### Monitoring Network

The initial three-year program (1977–1979) called for the establishment of approximately 300 monitoring stations on rivers, lakes and aquifers in a limited number of countries. This was subsequently to be expanded to approximately 800–1200 stations, or about one station for every 4–5 million inhabitants.

Given the relatively limited number of monitoring stations contemplated for the GEMS/WATER program, the selection of the monitoring sites is governed by the following considerations:

1. major water resources used for water supplies of large population centers for drinking, irrigation, livestock, fisheries, food and health-related industries and body-contact recreation;
2. major international rivers and lakes near boundary points;
3. major rivers discharging into oceans or seas; and
4. major rivers and lakes in presently remote areas for establishing "baseline levels" and monitoring any changes at relatively longer intervals of time (e.g., once a year or even once every few years).

In all of the aforementioned cases, advantage is taken of the already existing stations and laboratories, providing international assistance if and where required, and to the extent of the available resources, to enable the selected monitoring stations and laboratories to fulfill their "global" function. The establishment of new monitoring stations where none exists will be considered within the context of criteria set earlier in areas where presently important rivers, lakes or aquifers are not monitored for water quality. In any event, the recognition of existing monitoring stations and laboratories for the GEMS/WATER purposes or establishment of new stations is being made in full consultation with the national authorities and with their ultimate approval.

**Operating Levels**

The GEMS/WATER program, like other GEMS programs operates at three levels, national, regional and global. In general, each participating country designates a "national center" to advise on the selection of sampling sites and gauging stations and of analytical parameters to be monitored as well as to transmit national water quality data to the "regional center."

The regional center coordinates the inputs from national centers, in terms of the final selection of the sampling sites for the whole region, bearing in mind the regional significance of the proposed sites and the overall capability of the system. It also processes data received from national centers on manually filled forms into a computer-interpretable form (punched cards or magnetic tape) for transmittal to the global center. Moreover, the regional center has specific responsibilities for the conduct of the regional data quality assurance program and organization of periodic training courses.

The global center coordinates the development of an appropriate data storage and retrieval system and a data quality assurance program; prepares and distributes the necessary documentation, including operators' manuals; processes and stores data received from the regional centers; issues comprehensive annual reports comprising data from all monitoring stations and showing analytical results, flow and mass flow. Moreover, at request, the global center issues periodic reports providing profiles of selected parameters in selected geographic areas. The Canada Centre for Inland Waters in Burlington, Ontario, has agreed to initially assume the responsibilities of the global center.

**Analytical Determinands**

The water quality determinands to be measured fall into three distinct categories.

*Basic Determinands*

In this group are included 13 parameters which are considered important for the general assessment of water quality. They are to be measured at all monitoring sites. The monitoring of these determinands is relatively simple and does not require expensive instrumentation. Among them are: temperature, pH, biochemical oxygen demand (BOD), conductivity, dissolved oxygen, chloride, suspended solids and fecal coliforms.

*Determinands of Global Significance*

In this group are comprised such persistent and toxic pollutants as cadmium, mercury, lead and organohalogenated compounds. They are characterized by their transportability into far away places and by bioaccumulation.

*Optional Determinands*

In this category are currently listed 37 analytical parameters all of local significance. Among them are: total organic carbon, chemical oxygen demand, anionic tensides (MBAS), nonionic tensides, chromium (total and hexavalent), arsenic, boron, selenium, cyanide, phenols and fecal streptococci.

Not all parameters are meant to be measured in each of the three media (rivers, lakes and aquifers). As an example, chlorophyll-a is to be measured in lakes only and not in rivers or aquifers; suspended solids are to be measured in rivers and not in aquifers or lakes, and fluoride is to be monitored in aquifers and not in rivers or lakes.

Analytical parameters in the second category (global significance) are to be measured as "total content in water," "dissolved content in water," "content in bottom sediment" and "concentration in biota."

The implementation of the analytical portion of monitoring will be gradual (starting with the analysis of one or a few analytical parameters) and concomitant with the application of an international "analytical quality assurance" program.

**Analytical Quality Control**

The accuracy and comparability of results obtained within the global network are the cornerstones on which rests the GEMS edifice. Ideally, the application of identical analytical methodologies and instrumentation by all participating laboratories preceded by adequate and uniform training of laboratory personnel would greatly facilitate the achievement of the desired standards of accuracy and comparability worldwide. However, even then there would have been a need for running in parallel an analytical quality assurance program. Experience shows that even within the same corporate or governmental organizations, applying the same methodologies and using instruments of the same make, inadvertent problems do creep in: a minor malfunction of the instrument or a slip in method application that is being transferred from one operator to another. Problems are, of course, much greater when a large number of laboratories participate in a cooperative program employing instruments and applying methodologies of different performance characteristics, and moreover, having operators with widely different backgrounds in training and education.

It is in consideration of the aforementioned reasons that a program of inter- and intralaboratory quality control has been made an integral part of the GEMS/WATER project. The program, in a nutshell, consists of periodic analysis of (1) a standard solution of known concentration; (2) duplicate portions of the same sample, and (3) duplicate portions of a sample to which known amounts of a "determinand" have been added. If the results do not meet the desired standards of accuracy and precision, the causes of the problem have to be identified and corrected.

The agreed targets of accuracy and comparability within the GEMS/WATER system have been set at either 20% of the concentration or value of the determinand, or a fixed target based on actual concentration of each determinand, whichever is greater. Furthermore, the maximum tolerable error should be divided equally between precision (random error) and bias, using the 95% statistical confidence level for assessing random error.

## Data Storage and Retrieval

Based on biweekly sampling and an average analysis of 20 parameters per sample, 400 monitoring stations will generate when fully operational some 200,000 data points per year in the initial period to subsequently increase to over 600,000 data points as the network expands to about 1200 monitoring stations.

To handle this amount of data the WHO Collaborating Centre on Surface and Ground Water Quality has developed a computerized system named GLOWDAT (GLObal Water DATa).

GLOWDAT is a development of proven storage/retrieval techniques based on extensive use of the Canadian NAQUADAT system (NAtional water QUAlity DATa bank) further enhanced by recent database technology.

On completion of the analysis, the operator enters the data onto coded data forms. The laboratory adds other pertinent hydrographic information and forwards the forms to the regional center (or directly to the global center if there is still no regional center in the area). Each regional center is responsible for collating and keypunching the data and forwarding at quarterly intervals to the global center both the source data forms and the corresponding sets of 80-column punch cards. As a safety measure, the forms and the punch cards are sent separately. The data are processed at the global center and appropriate computer-printed reports are returned to the regional centers.

## Validity of Data

The establishment of a global network of monitoring stations averaging no more than one station for every 10–12 million inhabitants in the initial period, and raised subsequently to one station for every 4–5 million inhabitants, does not seem much. Questions have been raised, in fact, as to the value of such a scheme. How can possibly a single monitoring point in a country of 5–10 million inhabitants be indicative of the state of pollution of that country's rivers, lakes and aquifers? The answer is simple—it cannot. However, then it never was meant to be that. The 400 or 800 monitoring stations placed in carefully selected sites around the globe are to provide us with a statistical indication of the long-term trends in water pollution on a global basis.

Such a statistical approach is as valid as that of an election poll which, through interviews of a carefully selected population sample of 3000 individuals in a country of 220 million, can predict reasonably well how 80 million people are going to vote. Statistically, there is for every borderline individual that makes a last minute switch from Party A to Party B another borderline individual (or an equivalent proportion) that makes a last minute switch from Party B to Party A. Likewise, it is to be expected that a "borderline" monitoring station that erroneously shows slight improvement in water condition rather than slight deterioration should be statistically cancelled out by another "borderline" station showing erroneously the reverse conditions. In conclusion, GEMS/WATER is being created to monitor freshwater resources on our planet as a whole using comparable methodologies of sampling and measurement and achieving comparable accuracy. The lonely monitoring station in a country of 5–10 million is no substitute for the more elaborate monitoring network which that country requires.

There are, however, some tangible benefits that countries, particularly developing countries, are expected to derive within a short span of time from the establishment of one or a few global monitoring stations on their territories. The laboratory staff of the designated national laboratory will be exposed to and benefit from international training provided by experienced teachers of worldwide reputation. Through participation in international, interlaboratory quality assurance programs, the operators will achieve and maintain high standards of performance.

Such national laboratories participating in the GEMS/WATER program will eventually serve as model laboratories in their own countries, provide training to chemists and technologists from other laboratories in the country, and ultimately initiate and lead a national interlaboratory quality assurance program.

## ACCOMPLISHMENTS

### Monitoring Stations

As shown in Table II, by the end of 1979 a total of 388 water quality monitoring stations had been designated in 62 countries, compared with the targeted total of 300. The distribution of these stations is: 227 on rivers, 74 on lakes and 87 on aquifers. The term "designation" implies the formal acceptance by the country concerned to contribute water quality data from that station to GLOWDAT.

Table III shows the breakdown of monitoring stations by "impact" or "baseline." An impact stations is one located on a river, lake or aquifer subject to industrial and/or urban waste disposal, and/or agricultural runoff. A baseline station is one located in a hydrologic basin where there are no upstream anthropogenic emissions of pollutants arising from agricultural, industrial or urban activities.

**Table II. Georgraphic Distribution of GEMS/WATER Monitoring Stations**

| | | Number of Stations | | | |
|---|---|---|---|---|---|
| Region | Participating Countries | Rivers | Lakes and Reservoirs | Groundwater | Total |
| Africa | 13 | 25 | 11 | 5 | 41 |
| America | 13 | 67 | 21 | 15 | 103 |
| Eastern Mediterranean | 11 | 32 | 15 | 33 | 80 |
| Europe | 11 | 34 | 15 | 9 | 58 |
| Southeast Asia | 5 | 37 | 3 | 20 | 60 |
| Western Pacific | 9 | 32 | 9 | 5 | 46 |
| Total | 62 | 227 | 74 | 87 | 388 |

Table III. Breakdown of Monitoring Sites as "Impact" and "Baseline"

| Water Body | Impact Stations | Baseline Stations | Unclassified |
|---|---|---|---|
| River | 169 | 43 | 15 |
| Lake/Reservoir | 54 | 12 | 8 |
| Groundwater | 51 | 12 | 24 |
| Total | 274 | 67 | 47 |

## Pilot Project

During the second half of 1979, a GEMS/WATER pilot project was implemented involving reporting of water quality data from two or three more advanced countries in each of the six world regions. The purpose of the pilot project was to identify any potential problems in the program and correct them before initiating worldwide reporting of data from all the designated stations. The pilot project has served the intended purposes and, as of January 1980, the GEMS/WATER project became worldwide with assured participation from 62 countries.

## Training Courses

By the end of 1980, five regional training courses in all aspects of GEMS/WATER program were held in the following locations: Manila, Phillipines, for the benefit of the countries of the Western Pacific region; Lima, Peru, for the countries of Latin America; Nagpur, India, for the countries of the Southeast Asia region; Alexandria, Egypt, for the countries of the Eastern Mediterranean region; and Beijing, China for the exclusive benefit of the participating organizations of the People's Republic of China.

In November 1978 a workshop on sediment analysis was held in Budapest, Hungary, and in December 1979 a pilot project on GEMS/WATER analytical quality control (AQC) was held in Mexico City. In January 1981 a GEMS/WATER training course was held in Dakar, Senegal for the benefit of the French-speaking countries of Africa, while another course is scheduled in Nairobi, Kenya for the benefit of the English-speaking countries of Africa.

Subsequent to the successful completion of the AQC pilot project course in Mexico City, similar courses in AQC are being planned for all other regions of the world within the next two years.

# SECTION 8

# COMPUTERIZED DATA

# CHAPTER 32

# THE DISTRIBUTION REGISTER OF ORGANIC POLLUTANTS IN WATER–WATERDROP

**Bonnie L. Carson, John E. Going, Viorica Lopez-Avila, Joy L. McCann, Christopher J. Cole and Joel L. Holt**

Midwest Research Institute
Kansas City, Missouri

The initial effort of compiling a comprehensive list of organic pollutants found in water was made at the U.S. Environmental Protection Agency (EPA)/ORD Environmental Research Laboratory, Athens, GA. The product of this effort is known as the Shackelford-Keith List [1]. Plans were being made when it was published in December 1976 to fund a contractor to produce a database called WaterDROP that would include all the many fields of information not available in the Shackelford-Keith List.

The Shackelford-Keith List, which was limited to survey-type analyses, contained as of 1976, 5700 purported occurrences of 1259 unique compounds. (The June 1977 update contained 6944 occurrences of 1282 compounds. Since Chemical Abstracts Service Registry Numbers (CASRN) were not used, some fraction of these are probably duplicates. For example, both isopropanol and 2-propanol are listed as unique compounds. Most of the 160 references used for this list are not readily obtainable in hard copy. Eight references account for approx. 2700 entries; 17 references for approx. 1300 entries; and 137 references for the rest. Thus, about 67% of the entries in the 1976 list could be obtained from 25 survey references, only six of which were obtainable in hard copy. Cross-references between this list, a list compiled by Junk and Stanley (bibliography of about 80 references) [3], and the European list called the CEC Cost-Project 64B (about 300 references) [4] were so entwined, that the Shackelford-Keith List was counting occurrences up to three times for the same set of data (from the other two lists as well as from the original published literature references).

Table I shows the seven original output formats of the Shackelford-Keith List. Table II shows the initial goals for WaterDROP. This chapter attempts to show how well Midwest Research Institute (MRI) and Fein-Marquart Associates, Inc. (FMA), have met these goals and far exceeded them.

In early 1978, the prime contract for WaterDROP was awarded to FMA, a software firm in Baltimore, MD, which has received funding to design, develop, implement and install the Chemical Information System (CIS) from EPA and the National Institutes of Health (NIH). In August 1978 MRI, as a subcontractor to FMA, was engaged to review systematically the existing data and assist in the development of a highly detailed and searchable database. The resulting database is "WaterDROP," which is publicly available as a component of the NIH/EPA Chemical Information System.

The first goal of our program was to develop a data sheet that would satisfy both the information requirements of the EPA and the CIS software already developed. Useful data fields were quickly selected from the Shackelford-Keith List, additional requirements desired by A. Alford, the EPA Project Officer, and the experience of the MRI staff. However, making the data entry process compatible with the software and hardware was more difficult.

The initial data sheet defined the major types of data that would be entered into WaterDROP and served as the initial base for definition. As more specific entries were added, a single-page format could no longer be used. The

**Table I. Shackelford-Keith List Output**

1. Frequency of occurrence of each compound listed alphabetically
2. List of water types and the number of times each occurred
3. List of geographic sources of the pollutants and the number of times each was used
4. Complete list of all 5700 entries alphabetically by compound name (includes only geographical location, water type, date of sampling or date of publication, and frequency of occurrence)
5. Complete list, but this time sorted by alphabetized water type; compounds found in each water type are subalphabetized
6. Complete list sorted by geographic location or by abbreviated literature reference
7. Somewhat later output could also show the distribution among 112 functional group classes by water type

**Table II. Goals for WaterDROP Set by Shackelford and Keith [1] and Garrison et al. [2]**

Ease of searching and updating

Emphasis on compound identities, concentrations, locations and water types

Fields may include:

| | | |
|---|---|---|
| Water type | Chemical name | Literature reference |
| Sampling date and location | Synonyms | Source/lab reference |
| Concentration | Molecular formula and weight | WaterDROP entry date |
| Identification method | CAS registry number | Biological effects |
| Indication of confirmation | Wiswesser line notation | Taste and odor threshold |
| Structure | | |
| Chemical Class | | |

Types of searches
- By country, state or city
- By water types:
  - Groundwater
  - Surface water (and subclasses such as river, lake or sea)
  - Finished drinking water
  - Municipal wastewater
  - Process waters
  - Industrial wastewater (subclassified by SIC number)

most serious problem with this format was the inconvenience created by the need for multiple entries of compounds, concentrations, sampling locations and water types of any single article. Also, this type of article occurs more frequently in the literature than articles with single entries for each data type. Consequently, this type of format had to be dropped in favor of a more flexible format for data input.

The second data sheet was designed to handle literature with multiple compounds, sample locations and water types with minimum data entry, but the FMA-specified coding method to ensure proper ordering of the information by the computer was too complex to handle conveniently.

The third data sheet was developed in close conjunction with FMA to assure format compatibility between MRI and the CIS. A switch to coding on IBM sheets was also initiated at this time. MRI suggested and FMA approved a much improved system for globally coding information that would be frequently reiterated by the computer. The global code, which is alphanumeric, also ensures proper ordering of the data. The final data sheet format was made after several meetings with various EPA groups. Changes were made to incorporate their ideas into the data sheet. As originally designed, the data sheet has proved remarkably flexible in accommodating all the kinds of data that we have encountered in the literature.

Since the design of the data sheet, the role of MRI has included document selection, acquisition, rating and analysis, and data entry on IBM coding forms. Data entry was also performed by Systems Research Company (SRC), Philadelphia, PA, for a number of months, and MRI directed this activity. Support for this work, which chiefly included all the U.S. and Canadian drinking water literature that we could identify and acquire, was provided by the Health Effects Research Laboratory of EPA at Cincinnati, OH. R. D. Lingg was the project monitor for the SRC contract and provided many original publications and private communications.

Almost every document MRI processes has been quite complex, requiring a minimum of several hours and sometimes as long as weeks to complete data entry. Since document analysis and data entry were proceeding at a rate of 10–15 documents per month, it was soon realized that we would have to postpone our plans for soliciting unpublished data from EPA laboratories and their contractors.

Prior to July 1979 we processed mainly documents provided by W. Shackelford, about half of which had been used to create the Shackelford-Keith List. After July 1879 we processed mainly documents published since 1978. Top foreign articles, which seem to be produced at about the same rate as U.S. and Canadian articles, are filed for future acquisition and processing. It was felt that a database that could initially promise some degree of completeness of at least the most recent U.S. and Canadian published literature would be better than one that added new data unsystematically.

Data entry at MRI has been performed by J. L. McCann, C. J. Cole and J. L. Holt. All have contributed creatively to handling new data entry situations as they arise. McCann is responsible for the original idea for the alphanumeric global coding system, which greatly simplified the FMA requirement for uniquely identifying each line of data MRI submits. (Some documents have contained more than a thousand such lines.)

Confidence rating is a unique and sophisticated aspect of WaterDROP and will be described more fully below. M. Marcus originally was in charge of this task, but since May 1979, V. Lopez-Avila has overseen the confidence rating of the compound identifications. B. L. Carson, of the MRI Chemical Impact Assessment Section, has served as principal investigator and coordinator of all the ongoing MRI tasks. J. E. Going, Head of the MRI Environmental Analysis Section, is project leader.

At FMA the data are keyboarded, computer edited and , after MRI has made required corrections, loaded on the computer.

## DOCUMENT SELECTION AND CONFIDENCE RATING ASSIGNMENT

Initial data entry efforts were devoted largely to the original documents used for producing the Shackelford-Keith List published in 1976. These documents represent the majority of the data currently online. Most documents processed in the last 12 months, however, were published from 1978 through 1980. These are selected by regular review of *Environmental Science and Technology*, Chemical Abstracts (CA) Selects titles, *Pollution Monitoring*, *Environmental Pollution* and *Gas Chromatography-Mass Spectrometry Abstracts*.

A highly unique factor of WaterDROP is the Confidence Rating entry, which is an evaluation of the identification of the compound. The analytical method used to establish the identity is given a rating of "Confirmed," "Probable," "Unconfirmed" or "Unknown" by members of the MRI Analytical Chemistry Department. Table III depicts the Confidence Rating Sheet.

Analytical chemists in the Environmental Analysis Section code the isolation and analytical methods used and assign the confidence rating for each compound in the document. A "confirmed" rating is given to those identifications made by gas chromatography/mass spectrometry (GC/MS) using authentic standards for verification or computer matches with high confidence levels. Any other analysis procedure for identification must have substantial experimental evidence or be an established standard method to receive a confirmed rating.

An example of a method other than GC/MS that receives confirmed ratings is high-pressure liquid chromatography with fluorescence detection for the analysis of polycyclic aromatic hydrocarbons.

The "probable" confidence rating is assigned to GC/MS analyses with moderate confidence levels, to chromatographic results with nonselective detectors and multiple chromatographic systems, and to nonchromatographic procedures with considerable experimental evaluation. The "unconfirmed" rating is assigned to GC/MS results with low confidence levels, chromatographic measurements with nonselective detectors and nonchromatographic procedures without experimental verification. The "unknown" confidence level is assigned if there are insufficient experimental data to assign any of the other ratings. To assure consistent ratings by MRI staff, a training session was held to define in detail the criteria necessary for the rating levels, and a set of example literature was provided covering the most common types of data. Data sheets were checked in a random manner to assure that the ratings were consistent.

**Table III. WaterDROP Confidence Rating**

One of four ratings may be assigned to compound identification:

Confirmed
Probable
Unconfirmed
Unknown

Confirmed:

1. Identification made by GC/MS using authentic standards for retention time and mass match;
2. Identification made by GC/MS with computer matches of high confidence levels
   MSSS > 0.8;
   Finnigan (old software) fit > 0.8;
   Finnigan (present software) fit > 0.9; purity > 0.7;
3. Identification made by GC/MS with independent method, e.g., NMR or IR; and
4. Non-GC/MS identifications that have substantial experimental evidence or an established standard method.

Probable:

1. Identifications made by GC/MS with moderate computer matches
   MSSS, 0.4–0.8,
   Finnigan (old), 0.4–0.8,
   Finnigan (new) fit, 0.6–0.8; purity, 0.5–0.7;
2. Identification by GC/MS with manual interpretation of spectra without reference to standard spectra;
3. GC/selective element detectors retention time match with authentic standards and relatively clean samples;
4. GC nonselective detector on two chromatographic systems with authentic standards and clean samples; and
5. Nonchromatographic procedures with considerable experimental evaluation.

Unconfirmed:

1. Identification made by GC/MS with low computer matches;
2. Chromatographic results with nonselctive detectors; and
3. Nonchromatographic procedures without experimental verification.

Unknown: Insufficient experimental data to assign any other confidence level.

## THE DATA SHEET AND DOCUMENT ANALYSIS

Table IV is the WaterDROP Data Sheet, which serves as the guide for document analysis and data entry. (A more detailed list of data entry instructions is being developed by FMA and MRI, but this is the sheet we have used to the present.)

**Table IV. WaterDROP Data Sheet**

Data Source

10. First author, editor or individual to contact (last name first)
11. Laboratory, code, organization name and address

Published

12. Journal: coden, volume, issue, date, pages
13. Other: coden or title
14. Other: publisher, place, date, pages
15. MRI source library: national Union Catalog of Library of Congress code

Unpublished

16. Project report: title
17. Project report: project or contract number, sponsoring agency
18. Meeting: title of meeting, title of paper
19. Meeting: place, time
20. Private communication

Other sample types included

21. Gas
22. Soil
23. Plants
24. Animals, excreta
25. Other [free text]

Water Type Information

26. Drinking, unfinished (please indicate source below and treatment method on line 49)
27. Drinking, finished: plant name if different from that of city
28. Tapwater
29. Surface, lake, impoundment, pond, reservoir or freshwater marsh: name
30. Surface, river, stream, creek, brook, bayou or seaway: name
31. Surface, ocean, sea, estuary, fjord, gulf bay, harbor or strait(s): name
32. Surface, canal, channel, aqueduct or drainage ditch: name
33. Groundwater: aquifer name (if known)
34. Groundwater, deep well (greater than 2000 ft)
35. Groundwater, shallow well (2000 ft or less)
36. Precipitation or runoff: type
37. Solids, sediments (please indicate source above)
38. Solids, suspended particulates (please indicate source above)
39. Solids, sewage sludge: plant name if different from that of city
40. Solids, industrial sludge (please fill in line 69 and the associated liquid in lines 41, 42 or 43)
41. Industrial discharge, final effluent (please indicate receiving water body above)
42. Industrial, untreated
43. Industrial, midtreatment (please describe treatment, line 49)
44. Municipal discharge, final effluent: plant name if different from that of city (please indicate receiving water body above)
45. Municipal sewage, untreated liquids: plant name if different from that of city

**Table IV, continued**

46. Municipal sewage, midtreatment: plant name if different from that of city
47. Deep-well injection (please fill in lines 68 and 69 below)
48. Other (please specify)
49. Treatment system [free text entry]

Sampling Information

Sampling and Isolation Methods

50. Grab sampling
51. Composite
52. Accumulator (resin, carbon)
53. Volatile organic analysis (VOA)
54. Solvent extraction
55. Other (please specify)

Sampling Location and Dates

56. Monitoring data (2 or more sampling dates), starting date of sampling: month/day/year (enter ? where any part is unknown)
57. Monitoring data, ending date of sampling: month/day/year (enter ? where any part is unknown)
58. Survey data, sampling date: month/day/year (enter ? where any part is unknown)
59. Latitude (unnecessary if within a city)
60. Longitude (unnecessary if within a city)
61. Radius of uncertainty, km
62. Zip code
63. Other: industrial plant name, street address, etc. [free text]
64. City
65. County
66. State, province, district (use two-letter state code)
67. Country
68. Source of contaminant is identified in the document or speculated on
69. Industry type, SIC number

Identification Technique Information

| Identification Technique | Separation Technique[a] | Specifics[b] |
|---|---|---|
| 70. Mass spectrometry | GC HPLC TLC PC None | EI CI NI FD |
| 71. Flame ionization detector | GC HPLC TLC PC None | FI SIM HR Other |
| 72. Electron capture | GC HPLC TLC PC None | |
| 73. Flame photometric detector | GC HPLC TLC PC None | |
| 74. Alkali bead or flame detector | GC HPLC TLC PC None | |
| 75. Electrolyic conductivity | GC HPLC TLC PC None | |
| 76. Infrared | GC HPLC TLC PC None | |
| 77. Ultraviolet | GC HPLC TLC PC None | |
| 78. Nuclear magnetic resonance | GC HPLC TLC PC None | Proton $^{13}C$ |
| 79. Fluorescence | GC HPLC TLC PC None | |

Table IV, continued

| Identification Technique | Separation Technique[a] | Specifics[b] |
|---|---|---|
| 80. Thermal conductivity detector | GC HPLC TLC PC None | |
| 81. Colorimetric | GC HPLC TLC PC None | |
| 82. Other (please specify) | GC HPLC TLC PC None | |
| Confidence Rating (to be completed by MRI staff) | | |
| 83. Confirmed | | |
| 84. Probable | | |
| 85. Unconfirmed | | |
| 86. Unknown | | |
| Compound Information | | |
| Compound Identity | | |
| 87. Name (please indicate all necessary stereochemical information unless the CASRN is given) | | |
| 88. CASRN (if known) | | |
| 89. Wiswesser line notation (if known) | | |
| 90. Formula (only if lines 87 and/or 88 or 89 are not given) | | |
| 91. Structure (necessary only if lines 87, 88, 89 and/or 90 are not given; please use separate sheet and label with roman numeral you give here) | | |
| 92. Water solubility (if known) | | |
| 93. Boiling point (if known) | | |
| Concentration (please indicate detection limit (DL) and/or detection frequency (DF) where appropriate in lines 94–97) | | |
| 94. Water: one or average value reported, μg/L | | |
| 95. Water: range of values when detected, μg/L | | |
| 96. Solids: one or average value reported, μg/g | | |
| 97. Solids: range of values when detected, μg/g | | |
| 98. Trace | | |
| 99. Not reported | | |
| 100. Not detected | | |
| 104. Water (gases in): one or average value reported, μL/L | | |
| 105. Water (gases in): one or average value reported, μL/L | | |

[a]GC = gas chromatography; HPLC = high-performance liquid chromatography; TLC = thin-layer chromatography; PC = paper chromatography.
[b]EI = electron ionization; FD = field desorption; HR = high-resolution; CI = chemical ionization; FI = field ionization; NI = negative ion; SIM = selected ion monitoring.

Altogether there are about 90 fields. Fields 10 to 25 describe the data source (published, unpublished or private communication). Fields 26 to 49 list the water types and treatment fields. An attempt was made to segregate impounded freshwater in field 29, flowing freshwater in fields 30 (natural water bodies) and 32 (artificial water bodies), and seawater in field 31. At some future time, steps should be made to extract freshwater bays and harbors (such as those of the Great Lakes) and place them into field 29.

Fields 50 to 69 describe the sampling and isolation methods and dates, the geographical location, and the industry type, if applicable. Information for fields 70 to 86 is provided to the document analyst by the analytical chemist; these fields include the identification technique(s) used for each compound and the confidence rating.

Fields 87 to 105 describe the compound's identity and its concentration. Because of the ability to use the Structure and Nomenclature Search System (SANSS) of the CIS, MRI does not supply CASRN or a structure. The only time we envisioned our drawing a structure was if the author did and no name could be devised for the compound. This has yet to occur. We also have never encountered Wiswesser line notations in the original literature.

## WaterDROP ONLINE AND THE CIS

Currently, the WaterDROP data entered by MRI from 213 documents include 26,800 occurrences of about 1000 organic compounds in drinking water, natural water bodies, industrial and sewage effluents, aqueous process streams, and associated suspended particulates and sediments. Approximately half of these data have been available online from the NIH/EPA Chemical Information System since March 1980. By May 1980 WaterDROP was the eighth most used database in the CIS system, with mjaor users including the National Bureau of Standards.

Figure 1 depicts the structure of the CIS collection of computerized data storage and retrieval modules for chemical information. Each of these modules or databases is a self-contained or "stand-alone" system dealing with a particular aspect of chemistry. However, since all the databases have been prepared according to a standard set of CIS guidelines, they can share the same utility software and communicate among each other. In this way, it is relatively easy to conduct composite searches dealing with these various aspects of chemistry, and similarly to display, in association with retrieved compounds, information stored in the databases associated with the various modules.

The modules of the CIS communicate among themselves by stored lists of Registry Numbers, for the most part supplied by Chemical Abstracts Service. Interaction of WaterDROP with SANSS will be discussed in the section on Applications.

The online database contains 10,600 citations for 869 unique compounds, that is, complete records of occurrences of organic compounds with their

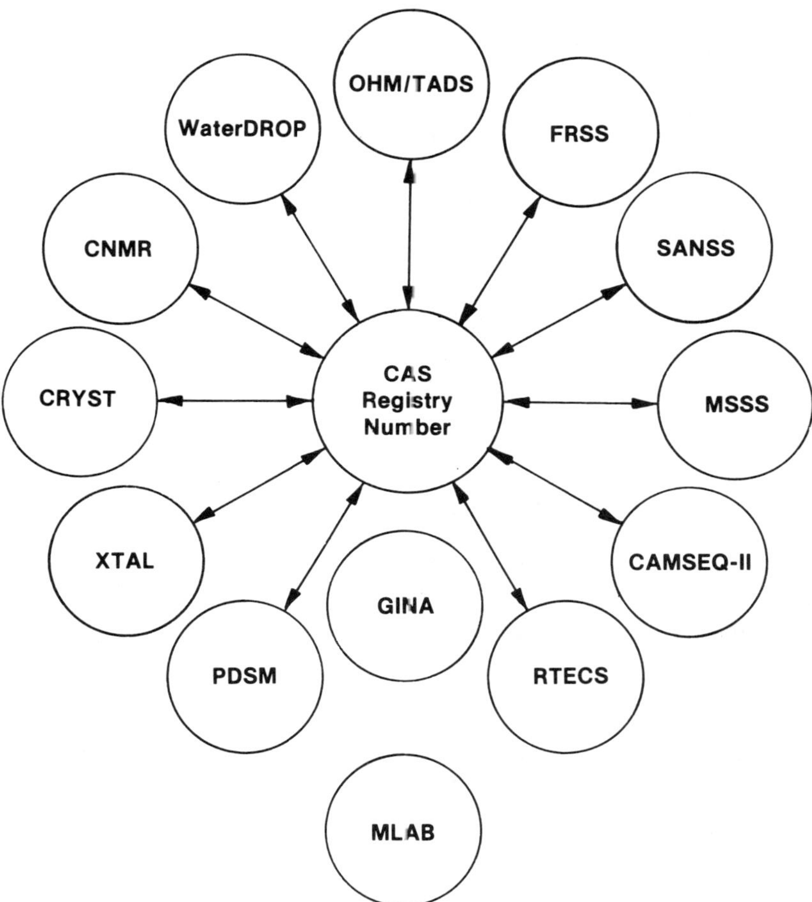


**Figure 1.** NIH/EPA chemical information system databases. WaterDROP = Distribution Register of Organic Pollutants in Water; OHM/TADS = Oil and Hazardous Materials/ Technical Assistance Data System; FRSS = Federal Register Search System; SANSS = Structure and Nomenclature Search System; MSSS = Mass Spectral Search System; CAMSEQ–II = Conformation Analysis of Molecules in Solution by Empirical and Quantum–Mechanical Techniques System; RTECS = Acute Toxicity Data from the NIOSH Registry of Toxic Effects of Chemical Substances; GINA = Graphical Interactive NMR Analysis; MLAB = Modeling Laboratory; PDSM = X-Ray Powder Diffraction; XTAL = Single Crystal Reduction and Search System; CRYST = Cambridge X-Ray Crystallography Data Base; CNMR = Carbon-13 Nuclear Magnetic Resonance Search System.

concentrations and all the associated data. If a document contains 988 occurrences (one does), you can get a complete record printed out for each one. However, since retrieved data may be for only one citation from this particular document, the citation must stand alone and contain all the information needed to identify the location, sampling and identification methods, etc.

Table V shows an analysis of the online database, which currently contains data from 65 published references and theses, 19 government reports and 10 personal communications. The average number of unique compounds (i.e., those assigned CASRN) per document are 73, 179 and 28 for the three document types, respectively, with the average numbers of citations per document (91, 235 and 34, respectively) being slightly higher. The overall averages are 92 compounds per document and 117 citations per document.

## APPLICATIONS

The usefulness of WaterDROP is in the online retrieval of information in almost limitless combinations using the searching routines and in the immediate availability of other CIS components. Not only are searches possible using almost all of the fields listed in the data sheet, any nontrivial word appearing in a WaterDROP citation can serve as a search term. Certain fields are numerical and are so indexed by the software that the user can search by

**Table V. Analysis of the Online Database (March–August 1980)**

| Range | No. of Documents with This No. of Unique Compounds (i.e., CASRN) | | | | No. of Documents with This No. of Unique Compounds (Occurrences) | | |
|---|---|---|---|---|---|---|---|
| | J[a] | G[b] | P[c] | | J[a] | G[b] | P[c] |
| 0[d] | 8 | – | – | | – | – | – |
| 1–99 | 48 | 6 | 6 | | 53 | 5 | 6 |
| 100–399 | 6 | 11 | 0 | | 7 | 11 | 0 |
| 400–699 | 0 | 0 | 0 | | 1 | 1 | 0 |
| 700–999 | 3 | 2 | 0 | | 4 | 2 | 0 |
| Total Documents | 65 | 19 | 6 | | 65 | 19 | 6 |
| Compounds/Document | 73 | 179 | 28 | Citations/Document | 91 | 235 | 34 |
| Overall Avg | | 92 | | | | 117 | |

[a]J = journal, book or thesis.
[b]G = government reports.
[c]P = personal communication.
[d]WaterDROP compounds that have not been assigned CASRN include Aroclors, humic acids, endotoxin and polysiloxanes (silicones).

ranges that he has constructed to retrieve all citations that fall within those ranges, e.g., all compounds found in surface rivers having boiling points between 200 and 300°C and concentrations greater than 5 μg/L. When the fields (called keywords at this point in the instruction manual for Water-DROP), which are represented by two- to five-letter mnemonics, such as SFRIV for surface rivers, are followed by a slash, one can limit the search to whatever term follows it. For example, the keyword SFRIV should retrieve all citations on samples taken from rivers, but SFRIV (now a field according to the manual) followed by a slash and OHIO retrieves citations only on samples collected from the Ohio River. A search, however, using only the word OHIO without the field mnemonic will also retrieve any mention in the text of Ohio Street or the state of Ohio as the sampling location as well as the Ohio River.

The logical (Boolean) operations AND, OR and NOT, can also be applied to searches to further define specific data desired. For example, a Boolean search on MAT/CHLOROFORM AND SFRIV/MISSISSIPPI AND CONCW/> 10 will retrieve only the data for chloroform found in the Mississippi River with concentrations greater than 10 μg/L. Table VI gives an example of the use of logical operations.

Thus, the final count (47) represents the total number of occurrences of compounds in the Ohio River at Cincinnati, OH.

For another example of the same type, we can similarly use the "AND" command to link files of the number of times (8093) solvent extraction (SOLV) has been used together with VOA (volatile organic analysis) (1218): 4; whereas the "OR" command gives 9307 as a result of adding all occurrences of SOLV or of VOA (8093 + 1218) and subtracting the number of occurrences of SOLV and VOA together: 9311-4 = 9307 to avoid a double count on both keywords.

More elaborate searches are possible, since every file created as a result of combining fields can subsequently be combined with more such files. We

**Table VI. Use of Logical Operations Online**

| | |
|---|---|
| OPTION ?<br>FILE 1 | SFRIV/OHIO<br>COUNT: 214 |
| OPTION ?<br>FILE 2 | CITY/CINCINNATI<br>COUNT: 320 |
| OPTION ?<br>FILE 3 | #1 or #2<br>COUNT: 487 |
| OPTION ?<br>FILE 4 | #1 AND #2<br>COUNT: 47 |

might ask "how many times has chloroform been detected in finished drinking water at concentrations greater than 5 $\mu$g/L?" (answer, 18) or "how many times have phthalate esters been found at concentrations up to 5 $\mu$g/L in surface rivers?" (324) or "what phthalate derivatives were found in Mississippi River sediments between August 1973 and March 1976, were grab samples, and were identified and confirmed by GC/MS?"

In the following illustrations of applications, retrieval methodology will generally not be explained except where there are problems not covered by the user's manual.

The simplest type of search is simply to enter the keyword mnemonic such as DWFIN in response to the computer terminal's query OPTION? Whereupon the terminal types out File: $\chi$ Count: Z. This is how Table VII was put together (except for Column 4, which was calculated offline).

In Table VII, the total number of occurrences by water type should give the total number of citations in the database (10,600), not 13,085. Unfortu-

**Table VII. Water Type**

| | | No. of Entries | Percent of Total |
|---|---|---|---|
| DWFIN | Drinking, finished | 810 | 6.2 |
| DWUNF | Drinking, unfinished | 50 | 0.4 |
| TAPW | Tapwater | 198 | 1.5 |
| SFLAK | Surface: lake, impoundment, pond, reservoir or freshwater marsh | 1,674 | 12.8 |
| SFRIV | Surface: river, stream, creek, brook, bayou or seaway | 5,970 | 45.6 |
| SFOCE | Surface, ocean, sea, estuary, fjord, gulf, bay, harbor or strait(s) | 444 | 3.4 |
| SFCAN | Surface: canal, channel, aqueduct or drainage ditch | 458 | 3.5 |
| GNDAQ | Groundwater, aquifer | 0 | |
| GNDDE | Groundwater, deep well (greater than 2,000 ft) | 15 | 0.11 |
| GNDSH | Groundwater, shallow well (2,000 ft or less) | 193 | 1.5 |
| PRECP | Precipitation or runoff | 1 | |
| SEDIM | Solids, sediments | 507 | 3.9 |
| PARTI | Solids, suspended particulates | 8 | 0.06 |
| SLUDG | Solids, sewage sludge | 12 | 0.09 |
| INDSL | Solids, industrial sludge | 1 | |
| INDFI | Industrial discharge, final effluent | 701 | 5.4 |
| INDUN | Industrial, untreated | 311 | 2.4 |
| INDMI | Industrial, midtreatment | 418 | 3.2 |
| MUNFI | Municipal discharge, final effluent | 869 | 6.6 |
| MUNUM | Municipal sewage, untreated liquids | 94 | 0.7 |
| MUNMI | Municipal sewage, midtreatment | 351 | 2.7 |
| DEEPW | Deep-well injection | 0 | |
| | Total | 13,085 | |

nately, the current software also counts entries wherein a surface water is listed as a source or recipient of another water type.

Since the database has not yet been programmed to recognize this problem, the following procedure can be used. This information should be included in future versions of the WaterDROP User's Manual.

| | |
|---|---|
| Option ? | SFRIV/MISSISSIPPI |
| File: 1 | COUNT: 845 |
| Option ? | # 1 NOT (MISSISSIPPIRIVERSOURCE OR MISSISSIPPI-RIVERRECIPIENT OR SEDIM) |
| File: 2 | COUNT: 612 |

There are three output formatting options available to the user. Format 1 gives only the unique citation number, the name, some synonyms and the CASRN (Figure 2). Format 2 gives the full citation (Figure 3), and format 4 (not shown) can be specified by the user. Figure 3 illustrates a real citation for a compound in Mississippi River water, but Figures 4 and 5 illustrate citations where the river was the recipient of a final industrial discharge and where it was the source of a finished drinking water. The sequence in which the data are displayed is undergoing improvements for future versions of WaterDROP.

There are 869 unique compounds in the online database. By merging with SANSS (by methods which will have to be explained in future editions of the User's Manual), we can determine this number or the number of unique compounds in any water type and the total number of their occurrences. Table VIII illustrates such retrieved data.

```
OPTION ?                        TYPE 6/1/1        ENTRY NO.

                            FILE NO.       FORMAT NO.
                            FOR SFRIV


                          FILE 6:   ENTRY 1
CITATION NO.:  4003
CAMPHOR (76-22-2)
1,7,7-TRIMETHYLBICYCLO(2.2.1)HEPTAN-2-ONE
2-CAMPHANONE
```

**Figure 2.** WaterDROP output format 1 sample output.

```
OPTION?                    TYPE 6/2/1     SAME ENTRY NO.

                             DIFFERENT (FULL) FORMAT


                     FILE 6:  ENTRY 1
CITATION NO.:  4003
CAMPHOR (76-22-2)
1,7,7-TRIMETHYLBICYCLO(2.2.1)HEPTAN-2-ONE
2-CAMPHANONE

SURFACE WATER FROM LOWER MISSISSIPPI RIVER; FROM BELOW
DISCHARGE POINT; COMPOSITE SAMPLE; SURVEY DONE:  7-7-72;
INDUSTRY TYPE:  2621; ANALYSIS METHOD:  GCMS SPECIFIC
METHOD:  UNSPECIFIED, SPECIFIC METHOD; IDENTIFICATION
CONFIRMED; CONC. NOT REPORTED; WATER SOLUBILITY:  0.125 g/
100 ml AT 25 DEG. C (COLLOIDAL SOLN.); BOILING POINT:  204
DEG C.

LABORATORY CODE:  0005; CODEN:  EPA6-
```

**Figure 3.** WaterDROP output format 2 sample output.

```
OPTION ?                   TYPE 5/2/1

                             FILE NO. FOR
                             MISSISSIPPIRIVERRECIPIENT


                     FILE 5:  ENTRY 1
CITATION NO.:  6001
BIS(2-ETHYLHEXYL) AZELATE (103-24-2)

SURFACE WATER FROM MISSISSIPPI RIVER (RECIPIENT);  FINAL
INDUSTRIAL DISCHARGE WATER; COMPOSITE SAMPLE; SOLVENT EX-
TRACTION (CHLOROFORM) SURVEY DONE:  7-7-73; ST. FRANCISVILLE,
LOUISIANA, USA; INDUSTRY TYPE:  PAPERMILL #1 SIC 2621;
ANALYSIS METHOD:  GCMS SPECIFIC METHOD:  EI, SPECIFIC METHOD;
IDENTIFICATION PROBABLE; CONC. NOT REPORTED

LABORATORY CODE:  0005; CODEN:  EPA1-
```

**Figure 4.** WaterDROP output for a river as the recipient of a final industrial discharge.

```
OPTION ?                    TYPE 4/2/1

                              FILE NO. FOR
                              MISSISSIPPIRIVERSOURCE


FILE 4:  ENTRY 1
CITATION NO.:  19003
O-NITROCHLOROBENZENE (88-73-3)

FINISHED DRINKING WATER; SURFACE WATER FROM MISSISSIPPI
RIVER (SOURCE); ACCUMULATOR SAMPLE (CARBON) SOLVENT EX-
TRACTION; BELOW ST. LOUIS, BUT OTHERWISE UNSPECIFIED;
USA; ANALYSIS METHOD:  UNSPECIFIED; PROBABILITY OF COR-
RECT IDENTIFICATION UNKNOWN; WATER SOLUBILITY:  INSOLUBLE;
BOILING POINT:  245-246 DEG. C; CONC. NOT REPORTED

LABORATORY CODE:  0018; CODEN:  CEPSAB 59,26 (1963)
```

**Figure 5.** WaterDROP output for a river as the source of a finished drinking water.

**Table VIII. Total Number of Unique Compounds per Total Number of Citations with CASRN**

| | |
|---|---|
| Surface Rivers | 304/4850 |
| Finished Drinking Waters | 104/683 |
| Final Industrial Effluents | 215/506 |
| Entire Database | 869/8305 |

Some very interesting searches can be performed rather simply by merging structure fragment codes from SANSS, converted to CASRN, with WaterDROP files. For example, merging CAS/(insert fragment file number from SANSS) and a file number from WaterDROP representing all the occurrences in a water type will give (rather slowly) the number of occurrences of compounds containing that fragment in that water type. Table IX illustrates retrievals of this type for alicyclic esters and for compounds containing a benzene, naphthalene or anthracene nucleus. Table X shows the 51 unique compounds containing the benzene nucleus that were found in finished drinking water.

Table IX. Results of Some Searches Merging SANSS and WaterDROP

| | No. of Unique Compounds | | | | No. of WaterDROP Citations with Those Compounds | | | |
|---|---|---|---|---|---|---|---|---|
| | Whole Data-base | SFRIV | DWFIN | INDFI | Whole Data-base | SFRIV | DWFIN | INDFI |
| Acyclic Esters | 34 | 14 | 7 | 10 | 520 | 341 | 30 | 14 |
| Benzene Nucleus | 276 | 99 | 51 | 61 | 2970 | 1840 | 239 | 135 |
| Naphthalene Nucleus | 39 | 7 | 4 | 8 | 115 | 21 | 9 | 31 |
| Anthracene Nucleus | 4 | 1 | 0 | 0 | 14 | 4 | 0 | 0 |

Table X. Unique Compounds Containing the Benzene Nucleus in Finished Drinking Waters

| | CAS Registry No. |
|---|---|
| 1,1-(2,2,2-Trichloroethylidene)-*bis*-[4-chlorobenzene] | 50-29-3 |
| 2,2′-Methylene-*bis*[3,4,6-trichlorophenol] | 70-30-4 |
| Benzene | 71-43-2 |
| 1,1′-(2,2,2-Trichloroethylidene)-*bis*-[4-methoxybenzene] | 72-43-5 |
| 1,1′-(Dichloroethenylidene)-*bis*-[4-chlorobenzene] | 72-55-9 |
| 1,2-Benzenedicarboxylic Acid, Diethyl Ester | 84-66-2 |
| 1,2-Benzenedicarboxylic Acid, *bis*-(2-Methylpropyl) Ester | 84-69-5 |
| 1,2-Benzenedicarboxylic Acid, Dibutyl Ester | 84-74-2 |
| 1,2-Benzenedicarboxylic Acid, Dihexyl Ester | 84-75-3 |
| 1,2-Benzenedicarboxylic Acid, Butyl Phenylmethyl Ester | 85-68-7 |
| 1,2-Benzenedicarboxylic Acid, 2-Butoxy-2-Oxoethyl Butyl Ester | 85-70-1 |
| Pentachlorophenol | 87-86-5 |
| 3-(1,1-Dimethylethyl)-4-methoxyphenol | 88-32-4 |
| 1-Chloro-2-nitrobenzene | 88-73-3 |
| 2-Methoxyphenol | 90-05-1 |
| 1,2-Dimethoxybenzene | 91-16-7 |
| Benzoic Acid, Methyl Ester | 93-58-3 |
| 1,2-Dimethylbenzene | 95-47-6 |
| 1-Phenylethanone | 98-86-2 |
| Nitrobenzene | 98-95-3 |
| Ethylbenzene | 100-41-4 |
| Ethenylbenzene | 100-42-5 |
| Benzaldehyde | 100-52-7 |
| Propylbenzene | 103-65-1 |
| Butylbenzene | 104-51-8 |
| Bromobenzene | 108-86-1 |

**Table X, continued**

| | CAS Registry No. |
|---|---|
| Methylbenzene | 108-88-3 |
| Chlorobenzene | 108-90-7 |
| Phosphoric Acid, Triphenyl Ester | 115-86-6 |
| 1,2-Benzenedicarboxylic Acid, *bis*-(2-Ethylhexyl) Ester | 117-81-7 |
| Hexachlorobenzene | 118-74-1 |
| 2-(1,1-Dimethylethyl)-4-methoxyphenol | 121-00-6 |
| 1-Chloro-3-nitrobenzene | 121-73-3 |
| 2,6-*bis*-(1,1-Dimethylethyl)-4-methylphenol | 128-37-0 |
| 1,2-Benzenedicarboxylic Acid, Dimethyl Ester | 131-11-3 |
| 1,2-Benzenedicarboxylic Acid, Dipropyl Ester | 131-16-8 |
| 2,6-*bis*-(1,1-Dimethylethyl)-4-methoxyphenol | 489-01-0 |
| 1,3-Dichlorobenzene | 541-73-1 |
| 2-Methyl-1,3-dinitrobenzene | 606-20-2 |
| 1-Ethyl-2-methylbenzene | 611-14-3 |
| 1-Ethyl-3-methylbenzene | 620-14-4 |
| 2-Methyl-1,1′-biphenyl | 643-58-3 |
| 3,5-*bis*-(1,1-Dimethylethyl)-4-hydroxybenzoic Acid | 1421-49-4 |
| 2,3,5,6-Tetrachloro-1,4-benzenedicarboxylic Acid, Dimethyl Ester | 1861-32-1 |
| 2-(1,1-Dimethylethyl)-4-methyl Phenol | 2409-55-4 |
| 2,6-*bis*-(1,1-Dimethylethyl)-4-ethylphenol | 4130-42-1 |
| 2-Chloro-N-(2,6-diethylphenyl)-N-(methoxymethyl)-acetamide | 15972-60-8 |
| N-(Butoxymethyl)-2-chloro-N-(2,6-diethylphenyl)-acetamide | 23184-66-9 |
| Dichlorobenzene | 25321-22-6 |
| Dibromobenzene | 26249-12-7 |
| Bromochlorobenzene | 28906-38-9 |

We have not yet mastered all the ways that SANSS can be used to retrieve information in classes of compounds. Table XI shows more attempts by using certain coded structural fragments from SANSS.

Some other kinds of very simple searches are shown in Table XII, which depicts the number of occurrences of various categories of sampling and location information. Table XIII shows the distribution of confidence ratings.

It is not too much more complicated to generate Table XIV, but since more than one identification technique has been used in some citations, the total exceeds the total number of citations. For example, IR has seldom been used alone (actually, about 50 times compared to a total number of over 1000 times in the entire database).

A somewhat more complex search (Tables XV and XVI) tried to find the chronological distribution of identification techniques (MS vs non-MS meth-

**Table XI. Number of Unique Compounds Belonging to Specific Classes in WaterDROP (Found by Merging WaterDROP with SANSS)**

| | Entire Database | Surface Rivers | Finished Drinking Water | Final Industrial Effluent |
|---|---|---|---|---|
| Certain Saturated Halocarbons | 51 | 25 | 24 | 4 |
| $\sim C(X)_2-X$ (e.g., chloroform) | 17 | 12 | 11 | 2 |
| Primary Halide | 20 | 7 | 8 | 2 |
| Secondary Halide | 5 | 1 | 1 | 0 |
| Tertiary Halide | 0 | 0 | 0 | 0 |
| Geminate Dihalide, Secondary Carbon | 1 | 0 | 0 | 0 |
| Geminate Dihalide, Primary Carbon | 8 | 5 | 4 | 2 |
| Certain Ketones and Aldehydes | 72 | 19 | 9 | 11 |
| Acycylic Aldehydes | 8 | 6 | 3 | 0 |
| Ketones Wherein the Carbonyl Is Not Attached to a Ring | 20 | 5 | 3 | 3 |
| Cyclic Ketones | 44 | 8 | 3 | 8 |

**Table XII. Sampling and Location Information in WaterDROP**

| Category | Occurrences |
|---|---|
| Sampling Methods | |
| Grab | 7028 |
| Composite | 692 |
| Accumulator (Resin and Carbon) | 1954 |
| Volatile Organic Analysis | 1218 |
| Solvent Extraction | 8123 |
| Dates | |
| Monitoring Start/End | 4593 |
| Survey | 5910 |
| Locations | |
| City | 8278 |
| County | 451 |
| State | 9445 |
| Country | 9675 |
| Source of Contaminant | 620 |

ods) and their confidence ratings. Again, the total exceeded 10,600. The reason in this instance is that when a paper has given only a range of years for the sampling date, the citations (or records) for each compound identified will be counted as hits in each and every year of the range.

**Table XIII. Distribution of Confidence Rating Assignments**

| | |
|---|---|
| Confirmed | 1097 |
| Probable | 7177 |
| Unconfirmed | 1030 |
| Unknown | 485 |

**Table XIV. Distribution of Identification Methods Used for WaterDROP Citations (as of July 30, 1980)**

| Method | No. of Times Used |
|---|---|
| Mass Spectrometry | 87 |
| Gas Chromatography/Mass Spectrometry | 6117 |
| Mass Spectrometry Subtotal | 6207 (58.5%)[a] |
| Flame Ionization Detector | 1064 |
| Electron Capture | 3029 |
| Flame Photometric Detector | 21 |
| Alkali Bead or Flame Detector | 0 |
| Electrolytic Conductivity | 0 |
| Microcoulometric Titration (which should be moved from "Other" to this field) | 2650 |
| Infrared Spectrometry | 1063 |
| Ultraviolet Spectrometry | 65 |
| Nuclear Magnetic Resonance | 6 |
| Fluorescence | 2 |
| Thermal Conductivity Detector | 37 |
| Colorimetry | 65 |

[a]94.2% since 1970.

**Table XV. Identification Using Mass Spectrometry**

| Date | Confidence Rating | | |
|---|---|---|---|
| | Confirmed | Probable | Unconfirmed |
| Before 1970 | 43 | 84 | 7 |
| 1970 | 16 | 122 | 2 |
| 1971 | 92 | 183 | 2 |
| 1972 | 327 | 418 | 90 |
| 1973 | 256 | 310 | 8 |
| 1974 | 652 | 618 | 9 |
| 1975 | 222 | 2422 | 18 |
| 1976 | 10 | 2165 | 0 |
| 1977 | 136 | 12 | 0 |

Table XVI. Identification Not Using Mass Spectrometry

| Date | Confidence Rating | | |
|---|---|---|---|
| | Confirmed | Probable | Unconfirmed |
| Before 1970 | 2 | 2961 | 652 |
| 1970 | 1 | 0 | 7 |
| 1971 | 0 | 93 | 1 |
| 1972 | 0 | 18 | 170 |
| 1973 | 0 | 54 | 0 |
| 1974 | 2 | 80 | 0 |
| 1975 | 10 | 0 | 0 |
| 1976 | 0 | 26 | 0 |
| 1977 | 0 | 0 | 0 |

Figures 6 through 11 depict the frequencies of occurrence of dieldrin, heptachlor, chloroform, trichloroethylene, toluene and phthalate esters in surface rivers and finished drinking water. It would be misleading from these histograms to conclude that phthalate esters occur more frequently in finished drinking waters than does chloroform. Two important studies that include chloroform data are not included. The National Organic Reconnaissance Survey (NORS) data on drinking waters have been processed but are not online. We have not yet received permission from EPA to include the National Organic Monitoring Survey (NOMS) data, but we believe we have already processed all the other major compilations of organics in U.S. drinking waters.

For people who do not have access to a terminal, a hard copy of the data in a few formats can be obtained. A paper version for the present online database is already several inches thick, and future plans call for microfiche versions to be able to accommodate all of the data.

## ACKNOWLEDGMENT

This work was supported by a subcontract with Fein-Marquart Associates, Inc., whose prime Contract No. 68-03-2711 is with the U.S. Environmental Protection Agency.

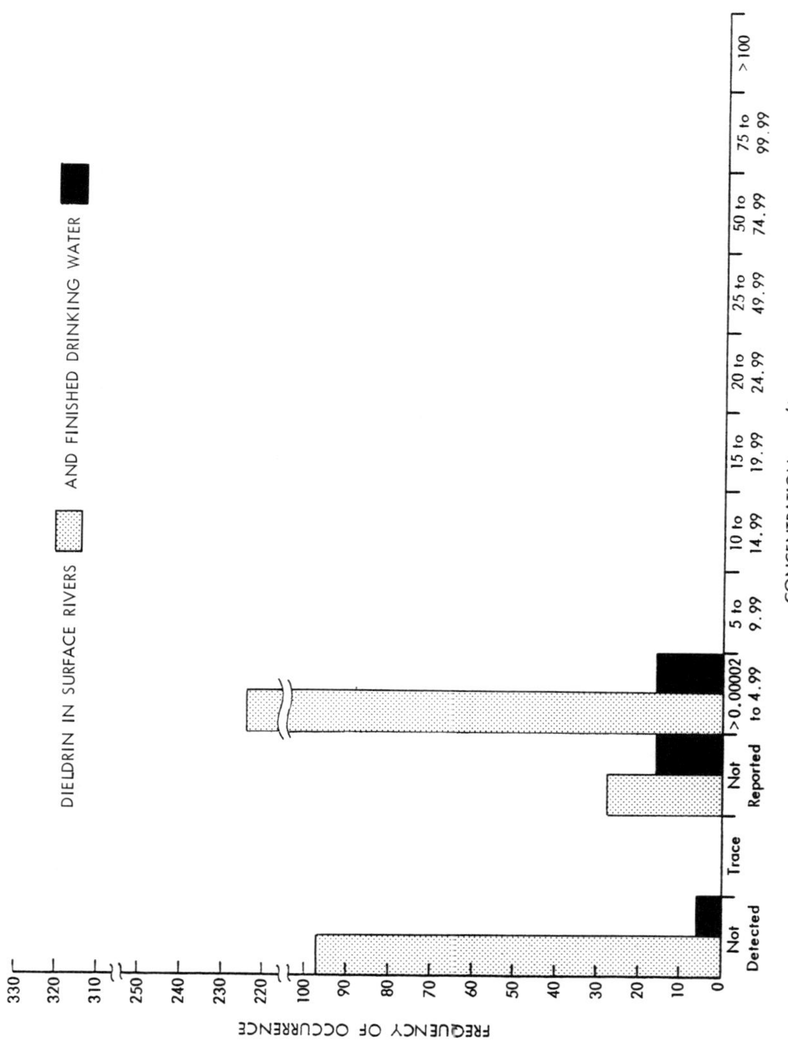


**Figure 6.** Frequency of occurrence of dieldrin in surface rivers and finished drinking water.

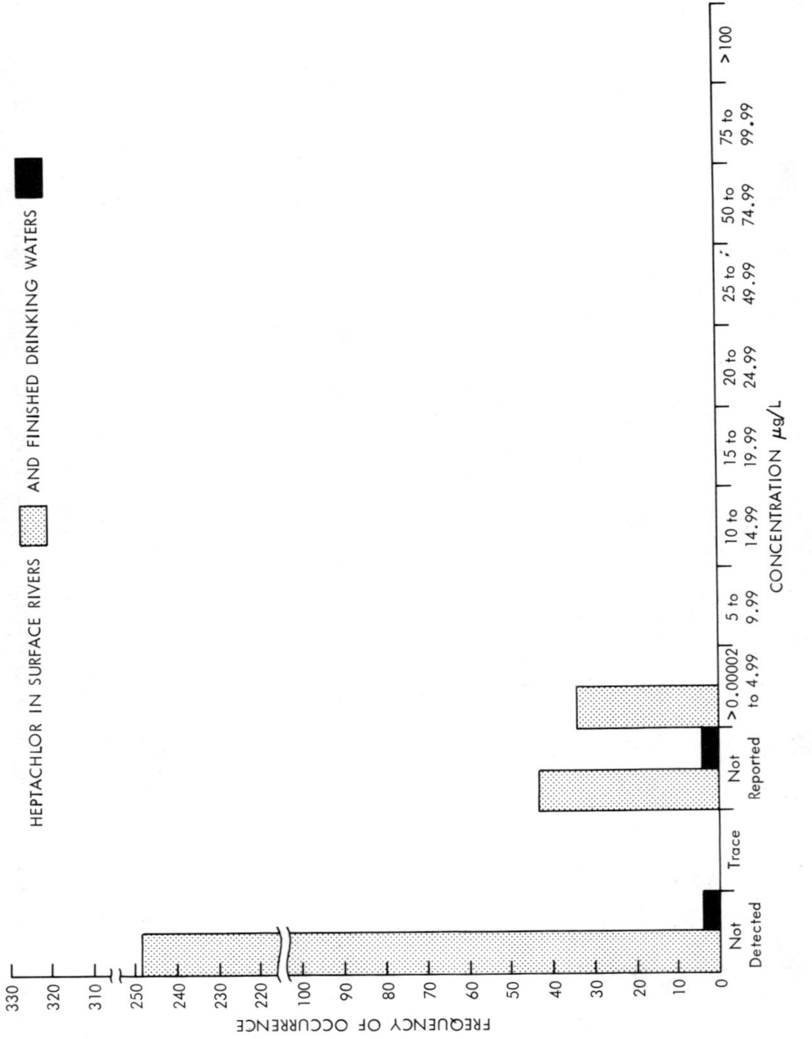


**Figure 7.** Frequency of occurrence of heptachlor in surface rivers and finished drinking water.

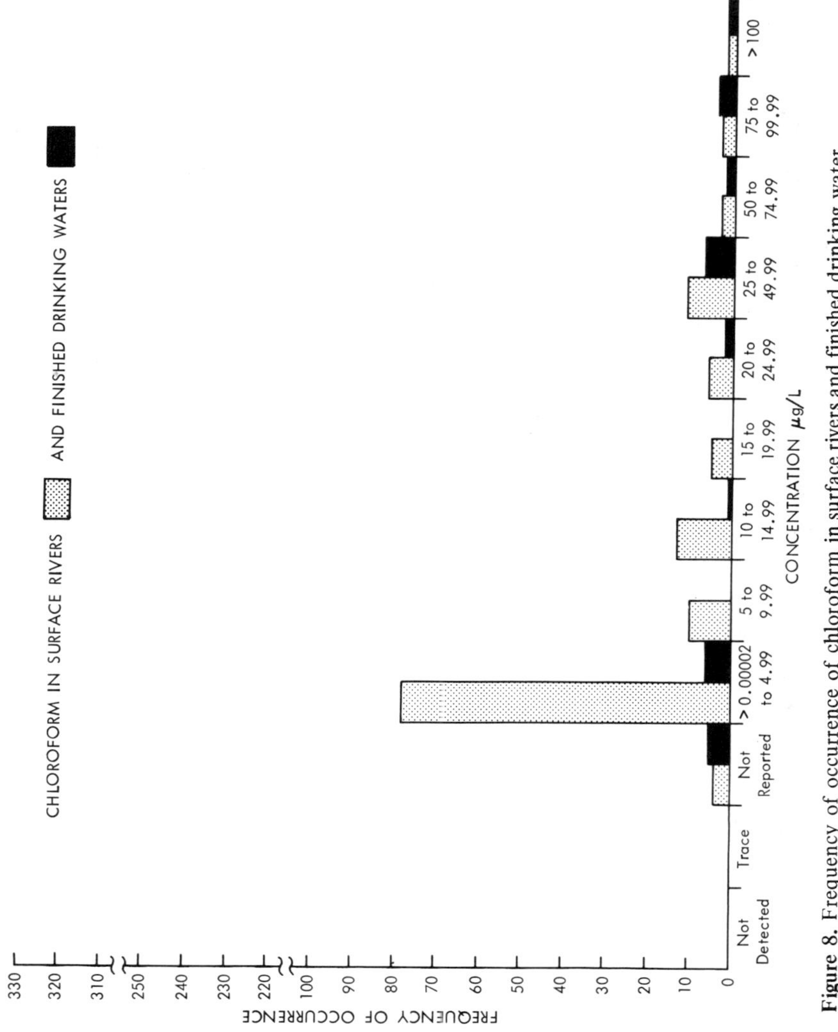


**Figure 8.** Frequency of occurrence of chloroform in surface rivers and finished drinking water.

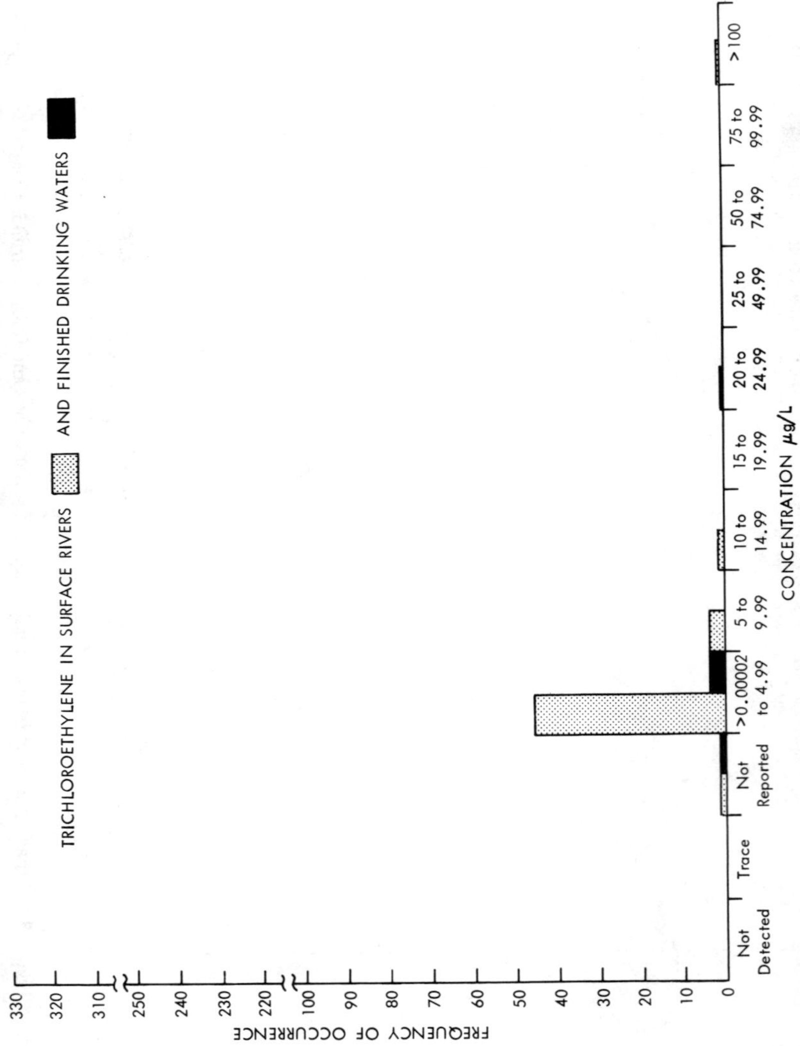


**Figure 9.** Frequency of occurrence of trichloroethylene in surface rivers and finished drinking water.

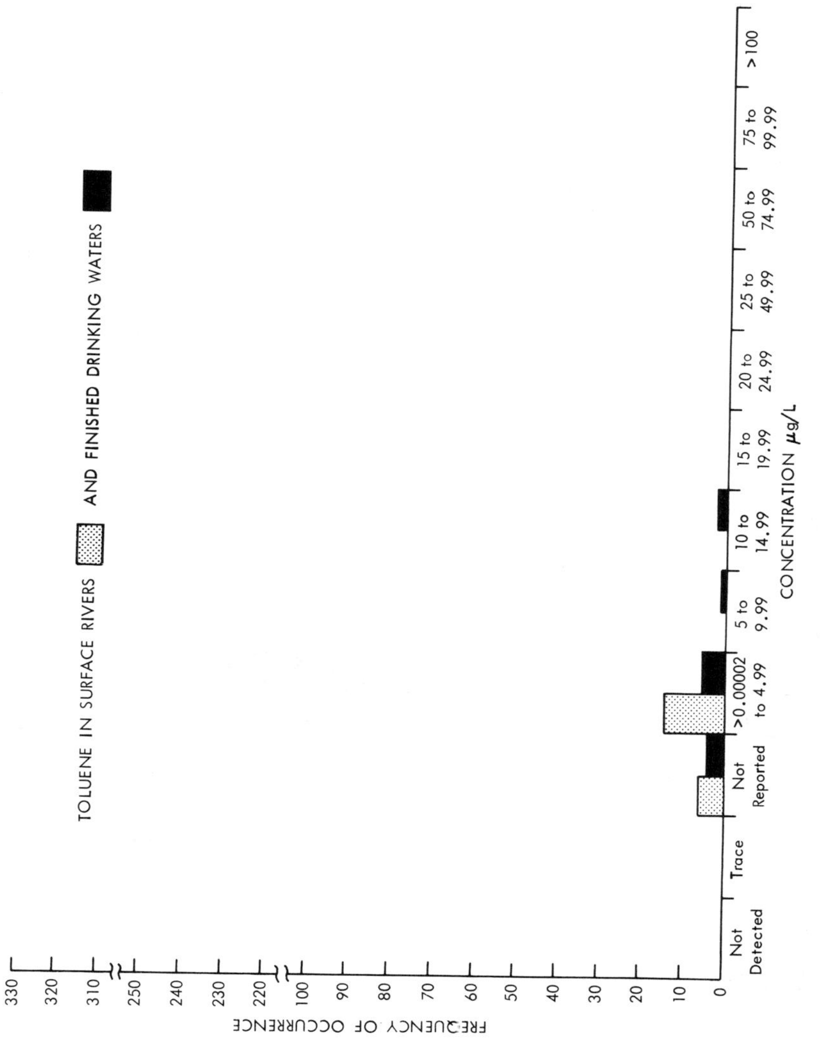


**Figure 10.** Frequency of occurrence of toluene in surface rivers and finished drinking water.

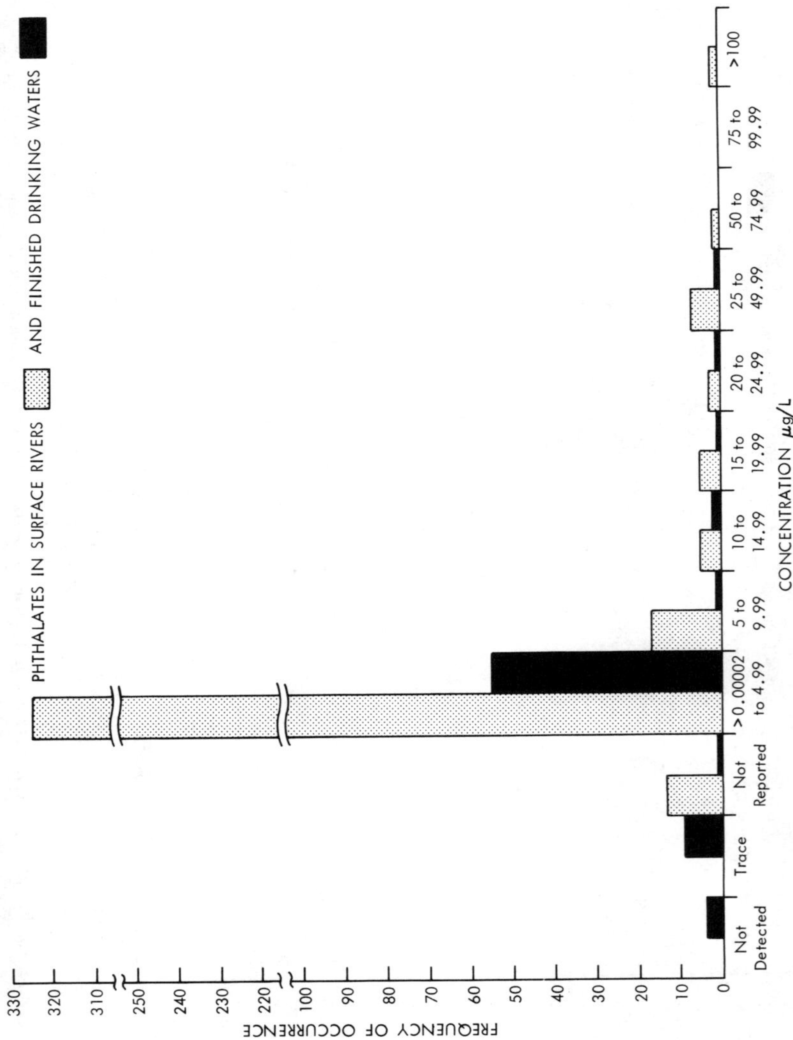


**Figure 11.** Frequency of occurrence of phthalates in surface rivers and finished drinking water.

## REFERENCES

1. Shackelford, W. M., and L. H. Keith. "Frequency of Organic Compounds Identified in Water," U.S. EPA Report EPA-600/4-76-062, PB-265 470, National Technical Information Service, Springfield, VA (1976).
2. Garrison, A. W., L. H. Keith and W. M. Shackelford. "Occurrence, Registry, and Classification of Organic Pollutants in Water, with Development of a Master Scheme for Their Analysis," in *Aquatic Pollutants: Transformation and Biological Effects*, C. Hutzinger, I. H. van Lelyveld and B. C. J. Zoeteman, Eds. (Oxford: Pergamon Press, 1978).
3. Junk, G. A., and S. E. Stanley. "Organics in Drinking Water. Part I. Listing of Identified Chemicals," prepared for the U.S. Energy Research and Development Administration, IS-3671, Distribution Category UC4, National Technical Information Service, Springfield, VA (1975).
4. Commission of the European Communities, Cost Project 64B, Analysis of Organic Micropollutants in Water Management Committee. "A Comprehensive List of Organic Pollutants Which Have Been Identified in Various Surface Waters, Effluent Discharges, Aquatic Animals and Plants, and Bottom Sediments," Water Research Centre, Stevenage Laboratory, Stevenage, Hertfordshire, England (1975).

## CHAPTER 33

# COMPUTER SURVEY OF GAS CHROMATOGRAPHY/ MASS SPECTROMETRY DATA ACQUIRED IN THE U.S. ENVIRONMENTAL PROTECTION AGENCY SCREENING ANALYSIS: SYSTEM AND RESULTS

**W. M. Shackelford and D. M. Cline**

U.S. Environmental Protection Agency
Environmental Research Laboratory
Athens, Georgia

**L. Burchfield, L. Faas,**
**G. Kurth and A. D. Sauter***

Computer Science Corporation
Falls Church, Virginia

In June 1976 the U.S. Environmental Protection Agency (EPA), as a result of court action by several environmental action groups, was directed by a Consent Decree from the U.S. District Court in the District of Columbia to assess the wastewater of 21 industrial categories for 65 chemical substances and prescribe the best available treatment (BAT) for the effluent. To begin the task, a scheme for analysis of the wastewaters for the 65 substances had to be designed.

---

*Present address: U.S. Environmental Protection Agency, Environmental Systems Laboratory, Las Vegas, NV.

Although some of the 65 substances were unique chemical compounds, many included whole classes of compounds, e.g., polynuclear aromatic hydrocarbons (PAH). Realizing that these classes of compounds could contain literally hundreds of individual members, EPA included for analysis only those members that had previously been identified a significant number of times, were produced in quantity by industry and were available as analytical standards. The now familiar 129-compound priority pollutant list was the result of this work. Details of the list-resolving procedure may be found elsewhere [1,2].

Even though the list of 129 specific substances made the analysis task manageable, the plaintiffs in the court action were concerned that some members of the chemical classes not on the 129-compound list would be missed in the analysis procedure. Because it was generally agreed that computerized gas chromatography/mass spectrometry (GC/MS) would be the analysis tool of choice, the advantage of saving all raw GC/MS data for later processing to look for compounds other than the priority pollutants became obvious. State-of-the-art GC/MS instrumentation includes a computer system; thus, the data could be saved in computer-readable format for later study. Because magnetic tape is the cheapest mass storage medium, it was chosen for saving all GC/MS data from sample analysis.

Initial analysis of each sample at the laboratories operating under EPA contract was to be directed only toward compounds among the 129 priority pollutants. Although EPA could have contracted for a general survey of all compounds in each sample, a number of limiting factors precluded this approach:

1. Cost of a general survey was estimated at $2000 per sample vs $700 for the limited analysis.
2. Time was extremely important. Although the data acquisition times for general survey and specific analysis are the same, data evalution times could be 5–10 times longer for survey analysis (if only computer matching is required for identification). Decreasing the number of samples per unit time by a factor of 5–10 would have played havoc with the court-ordered deadlines.
3. Management of the large volumes of unconfirmed data would have required a massive secondary effort to confirm and collate the results of the survey analysis.

By requiring that all data from each GC/MS acquisition be sent to a central location for survey processing, EPA could assure proper management of non-priority pollutant data while at the same time obtain timely response directly from contractor laboratories on the priority pollutants for a reasonable cost. Because all parties involved in the Consent Decree had agreed that the non-priority pollutant data were of less immediate need, no part of the spirit of the Consent Decree was sacrificed; yet provision was made for assessing all the data for compounds other than the 129.

A second provision for possible later analysis of the sample was the requirement for the analysis laboratories to supply each sample extract along with the GC/MS data. Thus, should some compound be identified tentatively in the GC/MS data, it could be confirmed by reanalysis of the corresponding extract. Also, reoccurring components that could not be identified from their mass spectra could possibly be identified using another analysis technique on the saved extract.

The Athens, GA, Environmental Research Laboratory had participated in the formation of the priority pollutant list [1] and with the development of the analysis scheme [2]. Responsibility for developing a program for studying the GC/MS data was given to the Athens laboratory in August 1978. A second program to confirm any tentative mass spectra identifications began in November 1979.

## TASK DESCRIPTION

The screening analysis phase of BAT review was expected to require the qualitative/semiquantitative analysis of about 4000 samples. Each sample analysis could be expected to involve GC/MS data acquisition for at least five fractions: a volatile organics analysis (VOA), a VOA blank, an extractable base/neutral (B/N), an extractable acid (ACI) and a direct aqueous injection (DAI). Other blank, standard and pesticide confirmation runs also could be expected. All calculations for the task were based on 20,000 GC/MS runs, i.e., 4000 samples x 5 fractions. When it is considered that each GC/MS run can be expected to contain some 500–1000 individual spectra, the magnitude of the task of evaluating this data is evident.

Implicit in this task was the development of a computer system that could evaluate these data in a manner comparable to a human using computer-aided spectrum extraction and spectra matching to identify tentatively all sample components. An additional goal of the task was to point out those spectra that did not match any spectrum in the reference library yet are seen in multiple GC/MS runs. Thus, a library of compounds tentatively identified in each industrial category as well as a library of recurring but unidentified spectra were to be generated for use in effluent regulation. In addition, the data in these libraries will be studied in a subsequent project in which the saved extracts will be reanalyzed. Tentative identifications made in this project will be confirmed by comparison with standards, and recurring but unidentified spectra will be examined for ab initio determination of compound identity.

## EXPERIMENTAL

### System Description

The PDP 11/70-based GC/MS Data Survey System consists of computer hardware and software necessary to:

1. inventory all incoming magnetic tapes and sample extracts;
2. copy the data on each magnetic tape to a second tape in an internal use format and plot the reconstructed gas chromatogram;
3. retrieve data as necessary from tapes in batch mode;
4. extract the spectra of components in each GC/MS run from the background spectra in the run;
5. match the extracted spectra with a library of reference spectra;
6. check to see whether matched spectra have been seen before under the same circumstances;
7. check spectra that are not matched against their fellow unmatched spectra;
8. generate reports on the numbers of matched spectra by industry, fraction type, analytical laboratory, GC/MS run conditions, etc.;
9. provide graphics capability necessary to view the data from any run; and
10. search any run for specific compounds.

### Instrumentation

Initially, all work for this project used the computer system shown in Figure 1. Basically, it consists of a DEC PDP 11/70 with 256K words of memory and 2K words of cache memory, two 88-Mbyte disk drives (DEC RP04), two magnetic tape drives (DEC TU16), a Calcomp 936 plotter, and a Tektronics 4014 graphics display terminal. Recently, all run processing functions were transferred to a PDP 11/34 multiuser dedicated system with 128K words of memory and 2K words of cache memory, two 170-Mbyte disk drives (DEC RP06), and a Versatec 1200 printer/plotter. This includes spectrum extraction, library matching and collation of matched and unmatched spectra. A flowchart of these software components is shown in Figure 2. An average of 200 runs per week is processed through the system.

Data submitted to EPA for study was acquired by 18 laboratories using 5 different systems:

1. Finnigan 3200 MS, Varian 1400 GC, System Industries data system;
2. Finnigan 4000 MS, Finnigan 9600 GC, Finnigan-Incos data system;

3. Dupont Dimaspec MS, Finnigan 9600 GC, Finnigan-Incos data system;
4. Hewlett-Packard 5985 GC/MS/data system; and
5. Hewlett-Packard 5982 GC/MS/data system.

## Inventory System

To inventory and track the 20,000 GC/MS data runs and the estimated 12,000 extracts (B/N, ACI and pesticide fractions for each of 4000 samples), a database management system has been implemented. This system is the INFORM management program and is a well-known tool for database management. In INFORM are kept the GC/MS data run descriptors that allow the physical locating of each run and corresponding extract and also all the information known about the sample.

As each magnetic tape or extract is received at the Athens laboratory it is manually entered into the INFORM system. Important parameters that are entered for each data run are the tape on which it is found, the EPA sample number, an Athens laboratory run number, the fraction type and various GC/MS parameters. The corresponding data for the extract will have all of this information and additionally the precise location of the extract in a freezer.

During the inventory process, data are copied from the tape of the submitting contractor laboratory onto an Athens tape in a format that is both more space-efficient and damage-resistant. Thus, the original tape and a backup copy are saved. The backup copy, which has only the Athens laboratory number for identification of each run, is used for all data processing needs. Confidentiality of the data is maintained through the use of the backup copy so that descriptive data are not associated with the GC/MS data. Software that has access to both descriptive and GC/MS data is password-protected.

At the time of tape conversion from contractor format to Athens laboratory format, the data of each run are scanned, and a reconstructed gas chromatogram (RGC) is plotted. The RGC are then bound in volumes to serve as references at the time runs are submitted for analysis. Inspection of the RGC by a chemist may result in discarding the corresponding run because of obvious flaws such as absence of peaks or premature end of data.

When data are to be processed, the chemist identifies the runs that have passed visual inspection for processing. Software then retrieves the designated runs from the magnetic tape and prepares each in turn for processing by the analytical system. The inventory system is reapplied when the run has been processed and descriptors contained in INFORM are necessary for reporting.

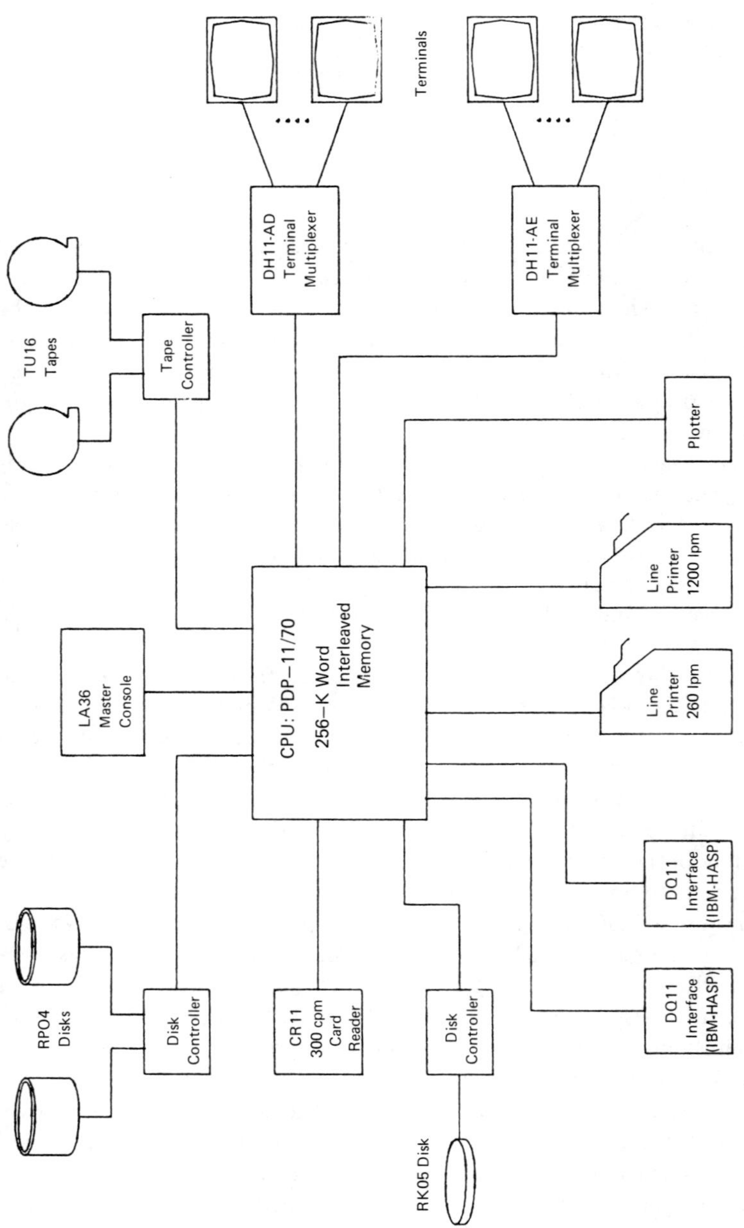


**Figure 1.** Athens, GA, EPA computer hardware configuration.

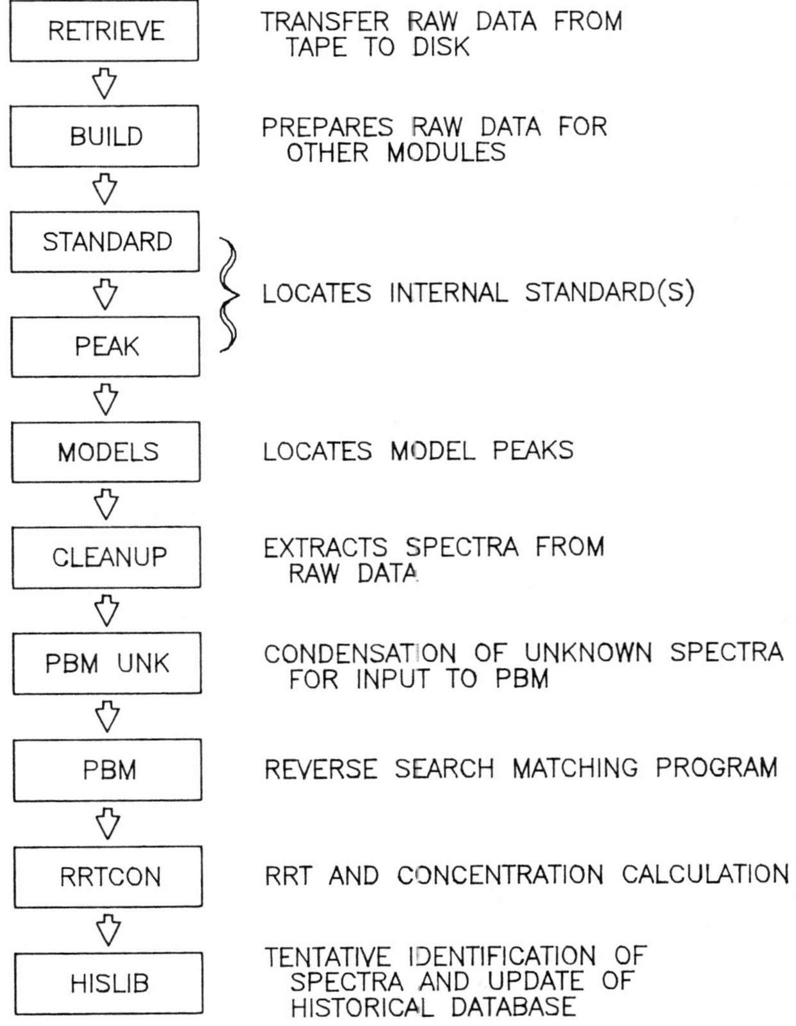


**Figure 2.** Flowchart of software components for run processing.

## Analytical System

The analytical system consists of four main parts: the internal standard locator, PEAK; the peak or spectrum extractor, CLEANUP; the spectrum matching system, PBM; and the result collator, HISLIB.* Chemists have opportunities at various points during the process to make decisions that can end processing or affect the further processing of any given component spectrum. The ideal data analysis, however, proceeds with minimal operator intervention. Only when the analytical system is presented with decisions that it is unqualified to make does the chemist intervene.

The program PEAK was developed to assure the identification of internal standard location in each run. Because all subsequent processing of the data requires knowledge of the internal standard in the run, it was imperative that software be available that would unambiguously define the location and area of the internal standard peak in each data run. The algorithm is discussed by Sauter et al. [6].

CLEANUP is a system of programs developed at Stanford University [4] that finds and extracts the spectra of components in the GC/MS data run. Successive 16-scan windows are searched for ion peaks that have two ascending points, a maximum and two descending points. When an ion peak is found, successive ions from mass 40 to 400 are checked to see whether any maximize within a distance of ±1 scan number of the first found peak. When eight or more such masses maximize simultaneously, a component peak is said to be detected. In this case all the masses maximizing at this point are collected, their areas are normalized to the largest mass of the group, and they are passed along to the next phase of the analysis as a mass spectrum.

CLEANUP involves a number of checks to ensure that such artifacts as column bleed, noise spikes and background are not chosen as sample components. Criteria are input at the start of processing to ensure that only ion peaks of a defined sharpness will be considered. This will normally eliminate peaks caused by column bleed, which usually shows up in the form of broad peaks. Noise spikes, which are generally of only one- or two-scan duration, are guarded against by requiring a minimum of four scans in the ion peaks. Instrumental background noise caused by pump oil or other contaminants normally does not peak during a run; therefore, it does not interfere with the CLEANUP process.

---

*HISLIB is the name used by Smith et al. [3] to describe their historical library software. The Athens laboratory-developed historical library uses a similar concept to that of Smith et al., but does not rely on any of their software. CLEANUP and PBM are names of computer programs reported in Dromey et al. [4] and Pesyna et al. [5], respectively, and are used with slight modification only.

A preprocessor to CLEANUP called MODELS searches the data for "model" peaks with four ascending and four descending points. Once the model peaks are found, they are used to define the peak shape that a component should have in the region close to the model. A polynomial curve-fitting technique is used to redefine the shape of the peak of interest. As one would expect, the model peaks toward the end of the GC/MS run are broader than the models toward the front. Thus, the "real-life" conditions of peak shape are taken into account by CLEANUP.

Once a spectrum has been identified by CLEANUP, the area of the base peak is calculated. This is later used to calculate the concentration of the component in the sample by ratioing it with the area of the base peak of the internal standard.

The spectra extracted by CLEANUP are then passed to PBM, a library matching program developed at Cornell University under an EPA grant [5]. PBM (probability-based matching) employs a reverse-search technique to compare a reference library of condensed spectra to a similarly condensed unknown spectrum. The algorithm, which is discussed in detail elsewhere [5,7], reports several match parameters by which to evaluate the quality of the matches. An overall match confidence or K value is calculated and compared to $K_{max}$, where $K_{max}$ is the value that can be achieved if all intensities fall within proper windows. These parameters are compared by calculation of $\Delta K$ where

$$\Delta K = K_{max} - K \tag{1}$$

A measure of spectrum contamination is calculated which represents the scaled intensities of ions in the unknown spectrum that are not matched in the reference spectrum and of matched ions in the unknown spectrum that fall outside the intensity tolerances used by PBM. The ratio of the above values to the sum of all scaled intensities for the ions studied (usually the 10 largest ions in the condensed spectrum) is termed "percent contamination." If percent contamination is large, more than one component may be present in the unknown spectrum.

Because the presence of the molecular ion is often helpful in differentiating among the spectra of compounds in a homologous series, PBM notes the presence of a matched molecular ion with a flag (a "+" printed with the K value).

A good match, then, has a large K and a small $\Delta K$, a low value for percent contamination, and a "+" printed with the K value. In this work, the ratio $K/(K + \Delta K)$ is calculated. Values of this ratio that are greater than 0.45 have been found empirically to indicate reasonable matches.

The 10 best PBM matches for each unknown spectrum are saved and passed to the historical library program. This algorithm looks at the relative retention time (RRT) of each component and the spectrum matches that pass certain match quality criteria ($K/(K + \Delta K) > 0.45$, contamination less than 60%) and attempts to match these against entries in the historical library. Each entry in the historical library contains an RRT window, the Chemical Abstracts Service Registry Number (CASRN) and the applicable GC column and internal standard. Therefore, when a match from the spectrum library matches the CASRN and the RRT falls within the window specified in HISLIB, a tentative identification is made. This tentative identification is added to the historical library for reporting purposes.

Of the remaining spectra, those that contain at least one PBM match for which $K/(K + \Delta K) > 0.40$ and percent contamination is less than 70% are passed to the chemist for inspection, and the rest are passed to the MISS library. The results of the chemist's examination may either cause a new entry to be made in the historical library or result in discarding of the PBM results. All new entries are confirmed by viewing the spectrum passed by CLEANUP in comparison with library matches suggested by PBM. This is done using a Tektronix 4014 graphics terminal.

These discarded spectra now join those from previous runs that did not match any spectrum in the library of reference spectra. The eight most intense ions for each spectrum are saved, and these unmatched spectra are put in the MISS file in order of their RRT. Each spectrum is matched against the spectra in the MISS file that have similar RRT values. If an ion of 75-100% relative intensity (RI) is present that agrees with the base peak in the MISS library and six of the remaining ions agree within broad window ranges, a match occurs. In this way unidentified spectra that appear to recur are cataloged. Figure 3 shows the logic used in taking a spectrum through the HISLIB procedure.

## Reporting

Reporting is accomplished in two ways. The first system is a series of hard copy outputs that describe the flow of data through the total system and the results generated from the data. The contents of the historical library can be printed out either in totality or as a listing of unique entries. The data may be sorted by parameters such as CASRN, RRT, GC column, analysis laboratory, industrial category or relative concentration.

A second method for reporting is a graphics system that allows the chemist to recall data and plot it in various ways. For instance the raw data for a spectrum, the cleaned-up spectrum and the reference spectrum can all be

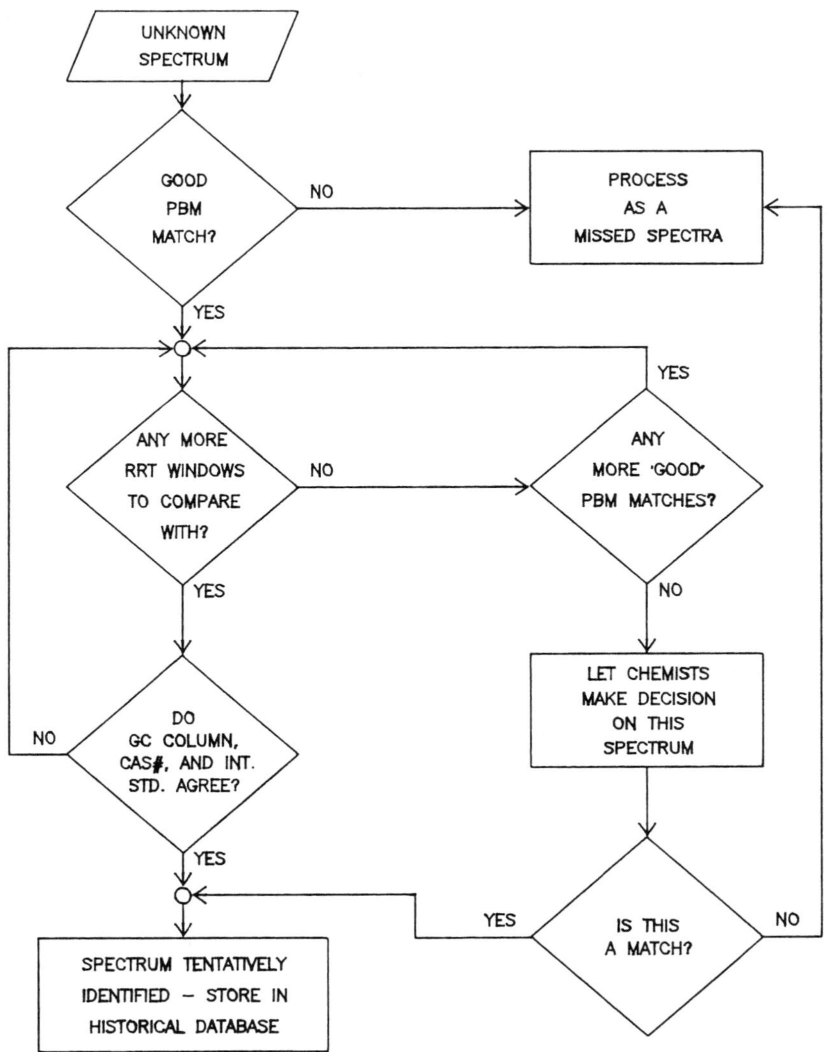


**Figure 3.** Logic used in processing a spectrum through HISLIB.

plotted on the same screen simultaneously. The extracted ion current profile (EICP) for any ion can be plotted between any scan limits. Multiple EICP plots can be displayed on the screen. The graphics system is used by the chemists to evaluate ambiguous results from the computer analysis.

## RESULTS AND DISCUSSION

### Spectrum Extraction

The extraction of information containing spectra from the mass of spectra in a GC/MS run is the key to a successful automated system for GC/MS data analysis. Figure 4 shows an RGC of a group of 11 phenols. The scans 213 and 215 can be seen to be on opposing sides of an apparent single-component peak. Manual subtraction of a baseline spectrum (e.g., 208 or 220) from spectrum 214 results in a spectrum that is not recognizable as any of the components injected. Using CLEANUP to find spectra, however, scans 213 and 215 reveal that the peak is actually the sum of two components. Figure 5 shows the resultant spectra of 213 and 215. Also depicted are spectra from the reference library that establish the identity of the two components. Obviously, this example is an ideal case because standards were used with no interferences, but it serves to illustrate the ability of CLEANUP to separate components eluting within two scans of each other.

A second example, Figure 6, shows an actual acid extract from an industrial effluent. Note that spectra were extracted and tentatively identified for 10 components as well as the internal standard, $d_{10}$-anthracene. Table I shows a comparison of component finding ability for manual vs automated techniques. Note that the automated system in each case discerned more components than the manual interpreter. Identification by spectrum matching and manual confirmation are given also. It is interesting to note in these cases that although both manual and automated methods identified the same components (ID), the automated method, by virtue of more discerned components, identified additional components (ID in parentheses).

The data shown thus far indicate that for the systems studied, automated techniques are at least equal to manual techniques for pointing out components in the run and identifying them by spectrum matching with a reference library. Before becoming too optimistic, however, it is best to look at cases where spectrum extraction and identification are not so clear cut as in those given above. It can be seen from Table I that despite the fact that more peaks are found with the automated method, the ratio of identifications to peaks has decreased. In fact, as the number of components in a run increases, identification becomes more difficult even though the automated system is able apparently to deliver a spectrum for each component. Figure 7 shows the relationship of identifications made to components indicated.

One of the reasons for loss in identification efficiency can be laid at the doorstep of CLEANUP. The components that are found that are missed by manual techniques are generally either very small peaks or shoulders on other

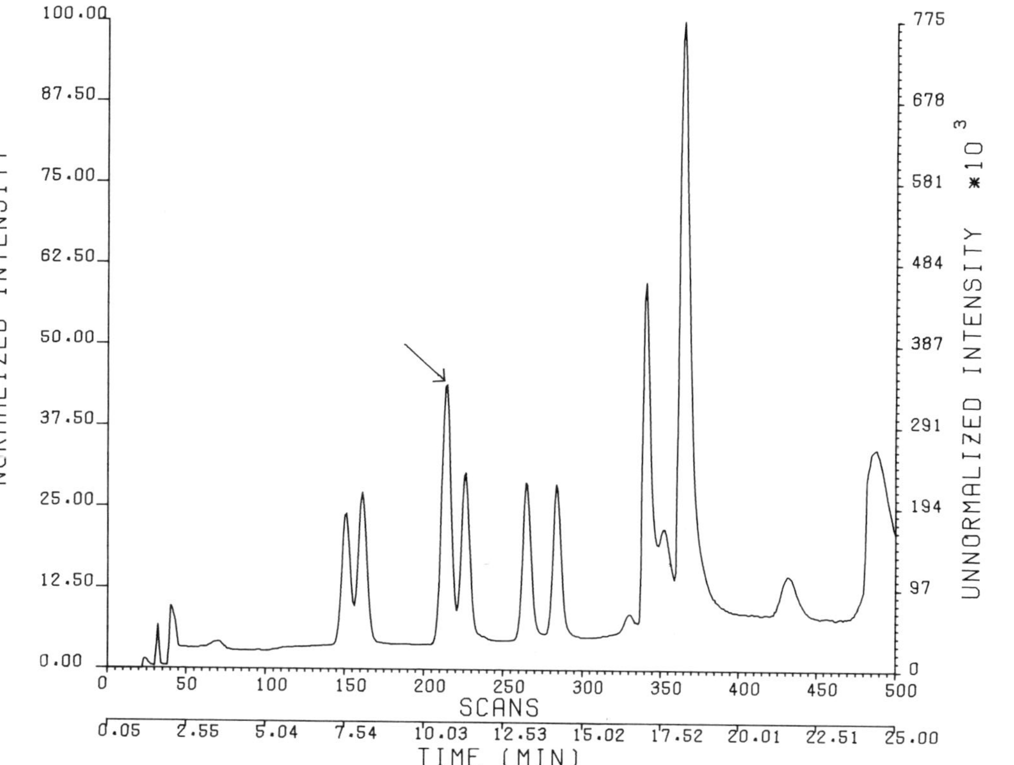


**Figure 4.** RGC of 11-component phenol standard. Arrow indicates apparent single component peak that is actually the sum of two components.

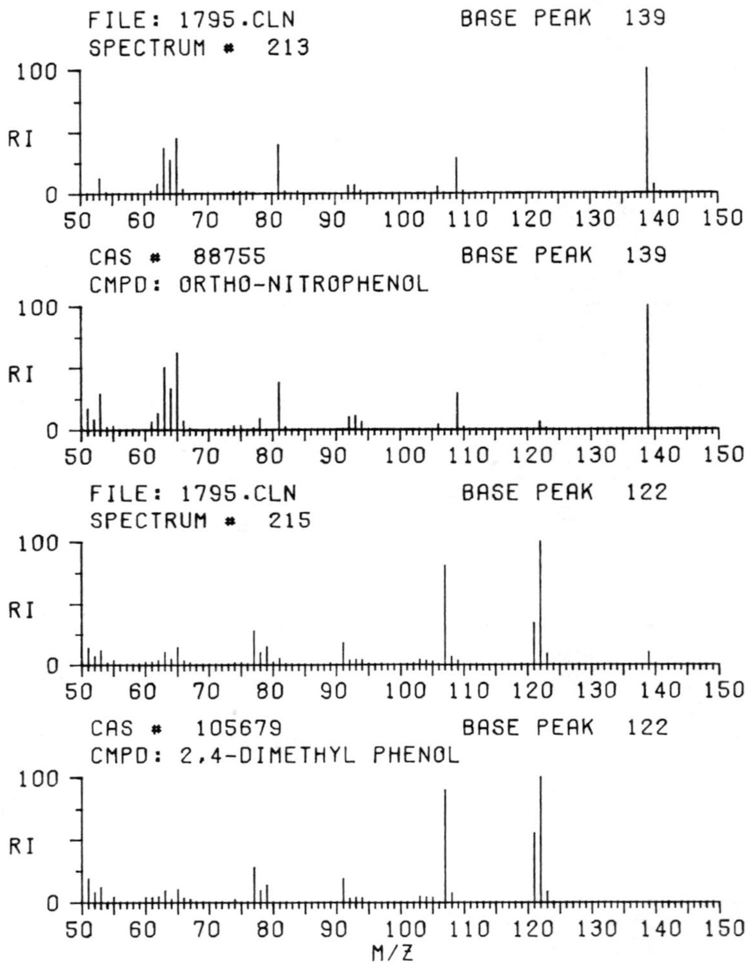


**Figure 5.** Resultant spectra from scans 213 and 215 of Figure 4 compared to matching library spectra.

peaks. CLEANUP then has a difficult time extracting a good spectrum. A good reason is that the general background matrix of components is due to long-chain substituted alkyls that are not identified by PBM with any degree of certainty.

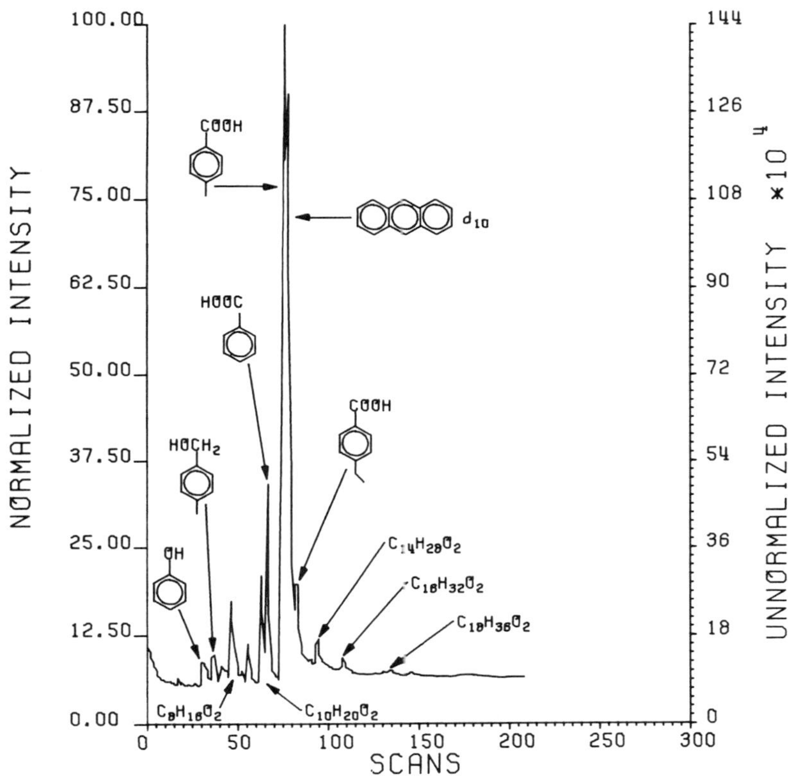


**Figure 6.** RGC from an actual extract of an industrial effluent with tentative identifications indicated.

**Table I. Comparison of Automated vs Manual Peak Extraction**

| Manual | | | Cleanup-PBM | | |
|---|---|---|---|---|---|
| Peaks | ID | % ID | Peaks | ID | % ID |
| 30 | 18 | 60 | 46 | 18 (6)[a] | 52 |
| 18 | 7 | 39 | 44 | 7 (4) | 25 |
| 31 | 11 | 35 | 41 | 11 (2) | 32 |
| 26 | 10 | 38 | 43 | 10 (2) | 28 |

[a]Numbers in parentheses are additional components identified by the automated method.

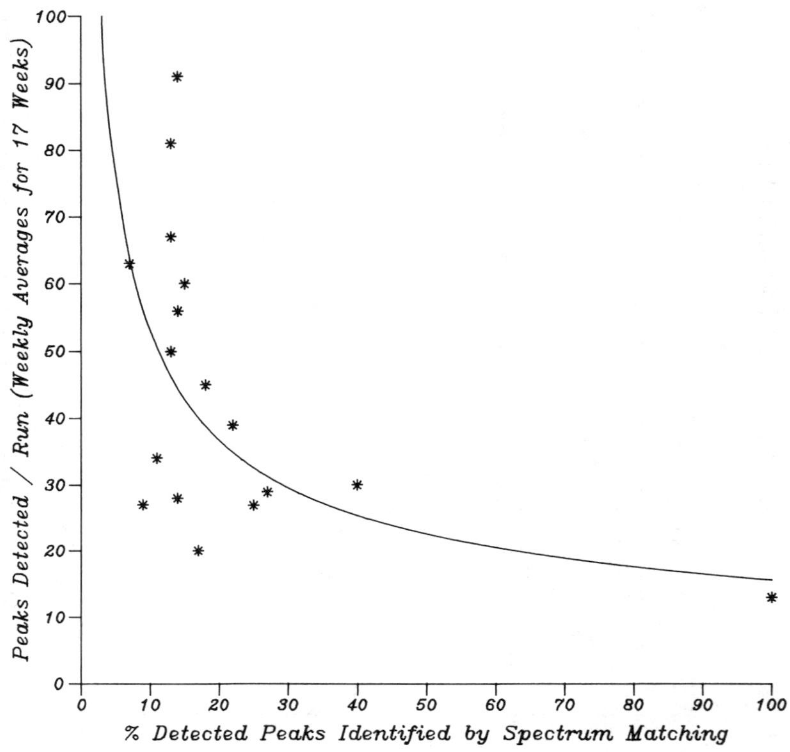


**Figure 7.** Relationship of identifications made by spectrum matching to the number of components extracted by CLEANUP.

CLEANUP requires each potentially detected ion to pass a number of tests before the ion is considered to be detected. As mentioned earlier, each peak must have two rising points, a maximum and two falling points. The ratio of the slope of these points to the peak width must exceed a user-defined value. The area under the described curve must exceed a second user-defined value after the local background has been subtracted. These criteria offer no great hurdle for normally chromatographed components with few interferences. Problems develop, however, in the case of more convoluted runs.

The first problem is the loss of weak-intensity ions that are important for identification. These ions get lost in the background because of insufficient peak area or by not having two rising and falling points. The same problems occur when several peaks overlap that have ions in common. Because the MS sees only the sum of the ions in common, no distinct peaks are seen for these components. Also, when local background is subtracted, these ions may be lost or their areas severely miscalculated.

A second problem is that as the end of the GC temperature program is reached, the widths of the eluted peaks become greater. As these widths increase, a point is reached at which the peaks no longer have acceptable slope ratios and are eliminated. Finally, noisy data peaks may be eliminated due to noise spikes on either the ascending or descending side of the peak.

## Spectrum Matching

The spectrum matching portion of the data analysis system has undergone the least modification. PBM has been evaluated in the literature [5], and has been used in the Athens laboratory for several years. Before the start of this project, PBM was operational on an in-house PDP 11/70.

Selection of a database of reference spectra for use with PBM involved no small problem. Three databases were available: the Wiley collection, the National Bureau of Standards collection and the EPA master collection. The Wiley library contains 30,476 spectra of 30,476 compounds; the NBS library 25,025 spectra of 25,025 compounds; and the EPA master list contained about 40,000 spectra of approximately 32,000 compounds, and is the master list of spectra from which the NBS library was taken.

Because the GC/MS data used in this study came from a great variety of sources it was thought that "duplicate" spectra, i.e., multiple spectra of the same compound that differ slightly due to run conditions, in the database would be of some help in the matching process.

Table II compares the matching ability of two databases on the same spectra. As can be seen, Database II (the EPA master database) enjoys a dis-

**Table II. Comparison of PBM Matching Ability for Two Databases**

| Database I[a] (K value, missing masses) | Database II[b] (K value, missing masses) | Manual ID |
|---|---|---|
| Toluene (75+)[c] | Toluene (75+) | Toluene |
| 7-Oxabicyclo-2,2,1-heptane (49+) | 2-Cyclohexene-1-ol (76+) | 2-Cyclohexene-1-ol |
| Phthalide (56, –2) | Methyl benzoate (69+) | Methyl benzoate |
| Hexacosanoic Acid (102, –3) | Octadecanoic acid (105+) | Octadecanoic acid |

[a]Database I is NBS library.

[b]Database II is EPA master database.

[c]+ indicates that the molecular ion was matched within the proper intensity tolerance.

tinct advantage over Database I (the NBS library) for the cases mentioned in the table. Comparison of the matches suggested by the two databases with the manual identification shows the superior ability of Database II. Data generated using the Wiley library showed shortcomings similar to those of the NBS library.

In cases where the identical matched spectrum occurred in all the libraries, as was generally the case, no problems were observed. In some cases when spectrum extraction or run conditions slightly affected the spectrum, only if there were duplicate spectra did a match occur. Thus, the EPA master library was chosen for use in this work.

The reverse search approach of PBM has also proven useful when analyzing environmental samples with an automated system. Figure 8 is an example of a mixed spectrum obtained by CLEANUP from a VOA standard run. Although the two compounds (*cis*-1,3-dichloropropene and 1,1,2-trichloroethane) are not resolved, PBM was able to match both compounds in this spectrum. This ability of a reverse search to recognize components of mixed spectra is clearly an advantage.

## Result Collator

The combination of MS data with retention indices provides a powerful tool in the automated extraction and identification of components in GC/MS data. The use of retention indices coupled with MS data for identification purposes was reported by Nau and Biemann [8] and Sweeley et al. [9]. Subsequently, automated systems in several laboratories [3,10,11] have been reported that rely on retention data as well as MS data for computer-aided identification of a known set of organic compounds. Table III shows RRT variation vs K variation for several compounds. Although the match quality parameters vary from 45 to 100, the RRT variation is only 0.01-0.03. RRT are particularly important in the identification of compounds such as alkanes or alcohols, which have little or no molecular ion and exhibit highly similar spectra. This is shown in Figure 9, which compares spectra and RRT of a $C_{13}$ and a $C_{14}$ alkane.

In this project, RRT have been such a sensitive indicator that we have been able to distinguish contractor laboratories that have not followed the prescribed temperature-programming conditions. Table IV shows a comparison of mean RRT for two laboratories with the mean RRT observed for all other laboratories for the same compounds. Based on these discrepancies, it was discovered that these laboratories were indeed not temperature-programming in the same manner as the others. As noted in the table, a linear correction factor is applied to the data from these laboratories to facilitate automatic spectra matching by HISLIB.

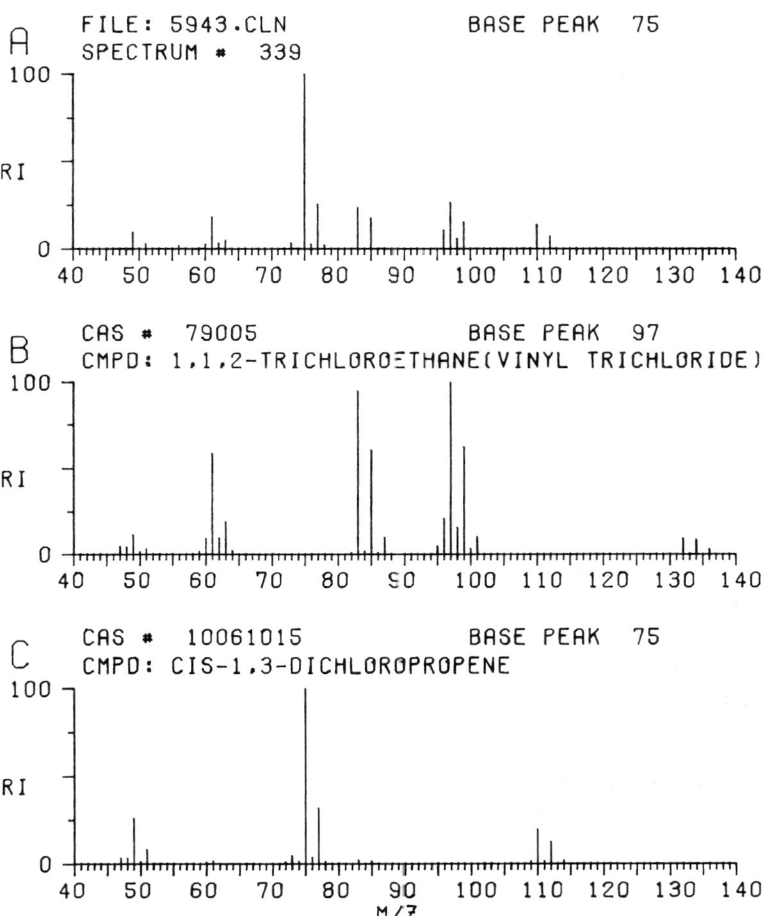


**Figure 8.** (A) Mixed spectrum extracted by CLEANUP from a VOA standard run. (B,C) Library spectra indicated as matched by PBM.

Before actual runs were processed through the system, the historical library was initialized with data from standard runs. Several ACI, B/N and VOA priority pollutant standard runs from the various contract laboratories were processed, and the compounds were entered in the historical library after being inspected visually and checked against the protocol for RRT consistency. Whenever possible, mean RRT values were calculated. All entries from standard runs were flagged so they would not be counted in the frequency of occurrence statistics.

Table III. Comparison for RRT vs K Variation for Selected Matches

| Compound | Range of RRT | Range of K |
|---|---|---|
| Dioctylphthalate | 0.03 | 45–100 |
| Phthalide | 0.01 | 55–77 |
| Toluic Acid | 0.03 | 48–85 |

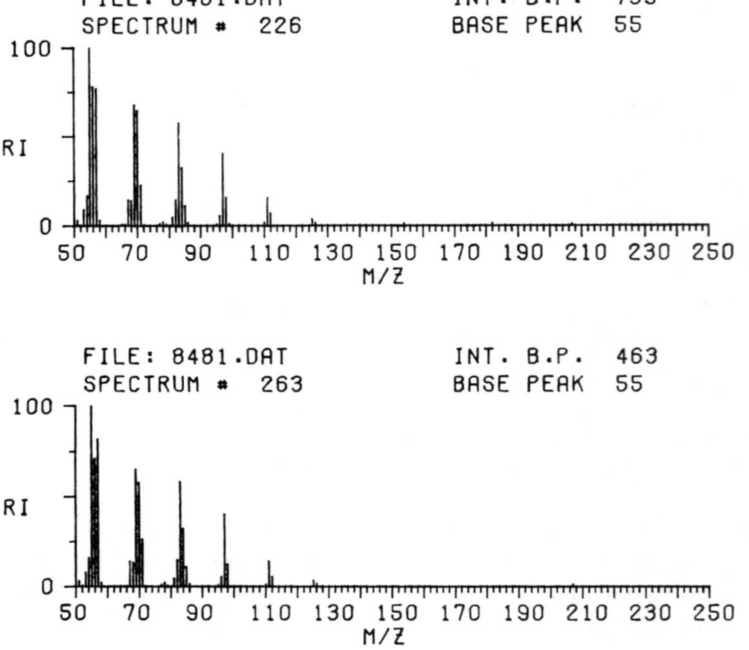


**Figure 9.** Comparison of spectra for two alkanes. Top spectrum (226) is tridecane (RRT = 0.510). Bottom spectrum (263) is tetradecane (RRT = 0.591).

Table IV. Mean RRT for Two Laboratories before and after Correction, Compared with Mean RRT for Other Labs for B/N Run on 1% SP-2250

| Compound | RRT before Correction | RRT after Correction[a] | RRT for all other labs |
|---|---|---|---|
| Xylenes | 0.041 | 0.137 | 0.136 |
| 2-*n*-Butoxyethanol | 0.090 | 0.180 | 0.166 |
| $C_3$ Benzenes | 0.111 | 0.200 | 0.208 |
| Naphthalene | 0.375[b] | 0.438 | 0.457 |
| Methyl Naphthalenes | 0.491 | 0.542 | 0.553 |
| $C_2$ Naphthalenes | 0.612 | 0.651 | 0.663 |
| Fluorene | 0.796[b] | 0.816 | 0.838 |

[a]Corrected using: $RRT_{corrected} = RRT_{orig} + 0.1(1 - RRT_{orig})$; for $RRT_{orig} < 1$.
[b]One observation only.

Table V. Homologous Series Standards

| | |
|---|---|
| $C_6$-$C_{19}$ *n*-Alkanes | Dicarboxylic Acids |
| $C_{19}$-$C_{40}$ *n*-Alkanes | Dimethyl Esters of Dicarboxylic Acids |
| $C_6$-$C_{10}$ Alkenes | $C_3$-$C_{18}$ Fatty Acids |
| $C_8$-$C_{22}$ Alkenes | $C_3$-$C_{12}$ Glycols |
| $C_4$-$C_{22}$ *n*-Alcohols | $C_3$-$C_{10}$ Glycol Ethers |
| $C_3$-$C_{16}$ Aldehydes | Low-boiling Esters |
| $C_4$-$C_{14}$ Primary Amines | $C_3$-$C_{19}$ Methyl Ketones |
| $C_4$-$C_{18}$ Secondary and Tertiary Amines | Phenols |
| Benzenoid Hydrocarbons | Phthalate Esters |

As discussed above, several compound classes are difficult to distinguish without RRT data. Therefore, standards were obtained for the homologous series compound classes listed in Table V. These standards were run on the various GC columns, and the compounds were entered to the historical library.

These standard runs, then, furnished a starting point for construction of the library. Compounds have been added to the library over the course of the project with nearly 950 unique compounds tentatively identified to date. The identification rate (the percent of spectra extracted by CLEANUP that have been tentatively identified) has increased as a function of the number of compounds in the library, and has reached a plateau at 23%. This is demonstrated in Table VI, which compares project status as of December 31, 1979, with the results to date. At present, nearly 100,000 spectra have been extracted by CLEANUP, of which approximately 23,000 have been tentatively identified.

Table VI. Project Status

| | As of | |
|---|---|---|
| | 12/31/79 | 07/18/80 |
| Unique Compounds in HISLIB | 620 | 945 |
| Spectra Tentatively Identified | 8,117 | 22,978 |
| Spectra Extracted by CLEANUP | 45,574 | 99,894 |
| Identification Efficiency | 17.8% | 23.0% |

Table VII. Compounds Most Frequently Found (as of 12/31/79)

| HISLIB Frequency | Compound |
|---|---|
| 266 | 1-(2-Butoxyethoxy) ethanol |
| 180 | Cresols |
| 168 | Phenol[a] |
| 152 | Toluene |
| 145 | Tetrachloroethylene[a] |
| 134 | Benzene[a] |
| 120 | Benzoic acid |
| 113 | Dichloromethane[a] |
| 102 | Naphthalene[a] |
| 92 | *p*-Xylene |
| 91 | Ethyl benzene[a] |
| 88 | 1-Methyl naphthalene |
| 85 | Octadecane |
| 81 | Di-*n*-octyl-phthalate[a] |
| 80 | Hexadecane |
| 75 | 2-Methyl naphthalene |
| 75 | 2-Butoxyethanol |
| 75 | 1-Methyl-2-ethyl benzene |

[a]Priority pollutant.

A comparison of the most frequently occurring compounds as of December 31, 1979, in Table VII with the present list in Table VIII shows little compound variation with some shift in the order of frequencies. Duplicate occurrences of compounds in the ACI, B/N and VOA fractions of the same sample have been eliminated prior to generation of these lists. The extracting solvent, dichloromethane, is never counted in ACI or B/N runs. Background contaminants found in blank runs, however, have not been subtracted from these statistics.

**Table VIII. 25 Most Frequently Found Compounds (as of 7/18/80)**

| HISLIB Frequency | Compound |
|---|---|
| 489 | 2-(2-Butoxyethoxy) ethanol |
| 485 | Tetrachloroethylene[a] |
| 436 | Cresols |
| 414 | Xylenes |
| 406 | Toluene |
| 349 | Dichloromethane[a] |
| 335 | Phenol[a] |
| 302 | Benzene[a] |
| 274 | Benzoic acid |
| 263 | Ethyl benzene[a] |
| 262 | Octadecane |
| 256 | Hexadecane |
| 253 | Trichloroethylene[a] |
| 230 | Naphthalene[a] |
| 214 | 1-Methyl naphthalene |
| 210 | Butyl phthalate[a] |
| 206 | Pentadecane |
| 194 | *bis* (2-Ethylhexyl) phthalate[a] |
| 188 | Di-*n*-octyl-phthalate[a] |
| 179 | 1,1,1-Trichloroethane[a] |
| 175 | 2-Methylnaphthalene[a] |
| 174 | Chloroform[a] |
| 151 | Diethyl *o*-phthalate[a] |
| 151 | 2-Butoxyethanol (butyl cellosolve) |
| 145 | Alpha-terpineol |

[a]Priority pollutant.

The increase in the number of unique library compounds over the last seven months is largely due to the processing of runs from a greater variety of industrial categories. Profiles of samples analyzed by industrial categories as of December 31, 1979, and July 18, 1980, are shown in Tables IX and X, respectively. Comparison of these tables shows that while "publicly owned treatment works" (POTW) is still the most frequently analyzed industry, other categories such as timber products, organics and plastics, and nonferrous metals have increased considerably. Examples of a few "industry-specific" compounds are listed in Table XI, which shows the total number of HISLIB entries and the percentage that have been tentatively identified in a given industry.

**Table IX. Profile by Industrial Category for Samples Analyzed as of 12/31/79**

| No. of Runs | Industrial Category |
|---|---|
| 544 | Publicly owned treatment works |
| 430 | Standards and blanks |
| 277 | Pulp & paper |
| 190 | Organic chemicals |
| 177 | Coal mining |
| 173 | Paint & ink |
| 167 | Inorganic chemicals |
| 146 | Auto & other laundries |
| 87 | Pharmaceuticals |
| 83 | Electronics |
| 72 | Mechanical products |
| 60 | Rum industry |
| 55 | Transportation equipment |
| 54 | Printing & publishing |
| 50 | Rubber processing |
| 46 | Industry unknown |
| 43 | Plastics & synthetics |
| 43 | Pesticides manufacturing |
| 36 | Photographic industries |
| 34 | Leather tanning |
| 33 | Textile mills |
| 33 | Porcelain enameling |
| 27 | Foundries |
| 24 | Plastics manufacturing |
| 23 | Organics & plastics |
| 15 | Soaps & detergents |
| 14 | Amusements & athletic goods |
| 14 | Ore mining |
| 8 | Petroleum refining |
| 4 | Steam electric |
| 2 | Iron & steel manufacturing |
| 1 | Battery manufacturing |
| 1 | Timber products |

The MISS library collates results for unidentified spectra that appear to recur within narrow RRT windows (±0.1 RRT unit). To date, over 72,000 such unidentified spectra have been entered into the MISS library. There are 254 unidentified components that have been found with a frequency of 15 or greater. These represent 10% of the total spectra cataloged in the MISS library. A sample entry is shown in Figure 10. This unknown has occurred 11

Table X. Profile by Industrial Category for Samples Analyzed as of 7/18/80

| No. of Runs | Industrial Category |
|---|---|
| 1623 | Publicly owned treatment works |
| 1325 | Standards & blanks |
| 463 | Pulp & paper |
| 426 | Inorganic chemicals |
| 423 | Electronics |
| 383 | Mechanical products |
| 366 | Organics & plastics |
| 362 | Paint & ink |
| 342 | Foundries |
| 334 | Organic chemicals |
| 316 | Coal mining |
| 312 | Auto & other laundries |
| 181 | Timber products |
| 162 | Pharmaceuticals |
| 153 | Nonferrous metals |
| 151 | Petroleum refining |
| 132 | Textile mills |
| 124 | Pesticides manufacturing |
| 107 | Printing & publishing |
| 104 | Iron & steel manufacturing |
| 102 | Plastics & synthetics |
| 96 | Rubber processing |
| 93 | Ore mining |
| 92 | Paving & roofing |
| 78 | Transporation equipment |
| 73 | Industry unknown |
| 71 | Rum industry |
| 63 | Steam electric |
| 50 | Photographic industry |
| 35 | Leather tanning |
| 33 | Porcelain enameling |
| 32 | Soaps & detergents |
| 24 | Plastics manufacturing |
| 16 | Amusements & athletic goods |
| 6 | Explosives |
| 1 | Adhesive & sealants |
| 1 | Battery manufacturing |

times in acid extracts run on 1% SP-1240 DA and is thought to be $D_8$ anthraquinone from degradation of the internal standard $D_{10}$ anthracene. This spectrum has been observed occasionally when manually locating internal standards, and because no deuterated compounds exist in the reference library, has been passed to the MISS library where recurrences have been collected.

Table XI. Selected HISLIB Results

| Compound | Total HISLIB Entries | % of Entries Found in Industrial Category | Industry |
|---|---|---|---|
| Dimethyl Sulfone | 16 | 100 | Pulp & paper |
| Caffeine | 64 | 80 | POTW |
| 2-Furaldehyde | 11 | 45 | Rum |
| Xanthene | 10 | 80 | Coal mining |
| *p*-Toluenesulfonamide | 12 | 67 | Printing & publishing |
| Triphenylphosphate | 26 | 27 | Nonferrous metals |
| | | 42 | Foundries |
| Syringaldehyde | 30 | 87 | Pulp & paper |
| Acetosyringone | 17 | 65 | Pulp & paper |
| | | 35 | Rum |
| Acridone | 11 | 55 | Timber products |
| Dibenzofuran | 33 | 36 | Coal mining |
| | | 36 | Timber products |
| Diphenyl Ether | 26 | 38 | Organic chemicals |
| | | 15 | Organics & plastics |
| Veratryl Alcohol | 5 | 100 | Nonferrous metals |

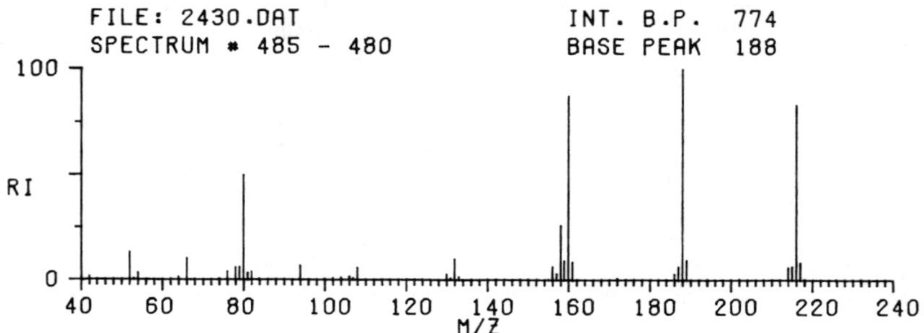


**Figure 10.** Sample spectrum from MISS library thought to be $D_8$ anthraquinone from degradation of internal standard $D_{10}$ anthracene. This unknown has occurred 11 times in acid extracts run of 1% SP-1240 DA.

## SUMMARY

Automated survey techniques for processing GC/MS data appear to work well on most of the data encountered in this work. Whereas the sensitivity of the CLEANUP-PBM package is not as great as that of a reverse search for specific ions, it is sufficient for the tentative identification of the major components in a sample. CLEANUP-PBM is cost-effective when compared to an operator's finding peaks, subtracting background and matching the spectrum manually or using a computer search.

Use of a historical library to catalog data collected over long periods of time aids identification by adding the dimension of GC retention data. Confidence in a tentative identification is heightened if there are collaborating GC retention data. This combination of spectral and retention data can be used effectively to catalog and highlight recurring unidentified substances for future study.

Two areas are the subject of continuing study. The proper compensation of background by CLEANUP is of concern since errors in intensity calculations reduce the chance that PBM will find a match for the spectrum. Studies underway indicate that background compensation similar to that used in PEAK would be effective in CLEANUP. Implementing such a routine is under study.

A second area of concern are the various parameters that must be set for CLEANUP. These parameters that pertain to peak shape and minimum area are inflexible during a given data run. Thus, changing chromatographic conditions that affect peak size and shape may cause CLEANUP to miss good data. The automatic setting of these parameters during data processing is desirable and is presently under study.

## ACKNOWLEDGMENTS

We would like to thank Bruce Hodge for his work in bringing CLEANUP up on our computers, Bruce Bartell for writing the graphics software and David Bradshaw for his work in evaluating data.

## DISCLAIMER

Mention of trade names or commercial products does not constitute endorsement or recommendation for use by the U.S. Environmental Protection Agency.

## REFERENCES


1. Shackelford, W. M., and L. H. Keith. "Evolution of the Priority Pollutant List from the Consent Decree," in *Proceedings of the Second Open Forum on Management of Petroleum Wastewater*, U.S. EPA Report EPA-600/2-78-058, Ada, OK (1978), pp. 103-109.
2. Keith, L. H., and W. A. Telliard. "Priority Pollutants. I. A Perspective View," *Environ. Sci. Technol.* 13(4):416-423 (1979).
3. Smith, D. H., M. Achenbach, W. J. Yeager, P. J. Anderson, L. Fitch and T. C. Rindfleisch. "Quantitative Comparison of Combined Gas Chromatography/Mass Spectrometric Profiles of Complex Mixtures," *Anal. Chem.* 49(11):1623-1632 (1977).
4. Dromey, R. G., J. Stefik, T. C. Rindfleisch and A. M. Duffield. "Extraction of Mass Spectra Free of Background and Neighboring Component Contributions from Gas Chromatography/Mass Spectrometry Data," *Anal. Chem.* 48(9):1368-1375 (1976).
5. Pesyna, G. M., R. Venkataraghavan, H. E. Dayringer and F. W. McLafferty. "Probability Based Matching System Using a Large Collection of Reference Mass Spectra," *Anal. Chem.* 48(9):1362-1368 (1976). EPA Research Report No. EPA-600/4-76-046 (1976).
6. Sauter, A. D., L. F. Faas, G. R. Kurth, W. M. Shackelford and D. M. Cline. "Applications of a Data-Adaptive Background Subtraction Technique to the Gas Chromatographic/Mass Spectrometric Analysis of Wastewater Samples," Chapter 35, this volume.
7. Pesyna, G. M., F. W. McLafferty, R. Venkataraghavan and H. E. Dayringer. "Statistical Occurrence of Mass and Abundance Values in Mass Spectra," *Anal. Chem.* 47(7):1161-1164 (1975).
8. Nau, H., and K. Biemann. "Computer-Assisted Assignments of Retention Indices in Gas Chromatography-Mass Spectrometry and Its Application to Mixtures of Biological Origin," *Anal. Chem.* 46(3):426-434 (1974).
9. Sweeley, C. C., N. D. Young, J. F. Holland and S. C. Gates. "Rapid Computerized Identification of Compounds in Complex Biological Mixtures by Gas Chromatography-Mass Spectrometry," *J. Chromatog.* 99:507-517 (1974).
10. Tsuchiya, Y., J. Boulanger and K. Sumi. "Identification of Chemical Compounds by Computer Matching of Low Resolution Mass Spectra and Retention Indices," *Chromatographia* 10(3):154-156 (1977).
11. Blaisdell, B. E. "Automatic Computer Construction, Maintenance, and Use of Specialized Joint Libraries of Mass Spectra and Retention Indices from Gas Chromatography-Mass Spectrometry Systems," *Anal. Chem.* 49(1):180-186 (1977).

# CHAPTER 34

# OVERCOMING BOTTLENECKS IN ENVIRONMENTAL SAMPLE ANALYSIS

**R. E. Finnigan, M. S. Story**
Finnigan Corporation
Sunnyvale, California

**Donald F. Hunt**
University of Virginia
Charlottesville, Virginia

In 1971 the U.S. Environmental Protection Agency (EPA) selected a computerized gas chromatography/mass spectrometry system as its principal analytical tool for identification and analysis of organic pollutants in water. Following adoption of GC/MS by EPA labs, this technique was adopted as the principal tool for analysis of drinking water, effluent streams from industrial plants, and organics in air and hazardous wastes. By the end of the 1970s, most major industrial companies in the United States and in western Europe had adopted GC/MS for analysis of their own effluents. Much of this increased use of GC/MS was fostered by the regulations formulated to implement the Clean Water Act and other environmental legislation. Current legislation includes National Pollutant Discharge Elimination System (NPDES) permit requirements and other regulations that essentially require GC/MS analysis of effluents prior to issuance of a discharge permit and, in most instances, for follow-on self monitoring. GC/MS is ideally suited to perform the complex environmental analyses required by law, to provide data that are acceptable as legal evidence, and to do these tasks in a cost-effective and highly productive manner.

Despite its success, GC/MS has encountered several significant "bottlenecks" limiting its even broader acceptance for environmental sample analysis:

1. Complexity of GC/MS systems requires qualified technical personnel to operate and maintain them.
2. High initial cost of GC/MS with associated data systems implies a higher cost for sample analysis when compared to other methods such as GC-only.
3. GC/MS has encountered limitations when analyzing complex environmental matrices such as sludges and soils, and very low level complex isometric pollutants such as TCDD.

## COMPLEX GC/MS SYSTEMS REQUIRE TRAINED OPERATORS

Recent publications [1], articles [2] and industry comments to the "Guidelines Establishing Test Procedures for the Analysis of Pollutants" proposed by the EPA [3], have alluded to the critical shortage of qualified chemists and technicians experienced in the application of GC/MS methodology to environmental analysis. It is implied that because of this shortage, it is not reasonable for EPA to require GC/MS methodology for regulatory programs being implemented under the Clean Water Act, even though most industrial companies, publicly own treatment works (POTW), etc., agree that GC/MS is the only proven method available and the most cost-effective approach as well.

It is true that the supply of chemists and technicians trained in GC/MS theory and practice is quite tight at the present. This situation has resulted from the explosive growth of the GC/MS field during the past decade coupled with the absence or shortage of educational programs in the colleges and universities in this technical area, particularly those courses providing "hands-on" training.

We have implemented two actions which we feel are effectively overcoming this bottleneck:

1. establishment of a training institute to train chemists and technicians in all aspects of GC/MS methodology; and
2. introduction of a highly automated GC/MS system dedicated to analysis of organics in water which minimizes the technical expertise required to perform complex environmental analyses.

## Training Institute

The Finnigan Institute was established at Cincinnati, OH in 1978 for the purpose of providing education and training in GC/MS and allied fields. The Finnigan Institute holds a series of courses throughout the year, taught by a professional staff under the direction of a respected mass spectroscopist and teacher, on the following subjects:

- GC/MS theory and practice (basic and advanced)
- data system theory and practice (basic and advanced)
- GC/MS/DS maintenance training
- organics in water analysis
- LC/MS theory and practice
- mass spectral interpretation (basic and advanced)
- derivatization techniques for GC/MS
- GC/MS for laboratory managers
- atomic absorption theory and practice
- GC and LC theory and practice

The enrollment in each class is purposely limited so that there will be no more than two students per instrument, allowing the students to spend at least 50% of their time working in the laboratory.

The institute has already graduated more than 1500 students in the courses listed during the first two years of its life while establishing a reputation for excellence in hands-on instrumental training. Through its intensive courses, the institute has been successful in bringing new organizations entering the GC/MS field up to the state-of-the-art in relatively short periods of time. The institute has the staff, facilities and GC/MS instruments to train at least twice as many students as are currently enrolled without sacrificing the quality of its instruction. By carrying out its pioneering role in the practical training and education of chemists and technicians in GC/MS theory and practice, the institute is producing a supply of trained GC/MS chemists, operators and technicians to satisfy the expanding needs of industrial and government laboratories.

## Organics in Water Analysis

The organics in water analyzer (OWA) is a fully automated compact GC/MS system designed specifically for routine and repetitive analysis of volatile and dissolved organic pollutants in water. System reliability is enhanced significantly by use of a turbomolecular pump to provide the vacuum for the mass spectrometer.

Automation of the OWA is accomplished using a dedicated minicomputer which performs the following functions:

1. automatically diagnoses the system to establish good working conditions and "tunes" the mass spectrometer using preestablished criteria and standards; no operator knowledge is required;
2. programs the liquid sample concentrator which purges and concentrates volatile organics from water samples prior to introduction into GC;
3. programs the GC parameters to preset values and injects sample using optional autosampler;
4. programs the mass spectrometer over the specified range and at the programmed scan rate following sample injection, continuing until the GC run is complete;
5. acquires mass spectral data and computes total ion current (equivalent to flame ionization detection (FID) output using a gas chromatograph), which are presented to the operator (if instrument is attended) and stored on hard disk for data processing;
6. processes acquired data, identifying all unknown compounds for which there are spectra contained in its library (this can be accomplished in parallel with acquisition using priority-interrupt, foreground/background features without operator interaction);
7. using appropriate internal and/or external standards, quantifies the peaks identified; and
8. outputs results of analyses on hard copy in the format specified by EPA or the user and the original data is then archively stored on a disc or on nine-track tape for later, more detailed analysis or for legal purposes.

User experience with the OWA over the past 24 months has shown that a bachelor-level chemist or a senior technician without prior GC/MS experience can become qualified to run priority pollutant analyses using the OWA after several weeks of training at the Finnigan Institute, followed by a short period of on-the-job training.

## HIGH INITIAL COST OF GC/MS IMPLIES HIGH COST FOR SAMPLE ANALYSIS

Until very recently, a computerized GC/MS research-grade system capable of performing priority pollutant analysis of water samples sold for $150,000 or more. This did not include the cost of the laboratory facility, utilities nor the technical staff required to operate and maintain the instrument. It is difficult for many organizations, particularly medium- and small-size industrial companies, to consider such an undertaking even if GC/MS is perceived as the only instrument which can provide a reliable and quantitatively accurate effluent analysis. This high initial cost has also been the primary motivation for the EPA to seek lower-cost alternatives such as gas chromatography alone.

In contrast, a gas chromatograph sells for $10,000–15,000, and when coupled to a data system will sell for approximately $40,000; thus, on first look, it would appear to be a more cost-effective solution for the organization contemplating analysis of their own effluents to comply with mandates of the Consolidated Permit Regulations under the Clean Water Act.

Unfortunately, it turns out to be an overwhelming task to attempt the analysis of the 114 organic pollutants on the EPA priority pollutant list by gas chromatography with conventional detectors. The job may take several operators, using as many as four GC, an HPLC, many selective detectors and a laboratory data system, several days to match the retention time of each unknown peak with the various possible candidate compounds iteratively. Moreover, even when an identification is considered made, the possibility of a false positive is great. In a recent survey [4], of chemists, laboratory managers, and officials in industry and government, GC was viewed as "unreliable" or "unsuitable" for priority pollutant analysis, whereas GC/MS was perceived as "very reliable."

The survey also determined the total charges for performing an analysis of an industrial effluent for the 114 priority pollutants by GC and GC/MS. When using GC alone, this charge was found to be $557 (1978 dollars) per analysis, while when GC/MS was used the charge per analysis was only $328.

As more compounds are monitored, there is little change in the time or cost to perform the analysis by GC/MS. The time and cost required by GC increases, however, when the number of compounds increases. Moreover, the practical limit of the number of compounds that can be monitored by GC is limited by the number of interfering compounds present. In the complex effluents of chemical plants, petroleum refineries and POTW, the effluents cannot be analyzed by GC alone.

The charges shown above for GC and GC/MS were those reported by industrial companies and contract service laboratories, each of whom had a large enough sample load to justify one-shift operation or more. They include the amortized cost of the capital equipment, facility costs and the loaded costs of the technicians and operators required to prepare the samples for analysis and to run them. The reason that GC-only analysis is far more costly than GC/MS is because of the much greater time required for sample extraction, cleanup and instrument operation, all of which are technician/operator-intensive. The total technican/operator costs reported using GC only were almost three times the amounts which were reported when using GC/MS.

Because of the lower cost per analysis, reliability of data and ease of use, GC/MS has rapidly become the standard analytical technique for analysis organics in water by industrial, contract service and governmental laboratories.

Many industrial companies do not and will not have a large enough sample load to justify purchase of their own GC/MS system and setting up their own environmental laboratory. These organizations will find it far more economical and timely to send their samples to a qualified contract service laboratory which has GC/MS capability. In the recently issued Consolidated Permit Regulations [5], EPA states there are approximately 66 commercial environmental testing laboratories which have a total of 129 GC/MS systems capable of performing priority pollutant analysis of customer samples. The prices charged per sample by these laboratories are very competitive, with sizable reductions for multiple samples. They are very familiar with EPA requirements and will generally provide the results in a standard format approved by the EPA. Furthermore, they will normally store the raw data on each sample analyzed for a reasonable period of time, so that in the event there are questions concerning the effluent analysis requiring further data reduction or processing, there is no need to retake effluent samples and redo the analysis. This can save a considerable amount of time and money, since the gathering of an effluent sample alone is estimated by EPA to cost approximately $1550 per outfall [5].

## GC/MS HAS LIMITATIONS WHEN ANALYZING VERY COMPLEX ENVIRONMENTAL SAMPLES

When one requires the detection and quantification of specific chemicals, such as priority pollutants or hazardous substances, at low levels in complex environmental or biological matrices, this provides a difficult challenge to GC/MS. In such instances, at the minimum, an extensive sample extraction, cleanup and concentration procedure is required which can often take 8-12 hours of time by an experienced technician, before the sample is ready to be analyzed by capillary column GC/MS. With some matrices, this cleanup can remove some or all of the compounds of interest along with the interfering contaminants. With such samples, what is needed is an analysis technique which does not require this tedious and time-consuming sample preparation.

Recently, it has been shown that tandem mass spectrometers (MS/MS), consisting of an ion source and multiple mass analyzers with a collision gas chamber between them, can be used to solve such complex analysis problems.

In the MS/MS methodology, ions are generated from all components in a complex mixture simultaneously. The entire mixture is put into the spectrometer with no prior chromatographic separation. After ionization of the entire mixture, an ion that is characteristic of a specific component is allowed to pass through the first stage of mass analysis. This is analogous to the GC mixture separation in GC/MS analysis. These ions are then purposely caused to fragment in a collision chamber; then in the second stage of mass analysis, all ions are scanned as in GC/MS analysis. This scanning identifies the ion(s) and the compound selected in the first stage of mass analysis, just as in GC/MS.

The advantages of this approach are manifold. A major breakthrough that this new technology brings to environmental analysis is the time required to prepare and run the sample. In most cases, no prior sample workup is necessary. We have successfully analyzed whole plant material for pesticide residues by putting the plant material directly into the mass spectrometer via the solid probe. We have then run the complete MS/MS analysis in less than two minutes. Because there were no chemical extractions, derivatizations or long GC runs, we accomplished considerable time savings over GC/MS analysis. These time savings suggest that this technique would be ideal for screening large numbers of samples for confirmation of the presence or absence of a number of suspected compounds.

Much larger and more polar molecules can be analyzed by MS/MS than by GC/MS because the new technique allows heating at lower temperatures for a much shorter period of time than with GC analysis. There is also no problem of separating the sample from the carrier gas or solvent, as is the case with chromatography. The technique has shown itself to be unique in its capability for functional group analysis.

A commercially available MS/MS system which utilizes a triple stage quadrupole mass spectrometer (TSQ) has been applied to the detection of specific priority pollutants in a complex environmental matrix. This instrument is similar to that described by Yost and Enke [6] and is constructed using three standard Finnigan quadrupole mass filters arranged in tandem (Figure 1). All of these components are mounted on a single flange. The total assembly is approximately 24 inches in length and is seen in Figure 2. Figure 3 shows the vacuum manifold which has been modified to incorporate the longer quadrupole assembly with more feedthroughs and to support two remote rf coils; it uses the standard Model 4000 spectrometer vacuum and inlet system. Figure 4 shows the overall MS/MS system (including data system), and Figure 5 shows the special electronics consoles which provide control of the individual quadrupoles and the multiple lenses between them. The MS/MS data system uses standard Finnigan/Incos hardware but contains powerful software which provides for integrated control of each quadrupole and subsequent data processing, allowing entire experiments to be performed easily.

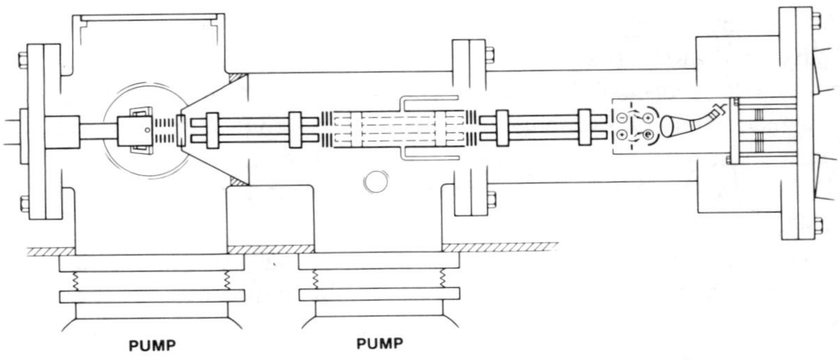


**Figure 1.** Triple-stage quadrupole mass spectrometer analyzer cross section.

In normal MS/MS operation, the rf/dc voltage on the first quadrupole, $Q_1$, is tuned to transmit the ion(s) of interest into the second quadrupole, $Q_2$. Only rf voltage is applied to $Q_2$, so in this configuration, $Q_2$ functions as an ion focusing device and collision chamber and transmits ions of all m/z. A collision gas pressure of 1-6 x $10^{-3}$ torr in $Q_2$ is required to optimize formation of fragment ions. Mass analysis of the resulting fragment ions is accomplished by scanning the rf and dc potentials in a normal manner on $Q_3$. A conversion dynode electron multiplier is employed for detection of either positive or negative ions.

To illustrate the fast screening and high specificity of this technique, a prototype TSQ mass spectrometer of the type described above was used in the laboratory of one of the authors at the University of Virginia. This analysis was performed in an effort to evaluate the specificity of the technique for the analysis of priority pollutants in sludge without prior extraction or chromatographic separation of the components [7]. *p*-Nitrophenol, 2,4-dinitrophenol and dioctyl phthalate were spiked into a sample sludge at the 100-ppb level. After the sample had been freeze dried, 5 mg of the solid residue containing approx. 10 ng of each component was inserted onto a solids probe directly into the ion source, which was operating in the chemical ionization negative ion mode, in order to detect the two phenols in the sludge sample. $Q_1$ was set to pass ions in a small window around the

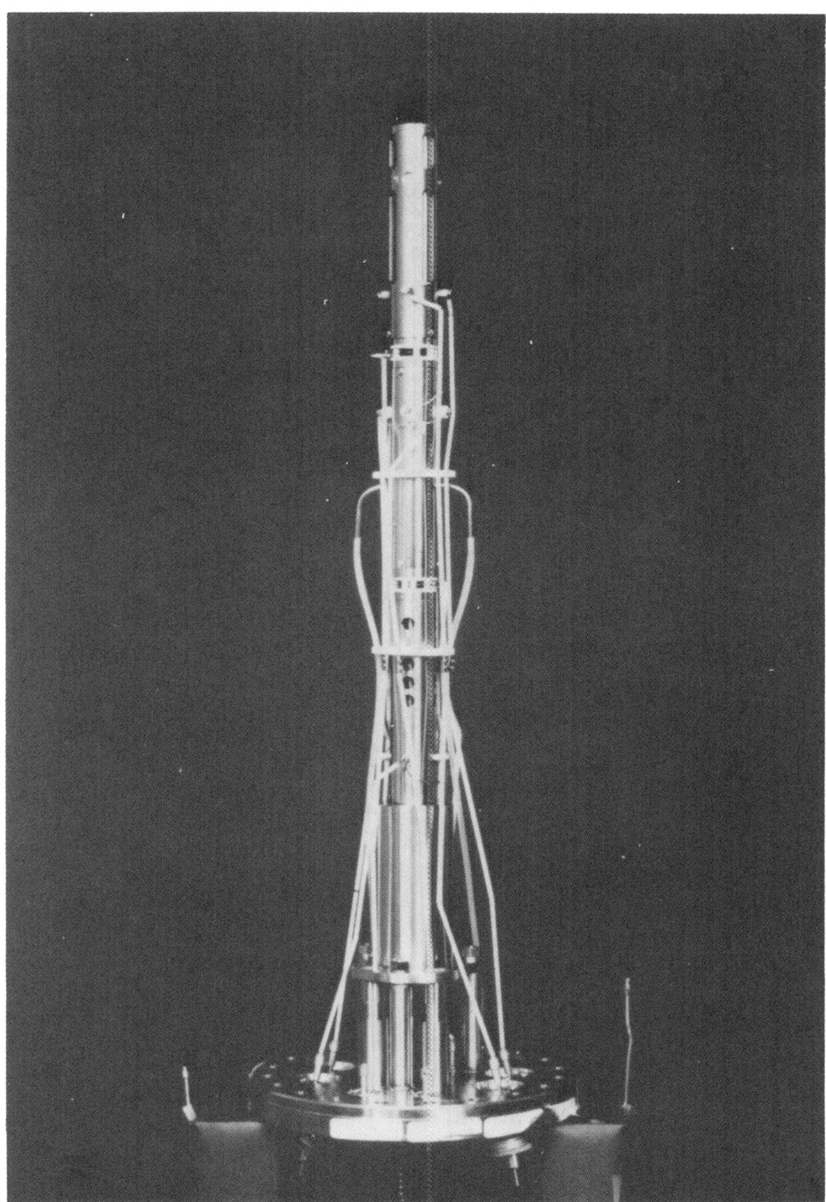

**Figure 2.** Triple-stage quadrupole mass spectrometer analyzer mounted on flange assembly.

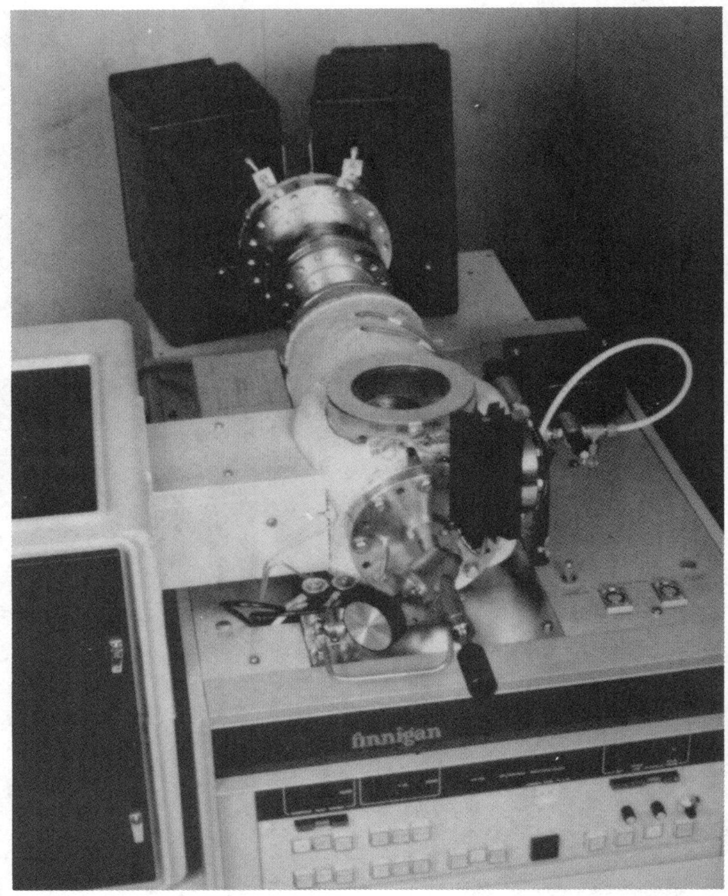


**Figure 3.** Triple-stage quadrupole mass spectrometer analyzer mounted in vacuum manifold of Model 4000. Two rf coils are seen mounted to the flange at the rear of the vacuum manifold.

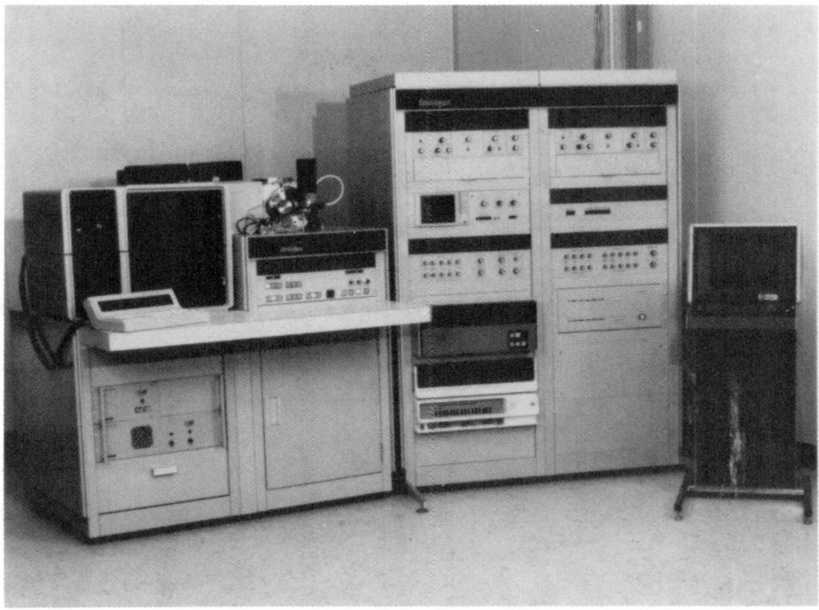

**Figure 4.** Triple-stage quadrupole MS/MS system. Gas chromatograph is seen on left, interfaced to vacuum manifold. MS/MS electronics and data system are seen in console at right of photo.

(M-1) ions at m/z 183 for dinitrophenol and m/z 138 for the nitrophenol, allowing these ions to pass through $Q_1$ and undergo CAD with nitrogen in $Q_2$, while $Q_3$ was scanned continuously. Sample identification was made by comparing the resulting spectra with those of pure standards. No nitrophenols were detected in the unspiked sludge samples. After repeating the above experiment several times, it was concluded that the detection limit for the two phenols in sludge using the prototype TSQ instrument is about 10 ppb. Significant improvement in sensitivity could undoubtedly be achieved by employing selected ion monitoring techniques in $Q_3$, but this was not done in this study.

**Figure 5.** MS/MS special electronics chassis which provides control of $Q_1$, $Q_2$, $Q_3$ and lenses between quadrupoles.

Analysis of dioctyl phthalate in sludge was similarly accomplished using the TSQ in the positive ion CI mode. Unspiked sludge was found to contain approx. 100 ppb of dioctyl phthalate based on measurement in the spiked sludge sample at the 200-ppb level.

We next attempted analysis of polycyclic aromatic hydrocarbons (PAH) in industrial sludge. The PAH investigated were pyrene, chrysene, benzo[a]-pyrene and fluorene. The procedure used was to take 3 mL of raw industrial sludge and freeze-dry to 150 mg of fibrous material. One 20-mg portion of this was spiked with 500 $\mu$g of each of the PAH corresponding to 1 ppm. Another 20-mg sample was spiked at the 5-ppm level. This mixture was then extracted with 1.5 mL of methylene chloride on a Millipore filter and dried. It was then brought up into 50 $\mu$L of methylene chloride, and a 10-$\mu$L aliquot was dried on a glass probe. The residue was nitrated in a $N_2O_4$ gas stream for 2-3 sec and put into the solid probe inlet of the mass spectrometer. The sample preparation procedure is summarized in Figure 6.

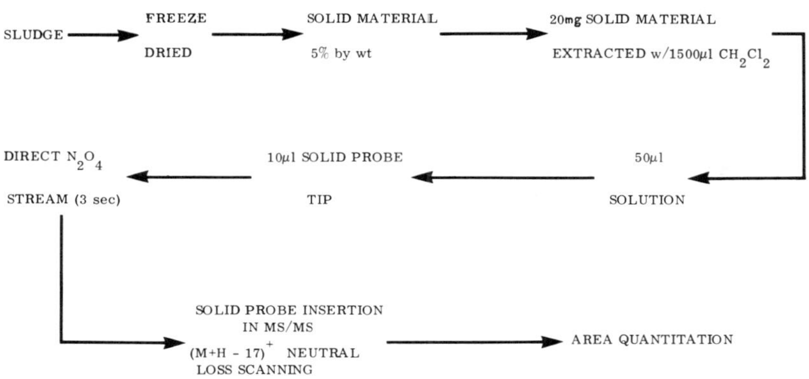


**Figure 6.** Sample preparation procedure for PAH in industrial sludge.

Under collisional activation, all of these compounds lose 12 amu; this characteristic is exploited by scanning $Q_1$ 17 mass units ahead of $Q_3$. This means there will not be any ions detected unless they are a result of this loss. The ion intensities for each of these compounds were summed and the resulting areas are reported in Table I. Figure 7 shows for comparison purposes the spectra which were obtained when analyzing pyrene by conventional positive ion CI (methane reagent gas) and by TSQ as described above using nitrogen as the collision gas.

The attended analysis time required for this work was about 20 min. Of this, the instrument time was 2-3 min and the rest was in data workup. The extractions were done in several minutes. The freeze-drying required 1-2 hr, but this was unattended time. In contrast, the current accepted methodology is approximately 10 hr, including 1 hr of instrument time for a capillary column GC/MS run.

Quantitative accuracy in the above analyses of PAH in sludge was determined to be better than a factor of two at the 1 ppm level. The measurements could be easily extended to the 100-ppb level.

Table I. Ion Intensities for Analysis of PAH by MS/MS

| | 100 ng PAH | 20 μL Sludge, 1 ppm PAH | 20 μL Sludge, 5 ppm PAH | 20 μL Sludge Blank | 1 ppm Sample Less Blank | 5 ppm Sample Less Blank | 1 ppm Sample Less Blank x 5 |
|---|---|---|---|---|---|---|---|
| Pyrene | 62.250 | 109,055 | 543,592 | 12,584 | 96,471 | 531,008 | 482,355 |
| Chrysene | 20,760 | 24,584 | 101,219 | 3,064 | 24,584 | 98,155 | 107,600 |
| Benzo[a]pyrene | 6,100 | 20,373 | 84,861 | 6,321 | 14,052 | 78,540 | 70,275 |
| Fluorene | 37,755 | 166,149 | 298,273 | 74,683 | 91,466 | 223,590 | 457,330 |

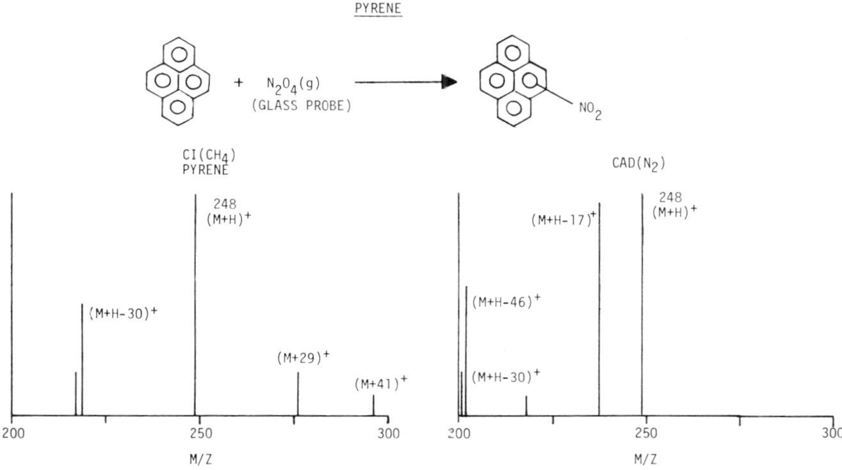


**Figure 7.** Spectra obtained for pyrene when using positive ion CI (left) and collisionally activated decomposition MS/MS (right).

In summary, when compared to GC/MS, MS/MS provides a technique which is faster, often shows higher specificity, is less destructive to the sample and is capable of analyzing some samples not amenable to GC/MS analysis.

## REFERENCES


1. Kagel, R. O., R. H. Stehl and W. B. Crummett. "Priority Pollutant Analysis: Problems and Opportunities," *Anal. Chem.* 51:223A (1979).
2. Finnigan, R. E. "The Role of GC/MS in Environmental Sample Analysis," Proceedings of the 42nd Annual Meeting of the American Council of Independent Laboratories (1979).
3. "Guidelines Establishing Test Procedures for the Analysis of Priority Pollutants," *Federal Register* 44(233) (1979).

4. Finnigan, R. E., D. W. Hoyt and D. E. Smith. "Priority Pollutants. II. Cost Effective Analysis," *Environ. Sci. Technol.* Vol. 13 (1979).
5. "Consolidated Permit Regulations," *Federal Register* 45(98):33542.
6. Yost, R. A., and C. G. Enke. "Selected Ion Fragmentation with a Tandem Quadrupole Mass Spectrometer," *J. Am. Chem. Soc.* 100:2274 (1978).
7. Hunt, D. F., J. Shabanowitz and A. B. Girodani. "Collision Activated Decompositions of Negative Ions in Mixture Analysis with a Triple Quadrupole Mass Spectrometer," *Anal. Chem.* 52:386 (1980).

## CHAPTER 35

# APPLICATIONS OF A DATA-ADAPTIVE BACKGROUND SUBTRACTION TECHNIQUE TO THE GAS CHROMATOGRAPHIC/MASS SPECTROMETRIC ANALYSIS OF WASTEWATER SUPPLIES

**A. D. Sauter,* L. Faas and G. R. Kurth**

Computer Sciences Corporation
Falls Church, Virginia

**W. M. Shackelford and D. M. Cline**

U.S. Environmental Protection Agency
Environmental Research Laboratory
Athens, Georgia

In recent years gas chromatography/mass spectrometry (GC/MS) has become a routine tool for the analysis of wastewater for organic compounds. The U.S. Environmental Protection Agency (EPA), in its recent study of industrial waste treatment technology, relied heavily on GC/MS techniques to find and quantify organic priority pollutants in industrial wastewater samples [1-3]. In this study, compounds of interest were identified through mass spectrometric (MS) and retention time data. Internal standards were used in these analyses to provide reference data for quantification and identification.

*Present address: U.S. Environmental Protection Agency, Environmental Monitoring and Support Laboratory, Las Vegas, NV.

Use of retention indices coupled with MS data for identification purposes was reported by Nau and Biemann [4] and Sweeley et al. [5]. Subsequently, automated systems in several laboratories [6-9] have been reported which rely on retention data as well as MS data for computer-aided identification of a known set of organic compounds.

As noted by Nau and Biemann [4], the use of internal standards for qualitative and quantitative purposes within an analysis necessitates that the internal standard is able to be located unambiguously and its intensity is able to be measured accurately. In analyzing highly predictable matrices, one can choose internal standards that have few interferences from the background and are present in sufficient quantity to be measured adequately. In cases where the internal standard has matrix interference, problems arise quickly.

This work resulted from problems encountered in a computer survey [9] of thousands of GC/MS analyses acquired in 18 different laboratories on a variety of GC/MS instrumentation under similar conditions. Common among all analyses were internal standards that could be used to compare the retention data of the various compounds and to provide a means for quantification. Here, as in the systems cited above, finding the internal standard was an indispensible step in the analysis. Quantification and identification depended on the ability of an operator or a computer to accurately gather MS data and retention data in the presence of varying degrees of interference.

In the initial phases of the computer survey, it became obvious that locating the internal standards was beyond the capability of the automated system in some cases. For example, the sample analysis procedure called for adding the internal standard at a concentration of 20 μg/mL regardless of sample complexity. In cases where 20 μg/mL was an extremely minor concentration, the computer system lacked the sensitivity to extract a spectrum that the spectrum matching program could identify. Consequently, work in this case had to cease until the internal standard could be located manually. Also, the superpositioning of ion signals from the internal standard on either a noisy baseline or a sharply rising or falling baseline made spectrum extraction and subsequent matching haphazard and made the computer survey in this case impossible without manual assistance.

To remedy this situation, a preprocessor to the survey system was constructed to look specifically for the internal standard. The ion current profiles of key internal standard ions were examined for simultaneous peaking and at such a point, provided the relative abundances met certain criteria, the internal standard was said to be located. Although this remedy eliminated many problems associated with sensitivity, baseline problems continued to hamper peak recognition and relative abundance calculation.

An automated data-adaptive solution to some of these problems has been published by Summons et al. [10] and was used to form the basis of our approach. In the earlier work, data were collected under carefully controlled conditions in one laboratory. Because our data were received from many sources and acquired from a variety of sample matrices, the work by Summons et al. [10] was modified to make the system even more flexible and data-adaptive.

These modifications, as well as the performance of the automated system on field data, are presented for several cases of data complexity. The applicability of this technique to reverse search analysis is also demonstrated.

## EXPERIMENTAL

### Instrumentation

All data were acquired on commercially available integrated GC/MS/data systems. As shown in Table I, 18 different systems in 14 laboratories were involved in the study.

Computer programs were run on a Digital Electronics Corporation (DEC) PDP 11/70 computer with 256K words of memory, 2K words of cache memory, two 88-Mbyte disk drives (DEC RP04), two 9-track magnetic tape drives (DEC TU16), a Calcomp 936 plotter and a Tektronics 4014 display terminal. Statistical analyses were performed using the Biomedical Computer Program P-Series package [11].

**Table I. Instrumentation for GC/MS Survey with Contributing Laboratories**

1. Finnigan 3200 MS, Varian 1200 GC, System Industries data system–Region IV (EPA), Region II (EPA), Region VII (EPA)
2. Finnigan 4000 MS, Finnigan 9600 GC, Finnigan-Incos data system–Region II (EPA), Region IV (EPA), Region VII (EPA), Carborundum Co., West Coast Technical Service, Jacobs Engineering, Analytical Research Laboratories, Inc.
3. Dupont Dimaspec MS, Finnigan 9600 GC, Finnigan-Incos data system–Systems, Science, and Software; TRW
4. Hewlett-Packard 5985 GC/MS/data system–Radian Corp., Gulf South Research Institute, Envirodyne Engineers
5. Hewlett-Packard 5982 GC/MS/data system–Radian Corp., Foremost Foods, Energy Research Corp.

## Laboratory Procedures

Data submitted to this laboratory for study consisted of raw GC/MS data acquired in a survey of industrial wastewater. The procedures used for analysis were recently published [3]. Purgeable, base/neutral and acid extractable fractions were analyzed by GC/MS for each sample taken in the field. One internal standard, anthracene-$d_{10}$, was added to the final extract of the acid and base/neutral fractions. Two internal standards, bromochloromethane and 1,4-dichlorobutane, were added to the purgeable fraction. During the whole of the study, one GC column liquid phase was used for purgeable analysis (0.2% Carbowax 1500), two liquid phases were used for acids analysis (Tenax GC and 1% SP-1240 DA); and three liquid phases were used for base/neutrals analysis (1% SP-2250, 3% SP-2250, 3% SP-2250 DB). Column performance was assured by the chromatography of test compounds (benzidine, pentachlorophenol) and GC/MS performance was required to meet specifications set forth by Eichelberger et al. [12].

## Data Analysis Procedure

A complete description of the data analysis system and procedures for analysis can be found elsewhere [9]. The procedure used for specific ion searches, whether for internal standard location, target compound location or compound class searching, involves both manual and automated operations. The amount of manual intervention called for depends on the degree to which the algorithms are tailored to each particular sample run. In routine cases, parameters were chosen that had been shown to give generally good results for the widest variety of run types. For special cases, such as a class specific search for polychlorinated biphenyl moieties, program parameters are varied until the best fit of the data is obtained.

## Background Subtraction

The background correction technique as reported by Summons et al. [10] utilizes a "piecewise" least-squares fit approach to estimate the local background in spectrum number ranges. In our work, extracted ion current profiles (EICP) are divided into sequences of N intensity values. Each sequence is further divided into intervals. The number of intervals (NUMINT) and the number of spectra in an interval (NUMSCAN) are user-defined variables employed in this technique. The product of these variables is the sequence size.

A program operates on raw data as follows. For each EICP sequence, the lowest intensity in each interval is determined. A least-squares line is calculated for these points within each sequence. Next all intensity values that fall above this first least-squares line are eliminated, and a new line is calculated with the remaining data points. In this way points that are not on the base line such as those between unresolved multiplets are eliminated from background consideration.

Summons et al. [10] employed a spline function to estimate the background between adjacent sequence midpoints:

$$f(\epsilon) = 1 - 3\epsilon^2 + 2\epsilon^3 \tag{1}$$

This joining function is evaluated at each spectrum between the midpoints of adjacent sequences by:

$$\epsilon = \frac{t - t_{mid}(i)}{t_{mid}(i+1) - t_{mid}(i)} \tag{2}$$

$\epsilon$ is the scan number of interest in this range and $t_{mid}(i)$ and $t_{mid}(i+1)$ are the midpoint scan or spectrum numbers of the respective sequences $S_I$ and $S_{I+1}$.

Figure 1 is a graphic representation of Equations 1 and 2. The background intensity is then calculated by:

$$B(t) = f(\epsilon)\, P_i\,(t) + [1 - f(\epsilon)]\, P_{i+1}\,(t) \tag{3}$$

where $P_i(t)$ and $P_{(i+1)}$ (t) are the second least-square approximations for sequence I and I + 1.

The background intensities [B(t)] are then subtracted from the raw data. Simple and complex EICP studied initially gave results that were encouraging because the ion current removed in most cases agreed well with manually estimated background values.

An example of a simple case is shown in Figure 2 with both raw and background-corrected EICP presented. The peak height was estimated manually for the raw data at approximately 1990 intensity units. The peak height resulting after automated background subtraction (and before thresholding) was 2030 intensity units. The agreement between manual and automatic background subtraction techniques for this simple EICP is representative of our observations. This example serves to demonstrate that the technique produces results that are consistent with manual compensation. The power of this approach becomes apparent when complex EICP are studied.

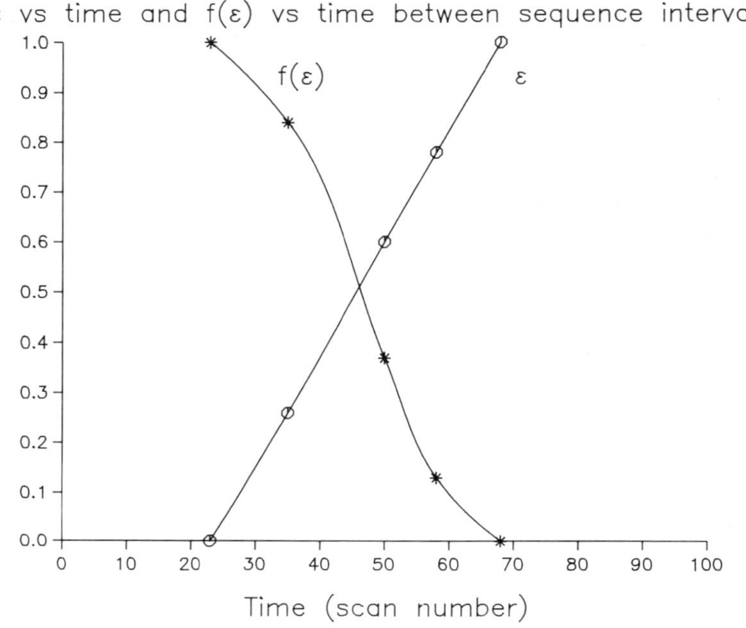


**Figure 1.** Plot of f ($\epsilon$) vs time for Equation 1 and $\epsilon$ vs time for Equation 2.

Figure 3 shows a complex EICP. By inspection, it is apparent that the relative intensities of peak maxima were preserved along with EICP fine structure. To demonstrate that manual and automatic background-corrected intensities are in agreement, manual peak height measurements are compared to background corrected data in Table II for the eight peaks labeled in Figure 3. The manual data represents the average of four chemists' estimations of peak heights above background. The automated background compensation routine generally falls well within one standard deviation of the corresponding manual value. The apparent positive bias of the automated values is probably due to the fact that the background line is constructed through peak minima.

Table II indicates that the automated background subtraction technique can produce results consistent with manual background corrected estimation of peak heights. The EICP in Figure 3 contains resolved and unresolved peaks of varying peak shapes and consequently represents a difficult test. Thus, it

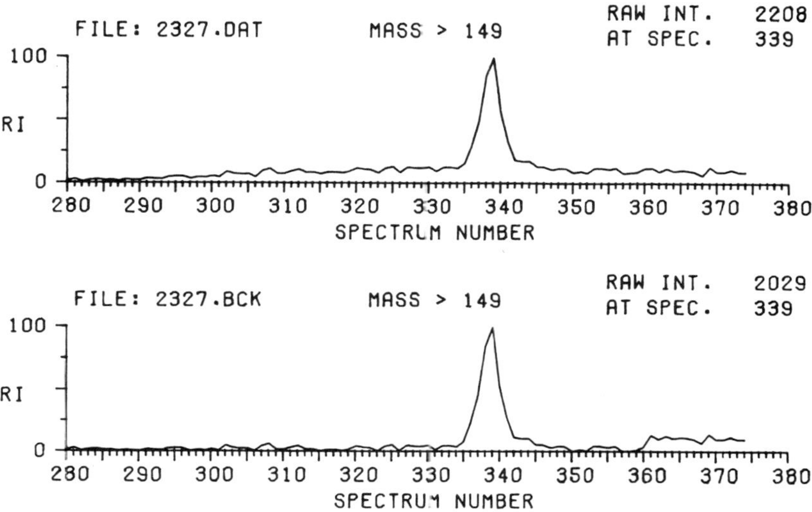


**Figure 2.** Background subtraction for the simple case of an isolated large peak. The upper plot depicts raw data.

**Table II. Calculated Peak Heights**

| Peak Number[a] | Peak Height: Manual[b] | Peak Height: Automatic | Δ |
|---|---|---|---|
| 1 | 1050 ± 36 | 1045 | -5 |
| 2 | 1932 ± 49 | 1947 | +15 |
| 3 | 1902 ± 56 | 1909 | +7 |
| 4 | 1621 ± 35 | 1636 | +15 |
| 5 | 261 ± 31 | 273 | +12 |
| 6 | 1462 ± 62 | 1462 | 0 |
| 7 | 597 ± 19 | 591 | -6 |
| 8 | 1253 ± 15 | 1268 | +15 |

[a]Refer to Figure 3.
[b]Average of observations by four chemists.

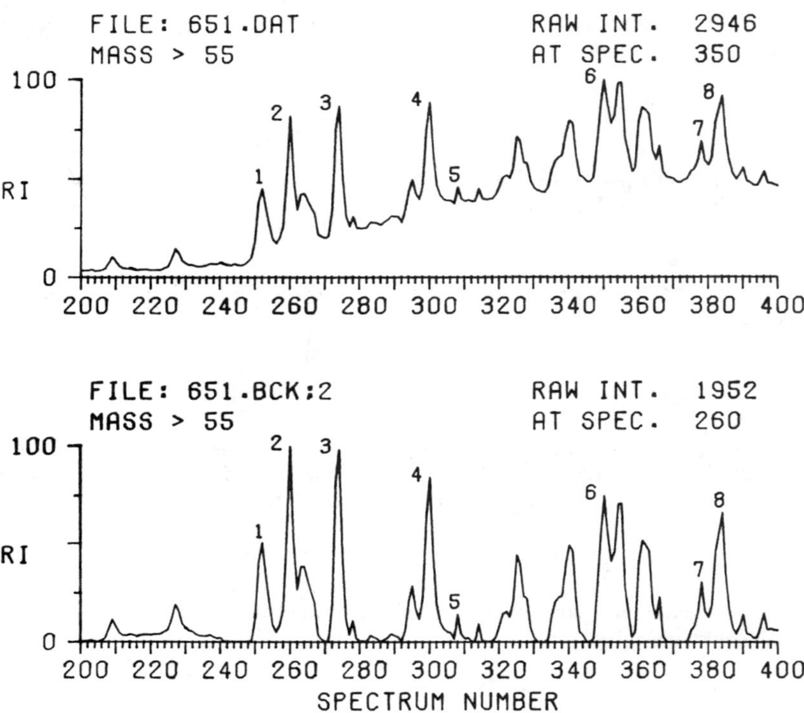


**Figure 3.** Background subtraction for a complex case. The upper plot depicts raw data.

can be seen that the automated background subtraction technique can accurately compensate for the fraction of the ion current due to "background" for major peaks in a given complex EICP. Minor peaks in clusters have not undergone background correction in many cases, and as such, represent a limitation of the background correction technique as applied to data that we encountered in our work. Oftentimes in practice this does not represent a major limitation because identification of species of relatively low concentration (in the work studied [9] less than 10 ppb) are of limited interest. Adjusting the user-defined variables can compensate to a degree for this property of the background correction program.

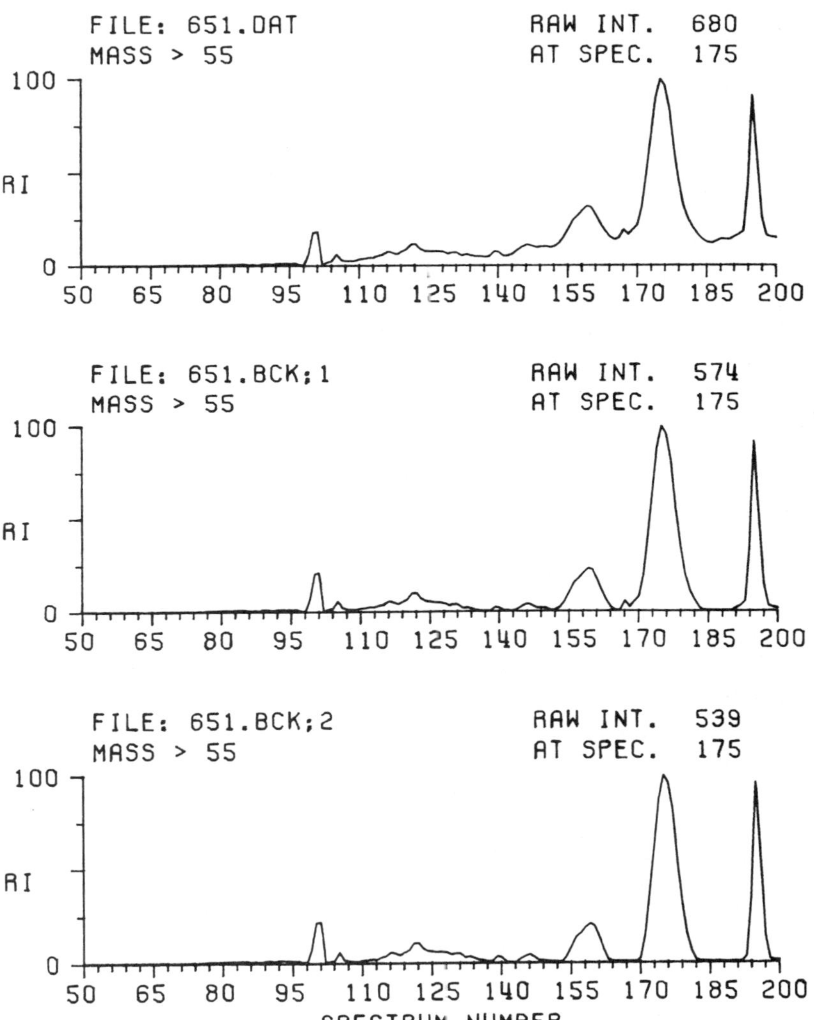


**Figure 4.** Effect of NUMSCAN/NUMINT on background subtract: (top) uncorrected data; (middle) data corrected with NUMSCAN/NUMINT set at 5/9; (bottom) NUMSCAN/NUMINT set to 5/15.

The otpimum value of the user-defined variables (NUMSCAN and NUMINT) is peak geometry-dependent. For example, when NUMSCAN is lowered to 3 (spectra/interval) and NUMINT is raised to 15 (intervals/sequence) maintaining a constant sequence size (45 spectra), the broad peak (Scan 176) in Figure 4 is too aggressively subtracted. The area for this peak is only 90% of the value attained when NUMSCAN and NUMINT are set to 5 and 9, respectively, and only 92% of the chemist estimated value. Some sharp peaks (shown in Figure 3), however, showed an equivalent or improved agreement with the manually estimated peak height with these values of the user-defined parameters.

Another example of the effect of the user-defined variables NUMINT and NUMSCAN is illustrated below. Figure 5 shows the background correction technique applied with default values for NUMSCAN and NUMINT (5 and 9). The manually estimated peak height (A) for the raw data was 570 intensity units. Figure 4B has a chopped appearance, and the peak height (474) is considerably lower than the manually estimated value. By changing NUMSCAN and NUMINT to values of 10 and 8, the resultant peak height was increased to 575 intensity units (5C), which is in better agreement with the manual peak height estimation. This observation served to illustrate that peak width considerations can affect the results of this technique even for simple EICP. Because EICP can contain peaks or clusters of varied geometry and complexity, it is unlikely that a single, fixed value for the user-defined variables NUMSCAN/NUMINT will be adequate for all peaks in a complex EICP. Because our initial application of the background subtraction technique was in the location of internal standards, NUMSCAN/NUMINT values were chosen that gave background corrected peak heights in agreement with manually estimated peak heights for major fragments in the mass spectrum of the internal standards in simple EICP. For data collected on GC columns except Tenax GC, the NUMSCAN/NUMINT values were set to 5/9. For Tenax GC values of NUMSCAN/NUMINT were set at 10/8.

## Noise Thresholding

After the best background line has been removed from the raw data, detection of the peak representing the component is initiated. In the work of Summons et al. [10], an intensity threshold was set to compensate for noise peaks that were above the corrected background line. The intensity threshold was set by examining the distribution of intensities within the background corrected data set. A data-adaptive threshold was calculated by

$$\text{threshold} = \text{mode} + 2.5 \times (\text{standard deviation from the mode}) \quad (4)$$

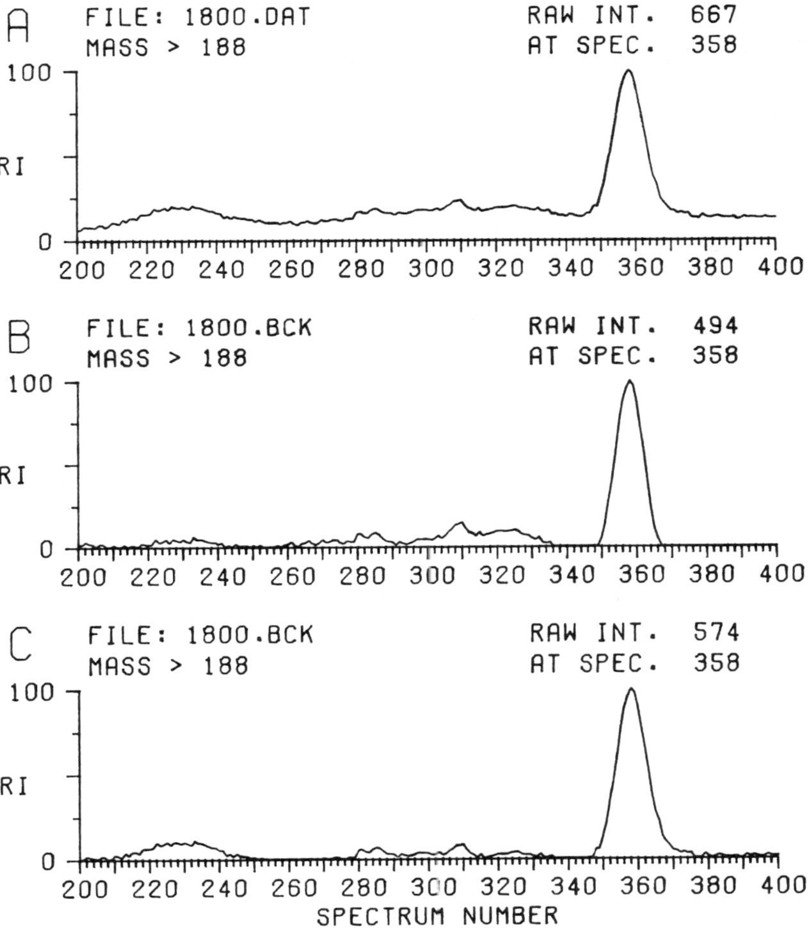


**Figure 5.** Effect of NUMSCAN/NUMINT on background subtract: (top) uncorrected; (middle) corrected with NUMSCAN/NUMINT set at 5/9; (bottom) corrected with NUMSCAN/NUMINT set at 10/8.

and appeared to give excellent noise compensation. In the present application, however, this approach to thresholding gave high values for the threshold that resulted in some peaks of interest falling below the threshold value. Our data did not exhibit good noise statistics because the usual mode of acquiring data involves setting an instrumental threshold for the noise. Thus, the noise looks like spikes on a flat baseline rather than a sinusoidal variance about some mode value. Although the peaks of the noise spikes may vary in height, the minima are invariably equal to zero. As the uncorrected baseline rises above the abcissa, use of the least-squares technique puts some minima above the calculated basline and some below it. The variance of these points about the calculated baseline not only involves instrument noise but also usable data. Thus, in our work the standard deviation from the mode value was too great to use for a signal threshold.

Intensity distributions of background-corrected EICP were examined to derive an alternative threshold-setting procedure. A typical ranged intensity distribution for one of the columns tested is shown in Figure 6 (raw and background-corrected EICP are also shown in Figure 6). This ranged intensity distribution is representative of the positively skewed distributions that resulted after background subtraction of simple EICP. The degree to which the raw data influenced these background corrected intensity distributions was not delineated; but it was recognized that the raw data, the algorithm employed and other factors could predispose the skew of the distribution resulting from background correction. For example, when many zero intensity values occur in the raw data of any EICP, the background-corrected EICP intensity distribution is likely to have a mode of zero intensity units. The degree to which these distributions represent the effect of the background correction technique as opposed to the raw data characteristics prior to background subtraction is a question that we did not address quantitatively. Empirically, we found that simple EICP yielded a positively skewed distribution with a zero value mode.

Other distribution types were noted also. For example, when ion current fluctuations around a given discrete value (above zero) were apparent in the background-corrected EICP, the distribution had a higher degree of symmetry and was generally less peaked and gave modes at greater than zero intensity units. A partial EICP and the total intensity distribution resulting after background correction is shown in Figure 7 for a case which could be considered representative of random noise characteristics.

It became apparent that a truly useful solution in our initial application, the location of the internal standard, had to be capable of handling positively skewed distributions of different types with occasional nonunique modes. Complex EICP were investigated first.

Complex EICP can be viewed as consisting of a mixture of data and background noise intensities. The relative proportion of each indicates the degree of complexity for the m/z value in the data file in question. For the threshold to be established in complex data files, a mechanism

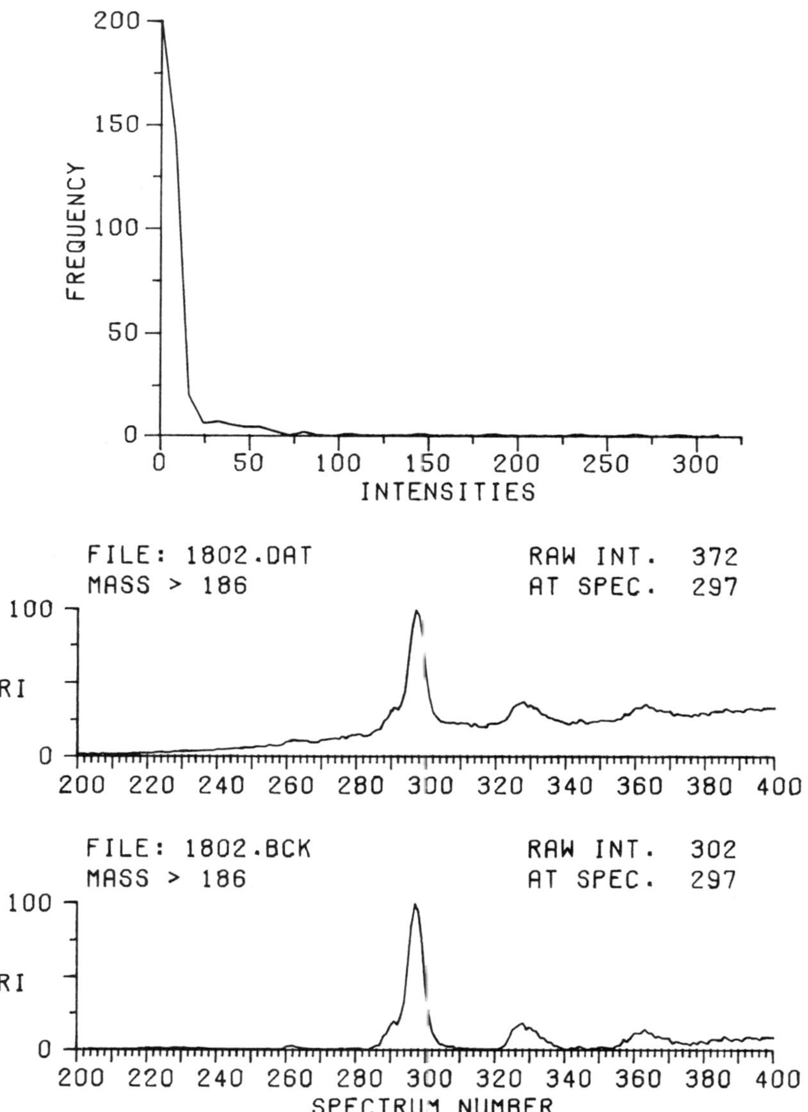


**Figure 6.** Plot of ranged intensity distribution for mass 186 after background correction. Chromatograms show uncorrected (top) and background corrected (bottom) data.

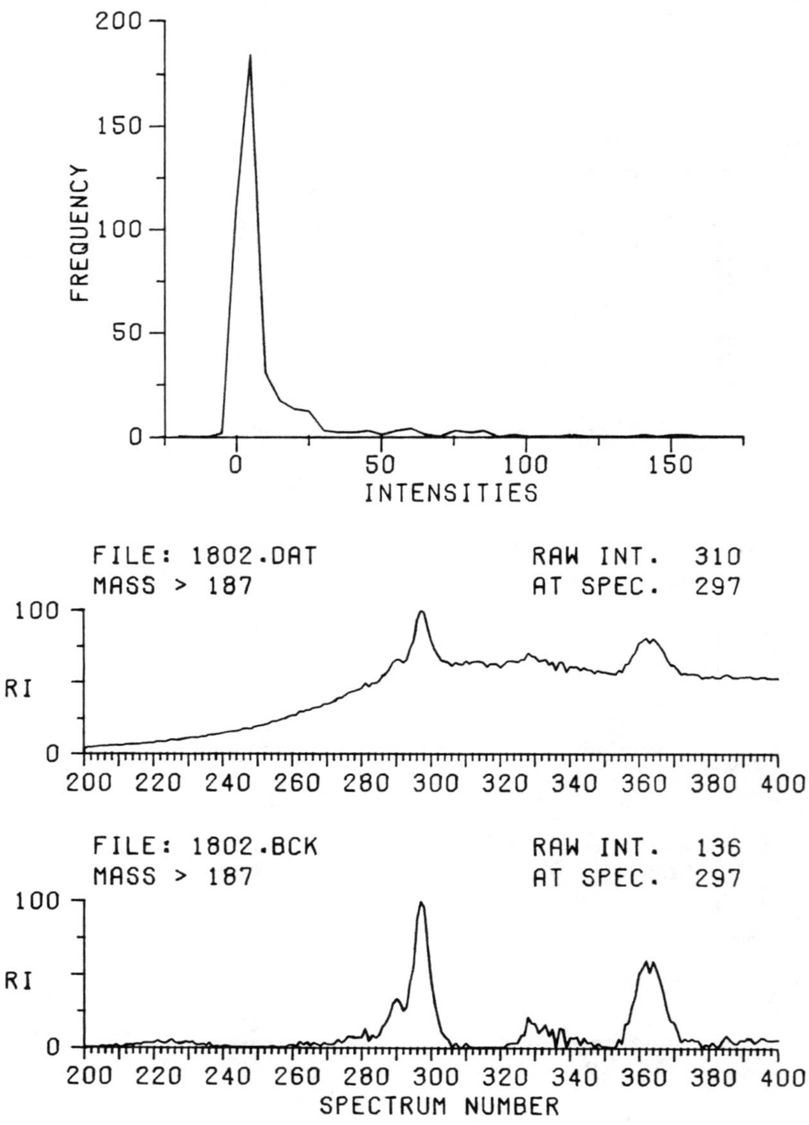


**Figure 7.** Plot of ranged intensities for normal distribution. Uncorrected and corrected data shown in lower plots.

is required to separate data from background noise. By examining lists of intensity distributions for background corrected EICP, it became apparent that discrete intensity values with a low frequency of occurrence were predominantly data. Even for EICP with regularly spaced peaks, this trait was apparent, as can be seen in Figure 8.

In Figure 8, a listing of discrete background intensity values and their frequency of occurrence for the entire EICP is represented. Two distinct groups of intensity value/frequency pairs can be observed in this list: high intensity/low frequency and low intensity/high frequency. Typically, any positive intensity value with a frequency of occurrence of two or less was a value that we would have assigned to the data region in manual integration. This was noted with few exceptions for a variety of complex EICP, and subsequently the intensity frequency was utilized to separate background and data by eliminating intensity values with frequency of 2 or less from the background-corrected intensity distribution. It remained to determine the exact intensity level that would define the noise level in the frequency filtered distribution.

Ideally, this technique had to handle different distributions successfully because different distributions were observed in our data. For this reason, we decided to set the threshold independent of assumptions about the distribution type. By applying the EICP peak-finding program to EICP with easily observed internal standards, minimum confusion was produced when the threshold was set at the 99th percentile of intensities in the frequency filtered background-corrected EICP. In the case of a Gaussian distribution, our procedure for threshold determination results in virtually identical results with that of Equation 4; however, our approach is flexible because the distribution is not required to be of any one type.

## Peak Detection

A rather simplistic peak detection algorithm is employed to point to the location of the sought components. Our experience has been that the conditions for internal standard location are rigorous enough, however, to preclude false positive identifications. The algorithm basically looks for a peak in the EICP that is above the threshold value. If such a feature is found, a second EICP characteristic of the component of interest is examined for a corresponding peak with maximum intensity ±2 scans from the first peak maximum. The more EICP that are examined, the more rigorous is the identification procedure. If simultaneous peaking is not observed (±2 scans), subsequent features in the EICP are examined until the end of the EICP is reached. Our procedure utilizes eight ions for identification of the internal standards. As few as two ions, provided they are judiciously chosen, have

INTENSITY / FREQUENCY

| | | | | | | | |
|---|---|---|---|---|---|---|---|
| -14. | 1 | 21. | 2 | 53. | 2 | 148. | 1 |
| -12. | 1 | 22. | 6 | 54. | 1 | 149. | 1 |
| -10. | 1 | 23. | 1 | 56. | 1 | 153. | 2 |
| -8. | 1 | 24. | 5 | 58. | 1 | 154. | 2 |
| -5. | 1 | 25. | 8 | 60. | 1 | 162. | 1 |
| -4. | 1 | 26. | 5 | 61. | 2 | 164. | 1 |
| -3. | 1 | 27. | 4 | 62. | 2 | 172. | 1 |
| -2. | 3 | 28. | 3 | 63. | 2 | 176. | 1 |
| -1. | 6 | 29. | 5 | 64. | 1 | 177. | 1 |
| 0. | 350 | 30. | 9 | 68. | 2 | 182. | 2 |
| 1. | 28 | 31. | 3 | 69. | 1 | 192. | 1 |
| 2. | 7 | 32. | 7 | 70. | 1 | 195. | 1 |
| 3. | 17 | 34. | 5 | 71. | 2 | 198. | 1 |
| 4. | 19 | 35. | 9 | 79. | 1 | 211. | 1 |
| 5. | 12 | 36. | 3 | 81. | 1 | 215. | 1 |
| 6. | 10 | 37. | 2 | 83. | 1 | 223. | 1 |
| 7. | 7 | 38. | 1 | 84. | 2 | 248. | 1 |
| 8. | 5 | 39. | 4 | 88. | 1 | 261. | 1 |
| 9. | 3 | 40. | 4 | 91. | 1 | 262. | 1 |
| 10. | 2 | 41. | 2 | 94. | 1 | 263. | 1 |
| 11. | 1 | 42. | 2 | 100. | 1 | 292. | 1 |
| 12. | 2 | 43. | 5 | 101. | 1 | 322. | 1 |
| 13. | 7 | 44. | 2 | 105. | 1 | 335. | 1 |
| 14. | 2 | 45. | 1 | 107. | 1 | 369. | 1 |
| 15. | 6 | 46. | 4 | 112. | 1 | 372. | 1 |
| 16. | 14 | 47. | 2 | 134. | 1 | 392. | 1 |
| 17. | 1 | 48. | 4 | 136. | 1 | 462. | 1 |
| 18. | 1 | 49. | 1 | 137. | 1 | | |
| 19. | 6 | 50. | 2 | 139. | 1 | | |
| 20. | 4 | 51. | 2 | 147. | 1 | | |

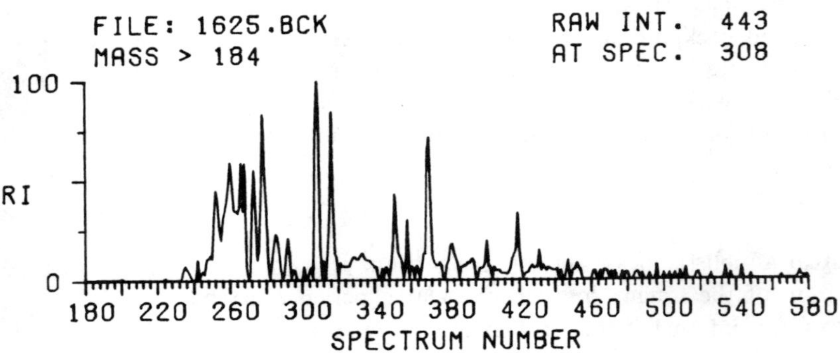


**Figure 8.** Background corrected EICP and frequency distribution for complex case.

been shown to be effective for compound class recognition. The component location software is capable of examining the data for features in 30 EICP concurrently.

## RESULTS AND DISCUSSION

### Internal Standard Location

The automated GC/MS data survey system (CLEANUP/PBM) [9] in initial tests was unable to find the internal standard in about 50% of the data. The initial test of this approach, then, was to locate the internal standard in the mass of GC/MS data for each sample. The background correction and thresholding technique was applied to selected EICP belonging to the mass spectrum of the internal standard for the acid and base/neutral fractions, anthracene-$d_{10}$. Initially, 12 data files from the same contract laboratory were chosen that used the same GC columns (1% SP-1240DA). Six of these files were examples of cases in which CLEANUP/PBM had failed to find the internal standard. On application of the preprocessor to all 12 data files, results were obtained that agreed with CLEANUP/PBM. For the data files in which no internal standard was located by either technique, manual examination of the files in question also found no internal standard.

It was hoped that this test would demonstrate that the preprocessor could locate the internal standard when CLEANUP/PBM had failed, but because these files did not have internal standards another test was required. At the same time, however, it is important to note that the preprocessor showed excellent resistance to false positive. As importantly, the ability to "flag" data files with no internal standard was demonstrated.

To demonstrate the ability of the approach to locate the internal standard where CLEANUP/PBM had failed, hundreds of such data files were run through the preprocessor. The preprocessor located anthracene-$d_{10}$ in 90% of these data files in agreement with their manually assigned retention time, but 5% could not be found at all. The reason that the internal standard was not located in some cases by the preprocessor was apparently due to the low absolute intensity values of the low abundance, low mass fragments. As a result, peaks were not found for these masses and the preprocessor algorithms rejected these runs.

In Figure 9, possible pitfalls for automated identification of anthracene-$d_{10}$ by specific mass search is shown. Masses 188 and 186 are both seen at scans 366 and 485. An examination of the relative intensities indicates that at both scans anthracene-$d_{10}$ is indicated. Examination of the spectra shows

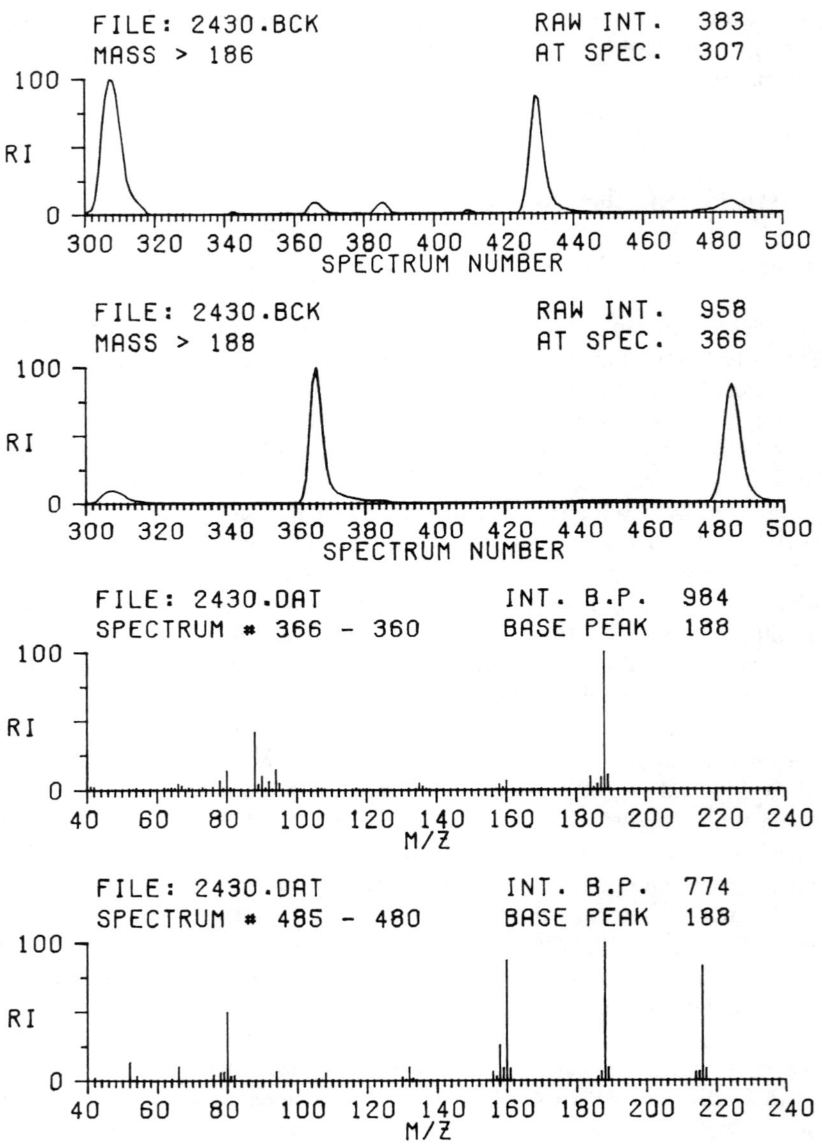


**Figure 9.** EICP plots for mass 188 and 186 with spectra for anthracene-$d_{10}$ and anthraquinone-$d_8$.

that the peak at 485 is due to anthraquinone-$d_8$ while anthracene-$d_{10}$ is at 366. Thus, either retention data or additional masses are necessary to give a high probability identification. In the case of finding an internal standard, only additional masses are appropriate when dealing with a wide variety of data sources.

Over the course of the project, the anthracene-$d_{10}$ recognition rate increased approximately 90-95% overall by utilizing the preprocessor coupled with the CLEANUP/PBM modules in the system. In general, manual examination of the file in question is warranted only on about 5-10% of all data at this time.

In the purgeable fraction, both bromochloromethane and 1,4-dichlorobutane were used as internal standards. Both compounds are plagued by interferences in many sample matrices. Problems arising in location of the internal standard bromochloromethane after background removal are shown in Figure 10. In the topmost frame m/z 49 is depicted. This ion is essential in the identification of bromochloromethane, but in this illustration its occurrence at spectrum 130 is dwarfed by the huge peak centered about spectrum 100. In fact use of the threshold (mode + 2.5 $\sigma$ mode) eliminates the peak at 130. Use of the frequency filtering technique for determining threshold (99th percentile of the intensities occurring 3 or more times) preserves this peak in the presence of a much larger one. In the lower frames of Figure 10, the importance of choosing internal standards that have little interference from sample components is shown. Mass 128 is the molecular ion for 1,4-dichlorobutane-$^{35}$Cl, $^{37}$Cl. The low abundance of m/z 128 requires that other ions be used for identification. Ions at m/z 90 and m/z 92 can be used in many cases because the isotopic abundance ratios (for $C_4H_7{}^{35}Cl^+$ and $C_4H_7{}^{37}Cl^+$) should give strong indications for detection. In Figure 10 it can be seen that m/z 128 is barely discernible at scan 326, and m/z 90 and 92 are completely obscured by toluene. The most intense ion of the 1,4-dichlorobutane spectrum, 55, is of little use due to an extremely convoluted EICP as shown in Figure 3. In this case the automated preprocessor gives negative results—as it should.

## Location of Chlorinated Biphenyls

Another application of a compound specific presearch was in detecting the presence of chlorinated bipheyls. A data file in which the CLEANUP/

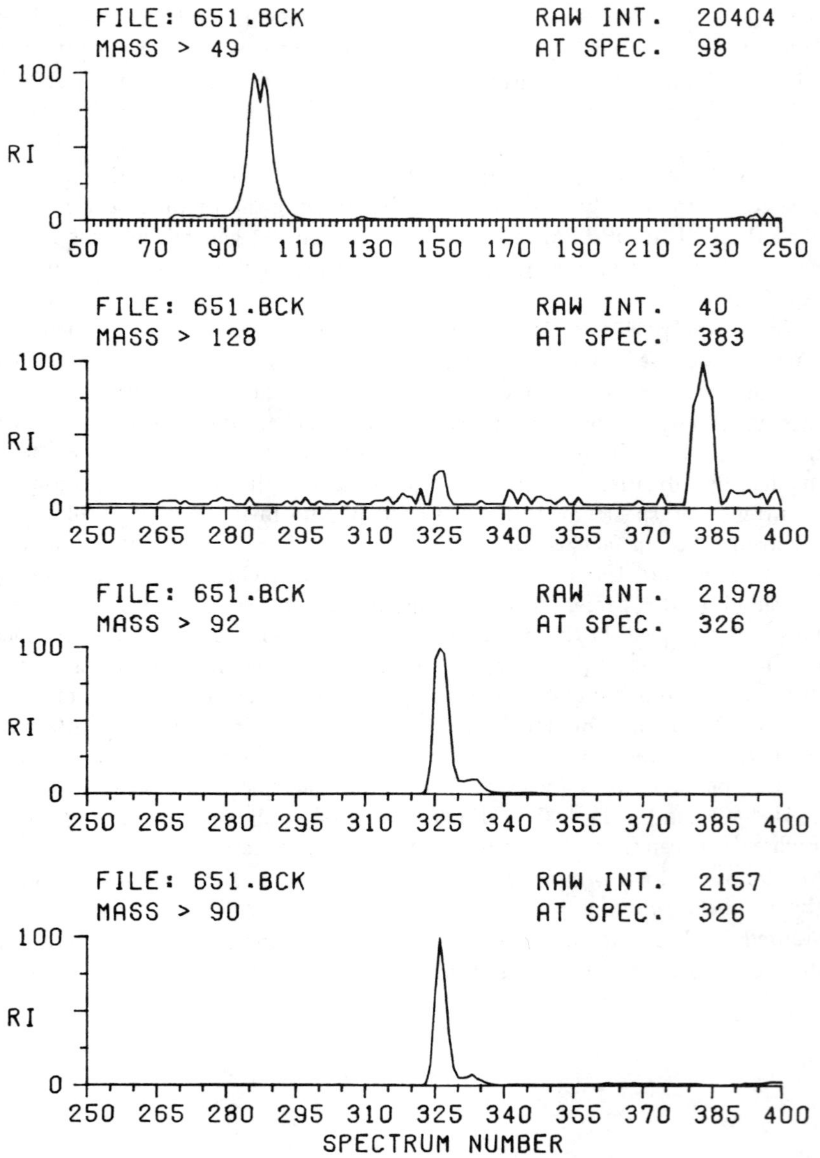


**Figure 10.** Illustration of problems in finding bromochloromethane (top) and 1, 4-dichlorobutane (lower three plots).

PBM programs had detected a tetrachlorobiphenyl species was selected for initial study. Utilizing the four most intense masses in a reference mass spectrum (m/z 292, 220, 290 and 294), we applied background correction with intensity threshold determination and the EICP peak-finding routine to determine whether we could locate the same tetrachlorobiphenyl species that CLEANUP/PBM had found. Our approach located the compound at the spectrum number in agreement with CLEANUP/PBM. The background-corrected EICP had relative intensities for the selected m/z values in good agreement with the standard reference spectrum matched by PBM as shown in Table III.

Moreover, three additional eluents were detected in this data file with similar mass spectral characteristics at nearby retention times. Subsequent manual examination of these spectra verified the 4 and 2 chlorine cluster in the m/z 290 and 220 mass range. In summary, our approach agreed with CLEANUP/PBM for the location of the tetrachlorobiphenyl based on EICP of selected mass. As importantly, additional eluents with similar mass spectral properties to tetrachlorobiphenyl were indicated for further study. This application demonstrated the usefulness of the preprocessor in extracting selected scans in complex data files for manual study. A similar approach was employed to search 65 GC/MS data files for the presence of selected chlorinated biphenyls.

In this application, we sought to provide, where possible, GC/MS confirmation to GC (electron capture) results that had indicated the presence of chlorinated biphenyls in effluents of a given industrial category. To this end, 65 GC/MS data files (greater than 25,000 mass spectra) were preprocessed in the fashion outlined above for mono-, di-, tri- and tetrachlorobiphenyl. The peak-finding algorithm required that three of four characteristic masses exhibit themselves in the thresholded, background corrected data within ±1 scan of each other. This presearch resulted in condensing the 25,000

**Table III. Relative Intensity Agreement**

| Tetrachlorobiphenyl Library Entry No. 13110 CASRN 11097691 | | Sample Data Relative Intensities after Background Correction | |
|---|---|---|---|
| m/z | RI (%) | m/z | RI (%) |
| 292 | 100 | 292 | 100 |
| 220 | 83 | 220 | 86 |
| 290 | 79 | 290 | 77 |
| 294 | 48 | 294 | 53 |

mass spectra to 35 mass spectra. For all but two of these spectra, manual examination determined that chlorinated biphenyls were *not* present. In the two data files remaining, chlorinated aromatics were present, but the eluents were not the chlorinated biphenyls. Manual study showed that these eluents were probable di- and trichloro nitrogen-containing aromatic species.

All data files in question were also studied manually to confirm the preprocessor results. The results of manual examination (plotting EICP base peaks for selected m/z values for selected chlorinated biphenyls) agreed with the presearch in that the compounds searched for were not indicated in the data.

The usefulness of this technique in examining large volumes of GC/MS data is apparent. We do not mean to imply, however, that this approach should be employed unequivocally to determined presence/absence of organic compounds in any data file without prior knowledge of retention data. Nor do we intend to imply that this type of EICP presearch can replace more in-depth manual or computer-based data analysis for survey analysis. Rather, we have demonstrated that this simplistic approach can be used to condense large volumes of GC/MS data to a humanly manageable fraction of the total data analysis task. The risk involved in misassignment is dependent on the "uniqueness" of the m/z values selected, so that this technique should always be used with caution and with other confirmatory technqiues. In practice, when "unique" ions are prevalent in the mass spectrum of interest, this approach can be employed in large volumes of GC/MS data to highlight spectra for further examination. The fact that this approach has been applied successfully to large volumes of GC/MS data generated by a variety of analytical laboratories utilizing different GC/MS instrumentation and chromatographic configurations is noteworthy.

## DISCLAIMER

Mention of trade names or commercial products does not constitute endorsement or recommendation for use by the U.S. Environmental Protection Agency.

## REFERENCES

1. Shackelford, W. M., and L. H. Keith. "Evolution of the Priority Pollutant List from the Consent Decree," in *Proceedings of the Second Open Forum on Management of Petroleum Wastewater*, U.S. EPA Report EPA-600/2-78-058, Ada, OK (1978), pp. 103-109.

2. Keith, L. H., and W. A. Telliard. "Priority Pollutants. I. A Prospective View," *Environ. Sci. Technol.* 13(4):416-423, 1979.
3. "Guidelines Establishing Test Procedures for the Analysis of Pollutants; Proposed Regulations," *Federal Register* 44(233):69464-69575 (1979).
4. Nau, H., and K. Biemann. "Computer-Assisted Assignments of Retention Indices in Gas Chromatography-Mass Spectrometry and Its Application to Mixtures of Biological Origin," *Anal. Chem.* 46(3):426-434 (1974).
5. Sweeley, C. C., N. D. Young, J. F. Holland and S. C. Gates." Rapid Computerized Identification of Compounds in Complex Biological Mixtures by Gas Chromatography-Mass Spectrometry," *J. Chromatog.* 99:507-517 (1974).
6. Smith, D. H., M. Achenbach, W. J. Yaeger, P. J. Anderson, W. L. Fitch and T. C. Rindfleisch. "Quantitative Comparison of Combined Gas Chromatography/Mass Spectrometric Profiles of Complex Mixtures," *Anal. Chem.* 49(11):1623-1632 (1977).
7. Tsuchiya, Y., J. Boulanger and K. Sumi. "Identification of Chemical Compounds by Computer Matching of Low Resolution Mass Spectra and Retention Indices," *Chromatographia* 10(3):154-156 (1977).
8. Blaisdell, B. E. "Automatic Computer Construction, Maintenance, and Use of Specialized Joint Libraries of Mass Spectra and Retention from Gas Chromatography-Mass Spectrometry Systems," *Anal. Chem.* 49(1):180-186 (1977).
9. Shackelford, W. M., D. M. Cline, L. O. Burchfield, L. Faas, G. R. Kurth and A. D. Sauter. "Computer Survey of Gas Chromatography/Mass Spectrometry Data Acquired in the U.S. Environmental Protection Agency Priority Pollutant Screening Analysis: System and Results," Chapter 33, this volume.
10. Summons, R. E., W. E. Periera, W. E. Reynolds, T. C. Rindfleisch and A. M. Duffield. "Analysis of Twelve Amino Acids in Biological Fluids by Mass Fragmentography," *Anal. Chem.* 46(4):582-587 (1974).
11. *Biomedical Computer Program, P-Series Package* (University of California Press, 1979).
12. Eichelberger, J. W., Harris, L. E. and W. L. Budde. "Reference Compound to Calibrate Ion Abundance Measurement in Gas Chromatography-Mass Spectrometry Systems," *Anal. Chem.* 47:955-1000 (1975).

# SECTION 9

# GROB CLOSED-LOOP STRIPPING

# CHAPTER 36

# COMPARISON OF GROB CLOSED-LOOP STRIPPING ANALYSIS WITH OTHER TRACE ORGANIC METHODS

**R. G. Melton, W. E. Coleman, R. W. Slater, F. C. Kopfler**
**W. K. Allen, T. A. Aurand, D. E. Mitchell and S. J. Voto**

U.S. Environmental Protection Agency
Health Effects Research Laboratory
Cincinnati, Ohio

**S. V. Lucas and S. C. Watson**

Battelle Columbus Laboratories
Columbus, Ohio

The Exposure Evaluation Branch of the Health Effects Research Laboratory (HERL), U.S. Environmental Protection Agency (EPA) Cincinnati, OH is responsible for validating sensitive and reproducible organic analysis procedures which are used in our research to determine the health effects of chemical contaminants of drinking water. The data presented were obtained in January 1980, when HERL was evaluating different procedures (lyophilization, reverse osmosis (RO) and XAD-2 resin adsorption) for concentrating organics in drinking water. The resulting concentrated organics are used by HERL for biological toxicity testing. XAD-2 resin was used in this situation to concentrate organics for biological testing purposes and not as an analytical procedure.

Since there continues to be a great deal of interest among environmental chemists concerning the comprehensive analysis of purgeable organics in drinking water, we have compared some of our January 1980 Grob closed-loop stripping analysis (CLSA) data with data from several more conventional methods of organics analysis. For researchers studying the health effects and use of alternative disinfectants, such as chlorine dioxide, chloramines and ozone, or the use of granular activated carbon (GAC) in the treatment of drinking water, simple packed-column gas chromatographic/flame ionization detector (GC/FID) chromatograms of organic components greater than 100 ng/L in concentration do not provide data on which decisions can be based. Instead, state-of-the-art methods that use internal standards (IS) spiked in water samples, a high degree of organic concentration, high-resolution capillary column separations, reproducible gas chromatography/mass spectrometry (GC/MS) measurements, and sophisticated computerized quantification methods are required.

Since the presentation of our papers [1,2] five years ago on the Bellar purge-and-trap (P&T) gas chromatography/mass spectrometry/data system (GC/MS/DS) analysis of drinking water, we have followed with great interest new developments in the methodology of purgeable organic analysis. Even though chemists worldwide have learned a great deal in the past five years about comprehensive analysis of volatile organics in water, there is no consensus at this time as to the optimum method or methods for comprehensive analysis of purgeable organics in drinking water. For example, the literature indicates that most European environmental chemists would recommend Grob CLSA (a P&T method) with wall-coated open tubular (WCOT) capillary GC/MS as the best method, whereas most environmental chemists in North America would probably recommend some alternative method. We became interested in applying comprehensive capillary GC/MS/DS methodology to our health effects research objectives soon after the development of the CLSA method by Grob in 1973. Progress in using Grob CLSA was slow in our laboratory between 1975 and 1977 until WCOT capillary column technqiues were learned. In the past two years we have measured approximately 500 unique purgeable organics using Grob capillary GC/MS/DS CLSA. We published a preliminary report [3] in December, 1979, detailing some of our CLSA results and applications. The present report reviews briefly U.S. P&T methods and presents some comparative analytical data of surface water samples (Cincinnati tapwater before and after GAC treatment) using the following four methods of analysis:

- Method A Bellar purge and trap analysis (EPA Method 601)
- Method B Grob capillary GC/MS/DS CLSA
- Method C Batch liquid-liquid extraction (BLLE) analysis using a modified Master Analytical Scheme (MAS) procedure
- Method D XAD-2 adsorption–ethyl ether elution method (XAD-EEE)

The authors wish to point out at the outset that even though the subsequent data indicate the presence of many organics in water from the Cincinnati waterworks (CWW), these specific Cincinnati drinking water samples are less contaminated than most tapwater samples that we have analyzed from other locations. For example, the average concentration of Grob CLSA purgeable organics (other than trihalomethanes) in water samples from CWW was 9.2 ng/L. Grob capillary GC/MS/DS CLSA is an extremely sensitive method of trace organic analysis. In fact, the lower GC/MS detection limit of Grob CLSA for more than 200 organics is 1-10 ng/L. Therefore, the reader should bear in mind that very reproducible chemical data of trace levels of volatile organics in relatively "clean" drinking water samples are being presented. Secondly, not all laboratories require purgeable organic analytical methods that are as sensitive and comprehensive as Grob CLSA. Certainly, research laboratories that are generating chemical data on which important decisions concerning the choice of drinking water treatment processes, such as research on the use of GAC, alternative disinfectants, filtration techniques, and the health effects of such water treatment processes should use state-of-the-art comprehensive analytical methods such as those which are proposed in the EPA MAS [4-10]. However, most laboratories are not equipped with good state-of-the-art capillary GC/MS/DS hardware and software, and the capital investment of comprehensive capillary GC/MS/DS methods should be put in perspective with the required objectives of each laboratory. Environmental scientists also realize that the cost per organic compound analyzed is constantly decreasing due to major improvements in analytical methods and laboratory hardware and software. Five years ago [2] we identified the presence of 60 purgeable organics in Miami tapwater using an "exotic" instrument (GC/MS) and the Bellar P&T method. Today, this same analysis (EPA Method 624) is no longer considered "exotic." In fact, it is now being used by several U.S. waterworks laboratories. Perhaps five years from now, Grob capillary GC/MS/DS CLSA and other comprehensive trace organic procedures will be "affordable" to more environmental and drinking water laboratories. The Grob CLSA data for 292 organics in this chapter were produced by our laboratory group and further illustrate the application of Grob CLSA in drinking water treatment research and in the determination of the health effects of drinking water treatment processes.

Coleman et al. [11] present a discussion of the use of GC/MS/DS and internal standards for long-term quantification studies. DeMarco et al. [12] discuss chemical data obtained over a four-month period on the effect of full-scale GAC contactors (1-mgd) at CWW. The Grob CLSA results presented by DeMarco et al. were conducted by the Exposure Evaluation Branch of HERL.

## HISTORICAL BACKGROUND OF GROB CLSA

In 1973 in Zurich, Switzerland, Grob [13] reported on CLSA for the measurement of purgeable, intermediate-molecular-weight organics in drinking water at the part-per-trillion (nanogram-per-liter) level. Later, in 1974, Bellar and co-workers [14,15] reported a P&T method for the analysis of purgeable volatile organics at the part-per-billion (microgram-per-liter) level. U.S. water analysis laboratories quickly adopted the Bellar P&T method [1,2,16] using packed GC columns, whereas Western European laboratories adopted the Grob CLSA method, which uses WCOT capillary columns. The primary reason for slow adoption of the Grob CLSA in the U.S. was the slow acceptance of state-of-the-art WCOT glass capillary column technology and capillary column hardware by U.S. manufacturers. Presently, U.S. laboratories remain behind our Western European counterparts in the use of capillary GC for the separation of environmental pollutants. Comprehensive organic analytical procedures, such as Grob CLSA and GC procedures in the MAS, require the use of high-resolution capillary column separations. Fortunately, U.S. manufacturers and environmental laboratories are beginning to catch up with our Western European counterparts. For this reason, the use of comprehensive trace organic methods in the United States can now realistically be proposed.

Grob CLSA utilizes 1.5 mg of activated carbon as a trapping adsorbent [17]. Activated carbon has been used in the past to monitor organic pollutants in air. For example, White et al., of the National Institute for Occupational Safety and Health (NIOSH), reported [18] in 1970 a standard method to measure selected solvent vapors in industrial atmospheres. The NIOSH method involves passing a standard 10-L volume of industrial room air through a standardized adsorption tube that is packed with activated carbon. After capping the tube and shipping it back to the laboratory, the activated carbon is removed from the tube and placed in a clean vial. One milliliter of carbon disulfide ($CS_2$) is added, and the resulting solution analyzed by GC/FID. All phases of this method have been standardized, and the equipment is readily available.

Grob thoroughly discussed the design and development of the CLSA procedure in his first CLSA paper [13]. Like the NIOSH air analysis method [18], Grob uses $CS_2$ to elute the organics from the activated carbon, and gas chromatography to separate the organics in the eluent. Like the Bellar P&T method [14], the CLSA method is a vapor-phase P&T stripping technique in which those compounds with appreciable vapor pressure over water are removed from the sample by purging it with a large volume of gas and by passing the stripping gas through an adsorption tube. Unlike other vapor-phase procedures, Grob has achieved nearly a $10^6$-fold concentration of most low- and intermediate-molecular-weight organics by using a closed-loop design where 0.5 L of stripping gas is recycled continuously through the water sample, and the adsorption trap is extracted with 12 $\mu$L of $CS_2$. Quantification is achieved by spiking the initial water sample with a series of internal and reference standards, stripping at 30°C for two hours and chromatographing the $CS_2$ extract on a WCOT capillary column. Grob reported the capillary GC/MS identification of 62 organics in samples of Lake Zurich water and Zurich potable water (approx. 60% comes from Lake Zurich) in his initial CLSA paper.

Grob's second paper [19] on CLSA was dedicated primarily to the application of CLSA to raw and finished drinking water in the area of Zurich. Using capillary GC/MS for identification, Grob and Grob [19] reported the occurrence of 136 organics in area water at the low-ng/L range and demonstrated that automobile gasoline was the major pollutant in Lake Zurich. In these first two CLSA papers [13,19] Grob identified 29 unique alkanes and 34 alkyl-substituted benzenes in Zurich raw and finished drinking water.

In 1975 Grob et al. [20] compared CLSA with a new trace organic analysis technique, rapid liquid extraction. Grob et al. point out the complementary nature of the two procedures. CLSA is very sensitive for low- and intermediate-molecular-weight nonpolar organics, whereas rapid solvent extraction is the method of choice for heavier compounds. They also point out something that many environmental laboratories have recently rediscovered; solvent extracts of water heavily stress capillary GC columns, because nonvolatile components in the extracts shorten column life. In contrast, CLSA extracts contain GC-volatile substances so that capillary columns may be used over a very long period of time without any loss of column performance.

In 1976 Grob and Zurcher [17] improved and standardized the CLSA procedure when they realized that many water research laboratories (mostly European) were already using the procedure routinely to study source pollution and drinking water treatment techniques, such as the use of bank filtration, activated carbon adsorption and alternative disinfection. Zurcher and Grob clearly point out the major limitations of CLSA, such as the limited intermediate volatility and molecular weight range of substances that are readily measured by the method. Most laboratories that are using CLSA to measure low-level organics in water are following this standardized method and have made only slight modifications of it.

We have recorded a total of 16 additional references from 7 different laboratories (5 European) that have used Grob CLSA to measure 192 unique organics in water. However, a brief tour of European drinking water laboratories indicates that CLSA is being used daily in many waterworks. Stieglitz et al. [21] in West Germany published in 1976 an early comprehensive applications paper. They used capillary GC/MS CLSA exclusively to measure 103 organics at three different water utilities on the River Rhine. Their data clearly show some of the effects on the raw water of different treatment techniques, such as bank filtration, chlorination and ozonation. Stieglitz et al. modified the CLSA method of Grob and Zurcher to analyze water from the heavily contaminated River Rhine. Two-liter samples were stripped at pH 3 in two different stages. After 15 min of stripping, the first activated carbon filter was removed from the loop and a new filter was inserted. Stripping was then continued for an additional 2 hr and 45 min. Each filter was extracted separately, and the eluents of filters I and II were combined prior to capillary GC/MS analysis. Stieglitz reported a relative standard deviation of 10–15% for 24 organics at the 100-ng/L level, and an average GC/MS detection limit of 0.2 ng/L.

Starting in 1976, Giger, in Dubendorf, Switzerland, began publishing a series of comprehensive application papers using Grob CLSA. Zurcher and Giger [22] reported the occurrence of 70 organics at different points on the Glatt River using capillary GC/MS. Giger et al. [23] reported capillary GC/MS analysis of trace organics using methylene chloride solvent extraction and Grob CLSA. In 1978 Giger et al. [24] applied Grob CLSA to trace the source of chlorinated volatile hydrocarbons in ground- and lakewaters in the Zurich area. Tetrachloroethylene, the most dominant chlorinated compound, was shown to originate from tertiary treated sewage and ground spills. Giger et al. clearly demonstrated that Grob CLSA is an excellent method to trace the source of chlorinated hydrocarbons and substituted aromatic hydrocarbons from industrial point sources. In 1979 Schwarzenbach et al. [25] used Grob CLSA to determine the distribution of tetrachloroethylene and 1,4-dichlorobenzene in Lake Zurich at various depths over a 12-month period. One-liter samples were stripped at 30°C for 90 min, and quantification was done by capillary GC/FID peak height measurements. Duplicate measurements over the one-year study at the 5- to 70-ng/L range had relative standard deviations of less than 10%, except at the thermocline depth of the lake, where concentration gradients were greatest. Using Grob CLSA data, Schwarzenbach et al. were able to conduct an accurate mass balance for 1,4-dichlorobenzene into and out of Lake Zurich. Sewage treatment plants introduced 62 kg/year of 1,4-dichlorobenzene to the lake, whereas the Zurich water utilities transferred 1 kg/year out of the lake. In 1977 Giger proposed [26] the use of Grob CLSA to measure volatile organics in the marine environment. In 1978 Schwarzenbach et al. [27] and Max Blumer conducted an extensive analysis of volatile organics in coastal seawater using Grob CLSA. Since most volatile organics in seawater are present below the 10-ng/kg range, Schwarzenbach et al. stripped 5-L samples at 35°C to have a higher concentration of organics for GC/MS analysis. Reproducibility for 20 selected organics in seawater samples was ±15–30%.

Reinhard and co-workers published many papers [28-32] using Grob CLSA as one of three analytical methods to assess advanced wastewater treatment processes and the transport of organics from groundwater injection wells. As early as 1976, Reinhard and co-workers chose the following analytical methods due to the complexity of the organics in biologically treated municipal wastewater at Water Factory 21 in California. Bellar P&T analysis using a packed column GC/Hall detector system was used for haloforms and halogenated compounds with one and two carbons. Grob GC/MS CLSA was used to measure compounds of medium volatility and low water-solubility. Solvent extraction with two different solvents was used to measure compounds of lower volatility and higher water solubility. Capillary separations were required except for Bellar P&T samples, and GC/MS was used to confirm all identifications. The data of Reinhard and co-workers clearly indicate the complexity of environmental water samples and the need for high-resolution capillary separations.

In December 1979 Coleman et al. [3] published a paper on Grob capillary GC/MS/DS CLSA of drinking water samples. Coleman et al. reported the use of GC/MS computer procedures to automatically quantify purgeable organics in Grob CLSA data files using internal standards spiked in water samples, a computer library of 215 reference standards with narrow relative retention time windows, reverse mass spectrum library searches, and relative response factors for the 215 standards based on single mass spectral ions. This procedure permits a laboratory to quantify three drinking water samples within a 24-hour period for 215 reference purgeable organics using 12 person-hours of time. The resulting Grob CLSA data were reported to have correctly identified 80% of the 215 reported compounds with quantitative accuracy to within ±25% for most solvent-type organics in the 50-ng/L range. Coleman reported a GC/MS detection limit of 1–10 ng/L for most of the 300–400 organics which are identifiable by the method.

## HISTORICAL BACKGROUND OF U.S. P&T METHODS

A number of P&T methods have recently been standardized by EPA. The authors will attempt to illustrate the design differences between Grob CLSA and these newer EPA P&T methods. On December 3, 1979, the EPA published a set of proposed chemical methods for the analysis of pollutants [33]. The use of these proposed methods would be required for filing applications under the National Pollutant Discharge Elimiation System, state certifications, compliance monitoring under the Clean Water act and analyses of 113 organic toxic pollutants (priority pollutants) under a Settlement Agreement [34] (*Natural Resources Defense Council, Inc., et al. v. Train*) and under Section 304h of the Clean Water Act of 1977. The following analytical methods were proposed [33] for the analyses of organic pollutants in water:

- 601: purgeable halocarbons using packed GC/Hall
- 602: purgeable aromatics using packed GC/photoionization
- 603: acrolein/acrylonitrile using packed GC/FID
- 604: phenols
- 605: benzidines
- 606: phthalate esters
- 607: nitrosamines
- 608: organochlorine pesticides and polychlorinated biphenyls (PCB)
- 609: nitroaromatics and isophorone
- 610: polynuclear aromatic hydrocarbons
- 611: haloethers
- 612: chlorinated hydrocarbons
- 613: 2,3,7,8-tetrachlorodibenzo-*p*-dioxin
- 624: purgeables using packed GC/MS
- 625: base/neutrals, acids and pesticides using packed GC/MS

The above 15 methods are designed for the analyses of 113 specific "consent decree" organics. Methods 601, 602, 603 and 624 are all Bellar P&T methods. All four P&T methods require the use of packed GC columns and different GC detectors. EMSL recently reported [35,36] the following two additional P&T methods:

- Method 502 purgeable halogenated chemical indicators of industrial contamination using packed GC/Hall
- Method 503 purgeable aromatic chemical indicators of industrial contamination using packed GC/photoionization

Methods 502 and 503 are identical to Methods 601 and 602, respectively. The only difference in Methods 601 and 502, and in Methods 602 and 503 is that Methods 601 and 602 are limited to 113 "consent decree" organics. Methods 502 and 503 were developed by EMSL for the EPA Office of Drinking Water (Washington, DC) to measure a broad spectrum of purgeable chemical indicators of industrial contamination of drinking water. EPA Method 601 will measure 29 "consent decree" organics, whereas Method 502 will measure 48 halogenated purgeable organics (chloromethane to 1,4-dichlorobenzene) at concentrations between 0.1 and 50 $\mu$g/L. Like EPA

Method 601, Method 502 requires a total analysis time of 1 hr and uses a packed-column GC/Hall instrument system. Method 503 [36], like Method 602, is designed to measure aromatic purgeable organics with a packed-column GC/photoionization instrument system. Method 602 measures seven "consent decree" aromatics, whereas Method 503 is capable of measuring 33 purgeable aromatic organics over a concentration range of 0.05–0.5 μg/L. The combined use of Bellar P&T Methods 502 and 503 will measure 81 unique purgeable organics in drinking or raw source water with a lower limit of detection of at least 0.1 μg/L.

Methods 601 to 625 are designed for analysis of 113 specific organics. These methods were not intended to be comprehensive methods for the in-depth analysis of a broad range of organics in water. To develop a comprehensive MAS, EPA (Environmental Research Laboratory, Athens, GA) awarded a competitive contract to Research Triangle Institute (RTI) in 1978 [4-10]. This research effort was designed by EPA to ensure the use of a minimum number of organic analysis procedures to analyze a very broad spectrum of organics in water. Consequently, EPA required the use of high-resolution chromatography separations and broad spectrum chromatography detectors such as state-of-the-art MS/DS hardware and software. The lower detection (LD) limits for the analysis of drinking water using MAS procedures is 0.1 μg/L [5]. For the analysis of "extractable" organics at the 1-μg/L range, the MAS recommends a BLLE procedure using methylene chloride to stirbar-extract one liter of water [7]. For the same "extractables" in cleaner water, such as drinking water, the MAS recommends passing three liters of water through a XAD-4 resin sorbent column and eluting adsorbed organics with ethyl ether solvent [7]. XAD-4 and XAD-2 resins differ only in pore size. Both resins have the same polymeric chemical composition and have similar sorptive characteristics. The MAS XAD-4 procedure is similar to the procedure described by Junk et al. [37] and the XAD-EEE procedure described in this report. These three XAD procedures differ primarily in volume of drinking water used and the adjusted pH of water that is passed through the sorbent column.

For the comprehensive analysis of purgeable organics for the MAS, RTI adopted the use of the P&T capillary GC/MS/DS procedure that was previously developed by RTI and outlined in Figure 1. This procedure was intended to cover a spectrum of purgeable organics from the very volatile gases (chloromethane and vinyl chloride), such as EPA Method 601 measures, to intermediate-molecular-weight purgeable organics. The lower limit of detection of the current MAS P&T procedure is 0.1 μg/L for drinking water [5]; thus, according to the designers of the MAS [8], "the MAS P&T procedure does not present competition with Grob capillary GC/MS CLSA for the measurement of purgeables in drinking water at the part-per-trillion level." However, since both P&T methods are intended to provide comprehensive research information on the level of purgeable organics in drinking water, the methods should be compared for differences in design and experimental performance. Such comparative information is important to chemists who must decide which P&T method (or methods) will provide the best and most cost-effective analytical data. RTI has not reported research or data on the use of Grob CLSA for low- to intermediate-molecular-weight nonpolar organics, for which the CLSA method was designed. Instead, RTI attempted to extend Grob CLSA for the analysis of water-soluble low-molecular-weight organics (volatile intractables), such as methanol and acetone. Not surprisingly, the method failed for this group of organics [5].

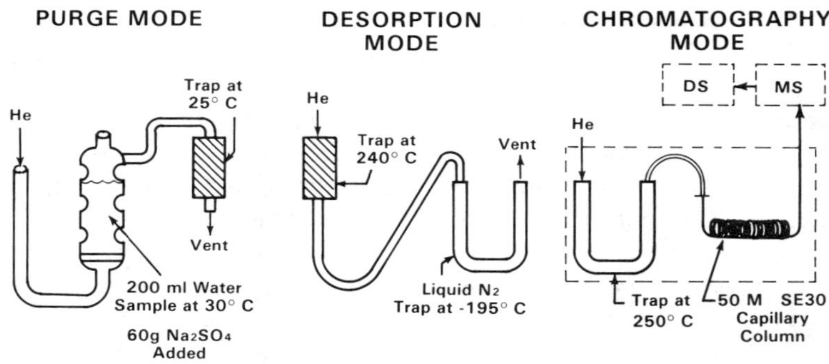


**Figure 1.** Schematic of the EPA MAS P&T method (RTI method).

## HERL SCHEME FOR TRACE ORGANIC ANALYSIS

During the past two years, HERL used the following three methods for the organic analysis of drinking water:

- Method A–Bellar P&T GC/Hall detector analysis (EPA Method 601)
- Method B–Grob capillary GC/MS/DS CLSA
- Method C–liquid-liquid extraction (BLLE) of 10-L samples using methylene chloride and capillary GC/MS/DS analysis

The reasons why we chose the combination of Bellar P&T (601) and Grob GC/MS CLSA to analyze purgeable organics are outlined in Figure 2. Overall, we have found that the combination of Methods 601 and CLSA provides a

Method A - Analysis of 5 ml Water Samples using EPA Method 601 (Bellar Purge and Trap Analysis) for Quantification of 23 Halogenated Low Molecular Weight (Chloromethane through Bromoform) Organics.

5 ml Sample

ADVANTAGES:

1. Low Cost - Packed GC Columns/Electro-Conductivity Detector
2. Fast - 20 minutes for Bromoform to Elute.
3. Well Researched and Accepted Method.
4. Good Quality Control Procedures.

DISADVANTAGES:

1. Very Few, Considering Cost/Organic

Method B - Analysis of 1-L Water Samples using Grob CLSA for MS Quantification of over 215 Organics and Qualitative Identification of over 400 Organics.

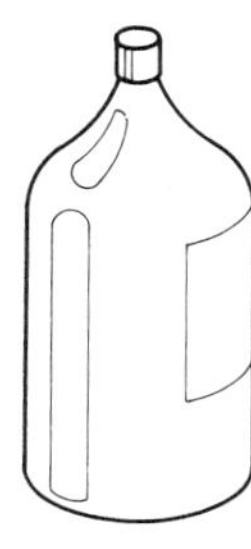

1 Liter Sample

ADVANTAGES:

1. Good MS Sensitivity - Detection Limit of 1 to 10 ng/l.
2. Good Reproducibility - Internal Standards are spiked in Water Prior to Purging; Accuracy of ± 25% for most Solvent Type Organics at the 50 ng/l Level.
3. Excellent Method for Control of Unit Processes such as;
   A) Use of Granular Activated Carbon.
   B) Disinfection with Ozone, Chlorine, Chlorine Dioxide, and Chloramines.
   C) Source Contamination of Drinking Water Supplies due to Industrial Spills and/or Discharges.

DISADVANTAGES:

1. Expensive
   A) Cost per capillary GC/MS/DS CLSA - $460, or $460 ÷ 215 Organics = $2 per Organic.
   B) Cost per capillary/FID CLSA - $80
2. Sample Matrix Interferences.
3. Activated Carbon Trap may become Overloaded with Organics in Industrial Effluents

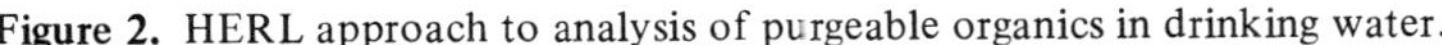

**Figure 2.** HERL approach to analysis of purgeable organics in drinking water.

comprehensive, broad-spectrum, cost-effective, quantitative analysis of trace levels of purgeable organics in drinking water. Method 601 (502) and CLSA are diagrammed simply in Figures 3 and 4. It is clear from these figures that the desorption modes of Bellar P&T analysis and Grob CLSA are distinctively different. Bellar P&T depends on thermal desorption of organics from the trapping material, whereas Grob CLSA depends on $CS_2$ solvent extraction of organics from the surface of the activated carbon. It is this basic difference in method of desorption of organics from the trapping material that makes Bellar P&T Method 601 (502) and CLSA complementary in the spectrum of organics analyzed. The gaseous-type purgeable organics, which are covered up by the $CS_2$ solvent in the Grob CLSA, are readily quantified by the cost-effective Bellar P&T method using a packed-column GC/HALL instrument system, whereas the Grob CLSA provides a very cost-effective, quantitative analysis of purgeable organics which elute after benzene and bromoform and which require the use of high-resolution capillary columns. The data presented in this report illustrate this important principle of complementary analysis. In addition, results obtained by using Method 601, CLSA and BLLE, above, will be compared with the XAD-2 adsorption method (XAD-EEE) of Junk et al. [37].

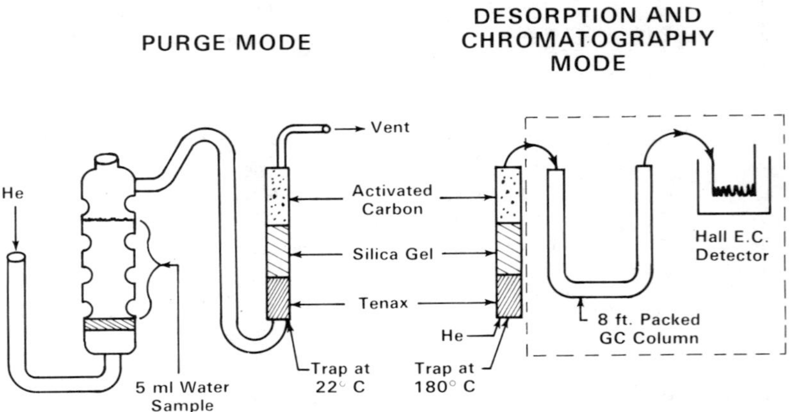


**Figure 3.** Schematic of EPA Methods 601 and 502 (Bellar P&T method).

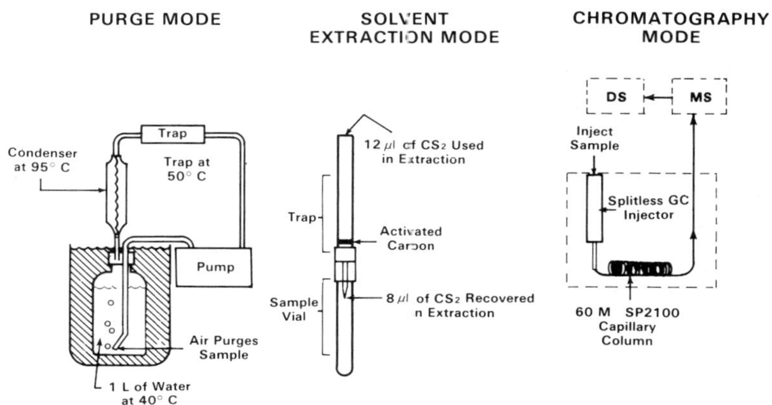


**Figure 4.** Schematic of Grob CLSA method as modified by HERL on August 1, 1980.

## EXPERIMENTAL

### Source of Water Samples

Drinking water samples were obtained from the Cincinnati Waterworks (CWW) on January 14, 1980 (GAC contactor A), and January 28, 1980 (GAC contactor D), at sampling points into (influent) and out of (effluent) 1-mgd GAC columns (contactors) that had been on line for seven and two weeks, respectively. These same GAC contactors at CWW are described in more detail by DeMarco et al. [12]. Method 601, CLSA and BLLE were applied to influent and effluent water samples from GAC contactor D (Figure 5). Analytical results of samples XAD-inf, XAD-eff and XAD-EEE were obtained from contactor A GAC-inf water on January 14, 1980. All water samples were preserved at collection with 10 mg/L of mercuric chloride and 20 mg/L of sodium sulfite. The data, however, indicate a possible problem with the use of mercuric chloride as a preservative (see Results). For the XAD-2 concentration experiments, five gallons of GAC-inf water (see Figure 5) were brought back to HERL for concentration. Bellar P&T analyses and CLSA were conducted by HERL. Water samples and reagent water samples for BLLE were shipped to Battelle–Columbus Laboratories (EPA Contract 68-03-2548) for analysis. Battelle also carried out the capillary GC/MS/DS analysis of XAD-EEE extracts. Reagent water, prepared by passing distilled water through a Millipore Super Q water purification system (all three cartridge housing units were filled with activated carbon cartridge filters), was concurrently analyzed by the same four methods. All analytical data reported in this paper have been corrected for methodology artifacts.

**Bellar P&T Analysis**

Purgeable, low-molecular-weight organohalides were analyzed using Method 601 (Figure 3), except that the purge-and-trap device described by Bellar and Lichtenberg [14] in 1974 was used, and the trap was packed with 60/80 mesh Tenax GC. This packing material is a deviation from the combination of Tenax GC, silica gel and activated carbon as is specified in Methods 601 and 502. Compounds such as chloromethane would not have been trapped appreciably by the sole use of Tenax GC at room temperature. Future Bellar P&T analyses from HERL will be conducted using the above combination packing material.

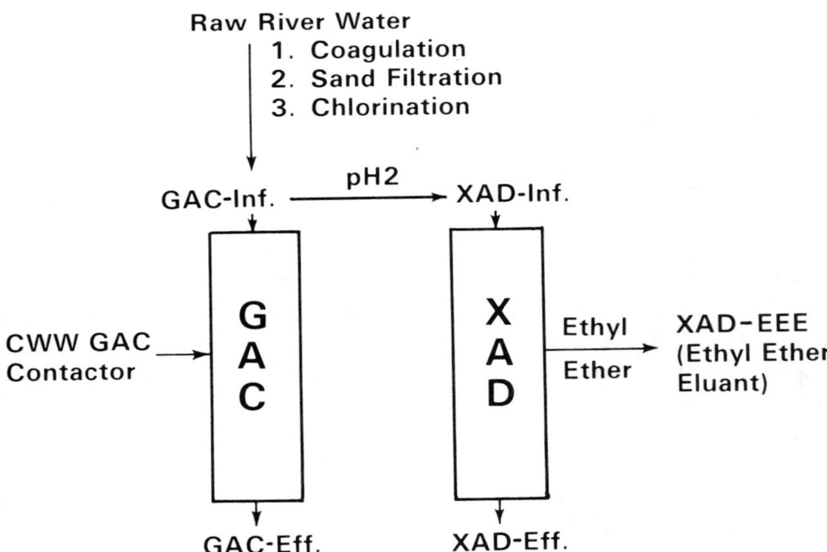


**Figure 5.** Water samples used to compare four analytical procedures. (1) GAC-inf was CWW water before passage through a 1-mgd GAC contactor at CWW (GAC contactor A samples were collected on January 14, 1980; GAC contactor D samples were collected on January 28, 1980); (2) GAC-eff water was collected after passage through the 1-mgd GAC contactor at CWW; (3) XAD-inf water was GAC-inf water which was brought back to HERL, before adsorption on a 37-$cm^3$ XAD-2 analytical column; (4) XAD-eff was water collected after passage of XAD-inf water through the 37-$cm^3$ XAD-2 column.

Bellar P&T samples were chromatographed according to the following conditions:

- injector temperature: 150°C;
- initial column temperature: 28°C
- temperature program sequence: (1) Heat column from 28 to 60°C at 40°C/min; (2) hold for 1 min; (3) heat from 60 to 160°C at 8°C/min; (4) hold at 160°C; and
- GC column: 0.2% Carbowax 1500 on Carbopak C (80/100 mesh) packed in 9-ft x 2-mm i.d. glass column.

## Grob Capillary GC/MS/DS CLSA

The experimental method described by Grob and Zurcher [17] was followed using the following minor modifications:

1. Water samples were collected in one-gallon screw-capped bottles.
2. Samples were analyzed in the above shipment bottles by decanting sample water down to the one-gallon mark, adding five internal standards (52 ng/L each of chlorohexane, chlorooctane, chlorododecane and chlorohexadecane and 260 ng/L of chlorooctadecane) dissolved in 0.6 $\mu$L of acetone to the sample, and then purging the sample for two hours at 30°C.
3. The filter holder (trap) was maintained at 40°C and the heat exchanger at 80°C.

Details of HERL modifications to the Grob and Zurcher CLSA method are given by Coleman and co-workers [3,11]. The schematic in Figure 4 reflects changes in HERL CLSA since the CWW samples in this publication were analyzed in January 1980. These August 1, 1980, modifications are designed to promote the CLSA of a broader and higher-molecular-weight spectrum of purgeable organics by maintaining a higher purge temperature and a heated all-glass system from the sample bottle to the trap. Ultrapure $CS_2$ from Matheson, Coleman, and Bell Chemical Company (Cincinnati, OH) or from Tedia Chemical Company (Fairfield, OH) was used without additional redistillation or cleanup.

Grob CLSA samples were separated by capillary GC according to conditions described by Coleman et al. [11]. Briefly, CLSA carbon extracts were injected (splitless) at 20°C with a capillary column flow velocity of 25 cm/sec. When the $CS_2$ begins to elute, the SP 2100 capillary column was heated at a rate of 2°C/min to a maximum temperature of 250°C. Data acquisition on a Finnigan-Incos GC/MS/DS was begun after the $CS_2$ finished eluting from the capillary column. The mass spectrometer was scanned at a rate of 14–450 amu/2 sec. Further GC/MS/DS details are described by Coleman et al. [11].

**BLLE Capillary GC/MS/DS Analysis**

The batch methylene chloride extraction method that is briefly outlined in the MAS [5,8] was used. If EPA Method 625 had been used, the GC/MS detection limit of 10 $\mu$g/L would have been unacceptable for the measurement of organics in these drinking water samples from CWW. The BLLE procedure below requires that 10-L water samples be collected in three 1-gal sample bottles and spiked with a series of deuterated internal standards at a concentration of 0.2 $\mu$g/L prior to stirbar extraction with methylene chloride. The following experimental details are provided because it is extremely difficult to achieve acceptable sensitivity and artifact levels for the BLLE analysis of trace-level organics in drinking water:

*Solvent Preparation*

One-gallon batches of Burdick and Jackson "distilled in glass" methylene chloride were redistilled in a 5-L flask equipped with a 60- x 1.8-cm i.d. column packed with medium size glass helices. The receiver was the original one-gallon solvent bottle which was preflushed with ultrahigh-purity $N_2$ (Matheson, 99.99 1/6). A positive pressure of $N_2$ was maintained throughout the distillation using a bubbler chamber. Methylene chloride was distilled at a rate of 1.4–1.8 mL/min. The first and the last 300 mL of solvent were discarded. After distillation, methylene chloride was stored under $N_2$ and used within three days for BLLE.

*Sample Extraction*

Ten liters of drinking water were extracted in the original one-gallon sample bottles by first removing all but 3.3 L of sample water from each of three sample bottles, then adding 33 $\mu$L of a mixture of deuterated internal standards (0.2 $\mu$g/L) [38], adding a 3-in. Teflon® stirring bar, stirring the water sample at maximum stable speed and adding concentrated sulfuric acid until the acidity was lowered to pH 2–2.5. Three 40-min solvent extractions were made using 250, 100 and 100 mL of redistilled methyline chloride. Solvent was removed after each extraction using an all glass and Teflon pipet-type device and about 5 in. Hg vacuum.

*Solvent Evaporation*

The stir-extraction, above, of 10 L of drinking water in three sample bottles yields approximately 1200 mL of methylene chloride. Two Kuderna-Danish (KD) apparatuses were used to concentrate the solvent to a volume of about 4.5 mL which was fractionated into an acids fraction (derivitized with diazomethane) and a neutrals fraction according to the procedure described by Lucas et al. [38]. BLLE samples were chromatographed on a 40-m x 0.25-mm i.d. SP1000 WCOT capillary column (prepared by Battelle) according to the following conditions:

- injector temperature: 250°C
- initial oven temperature (hold): 50°C (6 min)
- temperature program rate: 2°C/min
- upper temperature limit: 225°C
- injection volume: 2 $\mu$L sample + 1 $\mu$L heptane
- transfer line temperature: 250°C

## XAD-EEE Capillary GC/MS/DS Analysis

The grab sample method described by Junk et al. [37] was used for the concentration of organics with the following modifications:

1. XAD-2 resin sufficient for both CWW GAC contactor A and D experiments (see Figure 5) was cleaned up by consecutive 24-hr soxhlet extractions with methanol, acetonitrile, ethyl ether and methanol. Clean resin was stored wet under methanol until prior to packing columns.
2. Two columns were set up; one column for 10 L of CWW XAD-inf water, and the second for 10 L of Super Q reagent water. Each column was 2.7 cm in diameter by 6.5 cm in height. A silanized glass wool plug was placed on the bottom of the empty column.
3. The resin was removed from the methanol storage bath and slurried into a beaker of Super Q reagent water. The resin was then rinsed four times with Super Q reagent water. Next the resin (37 $cm^3$) in the beaker was slurried into the glass column, which was filled with reagent water. Silanized glass wool was placed on top of the resin which was always kept wet with reagent water. The column resin was then rinsed with 1 L of Super Q reagent water.
4. Ten liters each of CWW sample and reagent water were adjusted to pH 2 with 20 mL of 12 *N* sulfuric acid.
5. Each acidified sample water and reagent water were passed through each respective XAD-2 column at a flowrate of approximately 28 mL/min. Each column was immediately rinsed with 200 mL of pH 2

reagent water. Sample water and reagent water that passed through each respective XAD column were labeled XAD-eff and were later analyzed by HERL using EPA Method 601 and Grob CLSA, and by Battelle–Columbus Laboratories using BLLE.

6. Three bed volumes of freshly redistilled ethyl ether were used to elute the adsorbed organics from the XAD-2 resin. This ethyl ether eluent was labeled XAD-EEE.
7. The sodium sulfate drying procedure of Junk et al. [37] was used.
8. The ethyl ether eluent was evaporated to 1 mL (KD) and shipped to Battelle–Columbus for capillary GC/MS/DS analysis under EPA Contract 68-03-2548.

All capillary GC/MS/DS parameters for the analysis of XAD-EEE samples were the same as previously described for BLLE samples.

## RESULTS

Chromatograms using Method 601, Grob CLSA and BLLE are presented for GAC-inf and GAC-eff water only due to space limitations. Unfortunately, the XAD-2 ethyl ether eluent (XAD-EEE) samples from CWW GAC contactor D were heavily contaminated with chemical artifacts from the XAD-2 resin. Since we had good samples and data files using all four methods taken at the same points at GAC contactor A at the CWW on January 14, 1980 (two weeks prior to CWW contactor D samples), we have presented, instead, analytical results on the XAD-inf, XAD-eff and XAD-EEE samples from contactor D GAC-in water to contactor A water. Basically, this change from Contactor D GAC-inf water to contactor A GAC-inf water is simply a difference in sampling CWW raw Ohio River water on dates differing by two weeks. Comparison of CWW GAC-inf water samples on January 14, 1980, and January 28, 1980, using Grob CLSA and BLLE analyses shows that water samples on these two dates are quite similar, except that there was a slightly higher level of alkyl-substituted benzenes in the January 28, 1980, water samples.

Total organic carbon (TOC) measurements of combined volatile and nonvolatile organics were determined on GAC-inf and GAC-eff water from CWW. On January 14, 1980, the TOC of GAC-inf to contactor A was 1.9 mg/L and of GAC-eff water was 1.2 mg/L. This represents a removal of 37% TOC by the GAC (7 weeks old) in contactor A. The TOC of contactor D water on January 28, 1980 (GAC was in use for two weeks), showed a corresponding reduction from 1.6 to 0.2 mg/L, or a removal of 87% TOC.

Results of Bellar P&T analysis (Method 601) are presented in Table I, and representative chromatograms in Figure 6. Only six halogenated organics were detected due to the low level of purgeable organics in GAC-inf water.

All six compounds were Consent Decree organics with an average concentration of 16 μg/L. No organics were detected in the corresponding GAC-eff water. This would represent a 100% removal by GAC contactor D at CWW. According to EPA Method 601, an additional 23 (29 - 6) organics would have been detected in GAC-inf water if present in concentrations above 0.06 μg/L. EMSL Method 502 would have detected an additional 42 (48 - 6) halogenated organics if they had been present above 0.1 μg/L. Table I also shows the effect of the 37-$cm^3$ XAD-2 analytical column in removing halogenated purgeable organics from XAD-inf water. Overall, Table I indicates that the 1-mgd GAC contactor D at CWW was more effective in removing organics than was the small XAD-2 analytical column.

Results of Grob capillary GC/MS/DS CLSA are presented in Table II. Grob CLSA detected 107 purgeable organics in GAC-inf water (contactor D). Quantitative results of organics listed in Table II that have the designation "S" under "Quan. Method" are based on actual relative response factors of reference compounds, as compared to the internal standard, chlorododecane, which was initially spiked in each water sample at 52 ng/L. A total ion current area relative response factor of one (chlorododecane, IS) is assumed for the compounds which do not have the "S" designation in Table I. As mentioned earlier, the average concentration of the nontrihalomethane organics in GAC-inf water according to CLSA was 9.2 ng/L. The average concentration of these same organics in GAC-eff water was 1.8 ng/L. If the MAS P&T procedure had been used on GAC-inf water (contactor D), probably only five compounds (four trihalomethanes and 1,1,1-trichloroethane) would have been detected. This prediction is based on the MAS GC/MS lower detection limit of 0.1 μg/L [5]. The 80% removal of the CLSA organics in GAC-eff water by CWW GAC contactor D shows surprisingly good agreement to the 87% removal based on TOC cited above. Since most of the TOC material is probably of humic origin (and therefore not accessible to Grob CLSA), these data seem to indicate that GAC contactor D is removing the same percentages of purgeable organics and humic material. The XAD-inf and XAD-eff data in Table II indicate that the 37-$cm^3$ XAD-2 analytical column removed 79% of these same Grob CLSA organics. Accordingly, CWW GAC contactor D and the XAD-2 analytical column are doing similar jobs (80%, 79%) in removing the organics which can be measured by Grob CLSA in GAC-inf and XAD-inf water. The ability of Grob CLSA to measure directly the effect of XAD resin as a unit process is clearly illustrated in Table II. Grob CLSA is also an excellent method to determine if XAD-2 resin is adequately cleaned-up (see Figure 5) for analytical use as an adsorbent by measuring purgeable organics in reagent water before and after passage through an XAD column.

Table I. Results of Bellar P&T Analysis

| | Retention Time (min) | GAC Contactor D Water[a] | | | GAC Contactor A Water[a] | | | |
|---|---|---|---|---|---|---|---|---|
| | | GAC-inf ($\mu$g/L) | GAC-eff ($\mu$g/L) | GAC-eff (% Removed[b]) | GAC-inf ($\mu$g/L) | XAD-inf ($\mu$g/L) | XAD-eff ($\mu$g/L) | XAD-eff (% Removed[c]) |
| Methylene Chloride | 4.8 | ND[d] | ND | | ND | ND | ND | |
| Chloroform | 8.9 | 56 | ND | 100 | 65 | 23 | 3.4 | 85 |
| 1,2-Dichloroethane | 9.5 | ND | ND | | ND | ND | ND | |
| 1,1,1-Trichloroethane | 10.4 | 0.4 | ND | 100 | 1.9 | ND | 1.1 | |
| Tetrachloromethane | 10.7 | ND | ND | | ND | ND | TD | |
| Bromodichloromethane | 11.7 | 18 | ND | 100 | 83 | 10.9 | 0.6 | 94 |
| Trichloroethane | 13.3 | TD[e] | ND | | TD | 0.1 | 0.1 | 0 |
| Chlorodibromomethane | 14.4 | 5.8 | ND | 100 | 8.0 | 5.0 | 0.4 | 92 |
| Bromoform | 17.1 | 0.2 | ND | 100 | ND | ND | | |
| Tetrachloroethene | 19.1 | ND | ND | | ND | ND | ND | |

[a] Water samples described in Figure 5.
[b] % Removed by GAC in contactor D.
[c] % Removed by XAD-2 resin in analytical column (see Figure 5).
[d] ND = not detected.
[e] TD = trace detected.

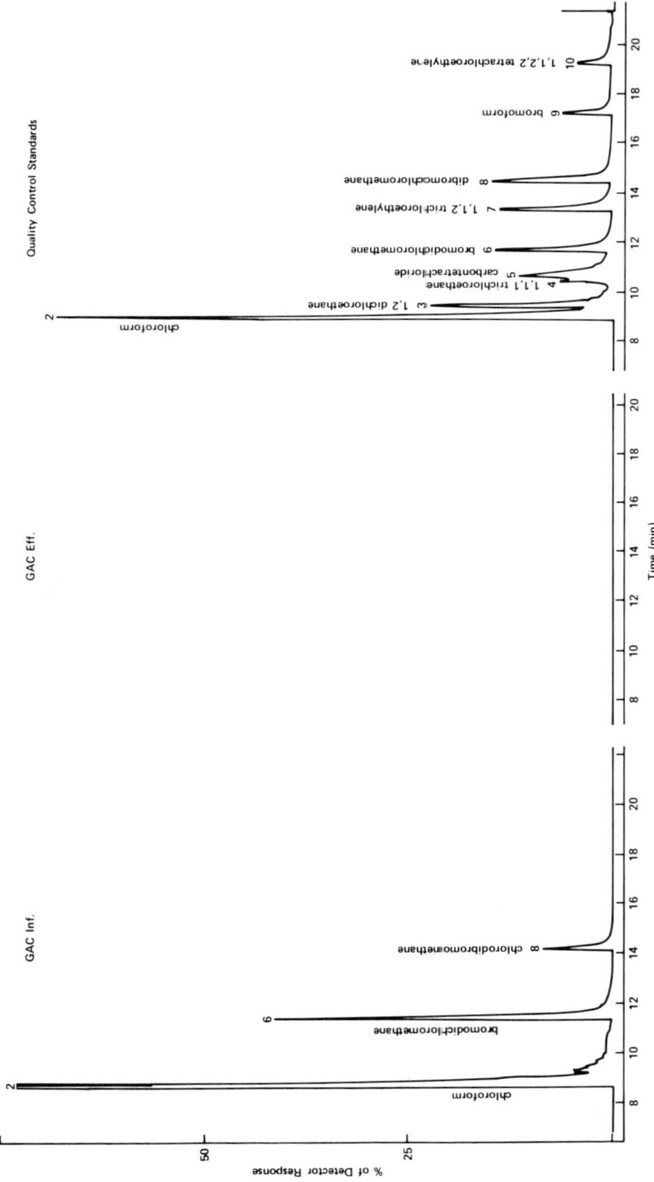


**Figure 6.** Chromatographic results of Bellar P&T analysis.

Table II. Results of Grob Capillary GC/MS/DS CLSA of GAC-inf, GAC-eff, XAD-inf and XAD-eff Samples

| | | | Contactor D Water[c] | | | Contactor A Water[c] | | | |
|---|---|---|---|---|---|---|---|---|---|
| Compound | RRT[a] | Quan Method[b] | GAC-inf (ng/L) | GAC-eff (ng/L) | % Removal by GAC | GAC-inf (ng/L) | XAD-inf (ng/L) | XAD-eff (ng/L) | % Removal by XAD |
| 2-Butanone[d] | 0.134 | | | | | | | | |
| 3-Methylpentane[d] | 0.136 | | | | | | | | |
| Diisopropylether | 0.146 | | 141 RAU[e] | 0 RAU | 100 | 460 RAU | 145 RAU | 0 RAU | 100 |
| Chloroform | 0.149 | S | 10 μg/L | 0.60 μg/L | 94 | 12 μg/L | 5 μg/L | 0.40 μg/L | 91 |
| Methylcyclopentane | 0.161 | | NQ[f] | | | | | NQ | |
| 1,1,1-Trichloroethane | 0.167 | S | 8 | 1 | 88 | 4 | 2 | 0 | 100 |
| 1-Chlorobutane[d] | 0.171 | | | | | | | | |
| 2-Pentanone | 0.175 | | | | | | | NQ | |
| Benzene | 0.180 | S | 86 | 8 | 91 | 53 | 57 | 4 | 93 |
| Carbon Tetrachloride | 0.184 | S | 14 | 4 | 71 | 8 | 6 | 3 | 50 |
| Cyclohexane | 0.186 | | 4 RAU | 0 RAU | 100 | 8 RAU | 7 RAU | 0 RAU | 100 |
| $C_7$ Alkane Isomer[d] | 0.198 | | | | | | | | |
| Cyclohexene[d] | 0.201 | | | | | | | | |
| Methylpropenoic Acid, Methyl Ester Isomer | 0.207 | | | | | | | NQ | |
| 1,2-Dichloropropane | 0.208 | | | | | 2 RAU | | | |
| Trichloroethene | 0.215 | S | 57 | 3 | 95 | 6 | 7 | 2 | 71 |
| Bromodichloromethane | 0.220 | S | 16 μg/L | 0.01 μg/L | 100 | 13 μg/L | 6.2 μg/L | 0.03 μg/L | 100 |
| Methylpropenoic Acid, Methyl Ester Isomer | 0.230 | | 2 RAU | 0 RAU | 100 | | | | |
| Heptane | 0.233 | | 5 RAU | 0 RAU | 100 | | | | |
| 1-Bromo-2-chloroethane[d] | 0.234 | | | | | | | | |
| 5,5-Dimethyl-2-hexene[d] | 0.240 | | | | | | | | |
| Methylcyclohexane | 0.248 | | 4 RAU | 0 RAU | 100 | 2 RAU | 3 RAU | 0 RAU | 100 |
| 4-Methyl-2-pentanone | 0.254 | | 64 RAU | 0 RAU | 100 | 16 RAU | 7 RAU | 0 RAU | 100 |
| Dichloromethylbutane Isomer | 0.276 | | 20 RAU | 0 RAU | 100 | | 5 RAU | 0 RAU | 100 |
| 4-Octanone[d] | 0.277 | | | | | | | | |
| 2,3,4-Trimethylpentane[d] | 0.284 | | | | | | | | |
| 2-Bromo-1-chloropropane[d] | 0.287 | | | | | | | | |

| | | | | | | | | | |
|---|---|---|---|---|---|---|---|---|---|
| Toluene | 0.288 | S | 32 | 9 | 72 | 19 | 32 | 59 | -84 |
| 4-Methyl-2-pentanol[d] | 0.289 | | | | | | | | |
| 1,3-Dichloropropane[d] | 0.299 | | | | | | | | |
| 2-Methylthiophene[d] | 0.299 | S | 0 | 0 | | 0 | 0 | 0 | |
| Butyl Acetate Isomer | 0.300 | | 4 RAU | 0 RAU | 100 | | | | |
| 3-Hexanone[d] | 0.304 | | | | | | | | |
| 2-Ethyl-4-methyl-1,3-dioxolane | 0.310 | | 5 RAU | 0 RAU | 100 | | | | |
| Dibromochloromethane | 0.313 | S | 6.2 μg/L | 0 | 100 | 5.7 μg/L | 2.9 μg/L | 0.01 μg/L | 100 |
| Hexanal | 0.319 | S | 6 | 4 | 33 | 0 | 6 | 0 | 100 |
| Ethylmethyl-1,3-dioxolane Isomer[d] | 0.320 | | | | | | | | |
| Trimethylcyclopentane Isomer[d] | 0.331 | | | | | | | | |
| Tetrachloroethene | 0.338 | S | 18 | 3 | 83 | 14 | 20 | 3 | 85 |
| Dichloroiodomethane | 0.345 | S | 9 | 0 | 100 | 47 | 15 | 0 | 100 |
| Octane | 0.345 | | | NQ | | | | | |
| Butyl Acetate Isomer[d] | 0.346 | | | | | | | | |
| Diethyltetrahydrofuran Isomer[d] | 0.363 | | | | | | | | |
| 1,1,1-Trichloro-2-propanone | 0.369 | | 34 RAU | 0 RAU | 100 | 1 RAU | 8 RAU | 0 RAU | 100 |
| Chlorobenzene | 0.375 | S | 14 | 2 | 86 | 10 | 14 | 1 | 93 |
| Dichloro-3-pentanone Isomer | 0.382 | | 13 RAU | 0 RAU | 100 | | 5 RAU | 0 RAU | 100 |
| Ethylbenzene | 0.398 | S | 24 | 4 | 83 | 2 | 4 | 2 | 50 |
| 1,3-Dimethylbenzene | 0.409 | S | 8 | 6 | 25 | 7 | 13 | 7 | 46 |
| 1,4-Dimethylbenzene | 0.410 | S | NQ | NQ | | NQ | NQ | NQ | |
| Bromoform | 0.415 | S | 0.51 μg/L | 0 | 100 | 0.66 μg/L | 0.38 μg/L | 0 | 100 |
| 3-Heptanone | 0.428 | | | | | NQ | | | |
| Trimethylcyclohexane Isomer[d] | 0.430 | | | | | | | | |
| Styrene | 0.430 | S | 0 | 1 | $-\infty$ | 2 | 2 | 0 | 100 |
| 1,2-Dimethylbenzene | 0.434 | S | 5 | 3 | 40 | 4 | 6 | 3 | 50 |
| Dibutylether | 0.436 | | 2 RAU | 0 RAU | 100 | NQ | 1 RAU | 0 RAU | 100 |
| Heptanal | 0.439 | | 2 RAU | 2 RAU | 0 | 2 RAU | 4 RAU | 1 RAU | 75 |
| Ethylmethylcyclohexane Isomer | 0.442 | | NQ | NQ | | | | | |
| Ethylmethylcyclohexane Isomer[d] | 0.445 | | | | | | | | |

**Table II, continued**

| Compound | RRT[a] | Quan Method[b] | Contactor D Water[c] GAC-inf (ng/L) | Contactor D Water[c] GAC-eff (ng/L) | Contactor D Water[c] % Removal by GAC | Contactor A Water[c] GAC-inf (ng/L) | Contactor A Water[c] XAD-inf (ng/L) | Contactor A Water[c] XAD-eff (ng/L) | Contactor A Water[c] % Removal by XAD |
|---|---|---|---|---|---|---|---|---|---|
| Bromochloroiodomethane + Bromotrichloroethene | 0.448 | S | 3 | 0 | 100 | 6 | 5 | 0 | 100 |
| Dimethylpentanal Isomer[d] | 0.448 | | | | | | | | |
| 1-Nonene | 0.449 | | | NQ | | | | | |
| Methylpropylcyclopentane Isomer[d] | 0.450 | | | | | | | | |
| 1,2,3-Trichloropropane[d] | 0.451 | | | | | | | | |
| 1,1,2,2-Tetrachloroethane | 0.453 | S | 7 | 7 | 0 | 5 | 7 | 4 | 43 |
| Methoxybenzene or Phenylhydrazine[d] | 0.457 | | | | | | | | |
| Trimethylcyclohexane Isomer | 0.458 | | | NQ | | | | | |
| Benzonitrile[d] | 0.461 | | | | | | | | |
| Trimethylcyclohexane Isomer[d] | 0.463 | | | | | | | | |
| $C_3$ Cyclohexane Isomer[d] | 0.467 | | | | | | | | |
| $C_4$-$C_5$ Tetrahydrofuran Isomer | 0.470 | | | | | NQ | | | |
| Isopropylbenzene | 0.473 | S | 1 | 0 | 100 | 1 | 2 | 0 | 100 |
| Methyloctahydropentalene Isomer[d] | 0.474 | | | | | | | | |
| Isopropylcyclohexane[d] | 0.478 | | | | | | | | |
| Trichloro-2-butanone Isomer | 0.479 | | 9 RAU | 0 RAU | 100 | | | | |
| Bromochloro-3-pentanone Isomer | 0.481 | | 9 RAU | 0 RAU | 100 | 3 RAU | 9 RAU | 0 | 100 |
| Ethylmethylcyclohexane Isomer[d] | 0.488 | | | | | | | | |
| Propylcyclohexane | 0.490 | | NQ | | | | NQ | | |
| Chlorotoluene Isomer | 0.498 | | NQ | | | NQ | 1 | 0 | 100 |
| 2-Ethylhexanal | 0.502 | | NQ | | | | | | |
| Propylbenzene | 0.506 | S | 3 | 2 | 33 | 2 | 4 | 2 | 50 |

| | | | | | | | | | |
|---|---|---|---|---|---|---|---|---|---|
| Octahydroindene | 0.511 | | | NQ | | | NQ | | |
| 1-Ethyl-4-methylbenzene | 0.515 | S | 2 | 3 | -50 | 2 | 10 | 5 | 50 |
| 1-Ethyl-3-methylbenzene | 0.517 | | 2 RAU | 3 RAU | -50 | 2 RAU | 5 RAU | 2 RAU | 60 |
| Dimethylcyclooctane or Tetramethyl Hexene Isomer | 0.519 | | | | | | 2 RAU | 0 RAU | 100 |
| 1,3,5-Trimethylbenzene | 0.524 | | 0 RAU | 3 RAU | $-\infty$ | 1 RAU | 7 RAU | 2 RAU | 71 |
| Pentachloroethane | 0.522 | S | NQ | | | | | | |
| 2,2,4,4-Tetramethyl-3-pentanone | 0.527 | | NQ | | | 11 | 13 | 0 | 100 |
| 1-Ethyl-2-methylbenzene | 0.534 | S | 1 | 1 | 0 | 2 | 2 | 1 | 50 |
| 3-Ethyl-2,4-dimethylpentane | 0.546 | | | NQ | | | 2 RAU | 0 RAU | 100 |
| 1,2,4-Trimethylbenzene | 0.551 | S | 3 | 4 | -33 | 2 | 5 | 3 | 40 |
| Octanal | 0.555 | | 5 RAU | 7 RAU | -40 | NQ | 0 RAU | 4 RAU | $-\infty$ |
| 1,3-Dichlorobenzene | 0.556 | S | 5 | 0 | 100 | 23 | 36 | 0 | 100 |
| Dimethylheptanal Isomer[d] | 0.561 | | | | | | | | |
| 1,4-Dichlorobenzene | 0.562 | S | 18 | 2 | 89 | 27 | 41 | 2 | 95 |
| Methylisopropylcyclohexane Isomer[d] | 0.566 | | | | | | | | |
| (2-Methylpropyl)benzene[d] | 0.567 | | | | | | | | |
| (1 Methylpropyl)benzene[d] | 0.571 | | | | | | | | |
| Decane | 0.576 | | NQ | NQ | | | NQ | NQ | |
| 1-Methyl-4-propyl-7-oxabicyclo[2.2.1]heptane | 0.577 | | 4 RAU | 0 RAU | 100 | 3 RAU | 2 RAU | 0 RAU | 100 |
| 1,2,3-Trimethylbenzene | 0.580 | S | 1 | 1 | 0 | 1 | 2 | 1 | 50 |
| Methylisopropylbenzene Isomer | 0.583 | | | NQ | | | NQ | NQ | |
| 1,2-Dichlorobenzene + Methylisopropylbenzene Isomer | 0.585 | S | 17 | 1 | 94 | 24 | 33 | 1 | 97 |
| Trimethylcyclohexanone Isomer[d] | 0.587 | | | | | | | | |
| Indan | 0.590 | S | 1 | 0 | 100 | 0 | 1 | 0 | 100 |
| 1,3,3-Trimethyl-2-oxabicyclo[2.2.2]octane | 0.593 | | NQ | | | NQ | | | |
| (1-Methylpropyl)cyclohexane | 0.595 | | NQ | NQ | | | 1 RAU | 0 RAU | 100 |
| Methylisopropylbenzene Isomer[d] | 0.599 | | | | | | | | |
| Indene | 0.599 | S | | NQ | | | | NQ | |

**Table II, continued**

| Compound | RRT[a] | Quan Method[b] | Contactor D Water[c] GAC-inf (ng/L) | Contactor D Water[c] GAC-eff (ng/L) | Contactor D Water[c] % Removal by GAC | Contactor A Water[c] GAC-inf (ng/L) | Contactor A Water[c] XAD-inf (ng/L) | Contactor A Water[c] XAD-eff (ng/L) | Contactor A Water[c] % Removal by XAD |
|---|---|---|---|---|---|---|---|---|---|
| Ethyldimethylbenzene Isomer[d] | 0.599 | | | | | | | | |
| Butylcyclohexane[d] | 0.604 | | | | | | | | |
| 2,2-Oxy-*bis*-[1-chloro] propane[d] | 0.606 | | | | | | | | |
| Pentylcyclopentane[d] | 0.609 | | | | | | | | |
| 1,3-Diethylbenzene | 0.611 | S | 1 | 1 | 0 | 1 | 1 | 0 | 100 |
| Methylpropylbenzene Isomer | 0.614 | | 3 RAU | 3 RAU | 0 | 1 RAU | 5 RAU | 2 RAU | 60 |
| 1,4-Diethylbenzene | 0.618 | S | 1 | 0 | 100 | NQ | 3 | 0 | 100 |
| *n*-Butylbenzene | 0.618 | S | 0 | 1 | $-\infty$ | 0 | 1 | 0 | 100 |
| 5-Ethyl-1,3-dimethylbenzene | 0.621 | S | 0 | 1 | $-\infty$ | 1 | 2 | 1 | 50 |
| Decahydronaphthalene | 0.622 | | 2 RAU | 0 RAU | 100 | | | | |
| Methylpropylbenzene isomer | 0.629 | | NQ | NQ | | NQ | NQ | NQ | |
| $C_4$ Cyclohexane Isomer[d] | 0.632 | | | | | | | | |
| Hexachloroethane | 0.634 | S | 8 | 1 | 88 | 7 | 5 | 1 | 80 |
| 2-Ethyl-1,4-dimethylbenzene | 0.640 | S | 2 | 1 | 50 | NQ | 4 | 0 | 100 |
| 4-Ethyl-1,3-dimethylbenzene | 0.642 | S | NQ | NQ | | NQ | 2 | 0 | 100 |
| *d*-Fenchone[d] | 0.645 | S | 0 | 0 | | 0 | 0 | 0 | |
| Ethylstyrene Isomer | 0.646 | | 1 RAU | 0 RAU | 100 | NQ | 1 RAU | 0 RAU | 100 |
| 4-Ethyl-1,2-dimethylbenzene | 0.648 | S | 1 | 1 | 0 | 1 | 2 | 1 | 50 |
| 2-Ethyl-1,3-dimethylbenzene | 0.654 | S | NQ | | | | | | |
| Ethylisopropylbenzene Isomer[d] | 0.657 | | | | | | | | |
| 1,1-Dimethylindan + $C_4$ cyclohexane Isomer[d] | 0.659 | S | | | | | | | |
| Nonanal | 0.662 | | 8 RAU | 24 RAU | -200 | 1 RAU | 23 RAU | 9 RAU | 61 |
| 3-Ethyl-1,2dimethylbenzene | 0.668 | S | 1 | 1 | 0 | 0 | 1 | 0 | 100 |
| $C_5$ Benzene Isomer[d] | 0.669 | | | | | | | | |
| $C_5$ Benzene Isomer[d] | 0.673 | | | | | | | | |
| $C_5$ Benzene Isomer[d] | 0.675 | | | | | | | | |
| Undecane | 0.679 | | | NQ | | | 0 RAU | 3 RAU | $-\infty$ |

| | | | | | | | | | |
|---|---|---|---|---|---|---|---|---|---|
| 1,2,4,5-Tetramethylbenzene | 0.681 | | 4 RAU | 3 RAU | 25 | NQ | 8 RAU | 0 RAU | 100 |
| 1,2,3,5-Tetramethylbenzene | 0.684 | S | 1 | 1 | 0 | NQ | 2 | 1 | 50 |
| (3-Methylbutyl)benzene[d] | 0.687 | | | | | | | | |
| Dimethylindan Isomer | 0.689 | | | | | NQ | NQ | NQ | |
| 1,3,5-Trichlorobenzene | 0.691 | S | 0 | 0 | | | NQ | | |
| $C_5$ Benzene Isomer[d] | 0.696 | | | | | | | | |
| 2-Methyldecahydronaphthalene | 0.698 | | 1 RAU | 0 RAU | 100 | 1 RAU | 3 RAU | 0 RAU | 100 |
| Methylindan Isomer | 0.699 | | NQ | NQ | | NQ | 5 RAU | 0 RAU | 100 |
| $C_5$ Benzene Isomer | 0.702 | | | NQ | | NQ | | | |
| Dimethylindan Isomer | 0.705 | | | | | | NQ | | |
| 1,3-Diethyl-5-methylbenzene | 0.707 | S | | NQ | | | | | |
| Methylindan or $C_2$ Sytrene Isomer | 0.709 | | | NQ | | | | | |
| $C_5$ Benzene Isomer[d] | 0.710 | | | | | | | | |
| 1,2,3,4-Tetramethylbenzene | 0.714 | S | 1 | 0 | 100 | NQ | 1 | 0 | 100 |
| *p*-Isobutyltoluene | 0.718 | | | NQ | | | | | |
| Tetrahydronaphthalene | 0.718 | S | 40 | 0 | 100 | NQ | NQ | | |
| Diethylmethylbenzene Isomer | 0.721 | | | | | NQ | | | |
| *n*-Pentylbenzene | 0.722 | S | | NQ | | | | | |
| $C_5$ Benzene Isomer[d] | 0.722 | | | | | | | | |
| (1,1-Dimethylpropyl)benzene | 0.725 | | | NQ | | | | | |
| $C_5$ Benzene Isomer[d] | 0.729 | | | | | | | | |
| 1,2,4-Trichlorobenzene + $C_5$ Benzene Isomer | 0.731 | S | 2 | 0 | 100 | NQ | 4 | 0 | 100 |
| Naphthalene | 0.738 | S | NQ | NQ | | NQ | NQ | NQ | |
| Dimethylindan or Methylbutenylbenzene Isomer[d] | 0.744 | | | | | | | | |
| $C_2$ Indan Isomer + a Siloxane | 0.748 | | | | | | 7 | 0 | 100 |
| $C_2$ Indan + $C_6$ Benzene Isomers[d] | 0.749 | | | | | | | | |
| $C_2$ Indan Isomer[d] | 0.753 | | | | | | | | |
| Ethyltrimethylbenzene Isomer | 0.756 | | | NQ | | | | | |
| $C_2$ Indan Isomer | 0.756 | | 1 RAU | 4 RAU | -300 | 1 | | | |
| 3-Ethyl-1,2,4-trimethylbenzene[d] | 0.760 | | | | | | | | |
| Decanol | 0.762 | | 17 RAU | 73 RAU | -330 | 8 RAU | 39 RAU | 17 RAU | 56 |

**Table II, continued**

| Compound | RRT[a] | Quan Method[b] | Contactor D Water[c] GAC-inf (ng/L) | Contactor D Water[c] GAC-eff (ng/L) | Contactor D Water[c] % Removal by GAC | Contactor A Water[c] GAC-inf (ng/L) | Contactor A Water[c] XAD-inf (ng/L) | Contactor A Water[c] XAD-eff (ng/L) | Contactor A Water[c] % Removal by XAD |
|---|---|---|---|---|---|---|---|---|---|
| 1,2,3-Trichlorobenzene | 0.763 | S | NQ | | | 1 | | NQ | |
| $C_3$ Indan Isomer | 0.764 | | 1 RAU | 0 RAU | 100 | 1 RAU | NQ | | |
| $C_6$ Benzene Isomer | 0.768 | | | | | NQ | | | |
| 1-Ethyl-2,3,5-trimethylbenzene | 0.770 | | | NQ | | | | | |
| $C_3$ Indan or $C_2$ THN[g] Isomer | 0.772 | | 1 RAU | 0 RAU | 100 | 2 RAU | 3 RAU | 0 RAU | 100 |
| Dodecane | 0.775 | | 11 RAU | 0 RAU | 100 | | 5 RAU | 0 RAU | 100 |
| $C_5$ Benzene Isomer[d] | 0.778 | | | | | | | | |
| Methyl THN Isomer[d] | 0.780 | | | | | | | | |
| Hexachloro-1,3-butadiene | 0.780 | S | 0 | 0 | | 0 | NQ | 0 | |
| $C_6$ Benzene Isomer | 0.786 | | 2 RAU | 0 RAU | 100 | NQ | | | |
| $C_6$ Benzene Isomer[d] | 0.788 | | | | | | | | |
| Methyl THN Isomer[d] | 0.790 | | | | | | | | |
| $C_5$ Benzene Isomer[d] | 0.792 | | | | | | | | |
| $C_6$ Benzene + $C_2$ THN or $C_3$ Indan Isomers | 0.797 | | 2 RAU | 0 RAU | 100 | NQ | | | |
| $C_6$ Benzene Isomer[d] | 0.799 | | | | | | | | |
| $C_6$ Benzene Isomer[d] | 0.802 | | | | | | | | |
| Dimethylindan or Methyl THN Isomer | 0.805 | | NQ | NQ | | | | | |
| $C_6$ Benzene Isomer | 0.808 | | | | | NQ | | | |
| Methyl THN Isomer or 4,7-Dimethylindan | 0.818 | S | 1 | 0 | 100 | | 2 | 0 | 100 |
| $C_6$ Benzene Isomer[d] | 0.821 | | | | | | | | |
| $C_6$ Benzene Isomer | 0.824 | | | | | NQ | | | |
| $C_7$ Benzene Isomer | 0.828 | | | | | 1 RAU | | | |
| $C_9$-$C_{12}$ Aldehyde Isomer | 0.829 | | 0 RAU | 2 RAU | $-\infty$ | | | | |
| $C_6$ Benzene Isomer | 0.833 | | | NQ | | 1 RAU | | | |
| Pentamethylbenzene | 0.835 | S | | NQ | | | | | |
| $C_6$ Benzene Isomer | 0.838 | | | NQ | | | | | |
| Dimethyl THN Isomer[d] | 0.841 | | | | | | | | |
| | [illegible] | S | NQ | NQ | | | NQ | NQ | |

| | | | | | | | | | |
|---|---|---|---|---|---|---|---|---|---|
| $C_3$ Indan or $C_2$ THN Isomer | 0.844 | | | | | NQ | 1 RAU | 0 RAU | 100 |
| $C_3$ Indan or $C_2$ THN Isomer | 0.847 | | | | | 3 RAU | 4 RAU | 0 RAU | 100 |
| $C_3$ Indan or $C_2$ THN Isomer | 0.849 | | | | | NQ | | | |
| $C_3$ Indan or $C_2$ THN Isomer[d] | 0.854 | | | | | | | | |
| 1-Methylnaphthalene + Undecanol | 0.855 | S | NQ | NQ | | | 0 | 2 | −∞ |
| | | | (1) RAU | (8) RAU | (−700) | | | | |
| Tridecane | 0.866 | | | NQ | | | 3 RAU | 0 RAU | 100 |
| Dimethyl THN Isomer | 0.868 | | | | | | 5 RAU | 0 RAU | 100 |
| $C_6$ Benzene Isomer[d] | 0.871 | | | | | | | | |
| Dimethyl THN Isomer | 0.873 | | | | | | 5 RAU | 0 RAU | 100 |
| Trimethylindan Isomer | 0.875 | | | | | | 4 RAU | 0 RAU | 100 |
| $C_3$ THN Isomer | 0.887 | | | | | | 6 RAU | 0 RAU | 100 |
| Dimethyl THN Isomer[d] | 0.890 | | | | | | | | |
| $C_7$ Benzene Isomer | 0.890 | | | | | | 4 RAU | 0 RAU | 100 |
| Ethyl THN or Trimethylindan Isomer[d] | 0.892 | | | | | | | | |
| Dimethyl THN Isomer | 0.896 | | | | | | 1 RAU | 0 RAU | 100 |
| Ethyl THN Isomer[d] | 0.900 | | | | | | | | |
| Trimethyl THN or $C_4$ Indan Isomer | 0.901 | | | | | | 7 RAU | 0 RAU | 100 |
| $C_7$ Benzene Isomer[d] | 0.903 | | | | | | | | |
| $C_3$ THN Isomer + Siloxane | 0.907 | | | | | | NQ | | |
| 1,1-Biphenyl[d] | 0.914 | | | | | | | | |
| Trimethyl THN Isomer | 0.920 | | | | | | 2 RAU | 0 RAU | 100 |
| $C_7$ Benzene Isomer | 0.924 | | | | | | 1 RAU | 0 RAU | 100 |
| Dimethyl THN Isomer | 0.929 | | | | | | NQ | | |
| Diphenylether | 0.932 | | 14 RAU | 0 RAU | 100 | | 3 RAU | 0 RAU | 100 |
| $C_2$ Napthalene Isomer | 0.938 | | | NQ | | | NQ | | |
| $C_7$ Benzene Isomer | 0.940 | | | | | | 1 RAU | 0 RAU | 100 |
| Dodecanal | 0.943 | | 0 RAU | 5 RAU | −∞ | | 6 RAU | 2 RAU | 67 |
| $C_3$ THN Isomer | 0.946 | | | | | | 1 RAU | 0 RAU | 100 |
| $C_3$ THN Isomer[d] | 0.950 | | | | | | | | |
| Dimethylnaphthalene Isomer | 0.951 | | | NQ | | | | | |
| Tetradecane | 0.953 | | 0 RAU | 3 RAU | −∞ | | 3 RAU | 0 RAU | 100 |

**Table II, continued**

| Compound | RRT[a] | Quan Method[b] | Contactor D Water[c] GAC-inf (ng/L) | Contactor D Water[c] GAC-eff (ng/L) | Contactor D Water[c] % Removal by GAC | Contactor A Water[c] GAC-inf (ng/L) | Contactor A Water[c] XAD-inf (ng/L) | Contactor A Water[c] XAD-eff (ng/L) | Contactor A Water[c] % Removal by XAD |
|---|---|---|---|---|---|---|---|---|---|
| $C_3$ THN or $C_4$ Indan Isomer[d] | 0.954 | | | | | | | | |
| $C_5$ THN or $C_6$ Indan Isomer | 0.957 | | | | | | 1 RAU | 0 RAU | 100 |
| $C_4$ Indan or $C_3$ THN Isomer | 0.958 | | | | | | 3 RAU | 0 RAU | 100 |
| $C_4$ Indan or $C_3$ THN Isomer[d] | 0.962 | | | | | | | | |
| $C_4$ Indan or $C_3$ THN Isomer[d] | 0.966 | | | | | | | | |
| Dimethylnaphthalene Isomer[d] | 0.967 | | | | | | | | |
| $C_4$ Indan or $C_3$ THN Isomer[d] | 0.971 | | | | | | | | |
| $C_4$ Indan or $C_3$ THN Isomer | 0.974 | | | | | | 1 RAU | 0 RAU | 100 |
| Trimethyl THN Isomer | 0.977 | | 4 RAU | 0 RAU | 100 | | 9 RAU | 0 RAU | 100 |
| 5,9-Undecadien-2-one,6,10-dimethyl | 0.977 | | | NQ | | | | | |
| $C_3$ THN or $C_4$ Indan Isomer[d] | 0.982 | | | | | | | | |
| $C_4$ Dihydronaphthalene Isomer | 0.984 | | | | | 1 RAU | 3 RAU | 0 RAU | 100 |
| 2,6-bis(1,1-Dimethylether)2,5-cyclohexadiene-1,4-dione | 0.991 | | 10 RAU | 0 RAU | 100 | NQ | 24 RAU | 0 RAU | 100 |
| Trimethyldihydronaphthalene Isomer | 0.992 | | | | | | 1 RAU | 0 RAU | 100 |
| $C_3$ THN Isomer | 0.997 | | NQ | | | | | | |
| 1-Chlorododecane, IS | 1.000 | S | 52 | 52 | | 52 | 52 | 52 | |
| Diphenylmethane[d] | 1.002 | | | | | | | | |
| $C_2$ Biphenyl Isomer[d] | 1.010 | | | | | | | | |
| $C_2$ Biphenyl Isomer[d] | 1.016 | | | | | | | | |
| $C_2$ Biphenyl Isomer[d] | 1.020 | | | | | | | | |
| Tridecanal | 1.027 | | | NQ | | | NQ | NQ | |
| Hexylindan Isomer[d] | 1.029 | | | | | | | | |
| Pentadecane | 1.034 | | 0 RAU | 5 RAU | $-\infty$ | | 9 RAU | 0 RAU | 100 |
| $C_9$ Benzene Isomer | 1.037 | | | | | | | | |
| Pentachlorobenzene[d] | 1.037 | S | 0 | 0 | | 0 | 0 | 0 | |
| Tetramethylindan Isomer[d] | 1.041 | | | | | | | | |

| | | | | | | | | | |
|---|---|---|---|---|---|---|---|---|---|
| $C_9$ Benzene + $C_3$ THN or $C_4$ Indan Isomers[d] | 1.066 | | | | | | | | |
| Trimethylnaphthalene isomer[d] | 1.066 | | | | | | | | |
| $C_2$ Biphenyl Isomer[d] | 1.079 | | | | | | | | |
| Diethylphthalate | 1.080 | S | 1 | 0 | 100 | 1 | | | |
| $C_2$ Biphenyl Isomer[d] | 1.085 | | | | | | | | |
| $C_3$ Biphenyl Isomer | 1.089 | | | | | | | | |
| Dimethylbiphenyl Isomer[d] | 1.093 | | | | | | NQ | | |
| $C_3$ Biphenyl Isomer | 1.095 | | 0 RAU | 2 RAU | $-\infty$ | | 1 RAU | 0 RAU | 100 |
| 2,2,4-Trimethylpenta-1,3-diol Diisobutyrate | 1.097 | | 3 RAU | 5 RAU | -67 | | 1 RAU | 0 RAU | 100 |
| $C_3$ Biphenyl Isomer[d] | 1.100 | | | | | | | | |
| $C_3$ Biphenyl Isomer | 1.103 | | | | | | 1 RAU | 0 RAU | 100 |
| Tetradecanal | 1.106 | | | | | | | NQ | |
| 1,2-Diphenylhydrazine | 1.108 | S | NQ | NQ | | 0.20 | 1 | 0 | 100 |
| Hexadecane | 1.110 | | 2 RAU | 4 RAU | -100 | 1 RAU | 3 RAU | 0 RAU | 100 |
| $C_3$ Biphenyl Isomer | 1.113 | | 0 RAU | 1 RAU | $-\infty$ | 1 RAU | NQ RAU | NQ RAU | |
| $C_3$ Biphenyl Isomer | 1.115 | | 0 RAU | 3 RAU | $-\infty$ | | 3 RAU | 0 RAU | 100 |
| Diethylbiphenyl Isomer[d] | 1.119 | | | | | | | | |
| $C_6$ Indan Isomer[d] | 1.125 | | | | | | | | |
| Phthalate Isomer[d] | 1.130 | | | | | | | | |
| $C_3$ Biphenyl Isomer | 1.136 | | | NQ | | | 2 RAU | 0 RAU | 100 |
| 2,5-*bis*-(1,1-Dimethylpropyl)-2,5-cyclohexadiene-1,4-dione | 1.141 | | NQ | NQ | | | | | |
| $C_5$ Biphenyl Isomer | 1.146 | | | | | | 7 RAU | 0 RAU | 100 |
| Diethylbiphenyl Isomer | 1.150 | | | | | 3 RAU | | | |
| 1-Chlorotetradecane | 1.157 | | NQ | NQ | | NQ | 0 RAU | 1 RAU | $-\infty$ |
| $C_4$ Biphenyl Isomer[d] | 1.160 | | | | | | | | |
| 1,1,3-Trimethyl-3-phenylindan | 1.181 | | 2 RAU | 6 RAU | -200 | 1 RAU | 6 RAU | 0 RAU | 100 |
| Heptadecane | 1.183 | | 2 RAU | 3 RAU | -50 | 1 RAU | 3 RAU | 1 RAU | 67 |
| Diethylbiphenyl Isomer[d] | 1.187 | | | | | | | | |
| Alkane Isomer | 1.191 | | | NQ | | | | | |
| Dibutylphthalate Isomer | 1.226 | | | NQ | | | | | |
| A Siloxane[d] | 1.227 | | | | | | | | |
| Octadecane | 1.254 | | NQ | NQ | | NQ | 1 RAU | 0 RAU | 100 |

**Table II, continued**

| Compound | RRT[a] | Quan Method[b] | Contactor D Water[c] GAC-inf (ng/L) | Contactor D Water[c] GAC-eff (ng/L) | Contactor D Water[c] % Removal by GAC | Contactor A Water[c] GAC-inf (ng/L) | Contactor A Water[c] XAD-inf (ng/L) | Contactor A Water[c] XAD-eff (ng/L) | Contactor A Water[c] % Removal by XAD |
|---|---|---|---|---|---|---|---|---|---|
| Alkane Isomer | 1.263 | | | NQ | | | | | |
| Phthalate + a Siloxane[d] | 1.274 | | | | | | | | |
| 1,1-*bis*-(Ethylphenyl)ethane[d] | 1.277 | | | | | | | | |
| Nonadecane | 1.321 | | NQ | NQ | | | | | NQ |
| Dibutylphthalate Isomer | 1.335 | | NQ | NQ | | NQ | NQ | | NQ |
| 4-Phenylbicyclohexyl[d] | 1.366 | | | | | | | | |
| Eicosane[d] | 1.385 | | | | | | | | |
| Benzylbutylphthalate | 1.563 | S | NQ | NQ | | | | | NQ |
| Dioctylphthalate | 1.683 | | NQ | | | NQ | NQ | | NQ |

[a] RRT = relative retention time, where RRT of chlorododecane = 1.000.

[b] Method of quantification; all quantification values reported are in ng/L unless otherwise noted. "S" indicates that a standard was purchased and the corresponding experimental relative response factor to that of chlorododecane, IS, was determined.

[c] See Figure 5 for an explanation of sample origin.

[d] This compound was detected in CWW contactor A GAC-inf water on February 20, 1980, but was not detected in any January 14 and 28, 1980 CWW samples. Compound is included to provide additional relative retention time.

[e] RAU (relative area unit) = $\frac{\text{Total Ion Current Area (UNK)}}{\text{Total Ion Current Area (IS)}} \times 52$; where UNK = Unknown Compound, IS = Chlorododecane

[f] NQ = Organic was detected but was not quantified.

[g] THN = Tetrahydronapthalene.

The chromatograms in Figure 7 are arranged to illustrate the Grob CLSA differences between GAC-inf and GAC-eff water. Note that the levels of internal standards (IS) in Figure 7 are the same. Also, that two of the first detectable organics that elute after the $CS_2$ solvent are isopropyl ether and chloroform. If more volatile nonpolar organics had been present, such as chloromethane, vinyl chloride and methylene chloride, they would have been covered up by the $CS_2$ solvent peak. These more volatile organics would have been detected, however, by the Bellar P&T analysis (Method 601) of these same CWW samples (see Table I and Figure 6) if we had used a trap packed with Tenax, silica gel and activated carbon.

Results of BLLE capillary GC/MS/DS analysis of GAC-inf and GAC-eff water are presented in Tables III and IV, which correlate with the chromatograms shown in Figures 8 and 9. Respectively, 51 and 38 organic compounds were identified in the neutral and methylated acid fractions of GAC-inf water. These BLLE samples are notably low in solvent artifacts due to the elaborate methylene chloride purification steps employed. Meticulously clean solvents, reagents and glassware are necessary for reproducible GC/MS analysis of BLLE samples of trace organics in drinking water. Figures 8 and 9 clearly illustrate the high level of artifact-free performance which has been achieved. The peaks labeled "IS" are deuterated internal standards (0.2 $\mu$g/L level) which were added to the water before extraction. The peak marked "IS HEB" is hexaethylbenzene, an internal standard, which was also added at the 0.2-$\mu$g/L level prior to GC/MS analysis. Artifacts which are due to aqueous extraction or extract fractionation are indicated by special symbols in the figures. Divinylmercury ($C_4H_6Hg$) was present in GAC-inf water according to BLLE (methylated acid fraction). Divinyl mercury also appeared in some of the BLLE blanks of reagent water. The presence of divinylmercury may be due to a chemical reaction between the preservative (mercuric chloride) and certain organics in the water samples.

The contamination of the XAD-EEE extract from the contactor D experiment by XAD-2 resin was so severe that the resulting GC/MS data files were of no value. These artifacts are quite typical of what we have seen on a number of occasions when analyzing XAD-2-generated organic concentrates. Since XAD-2 and XAD-4 resins are widely used for the concentration of organics from water (the MAS "extractables" method uses XAD-4 to concentrate organics from 3 L of drinking water), and since we are not aware of literature documentation of the specific contaminants one generally encounters, the abbreviated listing in Table V may be of some use to the reader. Capillary GC/MS/DS analysis of the XAD-EEE extract from the contactor A experiment is presented in Tables VI and VII. A total of 58 organic compounds were identified in the neutral and methylated acid fractions of this XAD-EEE extract. Columns labeled XAD-inf and XAD-eff

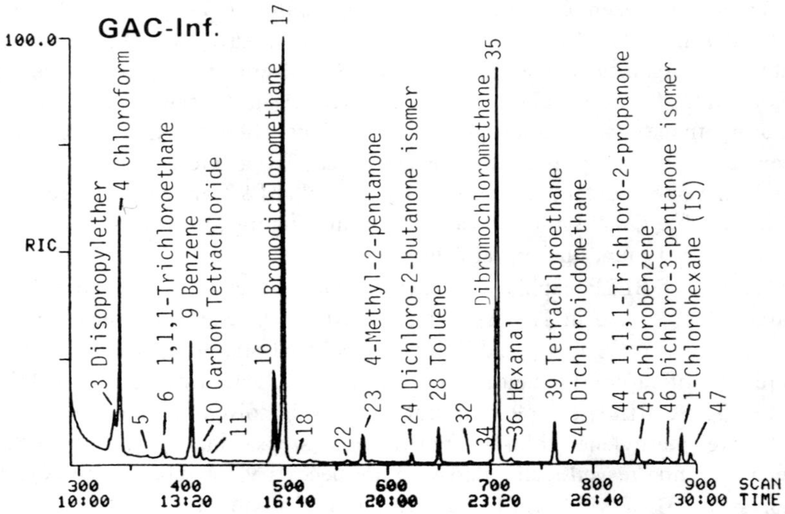


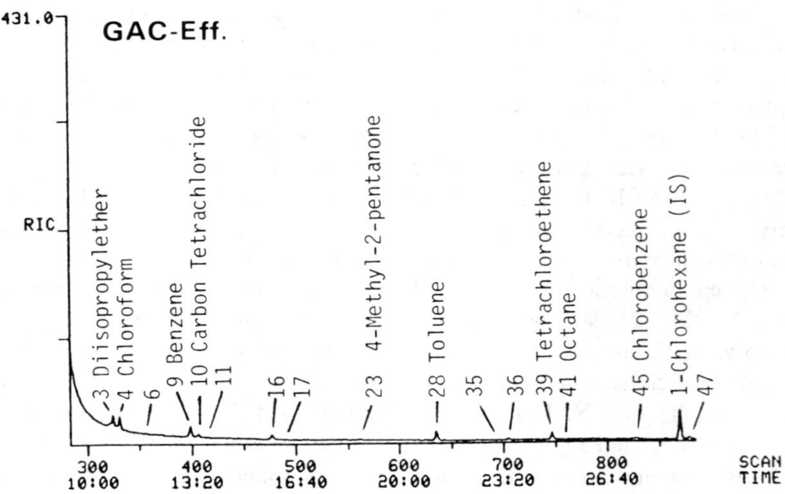


**Figure 7.** Chromatographic results of GAC-inf and GAC-eff water using Grob CLSA.

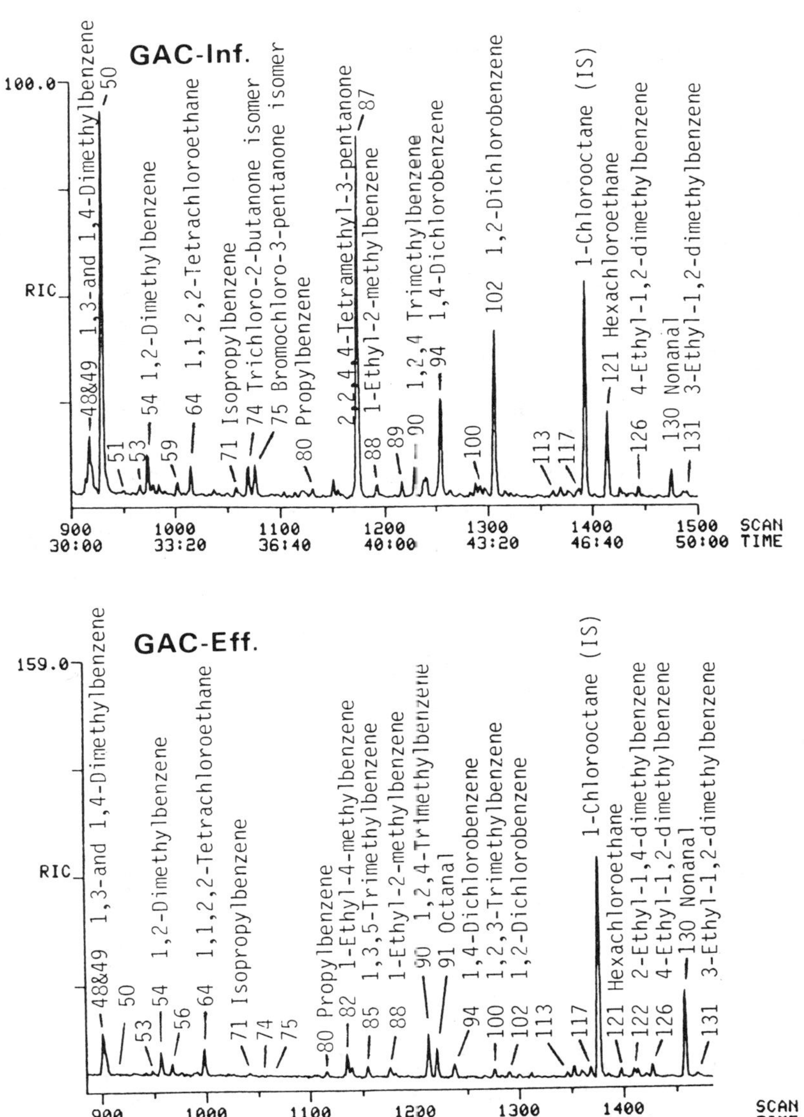


**Figure 7, continued**

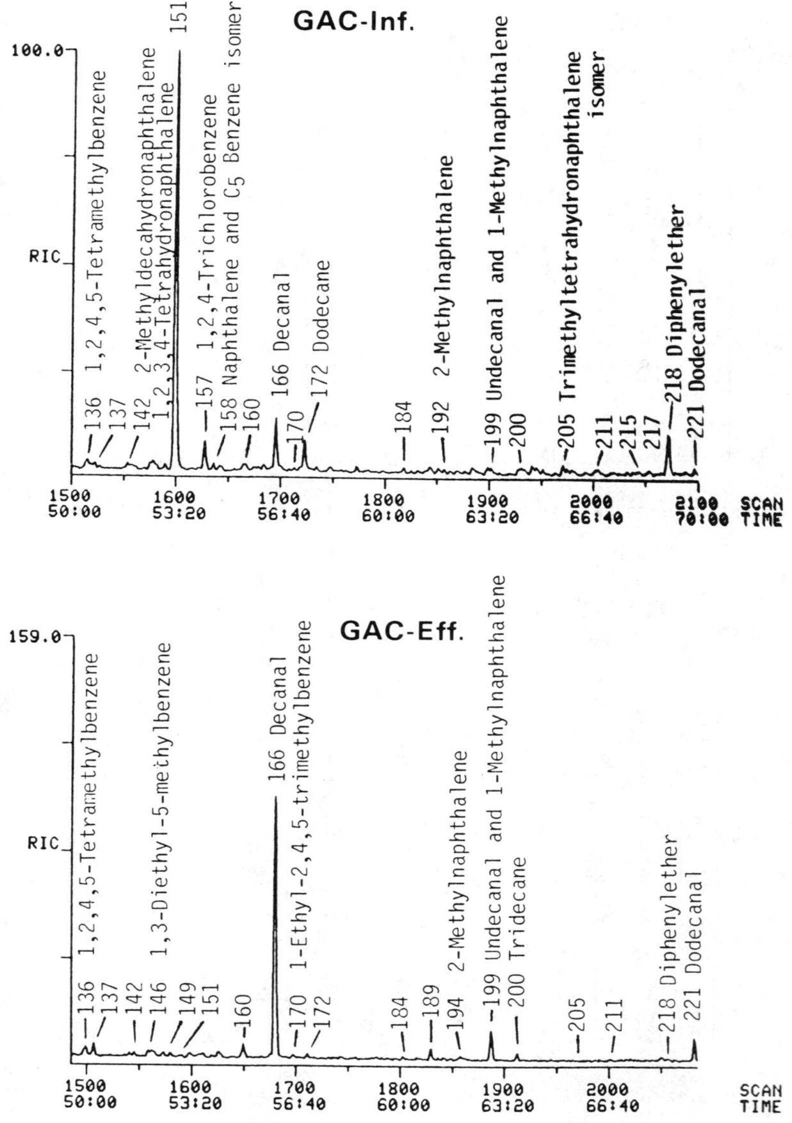


**Figure 7, continued**

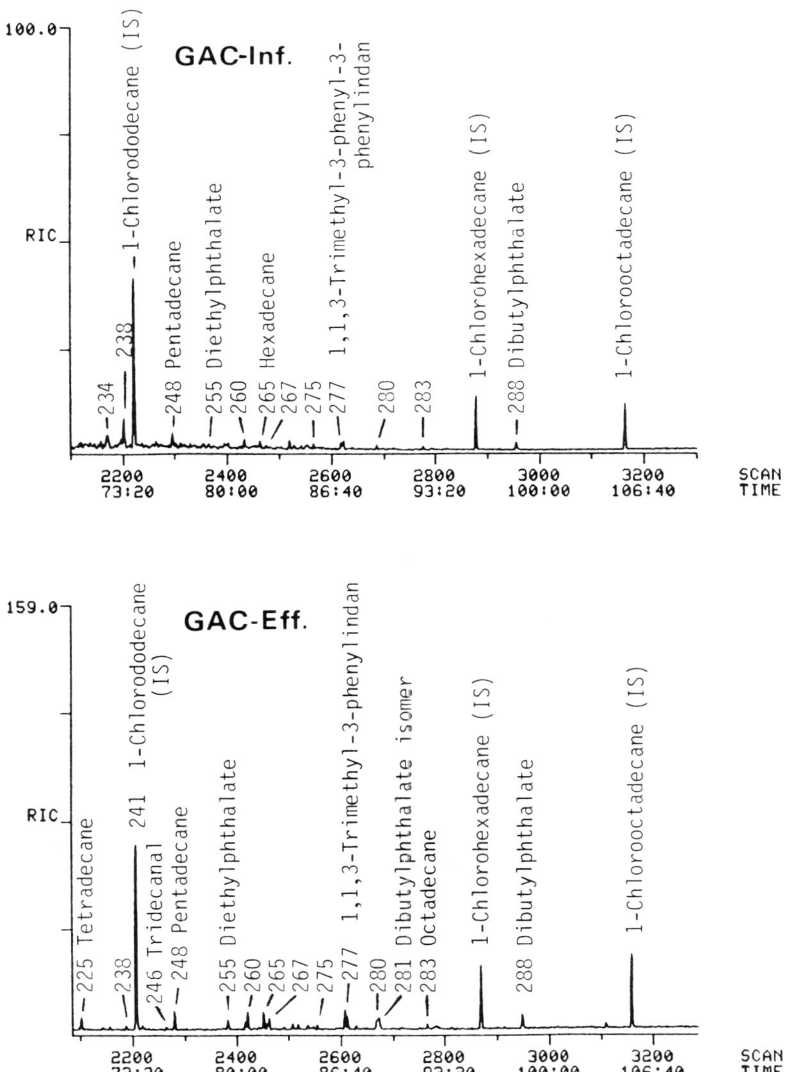


Figure 7, continued

**Table III. Results of BLLE Analysis of GAC-inf and GAC-eff Samples[a]–Neutral Fraction, GAC Contactor D Water**

| Sequence Number[b] | Relative Retention Time[c] | Compound Name | Formula | Relative Amount Detected[d] | | Percent Removal |
|---|---|---|---|---|---|---|
| | | | | GAC-inf | GAC-eff | |
| 1 | 0.14 | 2-Ethyl-4-methyl-1,3-dioxlane | $C_6H_{12}O_2$ | 25 | | 100 |
| 2 | 0.23 | 4-Methyl-3-penten-2-one | $C_6H_{10}O$ | 8 | | 100 |
| 3 | 0.23 | 1-Chloro-2,4-hexadiene | $C_6H_9Cl$ | 36 | | 100 |
| 4 | 0.25 | Bromodichloromethane | $CHBrCl_2$ | 420 | | 100 |
| 5 | 0.26 | 2,6-Dimethyl-4-heptanone | $C_9H_{18}O$ | 44 | | 100 |
| 6 | 0.30 | Chlorobenzene | $C_6H_5Cl$ | 9 | | 100 |
| 7 | 0.31 | 1,1-Dichlorocyclohexane | $C_6H_{10}Cl_2$ | 26 | | 100 |
| 8 | 0.39 | Dibromochloromethane | $CHBr_2Cl$ | 229 | | 100 |
| 9 | 0.40 | Cycloheptanone | $C_7H_{12}O$ | 7 | | 100 |
| 10 | 0.46 | 2-Methyl-2-cyclopenten-1-one | $C_6H_8O$ | 7 | | 100 |
| 11 | 0.47 | Dichloroacetonitrile | $C_2HNCl_2$ | 2 | | 100 |
| 12 | 0.47 | 1,1,3-Trichloro-1-propene | $C_3H_3Cl_3$ | 3 | | 100 |
| 13 | 0.51 | *m*-Dichlorobenzene | $C_6H_4Cl_2$ | 3 | | 100 |
| 14 | 0.53 | Hexachloroethane | $C_2Cl_6$ | 9 | | 100 |
| 15 | 0.54 | Bromoform | $CHBr_3$ | 75 | | 100 |
| 16 | 0.55 | 1,2-Dichlorocyclohexane | $C_6H_{10}Cl_2$ | 168 | | 100 |
| 17 | 0.57 | Alcohol | | 14 | | 100 |
| 18 | 0.59 | *o*-Dichlorobenzene | $C_6H_4Cl_2$ | 19 | | 100 |
| 19 | 0.62 | Tetralin | $C_{10}H_{12}$ | 49 | | 100 |
| 20 | 0.63 | Di(2-chloroethyl) ether | $C_4H_8OCl_2$ | 5 | | 100 |
| 21 | 0.64 | 3-Propylcyclopentene | $C_8H_{14}$ | 4 | | 100 |
| 22 | 0.66 | 1-Bromo-2-chlorocyclohexane | $C_6H_{10}CLBr$ | 29 | | 100 |
| 23 | 0.68 | Fenchyl alcohol | $C_{10}H_{18}O$ | 5 | | 100 |
| 24 | 0.69 | Isophorone | $C_9H_{14}O$ | 18 | | 100 |
| 25 | 0.70 | 4-Hydroxy-4-methylcyclohexanone | $C_7H_{12}O_2$ | 3 | | 100 |
| 26 | 0.70 | Benzonitrile | $C_7H_5N$ | 16 | | 100 |

| | | | | | | |
|---|---|---|---|---|---|---|
| 27 | 0.76 | Triethyl phosphate | $C_6H_{15}O_4P$ | 6 | | 100 |
| 28 | 0.83 | Nitrobenzene | $C_6H_5O_2N$ | 30 | | 100 |
| 29 | 0.83 | Naphthalene | $C_{10}H_8$ | 8 | | 100 |
| 30 | 0.85 | 2-Phenyl-2-propanol | $C_9H_{12}O$ | 7 | | 100 |
| 31 | 0.87 | *o*-Nitrotoluene | $C_7H_7O_2N$ | 18 | | 100 |
| 32 | 0.89 | Tripropylene glycol methyl ether | $C_{10}H_{22}O_4$ | 5 | | 100 |
| 33 | 0.89 | | | 9 | | 100 |
| 34 | 0.90 | | | 7 | | 100 |
| 35 | 0.90 | | | 6 | | 100 |
| 36 | 1.01 | Benzylcyanide | $C_8H_7N$ | 14 | | 100 |
| 37 | 1.03 | Benzthiazole | $C_7H_5NS$ | | 6 | |
| 38 | 1.08 | Indanone plus phenyl ether | $C_9H_8O$ | | | |
| | | | $C_{12}H_{10}O$ | 37 | | 100 |
| 39 | 1.10 | Dipropylene glycol 2-propenyl ether | $C_9H_{18}O_3$ | 5 | | 100 |
| 40 | 1.14 | 3,4-Dihydronaphthalen-1-one | $C_{10}H_{10}O$ | 31 | | 100 |
| 41 | 1.17 | Tributyl phosphate | $C_{12}H_{27}O_4P$ | 38 | | 100 |
| 42 | 1.22 | Trimethyl isocyanurate | $C_6H_9O_3N_3$ | 196 | 1 | 99 |
| 43 | 1.23 | Tetrahydro-1-naphthalenol | $C_{10}H_{12}O$ | 5 | | 100 |
| 44 | 1.24 | Sulfolane | $C_4H_8O_2S$ | 5 | | 100 |
| 45 | 1.29 | N-(4-Chlorophenyl)acetamide | $C_8H_8ONCl$ | 9 | | 100 |
| 46 | 1.33 | Tetrahydro-trimethyl benzofuranone | $C_{11}H_{16}O_2$ | 8 | | 100 |
| 47 | 1.33 | Phthalide | $C_8H_6O_2$ | 4 | | 100 |
| 48 | 1.44 | 2,4-Dinitrotoluene | $C_7H_6O_4N_2$ | 7 | | 100 |
| 49 | 1.49 | N-Phenylacetamide | $C_8H_9ON$ | 42 | | 100 |
| 50 | 1.66 | Fluorenone | $C_{13}H_8O$ | 1 | | 100 |

[a] See Figure 5 for explanation of sample code names.

[b] Numbers coorespond to the GC peaks, as labeled in Figure 8.

[c] Relative to the internal standard, hexaethylbenzene.

[d] Expressed as a peak height percentage of the internal standard, hexaethylbenzene, 0.2 ppb.

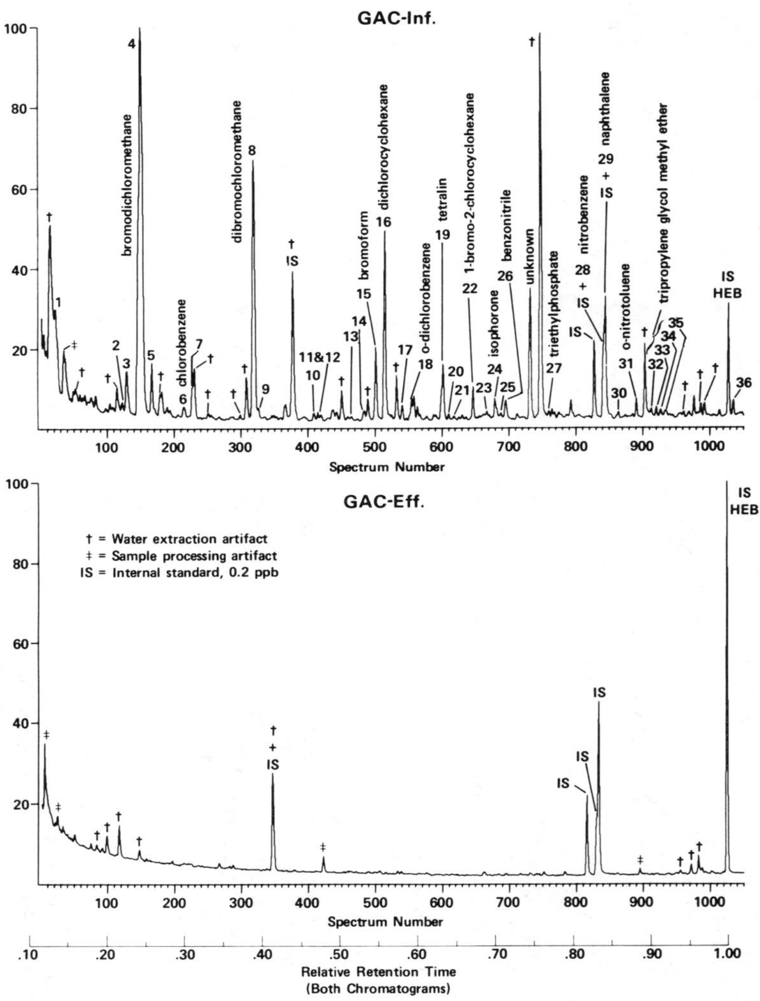


**Figure 8.** Chromatographic results of BLLE analysis of GAC-inf and GAC-eff. Neutral fraction.

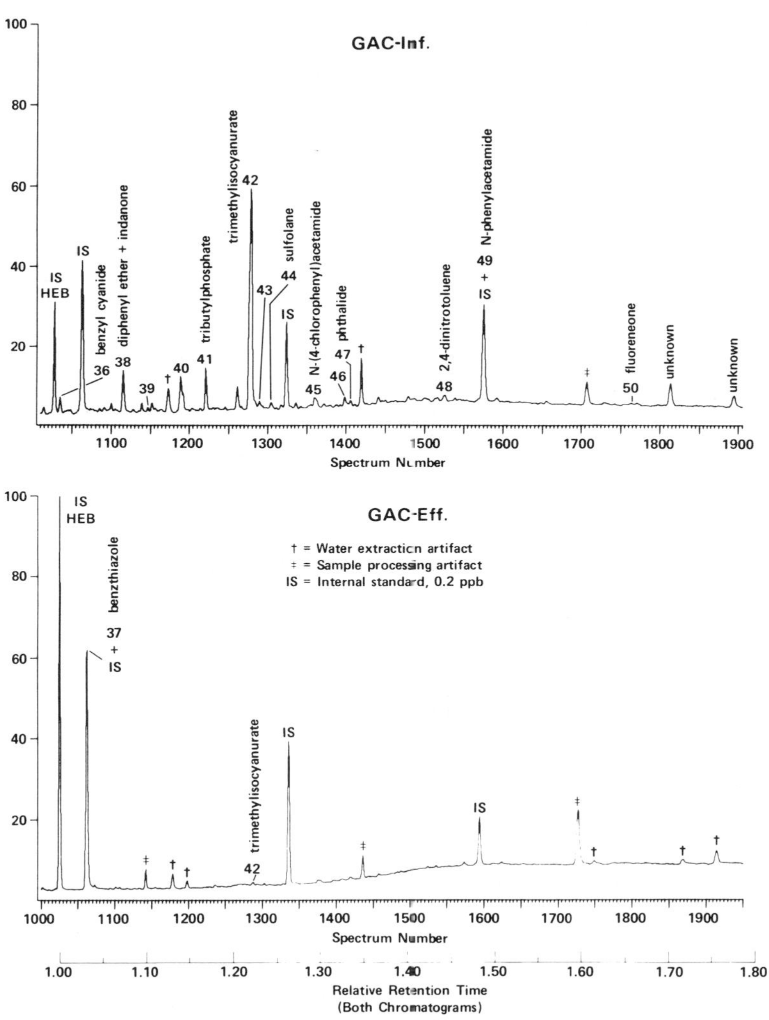


Figure 8, continued

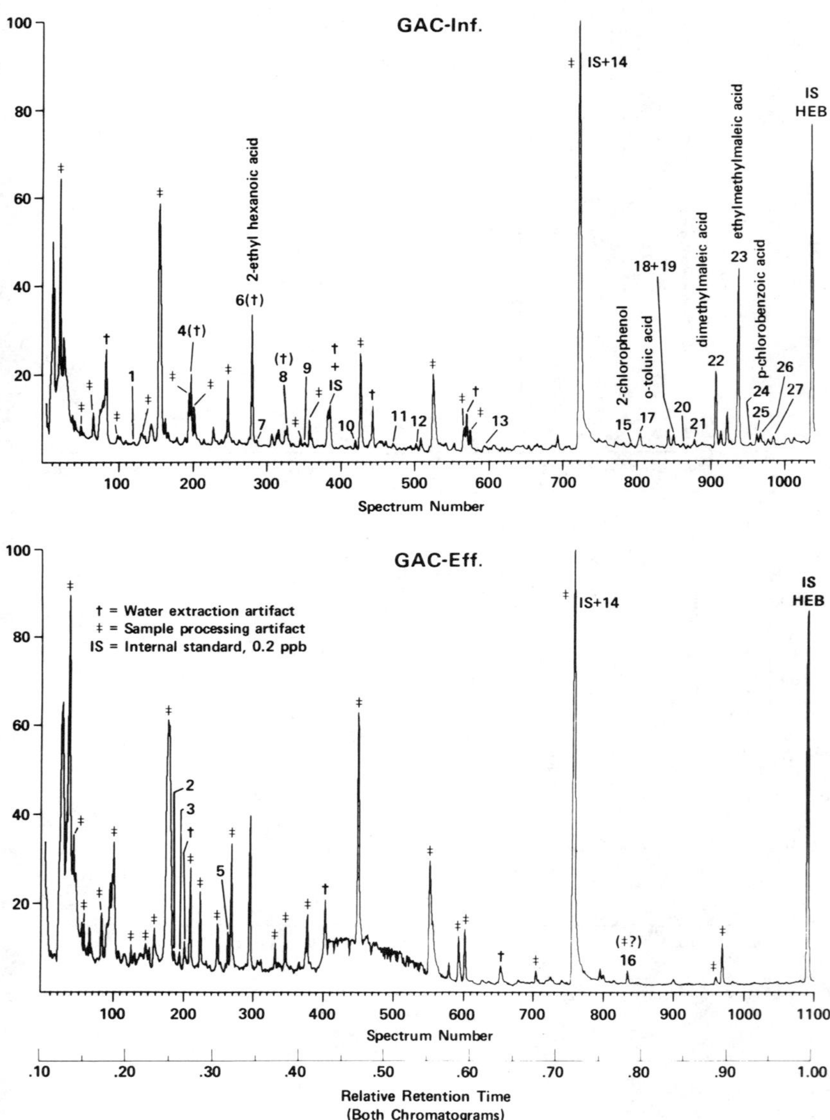


**Figure 9.** Chromatographic results of BLLE analysis of GAC-inf and GAC-eff. Methylated acid fraction.

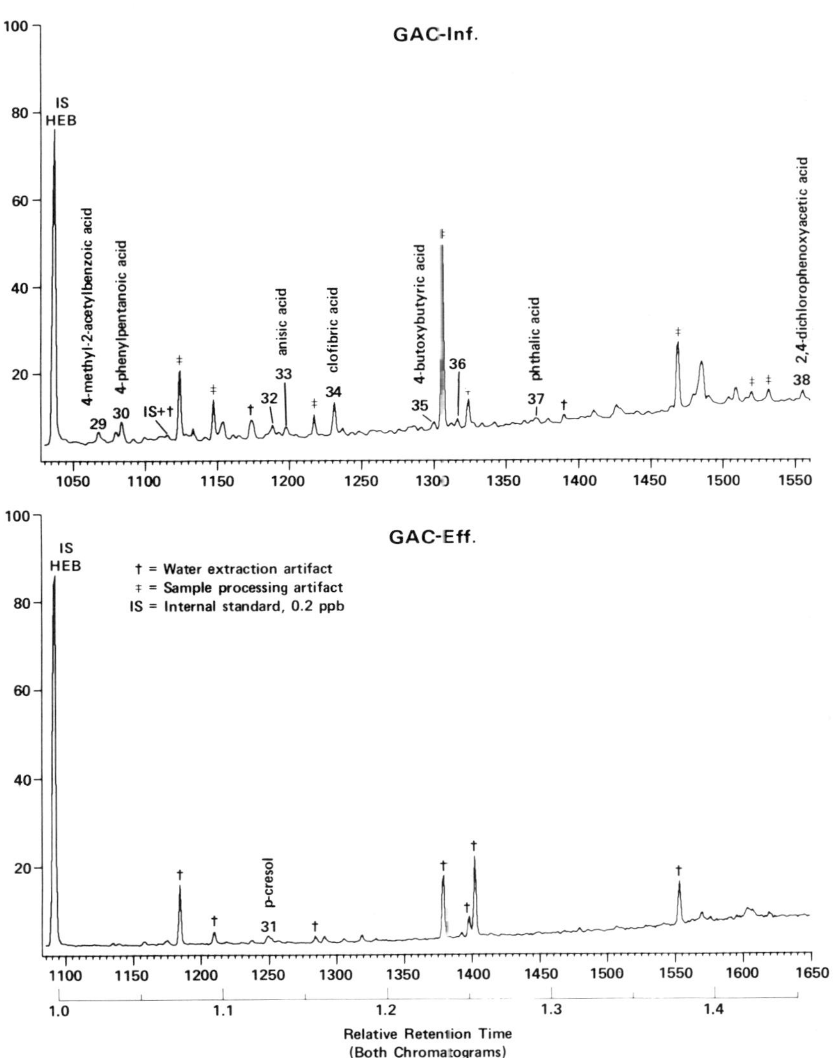


Figure 9, continued

Table IV. Results of BLLE Analysis of GAC-inf and GAC-eff Samples[a]–Methylated Acid Fraction, GAC Contactor D Water

| Sequence Number[b] | Relative Retention Time[c] | Compound Name[d] | Formula | Relative Amount Detected[e] | | Percent Removal |
|---|---|---|---|---|---|---|
| | | | | GAC-Inf. | GAC-Eff. | |
| 1 | 0.21 | Neoheptanoic acid, ME | $C_7H_{14}O_2$ | 1 | | 100 |
| 2 | 0.25 | 1-Butanol | $C_4H_{10}O_1$ | | 3 | $-\infty$ |
| 3 | 0.26 | Dibutyl sulfite | $C_8H_{18}O_3S_1$ | | 5 | $-\infty$ |
| 4 | 0.28 | Hexanoic acid, ME | $C_6H_{12}O_2$ | 20 | | 100 |
| 5 | 0.31 | 3-Methoxy-3-methyl-2-butanone | $C_6H_{12}O_2$ | | 10 | $-\infty$ |
| 6 | 0.35 | 2-Ethylhexanoic acid, ME | $C_8H_{16}O_2$ | 40 | | 100 |
| 7 | 0.35 | 2-Methyloctanoic acid, ME | $C_9H_{18}O_2$ | 1.5 | | 100 |
| 8 | 0.38 | Heptanoic acid, ME | $C_7H_{14}O_2$ | 5 | | 100 |
| 9 | 0.41 | Divinylmercury (artifact?) | $C_4H_6Hg$ | 1 | | – |
| 10 | 0.46 | Dichloroacetic acid, ME | $C_2H_2O_2Cl_2$ | 3 | | 100 |
| 11 | 0.51 | *m*-Dichlorobenzene | $C_6H_4Cl_2$ | 2 | | 100 |
| 12 | 0.53 | 3,6-Dimethyloctanoic acid, ME | $C_{10}H_{20}O_2$ | 2 | | 100 |
| 13 | 0.61 | 1,3,5-Trichlorobenzene | $C_6H_3Cl_3$ | 1.5 | | 100 |
| 14 | 0.72 | Benzoic acid, ME | $C_7H_6O_2$ | 102 | 5 | 95 |
| 15 | 0.77 | 2-Chlorophenol, ME | $C_6H_5Cl$ | 0.05 | | 100 |
| 16 | 0.78 | A glycol ether | | | 3.5 | $-\infty$ |
| 17 | 0.79 | *o*-Toluic acid, ME | $C_8H_8O_2$ | 5 | | 100 |
| 18 | 0.82 | *m*-Toluic acid, ME | $C_8H_8O_2$ | 2 | | 100 |
| 19 | 0.82 | 2-Phenylpropnaoic acid, ME | $C_9H_{10}O_2$ | 2 | | 100 |
| 20 | 0.83 | *p*-Toluic acid, ME | $C_8H_8O_2$ | 1.5 | | 100 |
| 21 | 0.85 | Phenylacetic acid, ME | $C_8H_8O_2$ | 2.5 | | 100 |
| 22 | 0.87 | Dimethyl maleic acid, di-ME | $C_6H_8O_4$ | 24 | | 100 |
| 23 | 0.90 | Ethylmethyl maleic acid, di-ME | $C_7H_{10}O_4$ | 55 | | 100 |
| 24 | 0.91 | *p*-Chlorobenzoic acid, ME | $C_7H_5O_2Cl$ | 1 | | 100 |

| | | | | | | |
|---|---|---|---|---|---|---|
| 25 | 0.92 | 3(*p*-Tolyl)propionic acid, ME | $C_{10}H_{12}O_2$ | 4 | | 100 |
| 26 | 0.92 | Hydrocinnamic acid, ME | $C_9H_{10}O_2$ | 3.5 | | 100 |
| 27 | 0.94 | 2,4-Dimethylbenzoic acid, ME | $C_9H_{10}O_2$ | 3 | | 100 |
| 28 | 0.96 | 3-Phenylpentanoic acid, ME | $C_{11}H_{14}O_2$ | 2 | | 100 |
| 29 | 1.03 | 4-Methyl-2-acetylbenzoic acid, ME | $C_{10}H_{10}O_3$ | 3.5 | | 100 |
| 30 | 1.04 | 4-Phenylpentanoic acid, ME | $C_{11}H_{14}O_2$ | 7 | | 100 |
| 31 | 1.13 | *p*-Cresol | $C_7H_8O_1$ | | 2 | $-\infty$ |
| 32 | 1.13 | A substituted naphthalene carboxylic acid, ME | | 4 | | 100 |
| 33 | 1.14 | Anisic acid, ME | $C_8H_8O_3$ | 3 | | 100 |
| 34 | 1.17 | Clofibric acid, ME | $C_{10}H_{11}O_3Cl$ | 10 | | 100 |
| 35 | 1.23 | 4-Butoxybutyric acid, ME | $C_8H_{16}O_3$ | 2 | | 100 |
| 36 | 1.25 | Isomer of clofibric acid, ME | $C_{10}H_{11}O_3Cl$ | 2.5 | | 100 |
| 37 | 1.30 | Phthalic acid, di-ME | $C_8H_6O_4$ | 1.5 | | 100 |
| 38 | 1.47 | 2,4-Dichlorophenoxyacetic acid, ME | $C_8H_6O_3Cl_2$ | 3 | | 100 |

[a] See Figure 5 for explanation of sample code names.
[b] Numbers correspond to the GC peaks, as labeled in Figure 9.
[c] Relative to the internal standard, hexaethylbenzene.
[d] ME indicates the compound was detected as the methyl ester or ether. The formula shown is that for the free acid.
[e] Expressed as a peak height percentage of the internal standard, hexaethylbenzene, 0.2 ppb.

**Table V. Artifact Contaminants from XAD-2 Resin in the XAD-EEE Sample[a]**

| Compound[b] | Relative Amount[c] | Compound[b] | Relative Amount[c] |
|---|---|---|---|
| *p*-Xylene | 890 | Methyl(1-ethylpropyl)benzene | 170 |
| *m*-Xylene | 5170 | 4-Ethyl styrene | 5160 |
| Cumene | 250 | 3-Ethyl styrene | 3280 |
| *o*-Xylene | 5600 | 5-Methyl indan | 3040 |
| A propyl benzene isomer | 200 | A methyl indan isomer | 180 |
| *p*-Ethyltoluene | 420 | A dimethyl indan isomer | 450 |
| Mesitylene | 150 | A methyl indan isomer | 250 |
| Styrene | 690 | Tetralin | 360 |
| 1,2,4-Trimethylbenzene | 130 | Divinylbenzene isomer | 790 |
| *m*-Diethylbenzene | 1820 | Divinylbenzene isomer | 630 |
| *p*-Diethylbenzene | 1710 | 2-Pentenylbenzene | 240 |
| *o*-Diethylbenzene | 720 | 1,1a,6,6a-Tetrahydrocycloprop[a]indene | 170 |
| An ethyl cumene isomer | 1000 | Methylbenzoate | 1720 |
| *t*-Pentylbenzene | 70 | Acetophenone | 90 |
| *p*-Ethyl cumene | 390 | *o*-Ethylbenzaldehyde | 120 |
| A propyl xylene isomer | 1040 | Naphthalene | 1390 |
| 3-Phenylpentane | 350 | Methyl *m*-ethylbenzoate | 50 |
| A methylindan isomer plus a C-6 benzene | 1560 | *p*-Ethylacetophenone | 50 |
| A methyl styrene plus a C-5 benzene | 80 | 2-Methylnaphthalene | 70 |
| 2-Ethyl styrene | 400 | 1-Methylnaphthalene | 40 |

[a] See Figure 5 for an explanation of sample code names.
[b] Listed in GC retention order.
[c] Expressed as the GC peak height percentage of the internal standard, hexaethylbenzene, added at the 0.2 ppb level.

Table VI. Results of Analysis of XAD-inf, XAD-eff and XAD-EEE Samples[a]–Neutral Fraction, GAC Contactor A Water

| | | | XAD-Inf. | XAD-Eff. | | XAD-EEE | |
|---|---|---|---|---|---|---|---|
| Number | RRT[b] | Compound Name | Relative Amount[c] | Relative Amount[c] | Percent Removal[d] | Relative Amount[c] | Percent Recovery[d] |
| 1 | 0.15 | Bromodichloromethane | 3 | | 100 | | 0 |
| 2 | 0.34 | Dibromochloromethane | 27 | | 100 | 14 | 52 |
| 3 | 0.39 | *p*-Methylpropylbenzene | | | | 2 | ∞ |
| 4 | 0.42 | 1,3-Dimethyl-5-ethyl benzene | | | | 3 | ∞ |
| 5 | 0.44 | (Dichloromethyl)naphthalene | 1 | | 100 | | 0 |
| 6 | 0.45 | 1,4-Dimethyl-2-ethylbenzene | | | | 2 | ∞ |
| 7 | 0.45 | 1,3-Dimethyl-4-ethylbenzene | | | | 2 | ∞ |
| 8 | 0.46 | 1,2-Dimethyl-4-ethylbenzene | | | | 4 | ∞ |
| 9 | 0.48 | Ethyl dichloroacetate | | | | 1 | ∞ |
| 10 | 0.48 | *m*-Dichlorobenzene | 1 | | 100 | | 0 |
| 11 | 0.52 | *p*-Dichlorobenzene | 3 | | 100 | | 0 |
| 12 | 0.52 | Bromoform | 12 | | 100 | 8 | 75 |
| 13 | 0.57 | *o*-Dichlorobenzene | 5 | | 100 | 60 | 1200 |
| 14 | 0.62 | Di(2-chloroethyl) ether | 1 | | 100 | | 0 |
| 15 | 0.68 | 1,1,2,3-Tetrachloropropane | 1 | | 100 | | 0 |
| 16 | 0.69 | Benzonitrile | 3 | | 100 | 2 | 66 |
| 17 | 0.72 | Methyl benzoate | | 4 | −∞ | 233 | ∞ |
| 18 | 0.76 | Ethyl benzoate | | | | 3 | ∞ |
| 19 | 0.80 | *o*-Ethylbenzaldehyde | | | | 13 | ∞ |
| 20 | 0.80 | Phenyl ethyl ketone | 5 | | 100 | | 0 |
| 21 | 0.82 | Nitrobenzene | 4 | | 100 | | 0 |
| 22 | 0.82 | Naphthalene | 5 | 3 | 40 | 11 | 220 |
| 23 | 0.86 | *o*-Nitrotoluene | 8 | | 100 | 7 | 88 |
| 24 | 0.90 | Methyl *m*-ethylbenzoate | | | | 13 | ∞ |
| 25 | 0.92 | Methyl *p*-ethylbenzoate | | | | 5 | ∞ |

Table VI, continued

| | | | XAD-Inf. | XAD-Eff. | | XAD-EEE | |
|---|---|---|---|---|---|---|---|
| Number | RRT[b] | Compound Name | Relative Amount[c] | Relative Amount[c] | Percent Removal[d] | Relative Amount[c] | Percent Recovery[d] |
| 26 | 0.93 | *p*-Nitrotoluene | 1.5 | | 100 | | 0 |
| 27 | 1.01 | Benzyl cyanide | 4 | | 100 | | 0 |
| 28 | 1.02 | 2-Chloroaniline | 1 | | 100 | | 0 |
| 29 | 1.04 | 2,4-Dichloro-1-nitrobenzene | 0.5 | | 100 | | 0 |
| 30 | 1.08 | Diphenyl ether | 0.5 | | 100 | | 0 |
| 31 | 1.11 | Hexachloropentane | 15 | 0.5 | 97 | | 0 |
| 32 | 1.27 | Ethyl palmitate | | | | 2 | ∞ |
| 33 | 1.37 | 2,6-Dinitrotoluene | 5 | | 100 | 3 | 60 |
| 34 | 1.44 | 2,4-Dinitrotoluene | 9 | | 100 | 8 | 89 |

[a] See Figure 5 for an explanation of sample code names.
[b] RRT = relative retention time. Relative to the internal standard, hexaethylbenzene.
[c] Expressed as a peak height percentage of the internal standard, hexaethylbenzene, 0.2 ppb.
[d] Percent removal or recovery, relative to the original water: XAD-inf.

Table VII. Results of Analysis of XAD-inf, XAD-eff and XAD-EEE Samples[a]–Methylated Acid Fraction, GAC Contactor A Water

| Number | RRT[b] | Compound Name[c] | XAD-Inf. | XAD-Eff. | | XAD-EEE | |
|---|---|---|---|---|---|---|---|
| | | | Relative Amount[d] | Relative Amount[d] | Percent Removal[e] | Relative Amount[d] | Percent Recovery[e] |
| 1 | 0.19 | Valeric acid, ME | | | | 7 | ∞ |
| 2 | 0.19 | Glycol ether | | | | 6 | ∞ |
| 3 | 0.27 | Formate | | | | 30 | ∞ |
| 4 | 0.30 | An ether | | | | 30 | ∞ |
| 5 | 0.38 | A fatty acid, ME | | | | 15 | ∞ |
| 6 | 0.41 | 1-Chloro-2-propanol | | | | 9 | ∞ |
| 7 | 0.46 | Dichloroacetic acid, ME | 15 | | 100 | 16 | 106 |
| 8 | 0.48 | 2,2′-bis-1,3-Dioxolane | | | | 9 | ∞ |
| 9 | 0.58 | 2,3,3-Trichloroacrylic acid, ME | | | | 6 | ∞ |
| 10 | 0.61 | An oxo-fatty acid, ME | | | | 2 | ∞ |
| 11 | 0.65 | An oxo-fatty acid, ME | | | | 4 | ∞ |
| 12 | 0.65 | An oxo-fatty acid, ME | | | | 4 | ∞ |
| 13 | 0.66 | Levulinic acid, ME | 6 | | 100 | | 0 |
| 14 | 0.67 | A 2,3-dimethyl fatty acid, ME | | | | 9 | ∞ |
| 15 | 0.69 | Capric acid, ME | | | | 19 | ∞ |
| 16 | 0.70 | 2,2-Dichloroethanol | 24 | | 100 | | 0 |
| 17 | 0.72 | Benzoic acid, ME | | 120 | −∞ | 200 | ∞ |
| 18 | 0.82 | *m*-Toluic acid, ME | | | | 3 | ∞ |
| 19 | 0.82 | 2-Phenylpropionic acid, ME | | | | 4 | ∞ |
| 20 | 0.85 | Phenylacetic acid, ME | 4 | | 100 | 13 | 320 |
| 21 | 0.86 | Salicylic acid, ME | 3 | | 100 | 6 | 200 |
| 22 | 0.87 | Dimethylmaleic acid, di-ME | 40 | 20 | 50 | 22 | 55 |
| 23 | 0.89 | Lauric acid, ME | | | | 74 | ∞ |

Table VII, continued

| Number | RRT[b] | Compound Name[c] | XAD-Inf. | XAD-Eff. | | XAD-EEE | |
|---|---|---|---|---|---|---|---|
| | | | Relative Amount[d] | Relative Amount[d] | Percent Removal[e] | Relative Amount[d] | Percent Recovery[e] |
| 24 | 0.90 | *m*-Ethylbenzoic acid, ME | | | | 10 | ∞ |
| 25 | 0.90 | Ethylmethylmaleic acid, di-ME | 68 | 30 | 44 | 96 | 140 |
| 26 | 0.91 | Tetrachlorobutenoic acid, ME | | | | 6 | ∞ |
| 27 | 0.92 | *p*-Ethylbenzoic acid, ME | | | | 13 | ∞ |
| 28 | 0.92 | Hydrocinnamic acid, ME | 4 | | 100 | 7 | 175 |
| 29 | 0.94 | 3,5-Dimethylbenzoic acid, ME | | | | 7 | ∞ |
| 30 | 0.94 | *o*-Methylphenylacetic acid, ME | 3 | | 100 | | 0 |
| 31 | 1.02 | *o*-Chloroaniline | 3 | | 100 | | 0 |
| 32 | 1.02 | 4-Methyl-2-acetylbenzoic acid, ME | | | | 5 | ∞ |
| 33 | 1.04 | Isomyristic acid, ME | 4 | 1 | 25 | | 0 |
| 34 | 1.04 | α-Phenyl-*t*-butyric acid, ME | 2 | | 100 | | 0 |
| 35 | 1.09 | Suberic acid, di-ME | | | | 6 | ∞ |
| 36 | 1.10 | Phenoxyacetic acid, ME | | | | 11 | ∞ |
| 37 | 1.10 | 3-Hydroxybenzisothiazole, ME | | 5 | −∞ | | |
| 38 | 1.11 | 4(1,5-Dimethyl-3-oxohexylcyclohexane carboxylic acid, ME | | | | 8 | ∞ |
| 39 | 1.13 | Isopentadecanoic acid, ME | 14 | | 100 | | 0 |
| 40 | 1.14 | Anisic acid, ME | 5 | | 100 | 8 | 160 |
| 41 | 1.17 | Clofibric acid, ME | 11 | | 100 | 9 | 82 |
| 42 | 1.18 | Azelaic acid, di-ME | | | | 30 | ∞ |
| 43 | 1.27 | N-Hydroxy phthalimide | | | | 6 | ∞ |
| 44 | 1.29 | Phthalic acid, di-ME | | | | 8 | ∞ |

Table VII, continued

| Number | RRT[b] | Compound Name[c] | XAD-Inf. | XAD-Eff. | | XAD-EEE | |
|---|---|---|---|---|---|---|---|
| | | | Relative Amount[d] | Relative Amount[d] | Percent Removal[e] | Relative Amount[d] | Percent Recovery[e] |
| 45 | 1.31 | Anteisoheptadecanoic acid, ME | 6 | | 100 | | 0 |
| 46 | 1.35 | Heptadecenoic acid, ME | 16 | | 100 | | 0 |
| 47 | 1.46 | Linoleic acid, ME | 28 | | 100 | | 0 |
| 48 | 1.47 | 2,4-Dichlorophenoxyacetic acid, ME | 6 | | 100 | 5 | 83 |
| 49 | 1.49 | A derivative of N,N-dimethyl urea | | | | 2 | ∞ |
| 50 | 1.48 | 4-Ethoxyethyl aniline | | 10 | −∞ | | |
| 51 | 1.49 | 1,4-Benzothiazin-2-one | 22 | | 100 | | 0 |
| 52 | 1.49 | N-Phenylacetamide | | 3 | −∞ | | |

[a] See Figure 5 for an explanation of sample code names.
[b] RRT = relative retention time. Relative to the internal standard, hexaethylbenzene.
[c] ME = methyl ester or methyl ether.
[d] Expressed as a peak height percentage of the internal standard, hexaethylbenzene, 0.2 ppb.
[e] Percent removal or recovery, relative to the original water: XAD-Inf.

in Tables VI and VII are the results of the BLLE analysis of contactor A drinking water (January 14, 1980) before and after passage through the XAD-2 analytical column.

## DISCUSSION

The data presented above on these and other water samples will help HERL determine the authenticity of organic concentrates derived from RO, lyophilization and XAD adsorption such as those which were produced from the same GAC-inf water on January 14 and 28, 1980. Chemical analysis of the RO concentrates, T4C and T4X, from 7500 L of GAC-inf drinking water (contactor A) is presented elsewhere [38]. A total of 210 organics (16 consent decreee organics) is reported in these RO concentrates from GAC-inf water to contactor A (January 14, 1980). Many of the same organics measured in RO concentrates T4C and T4X [38] are also reported in Table I (Bellar P&T), Table II (Grob CLSA), and Tables VI and VII (BLLE and XAD-EEE). For example, the herbicide, 2,4-dichlorophenoxyacetic acid, and the antilipidemic drug metabolite, clofibric acid, were measured by BLLE and XAD-EEE analysis in drinking water before (Table VII) and after RO concentration (Table V [38]). For our health effects research, it is clear to us that we will not be successful in producing representative organic concentrates of water for biological screening tests, if we do know how to conduct state-of-the-art organic analysis of the "starting material"—drinking water.

The contamination problem that we encountered with XAD-EEE sample from contactor D water on January 28, 1980, and not from contactor A water two weeks earlier has been a consistent problem in our use of XAD-2 resin over the past six years. The Grob CLSA data of XAD-eff water from contactor A and D experiments indicate that the XAD-eff water was not contaminated by XAD-2 resin (see Table II). Therefore, we have obviously contaminated the XAD-EEE sample from contactor D during the ethyl ether elution step, even though the same procedure was used for both the January 14 and 28, 1980, experiments. Hopefully the XAD-4 resin adsorption method described in the MAS for the measurement of "extractable" organics in drinking water will be designed to prevent absolutely such contamination from XAD resin during the ethyl ether elution of adsorbed organics.

A total of 215 organic compounds have been purchased as authentic standards and analyzed by Grob CLSA. For those compounds listed in Table VIII, experimental response factors and chromatographic behavior have been determined so that all CLSA data files can be automatically searched for these 215 compounds using reverse library search software.

Table VIII. List of 215 Reference Compounds Which Are Measured in Each Grob CLSA Sample

| Compound | Mol Wt[a] | Formula | RRT[b] |
|---|---|---|---|
| 1,1-Dichloroethane | 98 | $C_2H_4Cl_2$ | 0.137 |
| Bromochloromethane | 128 | $CH_2ClBr$ | 0.153 |
| Chloroform | 118 | $CHCL_3$ | 0.155 |
| 1,2-Dichloroethane | 98 | $C_2H_4Cl_2$ | 0.170 |
| 1,1,1-Trichloroethane | 132 | $C_2H_3Cl_3$ | 0.173 |
| Benzene | 78 | $C_6H_6$ | 0.185 |
| Carbon Tetrachloride | 152 | $CCl_4$ | 0.188 |
| Dibromomethane | 172 | $CH_2Br_2$ | 0.215 |
| Trichloroethene | 130 | $C_2HCl_3$ | 0.219 |
| Bromodichloromethane | 162 | $CHCl_2Br$ | 0.223 |
| N-Nitrosodimethylamine | 74 | $C_2H_6ON_2$ | 0.260 |
| Pyridine | 79 | $C_5H_5N$ | 0.260 |
| Bromotrichloromethane | 196 | $CCl_3Br$ | 0.284 |
| Toluene | 92 | $C_7H_8$ | 0.288 |
| 2-Methylthiophene | 98 | $C_5H_6S$ | 0.295 |
| Dibromochloromethane | 206 | $CHClBr_2$ | 0.315 |
| Hexanal | 100 | $C_6H_{12}O$ | 0.323 |
| 1,2,2-Trichloropropane | 146 | $C_3H_5Cl_3$ | 0.323 |
| Tetrachloroethene | 154 | $C_2Cl_4$ | 0.339 |
| Dichloroiodomethane | 210 | $CHCl_2I$ | 0.349 |
| 1,1,2-Trichloropropane | 146 | $C_3H_5Cl_3$ | 0.367 |
| 4-Hydroxy-4-methyl-2-pentanone | 116 | $C_6H_{12}O_2$ | 0.368 |
| Chlorobenzene | 112 | $C_6H_5Cl$ | 0.379 |
| Dibromodichloromethane | 240 | $CCl_2Br_2$ | 0.391 |
| 1-Chlorohexane | 120 | $C_6H_{13}Cl$ | 0.396 |
| Ethylbenzene | 106 | $C_8H_{10}$ | 0.400 |
| *m*-Xylene | 106 | $C_8H_{10}$ | 0.411 |
| *p*-Xylene | 106 | $C_8H_{10}$ | 0.411 |
| Bromoform | 250 | $CHBr_3$ | 0.417 |
| Styrene | 104 | $C_8H_8$ | 0.425 |
| *o*-Xylene | 106 | $C_8H_{10}$ | 0.435 |
| 1,2,3-Trichloropropane | 146 | $C_3H_5Cl_3$ | 0.442 |
| Bromotrichloroethene | 208 | $C_2Cl_3Br$ | 0.446 |
| 1,1,2,2-Tetrachloroethane | 166 | $C_2H_2Cl_4$ | 0.446 |
| Isopropylbenzene | 120 | $C_9H_{12}$ | 0.475 |
| 2-Chlorotoluene | 126 | $C_7H_7Cl$ | 0.499 |
| 3-Chlorotoluene | 126 | $C_7H_7Cl$ | 0.501 |
| 4-Chlorotoluene | 126 | $C_7H_7Cl$ | 0.504 |
| *n*-Propylbenzene | 120 | $C_9H_{12}$ | 0.507 |
| Bromocyclohexane | 162 | $C_6H_{11}Br$ | 0.512 |
| 1-Ethyl-4-methylbenzene | 120 | $C_9H_{12}$ | 0.514 |
| 1-Ethyl-3-methylbenzene | 120 | $C_9H_{12}$ | 0.515 |
| Pentachloroethane | 200 | $C_2HCl_5$ | 0.517 |
| Benzonitrile | 103 | $C_7H_5N$ | 0.518 |

**Table VIII, continued**

| Compound | Mol Wt[a] | Formula | RRT[b] |
|---|---|---|---|
| 1,3,5-Trimethylbenzene | 120 | $C_9H_{12}$ | 0.523 |
| *bis*-(2-Chloroethyl)ether | 142 | $C_4H_8OCl_2$ | 0.525 |
| a-Methylstyrene | 118 | $C_9H_{10}$ | 0.530 |
| 1-Ethyl-2-methylbenzene | 120 | $C_9H_{12}$ | 0.533 |
| (1,1-Dimethylethyl)benzene | 134 | $C_{10}H_{14}$ | 0.546 |
| Bromochloroiodomethane | 254 | CHClBrI | 0.550 |
| 1,2,4-Trimethylbenzene | 120 | $C_9H_{12}$ | 0.550 |
| 1,3-Dichlorobenzene | 146 | $C_6H_4Cl_2$ | 0.556 |
| 1,4-Dichlorobenzene | 146 | $C_6H_4Cl_2$ | 0.560 |
| a-Chlorotoluene | 126 | $C_7H_7Cl$ | 0.560 |
| (2-Methylpropyl)benzene | 134 | $C_{10}H_{14}$ | 0.567 |
| (1-Methylpropyl)benzene | 134 | $C_{10}H_{14}$ | 0.570 |
| 1,2,3-Trimethylbenzene | 120 | $C_9H_{12}$ | 0.579 |
| 1,2-Dichlorobenzene | 146 | $C_6H_4Cl_2$ | 0.585 |
| Indan | 118 | $C_9H_{10}$ | 0.589 |
| Indene | 116 | $C_9H_8$ | 0.597 |
| 1-Phenylethanone | 120 | $C_8H_8O$ | 0.608 |
| 1,3-Diethylbenzene | 134 | $C_{10}H_{14}$ | 0.610 |
| 1,4-Diethylbenzene | 134 | $C_{10}H_{14}$ | 0.615 |
| *n*-Butylbenzene | 134 | $C_{10}H_{14}$ | 0.617 |
| 2-Chloro-*p*-xylene | 140 | $C_8H_9Cl$ | 0.617 |
| N-Nitroso-di-*n*-propylamine | 130 | $C_6H_{14}ON_2$ | 0.618 |
| 5-Ethyl-1,3-dimethylbenzene | 134 | $C_{10}H_{14}$ | 0.619 |
| 2-Chlorostyrene | 138 | $C_8H_7Cl$ | 0.621 |
| 1,2-Diethylbenzene | 134 | $C_{10}H_{14}$ | 0.621 |
| 1-Chlorooctane | 148 | $C_8H_{17}Cl$ | 0.623 |
| 3-Chlorostyrene | 138 | $C_8H_7Cl$ | 0.624 |
| 4-Chlorostyrene | 138 | $C_8H_7Cl$ | 0.627 |
| Phenyl-2-butene | 132 | $C_{10}H_{12}$ | 0.627 |
| 2,6-Dimethylstyrene | 132 | $C_{10}H_{12}$ | 0.633 |
| Hexachloroethane | 234 | $C_2Cl_6$ | 0.634 |
| 2-Ethyl-1,4-dimethylbenzene | 134 | $C_{10}H_{14}$ | 0.639 |
| 1,1-Dimethylindene | 144 | $C_{11}H_{12}$ | 0.642 |
| 4-Ethyl-1,3-dimethylbenzene | 134 | $C_{10}H_{14}$ | 0.642 |
| *d*-Fenchone | 152 | $C_{10}H_{16}O$ | 0.644 |
| 4-Chloro-1,2-dimethylbenzene | 140 | $C_8H_9Cl$ | 0.645 |
| 4-Ethyl-1,2-dimethylbenzene | 134 | $C_{10}H_{14}$ | 0.647 |
| 2-Ethyl-1,3-dimethylbenzene | 134 | $C_{10}H_{14}$ | 0.651 |
| (1,1-Dimethylpropyl)benzene | 148 | $C_{11}H_{16}$ | 0.653 |
| 4-Ethylstyrene | 132 | $C_{10}H_{12}$ | 0.653 |
| 1,1-Dimethylindan | 146 | $C_{11}H_{14}$ | 0.662 |
| 3-Ethyl-1,2-dimethylbenzene | 134 | $C_{10}H_{14}$ | 0.666 |
| a-Chloro-*m*-xylene | 140 | $C_8H_9Cl$ | 0.667 |
| Isophorone | 138 | $C_9H_{14}O$ | 0.667 |

Table VIII, continued

| Compound | Mol Wt[a] | Formula | RRT[b] |
|---|---|---|---|
| a-Chloro-*o*-xylene | 140 | $C_8H_9Cl$ | 0.670 |
| a-Chloro-*p*-xylene | 140 | $C_8H_9Cl$ | 0.670 |
| 2,4-Dichlorotoluene | 160 | $C_7H_6Cl_2$ | 0.675 |
| 2,5-Dichlorotoluene | 160 | $C_7H_6Cl_2$ | 0.676 |
| 2,6-Dichlorotoluene | 160 | $C_6H_6Cl_2$ | 0.679 |
| 1,1,2,3,3-Pentachloropropane | 214 | $C_3H_3Cl_5$ | 0.679 |
| 5-Isopropyl-1,3-dimethylbenzene | 148 | $C_{11}H_{16}$ | 0.680 |
| *o*-Chloroaniline | 127 | $C_6H_6NCl$ | 0.681 |
| 1,2,3,5-Tetramethylbenzene | 134 | $C_{10}H_{14}$ | 0.681 |
| 1,3,5-Trichlorobenzene | 180 | $C_6H_3Cl_3$ | 0.688 |
| *d*-Camphor | 152 | $C_{10}H_{16}O$ | 0.694 |
| Isoborneol | 154 | $C_{10}H_{18}O$ | 0.696 |
| *p*-Methylphenol | 108 | $C_7H_8O$ | 0.699 |
| 3,4-Dichlorotoluene | 160 | $C_7H_6Cl_2$ | 0.702 |
| 1,3-Diethyl-5-methylbenzene | 148 | $C_{11}H_{16}$ | 0.704 |
| *bis*-(2-Chloroethoxy)methane | 172 | $C_5H_{10}O_2Cl_2$ | 0.706 |
| Menthone | 154 | $C_{10}H_{18}O$ | 0.706 |
| 1,2,3,4-Tetramethylbenzene | 134 | $C_{10}H_{14}$ | 0.713 |
| 1,2-Dihydronaphthalene | 130 | $C_{10}H_{10}$ | 0.717 |
| 1,3-Diisopropylbenzene | 162 | $C_{12}H_{18}$ | 0.717 |
| 1,2,3,4-Tetrahydronaphthalene | 132 | $C_{10}H_{12}$ | 0.718 |
| *n*-Pentylbenzene | 148 | $C_{11}H_{16}$ | 0.720 |
| Borneol | 154 | $C_{11}H_{18}O$ | 0.726 |
| 1,2,4-Trichlorobenzene | 180 | $C_6H_3Cl_3$ | 0.729 |
| a,2-Dichlorotoluene | 160 | $C_7H_6Cl_2$ | 0.730 |
| 1,4-Diisopropylbenzene | 162 | $C_{12}H_{18}$ | 0.734 |
| Naphthalene | 128 | $C_{10}H_8$ | 0.737 |
| 1-tert-Butyl-3,5-dimethylbenzene | 162 | $C_{12}H_{18}$ | 0.741 |
| a,3-Dichlorotoluene | 160 | $C_7H_6Cl_2$ | 0.742 |
| a,4-Dichlorotoluene | 160 | $C_7H_6Cl_2$ | 0.743 |
| *m*-Chloroaniline | 127 | $C_6H_6NCl$ | 0.751 |
| *p*-Chloroaniline | 127 | $C_6H_6NCl$ | 0.755 |
| 1,2,3-Trichlorobenzene | 180 | $C_6H_3Cl_3$ | 0.761 |
| 2-Methylpentylbenzene | 162 | $C_{12}H_{18}$ | 0.766 |
| 2,6-Dichlorostyrene | 172 | $C_8H_6Cl_2$ | 0.769 |
| a,a,a-Trichlorotoluene | 194 | $C_7H_5Cl_3$ | 0.773 |
| 2,5-Dichlorostyrene | 172 | $C_8H_6Cl_2$ | 0.777 |
| Hexachloro-1,3-butadiene | 258 | $C_4Cl_6$ | 0.779 |
| 1,3,5-Triethylbenzene | 162 | $C_{12}H_{18}$ | 0.782 |
| 2,5-Dichloro-*p*-xylene | 174 | $C_8H_8Cl_2$ | 0.784 |
| 3,4-Dichlorostyrene | 172 | $C_8H_6Cl_2$ | 0.804 |
| 4,7-Dimethylindan | 146 | $C_{11}H_{14}$ | 0.817 |
| *n*-Hexylbenzene | 162 | $C_{12}H_{18}$ | 0.817 |
| Pentamethylbenzene | 148 | $C_{11}H_{16}$ | 0.834 |
| 2-Methylnaphthalene | 142 | $C_{11}H_{10}$ | 0.839 |

**Table VIII, continued**

| Compound | Mol Wt[a] | Formula | RRT[b] |
|---|---|---|---|
| 5-Methyltetrahydronaphthalene | 146 | $C_{11}H_{14}$ | 0.840 |
| 2,4,5-Trichlorotoluene | 194 | $C_7H_5Cl_3$ | 0.841 |
| 2,3,6-Trichlorotoluene | 194 | $C_7H_5Cl_3$ | 0.851 |
| 1-Methylnaphthalene | 142 | $C_{11}H_{10}$ | 0.855 |
| a,a′-Dichloro-*o*-xylene | 174 | $C_8H_8Cl_2$ | 0.859 |
| Cyclohexylbenzene | 160 | $C_{12}H_{16}$ | 0.867 |
| 2,6-Dimethyltetrahydronaphthalene | 160 | $C_{12}H_{16}$ | 0.867 |
| 2,5-Dichloroaniline | 161 | $C_6H_5NCl_2$ | 0.868 |
| 1,2,3,5-Tetrachlorobenzene | 214 | $C_6H_2Cl_4$ | 0.870 |
| 1,2,4,5-Tetrachlorobenzene | 214 | $C_6H_2Cl_4$ | 0.870 |
| a,2,4-Trichlorotoluene | 194 | $C_7H_5Cl_3$ | 0.877 |
| a,2,6-Trichlorotoluene | 194 | $C_7H_5Cl_3$ | 0.880 |
| Hexachloro-1,3-cyclopentadiene | 370 | $C_5Cl_6$ | 0.884 |
| 1,8-Dimethyltetrahydronaphthalene | 160 | $C_{12}H_{16}$ | 0.886 |
| a,a′-Dichloro-*m*-xylene | 174 | $C_8H_8Cl_2$ | 0.896 |
| a,a′-Dichloro-*p*-xylene | 174 | $C_8H_8Cl_2$ | 0.904 |
| Butylbenzoate | 178 | $C_{11}H_{14}O_2$ | 0.908 |
| 2-Chloronaphthalene | 162 | $C_{10}H_7Cl$ | 0.909 |
| *n*-Heptylbenzene | 176 | $C_{13}H_{20}$ | 0.909 |
| 1,2,3,4-Tetrachlorobenzene | 214 | $C_6H_2Cl_4$ | 0.910 |
| a,3,4-Trichlorotoluene | 194 | $C_7H_5Cl_3$ | 0.914 |
| 2-Ethylnaphthalene | 156 | $C_{12}H_{12}$ | 0.928 |
| 5,7-Dimethyltetrahydronaphthalene | 160 | $C_{12}H_{16}$ | 0.929 |
| 1-Ethylnaphthalene | 156 | $C_{12}H_{12}$ | 0.929 |
| 2-Methylbiphenyl | 168 | $C_{13}H_{12}$ | 0.932 |
| 1-Phenyl-1-cyclohexene | 158 | $C_{12}H_{14}$ | 0.933 |
| 2,4,6-Trichloroaniline | 195 | $C_6H_4NCl_3$ | 0.935 |
| 2,6-Dimethylnaphthalene | 156 | $C_{12}H_{12}$ | 0.938 |
| 1,3-Dimethylnaphthalene | 156 | $C_{12}H_{12}$ | 0.950 |
| 1,6-Dimethylnaphthalene | 156 | $C_{12}H_{12}$ | 0.953 |
| Diphenylmethane | 168 | $C_{13}H_{12}$ | 0.955 |
| 1,4-Dimethylnaphthalene | 156 | $C_{12}H_{12}$ | 0.965 |
| 2,3-Dimethylnaphthalene | 156 | $C_{12}H_{12}$ | 0.966 |
| 2-Methoxynaphthalene | 158 | $C_{11}H_{10}O$ | 0.970 |
| 1,2-Dimethylnaphthalene | 156 | $C_{12}H_{12}$ | 0.977 |
| 2-Isopropylnaphthalene | 170 | $C_{13}H_{14}$ | 0.981 |
| Hexamethylbenzene | 162 | $C_{12}H_{18}$ | 0.982 |
| *n*-Octylbenzene | 190 | $C_{14}H_{22}$ | 0.996 |
| Acenaphthene | 154 | $C_{12}H_{10}$ | 0.999 |
| 1-Chlorododecane | 204 | $C_{12}H_{25}Cl$ | 1.000 |
| 2,7-Dimethyltetrahydronaphthalene | 160 | $C_{12}H_{16}$ | 1.000 |
| Pentachloropyridine | 249 | $C_5NCl_5$ | 1.015 |
| 2,4-Dinitrotoluene | 182 | $C_7H_6O_4N_2$ | 1.021 |
| Pentachlorobenzene | 248 | $C_6HCl_5$ | 1.031 |
| 2,4,5-Trichloroaniline | 195 | $C_6H_4NCl$ | 1.041 |

**Table VIII, continued**

| Compound | Mol Wt[a] | Formula | RRT[b] |
|---|---|---|---|
| 2,3,4-Trichloroaniline | 195 | $C_6H_4NCl_3$ | 1.064 |
| 2,3,5-Trimethylnaphthalene | 170 | $C_{13}H_{14}$ | 1.067 |
| Fluorene | 166 | $C_{13}H_{10}$ | 1.077 |
| Diethyl Phthalate | 222 | $C_{12}H_{14}O_4$ | 1.079 |
| *n*-Nonylbenzene | 204 | $C_{15}H_{24}$ | 1.079 |
| 4-Chlorophenyl Phenyl Ether | 204 | $C_{12}H_9OCl$ | 1.084 |
| 1,2-Diphenylhydrazine | 184 | $C_{12}H_{12}N_2$ | 1.104 |
| 2,4,5,6-Tetrachloro-*m*-xylene | 242 | $C_8H_6Cl_4$ | 1.111 |
| 3,4,5-Trichloroaniline | 195 | $C_6H_4NCl_3$ | 1.128 |
| BHC Isomer | 288 | $C_6H_6Cl_6$ | 1.153 |
| *n*-Decylbenzene | 218 | $C_{16}H_{26}$ | 1.158 |
| Hexachlorobenzene | 282 | $C_6Cl_6$ | 1.176 |
| Lindane | 288 | $C_6H_6Cl_6$ | 1.180 |
| BHC Isomer | 288 | $C_6H_6Cl_6$ | 1.197 |
| Phenanthrene | 178 | $C_{14}H_{10}$ | 1.221 |
| a,a,a,a′,a′,a′-Hexachloro-*p*-xylene | 310 | $C_8H_4Cl_6$ | 1.223 |
| Anthracene | 178 | $C_{14}H_{10}$ | 1.228 |
| 2,4,5-Trichlorobiphenyl | 256 | $C_{12}H_7Cl_3$ | 1.268 |
| 1-Chlorohexadecane | 160 | $C_{16}H_{33}Cl$ | 1.301 |
| Heptachlor | 370 | $C_{10}H_5Cl_7$ | 1.309 |
| Aldrin | 362 | $C_{12}H_8Cl_6$ | 1.355 |
| 2,3′,4,5′-Tetrachlorobiphenyl | 290 | $C_{12}H_6Cl_4$ | 1.370 |
| a,a,2,4,6-Hexachloro-*m*-xylene | 310 | $C_8H_4Cl_6$ | 1.383 |
| Heptachlor Epoxide | 386 | $C_{10}H_5OCl_7$ | 1.399 |
| Fluoranthene | 202 | $C_{16}H_{10}$ | 1.404 |
| 1-Chlorooctadecane | 288 | $C_{18}H_{37}Cl$ | 1.430 |
| 2,2′,4,4′,6,6′-Hexachlorobiphenyl | 358 | $C_{12}H_4Cl_6$ | 1.433 |
| Pyrene | 202 | $C_{16}H_{10}$ | 1.434 |
| 2,2′,4,5,5′-Pentachlorobiphenyl | 324 | $C_{12}H_5Cl_5$ | 1.442 |
| p,p′-DDE | 316 | $C_{14}H_8Cl_4$ | 1.472 |
| Dieldrin | 378 | $C_{12}H_8OCl_6$ | 1.476 |
| Endrin | 378 | $C_{12}H_8OCl_6$ | 1.499 |
| p,p′-DDD | 318 | $C_{14}H_{10}Cl_4$ | 1.518 |
| Benzylbutylphthalate | 312 | $C_{19}H_{20}O_4$ | 1.559 |
| p,p′-DDT | 352 | $C_{14}H_9Cl_5$ | 1.565 |
| Benz[a]anthracene | 228 | $C_{18}H_{12}$ | 1.628 |
| Chrysene | 228 | $C_{18}H_{12}$ | 1.628 |

[a] Molecular weights are calculated on the basis of the integral masses of the most abundant isotopes.

[b] RRT = relative retention time as compared to that of chlorododecane, IS.

For the purposes of this chapter, on the comparison of Grob CLSA with Bellar P&T, BLLE and XAD-EEE, an attempt has been made to summarize the comparative differences of the four selected methods in Tables IX and X using one water sample; GAC-inf water from contactor D (contactor A for XAD-EEE). Reverse library computer searching for the 215 organics was performed automatically on the CLSA GC/MS data file of GAC-inf water, and 64 organics were detected and quantified by the Incos data system. The method described by Coleman et al. [11] was used. Of the 215 organics, 171 were not found. This negative information is very valuable in that several of the 171 organics not detected in GAC-inf water are toxic. For example, Coleman reports [11] that 2,2′,4,4′,6,6′-hexachlorobiphenyl ( a PCB isomer of molecular weight 358), one of the 171 organics not detected, can be measured in drinking water by Grob GC/MS CLSA at a concentration of 2 ng/L. The standard deviation for the measurement of this PCB isomer at 6.2 ng/L (16 replicates, 59% recovery efficiency) was ±1.1 ng/L. For drinking water treatment researchers and toxicologists, this type of reproducibility and sensitivity is important. However, the drinking water consumers in Cincinnati are perhaps the most gratified group over the low GC/MS detection limits of Grob CLSA, since they probably dislike drinking PCB isomers. If this PCB isomer were detected in GAC-inf water at 2 ng/L the combined concentration of all Aroclor PCB isomers in the drinking water would have been dramatically higher than 2 ng/L. Unfortunately, we have detected PCB isomers on previous occasions in several drinking water samples from other major cities using Grob GC/MS CLSA. The analogous limits of detection pertaining to BLLE and XAD-EEE are not available, since it is very difficult and time-consuming to obtain quantitative data using BLLE and XAD adsorption.

TOC measurements indicate that CWW GAC contactor D was 87% effective in removal of organics, whereas CLSA indicates that contactor D was 80% effective in removal of purgeable organics. CLSA also indicates that the XAD-2 analytical column was 79% effective in removal of purgeable organics.

Table IX provides us with information on overlap between the four methods. For example, dibromochloromethane and bromoform were detected by all four methods. However, only Bellar P&T and Grob CLSA provided quantitative results [11,33]. Of the 12 carboxylic acids (including 2,4-dichlorophenoxyacetic acid) that were identified by BLLE and XAD-EEE analyses, none were observed in Grob CLSA data. BLLE and XAD-EEE analyses detected the presence of 3-nitrotoluene and 2,4-dinitrotoluene in GAC-inf water, but Grob CLSA did not detect these important compounds. Surprisingly, BLLE missed four isomers of ethyldimethylbenzene that Grob CLSA and XAD-EEE analyses detected. Perhaps these alkylated benzene

Table IX. Organics Detected in GAC-inf Sample by More Than One Analytical Method

| | GAC-inf Contactor D | | | | GAC-inf Contactor A | |
|---|---|---|---|---|---|---|
| | | | BLLE | | XAD-EEE | |
| | Bellar[a] P&T ($\mu$g/L) | Grob[b] CLSA (ng/L) | Neutrals (RS)[c] | Acids (RS) | Neutrals (RS) | Acids (RS) |
| Chloroform | 56 | 10 $\mu$g/L | | | | |
| 1,1,1-Trichloroethane | 400 | 8 | | | | |
| Trichloroethene | TD[d] | 57 | | | | |
| Bromoform | 0.2 | 0.51 $\mu$g/L | 75 | | 8 | |
| 2-Ethyl-4-methyl-1,3-dioxolane | | 5 RAU[e] | 25 | | | |
| Bromodichloromethane | 18 | 16 $\mu$g/L | 420 | | | |
| Chlorobenzene | | 14 | 9 | | | |
| 1,3,5-Trimethylbenzene | | 0 RAU | 5 | | | |
| 1,2,4-Trimethylbenzene | | 3 | 5 | | | |
| Dibromochloromethane | 5.8 | 6.2 $\mu$g/L | 229 | | 14 | |
| 1,3-Dichlorobenzene | | 5 | 3 | 2 | | |
| Hexachloroethane | | 8 | 9 | | | |
| 1,2-Dichlorobenzene | | 17 | 19 | | 60 | |
| Tetrahydronaphthalene | | 40 | 49 | | | |
| Naphthalene | | NQ | 8 | | 11 | |
| Phenylether | | 14 RAU | 37 | | | |
| Benzoic Acid | | | | 102 | | 200 |
| Phenylpropanoic Acid | | | | 2 | | 4 |
| *m*-Toluic Acid | | | | 1.5 | | 3 |
| Phenylacetic Acid | | | | 2.5 | | 13 |
| Dimethylmaleic Acid | | | | 24 | | 22 |
| Ethylmethylmaleic Acid | | | | 55 | | 96 |

**Table IX, continued**

| | GAC-inf Contactor D | | | | GAC-inf Contactor A | |
|---|---|---|---|---|---|---|
| | | | BLLE | | XAD-EEE | |
| | Bellar[a] P&T (μg/L) | Grob[b] CLSA (ng/L) | Neutrals (RS)[c] | Acids (RS) | Neutrals (RS) | Acids (RS) |
| Hydrocinnamic Acid | | | | 3.5 | | 7 |
| 4-Methyl-2-acetylbenzoic Acid | | | | 3.5 | | 5 |
| Anisic Acid | | | | 3 | | 8 |
| Clofibric Acid | | | | 10 | | 9 |
| Phthalic Acid | | | | 1.5 | | 8 |
| 2,4-Dichlorophenoxyacetic Acid | | | | 3 | | 5 |
| Methylpropylbenzene Isomer | | 3 RAU | | | 2 | |
| 5-Ethyl-1,3-dimethylbenzene | | 0 | | | 3 | |
| 2-Ethyl-1,4-dimethylbenzene | | 2 | | | 2 | |
| 4-Ethyl-1,3-dimethylbenzene | | NQ | | | 2 | |
| 4-Ethyl-1,2-dimethylbenzene | | 1 | | | 4 | |
| Benzonitrile | | | 16 | | 2 | |
| *o*-Nitrotoluene | | | 18 | | 7 | |
| 2,4-Dinitrotoluene | | | 7 | | 8 | |

[a] Standards were obtained and quantification is based on an experimental response factor.

[b] Standards were obtained and quantification is based on an experimental relative response factor to that of chlorododecane, IS.

[c] RS = relative size; GC/MS peak height compared to that of hexaethylbenzene, IS, 0.2 ppb.

[d] TD = trace detected.

[e] $$\text{RAU (relative area unit)} = \frac{\text{total ion current area (UNK)}}{\text{total ion current area (IS)}} \times 52$$; where UNK = unknown compound; IS = chlorodecane.

Table X. Comparison[a] of Analysis Results of Four Methods Using Organic Functional Groups and EPA Lists of Toxic Compounds

| | Bellar P&T | | Grob CLSA | | BLLE | | XAD-XEE[b] | |
|---|---|---|---|---|---|---|---|---|
| | No. | (%) | No. | (%) | No. | (%) | No. | (%) |
| Broad Categories | | | | | | | | |
| Aliphatic Hydrocarbon | | | 26 | 24.3 | 6 | 6.7 | | |
| Aromatic Hydrocarbon | | | 31 | 29.0 | 4 | 4.4 | 6 | 10.3 |
| Halogenated Organic | 6 | 100.0 | 26 | 24.3 | 16 | 17.8 | 10 | 17.2 |
| Nitrogen Compound | | | 1 | 0.9 | 8 | 8.9 | 6 | 10.3 |
| Oxygen Compound | | | 23 | 21.5 | 52 | 57.8 | 36 | 62.1 |
| Sulfur Compound | | | | | 1 | 1.1 | | |
| Phosphorus Compound | | | | | 2 | 2.2 | | |
| Mercury Compound | | | | | 1 | 1.1 | | |
| Total Compounds Detected | 6 | 100.0 | 107 | 100.0 | 90 | 100.0 | 58 | 100.0 |
| Specific Categories | | | | | | | | |
| Alkane | | | 6 | 5.6 | | | | |
| Alkene, Alkyne | | | | | | | | |
| Alicyclic Hydrocarbon | | | 6 | 5.6 | 2 | 2.2 | | |
| Benzene Hydrocarbon | | | 31 | 29.0 | 3 | 3.3 | 5 | 8.6 |
| Indeno Hydrocarbon | | | 7 | 6.5 | 1 | 1.1 | | |
| Biphenyl Hydrocarbon | | | | | | | | |
| Naphtheno Hydrocarbon | | | 4 | 3.7 | 1 | 1.1 | 1 | 1.7 |
| Polyhydrofuran | | | | | 1 | 1.1 | | |
| Aliphatic Mercury | | | | | 1 | 1.1 | | |
| Polyhydronaphthalene | | | 3 | 2.8 | 2 | 2.2 | | |
| Alcohols | | | | | 5 | 5.6 | 1 | 1.7 |
| Glycols | | | | | 5 | 5.7 | | |
| Amines | | | | | | | | |

Table X, continued

| | Bellar P&T | | Grob CLSA | | BLLE | | XAD-XEE[b] | |
|---|---|---|---|---|---|---|---|---|
| | No. | (%) | No. | (%) | No. | (%) | No. | (%) |
| Phenols | | | | | | | | |
| Aldehydes | | | 6 | 5.6 | | | 1 | 1.7 |
| Ketones | | | 2 | 1.9 | 5 | 5.6 | | |
| Quinones | | | 2 | 1.9 | | | | |
| Aliphatic Esters | | | 2 | 1.9 | | | 3 | 5.2 |
| Aromatic Esters | | | 4 | 3.7 | 1 | 1.1 | 4 | 6.9 |
| Ethers | | | 3 | 2.8 | 1 | 1.1 | 2 | 3.4 |
| Halogenated Ethers | | | | | 1 | 1.1 | | |
| Aliphatic Carboxylic Acids | | | 1 | 1.0 | 15 | 16.7 | 16 | 27.6 |
| Aromatic Carboxylic Acids | | | | | 18 | 20.0 | 15 | 25.9 |
| Amides | | | | | | | 1 | 1.7 |
| Nitriles | | | | | 2 | 2.2 | 1 | 1.7 |
| Cyclic Oxygen | | | 3 | 2.8 | 2 | 2.2 | | |
| Basic Nitrogen | | | 1 | 1.0 | 2 | 2.2 | 1 | 1.7 |
| Aromatic Nitro | | | | | | | 4 | 6.9 |
| Thiophenes | | | | | | | | |
| Halogenated Aliphatic | 6 | 100.0 | 15 | 14.0 | 10 | 11.1 | 2 | 3.4 |
| Halogenated Aromatic | | | 7 | 6.5 | 4 | 4.4 | 1 | 1.7 |
| Halogenated Ketones | | | 4 | 3.7 | | | | |
| Halogenated Phenols | | | | | 1 | 1.1 | | |
| Halogenaed Amides | | | | | | | | |
| Halogenated PCB & Pesticides | | | | | | | | |
| Phosphates | | | | | 2 | 2.2 | | |

| | | | | | | | | |
|---|---|---|---|---|---|---|---|---|
| Percent of Total Number (183) of Unique Organics | | 3 | | 58 | | 49 | | 32 |
| Consent Decree [28] Organics | | | | | | | | |
| Number | 6 | | 23 | | 11 | | 6 | |
| Percent[c] | | 5 | | 20 | | 10 | | 5 |
| EPA Chemical Indicators of Industrial Pollution in Drinking Water [34] | | | | | | | | |
| Number | 2 | | 18 | | 10 | | 5 | |
| Percent[d] | | 3 | | 29 | | 16 | | 8 |

[a] Comparison based on GAC-inf from contactor D.
[b] XAD-inf water from contactor A was used.
[c] There are 113 Consent Decree organics.
[d] There are 62 EPA Chemical Indicators.

isomers were lost during the evaporation of 1200 mL of BLLE methylene chloride down to 0.5-mL volume. BLLE and XAD-EEE analyses produced similar GC peak heights for most methylated acids and nitrotoluenes.

Table X provides a greater depth of comparative physical-chemical information than any other table or figure. Overall, 183 different organics were detected in GAC-inf water by all four methods. Six organics were detected by Bellar P&T, 107 by Grob CLSA, 90 by BLLE and 58 by XAD-EEE. As compared to 183 total organics, 3% were detected by Bellar P&T, 58% by CLSA, 49% by BLLE and 32% by XAD-EEE. Of the Consent Decree [33, 34] organics, Bellar P&T detected 5%, Grob CLSA detected 20%, BLLE detected 10% and XAD-EEE detected 5%. The EPA Office of Drinking Water published [37] a list of "chemical indicators of industrial pollution" as a yardstick to determine if a drinking water supply would be required to use GAC to remove toxic pollutants from potable water. Of the 62 organics or organic classes on this list, Bellar P&T detected 3%, Grob CLSA detected 27%, BLLE detected 16% and XAD-EEE detected 8%. In summary, for these samples of drinking water, Grob CLSA has resulted in the quantification of a larger number and higher percentage of the organics that EPA is currently monitoring than the three other methods. This summary statement would not be accurate if the concentration of Consent Decree organics and "chemical indicators of industrial pollution" had been greater than 40 ng/L. At concentrations greater than 40 ng/L, Method 601 would have detected 28 (26% of 113) Consent Decree organics and EMSL Method 502 would have detected 43 (69% of 62) "chemical indicators of industrial pollution." At concentrations greater than 0.1 $\mu$g/L, the MAS P&T and XAD adsorption procedures should have detected a majority of the Consent Decree organics and "chemical indicators of industrial pollution."

The above statistics do not provide an overview of the physical-chemical differences of the four methods. Table X indicates that Grob CLSA quantified more aliphatic hydrocarbons and aromatic hydrocarbons than the other three methods combined. However, CLSA detected a lower number of nitrogen compounds and oxygenated compounds than either BLLE or XAD-EEE. Concerning specific functional groups, Grob CLSA detected a greater number of alkanes, alicyclic hydrocarbons, alkylated benzenes, indeno hydrocarbons, naptheno hydrocarbons, aldehydes, quinones, aliphatic esters, ethers, oxygen-containing heterocycles, halogenated aliphatics, halogenated aromatics and halogenated ketones. However, BLLE detected a greater number of water-soluble compounds such as alcohols, glycols, ketones, halogenated ethers, aromatic carboxylic acids, amides, nitriles, halogenated phenols and phosphates. Table X also indicates that aliphatic carboxylic acids (fatty acids) were equally well detected by BLLE and XAD-EEE. Overall, XAD-EEE analysis did not perform as well as BLLE analysis. In summary, Table X indicates that more toxic or potentially toxic species may be quantified by Grob CLSA than by the other three methods, but that Bellar P&T, Grob CLSA, and BLLE have optimal performance for different classes of organics. Thus, it is clear that Bellar P&T (Method 601), Grob CLSA and BLLE are important complementary methods. For this reason, HERL will continue to require the use of all three methods for health effects research water samples.

The physical-chemical data in Table X also provide valuable information about the optimum choice of liquid phases for the GC separation of organics in CLSA, BLLE and XAD-EEE extracts. Satisfactory separation results can be obtained for the nonpolar organics in CLSA extracts using both nonpolar (methyl silicone) and polar GC liquid phases. The predominance of oxygenated polar organics in BLLE and XAD-EEE extracts require the use of polar liquid GC phases for optimum separation results. Chemists should not forget that splitless injection of solvent extracts on WCOT capillary columns requires the GC liquid phase be a liquid (not a solid) at the temperature needed to achieve the correct solvent effect performance. For example, the use of $CS_2$ as a CLSA extraction solvent requires a GC oven temperature of 20°C or less for a satisfactory solvent effect. Thus, the capillary column liquid phase must also be a liquid at 20°C. Consequently, the use of SP1000 or Carbowax 20M liquid phases for the capillary splitless injection of $CS_2$ extracts would be unsatisfactory because these polar phases are a semisolid at 20°C. Unfortunately, the operating temperature range (minimum *and* maximum temperatures) of most commercially available polar capillary columns is unacceptable for the splitless injection of CLSA, BLLE and XAD-EEE extracts. This limitation of too high of a minimum temperature may be overcome by using a solvent exchange step (with a higher-boiling solvent) or by adding a higher-boiling solvent to a CLSA, BLLE or XAD-EEE extract prior to splitless injection. Both of these approaches lead to complete masking and/or partial loss of many early-eluting components.

Cost/benefit (number of organics measured) is an important consideration when comparing methods. However, we have not been able to devise a fair way to make this type of comparison for BLLE and XAD-EEE analyses. The data in this report are not quantitative, and the limits of detection are unknown for these two methods; therefore, it is difficult to determine a fair basis for comparison with Bellar P&T and Grob CLSA. Precision and accuracy data have been reported previously for Method 601 [33] and Grob CLSA [11]. However, some cost information on Bellar P&T and Grob CLSA can be provided. For these calculations, the appropriate cost for the analysis of GAC-inf water by Bellar P&T (Method 601 or 502) is \$85 and by Grob CLSA is \$460. Even though only six organics were detected using Bellar P&T analysis, 48 halogenated organics above 0.1 $\mu$g/L according to EMSL Method

502 could have been detected. A total of 107 organics were detected by Grob CLSA, plus 171 organics (215 - 64) were not present above our CLSA limits of detection. Therefore, Grob CLSA could have detected 278 organics in GAC-inf water. Following this logic, the average cost to quantify an organic by EMSL Method 502 and by Grob CLSA is approximately $2. Even though these figures would indicate that Bellar P&T Method 502 and Grob CLSA have a similar cost/benefit ratio, the methods are not similar in the complexity of instruments required to perform an analysis. However, in considering the complementary nature and cost/benefit figures of Bellar P&T analysis (Methods 601 or 502) and Grob CLSA, both methods are used for important health effects research water samples, especially for studying a water treatment unit process such as GAC adsorption or ozone disinfection.

To illustrate further the effectiveness of CLSA to monitor the fate of purgeable organics in water, the Grob CLSA chromatograms (Figure 10) of a CWW sample before and after ozone treatment (Ozone-inf and Ozone-eff water, respectively) are presented. The predominant oxidation of specific, trace-level, alkyl- and halogen-substituted benzenes in this drinking water sample would not have been apparent using the other three methods or MAS procedures. Ozonolysis water treatment experiments conducted by the EPA Drinking Water Research Division (Cincinnati, OH) and analytical Grob CLSA conducted by HERL indicate that the reduction of purgeable organics in Figure 10 is due to chemical oxidation and not gas-phase stripping (ozone-oxygen). The apparent chemical oxidation of these organics indicates a possible reduction in toxic organics in ozonated drinking water. However, this is not to imply that the reduced amounts of purgeable halogenated and aromatic compounds detected in this ozonated water by Grob CLSA will provide evidence of a reduction in long-term health effects. Such a determination would also require the comprehensive measurement of ozone reaction products and the toxicological effects of ozonated drinking water. Grob capillary GC/MS/DS CLSA, however, does provide highly reproducible, quantitative information of many unit process effects (GAC, ozonation, etc.) for one small group of compounds in drinking water—purgeable organics.

## CONCLUSIONS

### Future Comprehensive Analytical Scheme for Purgeable Organics

The analysis of purgeable organics will continue to be important in future years, because many industrial pollutants are readily measured by P&T procedures. State-of-the-art knowledge of comprehensive purgeable analytical

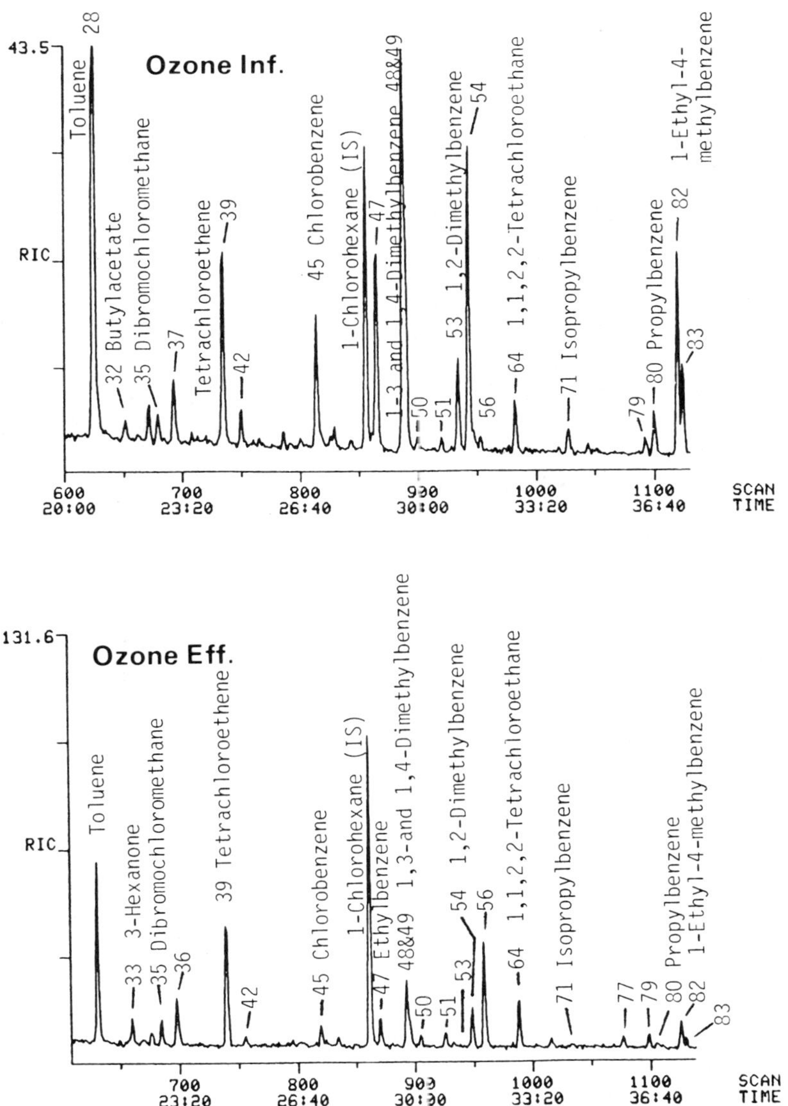


**Figure 10.** Chromatographic Grob CLSA results of CWW raw, settled water (ozone-inf) and ozone-treated water (ozone-eff).

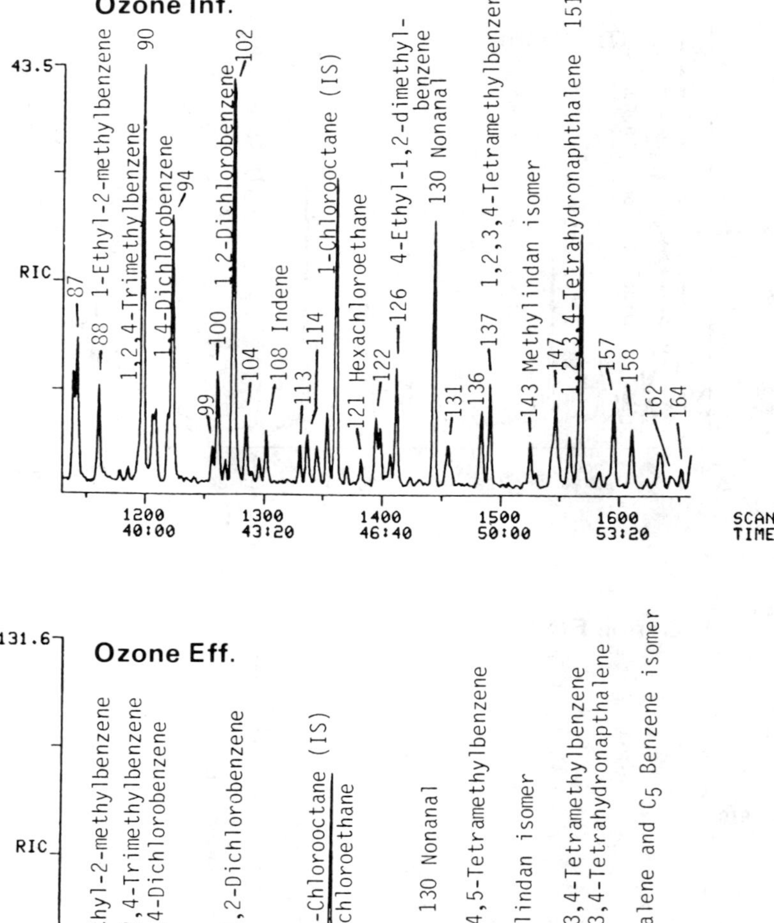


Figure 10, continued

methods has reached a sufficiently high level that allows environmental chemists to design superior analytical schemes for the comprehensive analyses of purgeables in drinking water. This comprehensive scheme, in our estimation, should consider the following requirements which are evident in the data presented:

1. Bellar P&T methods, such as EPA Methods 601 [33] and 502 [35], provide adequate sensitivity, GC resolution, detector specificity and overall method reproducibility to cost-effectively quantify low-molecular-weight halogenated purgeable organics in drinking water. Furthermore, gaseous-type halogenated organics such as chloromethane and vinyl chloride are readily measured by the "combination" trap (Tenax-silica gel-activated carbon) [33,35], which is required in Methods 601 and 502. The above advantages and data indicate that EPA Methods 601 and 502 will continue in the future to be widely used by U.S. drinking water laboratories. In addition, our data indicate that if Grob CLSA is used to measure purgeable organics in water, the Bellar P&T Method 601 (or 502) should also be used to measure low-molecular-weight halogenated organics (e. g., chloromethane, vinyl chloride and chloroform) which are not amenable to Grob CLSA.

2. High-resolution capillary columns are required to separate the hundreds of alkyl- and halogen-substituted aromatic compounds which are often present in drinking water (see Table II and Figure 10). Future data may show that, due to the complexity of environmental water samples containing substituted aromatic isomers, it is extremely difficult to measure purgeable substituted aromatics using packed GC columns and a photoionization detector such as are required in EPA Methods 602 [33] and 503 [36]. Instead, capillary GC/MS/DS analysis such as Grob CLSA or the MAS RTI P&T procedure will be necessary for these substituted aromatic compounds.

3. If future researchers find that statement 1, above, is correct, the required capillary GC/MS/DS procedures described in statement 2 should be developed to measure quantitatively a broad spectrum of purgeable organics from benzene, toluene and isomers of dichlorobenzene to as high a molecular weight range as is practical. Higher purging temperatures and optimized trapping materials should be developed and used. There seems to be a misconception today in the thinking of some environmental chemists that BLLE or XAD-EEE will cost-effectively measure low levels (1-100 ng/L) of purgeable organics such as alkyl- and halogen-substituted indans, tetrahydronaphthalenes and biphenyls. These organics are difficult to measure quantitatively in drinking water by BLLE and XAD-EEE due to interferences from concentrated solvent impurities, losses of these organics during KD evaporation and the overall insensitivity of BLLE and XAD-EEE. Therefore, the use of BLLE

or XAD-EEE for these purgeable organics is very difficult and costly. A comprehensive capillary GC/MS P&T procedure, such as Grob CLSA or the MAS P&T procedure, should be optimized to measure quantitatively trace amounts of these higher-molecular-weight aromatic compounds in drinking water.

4. The MAS RTI P&T method attempts to use one P&T procedure to achieve the comprehensive, combined results of both Method 601 (or 502) and Grob capillary GC/MS/DS CLSA. From a design viewpoint, the MAS procedure may poorly measure highly volatile compounds (such as chloromethane and vinyl chloride) which Methods 601 and 502 readily measure. This is due to the use of a removable-type Tenax (only) cartridge at room temperature. Furthermore, it is not certain that the MAS P&T procedure will measure the organics listed in statement 3 which are readily measurable by Grob CLSA. There is a need to systematically compare EPA Method 601 (502) and Grob capillary GC/MS/DS CLSA to the MAS RTI P&T procedure and to optimize a capillary GC/MS P&T procedure that meets the basic requirements of statement 3. Such a comparison should also include BLLE, EPA Methods 602 and 503 and the MAS XAD-4 procedure. The cost effectiveness of each method should be computed.

## Future Use of Grob CLSA

Even though Grob capillary GC/MS/DS CLSA is one of the first operational and feasible comprehensive purgeable analytical methods for drinking water, the subsequent large number of purgeable organics measured by CLSA comprise only a very small weight of the total mass of organic material present in potable water. The data, however, indicate that the use of Grob capillary GC/MS/DS CLSA and Bellar P&T analysis (Method 601 or 502) provide a feasible and useful approach for studying trace-level amounts of a surprisingly wide range of purgeable organics in drinking water. Some of the advantages of the CLSA method are:

1. The method is simple, straightforward and rapid. A water sample can go from cold storage to GC/MS injection of the CLSA filter extract in about 2.5 hr. Sophisticated, electronic "black boxes" are not required for CLSA. No alteration of a standard capillary GC/MS/DS (equipped with a Grob-type splitless injector) is required. Overall, CLSA equipment costs about $1300.
2. The method is ultrasensitive. Good mass spectra can be obtained on most purgeable compounds in the 1- to 10-ng/L range. The GC/MS detection limit of some PCB isomers is 2 ng/L.
3. Blanks are extremely clean. The sample is extracted from the activated carbon filter with only 12 $\mu$L of carbon disulfide.

4. Glass or fused-silica capillary columns coated with methyl silicone liquid phases can be used with CLSA carbon filter extracts for a year or more without developing sample-induced degradation of column performance. This is in direct contrast to the injection of BLLE and XAD-EEE extracts on similar capillary columns.
5. The method is relatively trouble-free. The most frequent problem (about once every three months at HERL) has been contamination of the closed-loop with high-molecular-weight organics from heavily contaminated water samples. Muffling the glass and metal components of the loop at 450°C for 2 hr corrects this problem.
6. On a per-compound-quantified basis, the method is highly cost-effective. Once initiated, sample purging, GC/MS data acquisition and automatic computerized quantification methods proceed virtually unattended. Solvent extraction of the activated carbon trap requires approximately 10 min. Overall, three drinking water samples can be quantified automatically for 215 organics in 24-hr period using 12 person-hours of labor.

Disadvantages of the CLSA method are:

1. The range of compounds effectively measured by Grob CLSA is somewhat limited. (1) Highly volatile compounds such as chloromethane, vinyl chloride, methylene chloride and chloroform are poorly recovered and/or covered up by the $CS_2$ extraction solvent. Therefore, Bellar P&T Method 601 (or 502) should also be used with CLSA to provide a comprehensive analysis of purgeable organics. (2) Moderate and highly polar or ionizable organic species are either poorly purged or not recovered at all.
2. Highly contaminated samples, such as industrial effluents, may overload the 1.5-mg activated carbon trap and contaminate the closed-loop. A larger capacity 5.0-mg trap, however, is now available from Bender and Holbein Company in Zurich.
3. The method requires the development of new laboratory skills (extraction of the trap) before the method can be implemented effectively. This may account for the surprisingly slow acceptance of Grob CLSA in the United States as compared to Europe. While the CLSA procedure can become highly routine in the hands of competent technicians, the method is demanding of careful and consistent manipulations.
4. There is currently no commercial U.S. supplier of the Grob-designed activated carbon filters or filter holders. In addition, there is no worldwide supplier of an integrated, self-contained CLSA apparatus. CLSA components must be purchased from a number of suppliers. However, this disadvantage may no longer exist. Grob recently announced that Brechbuhler AG, Schlieren, Switzerland, will manufacture and sell CLSA components in Europe, and through distributors in Japan and the United States. (Liquid Partition Systems, Inc., Ashland, MA, is the U.S. distributor.)
5. Grob CLSA has not been researched comprehensively to optimize analytical conditions and apparatus since Grob and Zurcher [12] standardized the procedure in 1976. It is suggested that Grob CLSA be optimized to measure purgeable compounds from benzene, toluene and dichlorobenzenes to as high a molecular weight range as is practical.

Additional features of the CLSA method include:

1. Grob CLSA is especially suitable for automatic quantification procedures using state-of-the-art GC/MS/DS. The method provides highly reproducible relative retention time data ($\pm 0.2\%$) and clean mass spectra, which are important for highly successful automatic GC/MS/DS software procedures.
2. The five internal standards added before sample purging greatly facilitate the requirement for good quality control and the monitoring of recovery efficiencies.

This report shows that many compounds which are not amenable to Bellar P&T and Grob CLSA (due to low stripping efficiencies caused by polarity, ionizability and/or nonvolatility) can be analyzed effectively using a large sample volume (10 L), BLLE and ultraclean laboratory techniques. The BLLE procedure has produced extremely clean blanks and an apparent GC/MS detection limit of 5-50 ng/L for a wide variety of extractable organic compounds.

The XAD-2 ethyl ether elution (XAD-EEE) procedure (10 L water sample) did not perform nearly as well as the BLLE procedure. This XAD adsorption procedure has also been shown to erratically contaminate extracts during the ethyl ether elution phase of the procedure. Consequently, the use of XAD adsorption procedures is not recommended (especially for the measurement of alkyl-substituted aromatic compounds in drinking water) unless this type of contamination can absolutely be prevented. The use of XAD-2 or XAD-4 resin to measure alkyl-substituted benzenes, indanes, tetrahydronapthalenes and napthalenes is especially difficult due to the contamination artifacts from XAD-2 and XAD-4 resin documented in this report (see Table V). A P&T method such as Grob CLSA is superior for this group of substituted aromatics.

This report presents experimental data which may be useful to the many research laboratories that are developing and using comprehensive analytical methods for the measurement of organics in drinking water. Development of comprehensive analytical methods is expensive. It is especially important in this coming era of diminishing monetary support for drinking water research that comprehensive methods of organic analysis be (1) broad spectrum, (2) sensitive, (3) cost-effective, and (4) scientifically sound. It is hoped that this report will stimulate renewed thinking along these lines.

The data presented suggest that the combined use of Bellar P&T Method 601 (502) and Grob capillary GC/MS CLSA (two cost-effective P&T methods) measures a number of organics (at concentrations between 10 and 100 ng/L) that the MAS P&T method and MAS XAD-4 adsorption method cannot measure [8]. The data also suggest that the combined use of Bellar

P&T Method 601 (502) and the described Grob CLSA and BLLE procedure will measure a considerable greater number of organics (both toxic and nontoxic) in drinking water than the combined use of the MAS P&T and MAS XAD-EEE procedures [4-10].

Experimental data on samples of drinking water from CWW indicate that twice as many Consent Decree organics [33] and "chemical indicators of industrial pollution" [39] were detected and quantified by Grob CLSA than by Bellar P&T, BLLE and XAD-EEE analysis. Furthermore, Grob CLSA produced this superior analysis at a low cost-per-compound-analyzed figure. The comparative data presented verify the words used by Grob, Reinhard, Stieglitz and Piet to describe Grob CLSA: "The method works." This is the primary reason why Grob CLSA continues to increase in worldwide popularity.

This report has attempted to present a brief review of U.S. P&T methods and to point out design differences of each method. Interested North American environmental chemists may find the detailed historical review of Grob CLSA particularly useful. Finally, experimental data from our health effects research have been presented to illustrate some basic differences between Grob CLSA, Bellar P&T (Method 601), BLLE and XAD-EEE analyses. Even though the experimental data from these CWW samples do not validate or invalidate any of the four tested methods, the data demonstrate some important differences in the analytical performance which one might expect of each of these methods. Overall, the data indicate that the combined use of Bellar P&T (EPA Methods 601 or 502), Grob capillary GC/MS/DS CLSA, and capillary GC/MS/DS BLLE analyses provide useful information on drinking water treatment unit processes such as the purification of water with 1-mgd GAC contactors.

## ACKNOWLEDGMENTS

We appreciate the outstanding cooperation of Mr. Richard Miller, Director, and Mr. Dave Hartman, Chemist, of the Cincinnati Waterworks in providing samples and physical facilities to conduct health effects research experiments. Their continuous cooperation over the years has been a service to drinking water consumers throughout the United States. The authors wish to thank Prof. K. Grob, M. Reinhard, L. Stieglitz and G. Piet for their valuable advice on the use of Grob CLSA, and the many individuals that diligently worked with us between 1975 and 1977 until we could produce superior WCOT capillary GC separations. We appreciate the valuable assistance over the past six years of Mr. Tom Bellar, EMSL, for his constant advice and help on the analysis of purgeable organics in water. We also thank Ms. Verna

Tilford, Ms. Melda Hatfield, Ms. Nancy Koopman, Ms. Deborah Dean and Mr. Lon Winchester for typing; Mr. Rob Brown, EPA, CERI, for graphical reproductions; Mrs. Marta Richards, Mrs. Jean Munch and Mrs. Dot Reynolds for proofreading; and Mrs. Judi Olsen for editorial assistance. At Battelle Columbus Labs, Ms. Vanessa Goff, Ms. Denise Contos, Mr. Tim Hayes and Mr. Dan Aichele have made significant contributions in this work.

## DISCLAIMER

This report has been reviewed by the Health Effects Research Laboratory, U.S. EPA, and approved for publication. Approval does not signify that the contents necessarily reflect the veiws and policies of the U.S. EPA, nor does mention of trade names or commercial products constitute endorsement or recommendation for use.

## REFERENCES

1. Kopfler, F. C., R. G. Melton, R. D. Lingg and W. E. Coleman. "GC/MS Determination of Volatiles for the National Organics Reconnaissance Survey (NORS) of Drinking Water," in *Identification and Analsis of Organic Pollutants in Water*, L. H. Keith, Ed. (Ann Arbor, MI: Ann Arbor Science Publishers, Inc., 1976).
2. Coleman, W. E., R. D. Lingg, R. G. Melton and F. C. Kopfler. "The Occurrence of Volatile Organics in Five Drinking Water Supplies Using Gas Chromatography/Mass Spectrometry," in *Identification and Analysis of Organic Pollutants in Water*, L. H. Keith, Ed. (Ann Arbor, MI: Ann Arbor Science Publishers, Inc., 1976).
3. Coleman, W. E., R. G. Melton, R. W. Slater, F. C. Kopfler, S. J. Voto, W. K. Allen and T. A. Aurand. "Determination of Organic Contaminants by the Grob Closed-Loop-Stripping Technique," Proceedings of the American Water Works Association Technology Conference VII, Philadelphia, PA (1979).
4. Bursey, J. T. et al. "Master Scheme for the Analysis of Organic Compounds in Water. Part I. State-of-the-Art Review of Analytical Operations," U.S. EPA, Contract No. 69-03-2704, Environmental Research Laboratory, Athens, GA (1979).
5. Pellizzari, E. D. et al. "Master Scheme for the Analysis of Organic Compounds in Water, Preliminary Draft Report. Part III. Experimental Development and Results," U.S. EPA, Contract No. 68-03-2704, Environmental Research Laboratory, Athens, GA (1980).
6. Michael, L. C., R. Wiseman, L. S. Sheldon, J. Bursey, K. E. Tomer, T. A. Scott, R. Coney, A. W. Garrison, J. Gebhart and J. F. Ryan. "Quality of Master Analytical Scheme Data: Purgeables and Volatile Organic Acids," in *Advances in the Identification and Analysis of*

*Organc Pollutants in Water, Vol. 1*, L. H. Keith, Ed. (Ann Arbor, MI: Ann Arbor Science Publishers, Inc., 1981), Chapter 6.

7. Garrison, A. W., A. L. Alford, J. S. Craig, J. J. Ellington, A. F. Haeberer, J. M. McGuire, J. D. Pope, W. M. Shackelford, E. D. Pellizzari and J. E. Gebhart. "The Master Analytical Scheme: An Overview of Interim Procedures," in *Advances in the Identification and Analysis of Organic Pollutants in Water, Vol. 1*, L. H. Keith, Ed. (Ann Arbor, MI: Ann Arbor Science Publishers, Inc., 1981), Chapter 2.
8. Gebhart, J. E., J. F. Ryan, R. D. Cos, E. D. Pellizzari, L. C. Michael and L. S. Sheldon. "The Master Analytical Scheme: Deveopment of Effective Techniques for Isolation and Concentration of Organics in Water," in *Advances in the Identification and Analysis of Organic Pollutants in Water, Vol. 1*, L. H. Keith, Ed. (Ann Arbor, MI: Ann Arbor Science Publishers, Inc., 1981), Chapter 3.
9. Tomer, K. E., J. Bursey, L. C. Michael, L. S. Sheldon, E. D. Pellizzari, A. L. Alford, J. D. Pope and A. W. Garrison. "Quantitative Aspects of the Master Analytical Scheme for Organics in Water," in *Advances in the Identification and Analysis of Organics in Water, Vol. 1*, L. H. Keith, Ed. (Ann Arbor, MI: Ann Arbor Science Publishers, Inc., 1981), Chapter 5.
10. Ryan, J. F., J. E. Gebhart, L. C. Rando, D. L. Perry, K. E. Tomer, E. D. Pellizzari and J. T. Bursey. "The Master Analytical Scheme: An Assessment of Factors Influencing Precision and Accuracy of Gas Chromatography/Mass Spectrometry Data," in *Advances in the Identification and Analysis of Organic Pollutants in Water, Vol. 1*, L. H. Keith, Ed. (Ann Arbor, MI: Ann Arbor Science Publishers, Inc., 1981), Chapter 4.
11. Coleman, W. E., W. K. Allen, R. W. Slater, S. J. Voto, R. G. Melton, F. C. Kopfler and T. A. Aurand. "Automatic Quantification and Statistical Evaluation of Organic Contaminants Using a Computerized Glass Capillary Gas Chromatography/Mass Spectrometry System and Grob Closed-Loop Stripping," Chapter 37, this volume.
12. DeMarco, J., A. A. Stevens and D. J. Hartman. "Application of Organic Analysis for Evaluation of Granular Activated Carbon Performance in Drinking Water Treatment," Chapter 47, this volume.
13. Grob, K. "Organic Substances in Potable Water and in Its Precursors. Part I. Methods for Their Determination by Gas-Liquid Chromatography," *J. Chromatog.* 84:255 (1973).
14. Bellar, T. A., and J. J. Lichtenberg. "The Determination of Volatile Organic Compounds at the Microgram per Liter Level in Water by Gas Chromatography," *J. Am. Water Works Assoc.* 66:739 (1974).
15. Bellar, T. A., J. J. Lichtenberg and R. C. Kroner. "The Occurrence of Organohalides in Chlorinated Drinking Waters," *J. Am. Water Works Assoc.* 66:703 (1974).
16. Lingg, R. D., R. G. Melton, F. C Kopfler, W. E. Coleman and D. E. Mitchell. "Quantitative Analysis of Volatile Organic Compounds by GC/MS," *J. Am. Water Works Assoc.* 69:605 (1977).
17. Grob, K., and F. Zurcher. "Stripping of Trace Organic Substances from Water, Equipment and Procedure," *J. Chromatog.* 117:285 (1976).

18. White, L. D., D. G. Taylor, P. A. Mauer and R. E. Kupel. "A Convenient Optimized Method for the Analysis of Selected Solvent Vapors in the Industrial Atmosphere," *J. Am. Ind. Hyg. Assoc.* 31:225 (1970).
19. Grob, K., and G. Grob. "Organic Substances in Potable Water and in Its Precursors. Part II. Applications in the Area of Zurich," *J. Chromatog.* 90:303 (1974).
20. Grob, K., K. Grob, Jr. and G. Grob. "Organic Substances in Potable Water and in Its Precursors. Part III. The Closed-Loop-Stripping Procedures Compared with Rapid Liquid Extraction," *J. Chromatog.* 106:299 (1975).
21. Stieglitz, L., W. Roth, W. Kuhn and W. Leger. "The Behavior of Organohalides in the Treatment of Drinking Water," *Vom Wasser* 47:347 (1976).
22. Zurcher, F., and W. Giger. "The Study of Volatile Organic Compounds in the Glatt River," *Vom Wasser* 47:37 (1976).
23. Giger, W., M. Reinhard, C. Schaffner and F. Zurcher. "Analyses of Organic Constituents in Water by High-Resolution Gas Chromatography in Combination with Specific Detection and Computer-Assisted Mass Spectrometry," in *Identification and Analysis of Organic Pollutants in Water*, L. H. Keith, Ed. (Ann Arbor, MI: Ann Arbor Science Publishers, Inc., 1976), pp. 433-452.
24. Giger, W., E. Molanr and S. Wakeham. "Volatile Chlorinated Hydrocarbons in Ground and Lake Water," in *Aquatic Pollutants*, O. Hutzinger, Ed. (Oxford: Pergamon Press, 1978).
25. Schwarzenbach, R. P., E. Molnar–Kubica, W. Giger and S. G. Wakeman. "Distribution, Residence Time, and Fluxes of Tetrachloroethylene and 1,4-Dichlorobenzene in Lake Zurich, Switzerland," *Environ. Sci. Technol.* 13:1367 (1979).
26. Giger, W. "Inventory of Organic Gases and Volatiles in the Marine Environment," *Marine Chem.* 5:429 (1977).
27. Schwarzenbach, R. P., R. H. Bromund, P. M. Gschwend, and O. C. Zafiriou. "Volatile Organic Compounds in Coastal Seawater," *Org. Geochem.* 1:93 (1978).
28. McCarty, P. L., M. Reinhard and D. G. Argo. "Organics Removal by Advanced Wastewater Treatment," Proc. 97th AWWA Annual Conf., Anaheim, CA, May 1977.
29. Reinhard, M., C. J. Dolce, P. L. McCarty and D. G. Argo. "Trace Organics Removal by Advanced Waste Treatment," *Proc. Am. Soc. Civil Eng.* 105:14760 (1979).
30. Reinhard, M., J. E. Schneier, T. Everhart and J. Graydon. "Specific Compound Analysis by Gas Chromatography and Mass Spectroscopy in Advanced Treated Waters," NATO/GCMS Conf. of Practical Application of Adsorption Techniques, Reston, VA, May 1979.
31. McCarty, P. L., and M. Reinhard. "Statistical Evaluation of Trace Organics Removal by Advanced Wastewater Treatment," Annual Conf. of the Water Pollution Control Federation, Anaheim, CA, October 1979.
32. McCarty, P. L., D. Argo and M. Reinhard. "Operational Experiences with Activated Carbon Adsorbers at Water Factory 21," *J. Am. Water Works Assoc.* 71:683 (Nov. 1979).

33. "Guidelines Establishing Test Procedures for the Analysis of Pollutants," *Federal Register* 44:69464 (1979).
34. *Natural Resource Defense Council, Inc., et al.* v. *Train*, 8 ERC 2120 (D.D.C. 1976).
35. "The Analysis of Halogenated Chemical Indicators of Industrial Contamination in Water by the Purge and Trap Method, Method 502," Environmental Monitoring and Support Laboratory, Cincinnati, OH (1978).
36. "The Analysis of Aromatic Chemical Indicators of Industrial Contamination in Water by the Purge and Trap Method, Method 503," Environmental Monitoring and Support Laboratory, Cincinnati, OH (1980).
37. Junk, G. A., J. J. Richard, J. S. Fritz and H. J. Svec. "Resin Sorption Methods for Monitoring Selected Contaminants in Water," in *Identification and Analysis of Organic Pollutants in Water*, L. H. Keith, Ed. (Ann Arbor, MI: Ann Arbor Science Publishers, Inc., 1976), pp. 135-154.
38. Lin, D. C. K., R. G. Melton, F. C. Kopfler and S. V. Lucas. "Gas Capillary Gas Chromatographic/Mass Spectrometric Analysis of Organic Concentrates from Drinking and Advanced Waste Treatment Water," Chapter 46, this volume.
39. "Interim Primary Drinking Water Regulations," *Federal Register* 43:5773 (1978).

CHAPTER 37

# AUTOMATIC QUANTIFICATION AND STATISTICAL EVALUATION OF ORGANIC CONTAMINANTS USING A COMPUTERIZED GLASS CAPILLARY GAS CHROMATOGRAPHY/MASS SPECTROMETRY SYSTEM AND GROB CLOSED-LOOP STRIPPING

**W. Emile Coleman, W. K. Allen, R. W. Slater, S. J. Voto, R. G. Melton, F. C. Kopfler and T. A. Aurand**

U.S. Environmental Protection Agency
Health Effects Research Laboratory
Epidemiology Division
Exposure Evaluation Branch
Cincinnati, Ohio

The Exposure Evaluation Branch (EEB) of the Health Effects Research Laboratory (HERL) in Cincinnati, OH, is responsible for validating sensitive methods for qualitative and quantitative measurements of organic contaminants in drinking water and other environmental media, mainly in support of toxicological and epidemiological studies. Grob closed-loop stripping analysis (CLSA) is currently being used by HERL to measure a wide range and large number of organic chemical contaminants (typically 200-300 compounds in a water supply) in U.S. drinking waters. Equipment, procedure, applications and validation of CLSA were discussed previously [1-3].

The large number of water samples analyzed by EEB and the presence of large numbers of organic contaminants in water samples demanded that we develop computerized automatic procedures to expedite the reporting of qualitative and quantitative results of analyses. Glass capillary gas chromatography/mass spectrometry (GC/MS) quantification is difficult because MS sensitivity and stability fluctuate daily. Incorporating internal standards as references for quantification, however, minimizes the effects of these variations. Although comparing total-ions quantification (TIQ) and single-ion quantification (SIQ) has indicated [1] that TIQ was as accurate and precise as SIQ when GC peaks were well resolved because of clean samples, and glass capillary GC, SIQ must be used to quantify the many overlapping GC peaks found in environmental samples. SIQ results of 22 compounds, most of which commonly occur in drinking water in the United States and are detected by CLSA are discussed in this chapter. Ten repetitive, direct injections of a 22-compound standard mixture and 16 repetitive purges of water spiked with the standard mixture at the 10- to 100-ng/L (ppt) level were done over a 4-week period. Computerized automatic quantification procedures, employing an in-house–developed computer library of standard compounds to reverse search a GC/MS data file for spectra matching, provided qualitative and quantitative analyses. The in-house library contained accurate spectra, relative retention times (RRT), response factors (RF) and selected quantification masses for 215 compounds. Statistical computations (SC) of the results emphasize the importance of maintaining reproducible GC/MS response factors.

## EXPERIMENTAL

To validate CLSA as a method for concentrating organics from water samples and subsequently to provide qualitative and quantitative analyses of the Grob carbon filter extracts by glass capillary GC/MS/DS, ten repetitive glass capillary GC/MS/DS analyses of the direct injections of a 22-compound standard mixture were performed to establish sets of average RRT and GC/MS RF values for each compound. These values were stored in a computerized quantification library.

Sixteen repetitive CLSA and glass capillary GC/MS/DS analyses of water spiked with a 22-compound standard mixture at the 10- to 100-ng/L (ppt) level were done. The RRT and RF data established from the direct injection experiment were used to quantify and determine the recovery efficiency of each of the 22 compounds from water by CLSA. The recovery data reported in Table III are the results of 16 purges of the same standard solution spiked in a different gallon of Milli-Q® water prior to each analysis. The recovery for each compound was determined by comparing the computer-quantified amounts from the purged samples to the calculated amounts in the spike, according to the formula:

$$\% \text{ recovery} = \frac{\text{amount found (ng/L)}}{\text{amount spiked (ng/L)}} \times 100$$

## Apparatus

### *Reagents*

All solvents and chemicals used for the preparation of standard solutions were reagent grade or of the highest purity obtainable from renowned chemical suppliers. The compounds used in the studies are listed in Tables I to III.

The direct injection standards for GC/MS, RRT and RF determinations were prepared in carbon disulfide ($CS_2$). The standards for CLSA, which were spiked into Milli-Q water, were prepared in acetone. Five internal standards (IS), 1-chlorohexane, 1-chlorooctane, 1-chlorododecane, 1-chlorohexadecane and 1-chlorooctadecane, were added to each direct injection standard solution and to each water sample prior to CLSA. These IS are not normally seen in drinking water.

*Direct Injection Standards (DIS).* A stock solution of standards in $CS_2$ was prepared to achieve a concentration of about 250 ng/$\mu$L per component. Typically a 1 to 10 dilution of the stock solution was used to provide a concentration of 25 ng/$\mu$L for injection into the GC/MS. Two microliters were used for glass capillary GC/MS analyses.

*CLSA Standards.* These standards, prepared in acetone, were usually made four times the level of the $CS_2$ standards to achieve a concentration of 50 ng/$\mu$L when added to the 1-gal ($\approx$4 L) water samples. A stock solution having a concentration of about 1000 ng/$\mu$L of each component was prepared. A 1 to 10 dilution of the stock provided a solution containing 100 ng/$\mu$L per component for spiking the water. Two microliters (200 ng/compound) of the latter solution, when added to the water, effected a concentration of 50 ng/L of each compound. Assuming 100% recovery of all compounds in the $CS_2$ extract of the carbon filter after CLSA, one-fourth of the final extract volume would represent the concentration of compounds in 1 L of water. We typically used 2 $\mu$L of the 8-$\mu$L $CS_2$ extract for glass capillary GC/MS analysis. The level of components in the 2-$\mu$L extract at 100% recovery should be equivalent to the concentration of components in 2 $\mu$L of the DIS mixture in $CS_2$.

*Internal Standards.* The IS for direct injection were prepared in the same manner as the DIS in $CS_2$. Prior to glass capillary GC/MS analysis, 1 ml of IS stock solution and 1 mL of DIS stock solution were added to a common 10-mL volumetric flask, and the volume was adjusted to the 10-mL mark with $CS_2$. This final solution of standards and internal standards served as the sample for the repetitive direct injection study. Incorporation of the internal standards in the DIS solution allowed us to monitor variations in GC/MS conditions by observing response factor profiles relative to documented profiles obtained over months of operation. During this study, the GC/MS re-

**Table I. Statistical Computations of Repetitive Direct Injection Study with Quantification Based on a Single Ion**

| Compound | Quantification Mass (m/e) | Relative Retention Time[a] | $\pm \sigma$ | Amount[a] (ng) | $\pm \sigma$ | Response Factor[a] | $\pm \sigma$ |
|---|---|---|---|---|---|---|---|
| 2–Methylthiophene | 97 | 0.287 | 0.002 | 50.7 | 4.3 | 4.366 | 0.367 |
| Dibromochloromethane | 129 | 0.311 | 0.002 | 122.5 | 12.0 | 2.333 | 0.077 |
| Styrene | 104 | 0.425 | 0.001 | 45.5 | 3.5 | 4.530 | 0.347 |
| Isopropylbenzene | 105 | 0.467 | 0.001 | 43.2 | 3.7 | 5.871 | 0.483 |
| 2–Chlorotoluene | 91 | 0.493 | 0.001 | 54.0 | 3.1 | 4.217 | 0.241 |
| *bis*-(2–Chloroethyl)ether | 93 | 0.525 | 0.001 | 61.0 | 3.1 | 2.869 | 0.144 |
| α-Methylstyrene | 118 | 0.530 | 0.001 | 45.0 | 3.7 | 3.645 | 0.300 |
| 1,4–Dichlorobenzene | 146 | 0.557 | 0.001 | 50.0 | 4.6 | 4.904 | 0.447 |
| 2–Ethyl–1,4–dimethylbenzene | 119 | 0.635 | 0.001 | 42.8 | 3.4 | 7.610 | 0.595 |
| 4–Chloro–*o*–xylene | 105 | 0.642 | 0.001 | 53.6 | 3.5 | 4.298 | 0.280 |
| 1,1–Dimethylindan | 131 | 0.662 | 0.000 | 42.0 | 3.6 | 8.187 | 0.701 |
| *p*–Methylphenol | 107 | 0.669 | 0.001 | 51.0 | 3.7 | 6.386 | 0.463 |
| Tetrahydronaphthalene | 104 | 0.714 | 0.000 | 48.6 | 3.2 | 3.837 | 0.250 |
| 1,2,4–Trichlorobenzene | 180 | 0.728 | 0.001 | 72.6 | 6.5 | 3.560 | 0.321 |
| Hexachloro–1,3–butadiene | 225 | 0.776 | 0.000 | 84.0 | 9.6 | 1.332 | 0.152 |
| 2–Methylbiphenyl | 168 | 0.932 | 0.000 | 50.9 | 4.9 | 3.907 | 0.373 |
| 1,6–Dimethylnaphthalene | 156 | 0.954 | 0.000 | 50.7 | 4.9 | 3.741 | 0.335 |
| 2–Isopropylnaphthalene | 155 | 0.982 | 0.000 | 47.8 | 4.6 | 8.034 | 0.769 |
| Pentachlorobenzene | 250 | 1.032 | 0.000 | 50.0 | 6.0 | 1.887 | 0.227 |
| Hexachlorobenzene | 284 | 1.178 | 0.001 | 50.0 | 6.5 | 0.921 | 0.120 |
| 2,2′,4,4′,6,6′–Hexachlorobiphenyl | 145 | 1.438 | 0.002 | 50.0 | 4.9 | 0.778 | 0.076 |
| 2,2′,4,5,5′–Pentachlorobiphenyl | 254 | 1.447 | 0.002 | 50.0 | 7.1 | 0.500 | 0.071 |
| 1–Chlorododecane (IS) | 91 | 1.000 | 0.000 | 52.0 | 0.0 | 1.000 | 0.000 |

[a] Average value ± standard deviation ($\sigma$) based on 10 injections.

Table II. Variations of GC/MS RF of Grob IS Caused by a Change in Ion Source Pressure[a]

| Internal Standards (52 ng each) | Library Value | Response Factors[b]: Helium Head Pressure (psi) 20 | 21 | 22 |
|---|---|---|---|---|
| 1-Chlorohexane | 2.207 | 2.242 (1.02)[c] | 1.941 (0.88) | 1.677 (0.76) |
| 1-Chlorooctane | 1.965 | 2.081 (1.06) | 1.865 (0.95) | 1.681 (0.86) |
| 1-Chlorododecane | 1.000 | 1.000 (1.00) | 1.000 (1.00) | 1.000 (1.00) |
| 1-Chlorohexadecane | 0.380 | 0.377 (0.99) | 0.388 (0.89) | 0.342 (0.90) |
| 1-Chlorooctadecane | 0.037 | 0.039 (1.06) | 0.052 (1.40) | 0.063 (1.69) |

[a]Ion source potentials and multiplier voltages held constant.
[b]Relative to m/e 91 of 1-chlorododecane and m/e 91 of other IS.
[c]Numbers in parentheses represent the ratio of GC/MS RF values of IS to the quantification library RF values.

sponse factors for the IS relative to 1-chlorododecane did not vary more than 10%.

The IS for CLSA, which were also spiked into the water samples, not only provided quantitative information, but the chromatogram of the IS also provided a profile from which we were able to detect discrepancies in the CLSA purging apparatus or in the extraction procedures. These standards were prepared in acetone at a concentration of 347 $\mu$g/mL each, with the exception of 1-chlorooctadecane which was 1733 $\mu$g/mL. When 0.6 $\mu$L of IS solution was added to 1 gal of water, the concentrations of the $C_6$, $C_8$, $C_{12}$ and $C_{16}$ IS were 52 ng/L each, and the concentration of the $C_{18}$ IS was 260 ng/L. The latter amount was added because of documented poor recovery efficiency of 1-chlorooctadecane at the 10- to 100-ng/L level.

## *Closed-Loop-Stripping Apparatus*

The CLSA was the same as that reported recently by Coleman et al. [1] and Melton et al. [2]. Sampling, stripping and carbon filter extraction procedures were discussed in detail by Coleman et al. [1]. All samples for CLSA were performed with the 1-gal stripping apparatus. CLSA was initiated by filling a sample container to capacity with Milli-Q water and decanting a small amount of the water down to the "1-gallon" mark, adding 2 $\mu$L of the acetone standard solution and 0.6 $\mu$L of the internal standard solution, and assem-

Table III. Statistical Computations of Repetitive Purge Study with Quantification Based on a Single Ion

| Compound | Standard Amount (ng) | Recovered Amount[a] | ± σ | Range | Recovery Efficiency[a] (%) | ± σ |
|---|---|---|---|---|---|---|
| 2-Methylthiophene | 20.1 | 11.1 | 3.4 | 6.5–18.6 | 55.3 | 17.0 |
| Dibromochloromethane | 44.5 | 7.0 | 5.0 | 2.5–18.4 | 15.7 | 11.2 |
| Styrene | 15.4 | 8.3 | 1.8 | 5.9–11.1 | 53.7 | 11.8 |
| Isopropylbenzene | 14.9 | 10.8 | 2.9 | 7.5–16.8 | 72.2 | 19.4 |
| 2-Chlorotoluene | 19.9 | 10.8 | 2.4 | 8.3–15.4 | 54.3 | 12.1 |
| *bis*-(2-Chloroethyl)ether | 21.5 | 0.1 | 0.1 | 0.0–0.2 | 0.4 | 0.3 |
| α-Methylstyrene | 14.3 | 8.1 | 1.8 | 6.0–11.7 | 56.4 | 12.8 |
| 1,4-Dichlorobenzene | 20.2 | 14.4 | 3.8 | 9.1–20.8 | 71.2 | 18.9 |
| 2-Ethyl-1,4-dimethylbenzene | 13.6 | 10.4 | 2.5 | 7.5–15.1 | 76.6 | 18.6 |
| 4-Chloro-*o*-xylene | 17.8 | 9.4 | 2.2 | 6.9–13.7 | 52.5 | 12.3 |
| 1,1-Dimethylindan | 13.0 | 9.5 | 2.4 | 6.6–14.0 | 73.4 | 18.8 |
| *p*-Methylphenol | 20.7 | 0.1 | 0.2 | 0.0–1.0 | 0.3 | 1.1 |
| Tetrahydronaphthalene | 17.0 | 9.1 | 2.1 | 6.5–13.3 | 53.6 | 12.4 |
| 1,2,4-Trichlorobenzene | 18.8 | 10.4 | 2.6 | 6.6–14.3 | 55.1 | 14.0 |
| Hexachloro-1,3-butadiene | 21.9 | 19.6 | 5.6 | 12.9–31.2 | 89.4 | 25.7 |
| 2-Methylbiphenyl | 14.1 | 7.3 | 2.1 | 4.4–10.2 | 51.9 | 14.7 |
| 1,6-Dimethylnaphthalene | 13.9 | 1.7 | 0.6 | 0.7–3.1 | 12.3 | 4.5 |
| 2-Isopropylnaphthalene | 13.3 | 6.5 | 1.7 | 4.1–9.0 | 49.1 | 13.1 |
| Pentachlorobenzene | 8.3 | 2.2 | 0.8 | 1.2–4.4 | 26.0 | 9.8 |
| Hexachlorobenzene | 14.0 | 1.7 | 1.1 | 0.0–4.2 | 11.9 | 7.8 |
| 2,2′,4,4′,6,6′-Hexachlorobiphenyl | 6.2 | 3.6 | 1.1 | 1.6–5.5 | 58.7 | 17.9 |
| 2,2′,4,5,5′-Pentachlorobiphenyl | 21.4 | 4.0 | 1.6 | 1.4–6.7 | 18.8 | 7.5 |
| 1-Chlorododecane | 52.0 | 41.9 | 15.4 | 19.4–70.5 | 80.6 | 29.5 |

[a]Average value ± standard deviation (σ) based on 16 purging analyses.

bling the purging apparatus. The sample was stripped of organics by recirculating the headspace through the carbon filter for 2 hr.

### *Computerized Glass Capillary GC/MS System*

All glass capillary GC/MS/DS analyses, structural identifications and quantification results for studies reported in this chapter were performed on a Finnigan Model 3300 quadrupole mass spectrometer equipped with an Incos Model 2300 data system computer. A Finnigan Model 9500 gas chromatograph equipped with a Grob-designed splitless injector [1,4], and a 60-m x 0.25-mm i.d. wall-coated open tubular (WCOT) glass capillary SP 2100 column was interfaced to the mass spectrometer with a glass-lined stainless steel dual-transfer line held at 230°C.

Two microliters of sample in $CS_2$ were injected with the split closed [5]. The split was opened (20:1 ratio) after 30 sec. The hot needle technique [1,6] was used for injection of all samples and standards. The helium carrier gas flowrate was about 3 mL/min at 25°C; the linear velocity through the GC column was 25 cm/sec with a helium head pressure of 20 psi (no flow controller). The injector temperature was maintained at 260°C. The column temperature program was: 20°C isothermal (cooled with liquid nitrogen) for approximately 8 min, then 2°C/min to 250°C. Mass spectra were acquired at the rate of one per 2 sec from 14 to 450 amu at 70 eV and 90°C source temperature. Source pressure was approximately 5 x $10^{-6}$ torr. For a typical glass capillary GC/MS/DS analysis of this type, 3900 scans were recorded (130 min from injection to termination). Confirmation of identifications was made by comparison of GC and GC/MS properties with those of authentic standards or library spectra from the HERL quantification library of 215 spectra of standard compounds. This spectral library represented compounds that were typically detected by CLSA of various drinking water supplies.

## Quantification

Automatic computerized quantification procedures based on the internal standard method of quantification were refined and used to calculate the level of organics in the DIS and in the $CS_2$ extracts of the CLSA samples. A computer quantification library containing the 22 compounds and 1-chlorododecane as the IS was established.

Briefly, the computerized automatic quantification procedure performed these functions:

1. wrote a quantification library list (names, RRT, RF, etc.) from the established library;
2. searched for the IS in each data file based on its retention time (RT) in the library;
3. stored the scan number and RT of the IS for RRT and quantification reference when the IS was found;
4. calculated the search window (scan No. ±10 scans) based on RRT for each library list entry; and then performed a reverse search of the data file with each entry to match the library spectra within preset purity and fit limits, to one in the data file in the selected narrow search window;
5. quantified the designated GC/MS peak in the chromatogram based on a single selected quantification mass (m/e) and wrote its identity in a quantification list, when a match was obtained; and
6. sorted the final quantification list according to scan number and wrote the quantification report with concentration levels, RT, RRT and RT referenced to 1-chlorododecane, after library list entries were searched and quantified.

The reference concentration for 1-chlorododecane was 52 ng; it was also spiked into the water samples at 52 ng/L. The quantification formula used by the computer is:

$$\text{amount} = \frac{\text{area x IS amount}}{\text{IS area x RF(L)}}$$

$$\text{RF(R)} = \frac{\text{area x IS amount}}{\text{amount x IS area}}$$

where L = library RF
R = computer-calculated relative RF

Quantification by GC/MS on a single ion of a mass spectrum is extremely difficult because of fluctuations in GC/MS sensitivity and stability. The fact that the GC/MS response factors are based on a single ion in the mass spectrum of the IS (m/e 91 of 1-chlorododecane) relative to a single ion in the mass spectrum of the other compounds mandates that the relative intensities of mass ions in a given spectrum should remain consistent in the concentration range being analyzed. Therefore, it was necessary to maintain all GC/MS conditions as constant as possible over the four-week period of these experiments. Maintaining reproducible response factors was the most critical element for producing accurate quantification.

Prior to starting the study it was determined that after the GC conditions were optimized and after the MS resolution, source potentials and multiplier voltage were selected for maximum sensitivity and quality spectra, all samples would be run under these conditions. However, the MS would be calibrated daily.

Figure 1 demonstrates the way the ion source potentials were adjusted. This procedure served as a reference for checking and maintaining the relative intensities of 1-chlorododecane mass ions at its elution temperature from the GC column. A DIS containing 52 ng of 1-chlorododecane was injected into the GC/MS under identical GC conditions established for all samples. The elution of 1-chlorododecane from the GC column was visually monitored on the oscilloscope; the height of the m/e 91 peak in centimeters and the temperature of the GC column oven were recorded.

Following this procedure, the GC column oven was maintained at the elution temperature of the 1-chlorododecane; 1-chlorododecane was leaked into the direct inlet of the MS by means of a fine metering valve until the height of the m/e 91 peak was the same height as the m/e 91 peak observed

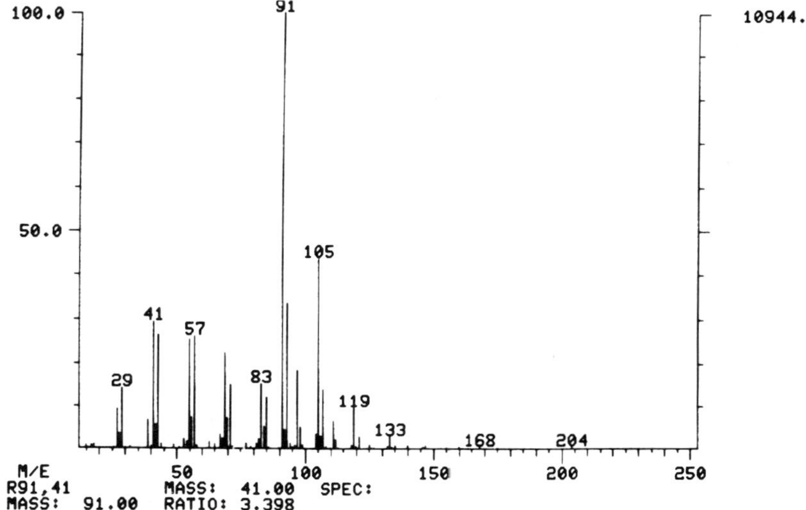


Figure 1. Mass spectrum of 1-chlorododecane showing the results of ion source adjustment with 1-chlorododecane leaked through the MS direct inlet. Ion energy, lens and extractor voltages were adjusted for a m/e 91/41 ratio of 3/1 with the GC column oven temperature maintained at 160°C.

on the oscilloscope from the GC injected 1-chlorododecane. Using the source potentials, the m/e 91/41 ion abundance ratio was arbitrarily adjusted for 3/1. The ion source pressure was also recorded. (Variations in ion source pressure have drastic effects on relative ion intensities of a mass spectrum and thereby affect individual response factors.) After this final adjustment, other than calibrating the MS daily, no other adjustments were performed unless the response factors of the internal standards varied by more than 15% of the quantification library values.

## Blanks

Since many organics in water can be measured at 5 ng/L or less with CLSA, it was necessary to incorporate procedural blanks into the analytical scheme. Prior to analyzing actual samples, filter blanks and CLSA blanks of low-organic (Milli-Q) water containing the same IS as discussed earlier were performed. When spiking water with low levels of organic chemicals for recovery studies, one must consider the background level of these chemicals in the spiking medium for correct quantification results.

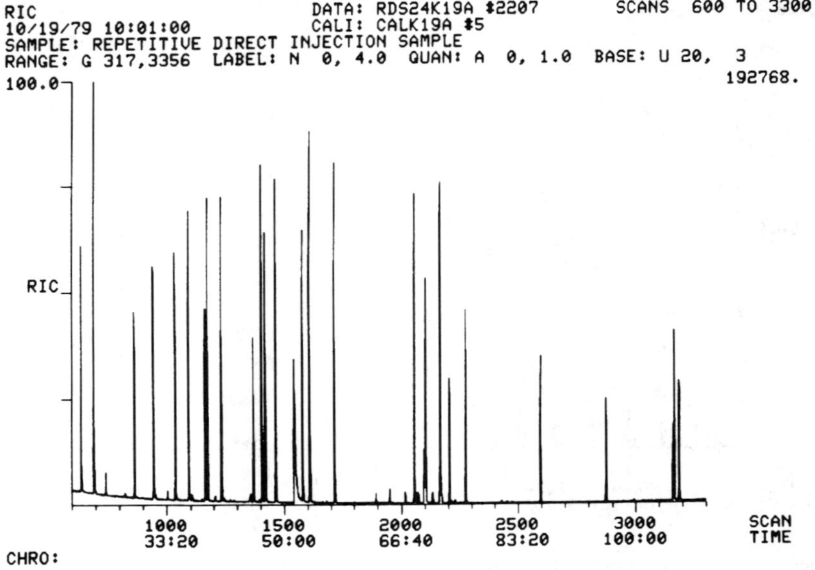


**Figure 2.** Reconstructed ion chromatogram of a DIS. 1-Chlorododecane (52 ng) is located at scan 2207.

Reconstructed ion chromatograms (RIC) of a DIS and of a CLSA standard spiked in Milli-Q water are shown in Figures 2 and 3, respectively. Figure 3 demonstrates the need for blank analyses, since there are obviously more compounds in the sample, than originally spiked into the water. Also, some of the compounds that were quantified did appear as contaminants in the water or in the solvents.

## RESULTS AND DISCUSSIONS

Statistical computations (SC) to determine standard deviations were performed on RRT, GC peak areas, RF, amounts and recovery efficiencies by CLSA. Table I shows the SC results of the repetitive direct injection study. RRT data show excellent reproducibility. Individual RRT values did not vary more than ±0.7% of the average RRT value. The individual RF obtained by SIQ of the listed masses in Table I varied from ±3.3% to ±14.2% of the average values. Compounds quantified on m/e ions 91-168 show RF variations of less than ±10% of the average RF values; those that were quantified on m/e ions 225-284 show RF variations of ±11.4-14.2% of the average RF values.

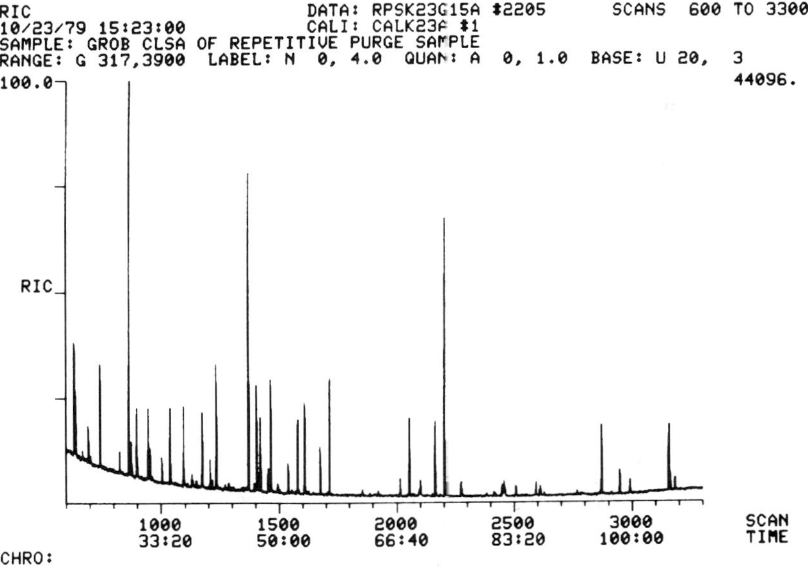


**Figure 3.** Reconstructed ion chromatogram of a CLSA standard spiked in Milli-Q water. This sample contained the same number of standard compounds, but one-fourth the amounts of the DIS in Figure 2. 1-Chlorododecane (52 ng/L) is located at scan 2205.

The latter group of compounds have many mass ions and an even distribution of ion abundance in their mass spectrum, thereby yielding lower RF and chances for greater errors for the RF values. Overall the RF varied on the average of ±8.6%.

SC on the GC peak areas are not included in a table. The peak areas (single ion peak areas) varied on the average of ±13%, whereas the amounts varied an average of ±9%. The sensitivity of the MS fluctuated daily and these variations in peak areas were expected. However, most importantly, the internal standard method of quantification compensated for these changes when the amounts were calculated by the computer based on the individual RF.

Figure 4 demonstrates the linearity of the computer calculated RF(R) for toluene and 1-chlorohexane with regard to concentrations in the 20- to 1000-ng range. The quantification library RF for toluene and 1-chlorohexane are based on 52 ng each vs 52 ng of 1-chlorododecane, and they remained constant. The computer calculated RF(R) varied as the concentration of the samples varied according to the quantification formulas discussed in the Quantification section. The computer calculated RF(R) should be propor-

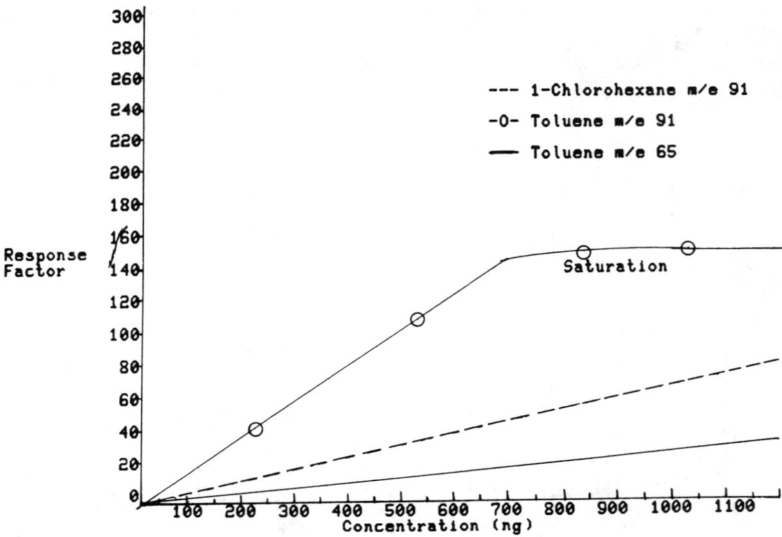


**Figure 4.** Toluene and 1-chlorohexane single ion computer calculated RF(R) versus con-concentration.

tional to the RF(L) in a concentration range as long as linearity of detector response is maintained. Even though toluene and 1-chlorohexane were both quantified on m/e 91, the figure shows that the RF(R) for toluene reached a plateau after 700 ng due to detector saturation, whereas the RF(R) for 1-chlorohexane maintained linearity. The ion abundance in the mass spectrum of toluene is mostly concentrated in two masses (m/e 91 and 92); but the ion abundance in the mass spectrum of 1-chlorohexane is distributed to m/e 91 and to other ions of fairly high intensities. Therefore, the ion contribution of m/e 91 of 1-chlorohexane is considerably less than that of toluene for the same concentrations of compounds. It can be seen from Figure 4 that the RF(R) determined for m/e 65 (a mass of lesser intensity) in toluene does maintain its linearlity in the 20- to 1000-ng range.

Table II shows the effect of ion source pressure variations on the GC/MS RF for the five Grob IS. It is evident from the data that such fluctuations, resulting in ion source pressure variations, could cause considerable errors in quantitative results.

The key elements for using computerized automatic quantification procedures are accurate library spectra acquired on the user's instrument for sample comparison, and reproducible RRT and RF as were shown in Table I.

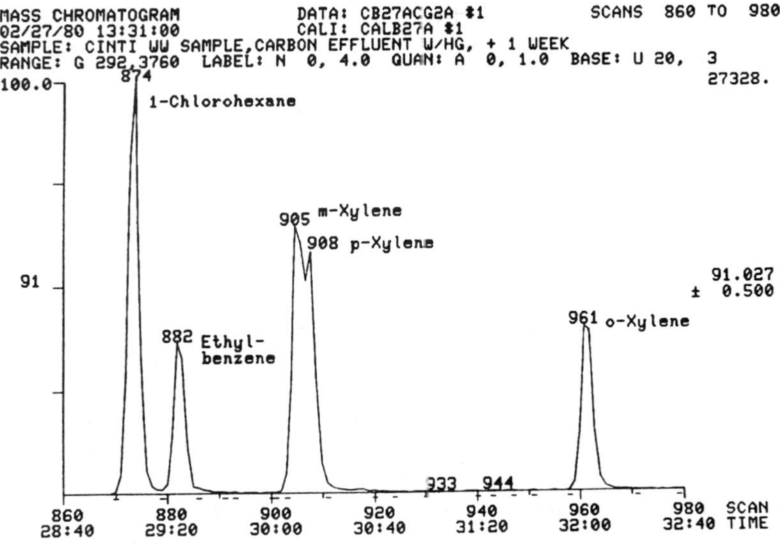


Figure 5. Mass chromatogram (m/e 91) of $C_2$ benzene isomers.

Figure 5 shows a cluster of $C_2$ benzene isomers within 100 scans of each other (200 sec). Since the mass spectra of the four isomers were almost identical, it was necessary to use very narrow search windows (±10 scans) relative to the scan selected on RRT by the computer search program to assign accurate identifications and quantities to each isomer. Such narrow search windows can be used only when RRT are reproducible within 2%. Water samples analyzed by CLSA at HERL have shown the presence of numerous substituted benzene isomers [2]. Their quantification was made possible by the high resolving power of glass capillary GC and automatic quantification procedures using narrow search windows.

Table III shows the recovery efficiency by CLSA for each compound and its standard deviation. The data indicate that all compounds do not have the same recoverability. The primary reason is because of differences in water solubility in water. *bis*-(2-Chloroethyl)ether was added to the water samples in this study because of its known toxicity and because of its occurrence in a few major U.S. water supplies. Detection of this compound in drinking water at low levels by CLSA means a much higher level of human exposure. The recoveries of pentachlorobiphenyl and hexachlorobiphenyl were admirable. Although we have not detected these compounds in drinking water using CLSA, their recoveries give some idea of the range (mol wt 358) and types of compounds measurable by CLSA.

## CONCLUSION

Statistical computations on RRT, GC peak areas, RF, amounts and recovery efficiencies by CLSA, have shown that glass capillary GC/MS analyses followed by computerized automatic quantification procedures employing MS SIQ based on the internal standard, can be used to quantify ng/L (ppt)-level results of CLSA within ±15% accuracy.

It was also demonstrated that reproducbile RRT and RF were mainly attributed to stable and reproducible glass capillary GC/MS/DS operating conditions, constant linear velocity through the GC column and constant ion source pressure. The data show that RF(L) should not only be reproducible, but RF(R) should be linear in the concentration range being analyzed.

## ACKNOWLEDGMENT

The authors thank Ms. Melda Hatfield for typing this manuscript.

## REFERENCES

1. Coleman, W. E., R. G. Melton, R. W. Slater, F. C. Kopfler, S. J. Voto, W. K. Allen and T. A. Aurand. "Determination of Organic Contaminants by the Grob Closed-Loop-Stripping Technique," *J. Am. Water Works Assoc.* 73(2):119 (1981).
2. Melton, R. G., W. E. Coleman, R. W. Slater, F. C. Kopfler, W. K. Allen, T. A. Aurand, D. E. Mitchell and S. J. Voto. "Comparison of Grob Closed-Loop Stripping Analysis with Other Trace Organic Methods," Chapter 36, this volume.
3. Grob, K., and F. Zurcher. "Stripping of Trace Organic Substances from Water. Equipment and Procedure," *J. Chromatog.* 117:285 (1976).
4. Grob, K., and K. Grob, Jr. "Splitless Injection and the Solvent Effect," *J. High Resolution Chromatog. Column Chromatog.* 1(1):57 (1978).
5. Grob, K., and K. Grob, Jr. "Isothermal Analysis on Capillary Columns Without Stream Splitting. The Role of the Solvent," *J. Chromatog.* 94:53 (1974).
6. Grob, K., and G. Grob. "Practical Capillary Gas Chromatography–Systematic Approach," *J. High Resolution Chromatog. Column Chromatog.* 2(3):109 (1979).

## CHAPTER 38

# DEVELOPMENT OF A CLOSED-LOOP STRIPPING TECHNIQUE FOR THE ANALYSIS OF TASTE- AND ODOR-CAUSING SUBSTANCES IN DRINKING WATER

**Stuart W. Krasner, Cordelia J. Hwang and Michael J. McGuire**

Water Quality Laboratory
The Metropolitan Water District
of Southern California
LaVerne, California

The majority of public complaints received by water utilities is normally associated with objectionable tastes and odors in drinking water. These problems are often difficult to treat and prevent. The causes are not always well understood, and the identification and quantification of odor-causing compounds at their threshold odor concentrations (low ng/L) has not been possible until now.

Some taste and odor occurrences have been traced to industrial effluent sources [1], or disinfection by-products such as chlorinated phenols [2]. However, natural causes due to microbiological sources are widespread and have been responsible for taste and odor incidents in The Metropolitan Water District of Southern California (Metropolitan) source-water reservoirs. Adams in his 1929 study of the Nile River, was the first to suggest that the problem of earthy odor and taste in drinking water was due to volatile products produced by microorganisms [3]. Silvey et al., starting in 1938, found actinomycetes to be responsible for the earthy, muddy, musty, woody and "potato-bin" odors common in the rivers, streams and reservoirs of the Southwest [4].

There have been several compounds identified with earthy, musty odors in water and soil. Gerber and Lechevalier isolated an earthy-smelling substance, geosmin, from a number of actinomycetes cultures [5]. The structure (Figure 1) has been elucidated by several complementary spectroscopic methods [6]. It has also been shown to be a metabolite of various blue-green algae [7,8]. Some actinomycetes produce a second highly odorous substance. This earthy-, musty- and camphorous-smelling compound has been identified as 2-methylisoborneol (MIB) [9,10], as shown in Figure 1. The literature indicates a worldwide distribution of geosmin and MIB. Both compounds have been detected throughout the United States and Europe, including lakes in Ohio [9] and surface waters of the Netherlands [11] and Japan [12].

Both geosmin and MIB are saturated tertiary alcohols and are thus resistant to oxidation. Typical water treatment processes using chlorine, ozone, chlorine dioxide or potassium permanganate are normally ineffective in destroying these odorants [13]. The steric configuration of the hydroxyl and methyl groups in these two compounds is believed to interact with receptors in the nose, imparting an earthy odor [14].

A third odorous metabolite of certain species of actinomycetes is 2-isopropyl-3-methoxy pyrazine [15] (see Figure 1). This musty, potato-bin-smelling compound is produced by certain microorganisms common in soil, and it could be a possible odor pollutant of rivers and lakes [16]. A similar smelling compound is 2-isobutyl-3-methoxy pyrazine, which was first isolated from bell peppers [17].

A fifth compound, 2,3,6-trichloroanisole (TCA) was included for consideration for several reasons. It has a "musty" odor and could thus be confused with these other compounds. Chlorinated anisoles have been detected in various California water systems [18,19], and they could be formed during a chlorination process.

Each of the five compounds discussed have strong odor characteristics, with threshold odor concentrations in water at low ng/L levels. Threshold odor concentrations reported in the literature vary considerably from one laboratory to another. Threshold values from 10 to 200 ng/L have been reported for geosmin in water [20,21]. Examination of the literature for the lowest reported threshold odor concentrations was made by the authors of this paper in order to set lower boundary goals for the analytical detection of these compounds. Table I lists these values plus characterizations of the odor quality of each compound.

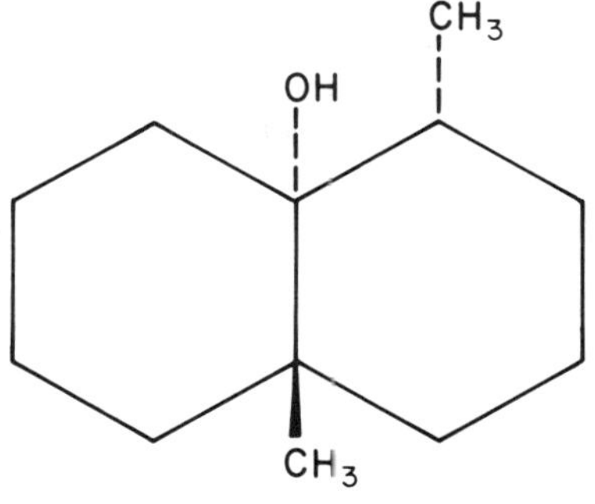


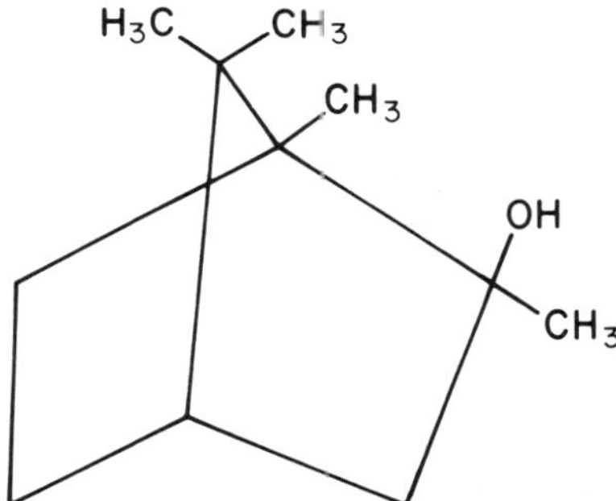


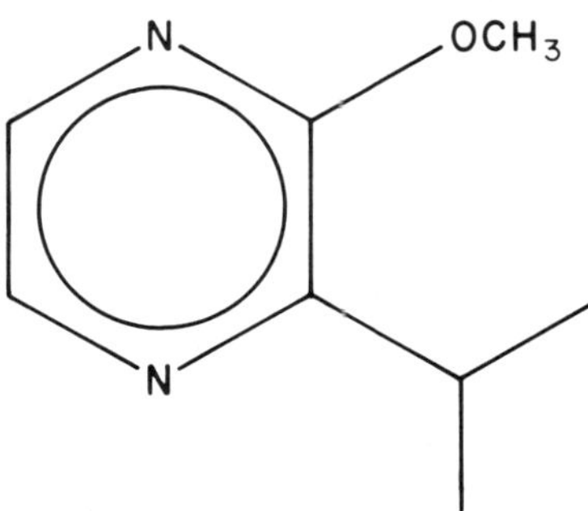


**Figure 1.** Molecular structures of microbiological metabolites.

Table I. Characteristics of Odor-Causing Compounds

| Compound | Lowest Reported Threshold Odor Concentration (ng/L) | Reference | Odor Characteristics |
|---|---|---|---|
| Geosmin | 10 | 21 | Earthy, musty |
| 2-Methylisoborneol | 29 | 22 | Earthy, musty, camphorous |
| 2-Isopropyl-3-methoxy Pyrazine | 2 | 23 | Earthy, musty, potato-bin |
| 2-Isobutyl-3-methoxy Pyrazine | 2 | 17 | Earthy, musty, bell pepper |
| 2,3,6-Trichloroanisole | 7 | 24 | Musty |

The analytical methods available for detection and quantification of these odorous compounds in water samples include sensory evaluation using trained human noses and instrumental techniques involving the gas chromatograph (GC). Odor panels have traditionally been used to determine the threshold odor number (TON), which is the reciprocal of the dilution factor needed to produce a just-detectable odor [25]. Persson has correlated sensory response with the logarithm of the concentration of geosmin and MIB [20]. While the human nose has been the most sensitive detector available, it is subjective, unable to discriminate specific odors in the presence of stronger ones, and suffers from fatigue after short exposure to geosmin or MIB.

The suitability of the GC with flame ionization detector (FID) or mass spectrometer (MS) as an instrument for analyzing odor compounds has long been accepted. The limitation has been the availability of satisfactory concentration procedures to provide the necessary sensitivities. Snoeyink and coworkers, in 1977, reported on the extraction of 1-L water samples with 35 mL of methylene chloride [26,27]. The detection limit of 100 ng/L by this method is well above the threshold odor concentrations of 29 ng/L for MIB and 10 ng/L for geosmin. The carbon adsorption method used by Rosen et al. in 1970 led to the recovery of geosmin and MIB from natural waters in Ohio [9]. A high concentration factor is achieved by this technique, but it is encumbered by the large sample size (3000 L), nonquantitative recoveries and the lengthy processing time of a week or more.

Grob developed a closed-loop stripping analysis (CLSA) technique in 1973, which combines a high concentration factor with a small sample size (generally one liter) and fast processing time (usually two hours) [28,29]. He has applied this to the analysis of alkanes up to eicosane and other semivolatile and intermediate-weight compounds in raw and finished drinking waters. Reinhard has subsequently applied it to the analysis for various priority pollutants in advanced wastewater treatment effluents [30]. This chapter presents the development and evaluation of the CLSA-GC/MS method of analysis for taste- and odor-causing organic substances in drinking water at detection levels at or below their respective threshold odor concentrations.

## EXPERIMENTAL METHODS

### Apparatus

The Grob CLSA apparatus and procedure were used with the modifications suggested by Reinhard [30,31] (Figure 2). The metal coils connected to the pump were replaced with flexible metal tubing, and the Sovirel-Rotulex joints connected to the stripping bottle were replaced with stainless steel quick-connects. Tall-form gas washing bottles modified with coarse cylindrical frits and fused glass-metal flexible tubing were used as stripping bottles for

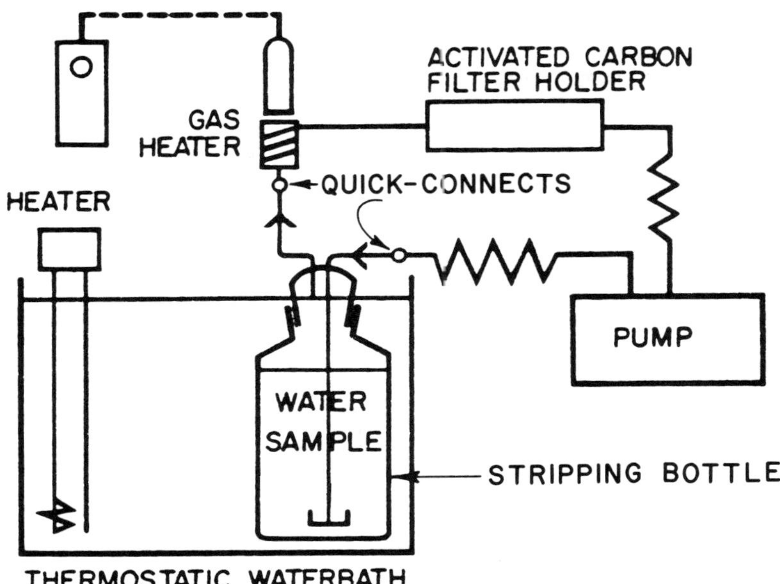


**Figure 2.** Schematic of closed-loop stripping apparatus.

125- and 500-mL samples. One-liter samples were analyzed in similarly fabricated bottles with effective height-to-diameter ratios of 1.7 and 3.3, the latter shown in Figure 3.

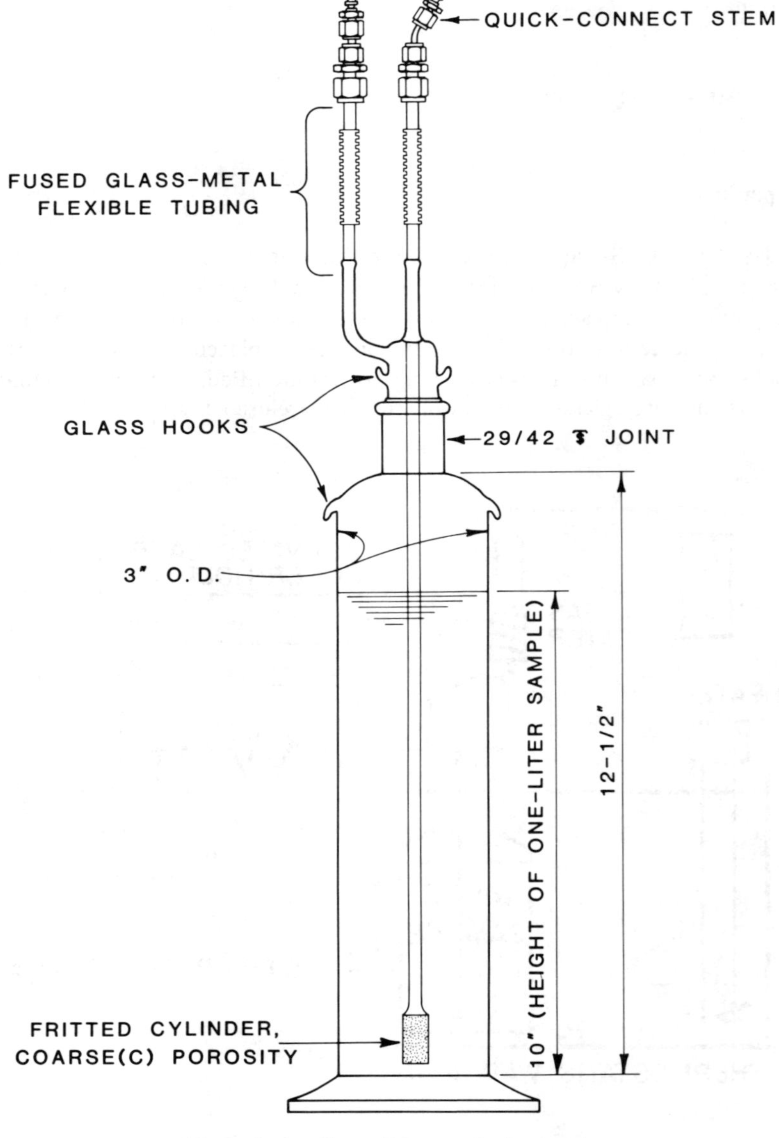


**Figure 3.** One-liter tall-form stripping bottle.

## Procedure

The internal standards 1-chlorooctane (Cl-$C_8$) and 1-chlorododecane (Cl-$C_{12}$) were added to the sample in the stripping bottle to give a concentration of 200 ng/L each. Samples were stripped for two or three hours by using the pump to recirculate the headspace air in the stripping bottle. The stripped organics were adsorbed on a 1.5-mg activated carbon (AC) filter. The water bath temperature was thermostatically maintained at 25°C, and the gas heater was kept at a temperature of 50°C.

After stripping, the AC filter was extracted with carbon disulfide ($CS_2$), as shown in Figure 4. A 2-$\mu$L aliquot of $CS_2$ was added to a vial, which was then connected to the filter by a piece of Teflon® (PTFE) tubing (5-mm i.d. flexible Teflon tubing, not the shrinkable Teflon used by Grob). A 10-$\mu$L aliquot of $CS_2$ was placed above the carbon, and the vial was warmed in the analyst's hand. The $CS_2$ was then alternately pulled and pushed through the carbon ten times by separating and reconnecting the filter and vial while still within the Teflon sleeve. With the filter and vial tightly butted, the vial was cooled with ice, drawing the $CS_2$ below the carbon and was shaken to complete the transfer. The extraction was continued with another 10 $\mu$L of $CS_2$, followed by a 5-$\mu$L aliquot, with approximately 20 $\mu$L total extract recovered.

Analyses were performed on a Finnigan Model 4023 GC/MS retrofitted with a Grob-designed split-splitless injector. Decafluorotriphenylphosphine (DFTPP) was analyzed to ensure that the MS was tuned to meet U.S. Environmental Protection Agency (EPA) specifications [32]. CLSA extracts were analyzed using the GC/MS parameters given in Table II. Identification and quantification of the five odor-causing compounds were incorporated into an automatic reverse-search computer program [33]; however, spot operator verification was performed frequently, especially on new samples with complex matrices.

## Materials

Geosmin (*trans*-1,10-dimethyl-*trans*-9-decalol) and 2-methylisoborneol were obtained from the EPA Water Supply Research Division, Cincinnati, OH. 2-Isopropyl-3-methoxy pyrazine, 2-isobutyl-3-methoxy pyrazine, 2-isobutyl-3-methoxy pyrazine and 2,3,6-trichloroanisole were purchased from K&K Labs, Division of ICN Pharmaceuticals, Inc. 1-Chlorooctane and 1-chlorododecane were purchased from Eastman Organic Chemicals. Carbon disulfide, Aldrich Gold Label, was used as received after GC/MS verification of purity.

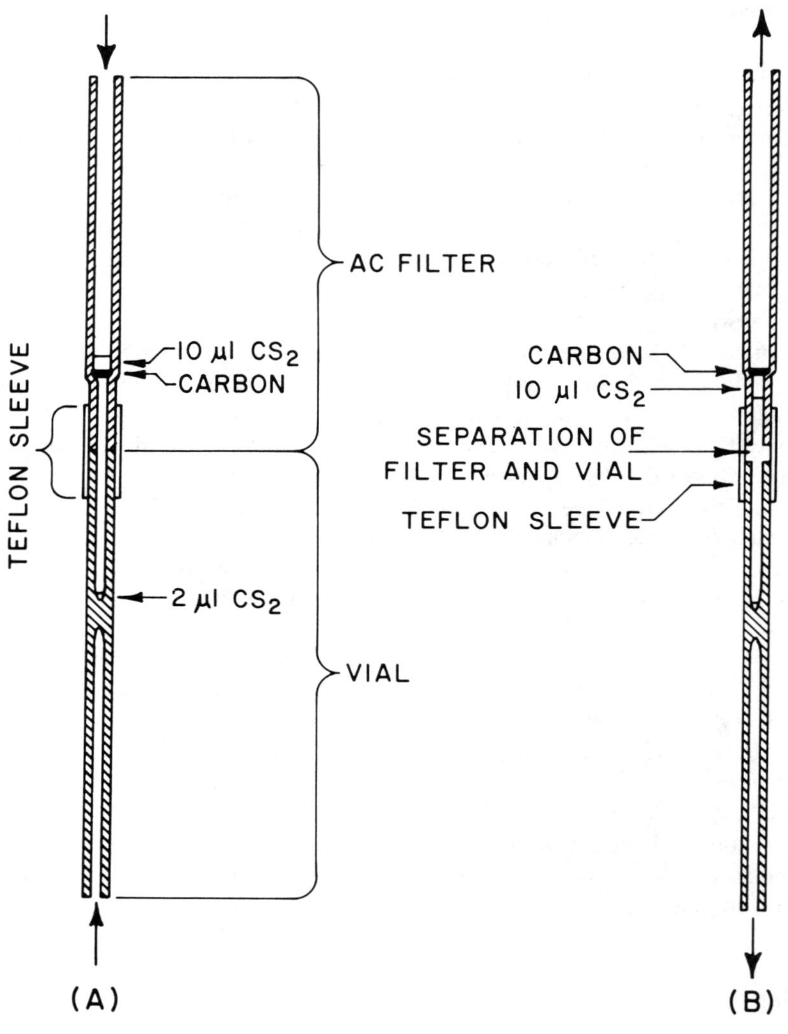


**Figure 4.** Extraction of filter.

Table II. GC/MS Parameters of CLSA Extracts

| | |
|---|---|
| Column | 30-m x 0.25-mm i.d. SE-52 fused-silica capillary column[a] |
| Column Temperature Program | Room temp, 8 min; 100–225°C at 7°C/min; 255°C 10 min |
| Carrier Gas | Helium |
| Linear Velocity on the Column | 29 cm/sec |
| Sample Size | Approx. 1.5 $\mu$L (splitless injection) |
| Injector Temperature | 200°C |
| Transfer Line Temperature | 200°C |
| Ionizer Temperature | 270°C |
| Source Pressure | Approx. 5 x $10^{-7}$ torr |
| Electron Energy | 70 eV |
| Mass Range Scanned | 41–75 amu, 77–400 amu[b] |
| Scan Time | 1.3 sec |

[a]J&W Scientific, Inc.

[b]Omitted 76 amu, the base peak for $CS_2$.

## DEVELOPMENT OF TECHNIQUE

### Optimization of CLSA Parameters

The ground-glass joint of the stripping bottle must be immersed below the surface in the water bath. Due to the shallow depth of the available water bath (12 inches) used during the initial phase of this study, the height of the 1-L bottle was limited, resulting in a "short-form" bottle. The effective height-to-diameter ratio for this 1-L bottle was only 1.7.

The stripping efficiency of the 1-L short-form bottle was optimized through a series of experiments where volume, stripping time and temperature were varied, as outlined in Figure 5. Initially, 500 mL of "organic-free" water (Millipore Corp. Super-Q water prestripped in CLSA apparatus) spiked with 200 ng each of the internal standards Cl-$C_8$ and Cl-$C_{12}$ was stripped in the 500-mL bottle for two hours, with a water bath temperature of 25°C. The GC/MS response ratio of Cl-$C_{12}$ to Cl-$C_8$ (using the 91-amu fragment peak areas) was 0.97, essentially the same response for both compounds. Then 1000 mL of water spiked with 200 ng of each of these two compounds was stripped in the 1-L bottle using the same conditions. The GC/MS response for Cl-$C_8$ was approximately the same as before, but the response for

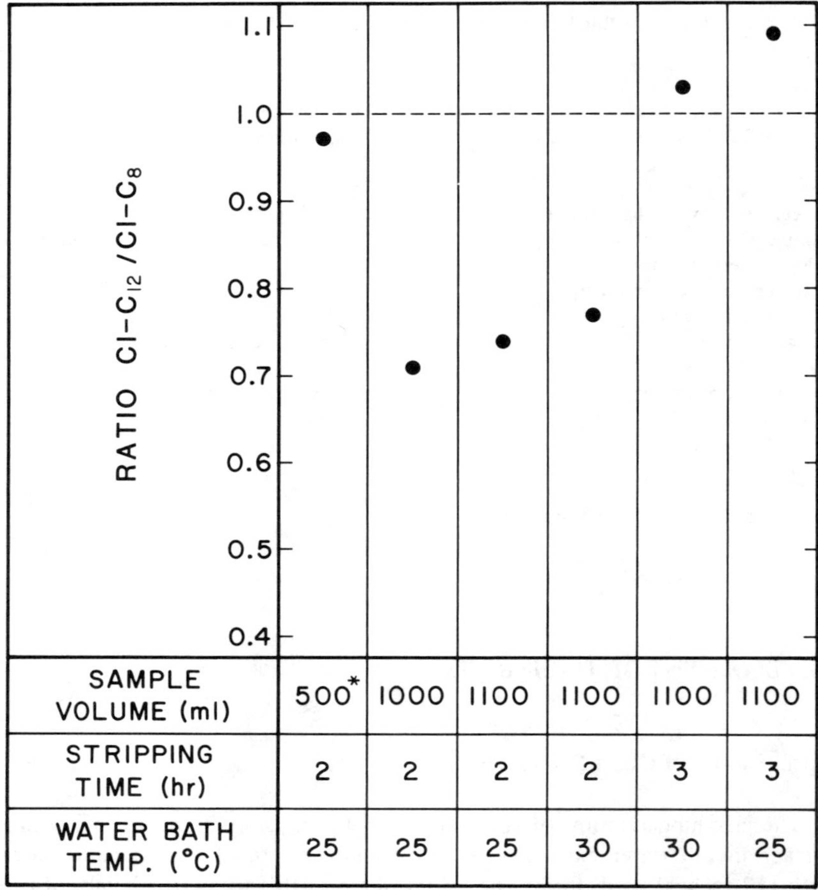


**Figure 5.** Optimization of CLSA parameters for one-liter short-form bottle. Comparison of GC/MS response ratios of the internal standards Cl-$C_8$ and Cl-$C_{12}$ (200 ng each).

Cl-$C_{12}$ was much lower, resulting in a ratio of only 0.71. The amount of headspace appeared to be too large, so the sample volume was increased to 1100 mL. This, however, did not result in any significant increase in the response ratio. The water bath temperature was increased to 30°C; there was still no significant improvement in the recovery of Cl-$C_{12}$. On increasing the stripping time to three hours, the response ratio increased to 1.03. One further experiment with a 1100-mL sample stripped for three hours was done in a water bath set at 25°C. Since the ratio was 1.09, it was decided to use these conditions for CLSA in the 1-L short-form bottle.

In another experiment, 1100 mL of water was spiked with 220 ng each of Cl-$C_8$ and Cl-$C_{12}$, plus 110 ng each of the five taste- and odor-causing compounds under analysis. This sample was stripped under standard conditions for three hours, after which the AC filter was removed. A fresh filter was placed in the system, and the sample was stripped further to see how much residual material remained. As indicated in Table III, the percent residuals for Cl-$C_8$, Cl-$C_{12}$ and TCA were relatively low (less than 10%), but higher than desired for the other four taste- and odor-causing compounds (32–38%).

Since the initial phase of methods development and sample analysis using the one-liter short-form bottle, a larger water bath (17.5 in. deep) (Corning No. 6945-13L cylindrical chromatography jar) was obtained. A new 1-L bottle was custom made, as shown in Figure 3, with an effective height-to-diameter ratio of 3.3, approximately twice the ratio for the short-form bottle. One liter of water was spiked with 200 ng each of Cl-$C_8$ and Cl-$C_{12}$, plus 20 ng each of the five taste- and odor-causing compounds. This sample was stripped in the 1-L "tall-form" bottle for three hours with a water bath temperature of 25°C. A second spiked water sample was similarly prepared and analyzed, except stripping was for only two hours. Since the GC/MS responses for all seven compounds were essentially the same for both runs, two-hour stripping was deemed sufficient for the tall-form bottle.

The level of residual material remaining after a two-hour stripping in the tall-form bottle was measured. As shown in Table III, the percent residuals for Cl-$C_8$, Cl-$C_{12}$, and TCA were still relatively low (15% or less), yet were still high for the other four compounds (34–40%).

Based on data developed during this study, the stripping efficiencies for these compounds are reproducible. Samples and calibration standards are run

**Table III. Percent Residuals after Initial Sampling**

| Compound | One-Liter Short-Form Bottle | One-Liter Tall-Form Bottle |
|---|---|---|
| 1-Chlorooctane | 4.5 | <1 |
| 2-Isopropyl-3-methoxy Pyrazine | 33 | 40 |
| 2-Isobutyl-3-methoxy Pyrazine | 32 | 39 |
| 2-Methylisoborneol | 38 | 40 |
| 1-Chlorododecane | 7.0 | 7.0 |
| 2,3,6-Trichloroanisole | 7.5 | 15 |
| Geosmin | 33 | 34 |

utilizing the same conditions, as given in Table IV, so that precise and accurate results can be obtained. Environmental samples containing 3-20 ng/L of geosmin were run in replicate. As indicated in Table V, the differences in replicates were only 1-3 ng/L. Table VI lists the results of a typical spiked sample. A treated water sample containing 8 ng/L MIB and 3 ng/L geosmin was spiked with all five taste- and odor-causing compounds at a 5-ng/L level each. Recoveries of the spiked material ranged from 89 to 119%.

Another estimate of precision was developed over a one-month period based on the analysis of 11 samples from the same reservoir. The variability of the geosmin results was very small and thought to be a valid estimate of the CLSA method precision. The mean of the 11 analyses was 4.4 ng/L geosmin with a standard deviation of 0.8 ng/L, resulting in a coefficient of variation of 18%.

**Table IV. Standard CLSA Parameters for Water Samples**

| Parameter | One-Liter Short-Form Bottle | One-Liter Tall-Form Bottle |
|---|---|---|
| Sample Volume (mL) | 1100 | 1000 |
| Stripping Time (hr) | 3 | 2 |
| Water Bath Temperature (°C) | 25 | 25 |

**Table V. Estimates of Precision**

| Sample | Geosmin Replicates (ng/L) | Differences (ng/L) |
|---|---|---|
| Raw Water | 17; 20 | 3 |
| Treated Water | 12; 15 | 3 |
| Raw Water | 4; 3 | 1 |
| Raw Water | 9; 6 | 3 |

**Table VI. Estimates of Accuracy**

| Compound | Percent Recovery of 5 ng/L Spike |
|---|---|
| 2-Isopropyl-3-methoxy Pyrazine | 89 |
| 2-Isobutyl-3-methoxy Pyrazine | 102 |
| 2-Methylisoborneol | 100 |
| 2,3,6-Trichloroanisole | 115 |
| Geosmin | 119 |

## CLSA of Nonaqueous Samples

After development of a CLSA technique for water samples, attention was directed toward the CLSA of sediments and laboratory-grown microbiological cultures to investigate the sources of the taste- and odor-causing compounds. Minimal manipulation of the sample was desired. Reservoir sediment samples were collected by scuba divers. A 125-mL aliquot of sediment and interstitial water was diluted to 500 mL with organic-free water and stripped according to conditions given in Table VII. AC filter extraction, GC/MS analysis and quantification paralleled the treatment of water samples.

Pure cultures of actinomycetes and unialgal cultures of blue-green algae were analyzed qualitatively using the parameters in Table VII. Liquid cultures were stripped directly. Where less than 125 mL of liquid culture was available, the volume was brought up to 125 mL with organic-free water. An actinomycete agar culture grown on the bottom of a flask was flooded with sterile phosphate buffer, and the resulting suspension was diluted to volume and stripped as noted before. It should be stressed that these analyses were only qualitative. A method for quantification of the geosmin and MIB produced in liquid culture has been presented elsewhere [34].

## Identification and Quantification

Automated search, identification and quantification were performed using parameters located in computer libraries developed on Metropolitan's GC/MS. The library entry for each compound consisted of relative retention time, quantification mass, response factor and reference spectrum. The first two parameters are listed in Table VIII. The two internal standards (IS) were used for referencing both retention times and responses for the five odor-causing compounds. The response factor for a compound (Z) was calculated from the CLSA of a calibration standard (cs) as follows:

$$\text{response factor (Z)} = \frac{\text{area compound (Z)}_{cs}}{\text{amount compound (Z)}_{cs}} \times \frac{\text{amount IS}_{cs}}{\text{area IS}_{cs}}$$

**Table VII. CLSA Parameters for Nonaqueous Samples**

| Parameter | Sediment Samples | Liquid Cultures |
|---|---|---|
| Stripping Bottle Size (mL) | 500 | 125 |
| Sample Volume (mL) | 125 | 125[a] |
| Organic-Free Water Dilution Volume (mL) | 375 | 0 |
| Stripping Time (hr) | 2 | 2 |
| Water Bath Temperature (°C) | 25 | 25 |

[a]Where <125 mL liquid culture available, volume brought up to 125 mL with organic-free water.

The retention times and response factors were updated frequently from the CLSA of water spiked with standards.

The reference spectrum for each compound was composed of 10-14 key masses (m/e) and their intensities. These were selected after the examination of chromatograms and spectra from standards and spiked natural waters to assure unambiguous identifications. One of these masses, usually the most intense one, was used as the quantification mass. Maximum sensitivity without interference was desired. For MIB the 107-amu mass was used for quantification rather than the base peak of 95 amu, since 95 amu was also frequently present in the spectrum of a closely eluting contaminant.

It should be noted that two different mass spectra have been reported for MIB in the literature. Medsker et al. showed 95 amu as the base peak (100%) with 107 amu at 21% [10], while Rosen et al. reported 107 amu as the base peak and 95 amu at 15% [9]. Over the course of a year's analyses for this compound at Metropolitan, spectral variations have been observed which appear to be dependent on the condition of the MS source and rods. In addition, Metropolitan rotated between the use of two different design versions of the ionizer source. The MIB spectrum was therefore monitored

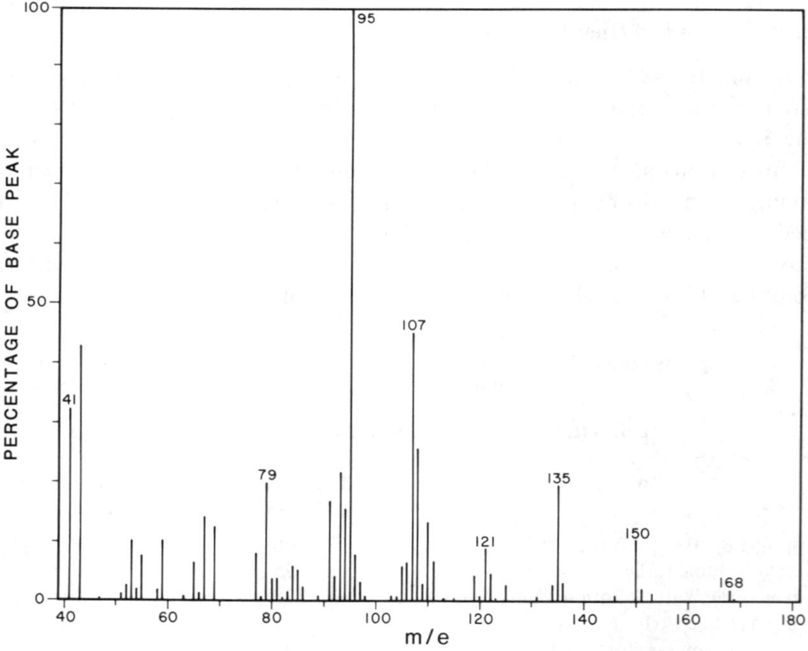


**Figure 6.** Mass spectrum of 2-methyliosborneol.

frequently and the reference library was modified accordingly. The most stable and common spectrum, which was achieved with a thoroughly cleaned, baked and pumped-out system, is given in Figure 6, with 95 amu as the base peak and 107 amu at 45%. Mass spectra for the other four compounds and the two internal standards did not vary significantly.

Unknown (u) samples were quantified as follows:

$$\text{amount compound } (Z)_u = \frac{\text{area compound } (Z)_u}{\text{response factor}} \times \frac{\text{amount IS}_u}{\text{area IS}_u}$$

The detection limits for these five compounds ranged from 2 to 5 ng/L, as shown in Table VIII. These values are at or below reported threshold odor concentrations. Figure 7 shows a typical calibration curve for CLSA. Concentrations of geosmin from 2 to 120 ng/L were analyzed and found to follow a linear response.

## APPLICATIONS

### Drinking Water Supplies

The CLSA technique was developed by Metropolitan to better deal with taste and odor problems in its own system. However, this technique was first used to assist two other utilities: the Los Angeles Department of Water and

**Table VIII. CLSA-GC/MS Analytical Parameters**

| No. | Compound | Relative Retention Time | Ref.[a] | Quant. Mass (amu) | Detect. Limit (ng/L) | Threshold Odor Conc (ng/L) |
|---|---|---|---|---|---|---|
| 1 | 1-Chlorooctane (IS) | 1.000[b] | 1 | 91 | | |
| 2 | 2-Isopropyl-3-methoxy-pyrazine | 1.034 | 1 | 137 | 2 | 2 |
| 3 | 2-Isobutyl-3-methoxy-pyrazine | 1.131 | 1 | 124 | 2 | 2 |
| 4 | 2-Methylisoborneol | 1.147 | 1 | 107 | 2 | 29 |
| 5 | 1-Chlorododecane (IS) | 1.000[c] | 5 | 91 | | |
| 6 | 2,3,6-Trichloroanisole | 0.925 | 5 | 210 | 5 | 7 |
| 7 | Geosmin | 0.959 | 5 | 112 | 2 | 10 |

[a]Reference to compound number of IS used for relative retention time and quantification calculations.

[b]Absolute retention time equals 13 min.

[c]Absolute retention time equals 19 min.

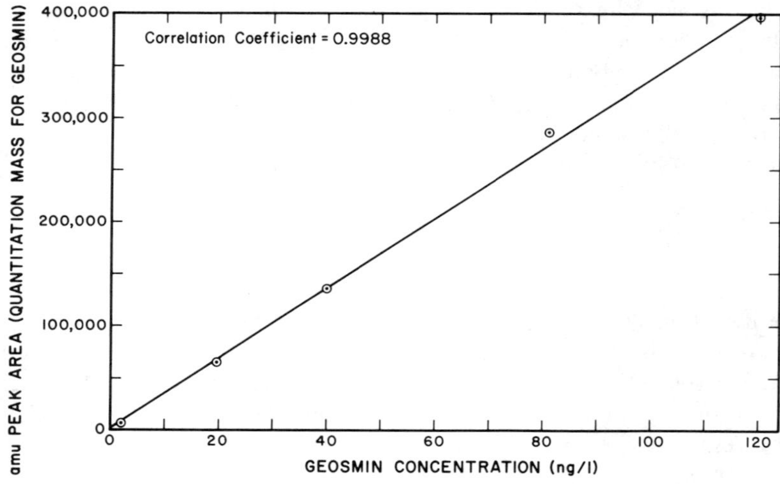


**Figure 7.** Calibration curve for geosmin, using CLSA.

**Table IX. Los Angeles Department of Water & Power Results, June 1980**

| Sample | TON | Odor Characteristics | Geosmin (ng/L) |
|---|---|---|---|
| Reservoir 1 | 8 | Musty, aromatic | <2 |
| Reservoir 2 | 24 | Aromatic, musty (geosmin-like) | 17; 20 |
| Distribution 2A | ⩾ 8 | Aromatic, musty (geosmin-like) | 9 |
| Distribution 2B | ⩾ 8 | Medicinal, aromatic, musty | 12; 15 |

Power (LADWP) in Southern California and Stockton–East Water District in Central California. The two utilities have completely different source waters.

In June 1980 LADWP received numerous complaints of earthy, musty odors in their drinking water. Samples from Reservoirs 1 and 2 and two distribution system locations serviced by Reservoir 2 were analyzed. The results are given in Table IX. All samples had TON of eight or greater. An exact TON could not be determined for the distribution samples because of limited sample volume. The character of the odors, however, was different. An aromatic, musty quality, which has now become associated with geosmin, predominated in Reservoir 2 and Distribution 2A samples, while Reservoir 1 had a weaker, slightly musty odor, and Distribution 2B had a predominantly medicinal odor. Geosmin, at or above the threshold odor concentration, was found by CLSA in Reservoir 2 and both of its distribution system samples and was undoubtedly the cause of the taste and odor problem. Based on this and other data, water was taken from Reservoir 1 instead of directly from Reservoir 2. Consumer complaints decreased significantly after this modification of their water supply delivery system.

The Stockton-East Water District's odor problem was not as well defined as that of LADWP. The initial samples in July 1980 had odors which were not predominantly earthy or musty, but rather grassy, septic and fishy. The results of sensory evaluations and CLSA are given in Table X. CR-1 reservoir effluent received TON of 70, 50 and 8 from successive panel members. These rapidly decreasing values were probably due to the extreme volatility of some odorant, not one of the five targeted compounds. Small amounts of geosmin, less than the threshold odor concentration, were found in the other two samples. A more extensive investigation was made with new samples collected in September 1980. The results of CLSA are given in Table XI. The odor problem appeared to have developed in the main stem of the Calaveras River (CR) and might have been aggravated by downstream impoundments and raw water reservoirs at the plant site. Based on these data, treatment process changes were implemented by the utility including the addition of powdered activated carbon.

Sediments from one of Metropolitan's reservoirs were analyzed by CLSA. Early summer 1980 results showed high levels of geosmin in lake sediment samples, but not in the overlying or surface waters. A couple of complaints were received in September 1980 regarding a musty odor in the water.

**Table X. Stockton-East Water District Results for July 1980 Samples**

| Sample | TON[a] | Odor Characteristics | Geosmin (ng/L) |
|---|---|---|---|
| CR-1: Reservoir Effluent | 70/50/8 | Grassy, swampy, musty | $<2$ |
| CR-5: River Intake Structure Downstream of CR-1 | 12/8 | Septic, musty, grassy | 4; 3 |
| Raw Water Reservoir at Plant | 8/12 | Fishy, aromatic | 9; 6 |

[a]Results of different panel members.

Table XI. Stockton-East Water District Results for September 1980 Samples

| Sample | Geosmin (ng/L) |
|---|---|
| CR-1: reservoir effluent | < 2 |
| CR-4: river sample downstream of CR-1 | 8 |
| CR-4A: river sample downstream of CR-4 | 7 |
| CR-5: river intake structure downstream of CR-4A | 14 |
| Plant effluent | 30 |

Table XII. Metropolitan's Results, September 1980

| | Geosmin (ng/L) | MIB (ng/L) |
|---|---|---|
| Source Waters | | |
| Colorado River water (CRW) | | |
| from aqueduct | <2 | <2 |
| after source water reservoir (CRW/R) | <2 | 16 |
| State Water Project water (SWP) | | |
| western branch (SWP/W) | <2 | <2 |
| eastern branch (SWP/E) | 6 | <2 |
| Water Treatment Plants | | |
| Plant 1 effluent[a] | 3 | 10 |
| Plant 2 effluent[a] | 3 | 8 |
| Plant 3 influent (100% SWP/W) | <2 | <2 |

[a]Plant influent = blend of 60% CRW/R + 40% SWP/E.

CLSA samples from Metropolitan source waters and water treatment plants were immediately performed. The results are listed in Table XII. Subsequent CLSA of Colorado River water leaving the source water reservoir indicated levels of MIB as high as 53 ng/L. Due to the low level of geosmin found in the eastern branch of the State Water Project system (6 ng/L) and the absence of geosmin and MIB in the western branch, deliveries of Colorado River water were substantially reduced and replaced by increased flows from both branches of the State Water Project system [35]. As a result of these changes, consumer complaints were greatly reduced.

## Cultures

The need to identify the microbiological and chemical causes of taste and odor in water has resulted in the coupling of the microbiologist's isolation and culturing techniques for microorganisms with the CLSA presented here. Actinomycetes and blue-green algae have been isolated from odorous source waters and sediments. Where the suspect microorganism cultures have demonstrated an odor potential, they have been analyzed by CLSA. A strain of *Oscillatoria curviceps* and one of *O. travis* have been identified as MIB producers. A species of actinomycete, *Anabaena* sp., and *Oscillatoria simplicissima* Gomont have been shown to produce geosmin. Figure 8 shows a chromatogram for the CLSA of *Oscillatoria simplicissima* Gomont in which geosmin is the most intense peak. In addition, because of the high concentration of compounds in the culture, many more peaks were visible than from the CLSA of natural water samples. Research is continuing in order to identify the other organic compounds detected on the chromatograms, which contribute to the overall odor of the samples or which may serve as fingerprints for particular microorganisms.

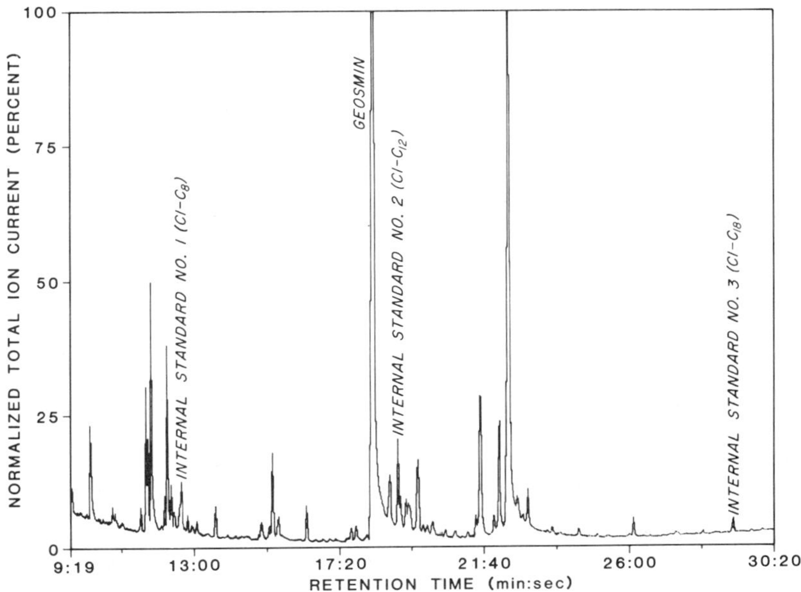


**Figure 8.** Chromatogram of CLSA: *Oscillatoria simplicissima* Gomont.

## CONCLUSION

A method has been developed for and applied to the determination of several taste- and odor-causing substances in water, sediments and cultures. Low detection limits have been coupled with manageable sample size and analysis time in the CLSA-GC/MS technique. This provides a valuable tool for extended surveys and regular monitoring of water supplies, identification of other taste- and odor-causing compounds, and evaluation of bench-scale and pilot-plant treatment processes at ng/L levels. The CLSA technique is now being used at Metropolitan in the understanding and management of taste and odor problems.

## ACKNOWLEDGMENTS

We appreciate the diligent work by Metropolitan's Microbiologist George Izaguirre in isolating and culturing the microorganisms used in this study. We are grateful to Martin Reinhard and Jim Graydon of Stanford University, Water Quality Control Research Laboratory, for assistance in helping us while we were first learning to use the closed-loop stripping technique.

## REFERENCES

1. Loy, E. W., Jr., D. W. Brown, J. H. M. Stephenson and J. A. Little. "Uses of Wastewater Discharge Compliance Monitoring Data," in *Identification and Analysis of Organic Pollutants in Water*, L. H. Keith, Ed. (Ann Arbor, MI: Ann Arbor Science Publishers, Inc., 1976), pp. 499-516.
2. Burttschell, R. H., A. A. Rosen, F. M. Middleton and M. B. Ettinger. "Chlorine Derivatives of Phenol Causing Taste and Odor," *J. Am. Water Works Assoc.* 51(2):205-214 (1959).
3. Adams, B. A. "Odors in Water of the Nile River," *Water Water Eng.* 31:309 (1929).
4. Silvey, J. K. G., J. C. Russel, D. R. Redden and W. C. McCormick. "Actinomycetes and Common Tastes and Odors," *J. Am. Water Works Assoc.* 42(11):1018-1026 (1950).
5. Gerber, N. N., and H. A. Lechevalier. "Geosmin, an Earthy-Smelling Substance Isolated from Actinomycetes," *Appl. Microbiol.* 13:935 (1965).
6. Gerber, N. N. "Geosmin, from Microorganisms is *trans*-1,10-Dimethyl-*trans*-9-decalol," *Tetrahedron Lett.* 25:2971 (1968).

7. Safferman, R. S., A. A. Rosen, C. I. Mashni and M. E. Morris. "Earthy-Smelling Substance from a Blue-Green Alga," *Environ. Sci. Technol.* 1:429 (1967).
8. Medsker, L. L., D. Jenkins and J. F. Thomas. "Odorous Compounds in Natural Waters. An Earthy-Smelling Compound Associated with Blue-Green Algae and Actinomycetes," *Environ. Sci. Technol.* 2(6):461-464 (1968).
9. Rosen, A. A., C. I. Mashni and R. S. Safferman. "Recent Development in the Chemistry of Odour in Water: The Cause of Earthy/Musty Odour," *Water Treat. Exam.* 19(2):106-119 (1970).
10. Medsker, L. L., D. Jenkins, J. F. Thomas and C. Koch. "Odorous Compounds in Natural Waters. 2-*exo*-Hydroxy-2-methylbornane, the Major Odorous Compound Produced by Several Actinomycetes," *Environ. Sci. Technol.* 3(5):476-477 (1969).
11. Piet, G. J., B. C. J. Zoeteman and A. J. A. Kraayeveld. "Earthy Smelling Substances in Surface Waters of the Netherlands," *Water Treat. Exam.* 21(4):281-286 (1972).
12. Tsuchiya, Y. "The Basic Research on the Causes of Odors in Water Studies on Odorous Materials in the Water of the Pond of Tairo at Miyakejima," *Tokyo Toritsu Eisei Kenkyusho Kenkyu Nempo* 25:445 (1974).
13. Gauntlett, R. B., and R. F. Packham. "The Removal of Organic Compounds in the Production of Potable Water," *Chem. Ind.* (London) (1973), p. 812.
14. Polak, E., D. Trotier and E. Baliguet. "Odor Similarities in Structurally Related Odorants," *Chemical Senses Flavour* 3(4):369-380 (1978).
15. Gerber, N. N. "Three Highly Odorous Metabolities from an Actinomycete. 2-Isopropyl-3-methoxypyrazine, Methylisoborneol, and Geosmin," *J. Chem. Ecol.* 3:475 (1977).
16. Gerber, N. N. "Volatile Substances from Actinomycetes: Their Role in the Odor Pollution of Water," *CRC Crit. Rev. Microbiol.* (November 1979), pp. 191-214.
17. Buttery, R. G., R. M. Seifert, D. G. Guadagni and L. C. Ling. "Characterization of Some Volatile Constituents of Bell Peppers," *J. Agric. Food Chem.* 17:1322 (1969).
18. Burlingame, A. L., B. J. Kimble, E. S. Scott, F. C. Walls, J. W. de Leeuw, B. W. de Lappe and R. W. Risebrough. "The Molecular Nature and Extreme Complexity of Trace Organic Constituents in Southern California Municipal Wastewater Effluents," in *Identification and Analysis of Organic Pollutants in Water*, L. H. Keith, Ed. (Ann Arbor, MI: Ann Arbor Science Publishers, Inc., 1976), pp. 557-585.
19. Reinhard, M., and P. L. McCarty. "Characterization of Organic Materials at Water Factory 21," in *Chemistry in Water Reuse, Vol. 2*, W. J. Cooper, Ed. (Ann Arbor, MI: Ann Arbor Science Publishers, Inc., 1981).
20. Persson, P.-E. "Sensory Properties and Analysis of Two Muddy Odour Compounds, Geosmin and 2-Methylisoborneol, in Water and Fish," *Water Res.* 14:1113-1118 (1980).

21. Cees, B., J. Zoeteman and G. J. Piet. "Cause and Identification of Taste and Odour Compounds in Water," *Sci. Total Environ.* 3:103-115 (1974).
22. Persson, P.-E. "Notes on Muddy Odour. III. Variability of Sensory Response to 2-Methylisoborneol," *Aqua Fennica* 9:48-52 (1979).
23. Seifert, R. M., R. G. Buttery, D. G. Guadagni, D. R. Black and J. G. Harris. "Synthesis of Some 2-Methoxy-3-alkylpyrazines with Strong Bell Pepper-Like Odors," *J. Agric. Food Chem.* 18:246 (1970).
24. Guadagni, D. G., and R. G. Buttery. "Odor Threshold of 2,3,6-Trichloroanisole in Water," *J. Food Sci.* 43:1346-1347 (1978).
25. *Standard Methods for the Examination of Water and Wastewater*, 14th ed. (New York: American Public Health Association, 1976).
26. Wood, N. F., and V. L. Snoeyink. "2-Methylisoborneol, Improved Synthesis and a Quantitative Gas Chromatographic Method for Trace Concentrations Producing Odor in Water," *J. Chromatog.* 132:405-420 (1977).
27. Herzing, D. R., V. L. Snoeyink and N. F. Wood. "Activated Carbon Adsorption of the Odorous Compounds 2-Methylisoborneol and Geosmin," *J. Am. Water Works Assoc.* 69(4):223-228 (1977).
28. Grob, K. "Organic Substances in Potable Water and in Its Precursor. Part I. Methods for Their Determination by Gas-Liquid Chromatography," *J. Chromatog.* 84:255-273 (1973).
29. Grob, K., and F. Zurcher. "Stripping of Trace Organic Substances from Water. Equipment and Procedure," *J. Chromatog.* 117:285-294 (1976).
30. Reinhard, M., J. E. Schreiner, T. Everhart and J. Graydon. "Specific Compound Analysis by Gas Chromatography and Mass Spectroscopy in Advanced Treated Waters," NATO/CCMS Conf. of Practical Application of Adsorption Techniques, Reston, VA, May 1, 1979.
31. Reinhard, M., and J. Graydon. Personal communication (1979).
32. Eichelberger, J. W., L. E. Harris and W. L. Budde. "Reference Compounds to Calibrate Ion Abundance Measurements in Gas Chromatography/Mass Spectrometry Systems," *Anal. Chem.* 47(7):995-1000 (1975).
33. Krasner, S. W., C. J. Hwang, R. S. Cohen and M. J. McGuire. "Development of a Volatile Organic Analysis Technique for the Orange–Los Angeles Reuse Study," in *Chemistry in Water Reuse, Vol. 2,* W. J. Cooper, Ed. (Ann Arbor, MI: Ann Arbor Science Publishers, Inc., 1981).
34. Izaguirra, G., C. J. Hwang, S. W. Krasner and M. J. McGuire. "Geosmin and 2-Methylisoborneol from Cyanobacteria in Three Water Supply Systems," paper presented at the 81st Annual Meeting of the American Society for Microbiology, Dallas, TX, March 3, 1981.
35. McGuire, M. J., S. W. Krasner, C. J. Hwang and G. Izaguirre. "Closed-Loop Stripping Analysis at the Parts-per-Trillion Level as a Tool for Solving Taste and Odor Problems," paper presented at the Eighth Annual AWWA Water Quality Technology Conference, Miami Beach, FL, December 9, 1980.

# SECTION 10

# SPECIALIZED PURGING TECHNIQUES

## CHAPTER 39

# A SCHEME FOR THE ROUTINE ANALYSIS OF PURGEABLE COMPOUNDS BY GAS CHROMATOGRAPHY/MASS SPECTROMETRY

**D. J. Munch, J. W. Munch, M. A. Feige,**
**E. M. Glick and H. J. Brass**

U. S. Environmental Protection Agency
Office of Drinking Water
Technical Support Division
Cincinnati, Ohio

The U.S. Environmental Protection Agency (EPA) is in the process of obtaining occurrence data in support of regulations for the control of organic contaminants in drinking water [1-4]. The EPA has recommended methods and procedures of analysis for certain of these contaminants [5-10]. Included are gas chromatographic (GC) methods for purgeable halocarbons and aromatic compounds using Hall electrolytic conductivity (HECD) and photoionization (PID) detectors, respectively. Using confirmatory techniques, a flame ionization detector (FID) can also be used to analyze for purgeable aromatic compounds [11].

This laboratory has developed a scheme for the routine gas chromatography/mass spectrometry (GC/MS) analysis of purgeable compounds in drinking water and sources of supply. This procedure uses purge-and-trap sample concentration followed by GC/MS analysis. Data reduction is performed by a computerized GC/MS data system. The quantification limits, precision and accuracy of these analyses have been demonstrated to be comparable to gas chromatographic procedures using electrolytic conductivity and FID [7-13].

## PROCEDURES

### Experimental

Compounds to be used for the preparation of standards were at least 95% pure. Single-component standard solutions for qualitative and quantitative analysis were prepared by filling a 10-mL volumetric flask to approximately 90% of capacity with "distilled in glass" methanol. The flask was allowed to stand open at room temperature for 10–15 min to permit evaporation of solvent from the neck. The flask was then weighed to ±0.1 mg, quickly removed from the balance, and the desired amount of pure compound was added using a microliter syringe. The flask was then reweighed, and the weight of added compound was determined. These stocks always contained greater than 2 mg of the standard component. Appropriate amounts of individual stock standards were then diluted in methanol to produce multicomponent working standards, such that 2 μL diluted into 25 mL of water produced the desired final concentrations [12].

Samples were then collected and stored in headspace free glass vials capped with Teflon®-backed silicone septa. It is suggested that samples be preserved with acid to pH 2 or with 10 mg/L mercuric chloride, and stored at approximately 4°C to inhibit organic compound degradation believed to be caused by biological action [8,14].

The plunger was removed, and samples were gently poured into the barrel of a 50-mL gas-tight syringe, equipped with a 9-in. 20-gauge needle. The volume was adjusted to 25 mL, the needle was removed, and 2 μL of an internal standard solution containing 18 ng/μL of fluorobenzene and 31 ng/μL of 2-bromo-1-chloropropane in methanol was added. This was followed by the addition of 2 μL of a standard containing 19 ng/μL pentafluorobromobenzene in methanol, used to assure that spectra generated met preestablished spectral quality characteristics [5,15]. The resulting solution was then purged at ambient temperature using a Tekmar LSC-1 liquid sample concentrator. The purge gas, helium, was first passed through a 5-Å molecular sieve scrubber. The purge flowrate and time were 40 mL/min and 11 min, respectively. Organics sparged by this procedure were trapped on a 9-in. x 0.105-in. i.d. stainless steel column packed with 2/3 Tenax™, 1/3 silica gel (Davidson grade 15). After purging, compounds were thermally desorbed at 180°C for 5 min, while backflushing the trap with helium at 25 mL/min.

The LSC-1 was interfaced to the gas chromatograph so that the GC carrier gas also served as the desorb gas. This permitted desorption to take place without disturbing the carrier gas flow. The LSC-1 interface was via a 0.105-in. i.d. stainless steel transfer line. Prior to installation, the transfer line was flamed to a glowing red color while being swept with a flow of oxygen gas to assure a chemically inactive surface. All sample transfer lines (from purging vessel to gas chromatograph) were heated to 135°C.

A purge desorb cycle was performed each day, before mass spectrometer calibration to condition the ion source of the GC/MS with water. This was required to maintain consistent instrument sensitivity. The LSC-1 adsorbent trap was baked at 250°C for 25 min between each run to remove any residual organic compounds remaining on the trap, and to reduce the amount of water entering the ion source from the previous analysis.

The gas chromatograph used was a Finnigan modified Varian Model 1400. Desorbed compounds from the LSC-1 were collected onto the head of a cool (40°C), 6-ft x 2-mm i.d. glass GC column packed with 0.1% SP1000 on Carbopack C. After desorption, the chromatographic column was temperature-programmed at 6°C/min to a final temperature of 220°C and held at that temperature until completion of the analysis. Total analysis time, beginning with desorption, was 40 min.

A single glass jet separator combined with a 2-mm i.d. glass transfer line was used to interface the gas chromatograph to a Finnigan Model 3200 quadrapole, 70-eV, electron impact, mass spectrometer. Separator and transfer line temperatures were held at 160 and 225°C, respectively. The mass spectrometer was scanned from 35 to 300 m/z at a rate of 2.95 sec per scan with a 0.05-sec settling time at the end of each scan. Data were acquired and stored, and automated data reduction was performed using a Finnigan-Incos Model 2300 GC/MS data system. Data were acquired for 40 min for each sample analysis.

## Data Reduction

### *Qualitative Analysis*

Automated data reduction was performed utilizing the Finnigan/Incos procedural language [16]. This language allows complex data system command stringing combined with certain standard computer programming functions. All qualitative and quantitative data reductions were performed unattended, overnight.

A reconstructed ion chromatogram (RIC) and spectrum map were displayed for the sample run, and the scan numbers of all chromatographic peaks defined by the data system were written into a scan list (Figure 1). To locate chromatographic peaks, the data system first subtracted 200 analog-to-digital units from the entire chromatogram to adjust for back-

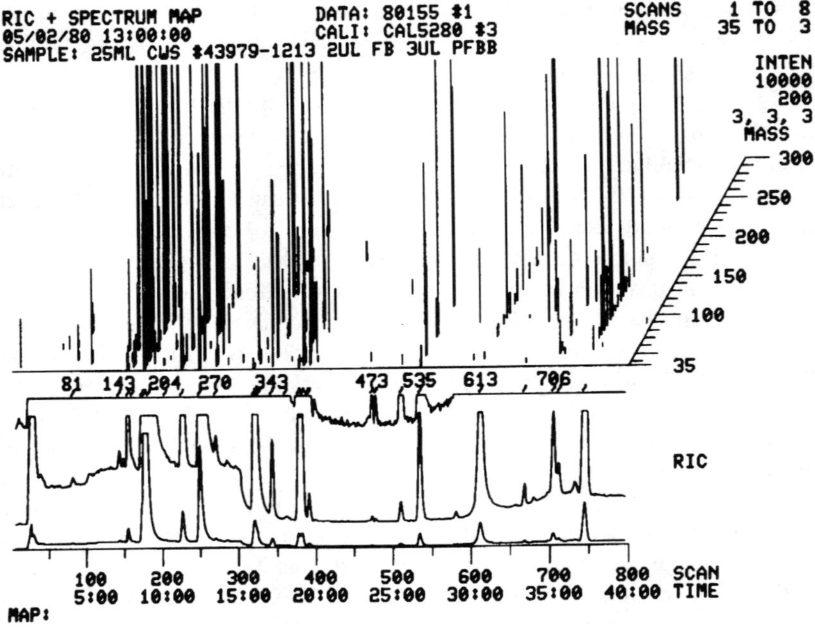


**Figure 1.** Reconstructed ion chromatogram (RIC) (bottom) and spectrum map (top) initially displayed in data workup. Ion intensity (below) and mass-to-charge ratio (above) are plotted against scan number and time.

ground noise. The chromatogram was then searched for any mass peaks greater than $3\sigma$ above the remaining background signal. Chromatographic peaks were then identified when the intensity of 3 such mass peaks maximized within a 3-scan (9-sec) window. After all peaks in the analysis were located, the data system retrieved a spectrum for each peak in the scan list. Each spectrum was enhanced by subtracting from it all ion peaks, with an intensity less than $2\sigma$ above the average intensity for each ion, within a 15-scan window about their corresponding chromatographic peak. A typical mass spectrum generated by the above data reduction procedures is shown in Figure 2.

This enhanced spectrum was then searched against the National Bureau of Standards Mass Spectral Library (27,409 compounds) and the spectra of the best six matches were displayed [16]. These six matches were composed of the three best matches of an entire unknown spectrum with the library (forward search) (Figure 3), and the three best matches of the library with

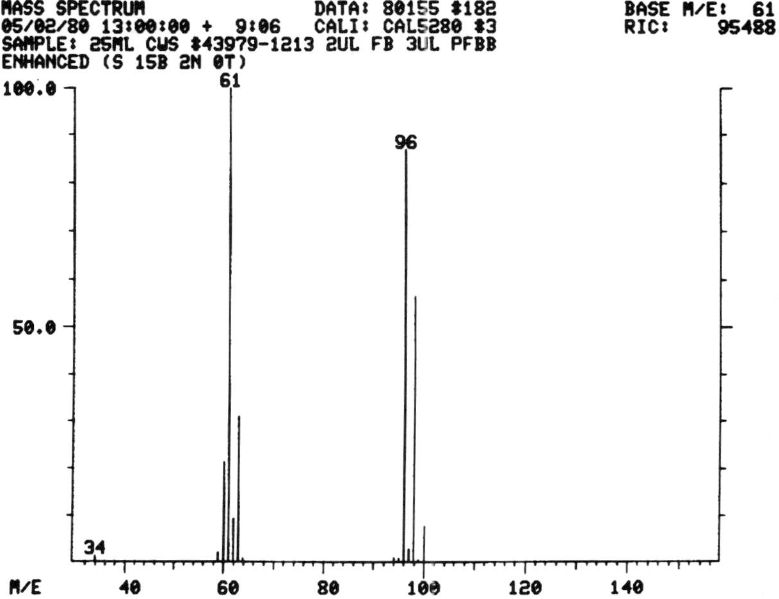


Figure 2. Enhanced spectrum generated during data workup. Ion intensity normalized to 100% is plotted against mass-to-charge ratio.

any portion of the unknown spectrum (reverse search). The forward search helped to identify spectra which resulted from single compounds, while the reverse search helped to identify spectra resulting from the simultaneous gas chromatographic elution of two or more compounds. Identifications were finalized only when spectra and retention times were compared to the analyses of standards.

Spectra which could not reasonably be identified were collected together in a separate spectral library. These library entries were searched against subsequent sample data within a 2.5-min window around the retention time that the spectra was initially discovered. In this way, unidentified spectra which recurred frequently were selected for more intensive interpretation.

## *Quantitative Analysis*

Electron multiplier voltages were adjusted so that the reference area of the fluorobenzene internal standard was maintained within ±10% of a previously

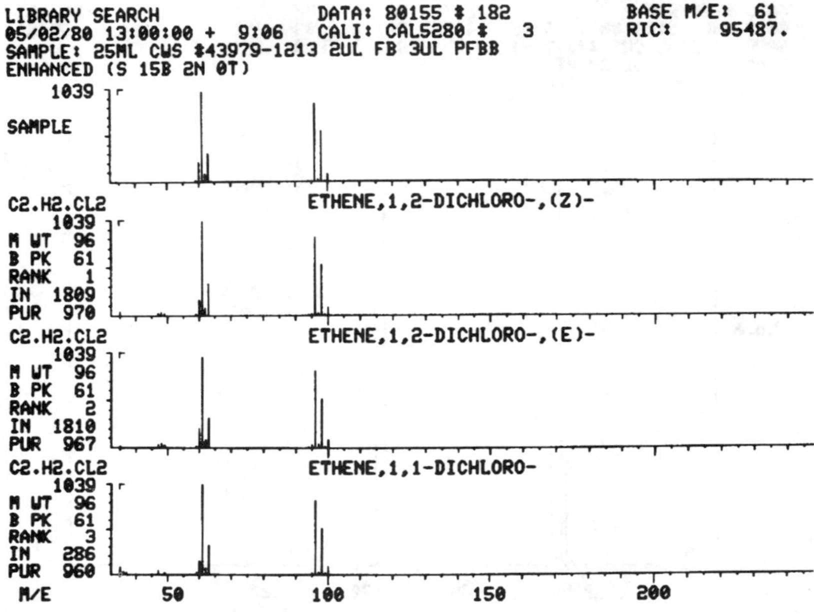


Figure 3. Output of NBS forward library search.

established value. This was done because changes in instrument sensitivity did not uniformly affect the compounds being determined. For example, a 50% increase in overall instrument sensitivity did not necessarily mean that a corresponding 50% increase in the responses of individual compounds occurred. These differences in compound responses were smaller for chemicals which were structurally similar. For this reason, 2-bromo-1-chloropropane was used as an internal standard for the measurement of halogenated aliphatic compounds, while fluorobenzene was used as an internal standard for the measurement of aromatic compounds.

As part of the analysis scheme, a library of compounds to be quantified was produced. The library spectra for the two internal standards, fluorobenzene and 2-bromo-1-chloropropane, and the spectral quality standard, pentafluorobromobenzene, were searched within 2.5-min retention time windows. If these standards could not be found, quantification was not performed.

Following the determination of the retention time of these internal standards, an extracted ion current profile (EICP) for the highest intensity ion for each compound in the library was performed (Figure 4). The area

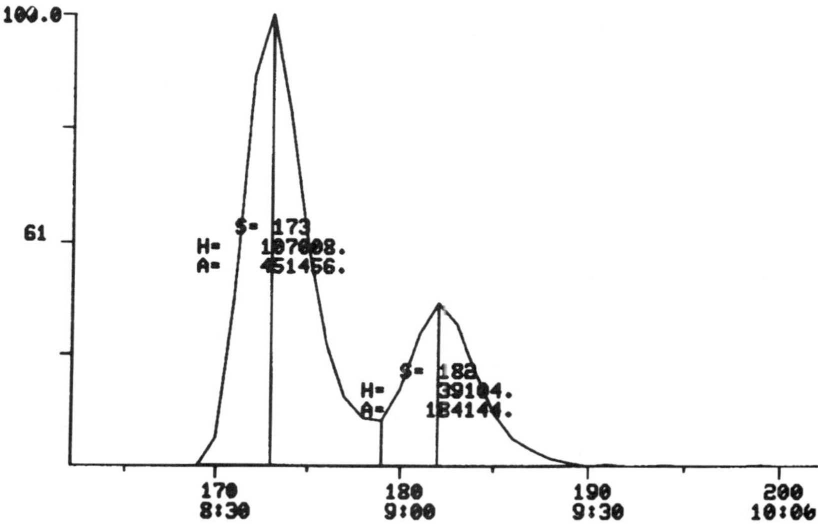


**Figure 4.** Extracted ion current profile for *trans*-1,2-dichloroethylene.

under each EICP peak selected was determined for each compound to be quantified by searching within a 1.5-min relative retention time window from the appropriate internal standard.

Quantification was accomplished by comparing the area under the EICP peak for the base ion of each compound against previously analyzed standards using the equation:

$$\frac{(\text{area}/\text{reference area}) - b}{m} = \text{amount}$$

| | |
|---|---|
| where area | = area of base peak, using the EICP for the compound to be quantified |
| reference area | = area of base peak, using the EICP of the internal standard |
| b | = intercept of the calibration curve |
| m | = slope of the calibration curve |

Calibration curves for each compound to be quantified were determined by plotting area/reference area against concentration in μg/L (Figure 5). Initially, standards at three different concentrations were each analyzed five times. Subsequently, each day that sample analyses were performed, a standard containing the compounds to be quantified was analyzed at one of the three concentrations used in the development of calibration curves. Once per month, the five standards most recently analyzed at each concentration were used to redevelop calibration curves. Every three months new standards were made and new calibration curves were established. Quantitative analysis was performed only if the concentration of a particular compound was within the concentration range of the standards analyzed for that compound.

## RESULTS AND DISCUSSION

The analytical scheme described permits the routine qualitative and quantitative analysis of 16 commonly occurring purgeable compounds over a concentration range of 0.1–0.2 to 5.0 μg/L. As needed, additional compounds can be analyzed by incorporating appropriate analytical standards.

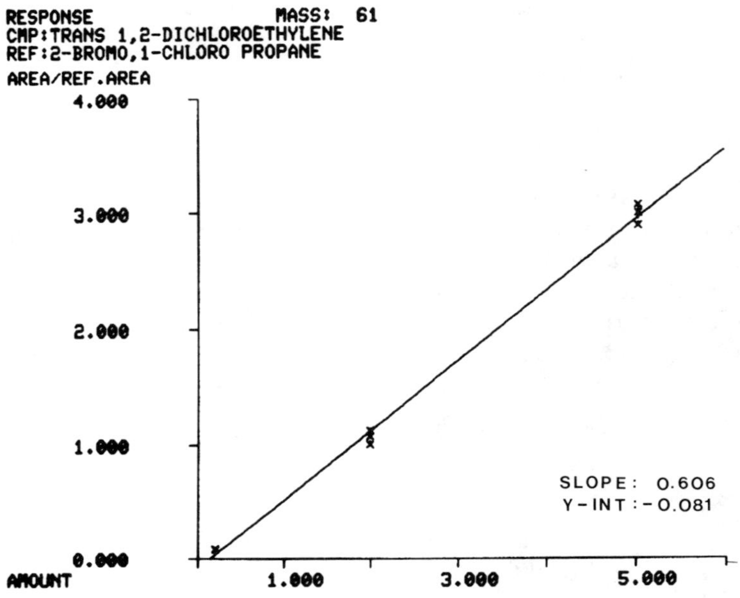


Figure 5. Typical calibration curve for *trans*-1,2-dichloroethylene.

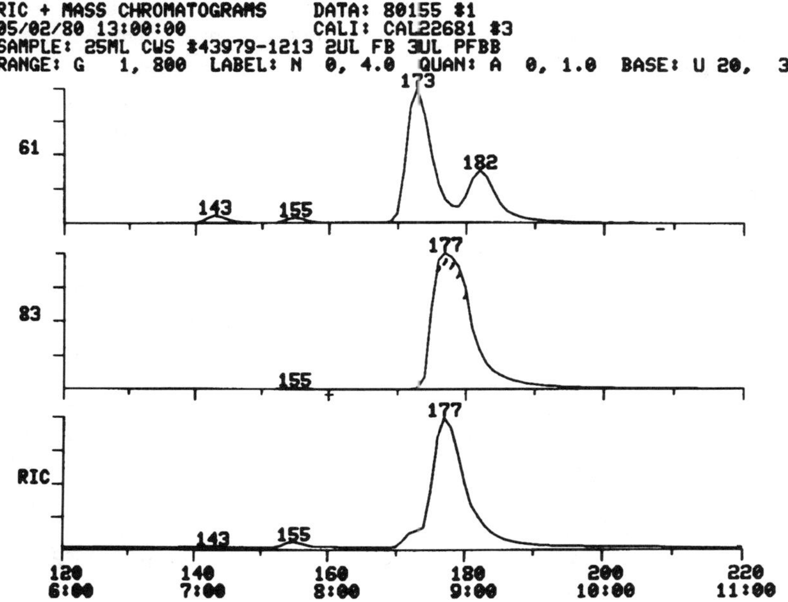


**Figure 6.** Extracted ion current profile from the analysis of a drinking water sample.

Software can be developed to extend the upper range of concentration. Since time-consuming data handling can be performed overnight, a greater number of samples are analyzed.

## Qualitative Analysis

The automated procedures easily located and "library-searched" all chromatographic peaks which eluted more than 3 scans (9 sec) apart. Compounds determined to be at or below their lower quantification limits using EICP were identified by this scheme, demonstrating its utility at low concentrations (generally 0.1-0.2 ppb or less). Even small peaks, which were not well resolved from larger peaks, and which appeared to coelute on an RIC, were satisfactorily identified. A portion of the chromatogram resulting from the analysis of a drinking water sample is shown in Figure 6. Additional data for this sample have been shown in Figures 1 to 4. The EICP for m/z 61 indicates the presence of 1,1-dichloroethylene (scan No. 143), *cis*-1,2-dichloroethylene (scan No. 173), and *trans*-1,2-dichloroethylene (scan No. 182). The large peak at scan No. 177 on both the RIC and the EICP for m/z 83 corresponds to chloroform. Even though chloroform nearly coelutes with two of the dichloroethylene isomers, and is present at a much higher concentration than either of them, the automated data reduction procedure was able to successfully identify and subsequently quantify these compounds. Interlaboratory comparisons of qualitative data acquired by both GC/HECD and GC/MS showed excellent correlation between the occurrences of halogenated compounds by both detectors.

Several steps were taken to ensure consistent data acquisition. The purge-and-trap procedure introduced a large amount of water into the GC/MS, reducing overall instrument sensitivity and affecting mass assignments. It was thus necessary to adopt procedures to minimize this problem, by keeping the amount of water in the ion source relatively constant during a work day. Water was introduced into the ion source at the beginning of each day by an initial purge/desorb cycle. After each analysis, the trapping column on the LSC-1 was baked to prevent an overload of the ion source with water. In this way, when instrument mass calibration was performed after an initial purge/desorb cycle, consistent instrument sensitivity and mass assignment accuracies were maintained. To demonstrate that proper mass assignment and selective ion abundances were maintained, pentafluorobromobenzene (PFBB) was added to all samples, standards and blanks. The masses and relative ion abundances of PFBB are known, and serve as a standard for consistent mass spectrometer operation [5,15].

## Quantitative Analysis

Following the quarterly preparation of standards and the development of calibration curves for each compound, an EPA Environmental Monitoring and Support Laboratory quality control (QC) sample was analyzed. Dilutions of separate concentrates containing halocarbon and aromatic compounds were combined and analyzed as a single sample. These analyses assured the accuracy of the overall analytical procedure. The data in Table I show that accurate and precise measurements could be made generally within 20% or 0.2 $\mu$g/L, at concentrations below 1 $\mu$g/L.

Once per week a sample from a set of spiked blanks was analyzed. The data obtained were used to routinely determine analytical precision. The results of recent analyses are shown in Table II. Also included in Table II are the m/z values used for quantification. Precision was comparable to that reported for purge-and-trap GC/HECD and GC/FID analyses at similar concentrations [8,11].

Data in Table III show results for the analysis of a drinking water sample by GC/MS and by the use of GC/HECD and GC/FID procedures. Both quali-

Table I. Quality Control Sample Analyses

| Compound | Expected Concentration (μg/L) | Mean[a] Concentration Measured (μg/L) | Error[b] | Relative Standard Deviation[c] |
|---|---|---|---|---|
| Chloroform | 2.7 | 2.7 | 0 | 14 |
| 1,2-Dichloroethane | 1.1 | 0.9 | -18 | 6.3 |
| 1,1,1-Trichloroethane | 0.4 | 0.4 | 0 | 0 |
| Carbon Tetrachloride | 0.5 | 0.4 | -20 | 11 |
| Bromodichloromethane | 0.5 | 0.6 | -20 | 9.7 |
| Dibromochloromethane | 0.7 | 0.5 | -29 | 17 |
| Trichloroethylene | 0.8 | 0.7 | -13 | 0 |
| Benzene | 0.9 | 1.0 | +11 | 12 |
| Bromoform | 0.6 | 0.6[d] | 0 | 0 |
| Tetrachloroethylene | 0.4 | 0.4 | 0 | 0 |
| Toluene | 0.7 | 1.0 | +43 | 7.3 |
| Ethyl Benzene | 0.5 | 0.5 | 0 | 11 |
| *m*-Xylene | 0.3 | 0.3 | 0 | 14 |
| *o*+*p*-Xylene[e] | 1.3 | 1.2 | -8 | 6.1 |

[a]Of seven analyses.

[b]Expressed as 100 times the difference between mean and expected concentrations, divided by the expected concentration.

[c]Of mean concentration.

[d]Found above the quantification limit of 0.5 μg/L in three of seven analyses.

[e]Summation of ortho and para isomers quantified as meta-xylene.

Table II. Results of Repetitive Spiked Blank Analyses

| Compound | Mass Used for Quantification | Mean Concentration[a] (μg/L) | Range[a] (μg/L) | Relative Standard Deviation[a] | Lower Quantification Limit (μg/L) |
|---|---|---|---|---|---|
| Chloroform[b] | 83 | 1.6 | 1.3–1.9 | 16 | 0.1 |
| t-1,2-Dichloroethylene[b] | 61 | 0.2 | 0.2–0.3 | 22 | 0.2 |
| 1,2-Dichloroethane[b] | 62 | 0.6 | 0.6–0.7 | 8.2 | 0.2 |
| 1,1,1-Trichloroethane[b] | 97 | 0.5 | 0.4–0.5 | 12 | 0.2 |
| Carbon Tetrachloride[b] | 117 | 0.9 | 0.9–1.0 | 5.5 | 0.2 |
| Bromodichloromethane[b] | 83 | 1.6 | 1.4–1.7 | 8.7 | 0.2 |
| Dibromochloromethane[b] | 129 | 1.6 | 1.4–1.8 | 11 | 0.2 |
| Trichloroethylene[b] | 95 | 0.6 | 0.5–0.6 | 7.0 | 0.2 |
| Benzene[c] | 78 | 1.3 | 1.1–1.4 | 10 | 0.2 |
| Bromoform[b] | 173 | 1.8 | 1.6–2.0 | 7.5 | 0.5 |
| Tetrachloroethylene[b] | 166 | 0.7 | 0.5–0.8 | 19 | 0.2 |
| Toluene[c] | 91 | 1.4 | 1.2–1.6 | 11 | 0.2 |
| Chlorobenzene[c] | 122 | 1.7 | 1.5–2.0 | 11 | 0.2 |
| Ethyl Benzene[c] | 91 | 1.6 | 1.3–1.7 | 8.9 | 0.2 |
| *m*-Xylene[c] | 91 | 1.5 | 1.3–1.6 | 8.4 | 0.2 |
| *p*-Dichlorobenzene[c] | 146 | 1.5 | 1.3–1.7 | 9.7 | 0.1 |

[a] Analyzed six times over an eight-week period.

[b] 2-Bromo-1-chloropropane was used as internal standard.

[c] Fluorobenzene used as internal standard.

**Table III. Quantitative Analysis Report: Sample 80155**

| Compound | Retention Time (min)[a] | Concentration (μg/L) | |
|---|---|---|---|
| | | GC/MS[b] | HECD or FID[c] |
| Chloroform | 8.84 | > 5.0 | 72 |
| $t$-1,2-Dichloroethylene | 9.06 | 2.4 | NA[d] |
| 1,2-Dichloroethane | 10.19 | 0.2 | < 0.5 |
| 1,1,1-Trichloroethane | 11.29 | 4.6 | 5.5 |
| Carbon Tetrachloride | – | < 0.2 | < 0.1 |
| Bromodichloromethane | 12.44 | > 5.0 | 22 |
| Dibromochloromethane | 15.99 | > 5.0 | 7.8 |
| Trichloroethylene | 16.19 | 0.2 | 0.3 |
| Benzene | 17.14 | 0.7 | 0.8 |
| Bromoform | 19.49 | 0.9 | 0.8 |
| Tetrachloroethylene | | < 0.2 | < 0.2 |
| Toluene | | < 0.2 | < 0.5 |
| Chlorobenzene | 26.74 | 1.2 | 1.5 |
| Ethyl Benzene | | < 0.2 | NA |
| $m$-Xylene | | < 0.2 | |
| | | | < 0.5 (m-, o+p-Xylene) |
| $o$+$p$-Xylene | 33.44 | 0.2 | |
| $p$-Dichlorobenzene | 35.29 | 0.7 | NA |

[a]For GC/MS analysis.

[b]At greater than 5.0 μg/L, compounds were identified but not quantified.

[c]Analyses using a HECD performed at TSD; those using an FID performed at SRI International.

[d]NA = not analyzed.

tative identifications and concentrations determined are in agreement. For each case where compounds were present at concentrations above 5 μg/L, qualitative agreement was observed. Where compounds were present between lower quantification limits and 5.0 μg/L, agreement was both qualitative and quantitative (within 20% or 0.2 μg/L). In two cases (1,2-dichloroethane and $o$+$p$-xylene) the MS quantification limits were lower than HECD and FID limits. Thus, these compounds were identified and quantified by GC/MS and not by these other detectors. While data in Table III represent the analysis of only one sample, they are typical of results obtained using the GC/MS scheme in a routine analytical mode.

Comparative data for drinking water samples are shown in Table IV. Analyses were performed by GC/HECD and GC/MS in our laboratory and by GC/FID at SRI International [11]. Since the upper quantification limit for all compounds analyzed by GC/MS was 5.0 μg/L, only compounds present at concentrations less than 5.0 μg/L were considered. The percent difference was calculated for each pair of values. The mean and median values, in percent, were calculated for the total number of pairs for each compound. Agreement was generally within 20%.

## CONCLUSIONS

Analysis by GC/MS for a broad range of purgeable compounds can be accomplished in a single gas chromatographic run at sensitivities comparable to GC/HECD and GC/FID procedures. Routinely, 16 compounds can be measured with precision and accuracy generally within 20% or 0.2 μg/L. Finally, the use of an automated data reduction procedure allows "round-the-clock" GC/MS operation, enabling it to be used as a routine analytical instrument for both qualitative and quantitative analysis.

## ACKNOWLEDGMENTS

The authors wish to express their appreciation to Robert F. Thomas and Michael J. Weisner for providing comparative data generated using a specific halogen detector and to Barbara A. Kingsley for providing comparative data using a flame ionization detector.

**Table IV. Comparative GC/MS Data[a]**

| Compound | No Cases | Mean % Diff[b] | Median % Diff[b] |
|---|---|---|---|
| 1,1,1-Trichloroethane | 6 | -3 | -2.7 |
| Dibromochloromethane | 16 | –19 | -17 |
| Tetrachloroethylene | 5 | -15 | -11 |
| Toluene | 12 | +16 | +24 |
| Xylenes | 16 | + 3.7 | + 5.8 |

[a]Compared to HECD or FID analyses. All concentrations between 0.2 and 5 μg/L.

[b]Percent difference for each data pair defined as 100 x [A – B/A + (B/2)]; where A = GC/MS value and B = HECD or FID value.

## REFERENCES

1. U.S. EPA. "National Interim Primary Drinking Water Regulations," *Federal Register* 40(248):59566-59588 (1975).
2. U.S.EPA. "Control of Organic Chemical Contaminants in Drinking Water," *Federal Register* 43(28):5755-5780(1978).
3. U.S. EPA. "Organic Chemical Contaminants; Control Options in Drinking Water," *Federal Register* 41(136):28991-28998 (1976).
4. U.S. EPA. "National Interim Primary Drinking Water Regulations: Control of Trihalomethanes in Drinking Water; Final Rule," *Federal Register* 44(231):68624-68707 (1979).
5. U.S. EPA. "Guidelines Establishing Test Procedures for the Analysis of Pollutants; Proposed Regulations," *Federal Register* 44(233):69464-69575 (1979).
6. Bellar, T. A., and J. J. Lichtenberg. "Determining Volatile Organics at the $\mu$g/l Level in Water by Gas Chromatography," *J. Am. Water Works Assoc.* 66:739 (1974).
7. "The Determination of Halogenated Chemical Indicators of Industrial Contamination in Water by Purge and Trap, Method 502.1," U. S. EPA Environmental Monitoring and Support Laboratory, Cincinnati, OH (1979).
8. "The Determination of Aromatic Chemical Indicators of Industrial Contamination in Water by Purge and Trap, Method 502.2," U.S. EPA Environmental Monitoring and Support Laboratory, Cincinnati, OH (1979).
9. "The Analysis of Trihalomethanes in Finished Waters by the Purge and Trap Method, Method 501.1," U.S. EPA Environmental Monitoring and Support Laboratory, Cincinnati, OH (1979).
10. Bellar, T. A., and J. J. Lichtenberg. "Semi-Automated Headspace Analysis of Drinking Waters and Industrial Waters for Purgeable Volatile Organic Compounds," in *Measurement of Organic Pollutants in Water and Wastewater*, C. E. Van Hall, Ed. (Denver, CO: American Society for Testing and Materials, 1979), p. 398.
11. Kingsley, B. A., C. Gin, W. R. Peiffer, D. F. Stivers, S. H. Allen, H. J. Brass, E. M. Glick and M. J. Weisner. "A Cooperative Quality Assurance Program for Monitoring Contract Laboratory Performance,"Chapter 45 this volume.
12. Brass, H. J., M. A. Feige, T. A. Halloran, J. W. Mello, D. J. Munch and R. F. Thomas. "The National Organic Monitoring Survey: Samplings and Analyses for Purgeable Organic Compounds," in *Drinking Water Enhancement Through Source Protection*, R. B. Pojasek, Ed. (Ann Arbor, MI: Ann Arbor Science Publishers, Inc., 1977), p. 398.
13. Munch, D. J., M. A. Feige and H. J. Brass. "The Analysis of Purgeable Compounds in the National Organic Monitoring Survey by Gas Chromatography/Mass Spectrometry," in *Proceedings of the American Water Works Association Water Quality Technology Conference* (Denver, CO: American Water Works Association, 1978).
14. Glick, E. M., D. J. Munch, J. W. Munch, M. A. Feige and H. J. Brass. "Decomposition of Unsaturated Compounds in Stored Water Samples," (in preparation).

15. Eichelberger, J. W., and W. L. Budde, U.S. EPA Environmental Monitoring and Support Laboratory, Cincinnati, OH. Personal communication.
16. *Finnigan Incos Data System MSDS Operators Manual, Revision 3,* (Sunnyvale, CA: Finnigan Corporation, 1978).

# CHAPTER 40

# ANALYSIS OF VOLATILE ORGANICS ON SEDIMENTS AND IN ASSOCIATED WATER

**Jackson Ellington**

U.S. Environmental Protection Agency
Environmental Research Laboratory
Analytical Chemistry Branch
Athens, Georgia

A wide variety of chemicals is introduced to surface waters directly from municipal and industrial effluents or indirectly in leachates from improperly designed landfills or corroded and broken containers. These dispersed leachates and effluents create a potential hazard to humans and the environment. The associated organic chemicals include compounds of diverse functionality, solubility, stability and water/particulate partition coefficients. Some of these pollutants are highly toxic and some are known mutagens.

The fate of organic chemicals in water includes ingestion by man and animals; loss to the atmosphere; transformation by hydrolysis, photolysis or other natural process; incorporation into aquatic organisms; and sorption to particulate matter in the water column and eventual deposition at the bottom of the water column. The origin of particulate matter is varied; this matter includes soils, suspended solids, sediments and sludges from sewage treatment plants and industry. In this chapter, initial attempts at desorbing organics from sediments prior to analysis by gas chromatography/mass spectrometry (GC/MS) are described. Methods of desorption include purging and solvent extraction.

Several methods for determining purgeable volatile organics in water are reported in the literature. These include headspace analysis [1-3], microsolvent extraction [4,5] and purge-and-trap [6]. Grob and Zurcher [7] reported a purge-and-trap system in which a small volume of air is recycled through a water sample, and the organic compounds trapped on a small charcoal filter. The entrapped organics are eluted from the charcoal with a small volume (<15 $\mu$L) of organic solvent (carbon disulfide, methylene chloride, hexane). This method was adapted to the analysis of volatile organics on sediment in the work report here.

The analysis of compounds of all volatility classes sorbed to sediments is desirable for reasons stated previously. Only a portion of this volatility range, however, is covered by the purge-and-trap method. The less volatile compounds that can be extracted into an organic solvent and are amenable to gas chromatographic analysis are addressed in a separate section of this chapter.

Initial experiments were performed with the conventional purge-and-trap system used with aqueous samples [6]. This method entailed purging with nitrogen, trapping the organics on Tenax® and thermal desorption onto a gas chromatographic column. The desirability of heating the sediment to increase desorption while limiting the volume of purge gas led to the use of the Grob closed-loop stripping analysis (CLSA) method [7].

## EXPERIMENTAL

### Samples

Sediment was collected from a northeast Georgia lake and river. The wet lake sediment was passed through a number 14 U.S. standard testing sieve, spread to 0.75-in. thickness in a flat pan and allowed to dry at room temperature in an organic vapor-free room. Sieving was necessary to prevent destruction of the Polytron Homogenizer head. The dried cake was pulverized in a mortar. River sediment was collected 20 ft above and below an Athens, GA, sewage treatment plant outfall. Samples were collected such that each one-liter bottle had approximately 200 mL of water above the collected sediment, and were stored at 18°C until analysis.

### Grob CLSA at Elevated Temperatures

The fully assembled system is shown in Figure 1. The ambient air trapped in the system when all openings are sealed is passed up through the sintered glass disk (near the bottom of the modified 250-mL flask), the sample, the

**Figure 1.** CLSA system.

fritted disk and the charcoal filter, and then through the pump (Metal Bellows Corporation MB-21) to start another cycle. Immersing the purge flask to the bottom of the 24/40 joint in a heated water bath maintains the desired temperature. The inline fritted disk before the filter holder is necessary because fine particles would be transported to the filter when dry samples are purged. For operation at elevated temperatures, the metal tubing is heated by a heat tape applied from the glass adapter to the fitting on the exit end of the filter holder. The temperature of the filter holder is determined by a thermometer inserted under the heat tape. The normal purge conditions were 1 hr with the bath at 65°C and the filter holder at 80°C. Figure 2 shows the disassembled filter holder with filter partially inserted, a filter connected to a sample vial by shrinkable Teflon® and a sample vial with stopper. The extraction of the filter [7] entails attachment of the filter to the sample vial with shrinkable Teflon, additon of 8-10 $\mu$L of solvent to the charcoal, cooling the sample vial side of the filter to draw solvent through the filter, then warming to force the solvent back through the filter. The cool-warm cycle is repeated at least three times, and on the last cool cycle, the assembled system is given a sharp downward sling to force the solvent into the sample vial. The procedure is then repeated with an additional 6-8 $\mu$L of solvent. The sample vial is then stoppered and stored at 0° C until analysis.

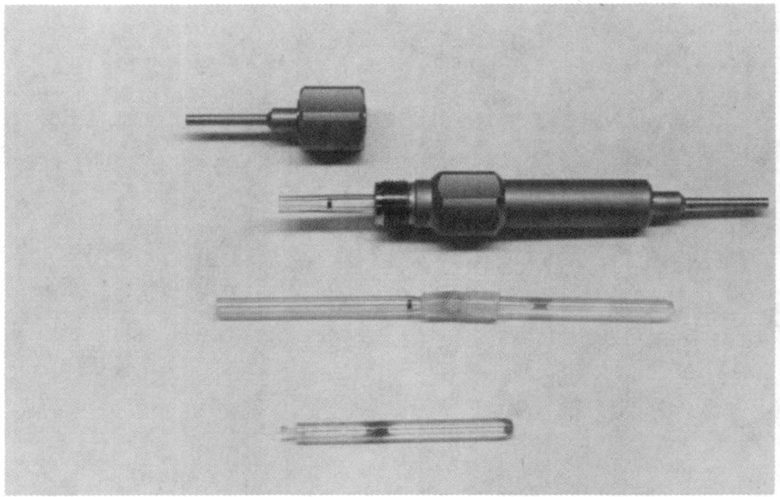

**Figure 2.** Filter holder, filter connected to sample vial and sample vial.

## Gas Chromatography

The Tracor 560 GC was equipped with a flame ionization detector (FID) and a Hall 700A electrolytic conductivity detector operated in the halogen mode. A 15-m, OV-101, 0.25-mm. i.d. glass wall-coated open tubular (WCOT) column was connected to the desired detector and the oven temperature was held at 20°C for 1 min then programmed at 4°C/min to 225°C. The difference in response time between the FID and the Hall 700A detector was determined by alternately connecting the column to each detector and injecting a homologous mixture of 1-chloroalkanes ($C_7$,$C_8$,$C_{10}$,$C_{12}$,$C_{14}$,$C_{16}$,$C_{18}$) followed by temperature programming.

The Hewlett-Packard 5840 GC was equipped with a 12-m SP-2100, 0.21-mm i.d. fused-silica WCOT column. Temperature program was 50°C for 1 min then 8°C/min to 200°C.

## Gas Chromatography/Mass Spectrometry

A Varian MAT-44 mass spectrometer was interfaced via a platinum-iridium transfer line to a factory-modified Varian 1400 GC. The GC was equipped with an 8-m, SP-2100, 0.22-mm i.d. glass WCOT column. The temperature program was 30°C for 1 min then 4°C/min to 175°C. Data from electron impact (EI) and chemical ionization (CI) runs were acquired, stored and manipulated on an online Spectro System MAT 200 computer. Tentative identifications, obtained from searching the spectral library supplied with the SS MAT 200 and manual matching with spectra in the Eight Peak Index of Mass Spectra, were confirmed by coinjection of standards.

### Oxidation of Radiolabeled Sediments

Sediment samples previously spiked with radiolabeled compounds were oxidized by a R. J. Harvey OX200 Biological Oxidizer. In a typical analysis, a 200-mg sample was inserted into the combustion chamber (900°C) of the oxidizer; the oxygen flow was maintained for 3.5 min followed by a 1-min nitrogen purge. The $^{14}C$-labeled carbon dioxide, formed in combustion, was trapped in 15 mL of Harvey Carbon 14 Cocktail. The disintegrations per minute (dpm) were then determined on a Packard Tri-Carb Scintillation Spectrometer Model 3380 operated in a 20-min duplicate scan mode. A conversion factor was determined for the oxidizer by triplicate oxidation of sediments containing known quantities of sorbed radiolabeled compounds. The conversion factor was used to relate the observed dpm after oxidation to the amount of radiolabeled compound sorbed to the sediment before oxidation.

### Scintillation Counting

A Packard Tri-Carb Liquid Scintillation Spectrometer Model 3380 was used in the $^{14}C$ study. Typically, an aliquot of the unknown solution containing the radiolabeled compound was added to 10 mL of Aquasol-2, and the solution counted in duplicate for 20 min. The average count was used for computations and the method of standard additions was used to compensate for instrument variation and quenching.

### Solvent Extractions

#### *Sonication*

In a typical analysis, 20 g of radiolabeled sediment was added to a 400-mL beaker followed by 200 mL of solvent. The sonicator horn was then immersed in the solvent and operated for 3 min in a 50% duty cycle pulsed extraction mode. The sediment was removed from the solvent by filtration on a Whatman No. 1 filter paper. The sediment on the paper was washed with two 20-mL aliquots of the extraction solvent. The eluent was transferred to a 250-mL flask, diluted to 250 mL with the extraction solvent and an aliquot was removed for determination of dpm. The sediment was dried, mixed and a 200-mg aliquot was oxidized, and the radiolabeled activity was compared to the value obtained before extraction.

### *Sonication-Homogenization*

A Brinkmann Polytron was used for the sonication and homogenization action. Typically, 20 g of radiolabeled sediment was added to the metal sonicator tube followed by 200 mL of solvent. The Polytron probe was then inserted into the sample for 30 sec at a setting of 7.5. The sediment was then separated from the solvent and both were analyzed as described above.

### *Soxhlet Extraction*

A Whatman extraction thimble was loaded with 10 g of sediment, and the solvent reservoir was charged with 80 mL of solvent and several boiling stones. The apparatus was assembled and reflux and siphoning (40/hr) continued for 16 hours.

### *Hypovial Extraction*

Two 5-mL hypovials were charged with 1 g each of 9-fluorenone-spiked and air-dried sediment. High-performance liquid chromatography (HPLC) water (1 mL) and 50/50 hexane/acetone (3 mL) were added to the first vial and only 3 mL of 50/50 hexane/acetone to the second vial. Both vials were vortexed for 1 min, sonicated for 15 min and left for 18 hr at room temperature. The vial containing 1 mL of water was centrifuged at 2500 rpm for 20 min to break an emulsion. Aliquots were removed from both vials and the extraction efficiency was determined.

## Sediment Spiking

The radiolabeled compounds (50 $\mu$Ci) obtained from California Bionuclear Corporation were dissolved in 100 mL of hexane and stored in a refrigerator at 0°C. Aqueous solutions were prepared by withdrawing an aliquot from the hexane solution at room temperature, adding the aliquot to a volumetric flask, evaporating the hexane, adding HPLC water to volume and sonicating for 15 min. Aliquots of the aqueous solution were withdrawn, sediment was added and the mixture was sonicated for 15 min. The water/sediment samples were then shaken overnight and transferred to metal centrifuge tubes; water was separated from sediment by centrifugation at 15,000 rpm for 30 min. The sediment was dried at room temperature and mixed by mortar and pestle, and the radiolabeled content was determined by oxidation. The aqueous layer was counted directly to determine the material balance.

## RESULTS AND DISCUSSION

### CLSA of Sediment Samples

Three lake sediment samples (50 g of sediment in 200 mL organic-free water) were purged at room temperature for 1 hr. The filters were extracted with carbon disulfide, methylene chloride or hexane. A comparison of the FID chromatograms obtained by injection on the Tracor 560 GC (under conditions given under Experimental) indicated carbon disulfide was more efficient at extracting one of the early eluting peaks as well as most of the compounds eluting from the column between 90 and 175° C. Carbon disulfide was the extracting solvent for the analyses presented in this chapter.

As a check of purging efficiency, 50 g of dry sediment was spiked with a homologous series ($C_7$,$C_8$,$C_{10}$,$C_{12}$,$C_{14}$,$C_{16}$,$C_{18}$) of chloroalkanes at 20 ppm; 200 mL of water was added to the sediment, and the sample was purged at room temperature for 1 hr. As expected, the purging efficiency decreased with increasing chain length. The 1-chlorooctadecane was not detected. A second spike sample (20 ppm) was purged at a bath temperature of 65°C and filter at 85°C. The drop in purging efficiency occurred between 1-chlorohexadecane and 1-chlorooctadecane. All subsequent analyses to be reported in this chapter, unless labeled otherwise, were performed at a bath temperature of 65°C and filter temperature of 85°C. The purge time was normally 1 hr. A system blank was run before each analysis.

The lake sediment sample (50 g in 200 mL of organic-free water) was purged and the carbon disulfide extract was analyzed on the Tracor 560 with detection by flame ionization and the 700A HECD operated in the halogen mode. Halogen-containing compounds were detected, but none were identified in the GC/MS analysis (Figure 3). The identifications in Figure 3 are based on computer matching of the mass spectra and coinjection of standards.

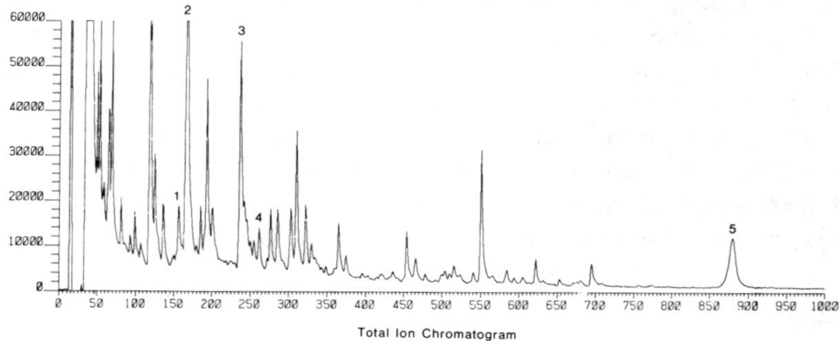


**Figure 3.** Total ion chromatogram from CLS purging of lake bottom sediment at 65°C. Peak identification: (1) benzaldehyde; (2) 2-octanone; (3) decane; (4) fenchone; (5) sulfur.

The river sediment (50 g wet in 200 mL of organic-free water) collected below the sewage outfall was analyzed (Figure 4) and only mass fragmentations typical of alkyl hydrocarbons were observed.

The river sediment above the outfall was not as complex. The total ion chromatogram (Figure 5) is surprisingly free of the early eluting volatile compounds seen in the previous samples. The two major peaks, labeled 1 and 2, eluted between 130 and 140° C. The HECD detected halogen in two peaks in this region of the chromatogram. The retention times, however, did not match those detected by FID and MS, and halogen content was not indicated in the MS spectra of peaks 1 and 2, shown in Figure 6. Perhaps the compounds detected by the HECD detector were formed by reaction of compounds 1 and 2 with residual chlorine in the water from the sewage treatment or perhaps there is no connection between the compounds. The halogen-containing compounds were not identified.

The similarities of peaks 1 and 2 are at once apparent in the two mass spectra, including common major mass ions and the same possible molecular ion at 173. The presence of nitrogen as indicated by the odd number molecular ion is not supported by the fragmentation at the lower masses, although the difference (173 - 143) could indicate the loss of NO. The m/e 55 and 56 ions suggested the presence of oxygen. The m/e 55 ion is typical of the $C_4H_7$ fragment from esters and the m/e 56 ion could be $C_4H_8$ from the rearrangement of a carbonyl compound. The ions at m/e 43, 71 and 85 are indicative of an alkyl hydrocarbon structure. Differences in the spectra occur at masses m/e 83, 98, 111, 113 and 161.

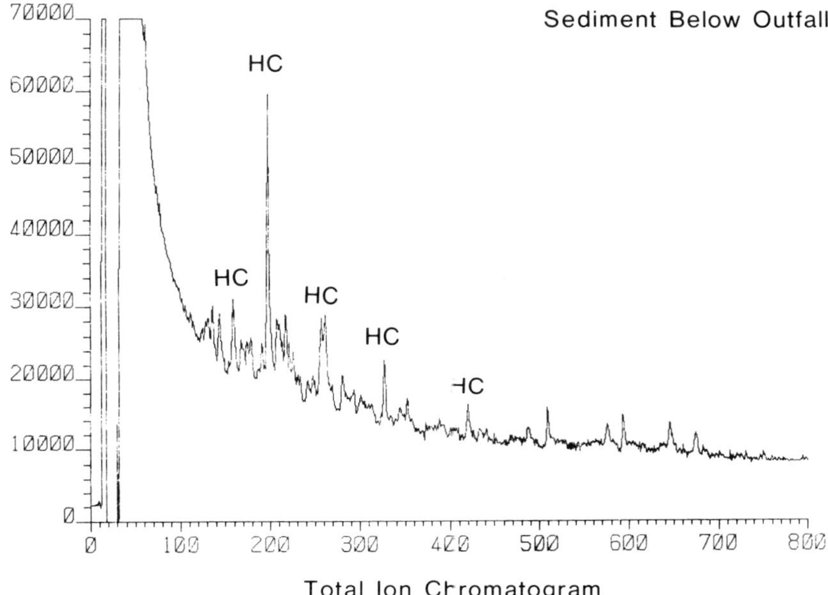


**Figure 4.** Total ion chromatogram from CLSA purging of river sediment below sewage outfall at 65°C. HC represents peaks identified as alkyl hydrocarbons.

The mass-to-charge intensities of the major masses of peaks 1 and 2 were normalized and the Eight Peak Index of Mass Spectra was searched for possible structures. Two candidate compounds were determined: 2,2,4-trimethylpentane-1,3-diol-isobutyrate and 2-ethylhexane-3-o1-1-nor-butyrate, both of molecular weight 216.

The identification of peaks 1 and 2 was investigated further because it seemed unlikely that two esters of the same molecular weight (216) would be present in the same sample at the apparent high concentrations. In an effort to find possible sources for the compounds the Distribution Register of Organic Pollutants in Water (WaterDROP) [8], which is part of EPA-NIH Chemical Information System (CIS) was searched. No responses were found for the esters, so the CIS Structure and Nomenclature Search System (SANSS) was searched for compounds having the molecular formula $C_{12}H_{24}O_3$.

Eighteen responses were obtained, including 2,2,4-trimethyl-1,3-pentanediol monoisobutyrate. A company trade name was also listed for this ester.

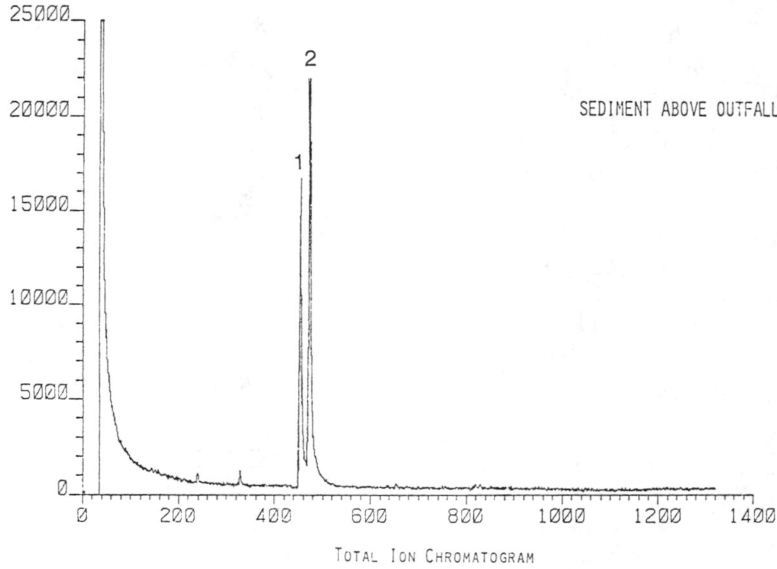


**Figure 5.** Total ion chromatogram from CLS purging of river sediment above sewage outfall at 65°C. Peaks 1 and 2 are identified in text.

The company supplied a sample of the monoester as well as the diisobutyl ester. The diisobutyl ester has been identified in the diethyl ether-soluble extract of a reverse osmosis concentrate of 400 gal of Cincinnati, OH, drinking water [9].

The monoester, hereafter referred to as TMI, was checked by GC and two peaks were detected in a 1:2 ratio, the same ratio as peaks 1 and 2 in the sediment sample. Coinjection of TMI and the sediment extract confirmed the same retention times. A sample of TMI was hydrolyzed with ethanolic potassium hydroxide and the acid/neutral fraction obtained after acidification was checked by GC. The presence of only isobutyric acid and 2,2,4-trimethyl-1,3-pentanediol in the hydrolyzate as confirmed by coinjection of standards was evidence that the commercial TMI was a mixture of two positional isomers. The isobutane chemical ionization spectra for TMI differed in that peak 1 had high intensity ions at m/e 112 and 130, whereas peak 2 had high intensity ions at m/e 111 and 129. Both had an M+2 at m/e 218, which is not unusual for alcohols. An ion at m/e 200 indicated loss of water. Similar CI spectra were obtained for the sediment peaks. The EI spectra of TMI peaks 1 and 2 had the same mass peaks in the same relative intensities as peaks 1 and 2 of

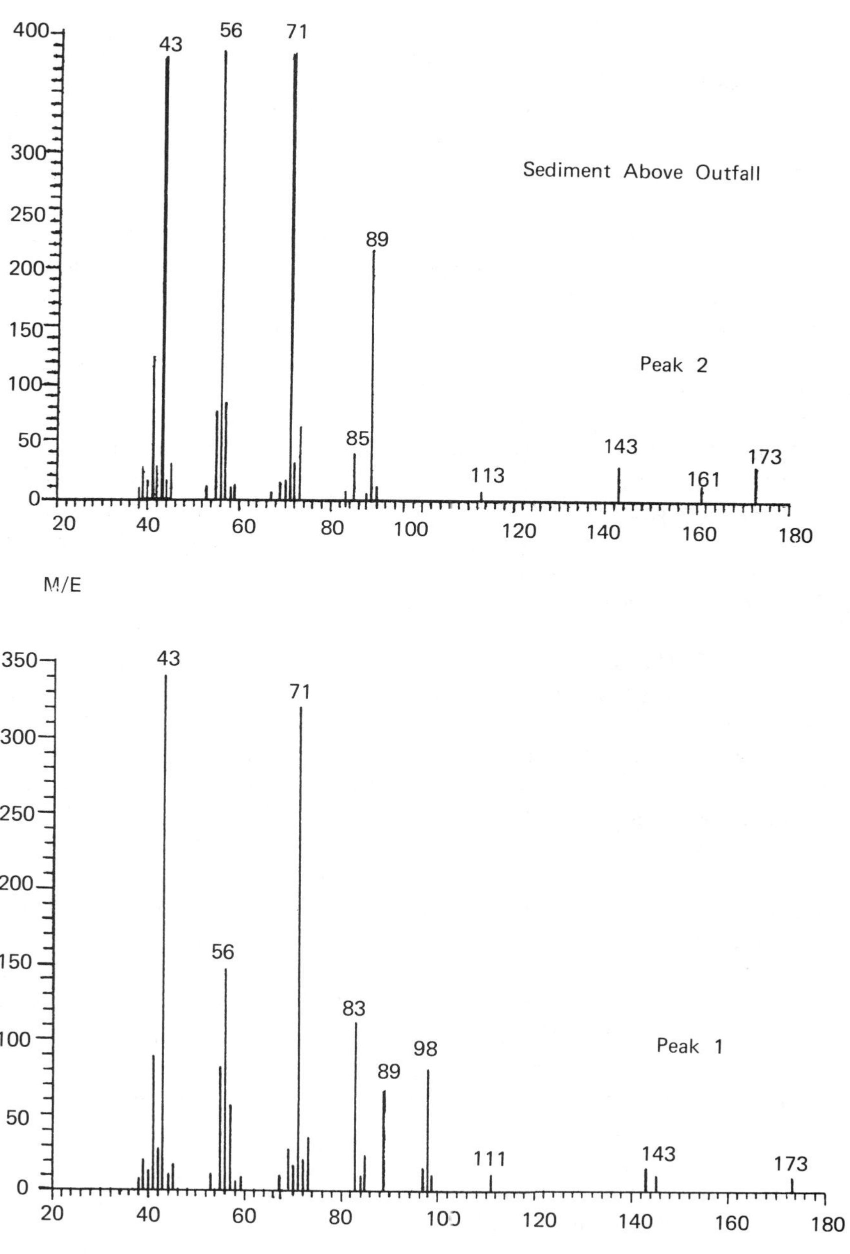


**Figure 6.** Mass spectra of peaks 1 and 2 from Figure 5.

the sediment sample. The results of analysis of TMI by hydrolysis, GC and GC/MS in the EI and CI modes when compared to analyses of peaks 1 and 2 in the sediment sample confirmed the presence of TMI.

The water associated with the sediment above the outfall was purged, after removal of the sediment by centrifugation, and peaks 1 and 2 were detected, as was fluorene.

The purging efficiency for TMI was checked by spiking a 100-mL water sample with TMI at 460 ppb and purging at a bath temperature of 65°C and filter at 80°C. The filter was extracted with carbon disulfide containing a known amount of decanol. A 32% recovery of TMI was calculated. To a second water sample spiked at the same 460-ppb level was added 50 g of lake sediment, and the sample was then purged as before. The recovery of TMI was reduced to 16% by the addition of sediment.

Figures 7, 8 and 9 are FID chromatograms obtained from purging, solvent-extracting and purging, respectively, of the solvent-extracted TMI contaminated sediment sample. The concentration of TMI on the sediment was determined by purging (Figure 7), and allowing for 16% recovery, was 3 ppm. Because solvent extraction in a soxhlet system is considered to be very efficient, the sediment was extracted for 16 hr with 50/50 hexane/acetone (Figure 8). The solvent-extracted sediment sample was then purged (Figure 9). The concentration of TMI obtained by totaling the soxhlet extraction and subsequent purge was 1.5 ppm. The water associated with the sediment contained TMI at 0.02 ppm.

Comparison of relative peak areas in Figures 7, 8 and 9 shows that TMI is present in Figure 8 (soxhlet extraction) in much lower relative concentrations to other compounds than in the purged samples (Figures 7 and 9). The area of peak 2 compared to the peak 3 area is 23:1 in Figure 7 and 0.27:1 in Figure 8, a 100-fold concentration increase in the purged sample. Conversely, the major peaks at 9.35-9.85 min in Figure 8 were only minor components of the purged sample.

The following uses of 2,2,4-trimethylpentanediol-1,3-monoisobutyrate were reported in the company literature: (1) a coalescing agent in latex paints, industrial latex finishes and water-base ink formulations; (2) a coalescing agent in poly(vinyl acetate) homopolymer and copolymer lattices; (3) an intermediate in chemical synthesis. Given these uses, the presence of TMI in an environmental sample is not surprising.

## Solvent Extraction of Sediment Samples

In addition to compounds sorbed to sediment that can be analyzed by a purge-and-trap method at either room or elevated temperatures, the extrac-

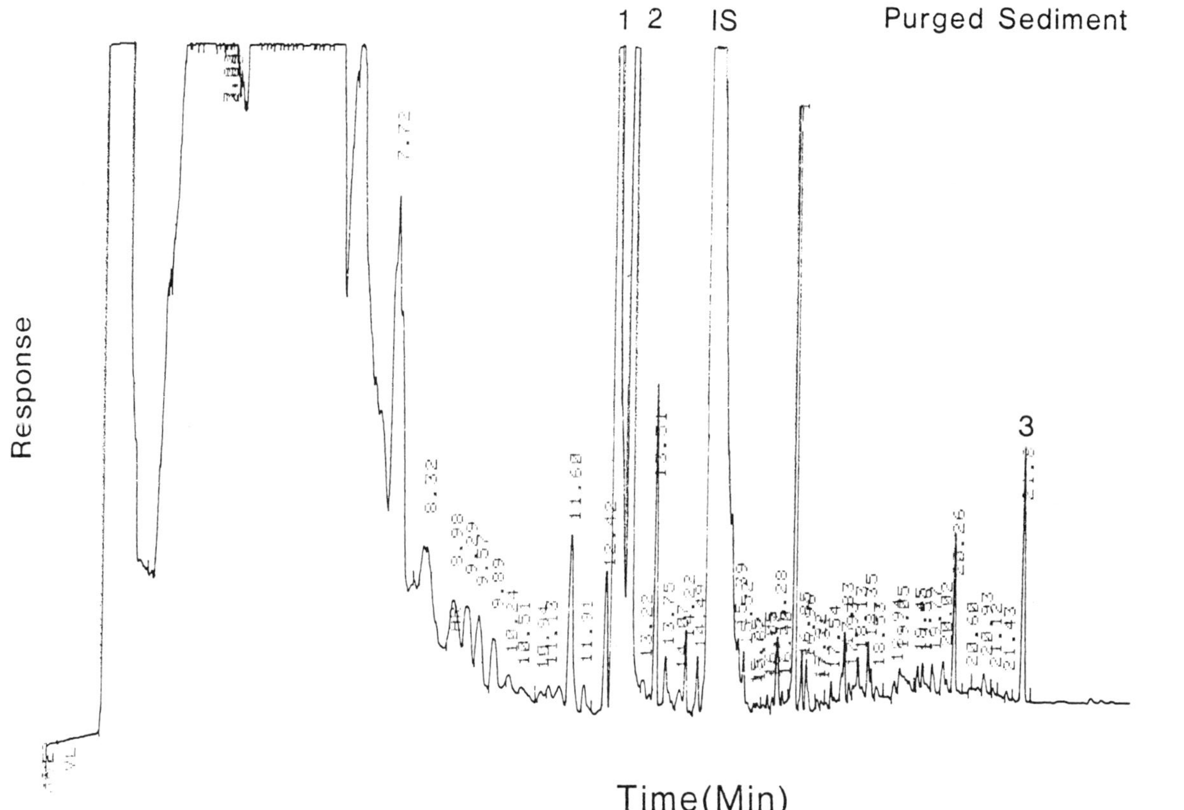


**Figure 7.** FID chromatogram from CLS purging of sediment above outfall at 65°C. Dodecanol (IS) was in the carbon disulfide extracting solvent. Peak identifications in text.

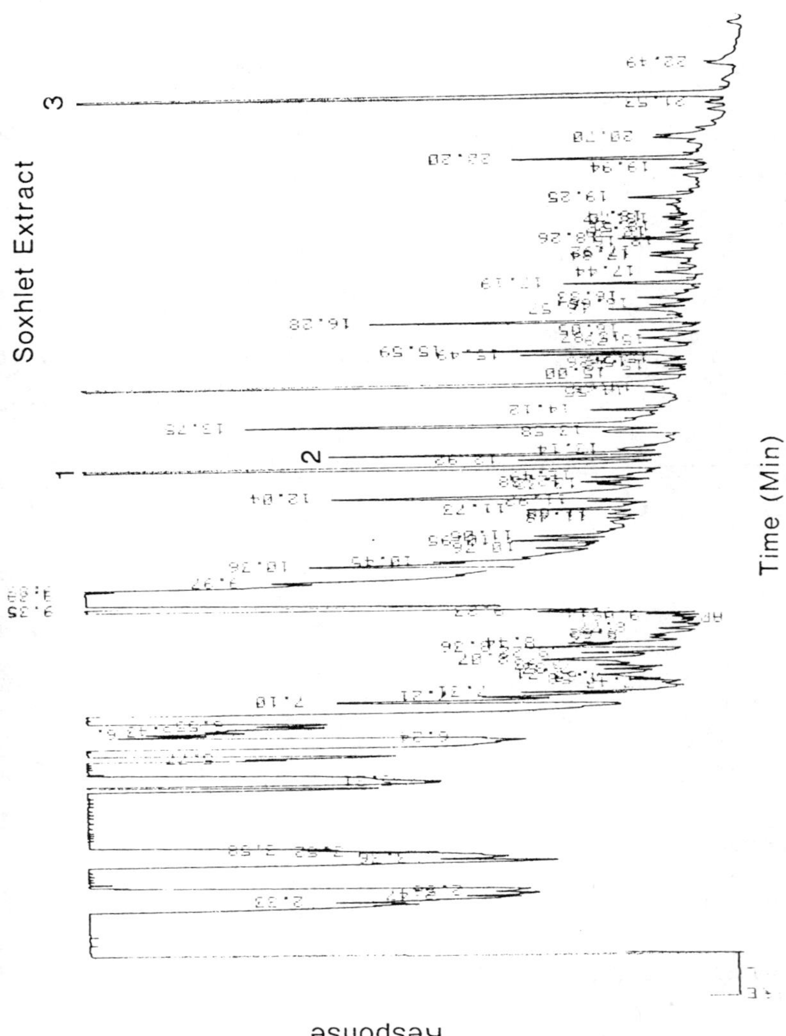


**Figure 8.** FID chromatogram from soxhlet extraction of sediment above outfall.

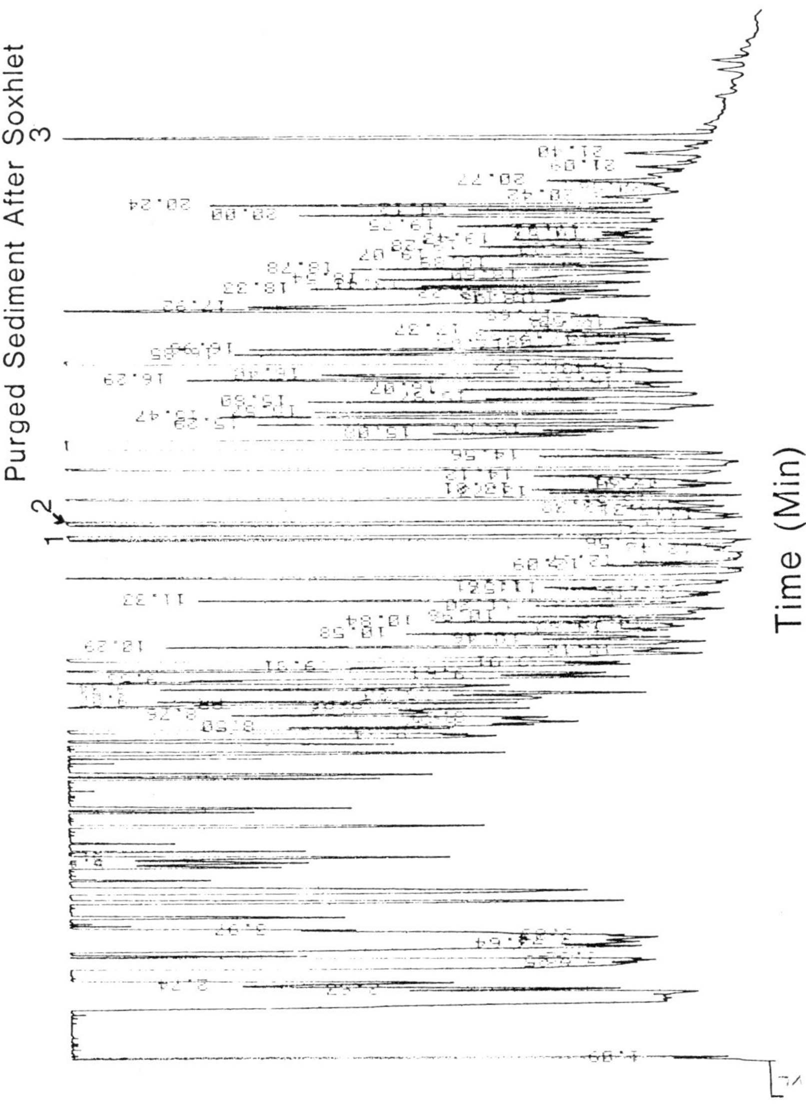


**Figure 9.** FID chromatogram from purging previously soxhlet extracted sediment (Figure 8).

tion of organics from sediments and their analysis by gas chromatography and GC/MS is of increasing importance. Radiolabeled compounds were utilized here to allow the use of real sediment samples in extraction studies while avoiding the interferences from previously sorbed organics. The radiolabeled compounds used were 9-fluorenone, dotriacontane and benzo[a]pyrene (B[a]P).

## Extraction Results and Discussion

The results in Table I are from an experiment to determine the completeness of sorption of 9-fluorenone to sediment at various sediment-in-water concentrations at two 9-fluorenone in water concentrations. At a sediment concentration of 250 $^{o}$/oo, the 9-fluorenone-C-14 at 250 ppb was almost completely sorbed from water. This sediment (94% 9-fluorenone-C-14 sorption) was solvent extracted by two methods (Table II). The sonication-only method (Heat Systems) was slightly more efficient than the sonication-homogenization (Polytron) method. In a study of extraction of chlordane from an environmental sample, sonication was equal to soxhlet extraction, but sonication-homogenization was only 75% as efficient [10].

**Table I. Adsorption of 9-Fluorenone to Sediment**

| Sediment in Water (ppm) | % Sorbed |
|---|---|
| 480 | 3.4[a] |
| 4,800 | 24.7[a] |
| 48,000 | 36.0[a] |
| 250,000 | 94.3[b] |

[a]9-Fluorenone concentration was 90 ppb.
[b]9-Fluorenone concentration was 250 ppb.

**Table II. Extraction of 9-Fluorenone with Hexane/Acetone (50/50)**

| 9-Fluorenone Determination | Sonication (% Extracted) | Sonication-Homogenization (% Extracted) |
|---|---|---|
| Counting Extracting Solvent | 83 | 69 |
| Oxidizing Sediment before and after Extracting | 83 | 70 |

$^{14}$C-B[a]P and dotriacontane were sorbed to sediment and extracted at two solvent polarities (Table III). The hexane/acetone (50/50) solvent was clearly superior for the extraction of the more polar B[a]P. The less polar dotriacontane is extracted by 100% hexane much more readily than the B[a]P, although the more polar hexane/acetone solvent is still significantly more efficient.

Table IV gives the results of an experiment based on a report by DeLeon et al. [11]. The extraction efficiencies, although not as good as sonication or sonication-homogenization, would be acceptable if a large number of samples had to be screened.

**Table III. Effect of Varying Solvent and Compound Polarities**

| Compound | Solvent | % Extracted |
|---|---|---|
| Benzo[a]pyrene | Hexane | 27 |
| | Hexane/Acetone (50/50) | 85 |
| Dotriacontane | Hexane | 74 |
| | Hexane/Acetone (50/50) | 96 |

**Table IV. Hypovial Extraction of 9-Fluorenone-Spiked Sediment**

| Solvent | % Extracted |
|---|---|
| Hexane/Acetone (50/50) 3 mL | 47 |
| Hexane/Acetone (50/50) 2 mL + 1 mL Water | 39 |

## CONCLUSIONS

The CLSA method offers several advantages over conventional purge-and-trap and microsolvent extraction techniques for desorption or organics from sediments. First, in a closed system, substances that break through the filter are automatically recycled. A second advantage is that a small volume of gas is constantly recycled through the filter, thus avoiding the large volumes of gas necessary in a conventional system to strip less volatile substances. Third, stripping methods, as opposed to extraction by solvent, are not as strongly influenced by matrix effects of the water, i.e., a high content of organic or inorganic particulate matter. Finally the CLS method is broader in scope; Grob [12] reports reasonable stripping efficiency for $C_{18}$ and lower substances at room temperature and for $C_{24}$ compounds at 80°C.

In this study, elevated temperature stripping of sediments gave a 100-fold concentration of TMI (Figure 7) compared to that obtained with soxhlet extraction. The TMI peaks were minor in the soxhlet extract and possibly would not have been identified. The low amount of TMI found in the associated water indicates a high partition from water to sediment. CLSA at elevated temperature, supplemented by solvent extraction for less volatile compounds, should be useful in screening sediments for a broad spectrum of organic compounds.

In addition, radiolabeled compounds were shown to be usedful in determining optimum extraction conditions (solvent, mechanical method).

## DISCLAIMER

Mention of trade names or commercial products does not constitute endorsement or recommendation for use by the U.S. Environmental Protection Agency.

## REFERENCES


1. Dietz, E. A., and K. F. Singley. *Anal. Chem.* 51:1809 (1979).
2. Kolb, B. *J. Chromatog.* 122:553 (1976).
3. Drozo, J., and J. Novak. *J. Chromatog.* 165:141 (1979).
4. Grob, K., K. Grob, Jr. and G. Grob. *J. Chromatog.* 106:299 (1975).
5. Thrun, K. E., and J. E. Oberholtzer. "Evaluation of the Microextraction Technique to Analyze Organics in Water," in *Advances in the Identification and Analysis of Organics in Water, Vol. 1*, L. H. Keith, Ed. (Ann Arbor, MI: Ann Arbor Science Publishers, Inc., 1981), Chapter 16.
6. Bellar, T. A., and J. J. Lichtenberg. *J. Am. Water Works Assoc.* 66:739 (1974).
7. Grob, K., and F. Zurcher. *J. Chromatog.* 117:285 (1976).
8. Carson, B., J. Going, V. Lopez, J. McCann and C. Cole. "The Distribution Register of Organic Pollutants in Water–WaterDROP," Chapter 32, this volume.
9. Coleman, W. E., R. G. Melton, F. C. Kopfler, K. A. Barone, T. A. Aurand and M. G. Jellison. *Environ. Sci. Technol.* 14:576 (1980).
10. Loy, W., Region IV, U.S. EPA, Athens, GA. Personal communication.
11. DeLeon, I. R., M. A. Maberry, E. G. Overton, C. K. Raschke, P. C. Remele, C. F. Steele, V. L. Warren and J. L. Laseter. *J. Chromatog. Sci.* 18:85 (1980).
12. Grob, K. *J. Chromatog.* 84:255 (1973).

## CHAPTER 41

# ORGANIC ANALYSES USING HIGH-TEMPERATURE PURGE-AND-TRAP TECHNIQUES

**R. L. Spraggins, R. G. Oldham,**
**C. L. Prescott and K. J. Baughman**

Radian Corporation
Austin, Texas

Studies at Radian Corporation have led to the development of useful techniques for analyzing soil, sludge and sediment samples. Analytical hardware has been built which permits the analysis of volatile organics from liquid, solid and mixed-phase samples. The methodology was based on purge-and-trap techniques first reported by Rook [1], Zlatkis et al. [2], Grob [3] and Mieure and Dietrich [4] in 1973.

The following year Bellar and Lichtenberg [5] refined the purge-and-trap technique. The U.S. Environmental Protection Agency (EPA) priority pollutant protocol used today for analysis of volatile organics in water is largely based on the work of these early investigators.

Over the past four years, the mass spectrometry group at Radian Corporation has been involved in wastewater analysis of volatile organics using the EPA priority pollutant protocol. The method involves purging volatiles from a water sample with an inert gas such as helium. The gas-phase organics from the water sample are then trapped from the purge gas on a solid sorbent system. The trapped organics, after concentration, are thermally desorbed onto an analytical system for analysis. The trap utilized for volatile priority pollutant analyses contains both six inches of Tenax-GC® (approximately 100 mg) and two inches of Davidson's silica gel (approximately 20 mg). Although this methodology works well for water sample, the procedure is not appropriate for semisolid and solid matrix samples.

The present analytical methodology was developed for analyzing semisolid and solid matrix samples including soil, sludge and sediment samples. The method is based on techniques similar to those of Bellar and Lichtenberg [5]. However, the samples are purged at an elevated temperature. The samples analyzed during this study included: water samples, soil from a petroleum landfarming site, solid waste samples from a chemical dumping area and an oily sludge from petroleum processing.

The effect of temperature on the purging efficiencies of volatiles from water samples was first reported by Kopfler et al [6]. In their study large water samples (140 and 500 mL) were purged at temperatures as high as 95°C. The effect of temperature on the purging efficiency of selected analytes from water was studied as a prelude to our analysis of solid matrix samples. Elevated purging temperatures increased the number of analytes that could be purged from water samples.

Two EPA laboratories have internally developed high-temperature purge-and-trap procedures for analyzing semisolid and solid matrix samples. Blazevich [7] published internal memos describing the purge-and-trap methodology shown in Figure 1. The purging chamber was heated to 50°C to increase analyte desorption from a water:sample mixture. This methodology was later used by EPA contractors for analyses of samples from the Love Canal area. Recently Speis [8] reported on another version of a high-temperature purge-and-trap device.

## HIGH-TEMPERATURE PURGE-AND-TRAP TECHNIQUES FOR SEMISOLID AND SOLID SAMPLES

Volatile organics from soil, sludge and sediment samples historically have been analyzed using static headspace techniques. This methodology is not very useful for environmental samples because detection limits are usually poor. The method is also very labor-intensive and, therefore, the cost of analysis is quite high.

The techniques developed in this study for analyzing volatiles from water, soil, sludge and sediment have been developed around "dynamic headspace" techniques. The methodology is based on high-temperature purging of vola-

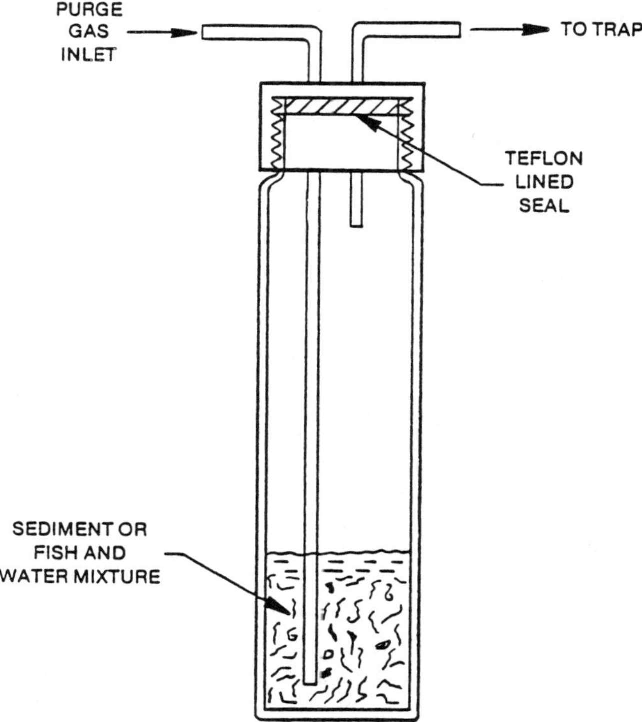


**Figure 1.** Modification of EPA purge Method 624.

tiles from a sample matrix with an inert purge gas. The volatiles are trapped from the purge gas on a solid sorbent, and then the trapped volatiles are thermally desorbed onto an analytical system for analysis. Detection limits are greatly improved compared to ordinary headspace analyses because analytes are concentrated on a solid sorbent prior to analyses.

Purging efficiency for selected analytes from water were studied at an elevated purging temperature. Water samples were spiked with various analytes at a level of 20 $\mu$g/L and were purged at 80°C. The apparatus used for these analyses was similar to the device used for purging solid samples which is described in the following section. Results from these purging studies are shown in Table I. Many more analytes could be analyzed by the purge-and-trap technique using elevated purging temperatures compared to the room-temperature conditions employed in conventional methodology. High-temperature purging was found to be even more important when analyzing solid matrix samples.

## ANALYTICAL HARDWARE

Hardware requirements for analyses of solids are more extensive than those needed for wastewater analyses. Two prototype purging chambers have been designed for these analyses (Figure 2). One is used primarily for solid samples. The other chamber is used for liquid and certain mixed-phase samples that can be introduced by syringe or pipet. The solids purging chamber is designed to be opened at a 24/40 standard taper joint for sample introduction and cleanup procedures. Purging chambers are made of glass and are equipped with a condenser at the gas exit of the chamber which reduces water levels in the exiting purge gases.

Samples are routinely heated during the purging phase of the analyses to effect increased desorption of analytes from sample matrices. The purge temperature necessary to effect complete desorption of an analyte is sample-dependent. Limited studies suggest that purging temperatures of 80–150°C or even higher may be required to completely desorb volatile analytes such as benzene from certain sample matrices. The prototype device uses an oil bath to heat the purging chamber.

**Table I. Purging Efficiency for Selected Compounds from Water at 80°C[a]**

| Compounds | Recovery Efficiency (%) | Coefficient of Variation (%) |
|---|---|---|
| Benzene | 99 | 7.9 |
| Toluene | 88 | 5.1 |
| Chlorobenzene | 84 | 5.1 |
| *m/p*-Xylene | 83 | 5.3 |
| *o*-Xylene | 82 | 5.4 |
| *m+p*-Dichlorobenzene | 90 | 15. |
| *o*-Dichlorobenzene | 78 | 10. |
| Nitrobenzene | 25 | 30. |
| 1,2,4-Trichlorobenzene | 58 | 6.7 |

[a]Analytical trap was Tenax-GC and silica gel; analytical column was Tenax-GC (6 in. x 2 mm i.d.).

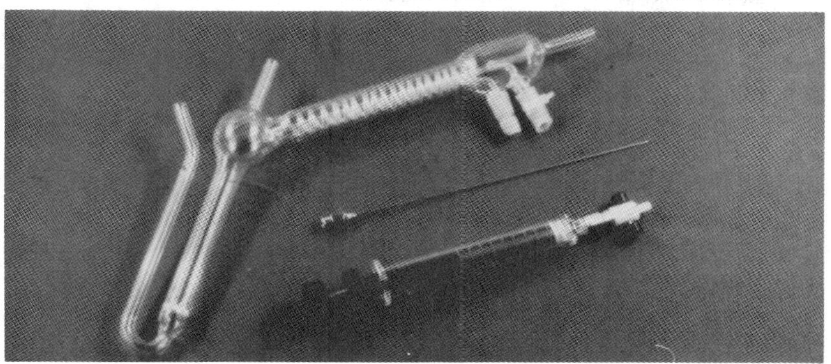

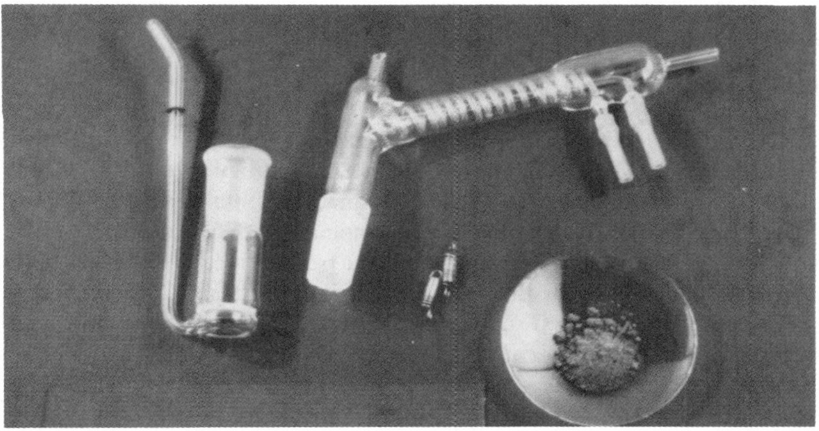

**Figure 2.** Liquid (top) and solids (bottom) purge assemblies.

Bellar's technique for wastewater analyses [5] uses room-temperature purge-and-trap procedures. Under these conditions, certain compounds with volatility equal to or greater than ethyl benzene can be analyzed quantitatively, but other compounds, e.g., the dichlorobenzenes, are not adequately purged. By heating the water samples to 80°C the dichlorobenzenes can be analyzed (see Table I). To avoid analyte condensation in the apparatus, valving and transfer lines are heated during analyses using elevated purging temperatures. All transfer lines are 1/8-in. i.d. glass-lined stainless steel. The lines and valves are first wrapped with Teflon® tape for electrical insulation. The heated zones are then wrapped with insulated nichrome wire and an outer insulation to prevent heat loss (Figure 3).

## ANALYTICAL PROCEDURES

The analytical procedures used were sample-dependent. For example, an oily sludge sample was dispersed on Celite prior to analysis to increase sample surface area, while other samples could be analyzed without prior sample preparation.

Selection of a representative sample is a major development problem when analyzing soil, sludge and sediment. In an attempt to minimize sample inconsistencies, a bulk soil sample from a petroleum landfarming facility was homogenized prior to analysis by lightly grinding it with a mortar and pestle. The bulk treated sample was then stored in a freezer in capped bottles until analysis. Reproducibility of the analytical method on this sample was found to be approximately 15%. Grinding procedures are not recommended for all samples, since volatile losses can occur during grinding.

Certain samples were also analyzed in replicate without any preparation. As expected, the reproducibility was poor compared to treated samples, and was for the most part dependent on the homogeneity of the sample.

The following two examples describe in detail procedures used for analyzing soil and sludge samples using the high-temperature purge-and-trap method.

### Analytical Procedure for Analyzing Soil Samples

A bulk soil sample obtained from a petroleum landfarming facility was first ground to a homogeneous mixture with a mortar and pestle. Care was taken not to pulverize the soil, but only to break the large clumps and to mix completely the sample. This preparation procedure is optional and can be used for any appropriate sample.

The bottom portion of the solids purging tube (see Figure 1) was tared on an analytical pan balance, and a 1-g soil sample was weighed into the purging vessel. The purging tube was then reassembled using a Teflon sleeve and springs to prevent leaks from occurring around the 24/40 tapered joint. The purge tube was then placed into the purging apparatus with Swagelok fittings.

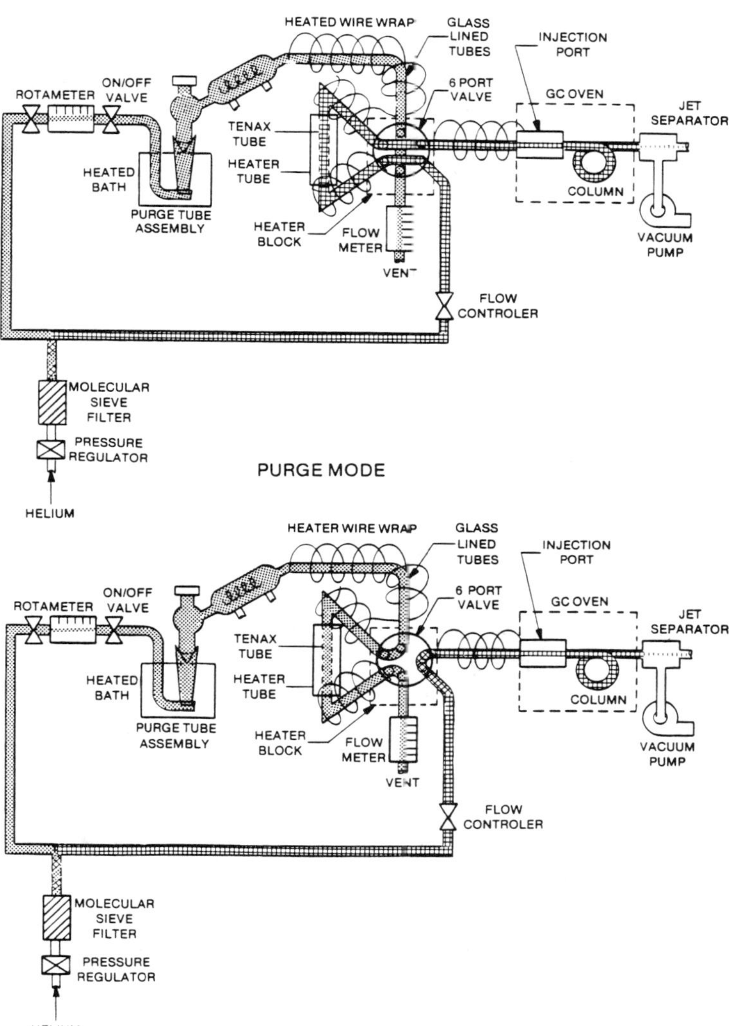


**Figure 3.** High-temperature purge-and-trap system GC/MS.

Bromochloromethane and 1,4-dichlorobutane (40 ng each) were spiked into the headspace of the purging chamber for use as internal standards. The sample was then heated in the purge chamber to 150°C in an oil bath and purged with helium gas at 10 mL/min for the first 10 min. The purge rate was then increased to 20 mL/min for an additional 20 min. The transfer lines and valves of the apparatus were then maintained at 150°C to prevent analyte condensation.

The purged analytes were trapped on an 1/8-in. glass-lined trap packed with 6 in. (approximately 100 mg) of Tenax-GC and 2 in. (approximately 20 mg) of Davidson 60/80 mesh silica gel. The analytical trap was maintained at room temperature during the purge cycle. The apparatus was then put into the desorb mode and the trap was desorbed at 250°C for 15 min with helium at 20 mL/min onto the head of an analytical column of a Hewlett-Packard 5982 GC/MS system. The 9-ft analytical column was packed with 0.2% Carbowax on Carbopak C and maintained at -50°C during the trap desorption. The column was then rapidly heated to an initial program temperature of 60°C. After holding for 4 min at 60°C, the temperature was programmed to 170°C at 8°C/min. The GC detector was a Hewlett-Packard 5982 repetitively scanned computer-controlled quadrupole mass spectrometer operated in the electron impact mode.

### Analytical Procedure for Analyzing Oily Sludge Samples

A 250-mg sample was weighed into the solids glass purging chamber containing 1.0 g of Celite. The sample was dispersed onto the Celite with the aid of a disposable pipet. The mixing end of the pipet was then broken off in the purging chamber and allowed to remain in the chamber during analysis. The purging chamber was then spiked with bromochloromethane and 1,4-dichlorobutane at 100 ng each. The sample was heated to 150°C and purged with helium at 10 mL/min for the first 10 min. The purge flow was then increased to 20 mL/min for 20 min. Apparatus transfer lines and valves were heated to 100°C for the analysis.

The purged volatiles were trapped on an analytical trap containing 6 in. (approximately 100 mg) of Tenax-GC and 2 in. (approximately 20 mg) of Davidson 60/80 mesh silica gel. The trapped volatiles were then desorbed at 250°C for 15 min onto the column of a Hewlett-Packard 5982 GC/MS system for analysis. The gas chromatography and mass spectrometry conditions were identical to those used for the soil sample described above.

## RESULTS OF HIGH-TEMPERATURE PURGE-AND-TRAP STUDIES

Two soil samples were analyzed using the described methodology without any prior sample preparation. The volatile organic priority pollutants found in these samples are shown in Tables II and III. Three replicate analyses were conducted for each soil sample. The replicate results are compared with sample results obtained using the EPA interim method for sediment and fish tissue (a static headspace analysis). Results in Table II and III point out two important analytical facts. The dynamic headspace, or high-temperature purge-and-trap, procedure was the more sensitive analytical method. A 1-g sample permitted priority pollutant detection at the 1-ppb level. Neither

**Table II. Volatile Priority Pollutants in Soil No. 1 Obtained from a Chemical Dump Site–Purge Temperature 80°C**

| Compounds Detected | Analyte Concentration (ppb) | | | |
|---|---|---|---|---|
| | By Headspace Analysis | By Modified Headspace | | |
| | | Analysis 1 | Analysis 2 | Analysis 3 |
| Methylene Chloride | ND[a] | 2.8 | ND | 2 |
| Chloroform | ND | ND | 1.8 | 5 |
| Benzene | 90 | 84 | 320 | 120 |
| Trichloroethylene | ND | 1 | 1 | 2.9 |
| Toluene | 110 | 150 | 240 | 250 |
| Tetrachloroethylene | ND | 14 | 24 | 16 |
| Chlorobenzene | ND | 3.7 | 4.3 | 6.9 |
| 1,1,2,2-Tetrachloroethane | ND | 28 | 100 | 160 |

[a]ND = not detected.

**Table III. Volatile Priority Pollutants in Soil No. 2 Obtained from a Chemical Dump Site–Purge Temperature 80°C**

| Compounds Detected | Analyte Concentration (ppb) | | | |
|---|---|---|---|---|
| | By Headspace Analysis | By Modified Headspace | | |
| | | Analysis 1 | Analysis 2 | Analysis 3 |
| Methylene Chloride | 360 | 40 | 15 | 20 |
| Chloroform | 55 | 5 | ND[a] | 1 |
| Benzene | 60 | 20 | 15 | 13 |
| Trichloroethylene | ND | 5 | 5 | 3.3 |
| Toluene | ND | 120 | 130 | 67 |
| Tetrachloroethylene | ND | 19 | 26 | 16 |
| Chlorobenzene | ND | 3.2 | 4.5 | 3.9 |
| 1,1,2,2-Tetrachloroethane | ND | 29 | 45 | 79 |

[a]ND = not detected.

analytical method was found to be very reproducible. A large number of static headspace analyses was run in our laboratory prior to the dynamic headspace work. Representative sample aliquots could not be obtained for either method without prior sample homogenization. By comparison, the replicate analyses shown in Table IV for soil sample No. 3 show good reproducibility. This sample was homogenized by grinding before sample aliquots were taken. The sample was also analyzed at several different purging temperatures. The results of these analyses are shown in Table V. These results demonstrate a binding or complexing property associated with the soil for volatile organic compounds. Few or no analyses were purged from the soil below 100°C. The binding effect this soil shows for light aromatic hydrocarbons could be a useful property for removing trace organics from ground- and wastewaters. A useful application for the high-temperature purge-and-trap analysis might be studying organic binding potentials for soil samples.

An oily sludge sample analyzed by the methodology described earlier was found to contain large amounts of volatile organics. The sample size had to

**Table IV. Volatile Aromatic Compounds from Land Farm Soil Sample No. 3– Purge Temperature 150°C**

| | Concentration (ppb) | | |
|---|---|---|---|
| Compounds Detected | Analysis 1 | Analysis 2 | Analysis 3 |
| Benzene | 140 | 160 | 190 |
| Toluene | 120 | 130 | 130 |
| Ethylbenzene | 160 | 170 | 180 |
| *o*-Xylene | 160 | 160 | 100 |
| *m+p*-Xylene | 300 | 300 | 330 |

**Table V. Volatile Aromatic Compounds from Land Farm Soil Sample No. 3 at Various Purge Temperatures**

| | Concentration (ppb) | | | |
|---|---|---|---|---|
| Compounds Detected | Analysis 1 50°C Purge | Analysis 2 80°C Purge | Analysis 3 120°C Purge | Analysis 4 150°C Purge |
| Benzene | 2.3 | 11 | 49 | 140 |
| Toluene | 2.8 | 7.6 | 36 | 120 |
| Ethylbenzene | 1.9 | 5.3 | 38 | 160 |
| *o*-Xylene | ND[a] | 7.1 | 47 | 160 |
| *m+p*-Xylene | ND | 16 | 80 | 300 |

[a]ND = not detected.

be reduced to 250 mg and the sample was dispersed on Celite to increase sample surface area. Only three aromtic analytes were measured. Benzene was measured at 6.0 ppm in the sample, while toluene and ethylbenzene were measured at 3.4 and 3.0 ppm, respectively. Two other sludge samples were diluted and analyzed by direct injection using GC/MS techniques. They were found to contain benzene at 700 and 1500 ppm, respectively. The volatile organics in these samples were too high for purge-and-trap analysis. It was impossible to weigh out a sample small enough to keep the desorbed volatile organics within the measurable range of the mass spectrometer.

A major advantage to using high-temperature purge-and-trap techniques is the expanded range of volatile compounds that can be analyzed (see Table I). Two soil samples from chemical dump sites were analyzed using an 80°C purging temperature, and the results of this work also demonstrate the expanded analysis range. The total results for these analyses are found in Tables VI and VII, while the volatile priority pollutants are reported in Tables II and III. Semivolatile analytes as large as dihydrophenanthrene were purged and trapped from these samples. The analytes were trapped on an 1/8-in. i.d. stainless steel glass-lined tube packed with 8 in. (approximately 120 mg) of Tenax GC.

Very polar compounds such as nitrobenzene and aniline were also detected using this methodology. A comparison of aniline and nitrobenzene results using the high-temperature purge-and-trap methodology with extraction techniques is found in Table VIII. Much more material was found by extraction. However, a purge temperature of only 80°C was used for these analyses. If higher purging temperatures were used, a higher recovery for these polar compounds might be obtained.

## CONCLUSION

The useful range for volatile organic analyses by the purge-and-trap technique is extended by high-temperature sample purging. Analytes considered only semivolatile, such as napthalene and acenapthene, can be analyzed by purge-and-trap techniques at elevated purging temperatures.

The detection limits for volatile organics in solid samples are improved using dynamic headspace techniques compared to static headspace analysis. Generally, a two-order-of-magnitude improvement is observed. The dynamic headspace technique will permit environmental monitoring of many volatile organics in semisolid and solid matrix samples at the 1-ppb level.

The development of this methodology is timely. The national environmental emphasis appears to be shifting from water matrix samples to monitoring studies of solid waste samples. Internally, the EPA has developed two alternative methods for determining volatile organics in these samples. A

Table VI. High-Temperature Purge-and-Trap of Soil No. 1 Taken from a Chemical Dump Site–Purge Temperature 80°C[a]

| Compound | Concentration (ppb) | | |
|---|---|---|---|
| | Analysis 1 | Analysis 2 | Analysis 3 |
| Methylene Chloride | 2.8 | NA[b] | 2.0 |
| Chloroform | NA | 1.8 | 5.0 |
| Benzene | 84 | 320 | 130 |
| Trichloroethylene | 1.0 | 1.0 | 2.9 |
| Toluene | 150 | 240 | 250 |
| Tetrachloroethylene | 14 | 24 | 16 |
| $\alpha,\alpha,\alpha$-Trichlorotoluene | 19 | 26 | 33 |
| Chlorobenzene | 3.7 | 4.3 | 6.9 |
| *m*+*p*-Xylene | 190 | 230 | 310 |
| *o*-Xylene | 100 | 110 | 150 |
| Styrene | 93 | 80 | 130 |
| 1,1,2,2-Tetrachloroethane | 28 | 100 | 160 |
| *n*-Decane | 170 | 110 | 230 |
| $C_3$-Alkylbenzenes | 70 | 150 | 160 |
| Benzaldehyde | NA | 48 | 56 |
| Indane/Methylstyrene | NA | 24 | 40 |
| Aniline | NA | 130 | 170 |
| $C_4$-Alkylbenzenes | NA | 36 | 58 |
| Indane | NA | 26 | 48 |
| Nitrobenzene | NA | 260 | 290 |
| $C_5$-Alkylbenzenes | NA | 32 | 44 |
| Methylindane | NA | 8.2 | 11 |
| $C_6$-Alkylbenzenes | NA | 96 | 140 |
| Naphthalene | NA | 38 | 63 |
| $C_7$-Alkylbenzenes | NA | 90 | 100 |
| 2-Methylnaphthalene | NA | 16 | 18 |
| 1-Methylnaphthalene | NA | 18 | 18 |
| $C_8$-Alkylbenzenes | NA | 60 | 60 |
| Biphenyl | NA | 28 | 23 |
| $C_9$-Alkylbenzenes | NA | 19 | 20 |
| $C_2$-Alkylnaphthalenes | NA | NA | 98 |
| Acenaphthylene | NA | 26 | 25 |
| Acenaphthene | NA | 7.6 | 8.3 |
| Dihydrophenanthrene | NA | 11 | 13 |

[a]Analytical trap was Tenax-GC.

[b]NA = not analyzed.

**Table VII. High-Temperature Purge-and-Trap of Soil No. 2 Taken from a Chemical Dump Site–Purge Temperature 80 °C[a]**

| Compound | Concentration (ppb) | | |
|---|---|---|---|
| | Analysis 1 | Analysis 2 | Analysis 3 |
| Methylene Chloride | 40 | 15 | 20 |
| Chloroform | 5.0 | NA[b] | 1.0 |
| Benzene | 20 | 15 | 13 |
| Trichloroethylene | 5.0 | 5.0 | 3.3 |
| Toluene | 120 | 130 | 67 |
| Tetrachloroethylene | 19 | 26 | 16 |
| $\alpha,\alpha,\alpha$-Trichlorotoluene | 39 | 38 | 16 |
| Chlorobenzene | 3.2 | 4.5 | 3.9 |
| *m+p*-Xylene | 94 | 130 | 96 |
| *o*-Xylene | 47 | 62 | 51 |
| Styrene | 15 | 17 | 17 |
| 1,1,2,2-Tetrachloroethane | 29 | 45 | 79 |
| *n*-Decane | 61 | 67 | 100 |
| $C_3$-Alkylbenzenes | 87 | 136 | 97 |
| Indane/Methylstryene | 7.6 | 15 | 10 |
| Aniline | 540 | 1100 | 670 |
| Indane/Methylstyrene | 3.6 | 5.4 | 3.5 |
| $C_4$-Alkylbenzenes | 140 | 280 | 170 |
| Indane | 3.2 | 6.9 | 4.4 |
| Nitrobenzene | 720 | 2100 | 1500 |
| $C_5$-Alkylbenzenes | 430 | 740 | 470 |
| Trichlorobenzene | 6.8 | 8.5 | 9.0 |
| Methylindane | 13 | 8.3 | 11 |
| $C_6$-Alkylbenzenes | 670 | 1200 | 700 |
| Naphthalene | 34 | 81 | 33 |
| $C_7$-Alkylbenzenes | 400 | 720 | 420 |
| 2-Methylnaphthalene | 34 | 71 | 47 |
| 1-Methylnaphthalene | 63 | 140 | 74 |
| $C_8$-Alkylbenzenes | 110 | 220 | 110 |
| Biphenyl | 18 | 75 | 49 |
| $C_9$-Alkylbenzenes | 16 | 40 | 16 |
| $C_3$-Alkylnaphthalene | 140 | 390 | 220 |
| Acenaphthylene | 2.7 | 29 | 14 |
| Acenaphthene | NA | 11 | 5.6 |
| Dihydrophenanthrene | NA | 4.0 | 2.1 |

[a]Analytical trap was Tenax-GC.
[b]NA = not analyzed.

**Table VIII. Comparison of Sample Results (ppb) for the High-Temperature Purge-and-Trap Method with Extraction Techniques for Aniline and Nitrobenzene–Purge Temperature 80°C[a]**

| | Soil No. 1 | | Soil No. 2 | |
|---|---|---|---|---|
| Analyte | Modified Headspace | Extraction Techniques | Modified Headspace | Extraction Techniques |
| Aniline | 150 | 3,700 | 770 | 11,000 |
| Nitrobenzene | 280 | ND[b] | 1,400 | 42,000 |

[a]Purging chamber = 80°C; transfer lines = 80°C; analytical trap = Tenax-GC; analytical column = Tenax-GC; purging time = 30 min; purge volume = 500 mL helium.

[b]ND = not detected.

comparison of the two EPA methods with the Radian Corporation method of analysis should be undertaken. Methods for the analysis of solid waste samples are currently under consideration and will have to be standardized before a comprehensive monitoring program can begin.

Our studies and the work of others suggest that several variables need to be considered in analyzing solid matrix samples. These variables include:

- surface of solid samples;
- water content of samples;
- nonvolatile sample component partitioning; and
- lack of homogeneity in sample analyte concentration.

These variables complicate sample analysis and result in high analytical variability. After analyzing numbers of solid matrix samples by both static and dynamic headspace methods, it appears that sampling and sample preparation prior to analysis is the key to good analytical reproducibility. We have shown analyte variability of greater than 100% in replicate samples without prior sample preparation using either methodology. However, replicate samples that received prior preparation in the form of grinding and mixing showed the same analytes can be measured with only about 15% variability.

Samples have been analyzed by the dynamic headspace technique with purging temperatures from 25 to 150°C. Traps packed with Tenax-GC or a combination of Tenax-GC and Davidson silica gel have been used for specific applications. Future studies will include purging temperatures as high as 200°C for specific applications. Published work involving static headspace techniques have been reported using sample analysis temperatures as high as

200°C without analyte degradation [9]. Other future studies will be centered around the solid sorbent used for analyte trapping. Analysis of semivolatile compounds, such as nitrobenzene and aniline, will be optimized by choosing the appropriate solid sorbent. The appropriate material will effectively trap selected analytes, but will readily give up the analyte on heating.

## ACKNOWLEDGMENTS

We wish to acknowledge the work of C. F. Mattox and J. E. Hudson, who helped process mass spectrometry data for this study.

## REFERENCES

1. Rook, J. J. *Water Treat. Exam.* 21:259 (1973).
2. Zlatkis, A., H. A. Lichtenstein and A. Tishbee. *Chromatographia* 6:67 (1973).
3. Grob, K. *J. Chromatog.* 84:255 (1973).
4. Mieure, J. P., and M. W. Dietrich. *J. Chromatog. Sci.* 11:559 (1973).
5. Bellar, T. A., and J. J. Lichtenberg. *J. Am. Water Works Assoc.* 66:703, 739 (1974).
6. Kopfler, F. C., R. G. Melton, R. D. Lingg and W. E. Coleman. "GC/MS Determination of Volatiles for the National Organics Reconnaisance Survey (NORS) of Drinking Water," in *Identification and Analysis of Organic Pollutants in Drinking Water*, L. H. Keith, Ed. (Ann Arbor, MI: Ann Arbor Science Publishers, Inc., 1976), pp. 87-104.
7. Blazevich, J. M. Personal communication (August 1980).
8. Speis, D. A. Paper presented at "Seminar/Workshop on Sampling and Analysis of Sediments, Bottoms Material, and Fish," U.S. EPA, EMSL, Cincinnati, OH, October 14-17, 1980.
9. O'Reilly, W. F. *Gov. Repts. Announcements (U.S.)* 74(20) (1974).

CHAPTER 42

# DEVELOPMENT OF METHODS FOR THE ANALYSIS OF PURGEABLE ORGANIC PRIORITY POLLUTANTS IN MUNICIPAL AND INDUSTRIAL WASTEWATER TREATMENT SLUDGES

**C. L. Haile, Y. A. Shan, L. S. Malone and R. V. Northcutt**

Midwest Research Institute
Kansas City, Missouri

The primary objective of this research program was to develop methods for the analysis of purgeable organic priority pollutants in sludges generated by the treatment of municipal and industrial wastewaters. Studies of emissions from wastewater treatment plants have historically focused on determining the concentrations of contaminants in influent and effluent wastewaters, and on their removal as the wastewater passed through the treatment system; Likely contaminant removal mechanisms include volatilization, chemical and/or biochemical degradation, and removal of solids, i.e., sedimentation. Many of the organic priority pollutants have very low aqueous solubilities and tend to associate with suspended solids (SS) in wastewater. Since solids removal is an important part of wastewater treatment and the solids are likely carriers of hydrophobic contaminants, sludge wasting may be a primary emission mode for organic priority pollutants.

This research was primarily directed toward applying and modifying the purge-and-trap gas chromatography/mass spectrometry (GC/MS) procedures specified by the U.S. Environmental Protection Agency (EPA) for industrial wastewater analyses [1]. The wastewater method was adapted from the original purge-and-trap GC procedure for volatile compounds in water reported by Bellar and Lichtenberg [2]. Volatile compounds that are sparingly soluble in water are sparged from a 5-mL sample aliquot with an inert gas stream. The analytes are adsorbed from the gas stream by a trap containing Tenax GC® and silica gel. The adsorbed compounds are then thermally desorbed into the inlet of a gas chromatograph interfaced with a mass spectrometer.

In response to an immediate need expressed by EPA for a purgeables method for municipal sewage treatment sludges, a preliminary purge-and-trap GC/MS method was developed by Midwest Research Institute (MRI) using sample dilution to control foaming [3]. Preliminary experiments with sludges from publicly owned treatment works (POTW) indicated that most, but not all, sludges could successfully be purged if the sample aliquot was diluted to approximately 0.5% (w/V) total solids with a 10-mL total volume. The recoveries for analytes spiked into POTW sludges and analyzed by this method are typical of those observed for highly polluted wastewaters. However, the precisions of the recovery determinations are frequently ±40–50% relative standard deviation. In addition, sample dilution decreases the effective method detection limit.

The experiments described below were designed to develop a purging procedure with improved recoveries, reproducibility and detection limits. A comparison of recoveries from spiked sludge and sludge supernatant was conducted to investigate the relative influences of dissolved solids (DS) and SS on volatiles purging. The feasibility of using salt or organic additives to improve purging reproducibility was investigated. Finally, alternative purging systems incorporating mechanical mixing were evaluated to enhance analyte transfer to the purge gas stream.

## EXPERIMENTAL

### Sample Handling and Spiking

All sludge samples were stored at 4°C with no headspace in 40-mL screw-cap vials sealed with TFE-lined septa. Vials of sludge were spiked by removing the cap, adding two to five 3.2-mm (1/8-in.) diam. stainless steel balls, injecting 10–20 $\mu$L of the spike solution (in methanol) under the surface of the sample and resealing with no headspace. Unless indicated, all samples were allowed to equilibrate overnight at 4°C (with tumbling) prior to analysis. The spiking and recovery experiments were conducted using a subset of the purgeable organic priority pollutants to simplify the method evaluations. The spiking compounds (Table I) were selected to represent the physical and chemical characteristics of the purgeable compounds. The solids contents of all sludge samples used for method evaluations were determined gravimetrically for triplicate 1.0-mL aliquots dried overnight at 110°C.

### Preliminary Purge-and-Trap GC/MS Method for POTW Sludge

Schematics of the purge-and-trap system in the sparge (or purge) and desorb modes are shown in Figures 1 and 2, respectively. The Tenax–silica gel trap used for industrial wastewater was modified as suggested by Bellar [4] to improve adsorption of the very volatile fluorochloromethanes. The packing of this trap is shown in Figure 3. The sample purging tube used by MRI for wastewater is shown in Figure 4. The tube is a chromatographic column with a medium porosity frit pressed into the TFE stopcock assembly.

An aliquot of the sludge containing 0.5 g total solids (TS) was transferred to the purge tube and diluted to 10 mL with volatile-free water. The diluted sludge aliquot was spiked with 9 $\mu$L of an internal standard spiking solution containing 20 ng/$\mu$L each of bromochloromethane, 2-bromo-1-chloropropane and 1,4-dichlorobutane in organic-free water. The purge tube was sealed into the purging system. The sludge was purged with prepurified nitrogen for 12 min at 40 mL/min. Following purging, the trap was rapidly heated to 180°C to thermally desorb the volatiles into the inlet of the gas chromatograph. The desorption condition (with the trap at 180°C) was maintained for 3 min. The column temperature program was initiated at the same time desorption was started. The GC column (2.4 m x 2 mm i.d. glass, packed with 0.2% Carbowax 1500 on 60/80 mesh Carbopack C) was held at 40°C for 3

**Table I. Representative Purgeable Organic Priority Pollutants Used for Method Development and Evaluation Experiments**

| |
|---|
| Benzene |
| Carbon Tetrachloride |
| Chlorobenzene |
| Chloroform |
| 1,2-Dichloroethane |
| 1,1-Dichloroethene |
| Ethyl Benzene |
| Tetrachloroethane |
| 1,1,1-Trichloroethene |
| Trichloroethane |
| Vinyl Chloride |

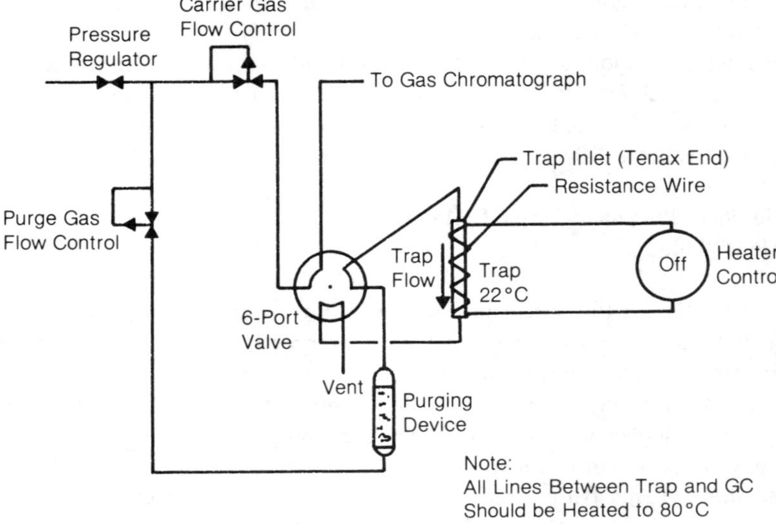


**Figure 1.** Schematic of the purge-and-trap system in the purge mode.

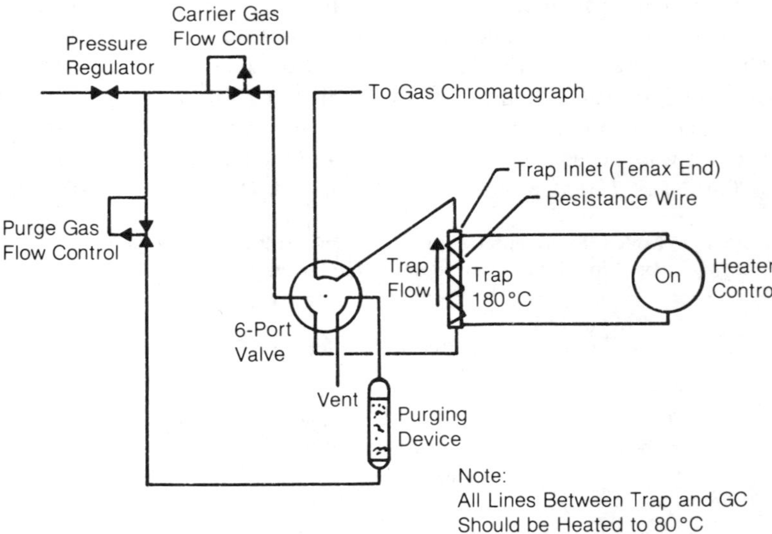


**Figure 2.** Schematic of the purge-and-trap system in the desorb mode.

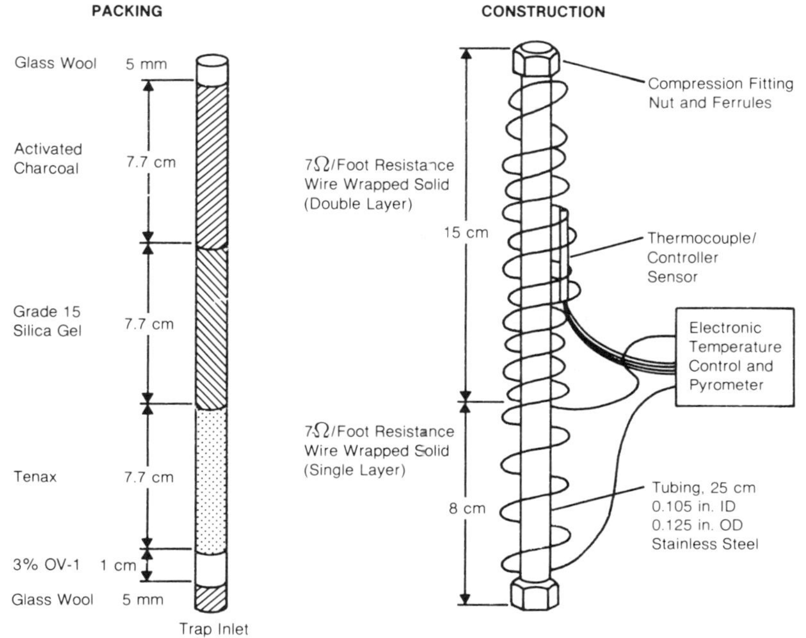


**Figure 3.** Trap packings and construction.

min, programmed to 170°C at 10°C/min, and then maintained at 170°C until after the elution time for ethyl benzene. The mass spectrometer was repetitively scanned over the range m/e 20–275 at 5 sec/scan.

The analytes were identified by the coincidence of peaks in extracted ion current plots (EICP) plotted for characteristic ions at the appropriate retention time and in the characteristic relative intensity ratios as described in the EPA industrial wastewater protocol. The analytes were quantified by comparing the height of the EICP peaks with those for mixed standards spiked into volatile-free water and were analyzed in the same manner as the diluted sludge aliquots. Concentrations of analytes were calculated using internal standard correction.

Some method evaluations with spiked clean water were conducted using purge-and-trap gas chromatography/flame ionization detection (GC/FID). Analytes were identified by retention times and were quantified using peak areas or heights.

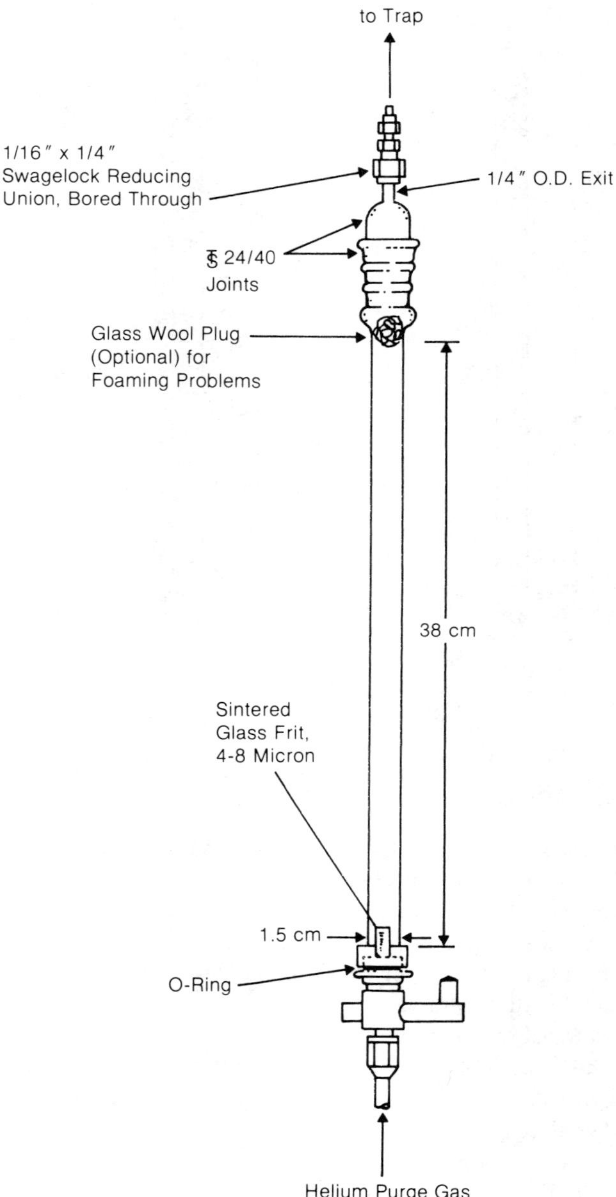


**Figure 4.** Sludge purging tube.

### Alternative Sludge Diluents

Saturated sodium chloride in volatile-free water and 10% ethylene glycol in volatile-free water were also used as sludge diluents.

### Alternative Purging Hardware for Sludge

Three alternative purge hardware systems were evaluated. There were stirred purging and dynamic headspace systems configured from a 100-mL round-bottom flask (Figures 5 and 6) and a stirred bottom frit tube designed by the authors (Figure 7).

## METHOD STUDIES

### Influences of DS and SS on Purging Efficiencies

The relative influences of DS and SS on the purging efficiencies of analytes from sludges were investigated by determining analyte recoveries from spiked sludge and spiked sludge supernatant. Aliquots of primary POTW sludge (4.8% solids) were centrifuged, and the supernatants were decanted. The supernatants and replicate aliquots of unfractionated sludge were spiked and analyzed by the preliminary POTW sludge method.

The results of the recovery determinations are shown in Table II. The recoveries for the spiked compounds from the sludge supernatant were both higher and less variable than from the unfractionated sludge for all compounds. Furthermore, the range of recoveries observed for the supernatant aliquots was similar to those typically observed for spiked clean water. Hence, it is likely that the presence of DS in the primary sludge does not significantly affect the purging efficiencies of the spiked compounds. Subsequent experiments were focused on examining procedures with a potential for enhancing the release of analytes from SS.

### Evaluation of the Addition of Salt or a High-Boiling Organic to Enhance Purging Efficiencies

The addition of salt or a high-boiling hydrophilic organic compound (with a very low tendency to purge from water) to sludge aliquots prior to purging was evaluated as a means of enhancing the purging efficiencies of volatile compounds, or at least improving the reproducibility of purging. The recov-

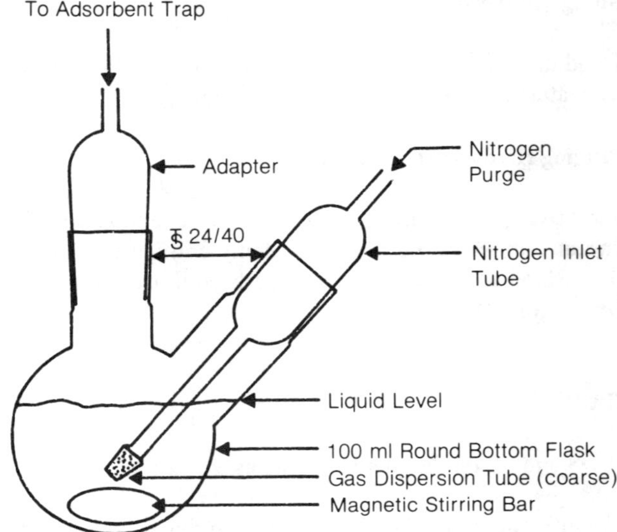


**Figure 5.** Stirred purging system using a round-bottom flask.

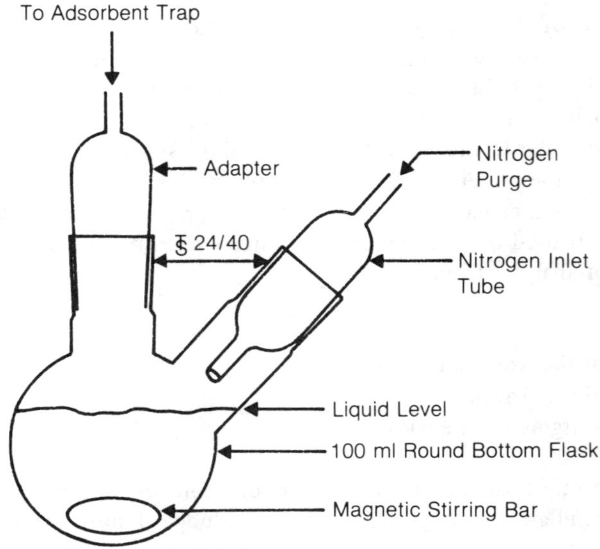


**Figure 6.** Stirred impinger system using a round-bottom flask.

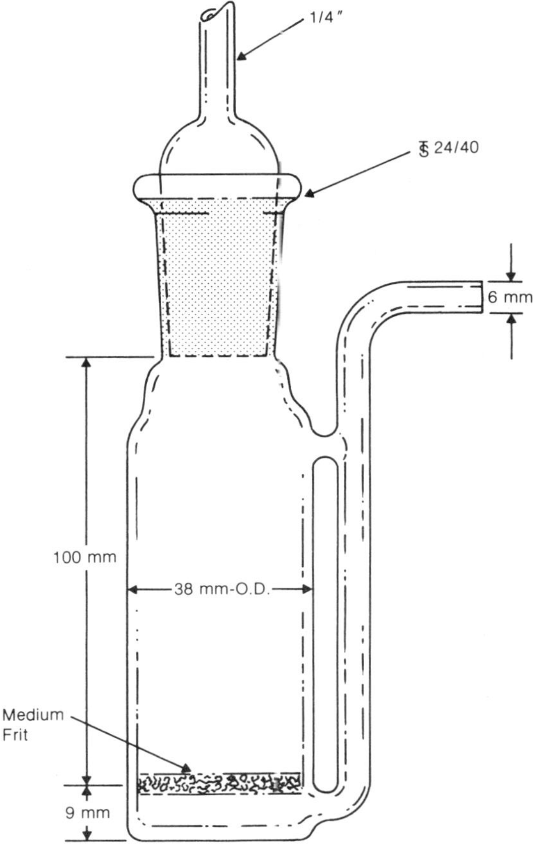


**Figure 7.** Bottom frit purge tube.

eries of compounds spiked into sludge and sludge supernatant indicated that interactions of the purgeable compounds with the SS may have a significant influence on their purging efficiencies. The addition of salt or a high-boiling organic was an attempt to alter favorably those interactions to provide a consistent matrix for purging. Adding salt may enhance the purging efficiencies for some compounds by decreasing their solubilities in the aqueous sludge matrix. However, increasing the polarity of the matrix may encourage stronger association of some compounds with the sludge solids and decrease their availability for purging. Alternatively, the addition of a high-boiling

Table II. Recoveries of Purgeable Compounds from Unspiked and Spiked Primary Sludge and Primary Sludge Supernatant by Purge-and-Trap GC/MS

| | | Sludge | | Sludge Supernatant | |
|---|---|---|---|---|---|
| Compound | Nominal Spike Level ($\mu$g/L) | Unfortified Concentration ($\mu$g/L) | Spike Recovery (%) | Unfortified Concentration ($\mu$g/L) | Spike Recovery (%) |
| Benzene | 250 | ND[a] | 58 ± 54[b] | ND | 130 ± 11 |
| Carbon Tetrachloride | 250 | ND | 47 ± 58 | ND | 110 ± 13 |
| Chloroform | 250 | ND | 37 ± 20 | ND | 120 ± 14 |
| 1,1-Dichloroethene | 250 | ND | 52 ± 33 | ND | 120 ± 20 |
| Tetrachloroethene | 250 | 10 | 72 ± 53 | ND | 90 ± 14 |
| Vinyl Chloride[c] | 250 | ~4100 | | | |
| 1,2-Dichloroethane | 250 | ND | 71 ± 44 | ND | 120 ± 13 |
| Trichloroethene | 250 | 6 | 53 ± 42 | ND | 110 ± 16 |
| 1,1,1-Trichloroethane | 250 | 59 | 83 ± 54 | ND | 120 ± 14 |
| Chlorobenzene | 250 | ND | 61 ± 44 | ND | 100 ± 5 |
| Ethyl Benzene | 250 | 325 | d | 27 | 95 ± 12 |

[a]ND = not detected.
[b]Recovery ± standard deviation for triplicate determinations.
[c]Present in sludge at very high concentrations beyond the method dynamic range. Not quantified in sludge fractions.
[d]Concentration in spiked sludge not determined because of peak saturation.

hydrophilic organic may lower the polarity of the matrix and encourage desorption of analytes from solids while tending to retard purging efficiencies. In either case, modification of the sludge matrix to a relatively consistent polarity may improve the reproducibility of recoveries, whatever the effects on the magnitude of recoveries.

The influences of salt or an organic on purging from aqueous media were investigated by analyzing spiked clean water by purge-and-trap GC/FID. Aliquots of volatile-free water (2.0 mL) were spiked and diluted to 10 mL prior to analysis with one of three diluents: (1) volatile-free water, (2) saturated sodium chloride in volatile-free water, or (3) 10% ethylene glycol in volatile-free water. The detector responses were compared to determine the influence of each diluent. Ethylene glycol was selected as the organic additive because of its low volatility, good aqueous solubility and general availability.

The results of these preliminary experiments indicated that the purging of most compounds does not appear to be significantly perturbed by the addition of salt or ethylene glycol. Chloroform and 1,2-dichloroethane exhibited a slightly higher response in water when salt was added. However, the responses for chloroform and 1,1,1-trichloroethane were slightly higher in water plus ethylene glycol. The sum of trichloroethene and benzene (coeluting) produced a response more than two times higher in the ethylene glycol–spiked solution. The addition of salt or ethylene glycol did not appear to decrease the response of any compound.

Recovery experiments were also conducted with primary POTW sludge (2% solids). Spiked and unspiked sludge aliquots were diluted with the salt or ethylene glycol solutions and analyzed by the preliminary purge-and-trap GC/MS methods. The results of these recovery measurements are shown in Tables II and IV. No specific advantages from adding salt or ethylene glycol were apparent. Recoveries for all compounds were good. Recoveries for carbon tetrachloride were significantly higher than typically observed using the preliminary POTW sludge protocol. However, both additives appeared to increase significantly the tendency of the sludge to foam during purging. Hence, the use of salt and ethylene glycol in sludge dilutions did not provide significant overall improvements from the preliminary POTW sludge method.

## Alternative Purging Hardware Systems

Alternative purging hardware systems which provide mechanical mixing during purging were evaluated for their potential to enhance desorption of analytes from the sludge solids to the supernatant and to improve transfer to the purge gas stream. All systems evaluated were designed to be used as direct replacements for the conventional purge tube in the purge-and-trap

**Table III. Recoveries of Purgeable Compounds by Purge-and-Trap GC/MS from Spiked Primary Sludge Diluted with a Saturated Salt Solution**

| Compound | Spike Level (ng) | Unspiked Sludge (ng)[a] | Spike Recovery[b] (%) |
|---|---|---|---|
| Benzene | 250 | 10.8;30.6 | 170 ± 48 |
| Carbon Tetrachloride | 250 | ND; ND[c] | 59 ± 22 |
| Chloroform | 250 | 16.9;12.0 | 180 ± 32 |
| 1,1-Dichloroethane | 250 | 15.4;20.5 | 220 ± 54 |
| Tetrachloroethene[d] | 250 | | |
| 1,2-Dichloroethane | 250 | 4.5;4.8 | 120 ± 22 |
| Trichloroethene | 250 | 322;364 | e |
| 1,1,1-Trichloroethane | 250 | ND; ND | 120 ± 37 |
| Chlorobenzene | 250 | 6.6;5.7 | 130 ± 15 |
| Ethyl Benzene | 250 | 33.6;75.0 | 97 ± 12 |

[a] 2.5-mL sludge aliquot.
[b] Mean recovery ± standard deviation for triplicate determinations.
[c] ND = not detected.
[d] Ion plots for all unspiked and spiked samples exhibited saturated peaks.
[e] Ion plots for spiked sludge samples exhibited saturated peaks.

**Table IV. Recoveries of Purgeable Compounds by Purge-and-Trap GC/MS from Spiked Primary Sludge Diluted with a 10% Ethylene Glycol Solution**

| Compound | Spike Level (ng) | Unspiked Sludge (ng)[a] | Spike Recovery[b] (%) |
|---|---|---|---|
| Benzene | 250 | ND; ND[c] | 100 ± 24 |
| Carbon Tetrachloride | 250 | ND; ND | 81 ± 12 |
| Chloroform | 250 | 8.7; 9.8 | 130 ± 30 |
| 1,1-Dichloroethene | 250 | 19.7; 10.0 | 130 ± 34 |
| Tetrachloroethene[d] | 250 | | |
| 1,2-Dichloroethane | 250 | ND; ND | 100 ± 30 |
| Trichloroethene | 250 | 215; 190 | e |
| 1,1,1-Trichloroethane | 250 | ND; ND | 130 ± 42 |
| Chlorobenzene | 250 | 7.7; 6.3 | 120 ± 6 |
| Ethyl Benzene | 250 | 36.5; 52.4 | 95 ± 12 |

[a] 2.5-mL sludge aliquot.
[b] Mean recovery ± standard deviation for triplicate determinations.
[c] ND = not detected.
[d] Ion plots for all unspiked and spiked samples exhibited saturated peaks.
[e] Ion plots for spiked sludge samples exhibited saturated peaks.

system employed in the preliminary POTW sludge method. Three hardware systems were evaluated. Two configurations were fabricated using a small round-bottom flask: stirred purging and stirred impinging. The third system was a stirred purge configuration using a specially designed stirred purge tube.

The two stirred systems illustrated in Figures 5 and 6 were evaluated by comparing the GC/FID responses for compounds spiked into water and analyzed by these systems with replicate spiked waters analyzed with the conventional purge method. Aliquots analyzed with the stirred systems were diluted to 50 mL prior to purging. The results of these experiments are shown in Table V. In general, the responses obtained from using stirred purging and stirred impinging were lower than those from the conventional purge tube. In particular, the internal standard responses were markedly lower. From these results, it appears that the geometry of the conventional purge tube, i.e., a small-diameter column with gas bubbling from a frit at the bottom to the surface of the sample, provides better transfer of purgeable organics from the sample to the gas stream, at least for clean water.

**Table V. Chromatographic Responses for Purgeables Spiked into Clean Water and Analyzed by Purge-and-Trap. Stirred Purge-and-Trap and Stirred Impinger Trap GC/MS**

| | Peak Height (mm), 100 ng Standard | | |
|---|---|---|---|
| Compound | Purge Tube | Stirred Purge | Stirred Dynamic Headspace |
| 1,1-Dichloroethene | 4.5; 4.5 | 5.0 | 5.0; 4.5 |
| Chloroform | 15.5; 13.0 | 12.0 | 10.5; 10.5 |
| 1,2-Dichloroethane | 11.5; 12.0 | 3.5 | 4.0; 5.0 |
| 1,1,1-Trichloroethane | 1.5; 1.5 | 1.5 | 2.0; 1.5 |
| Carbon Tetrachloride | 2.0; 2.0 | 0.5 | ND[a]; 0.5 |
| Trichloroethene | 17.5; 18.5 | 19.5 | 14.0; 14.5 |
| Benzene | 65.6; 71.0 | 63.0 | 51.0; 53.0 |
| Tetrachloroethene | 12.0; 14.0 | 17.0 | 14.0; 11.5 |
| Chlorobenzene | 54.0; 57.0 | 44.0 | 41.0; 36.0 |
| Ethyl Benzene | 22.0; 18.5 | 20.0 | 19.0; 18.0 |
| Bromochloromethane[b] | 26.0; 25.0 | 14.0 | 12.0; 8.0 |
| 2-Bromo-1-chloropropane[b] | 29.0; 29.0 | 16.0 | 15.0; 10.0 |
| 1,4-Dichlorobutane[b] | 46.0; 50.0 | 15.0 | 16.0; 18.0 |

[a]ND = not detected.
[b]Internal standard compounds.

To combine the advantages of the conventional purge tube geometry with mechanical agitation, a purge tube was designed and fabricated to allow the use of a magnetic stirrer. The diameter of the tube was increased to provide more cross-sectional area of sample headspace to reduce the potential for foaming. The modified purge tube, referred to hereafter as a "bottom frit tube," is shown in Figure 7.

The performance of the bottom frit tube with and without mechanical stirring was compared with that of the conventional purge tube for spiked water analyzed by purge-and-trap GC/FID. The responses observed for a few compounds analyzed in spiked water with the bottom frit tube without stirring were lower than those from the conventional purge tube. The bottom frit tube with stirring produced responses generally as high or higher than the conventional tube. These results likely reflect poorer contact of the sparging bubbles with the water in the unstirred bottom frit tube relative to the conventional purge tube. The purge flow (40 mL/min) produces only one or two streams of bubbles across the frit of the bottom frit tube. Since the 10-mL aliquot occupies a shorter portion of tube in the wider bottom frit apparatus, the residence time of a single bubble in the aliquot is less than in the conventional purge tube. Stirring the bottom frit tube causes the bubbles to swirl through the sample and increases the residence time of bubbles in the aliquot. Hence, stirring is necessary to obtain satisfactory purging using the bottom frit tube.

The performance of the bottom frit tube was further evaluated by determining the recoveries for compounds spiked into aliquots of primary POTW sludge (2% solids) immediately prior to analysis. The sludge aliquots were diluted and analyzed by the purge-and-trap GC/MS method using the stirred bottom frit tube in place of the conventional purge tube. The results of these recovery determinations are shown in Table VI. Good recovery data were obtained for all the compounds except carbon tetrachloride. This was likely due to the low sensitivity of carbon tetrachloride to GC/MS. The carbon tetrachloride response was too low to determine under the specific GC/MS operating conditions utilized. Recovery determinations were very reproducible. No foaming was observed with the bottom frit tube, whereas aliquots of the same sample previously analyzed with the conventional purge tube foamed profusely.

Since good data on accuracy and precision were obtained with the stirred bottom frit tube and there was no appreciable foaming, the system was evaluated for its capability to purge larger aliquots of sludge without troublesome foaming. Aliquots of 2.5, 5.0 and 10.0 mL of primary sludge (2% solids) were spiked with 250 ng of each of the purgeable spiking compounds just prior to analysis, diluted as required to 10 mL with volatile-free water, and analyzed by the preliminary POTW sludge GC/MS method. The recoveries of the

**Table VI. Recoveries of Purgeable Compounds from Spiked Sludge by Stirred Purging with a Bottom Frit Tube**

| Compound | Spike Level (ng) | Unspiked Sludge (ng)[a] | Spike Recovery[b] (%) |
|---|---|---|---|
| Benzene | 250 | ND; ND[c] | 113 ± 10 |
| Carbon Tetrachloride | 250 | ND; ND | 0 ± 0 |
| Chloroform | 250 | 8.6; 2.3 | 117 ± 8 |
| 1,1-Dichloroethene | 250 | 7.5; 6.0 | 123 ± 6 |
| Tetrachloroethene[d] | 250 | | |
| 1,2-Dichloroethane | 250 | ND; ND | 108 ± 4 |
| Trichloroethene | 250 | 217; 347 | e |
| 1,1,1-Trichloroethane | 250 | ND; ND | 106 ± 17 |
| Chlorobenzene[f] | 250 | 19; 13 | 145 ± 4 |
| Ethyl Benzene | 250 | 51; 45 | 101 ± 11 |

[a] 2.5-mL sludge aliquot.
[b] Mean recovery ± standard deviation for triplicate determinations.
[c] ND = not detected.
[d] Ion plots for all unspiked and spiked samples exhibited saturated peaks.
[e] Ion plots for spiked sludge samples exhibited saturated peaks.
[f] Peak interferences.

spiked purgeable compounds in different volumes of sludge are shown in Table VII. Recoveries for all compounds were good and did not appear to be strongly dependent on the volume of sludge purged. No foaming problems were encountered even with 10-mL aliquots. Evidently, the force imparted by the magnetic stirring bar swirling the sludge aliquot was an efficient foam dispersion mechanism.

To evaluate further the purging efficiencies afforded by the bottom frit tube and the practicality of purging larger aliquots of sludge, recoveries were determined for sludge samples allowed to equilibrate following spiking. In addition, the recoveries of purgeable compounds from spiked sludge using the stirred bottom frit tube were directly compared with those obtained using the conventional purge tube. Spiked and unspiked replicates of a primary POTW sludge (2.5% solids) were analyzed with the preliminary purge-and-trap GC/MS procedure using both the conventional purge tube and the bottom frit tube. That is, 2.0-mL aliquots were diluted to 10 mL prior to purging. In addition, the bottom frit tube was also used to analyze 5.0-mL aliquots diluted to 10 mL and 10-mL aliquots of undiluted sludge.

The results of these recovery measurements are shown in Table VIII. The recoveries were generally good. The recoveries did not appear to be dependent on the size of the aliquot purged even though the spike concentrations

**Table VII. Recovery of Spiked Purgeable Compounds from Spiked Sludge by Stirred Purging with a Bottom Frit Tube: Effect of Different Volumes of Sludge**

| Compound | Spike Level (ng) | 10-mL Aliquot Unspiked Sludge (ng) | Spike Recovery (%) | | |
|---|---|---|---|---|---|
| | | | 2.5 mL | 5 mL | 10 mL |
| Benzene | 250 | 49.2 | 133 | 142 | 148 |
| Carbon Tetrachloride | 250 | ND[a] | 77 | 68 | 55 |
| Chloroform | 250 | 43.6 | 101 | 123 | 122 |
| 1,1-Dichloroethene | 250 | 123 | 112 | 165 | 194 |
| Tetrachloroethene[b] | 250 | | | | |
| 1,2-Dichloroethane | 250 | ND | 102 | 114 | 152 |
| Trichloroethene[b] | 250 | | | | |
| 1,1,1-Trichloroethane | 250 | ND | 102 | 117 | 133 |
| Chlorobenzene | 250 | 19.0 | 135 | 108 | 61 |
| Ethyl Benzene[c] | 250 | 118 | | | |

[a]ND = not detected.
[b]Ion plots for all unspiked and spiked samples exhibited saturated peaks.
[c]Ion plots for all spiked sludge samples exhibited saturated peaks.

ranged from 125 μg/L in the 2.0-mL aliquots to 25 μg/L in the 10.0-mL aliquots. Recoveries of compounds in 10.0-mL aliquots of spiked sludge (at 25 μg/L) analyzed with the stirred bottom frit tube were very similar to those in 2.0-mL aliquots of spiked sludge (at 125 μg/L) using the conventional purge tube. In addition, no foaming was observed during stirred purging, even while purging 10.0-mL aliquots of undiluted sludge. Diluted aliquots of this sludge purged in the conventional tube foamed profusely. Two glass wool plugs were required in the top of the tube to control the foaming.

GC/MS chromatograms of unspiked and spiked sludge aliquots analyzed with the stirred bottom frit tube are shown in Figures 8 through 10. Interestingly, the levels of background material in the sludge chromatograms were fairly consistent. Background levels were not significantly higher in chromatograms of the larger sludge aliquots. However, the level of late-eluting, unresolved material did appear to increase during any single analysis day, regardless of the purging system or sample aliquot size. Presumably this was due to accumulation of materials on the GC column or the trap. The sludge background frequently obscured measurement of the response of two of the three internal standard compounds, especially bromochloromethane. Late-eluting materials also obscured measurement of responses from 1,4-dichlorobutane in some sludges. Only the second internal standard, 2-bromo-1-chloropropane, was consistently free from interferences. Although the compositions of

Table VIII. Recoveries of Purgeable Compounds from Spiked[a] Sludge by Stirred Bottom Frit Tube and Conventional Purge Tube

| Compound | Stirred Bottom Frit Tube | | | | | | Conventional Purge Tube | |
|---|---|---|---|---|---|---|---|---|
| | 2-mL Aliquot Sludge | | 5-mL Aliquot Sludge | | 10-mL Aliquot Sludge | | 2-mL Aliquot Sludge | |
| | Unspiked (ng) | Spike Recovery[b] (%) | Unspiked (ng) | Spike Recovery[b] (%) | Unspiked (ng) | Spike Recovery (%) | Unspiked (ng) | Spike Recovery (%) |
| Benzene | 9 | 118 ± 4 | 16 | 90 ± 4 | 56 | 121; 121 | 9 | 119; 112 |
| Carbon Tetrachloride | ND[c] | 84 ± 4 | ND | 43 ± 6 | ND | 87; 90 | ND | 64; 61 |
| Chloroform | ND | 112 ± 15 | ND | 124 ± 10 | ND | 116; 126 | ND | 112; 104 |
| 1,1-Dichloroethene | 1 | 113 ± 10 | ND | 98 ± 8 | ND | 150; 146 | 1 | 128; 100 |
| Tetrachloroethene | 9 | 145 ± 14 | 13 | 105 ± 6 | 22 | 88; 83 | 8 | 149; 117 |
| Vinyl Chloride | ND | 125 ± 18 | ND | 112 ± 7 | ND | 122; 120 | ND | 123; 96 |
| 1,2-Dichloroethane | ND | 109 ± 16 | ND | 121 ± 5 | ND | 121; 122 | ND | 124; 113 |
| Trichloroethlene | ND | 120 ± 20 | ND | 113 ± 6 | ND | 129; 128 | 2 | 126; 117 |
| 1,1,1-Trichloroethane | 38 | 115 ± 14 | 74 | 94 ± 7 | 204 | 122; 123 | 36 | 121; 114 |
| Chlorobenzene | ND | 132 ± 5 | 5 | 92 ± 4 | 5 | 88; 69 | 8 | 138; 108 |
| Ethyl Benzene | 40 | 138 ± 5 | 44 | 94 ± 4 | 99 | 80; 63 | 20 | 139; 104 |

[a] Spike level = 250 ng.
[b] Mean recovery ± standard deviation for triplicate determination.
[c] ND = not detected.

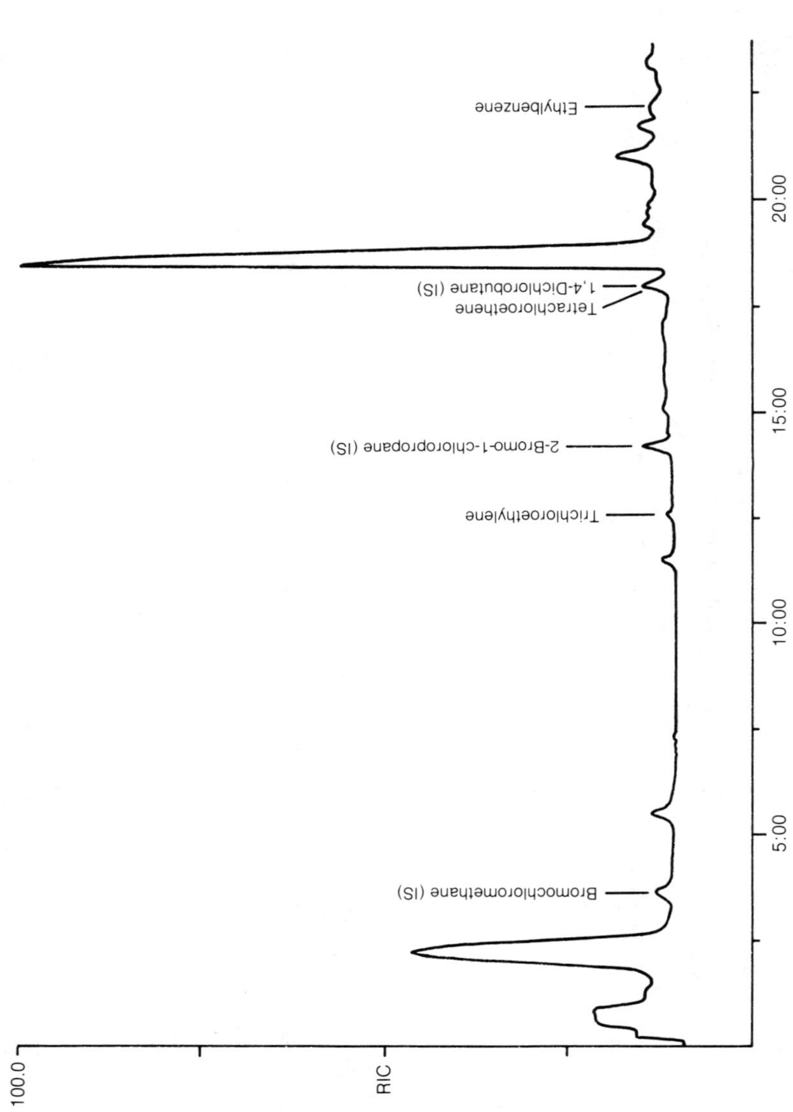


**Figure 8.** GC/MS chromatogram of a 2-mL aliquot of unspiked primary sludge analyzed with a stirred bottom frit tube.

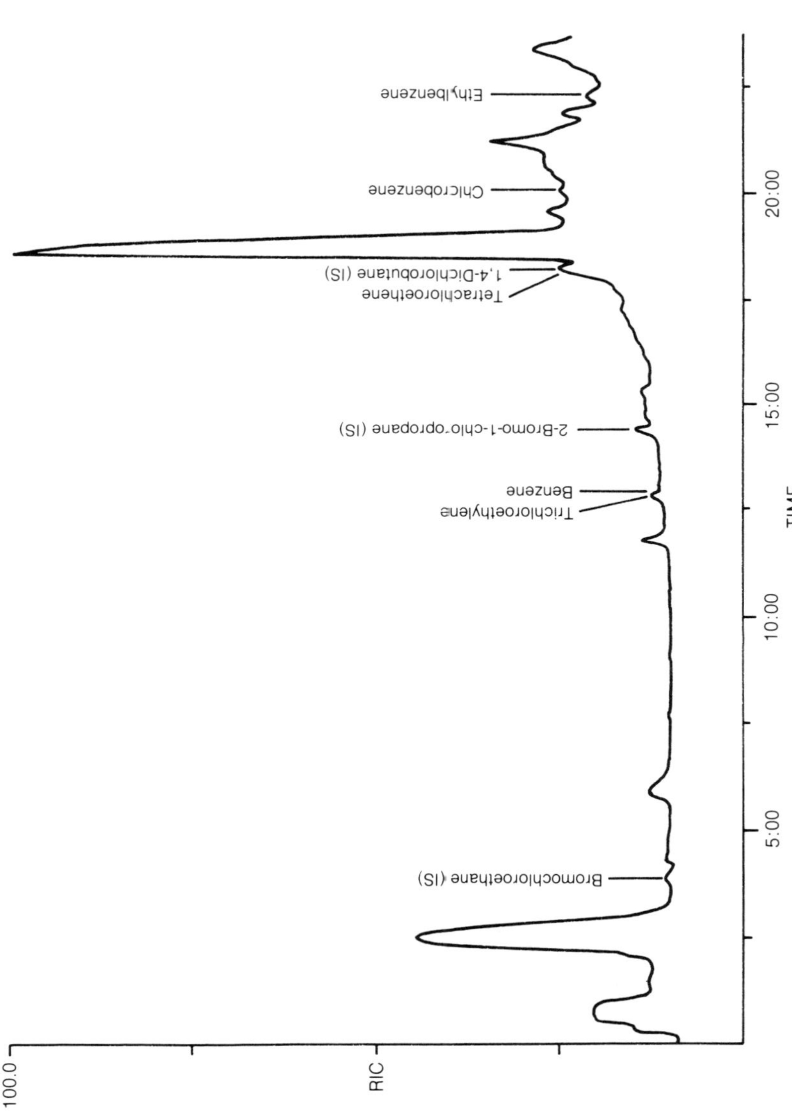


**Figure 9.** GC/MS chromatogram of a 5-mL aliquot of unspiked primary sludge analyzed with a stirred bottom frit tube.

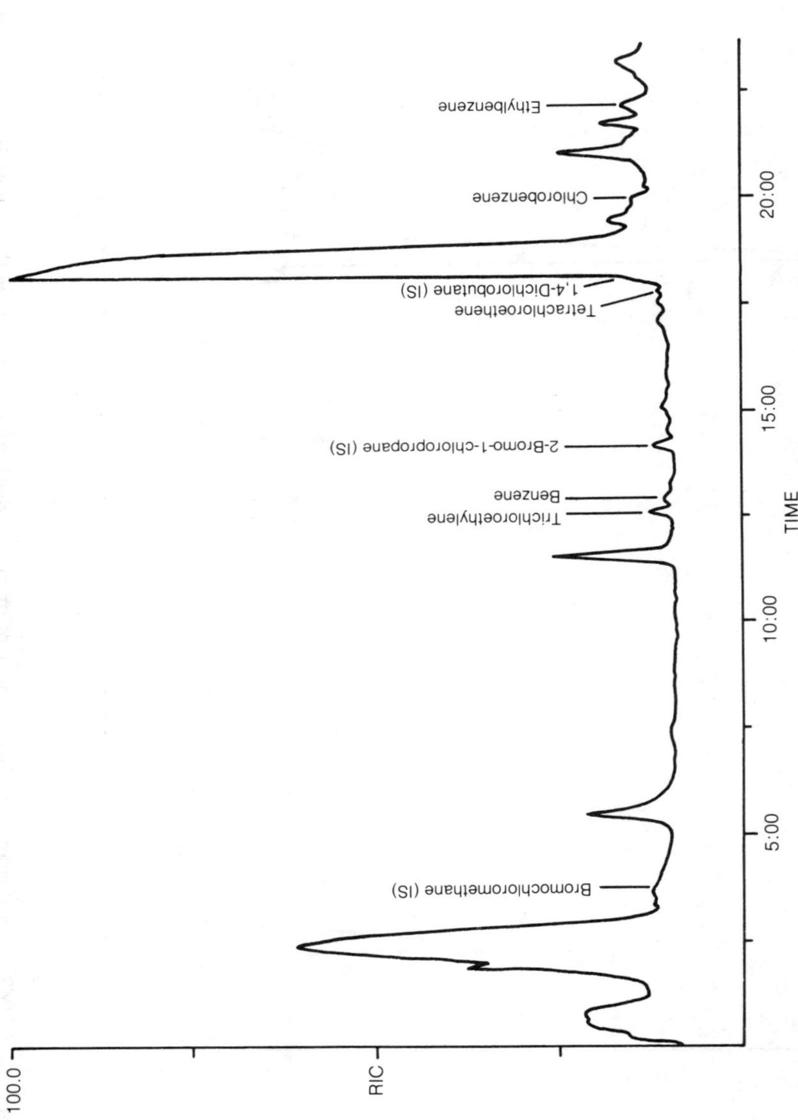


**Figure 10.** GC/MS chromatogram of a 10-mL aliquot of unspiked primary sludge analyzed with a stirred bottom frit tube.

sludges vary widely, the frequency of interferences with bromochloromethane and with 1,4-dichlorobutane indicates that the selection of alternate early- and late-eluting internal standards may be required for some samples.

## Selection of Methods for Purgeables in Sludge

The simple adaptation of the industrial wastewater screening method [1] for sludge analysis, i.e., dilution to approximately 0.5% solids prior to purging, developed for the preliminary POTW sludge method was not consistently successful in overcoming foaming problems during purging. Although recoveries observed for most compounds were good, recovery determinations were not very reproducible. In addition, sample dilution decreased the method detection limit. The stirred purge-and-trap GC/MS procedure using the MRI-designed bottom frit tube was demonstrated to provide significantly better reproducibility and better recoveries than were provided by the preliminary POTW method. In addition, the capability of the stirred bottom frit procedure to purge 10.0-mL aliquots of undiluted sludge without appreciable foaming provides a significant improvement in method detection limits. Hence, the stirred bottom frit purge system was selected for a revised method for the analysis of purgeables in municipal and industrial wastewater treatment sludges [5].

## Evaluation of the Precision and Accuracy of the Method

The precision and accuracy of the purgeables method were evaluated by determining recoveries for spiked compounds from five primary sludges. The subject sludges, three from POTW receiving various fractions of wastewater from residential and industrial sources and two from plants treating industrial wastewaters, were selected to provide a wide variety of sludge characteristics with which to test the performance of the method. The origin and characteristics of the five sludges are listed in Table IX. Recoveries were determined for the purgeable spiking compounds fortified into these aliquot sludges at three levels. The spiking levels were chosen to be 2, 20 and 200 times the individual detection limit typically observed for each compound in sludge. Undiluted 10-mL aliquots were analyzed for all sludges spiked at 2 and 20 times the detection limits. To ensure that the responses for all compounds were observed within the working range of the mass spectrometric detector, only 2.0-mL aliquots were analyzed for the sludge aliquots spiked at 200 times the detection limit.

**Table IX. Primary Wastewater Treatment Sludges Used in the Sludge Protocol Precision and Accuracy Spiking Study**

| Plant | Location | Principal Wastewater Inputs | % Solids in Primary Sludge |
|---|---|---|---|
| Platte County POTW | Suburban Kansas City, MO | Essentially 100% residential | 3.0 |
| Blue River POTW | Kansas City, MO | Residential and pretreated industrial | 7.7 |
| Kansas City, Kansas | Kansas City, KS | Large industrial | 9.5 |
| Industrial No. 1 | Southern United States | Petrochemical manufacturing | 1.9 |
| Industrial No. 2 | Eastern United States | Various small industrial concerns | 0.2[a] |

[a]This sludge was prepared by dilution from a sludge filter cake containing 37% solids.

The recoveries determined for purgeable compounds spiked into the five primary sludge samples are summarized in Table X. GC/MS chromatograms for unspiked and spiked (at 20 times the detection limits) sludge from the Blue River POTW are shown in Figures 11 and 12. Similarly derived chromatograms for sludges from the Industrial Plant No. 1 are shown in Figures 13 and 14. Although the recoveries observed were generally good, many recovery determinations were likely influenced by relatively high concentrations of the spiking compounds in the unspiked sludges. Several compounds were present in some of the sludges at levels greater than the spike level, and in some cases concentrations in the unspiked sludges were more than 10 times the spike level. Recoveries for the latter cases were not representative of the method precision and accuracy and were excluded from the table. Zero or very low recoveries were observed for carbon tetrachloride in all sludges except Industrial Plant No. 2. Carbon tetrachloride and, to some extent, chloroform are frequently troublesome compounds because of their poor sensitivity to mass spectrometric detection. Although the detection limits for these compounds were chosen based on considerable previous work, their detection limits in these particular sludges are evidently higher than is typical. Determinations of vinyl chloride, chlorobenzene and 1,1,1-trichloroethane were frequently obscured by coeluting interferences in the POTW sludges.

Nonetheless, in most cases the recoveries observed were both good and reproducible. Only 11% of all recovery determinations were less than 50%. More than 63% fell within the range of 50–150% recovery. An additional 25% of the recovery determinations exceeded 150%. The relative standard deviations (RSD) for triplicate recovery determinations were generally low. More than 90% of the RSD were 30% or less. Of those, 27% were less than 10% RSD. The method reproducibility was surprisingly good considering the difficulty in removing representative aliquots from a heterogeneous sample matrix.

## SUMMARY

The stirred bottom frit tube provided recoveries for compounds spiked into undiluted sludge and analyzed by purge-and-trap GC/MS that were as good as or better than the recoveries from dilution followed by purge-and-trap GC/MS with the conventional purge tube. The precision of recovery measurements was much better with the bottom frit tube, and no foaming problems were observed. In addition, analysis of undiluted 10-mL sludge aliquots provides a significant improvement in the method detection limit. Considering the variability of sludge characteristics, the heterogeneous nature of sludges and the detection limits achieved, the results of these precision and

## Table X. Results of Precision and Accuracy Evaluations for the Purgeables Method Applied to Municipal and Industrial Wastewater Treatment Primary Sludges

| Compound | Spike Conc (μg/L) | Platte County POTW Unspiked Conc (μg/L) | Platte County POTW Recovery (%) | Blue River POTW Unspiked Conc (μg/L) | Blue River POTW Recovery (%) | Kansas City POTW[a] Unspiked Conc (μg/L) | Kansas City POTW[a] Recovery (%) | Industrial No. 1. Unspiked Conc (μg/L) | Industrial No. 1. Recovery (%) | Industrial No. 2 Unspiked Conc (μg/L) | Industrial No. 2 Recovery (%) |
|---|---|---|---|---|---|---|---|---|---|---|---|
| Benzene | 1 | 3 ± 1[b] | 97 ± 20[b] | 4 ± 0.8 | 230 ± 320[c] | 6 ± 6 | 290 ± 80[c] | 0.7 ± 0.3 | 93 ± 25 | 16 ± 0.1 | d |
| | 10 | | 150 ± 30 | | 250 ± 50 | | 170 ± 29 | | 118 ± 22 | | 66 ± 9[c] |
| | 100[e] | | 120 ± 15 | | 140 ± 20 | | 150 ± 8 | | 114 ± 23 | | 97 ± 3 |
| Carbon Tetrachloride | 10 | ND[f] | 0 | ND | 0 | ND | 0 | ND | 0 | 252 ± 14 | d |
| | 100 | | 11 ± 4 | | 0 | | 0 | | 0 | | 85 ± 9[c] |
| | 1000 | | 0 | | 0 | | 0 | | 0 | | 87 ± 5 |
| Chloroform | 2 | 0.3 ± 0.3 | 120 ± 60 | ND | 0 | ND | 0 | 8 ± 0.4 | 80 ± 43[c] | 14 ± 0.6 | 82 ± 13[c] |
| | 20 | | 130 ± 15 | | 170 ± 39 | | 47 ± 7 | | 52 ± 16 | | 81 ± 19 |
| | 200 | | 62 ± 7 | | 160 ± 23 | | 110 ± 16 | | 77 ± 17 | | 82 ± 5 |
| 1,1-Dichloroethene | 5 | 0.3 ± 0.5 | 200 ± 27 | ND | 260 ± 57 | ND | 170 ± 30 | ND | 130 ± 31 | 33 ± 3 | 0[c] |
| | 50 | | 180 ± 26 | | 200 ± 34 | | 150 ± 24 | | 160 ± 30 | | 34 ± 24 |
| | 500 | | 130 ± 15 | | 130 ± 18 | | 100 ± 10 | | 130 ± 25 | | 95 ± 5 |
| Tetrachloroethene | 3 | 1 ± 0.5 | 160 ± 56 | 89 ± 10 | d | 200 ± 52 | d | ND | 170 ± 65 | 880 ± 110 | d |
| | 30 | | 180 ± 22 | | 150 ± 42[c] | | 29 ± 50[c] | | 170 ± 50 | | d |
| | 300 | | 127 ± 4 | | 140 ± 29 | | 140 ± 13 | | 200 ± 39 | | 52 ± 5[c] |
| Vinyl Chloride | 5 | ND | g | g | g | ND | d | ND | 120 ± 34 | ND | 0 |
| | 50 | | 160 ± 39 | | 170 ± 42 | | 100 ± 12 | | 160 ± 19 | | 76 ± 10 |
| | 500 | | 130 ± 16 | | 100 ± 9 | | 92 ± 2 | | 140 ± 24 | | 91 ± 8 |
| 1,2-Dichloroethane | 5 | ND | 140 ± 9 | ND | 240 ± 57 | ND | 150 ± 26 | ND | 81 ± 22 | 660 ± 110 | d |
| | 50 | | 120 ± 18 | | 170 ± 57 | | 87 ± 5 | | 130 ± 22 | | d |
| | 500 | | 130 ± 8 | | 110 ± 18 | | 95 ± 4 | | 130 ± 18 | | 86 ± 6[c] |
| Trichloroethene | 2 | 165 ± 33 | | 65 ± 11 | d | 13 ± 2 | d | ND | 170 ± 35 | 130 ± 8 | d |
| | 20 | | 130 ± 61 | | 280 ± 90 | | 150 ± 40 | | 160 ± 50 | | 67 ± 44[c] |
| | 200 | | 110 ± 13 | | 149 ± 20 | | 140 ± 12 | | 160 ± 39 | | 92 ± 6 |
| 1,1,1-Trichloroethane | 16 | 2 ± 3 | 140 ± 29 | ND | 91 ± 50 | ND | g | ND | 110 ± 17 | 1240 ± 52 | d |
| | 160 | | 120 ± 28 | | 190 ± 57 | | 91 ± 13 | | 180 ± 31 | | 67 ± 13[c] |
| | 1600 | | 130 ± 13 | | 170 ± 40 | | 120 ± 12 | | 110 ± 5 | | 100 ± 7 |
| Chlorobenzene | 2 | 0.4 ± 0.4 | 150 ± 60 | ND | g | g | g | ND | 250 ± 33 | 13 ± 0.5 | 85 ± 10 |
| | 20 | | 160 ± 11 | | 120 ± 19 | | 66 ± 7 | | 160 ± 40 | | 110 ± 6 |
| | 200 | | 120 ± 7 | | 120 ± 31 | | 130 ± 9 | | 180 ± 40 | | 97 ± 5 |
| Ethyl Benzene | 5 | 4 ± 2 | 100 ± 33 | 38 ± 2 | 160 ± 150 | 69 ± 24 | d | 1 ± 0.4 | 180 ± 21 | 1090 ± 190 | d |
| | 50 | | 150 ± 6 | | 110 ± 33 | | 47 ± 26[c] | | 190 ± 44 | | 88 ± 21[d] |
| | 500 | | 100 ± 8 | | 120 ± 27 | | 110 ± 9 | | 140 ± 16 | | 70 ± 7[c] |

[a]Kansas City, KS.
[b]Mean ± standard deviation for three determinations.
[c]Concentration determined in the unspiked sludge was higher than the spike level.
[d]Concentration determined in the unspiked sludge was 10 times higher than the spike level.
[e]Analyzed from a 2-mL aliquot diluted to 10 mL prior to purging.
[f]ND = not detected.
[g]Determination was not possible because of the presence of coeluting interferences.

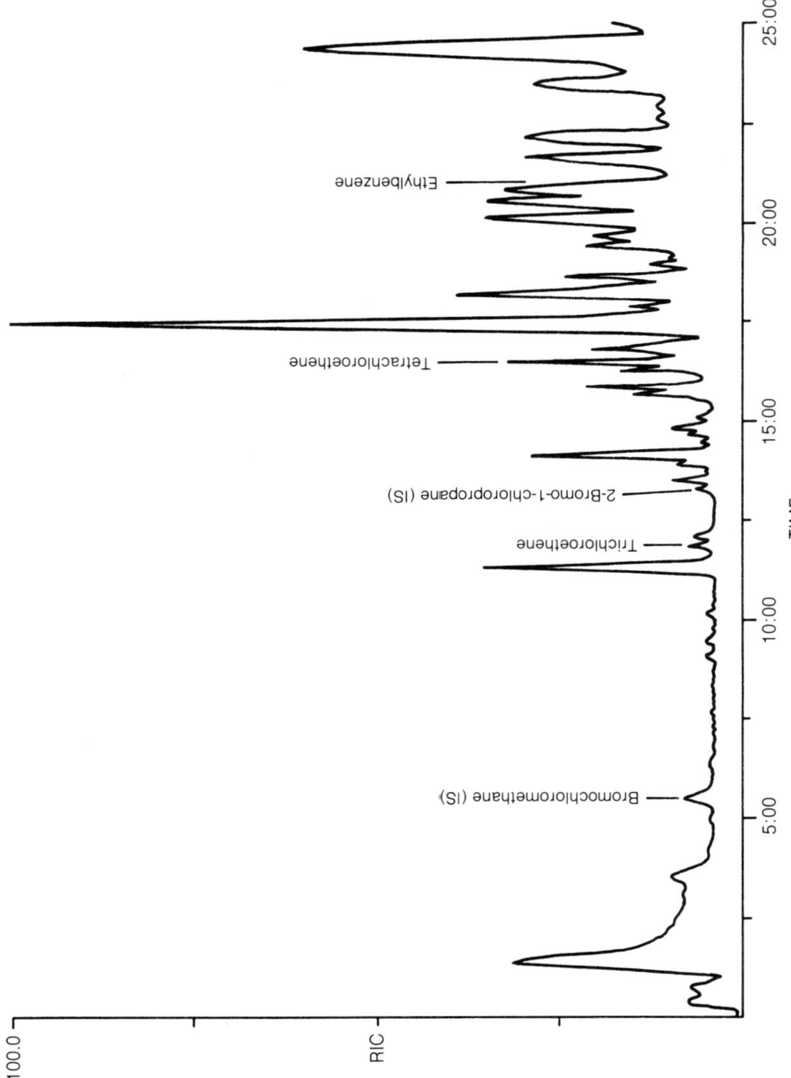


**Figure 11.** GC/MS chromatogram for purgeables in unspiked Blue River POTW primary sludge.

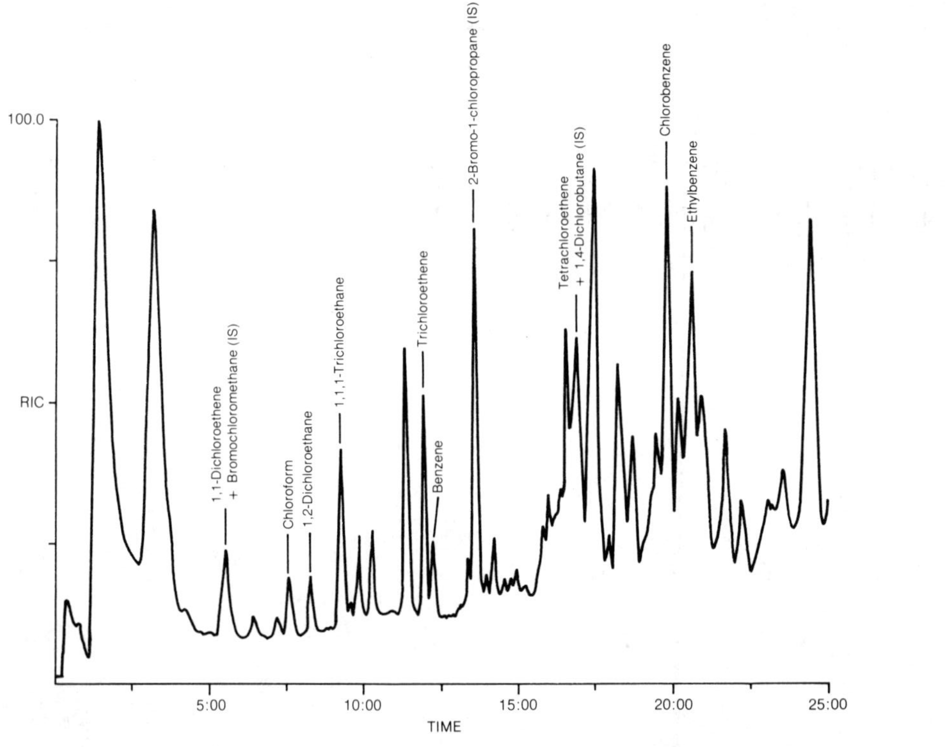


**Figure 12.** GC/MS chromatogram for purgeables in Blue River POTW primary sludge spiked at 20 times the detection limits for selected compounds.

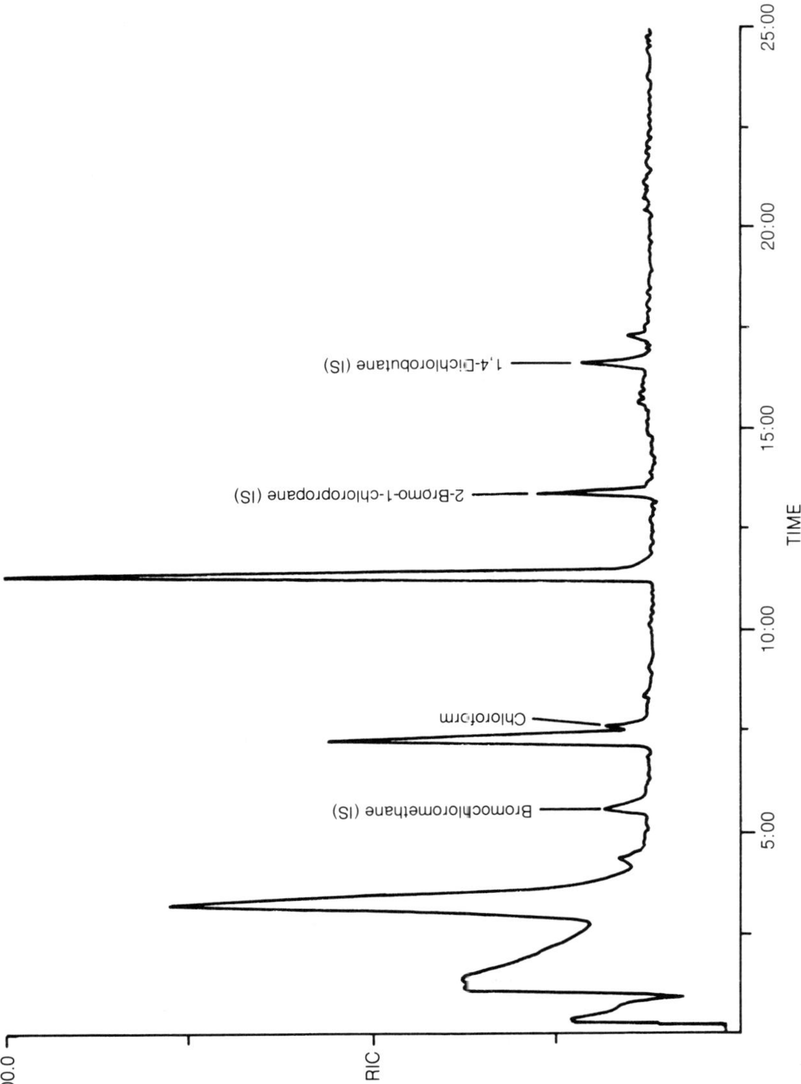


**Figure 13.** GC/MS chromatogram for purgeables in unspiked Industrial Plant No. 1 primary sludge.

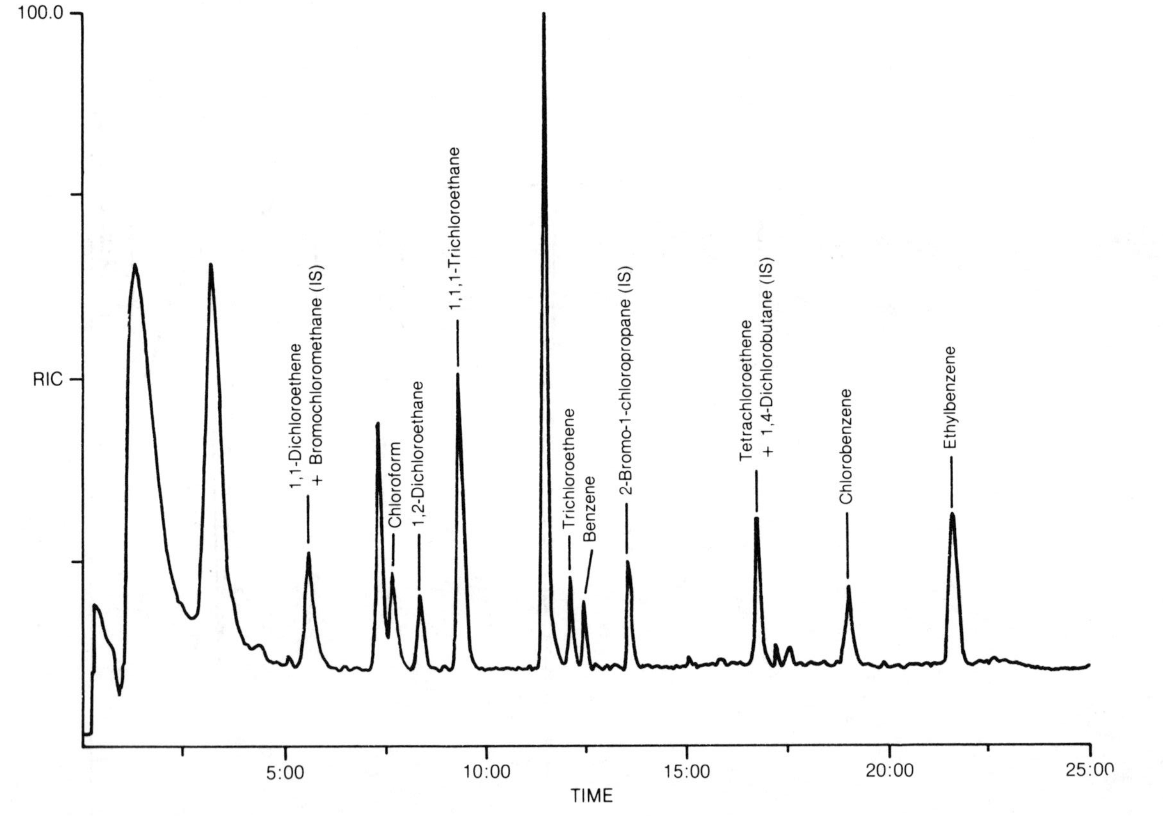


**Figure 14.** GC/MS chromatogram for purgeables in Industrial Plant No. 1 primary sludge spiked at 20 times the detection limits for selected compounds.

accuracy determinations demonstrate that the proposed method can reliably be applied to the analysis of purgeable compounds in municipal and industrial wastewater treatment sludges.

## ACKNOWLEDGMENTS

The authors gratefully acknowledge the assistance of Gil Radolovich, George Vaughn and Jon Onstot in the acquisition and retrieval of the GC/MS data. Support for this research program was provided by the U.S. Environmental Protection Agency, Contract No. 68-03-2695.

## REFERENCES

1. "Sampling and Analysis Procedures for Screening of Industrial Effluents for Priority Pollutants," U.S. EPA, Environmental Monitoring and Support Laboratory, Cincinnati, OH (1977).
2. Bellar, T. A., and J. J. Lichtenberg. "Determining Volatile Organics at Microgram-per-Liter Levels by Gas Chromatography," *J. Am. Waterworks Assoc.* 66:739-744 (1974).
3. Midwest Research Institute. "Analytical Protocol for Screening Publicly Owned Treatment Works (POTW) Sludges for Organic Priority Pollutants," Special Report No. 1 to EPA under Contract No. 68-03-2695 (1979).
4. Bellar, T. A. Personal communication (1980).
5. Midwest Research Institute. "Protocol for the Analysis of Purgeable Organic Priority Pollutants in Industrial and Municipal Wastewater Treatment Sludge," in *Development of Methods for the Analysis of Organic Priority Pollutants in Municipal and Industrial Wastewater Treatment Sludge*. Draft Final Report to EPA under Contract No. 68-03-2695 (1981).

CHAPTER 43

# DEVELOPMENT OF METHODS FOR THE ANALYSIS OF EXTRACTABLE ORGANIC PRIORITY POLLUTANTS IN MUNICIPAL AND INDUSTRIAL WASTEWATER TREATMENT SLUDGES

**V. Lopez-Avila, C. L. Haile,**
**P. R. Goddard, L. S. Malone, R. V. Northcutt,**
**D. R. Rose and R. L. Robson**

Midwest Research Institute
Kansas City, Missouri

The primary objective of this research program was the development of methods for the analysis of organic priority pollutants in sludges generated from the treatment of municipal and industrial wastewaters.

Although the term "sludge" is applied to any high-solids effluent from wastewater treatment, the characteristics of sludges are very diverse. Generally, sludges contain high levels of both dissolved and suspended organics. Sludges from publicly owned treatment works (POTW) (i.e., municipal sewage treatment plants) and industrial wastewater treatment plants receiving wastes of biological origin (e.g., process wastewaters from leather tanning) typically contain high levels of high-molecular-weight organics related to humic materials. Most of the organics in sludges from the treatment of synthetic organic process wastewaters have lower molecular weights. These sludges tend to have lower solids contents than POTW sludges. The characteristics of sludges also differ, depending on their origin within the treatment process. Primary sludges (from simple sedimentation) typically contain higher suspended solids (SS) levels and more dissolved organics than secondary sludges (from sedimentation following biological treatment). The solids in sludges from tertiary treatment processes may be primarily inorganic salts formed from chemical treatment additives. In addition, treatment of raw sludges by anaerobic or aerobic digestion or other processes practiced at some facilities significantly alters the nature of the sludge finally wasted. This tremendous diversity of sludge characteristics, coupled with the heterogeneous nature of sludges, presents a considerable challenge to precise and accurate determinations of trace levels of the organic priority pollutants in sludges.

This research program was conducted in two stages. In the first stage, a preliminary POTW sludge analysis protocol was developed from existing wastewater methods to satisfy fairly immediate U.S. Environmental Protection Agency (EPA) requirements for methods to be used in a survey of POTW wastewaters and sludges. Then, in the second stage, a systematic study of sludge analysis techniques was conducted, resulting in the development of a set of revised protocols.

The method development and evaluation experiments described in this chapter were conducted using a subset of the organic priority pollutants to simplify sample spiking and recovery determinations. These compounds were selected by EPA to represent the physical and chemical characteristics of most of the organic priority pollutants. The list of spiking compounds is shown in Table I.

**Table I. Representative Compounds for Phase I Study**

| |
|---|
| 1,4-Dichlorobenzene |
| Hexachloroethane |
| *bis*-(2-Chloroisopropyl) Ether |
| *bis*-(2-Chloroethyl) Ether |
| Acenaphthylene |
| 2,6-Dinitrotoluene |
| Fluoranthene |
| Benzidine |
| 3,3′-Dichlorobenzidine |
| *n*-Butylbenzyl Phthalate |
| *bis*-(2-Ethylhexyl) Phthalate |
| Benzo[a]pyrene |
| Phenol |
| 2,4-Dimethylphenol |
| 2,4-Dichlorophenol |
| Pentachlorophenol |

## DEVELOPMENT OF PRELIMINARY EXTRACTION METHODS FOR EXTRACTABLE COMPOUNDS

The development of a preliminary method for the analysis of base/neutral (B/N) and acid extractable compounds in POTW sludge was directed toward adapting industrial wastewater screening methods [1]. Under this protocol, wastewater samples are sequentially extracted with dichloromethane under basic and then acidic conditions. The B/N and acid extracts are individually concentrated and analyzed by packed-column gas chromatography/mass spectrometry (GC/MS) procedures.

The performance of these GC/MS procedures for analyses of B/N and acidic compounds has been well documented. Hence, the primary objective in adapting the wastewater methods was to develop procedures for sludge extraction and extract cleanup to provide extracts of sufficient quality for GC/MS analysis. Many of the extractable compounds were expected to associate strongly with the sludge solids. Hence, it was anticipated that the wastewater extraction method (simple liquid-liquid partitioning with dichloromethane) would not provide sufficient contact of the extracting solvent with the solids to allow efficient extraction of those compounds from sludges, and would be hindered by formation of emulsions. A procedure using a high-speed homogenizer probe to provide vigorous mixing and blending of the sludge aliquot with the extracting solvent was evaluated.

Unfortunately, this vigorous homogenization/centrifugation procedure also extracts large quantities of lipids, fatty acids and other high-molecular-weight compounds present in POTW sludges. These compounds can cause significant interferences during GC/MS analysis and necessitate extract cleanup. A preparative-scale, low-pressure, gel permeation chromatography (GPC) procedure was selected for removing these high-molecular-weight materials from sludge extracts. GPC techniques have been applied successfully to the removal of lipid materials from fish extracts prepared for analysis of chlorinated pesticides [2].

An analytical scheme for sludge similar to the wastewater procedure but utilizing homogenization/centrifugation extraction and GPC cleanup was proposed. This scheme (Figure 1) was evaluated with three POTW sludges. Details of the analytical procedure are given below.

### Experimental

The dual-pH extraction scheme was evaluated by determining recoveries for the extractable compounds spiked into primary and secondary sludge

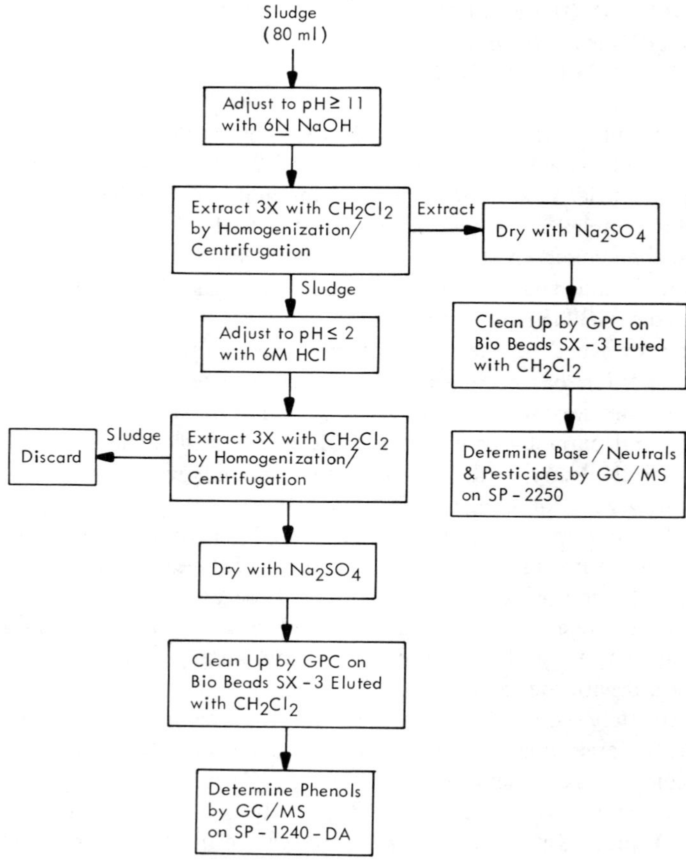


**Figure 1.** Dual-pH extraction scheme for extractable compounds in POTW sludges.

from the Belmont POTW (Indianapolis, IN) and combined sludge from the Muddy Creek POTW (Cincinnati, OH).

Each 320-mL sludge sample was divided into four 80-mL aliquots in four 200-mL glass centrifuge screw-cap tubes. A single sample for each sludge type was spiked with the extractable compounds in acetone to achieve a spike concentration of 310 $\mu$g/L. The tubes were sealed with TFE-lined screw caps and were held at 4°C overnight to allow equilibration of the spikes. Spiked and unspiked sludge aliquots were extracted with three portions of dichloromethane as follows. Each sludge aliquot was adjusted to pH 11 with 6 *N* sodium hydroxide. An 80-mL portion of dichloromethane was added to each centrifuge tube. The contents of the tube were homogenized with a Tekmar Tissumizer® blending probe for 1 min and then centrifuged at 3000 rpm for 30 min. This produced three phases consisting of a fairly clear aqueous layer on top, a dark solvent extract on the bottom, and a firm mat of solids at the solvent-water interface. The lower solvent layer was removed with a 50-mL pipette and the extraction was repeated twice. All extracts from the four tubes constituting a single sample were combined. These B/N sludge extracts were dried by passage through a short column of anhydrous sodium sulfate* prior to concentration to 5 to 25 mL in Kuderna-Danish (KD) evaporators. The lowest practicable volumes were achieved while avoiding excessive viscosity or solidification of the extracts. The sludge aliquots were adjusted to pH 1 with 6 *N* hydrochloric acid; the extraction, extract drying and extract concentration steps were repeated to produce the acidic sludge extracts.

The B/N and acidic extracts were cleaned by GPC on a 2.5-cm i.d. glass column packed with 50 g of Bio-Beads SX-3 and using dichloromethane as the eluent. The eluent flow was 5.0 mL/min with a column pressure of 350–700 mbars (5–10 psi). The column eluent was monitored with an ultraviolet (UV) detector (254 nm). The column was calibrated by chromatographing 1.0-g aliquots of corn oil and 100-$\mu$g portions of *bis*(2-ethylhexyl) phthalate and pentachlorophenol (PCP) in 5.0 mL of dichloromethane. The first 100 mL eluted from the column was discarded, the second 150 mL was collected as the cleaned extract, and the final 100 mL for each run was discarded. The "dump," "collect" and "wash" fractions were selected to provide 85% or better removal of the corn oil in the dump fraction, 85% or greater recovery of the phthalate, 100% recovery of the PCP in the collect recovery fraction and sufficient column washing (wash fraction) between extract injections to prevent carryover. The sludge extracts were chromatographed in one or more 5.0-mL injections. The combined cleaned extract fractions were concentrated to 1–5 mL for GC/MS analysis. Extracts that were still highly colored were cleaned by a second pass through the GPC column.

The B/N and acidic extracts were analyzed by GC/MS according to the procedures described in the industrial wastewater protocol. The GC column used for analyses of B/N extracts, a 1.8-m x 2-mm i.d. glass column packed with 1% SP-2250 on 100/120 mesh Supelcoport, was programmed from 50 to 260°C at 10°C/min following a 4-min initial hold. The column used for analysis of acidic extracts, a 1.2-m x 2-mm i.d. glass column packed with 1% SP-1240-DA on 100/120 mesh Supelcoport, was programmed from 85 to 200°C

*Cleaned by extracting with dichloromethane, drying and then heating to 650°C for 2 hr. The cleaned material was stored at 116°C until just prior to its use.

at 10°C/min following a 4-min initial hold. The final column temperature was maintained for 10 min. For both columns, helium at 30 mL/min was used as carrier gas. The base/neutral and acidic compounds were quantified by comparisons with standards prepared from the spiking solutions. Analyte concentrations were calculated using internal standard correction as follows:

$$(A/B)(B_{IS}/A_{IS})(V_E/V_I)(N/V_S)(1/F) = \mu g/L \text{ of sludge}$$

where A = area of peak in sample extract
B = area of peak in standard
$A_{IS}$ = area of internal standard peak in sample extract
$B_{IS}$ = area of internal standard peak in standard
$V_I$ = volume of extract injected (μL)
$V_E$ = total volume of extract (mL)
N = nanograms in standard
$V_S$ = volume of sludge extracted (L)
F = fraction of extract cleaned by GPC for analysis (e.g., 20 mL/25 mL = 0.80)

### Results and Discussion

The results of recovery determinations for sludges spiked with the extractable spiking compounds and analyzed with the dual-pH extraction scheme are shown in Table II. Although the recoveries observed for most compounds were fairly good, several compounds were not recovered from the primary sludge. These compounds were hexachloroethane, benzidene, 3,3′-dichlorobenzidene and PCP. Part of the loss of hexachloroethane can be attributed to volatilization during extract concentrations. Dilution of the primary sludge extracts to reduce the concentration of interfering coextractants also reduced the spike levels to, at or near the detection limit. The zero recoveries observed for *bis*(2-chloroethyl) ether and N-nitrosodimethylamine reflect, in part, their characteristically poor chromatography.

Overall, the combination of homogenization/centrifugation extraction and GPC extract cleanup provided fairly good recoveries for the B/N and acidic compounds spiked into POTW sludge. These procedures were employed by EPA in a nationwide survey of POTW emissions.

## DEVELOPMENT OF ALTERNATIVE METHODS FOR EXTRACTABLE COMPOUNDS

Although the preliminary methods provided reliable results for analyses of POTW sludges, this rigorous extraction procedure was time-consuming, labor-

Table II. Results of Evaluation of a Dual-pH Extraction Scheme (Basic Then Acidic) for the Analysis of Base/Neutrals and Acids from POTW Sludges

| Compound | Spike Level (μg/L) | Primary Sludge[a] Unfortified Conc (μg/L) | Primary Sludge[a] Spike Recovery (%) | Secondary Sludge Unfortified Conc (μg/L) | Secondary Sludge Spike Recovery (%) | Combined Sludge Unfortified Conc (μg/L) | Combined Sludge Spike Recovery (%) |
|---|---|---|---|---|---|---|---|
| 1,4-Dichlorobenzene | 310 | ND[b] | 150 | ND | 43 | ND | 78 |
| Hexachloroethane | 310 | ND | 0 | ND | 70 | ND | 150 |
| *bis*(2-Chloroisopropyl) Ether | 310 | ND | 110 | ND | 56 | ND | 100 |
| *bis*(2-Chloroethyl) Ether | 310 | ND | 0 | ND | 0 | ND | 0 |
| Acenaphthylene | 310 | ND | 170 | ND | 75 | ND | 110 |
| 2,6-Dinitrotoluene | 310 | ND | 94 | ND | 81 | ND | 110 |
| Phenanthrene[c] | 310 | 3600 | | ND | | 460 | |
| Fluoranthene | 310 | ND | 320 | ND | 59 | 180 | 140 |
| Benzidine | 310 | ND | 0 | ND | 39 | ND | 34 |
| 3,3′-Dichlorobenzidine | 310 | ND | 0 | ND | 67 | ND | 130 |
| *bis*(2-Ethylhexyl) Phthalate | 310 | 3500 | 68 | 420 | 66 | 630 | 330 |
| Benzo[a] pyrene | 310 | 94 | 270 | 7.7 | 80 | 74 | 150 |
| N-Nitrosodimethylamine | 310 | ND | 0 | ND | 0 | ND | 0 |
| Phenol | 310 | 400 | 180 | ND | 130 | 200 | 62 |
| 2,4-Dichlorophenol | 310 | ND | 130 | ND | 110 | ND | 27 |
| Pentachlorophenol | 310 | ND | 0 | ND | 80 | 180 | 100 |

[a]The primary sludge extracts were analyzed at a 1:5 dilution relative to other extracts because of high levels of interfering materials.

[b]ND = not detected.

[c]Phenanthrene was not included in the spikes.

intensive and not very selective. The large quantities of lipids and other high-molecular-weight biogenic compounds in many crude extracts exceeded the capacity of the GPC cleanup procedure. Although the GPC method was designed to remove 85% or greater of up to 1 g of high-molecular-weight biogenic compounds, multiple GPC runs were required for many extracts. The time and cleanup problems were partially reduced for the EPA survey of POTW emissions by extracting only an 80-mL sludge aliquot [3]. The experiments described below were designed to evaluate alternative procedures to allow more efficient and selective extraction, to evaluate alternative extract cleanup methods appropriate for the best extraction procedure and to develop capillary GC/MS methods to aid in the determination of the extractable compounds in complex sludge extracts.

## Alternative Extraction Techniques

Three alternative extraction procedures were evaluated for sludges in an attempt to develop a more efficient, selective and less time-consuming extraction method. The procedures evaluated include: continuous liquid-liquid extraction (CLLE), steam distillation and microextraction. The CLLE procedure has been used successfully by MRI for the analysis of a variety of wastewaters. Extractive steam distillation has been shown to be an efficient and selective procedure for concentrating chlorinated pesticides from water, soil, sediment and tissue samples. Successful liquid-liquid extraction of aqueous samples with a small volume of solvent was reported by Rhoades and Millar [4]. This method has been adapted to allow simple and rapid extraction of organic priority pollutants from a variety of industrial wastewaters [5-9].

## Experimental

The apparatus used for the CLLE experiments is shown in Figure 2. This apparatus was designed by MRI for use with heavier-than-water solvents. The sample (100-300 mL sludge containing 2.1% solids) was diluted with deionized water to 1-L volume, adjusted to pH 11 with 6 *N* sodium hydroxide and extracted for 10 or 20 hr with 300 mL dichloromethane. The extracts were concentrated to 1-2 mL and analyzed by GC/MS. No cleanup was performed.

The apparatus used for steam distillation, fashioned after the Kontes unit (Catalog No. K-523010), is shown in Figure 3. Aliquots of combined sludge (300 mL), one spiked and one unspiked, were basified to pH 11, and ex-

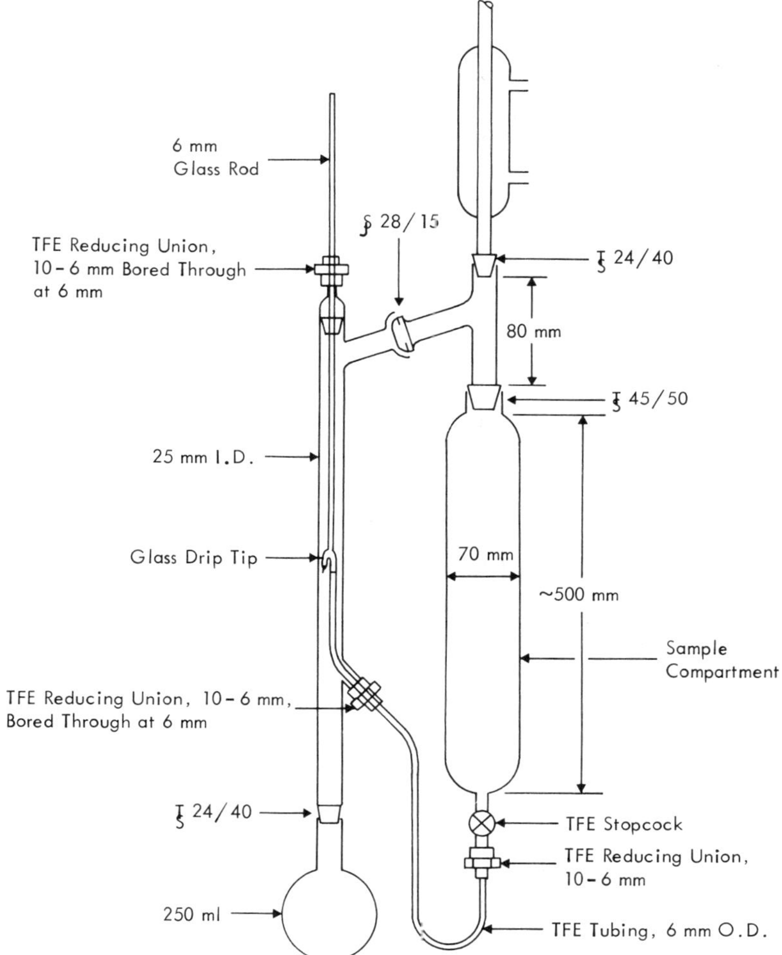


**Figure 2.** CLLE apparatus designed by MRI.

tracted for 6 hr with 300 mL dichloromethane. The extracts were concentrated and analyzed by GC/MS without any cleanup.

Microextraction procedures were evaluated using Babcock milk butterfat testing tubes, shaped much like volumetric flasks but designed to permit centrifugation; these tubes were used to assist solvent recovery from the

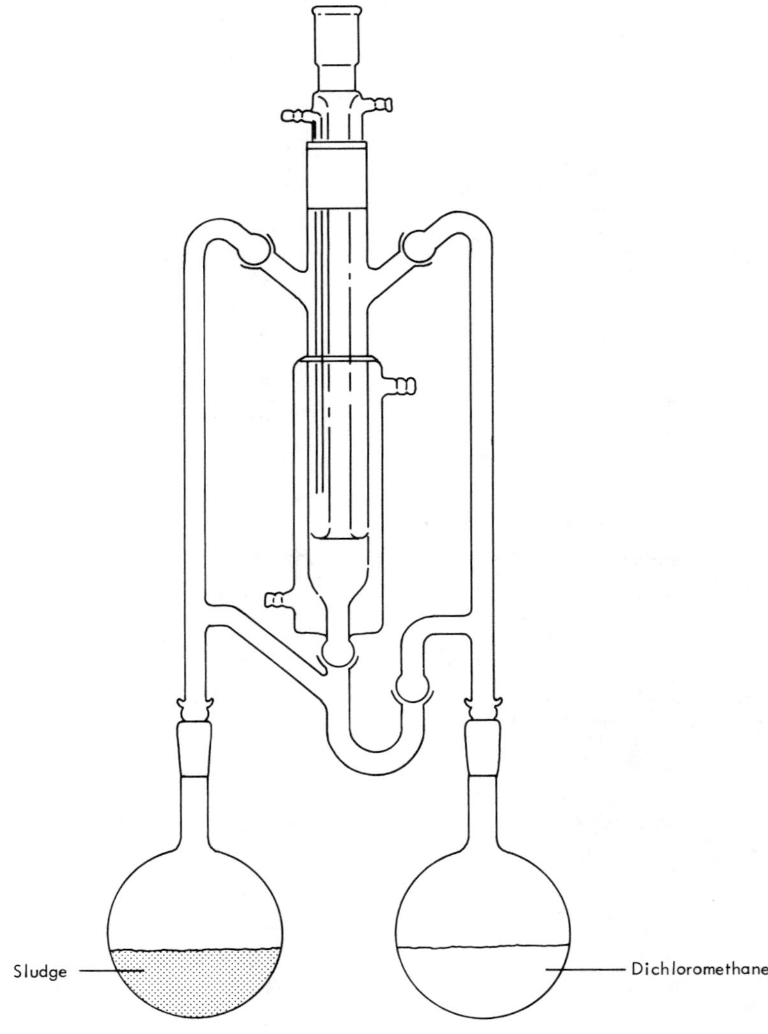


**Figure 3.** Continuous extractive steam distillation apparatus.

extracted sludge aliquots. Spiked and unspiked 30-mL aliquots of combined sludge were transferred to Babcock tubes. Separate aliquots were spiked with the B/N and acidic compounds nominally at 200 and 1000 $\mu$g/L. All sludge aliquots spiked at 1000 μg/L were allowed to equilibrate at 4°C overnight with tumbling. The aliquots were adjusted to pH 11 or 1, as appropriate, and were diluted with 15 mL of glass-distilled water that had been saturated with the extracting solvent. The diluted aliquots were extracted with 0.5 mL aliquots hexane or toluene or 1.0 mL of diisopropyl ether by hand-shaking for 2 min. The extracts were recovered by centrifuging at 1500 rpm for 15 min and removing the solvent with a 500-μL syringe. The volumes of extracts recovered were read from the syringe. All extractions were conducted in duplicate. The extracts were analyzed by the GC/MS procedures described above.

## Results and Discussion

The recoveries of the test compounds spiked into combined sludge aliquots and extracted by any of the three alternative extraction techniques, described above, are summarized in Table III.

Since a more detailed discussion of the results is given elsewhere [10], a few brief conclusions about these techniques will be given. As expected, each of these procedures achieved promising recoveries for concentrating spiked compounds from water. Recoveries from spiked combined POTW sludge however were generally lower. CLLE with dichloromethane proved ineffective in extracting polynuclear aromatic hydrocarbons (PAH) and similar hydrophobic compounds from sludge solids. Extractive steam distillation with dichloromethane was ineffective in extracting compounds with low vapor pressures. In addition, the steam distillate extracts from sludges contained high levels of coextracted aliphatic hydrocarbons. Although good recoveries were obtained for spiked water with microextraction and the procedure is exceptionally simple, physical recovery of the organic solvent extracts from sludge aliquots was a major hindrance. For most compound-solvent-salt combinations for which good recoveries were observed, most of the recovery was achieved in the first extraction. The results obtained for toluene and diisopropyl ether were similar to those reported for hexane in Table III. Single microextraction methods are more attractive for sludges, not only to simplify the extraction process, but also to avoid coextracted interferences that could be troublesome in composited multiple extracts. Extracting replicate sludge aliquots with multiple solvents and analyzing the resulting extracts separately are preferable to extracting the entire list of extractable compounds from the same sludge aliquot by multiple extraction and compositing the extracts.

None of the tested methods resulted in any significant improvements over the preliminary protocol. Hence, the homogenization/centrifugation method was selected for the revised sludge protocol. This selection necessi-

Table III. Recovery of Extractable Compounds from Spiked Sludge Samples by CLLE, Steam Distillation and Microextraction

| Compound | CLLE[a] | | | Microextraction | |
|---|---|---|---|---|---|
| | 100-mL Aliquot | 300-mL Aliquot | Steam Distillation[b] | Hexane, No Salt, Spike Level ~1300 μg/L Sludge | Hexane, No Salt, Spike Level ~5000 μg/L Sludge |
| 1,4-Dichlorobenzene | 33 | 57 | 103 | 51 | 48 |
| Hexachloroethane | 98 | 46 | 0 | 0 | 39 |
| *bis*-(2-Chloroisopropyl) Ether | c | c | 154 | 51 | c |
| *bis*-(2-Chloroethyl) Ether | c | c | 258 | 11 | c |
| Acenaphthylene | 88 | 35 | 97 | 52 | 93 |
| 2,6-Dinitrotoluene | 121 | 81 | 0 | 0 | 34 |
| Fluoranthene | 33 | 25 | 43 | 35 | 62 |
| Benzidine | 37 | 69 | 0 | 0 | c |
| 3,3′-Dichlorobenzidine | 68 | 43 | 0 | 12 | c |
| *n*-Butyl Benzyl Phthalate | 20 | 25 | 4 | 86 | 70 |
| *bis*-(2-Ethylhexyl) Phthalate | 57 | 6 | 17 | 24 | 63 |
| Benzo[a]pyrene | 20 | 26 | 3 | 10 | 26 |
| Phenol | 102 | c | d | 0 | c |
| 2,4-Dimethylphenol | 19 | c | 0 | 0 | c |
| 2,4-Dichlorophenol | 62 | c | 16 | 0 | c |
| Pentachlorophenol | 0 | c | 79 | 0 | c |

[a]Recovery was determined at 300-μg/L sludge spike level.

[b]Recovery was determined at 600-μg/L sludge spike level.

[c]These compounds were not spiked.

[d]Recovery not estimated since the concentration of phenol in the unspiked sludge was larger than the concentration found in the spiked sludge.

tated additional evaluation of extract cleanup procedures as alternatives or supplements to the GPC method used in the preliminary POTW protocol.

### Extract Cleanup Studies

Successful employment of a rigorous, nonselective extraction procedure for sludges, such as homogenization/centrifugation, necessitates the use of a very selective and efficient extract cleanup procedure to produce extracts of sufficient quality for reliable GC/MS determination. Two extract cleanup mechanisms were evaluated to meet these requirements, molecular size discrimination and polarity selection. Various GPC procedures were evaluated to determine the optimum molecular size fractionation procedure. The polarity-based cleanup method evaluated was adsorption chromatography on silica gel, Florisil and cesium silicate. The performance of adsorption procedures was evaluated for sludge extracts both with and without GPC precleaning.

### Experimental

Five GPC packings (Bio-Beads SX-2, SX-3, SX-4, SX-8 and Sephadex LH-20) were evaluated with three solvent systems. The solvents tested were dichloromethane, 15% cyclohexane in dichloromethane and 50% cyclohexane in dichloromethane. Each gel-solvent combination was evaluated by chromatographing solutions of corn oil, butyl benzyl phthalate and mixed phenols. Corn oil was used to represent the large biogenic materials in sludge extracts. Butylbenzyl phthalate is one of the largest extractable priority pollutants, and thus one of the first to elute. The mixed phenols solution contained phenol, 2,4-dichlorophenol and PCP. Elution patterns were obtained by monitoring the column eluent with a UV detector (254 nm).

Adsorption chromatography experiments were performed using 1% deactivated Florisil (60/100 mesh, Floridin) or 3% deactivated silica gel (Silica Woelm, 70/150 mesh, ICN). The elution scheme was:

- fraction I: 20 mL hexane
- fraction II: 50 mL 10% dichloromethane in hexane
- fraction III: 50 mL 50% dichloromethane in hexane
- fraction IV: 150 mL 5% acetone in dichloromethane

Cesium silicate was prepared according to procedures reported by DeWalle and Chian [11]. Acidic extracts (100-250 mL in volume) containing the phenols were passed through the cesium silicate, and the eluent (fraction I) was collected. The column was eluted with 20 mL methanol (fraction II). Since the cesium silicate has an appreciable solubility in methanol, the methanol eluents were concentrated to 4 mL and then were partitioned between 6 mL water (pH = 1) and 2 mL dichloromethane.

All the silica gel, Florisil and cesium silicate fractions were analyzed by GC/MS.

## Results and Discussion

The elution volumes and peak profiles data obtained for the various gels and solvent combinations indicate that the best overall separation is obtained with SX-3 gel when dichloromethane is the eluting solvent. A fraction cut selected at the valley between the corn oil and phthalate achieved 85% or greater elimination of the oil from the analytes, and 85% or greater recovery of the phthalate (based on peak areas). Furthermore, the selection of dichloromethane as the elution solvent is directly compatible with the homogenization/centrifugation sludge extracts. Elution with higher proportions of cyclohexane in dichloromethane resulted in poorer corn oil-phthalate separation as seen in Figure 4.

The recoveries of the test compounds spiked in blank extracts by GPC on Bio-Beads SX-3 eluted with dichloromethane are given in Table IV. The data indicate that recoveries were quantitative for all compounds with the exception of *bis*-(2-chloroethyl) ether and possibly benzidine and *bis*-(2-ethylhexyl) phthalate.

Recoveries by silica gel or Florisil chromatography were determined for the B/N compounds spiked into extracts prepared from 80- and 160-mL aliquots of combined POTW sludge. This study was designed to test both the recoveries and the cleanup capacities of the columns. In addition, recoveries were determined for extracts spiked following GPC cleanup to investigate the possible enhancement of the performance of adsorption chromatographic cleanup by removing high-molecular-weight biogenic compounds. Table V summarizes these data. The individual recoveries for each of the four fractions collected from the silica gel or Florisil columns are given elsewhere [10]. As the data indicate, the recoveries were similar for the silica gel and Florisil chromatography. Furthermore, cleanup by GPC prior to adsorption chromatography has no significant effect on the recovery of the spiked compounds. This is illustrated in Figures 5 through 9. Figures 5 through 8 show the GC/MS chromatograms of the individual fractions and the combined II through IV fractions, which were obtained by Florisil cleanup without any prior cleanup by GPC. For comparison, a GC/MS chromatogram of the composite of fractions II, III and IV for the same sludge that had been

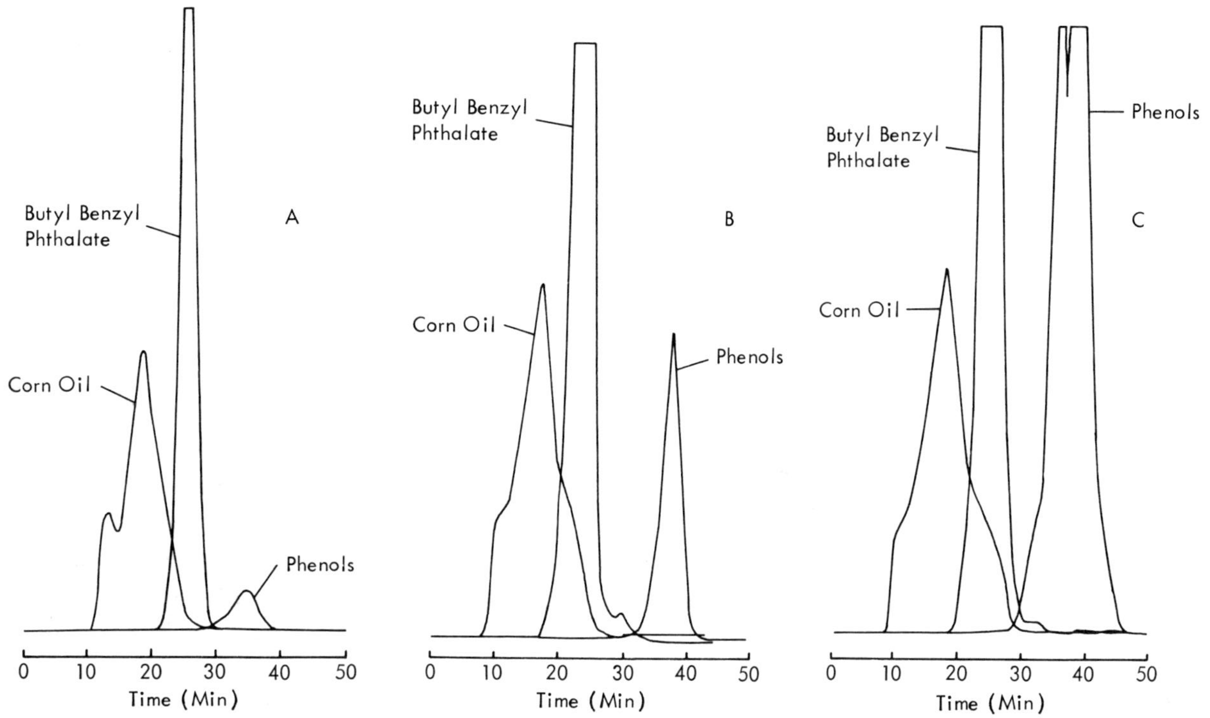

**Figure 4.** GPC chromatograms on Bio-Beads SX-3 eluted with (A) dichloromethane; (B) 15% cyclohexane in dichloromethane; and (C) 50% cyclohexane in dichloromethane.

**Table IV. Recoveries of Extractable Compounds by GPC Cleanup (SX-3, Dichloromethane)**

| Compound | Recovery (%) |
|---|---|
| 1,4-Dichlorobenzene | 69 |
| Hexachloroethane | 70 |
| *bis*(2-Chloroisopropyl) Ether | 63 |
| *bis*(2-Chloroethyl) Ether | 37 |
| Acenaphthylene | 77 |
| 2,6-Dinitrotoluene | 77 |
| Fluoranthrene | 87 |
| Benzidine | 52 |
| 3,3′-Dichlorobenzidine | 79 |
| *n*-Butyl Benzyl Phthalate | 81 |
| *bis*(2-Ethylhexyl) Phthalate | 57 |
| Benzo[a]pyrene | 79 |
| Phenol | 81 |
| 2,4-Dimethylphenol | 90 |
| 2,4-Dichlorophenol | 86 |
| Pentachlorophenol | 105 |

cleaned by GPC is shown in Figure 9. The background in this chromatogram is very similar to that for a replicate extract without GPC. The recoveries observed for extracts prepared from 80- and 160-mL sludge aliquots are comparable. However, the Florisil-cleaned extracts of 160-mL aliquots had to be analyzed at a larger final extract volume to avoid overloading the GC/MS detection system. Hence no real gain in method sensitivity was obtained by extracting larger sludge aliquots.

The results of recovery determinations for the phenolic compounds by cesium silicate chromatography are given in Table VI. Although no breakthrough of phenols was observed with spiked blanks, and quantitative recoveries were found for the extracts of 80-mL aliquots, significant amounts of the analytes eluted in fraction I when the 160-mL sludge aliquots were subjected to this type of cleanup. The breakthrough is probably due to the higher levels of coextracted materials blocking the adsorption sites on the cesium silicate. Since the results were no better for extracts that had been cleaned by GPC, the interferences are probably low-molecular-weight compounds. Hence, the cleanup capacity of cesium silicate may be somewhat limited.

Table V. Summary of Recoveries for the Extractable Spiking Compounds from Extracts of 80- and 160-mL Sludge Aliquots Cleaned with Florisil or Silica Gel

| Compound | Spike Level (μg) | | Average Recovery[a] (%) | | | | | | | |
|---|---|---|---|---|---|---|---|---|---|---|
| | | | Florisil | | | | Silica Gel | | | |
| | | | With GPC | | Without GPC | | With GPC | | Without GPC | |
| | 80 mL | 160 mL | 80 mL | 160 mL | 80 mL | 160 mL | 80 mL | 160 mL | 80 mL | 160 mL |
| 1,4-Dichlorobenzene | 47 | 94 | 92 | 142 | 68 | 88 | 73 | 109 | 62 | 95 |
| Hexachloroethane | 31 | 62 | 79 | 97 | 110 | 70 | 95 | 54 | 116 | 47 |
| *bis*-(2-Chloroisopropyl) Ether | 149 | 298 | 100 | 102 | 99 | 111 | 102 | 112 | 129 | 62 |
| *bis*-(2-Chloroethyl) Ether | 95 | 190 | 103 | 49 | 74 | 53 | 85 | 66 | 116 | 93 |
| Acenaphthylene | 29 | 57 | 97 | 148 | 96 | 109 | 87 | 72 | 91 | 93 |
| 2,6-Dinitrotoluene | 35 | 70 | 90 | 110 | 122 | 139 | 108 | 102 | 129 | 108 |
| Fluoranthene | 27 | 54 | 85 | 110 | 75 | 135 | 69 | 55 | 98 | 103 |
| Benzidine | 93 | 186 | 17 | 0 | 0 | 0 | 0 | 0 | 0 | 0 |
| 3,3-Dichlorobenzidine | 109 | 218 | 106 | 168 | 90 | 167 | 63 | 155 | 70 | 1 |
| *n*-Butyl Benzyl Phthalate | 45 | 90 | 170 | b | 108 | b | 148 | b | 103 | b |
| *bis*-(2-Ethylhexyl) Phthalate | 40 | 80 | 152 | b | 48 | b | 108 | b | 270 | b |
| Benzo[a] pyrene | 30 | 60 | 52 | 155 | 52 | 130 | 47 | 114 | 81 | 99 |

[a]Average of two determinations.

[b]Recoveries were not determined because the levels in the unspiked extracts were much higher than the spike level.

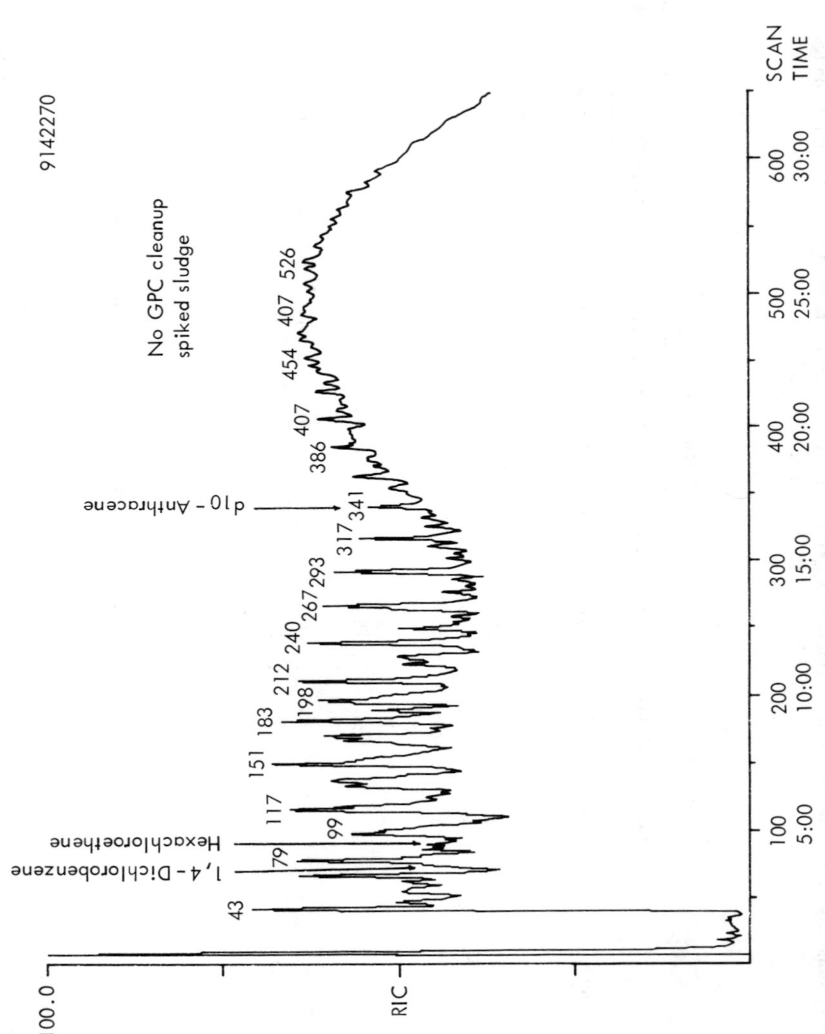


**Figure 5.** GC/MS chromatogram of fraction II obtained by Florisil cleanup of the spiked sludge (no GPC cleanup).

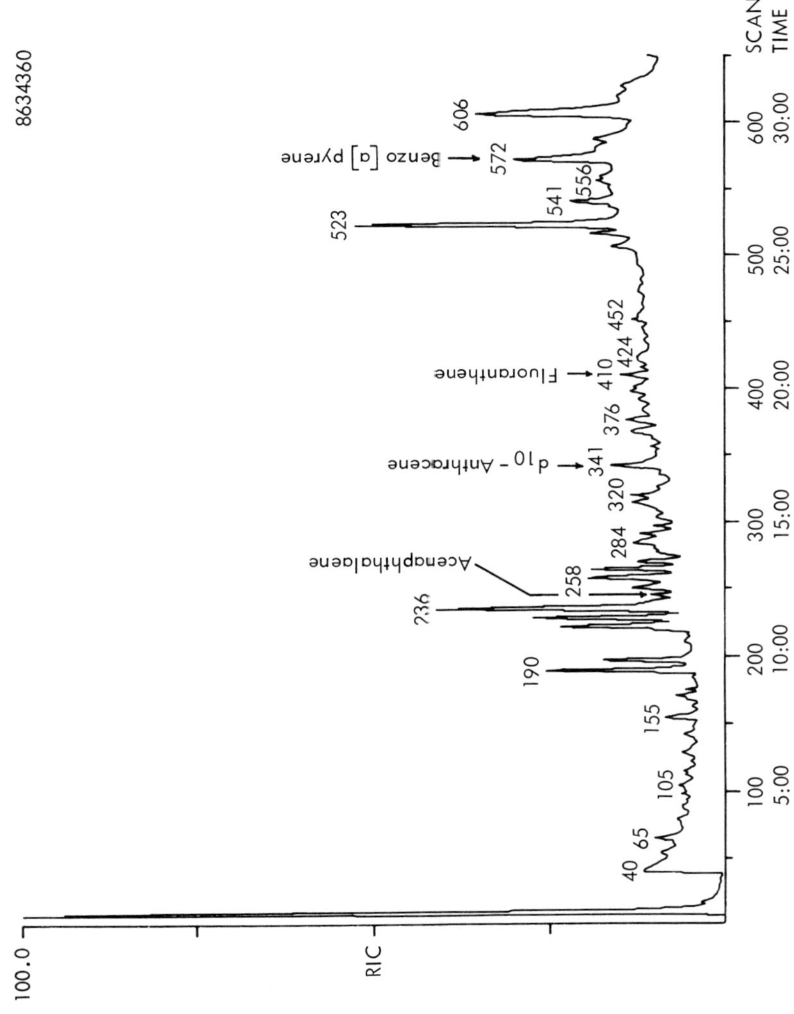


**Figure 6.** GC/MS chromatogram of fraction III obtained by Florisil cleanup of the spiked sludge (no GPC cleanup).

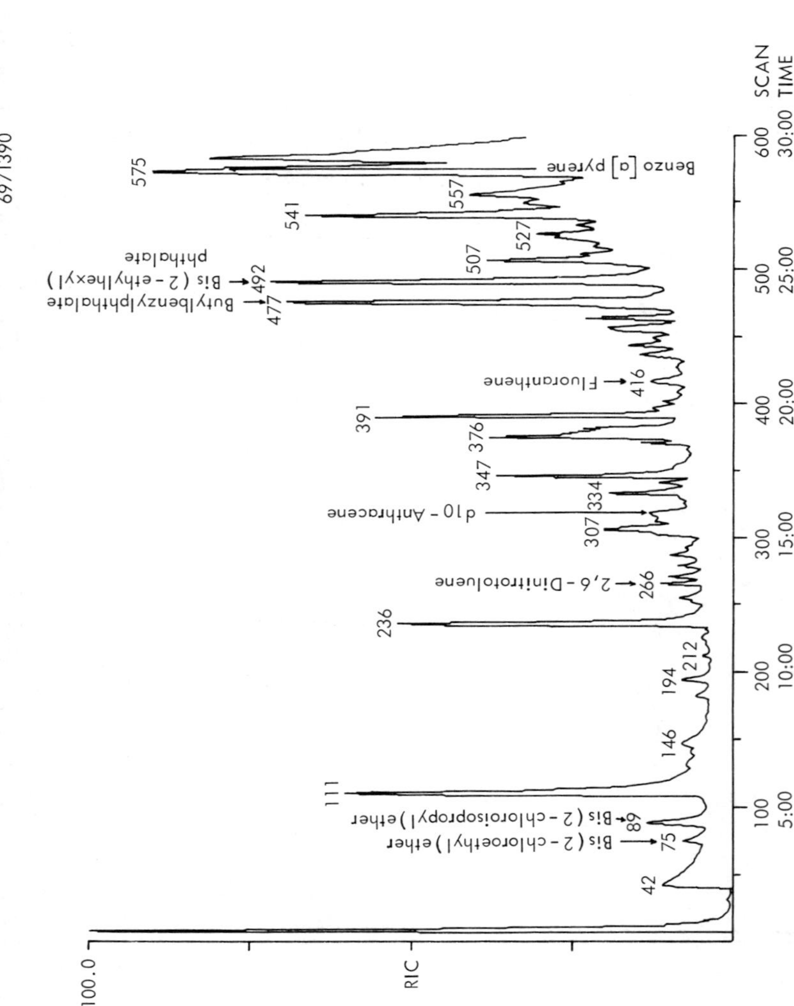


**Figure 7.** GC/MS chromatogram of fraction IV obtained by Florisil cleanup of the spiked sludge (no GPC cleanup).

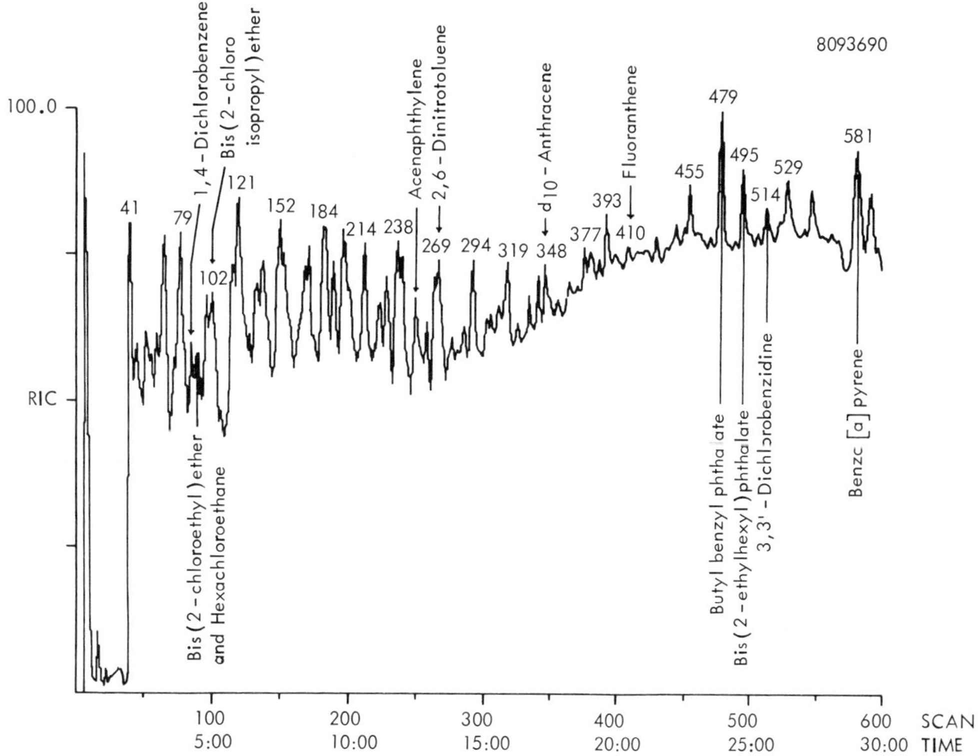

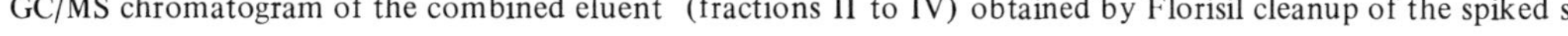


**Figure 8.** GC/MS chromatogram of the combined eluent (fractions II to IV) obtained by Florisil cleanup of the spiked sludge (no GPC clean-up).

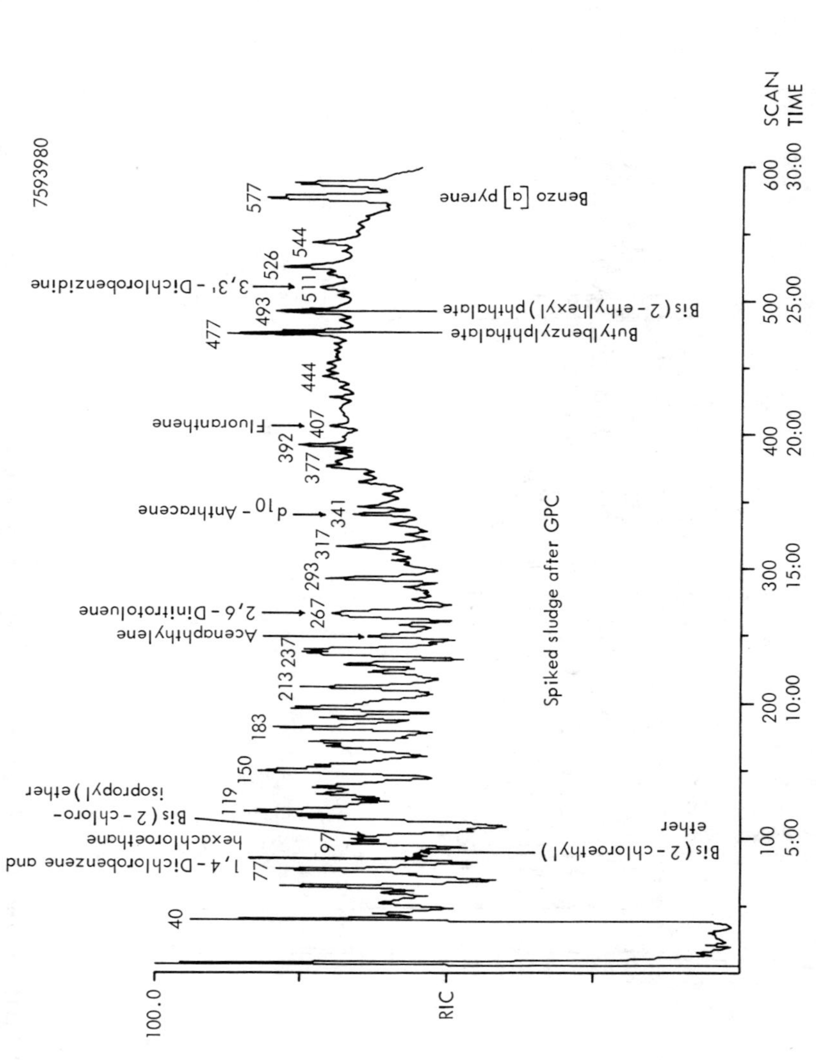


**Figure 9.** CH/MS chromatogram of the combined eluent (fractions II to IV) obtained by Florisil cleanup of the spiked sludge following GPC cleanup.

Table VI. Recoveries of Acidic Compounds from Spiked Sludge Extracts by Cesium Silicate Chromatography

| | 80-mL | | | 160-mL | | | | | | | | |
|---|---|---|---|---|---|---|---|---|---|---|---|---|
| | | | | | Without GPC | | | | With GPC | | | |
| | | | | | Unspiked Level | | Spike Recovery | | Unspiked Level | | Spike Recovery | |
| Compound | Spike Level (μg) | Unspiked Level (μg) | Recovery Fraction II[a] (%) | Spike Level (μg) | Fraction I[b] (μg) | Fraction II (μg) | Fraction I (%) | Fraction II (%) | Fraction I (μg) | Fraction II (μg) | Fraction I (%) | Fraction II (%) |
| Phenol | 48.4 | 8.4 | 68 | 76.2 | 18 | 85 | 150 | 62 | 33 | 88 | 120 | 150 |
| 2,4-Dimethylphenol | 35.4 | ND[c] | 75 | 49.2 | ND | ND | 68 | 43 | ND | ND | 63 | 33 |
| 2,4-Dichlorophenol | 32.8 | 2.8 | 89 | 53.1 | ND | ND | 57 | 23 | ND | ND | 51 | 39 |
| Pentachlorophenol | 30.4 | 9.8 | 120 | 45.6 | ND | 1.7 | 32 | 160 | ND | 0.5 | 0 | 180 |

[a]Eluted with 50 mL of methanol.

[b]The crude extract eluate.

[c]ND = not detected.

## GC/MS Procedures

Capillary GC/MS procedures were evaluated as an alternative to the packed column GC/MS methods described above. Four wall-coated open tubular (WCOT) capillary columns were evaluated using standard performance test mixtures and solutions of the B/N and acidic test compounds given in Table I. The details of these evaluations containing information on the height equivalent to a theoretical plate (HETP), peak asymmetries (AS), column adsorptivity and the acid/base character (pH) are given elsewhere [10]. Of the four columns tested, the SE-54 fused-silica column exhibited the best overall performance.

The suitability of the SE-54 fused-silica column for analyses of sludge extracts was evaluated by chromatographing the spiked sludge extracts prepared for the evaluation of adsorption chromatographic cleanup procedures. The better resolution and inertness of the SE-54 capillary column (relative to the packed column) provided GC/MS data for B/N extracts that were easier to interpret. Retention time and response factor data were generated for the entire list of B/N priority pollutants [10].

Although good chromatography was observed for the acidic compounds when standard solutions were chromatographed on all capillary columns, the phenols are likely to be very susceptible to adsorption on a partially active column. Analysis of acidic sludge extracts, cleaned by GPC only, would likely increase the adsorption character of capillary columns. In addition, the extracts for many sludges contain high levels of aliphatic hydrocarbons not removed by GPC that can hinder GC/MS analysis. Hence, capillary GC/MS may not be practicable for routine analyses of acidic sludge extracts.

## Selection of Methods for Extractable Compounds in Municipal and Industrial Wastewater Treatment Sludge

Although the methods described for the analysis of B/N and acidic compounds provided acceptable precision and accuracy for most compounds, some shortcomings were apparent. Homogenization/centrifugation extraction of 320-mL aliquots of sludge was both labor-intensive and time-consuming. The rigorous extraction also produced an extract enriched in interfering coextractants which challenged the capacity of the GPC cleanup. Unfortunately none of the alternative extraction procedures was found to provide recoveries for the broad range of the basic, neutral and acidic compounds that were comparable to those demonstrated for homogenization/centrifugation. Hence, the homogenization/centrifugation extraction was recommended for analyses of municipal and industrial sludge with one modification. Extraction of a single 80-mL aliquot of sludge was recommended to decrease the time required for extract preparation. Since extracts from 80-mL aliquots were analyzed at a smaller final volume, there was no real decrease in method sensitivity.

Both GPC and adsorption chromatography on Florisil and silica gel provided acceptable extract cleanup and recoveries for the B/N compounds. The Florisil and silica gel adsorbent methods were essentially equivalent, and removed aliphatic hydrocarbons and low- and high-molecular-weight polar interferences. Although the GPC method only removes mostly high-molecular-weight biogenic materials, it is easier to automate and more amenable to routine sludge analyses. GPC was the only cleanup procedure that provided good recoveries for the acidic compounds. Hence, GPC and adsorption chromatography on Florisil or silica gel were recommended as options for cleaning B/N extracts. GPC was recommended for cleanup of acidic extracts.

Although the packed-column GC/MS method described in the preliminary POTW protocol provided reliable analyses of sludge extracts, the SE-54 WCOT fused-silica capillary column was recommended as the preferred option for the GC/MS procedure for B/N extracts. However, the capillary column GC/MS method should only be used for extracts cleaned by chromatography on Florisil or silica gel. A comparison of chromatograms obtained for replicate POTW sludge B/N extracts by GPC cleanup and packed-column GC/MS with Florisil cleanup and capillary column GC/MS is shown in Figure 10. The packed column GC/MS method was recommended for analysis of acidic extracts. The level of cleanup provided by the GPC cleanup for acidic extracts may make it difficult to maintain the inactivity of the capillary column.

The revised analytical scheme for extractable priority pollutants in municipal and industrial wastewater treatment sludges is shown in Figure 11. The detailed analytical protocol is given elsewhere [10].

## EVALUATION OF THE PRECISION AND ACCURACY OF THE METHOD

The precision and accuracy of the extractables protocol [10] were evaluated by determining recoveries of the spiked compounds from five primary sludges. Recoveries were determined for the extractable compounds fortified into aliquots of these sludges at three spiking levels. The spiking levels were chosen to be 2, 20 and 200 times the individual detection limit typically observed for each compound in sludge. Two combinations of cleanup-analysis options were evaluated for the B/N compounds. B/N extracts were analyzed

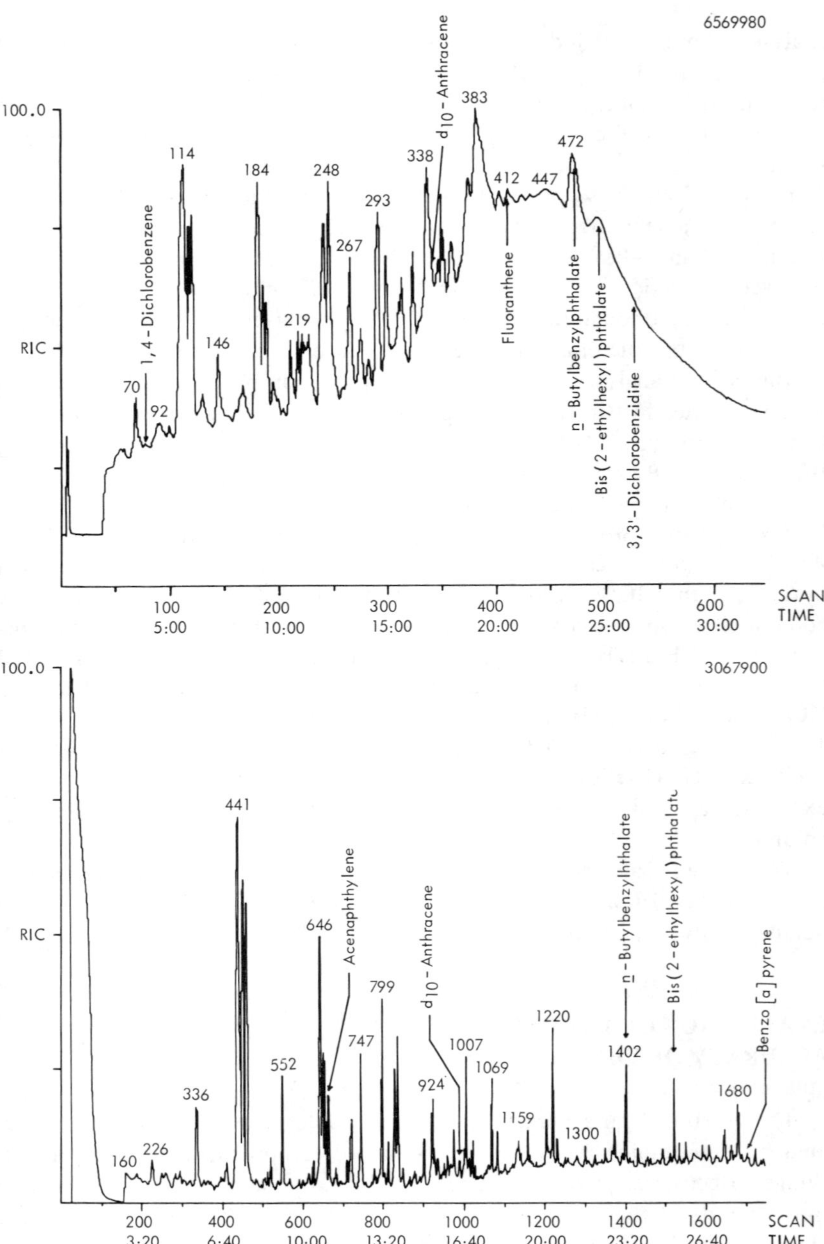


**Figure 10.** (top) GC/MS chromatogram for base/neutrals in unspiked Kansas City, KS, primary POTW sludge (GPC cleanup); (bottom) capillary column GC/MS chromatogram for base/neutrals in unspiked Kansas City, KS, primary POTW sludge (Florisil cleanup).

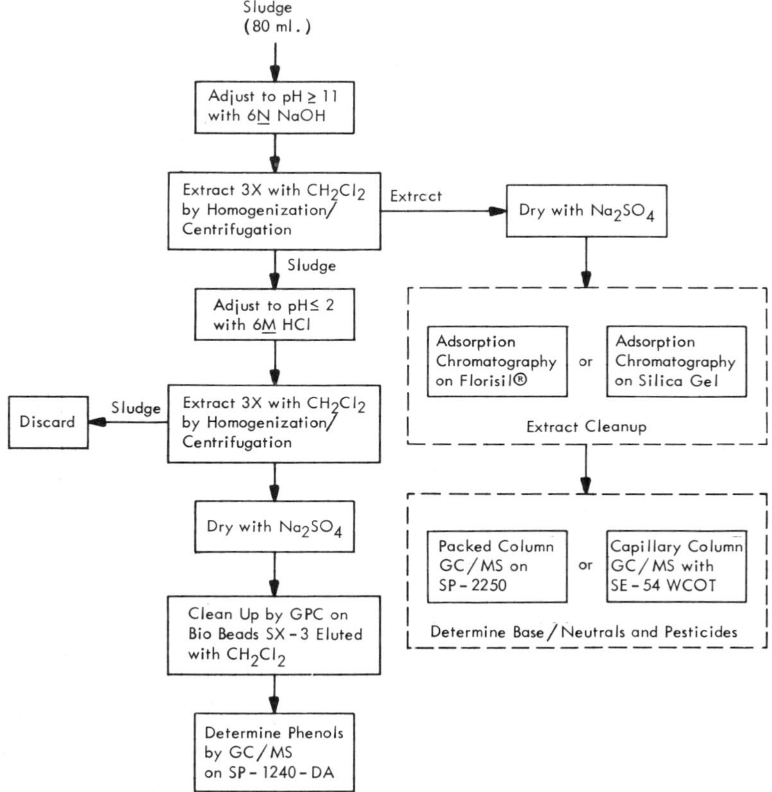


**Figure 11.** Scheme for analysis of extractable organics in sludge.

by packed-column GC/MS following GPC cleanup, and by capillary GC/MS following Florisil cleanup. All acidic sludge extracts were cleaned up by GPC and analyzed by packed-column GC/MS.

## Experimental

Eleven 80-mL aliquots of each of the five primary sludge samples were placed in 200-mL centrifuge bottles fitted with TFE-lined septa. Three sets of triplicate bottles were spiked with the extractable spiking compounds at

400, 4000 and 40,000 μg/L sludge. (Benzidine was spiked at 800, 8000 and 80,000 μg/L sludge.) The spikes were added in 2-3 mL of acetone solution. Immediately following the addition of the spikes, samples were homogenized for 30-45 sec and stored at 4°C overnight prior to analysis. The unspiked sludges were analyzed in duplicate and the spiked sludges were analyzed in triplicate as described above. Details of the analytical procedures are given elsewhere [10].

## Results and Discussion

The results of the precision and accuracy determinations for the B/N compounds obtained with GPC cleanup and packed-column GC/MS are shown in Table VII. Many of the recovery determinations were likely perturbed by high concentrations of the analytes in the unspiked sludge samples. Recoveries were not calculated for cases where the unspiked concentration was greater than 10 times the spike level. Although the recoveries varied somewhat from sludge to sludge for many compounds, all compounds were recovered from all five sludges spiked at 40,000 μg/L. Recoveries were generally lower for the polar compounds, including 2,6-dinitrotoluene, the benzidines and *bis*-(2-chloroethyl) ether. However, only 2,6-dinitrotoluene was not recovered from all sludges spiked at 400 μg/L. Of all recovery determinations, 49% fell within the range of 50-150%. An additional 19% were lower than 50% recovery, and only 8% of the recoveries were greater than 150%.

The accuracy of the method was somewhat dependent on the particular sludge samples. However, the precisions of the recovery determinations were good. The relative standard deviations for triplicate determinations were 30% or less for 95% of the measurements and 10% or less for 62% of the measurements.

The recoveries of B/N compounds from spiked sludge obtained with Florisil cleanup and capillary GC/MS are shown in Table VIII. Although these sludges are replicate samples of those spiked and analyzed with the packed-column GC/MS method, the levels determined in the unspiked samples by the capillary column GC/MS method were lower. This likely reflects losses during sample storage, as the samples were extracted for the capillary column GC/MS experiments 2-3 weeks later than for the packed-column GC/MS study. In general, the recoveries were similar to those observed with the packed-column GC/MS option. However, all compounds were recovered from all five sludges spiked at 400 μg/L. An investigation of the distribution of the recoveries shows that 48% were within the range of 50-150%, 28% were less than 50% and 13% were higher than 150% recovery. The reproducibility of the determinations was slightly poorer. Of the RSD values, 56% were 30% or

Table VII. Results of Precision and Accuracy Evaluations for the Extractables Method (Packed-Column GC/MS) Applied to the Analysis of Base/Neutral Compounds in Municipal and Industrial Wastewater Treatment Sludges

| | Platte County POTW | | | Blue River POTW | | | Kansas City, KS | | | Industrial No. 1 | | | Industrial No. 2 | | |
|---|---|---|---|---|---|---|---|---|---|---|---|---|---|---|---|
| Compound | Spike Conc (μg/L) | Unspiked[a] Sludge (μg/L) | Recovery (%) | Spike Conc (μg/L) | Unspiked[a] Conc (μg/L) | Recovery (%) | Spike Conc (μg/L) | Unspiked[a] Conc (μg/L) | Recovery (%) | Spike conc (μg/L) | Unspiked[a] Conc (μg/L) | Recovery (%) | Spike Conc (μg/L) | Unspiked[a] Conc (μg/L) | Recovery (%) |
| 1,4-Dichloro- | 417 | 45 | 70 ± 9[b] | 400 | 47 | 37 ± 21 | 400 | 21 | 11 ± 3 | 400 | ND | 51 ± 4 | 400 | 820 | c |
| benzene | 4,160 | | 87 ± 2 | 4,000 | | 77 ± 7 | 4,000 | | 93 ± 9 | 4,000 | | 73 ± 5 | 4,000 | | 88 ± 12 |
| | 41,610 | | 74 ± 8 | 39,990 | | 87 ± 7 | 39,990 | | 90 ± 9 | 39,900 | | 70 ± 4 | 40,000 | | 92 ± 9 |
| Hexachloro- | 441 | ND[d] | 39 ± 3 | 400 | ND | 0 | 400 | ND | 0 | 400 | ND | 17 ± 5 | 400 | ND | 0 |
| ethane | 4,400 | | 56 ± 7 | 4,000 | | 54 ± 3 | 4,000 | | 47 ± 8 | 4,000 | | 55 ± 9 | 4,000 | | 97 ± 6 |
| | 44,010 | | 53 ± 2 | 40,000 | | 59 ± 5 | 40,010 | | 57 ± 10 | 39,990 | | 51 ± 7 | 40,000 | | 77 ± 4 |
| *bis*(2-Chloro- | 408 | 62 | 35 ± 5 | 400 | ND | 100 ± 54 | 400 | ND | 0 | 400 | ND | 88 ± 6 | 400 | ND | 0 |
| ethyl) Ether | 4,070 | | 44 ± 2 | 4,000 | | 380 ± 37 | 4,000 | | 510 ± 123 | 4,000 | | 120 ± 14 | 4,000 | | 200 ± 4 |
| | 40,680 | | 27 ± 4 | 39,990 | | 280 ± 8 | 39,990 | | 410 ± 53 | 40,000 | | 95 ± 11 | 40,000 | | 130 ± 7 |
| Acenaph- | 411 | 1 | 77 ± 6 | 400 | 1,430 | 43,25[c] | 400 | ND | 0 | 400 | 6 | 65 ± 3 | 400 | ND | 200 ± 34 |
| thylene | 4,090 | | 84 ± 7 | 4,000 | | 42 ± 7 | 4,000 | | 82 ± 12 | 4,000 | | 89 ± 5 | 4,000 | | 130 ± 52 |
| | 40,000 | | 51 ± 2 | 39,900 | | 41 ± 2 | 40,000 | | 98 ± 10 | 39,000 | | 88 ± 3 | 40,000 | | 93 ± 2 |
| 2,6-Dinitro- | 407 | ND | 0 | 400 | ND | 0 | 400 | ND | 0 | 400 | ND | 0 | 400 | ND | 0 |
| toluene | 4,050 | | 0 | 4,000 | | 20 ± 1 | 4,000 | | 0 | 4,000 | | 59 ± 8 | 4,000 | | 150 ± 9 |
| | 40,550 | | 63 ± 4 | 40,040 | | 73 ± 3 | 40,040 | | 78 ± 20 | 39,990 | | 99 ± 14 | 40,000 | | 97 ± 5 |
| Fluoran- | 428 | 65 | 45 ± 5 | 400 | 8,330 | e | 400 | 124 | 16 ± 2 | 400 | 14 | 56 ± 8 | 400 | 670 | c |
| thene | 4,270 | | 58 ± 5 | 4,000 | | 59 ± 18 | 4,000 | | 52 ± 14 | 4,000 | | 82 ± 7 | 4,000 | | 120 ± 1 |
| | 40,000 | | 50 ± 4 | 39,530 | | 20 ± 2 | 40,000 | | 100 ± 6 | 39,920 | | 100 ± 14 | 40,000 | | 95 ± 4 |
| Benzidine | 789 | ND | 0 | 800 | ND | 0 | 800 | ND | 0 | 800 | ND | 38 ± 8 | 800 | ND | 0 |
| | 4,870 | | 20 ± 1 | 8,000 | | 0 | 8,000 | | 0 | 8,000 | | 82 ± 10 | 8,000 | | 120 ± 7 |
| | 80,000 | | 90 ± 7 | 79,080 | | 96 ± 12 | 80,000 | | 160 ± 13 | 80,000 | | 130 ± 6 | 70,000 | | 50 ± 2 |
| 3,3′-Di- | 415 | ND | 46 ± 6 | 402 | ND | 0 | 402 | 535 | c | 400 | ND | 0 | NA | ND | – |
| chloroben- | 4,130 | | 37 ± 2 | 4,020 | | 100 ± 14 | 4,020 | | 110 ± 13 | 4,000 | | 0 | NA | | – |
| zidine | NA[f] | | – | NA | | – | NA | | – | NA | | – | NA | | – |

| Compound | Platte County POTW | | | Blue River POTW | | | Kansas City, KS | | | Industrial No. 1 | | | Industrial No. 2 | | |
|---|---|---|---|---|---|---|---|---|---|---|---|---|---|---|---|
| | Spike Conc (μg/L) | Unspiked[a] Sludge (μg/L) | Recovery (%) | Spike Conc (μg/L) | Unspiked[a] Conc (μg/L) | Recovery (%) | Spike Conc (μg/L) | Unspiked[a] Conc (μg/L) | Recovery (%) | Spike conc (μg/L) | Unspiked[a] Conc (μg/L) | Recovery (%) | Spike Conc (μg/L) | Unspiked[a] Conc (μg/L) | Recovery (%) |
| *n*-Butyl | 452 | 544 | c | 400 | 3,090 | c | 400 | 5,950 | e | 400 | ND | 57 ± 8 | 400 | 12,500 | e |
| Benzyl | 4,500 | | 20 ± 1 | 4,000 | | 150 ± 17 | 4,000 | | 210 [b] 136[c] | 4,000 | | 54 ± 5 | 4,000 | | c |
| Phthalate | 40,000 | | 59 ± 2 | 40,000 | | 31 ± 2 | 40,000 | | 300 ± 30 | 40,000 | | 85 ± 7 | 40,000 | | 80 ± 4 |
| *bis*(2-Ethyl- | 444 | 652 | d | 400 | 2,850 | d | 400 | 1,790 | d | 400 | 85 | 30 ± 3 | 400 | 24,200 | d |
| hexyl) | 4,430 | | 22 ± 1 | 4,000 | | 120 ± 5 | 4,000 | | 190 ± 80 | 4,000 | | 42 ± 4 | 4,000 | | 230 ± 4[c] |
| Phthalate | 40,000 | | 47 ± 3 | 40,000 | | 28 ± 3 | 40,010 | | 180 ± 21 | 40,000 | | 53 ± 9 | 48,820 | | 200 ± 10[c] |
| Benzo[a]- | NA[f] | ND | – | 400 | 4,050 | e | 400 | ND | 0 | 400 | ND | 30 ± 1 | 400 | 308 | 0 |
| pyrene | NA | | – | 4,000 | | 290 ± 63 | 4,000 | | 80 ± 8 | 4,000 | | 33 ± 4 | 4,000 | | 140 ± 7 |
| | 40,550 | | 56 ± 3 | 40.010 | | 68 ± 10 | 40,010 | | 140 ± 19 | 40,080 | | 48 ± 4 | 40,000 | | 97 ± 6 |

[a]Mean of two determinations.
[b]Mean ± standard deviation for three determinations.
[c]Concentration determined in the unspiked sludge was higher than the spike level.
[d]ND = not detected.
[e]Concentration determined in the unspiked sludge was 10 times higher than the spike level.
[f]NA = this compound was not available when these samples were spiked.

Table VIII. Results of Precision and Accuracy Evaluations for the Extractables Method (Capillary Column GC/MS) Applied to the Analysis of Base/Neutral Compounds in Municipal and Industrial Wastewater Treatment Sludges

| | Platte County POTW | | | Blue River POTW | | | Kansas City, KS | | | Industrial No. 1 | | | Industrial No. 2 | | |
|---|---|---|---|---|---|---|---|---|---|---|---|---|---|---|---|
| Compound | Spike Conc (μg/L) | Unspiked[a] Sludge (μg/L) | Recovery (%) | Spike Conc (μg/L) | Unspiked[a] Conc (μg/L) | Recovery (%) | Spike Conc (μg/L) | Unspiked[a] Conc (μg/L) | Recovery (%) | Spike Conc (μg/L) | Unspiked[a] Conc (μg/L) | Recovery (%) | Spike Conc (μg/L) | Unspiked[a] Conc (μg/L) | Recovery (%) |
| 1,4-Dichloro-benzene | 417 | 21 | 52 ± 36[b] | 400 | ND[c] | 9 ± 2 | 400 | 6 | 21 ± 12 | 400 | ND | 24 ± 5 | 400 | 58 | 23 ± 12 |
| | 4,160 | | 100 ± 1 | 4,000 | | 35 ± 14 | 4,000 | | 41 ± 20 | 4,000 | | 160 ± 97 | 4,000 | | 57 ± 8 |
| | 39,990 | | 62 ± 28 | 39,900 | | 58 ± 8 | 39,990 | | 55 ± 18 | 39,990 | | 55 ± 8 | 40,000 | | 70 ± 19 |
| Hexachloro-ethane | 441 | ND | 21 ± 14 | 400 | ND | 3 ± 1 | 400 | ND | 0 | 400 | ND | 7 ± 10 | 400 | ND | 3 ± 2 |
| | 4,400 | | 22 ± 6 | 4,000 | | 10 ± 4 | 4,000 | | 15 ± 7 | 4,000 | | 161 ± 100 | 4,000 | | 75 ± 1 |
| | 40,010 | | 46 ± 12 | 40,010 | | 39 ± 6 | 40,010 | | 42 ± 6 | 39,990 | | 49 ± 21 | 40,000 | | 35 ± 5 |
| *bis*(2-Chloro-ethyl) Ether | 408 | ND | 78 ± 83 | 400 | ND | 24 ± 15 | 400 | ND | 45 ± 30 | 400 | ND | 38 ± 9 | 400 | ND | 0 |
| | 4,070 | | 140 ± 16 | 4,000 | | 69 ± 32 | 4,000 | | 130 ± 66 | 4,000 | | 130 ± 56 | 4,000 | | 63 ± 2 |
| | 39,990 | | 130 ± 81 | 39,990 | | 130 ± 66 | 39,990 | | 240 ± 66 | 40,000 | | 100 ± 8 | 40,000 | | 92 ± 19 |
| Acenaph-thylene | 411 | ND | 140 ± 72 | 400 | 34 | 18 ± 6 | 400 | 11 | 82 ± 46 | 400 | ND | 55 ± 5 | 400 | 730 | d |
| | 4,090 | | 230 ± 39 | 4,000 | | 53 ± 2 | 4,000 | | 150 ± 51 | 4,000 | | 110 ± 35 | 4,000 | | 93 ± 3 |
| | 40,000 | | 88 ± 26 | 40,000 | | 110 ± 4 | 40,000 | | 110 ± 9 | 39,990 | | 100 ± 2 | 40,000 | | 114 ± 27 |
| 2,6-Dinitro-toluene | 407 | ND | 43 ± 27 | 400 | ND | 7 ± 6 | 400 | ND | 0 | 400 | ND | 11 ± 13 | 400 | 1,090 | d |
| | 4,050 | | 5 ± 1 | 4,000 | | 4 ± 4 | 4,000 | | 16 ± 28 | 4,000 | | 160 ± 80 | 4,000 | | 62 ± 8 |
| | 40,040 | | 78 ± 5 | 40,040 | | 41 ± 43 | 40,040 | | 76 ± 21 | 39,990 | | 90 ± 3 | 40,000 | | 121 ± 32 |
| Fluoran-thene | 428 | 27 | 230 ± 122 | 400 | 750 | 7 ± 2[d] | 400 | 131 | 110 ± 69 | 400 | ND | 67 ± 3 | 400 | 125 | 0 |
| | 4,270 | | 190 ± 26 | 4,000 | | 120 ± 8 | 4,000 | | 120 ± 33 | 4,000 | | 160 ± 64 | 4,000 | | 121 ± 6 |
| | 40,000 | | 89 ± 23 | 40,000 | | 120 ± 10 | 40,000 | | 110 ± 4 | 39,920 | | 88 ± 19 | 40,000 | | 120 ± 25 |
| Benzidine | 789 | ND | 0 | 800 | ND | 0 | 800 | ND | 6 ± 10 | 800 | ND | 0 | 780 | ND | 0 |
| | 7,870 | | 1 ± 1 | 8,000 | | 11 ± 4 | 8,000 | | 1 ± 1 | 8,000 | | 240 ± 100 | 7,800 | | 35 ± 17 |
| | 80,000 | | 40 ± 11 | 80,000 | | 50 ± 10 | 80,000 | | 81 ± 9 | 80,010 | | 160 ± 6 | 70,000 | | 53 ± 4 |

| Compound | Platte County POTW | | | Blue River POTW | | | Kansas City, KS | | | Industrial No. 1 | | | Industrial No. 2 | | |
|---|---|---|---|---|---|---|---|---|---|---|---|---|---|---|---|
| | Spike Conc (μg/L) | Unspiked[a] Sludge (μg/L) | Recovery (%) | Spike Conc (μg/L) | Unspiked[a] Conc (μg/L) | Recovery (%) | Spike Conc (μg/L) | Unspiked[a] Conc (μg/L) | Recovery (%) | Spike Conc (μg/L) | Unspiked[a] Conc (μg/L) | Recovery (%) | Spike Conc (μg/L) | Unspiked[a] Conc (μg/L) | Recovery (%) |
| 3,3′-Di- | 415 | ND | 5 ± 8 | NA[e] | | | 402 | 20 | 6 ± 1 | NA | | – | NA | | – |
| chloro- | 4,130 | | 140 ± 32 | NA | | | NA | | | NA | | – | NA | | – |
| benzidine | NA | | – | NA | | – | NA | | – | NA | | – | NA | | – |
| *n*-Butyl | 452 | 770 | 290 ± 165 [d] | 400 | 2,380 | 80 ± 26 [d] | 400 | 5,230 | f | 400 | ND | 67 ± 8 | 400 | 9,930 | f |
| Benzyl | 4,510 | | 160 ± 26 | 4,000 | | 72 ± 20 | 4,000 | | 280 ± 266[d] | 4,000 | | 230 ± 36 | 4,000 | | d |
| Phthalate | 40,000 | | 120 ± 16 | 40,000 | | 120 ± 19 | 40,000 | | 310 ± 11 | 40,000 | | 72 ± 18 | 40,000 | | 140 ± 35 |
| *bis*(2-Ethylhexyl) | 444 | 3,500 | d | 400 | 3,380 | d | 400 | 1,800 | d | 400 | 54 | 58 ± 16 | 490 | 2,350 | d |
| Phthalate | 4,430 | | 150 ± 47 | 4,000 | | 18 ± 18 | 4,000 | | 200 ± 91 | 4,000 | | 140 ± 66 | 4,890 | | 170 ± 24 |
| | 40,010 | | 130 ± 57 | 40,010 | | 99 ± 11 | 40,010 | | 190 ± 16 | 40,000 | | 58 ± 13 | 48,820 | | 210 ± 62 |
| Benzo[a]- | 406 | 84 | 350 ± 202 | 400 | 540 | 63 ± 39 [d] | 400 | ND | 110 ± 51 | 400 | ND | 26 ± 2 | 400 | 910 | d |
| pyrene | 4,050 | | 100 ± 21 | 4,000 | | 46 ± 13 | 4,000 | | 120 ± 24 | 4,000 | | 170 ± 110 | 4,000 | | 57 ± 14 |
| | 40,010 | | 140 ± 25 | 40,010 | | 110 ± 29 | 40,010 | | 170 ± 13 | 40,080 | | 42 ± 13 | 40,000 | | 110 ± 22 |

[a]Mean of two determinations.

[b]Mean ± standard deviation for three determinations.

[c]ND = not detected.

[d]Concentration determined in the unspiked sludge was higher than the spike level.

[e]NA = this compound was not available when these samples were spiked.

[f]Concentration determined in the unspiked sludge was 10 times higher than the spike level.

less. Of these, 30% were less than 10% RSD. The poorer precision observed with the capillary column GC/MS method may reflect the influence repeated injections of sludge extracts on the performance of the capillary column. Concentrated samples can overload the column and degrade the column performance. The reproducibility of capillary column GC/MS determinations for extracts from the high level spikes was better than for other extracts. Of the extracts from samples spiked at 40,000 μg/L, 84% had RSD of less than 30%. columns evaluated, the fused-silica column coated with SE54 provided the best performance.

The recoveries determined for the acidic compounds are shown in Table IX. The recoveries observed were generally good and reproducible. No zero recoveries were observed. Although many of the spike levels were lower than the unspiked concentrations, 65% of all recoveries were in the range of 50-150%, 21% were less than 50% and 14% were greater than 150% recovery. The RSD values were low for triplicate determinations, 70% were 30% or less RSD and 40% were less than 10% RSD. Poor recoveries were determined for 2,4-dimethylphenol from the Platte County sludge samples. Recoveries observed for this compound during the method development experiments were frequently anomalously low. This may indicate that 2,4-dimethylphenol is susceptible to biodegradation in some sludges.

In view of the complexity and diversity of municipal and industrial wastewater treatment sludges, the precision and accuracy results presented here demonstrate that the protocol developed can reliably be applied to the analysis of the organic priority pollutants in sludge. The success of this protocol for the variety of extractable compounds for which it was developed and evaluated indicates that the methods included may also be useful for many nonpriority pollutant analytes.

## SUMMARY

Several approaches to the extraction of EPA priority pollutants from POTW sludges, including dual-pH homogenization/centrifugation, CLLE, steam distillation and microextraction, were evaluated. Quantitative recoveries were obtained only when POTW sludges were extracted with dichloromethane by the homogenization/centrifugation procedure. Since the homogenization/centrifugation procedure was not very selective, further development of extract cleanup procedures was conducted. Several alternative GPC systems, i.e., combinations of eluting solvent and packing, were evaluated. The best overall performance was achieved with Bio-Beads SX-3 eluted with dichloromethane. In addition to the GPC procedures, two essentially equivalent adsorption chromatographic cleanup procedures using silica gel and

| | Platte County POTW | | | Blue River POTW | | | Kansas City, KS | | | Industrial No. 1 | | | Industrial No. 2 | | |
|---|---|---|---|---|---|---|---|---|---|---|---|---|---|---|---|
| Compound | Spike Conc (μg/L) | Unspiked[a] Sludge (μg/L) | Recovery (%) | Spike Conc (μg/L) | Unspiked[a] Conc (μg/L) | Recovery (%) | Spike Conc (μg/L) | Unspiked[a] Conc (μg/L) | Recovery (%) | Spike Conc (μg/L) | Unspiked[a] Conc (μg/L) | Recovery (%) | Spike Conc (μg/L) | Unspiked[a] Conc (μg/L) | Recovery (%) |
| Phenol | 464 | 323 | 86 ± 14[b] | 400 | 160 | 67 ± 22 | 400 | 2,800 | c | 400 | 112 | 48 ± 9 | 400 | 800 | 101 ± 48 |
| | 4,630 | | 52 ± 8 | 4,000 | | 89 ± 8 | 4,000 | | 72 ± 71 | 4,000 | | 49 ± 6 | 4,000 | | 47,120 |
| | 46,280 | | 69 ± 6 | 40,000 | | 68 ± 15 | 40,000 | | 78 ± 14 | 40,000 | | 99,88 | 40,000 | | 57,66 |
| 2,4-Di-methyl-phenol | 412 | ND[d] | 7 ± 3 | 400 | ND | 94 ± 10 | 400 | ND | 174 ± 51 | 400 | ND | 88 ± 7 | 400 | 1,370 | d |
| | 4,110 | | 4 ± 1 | 4,000 | | 45 ± 2 | 4,000 | | 21 ± 9 | 4,000 | | 90 ± 11 | 4,000 | | 94,232 |
| | 41,080 | | 5 ± 3 | 40,000 | | 83 ± 68 | 40,000 | | 110,93 | 40,000 | | 190,190 | 40,000 | | 25,28 |
| 2,4-Di-chloro-phenol | 399 | ND | 140 ± 5 | 400 | 72 | 140 ± 7 | 400 | 1,780 | d | 400 | ND | 69 ± 9 | 400 | 114 | 380 ± 95 |
| | 3,970 | | 66 ± 13 | 4,000 | | 93 ± 7 | 4,000 | | 92 ± 64 | 4,000 | | 53 ± 3 | 4,000 | | 67,140 |
| | 39,750 | | 89 ± 8 | 40,000 | | 78 ± 7 | 40,000 | | 69 ± 17 | 40,000 | | 102,95 | 40,000 | | 68,79 |
| Penta-chloro-phenol | 401 | 323 | 130 ± 29 | 400 | 660 | 70 ± 10[d] | 400 | 1,120 | 230,230[d] | 400 | 229 | 41 ± 20 | 400 | 13,960 | e |
| | 4,000 | | 150 ± 27 | 4,000 | | 88 ± 6 | 4,000 | | 202 ± 75 | 4,000 | | 45 ± 3 | 4,000 | | d |
| | 40,010 | | 79 ± 101 | 39,970 | | 71 ± 12 | 39,970 | | 85 ± 47 | 40,000 | | 79,68 | 40,000 | | 39,53 |

[a]Mean for two determinations.

[b]Mean ± standard deviation for three determinations.

[c]Concentration determined in the unspiked sludge was higher than the spike level.

[d]ND = not detected.

[e]Concentration determined in the unspiked sludge was 10 times higher than the spike level.

Florisil were developed. These procedures were applicable only to the base/neutral compounds, as the acids were not separated from the polar interferences. Although the silica gel and Florisil procedures provided generally cleaner extracts than GPC, the GPC method can be automated to require much less labor. An adsorption chromatographic method using cesium silicate was evaluated for the acid compounds. However, the method recoveries were good only when moderate or low levels of coextractants were present. A capillary GC/MS method was developed for the base/neutral compounds to improve analysis of complex extracts. Of the four WCOT

A protocol for extractable compounds was developed based on the results of these method evaluations. The analytical scheme employed homogenization/centrifugation extraction, GPC cleanup and packed-column GC/MS for acidic extracts, and adsorption chromatographic cleanup and capillary column GC/MS for base/neutral extracts. GPC cleanup and packed-column GC/MS procedures were included as options for base/neutral extracts. This protocol was evaluated by analyzing aliquots of five primary sludges (three from POTW and two from industrial wastewater treatment plants) spiked with representative compounds at three levels. The precision and accuracy results indicated that these methods could be reliably applied to the analysis of acidic and base/neutral extractable compounds in sludge.

## ACKNOWLEDGMENTS

The assistance of Gil Radolovich, Jon Onstot and Margie Wickham in the acquisition and retrieval of the GC/MS data is gratefully acknowledged. Support for this research was provided by the U.S. Environmental Protection Agency, Contract No. 68-03-2695.

## REFERENCES

1. "Sampling and Analysis Procedures for Screening of Industrial Effluents for Priority Pollutants," U.S. EPA Environmental Monitoring and Support Laboratory, Cincinnati, OH (1977).
2. Stalling, D. L., R. C. Tindle and J. L. Johnson. "Cleanup of Pesticides and Polychlorinated Biphenyl Residues in Fish Extracts by Gel Permeation Chromatography," *J. Assoc. Off. Anal. Chem.* 55:32-38 (1972).
3. Hansen, E., Midwest Research Institute. Personal communication (1979).

4. Rhoades, J. W., and J. D. Millar. "Gas Chromatographic Method for Comparative Analysis of Fruit Flavors," *J. Agric. Food Chem.* 13:5-9 (1965).
5. Rhoades, J., and C. Nulton, Southwest Research Institute. Personal communication (1979).
6. Rhoades, J. W., and C. P. Nulton. "Microextraction as an Approach to Analysis for Priority Pollutants in Industrial Wastewater," in *Advances in the Identification and Analysis of Organic Pollutants in Water, Vol. 1,* L. H. Keith, Ed. (Ann Arbor, MI: Ann Arbor Science Publishers, Inc., 1981), Chapter 15.
7. Thrun, K. E., and J. E. Oberholtzer. "Evaluation of the Microextraction Technique to Analyze Organics in Water," in *Advances in the Identification and Analysis of Organic Pollutants in Water, Vol. 1*, L. H. Keith, Ed. (Ann Arbor, MI: Ann Arbor Science Publishers, Inc., 1981), Chapter 16.
8. Glaze, W. H., R. Rawley, J. L. Burleson, D. Mapel and D. R. Scott. "Further Optimization of the Pentane Liquid-Liquid Extraction Method for the Analysis of Trace Organic Compounds in Water," in *Advances in the Identification and Analysis of Organic Pollutants in Water, Vol. 1*, L. H. Keith, Ed. (Ann Arbor, MI: Ann Arbor Science Publishers, Inc., 1981), Chapter 17.
9. Junk, G. A., I. Ogawa and H. J. Svec. "Extraction of Organic Compounds from Water Using Small Amounts of Solvent," in *Advances in the Identification and Analysis of Organic Pollutants in Water, Vol. 1*, L. H. Keith, Ed. (Ann Arbor, MI: Ann Arbor Science Publishers, Inc., 1981), Chapter 18.
10. Midwest Research Institute. "Development of Methods for the Analysis of Organic Priority Pollutants in Municipal and Industrial Wastewater Treatment Sludge," Draft Final Report to EPA under Contract No. 68-03-2695 (1981).
11. DeWalle, F., and E. Chian. "Presence of Priority Pollutants in Sewage and Their Removal in Sewage Treatment Plants," First Annual Report to EPA, Grant No. R806102.

# SECTION 11

# DRINKING WATER ANALYSIS

# CHAPTER 44

# PROTOCOL FOR THE ANALYSIS OF A BROAD RANGE OF SPECIFIC ORGANIC COMPOUNDS IN DRINKING WATER

**P. A. Boland, B. A. Kingsley and D. F. Stivers**

SRI International
Menlo Park, California

**Irwin H. Pomerantz**

U.S. Environmental Protection Agency
Science and Technology Branch
Criteria and Standards Division
Office of Drinking Water
Washington, DC

An analytical protocol capable of application to raw and finished drinking water suitable for determining a large range of specific organic contaminants was developed. It was then used to obtain an occurrence database from a large number of U.S. water supplies for the specific compounds. The protocol was developed with minimum setup, labor and capital equipment requirements so that the technology could be transferred to other laboratories.

The protocol involves separation into fractions and analysis by gas chromatography (GC) as schematically represented in Figure 1. The analysis was segmented into three fractions: purgeables, base/neutral extractables and acid extractables. A list of compounds for which these procedures were evaluated is given in Table I. The quantification limits ranged from 0.1 to 8.0 ppb, and

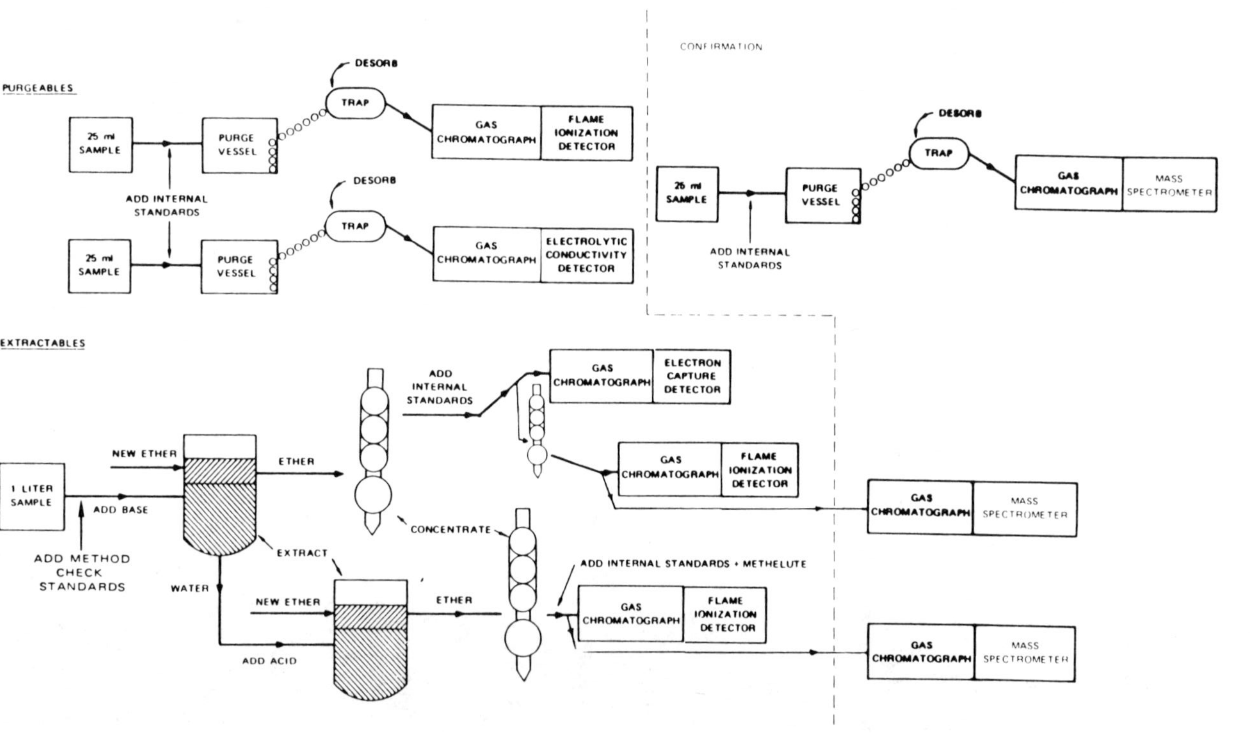


Figure 1. Protocol for screening organics in drinking water.

**Table I. Compounds Included in the National Screening Program for Organics in Drinking Water**

| Base/Neutral Extract Fraction | Acid Extract Fraction |
|---|---|
| Alachlor | Dicamba (Banvel D) |
| Atrazine | 2,4-Dichlorophenol |
| Benefin | 2,4-Dichlorophenoxyacetic acid (2,4-D) |
| Biphenyl | 2,4-Dimethylphenol |
| Bromacil | *o*-Methoxyphenol |
| 1,4-Bromochlorobenzene | Pentachlorophenol |
| Butachlor | Phenylacetic acid |
| *bis*(2-Chloroethyl) ether | 2,4,5-Trichlorophenol |
| Cyanazine | 2,4,5-Trichlorophenoxyacetic acid (2,4,5-T) |
| Diazinon | 2-(2,4,5-Trichlorophenoxy)propionic acid (Silvex) |
| Diphenylhydrazine | |
| Disulfoton | |
| Endrin | |
| Hexachlorobenzene | Purgeable Halocarbons |
| Hexachloro-1,3-butadiene | Bromobenzene |
| Hexachloroethane | Bromodichloromethane |
| Indene | Bromoform |
| Malathion | Carbon tetrachloride |
| Methoxychlor | Chlorobenzene |
| Methyl parathion | Chlorethane |
| Nitralin | Chloroform |
| Parathion | Chloromethane |
| Pentachloronitrobenzene (PCNB) | Dibromochloromethane |
| Phorate | *p*-Dichlorobenzene |
| Propachlor | 1,1-Dichloroethane |
| Propanil | 1,2-Dichloroethane |
| Propazine | 1,1-Dichloroethylene |
| Propyl benzene | *cis*-1,2-Dichlorethylene |
| Pyrene | *trans*-1,2-Dichloroethylene |
| Simazine | Dichloromethane |
| Treflan | 1,1,1,2-Tetrachloroethane |
| 1,2,4-Trichlorobenzene | Tetrachloroethylene |
| Purgeable Aromatics | 1,1,1-Trichloroethane |
| Benzene | 1,1,2-Trichloroethane |
| Styrene | Trichloroethylene |
| Toluene | Trichlorofluoromethane |
| *o*-, *m*- and *p*--Xylene | Vinyl chloride |

were compound-dependent. Gas chromatography/mass spectrometry (GC/MS) analysis was originally intended as an optional part of the protocol. It was used during this study to confirm the identities of the compounds that were indicated as present by the GC protocol and to determine the validity of the GC procedures.

## ANALYTICAL PROCEDURES

The purgeable fraction consists of two separate analyses: one for halocarbons and another for aromatic compounds. In both analyses the purgeable compounds are stripped from 25 mL of the raw or finished drinking water using a flow of helium. The compounds are trapped on a room-temperature Tenax/charcoal trap. The use of charcoal in the trap enhances the retention of the very low-molecular-weight halocarbons (e.g., methyl chloride and vinyl chloride). The compounds adsorbed on the Tenax/charcoal trap are thermally desorbed onto a cool GC column (0.2% Carbowax 1500 on Carbopack C), which is then temperature-programmed to its upper limit. Halocarbons are detected by an electrolytic conductivity detector and the aromatics by a flame ionization detector (FID). Identification and quantification are based on the addition of an internal standard (*n*-bromopentane) to the water before it is purged. Table II gives a step-by-step procedure for purge-and-trap analyses.

**Table II. Purgeable Analysis**

Halogenated

1. Pour 25 mL of sample into 30-mL gas-tight syringes.
2. Add *n*-bromopentane internal standard to the syringe.
3. Inject 25 mL into the sparging chamber.
4. Purge sample for 12 min at 40 mL/min flow of He onto a 2/3 Tenax: 1/3 charcoal trap (25 cm x 2.5 mm).
5. Desorb trap at 200°C for 4 min at 20 mL/min carrier gas flow onto a 0.2% Carbowax 1500 on Carbopack C Column (80/100 mesh, 8 ft x 2.5 mm), held at 10°C.
6. Chromatograph with 6-min hold, then at 6°C/min, 10–175°C into an electrolytic conductivity detector.

Nonhalogenated

1. Repeat steps 1–4 above with new water sample.
2. Desorb trap at 200°C for 4 min at 20 mL/min carrier gas flow onto a 0.2% Carbowax 1500 on Carbopack C column (80/100 mesh, 8 ft x 2.5 mm), held at 35°C.
3. Chromatograph with 4-min hold then at 6°C/min, 50–170°C into a FID.

Compounds of a molecular weight higher than that of, say, dichlorobenzene are not effectively determined by the purge-and-trap procedure described above. For these compounds, a liquid-liquid extraction (LLE) procedure was selected, using diethyl ether as the solvent and fractionating into base/neutral and acid fractions.

The procedure for the base/neutral fraction is outlined in Table III. The base/neutral compounds are extracted using diethyl ether from a 1-L water sample that has been made alkaline. The solvent should be distilled in glass within three days of use. The ether extract is concentrated using a Kuderna-Danish (KD) apparatus over a steam bath. The concentrate is analyzed using GC.

The GC analyses were performed on a glass capillary column that yields sharp, narrow chromatographic peaks that enhance sensitivity, separation and selectivity. The enhanced resolution gained by the capillary column was necessary for this work to obtain adequate selectivity for the individual compounds based on the elution (or retention) times. Detection of the eluting compounds in the initial analysis is made by an electron capture detector (ECD). Some compounds are not detected effectively, however, and a second analysis of the extract, using a FID, is necessary. No cleanup or fractionation procedures were necessary because of the relatively clean nature of drinking water and because of the enhanced resolution of capillary column chromatography.

**Table III. Extraction of Base/Neutral Fraction**

1. Transfer sample to a 1-L graduated cylinder from a 1-qt sample jar.
2. Record data, sample volume and pH.
3. Pour sample into separatory funnel.
4. Add *m*-bromophenol and 2,4-dibromoanaline internal standards.
5. Adjust pH to 11 with 10 *N* NaOH.
6. Extract with 150 mL of diethyl ether after rinsing sample vessel.
7. Drain off water layer into a 2-L Erlenmeyer flask.
8. Pass extract through sodium sulfate into a 1-mL KD receiver and 250-mL bulb.
9. Refill separatory funnel from the flask and then repeat steps 6–8 with two separate 50-mL diethyl ether portions.
10. Attach macro-Snyder column and concentrate to approximately 3 mL.
11. Remove KD bulb and macro column, then attach micro-Snyder column.
12. Concentrate to approximately 1 mL.
13. Add bromoalkane standards and analyze by GC using ECD.[a]
14. Concentrate to 400 $\mu$L with micro-Snyder column.
15. Analyze by GC using FID.[a]

[a]A SP2100 (OV-101) 30-m glass capillary column is used, programmed as follows: 30°C for 2 min, 7°C/min to 230°C.

The quality control and quantification/identification features of this procedure are based on the addition of two sets of internal standards. Just prior to extraction, known amounts of 2,4-dibromoaniline and *m*-bromophenol are added to the water sample. The amount of the compounds detected by the procedures was taken to represent the overall efficiency of the procedure, including the extraction, concentration and GC analysis.

A series of monobrominated straight-chain alkanes with 6–20 carbons is added to each concentrate before GC analysis. These compounds elute across the entire chromatogram and thereby serve as reference points in the elution sequence. The chromatographic peaks are identified based on a retention index, or the effective bromoalkane carbon number times 100. This increases the precision of the retention information by minimizing the effects of column flow, temperature program, column length and column aging, and therefore enhances the specificity in identifying individual compounds. Monobromoalkanes are used because they do not occur naturally in the environment and give good response on all the detectors used, including GC/MS. One of the bromoalkane compounds also serves as the quantification reference internal standard.

The water sample previously extracted under alkaline conditions is acidified, and extracted and concentrated as with the base/neutral extract. The bromoalkane series is also added as in the base/neutral fraction.

An aliquot of the extract is mixed with a methylating agent, MethElute. MethElute is an "on-column" methylating agent where the methylation takes place in the hot injection port of the GC instrument. The GC analysis utilizes a glass capillary column and FID. Table IV summarizes the extraction and analysis for the acid fraction.

**Table IV. Extraction of Acid Fraction**

1. Adjust pH of same water from base/neutral extraction to pH with 6 *N* HCl.
2. Extract as in base/neutral fraction (without rinsing sample vessel and using new sodium sulfate) with three 50-mL portions of diethyl ether.
3. Concentrate as in base/neutral fraction to approximately 400 μL.
4. Add bromoalkane standards.
5. Remove 50-μL aliquot and mix with 10 μL methElute derivatizing agent.
6. Analyze by GC using FID.[a]

[a]A SP2100 (OV-101) 30-m glass capillary column is used, programmed as follows: 30°C for 2 min, 7°C/min to 230°C.

The method check standard for this fraction is *m*-bromophenol, which was added along with the dibromoaniline to the water sample before the extraction. In addition to being a check on the extraction and concentration, it is also methylated, and therefore monitors this aspect of the procedure. The quantification limits are higher in the acid fraction, 4–8 ppb, because of a higher background contributed by the MethElute.

## ANALYTICAL RESULTS

The results of the GC/MS confirmation data indicate the confirmation rate is dependent on both fraction and compound. Table V shows the confirmation rate by fraction. The overall confirmation rate for the purgeable halocarbons was 86%. The majority of the cases where compounds were confirmed involved the trihalomethanes (THM) (chloroform, bromodichloromethane, dibromochloromethane and bromoform), and the confirmation rate for the THM was 91%. Most of the cases where confirmation was not possible were at or near the GC/MS detection limits.

The lower confirmation rate for the aromatic purgeables can be partially attributed to the coelution of trichloroethylene and dibromochloromethane with benzene. A sufficient concentration (greater than 5 ppb) of trichloroethylene gives a response on FID at the quantification limit for benzene. Eliminating these occurrences raises the confirmation rate to approximately 70%.

For the base/neutral extractable fraction, the overall confirmation rate was 40%. The confirmation rates for the most frequently observed compounds (observed by GC more than 5 times) varied greatly, as illustrated in Table VI. These 7 compounds account for 88% of the tentatively identified compounds. The most frequently occurring compounds were the commonly used herbicides atrazine, simazine, propazine and alachlor. The only other base/neutral compound with a confirmed identification was diazinon.

**Table V. Confirmation Rate by Fraction**

| | No. Tentatively Identified by GC | No. Confirmed by GC/MS | % Confirmed |
|---|---|---|---|
| Purgeable Halocarbons | 486 | 417 | 86 |
| THM | 325 | 297 | 91 |
| Purgeable Aromatics | 77 | 36 | 47 |
| Purgeable | 563 | 453 | 80 |
| Base/Neutral Extractables | 169 | 68 | 40 |
| Acid Extractables | 56 | 2 | 3.6 |

Table VI. Confirmation Rate for Base/Neutral Fraction

| Compound | No. Tentatively Identified by GC | No. Confirmed by GC/MS | % Confirmed |
|---|---|---|---|
| Alachlor | 8 | 7 | 88 |
| Atrazine | 62 | 44 | 71 |
| 1,4-Bromochlorobenzene | 7 | 0 | 0 |
| Disulfoton | 19 | 0 | 0 |
| Propazine | 6 | 3 | 50 |
| Simazine | 8 | 7 | 88 |
| 1,2,4-Trichlorobenzene | 38 | 4 | 11 |

For the acid extractable fraction, the two confirmations listed in Table V were for phenylacetic acid. We cannot account for the very low confirmation rate.

## CONCLUSIONS

In summary, the protocol confirmation data obtained by GC/MS for these specific compounds indicates that GC/MS confirmation should be performed on all fractions, except perhaps the purgeable halocarbon fraction, where the overall confirmation rate was 86%.

The usefulness of the protocol for other specific organics would require validation by GC/MS. The protocol as outlined, however, should be useful for "profiling" water samples without specific compound identification and quantification. In this mode, the relative type and molecular weights of compounds are indicated by the fraction and chromatographic retention times. For example, variations in influent and effluent for a particular part of the water purification process (e.g., chlorination, filtration) or for the overall treatment could be monitored, or daily variations at any point in the process could be monitored, yielding more detailed information than conventional bulk parameters such as TOC.

## ACKNOWLEDGMENTS

This work was supported by Contract No. 68-01-4666 with the U.S. Environmental Protection Agency (EPA). We are especially grateful to Dr. Joseph Cotruvo, Office of Drinking Water, EPA, for his valuable guidance and support.

# CHAPTER 45

# COOPERATIVE QUALITY ASSURANCE PROGRAM FOR MONITORING CONTRACT LABORATORY PERFORMANCE

**Barbara A. Kingsley, Christina Gin, William R. Peifer and Douglas F. Stivers**

SRI International
Menlo Park, California

**Scott H. Allen, Herbert J. Brass, Edward M. Glick and Michael J. Weisner**

U.S. Environmental Protection Agency
Office of Drinking Water
Technical Support Division
Cincinnati, Ohio

In the ongoing process of obtaining occurrence data in support of drinking water regulations, the U.S. Environmental Protection Agency (EPA) has conducted a Community Water Supply Survey [1-3]. One objective of this survey was to quantitatively analyze finished waters and sources of supply for purgeable halocarbons and aromatics, and total organic carbon (TOC). These analyses were performed by SRI International under contract to the EPA, Office of Drinking Water, Technical Support Division (TSD) [4]. A quality assurance protocol that was an integral part of the analytical program is described in this chapter.

In conjunction with the survey, 1189 field samples were analyzed, representing 452 water supplies serving populations from 25 to 100,000 persons.

Of these samples, 67% were from groundwater sources, 27% from surface waters and the remaining samples from other water source categories. Groundwater data are of particular interest at this time because of the growing concern about groundwater quality [5].

## ANALYTICAL PROGRAM

The analytical scheme specified in the contract is shown in Figure 1. Samples were collected in headspace-free, septum-sealed vials at the utilities and stored under refrigeration until analysis. Those intended for TOC analyses were acidified at the time of collection. Samples for purgeable analyses were collected both with and without a chlorine reducing agent (sodium sulfite). Analyses for purgeable aromatics and halocarbons and for TOC were performed for each sample. When the combined concentrations of the trihalomethanes (THM) exceeded 5ppb, a corresponding sample containing the chemical reducing agent (reduced sample) was also analyzed. Selection for the confirmatory analyses shown in this scheme is discussed as a part of the quality assurance (QA) program.

Purgeables were analyzed using the purge-and-trap procedure developed by Bellar and Lichtenberg, using 25-mL samples with 2-bromo-1-chloropropane as an internal standard [1-9]. The halocarbons shown in Figure 2 were detected with a Coulson electrolytic conductivity detector operated in the halogen-specific mode. This chromatogram was obtained by analysis of a 1-ppb standard of the compounds shown. Three gas chromatographs with automatic purge-and-trap devices were used for halocarbon analyses. After being stripped from the sample with a helium stream, volatiles were collected on a trap consisting of 2/3 Tenax®, 1/3 charcoal [10]. The primary chromatographic column (2.5 m x 2.1 mm i.d. stainless steel packed with 0.2% CW 1500 on Carbopack C) was held at 30°C during the 4-min desorption period, programmed at 4°C/min for 18 min, then at 8°C/min to a final temperature of 175°C. Compounds monitored include THM and many other purgeable halocarbons that have been found in drinking water. Concentrations above 0.5 ppb were reported for all compounds except bromoform, which was quantified at a 1-ppb limit. Other unknown peaks that occasionally occured were reported by relative retention time and relative area—that is, a quasi-concentration calculated by assuming a response equal to that of the internal standard.

The purgeable aromatics shown in Figure 3 were detected by flame ionization detection (FID). Two analytical systems were used for these analyses, one with an automatic purge-and-trap device and the other with an SRI-fabricated manual system [10]. The chromatogram shown was obtained by

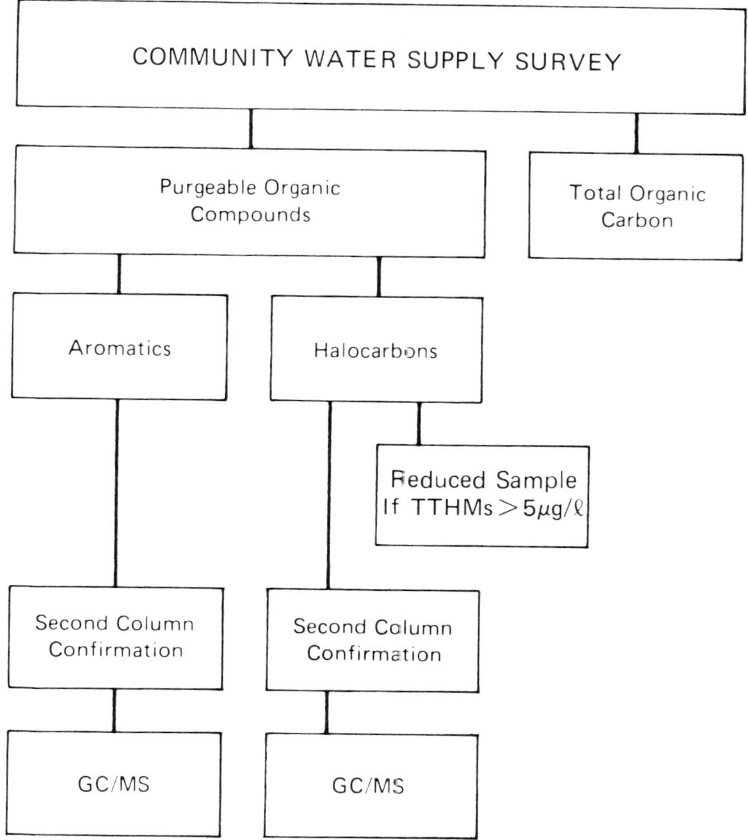


**Figure 1.** Analytical scheme for survey.

analysis of a 0.5 ppb standard of the purgeable aromatic compounds specified in the contract. Trichloroethylene was included at 3 ppb to show its potential for interference with benzene–in fact, these compounds were completely unresolved on the second analytical system and could have resulted in false identifications of benzene. The same type of gas chromatographic column used for the halocarbon analyses was held at 50° C during the 4-min desorption period, then programmed at 10° C/min to 150°C. Benzene, toluene and total xylenes (*o*-, *m*- and *p*-xylene) concentrations above 0.5 ppb were reported. In general, no attempt was made to identify other peaks.

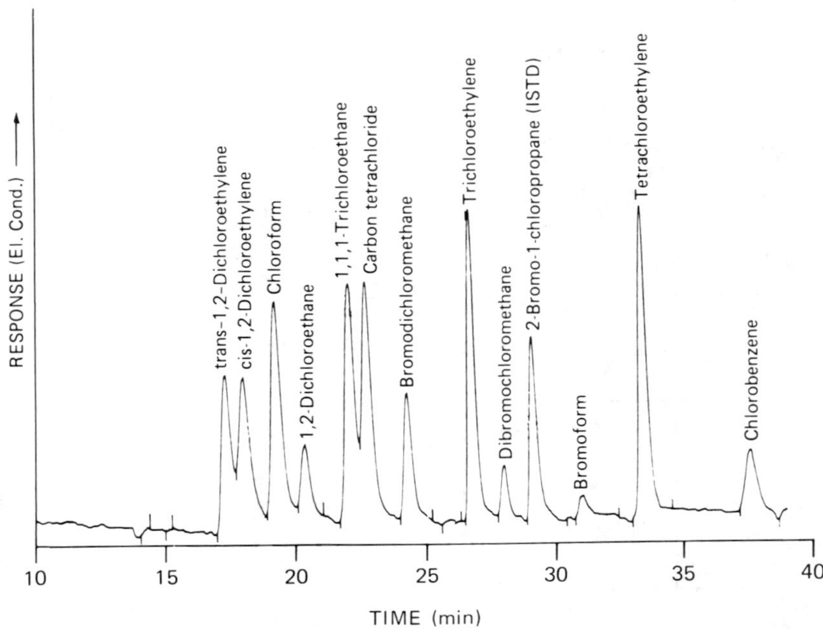


**Figure 2.** Analysis of purgeable halocarbon standard at 1 ppb using primary analytical column.

TOC analyses were performed using a Dohrmann DC-54 TOC analyzer with a modified sparger that allows the entire sample, including particulates, to be transferred into the ultraviolet (UV) reaction chamber [11,12].

## QA PROGRAM

In all, about 5000 analyses were performed using six analytical instruments over a one-year period. To allow monitoring and evaluation of these data, the five-part quality assurance protocol shown in Table I was established as a part of the program [13]. Examples of the results obtained from each part of this protocol will be presented and an assessment made regarding the validity of the data generated.

### Reference Standards

Concentrates containing purgeable halocarbons, aromatics or TOC standard mixtures were supplied as needed by the project officer. Reference standards were prepared by diluting these concentrates with blank water.

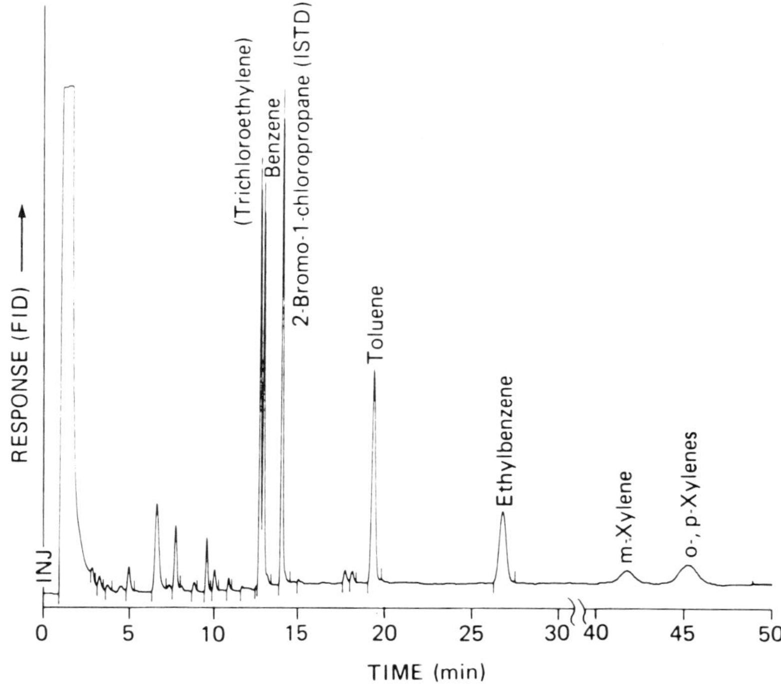


**Figure 3.** Analysis of purgeable aromatic standard at 0.5 ppb using the primary analytical column, showing interference of trichloroethylene at 3 ppb.

**Table I. QA Protocol**

| | |
|---|---|
| Reference Sample Analysis: | Twice per month in duplicate on each instrument |
| Aromatics: one concentration level | |
| Halocarbons and TOC: | two concentration levels |
| Duplicate Analyses: | First 10 samples of each type, 10% of remaining samples |
| Blind Sample Analyses: | 5% of samples |
| Split Sample Analyses: | 5% of samples |
| Confirmatory Analyses: | Secondary chromatographic column, 20%; GC/MS, ~6% |

Halocarbons in the reference standards were measured over a 13-month period using all 3 halocarbons systems. The expected concentration, and the mean, range and coefficient of variation (CV) of the concentrations determined using the primary analytical column are given in Table II. The protocol required precision and accuracy (error) to be 20% for concentrations above

Table II. Halocarbon Reference Standard Analyses–Primary Column

| | Expected Conc (ppb) | Low Level[a] Concentration Found (ppb) | | | | Expected Conc (ppb) | High Level[b] Concentration Found (ppb) | | | |
|---|---|---|---|---|---|---|---|---|---|---|
| | | Range | Mean | CV[c] | % Error[d] | | Range | Mean | CV[c] | % Error[d] |
| Chloroform | 8.2 | 6.0–10 | 7.5 | 14 | 8 | 34 | 24–37 | 30 | 10 | 12 |
| 1,2-Dichloroethane | 3.3 | 2.0–4.5 | 3.1 | 20 | 6 | 14 | 9.2–17 | 13 | 17 | 7 |
| 1,1,1-Trichloroethane | 1.3 | 0.70–1.4 | 1.1 | 20 | 15 | 5.6 | 2.8–5.5 | 4.3 | 14 | 23 |
| Carbon Tetrachloride | 1.5 | 0.87–1.8 | 1.3 | 17 | 13 | 6.2 | 3.6–8.8 | 5.5 | 18 | 11 |
| Bromodichloromethane | 1.4 | 1.2–2.4 | 1.8 | 13 | 29 | 6.0 | 5.4–9.2 | 6.7 | 14 | 12 |
| Trichloroethylene | 2.3 | 1.5–2.5 | 2.0 | 12 | 13 | 9.1 | 5.7–9.6 | 7.9 | 13 | 13 |
| Dibromochloromethane | 2.1 | 1.4–2.7 | 2.1 | 18 | <1 | 8.5 | 5.3–12 | 8.0 | 18 | 6 |
| Bromoform | 1.7 | 1.1–3.4 | 1.9 | 25 | 12 | 7.0 | 4.9–11 | 7.04 | 1.2 | <1 |
| Tetrachloroethylene | 1.1 | 0.68–1.4 | 1.1 | 20 | <1 | 4.4 | 2.8–5.5 | 4.3 | 16 | 2 |

[a]80 analyses.
[b]67 analyses.
[c]Coefficient of variation: 100 times the standard deviation divided by the mean value.
[d]Error expressed as 100 times the difference between the absolute value of the expected and mean found concentrations, divided b the expected concentration.

10 ppb and 40% for concentrations below that level [4]. These results satisfactorily met the stated performance criteria.

Reference standards were also analyzed for purgeable aromatics and TOC with results shown in Tables III and IV, respectively. Precision and accuracy requirements for the aromatics were the same as for the halocarbons—40%, since these were all below 10 ppb. Precision and accuracy requirements for total organic carbon were 20 and 10%, respectively, for concentrations below and above 0.3 ppm.

**Table III. Aromatic Reference Standard Analyses—Primary Column**

| Compound | Expected Conc (ppb) | Concentration Found (ppb)[a] Range | Mean | CV[b] | % Error[c] |
|---|---|---|---|---|---|
| Benzene | 8.7 | 5.6–11 | 7.4 | 15 | 15 |
| Toluene | 5.3 | 3.5–5.9 | 4.3 | 15 | 19 |
| Ethyl Benzene | 5.9 | 4.3–7.4 | 5.3 | 16 | 10 |
| Total Xylenes | 7.5 | 5.4–9.7 | 6.6 | 17 | 12 |

[a]57 analyses.
[b]Coefficient of variation: 100 times the standard deviation divided by the mean value.
[c]Error expressed as 100 times the absolute value of the difference between the expected and mean found concentrations, divided by the expected concentration.

**Table IV. TOC Reference Standard Analyses**

| | Expected Conc (ppm) | Concentration Found (ppm)[a] Range | Mean | CV[b] | % Error[c] |
|---|---|---|---|---|---|
| High Level[a] | 3.05 | 2.92-3.17 | 3.04 | 3 | <1 |
| Low Level[d] | 0.610 | 0.560–0.670 | 0.617 | 3 | 1 |

[a]38 analyses.
[b]Coefficient of variation: 100 times the standard deviation divided by the mean value.
[c]Error expressed as 100 times the absolute value of the difference between the expected and mean concentrations, divided by the expected concentration.
[d]36 analyses.

### Duplicate Analyses

Of all samples, 10% were analyzed in duplicate for halocarbons, aromatics and TOC including the first 10 samples for each type of analysis. Precision between duplicate values (calculated as 100 times the absolute value of their difference, divided by their average) was determined for each pair of data and the mean precision and standard deviation calculated for each parameter (compound or TOC). The results of these calculations are given in Table V for two concentration ranges.

### Split and Blind Analyses

Split samples were duplicates taken in the field or split by TSD and analyzed by both TSD and SRI. Blind samples were spiked standards prepared and analyzed by TSD and analyzed by SRI as samples. Together, splits and blinds represented about 10% of the samples analyzed in the survey.

The overall qualitative agreement between the two laboratories for combined split and blind samples analyzed for purgeable halocarbon and aromatic compounds is given in Table VI. The number of pairs where compounds were and were not detected above the quantification limts in both laboratories are shown, as well as the number of cases where an identification was made in one laboratory but not in the other. In general, disagreement occurred where compounds were observed at or near the quantification limits.

The results were also compared quantitatively. The plot shown in Figure 4 for all chloroform positives–splits and blinds, reduced and nonreduced sample values–shows the concentration found by TSD vs that found by SRI for concentrations from 0.5 to 250 ppb. The correlation coefficient ($R^2$) of 0.86 indicates a significant degree of scatter. The scatter was more pronounced for chloroform than for other compounds. This discrepancy may partially be explained by comparing the data obtained for split samples with those obtained for blinds (Table VII). Precision between TSD and SRI data was calculated for each pair by subtracting the TSD from the SRI value and dividing by their average. The range of individual precision values, as percents, are shown, as well as the mean precision calculated for n statistical pairs analyzed. The differences, reflected in both the range of precision values, are significantly lower for the blind than for the split samples. One possible explanation may be cross-labeling of reduced and nonreduced samples and of raw and finished samples at the time of sample collection.

Table V. Precision of Duplicate Analyses

| Compound | Concentration ≤10 ppb | | | Concentration >10 ppb | | |
|---|---|---|---|---|---|---|
| | No. of Duplicate Positives[a] | Mean %P[b] | $\sigma$[b] | No. of Duplicate Positives[a] | Mean %P[b] | $\sigma$[b] |
| Chloroform | 35 | 11 | 11 | 34 | 12 | 18 |
| Reduced[c] | 13 | 11 | 9 | 25 | 7.9 | 7.9 |
| Dibromochloromethane | 54 | 13 | 12 | 16 | 10 | 7.0 |
| Reduced[c] | 32 | 14 | 14 | 3 | 5.6 | 5.9 |
| Bromodichloromethane | 45 | 14 | 22 | 28 | 14 | 14 |
| Reduced[c] | 26 | 7.5 | 5.7 | 12 | 15 | 18 |
| Bromoform | 24 | 18 | 20 | 17 | 12 | 9.3 |
| Reduced[c] | 13 | 25 | 26 | 2 | <1 | |
| 1,2-Dichloroethane | 6 | 19 | 9.6 | 2 | 13 | 2.9 |
| Carbon Tetrachloride | 9 | 31 | 28 | 1 | <1 | |
| Chlorobenzene | 3 | 17 | 6.6 | 0 | | |
| 1,2-Dichloroethylene | 4 | 18 | 14 | 0 | | |
| Tetrachloroethylene | 7 | 18 | 14 | 0 | | |
| 1,1,1-Trichloroethane | 13 | 14 | 13 | 1 | <1 | |
| Trichloroethylene | 4 | 9.6 | 6.3 | 0 | | |
| Benzene | 3 | 8.1 | 11 | 0 | | |
| Toluene | 3 | 8.9 | 9.2 | 0 | | |
| Xylene | 0 | | | 3 | 6.5 | 6.6 |
| TOC | 8[d] | 13.3[d] | 11[d] | 114[e] | 4.1[e] | 3.5[e] |

[a]Number of times this compound found at or above the quantification limit in both analyses (separated into those cases with concentration ≤10 ppb and those >10 ppb).

[b]% Precision calculated for each pair as 100 times their difference divided by their average. The mean and standard deviation ($\sigma$) of these precision values were then calculated.

[c]Data obtained from samples containing a chemical reducing agent (sodium sulfite).

[d]Concentration ≤300 ppb.

[e]Concentration >300 ppb.

**Table VI. Qualitative Interlaboratory Comparison of Split and Blind Samples[a]**

| | Total Number of Pairs Analyzed | Number of Pairs | | |
|---|---|---|---|---|
| | | Negative | Positive | 1 Positive 1 Negative |
| *cis/trans*-1,2-Dichloroethylene | 79 | 75 | 4 | 0 |
| Chloroform | 79 | 27 | 49 | 3 |
| 1,2-Dichloroethane | 79 | 68 | 11 | 0 |
| 1,1,1-Trichloroethane | 78 | 65 | 10 | 3 |
| Carbon Tetrachloride | 79 | 65 | 14 | 0 |
| Bromodichloromethane | 79 | 32 | 47 | 0 |
| Trichloroethylene | 78 | 70 | 8 | 0 |
| Dibromochloromethane | 78 | 33 | 45 | 0 |
| Bromoform | 79 | 51 | 27 | 1 |
| Tetrachloroethylene | 79 | 69 | 10 | 0 |
| Chlorobenzene | 79 | 77 | 2 | 0 |
| Benzene | 70 | 53 | 13 | 4 |
| Toluene | 69 | 50 | 11 | 8 |
| Total Xylenes | 70 | 53 | 15 | 2 |

[a]Quantification limited to 0.5 ppb for all compounds except bromoform, which was quantified at 1 ppb.

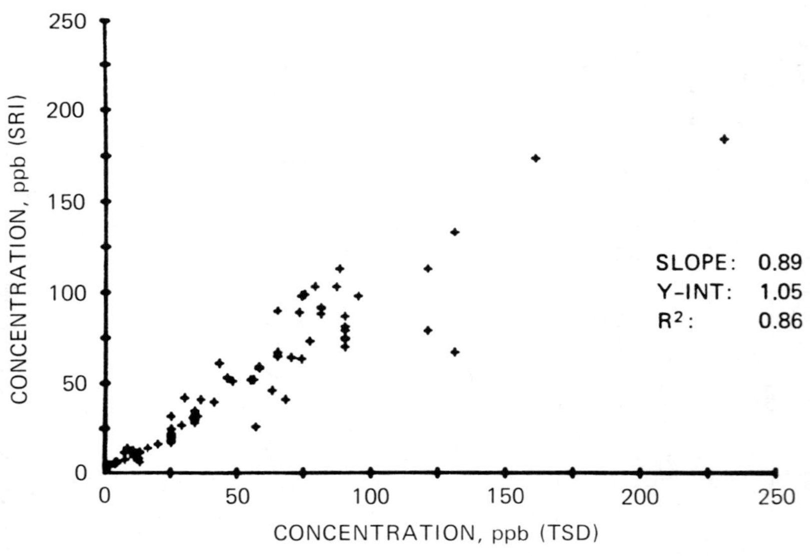


**Figure 4.** Interlaboratory comparison of split and blind sample analyses for chloroform.

Table VII. Comparison of the Precision of Split and Blind Sample Analyses for Chloroform[a]

| | n | Range of Precision Values | Mean Precision |
|---|---|---|---|
| Split Samples | 47 | -110 to +54 | -9.1 |
| Blind Samples | 42 | -53 to +11 | -18 |

[a]Precision calculated for each of n pairs of analyses as 100 times their difference divided by their average value.

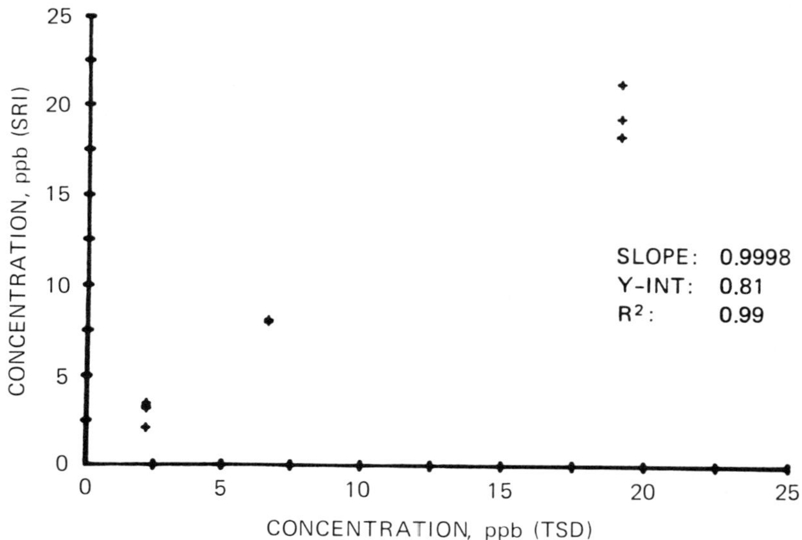


**Figure 5.** Interlaboratory comparison of split and blind sample analyses for 1,2-dichloroethane.

Plots similar to Figure 4 are shown in Figures 5 to 7 for 1,2-dichloroethane, total xylenes and TOC, respectively. The cumulative data for four representative compounds are shown in Table VIII, giving the range and mean percent precision values for n analyses in each laboratory. Because these compounds occurred infrequently there was no attempt to differentiate between split and blind samples.

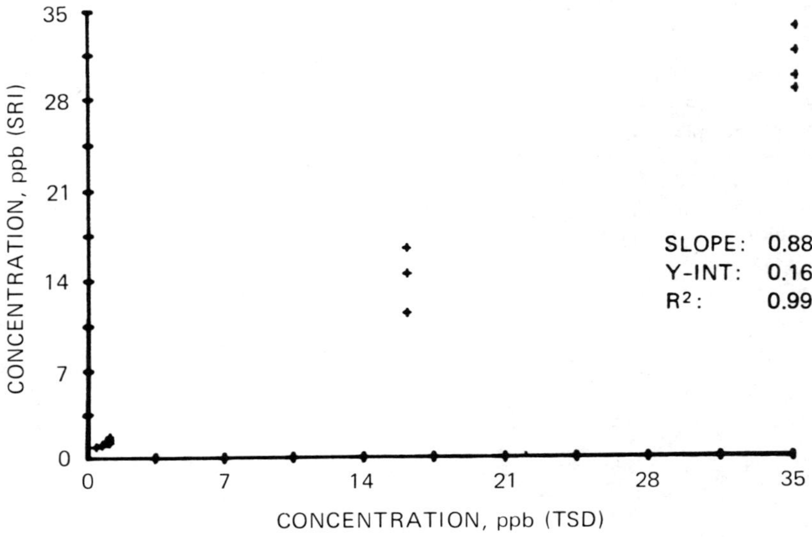


**Figure 6.** Interlaboratory comparison of split and blind sample analyses for total xylenes.

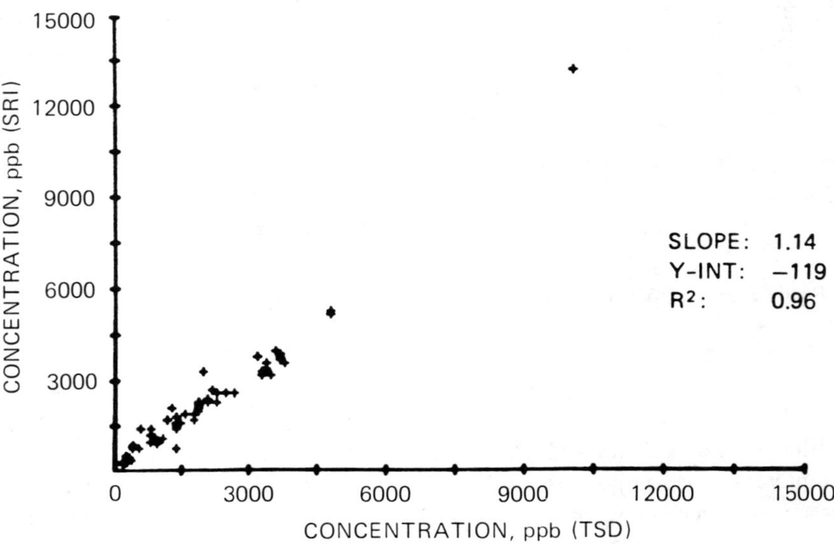


**Figure 7.** Interlaboratory comparison of split and blind sample analyses for total organic carbon.

**Table VIII. Precision of Split and Blind Sample Analyses[a]**

| | n | Mean Precision | Range of Precision Values |
|---|---|---|---|
| 1,2-Dichloroethane | 11 | +16 | −15 to +41 |
| Carbon Tetrachloride | 14 | −15 | −77 to +7 |
| Benzene | 14 | +17 | −9 to +78 |
| Total Xylenes | 16 | −3.7 | −37 to +54 |

[a]Precision calculated for each of n pairs of analyses as 100 times their difference divided by their average value.

Although there were no formal accuracy and precision requirements in the contract for split and blind analyses, it is interesting to look at the results obtained using the criteria applied to reference standard and duplicate analyses—taking into account the error propagated when two sets of data are compared [14]. Approximately 95% of all purgeable blinds agreed within the 20%/40% criteria for high and low concentrations (above and below 10 ppb). Split samples analyzed for THM did not agree as well, possibly reflecting sampling problems. Although 95% of the blind TOC samples agreed between laboratories within 40%, better precision had been anticipated. Overall analyses of split and blind samples provided a good external check of the data generated.

## Confirmatory Analyses

The fifth and final part of the QA protocol consisted of confirmatory analyses for halocarbons and aromatics using different analytical columns or gas chromatography/mass spectrometry (GC/MS). Confirmations, originally intended as qualitative, were also done quantitatively when possible to add an extra measure of confidence to the data generated. Selection criteria were as follows.

For second column analyses, 10 samples were initially selected at random—five ground samples and five surface supplies—and analyzed for both halocarbons and aromatics. All samples in which at least one compound other than THM was found in the primary analysis above 0.5 ppb were reanalyzed. In certain supplies samples were collected at up to five locations. If a compound (other than THM) was found in a sample from one location, samples collected at the other locations in the water system were reanalyzed when possible. Because of difficulties in detection of 1,2-dichloroethane in the presence of high chloroform concentrations, samples that contained chloroform at concentrations greater than about 60 ppb were reanalyzed to check for the presence of 1,2-dichloroethane [8]. Samples containing unidentified peaks were reanalyzed to obtain relative retention time data for comparison with data in the literature.

Samples for GC/MS analysis were selected from those in which at least one compound other than THM was found at or above a concentration of 1 ppb in the primary analysis. Selected samples containing unidentified compounds at concentrations apparently above 1 ppb were also analyzed.

The chromatogram shown in Figure 8 was obtained by analysis of a standard containing each of the indicated halocarbon compounds at 1 ppb using the confirmatory column (2.5 m by 2.1 mm i.d. stainless steel packed with *n*-octane on 100/120 Poracil C) held at 50°C during the 4-min desorption, then programmed to 140°C at 6°C/min. Several problems are readily apparent:

1. *cis*-1,2-Dichloroethylene and chloroform coeluted on this column and, thus, this isomer could not be confirmed when chloroform was present, usually in finished waters.
2. 1,1,1-Trichloroethane and trichloroethylene were not well resolved–quantification and identification of one of these compounds was questionable in the presence of the other.
3. The same problem occurred with tetrachloroethylene and bromodichloromethane. Unfortunately, this THM was present in many finished waters, often at significant concentrations.
4. Although not shown here, chlorobenzene and bromoform were also poorly resolved.

Confirmatory data for six halocarbons using the selection criteria are shown in Figure 9. These represent data from 127 analyses. The height for each compound represents the number of times that compound was seen in the primary analysis and for which alternative column confirmatory analyses were performed. The indicated shadings represent the number of times

1. the precision between the primary and confirmatory analyses were within 40%;
2. the compound was found to be present in the confirmatory analysis but with a precision outside a 40% limit (including those cases where quantification was not possible due to an interference–for example, where the compound was seen only as a shoulder on an interfering peak);
3. the interference was so great that nothing can be said about the presence or absence of the given compound; and
4. cases where confirmation should have been possible but the given compound was not seen on confirmatory analysis.

The problems anticipated by examination of the confirmatory column chromatogram are quite evident. For example, eight of the nine occurrences of 1,2-dichloroethylene could not be confirmed because chloroform was present. Interferences between 1,1,1-trichloroethane and trichloroethylene, and between bromodichloromethane and tetrachloroethylene are also reflected. Other cases "outside 40%" where there was no interferences were

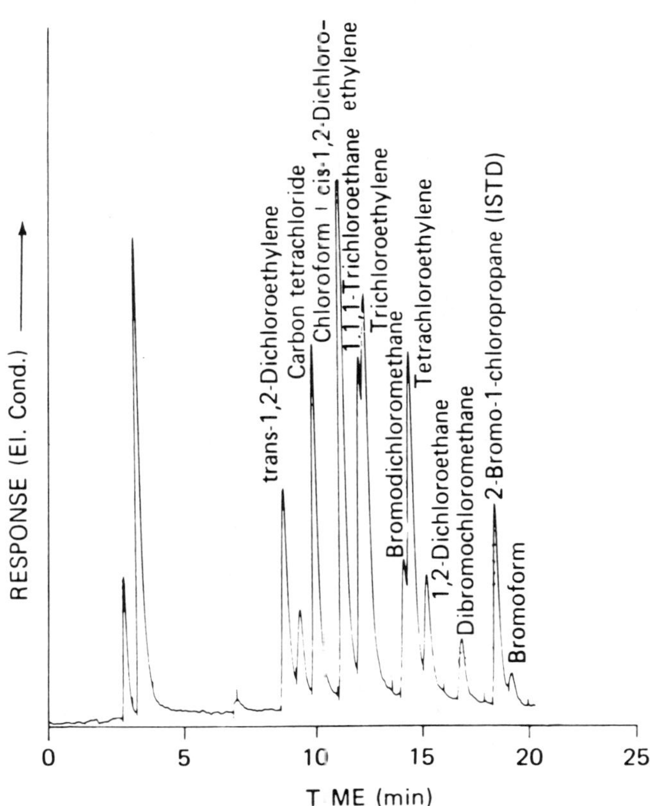


**Figure 8.** Analysis of purgeable halocarbon standard at 1 ppb using confirmatory analytical column.

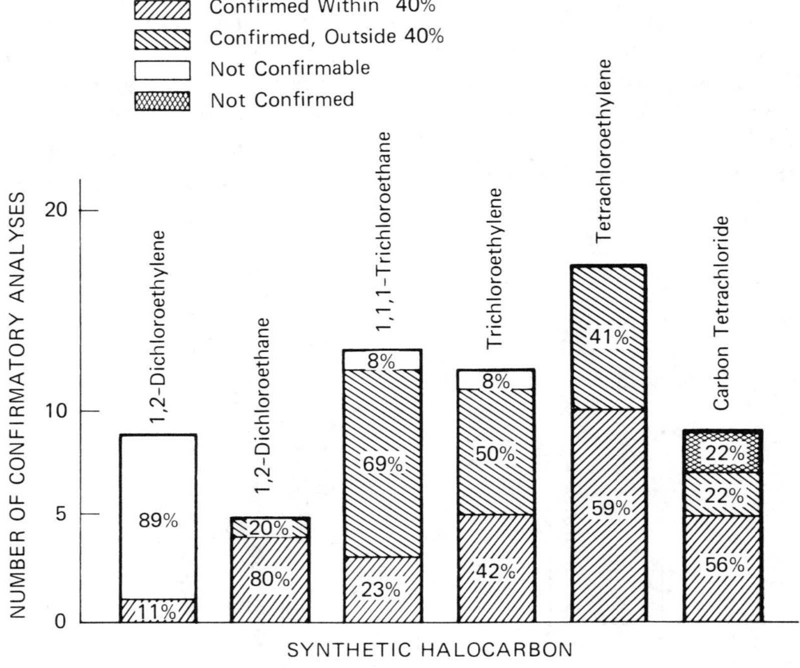


**Figure 9.** Second column confirmations for halocarbons.

generally at low concentrations—around 1 ppb—and most always agreed within a factor of two. The two cases of nonconfirmation for carbon tetrachloride were where this compound was measured in the primary analysis at about 0.6 ppb.

Confirmatory data for these same halocarbons using GC/MS are shown in Figure 10. All analyses were performed on a Finnigan 3200 GC/MS using the primary analytical column held at 25°C during the 4-min trap desorption, then programmed at 6°C/min to 175°C. No attempt was made to quantify below 1 ppb.

The total height for each compound represents the number of attempts to confirm. Note that 1,2-dichloroethylene, which was almost never confirmed using the second column because of the interference of the cis isomer with chloroform, was confirmed by GC/MS within 40% for all cases. Most confirmations for the other compounds that fell outside the 40% range were at concentrations near 1 ppb and agreed within a factor of 2. The two cases of nonconfirmation were near the 1-ppb level.

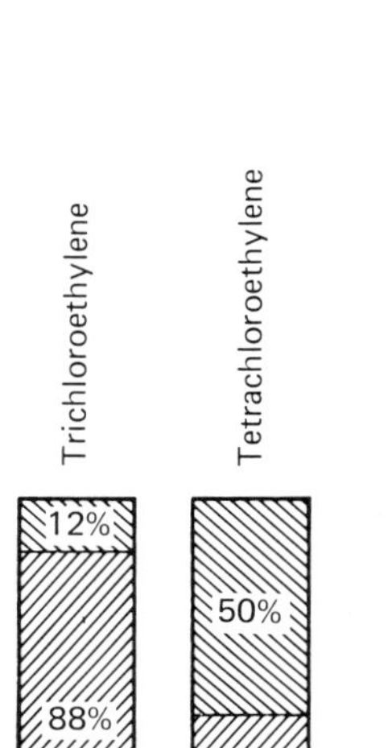


**Figure 10.** GC/MS confirmations for halocarbons.

Aromatic compounds were also confirmed in a similar manner. The chromatogram shown in Figure 11 was obtained using the confirmatory column (3.8 m by 2 mm i.d. glass packed with 5% SP1000/5% Bentone 34 on 100/120 Supelcoport) held at 45°C for 8 min, then programmed to 75°C at 10°C/min. The standard analyzed contained benzene, toluene and ethyl benzene at 0.5 ppb and each of the xylene isomers at 0.2 ppb. Trichloroethylene (3 ppb) was added to show the interference of this compound with benzene. Unfortunately, these two compounds coeluted on both the primary and confirmatory columns and careful cross-checking of halocarbon and aromatics data was necessary to interpret any identifications of benzene.

Confirmation data for 73 aromatics analyses are shown in Figure 12. Confirmations for benzene are deceptively high since many of these samples contained trichloroethylene. Almost all cases that fell outside the 40% range were at low levels. GC/MS analyses (Figure 13) were again restricted to occurrences above 1 ppb in the primary analyses. Here the problem with benzene is clarified: five of the nine attempts to confirm this compound failed. Trichloroethylene was present in all cases. One sample contained both benzene and trichloroethylene. Confirmation for the other aromatics appears quite satisfactory.

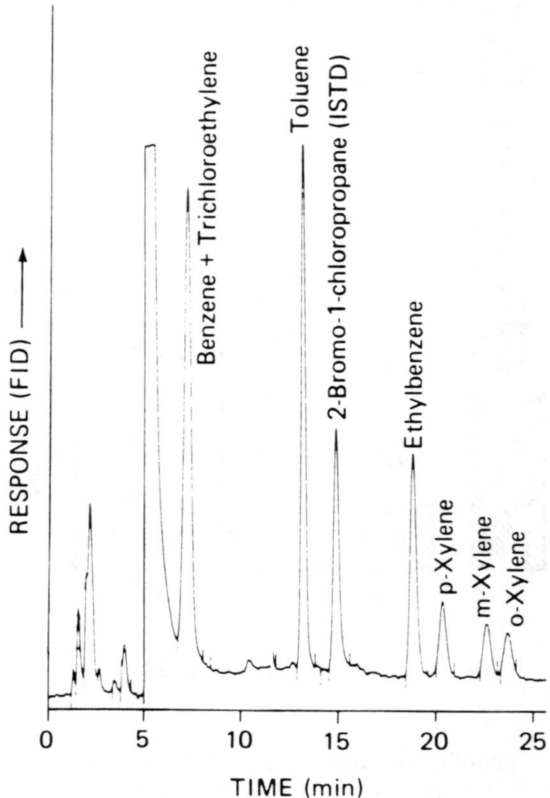


**Figure 11.** Analysis of purgeable aromatics standard at 0.5 ppb using confirmatory analytical column, showing interference of trichloroethylene at 3 ppb.

## SUMMARY

Reference standards, analyzed on a regular basis, allowed monitoring of overall performance with respect to accuracy and precision for a broad range of compounds and concentrations. They indicated when column replacement or other instrument maintenance was required and allowed comparison of results obtained with various instruments. They provided a clearly defined decision point for proceeding with analyses.

Duplicate analyses permitted monitoring of the precision within the laboratory, often between instruments, and were found to meet the quality assurance performance criteria.

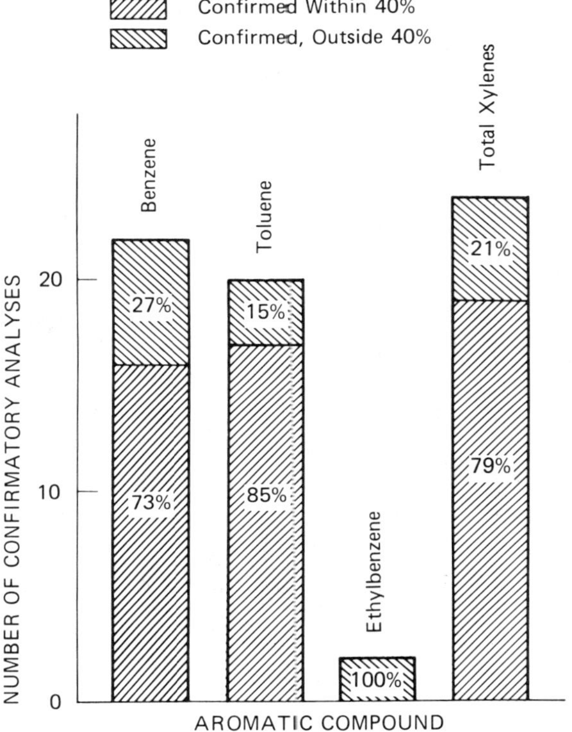


**Figure 12.** Second column confirmations for aromatics.

Analysis of split and blind samples provided valuable interlaboratory checks of the accuracy of the analytical data and some information about reproducibility between samples. Although no formal statistical requirements were specified for this part of the protocol, most of the comparative data agreed within 40%.

Confirmations, both alternative column and GC/MS, provided a high degree of confidence in the identifications made. The extra effort of making these measurements quantitative was very worthwhile.

Other subjects not covered here that would properly be considered parts of a QA program include daily analysis of standards generated within the analytical laboratory; monitoring of sample storage facilities; analysis of blanks; tracking samples from collection through analysis; and checking and correlating data.

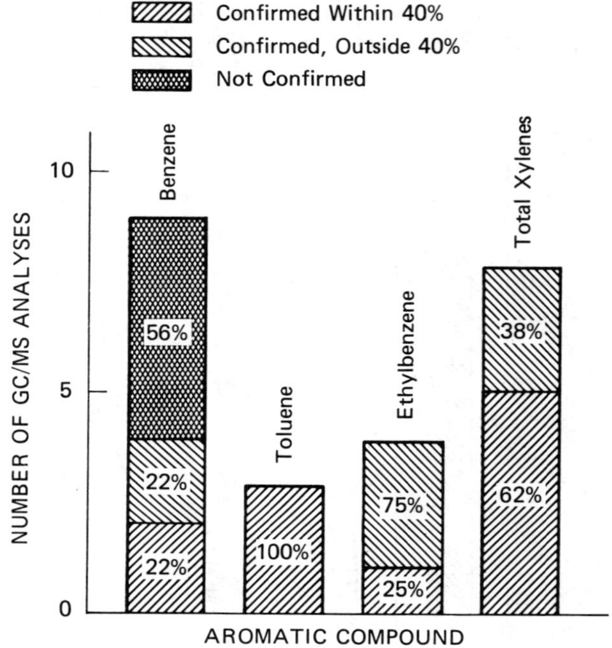


The main provisions of the quality assurance protocol were part of a cooperative venture. While this type of protocol is very time-consuming for both the contracting agency and the laboratory performing the analyses, it has provided the information essential for evaluation of the data generated in this large-scale survey. The protocol is suggested for possible incorporation in similar analytical efforts.

## ACKNOWLEDGMENTS

The authors wish to thank the project officer, Mr. Carl Hirth, Dr. Dale M. Coulson, Chairman of the Analytical and Inorganic Chemistry Department, and Ms. Patricia A. Boland of SRI International for their support and assistance in this work.

## REFERENCES

1. U.S. Environmental Protection Agency. "National Interim Primary Drinking Water Regulations," *Federal Register* 40(248):59566-59588 (1976).
2. U.S. Environmental Protection Agency. "Control of Organic Chemical Contaminants in Drinking Water," *Federal Register* 43(28):5756-5780 (1978).
3. U.S. Environmental Protection Agency. "National Interim Primary Drinking Water Regulations: Control of Trihalomethanes in Drinking Water; Final Rule," *Federal Register* 44(231):68624-68707 (1979).
4. "Analysis of Finished Water Quality of Community Water Systems," U.S. EPA Contract No. 68-03-2802 (1979).
5. "Proposed Ground Water Strategy," U.S. EPA Office of Drinking Water (1980).
6. Bellar, T. A., and J. J. Lichtenberg. "Determining Volatile Organics at the Microgram-per-Liter Level in Water by Gas Chromatography," *J. Am. Water Works Assoc.* 66:739 (1974).
7. Bellar, T. A., J. J. Lichtenberg and R. C. Kroner. "The Occurrence of Organohalides in Chlorinated Drinking Water," *J. Am. Water Works Assoc.* 66:703 (1974).
8. "The Determination of Halogenated Chemical Indicators of Industrial Contamination in Water by Purge and Trap, Method 502.1," U.S. EPA Environmental Monitoring and Support Laboratory, Cincinnati, OH (1979).
9. "The Determination of Aromatic Chemical Indicators of Industrial Contamination in Water by Purge and Trap, Method 502.2," U.S. EPA Environmental Monitoring and Support Laboratory, Cincinnati, OH (1979).
10. Boland, P. A., B. A. Kingsley, D. F. Stivers and I. H. Pomerantz. "Protocol for the Analysis of a Broad Range of Specific Organic Compounds in Drinking Water," Chapter 44, this volume.
11. "Total Organic Carbon, Low Level, Method," U.S. EPA Environmental Monitoring and Support Laboratory, Cinncinati, OH (1978).
12. "Dohrmann DC-54 Ultra Low Level Total Organic Carbon Analyzer System Equipment Manual," 2nd ed. (1978), pp. 2-14, 2-15.
13. American Chemical Society Committee on Environmental Improvement and Subcommittee on Environmental Analytical Chemistry. "Guidelines for Data Acquisition and Data Quality Evaluation in Environmental Chemistry," *Anal. Chem.* 52:2242 (1980).
14. Mandel, J. *The Statistical Analysis of Experimental Data* (New York: John Wiley & Sons, Inc., 1974), Chapter 4.

# CHAPTER 46

# GLASS CAPILLARY GAS CHROMATOGRAPHIC/MASS SPECTROMETRIC ANALYSIS OF ORGANIC CONCENTRATES FROM DRINKING AND ADVANCED WASTE TREATMENT WATERS

**Denis C. K. Lin**

Battelle Columbus Laboratories
Columbus, Ohio

**Robert G. Melton and Frederick C. Kopfler**

Health Effects Research Laboratory
U.S. Environmental Protection Agency
Cincinnati, Ohio

**Samuel V. Lucas***

Battelle Columbus Laboratories
Columbus, Ohio

Assessing health effects of trace (part-per-billion) levels of organic material in water is an important factor in establishing maximum permissible contamination levels for drinking water (DW) and finished water from advanced waste treatment (AWT) plants. Continued expansion of our chemical and manufacturing industries as well as an increasing population can be expected to challenge many DW resources in the near future. Certainly, for some areas of the United States, there is already some cause for concern. An obvious result of the ever-increasing demand on DW resources is the necessity for more effective wastewater reuse. Pilot AWT plants have been in operation to

*Author to whom correspondence should be addressed.

demonstrate the feasibility of direct wastewater reuse after AWT for high-quality requirement usage. One principal reuse mode is that of groundwater injection, either to prevent seawater intrusion into coastal aquifers or simply as aquifer replenishment. With respect to potential health effects, trace level organic contamination of these AWT reuse waters is obviously important for the ultimate protection of DW sources.

The Health Effects Research Laboratory (HERL), U.S. Environmental Protection Agency (EPA), Cincinnati, OH, is pursuing a research program to characterize the trace-level organic materials in DW and AWT water and to test the trace organic material concentrated from these waters for biological activity as an indication of potential health effects.

Large quantities (1500–15,000 L) of DW and AWT water have been concentrated, principally by reverse osmosis (RO) techniques, to yield gram-quantity amounts of concentrated organic material. Of each organic concentrate, 80% was reserved for biological testing. Of the remaining 20%, half was analyzed by glass capillary gas chromatography/mass spectrometry (GC/MS) for a detailed characterization of the organic material present. Thus, the sample aliquot analyzed represents between 150 and 1500 L of the original water for a theoretical concentration factor of 150,000:1 to 1,500,000:1. This extremely high concentration factor, coupled with fractionation of the sample into five separately analyzed organic polarity groups, has enabled characterization of the GC/MS -analyzable organic material of these relatively clean waters in greater detail than any other of which we are aware.

The compound identifications reported here have been generated at Battelle Columbus Laboratories under EPA Contract No. 68-04-2548. The scope of this research contract is limited to a detailed analysis of the organic concentrates. Ongoing work in the HERL laboratories as well as other contracted research projects have been involved in biological testing of the organic concentrates. The goal of the Battelle research effort in the overall program has been to characterize the organic materials present in sufficient detail to enable EPA to correlate the compound identification results with the biological test results. This correlation may suggest which of the organic compounds, or classes of compounds, have health effect significance. This chapter is limited to presenting and discussing some of the compound identification results obtained to date.

## PRODUCTION OF THE ORGANIC CONCENTRATE FROM WATER

While the principal emphasis of this report concerns our compound identification results, some discussion of the origins of the organic concentrates is germane.

Of the 16 concentrates reported here, 15 were prepared using RO technology. Of these 15, two (T1C and T1X, Table III) were prepared at HERL. The other 13 concentrates were prepared by Gulf South Research Institute. The remaining concentrate (T1Y, Table III) was prepared at HERL, using direct XAD-2 adsorption, with diethyl ether elution rather than RO.

The RO concentrate production methodology has been reported previously [1,2] and is only briefly summarized here. The water was processed batchwise at pH 5.5 through a cellulose acetate membrane. The permeate water was then processed (after pH adjustment to 10.0) through a second RO unit equipped with a DuPont Permasep® nylon fiber cartridge. The separate RO brines were recycled (with all cellulose acetate permeate being used as Permasep feed) until the brine volumes were reduced to about 38 L or until salt precipitation became a problem. The two RO brines were then sequentially extracted with pentane, methylene chloride and (after acidification) methylene chloride again. Thus, after solvent evaporation, six separate organic extracts were obtained. For some samples (Table III), the six solvent extracts were combined and analyzed as a composite concentrate (all DW concentrates). For other samples, one or more of the six individual solvent extracts was selected for analysis individually (as for some AWT concentrates).

For the DW concentrates, extraction of organics from the RO brine was carried one step further. Following solvent extraction, the RO brine was extracted with XAD-2 resin followed by ethanol elution to obtain additional organic material. While seven of these XAD-2 concentrates were produced, only two are included in the data presented here (concentrates T1X and T4X, Table III). Three of the five DW XAD-2 concentrates, which are not included in this report, were seriously contaminated with artifacts generally described as XAD "resin bleed"; the remaining two were less seriously contaminated. The XAD-2 artifact problem is discussed, and a list of the types of artifact compounds usually encountered is reported elsewhere [3].

The RO methodology for producing organic concentrates from large volumes of relatively clean water was, at best, an emerging technology when the concentrates were produced. Progress has been made since then to improve recovery and reduce artifact generation. HERL is currently sponsoring a program to characterize the performance of various organic concentration techniques from large volumes of clean water under carefully controlled conditions. The difficulty of obtaining an ideal RO process blank is formidable. Previous attempts have shown that RO processing of very large volumes of "high" quality reagent water produced a concentrate that was far from "blank" [1]. Moreover, the use of reagent water is inappropriate since the lack of dissolved salts might be expected to modify the contribution of the RO apparatus to the blank. For this reason, the Poplarville, MS

(V1C), concentrate has been used as a quasiblank. The Poplarville water originates from deep wells tapping an aquifer which is extraordinarily organic-free, yet contains dissolved salts at normal levels. More than 15,000 L of this water were used to prepare the Poplarville quasiblank concentrate. This volume is greater than the AWT volumes by a factor of 10, and greater than the DW volumes by a factor of 1.4–10 (Table IV). In addition, the Poplarville concentrate does represent true field sampling experimental conditions. Of more concern than the appropriateness of the blank is the lack of knowledge of the efficiency of RO membrane retention of relatively small organic molecules under conditions of extremely low concentration and very high volumes of processed water. (This issue is addressed in the Discussion section).

## WATER USED FOR CONCENTRATE PRODUCTION

The set of concentrates discussed in this report were selected as representative of the broad spectrum of concentrates analyzed. The 16 concentrates included encompass the following characteristics:

- Eleven were derived from DW, representing seven cities. (1) One DW sample was from a nearly organic-free source (Poplarville, MS) and served as a blank. (2) Two pairs of DW concentrates represented sequential solvent extraction followed by XAD-2 resin extraction of RO brine. The first pair was T1C and T1X from the Cincinnati sampling on October 17, 1978. The second pair was T4C and T4X from the January 14, 1980, sampling (see Tables III to V). Comparison of the solvent extract concentrates (T1C and T4C) with their respective XAD-2 concentrates (T1X and T4X) reflects on the effectiveness of these two different modes for removing organic material from the RO brine. In addition, one concentrate (T1Y) was a direct XAD-2 extract (i.e., RO was not used) of the split water sample which was also used for RO concentration of the T1C and T1X concentrates. Thus, comparison of the T1Y (direct XAD-2 extraction) results with the combined results of T1C and T1X (initial concentration by RO) enabled a comparison of the effectiveness of these two very different modes of organic compound removal from water. (3) The 11 DW concentrates represented the following types of source waters: ultraclean, deep wellwater; pristine mountain streamwater free of direct industrial and agricultural influence; contamination-vulnerable shallow wellwater; principally agricultural runoff waters; and surface water, highly reused and vulnerable to both agricultural and industrial contamination. (4) Concentrates from one DW sampling (T4C and T4X) were prepared from water on which Melton et al. [3] report results of Grob closed-loop stripping analysis (CLSA), purge-and-trap, liquid-liquid extraction (LLE) and direct XAD-2 adsorption/elution analysis.

- Five were derived from the final effluent water of pilot AWT plants. (1) Two AWT concentrates (B1M and B1N) were sequential solvent extracts of the same RO brine. (2) Three were composites of all solvent extracts of each respective RO brine. (3) One AWT plant sampled is the Orange County, CA "Water Factory 21" (code R1C, Tables IV and V) which has been the subject of previously published characterization by McCarty et al. [4].

The cities sampled for DW concentrates were chosen by EPA to provide a typical range in types of raw water and methods of treatment. Results of the analysis of volatile organics from DW sampled from these cities has previously been reported [5]. Details concerning the DW treatment plants at the time of sampling are available from HERL and are not discussed here. However note that none of the DW concentrates reported here are prepared from finished water treated by contact with activated carbon (AC). In contrast, all but one of the AWT plants (Table III) incorporated AC contact as the last step in the treatment scheme before final chlorination. Smith et al. [2] describe each of the AWT plants sampled and provide pre- and posttreatment data on suspended solids (SS), biochemical oxygen demand (BOD), chemical oxygen demand (COD), total organic carbon (TOC), ammonia, pH, hardness, inorganic anions and other standard water quality parameters.

## EXPERIMENTAL

The analytical scheme followed has been reported in detail elsewhere [6,7], and only a brief summary is included here. It is graphically presented in Figure 1.

Organic concentrate samples were split, and a quantity equal to 10% of the total original sample was analyzed. Organic acids and bases were extracted separately. The neutral components were partitioned on silica gel with four-step gradient elutions. The extracted bases were combined with one of the silica gel fractions, and the extracted acids were derivatized with diazomethane before analysis. The analytical scheme thus yielded the five fractions shown in Figure 1.

1. extracted, methylated acids;
2. hexane silica gel eluate (not analyzed by GC/MS);
3. hexane:benzene (1:1) silica gel eluate; aromatic fraction;
4. extracted bases plus methylene chloride silica gel eluate; medium polarity fraction; and
5. methanol silica gel eluate; high polarity fraction.

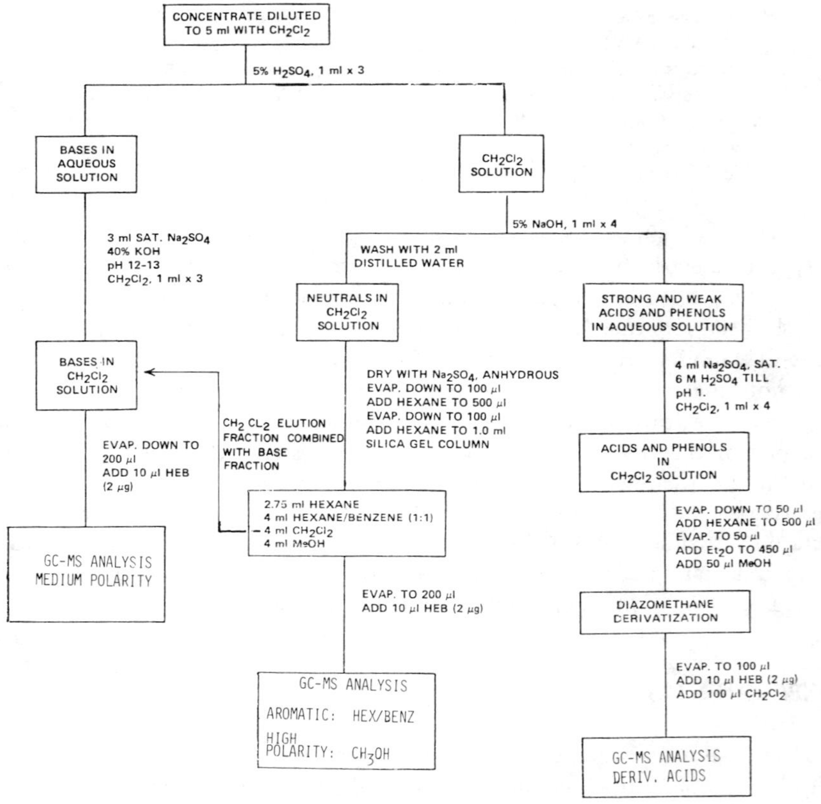


**Figure 1.** Scheme used for concentrate fractionation.

Thirteen deuterated internal standards were added before sample processing. This enabled both careful monitoring of sample partitioning as well as quantification of compounds of special interest. The deuterated internal standards are listed in Table I. A simultaneous and parallel blank sample was produced and analyzed along with each concentrate. A Finnigan 3200 mass spectrometer equipped with a Systems Industries System/150 data system was used for GC/MS analysis, and all data were archived on nine-track magnetic tape.

Glass capillary GC/MS analysis was performed on each of the four fractions of both sample and blank on two different GC liquid phases: SP1000 (a polar Carbowax 20M-nitroterephthalic acid phase, 40 m x 0.25 mm) and

Table I. Deuterium-Labeled Compounds Used as Internal Standards

| Internal Standards (mol wt) | Fraction in Which Internal Standard is Recovered[a] |
|---|---|
| Pyridine-$D_5$ (84) | Bases |
| Aniline-$D_5$ (98) | Bases |
| Benzoic Acid-$D_5$ (127) | Acids |
| Phenol-$D_5$ (99) | Acids |
| Eicosane-$D_{42}$ (324) | Nonaromatic hydrocarbon (not analyzed) |
| Bromobenzene-$D_5$ (167) | Aromatic |
| 1,4-Dibromobenzene-$D_4$ (238) | Aromatic |
| Naphthalene-$D_8$ (136) | Aromatic |
| Chrysene-$D_{12}$ (240) | Aromatic |
| Perylene-$D_{12}$ (264) | Aromatic |
| Nitrobenzene-$D_5$ (128) | Aromatic and medium polarity |
| Nitronaphthalene-$D_7$ (180) | Aromatic and medium polarity |
| Dimethylterephthalate-$D_4$ (198) | High polarity |

[a]See the fractionation scheme, Figure 1.

SP2100 (a nonpolar methyl silicone phase, 30 m x 0.25 mm). The unpartitioned concentrates of solvent extracted RO brine samples were also analyzed by GC/MS on the two columns with only deuterated internal standards added.

This procedure produced 18 GC/MS data files per concentrate (including those from the blank). The mass spectra (background subtracted) of all discernible GC peaks and shoulders (from both sample and blank) which were expected to be identifiable were processed through a Finnigan INCOS spectral matching routine which matched mass spectra against the latest NBS/NIH/EPA mass spectral library. Manual interpretation was performed as necessary. The GC/MS data of each fraction were carefully compared to those of the corresponding blank to ensure against reporting artifacts introduced by the analytical scheme.

An attempt was made to confirm all identifications based on mass spectral interpretation. Identified compounds available commercially at reasonable price were analyzed using glass capillary GC/MS conditions identical to those used for sample analysis. The resulting data provided standard mass spectra and GC retention data for comparison to those data obtained from concentrate analyses. If both mass spectra and GC retention data for an identified compound matched those obtained for the reference standard, the identification was designated as "confirmed"; otherwise it was designated as "tentative."

The GC/MS data of each sample fraction were also specifically searched by specially written software for the presence of the 53 compounds shown in Table II. All "hits" were validated manually, and quantification was performed using an appropriate deuterated internal standard via the response factor approach.

Spectral interpretation results and relevant descriptive data were entered into the Battelle central computer for storage and report generation. Such computerized data management allows highly flexible cross-correlation between sets of compound identification results for up to 27 concentrates. The following information relevant to each identification was entered into the computerized database:

**Table II. List of the 53 Compounds for Which GC/MS Data Are Searched**

| | |
|---|---|
| Chloroprene | N-Phenyl-1-naphthylamine |
| Aniline | Chrysene |
| Phenol[a] | 2,4-Dichlorophenoxyacetic acid[b] |
| Diethylnitrosamine | Dibromobenzene |
| Styrene | Benzo[a] pyrene |
| 4-Methylphenol | Benzo[b,k] fluoranthene |
| 2-Methyl Styrene | Hexachloro-1,3-butadiene |
| 2,4-Diamino Toluene | Pentachloroaniline |
| 2,4-Dimethylphenol | 2,4,5-Trichlorophenoxyacetic Acid[b] |
| Nitrobenzene | Hexachlorocyclopentadiene |
| 2-Chlorotoluene | Benzo[g,h,i] perylene |
| 2-Chloroaniline | Indeno[1,2,3-c,d] pyrene |
| 3-Nitroaniline | Pentachlorophenol[b] |
| 4-Chlorophenol[b] | Hexachlorobenzene |
| 2-Naphthylamine | Lindane |
| 1,4-Dichlorobenzene | Tetrachlorobiphenyls |
| *p*-Nitrophenol[b] | Pentachloronitrobenzene |
| 4-Chlorophenyl Methyl Ketone | Pentachlorobiphenyls |
| 2-Chloronaphthalene | Triphenyl Phosphate |
| Diphenylamine | DDT |
| 4-Phenylaniline | Hexachlorobiphenyls |
| 2,4-Dichlorophenol[b] | Aldrin |
| 1,2,4-Trichlorobenzene | Tri-(*m*)-cresyl phosphate |
| 2,4-Dichloronaphthalene | Heptachlor |
| Fluoranthene | Dieldrin |
| 2,4,6-Trichlorophenol[b] | Hexabromobiphenyl |
| Methylchlorophenoxyacetic Acid[b] | |

[a]Analyzed as the free acid and the methyl ether.

[b]Analyzed as the methyl ether or methyl ester.

- compound name, both systematic and common
- CASRN
- molecular weight
- molecular formula
- industrial source and use
- concentrate and fraction where the compound was identified
- GC phase on which the compound was identified
- GC/MS data file spectrum number and retention time
- quantification data, if available
- tentative or confirmed identification status
- relative size (compared to an internal standard) of the GC peak
- a general classification of type of compound
- specific functional groups contained in the molecule.

In addition, the following general information was also entered into the database:

- descriptive data on the concentrate
- residue weight data on the concentrate and its fractions
- GC/MS analytical conditions.

Extensive quality control measures employed for this work have been detailed elsewhere [6,7]. They will also appear in the final report on EPA contract No. 68-03-2548.

## RESULTS

The results of the analysis of 16 organic concentrates have been selected for presentation in this report. These 16 concentrates which represent about one-third of the total work performed under EPA Contract No. 68-03-2548, provide a broad spectrum of the kinds of concentrates analyzed.

The source of the DW and AWT water samples, the method of concentration used and the volume of water processed are presented in Table III. Table IV is a summary of the compound identification and residue weight analysis results. The residue weight analysis indicates the percentage of the original concentrate recovered into the four fractions analyzed by GC/MS.

The actual amount of organic material in each GC/MS-analyzed fraction can easily be approximated from the information in Table IV by using the original volume of water sampled, the concentration ($\mu$g/L, Table IV) which the sample represents in the original water and the fact that 10% of the original sample was used for GC/MS analysis. Typical fraction residue weights are 50 mg (acids), 0.6 mg (aromatic), 2 mg (medium polarity) and 30 mg (high polarity). The aliphatic and unsaturated hydrocarbon fraction, which is not listed in Table IV and is not analyzed by GC/MS, is generally about 0.3 mg.

## Table III. Origins and Description of 16 Organic Concentrates from DW and AWT Water

| Water Type | Three-Digit Code[a] | Location | Volume Sampled (L) | Method of Concentrate Production | Water Source |
|---|---|---|---|---|---|
| DW | T1C[b] | Cincinnati, OH | 1,510 | Combined solvent extracts of RO brine | Ohio River |
| | T1X | | | XAD-2 extraction/ethanol elution of RO brine after solvent extraction (see T1C) | |
| | T1Y | | | Direct XAD-2 extraction/ethyl ether | |
| | T4C[c] | Cincinnati, OH | 7,100 | Combined solvent extracts of RO brine | Ohio River |
| | T4X | | | XAD extraction/ethanol elution of RO brine after solvent extraction (see T4C) | |
| | M2C | Miami, FL | 2,280 | Combined solvent extracts of RO brine | Shallow well |
| | P2C | Philadelphia, PA | 6,620 | Combined solvent extracts of RO brine | Delaware River |
| | N2C | New Orleans, LA | 6,810 | Combined solvent extracts of RO brine | Mississippi River |
| | O2C | Ottumwa, IA | 6,450 | Combined solvent extracts of RO brine | Des Moines River |
| | S2C | Seattle, WA | 11,750 | Combined solvent extracts of RO brine | Pristine mountain stream |
| | V1C | Poplarville, MS | 15,100 | Combined solvent extracts of RO brine | Deep Well |
| AWT[d] | R1C | Orange County, CA | 1,500 | Combined solvent extracts of RO brine | Municipal sewage (30% industrial), 8 steps, including GAC |
| | E1C | Escondido, CA | 1,500 | Combined solvent extracts of RO brine | Municipal sewage (14% industrial), 3 steps, including RO but not GAC |
| | B2C | Blue Plains Washington, DC | 1,500 | Combined Solvent extracts of RO brine | Municipal sewage, 8 steps, including GAC |
| | B1M[e] | Blue Plains Washington, DC | 1,500 | Methylene chloride extraction of RO brine from cellulose acetate RO membrane | Municipal sewage, 5 steps, including GAC |
| | B1N | | | Methylene chloride extraction, after acidification, following the extraction of B1M | |

[a]Three-digit code is used in Table V.

[b]T1C, T1X and T1Y derived from the same water sample.

[c]T4C and T4X derived from the same water sample.

[d]AWT water volumes are approximate.

[e]B1M and B1N derived from the same water sample.

## Table IV. Pertinent Concentrate Information and Analysis Results Summary

| Water Type | Location | Code[a] Name | Date | Volume Sampled (L) | Concentration[b] (μg/L) | Residue Weight Analysis Fractions[c] (%) | | | | | Identification Results: Number of Compounds | | | | | | |
|---|---|---|---|---|---|---|---|---|---|---|---|---|---|---|---|---|---|
| | | | | | | | | | | | Status | | | | Special Compounds | | |
| | | | | | | Acid | Aromatic | Medium Polarity | High Polarity | Total % Recovered[d] | Confirmed | Tentative | Total | Unidentified[e] | Specific Search[f] | Consent Decree | Industrial Indicators[g] |
| DW | Cincinnati, OH | T1C[h] | 10/17/78 | 1,460 | 143 | 32 | 0.05 | 1.6 | 18 | 52 | 66 | 85 | 151 | 152; 53 | 6 | 7 | 9 |
| | | T1X | | | 397 | 11 | 0.0 | 0.2 | 1 | 12 | 35 | 42 | 77 | 69; 32 | 1 | 5 | 2 |
| | | T1Y | | | 527 | 23 | 0.2 | 1.3 | 16 | 41 | 126 | 117 | 243 | 121; 99 | 17 | 13 | 24 |
| | Cincinnati, OH | T4C[i] | 1/14/80 | 7,570 | 30 | 26 | 0.1 | 0.3 | 10 | 36 | 112 | 81 | 193 | 73; 64 | 16 | 13 | 17 |
| | | T4X | | | 261 | 21 | 0.1 | 0.3 | 1.2 | 23 | 106 | 99 | 205 | 64; 73 | 7 | 7 | 5 |
| | Miami, FL | M2C | 2/3/76 | 2,280 | 527 | 38 | 0.1 | 0.3 | 11 | 50 | 77 | 144 | 221 | 84; 42 | 6 | 12 | 13 |
| | Philadelphia, PA | P2C | 2/10/76 | 5,810 | 177 | 31 | 0.3 | 0.6 | 32 | 65 | 119 | 173 | 292 | 51; 46 | 10 | 11 | 18 |
| | New Orleans, LA | N2C | 1/14/76 | 6,620 | 217 | 31 | 0.5 | 5.0 | 15 | 53 | 92 | 146 | 238 | 55; 59 | 12 | 15 | 15 |
| | Ottumwa, IA | O2C | 9/10/76 | 5,450 | 139 | 39 | 0.1 | 0.5 | 17 | 57 | 83 | 149 | 232 | 52; 57 | 6 | 8 | 12 |
| | Seattle, WA | S2C | 11/5/76 | 11,752 | 37 | 30 | 0.4 | 1.5 | 20 | 53 | 109 | 281 | 390 | 68; 50 | 11 | 12 | 15 |
| | Poplarville, MS | V1C | 3/2/79 | 15,100 | 0.6 | 2.4 | 0.6 | 4.1 | 11 | 22 | 34 | 25 | 59 | 33; 41 | 4 | 9 | 9 |
| AWT | Orange County, CA | R1C | 1/27/76 | 1,500 | 104 | 6.6 | 0.7 | 0.5 | 28 | 37 | 79 | 47 | 126 | 99; 65 | 1 | 6 | 7 |
| | Escondido, CA | E1C | 7/8/75 | 1,500 | 18 | 2.3 | 0.7 | 0.2 | 2 | 23 | 32 | 24 | 55 | 64; 46 | 5 | 5 | 11 |
| | Blue Plains, | B2C | 5/31/75 | 1,500 | 46 | 12 | 1.1 | 1.4 | 63 | 79 | 65 | 59 | 124 | 82; 90 | 6 | 8 | 13 |
| | | B1M[j] | 9/20/74 | 1,500 | 23 | 26 | 0.2 | 1.0 | 13 | 41 | 80 | 61 | 141 | 159; 108 | 6 | 8 | 9 |
| | Washington, DC | B1N | | | 21 | 59 | 1.4 | 1.7 | 20 | 83 | 65 | 61 | 126 | 148; 72 | 5 | 5 | 3 |

[a]The three-digit code is used in Table V.

[b]The amount of organic material in the concentrate expressed as μg/L (ppb) in the original water sample.

[c]The amount of the recovered fractions, expressed as a percent of the original concentrate (residue basis). See Figure 1 for the fraction generation scheme. The aromatic, medium polarity and high polarity fractions are the hexane:benzene, methylene chloride and methanol silica gel eluates, respectively.

[d]Includes the recovered alkane/alkene fraction (hexane silica gel eluate, Figure 1), which is not analyzed by GC/MS.

[e]Unidentified components are listed separately for each GC phase. The first and second numbers are for the SP1000 and SP2100 GC phases, respectively. There is a high degree of redundancy in the two numbers shown.

[f]The 53 compounds for which the data are specifically searched are listed in Table II.

[g]The "Chemical Indicators of Industrial Pollution" list (Federal Interim Primary Drinking Water Regulations).

[h]T1C, T1X and T1Y derived from the same water sample.

[i]T4C and T4X derived from the same water sample.

[j]B1M and B1N derived from the same water sample.

A striking feature of the identification results shown in Table IV is that, while many compounds have been detected and identified in these 16 organic concentrates, relatively few of the 53 special-interest compounds have been found (see the "specific search" column in Table IV). Most special-interest compounds are found in significantly lower amounts relative to those for the other compounds identified. Indeed, most of the specific search compounds (Table II) identified and quantified would not have been found without extracted ion current profile data searching. Notable exceptions are the phthalates (and other plasticizers) and chlorinated phenols.

The extremely large bulk of identification data (implied by the numbers of identified compounds shown in Table IV) is stored and processed in Battelle's central computer facility. Computerized data processing software has been created to offer a wide range of flexibility in data processing. One option available for data processing is the nonredundant listing of all identified compounds (in decreasing order of amount detected) for up to 27 concentrates. The combined concentrate listing for the 16 concentrates which are the subjects of this report has been produced by this special software; the full listing for these 16 concentrates contains 1162 different organic compounds. Because of space limitations, a subset of 578 of these 1162 compounds has been selected for inclusion in this report (Table V). In selecting the compounds for Table V, priority has been given to compounds which:

- are on a special interest list (specific research, priority pollutant, indicator of industrial pollution and the EMIC database of compounds on which biological activity data are available)
- are present in relatively high concentration
- are confirmed identifications
- have been identified in three or more concentrates
- contain high-interest functional groups (halogen, nitrogen or sulfur heterocyclic, polynuclear aromatic hydrocarbon nucleus, etc.)
- are members of high-interest categories (pesticides, herbicides, drugs, biodegradation products, plasticizers, industrial solvents and intermediates, etc.)
- are in general indicative of the source of the original water.

In general, compounds with relatively low concentration, and/or relatively less-certain identification information, or compounds which are suspected artifacts have been omitted from Table V. In addition, many identifications of less commonly encountered fatty acids and related compounds (which are not expected to have high health significance) and aliphatic and olefinic hydrocarbons (which are probably a carryover from the hydrocarbon fraction not analyzed) have also been omitted from Table V.

Table V. Selected Compounds Which Have Been Identified in Eleven DW and Five AWT Concentrates

| Compound Name | Relative Size of the GC Peak[a] (Confirmed Identifications are Underlined) | | | | | | | | | | | | | | | |
|---|---|---|---|---|---|---|---|---|---|---|---|---|---|---|---|---|
| | T1C | T1X | T1Y | T4C | T4X | M2C | P2C | N2C | O2C | S2C | V1C | R1C | E1C | B2C | B1M | B1N |
| 1 Di-*n*-butyl phthalate (%, +, $)[b] | 5.0 | | 5.0 | | 0.6 | 8.7 | 8.1 | 8.3 | 4.0 | 2.5 | 2.0 | 8.1 | 7.0 | 8.1 | 6.0 | 5.0 |
| 2 Dioctyl phthalate (%, +, $) | | | 5.0 | | | | | 1.7 | | | | 8.0 | | 2.0 | | |
| 3 Palmitic acid (%) | 7.0 | | | 4.6 | 3.6 | 7.6 | 5.8 | 5.7 | 4.9 | 5.2 | | 6.1 | | 5.0 | 5.0 | 5.0 |
| 4 2-Ethyl-3-methylmaleic acid | 7.0 | 5.0 | | 5.6 | 4.6 | 5.6 | 4.8 | 6.7 | 6.9 | 4.2 | | 3.0 | | 4.0 | 6.0 | 5.0 |
| 5 2,3-Dimethylmaleic acid | 6.0 | | | 5.6 | 4.6 | 5.6 | 5.8 | 6.7 | 5.9 | 4.2 | | | | 4.0 | 5.0 | |
| 6 2-Ethylhexanoic acid | 5.0 | | 3.0 | 5.6 | 2.6 | 3.6 | 5.8 | 5.7 | 3.9 | 4.2 | -1.0 | 6.0 | | 4.0 | 4.0 | |
| 7 Dicamba (%) | 4.0 | | | 2.6 | 5.6 | | | 4.7 | | 2.2 | | | | | | |
| 8 *bis*(2-Ethylhexyl) adipate (%) | 4.0 | | | | | 3.6 | | 2.7 | 3.9 | | | 7.0 | | | 5.0 | |
| 9 Dioctyl adipate | | 2.0 | 5.0 | - 0.4 | 2.6 | 3.6 | 3.8 | | 1.9 | | | | 4.0 | 7.0 | | |
| 10 Dihexyl adipate | | | 4.0 | | | | | | | | | 7.0 | | | | 5.0 |
| 11 Cyclohexanol (%) | 5.3 | 3.0 | | 3.6 | | 6.9 | 5.9 | 6.8 | 5.0 | 5.3 | | | | | 4.0 | 3.0 |
| 12 Monobutyl phthalate (%) | | | | | | | 6.8 | 6.7 | | | | 6.0 | 3.0 | 5.0 | 3.0 | |
| 13 Cyclohexanone (%) | 5.0 | | | 2.6 | | 6.7 | 5.8 | 5.7 | 3.9 | 4.2 | | | | | 3.0 | 2.0 |
| 14 Benzoic acid (%) | 5.0 | 6.0 | 6.0 | 4.6 | 3.6 | 6.6 | 5.8 | 5.7 | 4.9 | 4.2 | | 5.0 | 5.0 | 5.0 | 5.0 | 5.0 |
| 15 Myristic acid (%) | 6.0 | | | 3.6 | | 6.6 | | 4.7 | 4.1 | 5.2 | 1.0 | 6.0 | | 5.0 | 4.0 | 5.0 |
| 16 Pentadecanoic acid | 4.0 | | | 2.6 | | 6.6 | | 3.7 | | 3.2 | 1.0 | 5.0 | | 3.0 | 3.0 | 4.0 |
| 17 Di-isobutyl phthalate (%, $) | 5.0 | | | | | 6.6 | 3.8 | 6.0 | | | | 5.1 | 3.0 | 5.0 | | |
| 18 2-Methylbutyric acid | | | 3.0 | 3.6 | 3.6 | 6.6 | 4.8 | 3.7 | 3.9 | 4.2 | | | | 3.0 | | 5.0 |
| 19 2,2,4,4-tetramethylpentanoic acid | | | 3.0 | 2.6 | | 6.6 | 3.8 | 3.7 | | | | 3.0 | | | 3.0 | |
| 20 Butyric acid (%) | | | | | 4.6 | 6.6 | 2.8 | 2.7 | | | | | | | | |
| 21 Margaric acid | | | | | | 6.6 | | 2.7 | | 3.2 | | 4.0 | | 3.0 | | 4.0 |
| 22 1,1-Dichloroacetone | | | | | | 6.6 | 4.8 | 5.7 | | | | | | | | |
| 23 3-Ethyl-4-methyl-maleic anhydride | | | | | | 6.6 | | 5.7 | 5.9 | 2.2 | | | | | | |
| 24 2-Hydroxy-3-methyl-2-cyclopenten-1-one | 4.0 | | 5.0 | | | | | | | 6.2 | | | | | | |
| 25 4-(Hexyloxy)-1-butanol | | | | | | | 2.8 | | 4.9 | 6.2 | | | | | | |
| 26 6-Bromo-2-hexanone | 4.1 | | 6.1 | | | 4.6 | | | | | | 3.0 | 3.0 | 4.0 | | |

Table V, continued

| Compound Name | Relative Size of the GC Peak[a] (Confirmed Identifications are Underlined) | | | | | | | | | | | | | | | |
|---|---|---|---|---|---|---|---|---|---|---|---|---|---|---|---|---|
| | T1C | T1X | T1Y | T4C | T4X | M2C | P2C | N2C | O2C | S2C | V1C | R1C | E1C | B2C | B1M | B1N |
| 27 Diisooctyl phthalate | | | | | 4.7 | | | | | | | | | | | |
| 28 Bromodichloromethane (%) | 1.0 | | 6.1 | | | 2.6 | 0.8 | 1.7 | | | | | | | | |
| 29 Bromoform (%) | 2.0 | | 6.0 | 0.6 | | 4.6 | | 3.7 | | | | | | | | |
| 30 Diethyl phthalate (%, +, $) | 5.0 | 3.0 | 6.0 | 1.6 | 3.6 | 4.6 | | 4.7 | 3.9 | | | 5.0 | 3.0 | 6.0 | | |
| 31 2,4,6-Trichlorophenol (%, *, +, $) | | | 6.0 | 3.6 | | 0.6 | 4.8 | | | 2.2 | -2.0 | | | 1.0 | | |
| 32 Amylene dichloride | | | 6.0 | 2.6 | | 4.6 | 2.8 | 4.7 | 2.9 | 3.2 | | | | | | |
| 33 9,10-Anthracenedione (%) | | | 6.0 | | | | | | | | | | | | | |
| 34 3-Methoxy-3-methyl-2-butanone | | | 6.0 | | | 3.6 | | 5.7 | | 4.2 | | | | | | |
| 35 Trichloropropane (%) | 6.0 | | | | | | | | | | | | | | | |
| 36 1-Methoxy-2-propanol | 6.0 | 6.0 | | | | | 4.8 | | 5.9 | | | | | 5.0 | | |
| 37 3-Hexanol | | 6.0 | | | | 5.6 | 2.8 | 4.7 | | 1.2 | | 4.0 | | 2.0 | 2.0 | 2.0 |
| 38 4-Chlorophenol (%, *, $) | 6.0 | | 3.0 | 2.6 | | | 3.8 | | 1.9 | | | | 2.0 | | | |
| 39 Phenylacetic acid (%) | 6.0 | 4.0 | 3.0 | 3.6 | 3.6 | 4.6 | 5.8 | 4.7 | 4.9 | 2.2 | | 3.0 | | 5.0 | 5.0 | 5.0 |
| 40 Salicylic acid (%) | 6.0 | 3.0 | 4.0 | 3.6 | 4.6 | 3.6 | 4.8 | 3.7 | 3.9 | | | | 3.0 | | 4.0 | 5.0 |
| 41 Caprylic acid (%) | 6.0 | | 3.0 | 3.6 | 1.6 | | 4.8 | 5.7 | 3.9 | 3.2 | | 6.0 | | 3.0 | 3.0 | 2.0 |
| 42 Pelargonic acid (%) | 6.0 | | | | | | 3.8 | 3.7 | 3.9 | 3.2 | | 6.0 | | 4.0 | | 2.0 |
| 43 Stearic acid (%) | 6.0 | | | 3.6 | | | | 3.7 | 4.9 | 5.2 | | 6.0 | | 4.0 | 4.0 | 5.0 |
| 44 1,3,5-Trimethylcyanuric acid | 6.0 | | 5.0 | 3.6 | | | | 4.7 | | | | | | | | |
| 45 2-(2-Butoxyethoxy)ethanol | 3.0 | 4.3 | | 3.6 | 2.7 | 5.6 | | 5.8 | 4.9 | 3.2 | | 6.0 | 6.0 | 5.0 | 1.0 | |
| 46 1-Methylbutylalcohol | 3.6 | | | | | 5.7 | 4.9 | 6.0 | 4.0 | 4.5 | | 3.0 | | | | |
| 47 3-Penten-2-ol | | | 4.0 | 3.0 | | 5.6 | 5.8 | 6.0 | 4.9 | | | | | | | 2.0 |
| 48 2-Ethyl-1-hexanol | 4.0 | 5.0 | | 4.0 | 4.6 | | | | 3.5 | 4.8 | -1.0 | 3.0 | | 3.2 | 3.6 | |
| 49 Phenoxyacetic acid | 4.0 | | | | 4.6 | | | | | | | | | | | |
| 50 Clofibric acid | 5.0 | | 5.0 | 4.6 | | 5.6 | 5.8 | 4.7 | 4.9 | | | | 4.0 | 3.0 | 3.0 | |
| 51 Isovaleric acid | 3.0 | | 4.0 | 2.6 | 4.6 | 3.6 | 4.8 | 3.7 | 4.9 | | | | | 4.0 | 4.0 | 4.0 |
| 52 Valeric acid (%) | 4.0 | 1.0 | | 2.6 | 4.6 | 5.6 | 2.8 | 4.7 | 3.9 | 3.2 | | 2.0 | | 3.0 | 4.0 | 2.0 |

| | | | | | | | | | | | | | | | | |
|---|---|---|---|---|---|---|---|---|---|---|---|---|---|---|---|---|
| 53 Caproic acid (%) | 5.0 | 4.0 | 2.0 | 4.6 | 4.6 | 3.6 | 4.8 | 4.7 | 4.9 | 3.2 | -1.0 | 5.0 | 3.0 | 3.0 | 5.0 | 3.0 |
| 54 *n*-Heptanoic acid (%) | 5.0 | 4.0 | 1.0 | 4.6 | 3.6 | 3.6 | 2.8 | 3.7 | 3.9 | 2.2 | | 6.0 | | 2.0 | | 2.0 |
| 55 Azelaic acid (%) | 5.0 | 4.0 | | 3.6 | 4.6 | 3.6 | 4.8 | 5.7 | 3.9 | | | 4.0 | | 4.0 | 3.0 | |
| 56 Capric acid (%) | 4.0 | | | 3.6 | | 4.6 | 4.8 | 5.7 | 3.9 | 4.2 | | 6.0 | 1.0 | 4.0 | | |
| 57 *bis*(2-Ethylhexyl) phthalate (%, +, $) | 3.0 | | | 4.6 | | 4.6 | 4.1 | 4.7 | 3.9 | | 2.6 | 6.0 | 2.0 | | 5.0 | |
| 58 Ethosuximide (%) | | 3.0 | | 2.6 | 4.6 | | | | | | | | | 4.0 | 5.0 | 4.0 |
| 59 Phthalic acid (%) | | 3.0 | | | 2.6 | 5.6 | 4.8 | 5.7 | 2.9 | | | 6.0 | | 5.0 | 3.0 | 2.0 |
| 60 2-Methoxy-4-methyl-2-pentenoic acid | | 4.0 | 4.0 | | 4.6 | 2.6 | | | 2.9 | | | | | | 3.0 | 4.0 |
| 61 *o*-Cresyl acetate | | | | | 3.6 | 6.0 | | | | | | | | 3.0 | | |
| 62 Oleic acid (%) | | | | | 3.6 | | | 5.7 | | 3.2 | | 4.0 | | | 5.0 | 6.0 |
| 63 Dicyclohexyl phthalate (%, $) | | | | | 2.6 | 2.0 | | | | | 3.0 | | 6.0 | 5.0 | | 5.0 |
| 64 2,4-Dimethoxybenzoic acid (%) | | | | | 4.6 | | | | | | | | | | | |
| 65 *n*-Butyl-isobutyl phthalate ($) | | | | | | 5.6 | 5.8 | 3.7 | | | | | 3.0 | 6.0 | | |
| 66 Tripropylene glycol, methyl ether (isomer) | | | | | | | 0.8 | | | | | 5.0 | | | 6.0 | |
| 67 Dioctyl sebacate | | | | | | | 2.8 | | 3.9 | | | | | | 6.0 | |
| 68 Dehydroabietic acid | | | | 3.6 | | | | 2.7 | 5.9 | 2.2 | 1.0 | 5.0 | 2.0 | 4.0 | 3.0 | |
| 69 *m*-Toluic acid | 4.0 | | 3.0 | 2.6 | 2.6 | | 5.8 | 4.7 | | 4.2 | | 4.0 | 2.0 | 3.0 | | |
| 70 *p*-Toluic acid | 4.0 | | 4.0 | 2.6 | | 2.6 | 5.8 | | | 4.2 | | 3.0 | 1.0 | 3.0 | | |
| 71 Isophorone (+) | | 3.0 | 1.0 | 2.6 | 2.7 | | 5.8 | | 1.9 | 1.2 | | | | | | |
| 72 2,6-Dimethylbenzoic acid | | | | | | | 5.8 | 1.7 | 2.9 | 4.2 | | | | | | |
| 73 1,5-Heptadiene-3,4-diol | | | | | | | 5.8 | 4.7 | | 5.2 | | | | | | |
| 74 3-Methyl-2-buten-1-ol | | | | | | 5.7 | | | 3.9 | 4.2 | | | | | | |
| 75 Benzoresorcinol | | | | | | | | 5.7 | | 1.2 | | | | | | |
| 76 Tributyl phosphate (%) | 3.0 | | 5.0 | 3.6 | | 4.6 | 4.8 | 5.7 | | 3.2 | 1.0 | 4.0 | 1.0 | 5.0 | | |
| 77 Suberic acid | | 3.0 | | | 3.6 | | | 5.7 | 2.9 | | | 2.0 | | 2.0 | | |
| 78 Lauric acid (%) | | 4.0 | | 3.6 | | | 2.8 | 5.7 | 3.9 | 5.2 | | | | 5.0 | 3.0 | 3.0 |
| 79 Atrazine (%) | | | | | | | | 5.7 | 2.9 | | | | | | | |
| 80 Dichloracetic acid (%) | 5.0 | 3.0 | 4.0 | | 3.6 | 5.6 | 2.8 | | | 3.2 | | | | | | |
| 81 Triethyl phosphate (%) | 3.0 | | 4.0 | 3.6 | | 5.6 | 3.8 | | 1.9 | | | | | | 3.0 | |
| 82 5,6,7,7A-Tetrahydro-4,4,7A-trimethyl-2-benzofuranone | 4.0 | 3.0 | | | 2.6 | 5.6 | 4.8 | 2.7 | 4.9 | | | | | | | |
| 83 *cis*-2-Methylcyclopentanol | | 5.0 | | | | 5.6 | | | | 0.2 | | | | | 1.0 | |

## Table V, continued

| Compound Name | Relative Size of the GC Peak[a] (Confirmed Identifications are Underlined) | | | | | | | | | | | | | | | |
|---|---|---|---|---|---|---|---|---|---|---|---|---|---|---|---|---|
| | T1C | T1X | T1Y | T4C | T4X | M2C | P2C | N2C | O2C | S2C | V1C | R1C | E1C | B2C | B1M | B1N |
| 84 3-Methyl-2-butenoic acid | | | 2.0 | | 2.6 | 5.6 | | | | 3.2 | | | | | | |
| 85 Tripropylene glycol, methyl ether | | | | 2.8 | 3.6 | 5.6 | | | | | | 5.0 | | | 5.0 | 4.0 |
| 86 Tripropylene glycol, methyl ether | | | | 2.6 | | 5.6 | 4.8 | | | | | | | | 5,0 | 5.0 |
| 87 6-Methyl-2-heptanone | | | | 0.6 | | 5.6 | 4.8 | | 3.9 | 2.2 | | | 2.0 | | | |
| 88 Amyl chloride | | | | | | 5.6 | 4.3 | | | | | | | | | |
| 89 3,5-Dimethylcyclohexanol | | | | | | 5.6 | | 1.7 | 2.9 | 2.2 | | | | | | |
| 90 *n*-Octanol (%) | | | | | | 5.6 | 4.8 | | | | | 3.0 | | | | |
| 91 1h-Imidazole-2-carboxaldehyde | | | | | | 5.6 | 3.8 | | | | -1.0 | | | | | |
| 92 Cyclohexyl formate | | | | | | 5.6 | 1.8 | | | 2.2 | | | | | 1.0 | 5.0 |
| 93 2-Methyl-3-buten-2-ol | | | | 2.6 | | 4.6 | 4.8 | 5.3 | 3.9 | | | | | | | |
| 94 Benzaldehyde (%) | 3.0 | 4.0 | 5.3 | 2.8 | 2.8 | 4.6 | 3.8 | 4.7 | 4.9 | 2.2 | | | 3.0 | | 2.0 | 1.0 |
| 95 Dipropylene glycol, methyl ether | | | | 2.7 | 3.8 | 1.6 | 2.8 | 4.7 | | | | | | | 5.0 | 1.0 |
| 96 Dipropylene glycol, methyl ether | | | | 1.7 | 3.8 | 3.7 | 2.8 | 4.7 | | | | 3.0 | | | 4.0 | 2.0 |
| 97 Diethylene glycol, butyl ether | | | 1.0 | | | | | 1.7 | 3.9 | 5.2 | | | | | | |
| 98 Diacetylbenzene | | | | | | | | | | 5.2 | | | | | | |
| 99 Acetophenone (%) | 2.0 | | 5.1 | | 3.6 | 4.6 | 4.9 | 3.7 | 2.9 | 2.2 | | | | | | |
| 100 6-Methyl-5-hepten-2-one | 3.1 | | | | | | 3.8 | | 2.9 | 2.2 | | | | | 5.1 | 2.0 |
| 101 2,5-Hexanediol | 3.0 | 3.0 | | | | | | | | | | | 5.1 | 2.0 | | |
| 102 N-Methyl-2-pyrolidone (%) | | 1.0 | | 3.7 | 3.6 | | | | | | | | | | 2.0 | 3.2 |
| 103 Caffeine (%) | | | | 3.7 | 2.8 | | 3.8 | 3.7 | | | | 3.0 | | | | |
| 104 3-Ethylstyrene | | | | | 3.7 | | 0.8 | | 4.9 | | | | | | | |
| 105 1,1-Diphenylethane | | | | | 3.7 | | | 1.7 | - 0.1 | | | | | | | |
| 106 1-(2-Butoxyethoxy)ethanol | | 4.0 | | -0.4 | | | 1.8 | | 3.9 | 2.2 | | 5.0 | 5.1 | | | |
| 107 2-Methoxy-3-methylcrotonic acid | 3.0 | 3.0 | 5.0 | | 3.6 | 2.6 | | | | | | | | | | 2.0 |
| 108 Dibromochloromethane (%) | 3.1 | | 5.0 | - 0.4 | | 4.6 | 2.8 | 4.7 | | 1.2 | | | | | | |
| 109 2-Cyclohexenone | 4.1 | | 5.0 | | | 4.6 | 4.8 | 4.7 | | | | 3.0 | | | | 2.0 |

| | | | | | | | | | | | | | | | | |
|---|---|---|---|---|---|---|---|---|---|---|---|---|---|---|---|---|
| 110 2,3,5,6-Tetrachloroterephthalic acid | | 5.0 | 5.0 | | 3.6 | | | | | | | | | | | |
| 111 *m*-Xylene (%, $) | | | 5.0 | 0.6 | | | | | 1.9 | 2.2 | | | | | | |
| 112 Styrene (%, *, $) | | | 5.0 | | | | - 0.2 | | 3.9 | 0.2 | | | | | | |
| 113 3-Chloro-2-methyl propene (%) | | | 5.0 | | | | | | | | | | | | | |
| 114 *o*-Dichlorobenzene (%, +, $) | | | 5.0 | | | 2.6 | 0.8 | | 0.9 | 2.2 | | | | | | |
| 115 Hexachloro-1,3-butadiene (%, *, +, $) | | | 5.0 | | | | | 0.7 | | | | | | | | |
| 116 *p*-Methylsulfonyltoluene | | | 5.0 | | | | | | | | | | | | | |
| 117 di-(2-Bromoethyl) ether | | | 5.0 | | | | | | | | | | | | | |
| 118 Sulfonyl-*bis*-(4-chlorobenzene) | | | 5.0 | | | | | | | | | | | | | |
| 119 Naphthalene (%, +) | | 5.0 | | 0.6 | 3.6 | | | | 4.9 | 2.2 | | | 1.0 | | | |
| 120 Phenol (%, *, +) | 5.0 | 4.0 | 3.0 | 3.6 | 3.6 | 2.6 | | 3.7 | 0.9 | | | | | 3.0 | 3.0 | 3.0 |
| 121 2,3,3-Trichloro-2-propenoic acid | 5.0 | 4.0 | 3.0 | | 3.6 | | 4.8 | | | | | | | | | 3.0 |
| 122 N-Hydroxyphthalamide | 5.0 | 4.0 | | | 3.6 | | | | | | | | | | | |
| 123 1,1,2,2-Tetrachloroethane (%, $) | 5.0 | 4.0 | | 0.6 | | 1.6 | | | | 0.2 | | | | 4.0 | | |
| 124 4-Ethylstyrene | | | 3.0 | | 3.6 | | | | 4.9 | | | | | | | |
| 125 Tripropylene glycol, methyl ether | | | | | | | | | | | | | | | | 5.0 |
| 126 Tripropylene glycol, methyl ether | | | | | | | | | | | | | | | | 5.0 |
| 127 4-Ketoisoheptanoic acid | 3.0 | 3.0 | | 2.6 | 3.6 | 3.6 | | 2.7 | | 2.2 | | | | | | 2.0 |
| 128 Levulinic acid | 2.0 | 4.0 | 3.0 | 3.6 | | 4.6 | 2.8 | 2.7 | 2.9 | 3.2 | | | | | | 2.0 |
| 129 Cyclic tetramethylene adipate | 4.0 | | | 3.6 | | | | | | | | | | | | |
| 130 1-Cyclopentylethanone | 4.0 | | | | | | | | 1.9 | 3.5 | | | | | | 5.0 |
| 131 *n*-Butylbenzene sulfonamide | 2.0 | | | | | | 3.8 | 4.7 | | | | 4.0 | | 3.0 | 5.0 | |
| 132 Tetrahydrotrimethylbenzofuranone | 3.0 | | | 3.6 | | | 3.8 | 4.7 | | 3.2 | | 3.0 | | | | |
| 133 Methyl *n*-amyl ketone | | 4.0 | | | | 3.6 | | 3.7 | | | | | | 5.0 | | |
| 134 2,3-Dimethylbutyric acid | | | 2.0 | | 3.6 | 2.6 | 1.8 | 3.7 | 1.9 | 3.2 | | | | | 2.0 | 1.0 |
| 135 2-Methylenebutyric acid | | | 1.0 | | 3.6 | | | | 1.9 | | | | | | 2.0 | |
| 136 *o*-Ethylbenzoic acid | | | 4.0 | | 3.6 | | | | | | | | | | | |
| 137 Anisic acid | | | 4.0 | 3.6 | 3.6 | | 3.8 | 4.7 | | 2.2 | | | 1.0 | 2.0 | | |
| 138 Ethylbenzaldehyde | | | 4.0 | | 3.6 | | | | 0.9 | | | | | | | |
| 139 Ethylene glycol monethyl ether (%) | | | | 3.6 | | | | | | | | | | 3.0 | | |

### Table V, continued

| Compound Name | Relative Size of the GC Peak[a] (Confirmed Identifications are Underlined) | | | | | | | | | | | | | | | |
|---|---|---|---|---|---|---|---|---|---|---|---|---|---|---|---|---|
| | T1C | T1X | T1Y | T4C | T4X | M2C | P2C | N2C | O2C | S2C | V1C | R1C | E1C | B2C | B1M | B1N |
| 140 Hydrocinnamic acid (%) | | | | 3.6 | 2.6 | | 4.8 | 3.7 | | | | | | | | |
| 141 3,4-Dimethylbenzoic acid | | | | 3.6 | | | 4.8 | | 3.9 | 3.2 | | 5.0 | | | | |
| 142 Tripropylene glycol, methyl ether | | | | 2.6 | 3.6 | 4.6 | 4.9 | | | | | 5.0 | | | 5.0 | 2.0 |
| 143 Tripropylene glycol, methyl ether | | | | 2.6 | 3.6 | 4.6 | 4.9 | | | | | | | | 5.0 | |
| 144 Benzyl alcohol (%) | | | | 2.6 | 3.6 | 4.6 | | 5.0 | | | | 4.0 | | 3.3 | 4.3 | 2.2 |
| 145 *tris*-2-Chloroethyl phosphate (%) | | | | 2.6 | | | | | | | | | | | 3.1 | 5.0 |
| 146 3,3-Dimethylbutyric acid | | | | | 3.6 | 3.6 | 2.8 | 2.7 | | 3.2 | | | | | | 2.0 |
| 147 1,2-Dihydro-3,6-pyridazinedione (%) | | | | | 3.6 | | | | | | | 3.0 | | | | |
| 148 3,4-Dihydroxybenzaldehyde | | | | | 3.6 | | | | | | | | | | | |
| 149 2,3-Dichloroacrylic acid | | | | | 3.6 | | | | | | | | | | | |
| 150 Toluene (%, $) | | | | | 3.6 | | 4.8 | | | | | | 4.0 | | | |
| 151 1,2,3-Trimethyl-4-propenylnaphthalene | | | | | 3.6 | 0.6 | | 1.7 | | | | 2.0 | | | | |
| 152 Tripropylene glycol, methyl ether | | | | | 3.6 | | | | | | | 2.0 | | | | |
| 153 *p*-Ethylbenzaldehyde | | | | | 3.6 | | | | 2.9 | | | | | | | |
| 154 Aniline (%, *) | | | | | 3.6 | | | | | | | | | | | |
| 155 Butyl benzyl phthalate (%, +, $) | | | | | | 4.6 | | 3.7 | | | | | | 5.0 | | |
| 156 *n*-Undecyclic acid (%) | | | | | | | | | 1.9 | | | 5.0 | | 3.0 | | |
| 157 Dioctyl acelate | | | | | | | | | | 1.2 | | | | 5.0 | | |
| 158 (2,5-Dimethylbenzene)butanoic acid | | | | | | | | | | | | 4.0 | 1.0 | 5.0 | | |
| 159 Triporpylene glycol, methyl ether | | | | | | | | | | | | 5.0 | | | 3.0 | |
| 160 *p*-Methoxy-*t*-butyl phenol (%) | | | | | | | | | | | | 5.0 | | | | |
| 161 3-Butyl-6-methyl-2,4-pyridinediol | | | | | | | | | | | | 5.0 | | | | |
| 162 *m*-Cresol (%) | | | | | | | | | | | | | | 2.0 | 5.0 | |
| 163 Tricresyl phosphate (isomer unknown) (*) | | | | | | | | | | | | | | 5.0 | | |
| 164 8,11-Octadecadienoic acid | | | | | | | | | | | | | | 5.0 | 5.0 | |
| 165 Acenaphthylene | | | | | | | | | | | | | | 5.0 | 2.0 | |

| | | | | | | | | | | | | | | | | | |
|---|---|---|---|---|---|---|---|---|---|---|---|---|---|---|---|---|---|
| 166 | Piperidinol | | | | | | | | | | | | | | 5.0 | | |
| 167 | Diethyl carbinol | | | | 1.8 | | 2.6 | 3.9 | 5.0 | 3.1 | 4.3 | | | | | | |
| 168 | 2,3-Dihydro-4-methylfuran | | | | 2.6 | | 1.6 | | | 4.9 | 2.5 | | | | | | |
| 169 | 2,3-Dimethylbenzoic acid | 4.0 | | | | | | 4.8 | 3.7 | 4.9 | 3.2 | | | | | | |
| 170 | (2,4,5-Trichlorophenoxy)acetic acid (%, *) | 1.0 | | 3.0 | - 0.4 | | | | 3.7 | 4.9 | | | | | | | |
| 171 | 3,4-Dimethyl-2,5-furandione | | | | | 2.6 | | | 3.7 | 4.9 | | | | | | 2.0 | |
| 172 | Diethylene glycol, monomethyl ether | | | | | | | | | 4.9 | 4.2 | | | | | | 3.0 |
| 173 | 3-Ethyl-4-methyl-1h-pyrrole-2,5-dione | | | | | | | | | 4.9 | | | | | | | |
| 174 | 1-(1-Methylethyl)-1h-pyrrole-2,5-dione | | | | | | | | | 4.9 | | | | | | | |
| 175 | 2,5-Hexanedione | 3.0 | 3.0 | | 0.6 | | 2.9 | 4.9 | | 3.9 | 4.2 | | 2.0 | 4.0 | | 1.0 | |
| 176 | 3-Methylpentan-3-ol | | 4.0 | 3.0 | | | | 4.9 | 2.7 | | 2.2 | | | | | | |
| 177 | 2,2,6-Trimethyl-1,4-cyclohexanedione | | 3.0 | | 2.6 | 2.7 | 4.6 | 4.8 | | 3.9 | 4.2 | | | | | 3.0 | |
| 178 | 2,4,6-Trimethylbenzoic acid | | | 2.0 | 2.6 | | 1.6 | 4.8 | | 2.9 | 2.2 | | | | | | |
| 179 | *o*-Toluic acid | | | | 2.6 | | 3.6 | 4.8 | 3.7 | 2.9 | 4.2 | | 3.0 | 2.0 | 2.0 | | |
| 180 | 3,5-Dimethylbenzoic acid | | | | 2.6 | | 2.6 | 4.8 | | | 3.2 | | 2.0 | | | | |
| 181 | Palmitoleic acid | | | | 2.6 | | | 4.8 | 4.7 | | 4.2 | 1.0 | | | | | |
| 182 | iso-Hexanoic acid | | | | | | 2.6 | 4.8 | 2.7 | 1.9 | | | 2.0 | | | | |
| 183 | Hexahydrobenzoic acid | | | | | | 3.6 | 4.8 | | | | | | | | | 3.0 |
| 184 | 2-Phenylisopropanol (%) | | | | | | 3.6 | 4.8 | 4.7 | | | | 3.0 | | | | |
| 185 | 4-Methylhexanoic acid | | | | | | | 4.8 | | 0.9 | 2.2 | | | | | | |
| 186 | 1-(2-Methoxy-1-methylethoxy)-2-propanol | | | | | | | 4.8 | 2.7 | | | | | | | 2.6 | |
| 187 | 2-(2-Methoxy-1-methylethoxy)-2-propanol | | | | | | | 4.8 | | | | | | | | | 4.0 |
| 188 | Fenchyl alcohol | | | | | | | 4.8 | | | | | | | | | |
| 189 | Endo-borneol (%) | | | | | | | 4.8 | | | | | | | | | |
| 190 | Phthalide | | | | | | | 4.8 | | | 2.8 | | | | | 2.0 | |
| 191 | Tripropylene glycol, methyl ether | | | | | | | 4.8 | | | | | | | | | 4.1 |
| 192 | Diisobutyl ketone | | | | | | | 4.8 | | | | | | | | | |
| 193 | 5,5-Dimethyldihydrofuranone | | | | 2.6 | | | 2.8 | | 2.9 | 4.8 | | | | | 3.0 | 4.0 |
| 194 | 5,5-Dimethyl-2-furanone | | | | 2.6 | | 4.6 | | 4.7 | | 3.2 | | | | | 3.0 | 2.0 |
| 195 | Pivalic acid | | | | | | | 2.8 | 4.7 | | 4.2 | | | | | | |
| 196 | Lauryl alcohol (%) | | | | | | | 2.8 | 4.7 | | 3.2 | | | | | | |

**Table V, continued**

| Compound Name | Relative Size of the GC Peak[a] (Confirmed Identifications are Underlined) | | | | | | | | | | | | | | | |
|---|---|---|---|---|---|---|---|---|---|---|---|---|---|---|---|---|
| | T1C | T1X | T1Y | T4C | T4X | M2C | P2C | N2C | O2C | S2C | V1C | R1C | E1C | B2C | B1M | B1N |
| 197 Trichloroacetic acid (%) | | | | | | | | 4.7 | | 4.2 | | | | | | |
| 198 Dichlorophenoxyacetic acid (isomer unknown) | | | | | | | | 4.7 | | | | | | | | |
| 199 Diethyl pentyl phosphate | | | | | | | | 4.7 | | | | | | | | |
| 200 N,N,4-Trimethylbenzenesulfonamide | | | | | | | | 4.7 | | | | | | | | |
| 201 1,4-Dioxane (%) | 3.0 | | | | | 4.6 | | 3.7 | 1.9 | | | 3.0 | | | | 2.0 |
| 202 1-Hepten-4-ol | | 3.0 | | | | 4.6 | | | | 3.2 | | | | | | |
| 203 N,N-Diethyl formamide | | | | 2.6 | 2.6 | 4.6 | | | 0.9 | 3.2 | | 3.0 | | | | |
| 204 2-Methylcyclopentanol | | | | | | 4.6 | 3.8 | | | | | | | 3.0 | | |
| 205 3,5-Dimethoxyphenol | | | | | | 4.6 | | | | | | | | | | |
| 206 Amobarbital (%) | | | | | | 4.6 | | | | | | | | | 3.0 | |
| 207 Pentabarbital (%) | | | | | | 4.6 | | | | | | | | | | |
| 208 Cyclohexyl bromide (%) | | | | | | 4.6 | 1.8 | | | 0.2 | | | | | | |
| 209 3,4-Epoxy-2-hexanone | | | | | | 4.6 | | | | 2.2 | 2.0 | | 3.0 | | | |
| 210 Dihexyl phthalate (%, +, $) | | | | | | 4.6 | | | | | | | | | | |
| 211 Cyclohexyl acetate (%) | | | | | | 4.6 | 0.8 | 0.7 | | 1.2 | | | | | | |
| 212 Methyl butyl ketone (%) | | | | | | 4.6 | 2.8 | | | 2.2 | | | | | | |
| 213 *p*-Cresol (%, *) | | | | - 0.4 | - 0.4 | | | | | | 0.2 | | | 4.2 | 4.0 | 3.0 |
| 214 Tetrahydro-1,1-dioxide-thiophene | | | | | | | | | | | | 4.2 | | | | |
| 215 2,4-Dimethylbenzoic acid | | | 3.0 | -2.6 | | | 3.8 | 1.7 | 2.9 | 4.2 | | | | | | |
| 216 Ethyl carbonate (%) | | | 3.0 | | | | 3.8 | | | 4.2 | | | | | 2.0 | |
| 217 5-Methylhexanoic acid | | | 2.0 | | | | 3.8 | 2.7 | 2.9 | 4.2 | | 2.0 | | | 2.0 | |
| 218 2-Butyl THF | | | | | | 2.6 | | | | 4.2 | | | | 2.0 | | |
| 219 5-Methyl-2-heptanone | | | | | | 3.6 | | | | 4.2 | | | | 2.0 | 1.0 | |
| 220 4(1h)-pyrimidinone (%) | | | | | | | | | | 4.2 | | | | | | |
| 221 1,1-Dichloro-2-hexanone | | | | | | | | | | 4.2 | | | | | | |
| 222 4,4-Dichloro-3-hexanone | | | | | | | | | | 4.2 | | | | | | |

| No. | Compound | | | | | | | | | | | | | | | |
|---|---|---|---|---|---|---|---|---|---|---|---|---|---|---|---|---|
| 223 | Dihydro-5-methylfuran-2-one | | | | | | | | | | 4.2 | | | | | 3.0 |
| 224 | 2,3-Dichloroaniline | 3.0 | | 4.0 | | 1.6 | | 1.8 | | | 1.2 | | | | | |
| 225 | *p*-Dichlorobenzene (%, *, +, $) | 1.0 | | 4.0 | -0.4 | | 1.6 | -0.2 | 3.7 | | - 0.8 | | | | | |
| 226 | Dimethyl phthalate (%, $) | 4.0 | | 4.0 | | | | | | | | | 4.0 | 2.0 | 4.0 | 4.0 |
| 227 | Benzyl cyanide (%) | 2.0 | | 4.0 | | | 1.6 | 3.8 | 3.7 | | 1.2 | | 4.0 | | 1.0 | |
| 228 | 3-Heptanol | | 3.0 | 4.0 | | | | | | | | | 3.0 | | 4.0 | 2.0 |
| 229 | 1-Bromo-4-ethylbenzene | | | 4.0 | | | | | | | | | | | | |
| 230 | 2,4,5-Trichloroaniline | | | 4.0 | | | | | | | 1.2 | | 2.0 | | | |
| 231 | Pentachloroaniline (*) | | | 4.0 | | | | | | | | | | | | |
| 232 | 1,2,2,3-Tetrachloropropane | | | 4.0 | 1.6 | | | | | | | | | | | |
| 233 | 1,2,4-Trichlorobenzene (%, *, +, $) | | | 4.0 | | | 1.6 | -0.2 | | | -0.8 | | | | | |
| 234 | 2,4-Dichloro-1-nitrobenzene | | | 4.0 | | | | | | | 0.2 | | | | | |
| 235 | Lindane (beta isomer) (%) | | | 4.0 | | | | | 2.7 | | | | | | | |
| 236 | Pentachloronitrobenzene (%, *) | | | 4.0 | | | | | | | | | | | | |
| 237 | 4-Tert-butyl-2-methylphenol | | | 4.0 | -0.4 | | | | | | 2.2 | | | | | |
| 238 | 2-Methyl-2-pentanol | | | 4.0 | | | 3.6 | | 3.7 | 0.9 | 2.2 | | | | | |
| 239 | Camphor (%) | | | 4.0 | | | | | | | 2.2 | | | | | |
| 240 | 2-Ketoisocaproic acid | | | 4.0 | | 2.6 | | | | | | | | | | |
| 241 | Pentylenetetrazole | | | 4.0 | | | | | | 2.9 | | | | | | |
| 242 | Succinic acid, dimethyl ester | | | 4.0 | | | | | | | | | | | | |
| 243 | Diethyl succinate (%) | | | 4.0 | | | | | | | | | | | | |
| 244 | 2,2-Dimethylglutaric acid | | | 4.0 | | | | | | | | | | | | |
| 245 | 1-(Methylphenyl)ethanone | | | 4.0 | | 1.6 | | | | | | | | | | |
| 246 | 4′-Ethylacetophenone (%) | | | 4.0 | | 2.6 | | | | | | | | | | |
| 247 | Methyl caproate | | | 4.0 | | | | | | | | | | | | |
| 248 | Methyl caprylate | | | 4.0 | | | | | | | | | | | | |
| 249 | Methyl pelargonate | | | 4.0 | | | | | | | | | | | | |
| 250 | Ethyl nonenoate | | | 4.0 | | | | | | | | | | | | |
| 251 | Methyl caprate | | | 4.0 | | | | | | | | | | | | |
| 252 | Benzonitrile | | | 4.0 | | | | 1.8 | | | | | | | | |
| 253 | Ethyl benzoate | | | 4.0 | | | | | | 1.9 | | | | | | |

Table V, continued

| Compound Name | T1C | T1X | T1Y | T4C | T4X | M2C | P2C | N2C | O2C | S2C | V1C | R1C | E1C | B2C | B1M | B1N |
|---|---|---|---|---|---|---|---|---|---|---|---|---|---|---|---|---|
| | Relative Size of the GC Peak[a] (Confirmed Identifications are Underlined) | | | | | | | | | | | | | | | |
| 254 Methyl laurate | | | 4.0 | | | | | | | | | | | | | |
| 255 Phenylpropionic acid, methyl ester | | | 4.0 | | | | | | | | | | | | | |
| 256 2,4,5-Trichlorophenoxy acetic acid, methyl ester | | | 4.0 | | | | | | | | | | | | | |
| 257 Dichloroethyl ether (%, +, $) | | | 4.0 | | | | | | | | | | | | | |
| 258 1,2-Dimethyl-4-nitrobenzene | | | 4.0 | | | | | | | | | | | | | |
| 259 Anteisopentadecanoic acid | | | 4.0 | | | | | | | | | | 3.0 | | | 3.0 |
| 260 2-Hexenoic acid | 4.0 | 4.0 | 3.0 | | | | | | | | | | | | | |
| 261 Dalapon (%) | | 4.0 | 4.0 | | 2.6 | | | | | 3.2 | | | | | | 3.0 |
| 262 *p*-Isopropyl benzoic acid | 4.0 | | | 2.6 | | | 1.8 | | | 3.2 | | | | 2.0 | | |
| 263 4,4-Dichlorobutenoic acid | 4.0 | | | | 2.6 | | 3.8 | | | | | | | | | 2.0 |
| 264 Anthranilic acid (%) | 4.0 | | | | | | | | | | | | | | 4.0 | |
| 265 Adipic acid, dihexyl ester | 4.0 | 2.0 | | | | | | | | | | 3.0 | | | | |
| 266 3-Methyl-2-cyclohexen-1-one | 2.3 | | | 2.6 | | | 0.8 | | | 2.2 | | | | | 4.0 | 2.0 |
| 267 2-Methoxy-3-methyl-2-pentenoic acid | 3.0 | 3.0 | 3.0 | | 2.6 | | | | | | | | | | 4.0 | 3.0 |
| 268 2,4-Dichlorophenoxyacetic acid (2,4-D) (%, *) | 1.0 | | 3.0 | 2.6 | | | | 3.7 | 3.9 | | | | | | | |
| 269 2-Butoxyethanol | 3.0 | 3.0 | | 0.6 | 0.6 | | | | | | | | 4.0 | | 4.0 | 3.0 |
| 270 Benzothiazole | 3.0 | | 3.0 | 2.6 | | | 3.8 | | | 3.2 | 1.0 | 3.0 | | | | |
| 271 Butylbenzyl phthalate | 3.0 | | | | | | | | | | | 4.0 | | | | |
| 272 α-Picoline | 2.0 | | | 2.6 | | | | | | | | | | | | |
| 273 2,4-Dichlorobenzoic acid | | 2.0 | | | | | | 2.7 | | 2.2 | | | | | 4.0 | 3.0 |
| 274 2-Methylnaphthalene (%) | | 2.0 | | | 1.6 | 0.6 | 0.8 | 1.7 | 0.9 | 0.2 | | 1.0 | | | 4.0 | |
| 275 3,3-Dichloro-2-propenoic acid | | | 2.0 | | 2.6 | | 1.8 | | | | | | | | | |
| 276 *m*-Chlorobenzoic acid | | | 3.0 | 2.6 | | | | | | | | 3.0 | | | | |
| 277 Biphenyl (%) | | | 3.0 | 0.6 | 2.6 | | -0.2 | 2.7 | 2.9 | 0.2 | | 1.0 | | | 1.0 | |
| 278 2-Phenylpropionic acid | | | | 2.6 | 2.6 | | | | | | | | | | | |
| 279 Di-*n*-propyl phthalate ($) | | | | 2.6 | | | | | | | | | | | | |

| No. | Compound | | | | | | | | | | | | | | | | |
|---|---|---|---|---|---|---|---|---|---|---|---|---|---|---|---|---|---|
| 280 | 2,5-Dimethylbenzoic acid | | | | 2.6 | | | 1.8 | | 2.9 | 3.2 | | 2.0 | | 4.0 | | |
| 281 | Ethylene tetrachloride (%) | | | | 2.6 | | | | | | | | | | | | |
| 282 | 3-Methyl-2-cyclopenten-1-one | | | | 2.6 | 1.6 | | | 2.7 | | 3.2 | | | | | | |
| 283 | N-Phenyl acetamide | | | | 2.6 | 2.6 | | | | | | | | | | | |
| 284 | N,N-Dimethylbenzylamine | | | | 2.6 | 1.6 | | | | | | | | | | | |
| 285 | Tetraethylene glycol dimethyl ether | | | | - 0.4 | 2.6 | | | | | | | | | | | 3.0 |
| 286 | 3-Hydroxybenzaldehyde | | | | | 2.6 | | | | | | | | | | | |
| 287 | Sebacic acid | | | | | 2.6 | | | | | | | | | | | |
| 288 | Saccharin (%) | | | | | 2.6 | | | | | | | | | | | 2.0 |
| 289 | Terpin (%) | 3.6 | 2.0 | | | 1.6 | | | | | | | | | | 3.2 | 4.0 |
| 290 | *o*-Anisic acid | | | | | | | 3.8 | | 3.9 | | | | | | | 4.0 |
| 291 | 3-Methoxybenzoic acid | | | | | | | 1.8 | 3.7 | | | | | | | | 4.0 |
| 292 | Diacetone alcohol | | | | | | | 3.8 | 4.0 | | 2.5 | | 4.0 | | | 2.0 | |
| 293 | Dicyclohexyladipate | | | | | | | 2.8 | | | | | | | | | 4.0 |
| 294 | 2,6-Dichlorobenzoic acid | | | | | | | | 3.7 | | | | | | 4.0 | | |
| 295 | Undecanedioic acid | | | | | | | | 3.7 | | | | 4.0 | | 3.0 | | |
| 296 | Piperidinone | | | | | | | | | 3.9 | | | 4.0 | | | | 2.0 |
| 297 | Trimethylnitroso urea (%) | | | | | | | | | | | | 4.0 | | | | |
| 298 | Butyl methyl phthalate ($) | | | | | | | | | | | | 4.0 | 3.0 | 4.0 | | |
| 299 | 3-Hydroxy-1,2-benzisothiazole | | | | | | | | | | | | | | 4.0 | | |
| 300 | 2-Hydroxydecanoic acid | | | | | | | | | | | | | | 4.0 | | |
| 301 | 2-Methyl-4-dibromomethylbenzoic acid | | | | | | | | | | | | | | 4.0 | | |
| 302 | *o*-Cresol (%) | | | | | | | | | | | | | | | 4.0 | |
| 303 | 2-Amino-5-chlorobenzoic acid | | | | | | | | | | | | | | | 4.0 | |
| 304 | *exo*-Borneol | | | | | | | | | | | | | | | 4.0 | |
| 305 | Pentachlorophenol (%, *, +, $) | 1.0 | | 2.0 | 1.6 | | | | 3.7 | 3.9 | 1.2 | | | 1.0 | 1.0 | 1.0 | 1.0 |
| 306 | 1,1,3-Trichloro-1-propene | 2.0 | | | 0.6 | | 3.6 | 3.8 | 3.7 | 3.9 | 1.2 | | | | | | |
| 307 | 3,3,3-Trichloropropene | | | | -0.4 | | | | | 3.9 | | | | | | | |
| 308 | 2,4,5-Trimethylbenzoic acid | | | | | | 2.6 | 3.8 | | 3.9 | 3.2 | | 1.0 | | | | |
| 309 | 3-Octen-2-one | | | | | | | 2.8 | | 3.9 | | | | | | | 2.0 |
| 310 | 9-Ketocapric acid | | | | | | | | | 3.9 | | -1.0 | 3.0 | 2.0 | 3.0 | | |

## Table V, continued

| | Compound Name | Relative Size of the GC Peak[a] (Confirmed Identifications are Underlined) | | | | | | | | | | | | | | | |
|---|---|---|---|---|---|---|---|---|---|---|---|---|---|---|---|---|---|
| | | T1C | T1X | T1Y | T4C | T4X | M2C | P2C | N2C | O2C | S2C | V1C | R1C | E1C | B2C | B1M | B1N |
| 311 | 2-Naphthoic acid (%) | | | | $\underline{1.6}$ | | | 3.8 | $\underline{2.7}$ | | $\underline{2.2}$ | | | | | $\underline{3.0}$ | |
| 312 | 2,4-Dichlorophenol (%, *, +, $) | | | | $\underline{-0.4}$ | | | $\underline{3.8}$ | $\underline{2.7}$ | | | | | | $\underline{1.0}$ | | |
| 313 | 1,2,3,3-Tetrachloro-1-propene | | | | | | 3.6 | 3.8 | | | 2.2 | | | | | | |
| 314 | 2-Methylvaleric acid | | | | | | 1.6 | $\underline{3.8}$ | $\underline{1.7}$ | $\underline{0.9}$ | | | | | | | |
| 315 | Cineole (%) | | | | | | | $\underline{3.8}$ | | | | | | | | $\underline{3.0}$ | |
| 316 | Fenchone | | | | | | | $\underline{3.8}$ | $\underline{3.7}$ | | | | | | | | |
| 317 | *p*-Isopropylbenzaldehyde (cuminaldehyde) | | | | | | | 3.8 | | 2.9 | 0.2 | | | | | | |
| 318 | 2,2,6-Trimethylcyclohexanone | $\underline{2.0}$ | | $\underline{3.6}$ | | $\underline{1.8}$ | 3.6 | $\underline{2.8}$ | $\underline{3.7}$ | $\underline{1.9}$ | $\underline{2.2}$ | | | | | | |
| 319 | *p*-Chlorobenzoic acid (%) | | | $\underline{3.0}$ | | | | 2.8 | 3.7 | | | | | $\underline{2.0}$ | | | |
| 320 | 3,4-Dichlorobenzoic acid | | | $\underline{3.0}$ | | | | | $\underline{3.7}$ | | 2.2 | | | | 2.0 | | |
| 321 | Dichloroisopropyl ether | | | 1.0 | | | | | 3.7 | | | | | | | | |
| 322 | Pentachlorocyclopropane | | | | 1.6 | | | | 3.7 | | | | | | | | |
| 323 | Butyl benzoate | | | | | | $\underline{2.6}$ | $\underline{0.8}$ | $\underline{3.7}$ | | | | $\underline{3.1}$ | | $\underline{2.0}$ | | |
| 324 | Chlorobromomethane (%) | | | | | | | | 3.7 | | | | | | | | |
| 325 | Trichlorophenoxyacetic acid (isomer unknown) | | | | | | | | $\underline{3.7}$ | | | | | | | | |
| 326 | Elaidic acid | | | | | | | | 3.7 | | 3.2 | $\underline{2.0}$ | | | | | |
| 327 | 3-Bromo-1,1-dichloropropane | | | | | | | | 3.7 | | | | | | | | |
| 328 | 3-Methylvaleric acid | | | | | | $\underline{3.6}$ | $\underline{2.8}$ | | 0.9 | | | | | | $\underline{2.0}$ | $\underline{1.0}$ |
| 329 | 3-Chloro-2-cyclopenten-1-one | | | | | | 3.6 | | | $\underline{1.9}$ | | | | | | | |
| 330 | 2,2,2-Trichloroethanol (%) | | | | | | | | | | $\underline{3.3}$ | | | | | $\underline{2.0}$ | |
| 331 | *p*-Xylene (%, $) | | | $\underline{3.3}$ | | | | | | $\underline{1.9}$ | $\underline{1.2}$ | | | | | | |
| 332 | 4-Acetylmorpholine | 3.3 | | | | | | | | | | | | | | | 1.0 |
| 333 | 3-Hydroxy-3-methyl-2-butanone | | | $\underline{3.0}$ | $\underline{1.6}$ | $\underline{1.8}$ | 2.6 | 0.8 | | 2.1 | 2.5 | | | | | | |
| 334 | *m*-Aminoacetophenone (%) | | | | | 1.8 | 2.6 | | | | | | | | | | |
| 335 | *o*-Chlorobenzoic acid | | | $\underline{3.0}$ | | | | | | | $\underline{3.2}$ | | | | $\underline{2.0}$ | | |

| | | | | | | | | | | | | | | | | | |
|---|---|---|---|---|---|---|---|---|---|---|---|---|---|---|---|---|---|
| 336 | 2,3,4,5,5-Pentachloro-2,4-pentadienoic acid | | | | | | | 1.8 | | | 3.2 | | | | | | |
| 337 | 5-Epideoxypodocarpic acid (+) | | | | | | | | 1.7 | 2.9 | 3.2 | | | | | | |
| 338 | 4-Amino-3,5-dichlorobenzoic acid | | | | | | | | | | 3.2 | | | | | | |
| 339 | 2,6-Di-*t*-butyl-*p*-benzoquinone | 2.0 | | | 1.7 | | | 1.8 | 2.7 | | 1.2 | -1.0 | 3.0 | | 2.0 | 2.0 | |
| 340 | Butylated hydroxy toluene (BHT) | | 1.0 | | 1.7 | | | 0.8 | 1.7 | | | | 2.0 | | | | 1.0 |
| 341 | Dibenzofuran (diphenylene oxide) | 1.0 | | 3.0 | | | | -0.2 | | | | | | | | | |
| 342 | 2,2-Dichlorobutyric acid | | 2.0 | 3.0 | | | | | | | | | | | | | 3.0 |
| 343 | Methyl palmitate | | 2.0 | 3.0 | | - 0.4 | | | | | | 2.0 | | | 2.0 | | |
| 344 | 4-Chlorophenyl acetate | | | 3.0 | | | 2.6 | | | | | | | | | | |
| 345 | 2,3,4,6-Tetrachlorophenol | | | 3.0 | | | | | | | | | | | | | |
| 346 | Chlorobenzene (%) | | | 3.0 | - 0.4 | | | | 0.7 | | - 0.8 | | | | | | |
| 347 | 1-Chloro-2-ethylbenzene | | | 3.0 | | | | | | | | | | | | | |
| 348 | Perchloroethane | | | 3.0 | | | | 0.8 | | | | | | | | | |
| 349 | 1-Bromo-2-ethylbenzene | | | 3.0 | | | | | | | | | | | | | |
| 350 | Chlorovinylbenzene | | | 3.0 | | | | | | | - 0.8 | | | | | | |
| 351 | 1,2,3-Trichlorobenzene ($) | | | 3.0 | | | | | | | | | | | | | |
| 352 | 3,5-Dibromotoluene | | | 3.0 | | | | | | | | | | | | | |
| 353 | (1,2-Dichloroethyl)benzene | | | 3.0 | | | | | | | | | | | | | |
| 354 | *m* Chloronitrobonzono (%) | | | 3.0 | | | | | 2.7 | | | | | | | | |
| 355 | 2,4-Dichloronaphthalene (*) | | | 3.0 | | | | | | | | | | | | | 1.0 |
| 356 | *m*-Dichlorobenzene (%, +, $) | | | 3.0 | | | | - 0.2 | | | | | | | | | |
| 357 | Nitrobenzene (%, *, +, $) | | | 3.0 | - 1.4 | | | | | | | | | | | | |
| 358 | Ethylene glycol, *bis*(2-chloroethyl)ether (%) | | | 3.0 | | | | 2.8 | | | | | | | | | |
| 359 | Dimethyl suberate | | | 3.0 | | | | | | | | | | | | | |
| 360 | Methyl succinate, dimethyl ester | | | 3.0 | | | | | | | | | | | | | |
| 361 | N-Methylethenamine | | | 3.0 | | | | | | | | | | | | | |
| 362 | 2-(2-Chloroethoxy)ethanol | | | 3.0 | | | | | | | | | | | | | |
| 363 | 2,5-Dimethylbenzaldehyde | | | 3.0 | | | | | | 2.9 | | | | | | | |
| 364 | Clofibrate (%) | | | 3.0 | | | | | | | | | | | | | |
| 365 | 9h-Fluoren-9-one (%) | | | 3.0 | | | | | | | 0.2 | | | | | | |
| 366 | 1-Methyl-2,4-dinitrobenzene (%) | | | 3.0 | -0.4 | | | | | | | | | | | | |

## Table V, continued

| Compound Name | Relative Size of the GC Peak[a] (Confirmed Identifications are Underlined) | | | | | | | | | | | | | | | |
|---|---|---|---|---|---|---|---|---|---|---|---|---|---|---|---|---|
| | T1C | T1X | T1Y | T4C | T4X | M2C | P2C | N2C | O2C | S2C | V1C | R1C | E1C | B2C | B1M | B1N |
| 367 2,4-Dichlorophenoxyacetic acid, methyl ester | | | 3.0 | | | | | | | | | | | | | |
| 368 Clofibric acid, methyl ester | | | 3.0 | | | | | | | | | | | | | |
| 369 *n*-Amyl alcohol (%) | 3.0 | 3.0 | | 1.6 | | | | | 1.9 | 1.2 | | | | | | 2.0 |
| 370 5-Hexen-2-ol | 2.0 | 3.0 | | | | 2.6 | | | | | | | | 3.0 | | |
| 371 Nicotine (%) | 2.0 | 3.0 | | −0.4 | 1.6 | | | | | | | | | | | |
| 372 Hexyl methyl ketone (%) | | 3.0 | | | 0.6 | | | | | 2.2 | | | | | | |
| 373 1-Hexanol | | 3.0 | | | | | | | | | 0.2 | | | | | |
| 374 2-Hydroxy-6-methylbenzoic acid | 3.0 | | | | | | | | 2.9 | | | | | | | |
| 375 1,2-Dichloroethane (%, $) | 3.0 | | 2.0 | | | | | | | | | | | | | |
| 376 2-Bromo-1,2-dichloropropane | 3.0 | | | | | | | 2.7 | | | | | | | | 2.0 |
| 377 1-Bromo-2,3-dichloropropane | 3.0 | | | | | | | 1.7 | | | | | | | | |
| 378 Tripropylene glycol | 3.0 | | | | | | | | | | | | | | | |
| 379 2,6-Lutidine (%) | 3.0 | | | 0.6 | | | | | | | | 2.0 | | | | |
| 380 2,4-Lutidine (%) | 3.0 | | | - 0.4 | | | 0.8 | | | | | | | | | |
| 381 2,3-Lutidine | 3.0 | | | - 0.4 | | | | | | | | 2.0 | | | | |
| 382 2,4,6-Collidine | 3.0 | | | | | | | | 0.9 | | | 2.0 | | | | |
| 383 2-Phenyl acrolein | 3.0 | | | | | | | | | | | | | | | |
| 384 1-Methylisoquinoline | 3.0 | | | | | | | | | | | | | | | |
| 385 2-3-Dimethylquinoline | 3.0 | | | | | | | | | | | | | | | |
| 386 2,4,6-Trichloroaniline | 2.0 | | | 1.6 | | | 1.8 | 2.7 | | 0.2 | -1.0 | | | | | |
| 387 Tetralin | | | 2.0 | | 1.6 | | - 0.2 | 1.7 | 1.9 | - 0.8 | | | | | | |
| 388 2,4-Dimethylphenol (xylenol)(%, +, *) | | | | - 0.4 | - 0.4 | | | | | | | | | | 3.0 | 1.0 |
| 389 Ethylbenzene ($) | | | | 1.6 | | | | | 1.9 | 2.5 | | | | | 1.0 | |
| 390 Phenyl ether (diphenyl ether) (%) | | | | 1.6 | | | 0,8 | 1.7 | | | | 1.0 | | | | |
| 391 2,2-Dimethylbutyric acid | | | | | 1.6 | | | 2.7 | | 2.2 | | | 3.0 | | | 2.0 |

| | | | | | | | | | | | | | |
|---|---|---|---|---|---|---|---|---|---|---|---|---|---|
| 392 | 2,4,6-Trimethylbenzaldehyde | | 1.6 | | | | 0.9 | | | | | | |
| 393 | Stilbene oxide (%) | | 1.6 | | | | | | | | | | |
| 394 | *m*-Anisaldehyde | | 1.6 | | | | | | | | | | |
| 395 | Dibromocyclohexene (unidentified isomer) | | 1.6 | | | | | | | | | | |
| 396 | Cinnamaldehyde (%) | | 1.6 | | | | | | | | | | |
| 397 | Tridecylic acid | | | | | | 1.9 | 2.2 | 3.0 | | | | |
| 398 | Dimethylsulfone | | | | | | | 2.2 | 3.0 | | | 2.0 | 2.6 |
| 399 | Didehydrogenated abietic acid | | | | | | | | | | 3.0 | 3.0 | |
| 400 | Isoindole-1,3-dione | | | | | | | | | 3.0 | | | |
| 401 | Arachidonic acid (%) | | | | | | | | | 3.0 | | | |
| 402 | Nonachlor | | | | | | | | | 3.0 | | | |
| 403 | 3-Ethoxy-1(3h)-isobenzofuranone | | | | | | | | | 3.0 | | | |
| 404 | 2-Isopropylthiophene | | | | | | | | | 3.0 | | | |
| 405 | 4-Bromobenzoic acid | | | | | | | | | | | 3.0 | |
| 406 | 2,4-Dibromobenzoic acid | | | | | | | | | | | 3.0 | |
| 407 | 4-Phenyl-4-oxo-butyric acid | | | | | | | | | | | 3.0 | |
| 408 | 4(2,5-Xylyl)butyric acid | | | | | | | | | | | 3.0 | |
| 409 | 2,5-Di(chloromethyl)-3-ethylbenzoic acid | | | | | | | | | | 3.0 | | |
| 410 | N-Ethyl-4-methyl-2-pentaneamine | | | | | | | | | | 3.0 | | |
| 411 | 4-Ethylphenol | | | | | | | | | | | 3.0 | |
| 412 | 1,2-Benzisothiazol-3(2h)-one | | | | | | | | | | | 3.0 | |
| 413 | 3-Dichloromethyl-4,6-di-*t*-butyl-*o*-benzoquinone | | | | | | | | | | | 3.0 | |
| 414 | 3,5-Dimethyldihydrofuranone | | | | | | | | | | | 1.0 | 3.0 |
| 415 | Lignoceric acid | | | | | | | | | | | | 3.0 |
| 416 | Hexahydro-2h-azepine-2-one | | | | | | | | | | | | 3.0 |
| 417 | Hexan-3-one | - 0.4 | | | 2.8 | | 2.9 | | | | | | |
| 418 | 2,3,4-Trimethylbenzoic acid | | | | 1.8 | | 2.9 | | | | | | |
| 419 | 2,3,6-Trimethylbenzoic acid | | | | | | 2.9 | | | | | | |
| 420 | Chloroacetone (%) | | | | | | 2.9 | | | | | | |
| 421 | Cyclohexyl chloride | | | 2.9 | | 1.7 | | 1.2 | | | | | |
| 422 | *o*-Xylene (%, $) | 0.6 | | | 2.8 | | | 2.2 | | | | | |

## Table V, continued

| | Compound Name | Relative Size of the GC Peak[a] (Confirmed Identifications are Underlined) | | | | | | | | | | | | | | | |
|---|---|---|---|---|---|---|---|---|---|---|---|---|---|---|---|---|---|
| | | T1C | T1X | T1Y | T4C | T4X | M2C | P2C | N2C | O2C | S2C | V1C | R1C | E1C | B2C | B1M | B1N |
| 423 | Iodocyclohexane | | | | | | 2.6 | 2.8 | 1.7 | 1.9 | 0.2 | | | | | | |
| 424 | *d*-Fenchone | | | | | | | 2.8 | | | - 0.8 | | | | | | |
| 425 | Menthone | | | | | | | 2.8 | | | | | | | | | |
| 426 | 1,6-Dimethyl-4-isopropylnaphthalene | | | | 0.6 | | 0.6 | 1.8 | 2.7 | 0.9 | 0.2 | | | | | | |
| 427 | 1,2,3-Trimethylbenzene | | | | - 0.4 | | 1.6 | 1.8 | 2.7 | | 1.2 | | | | | | |
| 428 | 2-Ethylbutyric acid | | | | | | 1.6 | | 2.7 | | | | | | | 2.0 | |
| 429 | Mesitylene (%) | | | | | | | 1.8 | 2.7 | | 0.2 | | | | | | |
| 430 | 3-Octanone | | | | | | | 0.8 | 2.7 | | 0.2 | | | | | | |
| 431 | 3,4-Dichlorophenoxyacetic acid (%) | | | | | | | | 2.7 | | | | | | | | |
| 432 | *o*-Chloronitrobenzene (%) | | | | | | | | 2.7 | | | | | | | | |
| 433 | 1-(Dichloromethyl)-4-ethylbenzene | | | | | | | | 2.7 | | | | | | | | |
| 434 | Triphenyl phosphate (%, *) | | | | | | | | 2.7 | | | | | | | | |
| 435 | 2′-Methylacetophenone | | | | | | | | 2.7 | | | | | | | | |
| 436 | 4-Isopropylacetophenone | | | | | | | | 2.7 | | 1.2 | | | | | | |
| 437 | Diamyl phthalate | | | | | | | | 2.7 | | | | | | | | |
| 438 | *o*-Chlorotoluene (%, *) | | | - 0.4 | - 0.4 | | 2.6 | | | | - 0.8 | | | 1.0 | | | |
| 439 | 1,1,2,3-Tetrachloropropane | | | | - 0.4 | | 2.6 | 0.8 | 1.7 | | 0.2 | | | | | | |
| 440 | Ethyl butyl ketone | | | | 0.6 | | 2.6 | | | | 1.2 | | | | | | |
| 441 | α-Chloropropionic acid | | | | | | 2.6 | | | | | | | | | | |
| 442 | Tigaldehyde | | | | | | 2.6 | 0.8 | | | 1.2 | | | | | | |
| 443 | Cetyl alcohol (%) | | | | 0.6 | | | 0.8 | | | 2.2 | | | | | | |
| 444 | *o*-Chloroaniline (%, *) | | | | - 0.4 | 0.6 | | | | | 2.2 | | | 1.0 | | | |
| 445 | *p-tert*-Butylphenol (%) | | | | | | | | | | 2.2 | | | | | | |
| 446 | 2,5-Dichloro-4-methylbenzoic acid | | | | | | | | | | 2.2 | | | | | | |
| 447 | Veratraldehyde | | | | | | | | | | 2.2 | | | | | | |
| 448 | 1-Naphthoic acid | | | | | | | | | | 2.2 | | | | | 2.0 | |

| Compound | | | | | | | | | | | | |
|---|---|---|---|---|---|---|---|---|---|---|---|---|
| 449 2-Chloro-3-methyl-2-butene | | | | | | | | 2.2 | | | | |
| 450 2,4-Dichloropentane | | | | | | | | 2.2 | | | | |
| 451 6-Methyl-3-(2h)-benzofuranone | | | | | | | | 2.2 | | | | |
| 452 Xanthene | | | | | | | | 2.2 | | | | |
| 453 2,6-Diisopropylphenol | | | | | | | | 2.2 | | | | |
| 454 α-Farnesal | | | | | | | | 2.2 | | | | |
| 455 2,4-Di-*tert*-butylphenol | | | | | | | | 2.2 | | | | |
| 456 2,3,4,5-Tetrachloroaniline | 1.0 | | 2.0 | | | 0.6 | | | | | | |
| 457 *o*-Chlorophenol ($) | | | 2.0 | | | | | | | | | |
| 458 2-Methyl-1-naphthalenol | | | 2.0 | | | | | | | | | |
| 459 2,3-Dichlorobutene | | | 2.0 | | | | | | | | | |
| 460 1-Bromo-2-chloro-2-butene | | | 2.0 | | | | | | | | | |
| 461 2,4-Dichloro-1-(chloromethyl)benzene | | | 2.0 | | | | | | | | | |
| 462 1,2-Dichloro-4-(chloromethyl)benzene | | | 2.0 | | | | | | | | | |
| 463 9-Fluorene (%, +) | | | 2.0 | | | | | | | 2.0 | 2.0 | |
| 464 *p*-Bromotoluene | | | 2.0 | | | | | | | | | |
| 465 1,2-Dichloro-3-nitrobenzene | | | 2.0 | | | | | | | | | |
| 466 2-Chloro-*p*-cymene | | | 2.0 | | | | | | | | | |
| 467 Lindane (%, *, $) | | | 2.0 | | | | 1.7 | 1.2 | | | | |
| 468 Fluoranthene (%, *, +, $) | | | 2.0 | -0.4 | -0.4 | 0.6 | | | -2.0 | | 1.0 | |
| 469 Methyl dichloroacetate | | | 2.0 | | | | | 1.2 | | | | 2.0 |
| 470 Urethane (ethylcarbamate) (%) | | | 2.0 | | 0.6 | | | | | | | |
| 471 4-Chlorophenylacetic acid, methyl ester | | | 2.0 | | | | | | | | | |
| 472 4-Chloro-2-nitroaniline | | | 2.0 | | | | | | | | | |
| 473 2,4-Dichlorophenoxyacetic acid ethyl ester | | | 2.0 | | | | | | | | | |
| 474 (1-Nitroethyl)benzene | | | 2.0 | | | | | | | | | |
| 475 Phenanthrene (%, +) | 1.0 | 2.0 | | | | 0.6 | 0.7 | 0.2 | -1.0 | 2.0 | 1.0 | |
| 476 γ-Picoline | 2.0 | | | -0.4 | | | | | | | | |
| 477 2,5-Lutidine | 2.0 | | | | | | | | | | | |
| 478 2,3,6-Trimethylpyridine | 2.0 | | | | | | | | | | | |
| 479 5-Ethyl-2-picoline | 2.0 | | | | | | | | | | | |

## Table V, continued

| | Compound Name | Relative Size of the GC Peak[a] (Confirmed Identifications are Underlined) | | | | | | | | | | | | | | | |
|---|---|---|---|---|---|---|---|---|---|---|---|---|---|---|---|---|---|
| | | T1C | T1X | T1Y | T4C | T4X | M2C | P2C | N2C | O2C | S2C | V1C | R1C | E1C | B2C | B1M | B1N |
| 480 | 2,3,4-Trimethylpyridine | 2.0 | | | | | | | | | | | | | | | |
| 481 | 2,4,5-Collidine | 2.0 | | | | | | | | | | | | | | | |
| 482 | 2-Ethyl-6-picoline | 2.0 | | | | | | | | | | | | | | | |
| 483 | Quinoline (%) | 2.0 | | | | | | | | | | | | | | | 1.0 |
| 484 | Quinaldine (%) | 2.0 | | | 1.6 | | | | | | | | | | | | |
| 485 | Dipropylene glycol | 2.0 | | | | 1.6 | | | | | | | | | | | |
| 486 | 2-(1-Methyl-2-piperidinyl)pyridine | 2.0 | | | | | | | | | | | | | | | |
| 487 | Triethylene glycol, monoethyl ether | 1.0 | | | | 1.6 | | 1.8 | | | | | | | | | 1.0 |
| 488 | 1-Methylnaphthalene (%) | | | 1.0 | | 1.6 | | 0.8 | 1.7 | -0.1 | -0.8 | | 1.0 | | | | |
| 489 | Cotinine | | | | | 1.6 | | | | | | | | | | | |
| 490 | Benzylamine | | | | | 1.6 | | | | | | | | | | | |
| 491 | 3,5-Dichlorophenol (%) | | | | | | | 1.8 | | | | | | | 2.0 | | |
| 492 | 4-Phenylbicyclohexane | | | | | | | | 1.7 | | 1.2 | | 2.0 | | | | |
| 493 | 2-Methylcyclopentanone | | | | | | | | | 1.9 | | | | | | 2.0 | 1.0 |
| 494 | Abietic acid | | | | | | | | | | | | | | 2.0 | | |
| 495 | 1,3,5-Trichloro-2-methoxybenzene | | | | | | | | | | | | 2.0 | | | | |
| 496 | N,4-Dimethylbenzenefulfonamide (%) | | | | | | | | | | | | 2.0 | | | | |
| 497 | α-Dichlorophenylacetic acid | | | | | | | | | | | | | 2.0 | | | |
| 498 | Phenobarbitol (%) | | | | | | | | | | | | | 2.0 | | | |
| 499 | Bromochlorobenzoic acid (isomer unidentified) | | | | | | | | | | | | | | 2.0 | | |
| 500 | Terephthalic acid | | | | | | | | | | | | | | 2.0 | | |
| 501 | Barbitol | | | | | | | | | | | | | | | | 2.0 |
| 502 | N-Methyl-2-piperidinone | | | | | | | | | | | | | | | | 2.0 |
| 503 | Pseudocumene | | | | -0.4 | | 1.6 | 0.8 | 1.7 | 1.9 | 1.2 | | | | | | |
| 504 | Chloroacetic acid (%) | | | | | | | | | 1.9 | | | | | | | 1.0 |
| 505 | Clofibric acid, *m*-chloro isomer | | | | | | | | | 1.9 | | | | | | | |

| | | | | | | | | | | | | |
|---|---|---|---|---|---|---|---|---|---|---|---|---|
| 506 1,1,2,3,3-Pentachloropropane | | | | -0.4 | | 1.6 | 1.8 | 1.7 | | 0.2 | | |
| 507 Dimethylformamide (%) | | | | | | 1.6 | 1.8 | | | 1.2 | | |
| 508 1,6-Dimethylnaphthalene | | | | | | | 1.8 | 1.7 | -0.1 | -0.8 | | 1.0 |
| 509 4-Ethylpyridine | | | | | | | 1.8 | | | | | |
| 510 1,3,6-Trimethyl-2,4(1h,3h)-pyrimidinedione (%) | | | | | | | 1.8 | | | | | |
| 511 Capraldehyde (%) | | | | | | | 1.8 | | | | | |
| 512 3′,4′-Dimethoxyacetophenone | | | | | | | 1.8 | | | 0.2 | | |
| 513 Diphenylamine (%, *) | | | | | -0.4 | | -0.2 | 1.7 | | 1.8 | | |
| 514 1,1,2-Trichloro-1-propene | 1.0 | | | | | 1.6 | | 1.7 | | | | |
| 515 1,4,6-Trimethylnaphthalene | | | | | | | -0.2 | 1.7 | | | | |
| 516 2,3,6-Trimethylnaphthalene | | | | | | | -0.2 | 1.7 | | -0.8 | | |
| 517 1,6,7-Trimethylnaphthalene | | | | | | | 0.8 | 1.7 | | | | |
| 518 *p*-Chloronitrobenzene | | | | | | | | 1.7 | | | | |
| 519 Chlorofluorobenzene (isomer unidentified) | | | | | | | | 1.7 | | | | |
| 520 1,1,3-Trichloro-2-methyl-1-propene | | | | | | | | 1.7 | | | | |
| 521 Pentachlorobenzene | | | | | | | | 1.7 | | | | |
| 522 Heptachlor (%, *, +, $) | | | | | | | | 1.7 | | | | |
| 523 Pentachlorobiphenyl + other PCB (*, +, $) | | | | | | | | | | 1.2 | | |
| 524 2,4-Diisopropylphenol | | | | | | | | | | 1.2 | | |
| 525 Dimethylurea | | | | | | | | | | 1.2 | | |
| 526 Benzyl methyl ketone | | | | | | | | | | 1.2 | | |
| 527 *o*-Isopropylacetophenone | | | | | | | | | | 1.2 | | |
| 528 Nonylphenol | | | | | | | | | | 1.2 | | |
| 529 1-(Bromomethyl)-4-methylbenzene | | | 1.0 | | | | | | | | | |
| 530 *p*-Chloroanisole | | | 1.0 | | | | | | | | | |
| 531 1,2,4,5-Tetrachlorobenzene | | | 1.0 | | | | | | | | | |
| 532 (1-Chloroethyl)dimethylbenzene | | | 1.0 | | | | | | | | | |
| 533 2-Bromo-4-methyl-1-isopropylbenzene | | | 1.0 | | | | | | | | | |
| 534 1-(Chloromethyl)naphthalene | | | 1.0 | | | | | | | | | |
| 535 Tetrachlorobiphenyl (*, $) | | | 1.0 | | | | | | | | | |
| 536 Pentachlorobiphenyl (*, $) | | | 1.0 | | | | | | | | | |

## Table V, continued

| Compound Name | Relative Size of the GC Peak[a] (Confirmed Identifications are Underlined) | | | | | | | | | | | | | | | |
|---|---|---|---|---|---|---|---|---|---|---|---|---|---|---|---|---|
| | T1C | T1X | T1Y | T4C | T4X | M2C | P2C | N2C | O2C | S2C | V1C | R1C | E1C | B2C | B1M | B1N |
| 537 1,1,2-Trichloropropane | 1.0 | | | | | | | | | -0.8 | | | | | | |
| 538 5,8-Dimethylquinoline | 1.0 | | | | | | | | | | | | | | | |
| 539 2,3,5-Collidine | 1.0 | | | | | | | | | | | | | | | |
| 540 4-Ethyl-2,6-dimethylpyridine | 1.0 | | | | | | | | | | | | | | | |
| 541 *m*-Toluquinone (%) | 1.0 | | | | | | | | | | | | | | | |
| 542 3-Ethyl-2,4,5-trimethyl-1h-pyrrole | 1.0 | | | | | | | | | | | | | | | |
| 543 2,4-Dimethylquinoline (%) | 1.0 | | | | | | | | | | | | | | | |
| 544 1,2-Benzisothiazole | | | | | | | | | | | 1.0 | | | | | |
| 545 Coumarone | | | | - 0.4 | | | | | | | | | | | | |
| 546 2,4-Dichloroaniline | | | | - 0.4 | | | 0.8 | | | | | | | | | |
| 547 3,4-Dichloroaniline (%) | | | | - 0.4 | | | 0.8 | | | | | | | | | |
| 548 Acenaphthalene (%, +) | | | | -0.4 | | | | | | - 0.8 | | | | | | |
| 549 2-Methylstyrene (*) | | | | - 0.4 | | | 0.2 | | | - 0.8 | | | | | | |
| 550 Hexachlorobenzene (%, *, +, $) | | | | | | | 0.2 | 0.7 | | | | | | | 1.0 | |
| 551 Trichlorobiphenyl (*) | | | | | | | | | | | | 1.0 | | | | |
| 552 *p*-Nitrophenol (%, *) | | | | | | | | | | | | | 1.0 | | | |
| 553 *m*-Cyclohexylphenol | | | | | | | | | | | | | 1.0 | 1.0 | | |
| 554 Bromodichloroaniline | | | | | | | | | | | | | | 1.0 | | |
| 555 Acetamide | | | | | | | | | | | | | | 1.0 | | |
| 556 *n*-Butylamine | | | | | | | | | | | | | | | | 1.0 |
| 557 Cumene (isopropylbenzene) (%, $) | | | | | | | 0.8 | | 0.9 | | | | | | | |
| 558 *n*-Propylbenzene ($) | | | | | | | -0.2 | | 0.9 | 0.2 | | | | | | |
| 559 1,1,3-Trichloro-1-butene | | | | | | | 0.8 | | | | | | | | | |
| 560 2-(Methylthio)benzothiazole | | | | | | | 0.8 | | | 0.2 | | | | | | |
| 561 Nonanal | | | | | | | 0.8 | | | | | | | | | |
| 562 DDE (%, +, $) | | | | | | | | 0.7 | | | | | | | | |

| | | | | | | | | | | | | | | | | |
|---|---|---|---|---|---|---|---|---|---|---|---|---|---|---|---|---|
| 563 3,3,3-Trichloro-2-methylpropene | | | | | | 0.6 | | | | -0.8 | | | | | | |
| 564 *bis*(Chlorophenyl)methane | | | | | | | | | | 0.2 | | | | | | |
| 565 Styrene glycol (%) | | | | | | | | | | 0.2 | | | | | | |
| 566 Benzyl chloride (%) | | | | | | | | | | -0.8 | | | | | | |
| 567 *o*-Nitrotoluene | | | | - 1.4 | | | | | | | | | | | | |
| 568 Pyrene (%, +) | | | | | | | | | | | 0.0 | | | | | |
| 569 1-Iodopentane | | | | | | | - 0.2 | | | | | | | | | |
| 570 2-Chloronaphthalene (%, *, +, $) | | | | | | | - 0.2 | | | | | | | | | |
| 571 4-Methylstyrene | | | | | | | | | | - 0.8 | | | | | | |
| 572 *m*-Terphenyl | | | | | | | | | | - 0.8 | | | | | | |
| 573 *o-tert*-Butylphenol | | | | | | | | | | - 0.8 | | | | | | |
| 574 *p*-(1,1,3,3-Tetramethylbutyl)phenol | | | | | | | | | | - 0.8 | | | | | | |
| 575 *p*-(1-Ethyl-1-methylhexyl)phenol | | | | | | | | | | - 0.8 | | | | | | |
| 576 *p*-(2,2,3,3-Tetramethylbutyl)phenol | | | | | | | | | | - 0.8 | | | | | | |
| 577 Dibromophenol | | | | | | | | | | | -1.0 | | | | | |
| 578 Butyl thiazole | | | | | | | | | | | -1.0 | | | | | |
| Total Number of Compounds | 127 | 60 | 176 | 143 | 118 | 142 | 193 | 177 | 136 | 208 | 28 | 101 | 50 | 91 | 111 | 97 |

[a]Symbols: * = a compound on the list of 53 compounds given in Table II; + = a priority pollutant; $ = a compound on the list of "Chemical Indicators of Industrial Pollution"; % = a compound in the EMIC database. See text for further explanation.

[b]The code identification can be found in Table III. See Table VI for explanation of Relative Size.

Four flags are used to designate compounds of special interest in Table V. These flags, appearing after the compound indicate the following:

- *–a compound on the list (Table II) of 53 compounds for which the GC/MS data are specifically searched
- +–a compound on the EPA list of priority pollutants
- $–a compound on the list of "Chemical Indicators of Industrial Pollution" of the Federal Interim Primary Drinking Water Regulations (February 9, 1978)
- %–a compound cataloged in the Oak Ridge National Laboratory computerized "EMIC" database of compounds on which bioactivity data, generally from the Ames test for mutagenicity, are available.

The compound names used in Table V are nonsystematic or common names rather than the systematic chemical name because of space limitations.

The numerical entry in Table V shows that a compound was identified in a particular concentrate and also indicates the relative size of the GC peak in the total ion chromatogram. When this parameter is underlined, the identification has been confirmed (see Experimental); if not underlined, the identification is tentative. The relative size parameter is a somewhat imprecise but extremely cost-effective method for estimating the amount of each identified component. It is a logarithmic scale spanning four orders of magnitude. Its assignment involves visually estimating the size of each GC peak in the total ion current chromatogram relative to that of hexaethylbenzene (designated HEB in Figure 1), an internal standard added immediately before GC/MS analysis. Thus, each GC peak corresponding to an identified compound is assigned a relative size (RS) value consisting of an integer from 1 to 9 on the basis shown in Table VI.

The computer report-generation software normalizes all RS values to a typical 1514-L sample by the following equation:

$$\text{RS (output)} = 2\log_{10}\frac{\text{antilog}_{10}\,[\text{RS(input)}/2]}{(\text{liters sampled})/(1514\text{ liters})} \quad (1)$$

Thus, computer output RS values are directly comparable from one concentrate report to another, regardless of the volume of water sampled. Generally, the output RS value can range between 0 and 9. However, when the sampled water volume exceeds 1514 L, it becomes possible for the RS (output) value (Equation 1) to become *negative* for the smallest RS (input) values since the scale is logarithmic. Thus, the RS value is designed to provide "ballpark" quantification for every water concentrate sample component identified. While it is tempting to assign the corresponding value in nanograms per liter

**Table VI. GC Peak Relative Size (RS) Values for Semiquantitative Estimation of Compound Concentrations**

| RS Value | GC Peak Size Relative to the Internal Standard[a] | Approximate Concentration[b] |
|---|---|---|
| 9 | Greater than 100X | >1300 ng/L (1.3 ppb) |
| 8 | 30 to 100 | |
| 7 | 10 to 30 | |
| 6 | 3 to 10 | |
| 5 | 1 to 3 | 13 ng/L (13 ppt) |
| 4 | 1/3 to 1 | |
| 3 | 1/10 to 1/3 | |
| 2 | 1/30 to 1/10 | |
| 1 | 1/100 to 1/30 | 0.13 ng/L (0.13 ppt) |

[a]The ranges are half orders of magnitude, so the RS value numbers span a "linear" log range. Technically, they should be 1–3.16, . . .31.6–100, etc., but for visual comparison purposes only one significant figure is more appropriate.

[b]Assuming the GC/MS response factor for the analyte to be equal to that of the internal standard and a 1514-L (400-gal) water sample. See text for other precautions concerning the use of the RS value.

of original water to each compound (as indicated in Table VI), such an indication could be misleading and easily misinterpreted. If the following conditions exist, the RS criteria would provide exact quantification: (1) the sample component and internal standard have the same response factor for mass spectrometric detection, (2) instrument response is linear through the entire 10,000-fold RS range, (3) the sample component is recovered quantitatively through the partitioning scheme and suffers the same relative chromatographic losses as the internal standard, and (4) the component is removed from the original water into the concentrate quantitatively. Clearly, the preceding conditions will never be met completely. Since detection of the internal standard is expected to be reproducible, one may directly compare RS values for the same component from one concentrate to another with relative confidence. While it is known that RO recovers nearly quantitatively many organic species (especially those of high molecular weight), it may be less efficient for others. In addition, the residue weight results of Table IV clearly show that significant amounts of organic material are not recovered in the partitioning scheme. These and other factors giving rise to the underestimation of concentration based on the RS values (as per Table VI) will generally outweigh other factors which cause overestimation. Thus, the correlation of concentration with RS value as indicated in Table VI should be taken as a lower bound only.

## DISCUSSION

### Reverse Osmosis Blank

Contribution of artifacts to the concentrate by the RO membrane or concentration apparatus did not seem to be an appreciable problem. An extremely clean and very large volume of water was used to prepare the Poplarville, MS, concentrate (V1C, Tables IV and V). This concentrate functioned as a quasi-RO "blank." The materials identified in this "blank" concentrate were both few (only 28 of the 578 entries of Table V) and in lower–often substantially lower–concentration than other concentrates. Most compounds identified in this "blank" concentrate were plasticizers, fatty acids or possible extraction solvent contaminants. Although the Poplarville sample cannot be considered a definitive blank in terms of either the type of water used or number of replicate samplings desirable, it is, nevertheless, an adequate blank since it represents an actual field sampling of an extremely large volume of extraordinarily clean water.

### Comparison of Direct XAD-2 Resin Removal of Organics with the Reverse Osmosis Method

Comparison of three sets of data (T1C, T1X, T1Y) presented in Table V suggests that the RO concentrate production method may have poor efficiency for some compounds. The T1C and T1X RO concentrates and the T1Y XAD-2 resin direct adsorption concentrate (see Table III) were produced from identical water samples (via a split stream). Thus, the identified compounds listed in Table V for these three concentrates offer some basis for comparison of these two fundamentally different methods. As one might expect, highly polar species, such as alcohols and ionized acids or bases, are recovered better by RO than by direct XAD-2 adsorption. For example, a large number of substituted pyridines and relatively basic anilines were identified in the RO solvent extract concentrate (T1C), but not in the XAD-2 direct adsorption concentrate (T1Y). This is undoubtedly due to acidification of the water before exposure to the XAD-2 resin so that the protonated species were not retained by the resin. The chlorobromoforms ($CHCl_2Br$, $CHClBr_2$ and $CHBr_3$; 28, 29 and 108, respectively, Table V) were detected in considerably higher concentration in the XAD-2 direct absorption concentrate (T1Y) than the RO concentrate (T1C) with the differences in detected amounts ranging from about 1 to 2.5 orders of magnitude. This difference may be due to losses on solvent evaporation, which are expected to be more severe with RO concentrates due to the much larger volumes of extraction solvent required. Nevertheless, poor retention by the RO membrane cannot be ruled out. One striking difference between the two sets of data is the large number of halogenated aromatics detected only in the XAD-2 resin direct absorption concentrate (T1Y) (Table V, entries 111-118, 229-259, 344-368, 456-474 and 529-536). Many of these compounds are relatively small, halogenated (especially chlorinated) aromatic or olefinic species. A possible explanation for these differences is that the RO membrane is either not retaining these materials sufficiently or is adsorbing them.

It should be emphasized that the difference in comparing the results of the T1C and T1X RO concentrates with those of the T1Y XAD-2 direct absorption concentrate cannot have an unambiguous bearing on the other results presented in Tables IV and V. First, these three concentrates were prepared by HERL, while all of the other concentrates were prepared under contract by Gulf South Research Institute. In addition, acidification of the flowing stream to pH 2 immediately prior to XAD-2 adsorption may have resulted in producing some of the chloro aromatics by reaction with the chlorine residual. Analysis of a blank elution of XAD-2 resin identical to that used for concentrate production, showed that most of the "extra" compounds and all of the chloro aromatics were not artifacts contributed by the XAD-2 resin. The most plausible explanation for these observations is that the RO membrane is not retaining many organic materials similar to those cited above. Ineed, Cabasso et al. [8] and Fang and Chain [9] indicated poor RO performance for organic compounds with molecular weights below 200.

## Comparison of Solvent Extraction and XAD-2 Extraction of RO Brine

The effectiveness of the two methods for organic compound removal from the RO brine (solvent extraction followed by XAD-2 resin adsorption/ethanol elution) can be compared using the results in Table V for the pair, T1C and T1X and the pair T4C and T4X. In each case, both Table IV and Table V show more identified compounds in the solvent extracted concentrates (T1C and T4C) than in the XAD-2 extracted concentrates (T1X and T4X). This result is not surprising, since solvent extraction is performed on the RO brine first. Note, however, that the higher effectiveness of XAD-2 for removing humic material results in considerably more material, by weight, in the T1X and T4X concentrates (see Recovery and Fractionation, below). Generally, compounds found in both the solvent and XAD-2 extract concentrates of the same RO brine occur at higher concentrations in the solvent extract sample.

The more nonpolar the compound, the higher will be the proportion in the solvent extract concentrate.

In addition, compounds identified only in the XAD-2 extract are usually extremely polar or ionizable. Conversely, highly nonpolar compounds found only in the XAD-2 extract, such as substituted benzenes, naphthalenes and tetralins, must be suspected as resin bleed artifacts. Since many of the compounds found in the XAD-2 extract but not in the solvent extract are of some interest, this extra extraction may be worthwhile for compound identification purposes. Moreover, for the purpose of biological testing, the goal is to obtain the most complete organic concentrate possible, and thus production of the XAD-2 RO brine extract is required. Counteracting these benefits is the difficulty of producing an artifact-free XAD-2 extract.

## Recovery and Fractionation

Certainly, the concentration of total organic material in the original water is higher (far higher in some cases) than that represented in Table IV. For example, humic material (usually present in relatively high concentrations in surface waters) will be recovered with good efficiency into the RO brine, but not as well into the solvent extract of the RO brine. Two pairs of concentrates, T1C & T1X and T4C & T4X (HERL RO brine solvent extracts and XAD-2 extracts of the October 17, 1978, and January 14, 1980, samplings, respectively) illustrate this result dramatically. The solvent extract concentrates, T1C and T4C, represent only 26 and 10%, respectively, of the total amount of organic material recovered from the respective RO brines. The remaining 74 and 90% were recovered from the RO brine by XAD-2 adsorption. These results are clearly due to the greater effectiveness of XAD-2 resin in removing humic material from RO brine. Indeed, the sum of the Table IV concentration values for T1C and T1X concentrates approximate that for concentrate T1Y, produced from the same water (by split stream) with direct XAD-2 resin extraction.

For the DW concentrates produced by solvent extraction of RO brine (excluding the quasiblank, Poplarville, MS), the recovery of organic material into the analyzed fractions of the partitioning scheme (Figure 1) ranged from 36 to 65%. Corresponding values for the two XAD-2 extracted concentrates (T1X and T4X) were substantially less at 12 and 23%, respectively. The significantly reduced recoveries into analyzed fractions for the XAD-2 extracted concentrates were attributable to the higher amounts of nonrecoverable humic material in them. Most of the material recovered into the derivitized acid fraction was humic in origin and resulted in the typical middle-to-late eluting chromatographic "hump" of unresolved components in the gas chromatogram. The amount of the humic "hump" in the acid fraction of solvent extract type concentrates was generally between 30 and 60% of the total chromatogram. Thus, the nonhumic material recovered into the analyzed fractions corresponded to only about 25-50% of the original solvent extract of the RO brine concentrate. This amount of material corresponded to a range of about 10-200 $\mu$g/L of nonhumic organic material in the original water.

Most of the material recovered into the analyzed fractions was highly polar, appearing in comparable amounts in the acid fraction and the higher polarity (methanol silica gel eluate) fraction. The material recovered into the aromatic and medium-polarity fraction was usually less than these two high-polarity fractions by one or two orders of magnitude.

Comparing the residue weight results for DW concentrates with those from AWT water concentrates shows that, on a per-liter of original water basis, far less organic material was recovered from the AWT water. The proportion of acidic material was also somewhat less for the AWT water concentrates. This result is undoubtedly due to use of RO processing (Escondido) or activated carbon contactors (Orange County and Blue Plains) as the last stage in the AWT pilot plant treatment. Either of these final treatment steps should perform well in removing humic and other relatively large organic species. This effect is also manifested in a lower percentage recovery of acidic materials for these AWT concentrates. By far, the AWT concentrate containing the fewest compounds and the lowest concentrations is that for Escondido, CA (E1C) which contained only 50 of the 578 compounds in Table V. This result may reflect more effective removal of organics by RO treatment than from the AC contactor treatment.

The two Blue Plains AWT concentrates (B1M and B1N) represent sequential solvent extraction of the same reverse osmosis brine (see Table III). For a comparison with other concentrates in Table IV, one must add the values shown under "Concentration ($\mu$g/L)" and take a weighted average of the percent recoveries into the analyzed fractions. Although the B1M and B1N concentrates were only 2 of the 6 RO solvent extracts for this sampling, they contained about 80% of the organic material recoverable by the solvent extractions from the RO brine.

The B1M and B1N concentrates were the unbuffered methylene chloride and the pH 2 methylene chloride extracts, respectively, from the cellulose acetate RO membrane. The pentane extract, first in the sequence, generally contained far less organic material than the two succeeding methylene chloride extracts. The three extracts from the nylon RO membrane, serving as a backup to the cellulose acetate membrane, usually did not contain significant amounts of organic material. Not surprisingly, the pH 2 methylene chloride extract, B1N, contained a significantly higher proportion of material in the acid fraction than the unbuffered extract, B1M. This result was also reflected in larger numbers of acidic compounds identified at higher concentration in B1N as compared to B1M (see Table V).

## Identified Compounds

The tabulation of compound identification results in Table IV represents a cost-effective attempt to identify every compound detected in the concentrates and their fractions. Background-subtracted mass spectra of each GC peak and shoulder with a reasonable signal-to-noise ratio have been computer-matched against the EPA/NBS/NSRDS mass spectral library of approximately 31,000 mass spectra. A limited attempt was made manually to identify all spectra for which the computer matching results were not definitive. Usually, manual interpretation did not go beyond identifying probable functional groups or assigning the substance to a general class of compounds. Unidentified mass spectra that were encountered frequently received greater attention in manual interpretation in proportion to their concentration and frequency of occurrence. The listing of the number of unidentified compounds in Table IV represents only those which have reasonably interpretable mass spectra. Since many of the unidentified mass spectra were encountered frequently, the numbers in Table IV reflect a relatively high degree of redundancy. They have been included to provide some perspective on how well Table V represents the organic material in the concentrates.

Only identifications for which the analyst was reasonably confident (80% or greater) were included in the computer-managed database. This minimum confidence level applied to all of the nonunderlined compounds listed in Table V. Compounds underlined in Table V are confirmed identifications, as described in the Experimental section. About 2100 identified compounds have been entered into the database through the course of this research program, and about 1100 reference compounds have been analyzed to generate data for identification confirmation. Of the 578 compounds selected for inclusion in Table V, 340 (60%) have confirmed identification for one or more of the concentrates in which they were reported.

The compounds in Table V are arranged in decreasing order of the relative size of the GC peak. For compounds found in more than one concentrate, the largest relative size value is used in sorting the list. Taking the first 200 compounds in Table V as characteristic of the material generally found at the highest concentrations, one finds that these 200 compounds can be classified as follows:

- 59 compounds are carboxylic acids (of these, 39 are aliphatic (mostly fatty acid-related) of which, 3 have chloro substitution; 16 of the compounds have an aromatic nucleus; 4 are related to maleic acid)
- 21 compounds are plasticizers (these include 10 phthalic acid diesters)

- 40 compounds probably originate from industrial sources or are household chemicals
- 23 compounds have a hydroxyl functional group (of these, 18 are nonaromatic and 5 are phenols)
- 15 compounds have possible uses as organic solvents (of these, 5 contain halogen substitution)
- 4 compounds are drugs, drug metabolites or food additives (caffeine, salicylic acid, clofibric acid and ethosuximide)
- 6 compounds are natural products
- 4 compounds are pesticides or herbicides (*tris*-β-chloroethyl phosphate, atrazine, 2,4,5-T and 2,4-D)
- 3 compounds are haloforms ($CHBr_3$, $CHBr_2Cl$ and $CHBrCl_2$)
- 2 compounds are polynuclear aromatic hydrocarbons (naphthalene and acenaphthylene).

By far, the predominant species at higher concentrations are the carboxylic acids (especially the fatty acids), plasticizers and polyglycol ethers. One rather surprising result is the detection of clofibric acid (50, Table V) in nearly every concentrate at relatively high levels. This compound is the free acid of the antilipidemic drug, clofibrate. The free acid of this drug is the major product of clofibrate metabolism. Since it is always seen at far higher concentrations than other indicators of water reuse (i.e., salicylic acid, caffeine, saccharin, etc.), it must be particularly resistant to further chemical degradation once it enters wastewater. Clofibrate is a prescription antilipidemic agent which is administered at the gram-per-day level to control the levels of blood lipids and cholesterol in individuals who may be predisposed toward atherosclerosis. It was widely prescribed during the time over which most of these concentrates were produced. The use of clofibrate has declined in recent years, so its sensitivity and reliability as a water reuse indicator may now be reduced [10]. Note, however, that the T4C concentrate, sampled in January 1980, did contain a relatively high concentration of clofibric acid.

Seattle (S2C) and Poplarville (V1C) are the only two cities which did not have clofibric acid in the sampled drinking water, as might be expected based on the raw water source (Table III). It is somewhat surprising, however, to see clofibrate in the Miami sample (M2C), since that raw water source is from ground- rather than surface water. Other water reuse indicators (i.e., amobarbitol, pentabarbitol, salicylic acid, dichlorobenzene and relatively high levels of plasticizers) found in the Miami concentrate (M2C) indicate there may be substantial infiltration of the drinking water aquifer with wastewater and/or leachate from solid waste sources.

Polyglycols and polyglycol ethers comprised a class of compounds identified at moderate to extremely high levels in nearly every concentrate. Oligomers of both polyethylene glycol and polypropylene glycol (found as the alcohol and/or ether terminated species) have tentatively been identified. No attempt was made to obtain high confidence identifications on the multitude of possible species. Generally, in concentrates where they were observed at high levels, two or three different series of oligomeric species were detected. These materials often represented 30-70% of the GC/MS-analyzable material in the high-polarity fraction (the methanol silica gel eluate, Figure 1). Since these materials are known to be highly resistant to biological degradation, it was not surprising that they were found at the highest levels in AWT and DW concentrates for which the water source was highly reused surface water. Apparently, these species are not well absorbed by granular activated carbon (GAC). Although these compounds have previously been reported as present in RO process blanks [1], three factors combine to indicate with high probability that they are not artifacts of the concentrate production process: (1) they are widely used in cosmetics, shampoos, ointments, fabric softeners, paints, packaging materials and a wide variety of industrial processes so that their presence in reuse water is expected; (2) they are resistant to biological degradation, so their concentrations are expected to be high relative to materials that undergo biodegradation; (3) they are found at the highest levels in concentrates from the most highly reused water.

Another group of compounds identified in nearly every concentrate was related to maleic acid. The materials were nearly always found at moderately high levels, with dimethyl and methylethyl maleic acids the most commonly encountered. At present, no reasonable explanation can be offered for this observation, and the possibility that these materials were artifacts cannot be ruled out.

Lindane, nonachlor, heptachlor and DDE (235/467, 402, 522 and 562, respectively, Table V) were the only pesticides or pesticide-related compounds identified in these 16 concentrates. They were detected in quite low levels and only in a few concentrates. Heptachlor and DDE were identified only in the New Orleans (N2C) concentrate; lindane was found in the Cincinnati (T1Y) and Seattle (S2C) samples. The detection of nonachlor in the otherwise extremely clean Escondido AWT concentrate (E1C) was so surprising that it should be suspected as an artifact. The agricultural herbicide, atrazine, was detected only in the Ottumwa (O2C) and New Orleans (N2C) concentrates. This result was not surprising considering the extensive agricultural drainage supplying these DW sources. On the other hand, the widely used phenoxy herbicides, 2,4,5-T and 2,4-D (170 and 268, respectively, Table V), have been detected at appreciable concentrations in the Cincinnati, New Orleans and Ottumwa samples.

Two other groups of compounds noteworthy for their low frequency of detection and relatively low concentrations found were the polynuclear aromatic hydrocarbons (PAH) and polychlorinated biphenyls (PCB). Acenaphyhylene, fluorene, fluoranthene, phenanthrene, napthalene and pyrene

(165, 463, 468, 475, 119 and 568, respectively, Table V) were the only PAH compounds identified. Fluoranthene, phenanthrene and naphthalene were found with the highest frequency (6, 8 and 6 concentrates, respectively). Acenaphthylene and naphthalene were the only PAH compounds found at moderate concentrations (maximum RS value of 5). Fluoranthene, phenanthrene and fluorene all have maximum RS values of 2. Pyrene, on the other hand, was found only in the quasiblank concentrate, Poplarville (V1C) at a very low level (RS value of 0.0). Thus, it may be an artifact. It is noteworthy that fluoranthene is the only one of the six PAH compounds on the specific search list (Table II) found in these concentrates. Furthermore, in five of the six identifications of fluoranthene, detection would not have been possible without the use of specific search data processing. The five PAH compounds of Table II that were not detected by the specific search were chrysene, benzo[a]pyrene, benzo[b,k]fluoranthene, benzo[g,h,i]perylene and indeno-[1,2,3-c,d]pyrene.

PCB compounds (523, 535, 536 and 551, Table V) were only found in 3 of the 16 concentrates. The concentrations of these PCB were so low that they could be found only by searching the GC/MS data files for the indicative ions. In all cases, the concentrations were far too low to reveal an Aroclor profile.

One clearly evident conclusion from Tables IV and V is that, at least with respect to the organic materials which can be analyzed by glass capillary GC/MS, the AWT concentrates are cleaner than the DW concentrates. Comparing only the concentrates consisting of the combination of the six solvent extracts of RO brine (T1C, T4C, M2C, P2C, N2C, O2C and S2C for DW, and R1C, E1C and B2C for AWT) one finds the average number of compounds identified (Table IV) for these seven DW concentrates to be 245 while the average for the three AWT concentrates is only 102. The corresponding average values for the more selective list in Table V are 160 compounds for the 7 DW concentrates and 81 compounds for the 3 AWT concentrates. In addition, there is a definite trend in Table V that compounds found in both DW and AWT concentrates are generally found in higher concentrations in the DW concentrates. For comparison, these values for the quasiblank concentrate, V1C, are 59 and 28 compounds listed in Tables IV and V, respectively, with nearly all RS values substantially lower than for the AWT and DW concentrates. While it may be somewhat surprising that AWT effluent RO concentrates may contain less GC/MS-analyzable organic material than finished DW RO concentrates, the difference is undoubtedly due to the use of GAC contact or RO (see Table III) in the final processing of the AWT plants. These final treatment steps were not used for the production of any of the finished DW samples from which concentrates were produced. Note, however, that the average TOC value for the posttreatment AWT waters was about a factor of three higher than the corresponding average TOC value for the posttreatment

DW waters, indicating that organic material not removed by GAC treatment is also not well recovered from the RO brine.

## SUMMARY AND CONCLUSIONS

The compound identification results presented here represent detailed analysis of the organic materials which can be isolated from extremely large volumes (1500-15,000 L) of relatively clean water from typical DW and AWT sources. These concentrates were extremely complex mixtures, and even the best possible glass capillary chromatography would be insufficient to separate all of the compounds for mass spectrometric identification. For this reason, partitioning of the concentrate into five chemical classes before analysis provided the critical step in allowing identification of some of the more important toxic organic species in the aromatic and medium-polarity groups. These two sample fractions usually represented only 0.5-5% of the original concentrate. Coleman et al. [11], using a partitioning scheme nearly identical to that used here, have reported the necessity of concentrate partitioning for the detection of a wide range of aromatic (including PCB) species at very low levels in RO concentrates of the same type reported here. Generally, only about 5-30% (by weight) of the organic material present in these concentrates, is accessible to identification using the glass capillary GC/MS analytical method employed. The balance of the material was humic in nature and, thus had to be removed before GC/MS analysis. In addition, removal for separate analysis of the extremely complex acid fraction was a significant aid in the analysis of the remaining neutral compounds. Derivitization of the acid fraction with diazomethane before GC/MS analysis was also an important step for optimal identification of the acidic compounds present.

The compound identification results included in this report cover 16 of the concentrates analyzed under the cited contract and comprise about one-third of the total effort. Through the entire research program, over 2100 compounds have been entered in the computer-managed database, and data for identification confirmation on more than 1100 reference compounds have been acquired. Analysis of the 16 representative concentrates covered in this report resulted in the identification of 1162 compounds, of which 578 were selected for presentation in Table V. Of these 578 identified compounds, 340 (60%) were confirmed identifications.

In general, the compounds found in highest concentrations were fatty acids and related compounds, phthalates, and polyglycol compounds. A wide variety of compounds probably originating from industrial sources have been identified at lower concentrations (generally in the middle to low part-per-trillion range). Compounds of relatively high interest, such as PCB, PAH, pesticides and herbicides were generally absent or present at quite low levels (often the sub-part-per-trillion range). The only distant exceptions to this observation were atrazine and the phenoxy herbicides 2,4,5-T and 2,4-D.

Comparison of some of the results presented suggests that the RO concentration method may not be effective for the recovery of relatively small, nonpolar organic species (i.e., substituted benzene compounds). On the other hand, one goal of this research program was to test the concentrated organic material for biological activity, and RO remains a superior method for the isolation of humic material which is the major organic constituent of surface waters. In addition, RO concentration does not present the artifact generation problem which can often be severe with the XAD-2 adsorption/solvent elution method.

In two cases, concentrate analysis results are related to previously reported work: (1) the concentrates T4C and T4X (Table III) were produced from the same water as that on which Melton et al. [3] have presented purge-and-trap, CLSA, direct XAD-2 adsorption/elution and LLE glass capillary GC/MS analysis. They designate this water as "granular activated carbon contactor A feed water." (2) The concentrate R1C (Table III) was produced from the Orange County, CA, AWT plant Water Factory 21. This production-scale AWT plant has received significant attention in the literature from McCarty et al. [4].

The analytical results presented here and elsewhere [3] indicate that many consumers of DW are exposed through this medium to a variety of toxic organic materials in the part-per-billion to sub-part-per-trillion concentration range. Certainly, these same individuals may have considerably higher exposures to these and similar toxic or carcinogenic organic compounds through the media of foods, beverages, tobacco smoke, drugs, cosmetics or other routes. Nevertheless, it is noteworthy that, for most DW consumers, exposure to the organic materials identified in this work and elsewhere [3] is not really optional. Thus, continued investigation of the nature and biological activity of the organic materials in water destined for human consumption can be expected to remain a primary focus of the Health Effects Research Laboratory of the U.S. Environmental Protection Agency.

## ACKNOWLEDGMENTS

This report was supported by the U.S. Environmental Protection Agency, Health Effects Research Laboratory, Cincinnati, Ohio (Robert G. Melton, Project Officer), under Contract No. 68-03-2548.

AWT and DW concentrates were produced under EPA Contracts 68-02-2090 and 63-03-2367, respectively, by Gulf South Research Institute.

The high quality achieved in this work could not have been possible without the excellent assistance of the following Battelle Columbus Laboratories staff: Daniel G. Aichele, Denise A. Contos, Vanessa R. Goff, Timothy L. Hayes, Roderick G. Warren and Susan C. Watson.

## DISCLAIMER

The use of a specific manufacturer's name in this report is for informational purposes only and does not imply endorsement by the U.S. Environmental Protection Agency.

## REFERENCES

1. Kopfler, F. C., W. E. Coleman, R. G. Melton, R. G. Tardiff, S. C. Lynch and J. K. Smith. "Extraction and Identification of Organic Micropollutants: Reverse Osmosis Method," *Ann. N.Y. Acad. Sci.* 298:20 (1977).
2. Smith, J. K., A. J. Englande, M. M. McKown and S. C. Lynch. "Characterization of Reusable Municipal Wastewater Effluents and Concentration of Organic Constituents," U.S. EPA Technology Series No. 600/2-78-016 (1978).
3. Melton, R. G., W. Coleman, R. W. Slater, F. C. Kopfler, W. K. Allen, T. A. Aurand, D. E. Mitchell, S. J. Voto, S. V. Lucas and S. C. Watson. "Comparison of Grob Closed-Loop Stripping Analysis with Other Trace Organic Methods," Chapter 36, this volume.
4. McCarty, P. L., D. Argo and M. Reinhard. "Operational Experiences with Activated Carbon Adsorbers at Water Factory 21," *J. Am. Water Works Assoc.* 71:683 (1979).
5. Coleman, W. E., R. D. Lingg, R. G. Melton and F. C, Kopfler. "The Occurrence of Volatile Organics in Five Drinking Water Supplies Using Gas Chromatography/Mass Spectrometry," in *Identification and Analysis of Organic Pollutants in Water*, L. H. Keith, Ed. (Ann Arbor, MI: Ann Arbor Science Publishers, Inc., 1976), p. 305.
6. Lin, D. C. K., R. L. Foltz, S. V. Lucas, B. A. Peterson, L. E. Slivon and R. G. Melton. "Glass Capillary Gas Chromatographic-Mass Spectrometric Analysis of Organics in Drinking Water Concentrates and Advanced Waste Treatment Concentrates, II," ASTM Special Technical Publication 686 (1979), pp. 68-84.
7. Concentrate Analysis Reports for EPA Contract No. 68-03-2548, U.S. EPA, Health Effects Research Laboratory, Cincinnati, OH.
8. Cabasso, I., C. S. Eyer, E. Klein and J. K. Smith. "Evaluation of Semipermeable Membranes for Concentration of Organic Contaminants in Drinking Water," U.S. EPA Report 670/1-75-0001, NTIS Accession PB-243 245/AS (1975).
9. Fang, H., and E. Chain. "Reverse Osmosis Separation of Polar Organic Compounds in Aqueous Solution," *Environ. Sci. Technol.* 10:364 (1976).
10. Lucas, J. B., Health Effects Research Laboratory, U.S. EPA, Cincinnati, OH. Personal communication.
11. Coleman, W. E., R. G. Melton, F. C. Kopfler, K. A. Barone, T. A. Aurand and M. G. Jellison. "Identification of Organic Compounds in a Mutagenic Extract of a Surface Drinking Water by a Computerized Gas Chromatography/Mass Spectrometry System (GC/MS/COMP)," *Environ. Sci. Technol.* 14:576 (1980).

# CHAPTER 47

# APPLICATION OF ORGANIC ANALYSIS FOR EVALUATION OF GRANULAR ACTIVATED CARBON PERFORMANCE IN DRINKING WATER TREATMENT

**J. DeMarco and A. A. Stevens**

U.S. Environmental Protection Agency
Cincinnati, Ohio

**D. J. Hartman**

Cincinnati Water Works
Cincinnati, Ohio

A growing number of organic pollutants in drinking water supplies are being identified as potentially harmful to health, and some requirement for removing these substances from drinking water appears to be likely [1,2]. Granular activated carbon (GAC) treatment for removal of organic pollutants from drinking water has received widespread attention because of the ability of GAC to remove a broad spectrum of organic substances. Thus, a GAC adsorption unit process seems especially applicable for use at locations where drinking water sources are subject to a wide variety and varying load of pollutants. The Cincinnati Water Works (CWW) site was considered to represent a type of system that might require GAC treatment. The water intake is located on the Ohio River below six major tributaries, one of which is the Kanawha River. Numerous large chemical plants that contribute to the vulnerability of the source water to contamination are located in the Kanawha River valley. As additional evidence of this vulnerability, 39 organic spills have been recorded by the utility laboratory during the past three years. Accordingly, the Cincinnati Water Works and the U.S. Environmental Protection Agency (EPA) are cooperating in a joint research project to assess the performance of a large-scale GAC adsorption system.

The principal research objectives of interest to utility personnel are the determination of the ability of GAC to improve their water quality, the best type of carbon to use, the most effective way to use the carbon and, always, the total cost to achieve the improved water quality. No single study at one location can provide a definitive answer that will apply to all locations because results from the use of GAC are highly site-specific. However, at any location the answer to the question of improved quality and most effective use must involve the application of chemical analyses of the water treated. Not all of the organic analyses used in assessing carbon performance described in this chapter are for use in normal utility assessments. The wide variety of analyses used reflects the complexity of a research effort to define the usefulness of GAC for water quality improvement with respect to measurable organic substances. Rather than simply looking for the small number of presently regulated contaminants included in the Interim Primary Drinking Water Regulations [3,4] or for the popular list of priority pollutants [5], an attempt has been made to maximize use of currently available organic analytical techniques. For the most part these techniques are nonstandard, in that they do not appear in standard methods texts nor are they always standard EPA analytical methods. This research approach provides broader assessment of the usefulness of carbon for water quality improvement for organic contaminants. Specifically, the purpose of performing the broader scope of analyses is threefold: (1) a database of the occurrence of numerous specific organic compounds is provided that is useful in defining whether a problem substance occurs with a sufficient degree of frequency and concentration to merit concern; (2) the degree of removal of many specific compounds achieved by a given unit process in an ambient environment is determined; and (3) the potential of using some easily measured general organic parameters as operating surrogate tools in lieu of the more specific and often more costly specific tests can be evaluated.

The research project includes the study of GAC systems in two modes. In the first mode the systems are acting as filter/adsorbers (sand replacement systems). These GAC beds receive coagulated and settled water, and the activated carbon acts both as a filter for carryover solids from the preceding processes and as an adsorbent material for dissolved organics. The second mode of GAC use under investigation employs carbon for the purpose of adsorption-only by using conventional sand filters prior to the adsorption process. The latter carbon mode is referred to as postfiltration adsorption, and the carbon units are referred to as adsorbers or contactors.

## PARAMETERS AND METHODS OF MEASUREMENT

### Total Organic Carbon (TOC)

Carbon is the necessary basic building block of all organic compounds. TOC is, therefore, the most basic measure of the presence or absence of organic material in a water sample. However, measures of the amount of organic material can be ambiguous because organic carbon makes up different fractions of the total weight of different compounds. Nevertheless, this measurement was included in this study as the most comprehensive available index of otherwise undefined organic material present.

The TOC sample was collected in a 25-mL screw-cap vial. An 8-mL aliquot of the sample is analyzed on a Dohrmann Envirotech DC-54 Ultra Low Level TOC Analyzer System. This system uses the principle of ultraviolet (UV) light-catalyzed acid persulfate oxidation to form $CO_2$ that is converted to methane by high-temperature catalytic reduction. The methane thus formed is measured by flame ionization detection (FID) and is proportional in amount to the organic carbon concentration in the original sample.

### Purgeable Halogenated Organics Methods

Three water samples were collected that were ultimately analyzed for halogenated organics. The purpose of these analyses was to measure:

1. Instantaneous trihalomethanes (THM), which is the THM concentration at the time of sampling [6]. This sample is analyzed for the four THM and, routinely at CWW, for eight other purgeable halogenated compounds: methylene chloride, carbon tetrachloride, 1,2-dichloroethane, trichloroethylene, tetrachloroethylene, 1,1,1-trichloroethane, chlorobenzene and *o*-dichlorobenzene.
2. Simulated distribution THM, which is an estimate of the THM concentration found in the distribution system and is one variation of the terminal THM described by Stevens and Symons [6]. For the simulated distribution determination, the sample was dosed with excess free chlorine and stored under controlled conditions. Chlorine dose, pH, temperature and storage or reaction time were controlled to match those present in the actual distributed water at the time of sampling. After the specified storage time, the THM concentration was measured and represented the simulated distribution THM. This measurement assisted in evaluating the effect of the experimental treatment on the actual quality of distributed product water.
3. Trihalomethane formation potential (THM FP), which is a practical measure of the precursor concentration, generally regarded as aquatic

humic materials, present at the time of sampling [6]. For this determination, the instantaneous THM sample concentration was subtracted from a terminal THM sample concentration. The terminal THM sample was dosed with excess free chlorine and treated similarly to the simulated distribution sample, except that the storage conditions were constant and represented the water conditions that maximize THM formation experienced in the Cincinnati distribution system.

## Procedures

The instantaneous THM sample was collected in the presence of sodium sulfite to chemically reduce chlorine residual to arrest the formation of THM after sample collection.

The simulated distribution THM procedure involved placing the sample in a 325-mL glass-stoppered amber bottle containing pH 8.2 boric acid–borax buffer and sufficient calcium hypochlorite solution to result in an initial free chlorine residual of 2.5 mg/L. The bottles were then stored in an insulated water bath for three days. Finished water from the water treatment plant continuously passed through the water bath to maintain the temperature of the samples at that of the plant finished water. After three days the THM reaction was stopped by collecting a sample in a muffled 40-mL screw-cap vial containing sodium sulfite. The sample was then analyzed to determine the simulated distribution THM concentrations.

The THM FP procedure involved placing a terminal THM sample in a 325-mL glass-stoppered amber bottle containing pH 9.5 borax-NaOH buffer and sufficient calcium hypochlorite solution to result in a free chlorine residual of 15 mg/L. The bottles were then stored at 29.4°C for 7 days, after which the THM reaction was stopped by collecting a sample in a 40-mL vial containing sodium sulfite. The THM FP was determined by subtracting the instantaneous THM concentration, determined earlier, from the THM concentration reached after storage in the above sample.

Purgeable organohalides were measured by the approved EPA purge-and-trap THM analytical procedure [7]. In summary, THM and similarly volatile compounds were stripped from a 5-mL water sample with a stream of helium. The organics were collected on a trap of Tenax GC, then desorbed by heating, and were collected on a cool gas chromatographic column. This process was facilitated by the use of a Tekmar LSC-1 liquid sample concentrator. The chromatographic column was a packed 0.4% Carbowax 1500 on Carbopack-A (80/100 mesh) (6 ft x 2 mm i.d. glass). The chromatogram was then developed by temperature programming, and a Hall model 700 electrolytic conductivity detector (Tracor) was used in the halogen-specific mode for detection and measurement.

## Organic Halogen (OX)

Organohalides represent a group of compounds of interest in the field of drinking water treatment research because many members of the group are suspected of causing adverse health effects [1,2,8]. Analysis of organohalides as a group parameter may be desirable because experience has shown that conventional purge-and-trap and solvent extraction based gas chromatographic techniques fail to include high-molecular-weight and very polar by-products of halogen disinfection [9-14]. Since organic halogen is a measure of all halogen-containing organic species, this test may be an important measure of the adequacy of drinking water treatment efforts to reduce the concentrations of potentially harmful disinfectant by-products in the consumers' water. The OX measurement does include THM and significant halogen-containing industrial or agricultural wastes that can be measured by gas chromatographic techniques, and interpretations of OX removal data are often preceded by a correction for these known factors.

The OX methods used in this study were for carbon-adsorbable organohalides described in detail by Dressman et al. [15]. In general, the organic material was adsorbed nearly quantitatively on 40 mg of activated carbon from a 100-mL sample to which $HNO_3$ had been added (to pH 2) to improve adsorption of organics. Sodium sulfite had also been added to reduce chemically the chlorine residual. Adsorption was facilitated by use of a minicolumn assembly and a commercially available (Dohrmann AD-2) adsorption module. Following the adsorption step, the activated carbon was washed with a nitrate solution to remove interference of chloride. Halide ions were formed by combustion of the sample in a controlled atmosphere and were measured by microcoulometric titration. This process was accomplished with either a MCTS-20 or MC-1 system (Dohrmann Envirotech). Results are expressed in $\mu g/L$ as chloride.

## Capillary Column GC/FID Profiles

Response profiles from capillary column gas chromatographic analysis of influent and effluent samples from operating GAC beds have been used to qualitatively judge adsorber performance.

The ultimate goal of any gas chromatographic technique is to obtain sufficient separation of individual organic compounds to make possible their identification and individual quantification. Capillary gas chromatographic columns used in the proper way together with mass spectrometric identification and quantification is bringing that goal closer. These techniques, coupled with special sample pretreatment, have been used in Switzerland [16],

Federal Republic of Germany [17] and the Netherlands [18] to give a quantitative assessment of relative water quality through various stages of treatment. In these cases, chromatographic peaks were identified, and the individual compounds were quantified and tabulated. The same techniques were also used in this study. Examination of the chromatograms themselves, however, leads to the conclusion that, even without identification of individual compounds, valid "impressions" of relative water quality can be perceived. Variations of this approach are also given with varying methods of interpretation in the literature [19,20]. Certain precautions in interpretation, as noted below, must be observed.

First, the stability of the detector used and chromatographic retention times must be assured or a suitable internal standard must be used to compensate for variations in responses from one run to the next.

Secondly, chromatograms should not be compared rigorously, peak by peak when identifications have not been made. Coeluting compounds would not be recognized as such, and because detector responses to different compounds differ, misleading "percent removals" could be calculated.

Thirdly, "total chromatographic areas" generally should not be compared numerically in a rigorous manner because of the significant natural variation of baseline drift, mostly caused by chromatographic column bleed or dominance of a few compounds present at higher concentration that would seriously affect the resulting interpretation. Furthermore, total areas subvert the purpose of chromatographic separation.

The important characteristics of a chromatogram are: peak resolution, sensitivity of individual compound detection and reliability of quantitative estimation. Without compound identifications, however, perfect quantitative interpretation of a chromatogram is impossible. Nevertheless, with the three precautions listed above in mind, useful impressions of relative water quality can be developed.

In this work organic profiles were generated by a liquid-liquid extraction (LLE) of 1-gal (3.8-L) samples adjusted to pH 2 with hydrochloric acid. Fresh volumes of 250, 100 and 100 mL of dichloromethane ($CH_2Cl_2$) were shaken consecutively with the sample in a 6-L separatory funnel, combined after separation, dried with anhydrous sodium sulfate, and reduced in volume by boiling to 0.5 mL in a Kuderna-Danish (KD) apparatus. A 1-$\mu$g quantity of anthracene (corresponding to 0.25 $\mu$g/L in the original sample) was added to the final volume after concentration as an internal standard. Blanks corresponding to each batch of redistilled solvent with internal standard added were analyzed. Analysis was by a microprocessor-controlled gas chromatograph (Hewlett-Packard model 5840) equipped with a splitless capillary injection system, a 30-m SP-2100 wall-coated open tubular (WCOT) glass capillary column and a FID.

## Specific Organics Other Than Purgeable Organohalides

Sample preparation by a variation of the closed-loop stripping analysis (CLSA) first described by Grob [10,16] followed by glass capillary gas chromatography/mass spectrometry (GC/MS) was considered to be a sensitive and convenient approach to the measurement of individual organic compounds in the intermediate volatility range. The exact method used in this study is described in detail by Coleman et al. [21]. One-gallon samples were purged at 30°C with their own headspace gas continuously recirculating for two hours. By use of an online activated carbon filter, the purged organics were adsorbed from the gas phase. Sorbed organics were later desorbed from the carbon with microliter quantities of carbon disulfide. Aliquots were then analyzed by GC/MS with results interpreted relative to the appropriate added internal standards.

## Sampling for All Organic Analyses

All glass sample bottles for this study were washed, rinsed, covered with aluminum foil and heated to 400°C for at least 0.5 hr in a muffle furnace. Vials and bottles with polytetrachloroethylene (TFE)-faced silicone rubber seals and plastic screw caps were used for the THM, OX and TOC samples, and 1-gal bottles employing plastic screw tops with TFE liners were used for the extractable organic GC/FID profile and Grob CLSA samples. Samples were returned to the laboratory and stored at 4°C until analysis or other treatment prior to analysis. At the time of the sampling, sufficient $Na_2SO_3$ was added to chemically reduce chlorine residual to prevent further formation of disinfectant by-products (except THM FP and simulated distribution THM samples). Mercuric chloride was added at approximately 10 mg/L in dry powder form to the 1-gal Grob CLSA and GC/FID profile samples to retard microbiological activity.
Special reagents included:

1. Chlorine dosing solution: 0.5 g of calcium hypochlorite is diluted to 500 mL with helium-stripped double distilled water. This results in approximately 500 mg/L of free available chlorine.
2. Chlorine neutralizing solution, 0.2 *N* sodium sulfite: 2.5 g $Na_2SO_3$ is diluted to 100 mL with helium-stripped double-distilled water; 0.5 mL is used for one 40-mL sample vial.
3. Boric acid/borax, pH 8.2 buffer: 3.1 g boric acid ($H_3BO_3$) and 0.7 g borax ($Na_2B_4O_7 \cdot 10H_2O$) are diluted to 1 L with helium-stripped double-distilled water; 15 mL of this buffer is added to each 325-mL sample bottle in the simulated distribution THM test.

4. Borax, pH 9.5 buffer: 4.8 g borax ($Na_2B_4O_7.10\ H_2O$) and 0.7 g NaOH are diluted to 1 L with helium-stripped double distilled water; 15 mL of this buffer is added to each 325-mL sample bottle in the THM FP test.

## FACILITIES DESCRIPTION

The study was conducted at the California Plant of the CWW, Cincinnati, OH. The normal water treatment process at the 235-mgd California plant consists of four sequential unit processes that include long-term presettling followed by coagulation, sedimentation and sand filtration (Figure 1). Facilities are available to apply chlorine, aluminum sulfate and powdered activated carbon to the raw water prior to presettling. The plant also has facilities for feeding chlorine, ferric sulfate, lime, soda ash and powdered activated carbon between the presettling basins and flocculation basins. Also, chlorine can be added between the sedimentation basin and sand filters as well as between the sand filters and clear wells.

Figure 2 is a schematic of the carbon units under investigation at CWW. Two separate 5-mgd sand filters have been converted to carbon filter/adsorbers (sand replacement systems). This conversion was effected by removing the 30 in. of filter sand and replacing it with 30 in. of 12 x 40 mesh carbon. A system of four 1-mgd carbon adsorbers (postfiltration adsorption system) that treat sand-filtered water has been also constructed. These two operating modes of carbon application provide a means of comparing the operation of a sand replacement carbon system (filter/adsorber) with that of a postfiltration carbon system (adsorbers). Table I lists some of the pertinent physical data that describe the carbon system now operating. The 1-mgd postfiltration contactors also have sample ports to allow comparison with the sand replacement systems at the same empty bed contact times (EBCT). Pilot carbon columns are also being operated parallel to the full-scale systems to assess their ability to simulate the large systems.

Figure 3 is a schematic diagram of the 550-lb/hr fluid-bed carbon reactivation system being used onsite. To date about 300 tons of carbon have been reactivated at the utility. Carbon is moved from the carbon adsorption systems to a spent carbon storage tank. The carbon is then dewatered to about 50% moisture and fed to the fluid-bed reactivation furnace. The carbon is reactivated to near virgin properties under a controlled combustion process, water-quenched to cool it, then placed in a regenerated carbon storage tank for subsequent transfer to the appropriate carbon adsorption system.

The data contained in this chapter will be confined primarily to the organic analyses performed on the postfiltration carbon contactors.

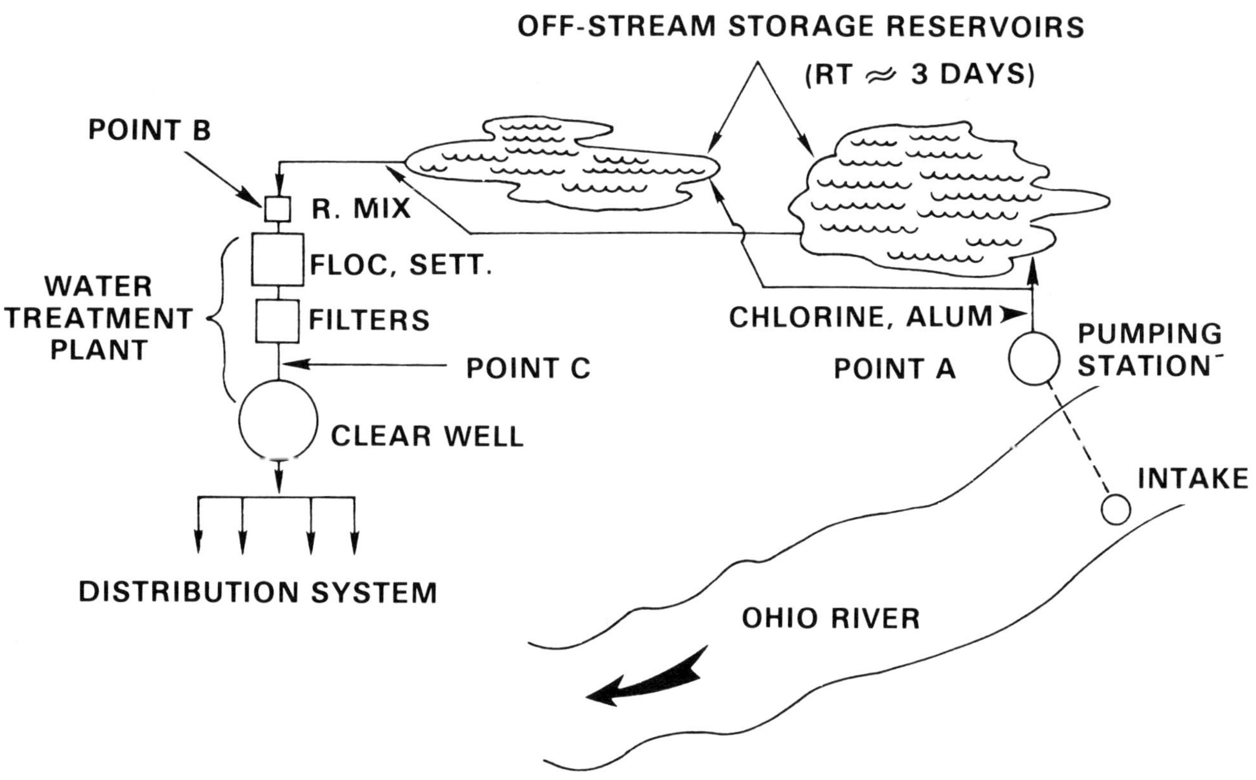


Figure 1. Schematic of Cincinnati waterworks.

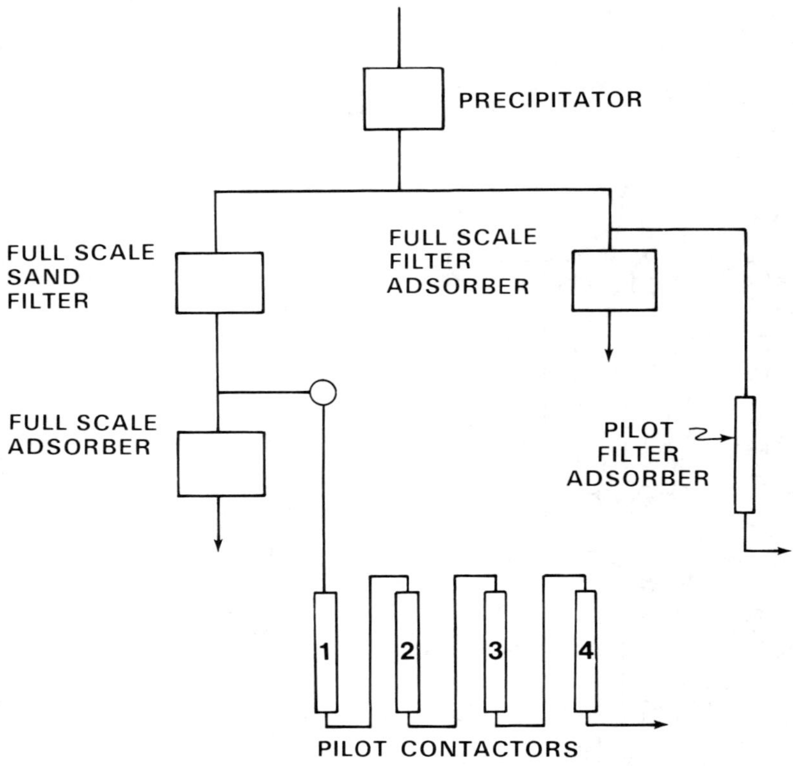


**Figure 2.** Schematic of carbon units.

**Table I. Physical Data for Carbon Systems**

| Carbon System | Flow | EBCT (min) | Carbon Depth (ft) | Carbon Weight (lb) | Hydraulic Loading (gpm/ft$^2$) |
|---|---|---|---|---|---|
| Postfiltration Contactor | 1 mgd | 16 | 14.8 | 42,500 | 6.94 |
| Sand Replacement | 5 mgd | 7.5 | 2.5 | 96,000 | 2.5 |
| Pilot Postfiltration Contactor | 920 gpd | 16 | 15.1 | 34.8 | 7 |
| Pilot Sand Replacement | 170 gpd | 7.8 | 2.6 | 3.1 | 2.5 |

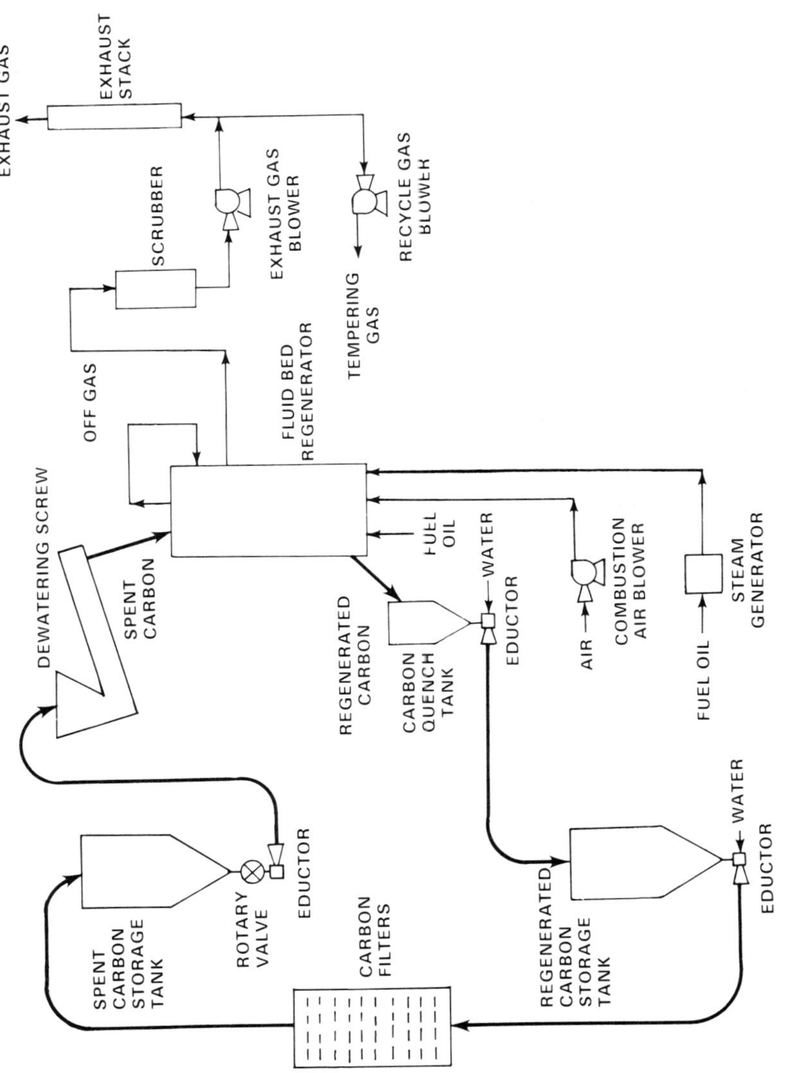


**Figure 3.** Carbon reactivation system.

## APPLICATION OF ANALYTICAL METHODS

How, then, can the various types of organic analyses effectively be used to judge the performance of GAC? Obviously, when a drinking water concentration of a specific organic substance is required to be controlled within given limits, there exists a definitive use of analytical methods to assess performance. However, drinking water regulations currently contain maximum contaminant levels (MCL) for only a few organics [3,4]. These are the pesticides endrin, lindane, methoxychlor, toxaphene, 2,4,-D and 2,4,5-TP; and total THM (chloroform, bromodichloromethane, dibromochloromethane and bromoform).

Although additional MCL are likely to be promulgated, the evaluation of technology should proceed without waiting for these possible future standards. Thus, all of the general and specific organic analyses currently available have some place in assessing treatment performance. The use of a single surrogate analysis to replace complex specific compound determinations will probably never be applicable as a sole measure of water quality, but a continued assessment of the role of surrogate analyses in technology evaluations is needed.

The six methods discussed in this chapter represent a combination of general and specific compound analyses that attempt to show the extent to which GAC reduces the organic content of drinking water. Table II presents an overview of the fraction of each measured parameter that passes into the drinking water after GAC treatment. As expected, the data indicate that a variety of adsorption performances exists, but in general the organic content of the drinking water was reduced.

More definitive evaluation of the data for this general objective and other more specific objectives are possible through interpretation of organic data in the form of simple breakthrough curves. These evaluations will be discussed along with interpretation problems posed by nonanalytical variability in results of analyses when attempting to answer the more specific problems, such as the length of GAC service life while maintaining a specific effluent criteria.

### Total Organic Carbon

Figure 4 is a detailed plot of the TOC data. The generation of such plots allows the evaluation of the ability of the GAC system to remove a variety of nonspecific organic substances present in most drinking water. The 0.2-mg/L concentration of TOC that was present in the initial effluent of the 15-ft deep carbon bed (16-min EBCT) was probably caused by the presence of a fraction

Table II. GAC Bed Performance

| Analytical Technique | Organic Fraction Passing (Ce/Ci) | | | | | |
|---|---|---|---|---|---|---|
| | Startup | 1 mo | 2 mo | 3 mo | 4 mo | 4.5 mo |
| TOC | 0.08 | 0.19 | 0.48 | 0.50 | 0.60 | 0.62 |
| THM FP | 0.02 | 0.13 | 0.24 | 0.42 | 0.52 | 0.46 |
| OX | a | 0.12 | a | a | a | 0.53 |
| Purge-and-Trap | | | | | | |
| Chloroform | 0.00[b] | 0.00 | 0.07 | 0.39 | 0.69 | 1.24 |
| Bromodichloromethane | 0.00 | 0.00 | 0.00 | 0.07 | 0.16 | 0.39 |
| Dibromochloromethane | 0.00 | 0.00 | 0.00 | 0.00 | 0.00 | 0.00 |
| CLSA | | | | | | |
| Carbon Tetrachloride | 0.00 | 0.06 | 0.07 | 0.00 | a | 0.00 |
| 1,2,4-Trimethylbenzene | 0.73 | 0.57 | 0.89 | 0.54 | a | 0.01 |
| Toluene | 0.62 | 0.48 | 1.03 | 0.41 | a | 0.00 |
| Ethylbenzene | 0.75 | 1.0 | 0.78 | 0.38 | a | 0.01 |
| 1,2-Dimethylbenzene | 0.60 | 0.42 | 0.83 | 0.41 | a | 0.01 |
| GC/FID Profiles[c] | | | | | | |

[a]No sample available.

[b]0.00 means effluent concentration was not detectable.

[c]Impression of no major increase in organic fraction passing with time.

of the TOC that is not carbon-adsorbable under the conditions tested. Results of studies of various long carbon contact time systems at other locations have shown similar patterns [22-24]. Although the TOC measurement does not measure specific organic compounds, it does serve to document some advantages and disadvantages of GAC. Note that the carbon achieved 90% removal [(1.6 - 0.2)/1.6] for about six weeks before rapid breakthrough of TOC took place and a steady-state plateau was established below the influent concentration (about 40% removal). These results (steady-state plateau) have supported speculation that biodegradation may also provide a useful asset in drinking water systems either with or without carbon. Such systems are now being investigated.

A negative view of carbon performance might be obtained if consideration is given that the 0.2 mg/L TOC present in the effluent at startup could represent a considerably larger concentration of a nonadsorbable organic substance or family of substances. The magnitude would depend on the carbon content of the compound. For example, a hypothetical vinyl chloride concentration based on 0.2 mg/L of carbon would be 0.53 mg/L of the compound.

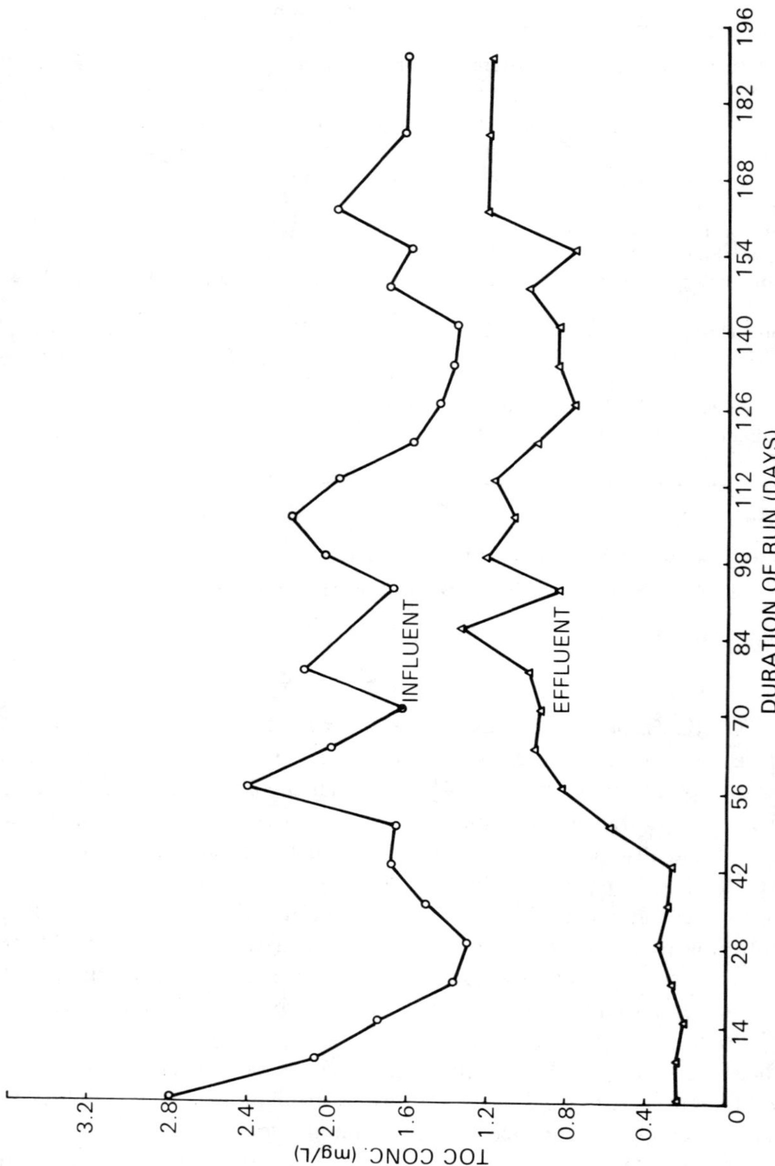


**Figure 4.** Carbon contactor performance for TOC.

Thus, this variability of composition of the mixture of materials measured by the TOC parameter can potentially cause considerable problems in assessing the relative safety and treatability of various waters. No manmade specific organic substances, however, have been reported in either influents or effluents of drinking water systems at concentrations that can account for more than a small fraction of the TOC generally present. Despite this question regarding the exact meaning of the absolute value of TOC measurements, especially in a regulatory role, the parameter can serve as an operating guide at a given location if proper relationships can be established with concentrations of other substances that may otherwise require strict regulation. The basic plot shown can also be used along with other data to calculate organic loadings on the carbon which are used in design considerations.

## Trihalomethanes

The concern regarding organics in drinking water has given rise to a variety of uses of the basic analytical method for measuring THM. The basic concepts and terminology for the three common parameters found to be useful in evaluating technology for THM reduction have been discussed above. Figure 5 shows the results of the simulated distribution system THM measurements for two sampling points in the CWW investigation. The influent concentrations to the carbon contactor represents the THM concentration for normal CWW sand-filtered effluent samples that were stored three days in contact with free chlorine and exposed to ambient distribution temperature and pH conditions as simulated in the bottle storage procedures previously described. The influent line, from day 14 on, indicates that the simulation process predicts that a consumer of finished water three days from the conventional treatment plant would receive a minimum THM conc of 39 μg/L, a maximum THM conc of 86 μg/L and an average THM conc of 61 μg/L.

Although not plotted, the tapwater THM concentration at an actual distribution system sampling point three days water travel time from the plant was a minimum of 34 μg/L, a maximum of 100 μg/L and average of 61 μg/L during the same time period. Repeated results, such as these at Cincinnati, along with those of Wood and DeMarco [22,23] in similar investigations at Miami, FL, have provided a growing confidence in the ability of this simulation procedure to predict actual distribution system concentrations. An important application of this procedure is found in evaluating the remaining data shown by the effluent line in Figure 5. Once the simulation procedure is accepted, the evaluation of the predicted water quality that a consumer would receive if the water was subjected to carbon treatment and then chlorinated and transported three days under ambient pH and temperature

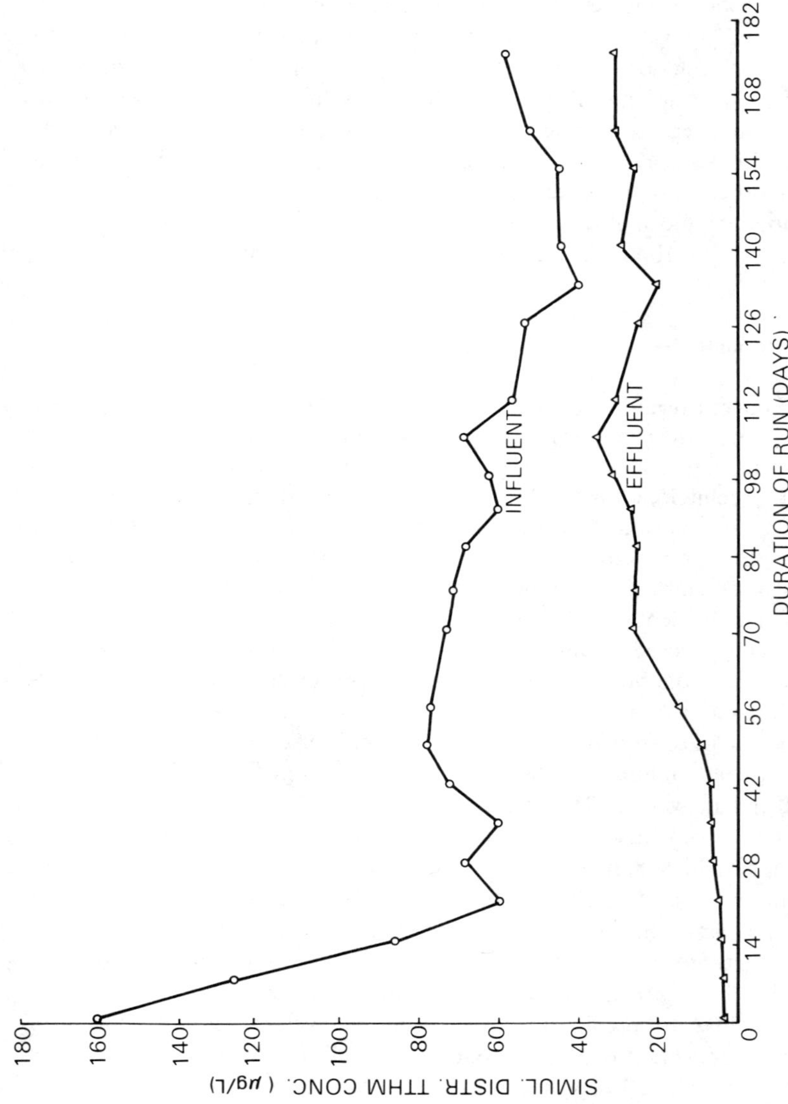


**Figure 5.** Carbon contactor performance for simulated distribution system total THM.

conditions can be compared to conventional treatment. For research, this acceptance is important because the entire plant flow rarely passes through the system being tested, rendering actual distribution system sampling impractical for assessing unit process performance. The influent and effluent plots of Figure 5 are a comparison of what the consumer might expect with conventional treatment (influent line) compared to carbon treatment (effluent line).

A general conclusion that GAC reduces simulated distribution THM can be supported by Figure 5. During the period studied, the use of GAC could be estimated to provide about 20 μg/L of total THM in the consumers' water as contrasted to 67 μg/L provided if GAC is not used. Although the simulated distribution THM concentrations are obviously reduced by GAC, associated costs with the inability to maintain high simulated distribution THM removals for longer than a few months has placed limits on GAC use for maintaining low tapwater THM concentrations in systems using free chlorine residuals.

By use of Figure 5, comparisons of the time that the GAC is most effective for adsorption of the mixture of THM and precursor substances can be made in the same way as previously described for TOC. Thus, the largest reduction in consumer exposure to THM would occur during the first six weeks of carbon use with breakthrough taking place at that time. A plateau exists starting at about ten weeks and continues through the six months shown. About 50% removal is achieved. An explanation of the steady-state plateau below the influent concentration will be provided in the subsequent discussion of THM FP and total THM. The method has application for assessing other treatment changes to reduce THM concentrations as well.

THM FP results for the finished water from the conventional treatment process (influent) and the GAC contactor system (effluent) are shown in Figure 6. The effluent steady-state plateau shows that about 57% of THM FP is removed from about the sixth week to the sixth month of operation. Figure 7 shows that at 140 days the GAC effluent instantaneous THM concentration equaled the influent concentration and that the adsorptive capacity of the GAC had been exhausted under the conditions present for instantaneous THM. Thus, the plateau that exists below the influent concentration of simulated distribution THM in Figure 5 is caused by the removal of THM FP (precursor) that was present in the mixture of THM FP plus instantaneous THM represented by the data in Figure 5.

Figures 8 to 10 show the variation of GAC adsorptive performance for substances within a chemically similar family. Application of the analytical technique at many locations has shown that, despite varying conditions, chloroform is the least well adsorbed of the THM, and that increasing bromine substitution results in increased adsorption. Since there is a variability

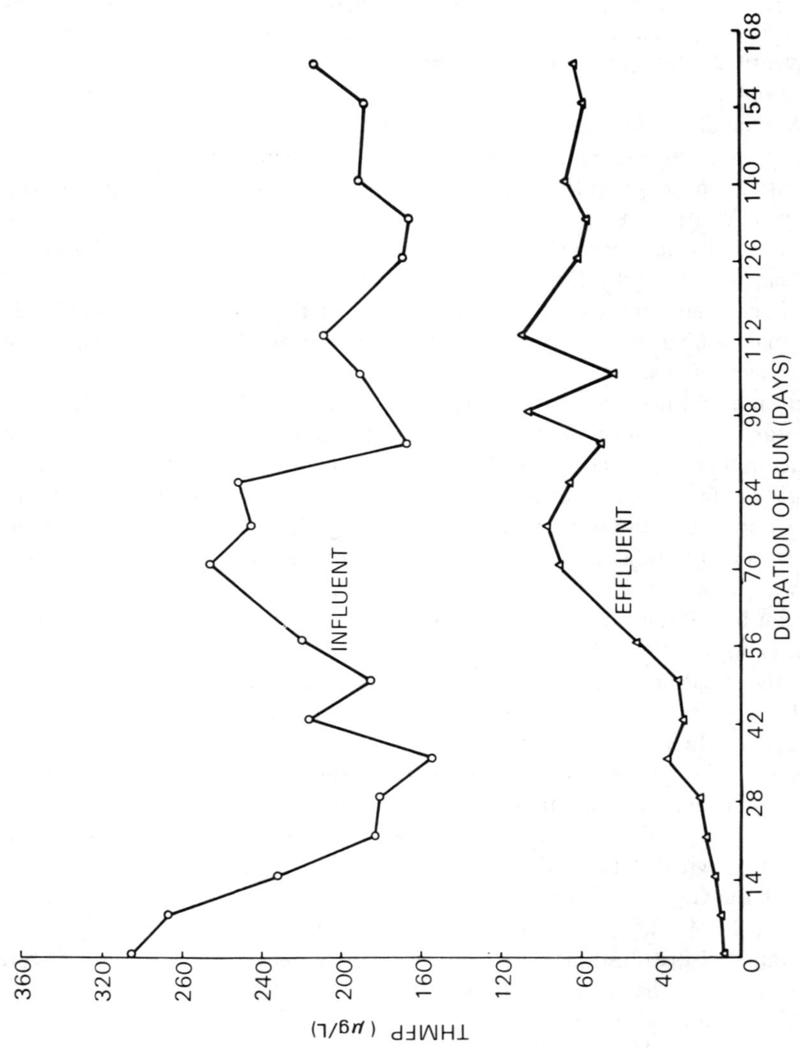


**Figure 6.** Carbon contactor performance for THM FP.

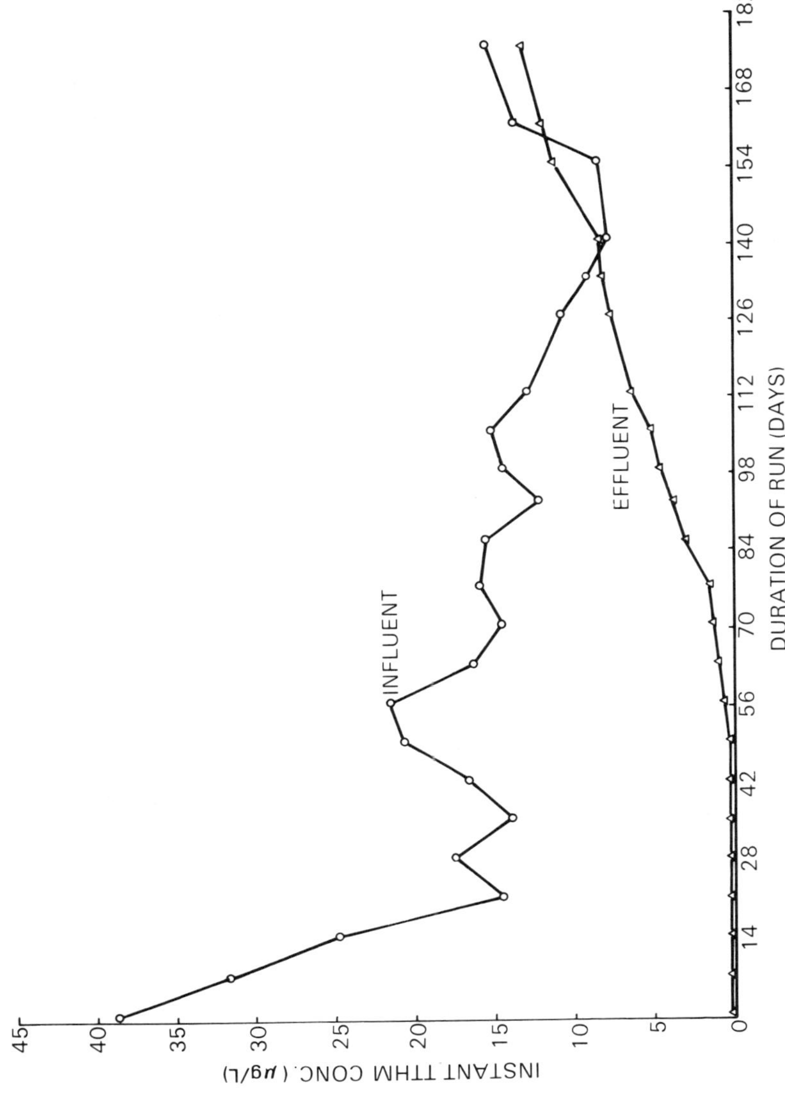


**Figure 7.** Carbon contactor performance for total THM.

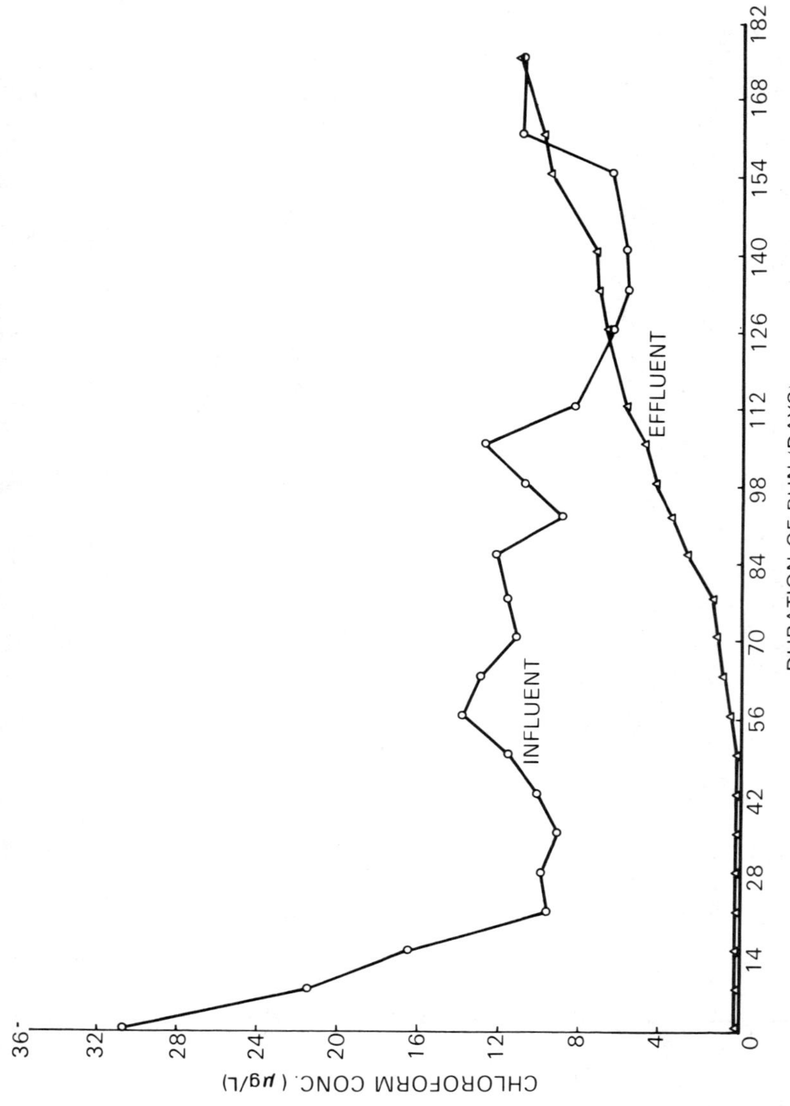


**Figure 8.** Carbon contactor performance for chloroform.

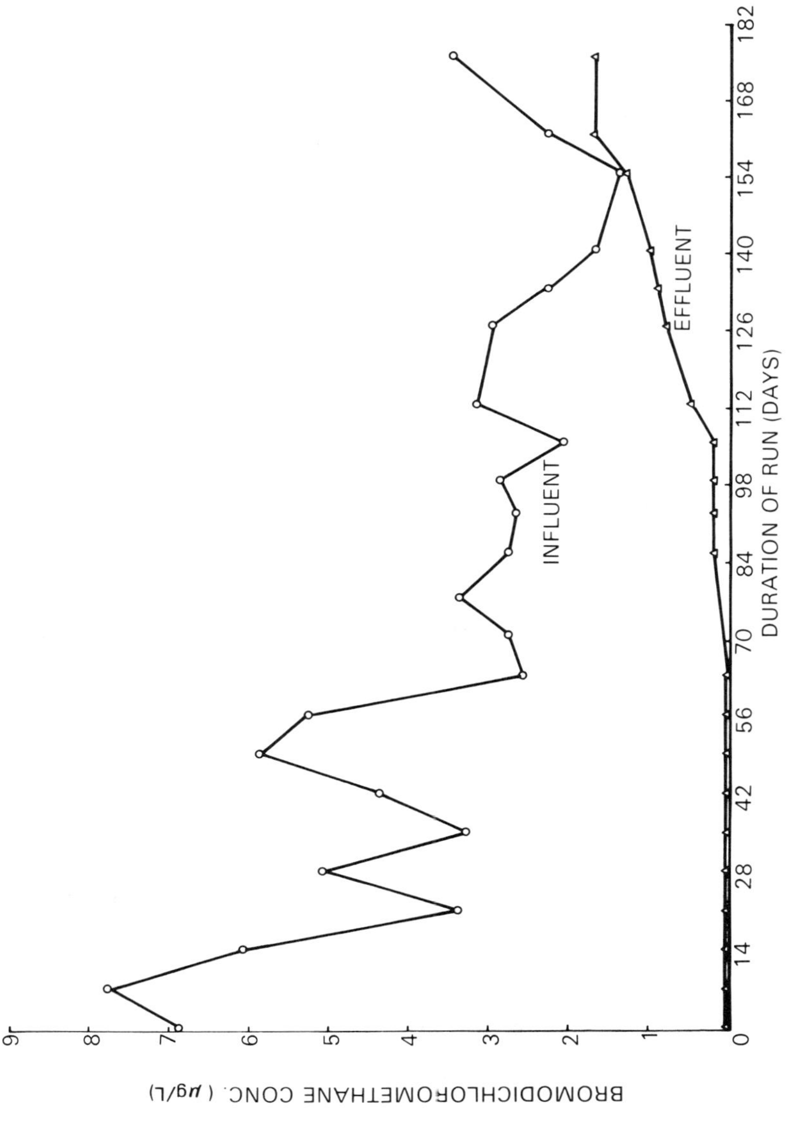

**Figure 9.** Carbon contactor performance for bromodichloromethane.

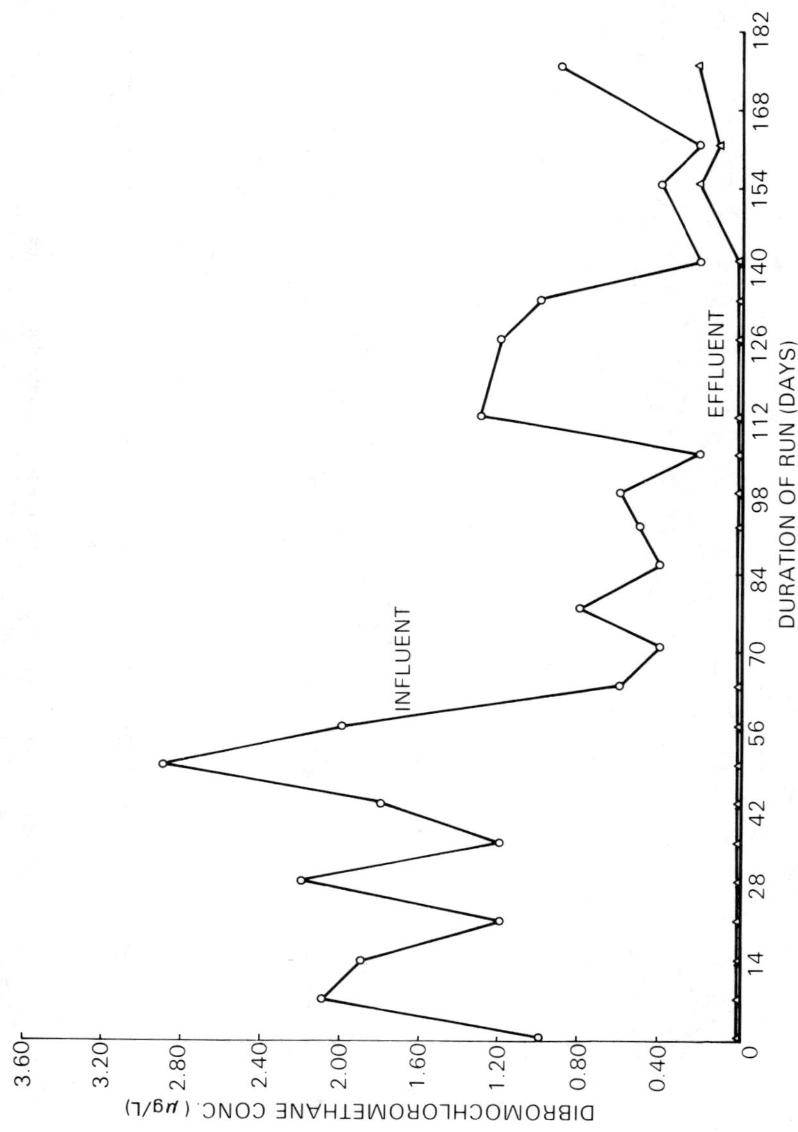


**Figure 10.** Carbon contactor performance for dibromochloromethane.

in the relative adsorbability for these four THM compounds, the design of a GAC adsorption system to consistently reduce an influent total THM concentration to a specified effluent total THM concentration would depend largely on the mixture of these four substances present at a given location and the constant appearance of given concentrations of the mixture of substances as well as many other factors. Note that equilibrium for this group of compounds was generally achieved as mentioned above, supporting the conclusion above that the plateaus observed for simulated distribution THM was a result of continued removal of precursors as measured by THM FP.

Although the various uses of THM analyses have been applied to provide a general description of the ability of GAC to successfully remove THM FP, the most common mixtures of THM, or mixtures of THM FP and THM, other considerations may lead to the conclusion that carbon adsorption is not necessarily a first-priority alternative to achieve this end.

## Organic Halogens

Few organic halogen data have been reported to date. The data in Table II show that during the first 4.5 months about 50% of the organohalide compounds designated by the OX measurement were removed. Note, although the THM substances comprise a portion of the OX concentration, the THM do not generally account for even as much as one-third of the OX measured in this study; the reported concentrations are not corrected for THM contribution to the total. Additional sample results are shown in Table III. The effluent OX concentration of contactor A exceeded the influent concentration at the seventh and eighth month. When the influent concentration increased rapidly at the ninth month, removal was again achieved. Contactor D was started three months after contactor A, and generally a received a higher concentration of OX earlier in its life cycle. The data show that OX are reduced in concentration by GAC, but that variation in effectiveness exists because of the changing influent conditions.

Achieving a specific effluent concentration level would provide different carbon life cycles for the two contactors. For example, if the effluent OX concentration were to be maintained at less than 40 $\mu$g/L, Table III data would indicate that contactor A could be used for about seven months, whereas contactor D would have a useful life of about five months.

Thus, the variation of influent concentrations and possibly chemical characteristics of OX materials to identical contactors was seen to cause significant change in expected adsorber life for a given target effluent. Therefore, careful definition of breakthrough curves under one set of conditions does not necessarily define the performance of even identical adsorber

**Table III. OX Removal by GAC (μg/L as Cl⁻)**

| Time (mo) | Contactor A | | Contactor D | |
|---|---|---|---|---|
| | Influent | Effluent | Influent | Effluent |
| 0 | NA[a] | NA | 51 | 3 |
| 1 | 57 | 7 | 43 | 5 |
| 2 | NA | NA | 46 | 8 |
| 3 | NA | NA | 76 | 10 |
| 4.5 | 43 | 23 | 44 | 16 |
| 5.5 | 46 | 24 | 150 | 44 |
| 6.5 | 76 | 31 | 174 | 77 |
| 7 | 44 | 49 | NA | NA |
| 8 | 27 | 43 | NA | NA |
| 9 | 150 | 89 | NA | NA |
| 10 | 175 | 103 | NA | NA |

[a]NA = no sample available.

systems with the present state of understanding of the meaning of this parameter and its natural controlling factors.

## GC/FID Profiles

Figures 11 to 14 demonstrate the use of GC/FID capillary column profiles to provide impressions of the overall ability of GAC to remove the spectrum of organics represented by the analytic method described. Although the following interpretations of Figures 11 to 14 are considered to be useful indicators of the GAC performance, the reader is reminded of the method

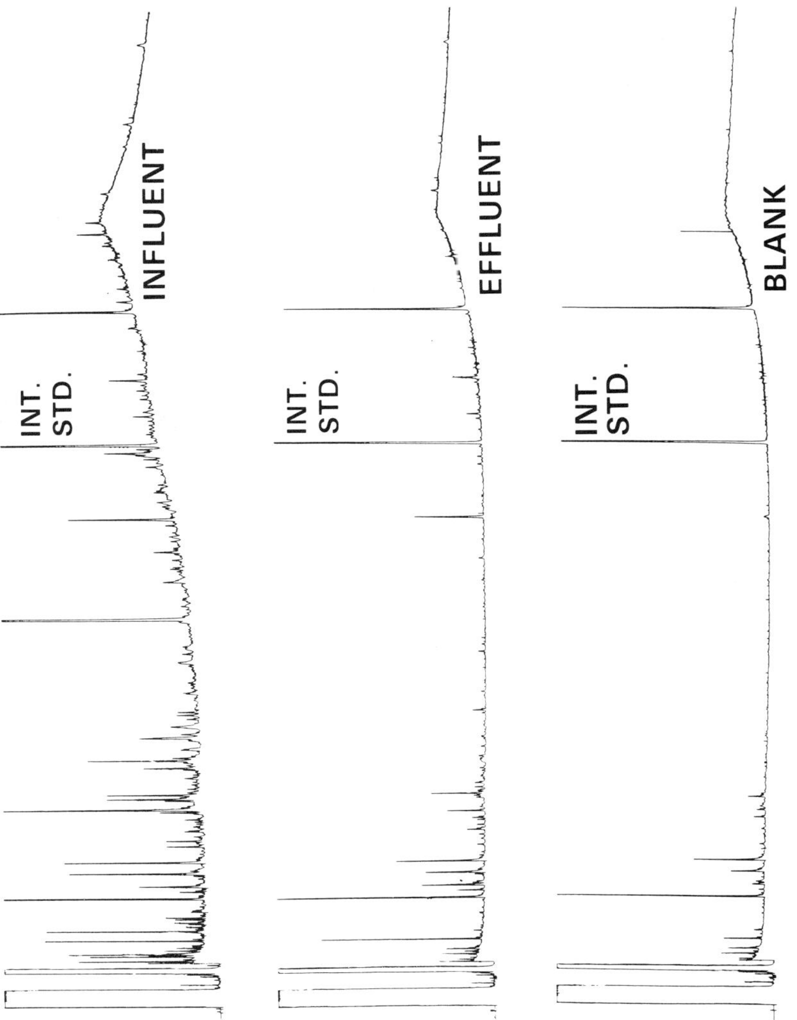


Figure 11. GC/FID capillary column profiles, contactor A—startup.

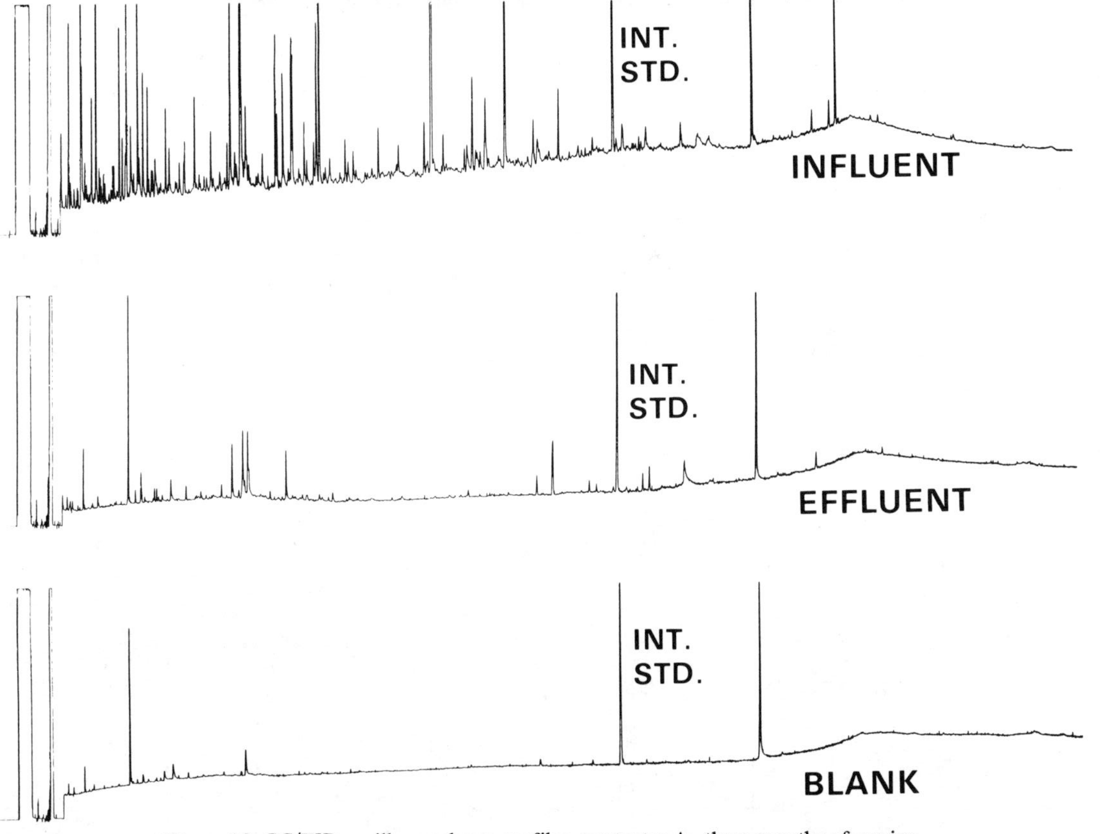


**Figure 12.** GC/FID capillary column profiles, contactor A—three months of service.

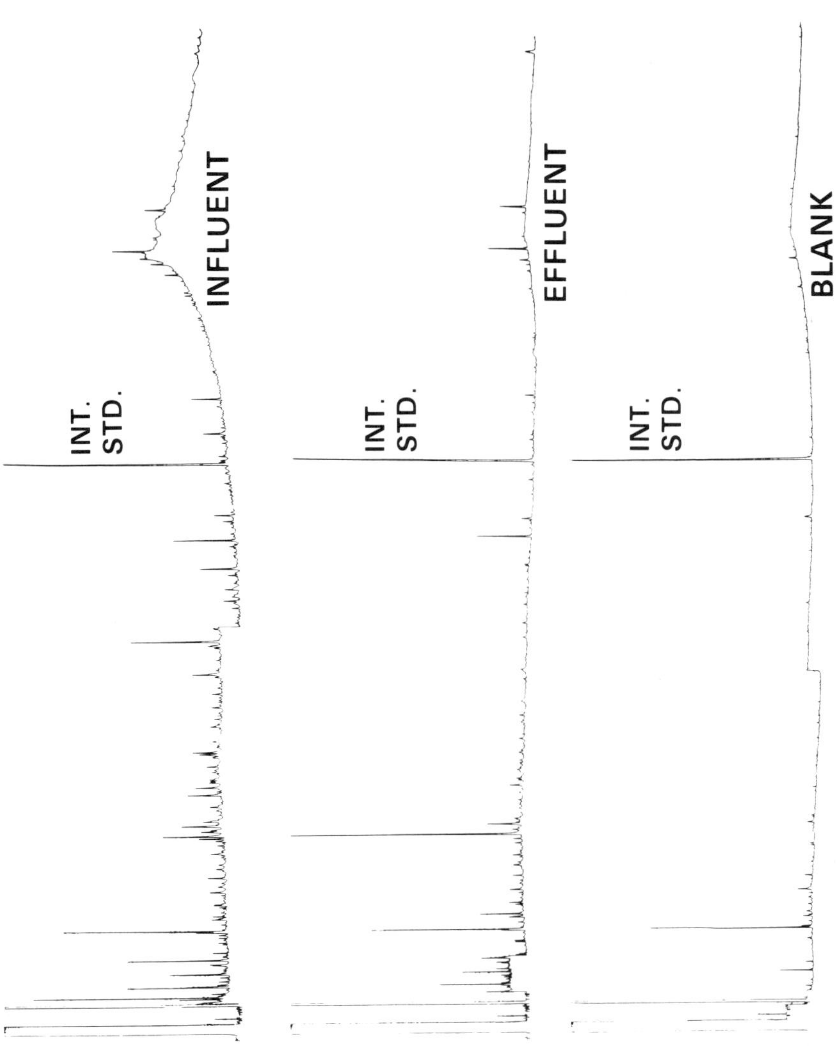


Figure 13. GC/FID capillary column profiles, contactor A—six months of service.

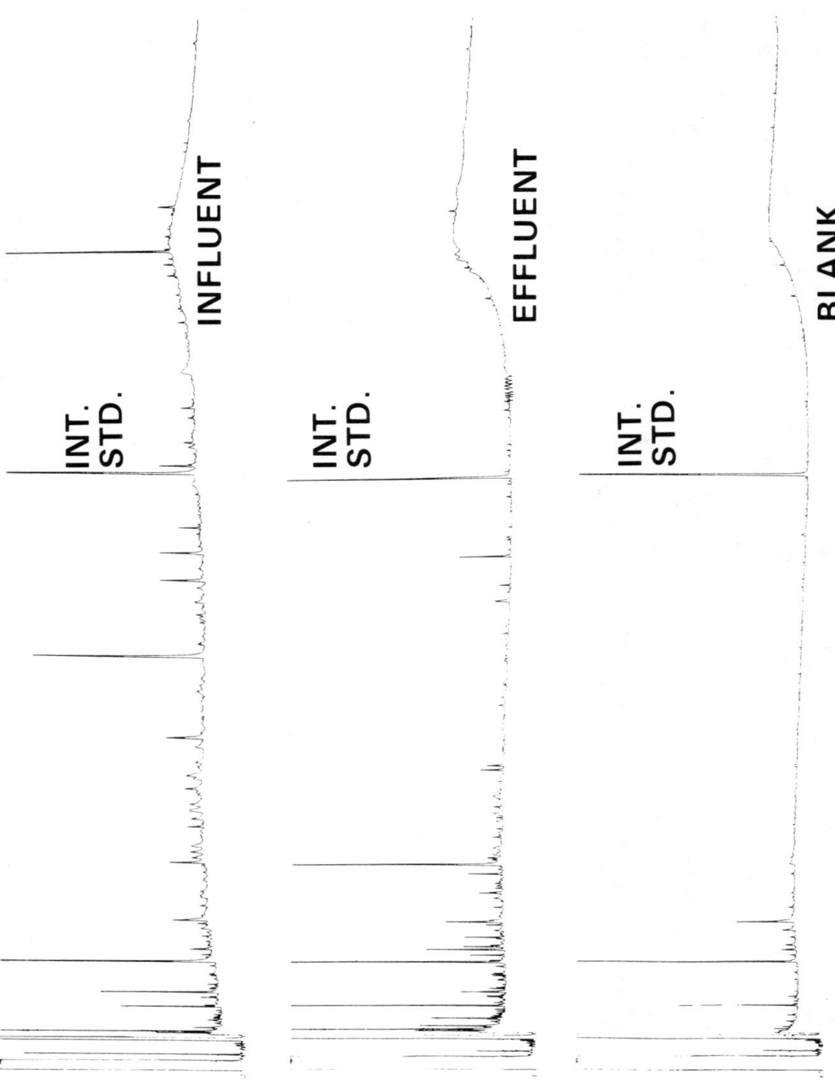


**Figure 14.** GC/FID capillary column profiles, contactor A—seven months of service.

deficiencies and precautions to be taken in interpretation given in the method description above.

Figure 11 presents the impression that the effluent of the GAC contactor at startup has few compounds, and those detected appear to have lower concentrations as seen by a lesser number of peaks and also lesser peak heights for peaks at the same retention times. Figure 11 also shows the general appearance of better GAC removal of organics after the first third of the chromatogram. The substances in the first third apparently have a lesser affinity for adsorption on carbon, more solubility in water or both.

Figure 12, third month of operation, provides the appearance of more substances present in the influent than were present in the influent shown in Figure 11; however, even with the increased appearance of more substances in the influent, the GAC effluent still appears to be of better quality than the influent.

Figure 13 shows an improvement in influent quality during the sixth month of GAC operation compared with the third month (Figure 12). The occurrence of at least two peaks in larger concentrations in the effluent than in the influent possibly indicates a desorption of these substances because of the decrease in the influent concentration.

Figure 14 shows the continued improvement of quality of the influent water to the GAC contactor. A greater number of substances also appear to have increased peak heights in the GAC effluent as compared to the GAC influent. The substances that appear to show a variable GAC performance for removal are still those that elute early on the chromatogram. The impression of desorption of previously adsorbed substances does pose a question about reactivation frequency for the GAC system. One must keep in mind, however, that the chromatograms represent impressions of relative influent/effluent concentration of detected substances and not an absolute indicator of water quality. While they serve to indicate effectiveness of treatment of the incoming water, they cannot alone indicate the requirement for the observed degree of treatment.

## Specific Organic Compounds

Table IV lists 25 of the approximately 160 compounds found by using the CLSA method as described by Coleman et al. [21]. All of the compounds listed were generally present in the low ng/L concentration range, except for the four commonly found THM. Of the other substances, benzene was found in the highest influent concentration with a range of 10-834 ng/L. Toluene, with a concentration range of 7-231 ng/L, and 1,2,4-trimethylbenzene, with a range of 2-127 ng/L, were the only others listed in Table IV that exceeded

Table IV. Example of Organics Found by CLSA Technique

| Compounds | Percent Removal[a] | | | | |
|---|---|---|---|---|---|
| | Startup | 1 mo | 2 mo | 3 mo | 4 mo |
| 1,1,1-Trichloroethane | 100 | 95 | 87 | 94 | 98 |
| Trichloroethene | b | 98 | 98 | 100 | 100 |
| Tetrachloroethene | 100 | 100 | 98 | 100 | 100 |
| Benzene | b | 100 | b | 100 | 100 |
| Carbon Tetrachloride | 100 | 94 | 93 | 100 | 100 |
| Hexachloroethane | 100 | 99 | 99 | 100 | 99 |
| Chloroform | 100 | 98 | 72 | 79 | 70 |
| Bromodichloromethane | 100 | 100 | 93 | 98 | 90 |
| Dibromochloromethane | 100 | 100 | 97 | 100 | 97 |
| Dichloroiodomethane | 100 | 100 | 100 | 100 | 100 |
| Bromoform | 100 | 100 | 100 | 100 | 98 |
| 1,2-Dichloropropane | b | 100 | b | b | b |
| 1,2-Dichlorobenzene | 100 | 100 | 99 | 100 | 99 |
| 1,2,4-Trichlorobenzene | 95 | 96 | 96 | 100 | 98 |
| Toluene | 38 | 52 | -3 | 59 | 100 |
| 1,2-Dimethylbenzene | 40 | 58 | 17 | 59 | 99 |
| Isopropylbenzene | 38 | 50 | 33 | 67 | 99 |
| Propylbenzene | 33 | 36 | 0 | 62 | 97 |
| 1,2,4-Trimethylbenzene | 27 | 43 | 11 | 46 | 99 |
| 1,2,3-Trimethylbenzene | 33 | 46 | 20 | 50 | 99 |
| Heptadecane | 35 | c | -177 | 100 | c |
| Octanal | c | c | - 77 | 39 | c |
| Heptanal | b | c | - 50 | b | c |
| Decanal | -150 | -683 | -177 | -17 | c |
| Nonanal | -350 | -328 | -152 | - 9 | -1271 |

[a]100% removal indicates not detected in effluent.
[b]Not detected in either influent or effluent.
[c]Effluent concentrations quantifiable, but influent concentrations not quantifiable.

100 ng/L. Most of the other substances listed were present in a concentration range of 1-70 ng/L. Also, only five of the substances were present in less than four of the five influent samples. These were heptanal, 1,2-dichloropropane, octanal, heptadecane and benzene.

Over the first four months (October to February 1980) of this study, many compounds continued to be removed at a high efficiency regardless of influent concentration: 1,1,1-trichloroethane (87-100%), benzene (100%), trichloroethene and tetrachloroethene (98-100%), carbon tetrachloride (93-100%), hexachloroethane (98-100%), and chlorobenzene (100%).

The alkyl benzenes varied in percent removal with concentration. For example toluene and 1,2-dimethylbenzene were poorly removed at about 20–60% when influent concentrations were at about 10 ng/L; these substances as well as others showed an increase to 99% removal when the influent concentrations increased greatly. Observations such as these should serve as a warning to carefully evaluate influent levels before making general conclusions about removal data.

Some *n*-aldehydes such as octanal, nonanal and decanal were normally found to be present in higher concentrations in the effluent from the GAC contactor.

As mentioned above, techniques such as CLSA coupled with GC/MS are potentially able to generate, in a routine manner, lists of a large number of compounds with their concentrations. This analytical approach is potentially the most useful when individual compounds of concern can be targeted. A perceived difficulty with application of the full output of this analytical approach is the manual conversion of the large database to a usable format, e.g., breakthrough curve plots, and manipulation of the data on a continuing basis in such a way as to make timely judgments regarding adjustments to the treatment process. This is increasingly important as the number of treatment targets and frequency of analysis is increased. This difficulty is, however, viewed as largely a data management problem that will be resolved through automatic data processing procedures as the analytical approach becomes more widely accepted and refined. However, success with this approach does not obviate the need for more general parameters as used in this study to characterize the performance of unit processes with regard to removals of the nonvolatile portion of the total organic load. This nonvolatile portion represents the majority of the organic material present and possibly the most significant factor.

## SUMMARY AND CONCLUSIONS

Analytical chemists have vastly improved methodology that can provide data for better defining what is meant by the often-stated phrase, "activated carbon is capable of removing a broad spectrum of organics." The application of these tools has only recently begun to receive attention in drinking water activities, and the optimum uses of them for GAC performance evaluation are yet to be established. Current nonanalytical deterrents to the optimum use of the methods applied in this study are (1) variable influent concentrations of individual substances; (2) variable composition within group parameters; (3) required subjective interpretations of some outputs; and (4) perceived difficulties in data management when high frequency of analysis or a large number of compounds are encountered. Most of these deterrents will always be present to complicate the use of the chemical methods when applied to field studies. Although much research information relative to definition of performance under given situations has been and still can be collected, the actual design and operation of carbon adsorption systems must still rely on development of specific targets for removal, i.e., performance criteria. Development of meaningful surrogate analyses, group parameters or practical specific analyses has not as yet been achieved in the practical sense.

The trend toward better understanding the organic quality of available water sources, the capabilities of conventional treatment and the advantages of improved treatment for removing organics has begun. Water supply utilities have begun to view modern analytical devices, such as gas chromatographs, as useful product water quality control tools and are beginning to apply them effectively for that purpose. The trend appears to be toward more effective use of them in the future as greater understanding of their use and application of results is developed.

We need to continue to define GAC capability by using the best tools now available, rather than limit our scope by rigorous lists. Also, the application of these data needs to continue to be focused on their end uses in real life situations.

## ACKNOWLEDGMENTS

We appreciate the cooperation and assistance of Mr. Richard Miller, Director of the Cincinnati Water Works, in providing the data presented from the cooperative research project. We also appreciate the analytic assistance provided by Mr. Dennis Seeger, Mr. Ronald Dressman and Mr. Emile Coleman of the EPA. Finally, we wish to thank Mrs. Patricia Pierson for typing the manuscript.

## REFERENCES

1. *Drinking Water and Health* (Washington, DC: National Academy of Sciences, 1977).
2. "Drinking Water and Health–Recommendations of the National Academy of Sciences," *Federal Register* 42:35764 (1977).

3. "National Interim Primary Drinking Water Regulations," *Federal Register* 40(248):59566-59588 (1975).
4. "National Interim Primary Drinking Water Regulations: Control of Trihalomethanes in Drinking Water," *Federal Register* 44(231):68624-68707 (1975).
5. "Publication of Toxic Pollutant List," *Federal Register* 43(21):4108-4109 (1978).
6. Stevens, A. A., and J. M. Symons. "Measurement of Trihalomethane and Precursor Concentration Changes," *J. Am. Water Works Assoc.* 69(10):546(1977).
7. "The Analysis of Trihalomethanes in Finished Water by the Purge and Trap Method," U.S. EPA Method 501.1, Environmental Monitoring and Support Laboratory, Cincinnati, OH (1977).
8. Cotruvo, J. A. "Regulatory Aspects of Potable Water Disinfection," in *Water Chlorination: Environmental Impact and Health Effects, Vol. 2*, R. L. Jolley, H. Gorchev and D. H. Hamilton, Jr., Eds. (Ann Arbor, MI: Ann Arbor Science Publishers, Inc., 1978).
9. Bellar, T., and J. Lichtenberg. "Determining Volatile Organics at the Microgram per Liter Level by Gas Chromatography," *J. Am. Water Works Assoc.* 66:734-744 (1974).
10. Grob, K., and F. Zurcher. "Stripping of Trace Organic Substances from Water–Equipment and Procedure," *J. Chromatog.* 117:285-294 (1976).
11. "Manual of Methods for Organic Pesticides in Water and Wastewater," U.S. EPA, NERC, Cincinnati, OH (1971).
12. Dressman, R. C., J. Fair and E. F. McFarren. "Determinative Method for Analyses of Aqueous Sample Extracts for bis(2-Chloro-)ethers and Dichlorobenzenes," *Environ. Sci. Technol.* 11:719-721 (1977).
13. Dressman, R. C., A. Stevens, J. Fair and B. L. Smith. "Comparison of Methods for Determination of Trihalomethanes in Drinking Water," *J. Am. Water Works Assoc.* 71:392-396 (1978).
14. "A Study of Exposure to Organics via Drinking Water in Eight U.S. Cities," U.S. EPA Health Effects Research Laboratory, Cincinnati, OH (1979).
15. Dressman, R. C., B. A. Najar and R. Redzikowski. "The Analysis of Organohalides (OX) in Water as a Group Parameter," in *Proceedings of the Seventh Water Quality Technology Conference* (Denver, CO: American Water Works Association, 1980).
16. Grob, K., and G. Grob. "Organic Substances in Potable Water and in Its Precursor. Part II. Applications in the Area of Zurich," *J. Chromatog.* 90:303 (1974).
17. Stieglitz, L., W. Roth, W. Kuhn and W. Leger. "The Behavior of Organohalides in the Treatment of Drinking Water," *Vom Wasser* 47:347 (1976).
18. Piet, G. J., and C. F. Morra. "Behavior of Micropollutants in River Water During Bank Filtration," in *Proceedings of the Conference on Oxidation Techniques in Drinking Water Treatment*, U.S. EPA Report EPA-570/9-79-020 (1978), p. 608.
19. Glaser, E. R., B. Silver and I. H. Suffet. "Computer Plots for the Comparison of Chromatographic Profiles," *J. Chromatog. Sci.* 15:22-28 (1977).

20. Van Rensburg, J. F. J., A. Hassett, S. Theron and S. G. Wiechers. "The Fate of Organic Micropollutants through an Integrated Wastewater Treatment Water Reclamation System," *Prog. Water Technol.* 12:537-552 (1980).
21. Coleman, W. E., R. G. Melton, R. W. Slater, F. C. Kopfler, S. J. Voto, W. K. Allen and T. A. Aurand. "Determination of Organic Contaminants by the Grob Closed-Loop Stripping Technique," in *Proceedings of the Seventh Water Quality Technology Conference* (Denver, CO: American Water Works Association, 1980).
22. Wood, P. R., and J. DeMarco. "Treatment of Groundwater with Granular Activated Carbon," *J. Am. Water Works Assoc.* 71(11):674 (1979).
23. Wood, P. R., and J. DeMarco. "Removing Total Organic Carbon and Trihalomethane Precursor Substances," in *Activated Carbon Adsorption of Organics from the Aqueous Phase, Vol. 2*, M. J. McGuire and I. H. Suffet, Eds. (Ann Arbor, MI: Ann Arbor Science Publishers, Inc., 1980), pp. 115-136.
24. DeMarco, J., and N. V. Brotdmann, Jr. "Prediction of Full Scale Plant Performance from Pilot Columns," in *Proceedings–Symposium on Practical Application of Adsorption Techniques in Drinking Water Treatment*, U.S. EPA (in press).

# CHAPTER 48

# DETECTION, IDENTIFICATION AND QUANTITATIVE ANALYSIS OF DIHALOACETONITRILES IN CHLORINATED NATURAL WATERS

**Michael L. Trehy**

City of West Palm Beach
Water Department
West Palm Beach, Florida

**Theodore I. Bieber**

Department of Chemistry
Florida Atlantic University
Boca Raton, Florida

Independent studies by Rook [1] and Bellar et al. [2] revealed the formation of trihalomethanes (THM) when natural waters are chlorinated to kill harmful bacteria. We have discovered that the chlorination of natural water is also responsible for the formation of another class of halogenated organic compounds, the dihaloacetonitriles (DHAN). Coleman identified dichloroacetonitrile in the tapwater at the National Institute of Environmental Health Sciences in Durham, NC, for McKinney et al. [3]. They did not quantify the dichloroacetonitrile, nor did they report whether the substance was an industrial pollutant or the result of chlorination of the raw water which originates in the Flat River. Stevens [4] reported that he had observed chloroacetonitrile derivatives in a finished tapwater in recent studies at the U.S. Environmental Protection Agency (EPA) Cincinnati, OH, laboratory.

We have detected the following DHAN in chlorinated ground- and surface water supplies of south Florida, but not in the water prior to chlorination: dichloroacetonitrile (**I**), bromochloroacetonitrile (**II**) and dibromoacetonitrile (**III**).

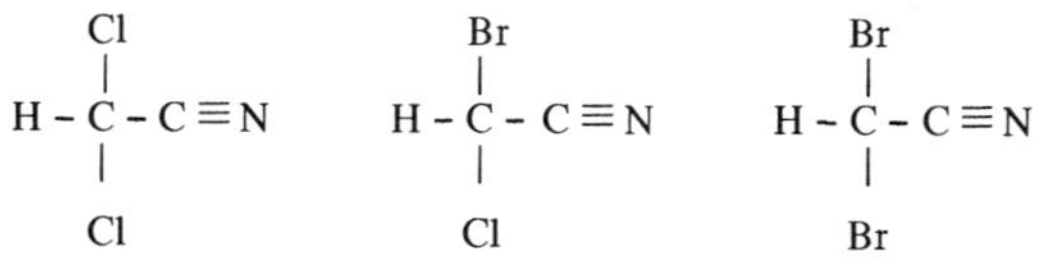

```
      Cl                Br                Br
      |                 |                 |
  H - C - C≡N       H - C - C≡N       H - C - C≡N
      |                 |                 |
      Cl                Cl                Br

     (I)               (II)              (III)
```

We have also observed that DHAN are not found if the finished water has a high pH (about 9 or above). The highest total concentration of DHAN so far found has been 42 $\mu g/L$ in a chlorinated wellwater supply. A reconstructed GC of this chlorinated water supply is shown in Figure 1.

The DHAN and their decomposition products thus represent a portion of the previously unidentified total organic chlorine reported by others [5].

The existence of DHAN in a public water supply could be of concern to public health because dichloroacetonitrile has been shown to be mutagenic in the Ames test [6]. Furthermore, two actual or proposed uses of dichloroacetonitrile, as an insecticide for grains [7], and as a biological growth inhibitor in cooling towers [8], are suggestive of considerable biological activity. Clearly, the effects on the human body of dichloroacetonitrile and the other members of the series need to be studied with regard to both short-term and long-term cumulative exposure.

The detection of DHAN in chlorinated water supplies by the liquid-liquid extraction (LLE) technique was the starting point of this investigation. The realization that DHAN were being formed by the chlorination of natural waters, but had seldom been reported, prompted us to determine if DHAN were being degraded before and/or during analysis. The headspace and purge-and-trap techniques were evaluated to determine their suitability for quantification of DHAN. A literature search helped to identify the hydrolysis products of DHAN that would be expected to be formed in drinking water as a result of the hydrolysis of DHAN.

Simultaneous to the discovery that DHAN were being formed by chlorination of natural waters, it was found that amino acids are potential precursors for both DHAN and THM. We therefore studied the halogenation of precursors for DHAN and THM. A literature search revealed possible pathways for the oxidation of amino acids by hypohalites. This led to the realization that many compounds, both halogenated and nonhalogenated, reported by

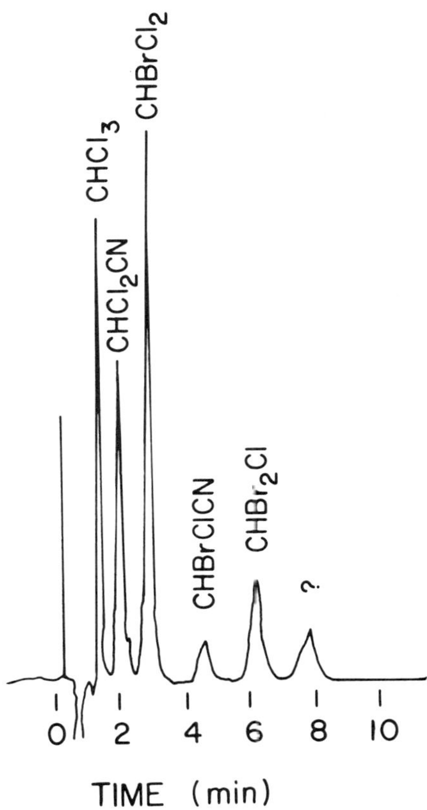


**Figure 1.** Reconstructed gas chromatogram of pentane extract of chlorinated wellwater. Squalane column.

others to be present in chlorinated water supplies could be accounted for on the basis that they are products resulting from the oxidation and halogenation of amino acids.

## CHROMATOGRAPHIC TECHNIQUES

### LLE Technique

A LLE procedure has been used extensively in this study for THM and DHAN analysis. This method is a modification of that developed by Henderson et al. [9]. A sample was collected in a 30-mL glass bottle containing a 1-in. Teflon®-coated magnetic stirring bar and 1 drop of 10% sodium thiosulfate solution to eliminate residual chlorine. The sample was sealed with a Teflon-coated septum. Six milliliters of water were replaced by 6 mL of nanograde pentane using two 10-mL syringes, one to supply the pentane and the other to receive the water. The sample was then extracted for 20 min while being stirred magnetically (a Magnestir was used at a setting of 3).

A 3-$\mu$L aliquot of pentane was then injected into a gas chromatograph (GC) containing a 6-ft x 2-mm i.d. glass column packed with squalane-coated (10% w/w) 60/80 mesh Chromosorb W/AW. A Varian series 1400 GC was used equipped with a constant-current tritiated scandium electron capture detector (ECD).

Nitrogen was used as a carrier gas with a flowrate of 30 mL/min. The GC operating temperatures were 100°C at the injection port, 59°C at the column and 150°C at the detector. A typical GC run requires 12 min for bromoform to elute.

## Headspace Technique

The headspace technique we used is a modification of that described by Kaiser and Oliver [10]. A water sample was collected in a sealed bottle containing two drops of 10% sodium thiosulfate. A 60-mL separatory funnel was completely filled with sample. A 4-mL quantity was then removed from the separatory funnel. The funnel was inverted and the air space was rapidly evacuated using a vacuum pump. The entire funnel was then submerged in a water bath at 44.5°C for 30 min. The sample was rapidly brought back to atmospheric pressure. A 25-$\mu$L sample of the headspace was injected with a gas-tight syringe into the GC containing a 6-ft x 2-mm i.d. glass column packed with Chromosorb 101 (60/80 mesh). The injection port was at 200°C, the column temperature at 160°C and the detector at 220°C. The squalane column previously described under the LLE technique was sometimes used in place of the Chromosorb 101 column to provide a comparison between the two columns.

## Purge-and-Trap Technique

The purge-and-trap technique developed by Bellar and Lichtenberg [11] involves purging the volatile organic compounds from a water sample by bubbling a gas, such as $N_2$ or He, through it and trapping the purged organics on a column packed with Tenax® at room temperature. The trapping column is then connected to the GC column, whereupon the trapping column is heated to 125–130°C (200°C in some cases) with a flow of $N_2$ or He to desorb the trapped organic compounds. As a result the organic compounds are transferred onto the GC column (which is at room temperature). At the completion of this process, the trapping column is disconnected and the GC is temperature-programmed.

The West Palm Beach Water Department does not have the facilities to perform this technique, but two samples were analyzed by EPA in Cincinnati, OH, using this technique [12].

## DETECTION AND IDENTIFICATION OF DHAN

Prior to July 1978 the headspace technique was used for the analysis of water samples for THM at the West Palm Beach water treatment plant. At that time the decision was made to switch to the LLE technique because of its greater sensitivity for less volatile components such as bromoform.

When unchlorinated water was analyzed, none of these unknown peaks were observed. Hence these peaks are not due to organic pollutants present in the raw water, but to compounds arising from the chlorination process. Two of the unknown peaks (A and B, where B has the longer retention time) were shown to be due to compounds belonging to the same haloorganic series. This was done by modifying the halogenating system. When sodium hypochlorite (10 mg/L) was used and the bromide ion content was that of the raw water, peak A was considerably larger than peak B (Figure 2). When sodium hypochlorite (10 mg/L) was used and the bromide ion content of the water was augmented by addition of KBr (2.5 mg/L), peak A was diminished and peak B was considerably enhanced, so that peak B was now larger than peak A (Figure 3). Furthermore, a new peak C, of still longer retention time than peak B had now appeared, the peak height of C approaching that of B. When the halogenating system consisted of sodium hypochlorite (10 mg/L), the original bromide content of the water, and a very much larger quantity of added KBr than in the previous experiment, peaks A and B were no longer present. Peak C, the sole remaining peak of the trio, was considerably enhanced (Figure 4). Inspection of Figures 2 to 4 also shows that the quartet of THM peaks underwent a change toward the more brominated members of the THM series, as the bromide ion content of the halogenating system was increased. All these halogenations were carried out at a pH of approximately 8.3.

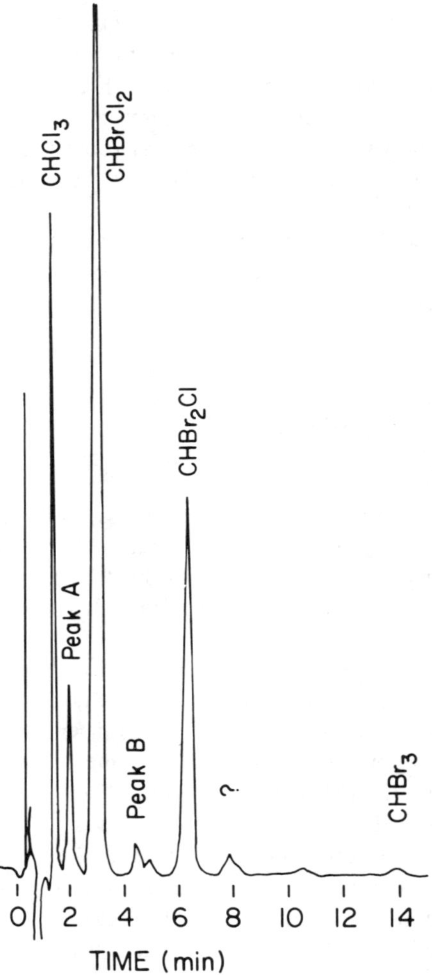


**Figure 2.** Reconstructed gas chromatogram of pentane extract of chlorinated lakewater. Squalane column.

The existence of three related peaks (A, B and C) makes it appear likely that these peaks are due to a certain dihalostructure, $QX_2$, where A is $QCl_2$, B is QBrCl, and C is $QBr_2$. If, as appears to be the case, there is only one QBrCl, then the relationship between the two halogen atoms in $QCl_2$ (as well as in $QBr_2$) must be either equivalence or enantiotopism. Otherwise, there would have been two chromatographically distinct QBrCl compounds.

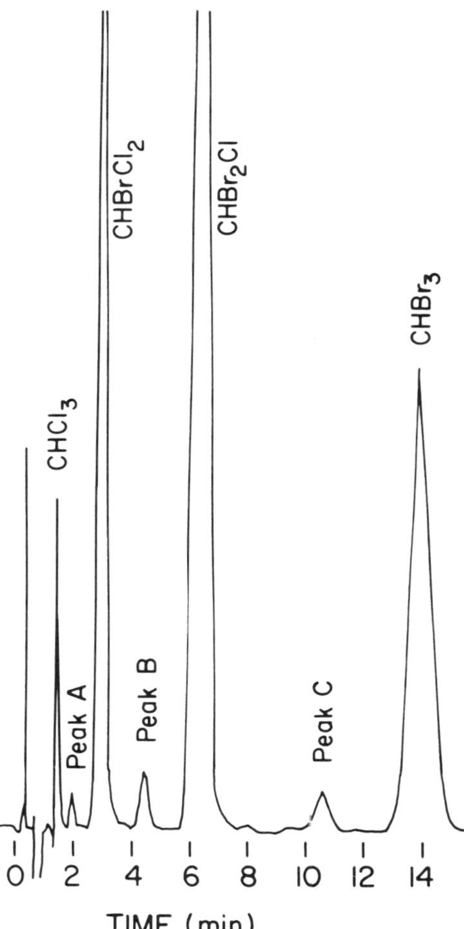


**Figure 3.** Reconstructed gas chromatogram of pentane extract of chlorinated lakewater with 2.5 mg/L KBr added. Squalane column.

Other water supplies in the area were analyzed to determine whether peaks A and B (and possibly C) were generally present. The unknowns were found in all supplies that had not been lime-softened. The concentrations of the unknowns varied considerably from supply to supply and did not correlate with the THM content, as can be seen from Figures 1, 2 and 5.

Numerous chlorinated organics were purchased to compare their retention times with those of the unknowns. Unfortunately, this approach did not lead to identification of the unknowns as none of the retention times matched.

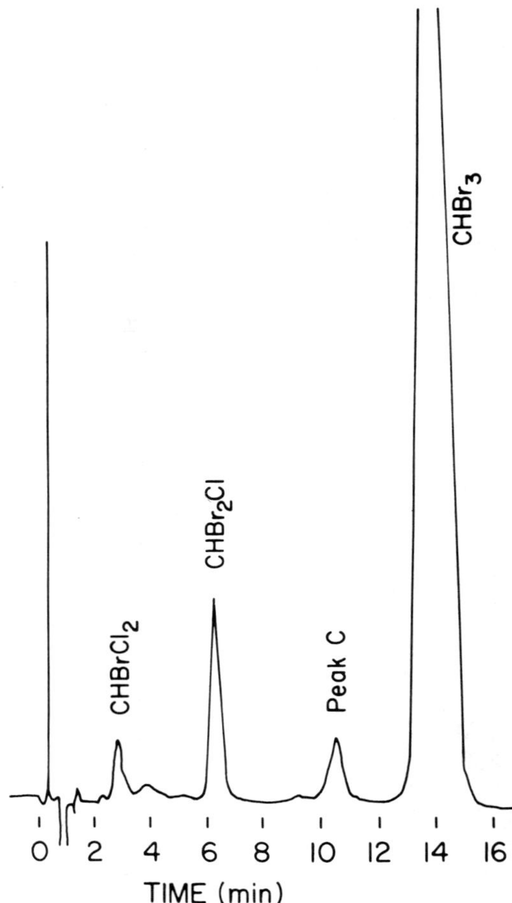


**Figure 4.** Reconstructed gas chromatogram of pentane extract of chlorinated lakewater with a large amount of KBr added. Squalane column.

Another approach was therefore taken. A search was made for substances which would yield the unknowns on chlorination, since the discovery of such substances would provide a clue to the structure of the unknowns. Furthermore, the unknowns could then be prepared in isolable quantities and identified by conventional methods.

Several compounds containing carbon, hydrogen and oxygen (ethanol, acetaldehyde, lactic acid, pyruvic acid, malic acid, oxalacetic acid and resorcinol) did not yield the unknowns on chlorination. On the other hand,

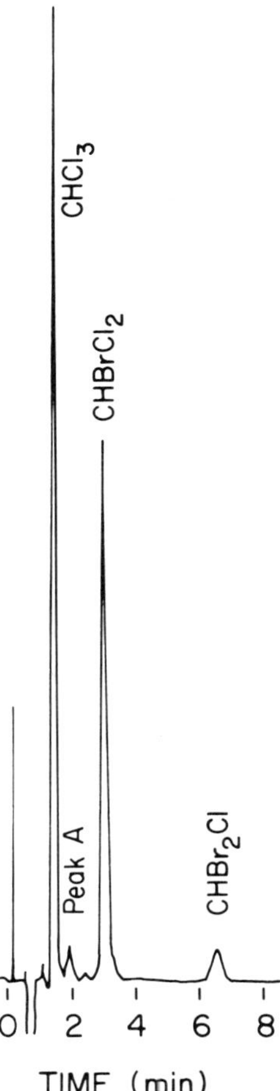


**Figure 5.** Reconstructed gas chromatogram of pentane extract of another chlorinated wellwater sample. Squalane column.

chlorination (at pH 7) of a sample of brilliant green bilebroth (Difco) did yield peak A. Nitrogen-containing components of the bile broth, such as peptones and amino acids, were then suspected to be the precursor substances. Various amino acids were therefore chlorinated to see whether one or more of them would produce peak A. Of 12 amino acids tried, 5 gave rise to peak A (aspartic acid, tryptophan, kynurenine, kynurenic acid and tyrosine).

Since aspartic acid (**IV**) produced peak A, it becomes especially significant that the closely related nitrogen-free compounds malic acid (**V**) and oxalacetic acid (**VI**) do not give rise to this peak.

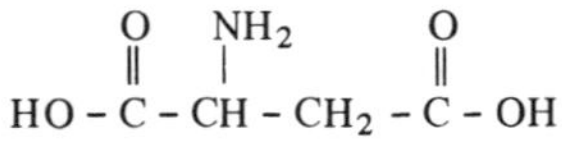

$$HO-\overset{\overset{O}{\|}}{C}-\overset{\overset{NH_2}{|}}{C}H-CH_2-\overset{\overset{O}{\|}}{C}-OH \qquad \text{(IV)}$$

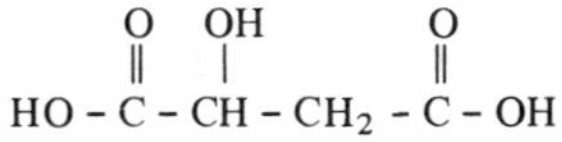

$$HO-\overset{\overset{O}{\|}}{C}-\overset{\overset{OH}{|}}{C}H-CH_2-\overset{\overset{O}{\|}}{C}-OH \qquad \text{(V)}$$

$$HO-\overset{\overset{O}{\|}}{C}-\overset{\overset{O}{\|}}{C}-CH_2-\overset{\overset{O}{\|}}{C}-OH \qquad \text{(VI)}$$

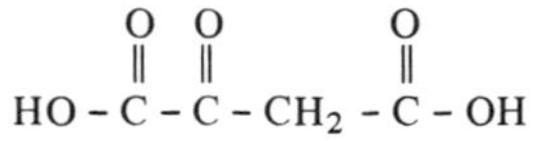

The results of these experiments are summarized in Tables I and II, not only with regard to the formation of peak A, but also with regard to chloroform formation. The chlorinations of Table I were carried out at pH 7-8 and those of Table II at pH 10-11.

The fact that aspartic acid, but not malic acid or oxalacetic acid gave rise to peak A indicates that the presence of a nitrogen atom is essential for the formation of the unknown and that in all likelihood, the unknown contains a nitrogen atom. All three unknowns (A, B and C) were subsequently generated from aspartic acid by varying the bromide ion content of the halogenating system. The respective retention times matched exactly those obtained on chlorination of natural water.

In view of this study, it was suspected that one or more of the amino acids which yield DHAN were present in the natural water of the area, possibly complexed to other substances. However, the natural waters have not been analyzed for the presence of any of these amino acids.

Aspartic acid was then brominated in the following manner to obtain a sufficient amount of compound C to permit its identification by conventional methods. Aspartic acid (10 g) was dissolved in 1 L of distilled water along with 10 g of disodium phosphate. To this solution 3.7 mL of bromine was added. The reaction was at first allowed to proceed without stirring, and as the reaction slowed, with stirring. The reaction rate can easily be followed by the evolution of gas. It was observed that the addition of disodium phosphate increased the rate of the reaction. The final pH was 3.3.

The gas that evolved during bromination was found to be more than 98% carbon dioxide. The brominated aspartic acid solution was allowed to stand for 3 days. The solution was then extracted with 25 mL of pentane. This particular pentane extract was discarded because it contained bromoform, peak C, and another, much smaller, unidentified peak having a shorter retention time than peak C. The pentane extraction was repeated with a larger quantity of pentane. This pentane extract was dried with anhydrous sodium sulfate and the pentane was then removed by distillation to yield a residue of approximately 1 mL of a high-boiling colorless liquid.

**Table I. Results of Chlorination of Potential Precursors for THM and DHAN at pH 7–8**

| | Chloroform | Peak A (Dichloroacetonitrile) |
|---|---|---|
| Glycine | +[a] | – |
| Alanine | – | – |
| β-Alanine | + | – |
| Serine | – | – |
| Cysteine | + | – |
| Proline | + | – |
| Aspartic Acid | + | +++ |
| Asparagine | + | – |
| Tyrosine | + | + |
| Tryptophan | + | ++ |
| Kynurenine | ++ | ++ |
| Kynurenic Acid | +++ | + |
| Ethanol | – | – |
| Lactic Acid | – | – |
| Pyruvic Acid | – | – |
| Malic Acid | – | – |
| Oxalacetic Acid | – | – |
| Resorcinol | +++ | – |

[a]Symbols: – = not detected; + = detected; ++ = moderate production; +++ = large production.

**Table II. Results of Chlorination of Potential Precursors for THM at pH 10–11**

| | Chloroform | Peak A (Dichloroacetonitrile) |
|---|---|---|
| Alanine | +[a] | – |
| Ethanol | + | – |
| Acetaldehyde | ++ | – |
| Lactic Acid | + | – |
| Pyruvic Acid | ++ | – |
| Malic Acid | + | – |
| Oxalacetic Acid | ++ | – |
| Resorcinol | +++ | – |

[a]Symbols: – = not detected; + = detected; ++ = moderate production; +++ = large production.

This liquid was subjected to mass spectroscopy. The mass spectrum (Figure 6A) revealed the liquid to be dibromoacetonitrile. The molecular ion peaks have m/e = 197, 199 and 201 in an approximate ratio of 1:2:1, as required by $CHBr_2CN$. The various ion fragments are also readily explained in terms of this molecule. A sample of dibromoacetonitrile prepared

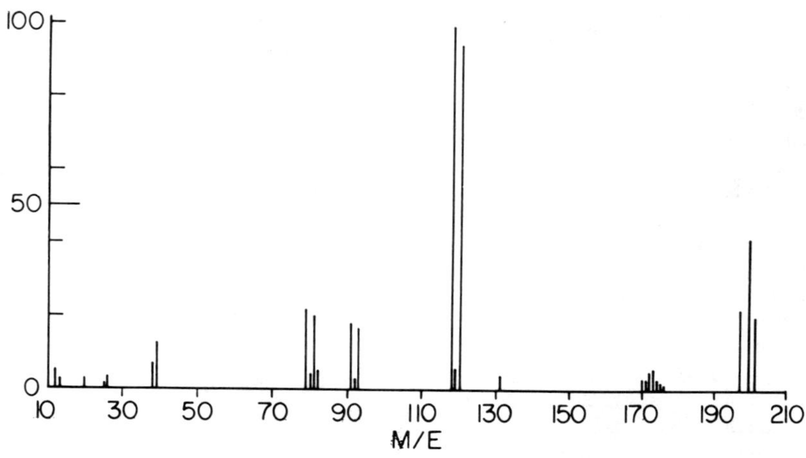


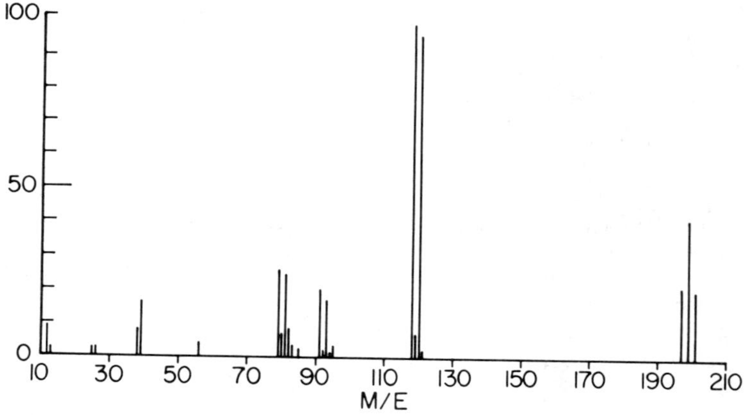


**Figure 6.** Reconstructed mass spectra of (A) brominated aspartic acid and (B) synthesized dibromoacetonitrile.

from cyanoacetic acid [13,14] gave the same mass spectrum (Figure 6B). Finally, the IR spectrum (Figure 7A) of the liquid obtained from aspartic acid resembles closely a published IR spectrum of dibromoacetonitrile (Sadtler IR 24255) [15], as well as the IR spectrum (Figure 7B) of dibromoacetonitrile prepared from cyanoacetic acid. The presence in our spectra of a medium-strength absorption near 6 $\mu$ indicates some carbonyl impurity.

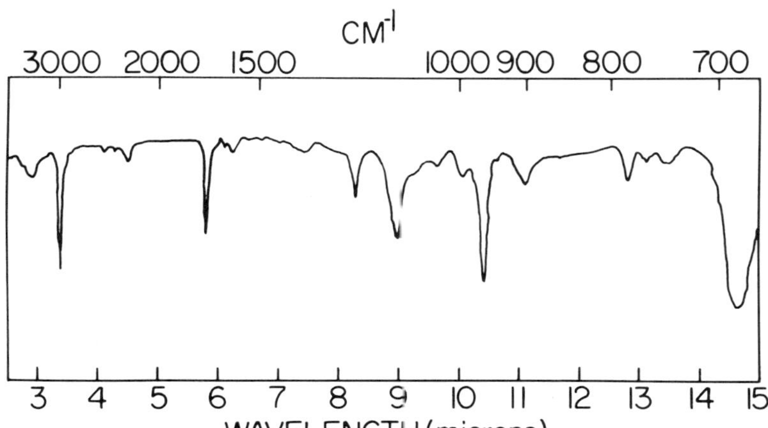


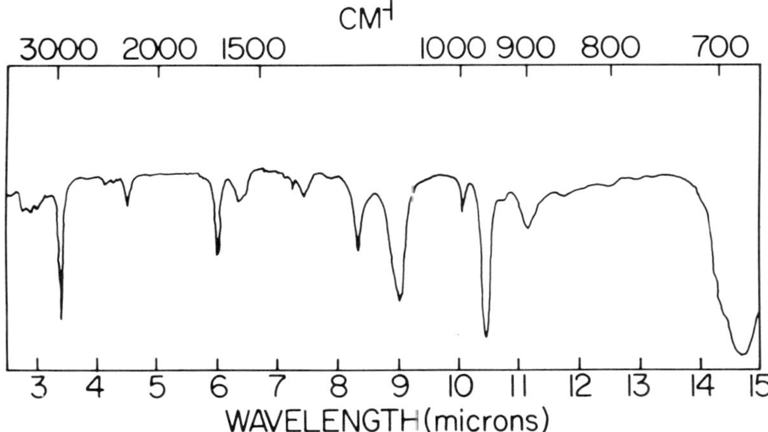


**Figure 7.** Reconstructed infrared spectra of (A) extract from bromine treatment of aspartic acid and (B) synthesized dibromoacetonitrile.

At this stage the three unknown peaks A, B and C observed after chlorinating various water supplies were suspected to be DHAN.

All three DHAN were synthesized by published procedures to compare their retention times and some chemical properties (reactivity with water and with thiosulfate ion) with the unknowns A, B and C found in the chlorinated water supplies. Dichloroacetonitrile was synthesized by reaction of cyanoacetic acid with N-chlorosuccinimide in a 1:2 molar ratio in water at room temperature [13,14]. Dibromoacetonitrile was made by a similar reaction using N-bromosuccinimide [13,14]. Bromochloroacetonitrile was synthesized by bromination of chloroacetonitrile with bromine on a steam bath [16]. In the course of this work, dichloroacetaldehyde was also prepared. This was made by heating dichloroacetaldehyde diethylacetal with benzoic anhydride to 170–180°C in the presence of concentrated sulfuric acid [17,18]. The dichloroacetaldehyde that is formed distills over and is collected. The physical properties of the DHAN are presented in Table III.

Synthesized samples of dichloroacetonitrile, bromochloroacetonitrile and dibromoacetonitrile had retention times exactly matching those of peaks A, B and C, respectively. The retention time of dichloroacetaldehyde is shorter than that for peak A.

Table IV lists the retention times for the various THM and DHAN on a 6-ft x 2-mm i.d. glass column packed with squalane-coated 60/80 mesh Chromosorb W/AW (10% w/w). The temperature of the column was 59°C and the nitrogen flowrate was 30 mL/min.

**Table III. Physical Properties of DHAN**

| Compound | Mol Wt | Density | Boiling Point (°C) |
|---|---|---|---|
| $CHCl_2CN$ | 109.9 | 1.369 | 110–112[a] |
| $CHBrClCN$ | 154.4 | 1.68 | 125–130[a] |
| $CHBr_2CN$ | 198.8 | 2.369 | 70–72[b] |

[a]At 760 mm Hg pressure.

[b]At 20 mm Hg pressure.

**Table IV. Retention Times for THM and DHAN on a 10% Squalane Column**

| Compound | $t_R$[a] | $t_R$ Compound[a] / $t_R$ $CHBrCl_2$ |
|---|---|---|
| $CHCl_3$ | 1 min 10 sec | 0.5 |
| $CHCl_2CHO$ | 1 min 30 sec | 0.65 |
| $CHCl_2CN$ | 1 min 41 sec | 0.70 |
| $CHBrCl_2$ | 2 min 19 sec | 1.00 |
| $CHBrClCN$ | 3 min 49 sec | 1.65 |
| $CHBr_2Cl$ | 5 min 7 sec | 2.21 |
| $CHBr_2CN$ | 8 min 52 sec | 3.83 |
| $CHBr_3$ | 11 min | 4.75 |

[a]Values uncorrected for retention time of air.

Samples of unchlorinated wellwater serving as the supply for a private utility in South Florida were sent to the Cincinnati EPA laboratory. The water samples were chlorinated at Cincinnati and analyzed for dichloroacetonitrile. Dichloroacetonitrile was detected using the purge-and-trap technique and confirmed by gas chromatography/mass spectrometry (GC/MS). The reconstructed gas chromatogram of the chlorinated wellwater sample is shown in Figure 8. The mass spectrum of peak No. 289, which they identified as dichloroacetonitrile, is shown in Figure 9.

Additional corroboration of the identity of peak A with dichloroacetonitrile was obtained by comparison of their hydrolysis rates. As is shown in Figure 10, the hydrolysis rates at pH 9.77 are identical.

In summary, peaks A, B and C observed in chlorinated drinking water by our LLE technique have unequivocally been identified as dichloroacetonitrile, bromochloroacetonitrile and dibromoacetonitrile, respectively. The method of identification included:

1. identity of the retention time for each synthesized DHAN with the retention time of the respective unknown in chlorinated water;

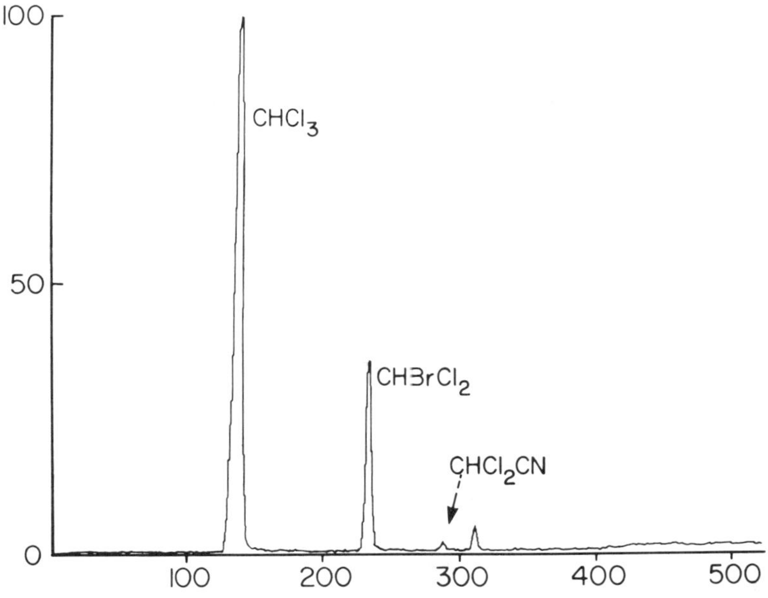


**Figure 8.** Reconstructed gas chromatogram of chlorinated wellwater by purge-and-trap technique.

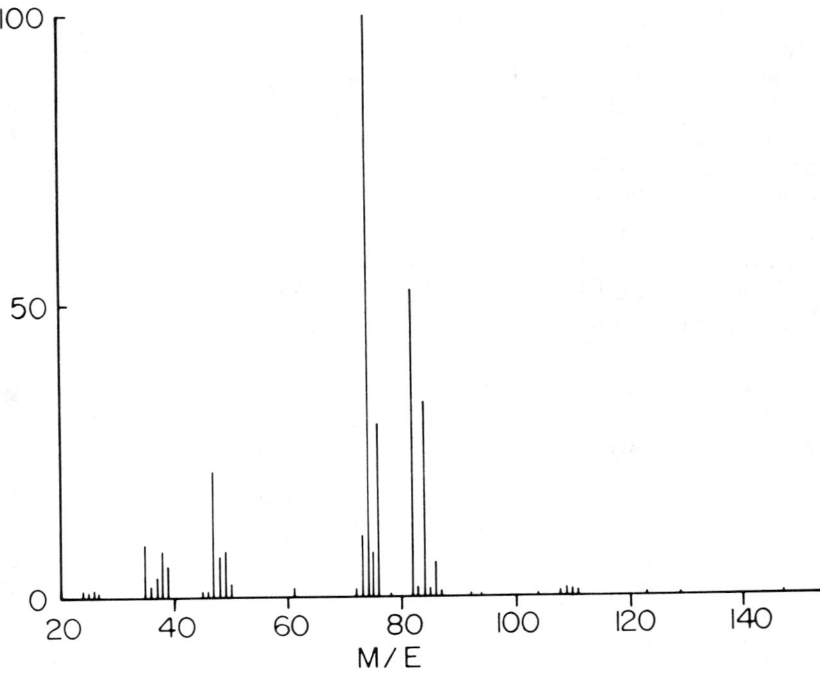


**Figure 9.** Reconstructed mass spectrum of peak No. 289 found in chlorinated wellwater by purge-and-trap technique.

2. observation of identical half-lives of disappearance due to hydrolysis for dichloroacetonitrile and peak A; and
3. confirmation of the presence of dichloroacetonitrile in a water sample following chlorination by GC/MS using the purge-and-trap technique.

## ANALYSIS FOR DHAN AND THM

A modification of the LLE method of Henderson et al. [9] has been used in this study. Several problem areas in the analysis for DHAN and THM need to be considered. The analysis for these compounds will first be discussed for spiked samples and then for chlorinated natural waters.

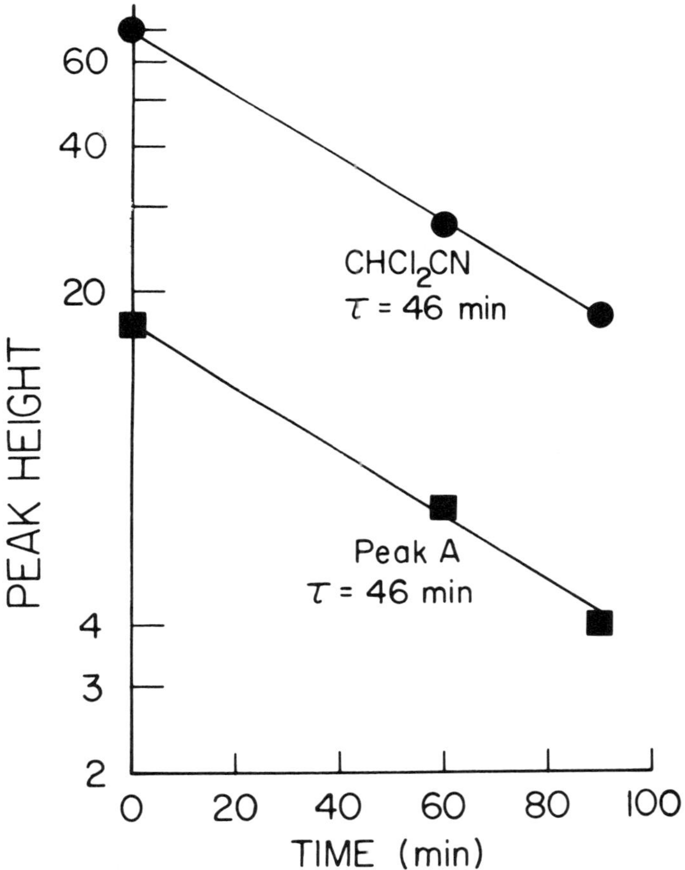


**Figure 10.** Rate of hydrolysis of peak A and dichloroacetonitrile at pH 9.77.

## Analysis of Spiked Samples

### *Sample Bottles*

Analysis for THM and DHAN requires collection of the sample in a 5-mL 30-mL septum bottle (purchased from Pierce) containing a 1-in. magnetic stirring bar. The variability in the volumes of these bottles was determined. The average volume of ten 50-mL bottles containing stirring bars was 58.05 mL, with a range from 57.6 (–0.8%) to 58.7 mL (+1.1%). The average volume of ten 30-mL bottles containing stirring bars was 36.33 mL with a range from 35.39 (-2.6%) to 36.9 mL (+1.6%).

*Preparation of Standards*

A methanol or acetone solution spiked with all the THM and DHAN is prepared as follows: each component (1.5-10 $\mu$L) is added, using a 10-$\mu$L syringe, to 100 mL of methanol or acetone. The accuracy of the syringe was found to be within 1% by weighing the syringe filled with chloroform and again after discharge of the chloroform. The spiked methanol or acetone solution must be prepared immediately before use because of the appreciable decomposition of the DHAN in these solvents. Thus, in 19 hr, approximately 30% of the dichloroacetonitrile and approximately 15% of the dibromoacetonitrile decomposes in either solvent.

The spiked methanol or acetone solution can be used to prepare standards in water. We have also used the acetone solution to prepare standards in pentane. One milliliter of the spiked solution is added to a volume of lakewater or pentane. That volume can be varied to obtain a number of standards.

While the water standards are fairly stable at pH 7 and in the absence of reducing agents, they were prepared daily. The pentane standards were observed to be stable for at least one month.

*Extraction Time*

The optimum time for the extraction of the THM and DHAN from water into pentane was determined by extracting identical 50-mL samples containing all components with 4 mL of pentane for various lengths of time. During extraction, the mixtures were magnetically stirred (setting of 3) while being kept in a constant temperature bath to counteract the heating effect of the magnetic stirrer. Graphs of peak heights vs time of extraction for the various components are shown in Figure 11 for the THM and in Figure 12 for the DHAN. It can be seen that after 20 min, equilibrium is reached for the various THM, while DHAN reach their maximum peak heights in just 10 min and then decrease slowly, probably due to hydrolysis. The optimum time for extraction of a sample containing both THM and DHAN is therefore 20 min.

*Column Effects*

DHAN are far more easily degraded than THM. In fact, DHAN are easily degraded by some column packings. Degradation of dichloroacetonitrile became very serious when the 10% squalane column normally used was replaced by a Chromosorb 101 column, resulting in a low sensitivity. The ratio of the

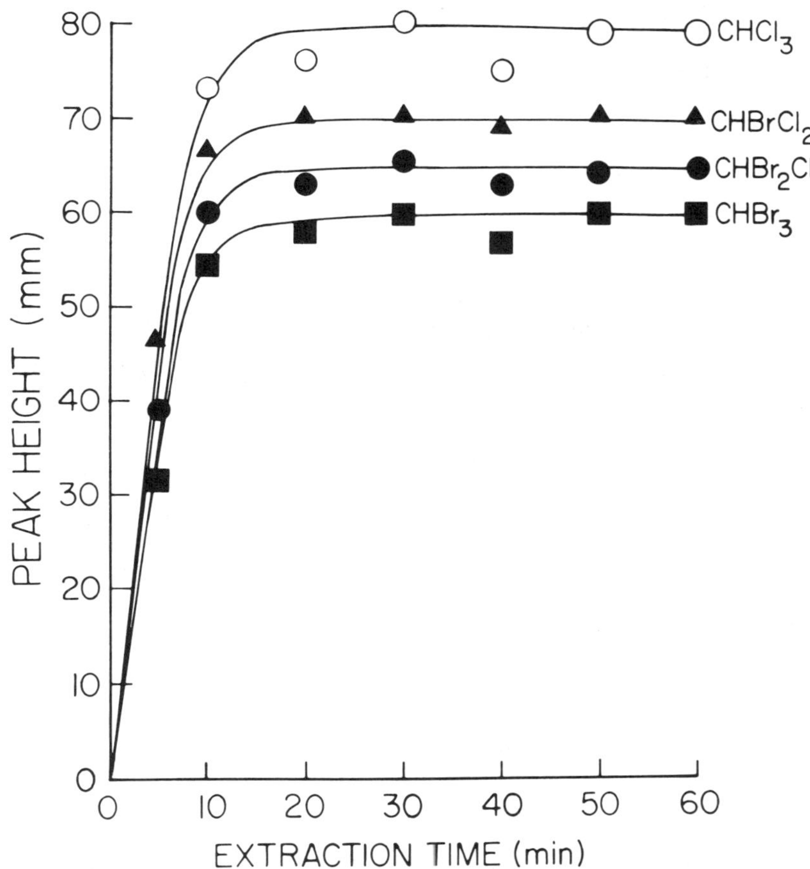


**Figure 11.** Peak heights vs extraction time for THM.

peak heights of dichloroacetonitrile to chloroform was much smaller when a Chromosorb 101 column was used rather than the 10% squalane column. Thus, care must be taken in the selection of a column to avoid degradation of DHAN during analysis.

*Detector Response*

We used a constant-current tritiated scandium ECD. To avoid errors it is important that the detector response (peak height) be linear with respect to the quantity of the injected component. If the quantity of that particular

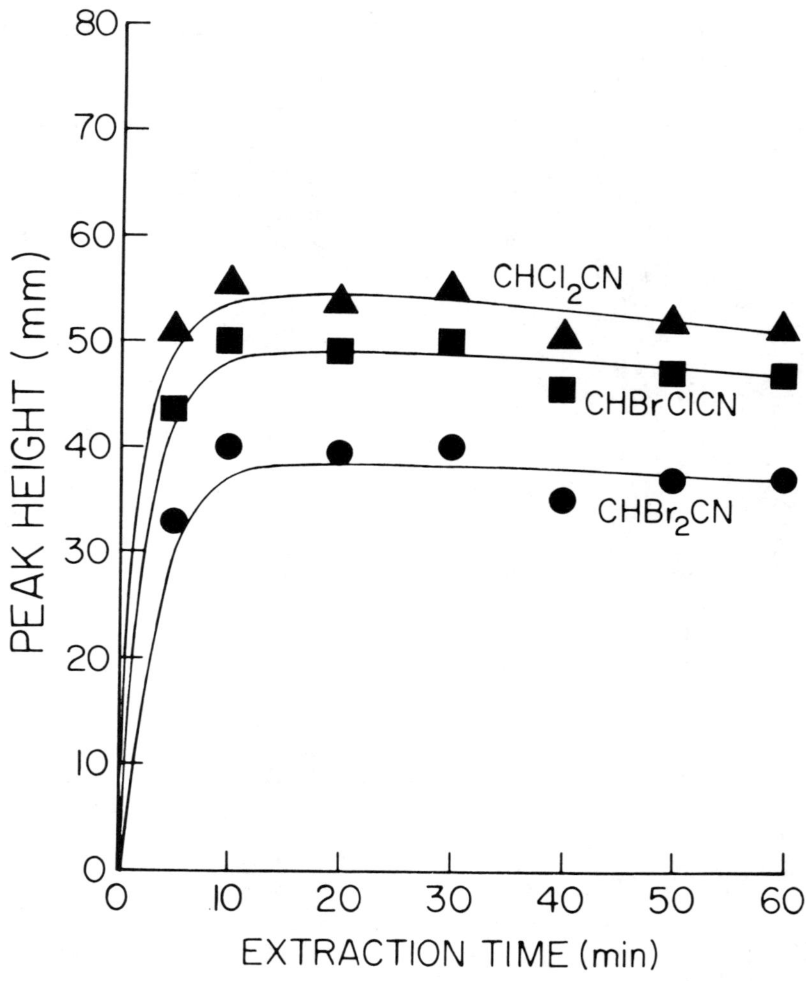


**Figure 12.** Peak heights vs extraction time for DHAN.

component exceeds a certain value for a given detector under given operating conditions, the response begins to level off, and becomes nonlinear [19]. With our detector we found that the response was approximately linear toward chloroform when less than 1.3 ng were injected. With larger amounts, the detector response becomes increasingly insensitive. With a chloroform concentration in the water of up to 90 $\mu$g/L one gets a nearly linear response if one uses a 30-mL bottle, extracts with 6 mL of pentane and injects 3 μL of the pentane extract. Using this particular pentane-to-water ratio and injecting 3 μL of pentane extract, a 90-μg/L concentration of chloroform in water corresponds to the injection of approximately 1.2 ng of chloroform. On the other hand, if one uses a 50-mL bottle, extracts with 4 mL of pentane and injects 5 μL of pentane extract, a chloroform concentration in water of 30-μg/L corresponds to the injection of approximately 1.6 ng of chloroform, so that one is in the nonlinear response range. Thus, accurate analysis requires the use of a higher pentane-to-water ratio to avoid near-saturation of the detector with the accompanying nonlinear response.

*Partition Coefficients*

The partition coefficients for the various THM and DHAN were determined in the following manner. A 30-mL bottle containing a 1-in. magnetic stirring bar was filled with water (36.33 mL) containing all the DHAN and THM in question. Using two syringes as previously described, 6 mL of the aqueous solution was replaced by 6 mL of the pentane. Extraction was carried out for 20 min, and 3 μL of the pentane extract were injected into the GC. The concentration of each component was determined from its peak height by comparison with four pentane standards. The percent recovery and partition coefficients for each component could then be calculated. The results are shown in Table V.

The percent recovery and hence the partition coefficients obtained for the DHAN may be subject to a slight correction because some hydrolysis is likely during extraction. Our results for the THM are in general agreement with those of Dressman et al. [20].

**Table V. Percent Recovery and Partition Coefficients for Extraction of THM and DHAN from Water to Pentane at a 5:1 Water-to-Pentane Volume Ratio**

| | Dose (μg/L) | Recovery (%) | Partition Coefficient |
|---|---|---|---|
| $CHCl_3$ | 74.2 | 91 | 51 |
| $CHBrCl_2$ | 29.7 | 90 | 45 |
| $CHBr_2Cl$ | 36.8 | 91 | 51 |
| $CHBr_3$ | 43.3 | 91 | 51 |
| $CHCl_2CN$ | 34.2 | 32 | 2.4 |
| $CHBrClCN$ | 35 | 39 | 3.2 |
| $CHBr_2CN$ | 59.2 | 40 | 3.4 |

We observed no significant differences in the percent recoveries when two pentane extractions were performed on a water sample in series. For example, if 90% recovery of the THM is obtained in the first extraction, an additional extraction results in the recovery of an additional 90% of remaining THM. This is, of course, as expected.

*Precision of Analysis*

Ten identical samples were collected in 30-mL bottles and extracted with 6 mL of pentane for 20 min. A 3-μL aliquot of each extract was injected into the GC. Peak heights were used to determine the concentration of the various THM and DHAN, and the standard deviations were calculated. For the concentrations indicated the standard deviations were: $CHCl_3$ 37.1 μg/L ± 3.0; $CHBrCl_2$ 29.7 μg/L ± 2.0; $CHBr_2Cl$ 36.8 μg/L ± 2.1; $CHBr_3$ 43.3 μg/L ± 2.6; $CHCl_2CN$ 36.0 μg/L ± 1.9; CHBrClCN 42 μg/L ± 2.4; $CHBr_2CN$ 59.2 μg/L ± 3.7.

*LLE vs Other Techniques in DHAN Analysis*

The LLE technique has several important advantages in analyzing for DHAN:

1. Elevated temperature in the presence of water is avoided, thus minimizing hydrolysis.
2. Since pentane extracts are injected, the chromatography occurs in the virtual absence of water.
3. The 10% squalane column does not degrade the DHAN.

The headspace technique requires the water samples to be heated to obtain sufficiently high vapor phase concentrations of THM and DHAN. Of the DHAN, we have detected only dichloroacetonitrile with the method. Hydrolysis during heating would make quantitative analysis very difficult.

The purge-and-trap technique would expose DHAN to water vapor at 180°C during the desorption stage. Even though the contact time would be brief, hydrolysis would surely take place. Furthermore, in the trapping column as well as in the GC column Tenax or Chromosorb 100 series column packings are used. The Chromosorb 101 column that we employed caused appreciable decomposition of the DHAN even when pentane extracts were injected.

### Analysis of Chlorinated Natural Waters

#### *Dechlorination*

Samples of chlorinated natural waters were dechlorinated with sodium thiosulfate, because, of the three dechlorinating agents that we investigated, it was the least reactive toward DHAN. However, since its reactivity with bromochloroacetonitrile and dibromoacetonitrile is appreciable, minimal amounts of it should be added.

Another possibility might be to quench the free chlorine with ammonium chloride in an appropriate buffer. One would expect formation of monochloramine, which does not appear to give rise to THM or DHAN. Reduction of bromine containing DHAN is not expected to occur in the presence of ammonium chloride or chloramine.

#### *Effects of Time and pH*

Destruction of DHAN by hydrolysis is appreciable at the pH frequently encountered in water treatment plants. As we have shown, hydrolysis of DHAN, especially of dichloroacetonitrile, proceeds to a significant degree, even at a pH between 7 and 8, if analysis is delayed for several days. For accurate analysis of DHAN it is therefore essential that a water sample be analyzed as soon as possible after collection. This will minimize hydrolysis of DHAN as well as their destruction by thiosulfate. With THM, on the other hand, a delay in analysis may result in higher values. This is because at pH 7–8, the hydrolysis of trihaloacetaldehydes and trihalomethyl ketones, which results in the release of haloforms, is a slow process. Only at pH 10 or above does this hydrolysis become fast.

## REACTION OF DHAN WITH THIOSULFATE, SULFITE AND FERROCYANIDE

Since the water samples are treated with sodium thiosulfate to remove residual chlorine prior to extraction with pentane, the possibility of reaction between the DHAN and thiosulfate has to be considered. Thiosulfate ion is known to react with alkyl halides to form Bunte salt anions [21], according to the following equation:

$$RX + S_2O_3^{2-} \rightarrow R\text{-}S\text{-}SO_3^- + X^-$$

If thiosulfate were to react rapidly with DHAN in this or some other manner, their analysis by the LLE method would be severely compromised.

To throw light on the situation we measured the disappearance of DHAN in very dilute solutions (30-100 $\mu$g/L) in the absence and presence of thiosulfate. Ten drops of 10% sodium thiosulfate were used per 60 mL. This represents a large excess of thiosulfate so that its concentration remains essentially constant in the course of a run. Thus, for a given DHAN at a given pH, we could obtain $\tau_w$, its half-life in the absence of thiosulfate, and $\tau_{w+s}$, its half-life in the presence of a given constant concentration of thiosulfate. The latter half-life is, of course, due to the disappearance of the DHAN by a superposition of hydrolysis and the thiosulfate reaction.

Figure 13 shows that for each DHAN the plot of the log of peak height vs time is linear not only in the absence of thiosulfate but also in its presence, and that the two lines thus obtained deviate from each other. The line obtained in the presence of thiosulfate has the larger slope, indicating that the rate of disappearance of the DHAN is increased by the presence of thiosulfate. This means that $\tau_{w+s}$ is shorter than $\tau_w$. Table VI lists $\tau_w$ and $\tau_{w+s}$ for each of the DHAN at pH 8.32.

To compare the reactivities of the various DHAN to thiosulfate, we compare their $\tau_w:\tau_{w+s}$ ratios. The larger this ratio is, the greater is the reactivity of a DHAN to thiosulfate. The same information can be obtained by comparing the deviations in slope caused by the presence of thiosulfate. The larger the deviation, the greater is the reactivity of a DHAN to thiosulfate.

Our work clearly shows that DHAN have the following order of reactivity to thiosulfate:

$$CHCl_2CN < CHBrClCN < CHBr_2CN$$

This order is consistent with nucleophilic attack by thiosulfate on the alpha-carbon and expulsion of halide ion [21]. This order is the reverse of that for nitrile hydrolysis in the same series of compounds.

**Table VI. Half-lives of DHAN at pH 8.32 in the Absence and Presence of Sodium Thiosulfate**

| | $\tau_w$(h) | $\tau_{w+s}$(h) | $\frac{\tau_w}{\tau_{w+s}}$ |
|---|---|---|---|
| $CHCl_2CN$ | 29 | 25 | 1.1 |
| CHBrClCN | 55 | 32 | 1.7 |
| $CHBr_2CN$ | 85 | 12 | 7 |

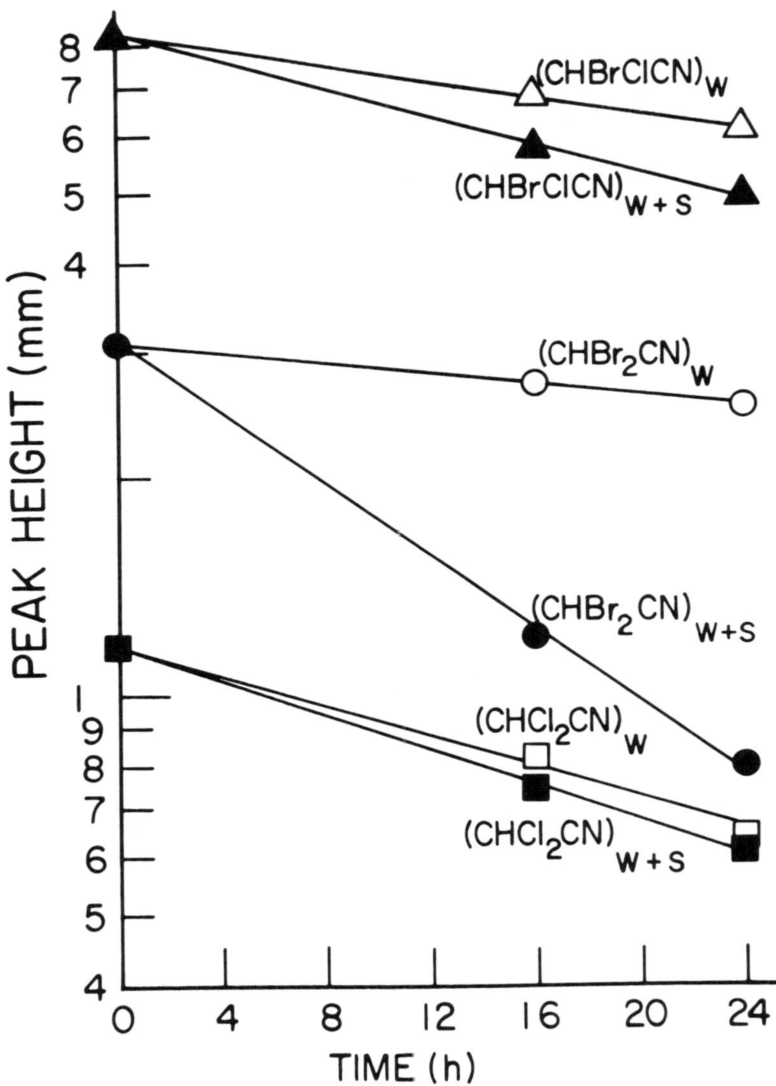


**Figure 13.** Hydrolysis rates for DHAN in the presence (w+s) and absence (w) of thiosulfate.

We have not determined the identity of the products resulting from the thiosulfate reaction and hence cannot distinguish between alternative types of nucleophilic attack. It is possible that thiosulfate, being the nucleophile, becomes attached to the carbon atom with formation of a Bunte salt. Or the attack may result in hydrogenation at the carbon atom, reduction (replacement of halide ion by a pair of electrons) being followed by addition of a proton.

We may write several alternative general reactions:

$$R\text{-}X + S_2O_3^{2-} \rightarrow R\text{-}S_2O_3^- + X^- \tag{1}$$

$$R\text{-}X + S_2O_3^{2-} + H_2O \rightarrow R\text{-}H + SO_4^{2-} + S + H^+ + X^- \tag{2}$$

$$R\text{-}X + 2\,S_2O_3^{2-} + H_2O \rightarrow R\text{-}H + S_4O_6^{2-} + OH^- + X^- \tag{3}$$

It will be noticed that in concentrated solutions the pH would remain unchanged in Reaction 1, would fall in Reaction 2 and would rise in Reaction 3. Furthermore, a precipitate of sulfur would form in Reaction 2.

To test these possibilities, 0.5 mL of dibromoacetonitrile was added to 100 mL of 10% sodium thiosulfate solution at room temperature. A portion of the resulting solution was used to fill completely a septum bottle which was immediately sealed. Another portion was used to monitor the pH with respect to time. The pH rose from 6 to 7.2 in the first few minutes and then decreased to pH 5 in about half an hour. Both the sealed solution and the open solution became turbid on standing.

These results do not lend themselves to ready explanation, but we lean to the view that the reaction is mainly hydrogenation to acetonitrile in the case of dibromoacetonitrile.

We did a similar study with dichloroacetonitrile and found that the pH did not change nor was a precipitate formed. The solution remained clear but turned yellowish after several days.

It is of interest that 2,2-dibromo-3-nitrilopropionamide is converted by "a number of reducing nucleophiles such as: $I^-$, $HS^-$, $HSO_3^-$, $S_2O_3^{2-}$ and $SO_3^{2-}$" to cyanoacetic acid [19].

$$\mathrm{Br_2C(CONH_2)(CN)} \longrightarrow \mathrm{H_2C(COOH)(CN)}$$

We also studied the effects of sodium sulfite and potassium ferrocyanide on the various DHAN, since both of these can serve in the removal of chlorine from chlorinated water.

A solution containing all of the DHAN was prepared (36 μg/L $CHCl_2CN$, 40 μg/L CHBrClCN, 59 μg/L $CHBr_2CN$) and the pH was adjusted to 7 using a phosphate buffer solution. To determine the relative rates of reaction of the ferrocyanide, thiosulfate and sulfites with DHAN, the bottles were divided into four groups. There was added to the first group 10 drops of 10% sodium thiosulfate, to the second group 1 drop of 10% potassium ferrocyanide and to the third group 1 drop of 10% sodium sulfite. The fourth group served as a control.

Sodium sulfite (actually a mixture of $SO_3^{2-}$ and $HSO_3^-$ at pH 7) was found to be the most reactive. In less than 40 min neither of the bromine containing DHAN could be detected. A 10% loss of dichloroacetonitrile was observed after 24 hr.

Potassium ferrocyanide was also observed to react with the bromine-containing DHAN, but not as rapidly as sodium sulfite. This reaction was not first-order under our conditions. No degradation of dichloroacetonitrile by potassium ferrocyanide was observed.

Of the three dechlorinating agents investigated, thiosulfate was the least reactive. Even though the concentration of the thiosulfate was ten times greater than that of either sulfite or ferrocyanide, the rate of disappearance of the bromine-containing DHAN was slowest in its presence. It attacks dichloroacetonitrile extremely slowly at pH 8.3. At pH 6.8, thiosulfate appears to be unreactive toward dichloroacetonitrile.

To summarize, the order of reactivity of the three dechlorinating agents to the DHAN is:

$$SO_3^{2-} > Fe(CN)_6^{4-} > S_2O_3^{2-}$$

## FORMATION OF DHAN

We consider candidate precursors for DHAN to be certain amino acids, polypeptides containing the corresponding amino acyl groups at the N-terminus, and humic and fulvic substances with amino acid moieties appended to the ring system. We have, indeed, shown that at least five amino acids, namely aspartic acid (**IV**), tryptophan (**VII**), kynurenine (**VIII**), kynurenic acid (**IX**) and tyrosine (**X**) yield DHAN on halogenation.

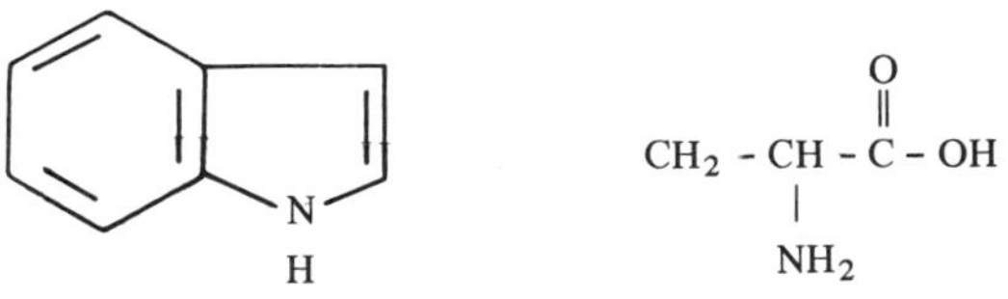

N
H

O
||
$CH_2$ - CH - C - OH
|
$NH_2$

**(VII)**

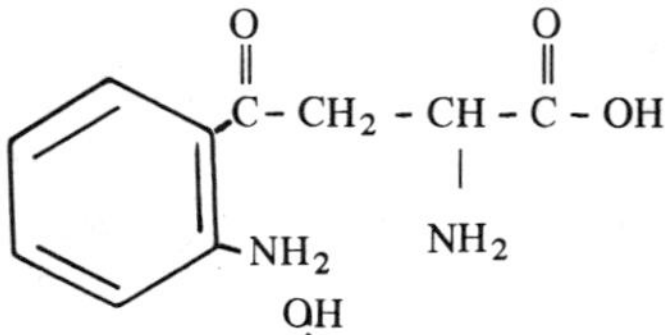

$$\text{(o-aminophenyl ring)}-\overset{O}{\overset{\|}{C}}-CH_2-\underset{NH_2}{\underset{|}{CH}}-\overset{O}{\overset{\|}{C}}-OH \quad (\text{ring substituent: } NH_2)$$ **(VIII)**

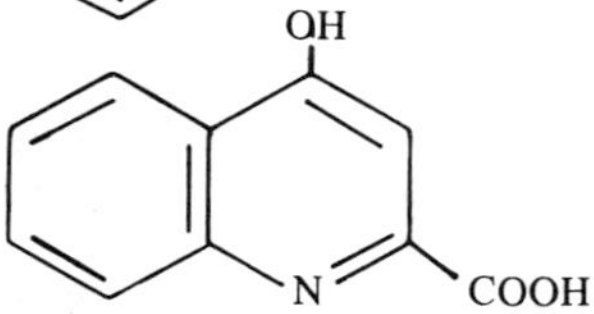

OH, N, COOH (quinoline ring structure) **(IX)**

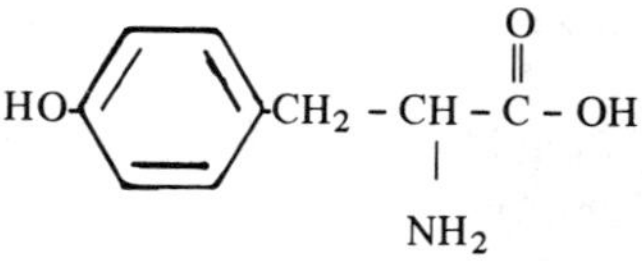

$$HO-\text{(benzene ring)}-CH_2-\underset{NH_2}{\underset{|}{CH}}-\overset{O}{\overset{\|}{C}}-OH$$ **(X)**

Moreover, we have found that humic acid (purchased from Aldrich Chemical Co.) yields on chlorination appreciable quantities of DHAN, besides THM. On the other hand, two porphyrins (etioporphyrin I and protohemin IX) gave no DHAN.

The results of our chlorination experiments with a variety of substrates are summarized in Tables I and II, not only with regard to dichloroacetonitrile formation, but also with regard to chloroform formation. Not all of the substrates contain nitrogen.

Past work on the halogenation of amino acids will now be reviewed briefly and brought into perspective.

The halogenation of amino acids was investigated by Langheld [22], using sodium hypochlorite, by Dakin [23,24], using chloramine-T, and by Goldschmidt et al. [25] and Friedman and Morgulis [26], using sodium hypobromite. Goldschmidt et al. also studied the reaction of polypeptides with sodium hypobromite [25].

Reaction of an amino acid with an equimolar amount of halogenating agent leads generally to aldehyde and carbon dioxide.

$$R-\underset{NH_2}{\underset{|}{CH}}-COOH \rightarrow R-\underset{O}{\underset{\|}{CH}} + CO_2$$

If the halogenating agent is used in excess, the decarboxylation yields not only aldehyde, but also nitrile, the nitrile-to-aldehyde ratio tending to decrease as the pH at which the reaction is run is increased.

$$\begin{array}{l} \text{R-CH-COOH} \longrightarrow \text{R-CHO} + CO_2 \\ \quad\;\; | \\ \quad\;\, NH_2 \longrightarrow \text{R-C}{\equiv}\text{N} + CO_2 \end{array}$$

Figures 14 and 15 account for the facts just cited. In Figure 15, as more alkaline reaction conditions are chosen for the halogenation, hydrolysis of the N-haloaldimine to aldehyde is presumed to undergo a larger rate increase than its dehydrohalogenation to nitrile. Thus, the nitrile to aldehyde product ratio is reduced with increasing pH (even though the formation of both products may be accelerated).

If the amino acid is, for example, alanine ($R{=}CH_3$), halogenation with excess reagent produces acetaldehyde and acetonitrile. The halogenation of either of these substances would require an elevated pH (above 9) and would in the case of acetaldehyde give rise to THM. It is not surprising, therefore, that unhalogenated aldehydes (such as acetaldehyde) and unhalogenated nitriles (such as acetonitrile) have been reported in chlorinated water supplies [27].

Although the halogenation of cyanoacetic acid is a known method for the synthesis of dihaloacetonitriles [13,14], their formation by the halogenation of aspartic acid has not been reported heretofore. Dichloroacetaldehyde was reported to result from the chlorination of aspartic acid with chloramine T near the neutral point [23] as well as by the reaction of aspartic acid with the $KMnO_4 + KBr + H_2SO_4$ system [28].

With the LLE method, employing pentane as the extracting solvent, we observed dichloroacetonitrile and a smaller quantity of chloroform after

$$\underset{\displaystyle NH_2}{\text{R–CH–COOH}} \xrightarrow{NaOX} \underset{\displaystyle NHX}{\text{R–CH–COOH}}$$

$$\downarrow -(CO_2 + HX)$$

$$\text{R–CH}{=}\text{NH}$$

$$\downarrow + H_2O$$

$$\text{R–CH}{=}\text{O} + NH_3$$

**Figure 14.** Use of an equivalent amount of halogenating agent.

$$\underset{\displaystyle NH_2}{R\!-\!CH\!-\!COOH} \rightarrow \underset{\displaystyle NX_2}{R\!-\!CH\!-\!COOH}$$

$$\downarrow -(CO_2 + HX)$$

$$\underset{\displaystyle NX}{\overset{}{R\!-\!CH}} \;(\text{R–CH}=\text{NX})$$

$$-HX \swarrow \qquad \searrow + H_2O$$

$$R\!-\!C\!\equiv\!N \qquad\qquad R\!-\!CHO + H_2NX$$

$$\downarrow \tfrac{1}{2}NaOX$$

$$\tfrac{1}{2}N_2$$

**Figure 15.** Use of an excess of halogenating agent.

chlorinating an aspartic acid solution (1 mg/L) with sodium hypochlorite at pH 7.8. The yield of dichloroacetonitrile after 30 min at room temperature was 7%. Although dichloroacetaldehyde is likely to be a significant product of the chlorination of aspartic acid, we have not been able to detect it by the LLE method, probably because it is too water-soluble to be extracted by pentane. We furthermore suspect the presence of dihaloacetaldehydes in chlorinated water supplies.

At a pH higher than 7, the dihaloacetaldehydes are expected to be halogenated to trihaloacetaldehyde (such as chloral), which, on hydrolytic cleavage, would produce THM. It is also possible that the decarboxylation of HOOC-$CX_2$-CHO (Figure 16) proceeds directly to the enol form of dihaloacetaldehyde. This enol form, unlike the carbonyl form, would be rapidly halogenated.

In the case of aspartic acid (R=$CH_2COOH$), halogenation is expected to yield malonic acid semialdehyde (**XI**) and cyanoacetic acid (**XII**). Each of these products contains a highly activated methylene group and should therefore be rapidly halogenated even at low pH. Halogenation would transform $CH_2$ to $CX_2$ and would be followed by decarboxylation, leading to dihaloacetaldehyde and DHAN, respectively (Figure 16).

We plan to determine the pH dependence of the chlorination of dichloroacetaldehyde. A method for detecting and quantifying dihaloacetaldehydes and trihaloacetaldehydes in chlorinated water supplies also needs to be developed.

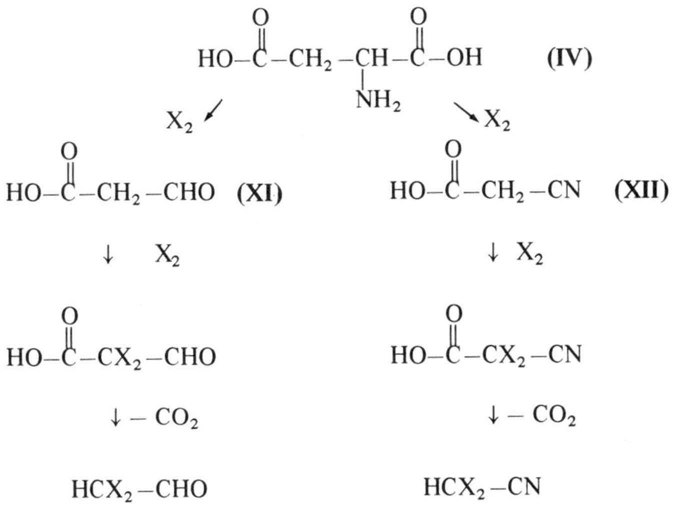

$$HO-\overset{\overset{O}{\|}}{C}-CH_2-\underset{\underset{NH_2}{|}}{CH}-\overset{\overset{O}{\|}}{C}-OH \qquad \textbf{(IV)}$$

$X_2 \swarrow \qquad \searrow X_2$

$$HO-\overset{\overset{O}{\|}}{C}-CH_2-CHO \ \textbf{(XI)} \qquad HO-\overset{\overset{O}{\|}}{C}-CH_2-CN \ \textbf{(XII)}$$

$\downarrow X_2 \qquad \downarrow X_2$

$$HO-\overset{\overset{O}{\|}}{C}-CX_2-CHO \qquad HO-\overset{\overset{O}{\|}}{C}-CX_2-CN$$

$\downarrow -CO_2 \qquad \downarrow -CO_2$

$$HCX_2-CHO \qquad HCX_2-CN$$

**Figure 16.** Halogenation of aspartic acid.

In addition to aspartic acid, tryptophan and tyrosine have been reported to give dibromoacetaldehyde by reaction with the $KMnO_4$ + KBr + $H_2SO_4$ system [28]. We therefore chlorinated dilute solutions (1 mg/L) of tryptophan and tyrosine at pH 7.8 to see whether chloroform and dichloroacetonitrile would be produced. We also extended this investigation to kynurenine and kynurenic acid, both of which are biochemically related to tryptophan.

Chlorination of tryptophan as well of kynurenine produced appreciable quantities of dichloroacetonitrile. Both also produced chloroform, kynurenine more so than tryptophan. Kynurenic acid gave a moderate quantity of chloroform and only a trace of dichloroacetonitrile. Tyrosine yielded small quantities of both chloroform and dichloroacetonitrile.

These results may be rationalized in terms of Figures 17 (for tryptophan and kynurenine), 18 (for kynurenic acid) and 19 (for tyrosine).

## ACKNOWLEDGMENTS

The mass spectra shown in Figure 6 were run by J. W. Louda at Florida Atlantic University.

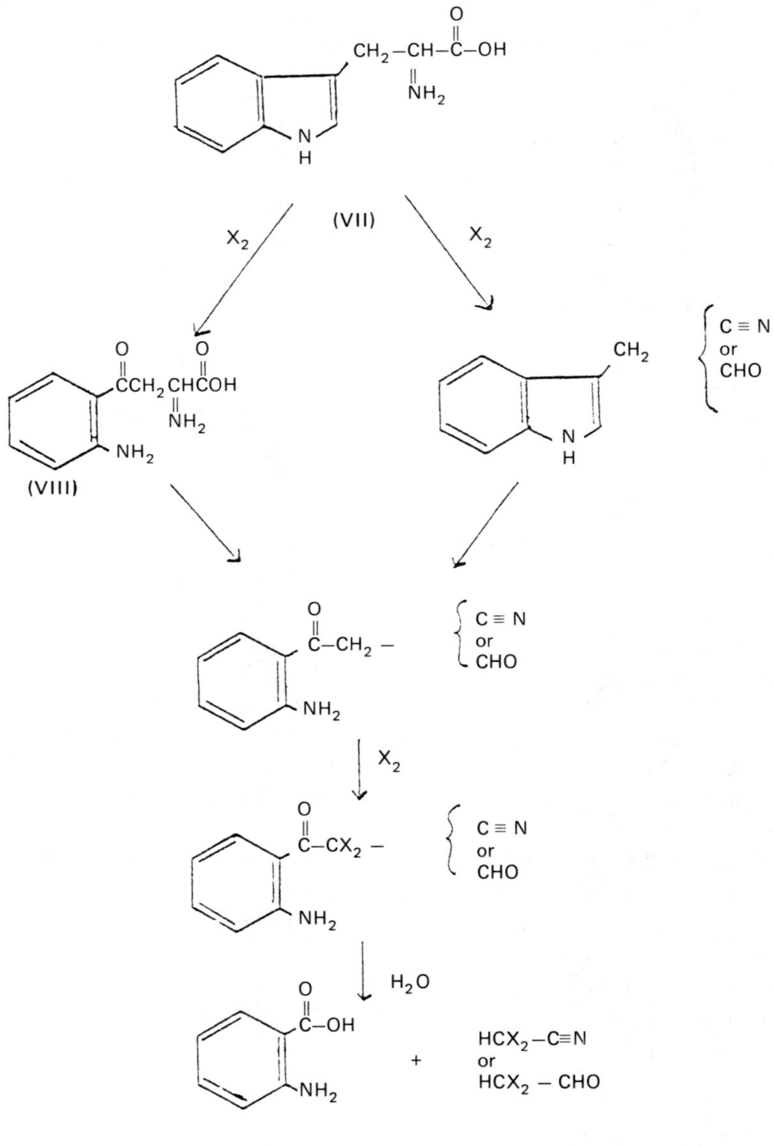

**Figure 17.** Halogenation of tryptophan and kynurenine.

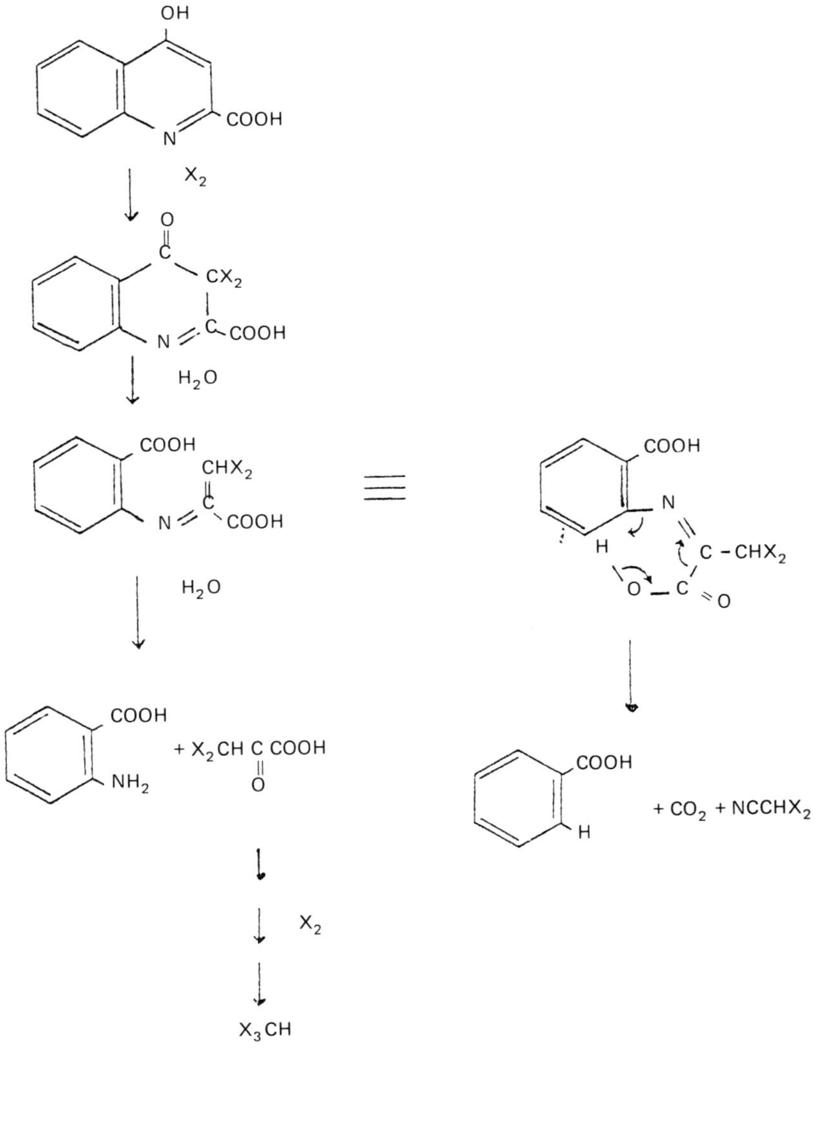

OH

COOH

N

$X_2$

O

C

$CX_2$

C COOH

N

$H_2O$

COOH

$CHX_2$

C

N COOH

≡

COOH

N

H

C - $CHX_2$

O C

O

$H_2O$

COOH

+ $X_2$CH C COOH

O

$NH_2$

COOH

+ $CO_2$ + $NCCHX_2$

H

$X_2$

$X_3CH$

**Figure 18.** Halogenation of kynurenic acid.

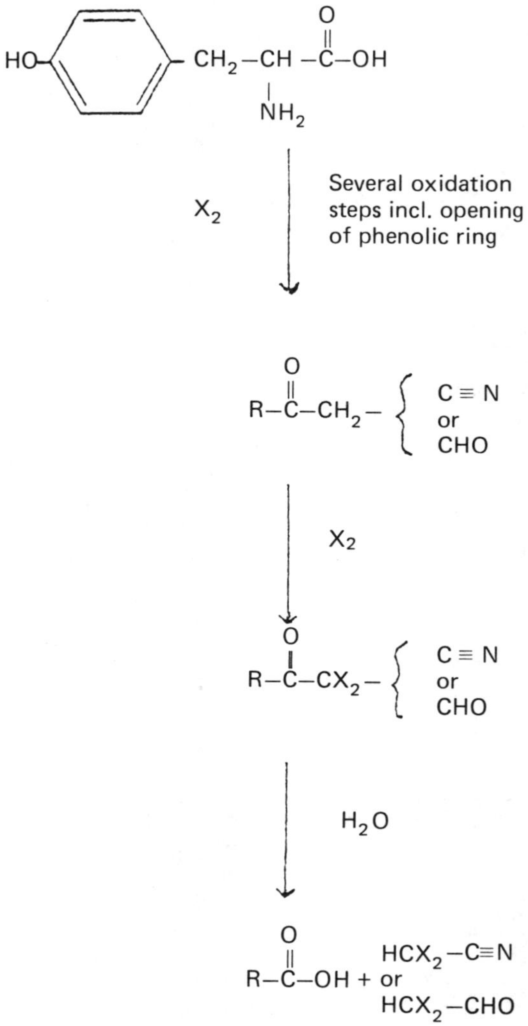

$$HO-C_6H_4-CH_2-CH(NH_2)-C(=O)-OH \quad (X)$$

$X_2$ — Several oxidation steps incl. opening of phenolic ring

$$R-C(=O)-CH_2-\begin{cases} C\equiv N \\ \text{or} \\ CHO \end{cases}$$

$X_2$

$$R-C(=O)-CX_2-\begin{cases} C\equiv N \\ \text{or} \\ CHO \end{cases}$$

$H_2O$

$$R-C(=O)-OH + \begin{matrix} HCX_2-C\equiv N \\ \text{or} \\ HCX_2-CHO \end{matrix}$$

**Figure 19.** Halogenation of tyrosine.

## REFERENCES


1. Rook, J. J. *Water Treat. Exam.* 23:234 (1974).
2. Bellar, T. A., J. J. Lichtenberg and R. C. Kroner. *J. Am. Water Works Assoc.* 66:703 (1974).
3. McKinney, J. D., R. R. Mauer, J. R. Hass and R. O. Thomas. In: *Identification and Analysis of Organic Pollutants in Water*, L. H. Keith, Ed. (Ann Arbor, MI: Ann Arbor Science Publishers, Inc., 1976), p. 417.
4. Stevens, A. A. Paper presented at the Conference on Oxidation Techniques in Drinking Water Treatment, Karlsruhe, Federal Republic of Germany, September 9-13, 1979.
5. Glaze, W. H., F. R. Peyton and R. Rawley. *Environ. Sci. Technol.* 11: 685 (1977).
6. Simmon, V. F., K. Kauhanen and R. G. Tardiff. *Prog. Gen. Toxicol.* 2: 249 (1977).
7. Cotton, R. T., and H. H. Walkden. *J. Econ. Entomol.* 30:529 (1946).
8. Matt, J. U.S. Patent 3,608,084 (1968).
9. Henderson, J. E., C. R. Peyton and W. H. Glaze. In: *Identification and Analysis of Organic Pollutants in Water*, L. H. Keith, Ed. (Ann Arbor, MI: Ann Arbor Science Publishers, Inc., 1976), p. 105.
10. Kaiser, L. E., and B. G. Oliver. *Anal. Chem.* 48:2207 (1976).
11. Bellar, T. A., and J. J. Lichtenberg. *J. Am. Water Works Assoc.* 66: 739 (1974).
12. Stevens, A. A., and D. Seeger, U.S. EPA, Cincinnati, OH. Personal communication (1979).
13. Wilt, J. W. *J. Org. Chem.* 21:920 (1956).
14. Rabjohn, N., Ed. *Organic Synthesis, Vol. IV* (New York: John Wiley & Sons, Inc., 1963), pp. 254-256.
15. "Spectrum 24255," in *Standard Infrared Spectra* (Philadelphia, Sadtler Research Laboratories, Inc., 1968).
16. Hechenbleikner, I. U.S. Patent 2,394,912 (1946).
17. Wohl, A., and H. Roth. *Ber. Deutsch. Chem. Ges.* 40:212 (1907).
18. Shirley, D. A. *Preparation of Organic Intermediates* (New York: John Wiley & Sons, Inc., 1951), p. 108.
19. Sherma, J., and M. Beroza. "Manual of Analytical Quality Control for Pesticides in Human and Environmental Media," U.S. EPA, Research Triangle Park, NC (1979), p. 114.
20. Dressman, R. C., A. A. Stevens, J. Fair and B. Smith. *J. Am. Water Works Assoc.* 71:392 (1979).
21. March, J. *Advanced Organic Chemistry*, 2nd ed. (New York: McGraw-Hill Book Co., 1977).
22. Langheld, K. *Ber. Deutsch. Chem. Ges.* 42:392 (1909).
23. Dakin, H. D. *Biochem. J.* 10:319 (1916).
24. Dakin, H. D. *Biochem. J.* 11:79 (1917).
25. Goldschmidt, S., E. Wiberg, F. Nagel and K. Martin. *Annal. Chem.* 456:1 (1927).
26. Friedman, A. H., and S. Morgulis. *J. Am. Chem. Soc.* 58:909 (1936).
27. Coleman, W. E., R. D. Lingg, R. G. Melton and F. C. Kopfler. In: *Identification and Analysis of Organic Pollutants in Water*, L. H. Keith, Ed. (Ann Arbor, MI: Ann Arbor Science Publishers, Inc., 1976), p. 305.
28. Suomalainen, H., and E. Arhimo. *Z. Anal. Chem.* 28:206 (1948).

# SECTION 12

# SURFACE WATER ANALYSES

# CHAPTER 49

# OXIDATIVE DEGRADATION OF AQUATIC HUMIC MATERIAL

**R. F. Christman, W. T. Liao,**
**D. S. Millington and J. D. Johnson**

Department of Environmental Sciences
and Engineering
University of North Carolina at Chapel Hill
Chapel Hill, North Carolina

The structural composition of the diverse natural organic materials in terrestrial streams is not known with a desirable level of scientific accuracy. Many of the classifications of aquatic organic material (humic acid, fulvic acid, plant extractives, etc.) are defined procedurally, and it is presumptive to conclude that these substances are characterized by any chemical or physical homogeneity in time or space. Nevertheless, it has been suggested that aquatic humic substances are responsible for the ubiquitous production of trihalomethanes (THM) in natural waters exposed to HOCl [1].

Given the growing list of organic compounds identified in natural water systems, it is reasonable to believe that natural product organic materials in terrestrial streams comprise a reactive pool of organic carbon that, when exposed to oxidants in treatment situations, produces materials in addition to THM. They, therefore, constitute an important background source or organic compounds against which other organic contaminants can be interpreted. In addition, many of these natural oxidative products may have human health consequences. Our objectives have been to isolate natural aquatic humic substances (humic and fulvic acids) from natural aquatic environments subject them to controlled chemical oxidative experiments and to identify their principal degradation products.

## HUMIC ISOLATION PROCEDURES

We have used two isolation procedures to obtain humic substances from Black Lake (an eastern North Carolina surface water), the first of which is shown in Figure 1. Samples of Black Lake water were filtered through a 10-$\mu$m filter and acidified to pH 2 with HCl to precipitate a combined humic-hymatomelanic acid (HA) fraction which was separated by settling and centrifugation. The precipitated pellet was redissolved in a HaOH solution, filtered, reprecipitated with HCl and recentrifuged. The resulting pellet was washed with HCl solution (pH 2) and distilled water prior to drying in a vacuum. The final humic acid was found to contain approximately 10% ash. The supernatant containing soluble fulvic acid (FA) at pH 2 was passed through a column containing XAD-2 resin under conditions outlined by Mantoura and Riley [2]. The column effluent was discarded. The adsorbed FA fraction was eluted from the resin with 0.2 *N* NaOH, filtered, reacidified to pH 5 and dialyzed against distilled water. A powdered fulvic acid with an ash content of 4.3% was obtained after freeze-drying.

More recently our humic materials have been prepared based on the procedure of Thurmann and Malcolm [3]. The procedure was slightly modified as shown in Figure 2. Black Lake water was filtered and acidified as before. After passing the sample through a column of XAD-8 polyacrylic ester with a flowrate of 1-2 column volumes per hour, the column was eluted with 0.1 *N* NaOH, and the dissolved organic concentration of collected eluent was adjusted to 500 ppm. The HA fraction was then separated from the solution as before, dialyzed against an AG MP-50 cation exchange resin in the hydrogen form (placed in the bottom of the dialysis container) and freeze-dried.

The supernatant containing the FA fraction was put on the XAD-8 column, and excess chloride was removed by washing with distilled water until the organic material began to elute. The FA was then eluted with 0.1 *N* NaOH followed by ion exchange on the AG MP-50 resin. The final desalted FA fraction was then freeze-dried. The elemental and ash analyses of the products of this procedure are shown in Table I.

## GEL CHROMATOGRAPHY

We have used an acrylic ester resin of intermediate polarity (Amberlite XAD-8) to help characterize the macromolecular properties of the unde-

Figure 1. Isolation of aquatic humic material with XAD-2.

graded humic materials. HA and FA are adsorbed from acidified solution at low pH on these resins, and are effectively desorbed in base. After cleaning, the gel bed was prepared in a 2.5-cm, 40-mL glass column and preconditioned with four bed volumes of the appropriate eluent. A 1-ml volume containing 1-5 mg of humic sample or standard compound was applied to the top of the bed with a pipet. The elution was performed at a flowrate of 1 mL/min, and effluent was collected continuously in 3-mL fractions with an automatic fraction collector. Elution profiles were determined by measurement of total organic carbon (TOC) with a Beckmann 915-A Carbon Analyzer and by measurement of optical density at 254 nm for the standard compounds. The void volume ($V_O$) of the column was determined using Blue Dextran 2000, and total effective pore volume ($V_t$) was determined using acetic acid, for which there was no absorption in basic eluent.

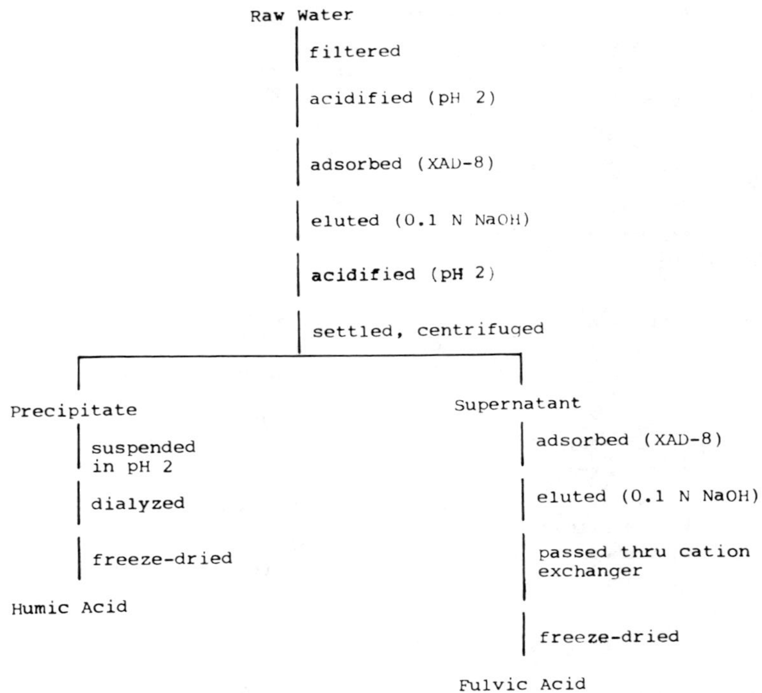


**Figure 2.** Isolation of aquatic humic material with XAD-8.

**Table I. Elemental Composition of Aquatic Humics (Thurmann Procedure)**

| | C | H | N | Ash |
|---|---|---|---|---|
| Humic Acid | 45.0 | 4.5 | 2.6 | 4.6 |
| Fulvic Acid | 46.7 | 3.5 | 1.3 | 1.0 |

We have found that higher pH values in the eluent tend to increase the fraction of the total humic carbon which elutes within $V_t$. It is attractive to assume that this is due to decreased adsorption that is not operative at pH 12. In addition to adsorptive effects on these resins, coulombic exclusion is apparently involved through the effect of salt concentrations on the small number of carboxyl groups within the resin matrix. We have observed that the effective pore size of the XAD-8 increases with increasing ionic strength of the eluent. Therefore our experiments were run at pH 11.5-12.0 with controlled ionic strength. The elution volumes for a selection of standard compounds are shown in Figure 3. The data in Figure 4 establish that humic acids isolated as described above are of apparently larger size and/or greater polarity than fulvic acids isolated by the same procedure. The effect of exposure to different relative concentrations of permanganate oxidant is apparent in Figures 5 and 6. In general, increased permanganate concentrations tend to reduce the apparent size and/or the relative polarity of both humic and fulvic acids such that the resulting elution volumes are quite close to those of the model compounds. Since the oxidation most probably increases polarity, it is attractive to assume that the effect is one of size reduction. Smaller relative concentrations of this oxidant tend to attack the larger size segments of the humic acid first converting them to size ranges more characteristic of fulvic acids. Finally these degradation experiments tend to attack all size fractions of the humic and fulvic materials.

## DEGRADATION EXPERIMENTS

Both HA and FA obtained from the above isolation procedure were subjected to chemical degradation study. The general degradation scheme is presented in Figure 7.

A 1-g sample of HA was first refluxed with 150 mL of 2 $N$ $H_2SO_4$ solution for 12 hr. After cooling, the reaction mixture was adjusted with sodium hydroxide to pH 2 and centrifuged. The supernatant containing acid hydrolysis products was freeze-dried and the residue was extracted with methylene chloride. Methyl esters of the extracted compounds were then formed using diazomethane in ether solution.

The precipitated HA fraction after acid hydrolysis was treated with 150 mL of 1.5 $N$ NaOH solution at 85°C under nitrogen for 4 hr. The reaction mixture was then made acidic (pH 2) with concentrated sulfuric acid and the base hydrolysis products were extracted and methylated as described above.

For oxidation with potassium permanganate, the HA after acid and base hydrolysis was dissolved in 150 mL of solution at pH 10. Measured amounts of potassium permanganate were then added to give carbon-to-potassium permanganate ratios by weight in the range of 1.5-3. The reactions were carried out at 65°C for 3-4 hr. No excess permanganate was found at the end of this period. After the pH value was adjusted to 12, reaction mixtures were filtered to remove manganese dioxide and then acidified to pH 2 with concen-

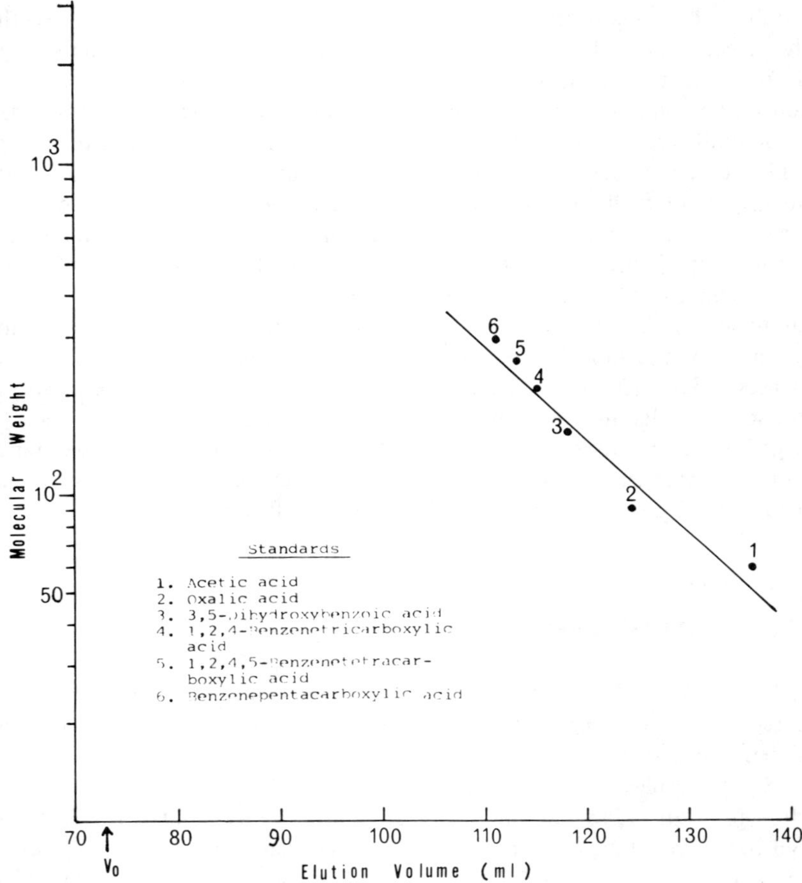


**Figure 3.** Standard elution volumes on XAD-3 at pH 12.

trated hydrochloric acid and divided into two aliquots. One aliquot was extracted with hexadecane for volatile compounds analysis. The other aliquot was freeze-dried, extracted with methylene chloride and methylated with diazomethane.

FA was directly subjected to permanganate oxidation. The workup procedures were then the same as above except the reaction had carbon-to-potassium permanganate ratios in the range of 1–2.

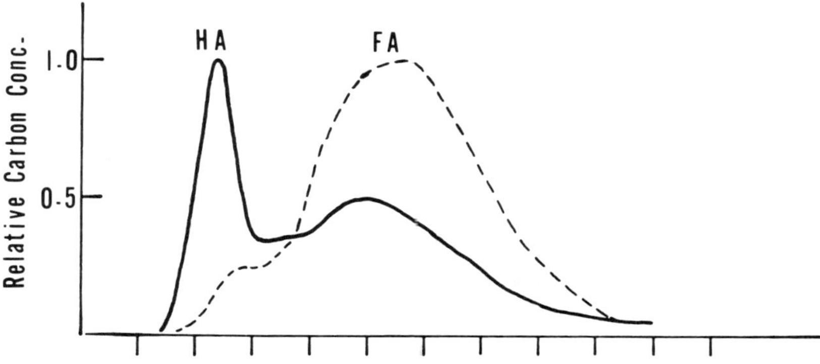


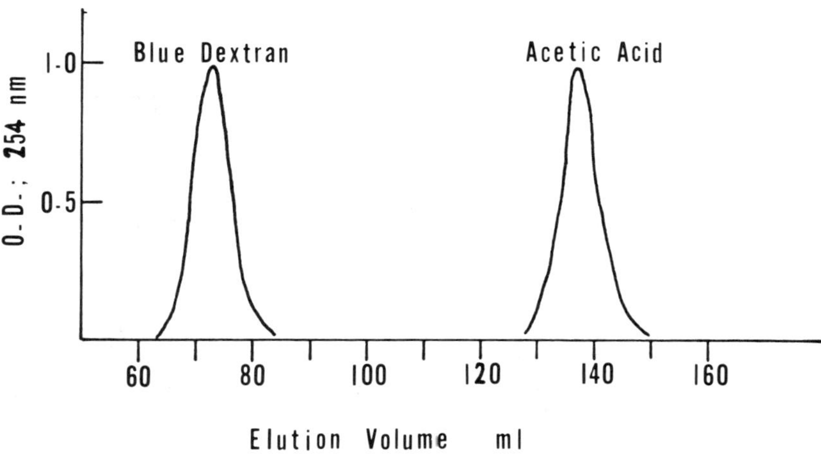


Figure 4. Elution of HA and FA on XAD-8 at pH 12.

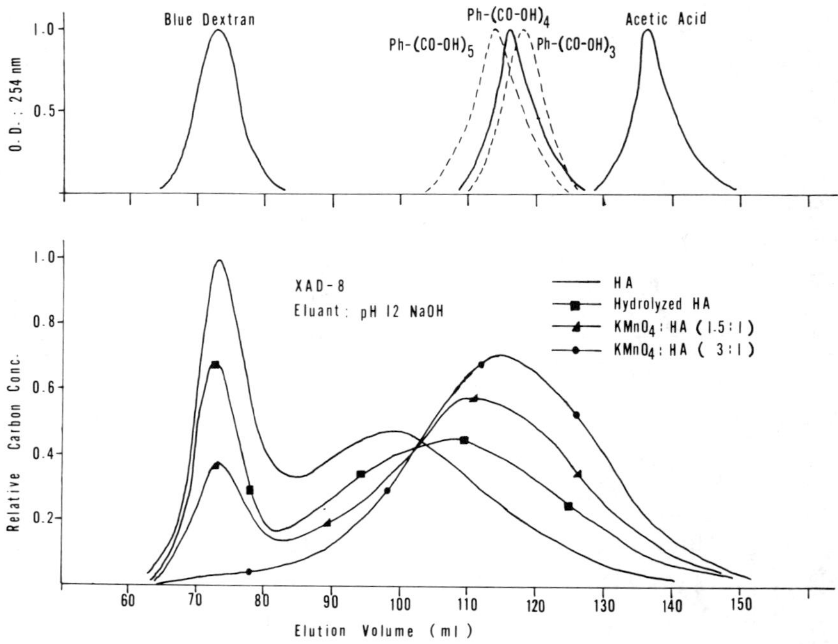


**Figure 5.** Effect of hydrolysis and oxidation on HA elution characteristics (XAD-8).

Carbon yields were monitored throughout the procedure by measuring TOC using a Beckmann carbon analyzer. Yields of chromatographable compounds were calculated by measurement of GC peak area using 1,2,4-benzenetricarboxylic acid methyl ester as an external standard (see Figure 7).

## GAS CHROMATOGRAPHY/MASS SPECTROMETRY (GC/MS) ANALYTICAL PROCEDURES

The GC/MS facility used in these studies consists of a Hewlett-Packard 5710-A gas chromatograph, interfaced to a VG-Micromass 7070F double-focusing mass spectrometer equipped with a combined electron impact/chemical ionization (EI/CI) source. Data reduction was performed by a VG-Datasystems 2035 F/B computer and storage on System Industries compatible 6.6-M-word dual-density cartridge disks. A Versatec 800A electrostatic printer/plotter was used to output the data.

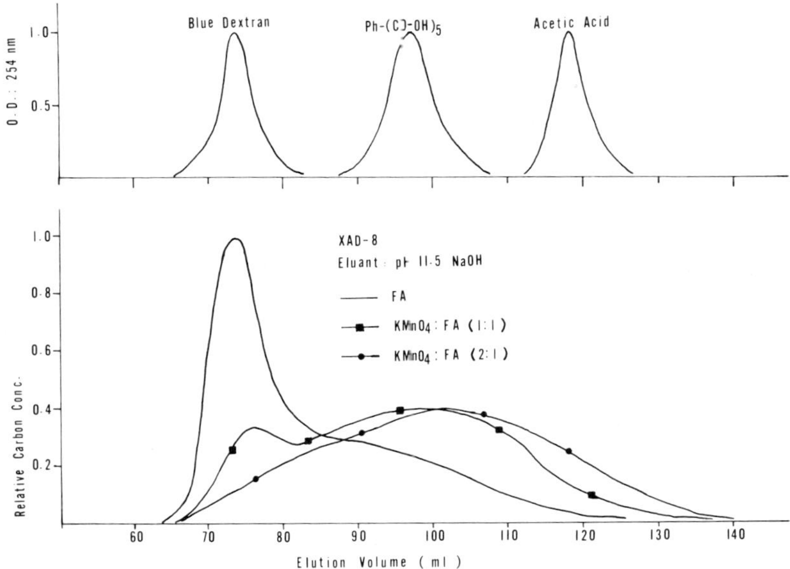


Figure 6. Effect of oxidation on FA elution characteristics (XAD-8).

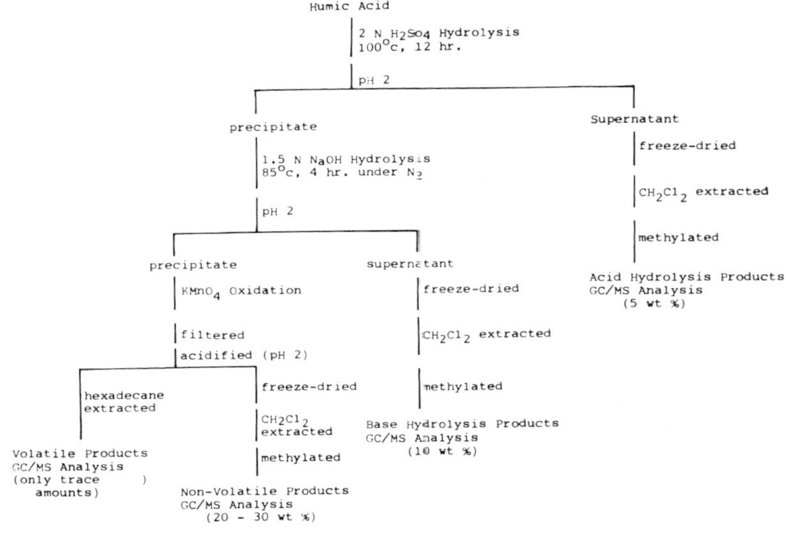


Figure 7. HA degradation procedure.

The gas chromatograph was fitted with a Grob-type split/splitless injector, supplied by GC[2] (chromatography) Ltd., Northwich, Chesire, U.K. The column used for most of the separations was a Hewlett-Packard 25-m OV-1 fused-silica capillary, which was easily inserted within the 0.7-mm i.d. GLT direct interface to the MS as far as the quartz inlet restriction, thus eliminating most of the potential interface problems such as dead-volume or active sites. At a head pressure of 4 $lb/in^2$, the helium flowrate through this column was about 1 mL/min at 60°C. Some analyses were also performed on a 50-m OV-17 Quadrex glass capillary column, supplied by Applied Science, Inc. The flowrate through this column was about 0.8 mL/min at a head pressure of 10 $lb/in.^2$ The program rate used on either column was usually 4°C/min after a 2-min delay from the injection temperature of 60°C, to a final temperature of 260°C.

The 7070F spectrometer is equpped with a Hall-probe regulated field-control system, enabling scan cycle times of around 1.2 sec. For convenience of data handling and compatibility with the present analytical problem, a scan cycle time of 2.3 sec was routinely employed (1-sec/decade, 1-sec interscan delay) over the mass range 450-20-450. For low-resolution accurate mass measurement, the cycle time was 3.0 sec (1.5-sec/decade, 1-sec interscan delay).

The standard operating conditions in EI mode were accelerating voltage, 4 kV; trap current, 100 $\mu$A; electron energy, 70 eV; source pressure, $2 \times 10^{-6}$ torr (1 mL/min helium flow); temperature, 200°C; resolution, 1000. In CI mode, the reactant gas used was isobutane and the operating conditions were filament current, 500 $\mu$A; electron energy, 100 eV; source pressure, approx. 0.3 torr; temperature, 180°; resolution, 1000. For accurate mass measurements, a low-resolution (~ 1500) technique employing tetraiodoethylene as internal standard was used enabling operation at full sensitivity. Details of the technique have been described previously [4].

Sample injection volumes were typically 1.5 $\mu$L, and split ratios of 5:1 to 10:1 were employed.

## DEGRADATION RESULTS AND DISCUSSION

The TIC chromatogram from the methylated products of an exhaustive permanganate oxidation of humic acid isolated by the original procedure is shown in Figure 8. The main compounds identified are straight (and some branched) chain fatty acid methyl esters, a series of aliphatic diesters with terminal carboxyl groups and a large number of one-ring aromatic esters. Interestingly, all the isomers of benzene mono-, di-, tri-, tetra-, penta- and hexacarboxylic acid methyl esters were identified. Most of these components

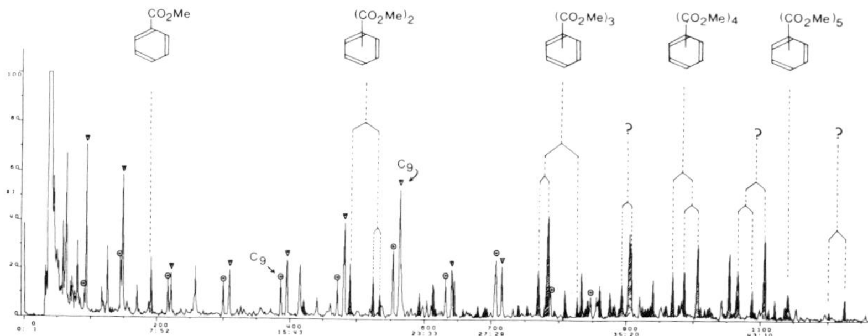


**Figure 8.** Total ion chromatogram of HA/$KMnO_4$ degradation products (OV-17).

were verified by comparison of GC retention and mass spectral data with those of authentic specimens.

In addition, another prominent series of aromatic esters, indicated by the "?" symbol in Figure 8, was observed, which have been tentatively interpreted as derivatives of mono $\alpha$-keto aromatic acids. For example, the lowest-molecular-weight member, whose EI and CI mass spectra are shown in Figure 9, is believed to be phthalonic acid dimethyl ester. The CI spectrum gave the most useful structural information here, including the molecular weight, whereas the EI spectrum contained only one prominent peak, corresponding to cleavage of the $\beta$-carbomethoxy group. Most of the aromatic esters do not exhibit molecular ion peaks in their EI spectra, hence the utility of CI as an interpretative aid in analysis of these mixtures.

It is interesting to note that phthalonic acid is produced in high yield by permanganate oxidation of naphthalene (Figure 10). If our mass spectral interpretations are correct, this constitutes evidence for the existence of fused aromatic rings in the humic acid macromolecule.

In general, the following criteria were applied in the structure identification of components in the complex mixtures produced by oxidative degradation of humic acid.

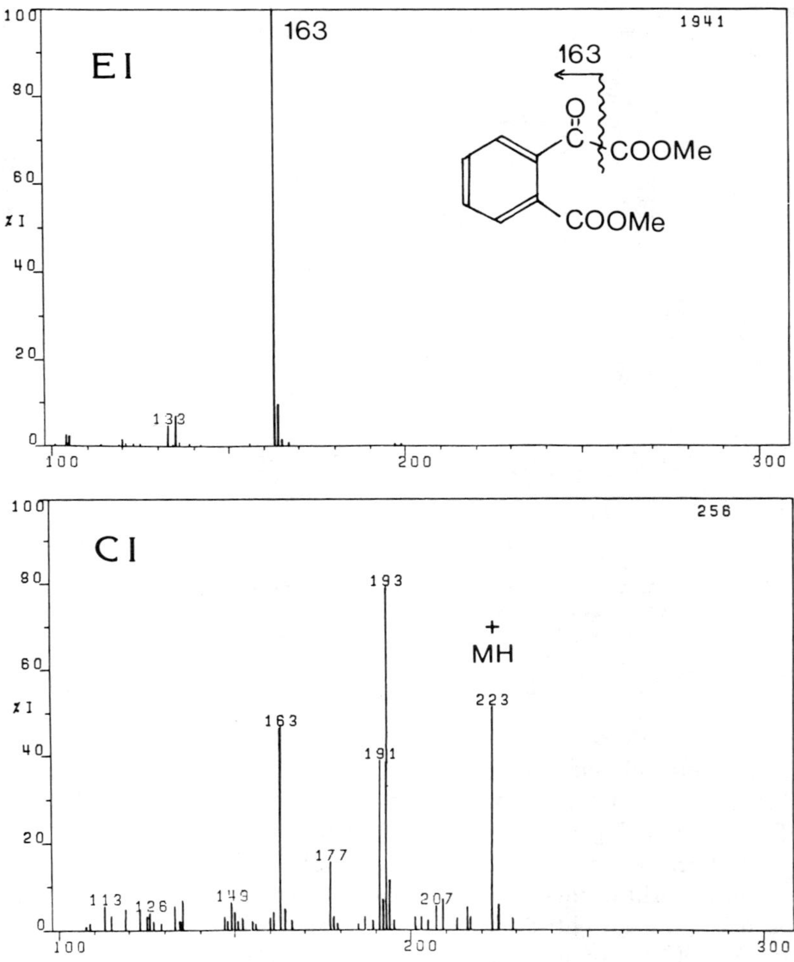


Figure 9. EI and CI mass spectra of HA/$KMnO_4$ degradation product.

1. the EI mass spectrum, giving a characteristic fragmentation pattern;
2. the CI mass spectrum, from which the molecular weight is deduced;
3. accurate mass measurements, which enables the assignment of elemental compositions to the major peaks in the EI spectrum, hence the molecular formula; and
4. comparison of GC retention and MS data with those of authentic compounds when available.

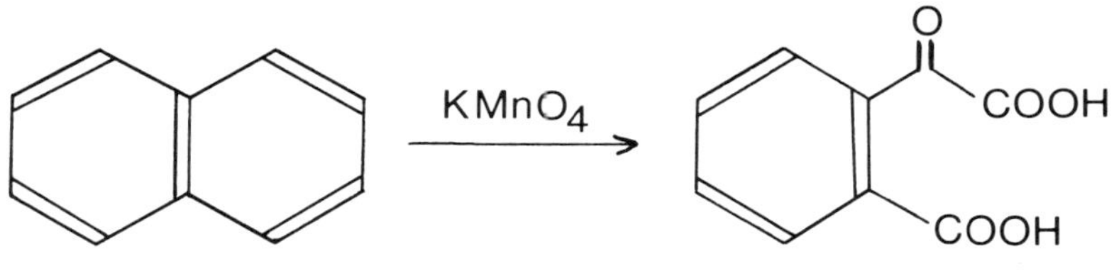

$$\xrightarrow{KMnO_4}$$

Figure 10. Possible source of phthalonic acid structures in degradation mixtures [5].

As a typical example of how the MS criteria were employed, Figure 11 shows EI, CI and accurate mass measurement data on one of the aliphatic dicarboxylic acid esters. The mass measurements obtained on the three most abundant EI peaks are correct to within 10 ppm of their true values. This was sufficient for unequivocal assignment of their elemental compositions based on the presence of C, H, N and O as possible elements of the molecule. From these and the CI spectrum, the molecular formula was established beyond doubt, and the EI fragmentation pattern interpreted as that of the diester shown. Finally, the EI spectrum and GC retention time were compared with those of an available authentic sample and the two sets of data did indeed correspond.

In Figure 12, the methylated oxidation products of the same humic acid sample using permanganate and HOCl are compared. The analyses shown here were performed on an OV-17 capillary column, which has better resolution than the OV-1 column for the lower-molecular-weight components. These data show that HOCl is an effective degradation agent for humic acid [6]. In general, similar aromatic compounds are produced as with permanganate, but more significantly, a high proportion of chlorinted aliphatic compounds are also produced, the most prominent being indicated in Figure 12 by arrows.

The major products have been identified as di- and trichloroacetic acid methyl esters, with dichlorosuccinic and both *cis*- and *trans*-dichlorofumaric acid methyl esters also prominent. A more complete summary of the compounds identified is given in Table II. The number and variety of chlorinated

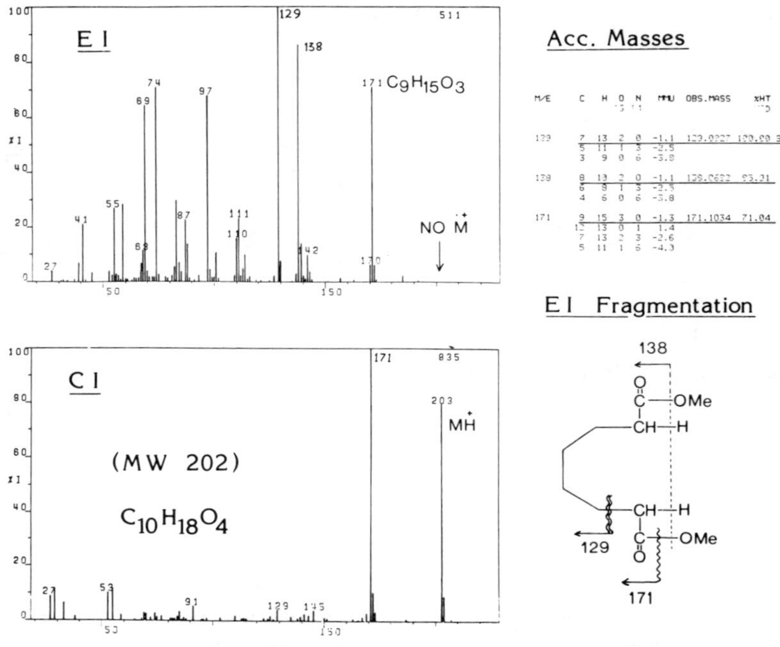


**Figure 11.** Example of structural assignment for dibasic acid degradation product.

products is noteworthy; however, no ring-chlorinated aromatic compounds were identified.

We would emphasize again that although many of the component identifications are tentative in that they have not been confirmed by comparison with authentic samples, the CI data and especially the accurate mass measurements were invaluable in making structural assignments with much greater confidence and for many more components than would otherwise have been possible.

These analytical procedures have also been applied to samples of humic and fulvic acid prepared by the new Thurmann procedure. Controlled oxidation of humic acid with permanganate (ratio 1:3) followed by extraction and methylation of the freeze-dried product gave the TIC chromatogram shown in Figure 13. Low-molecular-weight aliphatic compounds, some not yet identified, are produced in high yield. These would have been discriminated against by the extraction procedure described earlier and applied to the old humic acid sample (see Figure 8). In essence, the other oxidation products contain

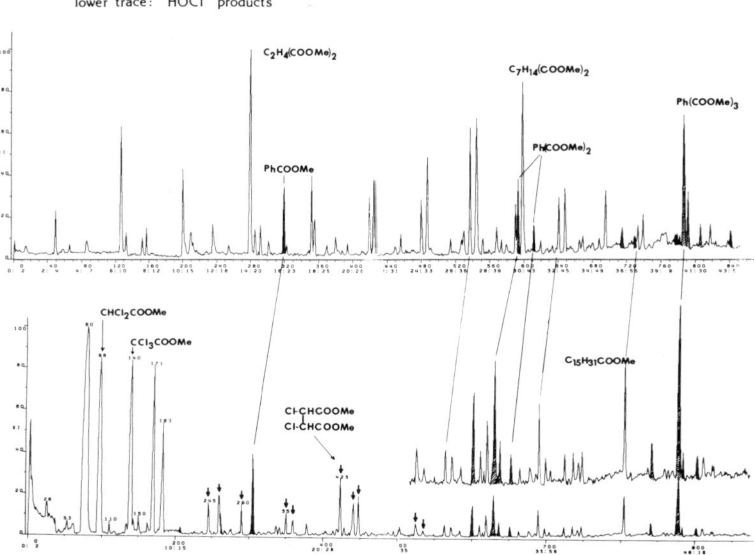


**Figure 12.** Comparison of HA degradation products with $KMnO_4$ and HOCl (OV-17).

**Table II. Humic Acid Chlorination: Ether Extract Methyl Esters (59-m OV-17 Capillary)**

| Proposed Structure | EI Scan | CI Scan | Accurate Masses Confirmed |
|---|---|---|---|
| $CH_2Cl{-}COOCH_3$ | 53 | 108 | Yes |
| $CH_3{-}\overset{O}{\overset{\|\|}{C}}{-}OC_2H_5$ (solvent) | 80 | – | Yes |
| $CHCl_2{-}COOCH_3$ | 96 | 189 | Yes |
| $H_3COOC{-}COOCH_3$ | 110 | 209 | Yes |
| Unknown (related to 143, 171, 183) | 134 | 251 | No |
| $CCl_3{-}COOCH_3$ | 137 | 260 | Yes |
| Unknown Isomer of 171 & 183 | 143 | 265 | No |

Table II, continued

| Proposed Structure | EI Scan | CI Scan | Accurate Masses Confirmed |
|---|---|---|---|
| $Cl_2C{=}CH{-}COOCH_3$ (Cl, Cl on C; H on C) (or) $+H_3C{-}C(=O){-}OCH_2Cl$; H(Cl)C=C(Cl)COOCH$_3$ | 150 | 274 | Yes |
| $H_3COOC{-}CH_2{-}COOCH_3$ | 162 | 299 | Yes |
| Unknown Methyl Ester (isomer of 183 & 143) | 171 | 310 | No |
| $C_2H_2Cl_3{-}COOCH_3$ (very tentative) | 176 | 325 | No |
| Unknown Isomer of 143 & 171 | 183 | 329 | No |
| $C_6H_5{-}COH$ (benzene ring with COH) | 207 | 363 | Yes |
| $C_3H_7CCl_2{-}COOCH_3$ | 245(246) | 425 | Yes |
| $H_3COOC{-}CH_2{-}CH_2{-}COOCH_3$ | 259 | 446 | Yes |
| $C_5H_7NO_2$ (ring: O, N–H, with $OCH_3$) tentative | 262 | 452 | Yes |
| $H_3C{-}C(Cl)(COOCH_3){-}COOCH_3$ (?) or most likely $H_2C(Cl){-}CH(COOCH_3){-}COOCH_3$ | 290 | 486 | Yes |

Table II, continued

| Proposed Structure | EI Scan | CI Scan | Accurate Masses Confirmed |
|---|---|---|---|
| $C_6H_5$-$COOCH_3$ (benzene ring with $COOCH_3$) | 305 | 507 | Yes |
| $H_3COOC$-$CH(C_2H_5)$-$COOCH_3$ (may be interchanged) | 337 | 555 | Yes |
| $H_3COOC$-$CH_2$-$CH(CH_3)$-$COOCH_3$ (may be interchanged) | 341 | 558 | Yes |
| $H_3COOC$-$CH_2$-$CHCl$-$COOCH_3$ + another methyl ester | 351 | 570 | Yes |
| $H_3COOC$-$CCl_2$-$COOCH_3$ | 360 | 581 | Yes |
| $H_3COOC$-$(CH_2)_3$-$COOCH_3$ | 378 | 610 | Yes |
| $H_3COOC$-$CCl_2$-$CH_2$-$COOCH_3$ | 423 | 667 | Yes |
| $H_3COOC$-$(CH_2)_4$-$COOCH_3$ | 428 | 674 | Yes |
| $H_3COOC$-$CCl$=$CCl$-$COOCH_3$ | 441 | 688 | Yes |
| (cis + trans) | 447 | 697 | Yes |
| $H_3COOC$-$(CH_2)_5$-$COOCH_3$ | 503 | 766 | Yes |
| $Cl$-$C_7H_6COOCH_3$ | 525 | 795 | Yes |

**Table II, continued**

| Proposed Structure | EI Scan | CI Scan | Accurate Masses Confirmed |
|---|---|---|---|
| | 535<br>(Isomer) | 809 | Yes |
| $C_{11}H_{23}$-$COOCH_3$ | 564 | 841 | Yes |
| $H_3COOC$-$(CH_2)_6$-$COOCH_3$ | 572 | 853 | Yes |
| $COOCH_3$<br>$[CH_3H_2Cl_2O]$ (?) (very tentative)<br>$COOCH_3$ | 584 | 870 | Yes |
| $C_{13}H_{10}O_2$(?) or $C_8H_{10}N_2O_4$ | 601 | 895 | Yes |
| 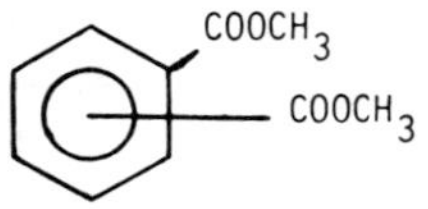 $COOCH_3$<br>$COOCH_3$ | 629 | 931 | Yes |
| 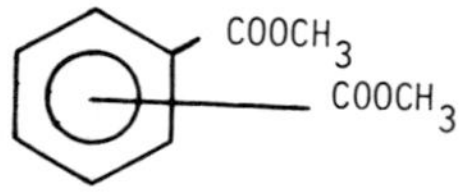 $COOCH_3$<br>$COOCH_3$ | 636 | 935<br>938 | Yes |
| 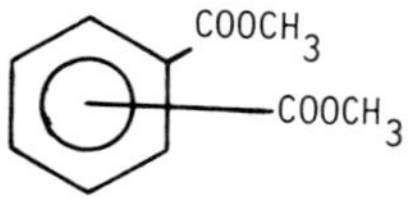 $COOCH_3$<br>$COOCH_3$ | 651 | 956 | Yes |
| $C_{13}H_{27}COOCH_3$ | 688 | 1001 | Yes |
| $H_3COOC$-$(CH_2)_8$-$COOCH_3$ | 698 | 1011 | Yes |

Table II, continued

| Proposed Structure | EI Scan | CI Scan | Accurate Masses Confirmed |
|---|---|---|---|
| 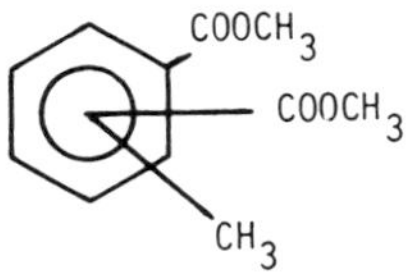 Benzene ring with $COOCH_3$, $COOCH_3$, $CH_3$ | 704 | 1022 | Yes |
| $C_{14}H_{29}COOCH_3$ | 724 | 1042 | Yes |
| | 730 (Isomer) | 1051 | Yes |
| | 747 (Isomer) | 1071 | Yes |
| $C_{15}H_{31}OCOOCH_3$ | 805 | 1143 | Yes |
| $C_{14}H_{12}O_2$ (possible artifact) | 840 | – | Yes |
| 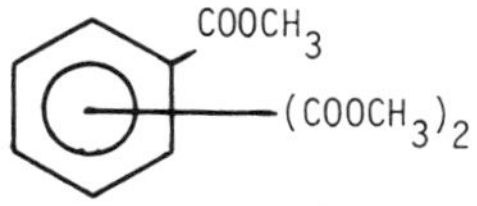 Benzene ring with $COOCH_3$, $(COOCH_3)_2$ | 878 | 1242 | Yes |
| 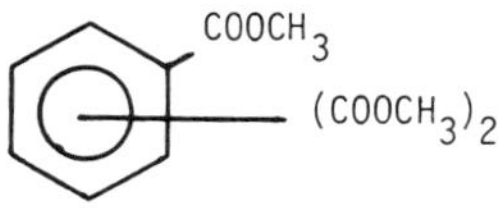 Benzene ring with $COOCH_3$, $(COOCH_3)_2$ | 902 | 1271 | Yes |
| 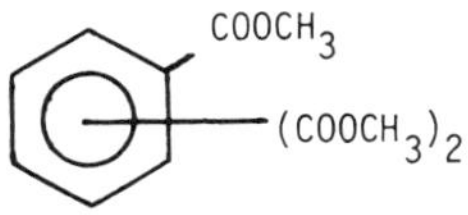 Benzene ring with $COOCH_3$, $(COOCH_3)_2$ | 883 | 1248 | Yes |
| $C_{17}H_{35}$-$COOCH_3$ | 910 | 1276 | Yes |

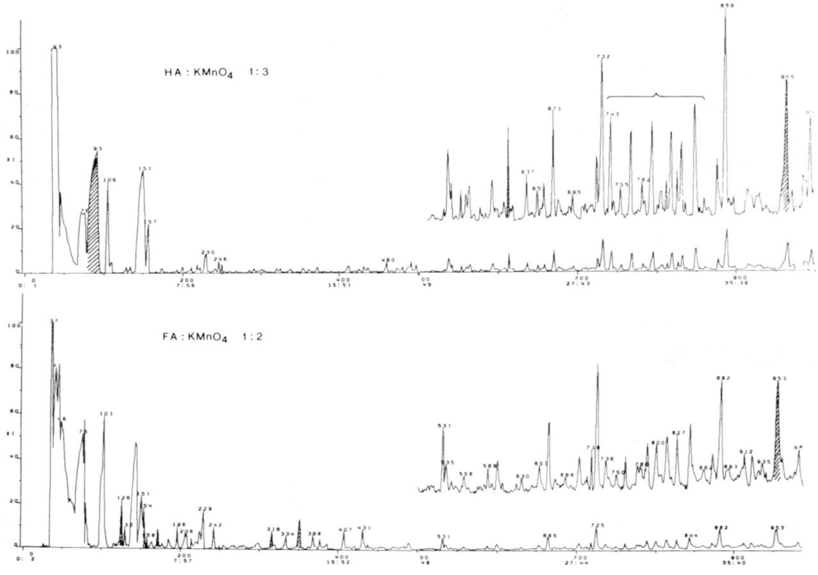


**Figure 13.** Comparison of $KMnO_4$ degradation products of HA and FA (OV-1).

the same components observed in the earlier experiment plus additional components believed to be less extensively degraded relatives, as judged from their mass spectra. The less stringent oxidation conditions employed would account for this observation.

Comparison of the TIC traces obtained form the excess permanganate oxidation of the new preparation of HA and FA (Figure 13) shows a similar distribution of product types, between low-molecular-weight oxidation products, other aliphatics and aromatics, but there are notable differences in the relative proportions of certain individual components. Owing to the incomplete status of the product identification, the significance of these results is not yet understood. Finally, the effect of increasing the permanganate-to-carbon ratio is apparently to shift the product distribution toward a preponderance of simpler products, especially one-ring aromatic acids and low-molecular-weight aliphatic acids. This is evident from the TIC chromatograms of both the fulvic and humic acid degradation products.

## ACKNOWLEDGMENT

This research was supported in part by EPA Research Grant No. R804430 from the Municipal Environmental Research Laboratory, Alan A. Stevens, Project Officer. ESE Publication No. 614.

## REFERENCES

1. Rook, J. J. "Haloforms in Drinking Water," *J. Am. Water Works Assoc.* 68:168-172 (1976).
2. Mantoura, R. F. C., and J. P. Riley. "The Analytical Concentration of Humic Substances from Natural Waters," *Anal. Chim. Acta* 76:97-106 (1975).
3. Thurmann, E. M., and R. L. Malcolm. Personal communication.
4. Powers, D., P. H. D'Arcy, J. C. Bill and M. J. Wallington. Proceedings of the Conference on Mass Spectrometry and Allied Topics, St. Louis, MO (1978), p. 480.
5. Gardner, J. H., and C. H. Naylor. *Org Syn. Coll.* 2:523 (1943).
6. Christman, R. F., J. D. Johnson, F. K. Pfaender, D. L. Norwood, M. Webb, J. R. Hass and M. H. Bobenrieth. "Chemical Identification of Aquatic Humic Chlorination Products," in *Water Chlorination: Environmental Impact and Health Effects, Vol. 3*, R. L. Jolley, W. A. Brungs and R. B. Cummings, Eds. (Ann Arbor, MI: Ann Arbor Science Publishers, Inc., 1980), pp. 75-84.

# CHAPTER 50

# LEVELS OF POLYCHLORINATED BIPHENYLS IN THE FORT EDWARD, NEW YORK, WATER SYSTEM

**M. Brinkman, K. Fogelman and J. Hoeflein**

Rensselaer Polytechnic Institute
Troy, New York

**T. Lindh**

University of Rochester
Rochester, New York

**M. Pastel***

State University of New York
at Albany
Albany, New York

**W. C. Trench and D. A. Aikens***

Rensselaer Polytechnic Institute
Troy, New York

Over the past decade, it has been established clearly that polychlorinated biphenyls (PCB) constitute a class of pollutants characterized by extreme environmental persistence [1-3] and severe toxicity to humans and animals [4-7]. Commercial PCB, which were marketed under the trade name Aroclor, are complex mixtures of polychlorinated derivatives of biphenyl having specified chlorine contents. Discovery of high concentrations of PCB in up-

*Authors to whom correspondence should be addressed.

state New York has resulted in widespread efforts to monitor and contain the distribution of PCB in this area [8,9]. Understandably, most studies have concerned sites of gross contamination, such as the Hudson River near the village of Fort Edward, NY, where an estimated 270,000 kg of PCB have entered the river [10]. On the other hand, the distribution of PCB in isolated surface waters close to this highly polluted section of the Hudson River has received little attention. We report here such a study of the public water supply system of the village of Fort Edward, NY, in June and July of 1978. In addition to defining the distribution of PCB in the sediment, water and biota of the water system, the results indicate that a substantial portion of the PCB burden of the water system originates from local sources and enters the water through the rainfall and surface runoff.

The Fort Edward water system, which is described in Figures 1 and 2, is located in a 290-ha watershed in the town of Moreau, Saratoga County [11]. The watershed lies approximately 0.5 km west of the Hudson River, and approximately 1 km upstream of the area of highest PCB contamination. The Caputo Disposal Site, a depository for transformers and capacitors containing PCB, is less than 0.5 km north of the northern perimeter of the watershed.

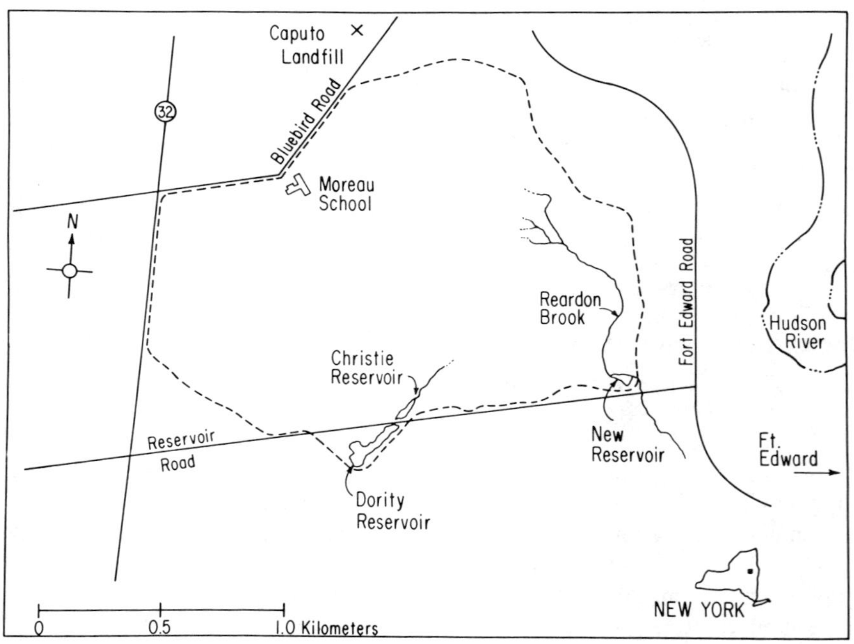


Figure 1. Map of Fort Edward, NY, water system and surrounding area.

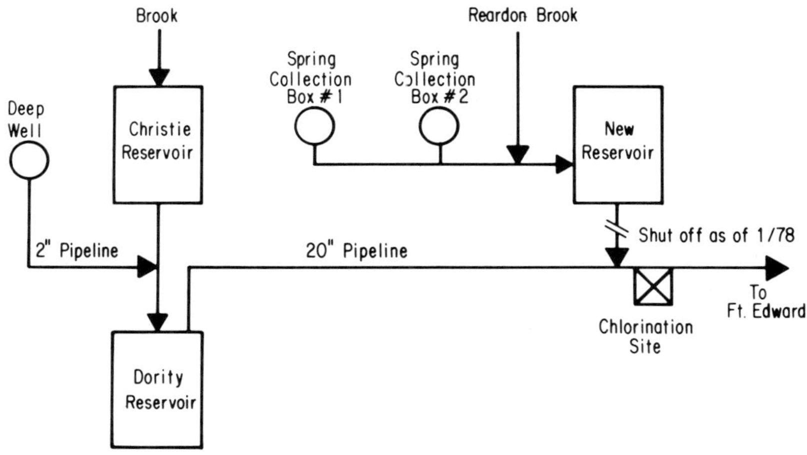


**Figure 2.** Schematic diagram of Fort Edward, NY, water supply and distribution system.

The three reservoirs comprising the impoundment area lie along the southern boundary of the watershed, the New Reservoir at the eastern end, and the Dority and Christie Reservoirs about 1 km west, near the midpoint. The New Reservoir, which was dredged of sediment in fall 1977, was not in service during the study. All three reservoirs are quite small. The Dority has a surface area of 8100 $m^2$, while those of the New and the Christie are only 1600 $m^2$ and 800 $m^2$, respectively. Figure 2 is a schematic representation of the water supply and distribution network that shows the major ground- and surface water sources and the interconnection of the reservoirs. Because the New Reservoir was not in service, water entered the distribution system only from the Dority Reservoir. Aside from the overflow from the Christie, the principal inputs to the Dority Reservoir are a 300-ft deep well and two springs. A small, unnamed brook feeds the Christie Reservoir, while Reardon Brook and two springs feed the New Reservoir, so that altogether seven discrete ground- and surface-water sources feed the water system.

## EXPERIMENTAL

Because details of the experimental methodology are reported elsewhere [12], only a brief outline is given here. Samples of water, sediments, aquatic vertebrates, macroinvertebrates and algae were collected, and the depth, temperature and dissolved oxygen (DO) levels of the water were measured.

Rainfall was monitored during the entire study using an automatic Wong sampler to collect rain. Samples were collected from this unit at weekly intervals and combined to yield a composite sample. Vertebrate samples were frozen, and all other samples were held at 4°C prior to analysis.

Methods of sample preparation and extraction were determined by the type of sample. Pesticide-grade solvents were used in all extractions and cleanup steps. For water, a 2-L sample was extracted three times with 30-mL portions of hexane, and the combined extract was passed through 1 g of anhydrous $Na_2SO_4$ and concentrated to 1.0 mL in a Kuderna-Danish (KD) appartus. For sediments and soils, 1000 g of wet material was mixed with 10 g of clean sea sand, placed in a preextracted cellulose thimble and extracted for 10 hr with 150 mL of 1:1 hexane-acetone in a soxhlet apparatus. Water in the extract was removed with a pipet and extracted with hexane. The combined extract was shaken with mercury to remove sulfur, concentrated to 5 mL in a KD apparatus, and placed on a column containing 10 g of 2% deactivated Florisil topped with 1 g of anhydrous $Na_2SO_4$. The column was eluted with 120 mL of hexane, and the first 100 mL of eluent was collected and concentrated to 1.0 mL in a KD apparatus. The dry weight of each sample was determined after heating in an oven at 110°C.

Biological samples were lyophilized prior to extraction. Aquatic vertebrates were boned and homogenized, and a 5.0-g portion (wet weight) was lyophilized. Because the wet weight of macroinvertebrate samples ranged from 0.2 to 7.2 g, and those of algae samples ranged from 6 to 15 g, the entire sample was lyophilized. After the dry weights had been recorded, the lyophilized samples were extracted with 150 mL of hexane for 2 hr in a soxhlet extractor, and the extract was treated as described for soils and sediments. The recoveries of PCB from fish oil spiked with Aroclor 1254 by this technique have been shown by Pastel et al. [13] to range from 66 to 95%.

PCB levels were determined by temperature-programmed gas chromatography (GC) of 2.0-$\mu$L samples on a Hewlett-Packard 5840 A chromatograph equipped with an automatic sampler, a 0.25-mm i.d. x 30-m capillary column with Apiezon L liquid phase, a $^{63}$Ni electron capture detector (ECD), a Grob injection system and a data microprocessor. The column temperature was held at 140°C for the first 5 min, and then increased at 3°C/min to 225°C.

The chromatograph was calibrated using an external standard containing 0.5 $\mu$g/mL each of Aroclors 1016, 1221, 1254 and 1260, yielding the chromatogram in Figure 3. For this chromatogram, Table I reports the Aroclor(s) and the compound associated with each of the 60 resolved peaks as determined by Bush and Snow [14], who combined gas chromatography/mass spectrometry (GC/MS) studies of pure compounds with the results of the earlier studies of Jensen and Sundström [15] and Sissons and Welti [16]. Each peak in a sample chromatogram which matched a peak in the calibration

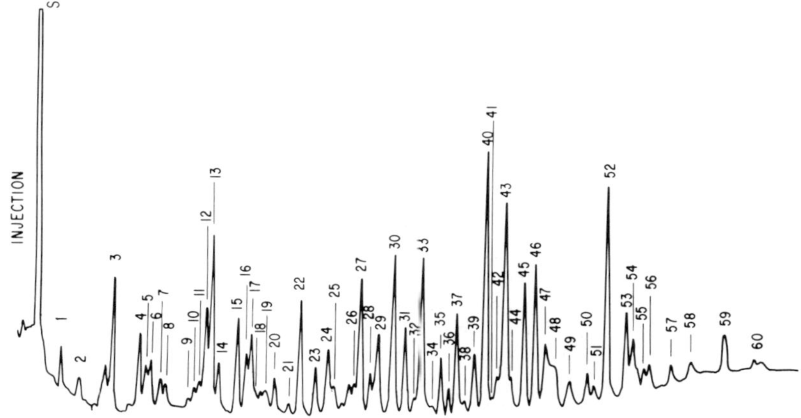

Figure 3. Gas chromatogram of calibration standard (0.5 μg/mL each of Aroclors 1016, 1221, 1254 and 1260). The peak numbers refer to Table I.

**Table I. Identities of Peaks in Chromatogram of Standard Aroclor Mixture**

| Peak No. | Constituent(s) | Aroclor(s) |
|---|---|---|
| 1 | 2-Chlorobiphenyl | 1221 |
| 2 | 4-Chlorobiphenyl and 2,2′-dichlorobiphenyl | 1016 & 1221 |
| 3 | 2,4′-Dichlorobiphenyl | 1016 & 1221 |
| 4 | 2,2′,5′-Trichlorobiphenyl | 1016 |
| 5 | 2,2′,4′-Trichlorobiphenyl | 1016 |
| 6 | 2,2′,3′-Trichlorobiphenyl and 3,2′,6′-trichlorobiphenyl | 1016 |
| 7 | 4,2′,6′-Trichlorobiphenyl | 1016 |
| 8 | Trichlorobiphenyl | 1221 |
| 9 | Trichlorobiphenyl | 1016 |
| 10 | Trichlorobiphenyl | 1016 |
| 11 | 3,2′,5′-Trichlorobiphenyl | 1016 |
| 12 | 3,2′,4′-Trichlorobiphenyl | 1016 |
| 13 | 2,4,4′-Trichlorobiphenyl | 1016 & 1221 |
| 14 | 2,3′,4′-Trichlorobiphenyl | 1016 |
| 15 | 2,5,2′,5′-Tetrachlorobiphenyl | 1016 & 1254 |
| 16 | 2,4,2′,5′-Tetrachlorobiphenyl | 1016 & 1254 |
| 17 | 2,3,2′,5′-Tetrachlorobiphenyl | 1016 & 1254 |
| 18 | 2,4,2′,4′-Tetrachlorobiphenyl | 1016 & 1254 |
| 19 | 2,3,2′,3′-Tetrachlorobiphenyl | 1016 & 1254 |
| 20 | Tetrachlorobiphenyl | 1016 & 1254 |
| 21 | Tetrachlorobiphenyl | 1016 |

Table I, continued

| Peak No. | Constituent(s) | Aroclor(s) |
|---|---|---|
| 22 | 2,5,2′,3′,6′-Pentachlorobiphenyl | 1254 & 1260 |
| 23 | 2,3,2′,3′,6′-Pentachlorobiphenyl | 1254 |
| 24 | 2,5,3′,4′-Tetrachlorobiphenyl | 1254 |
| 25 | 2,4,3′,4′-Tetrachlorobiphenyl | 1254 |
| 26 | 2,3,6,2′,3′,6′-Hexachlorobiphenyl | 1254 & 1260 |
| 27 | 2,5,2′,4′,5′-Pentachlorobiphenyl | 1254 & 1260 |
| 28 | 2,4,2′,4′,5′-Pentachlorobiphenyl | 1254 |
| 29 | 2,3,4,2′,5′-Pentachlorobiphenyl | 1254 & 1260 |
| 30 | 2,3,2′,3′,4′-Pentachlorobiphenyl | 1254 & 1260 |
| 31 | 2,3,5,6,2′,5′-Hexachlorobiphenyl | 1254 & 1260 |
| 32 | 2,3,2′,3′,5′,6′-Hexachlorobiphenyl | 1254 & 1260 |
| 33 | 2,3,4,2′,3′,6′-Hexachlorobiphenyl | 1254 & 1260 |
| 34 | Pentachlorobiphenyl | 1254 & 1260 |
| 35 | 2,3,6,2′,3′,5′,6′-Heptachlorobiphenyl | 1254 & 1260 |
| 36 | 3,4,2′,4′,5′-Pentachlorobiphenyl | 1254 & 1260 |
| 37 | 2,3,6,2′,3′,4′,6′-Heptachlorobiphenyl | 1254 & 1260 |
| 38 | 2,3,5,2′,4′,5′-Hexachlorobiphenyl | 1254 & 1260 |
| 39 | 2,4,2′,3′,4′-Pentachlorobiphenyl | 1254 & 1260 |
| 40 | 2,4,5,2′,4′,5′-Hexachlorobiphenyl | 1254 & 1260 |
| 41 | 2,3,4,2′,3′,5′-Hexachlorobiphenyl | 1254 & 1260 |
| 42 | 2,3,2′,3′,4′,5′-Hexachlorobiphenyl | 1254 & 1260 |
| 43 | 2,3,4,2′,4′,5′-Hexachlorobiphenyl | 1254 & 1260 |
| 44 | 2,3,5,2′,3′,5′,6′-Heptachlorobiphenyl | 1254 & 1260 |
| 45 | 2,3,4,2′,3′,4′-Hexachlorobiphenyl | 1254 & 1260 |
| 46 | 2,5,2′,3′,4′,5′,6′-Heptachlorobiphenyl | 1254 & 1260 |
| 47 | 2,3,4,2′,3′,4′,6′-Heptachlorobiphenyl | 1254 & 1260 |
| 48 | 2,4,5,3′,4′,5′-Hexachlorobiphenyl | 1254 & 1260 |
| 49 | 2,3,6,2′,3′,4′,5′,6′-Octachlorobiphenyl | 1254 & 1260 |
| 50 | 3,4,2′,3′,4′,5′-Hexachlorobiphenyl | 1254 & 1260 |
| 51 | 2,3,5,2′,3′,4′,5′-Heptachlorobiphenyl | 1260 |
| 52 | 2,4,5,2′,3′,4′,5′-Heptachlorobiphenyl | 1254 & 1260 |
| 53 | 2,3,4,2′,3′,4′,5′-Heptachlorobiphenyl | 1254 & 1260 |
| 54 | Heptachlorobiphenyl | 1260 |
| 55 | Heptachlorobiphenyl | 1260 |
| 56 | Heptachlorobiphenyl | 1260 |
| 57 | Heptachlorobiphenyl | 1260 |
| 58 | 2,3,4,2′,3′,4′,5′,6′-Octachlorobiphenyl | 1254 & 1260 |
| 59 | 3,4,5,2′,3′,4′,5′-Heptachlorobiphenyl | 1260 |
| 60 | 2,3,4,5,6,2′,3′,4′,5′-Nonachlorobiphenyl | 1260 |

chromatogram thus provided an independent estimate of the quantity of the corresponding Aroclor(s). The appropriate response factor for each peak was entered in the microprocessor, and the concentration of each Aroclor was taken as the average of the concentrations estimated by each of its peaks.

For an Aroclor to be considered present, it was necessary that five of its peaks be detected. This method, which is accurate to about 10% [14], has the advantage that it yields directly and simply the concentration of each Aroclor. The detection limit was 0.050 ng of Aroclor, which corresponds to a level of 25 ng/mL in the 2.0-$\mu$L injected sample.

Because GC peak ratios were consistent among samples of a given type, only representative samples of each type were analyzed by GC/MS to confirm the identities of PCB found by GC. A Finnigan GC/MS/Data System Series 9500/1015-D/6000 was used with a ⅛-in. by 5-ft glass column packed with 2% Apiezon L on Ultrabond support. The column temperature was 190°C, the ionization voltage was 70 eV, and the mass range scanned was 100-400 m/e.

## RESULTS AND DISCUSSION

Of the several Aroclors, only Aroclor 1016 (A-1016) and Aroclor 1254 (A-1254) were detected in our samples. Detection limits for each type of sample are calculated by considering the 25-ng/mL detection limit of the chromatograph and the volume or mass of the original sample relative to the 1.0-mL hexane extract. Thus, for a 2.0-L water sample, the detection limit is about 12 ng/L (25 ng/mL GC detection limit x 1.0 mL extract volume/ 2.0-L sample volume). Detection limits estimated in a similar manner on a dry-weight basis, for sediments, algae, vertebrates and macroinvertebrates are respectively, 5, 25, 25 and 200 ng/g (for example, for sediments; 25 ng/mL GC detection limit x 1.0 mL extract volume /5.0 g dry weight).

### Water

The water samples consisted of seven grab samples of the water from the three reservoirs, two from the distribution system (one from the chlorination site and one from a tap in the Village), and seven of the ground- and surface water sources feeding the reservoirs. For comparison, we also obtained three grab samples from a small private water system at Burden Lake approximately 100 km south of Fort Edward, and one grab sample from the Hudson River at Fort Edward. No samples contained detectable levels of common chlorinated pesticides, and A-1016 and A-1254 were the only Aroclors present in detectable quantities. Table II summarizes Aroclor levels in these water samples in terms of the median concentrations and the concentration ranges of A-1016, A-1254 and total PCB.

Table II. PCB Levels (ng/L) in Water

| Source | Samples | A-1016 | | A-1254 | | Total PCB | |
|---|---|---|---|---|---|---|---|
| | | Median | Range | Median | Range | Median | Range |
| Christie Reservoir | 1 | ND[a] | ND | ND | ND | ND | ND |
| Dority Reservoir | 4 | 99 | 70–130 | ND | ND | 99 | 70–130 |
| New Reservoir | 2 | | 110–120 | | ND–36 | | 110–160 |
| Distribution System | 2 | | 69–100 | ND | ND | | 69–100 |
| Ground- and Surface Water | 7 | ND | ND | ND | ND | ND | ND |
| Rain | 1 | 1300 | | ND | ND | 1300 | |
| Hudson River (Ft. Edward) | 1 | 360 | | 170 | | 530 | |
| Burden Lake, NY | 3 | ND | ND | ND | ND | ND | ND |

[a]ND = not detected, less than 12 ng/L.

### Impounded Water

Despite differences between total PCB levels of individual samples, it appears that with one exception, the total PCB level is relatively uniform throughout the reservoirs and the distribution system. Collectively, the nine samples from the three reservoirs and the distribution system exhibited total PCB levels ranging from less than the detection limit of 12 ng/L to 160 ng/L, with a median level of 100 ng/L. In all but one sample, the PCB levels were well above the detection limit, the exception being the sample from the Christie Reservoir, which did not exhibit a detectable level of PCB. Because this single result is consistent with the physical nature of the Christie Reservoir, it appears to be truly representatiave of the PCB level of the Christie Reservoir water. The Christie Reservoir is small and shallow, and could well be considered as an extension of the brook which feeds it. Other similar surface waters (below) do not exhibit detectable levels of PCB, so that the absence of PCB in the Christie water is not surprising. Despite the considerable variation of the total PCB levels from one sample to another within both reservoirs, the total PCB levels in the New Reservoir and the Dority Reservoir appeard to be comparable. Clearly, the median total PCB levels of 99 ng/L in the Dority Reservoir and 130 ng/L in the New Reservoir did not differ significantly ($p > 0.25$). The PCB levels of the finished water also appeared comparable to those of the water in the Dority Reservoir from which the water enters the treatment and distribution system ($p > 0.25$). As was true of the Dority water, the finished water did not exhibit detectable levels of A-1254, and the median A-1016 level of 85 ng/L in the finished water corresponded well to the median level of 99 ng/L in the Dority water. Although the PCB level in the sample taken at the chlorination site was approximately 30% higher than the PCB level in the sample drawn from a tap in the village, this difference is probably not significant ($p > 0.25$). It has been shown that chlorination temporarily increases the apparent PCB level in waters containing significant quantities of humic materials [17], but demonstrating the operation of such an effect would require 10-20 samples.

### Ground- and Surface Water Sources

Each of the seven discrete ground- and surface water sources feeding the reservoirs was represented by one of the seven samples denoted collectively in Table II as Ground- and Surface Waters. None of the seven exhibited detectable levels of A-1016 or A-1254, showing that these sources do not constitute a significant input of PCB. Even on the worst-case assumption that the A-1016 level in each source is barely below the detection limit of 12 ng/L, it still falls nearly an order of magnitude below the median A-1016 level of 100 ng/l in the reservoir water.

As might be expected, the water in the Hudson River at Fort Edward not only exhibits a considerably higher total PCB level than does the water in the Fort Edward reservoirs, it also contains substantial levels of A-1254. The total PCB level in the Hudson River sample was 530 ng/L, approximately five times as high as the median total PCB level in the Fort Edward Reservoir, and approximately 30% of it consisted of A-1254.

## Rainfall

In the 30-day composite rain sample, the A-1254 level was below the detection limit of 12 ng/L, but the A-1016 level was 1300 ng/L. This A-1016 level, which is more than tenfold higher than the median concentration of A-1016 in the reservoir water, suggests that rain represents a significant source of PCB input into the reservoirs. An estimate of the lower limit of the contribution of rain to the PCB burden of the reservoirs is the level of A-1016 in the water which would result from rain falling directly on the reservoir surface. Consider as an example, the Dority Reservoir, which has a volume of about $1.9 \times 10^4$ $m^3$, a surface area of about $8.1 \times 10^3$ $m^2$, and in which the mean residence time of the water is about 13 days [11]. In this period, the total precipitation was 1.9 cm, so that the rain falling on the surface of the reservoir provided enough A-1016 to increase the level in the impounded water by approximately 10 ng/L, or about 10% of the total A-1016 burden of the reservoir water. Because this figure neglects the contribution of runoff, it represents a lower limit for the contribution of rainfall to the PCB level of the impounded water. Thus, although its contribution is relatively small compared to the PCB level of the impounded water, rainfall appears to be the major source of PCB input. This result is consistent with the earlier studies of Harvey and Steinhauer [18], Lunde et al. [19], Murphy and Rzeszutko [20] and Eisenreich et al. [21], which indicated that in general rainfall is a significant and widespread source of PCB.

The high PCB level of the rainfall at the reservoir reflects the general elevation of PCB levels in the air in the Fort Edward area, which are far higher near Fort Edward than a places remote from sites of high PCB contamination. Hetling et al. [9] found that PCB levels in ambient air at sites in the village of Fort Edward ranged from 25 to 300 $ng/m^3$ and, at the Caputo Disposal Site, they were as high as 3200 $ng/m^3$. In contrast, they found that the PCB level in ambient air at Providence, RI, was only 9.4 $ng/m^3$ and over the western Atlantic Ocean it was only 0.1 $ng/m^3$.

Water samples from the small private water system at Burden Lake confirm that local sources are the major cause of the high PCB level in rainfall at Fort Edward. Neither A-1016 nor A-1254 was present at detectable levels in the three water samples from Burden Lake. Located approximately 100 km south of Fort Edward, Burden Lake is subject to the same general weather patterns as Fort Edward, so that if long-range atmospheric transport of PCB were important, the PCB levels at Fort Edward and at Burden Lake should be comparable. However, the median A-1016 level in Burden Lake water of less than 12 ng/L is approximately tenfold lower than the median value of 100 ng/L in the Fort Edward Reservoir water. Despite the small number of samples, a difference of this magnitude indicates clearly that local PCB sources constitute a major source, and perhaps the principal source, of the Aroclor burden in the Fort Edward reservoir system.

**Soil and Sediment**

The 24 soil and sediment samples consisted of 12 sediment samples from the 3 reservoirs, 6 sediment samples from the springs and brooks feeding the reservoirs and 5 samples of shoreline soil Table III summarizes the PCB levels in these samples in ng/g on a dry-weight basis, in terms of the median con-concentrations and concentration ranges of A-1016, A-1254 and total PCB.

Contamination of the reservoir sediments and the adjacent land by PCB appears to be widespread, with both A-1016 and A-1254 present at comparable levels. Four of the five soil samples and eight of the thirteen reservoir sediment samples exhibited detectable levels of A-1016, A-1254 or both, so that approximately two-thirds of the samples were contaminated. Among the entire group of 18 samples, 10 contained detectable levels of A-1016, and 9 contained detectable levels of A-1254; thus, the two Aroclors appeared with similar frequencies. Both the soils and the reservoir sediments exhibited sites with very high PCB levels, but these occurred infrequently. For example, if the 13 reservoir sediments are considered as a group, the total PCB level ranged up to 190 ng/g, but in 12 of the 13 samples, the total PCB level was 40 ng/g or less. The five soil samples exhibited a similar distribution of total PCB levels, with one sample containing 160 ng/g, and the remaining four containing less than 30 ng/g. It is not clear what causes localized concentrations of PCB in soils and sediments, but it is clear that individual samples can be very misleading, and a reliable assessment of the extent of PCB contamination of soils and sediments requires a large number of samples.

If the five groups of sediments and soil samples in Table III are classified by total PCB level, all but one of the groups of samples fall into two distinct categories in a manner which points out relationships that are otherwise not apparent. A convenient index of the total PCB level for this purpose in each group is the frequency with which the samples display detectable PCB levels,

Table III. PCB Levels (ng/g dry wt) in Sediment and Soil

| Source | Samples | A-1016 | | A-1254 | | Total PCB | |
|---|---|---|---|---|---|---|---|
| | | Median | Range | Median | Range | Median | Range |
| Brooks and Springs | 6 | ND[a] | ND | ND | ND | ND | ND |
| Dority Reservoir | 6 | 22 | ND–180 | 13 | ND–21 | 28 | ND–190 |
| Christie Reservoir | 2 | ND | ND | ND | ND | ND | ND |
| New Reservoir | 5 | ND | ND–29 | ND | ND–17 | ND | 17–30 |
| Soil | 5 | 8 | ND–66 | 11 | ND–89 | 28 | ND–160 |

[a]ND = not detected, less than 5 ng/g.

Table IV. PCB Levels (ng/g dry wt) in Biota

| Source | Samples | A-1016 | | A-1254 | | Total PCB | |
|---|---|---|---|---|---|---|---|
| | | Median | Range | Median | Range | Median | Range |
| Dority Reservoir Algae | 5 | ND[a] | ND | 36 | ND–120 | 36 | ND–120 |
| Dority Reservoir Macroinvertebrates | 3 | ND | ND | 1600 | ND–3300 | 1600 | ND–3300 |
| Christie Reservoir Macroinvertebrates | 3 | ND | ND | ND | ND–320 | ND | ND–320 |
| New Reservoir Macroinvertebrates | 8 | ND | ND–600 | ND | ND–3800 | 760 | ND–3800 |
| Dority Reservoir Vertebrates | 11 | ND | ND–260 | 290 | ND–850 | 930 | ND–1100 |
| New Reservoir Vertebrates | 2 | ND | ND | ND | ND | ND | ND |

[a]ND = not detected, less than detection limits of 25 ng/g for algae, 200 ng/g for macroinvertebrates and 25 ng/g for vertebrates.

i.e., the fraction of the samples containing detectable levels of PCB. Viewed in this manner, the Christie Reservoir sediments (0/2) and the sediments from the springs and brooks (1/6) form one group whose frequency of PCB contamination is low, whereas the shoreline soils (4/5) and the Dority sediments (5/6) form a second group whose frequency of PCB contamination is high. The low frequency of PCB contamination exhibited by the sediments from the Christie Reservoir and the springs and brooks is consistent with two aspects of these sites. First, as Table II shows, the Aroclor levels in the water fall below the detection limit. Second, at these sites the water velocity is relatively high, so that there is little opportunity for deposition of PCB-laden sediments from the surroundings. On the other hand, the high and similar frequencies of PCB contamination characteristic of the Dority sediments and shoreline soils suggest that an important source of the PCB in the Dority sediments is soil carried into the reservoir by runoff. Kenaga [22] has shown that soils strongly adsorb PCB, and the estimated rate of sediment accumulation in the Dority is 1.2 cm/yr [11]. Clearly, the input of sediment-borne PCB could be substantial.

The frequency of PCB contamination exhibited by the New Reservoir sediments (2/5) is interesting because it falls between that of the Christie Reservoir, springs and brooks on the one hand and that of the Dority Reservoir and soil on the other. Except for its smaller size, the physical characteristics of the New Reservoir are similar to those of the Dority, so that one would expect the sediments to exhibit comparable frequencies of PCB contamination. The New Reservoir had been dredged six months before the study, and we suspect that this caused the frequency of PCB contamination of the New Reservoir sediments to be lower than expected.

## Biota

In general, the biota exhibited much higher PCB levels that did the surrounding water or sediment. Of the 32 samples of algae, macroinvertebrates and vertebrates collected from the reservoirs, 20 contained detectable levels of PCBs. Table IV reports the PCB level for each category in terms of the median concentrations and the concentration ranges of A-1016, A-1254, and total PCB.

The biota exhibited a clear selectivity for A-1254 over A-1016. Of the 20 samples with detectable PCB levels, all contained A-1254, but only 9 contained A-1016. Further, in the eight samples containing both Aroclors, the levels of A-1254 were approximately two- to threefold greater than the levels of A-1016. This preferential uptake of A-1254 is consistent with the finding of Hansen et al. [23] that sheepshead minnows exposed to Aroclors in the laboratory incorporate the more highly chlorinated PCB faster than the less highly chlorinated PCB, and retain them longer. As a rule, the PCB level increases with the age of the organism. For example, total PCB levels of composite samples of tadpoles and frogs increased with age in the following manner: 0–1 years, 25 ng/g; 1–2 years, 230 ng/g; 2 years, 350 ng/g.

Of special interest are the macroinvertebrates, which although near the bottom of the food chain, exhibited by far the highest PCB levels. As a group, they exhibited a median total PCB level of 580 ng/g, which was approximately twice the median total PCB level of the vertebrates which feed on them. Macroinvertebrates inhabit the sediments, in which the PCB level is several-hundred-fold higher than that of the water, and their high PCB levels may result mainly from the high PCB levels of their local environment. The high concentrations of PCB in the macroinvertebrates are consistent with the laboratory uptake studies of Sanders and Chandler [24], and they support the suggestion of Simpson et al. [25] that these organisms may provide a simple means for monitoring PCB levels in aquatic systems. Increasing the sample size to several grams with mechanical collection devices should permit monitoring of PCB levels even in waters where the PCB levels are too low for direct detection.

## ACKNOWLEDGMENTS

Supported in part by NSF Grant SP178-03499. Valuable support and technical assistance were given by B. Bush, E. Legere, R. Narang, J. Snow, K. Simpson and S. Syrotynski of the New York State Department of Health; J. S. Kim of State University of New York at Albany; and E. Horn and J. Spagnoli of the New York State Department of Environmental Conservation.

## LITERATURE CITED

1. Gustafson, C. G. *Environ. Sci. Technol.* 4:814 (1970).
2. Hutzinger, O., S. Safe and V. Zitko. *Chemistry of Polychlorinated Biphenyls* (Cleveland, OH: CRC Press, 1974).
3. Furukawa, K., K. Tonomura and A. Kamibayashi. *Appl. Environ. Microbiol.* 35:223 (1978).
4. Bush, B., C. F. Tumasonis and F. D. Baker. *Arch. Environ. Contam. Toxicol.* 2:195 (1974).
5. Kuratsune, M., V. Matsuda and J. Nagayama. In *Proceedings of the National Conference on Polychlorianted Biphenyls*, EPA 56016-5-74 U.S. EPA, Washington, DC (1975).
6. Braunberg, R. C., R. L. Dailey, E. A. Brower, L. Kasza and A. M. Blaschka. *J. Toxicol. Environ. Health* 1:683 (1976).

7. Mayer, F. L., P. M. Mehrle and H. O. Sanders. *Arch. Environ. Contam. Toxicol.* 5:501 (1977).
8. Spagnoli, J., and L. C. Skinner. *Pestic. Monit. J.* 11:69 (1977).
9. Hetling, L., E. Horn and J. Tofflemire. "Summary of Hudson River PCB Study Results, Technical Paper 51," New York State Department of Environmental Conservation, Albany, NY (1978).
10. Sofaer, A. D. "Interim opinion and Order, File Number 2833" New York State Department of Environmental Conservation, Albany, NY (1976).
11. "Water Supply Study: Village of Fort Edward, New York," (Colonie, NY: Clough Associates, 1975).
12. Brinkman, M., K. Fogelman, J. Hoeflein, T. Lindh, M. Pastel, W. C. Trench and D. A. Aikens. "Levels of Polychlorinated Biphenyls in the Fort Edward Water Supply Ecosystem: a Baseline Approach," Rensselaer Polytechnic Institute, Troy, NY (1978).
13. Pastel, M., B. Bush and J. S. Kim. *Pestic. Monit. J.* 14:11 (1980).
14. Bush, B., and J. Snow. Paper presented at the Pittsburgh Conference on Analytical Chemistry and Applied Spectroscopy, Cleveland, OH (1978).
15. Jensen, S., and G. Sundström. *Ambio.* 3:70 (1974).
16. Sissons, D., and D. Welti. *J. Chromatog.* 60:15 (1971).
17. Billings, W. N., T. F. Bidleman and W. B Bornberg. *Bull. Environ. Contam. Toxicol.* 19:215 (1978).
18. Harvey, G. R., and W. G. Steinhauer. *Atmos. Environ.* 8:777 (1974).
19. Lunde, G., J. Gether, N. Gjos and M. Stobet-Lande. *Atmos. Environ.* 11:1007 (1977).
20. Murphy, T. J., and C. P. Rzeszutko. "Polychlorinated Biphenyls in Precipitation in the Lake Michigan Basin," EPA-600/3-78-071, U.S. Environmental Protection Agency, Washington, DC (1978).
21. Eisenreich, S. J., G. J. Hollod and T. C. Johnson. *Environ. Sci. Technol.* 13:569 (1979).
22. Kenaga, E. *Ecotoxicol. Environ. Safety* 4:26 (1980).
23. Hansen, D. J., S. C. Schimmel and J. Forester. *Trans. Am. Fish. Soc.* 104:584 (1975).
24. Sanders, H. O., and J. H. Chandler. *Bull. Environ. Contam. Toxicol.* 7:257 (1972).
25. Simpson, K., R. Mt. Pleasant and B. Bush. Paper presented at Kepone Symposium II, Easton, MD (1977).

# CHAPTER 51

# INDUSTRIAL ORGANIC COMPOUNDS IN THE NIAGARA RIVER WATERSHED

**Bertha L. Proctor, Vincent A. Elder and Ronald A. Hites**

School of Public and Environmental Affairs and
Department of Chemistry
Indiana University
Bloomington, Indiana

Love Canal is a 6-ha rectangular site located in the southeast corner of the city of Niagara Falls which has become the focus of international attention. Chemicals buried within the boundaries of this unfinished canal 30 years ago have now infiltrated several homes. Many different chemicals have been found in their basements, including some known or presumed carcinogens [1]. Unfortunately, Love Canal is only one of the waste disposal sites in western New York. An interagency task force (state and federal agencies) has identified 152 hazardous waste disposal sites in Erie and Niagara Counties, New York [1]. Because the Niagara River is an important drinking water source for several U.S. and Canadian cities and the major source of water for Lake Ontario [2], identifying the industrial organic compounds in the Niagara River-Lake Ontario system is imperative. This chapter reports on such a study of the Niagara River and three of its tributaries.

## STUDY AREA

The Niagara River is approximately 48 km long and connects Lake Erie with Lake Ontario (Figure 1). The Niagara River is between 0.4 and 2.3 km wide, and the flowrate averages 6000 $m^3$/sec [3]. The 102nd Street dumpsite forms a bay which is actually part of the Niagara River. The 102nd Street bay is 0.8 km long and 0.4 km wide. Four different dumpsites make up the shore line of the 102nd Street bay; it also receives runoff from the Love Canal area through a storm sewer. Bloody Run Creek is 200 m long and is little more than a drainage ditch for the Hyde Park landfill. Gill Creek runs through a highly industrialized complex near the Niagara River. The companies in this complex have several dumpsites located on their property [1].

## EXPERIMENTAL

With the cooperation of the New York State Department of Environmental Conservation and the U.S. Coast Guard (USCG), a series of samples were collected from the study area during June and November 1979 (Figure 1). Water samples were collected from midchannel in the Niagara River. Three of the stations were above Niagara Falls, and one station was below the falls. Riverwater samples were collected in a 4-L glass beaker securely attached to a 2-m rod, and then transferred to 20-L glass carboys. Water samples of 50 L volume were collected at each end of the river (stations 1 and 4), while 20-L water samples were collected at stations 2 and 3. An additional water sample was collected in a 3.8-L amber glass bottle at station 3 for volatile organic analysis (VOA). The 20-L glass carboys were kept in cardboard boxes to prevent photodegradation. Due to the high flow in the main channel of the Niagara River [3], no sediment samples were collected there.

Grab water samples were collected in 3.8-L amber glass bottles with Teflon®-lined caps at midstream in Gill Creek and Bloody Run Creek and at several places in the 102nd Street bay. Sediment samples were collected at each site with an Eckman dredge and stored in 1-L glass bottles with Teflon-lined lids.

Water samples for volatile analyses and sediment samples were packed in ice to slow biological degradation prior to analysis. Water samples for solvent extraction were immediately preserved for acidification to pH 2 with hydrochloric acid and by adding 250–500 mL of dichloromethane. Sample workup was begun as soon as possible after returning to the laboratory. Water samples were refrigerated until analysis. Sediment samples were kept frozen until analysis. Analytical techniques for concentration, separation and identification using solvent extraction, and gas chromatography/mass spectrometry (GC/MS) have been discussed in detail elsewhere [4].

**Figure 1.** Map of Niagara River and sampling sites.

Initial results indicated most samples were extremely complex, containing high levels of unresolved components along with mixtures of industrial and natural organic compounds. The unresolved complex organic interferences were removed from the extracts using a silica gel cleanup procedure. Extracts were evaporated to dryness and transferred to a 10 x 1-cm column packed with deactivated silica gel (1% water). The sample was then fractionated by successively eluting with 75 mL each of hexane, hexane with 10% dichloromethane, dichloromethane and methanol. To remove fatty acid interferences, fractionated extracts were dissolved in dichloromethane and extracted with aqueous NaOH (pH 9-11). Several sediment samples were passed through a colloidal copper column to remove elemental sulfur [4].

GC/MS analyses were run on most samples prior to each cleanup procedure to verify that sample integrity was maintained, that contaminants were not introduced and that major components were not lost or degraded. Blanks were also run for all extraction and cleanup procedures. The blanks showed no sigificant contamination.

Identification of the compounds in these samples was based on coincidence of gas chromatographic retention times and on equivalence of electron impact (EI) and (where possible) chemical ionization (CI) mass spectra with those of reference compounds. Those reference compounds not commercially available were synthesized in our laboratory.

It should be emphasized that this study was primarily qualitative in nature. Our principal goal was to identify compounds in the Niagara River and their probable sources rather than to measure their exact abundance. For this reason the concentration data are only semiquantitative. Compounds were quantified after silica gel fractionation by GC peak heights with a flame ionization detector (FID), and an approximate GC response factor was determined with methyl stearate.

## RESULTS AND DISCUSSION

The organic compounds found in the 102nd Street bay area, Bloody Run Creek, Gill Creek and the Niagara River are listed in Table I (compounds 1–90). The data include maximum concentrations for major compounds in the sediments. Water samples were not quantified. The structures for a select group of compounds are given in Figure 2. The sources of compounds in Table I can be classified into three general categories: (1) compounds which were known to be in the dump sites; (2) compounds with unknown sources; and (3) compounds which were by-products of industrial chemical processes and which were not known to be in the dumpsites.

Chlorinated benzenes and toluenes (No. 1–12, Table I) were found in almost all samples with the highest concentrations in the 102nd Street bay area. Mono-, di- and trichlorobenzenes (No. 1-3, Table I) were found in water from Bloody Run Creek and Gill Creek. Traces of chlorinated benzenes (No. 1, 4–6, Table I) were also detected in Niagara River water. Tetra-, penta- and hexachlorobenzenes are probably by-products of trichlorobenzene production [5].

**Table I. Organic Compounds Identified in the Niagara River Watershed (ppm)**

| | 102nd Street | Bloody Run Creek | Gill Creek | Niagara River |
|---|---|---|---|---|
| Chlorobenzenes | | | | |
| 1. Chlorobenzene | NA[a] | +[b] | + | + |
| 2. Dichlorobenzenes | + | + | + | –[c] |
| 3. Trichlorobenzenes | 40[d] | 8 | + | – |
| 4. Tetrachlorobenzenes | 200 | 25 | + | + |
| 5. Pentachlorobenzene | 100 | 10 | – | + |
| 6. Hexachlorobenzene | 8 | 10 | 30 | + |
| Chlorotoluenes | | | | |
| 7. Dichlorotoluenes | 20 | 90 | – | + |
| 8. Trichlorotoluenes | 100 | 50 | – | – |
| 9. Tetrachlorotoluenes | 40 | 10 | – | – |
| 10. Pentachlorotoluenes | 40 | 5 | – | – |
| 11. Hexachlorotoluenes | 40 | – | – | – |
| 12. Heptachlorotoluenes | 20 | – | – | – |
| Polycyclic Aromatic Hydrocarbons and Derivatives | | | | |
| 13. Methylnapththalenes | + | – | – | + |
| 14. $C_2$-Naphthalenes | + | – | – | + |
| 15. $C_3$-Naphthalenes | + | – | – | + |
| 16. Chloronaphthalene | 20 | – | – | – |
| 17. Dichloronaphthalene | 8 | – | – | – |
| 18. Trichloronaphthalene | 6 | – | – | – |
| 19. Biphenyl | 8 | – | – | – |
| 20. Anthracene (or phenanthrene) | – | + | – | – |
| 21. Chloroanthracene | – | + | – | – |
| 22. Dichloroanthracene | 20 | + | – | – |
| 23. Trichloroanthracene | – | + | – | – |
| 24. Pyrene | + | – | – | + |
| 25. Fluoranthene | + | – | – | + |
| Polychlorinated Biphenyls (PCB) | | | | |
| 26. Dichlorobiphenyls | – | – | 100 | + |
| 27. Trichlorobiphenyls | – | – | 300 | + |
| 28. Tetrachlorobiphenyls | – | 5 | 600 | + |
| 29. Pentachlorobiphenyls | – | – | 300 | + |
| 30. Hexachlorobiphenyls | – | – | 300 | – |
| Phenols and Derivatives | | | | |
| 31. Phenol | + | NA | – | NA |
| 32. Dichlorophenol | + | 2 | – | – |
| 33. Trichlorophenol | + | 5 | – | – |
| 34. *tert*-Butylphenol | – | – | – | + |
| 35. Di-*tert*-butylphenol | – | – | – | + |

Table I, continued

| | 102nd Street | Bloody Run Creek | Gill Creek | Niagara River |
|---|---|---|---|---|
| 36. Tetramethylbutylphenol | + | + | – | + |
| 37. 2-(2,4-Dichlorophenoxy)ethanol | + | – | – | – |
| 38. N-(4-Hydroxyphenol)acetamide | – | – | – | + |
| Cyclohexane Derivatives | | | | |
| 39. BHC (hexachlorocyclohexane) | 10 | + | – | + |
| 40. Hexamethylcyclohexane | – | + | – | – |
| 41. Cyclohexylcyclohexanol[e] | 30 | – | – | – |
| 42. 2-Cyclohexylcyclohexanone | 5 | – | – | – |
| 43. Phenylcyclohexane | + | – | – | – |
| Aliphatic Alcohols and Thiols | | | | |
| 44. Dodecanol | 5 | – | – | – |
| 45. Tetradecanol | 5 | – | – | – |
| 46. Hexadecanol | 4 | – | – | – |
| 47. Octadecanol | 3 | – | – | – |
| 48. 1-Dodecanethiol | 3 | – | – | – |
| Benzyl Derivatives | | | | |
| $C_7$-Benzyl Derivatives | | | | |
| 49. Benzyl alcohols (0–5 Cl)[e] | + | – | – | – |
| 50. Benzaldehydes (0–5 Cl)[e] | + | – | – | – |
| 51. Benzoic acids (0–5 Cl)[e] | + | – | – | – |
| 52. Benzamides (0–4 Cl)[e] | + | – | – | – |
| $C_{14}$-Benzyl Derivatives | | | | |
| 53. Methyldiphenylmethanes[e] | 90 | + | – | – |
| 54. Chloromethyldiphenylmethanes | 20 | + | – | – |
| 55. Dichloromethyldiphenylmethanes | – | + | – | – |
| 56. Trichloromethyldiphenylmethanes | 30 | + | – | – |
| 57. (Phenylmethyl)benzenemethanols[e] | 4 | – | – | – |
| 58. Methylbenzophenone[e] | 7 | + | – | – |
| 59. Phenylmethylbenzoic acid[e] | + | – | – | – |
| 60. Benzylbenzoic acid[e] | + | – | – | – |
| 61. Benzyl ether[e] | 70 | + | – | – |
| $C_{21}$-Benzyl Derivatives | | | | |
| 62. Methylbis(phenylmethyl)benzenes[e] | 30 | – | – | – |
| 63. Chloro(methyl)bis(phenylmethyl) benzenes | 2 | – | – | – |
| 64. Dichloro(methyl)bis(phenylmethyl)benzenes | 9 | – | – | – |
| 65. Trichloro(methyl)bis(phenylmethyl)benzenes | 5 | – | – | – |

Table I, continued

| | 102nd Street | Bloody Run Creek | Gill Creek | Niagara River |
|---|---|---|---|---|
| 66. Tetrachloro(methyl)bis(phenylmethyl)benzenes | 6 | – | – | – |
| 67. Pentachloro(methyl)bis(phenylmethyl)benzenes | 4 | – | – | – |
| 68. Dimethyltriphenylmethane[e] | 20 | + | – | – |
| $C_{28}$-Benzyl Derivative | | | | |
| 69. Methyltris(phenylmethyl) benzenes[e] | 6 | – | – | – |
| Fluorine-Containing Compounds | | | | |
| 70. Benzotrifluoride | – | + | – | – |
| 71. Chlorobenzotrifluoride[e] | – | + | – | – |
| 72. Dichlorobenzotrifluoride | – | + | – | – |
| 73. Chloro(difluoro,chloromethyl) benzene[e] | – | + | – | – |
| 74. Dichloro(trifluoromethyl)benzophenone[e] | – | 25 | – | – |
| 75. Dichloro(trifluoromethyl)-α,α-difluorodiphenylmethane[e] | – | 15 | – | – |
| Miscellaneous | | | | |
| 76. Mirex | – | + | – | – |
| 77. Phenothiazine[e] | – | – | 80 | – |
| 78. Chloro,hydroxyphenothiazine | – | 6 | – | – |
| 79. Phenyl ether | 20 | + | – | + |
| 80. DDE[1,1′-(dichloroethylidene) bis(4-chlorobenzene)] | – | + | – | – |
| 81. Methoxybenzophenone[e] | – | + | – | – |
| 82. Chloro,methoxybenzophenone | – | + | – | – |
| 83. Chloro,hydroxybenzophenone | – | 15 | – | – |
| 84. Aminoacetophenone[e] | + | – | – | – |
| 85. Benzothiazole | – | – | – | + |
| 86. Tetraethyleneglycol-di(2-ethylhexanoate) | – | – | – | + |
| 87. Trichloroethylene | NA | + | + | – |
| 88. Tetrachloroethylene | NA | + | + | – |
| 89. Hexachlorobutadiene | NA | + | + | – |
| 90. Furan | NA | – | + | – |

[a]NA = sample was not analyzed for this compound.

[b]+ = compound was detected in water or sediment, but was not quantified.

[c]– = compound was not detected in water or sediment.

[d]Maximum level (in ppm) of compound in sediment.

[e]For structure, see Figure 2.

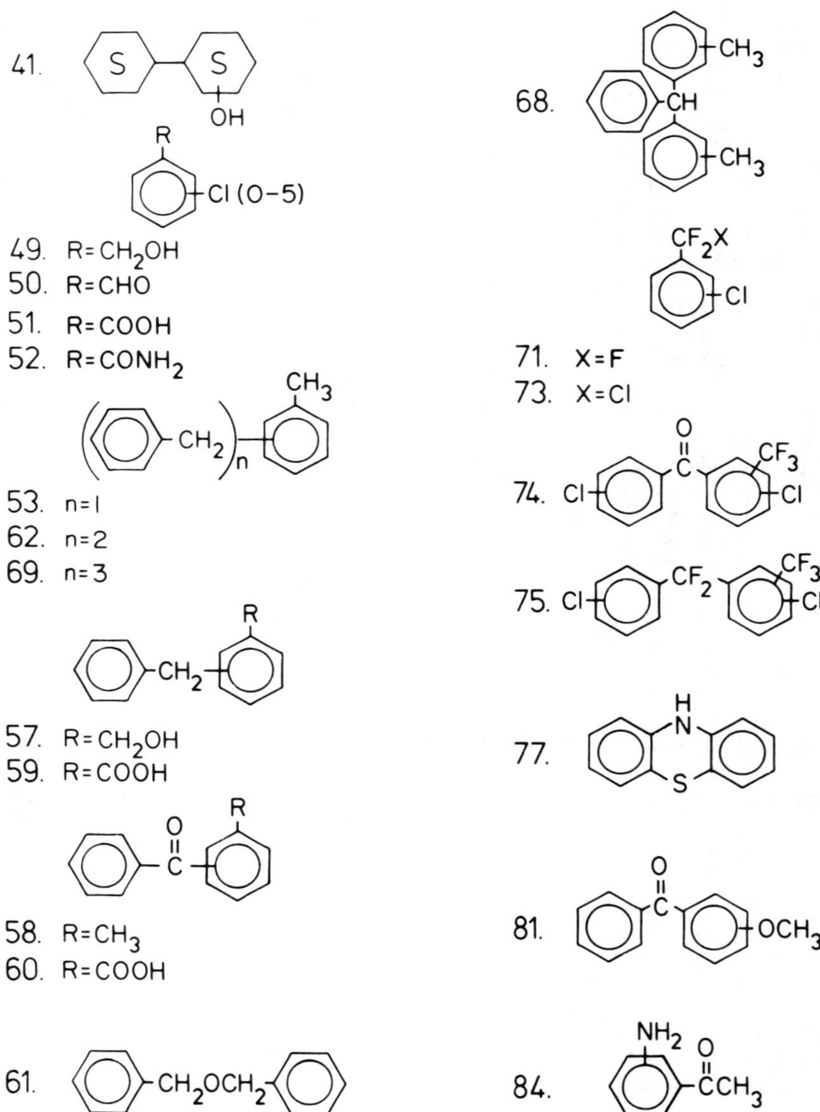


**Figure 2.** Structures of selected compounds.

The dumpsites contain large amounts of other compounds listed in Table I. BHC (No. 39), dichlorophenol (No. 32) and trichlorophenol (No. 33) were dumped in large amounts and were generally present in the sediment and water samples. Chloronaphthalenes (No. 16–18), dodecanol and other alcohols (No. 44–47), and dodecanethiol (No. 48) were detected in a heavily contaminated area of the 102nd Street bay. Mirex (No. 76) and another derivative of hexachlorocyclopentadiene were found in sediment of Bloody Run Creek. This is not unexpected because the Hyde Park dump contains 4100 metric tons of hexachlorocyclopentadiene wastes [1].

The sources of some compounds that occurred in major amounts have not been identified. Chloroanthracenes or chlorophenathrenes (No. 21-23) were identified in both 102nd Street bay and Bloody Run Creek sediments. Unusual cyclohexane derivatives (No. 41–43) were also detected in 102nd Street bay. Unusually high levels (avg 300 ppm) of polychlorinated biphenyls (PCB) (No. 26-30) and substantial levels (30 ppm) of phenothiazine (No. 77), a pesticide and intermediate in drug syntheses, were detected in Gill Creek. Normally phenothiazine is biodegradable [6], but microbial activity in the Gill Creek sediment may be limited by the high levels of toxic organic compounds. Phenothiazine did not appear on any lists of compounds dumped in western New York [1].

A series of unexpected compounds which were related to benzyl chloride and benzoyl chloride wastes were found at the 102nd Street bay and in Bloody Run Creek. We note that the Love Canal and Hyde Park sites contain 2900 and 8700 metric tons of these wastes, respectively [1]. The benzyl derivatives detected in the 102nd Street bay probably came from the Love Canal because the 102nd Street dump is not known to contain benzyl chloride wastes [1] and because the highest concentration of benzyl derivatives in the bay was directly in front of the storm sewer outfall from the Love Canal area. Benzyl-related compounds which were commercially produced in the Niagara Falls area include benzyl chloride, benzal chloride ($\alpha,\alpha$-dichlorotoluene), benzyl alcohol, benzyl thiocyanate and benzotrichloride [1]. Reaction of labile chlorines probably produced compounds No. 49–52 [5]:

$$C_6H_5CH_2Cl \xrightarrow{H_2O} C_6H_5CH_2OH \quad (49)$$

benzyl chloride

$$C_6H_5CHCl_2 \text{ (benzal chloride)} \xrightarrow{H_2O} C_6H_5CHO \quad (50)$$

$$C_6H_5CCl_3 \text{ (benzotrichloride)} \xrightarrow{H_2O} C_6H_5COOH \quad (51)$$

$$C_6H_5COCl \text{ (benzoyl chloride)} \xrightarrow{NH_3} C_6H_5CONH_2 \quad (52)$$

Each of these unsubstituted compounds (No. 49-52) was accompanied by chlorine-substituted analogs containing up to five chlorines (0-5 Cl, Figure 2). These chlorine atoms were probably derived from chlorine required for synthesis of benzyl chloride and other chlorinated derivatives. Benzyl ether (No. 61) was probably formed by the dimerization of benzyl alcohol [5]:

$$2\ C_6H_5CH_2OH \ (49) \xrightarrow{-H_2O} C_6H_5CH_2OCH_2C_6H_5 \ (61)$$

The $C_{14}$, $C_{21}$ and $C_{28}$ benzyl derivatives (No. 53-60, 62-69) were probably formed by combining benzyl groups. There were no records that the $C_{14}$, $C_{21}$ or $C_{28}$ derivatives had ever been produced commercially or dumped in the Niagara Falls area [1], but large amounts were detected at 102nd Street bay. Most of these benzyl derivatives have not been cataloged in the U.S. Environmental Protection Agency/National Institutes of Health (EPA/NIH) Mass Spectral Data Base [7]; thus, they were identified by interpretation of their mass spectra and confirmed by synthesis of reference standards.

Methyldiphenylmethane (No. 53) was the key to the identification of the benzyl derivatives (Figure 3A). This mass spectrum was initially thought to be that of ethyl biphenyl, diphenylethane or possible dimethylbiphenyl, because it was similar to the cataloged spectra of these compounds [7]. However, after reference standards of ethyl biphenyl, diphenylethane and 4,4′-dimethylbiphenyl were obtained, they had slightly different mass spectra and completely different GC retention times than the unknown. Methyldiphenylmethane (No. 53) was the only remaining compound with the formula $C_{14}H_{14}$ likely to fit the unknown mass spectrum.

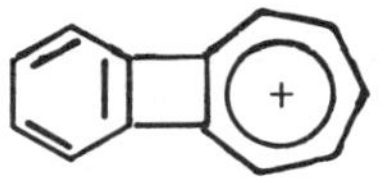

m/e 165 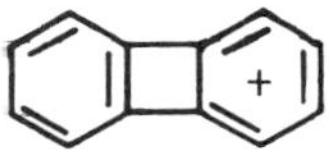 m/e 152

The characteristic ions at m/e 165 and 152 are produced by two phenyl rings joined directly with an aliphatic carbon attached to one of the rings or by two phenyl rings separated by one carbon. After methyldiphenylmethane (No. 53) was synthesized, it and the unknown had identical retention times and mass spectra. After this benzyl dimer was identified, other spectra were readily interpreted. The mass spectrum of chloromethyldiphenylmethane (No. 54) had the expected shifts due to the addition of chlorine to methyldiphenylmethane (No. 53) (Figure 3A,B).

There were two types of $C_{21}H_{20}$ benzyl derivatives. One type (No. 62) had two benzyl groups bonded to a phenyl ring, which gave intense fragment ions at m/e 91 and 181 and a low-intensity ion at m/e 257 from methyl loss (Figure 3C).

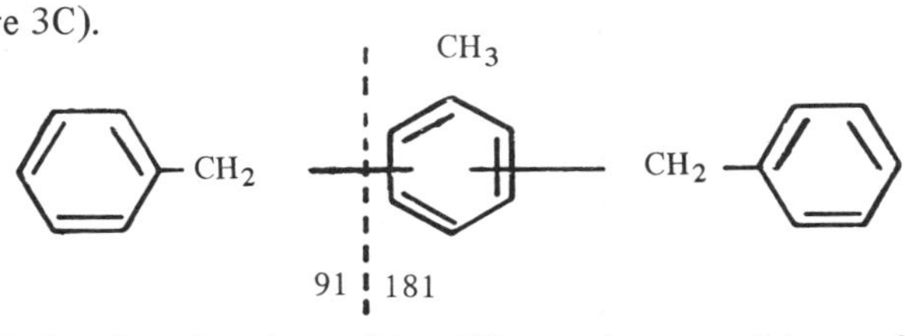

$CH_3$

$CH_2$ $CH_2$ (62)

91 181

Dimethyltriphenylmethane (No. 68) was the second type of $C_{21}H_{20}$ benzyl derivative, and had an intense ion at m/e 257 from methyl loss and low-intensity ions at m/e 91 and 181 (Figure 3D):

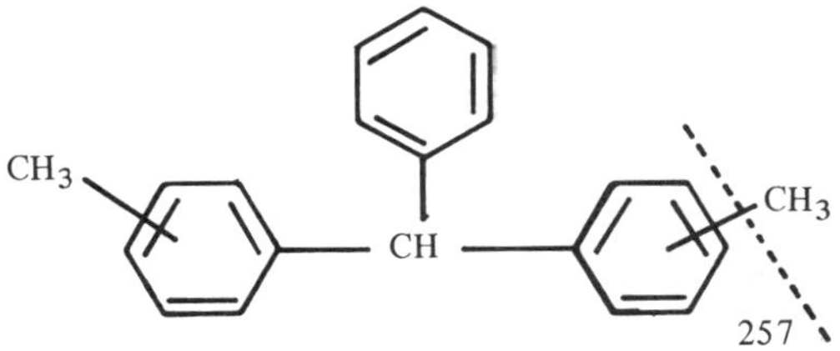

(68)

$CH_3$ CH $CH_3$

257

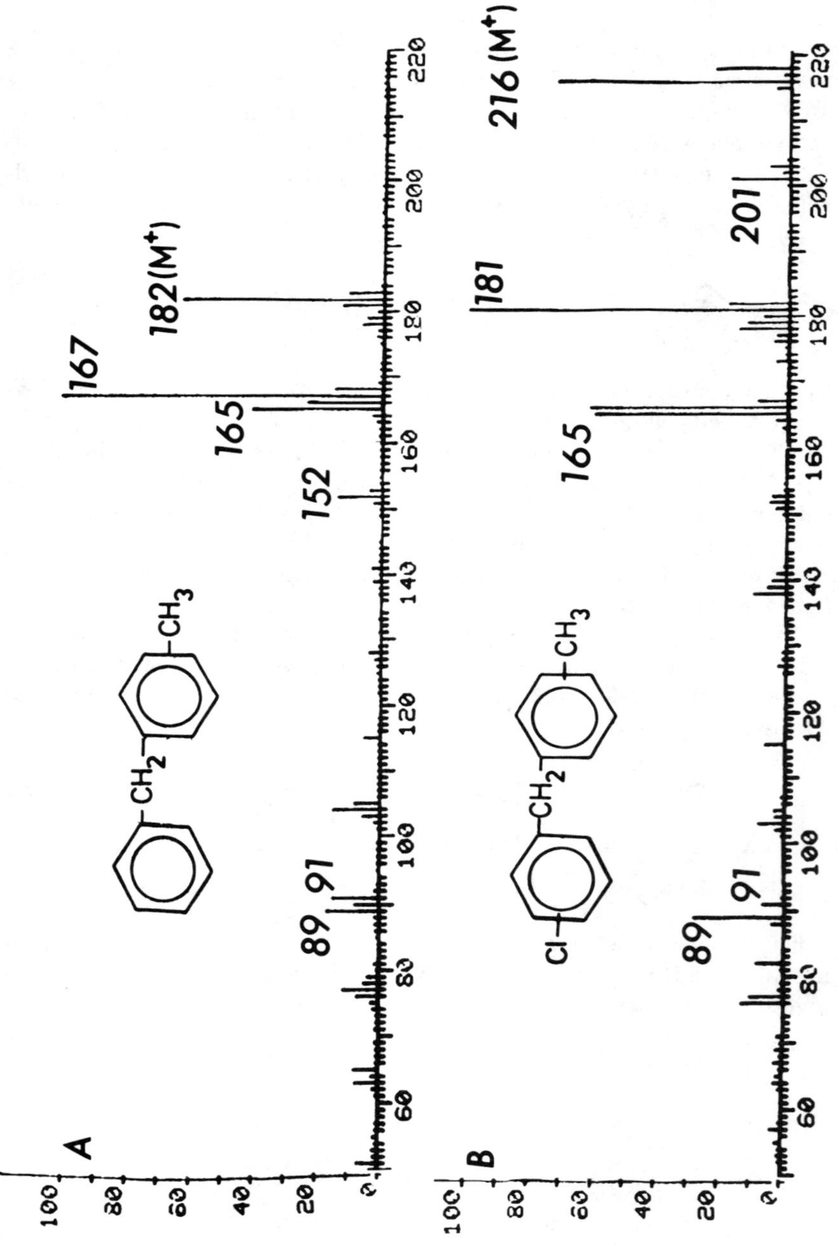
A
167
182(M+)
165
152
89
91
CH2
CH3
B
216(M+)
201
181
165
89
91
Cl
CH2
CH3

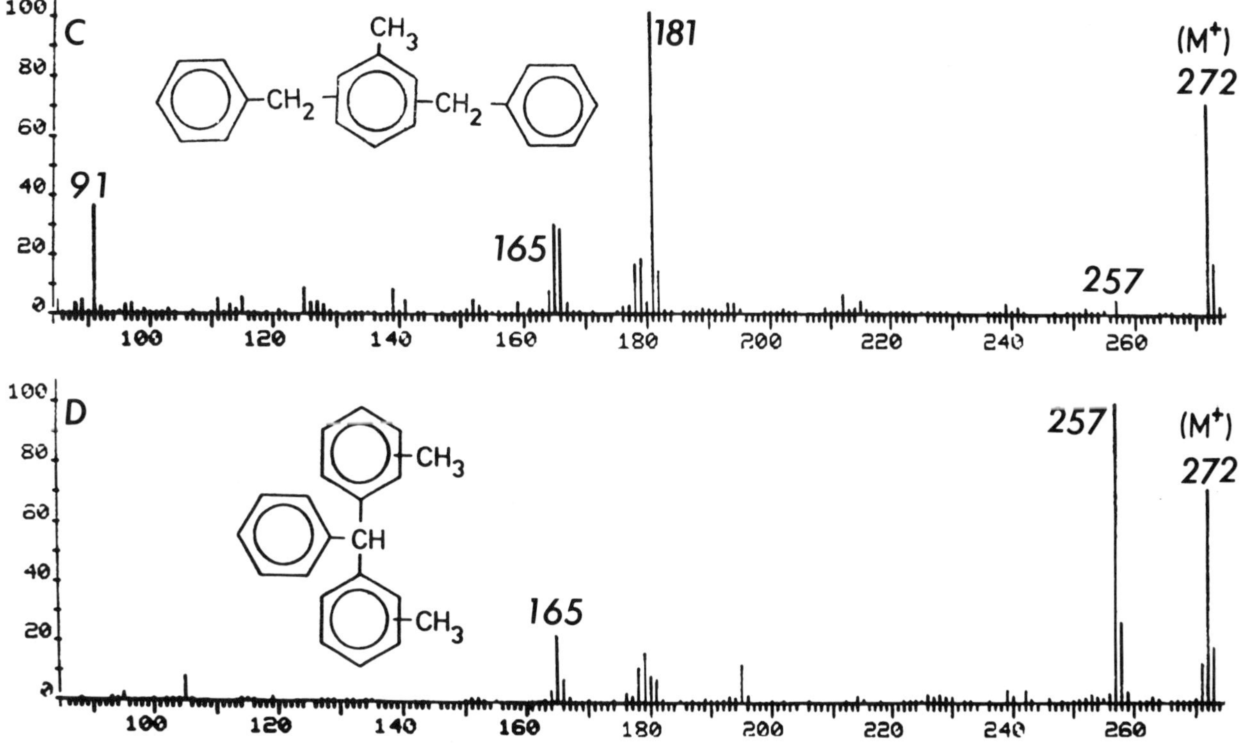


Figure 3. Electron impact mass spectra of benzyl derivatives: (A) methyldiphenylmethane (No. 53); (B) chloromethyldiphenylmethane (No. 54); (C) methylbis(phenylmethyl)benzene (No. 62); (D) dimethyltriphenylmethane (No. 68).

The $(M\text{-}15)^+$ base peak of dimethyltriphenylmethane (No. 68) is analogous to the $(M\text{-}15)^+$ base of methyldiphenylmethane (No. 53); compare Figures 3A and 3D.

Most of the $C_{14}$, $C_{21}$ and $C_{28}$-benzyl derivatives were probably by-products of the industrial production of benzyl chloride and related compounds. These are industrially produced by free-radical chlorination of toluene [5]:

$CH_3$ $\xrightarrow{Cl_2}$ $CH_2Cl$ $\xrightarrow{Cl_2}$ $CHCl_2$ $\xrightarrow{Cl_2}$ $CCl_3$

Benzyl chloride readily reacts with toluene at 0-20°C and atmospheric pressure by Friedel-Crafts alkylation; a reaction which requires a Lewis acid catalyst such as aluminum chloride or ferric chloride and anhydrous conditions [5,8]. In fact, our reference standard of methyldiphenylmethane (No. 53) was synthesized by Friedel-Crafts alkylation.

$CH_2Cl$ + $CH_3$ $\xrightarrow{AlCl_3}$ $CH_2$ $CH_3$

(53)

Figure 4 shows synthetic pathways which may explain the accidental synthesis of the other benzyl compounds during industrial benzyl chloride production. The chlorinated $C_{14}$-benzyl derivatives (No. 54-56) and chlorinated $C_{21}$-benzyl derivatives (No. 63-67) were probably produced by chlorination before or after the Friedel-Crafts reaction, since chlorine is used in the industrial synthesis of benzyl chloride [5]. The ferric chloride catalyst required for the Friedel-Crafts reaction was probably produced by the corrosion of process equipment with hydrochloric acid. (It is known that benzyl chloride cannot be stored in stainless steel barrels without a stabilizer that prevents the formation of ferric chloride [5].) After purification of benzyl chloride by distillation, the higher-boiling residue probably contained the $C_{14}$, $C_{21}$ and $C_{28}$ benzyl derivatives; wastes in the dumpsites were often described as "distillation residues, still bottoms, and tarry residues" [1].

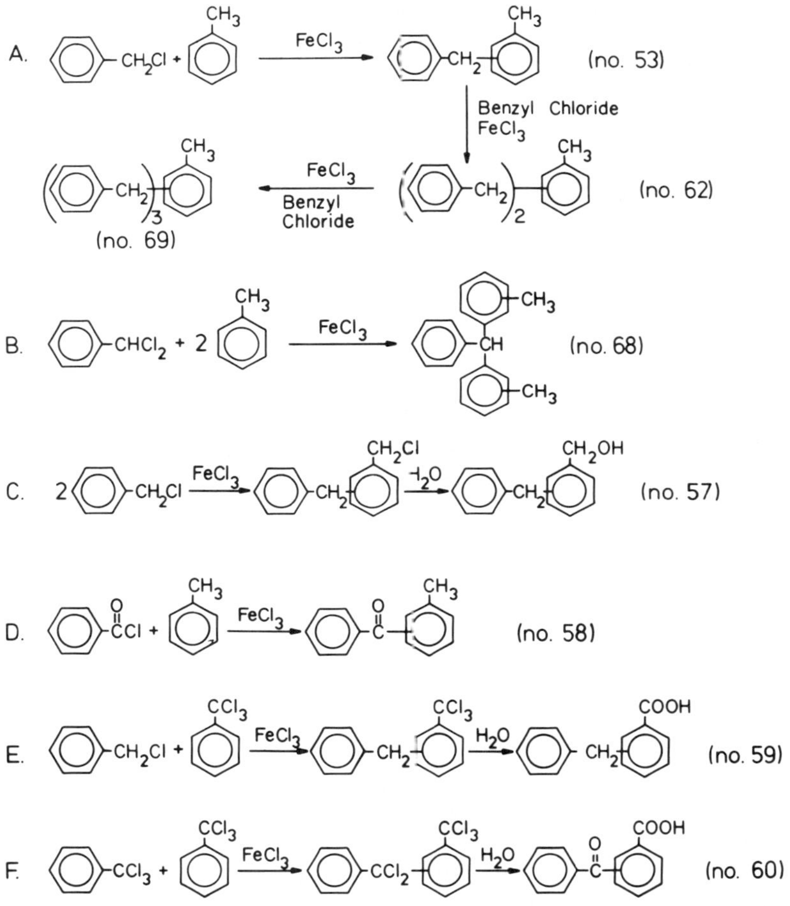

**Figure 4.** Possible syntheses of benzyl derivatives by the Friedel-Crafts reaction.

The benzyl derivatives probably could not be formed in the environment because the Friedel-Crafts reaction requires anhydrous conditions, but these compounds may have been formed when benzyl chloride wastes were stored in steel drums.

Several fluorinated compounds were detected in the Bloody Run Creek which drains the Hyde Park dump; this is not surprising since it is known to contain 7700 metric tons of benzotrifluoride derivatives [1]. Benzotrifluoride compounds No. 70-72 were detected in Bloody Run Creek and have also been reported in fish from the lower Niagara River [9]. Other fluorinated compounds were probably by-products of the industrial synthesis of 4-chlorobenzotrifluoride, which is industrially produced in the following way [5]:

$$Cl\text{-}C_6H_4\text{-}CH_3 \xrightarrow{Cl_2} Cl\text{-}C_6H_4\text{-}CCl_3 \xrightarrow{HF} Cl\text{-}C_6H_4\text{-}CF_3 \qquad (71)$$

Compound No. 73 was probably formed when the chlorines on the methyl group were not completely replaced with fluorines.

$$Cl\text{-}C_6H_4\text{-}CCl_3 \xrightarrow{HF} Cl\text{-}C_6H_4\text{-}CF_2Cl \qquad (73)$$

Compounds No. 74 and 75 were detected in major amounts in Bloody Run Creek, but neither compound has ever been listed in *Chemical Abstracts.* The EI mass spectrum of compound No. 74 had a small $(M\text{-}F)^+$ ion; however, an unexpected benefit of methane CI was a much larger $(M\text{-}F)^+$ ion; (Figure 5). Figure 6 is the electron impact mass spectrum of compound No. 75. Compounds 74 and 75 are clearly related to each other and are by-products of the industrial production of 4-chlorobenzotrifluoride. Compound 74

$$Cl\text{-}C_6H_4\text{-}CF_2\text{-}C_6H_3(CF_3)Cl \xrightarrow{H_2O} Cl\text{-}C_6H_4\text{-}CO\text{-}C_6H_3(CF_3)Cl$$

has been confirmed by synthesis. A compound corresponding to No. 75 but containing no chlorines on the aromatic ring has also been synthesized in our laboratory; the mass spectra of this synthesized compound and com-

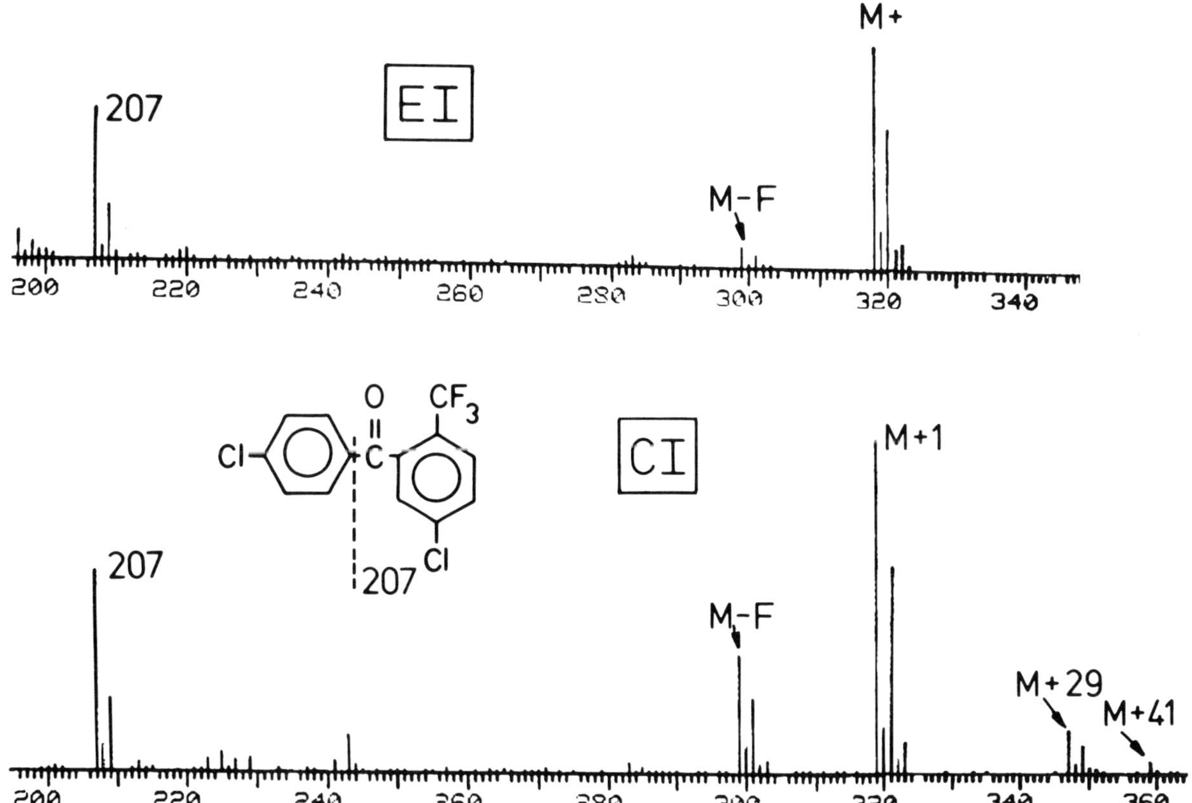


**Figure 5.** Electron impact and methane chemical ionization mass spectra of dichloro(trifluoromethyl)benzophenone (No. 74).

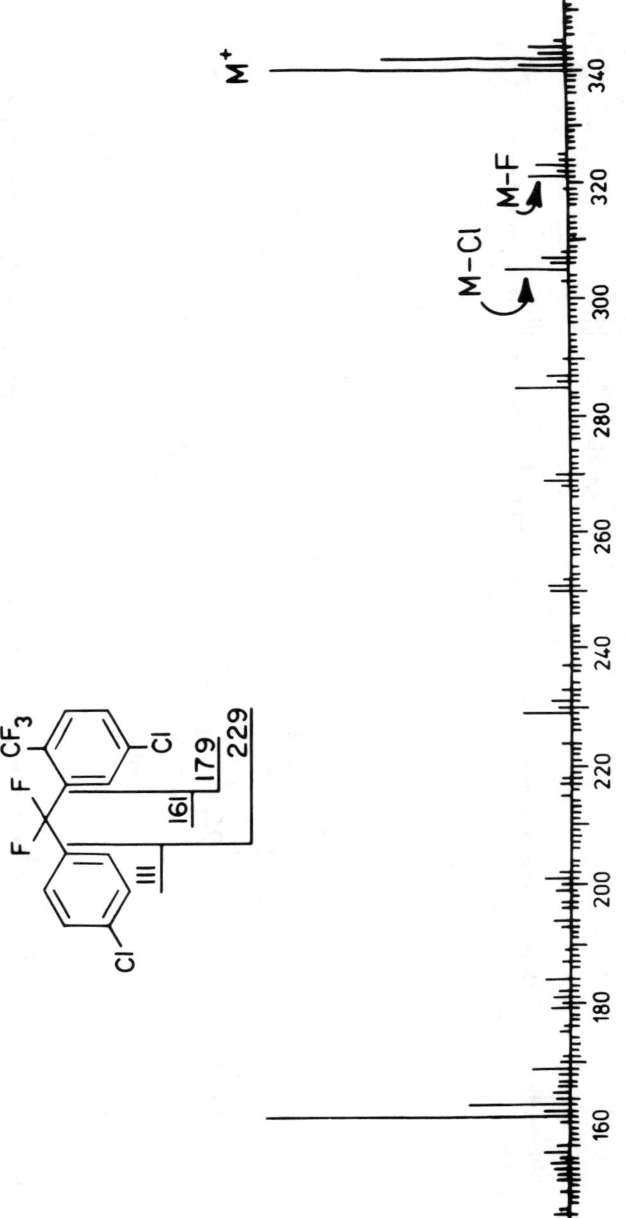


**Figure 6.** Electron impact mass spectrum of dichloro(trifluoromethyl)-α,α-difluorodiphenylmethane (No. 75).

pound No. 75 were similar. In addition to the fluorinated compounds listed in Table I, other fluorinated compounds have been detected by their intense $(M\text{-}19)^+$ ions in methane CI mass spectra, but they have not yet been identified.

We have attempted a rough mass balance calculation between the sediments and the contents of the known dumpsites in the area. No attempt was made to estimate removal of organic compounds through runoff or sediment transport. A uniform sediment depth was estimated at 10 cm, and the sediment density was estimated at 1.5 $g/cm^3$. In this way, the weight of the sediment in each area was obtained. This weight, when multiplied by the concentration (in ppm) gave us a total mass of each organic compound at each location. These total loadings are given in Table II. In all cases, these weights are much less than 0.1% of the estimated weight of these compounds deposited in the dumpsites [1]. Clearly, most of the organic compounds remain buried in the dumps. This is not to minimize the potential problems associated with the migration of these compounds out of the dumpsites.

**Table II. Total Loadings (in kg) of Various Compounds in the Niagara River Watershed**

| | 102nd Street | Bloody Run Creek | Gill Creek |
|---|---|---|---|
| Alcohols and Thiols | 20 | | |
| Benzyl Chloride Derivatives | 300 | | |
| Benzotrifluorides | | 3 | |
| Chlorobenzenes | 400 | 2 | 20 |
| Chlorotoluenes | 300 | 5 | |
| Hexachlorocyclohexane | 10 | | |
| PCB | | | 1000 |
| Phenothiazine | | | 60 |

## ACKNOWLEDGMENTS

We thank John Spagnoli, Director of the Buffalo Office, New York State Department of Environmental Conservation, and the U.S. Coast Guard, Fort Niagara, NY, for their assistance in sampling. Special thanks to Paul Foersch, Buffalo Office, New York State Department of Environmental Conservation, for acting as guide and helping to collect all samples. We extend our thanks and appreciation to the U.S. Environmental Protection Agency for supporting this research under grant No. R-806350.

## REFERENCES


1. "Draft Report on Hazardous Waste Disposal in Erie and Niagara Counties, New York," Interagency Task Force on Hazardous Wastes (1979).
2. Frey, D. G. *Limnology in North America* (Madison, WI: University of Wisconsin Press, 1963).
3. "United States Coast Pilot. Great Lakes: Lake Ontario, Erie, Huron, Michigan and Superior and St. Lawrence River," U.S. Department of Commerce, National Oceanic and Atmospheric Administration, Washington, DC (1979).
4. Jungclaus, G. A., V. Lopez-Avila and R. A. Hites. *Environ. Sci. Technol.* 12:88 (1978).
5. Kirk, R. E., D. F. Othmer, M. Grayson and P. Eckroth. *Kirk-Othmer Encyclopedia of Chemical Technology*, 3rd ed. (New York: John Wiley & Sons, Inc., 1980).
6. Coats, J. R., R. L. Metcalf, P.-Y. Lu, D. D. Brown, J. F. Williams and L. G. Hansen. *Environ. Health Persp.* 18:167 (1976).
7. Heller, S. R., and G. W. A. Milne. "EPA/NIH Mass Spectral Data Base," U.S. Department of Commerce, National Bureau of Standards, Washington, DC (1978).
8. Denny, R. C. *Named Organic Reactions* (New York: Plenum Press, 1969), p. 143.
9. Yurawecz, M. P. *J. Assoc. Off. Anal. Chem.* 62:36 (1979).

# SECTION 13

# INDUSTRIAL WASTEWATER ANALYSES

## CHAPTER 52

# CHLORINATED ORGANICS OF LOW AND HIGH RELATIVE MOLECULAR MASS IN PULP MILL BLEACHERY EFFLUENTS

**K. Lindström, J. Nordin and F. Österberg**

Swedish Forest Products Laboratory
Stockholm, Sweden

In Sweden, most chemical pulp is produced in the sulfate pulping process, whereby most of the original lignin is partially degraded and dissolved from the wood. The residual lignin, normally about 50 kg/metric ton of pulp, is removed in a subsequent bleaching process.

Utilizing chlorine-containing bleaching agents, such as chlorine (C), chlorine dioxide (D) and hypochlorite (H), lignin and other organic matter in the pulp are chlorinated.

To obtain pulp of good quality (high brightness and strength), it is necessary to wash out residual oxidized, degraded and chlorinated organic products by means of alkali extraction (E). In a conventional bleaching process, chlorination is the first of several stages. A typical six-stage bleaching sequence is: CEHDED, although a variety of modifications exists. An average mill produces 500-900 metric tons of bleached pulp daily.

Unlike the cooking process, the bleaching process is not closed, i.e., the bleaching effluents entering the receiving waters are more or less untreated.

From an environmental point of view, it is therefore of utmost importance to find out what categories and also, more specifically, what kind of compounds are formed in the bleaching process. Some of these compounds, particularly those of manmade origin (e.g., chlorinated compounds), may be hazardous to aquatic organisms in the receiving waters.

During the past four years much attention has been given by our institute to the formation of chlorinated organics from the bleaching process.

This chapter briefly reports on the characterization and determination of chlorinated organic matter of high and low relative molecular mass (mol wt) in spent bleach liquors (SBL) from a conventional bleaching process in relation to total organically bound chlorine (TOCl).

## RESULTS AND DISCUSSION

### Fractionation

The highest load of organic material occurs in the SBL of the prebleaching stage (C) and the first alkaline extraction stage ($E_1$) [1]. Therefore, most attention has been focused on the SBL of these two stages.

Due to differences in the chemical and physical properties of the chlorinated organic matter, it is practical from an analytical viewpoint to divide the SBL into four groups of chlorinated organics:

1. high-molecular-weight ($>1000$)
2. ether extractable (mol wt $<1000$)
3. extremely volatile (mol wt $<500$)
4. water-soluble (mol wt $<1000$)

The organic matter in group 1 can be separated from compounds of low relative molecular mass by means of ultrafiltration (UF) and further chemically characterized. Compounds in group 2 are suspected to be the most important, since they exhibit a less polar character than those in group 4, i.e., they are relatively lipophilic. Bioaccumulation in the fat of aquatic organisms, resulting in possible toxic effects may therefore occur [2,3].

The third group has been defined as extremely volatile compounds. It is best to separate this class of compounds from others although they may occur to a large extent in the ether extract, because concentration of the extract by means of evaporation may reduce the amounts of these compounds. Stripping the SBL with helium gas and passing the gas through a column with a solid sorbent traps most of the compounds in this group. Desorption can be carried out either thermally or by solvent extraction.

The least attention has been paid to compounds of group 4. Investigation of this class of compounds has not yet started.

## TOCl

A separate method for determining TOCl in SBL has been developed at our institute by Sjöström et al. [4]. The method is based on ultrafiltration, sorption on a crosslinked polymer, separation of inorganic from organic chlorine and Schöniger combustion of the organic residue followed by potentiometric titration with silver nitrate. Extremely volatile chlorinated organics, such as chloroform, are not included in the TOCl determination.

It appears that about 5 kg/metric ton of pulp, or 70% of the overall (about 7 kg/metric ton) TOCl from a bleaching sequence, is found in the SBL of the two first consecutive stages (C and $E_1$), although considerable variations of TOCl in the SBL have been observed. The SBL of the C and $E_1$ stages contain approximately 3 and 2 kg/metric ton, respectively.

Figure 1 illustrates the distribution of TOCl in the two SBL after fractionation (UF) in different relative molecular mass ranges. As much as about 30% of the chlorinated matter is constituted by components of low relative molecular mass (mol wt $< 1000$) in the C-stage SBL compared to only about 5% in the $E_1$-stage SBL. These results led us first to investigate the $E_1$-stage SBL for chlorinated organics of high relative molecular mass and the C-stage SBL for low relative molecular mass of preferably lipophilic character.

## Chlorinated Matter of High Relative Molecular Mass

The SBL of the C and $E_1$ stages have been investigated by Erickson and Dence for their high molecular mass matter after derivatization of the phenolic groups [5]. Degradation was carried out by potassium permanganate and sodium periodate in alkaline medium. The resulting carboxylic acids were derivatized with diazomethane and the corresponding methyl esters were determined by gas chromatography/mass spectrometry (GC/MS).

By slightly modifying the procedure utilizing ultrafiltration and deuterated dimethyl sulfate as the derivatizing reagent of the phenolic group (Figure 2) and using computerized glass-capillary column GC/MS and emission spectroscopy (ES) we have identified a number of chlorinated degradation products. A computer-reconstructed gas chromatogram is shown in Figure 3. The shaded peaks are the 14 most abundant chlorinated derivatives.

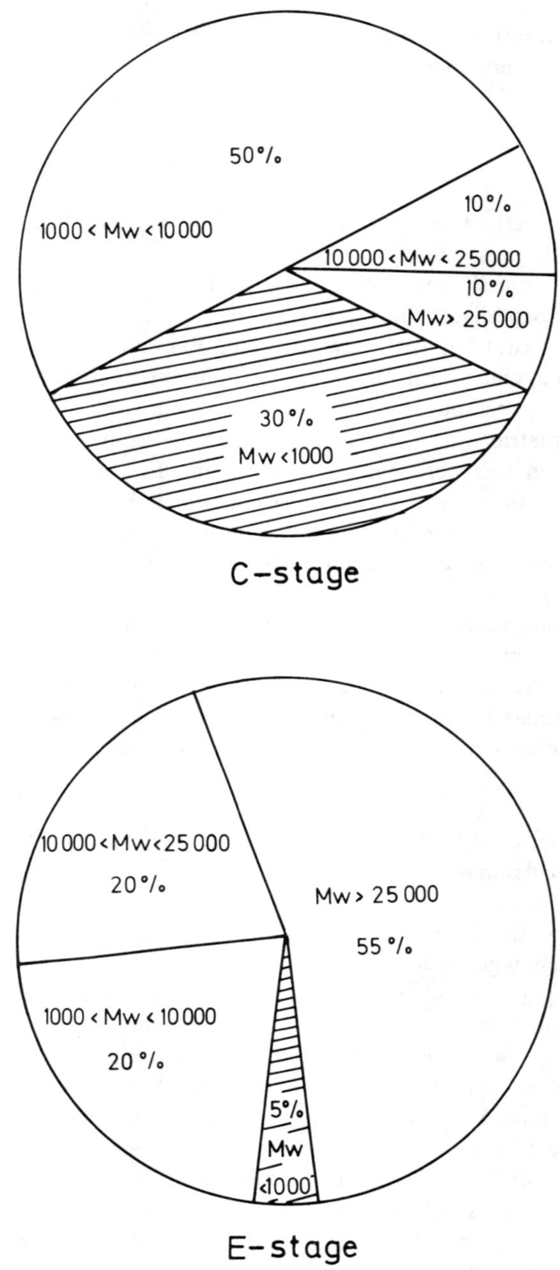


**Figure 1.** Distribution of TOCl in ultrafiltrated SBL of the C and E stages.

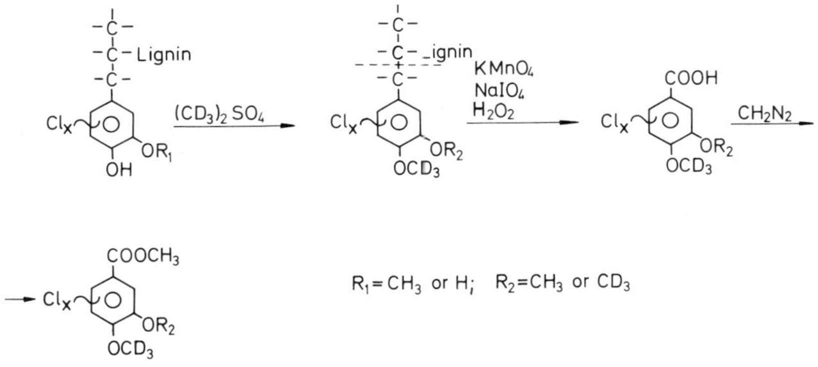

$R_1 = CH_3$ or H; $R_2 = CH_3$ or $CD_3$

Figure 2. Derivatization of the phenolic group, oxidative degradation and derivatization of the carboxylic group of ultrafiltrated (mol wt > 25,000) "chlorolignin."

GC in combination with ES gives the elementary composition of the molecules in the GC peak and thus facilitates the identification (Figure 4). Some weak sulfur signals (Figure 4) indicate the existence of organosulfur compounds.

About 40 different chlorinated degradation products have tentatively been identified in an ultrafilter SBL in the fraction of mol wt > 25,000. Figure 5 shows the 14 most abundant acids and their relative distribution. Mono and dibasic acids of guaiacolic, catecholic and phenolic type, containing one or two chlorine atoms in the aromatic nuclei predominate.

Derivatization of the phenolic groups with a deuterated methyl group results in a reduction of the number of GC peaks due to equal chromatographic properties of the guaiacolic and catecholic derivatives compared with, for example, ethylation. Yet, the proportions of guaiacolic and catecholic origin can easily be calculated from the corresponding computer limited reconstructed mass chromatograms. From the chlorine channel of the multiple plasma detector (Figure 4) the organically bound chlorine of the chlorinated degradation products can be calculated. The recovery of identified derivatives is only 10% of TOCl in the fraction investigated which corresponds to about 0.1 kg/metric ton of pulp, calculated as organically bound chlorine. The details of this work will be published in the near future [6].

The results obtained indicate that it is theoretically possible for chlorinated aromatics of low relative molecular mass to be formed from "chlorolignin" of high relative molecular mass in the receiving water. Degradation may be caused chemically, biochemically or physically (ultraviolet irradiation).

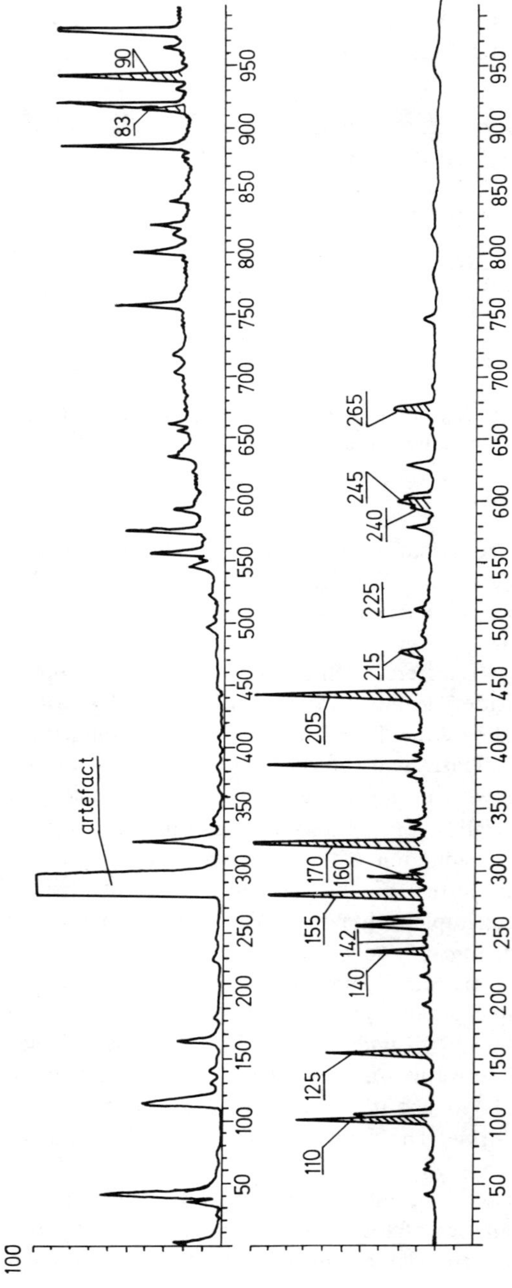


**Figure 3.** Computer-reconstructed total ion current gas chromatogram of methyl esters of degraded "chlorolignin" (mol wt >25,000). The shaded peaks constitute the 15 most abundant chlorinated compounds.

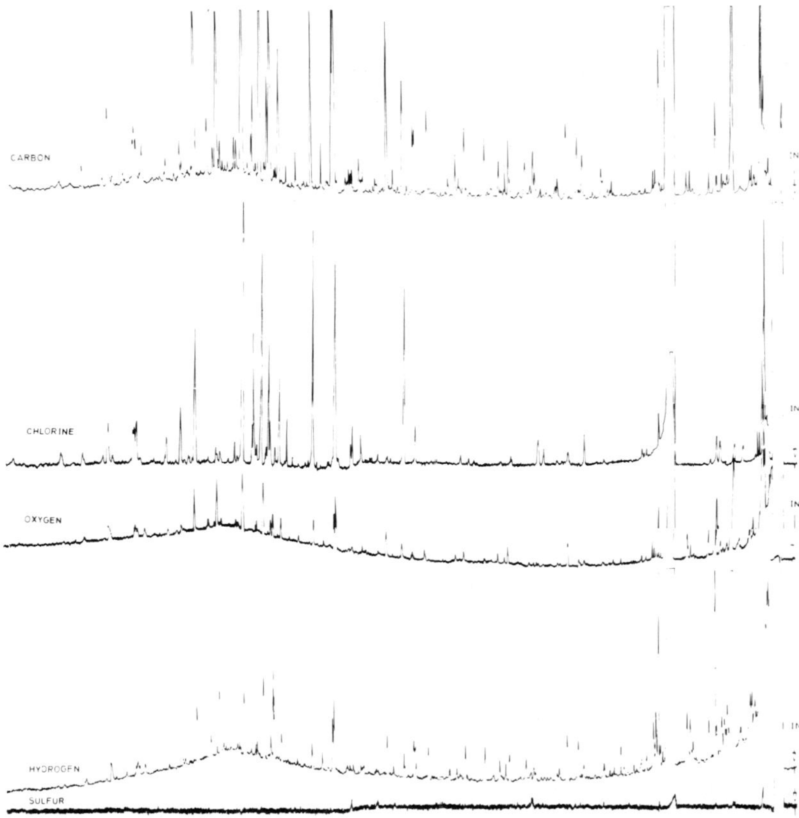


**Figure 4.** Multiple plasma detected gas chromatogram of the same sample shown in Figure 3.

## Ether-Extractable Chlorinated Organics of Low Relative Molecular Mass

In a bleachery effluent, three main classes of chlorinated compounds of low relative molecular mass can be expected: carboxylic acids, phenolics and neutrals. As shown above, the C-stage SBL contributes to most of the chlorinated organic matter of low relative molecular mass. Figure 6 illustrates schematically the workup procedure for determining TOCl of the three main classes of compounds in the final ether and ether/pentane (2:1) extracts.

| Degradation products | Peak no | R= H;% | R = $CH_3$;% | Relative distribution;% |
|---|---|---|---|---|
| 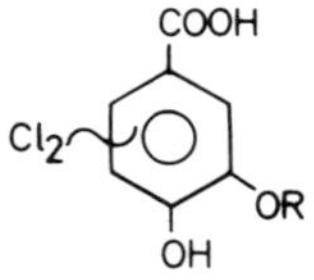 COOH, $Cl_2$, OR, OH | 110<br>125<br>142<br>160<br>170 | 65 | 35 | 46 |
| 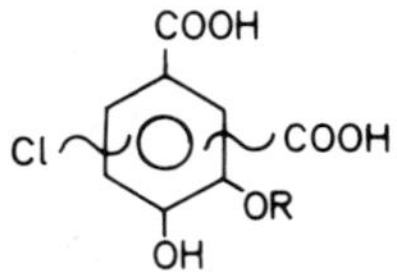 COOH, Cl, COOH, OR, OH | 205<br>245<br>265 | 74 | 26 | 20 |
| 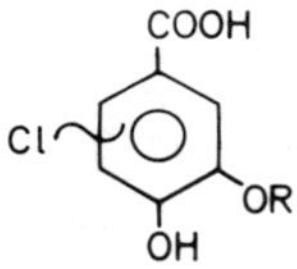 COOH, Cl, OR, OH | 140<br>155 | 35 | 65 | 19 |
| 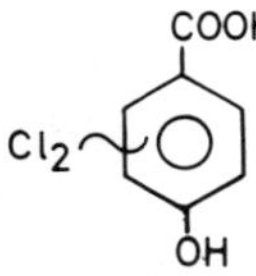 COOH, $Cl_2$, OH | 83<br>90 | | | 8 |
| 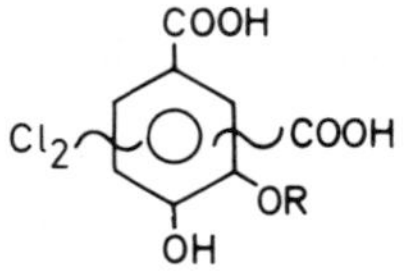 COOH, $Cl_2$, COOH, OR, OH | 215<br>225<br>240 | 83 | 17 | 7 |

**Figure 5.** The relative distribution of the 14 most abundant chlorinated acids obtained after ultrafiltration and degradation. The peak number refers to Figure 3.

*n*-Pentane was used to reduce the polarity of the extracting solvent in order to recover the most lipophilic components. Except for the ultrafiltration step, the workup procedure is similar to that described previously [7].

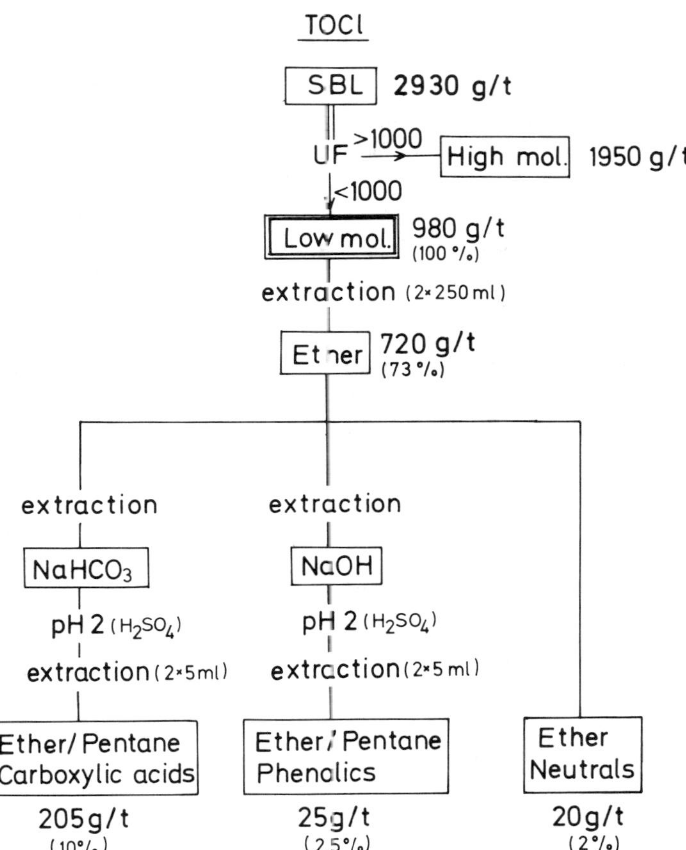


**Figure 6.** A schematic presentation of the workup procedure for TOCl determination of the acidic, phenolic and neutral extracts of C-stage SBL.

Although the values are referred to a C-stage SBL, the $E_1$-stage SBL exhibits about the same distribution of TOCl in the extracts. From a bioaccumulation viewpoint, the phenolics and neutrals should be more important than the acids. The occurrence of some chlorinated analogs in the SBL has previously been reported [8,9].

Based on the structure of native lignin [10] and the investigations carried out by Rogers and Keith [11] and Leach and Thakore [9], the phenolic fraction was the first to be examined, although the acidic fraction exhibited the highest concentration of chlorinated organic matter (TOCl).

## Chlorinated Phenolic Compounds

As previously reported, there are basically four kinds of chlorinated monomeric phenolic compounds (Figure 7) whose concentrations have been comprehensively investigated in the C- and $E_1$-stage SBL [7,12-14]. The predominating chlorophenolics are the chlorocatechols in the C stage and the chloroguaiacols in the $E_1$ stage. Considerable variation in concentration due to chlorine concentration and pH seems to be the most important factor during bleaching [13]. About 50% of the chlorinated phenols in the phenolic fraction have been calculated as TOCl.

Bioaccumulation of chlorophenolics in fish has been described previously [2,3]. As expected, the concentrations of chloroguaiacols exceed those of the chlorocatechols due to the more lipophilic character of the former.

3,4,5-Trichloroguaiacol [15] exhibited the highest concentration in the fish fat and not the 4,5,6-isomer stated by Landner et al. [2].

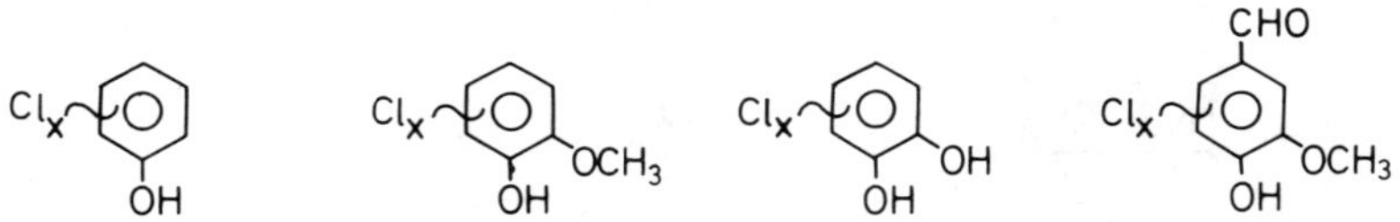

**Figure 7.** The four main types of monomeric chlorinated phenolic compounds found in SBL.

## Chlorinated Neutral Compounds

In an earlier study, some chlorinated neutral compounds from the C-stage SBL were identified [16]. Table I lists 45 different chlorinated organics and their concentrations found in C-stage SBL after various laboratory-scale bleaching conditions. The values in columns 1 and 2 were obtained after bleaching of carefully washed pulp of high and low concentrations of chlorine, respectively. In column 3, the values are given after bleaching of inadequately washed pulp at low chlorine concentration. The numbers and the concentrations of the chlorinated compounds increase considerably when using high chlorine concentration.

Chlorinated thiophenic compounds (Table I) cannot be detected when washing the pulp carefully, indicating that the corresponding nonchlorinated analogs are formed in the cooking process, where sodium sulfide and sodium hydroxide are used.

Table I. Recovery of Identified Chlorinated Neutral Compounds after Laboratory Bleaching

| Peak | Compound | Concentration (mg/metric ton of pulp) | | |
|---|---|---|---|---|
| | | 1[a] | 2[a] | 3[b] |
| 1 | Dichloroallene[c] | 50 | 20 | |
| 4 | Tetrachloroethene[d] | 20 | 10 | |
| 5 | Chlorobenzene[d] | | | |
| 6 | Trichloroallene[c] | 1100 | 550 | 30 |
| 7 | Dichlorobutenal[c] | 20 | | |
| 10 | Trichlorobutatriene[c] | 10 | | 15 |
| 11 | Trichlorobutatriene Isomer[c] | 10 | | |
| 12 | Tetrachloropropene[c] | 20 | | |
| 13 | Tetrachloropropene Isomer[c] | 70 | 10 | |
| 14 | Tetrachloropropene Isomer[c] | 90 | | 15 |
| 15 | Trichlorocyclopropanone[c] | 120 | | 30 |
| 16 | Tetrachlorobutene[c] | | | 5 |
| 17 | Dichlorobenzene[d] | 10 | | |
| 18 | Tetrachlorobutatriene[c] | 130 | 5 | 60 |
| 19 | Dichlorocyclopentenedione[c] | 5 | 280 | 30 |
| 20 | Pentachloropropene[c] | 510 | 25 | 30 |
| 21 | Trichlorothiophene[d] | | | 30 |
| 23 | Chlorobenzaldehyde[d] | 1060 | 30 | 30 |
| 24 | Chlorobenzaldehyde Isomer[d] | | 20 | 45 |
| 26 | Pentachlorobutadiene[c] | 380 | 5 | 60 |
| 27 | Pentachlorobutadiene Isomer[c] | 90 | | 90 |
| 30 | Trichlorocyclobutenone[c] | | 10 | |
| 31 | Dichloroformylthiophene[d] | | | 75 |
| 32 | Tetrachlorothiophene[d] | | | 90 |
| 33 | Dichloroacetylthiophene[d] | | | 60 |
| 34 | Dichloroacetylthiophene Isomer[d] | | | 120 |
| 35 | Dichloropripionylthiophene[d] | | | 45 |
| 36 | Trichloroformylthiphene[c] | | | 30 |
| 37 | Trichloroacetylthiphene[d] | | | 75 |
| 38 | Dichlorocymene-8-ol[c] | | | 45 |
| 39 | Dichlorobenzaldehyde[d] | 110 | | |
| 40 | Trichlorocyclopentenedione[c] | 70 | 10 | |
| 41 | Trichlorocyclopentenedione Isomer[c] | 150 | 40 | 30 |
| 42 | Trichlorocyclopentenedione Isomer[c] | 380 | 5 | |
| 43 | Tetrachlorobenzene[d] | 130 | | |
| 44 | Tetrachlorobenzene Isomer[d] | 220 | | |
| 45 | Pentachlorobenzene[d] | 30 | | |
| 49 | Hexachlorohexatriene[c] | 240 | 150 | |
| 50 | Hexachlorohexatriene Isomer[c] | 1960 | 220 | 45 |
| 51 | Hexachlorohexatriene Isomer[c] | 2310 | 15 | 60 |
| 52 | Hexachlorohexatriene Isomer[c] | 630 | | |

Table I, continued

| Peak | Compound | Concentration (mg/metric ton of pulp) | | |
|---|---|---|---|---|
| | | 1[a] | 2[a] | 3[b] |
| 53 | Hexachlorohexatriene Isomer[c] | 230 | | |
| 54 | Heptachlorohexatriene[c] | 85 | 5 | |
| 55 | Heptachlorohexatriene Isomer[c] | 110 | | |
| 56 | Trichlorotrimethoxybenzene[c] | | | 30 |

[a]Results obtained after bleaching of carefully washed pulp at high (Column 1) and low (Column 2) concentrations of chlorine. Bleaching conditions: unbleached pine kraft pulp, kappa number 30.1. Pulp consistency, 3.5%, 25°C, 30-min contact, 73.6 kg $Cl_2$/metric ton of pulp (Column 1) or 55.2 kg $Cl_2$/metric ton of pulp (Column 2).

[b]Results after bleaching of inadequately washed pulp at low chlorine concentration. Bleaching conditions: unbleached pine kraft pulp, kappa number 30.1, 3% pulp consistency, 56.0 kg $Cl_2$/metric ton of pulp, 20°C for 60 min.

[c]Mass spectral interpretation only.

[d]Mass spectra matched against mass spectra of reference compounds.

Quantitative determination was carried out by means of glass capillary column GC/MS using an external standard. The values in Table I are therefore very approximate due to interferences of nonchlorinated compounds in the gas chromatogram. Figure 8 shows a total ion current gas chromatogram of an ether extract. The numbered peaks of the corresponding compounds are listed in Table I.

A multiple plasma detected gas chomatogram (GC/ES) of an ether extract of a kraft pulp bleachery C-stage SBL is shown in Figure 9. Some of the compounds listed in Table I have been identified by matching the retention times of the peaks against those in a GS/MS total ion current chromatogram.

One of the most predominating peaks exhibits a very strong sulfur and chlorine signal. Identification by GC/MS shows fragment ions at m/e 83 (100%), 85 (66%), 87 (10%), 147 (3%) and 149 (2%). The compound, 1,1-dichlorodimethylsulfone, is reported by McKague and Walden to occur in bleaching effluents [17]. Mass spectra and retention times were identical to McKague and Walden's synthesized substance.

Only 2% of lipophilic chlorinated compounds of low relative molecular mass are found in the neutral fraction calculated as TOCl (see Figure 6).

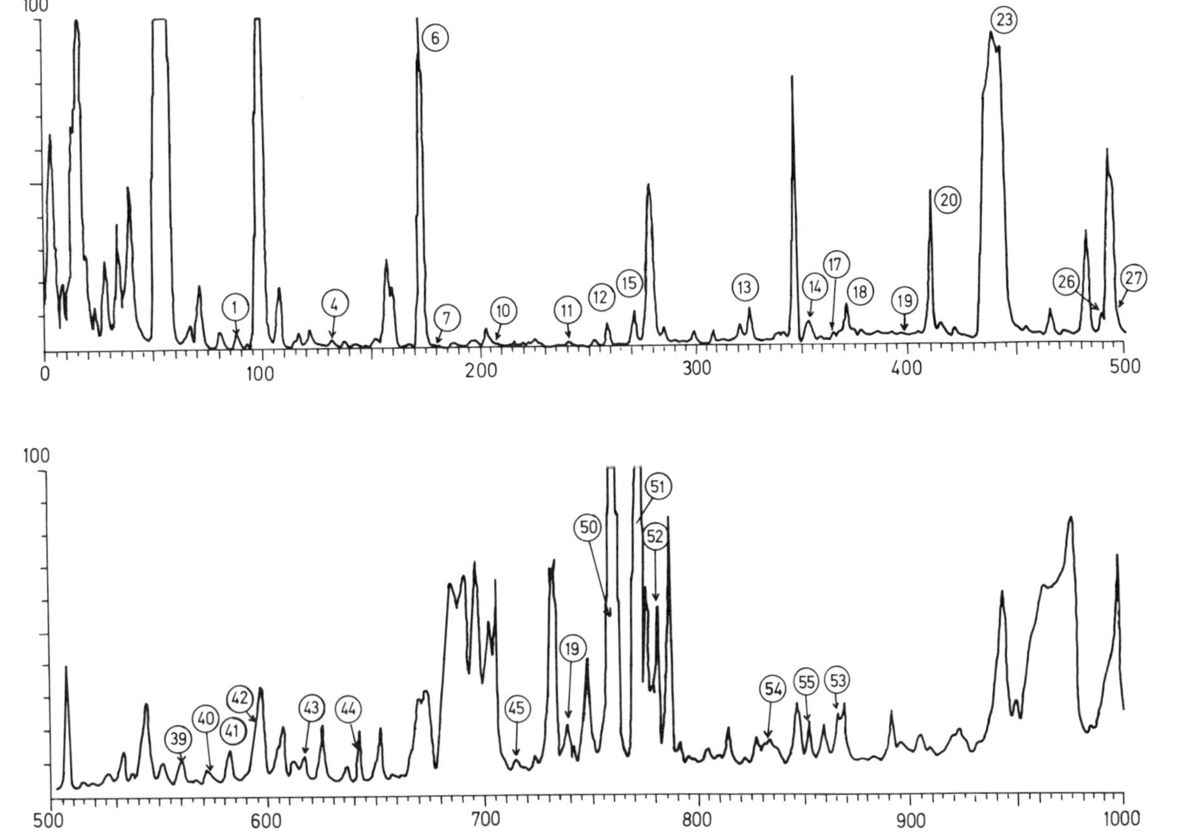


**Figure 8.** Computer-reconstructed total current ion gas chromatogram of the ether extract of the neutral fraction from a C-stage SBL of laboratory bleached pulp. The peak number refers to Table I.

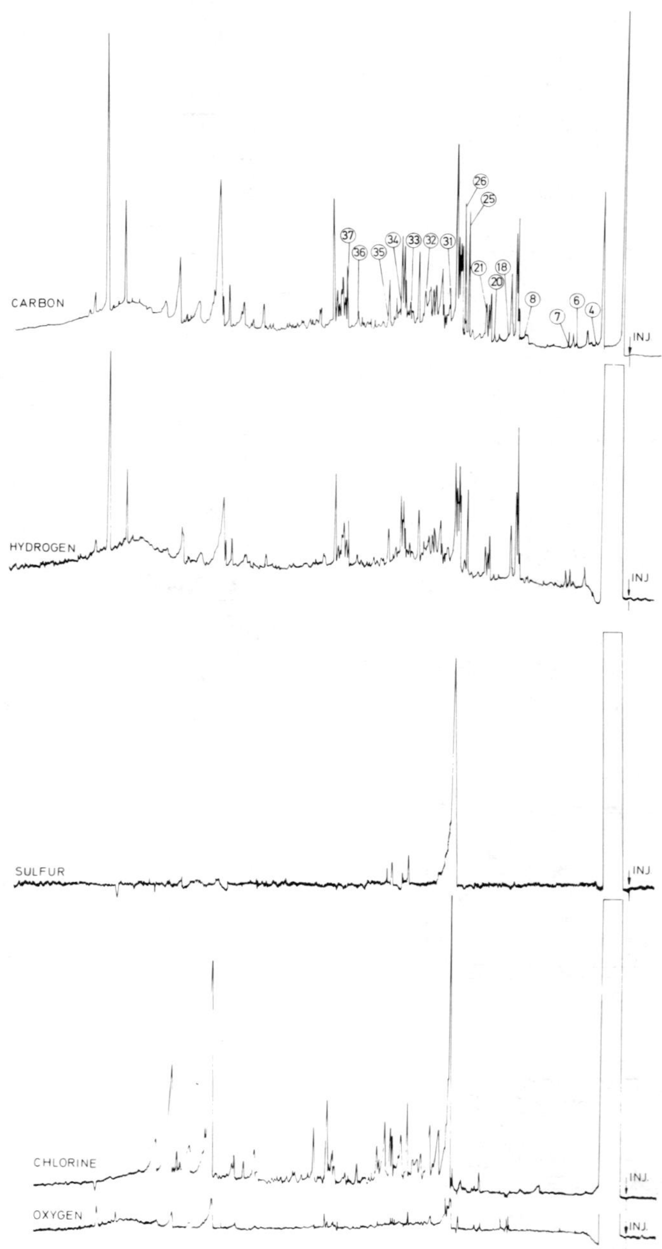


**Figure 9.** Multiple plasma detected gas chromatogram from an ether extract of the neutral fraction of a C-stage bleachery effluent. The peak number refers to Table I.

A rough calculation in the GC/ES chromatogram of the total chlorine signal of identified compounds gives about 4 g of organically bound chlorine, which is equivalent to about 20% so far identified.

The workup procedure was first used for the isolation of chemically relatively persistent chlorinated organics which might accumulate in the fat of aquatic organisms. Increased attention was, however, focused on the C-stage SBL when Ander et al. [18] reported on the mutagenicity of that spent liquor. Alkali as well as bicarbonate eliminated the mutagenicity. Recently Kringstad et al. reported a number of identified chlorinated neutrals by isolating parts of the extracted material on a sorption polymer followed by anion exchange of the eluate to eliminate the chlorophenolic compounds [19]. 2-Chloropropenal and 1,3-dichloroacetone exhibited strong mutagenicity. These compounds have not been identified in the extracts obtained with our procedure. They seem to be extremely alkali-labile.

Very little has been done to investigate bleachery effluents of bleached sulfite pulp. Among other compounds, chlorinated *p*-cymenes in relatively high concentrations have been reported to occur in sulfite bleachery effluents [20,21]. Bioaccumulation studies were therefore carried out [22]. By selected ion monitoring (SIM), two isomers of dichloro-*p*-cymene appeared to predominate in fish liver fat.

## Extremely Volatile Chlorinated Organics

By stripping the SBL with helium gas and trapping and volatile organics, one of the most abundant chlorinated compounds found is chloroform. Therefore, a separate analytical method has been developed to determine the chloroform concentration in various bleaching stages and under various bleaching conditions. The method is based on the addition of deuterated chloroform as internal standard to an acidified sample. Determination is carried out by glass capillary column GC and SIM. Figure 10 shows a selected ion chromatogram of a C-stage sample. The complete analytical time required is short (10–15 min), and the method is accurate and precise. Table II shows the results obtained after laboratory bleaching of pulp. The method is to be published in detail elsewhere [23].

As illustrated in Table II, chloroform constitutes as much as 17% of the chlorinated organic matter in the H stage. Since the TOCl method has, to date, been applied exclusively to SBL of C, E and D stages, there may be errors in the analytical performance of TOCl of this particular stage. The chloroform concentration does not affect the TOCl values materially except for the H stage. In such SBL, the TOCl determination has to be supplemented with a chloroform determination to get a true value of TOCl.

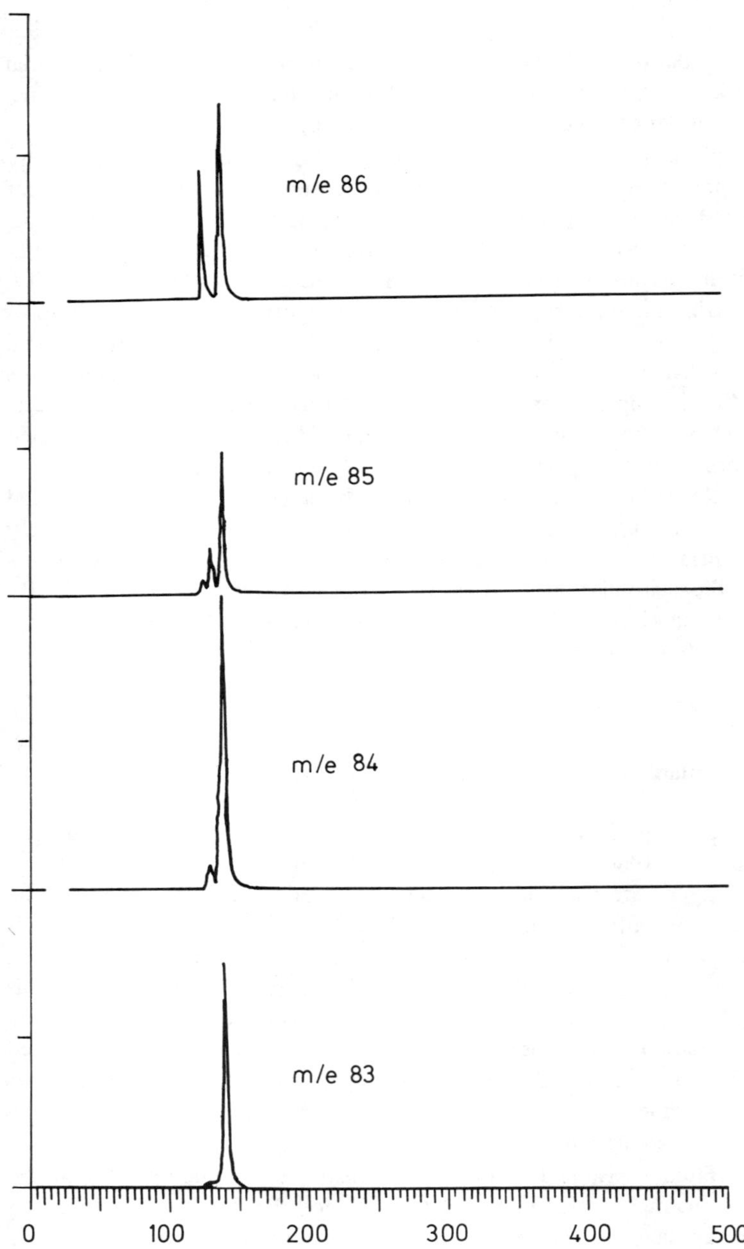


**Figure 10.** Selected ion monitored gas chromatogram for the determination of chloroform in SBL. Fragment ion m/e 83 and m/e 85 originate from chloroform and ions m/e 84 and m/e 86 from deuterated chloroform.

Table II. Relationship of Chloroform and TOCl in Different Bleaching Stages[a]

| Stage | TOCl (kg/metric ton) | $CHCl_3$[b] (kg/metric ton) | TOCl + $CHCl_3$[b] (kg/metric ton) | $CHCl_3$[b] (% of TOCl) |
|---|---|---|---|---|
| Chlorination[c] | 2.32 | 0.01 | 2.33 | 0.4 |
| Extraction[d] | 0.91 | 0.03 | 0.94 | 3.2 |
| Hypochlorite[e] | 0.72 | 0.15 | 0.87 | 17 |

[a]Unbleached pine kraft pulp, kappa number 35.1

[b]Calculated as Cl.

[c]Pulp consistency, 3% washed pulp, 7.0% $Cl_2$ (calculated on the pulp), 20°C, 60 min.

[d]Pulp consistency, 8% bleached pulp, 4% NaOH (calculated on the pulp), 20°C, 60 min.

[e]Pulp consistency, 6% extracted pulp, 4% available $Cl_2$, 40°C, 90 min.

Dichloromethane has also been identified, but its concentration has not been determined.

## EXPERIMENTAL

### Sampling

About 5 L of SBL were collected in glass bottles at the plant, sent to the laboratory and stored in the refrigerator at 4°C. The transportation of the sample usually took 2–3 days.

#### *Chlorinated Organic Matter of High Relative Molecular Mass*

One liter of SBL was ultrafiltrated through a Pellicon membrane disc (Millipore PSED, mol wt 25,000) in a 90-mm Hi-Flux Cell (Millipore xx 4209052) including a 5-L dispensing pressure vessel with an applied pressure of about 400 kPa. About 50 mL of the retentate was washed with 1 L distilled water. The retentate was freeze-dried, homogenized and submitted to oxidative degradation according to Erickson and Dence [5]. Hexadeuterated dimethyl sulfate was substituted for dimethyl sulfate.

*Ether-Extractable Compounds*

About 4 L of C-stage SBL were continuously extracted with 4 x 250 mL diethyl ether (purum) in four percolators holding 1.5 L each. After 48 hr, the ether extract was worked-up as previously described [7]. The neutral fraction of the ether extract was dried over anhydrous sodium sulfate, filtered and evaporated in a Kuderna-Danish evaporator and further concentrated in a modified micro-Snyder evaporator to a final volume of 500 $\mu$L.

Determination of TOCl of the different extracts was carried out as shown in Figure 6. One liter of SBL was ultrafiltered through an Amicon membrane disc (mol wt 1000) with the same equipment and under similar conditions described above.

*Extremely Volatile Compounds*

One liter of SBL was stripped with helium gas at 60°C for 10 min. The gas was allowed to pass through a column with activated carbon (charcoal tubes, SKC 107, Env. Compliance Corp.) for 10 min at a flowrate of 50 mL/min. The carbon was transferred to a 2-mL septum flask containing 0.5 mL of carbon disulfide. The flask was allowed to stand in an ultrasonic bath for 10 min, and 1-2 $\mu$L was injected on a GC/MS.

## Determination of Chloroform

A 5-mL aqueous sample was collected in a 10-mL test tube, and the pH was immediately adjusted (if necessary) with 0.1 mol/L sulfuric acid. A 1-mL aliquot of 0.42-mmol/L solution of deuterochloroform (Merck AG, deuteration grade 99%) in acetonitrile (Fluka AG p.a.), was added. The aqueous solution was shaken for 1 min. A 1-mL quantity of *n*-pentane (Merck AG p.a.) was added and the test tube shaken for 1 minute. After centrifugation for 3 min, most of the *n*-pentane was sucked off, 1 $\mu$L was injected on the GC/MS and the chloroform was determined by selected ion monitoring [23].

## Gas Chromatography/Mass Spectrometry

A Finnigan MS system (Model 3200 F) combined with an online computer (Model 6000) and a GC (Carlo Erba, Model 2900) were connected to the ion source by means of a platinum capillary interface. The columns used were BDS (degradation products), OV-17 and UCON (neutrals). Helium was used as carrier gas (0.6 mL/min). The GC oven was operated from 30 to 250°C at 2°C/min (neutrals) and from 50 to 210°C at 3°C/min (degradation products). The samples (1-5 $\mu$L) were injected at a split ratio of 1:30. Injector, capillary interface and ion source temperatures were 250, 270 and 60°C, respectively. Mass spectra were recorded every 2 sec at an electron energy of 70 eV.

## GC/ES

An Applied Chromatography System (Model MPD 850) was used, equipped with channels for simultaneous detection of carbon, hydrogen, chlorine, sulfur and oxygen. Nitrogen was used as purge gas (30 mL/min) for microwave plasma detection. The instrument was combined with a Hewlett-Packard GC (Model 5880A) and a Hewlett-Packard 3350 online Lab-System with a 21 MX computer. The glass capillary columns (OV-17 and BDS) were connected to the plasma head with a GLT-tubing interface. Helium was used as carrier gas (0.6 mL/min). The GC was operated as described under GC/MS. Two microliters of the sample was injected at a split ratio of 1:20. Injector, interface and plasma head temperatures were 250, 230 and 230°C, respectively. Microwaves were produced with a frequency of 2450 MHz and an output effect of 50 W. The detector was operated at a pressure of 665 Pa (5 torr).

## CONCLUSIONS

C-stage SBL exhibited the highest concentration of chlorinated organics of low relative molecular mass (mol wt $< 1000$). Chlorinated organics of high relative molecular mass (mol wt $> 1000$) are basically found in $E_1$-stage SBL.

Pulp washing, chlorine concentration and pH should be considered if the concentration of chlorinated organics of low relative molecular mass are to be reduced.

Of the TOCl in the fraction of low relative molecular mass of the C-stage SBL, 25% consists of relatively lipophilic chlorinated compounds.

The major extremely volatile chlorinated component is chloroform, and the H-stage SBL exhibits the highest concentration.

## ACKNOWLEDGMENTS

We are indebted to Mr. P. Lindblad at the Swedish Cellulose Co. (SCA) for producing laboratory spent bleach liquors and to Mr. J. Korostensky at GFL, Sweden, for carrying out the GC/ES analysis. Thanks are also due to Dr. K. Kringstad for review of the manuscript. Mrs. L. Johansson is acknowledged for excellent technical assistance.

## REFERENCES


1. Hardell, H. L., and F. de Sousa. *Sv. Papperstidn.* 80:110 (1977).
2. Landner, L., K. Lindström, M. Karlsson, J. Nordin and L. Sörensen. *Bull. Environ. Contam. Toxicol.* 18:663 (1977).
3. Renberg, L., O. Svanberg, B. E. Bengtsson and G. Sundström. *Chemosphere* 9:143 (1978).
4. Sjöström, L., R. Rådeström and K. Lindström (in preparation).
5. Erickson, M., and C. W. Dence. *Sv. Papperstidn.* 79:316 (1976).
6. Lindström, K., and F. Österberg (in preparation).
7. Lindström, K., and J. Nordin. *J. Chromatog.* 128:13 (1976).
8. Ota, M., W. B. Durst and C. W. Dence. *Tappi* 56:139 (1973).
9. Leach, J. M., and A. N. Thakore. *J. Fish Res. Board Can.* 32:1249 (1975).
10. Adler, E. *Svensk Kemisk Tidskr.* 80:279 (1968).
11. Rogers, I. H., and L. H. Keith. "Identification of Two Chlorinated Guaiacols in Kraft Bleaching Wastewaters," in *Identification and Analysis of Organic Pollutants in Water*, L. H. Keith, Ed. (Ann Arbor, MI: Ann Arbor Science Publishers, Inc., 1976), pp. 625-640.
12. Voss, R. H., J. T. Wearing, R. O. Mortimer, T. Kovaes and A. Wong. "Chlorinated Organics in Kraft Bleachery Effluent," paper presented at the Third International Congress on Industrial Waste Water and Wastes, Stockholm, Sweden, February 1980.
13. Voss, R. H., et al. "Effect of Pulp Chlorination Conditions on the Formation of Toxic Chlorinated Compounds," CPAR Project Report No. 828, Environment Canada, Ottawa, Ontario, Canada (1979).
14. Kachi, S., et al. "Low Molecular Weight Compounds Related to Toxicity of Beech Spent Bleaching Liquors," paper presented at the 1979 CPPA/TS Environmental Improvement Conference, Victoria, British Columbia, Canada, October 1979.
15. Lindström, K., and F. Österberg. *Can. J. Chem.* 58:815 (1980).
16. Lindström, K., and J. Nordin. *Sv. Papperstidn.* 81:55 (1978).
17. McKague, A. B., and C. C. Walden. "Identification of the Toxic Constituents in Kraft Mill Bleach Plant Effluents," CPAR Project No. 245, Environment Canada, Ottawa, Ontario, Canada (1979).
18. Ander, P., K. E. Erikson, M. C. Kolar, K. P. Kingstad, V. Rannung and C. Ramel. *Sv. Papperstidn.* 80:454 (1977).
19. Kringstad, K. P., P. O. Ljungquist, F. de Sousa and L. M. Stromberg. *Environ. Sci. Technol.* 15:562 (1981).
20. Bjorseth, A., G. Lunde and N. Gjos. *Acta Chem. Scand.* B31:979 (1977).
21. Eklund, G., B. Josefsson and A. Bjorseth. *J. Chromatog.* 150:161 (1978).
22. Landner, L., and K. Lindström. "Chemical and Biological Characterization of Effluents from a Magnetite Mill," IVL Report 27:3 (1978) (in Swedish).
23. Lindström, K., and J. Nordin (in preparation).

## CHAPTER 53

# A NOVEL GAS CHROMATOGRAPHIC METHOD FOR THE ANALYSIS OF CHLORINATED PHENOLICS IN PULP MILL EFFLUENTS

**R. H. Voss, J. T. Wearing and A. Wong**

Pulp and Paper Research Institute
of Canada
Pointe Claire, Quebec, Canada

In the production of bleached kraft pulp, a multistage (typically five to six stages) bleaching process is used to brighten the pulp by removing the residual lignin which still remains in the pulp after the chemical pulping process and which gives a brownish color to the pulp. Most of the delignification is achieved in the first two bleaching stages. The brown pulp is commonly treated with chlorine in the first (chlorination, C) stage of the bleaching process and subsequently extracted with alkali in the second (extraction, E) stage. Laboratory studies of spent liquors from the chlorination and alkali extraction stages in the prebleaching of softwood kraft pulp indicate (1-4) that about 10% by weight of the active chlorine charge applied to the pulp in the first bleaching stage will appear as organically bound chlorine in the spent liquors. At present, only a small fraction of the chlorinated organic material in kraft bleaching effluents, perhaps 1% of the total organically bound chlorine (TOCl), has been characterized with respect to specific chlorinated organic compounds. On a weight basis, the bulk of the chlorinated organic compounds which have been identified consists of chlorinated phenolics. Although the chlorinated phenolic compounds account for only a small part of the total chlorine-containing organic material in kraft bleaching effluents, their pollution potential is considered to be large in view of their acute toxicity to fish [5,6], bioaccumulation tendencies [7] and resistance to biodegradation [8,9]. Consequently, the availability of analytical methods for the determination of these compounds is essential to understanding and assessing their pollution impact.

This chapter describes a novel method that we have developed for the determination of chlorinated phenolics in kraft pulp mill effluents and natural waters receiving such wastewaters. Gas chromatography (GC) for the determination of low levels of phenolic compounds usually includes a derivitization step to improve the chromatographic response of these relatively polar compounds which would otherwise exhibit excessive peak tailing. Various derivitization reagents have been used for GC analysis of chlorinated phenolic compounds in pulp mill effluents; some of these are diazomethane [5,10,11], diazoethane [11-13], heptafluorobutyric anhydride [14] and silanizing reagents [15]. The analytical method that we have developed for the analysis of chlorinated phenolics of pulp mill origin involves the determination of the chlorinated phenolic compounds as their acetate derivatives by capillary GC with electron capture detection (GC/ECD). Several methods have previously been reported for the determination of low levels of chlorinated phenolics by derivitization with acetic anhydride followed by GC/ECD [16-18]. A feature common to each of these methods, however, was the isolation of the chlorinated phenolics from the aqueous solution by extraction with an organic solvent and then reextraction into an aqueous alkaline solution wherein the phenolics reacted with acetic anhydride. In our procedure, the preisolation step has been eliminated by forming the acetate derivatives in situ after buffering of the water sample with potassium carbonate. This direct-acetylation approach provides a significant reduction of sample manipulation and hence a more rapid method for the routine determination of chlorinated phenolics in pulp mill effluents and receiving waters. A similar approach for the determination of trace amounts of phenols in aqueous samples has recently been reported by Coutts et al. [19].

## EXPERIMENTAL

### Apparatus

#### *Gas Chromatograph*

Hewlett-Packard Model 5830A equipped with a $^{63}$Ni electron capture detector and a Hewlett-Packard Model 18835A split/splitless capillary column inlet system operated in the splitless mode.

*Gas Chromatography/Mass Spectrometry (GC/MS) System*

Hewlett-Packard Model 5985A GC/MS with computerized data processing. For analysis with packed GC columns, the gas chromatograph and quadrupole mass spectrometer were coupled via a glass jet separator.

*Columns*

Three types of columns were used: (1) a 1.8-m x 2-mm i.d. glass column packed with 3% SP2100 on 100/120 mesh Supelcoport (Supelco Inc., Bellefonte, PA); (2) a 15-m x 0.25-mm i.d. glass capillary column, wall-coated with SP2100 (AA grade, Supelco Inc., Bellefonte, PA); and (3) a 15-m x 0.25-mm i.d. fused-silica capillary column, wall-coated with SE-30 (J&W Scientific, Orangeville, CA).

## Instrumental Conditions

*GC Analysis*

For column 1, the column temperature regime was 150°C (hold for 4 min), then programmed at 6°C/min to 230°C; injector, 250°C; detector, 300°C; argon (95%) methane (5%) flowrate = 28 mL/min. For columns 2 and 3 a splitless injection technique was used. Column temperature was 45°C (hold for 1 min), then rapidly programmed at 15°C/min to 120°C followed immediately by 2°C/min temperature program to 170°C; injector, 210°C; detector, 250°C; attenuation, usually 1024-4096; inlet purge activation time, 0.6 min; carrier gas, helium at 9-10 psig; ECD auxiliary gas, argon (95%)/ methane (5%) with flowrate of 63 mL/min.

*GC/MS Analysis*

Column 1 was used. Column temperature was 140°C (3-min hold), then programmed to 190°C at 10°C/min; injector temperature, 230°C; helium carrier gas flowrate, 30 mL/min; jet-separator interface temperature, 250°C; ion source temperature, 200°C; electron energy, 70 eV; mass scan range, 40-340 amu; mass scan rate, 300 amu/sec.

## Reagents

2,6-Dibromophenol, which was used as an internal standard, was obtained from Eastman Kodak (Rochester, NY). Acetic anhydride (99+%, Matheson, Coleman & Bell) was distilled twice before use. A strong buffer solution containing 0.72 g/mL of reagent-grade potassium carbonate (Anachemia, Montreal, Quebec, Canada) was prepared in distilled water. Distilled-in-glass grade solvents (Burdick & Jackson) were used throughout. XAD-2 macroreticular resin (Rohm and Haas) was purified by sequential washing with acetone, methylene chloride and acetone.

## Reference Compounds

Authenic samples of chlorinated phenols (2,4-dichlorophenol, 2,4,6-trichlorophenol and 2,3,4,6-tetrachlorophenol) were obtained from commercial suppliers [Anachemia Chemicals Ltd. and RFR Corp. (tetrachlorophenol)]. Two types of reference compounds were prepared in the laboratory: (1) pure standard chlorinated phenolics and (2) "crude" synthetic products. The first group of reference compounds was used to confirm the identity of several of those compounds found by GC/MS analysis and to determine the ECD response factors for quantitative GC/ECD analysis. In some syntheses, a relatively simple mixture of chlorinated phenolics (differing in position and degree of chlorine substitution) was obtained as a final product. These mixtures proved to be a valuable aid in (1) confirming the presence of compounds identified in bleaching effluents by GC/MS and (2) relating the packed-column GC/MS peaks to their corresponding peaks in the capillary ECD gas chromatograms. Except for two compounds (monochlorosyringealdehyde and 3,4,5-trichlorocatechol) the pure compounds were prepared according to procedures described in the literature. In most cases, the methods used to prepare the crude product mixtures were variations of published procedures for the synthesis of the pure compounds. The reference compounds prepared for this study are listed in Table I along with information about the methods used for their synthesis. Pure standards of the four possible dichlorinated catechols were generously provided by M. Nazar (Department of Chemical Engineering, University of Toronto).

## Procedure

### *Analysis of Pulp Mill Effluents*

A schematic of the analytical procedure used for the determination of chlorinated phenolics in effluent samples based on our in situ acetylation technique is shown in Figure 1.

The volume of effluent sample used for analysis was either 50 mL or an aliquot of the sample made up to 50 mL with distilled water. Typically 10

Table I. Reference Compounds for GC and GC/MS Analysis

| Compound | Preparative Method |
|---|---|
| Pure Standards | |
| 3,5-Dichloro-4-hydroxybenzaldehyde | Chlorination of 4-hydroxybenzaldehyde in chloroform [20] |
| 4,5-Dichloroguaiacol | Reaction of guaiacol with sulfuryl chloride [21] |
| 6-Chlorovanillin | Method of Raiford and Lichty [22] |
| 4,5,6-Trichloroguaiacol | Chlorination of guaiacol in acetic acid [23] |
| 5,6-Dichlorovanillin | Method of Raiford and Lichty [22] |
| Monochlorosyringealdehyde | Chlorination of syringealdehyde in chloroform, m.p. 191–193°C |
| Tetrachloroguaiacol | Chlorination of guaiacol in acetic acid [23] |
| Tetrachlorocatechol | Chlorination of catechol with $HCl_{(aq)}/H_2O_2$ [24] |
| 3,4,5-Trichlorocatechol | Same as for tetrachlorocatechol |
| Mixtures | |
| Chlorinated syringols (contained all five possible chlorinated forms) | Chlorination of syringol with $HCl_{(aq)}/H_2O_2$ |
| Chlorinated guaiacols (contained four dichloroguaiacols, three trichloroguaiacols and tetrachloroguaiacol) | Chlorination of guaiacol with $HCl_{(aq)}/H_2O_2$ |
| Chlorinated catechols (contained all nine of the possible chlorinated forms of catechol) | Chlorination of guaiacol in water |

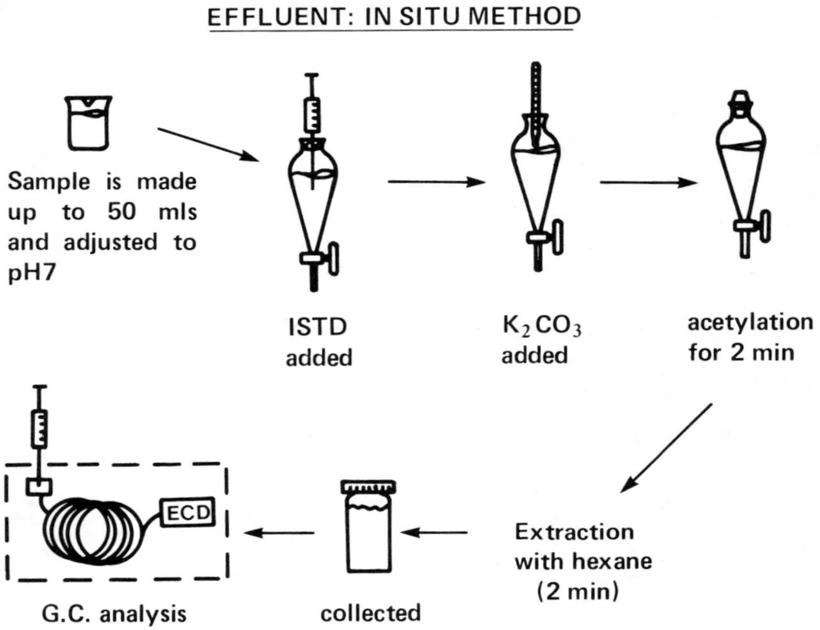

**Figure 1.** Schematic of the analytical procedure used to determine chlorinated phenolics in kraft bleaching effluents by in situ acetylation technique.

and 50 mL, respectivey, of softwood kraft bleaching (chlorination- and extraction-stage) and combined mill effluents (BKME) were used. Larger volumes of sample were used for hardwood bleaching effluents which contain smaller (approximately five times less) amounts of chlorinated phenolics. Chlorination-stage effluents were neutralized prior to analysis. Also, chlorination effluents containing a relatively large chlorine residual require pretreatment by sparging with nitrogen and/or addition of sodium thiosulfate to remove the excess chlorine which will interfere with the subsequent acetylation reaction.

The 50-mL sample was transferred to a 125-mL separatory funnel. The internal standard, 2,6-dibromophenol, was added to the sample with a 100-$\mu$L syringe (typically 70 $\mu$L of a 15.1-$\mu$g/mL solution of 2,6-dibromophenol in methanol or dimethylsulfoxide was used). A 1-mL quantity of strong $K_2CO_3$ buffer solution (0.72 g/mL) was then added such that the final concentration of $K_2CO_3$ in the sample was approx. 0.1 *M*. Once buffered, the sample was acetylated by addition of 1 mL of acetic anhydride. After vigorous shaking with frequent venting of the separatory funnel to permit the release of evolved carbon dioxide, the sample was allowed to stand for about 2 min and then extracted with 5 mL of *n*-hexane. Samples developing difficult-to-break emulsions on extraction with hexane are normally further centrifuged in 15-mL centrifuge tubes. Because the internal standard had been added to the effluent prior to acetylation, only a partial recovery of the hexane was needed. A 2-$\mu$L portion of the hexane extract was analyzed by splitless capillary GC/ECD. Quantitative information for the individual chlorinated phenolics was obtained from the ECD gas chromatograms by the use of the internal standard method.

### *Analysis of Receiving Waters by Preisolation with XAD-2 Resin*

Figure 2 shows a schematic of the analytical procedure used for the determination of chlorinated phenolics in receiving waters. A resin adsorption technique is used for preisolation and concentration of the phenolics. Typically a riverwater sample (1-10 L) was first adjusted to about pH 2 by addition of sulfuric acid. The chlorinated phenolics were isolated from the acidified water samples by adsorption on a porous polymeric resin, XAD-2. The acidified sample was passed by gravity flow at approximately 40 mL/min through a bed (10 cm long x 1.9 cm diameter) of XAD-2 resin (approx. 30 mL) held in a glass column. Following the adsorption step, the resin was washed sequentially with acetone (50 mL) first and then methylene chloride (150 mL). The acetone/methylene chloride eluate was collected in a 250-mL separatory funnel, separated from a small upper layer of water and then evaporated to dryness under reduced pressure with a rotary evaporator. After dissolving the residue in 50 mL of a 0.1 *M* potassium carbonate solution, the analysis proceeded according to the procedure described above for effluent samples.

### *Analysis of Receiving Waters by Direct Acetylation and Extract Concentration*

A schematic summary of the in situ acetylation method used for the analysis of chlorinated phenolics in receiving water samples is given in Figure 3. The method is very similar to that utilized for effluent samples with some modifications to concentrate the hexane extract. Up to the acetylation step, the method was identical to the procedure used for effluents except that the sample volume was larger (100 mL) and, consequently, twice the amount of

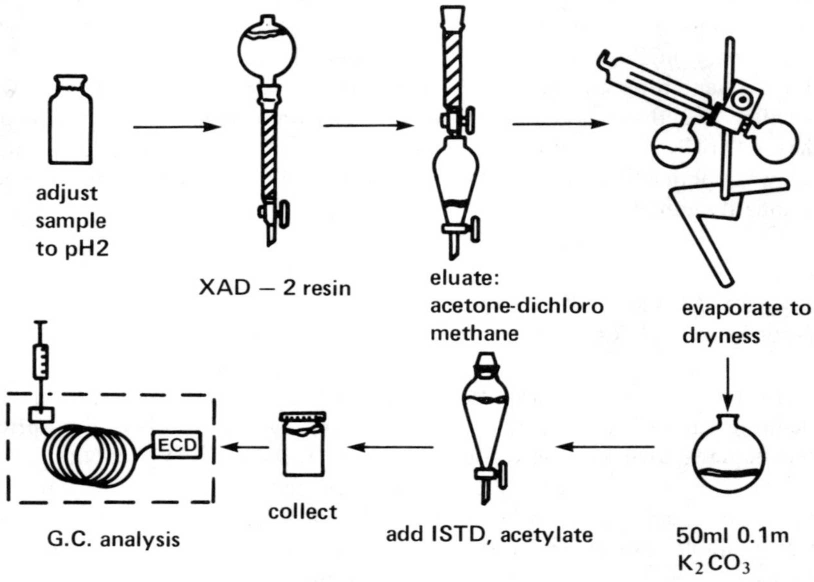


**Figure 2.** Schematic of XAD-2 preisolation/acetylation procedure for the determination of chlorinated phenolics in receiving waters.

strong $K_2CO_3$ buffer solution was added. Also, the amount of internal standard added was reduced according to the extent of effluent dilution in the receiving water. After acetylation, the sample was extracted with two 5-mL volumes of hexane. Unlike the case with effluent samples, special care was taken to recover all the hexane (10 mL) used for extraction. Prior to capillary GC/ECD analysis, the hexane extract was concentrated using a two-step procedure–a vacuum rotary evaporator was first used to reduce the volume to about 0.8 mL and then further concentration was accomplished by evaporation under a gentle stream of nitrogen. The concentration vessel used for the first step was a 50-mL round-bottom flask with a tapered glass finger joined to its bottom [25]; for the second concentration step, a 1-mL reactivial with a conical bottom was used. The degree of concentration of the hexane extract was adjusted according to the effluent dilution in the receiving water. For example, for an effluent concentration of 1% (v/v), the hexane extract would be concentrated to 100 $\mu$L.

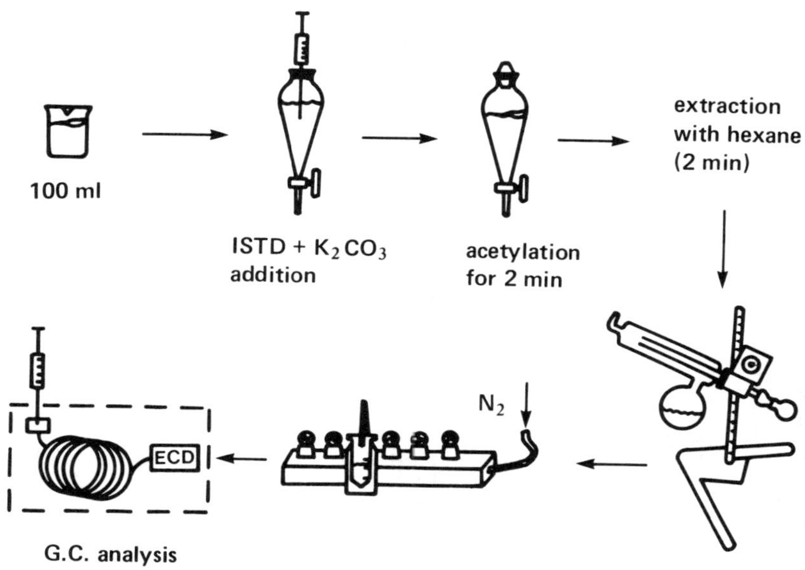


Figure 3. Schematic of in situ acetylation/concentration procedure for the analysis of chlorinated phenolics in receiving waters.

## RESULTS AND DISCUSSION

### Characterization of Chlorinated Phenolics in Kraft Bleachery Effluents by GC/MS

Spent liquors from laboratory bleaching of soft- and hardwood kraft pulps were used for GC/MS investigation. Background information for these effluent samples is given in Table II. The workup procedure involved isolation of the chlorinated phenolics from the bleaching liquors by adsorption on XAD-2 resin. An example of the total ion current profiles obtained by (packed column) GC/MS analysis of the acetylated effluent extracts is illustrated for the softwood bleaching effluent in Figure 4. Also shown in Figure 4 for comparison is the capillary GC/ECD chromatogram which was obtained by analysis of the GC/MS extract after appropriate dilution (1:100 with *n*-hexane containing 2,6-dibromophenyl acetate as a reference compound).

Table II. Bleaching Conditions for Effluent Samples Examined by GC/MS

| Sample No. | Pulp | Effluent Type[a] | Volume Used for GC/MS Work-Up[b] (L) | Chlorination Conditions[c] | | | NaOH on O.D. Pulp[d] (%) |
|---|---|---|---|---|---|---|---|
| | | | | $Cl_2$ Dosage (% of demand) | End pH | Temp (°C) | |
| 1 | Softwood | C-stage | 3.1 | 80 | 1.9 | 60 | 4 |
| 2 | Softwood | E-stage | 1.7 | 80 | 1.9 | 60 | 4 |
| 3 | Hardwood | C-stage | 1.6 | 120 | 1.8 | 25 | 2 |
| 4 | Hardwood | E-stage | 1.5 | 120 | 1.8 | 25 | 2 |

[a]Filtrates from each stage were collected from a pulp at 3.5% consistency.
[b]Acetate derivatives prepared for GC/MS analysis.
[c]Pulp consistency = 3.5% on o.d. pulp; reaction time: 45 min for 25° C and 6 min for 60° C.
[d]Apart from the NaOH applied, the conditions used for caustic extraction of the chlorinated pulp were the same: 60°C and 90 min.

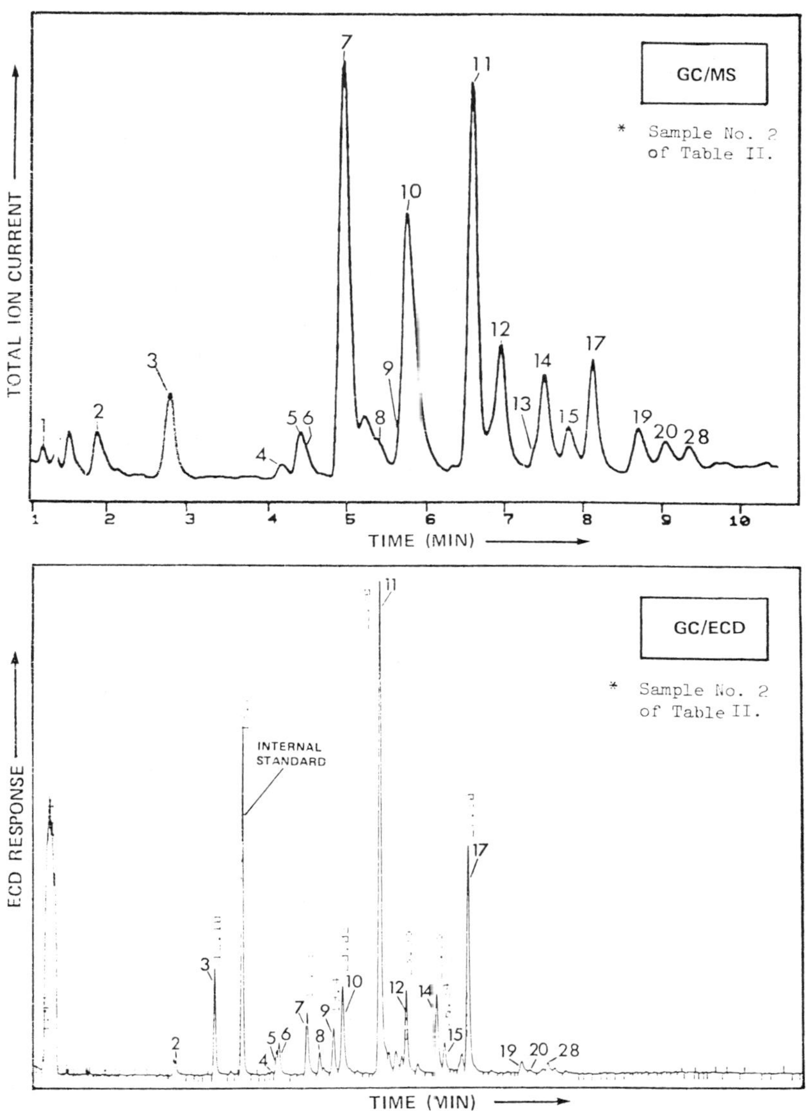


**Figure 4.** Total ion current profile and capillary GC/ECD trace of acetylated chlorinated phenolics isolated from softwood extraction-stage effluent. Numbers identifying peaks are keyed to Table III.

Capillary GC/ECD chromatograms for the GC/MS extracts obtained from hardwood chlorination- and alkali extraction-stage bleaching liquors are shown in Figure 5. Initial identifications of the chlorinated phenolics corresponding to peaks in the total ion current profiles were made primarily by interpreting the fragmentation patterns of the mass spectral data. (Representative mass spectra for the chlorinated phenolic compounds identified in the present study can be found in Voss et al. [26]). For example, the fragmentation behavior for trichloroguaiacol acetate is illustrated in some detail in Figure 6. The presence of three chlorine atoms in this compound can be deduced from the chlorine isotope distribution pattern found in the region of the molecular ion ($M^+ = 268$) and elsewhere in the spectrum. (Four ions, two mass units apart, with an intensity ratio of approximately 3:3:1:0.1, are indicative of the presence of three chlorines). A predominant ion for trichloroguaiacol acetate, and for other chlorinated guaiacols and chlorinated phenols, results in the loss of a ketene group ($CH_2CO$, m/e = 42) from the molecular ion. For chlorinated catechols, the most abundant ions correspond to the loss of two ketene groups from the parent molecular ion, i.e., $(M\text{-}84)^+$. For those chlorinated phenolic acetates which contain an aldehyde group, peaks of almost equal intensity are found at $A^+$ and $(A\text{-}1)^+$ where $A^+$ is the fragment ion formed after the loss of the ketene group(s). The m/e = 43 ion was present in all the mass spectra for the chlorinated phenolic acetates.

Wherever possible, the tentative assignments made from interpreted mass spectra were subsequently confirmed by running pure compounds through the GC/MS system. A packed column was used for GC/MS work, and a capillary column was used for the quantitative analyses. Consequently, the correlation of peaks identified in the total ion current profiles with those peaks in the capillary GC/ECD chromatogram could be tenuous, especially for the minor components for which no pure standards were available. This problem was largely overcome by using crude preparations of chlorinated phenolics (see Table I) for GC/MS analysis. As the crude mixtures were often much simpler to characterize and interpret than the relatively complex effluent samples, it was much easier to relate the peaks identified by packed-column MS to those peaks present in the corresponding capillary GC/ECD trace. This technique of qualitative analysis is illustrated in Figure 7 for the (acetylated) product obtained from the chlorination of guaiacol in aqueous solution. GC/MS of the ($Cl_2$(aq) + guaiacol) reaction product after acetylation showed it to be primarily a mixture of the nine possible chlorinated forms of catechol: monochlorocatechol (two isomers), dichlorocatechol (four isomers), trichlorocatechol (three isomers) and tetrachlorocatechol. Lesser amounts of di-, tri- and tetrachloroguaiacols were also detected. For qualitative analysis, this product mixture represents a considerable saving of time and effort required to synthesize each of the nine possible chlorinated forms of cate-

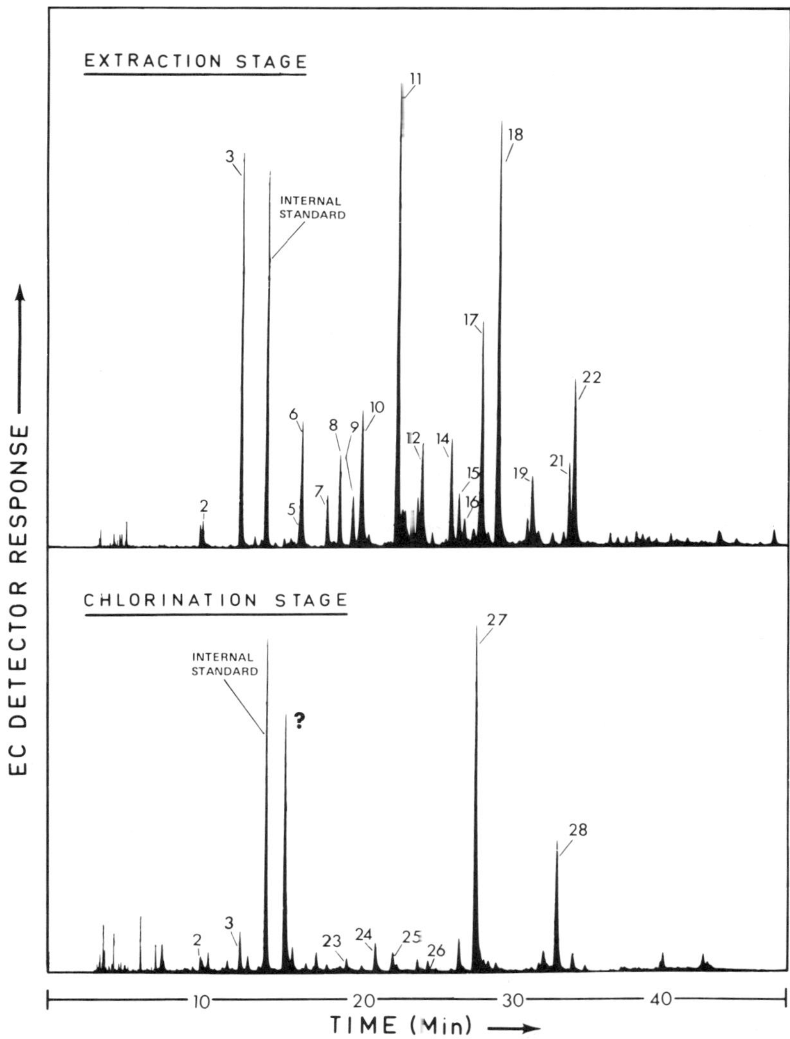


**Figure 5.** Capillary GC/ECD trace of acetate derivatives of chlorinated phenolics isolated from chlorination and caustic extraction stages of bleached hardwood pulp. Numbers identifying peaks are keyed to Table III.

chol. One obvious limitation is that the exact structures of the individual isomers are not defined. Cleavage of methyl aryl ether linkages (demethylation) resulting from reaction wtih chlorine is known to occur readily in aqueous solution [27].

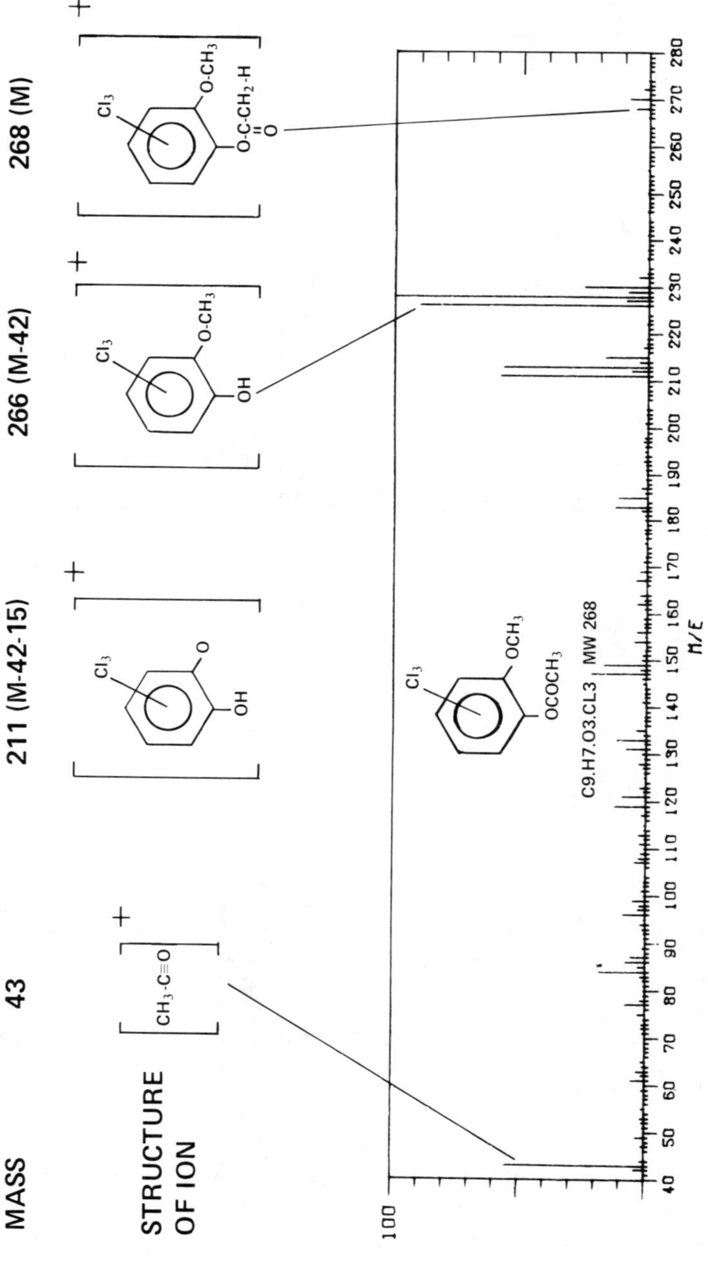


**Figure 6.** Mass spectral fragmentation pattern for 4,5,6-trichloroguaiacol-acetate.

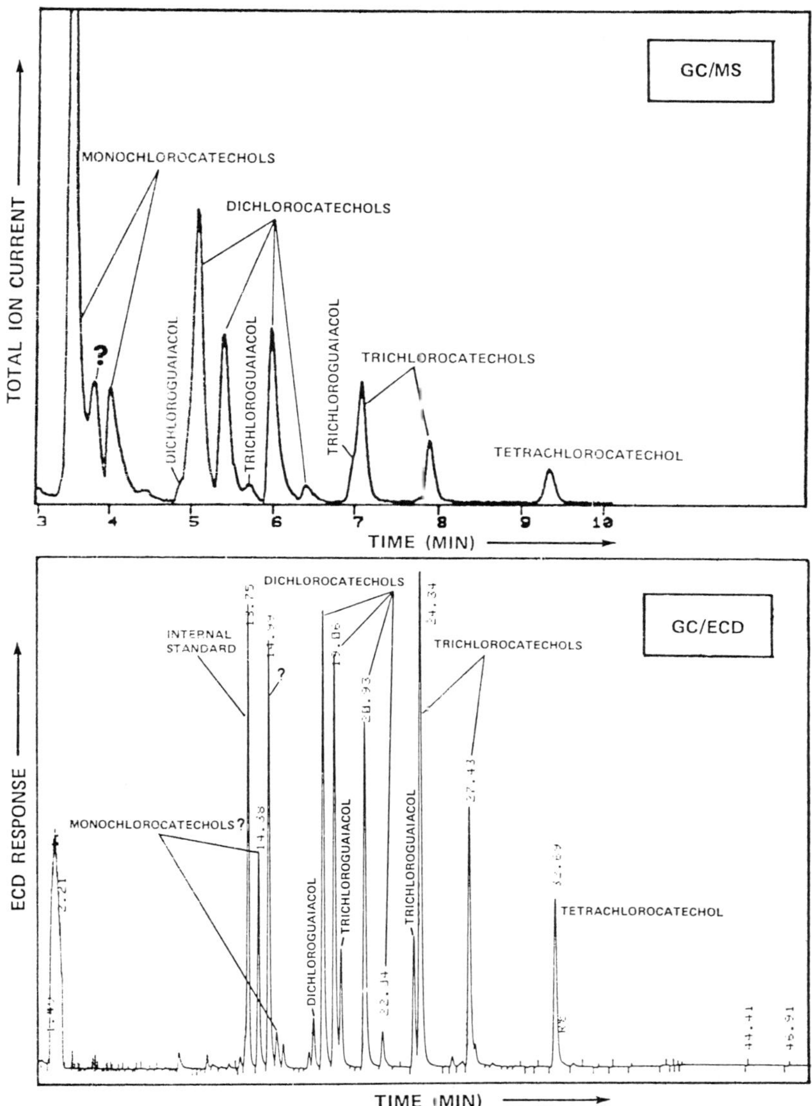


**Figure 7.** Total ion current profile and capillary GC/ECD trace of (acetylated) product from the chlorination of guaiacol in aqueous solution.

Table III gives a summary of the chlorinated phenolics which were identified in the acetylated extracts of laboratory-prepared effluents obtained from the bleaching of soft- and hardwood kraft pulps. The presence of trichloro-

**Table III. Chlorinated Phenolics Identified in Soft- and Hardwood Chlorination and Caustic Extraction Effluents**

| No.[a] | Compound | Softwood | | Hardwood | |
|---|---|---|---|---|---|
| | | C-stage | E-stage | C-stage | E-stage |
| 1 | Monochlorophenol | *[c] | * | * | * |
| 2 | 2,4-Dichlorophenol | ***[d] | *** | *** | *** |
| 3 | 2,4,6-Trichlorophenol | *** | *** | *** | *** |
| 4<br>5 | Dichloroguaiacols | | **[e] | | |
| 6 | 3,5-Dichloro-4-hydroxy-benzaldehyde | | *** | | *** |
| 7 | 4,5-Dichloroguaiacol | | *** | | *** |
| 8 | 2,3,4,6-Tetrachlorophenol | | *** | | *** |
| 9 | 3,4,6-Trichloroguaiacol | | ** | | ** |
| 10 | 6-Chlorovanillin | | *** | | *** |
| 11 | 3,4,5-Trichloroguaiacol | | ** | | ** |
| 12 | 4,5,6-Trichloroguaiacol | | *** | | *** |
| 13 | Monochloroprotocatechu-aldehyde | | * | | |
| 14 | 5,6-Dichlorovanillin | | *** | | *** |
| 15 | Dichlorovanillin | | * | | * |
| 16 | Monochlorosyringealdehyde | | | | *** |
| 17 | 3,4,5,6-Tetrachloroguaiacol | | *** | | *** |
| 18 | 3,4,5-Trichlorosyringol | | | | ** |
| 19<br>20 | Dichloroprotocatechualdehydes | | * | | * |
| 21 | Dichloroacetosyringone | | | | * |
| 22 | Dichlorosyringealdehyde | | | | * |
| 23 | 3,5-Dichlorocatechol | *** | | *** | |
| 24 | 3,4-Dichlorocatechol | *** | | *** | |
| 25 | 4,5-Dichlorocatechol | *** | | *** | |
| 26 | 3,4,6-Trichlorocatechol | | ** | | ** |
| 27 | 3,4,5-Trichlorocatechol | | *** | | *** |
| 28 | Tetrachlorocatechol | | *** | | *** |

[a]From Figures 4 and 5.
[b]Identified as the acetate derivative.
[c]Identification based on interpretation of mass spectrum.
[d]Identification confirmed by comparison with mass spectrum and GC retention time of a pure standard compound.
[e]Identification supported by coincidence of GC retention time with that for a compound identified in a "crude" product mixture.

guaiacols (compound No. 9, 11 and 12, Table III) and tetrachloroguaiacol (compound No. 17) in kraft mill caustic extraction effluents has previously been reported [5,10,11]. The chlorinated guaiacols were determined by GC after conversion to their methyl derivatives (i.e., tri- and tetrachloroveratroles) on methylation with diazomethane. As shown in Figure 8, there are four possible structural isomers of trichloroguaiacol. After reaction with diazomethane, two structures for the resulting trichloroveratrole are possible. To date, only the 3,4,5-trichloroveratrole species has been found by GC analysis [5,28,29]. The precursor of the 3,4,5-trichloroveratrole species has been demonstrated to be a mixture of 3,4,5- and 4,5,6-trichloroguaiacol (and not the corresponding catechol). This conclusion was reached by ethylating rather than methylating a bleach plant effluent extract [10,11] and actually isolating the trichloroguaiacols from the caustic extraction effluent [30,31]. However, the predominant isomer was not identified. Assignment of compound No. 12 (Table III) as the 4,5,6-isomer of trichloroguaiacol was confirmed by comparison with the GC retention time behavior and mass spectral fragmentation pattern of a pure sample of 4,5,6-trichloroguaiacol which was synthesized by the chlorination of guaiacol in acetic acid solution. Given that only 3,4,5-trichloroveratrole has been found [5,28] in the GC analysis of methylated kraft bleach plant extracts, and that one of the two possible precursors is compound No. 12, we were able to conclude that the other isomer, 3,4,5-trichloroguaiacol, also the predominant form of trichloroguaiacol, corresponds to compound No. 11 (Table III). The remaining trichloroguaiacol (compound No. 9) could have been either the 3,5,6- or the 3,4,6-isomer (see Figure 8). Its assignment as the 3,4,6- isomer was based on a

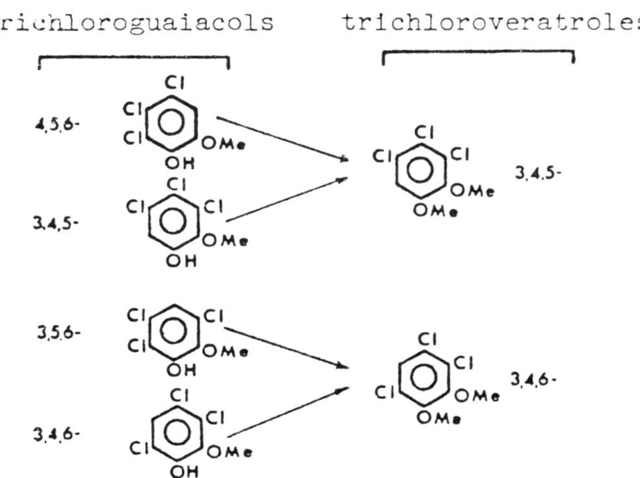


Figure 8. Possible structures of trichloroguaiacols and corresponding trichloroveratroles (adapted from Rogers [28]).

consideration of the possible reactions that may occur between lignin (Figure 9) and chlorine in the first bleaching stage. Clearly, electrophilic aromatic displacement of the side chain by chlorine (i.e., position No. 4) must take place to cleave the monomeric guaiacol from the lignin polymer.

Recently, Lindström and Nordin [12] used GC/MS analytical techniques to provide the first comprehensive characterization of the chlorinated phenolics present in spent bleach liquors, in which 14 different chlorinated phenolics were identified. These belonged to one of three main classes: chlorinated guaiacols, chlorinated catechols and chlorinated phenols. Although Servizi et al. [6] determined in 1968 the toxicity of two chlorinated catechols on the assumption that such compounds were present in kraft pulp bleach wastes, eight years had elapsed before their existence in bleach plant effluents was confirmed by Lindström and Nordin. Except for monochloropropiovanillone, all of the compounds reported by Lindström and Nordin were also detected in our samples. Most of the chlorinated phenolics identified in our study of spent chlorination and alkali extraction liquors from the bleaching of hard- and softwood kraft pulp can be grouped into essentially six major classes as shown in Figure 10. The chlorinated syringyl derivatives found in the hardwood alkaline bleach effluent were to be expected as the hardwood lignin polymer is known to consist of roughly equal proportions of guaiacyl and syringyl structural units. The chlorinated vanillins, syringols and syringealdehydes were newly identified compounds in our study. These same compounds have also been recently reported by Kachi et al [13].

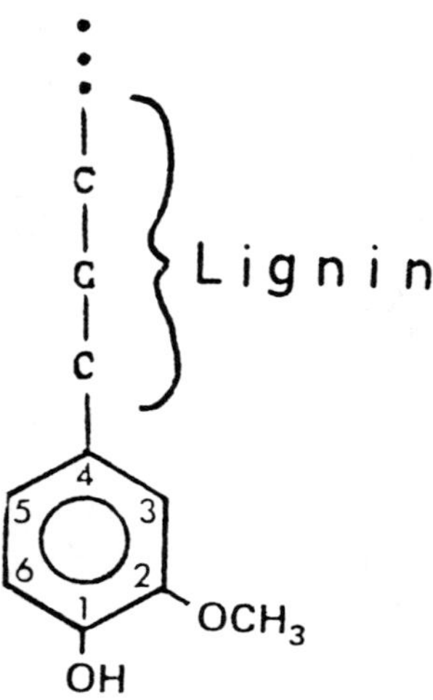


Figure 9. Softwood lignin unit.

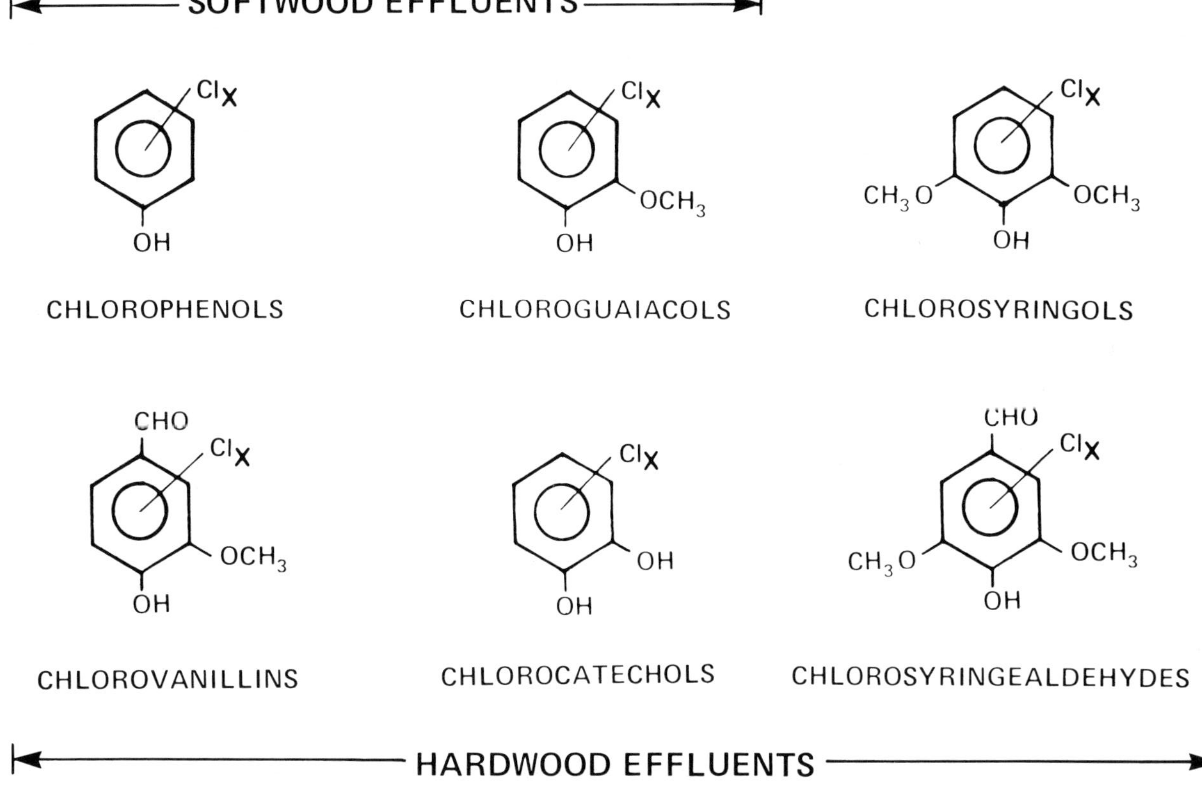


**Figure 10.** Major classes of chlorinated phenolics found in chlorination and caustic extraction effluents.

## GC Calibration

Prior to acetylation of a sample, a known amount of 2,6-dibromophenol was added as an internal standard. For an individual GC peak, its relative retention time and relative area to that of the internal standard were used for the identification and quantification of the peak, respectively. Concentrations of the individual chlorinated phenolics were calculated from the formula:

$$C_a\ (\mu g/L) = RRF_a \times \frac{Aa}{As} \times Ws \quad /V$$

where $C_a$ = concentration ($\mu$g/L) of chlorophenolic compound a in the sample
$RRF_a$ = ECD response factor for compound a relative to the response factor for the internal standard
Aa, As = peak areas for compound a and internal standard, respectively
Ws = weight ($\mu$g) of 2,6-dibromophenol internal standard
V = volume (L) of sample taken for analysis

Response factors and retention times of pure standard chlorinated phenolics relative to the internal standard were determined by acetylating aqueous solutions containing known amounts of each standard in addition to the internal standard and then injecting the hexane extracts of these solutions into the GC. GC/ECD response factors and retention times relative to the internal standard are listed in Table IV for those chlorinated phenolic compounds which were identified in kraft bleachery effluents.

## Method Evaluation

### *GC Column Selection*

Preliminary work on the development of the in situ acetylation procedure was done using a packed column (column 1) for GC analysis. However, a capillary column was soon incorporated in the analytical method because of its superior qualities with respect to both qualitative and quantitative analysis. An example of the superior resolution capabilities of a capillary column is illustrated in Figure 11, where 3,4,5-trichlorocatechol and tetrachloroguaiacol are shown to be baseline resolved on a capillary column (column 2), but badly overlapped on a packed column (column 1).

For the determination of chlorinated catechols as acetates, there is the possibility of irreversible adsorption of these compounds on glass capillary columns (Figure 12). Chlorinated phenyl acetates extracted from the total

Table IV. Relative Retention Times and Relative Response Factors for Chlorinated Phenolic Acetates[a]

| Parent Compound | Relative Retention Time | Relative Response Factor |
|---|---|---|
| 2,6-Dibromophenol[b] | 1.00 | 1.00 |
| 2,4-Dichlorophenol | 0.72 | 15.2 |
| 2,4,6-Trichlorophenol | 0.90 | 3.21 |
| 4,5-Dichloroguaiacol | 1.28 | 19.0 |
| 2,3,4,6-Tetrachlorophenol | 1.38 | 2.42 |
| 6-Chlorovanillin | 1.43 | 10.5 |
| 3,4,6-Trichloroguaiacol | 1.44 | 3.20[c] |
| 3,4-Dichlorocatechol | 1.50 | 10.9 |
| 4,5-Dichlorocatechol | 1.60 | 10.9 |
| 3,4,5-Trichloroguaiacol | 1.64 | 3.20[c] |
| 4,5,6-Trichloroguaiacol | 1.75 | 3.20 |
| 5,6-Dichlorovanillin | 1.87 | 4.02 |
| 3,4,5-Trichlorocatechol | 2.00 | 2.09 |
| 3,4,5,6-Tetrachloroguaiacol | 2.07 | 1.27 |
| 3,4,5,6-Tetrachlorocatechol | 2.42 | 1.89 |

[a]Fused-silica capillary column (column 3); GC conditions as given in experimental section.
[b]Internal standard.
[c]Response factor assumed to be the same as that for 4,5,6-trichloroguaiacol.

mill effluent of a bleached kraft pulp mill were injected onto an unused glass capillary column (column 2) and then the same extract was injected onto the same column after several days of use. From our experience, once the active sites responsible for the adsorption of the catechol acetates become saturated, the glass capillary column becomes usable for the determination of chlorinated catechols. However, more frequent calibration may be required for these columns. The lower surface activity reported for fused-silica capillary columns was certainly evident for the acetate derivatives of the chlorinated catechols. Large GC peaks were observed for the chlorinated catechols with the first injection onto a new fused-silica capillary column (column 3). Because of the advantages of the fused-silica capillary column with respect to analysis for chlorinated catechols, this type of column is favored over a glass capillary column.

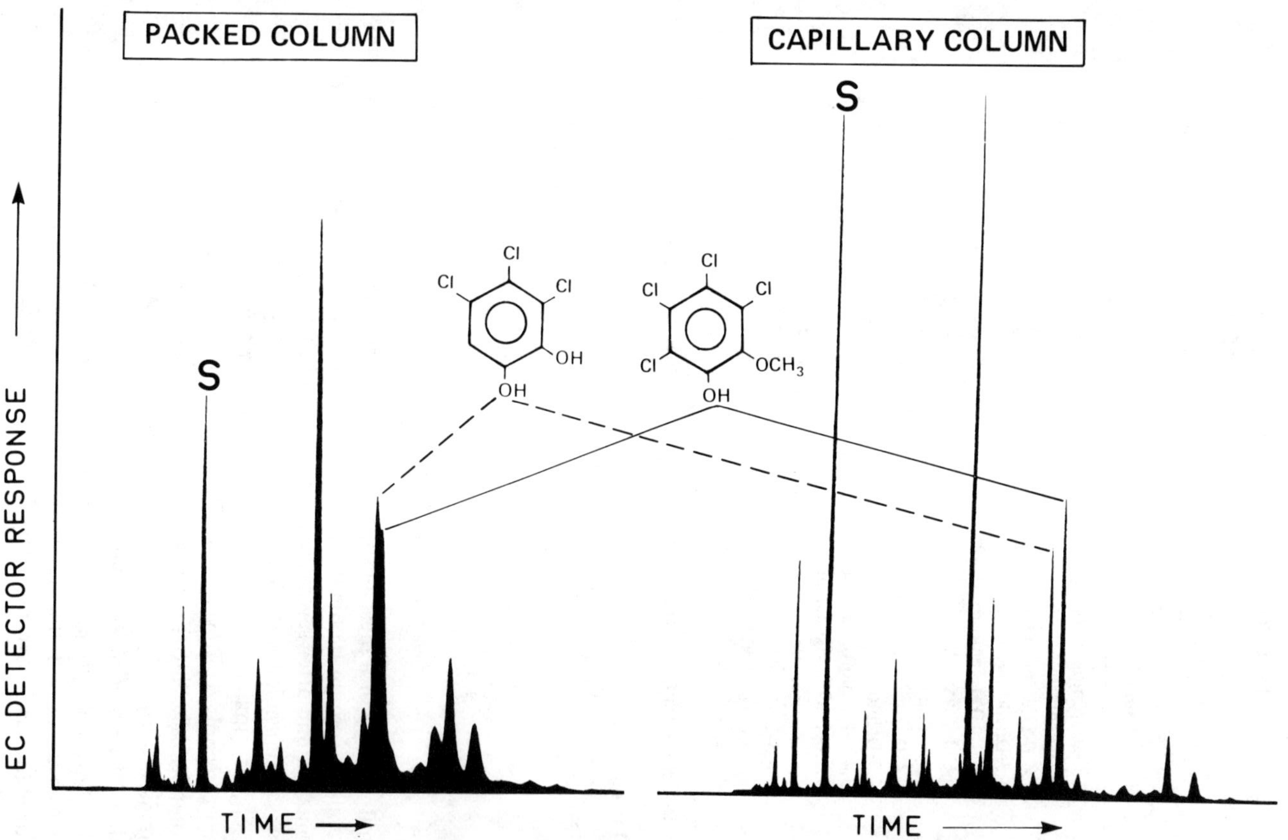


**Figure 11.** Comparative resolving powers of a capillary column (column 2) and a packed column (column 1) for chlorinated phenolics isolated from a sample of total mill effluent of a bleached kraft pulp mill.

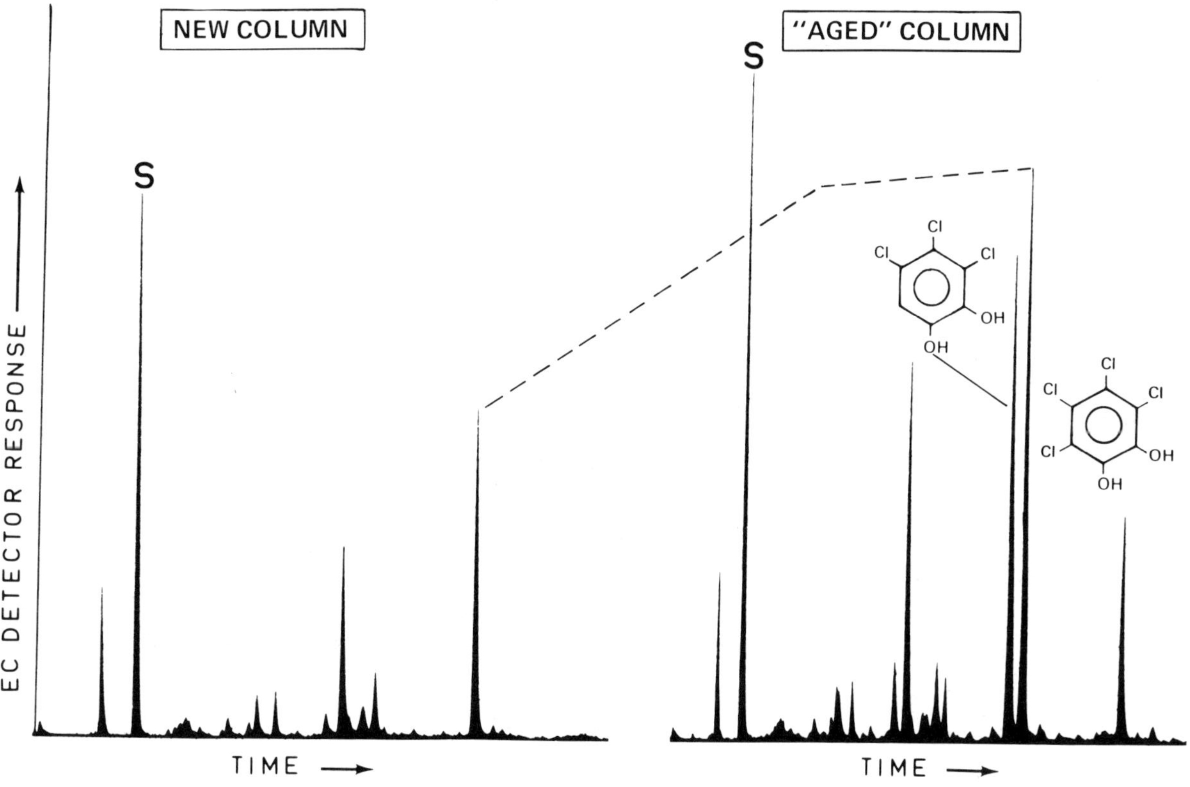


**Figure 12.** Irreversible adsorption effects for chlorinated catechols (as acetates on a glass capillary column (column 2).

*Precision of Analytical Method for Effluent Analyses*

The precision of the in situ acetylation method for chlorinated phenolics analysis in pulp mill effluents was determined by performing replicate analyses on the three types of effluents to which the method has been applied (C-stage effluent, E-stage effluent and BKME). The results in Table V indicate some loss of precision, particularly for chlorinated catechols, when BKME rather than individual effluents from the bleach plant (i.e., C and E) are examined. Also given in Table V is the precision found for the in situ analysis of a synthetic "downstream" riverwater sample (BKME 0.4% (v/v) in riverwater collected upstream from a bleached kraft pulp mill). For the determination of chlorinated phenolics at lower concentrations as in the riverwater sample, some decrease of precision is expected.

*Recovery Studies*

An indication of the recovery of chlorinated phenolics in pulp mill effluents by the direct acetylation method was determined by spiking various samples of extraction-stage effluent and combined mill effluent with known amounts of chlorinated phenolics. As shown for E-stage effluent in Table VI, a recovery of 80% or better was achieved for those compounds typically found in this type of effluent. Except for tetrachlorocatechol, a recovery of 68% and better was found for chlorinated phenolics added to biologically treated BKME (Table VII). We have no explanation for the much lower recovery of tetrachlorocatechol. However, from our experience the analytical results for the chlorinated catechols have tended to be more erratic than those for the other classes of chlorinated phenolics.

**Table V. Precision of in Situ Acetylation Method for Pulp Mill Effluent Analysis**

| | Relative Standard Deviation[a] (%) | | | | |
|---|---|---|---|---|---|
| Chlorinated Phenolic | C | E | BKME | 0.4% BKME | Mean Value |
| 2,4,6-Trichlorophenol | | 16 | 20 | 13 | 16 |
| 2,3,4,6-Tetrachlorophenol | | 7 | 5 | 15 | 9 |
| 3,4,5-Trichloroguaiacol | | 4 | 6 | 12 | 7 |
| 3,4,5,6-Tetrachloroguaiacol | | 7 | 5 | 17 | 10 |
| 3,4,5-Trichlorocatechol | 5 | | 13 | 25 | 14 |
| 3,4,5,6-Tetrachlorocatechol | 7 | | 26 | 45 | 26 |

[a]Six determinations for each effluent; all effluents from bleaching of softwood pulps.

Table VI. Recovery of Chlorinated Phenolics from E-Stage Effluent by the in Situ Method

| Chlorinated Phenolic | Added[a] (μg) | Found (μg) | Recovery[b] (%) |
|---|---|---|---|
| 2,4-Dichlorophenol | 4.8 | 4.5 | 93 |
| 2,4,6-Trichlorophenol | 0.67 | 0.58 | 86 |
| 4,5,6-Trichloroguaiacol | 0.98 | 0.94 | 96 |
| 3,4,5,6-Tetrachloroguaiacol | 0.61 | 0.49 | 80 |

[a]Amount added to 10 mL of effluent.
[b]Four determinations; softwood bleaching effluent.

Table VII. Recovery of Chlorinated Phenolics from Biotreated BKME by the in Situ Method

| Chlorinated Phenolic | Added[a] (μg) | Found (μg) | Recovery[b] (%) |
|---|---|---|---|
| 2,4-Dichlorophenol | 8.7 | 9.8 | 113 |
| 2,4,6-Trichlorophenol | 1.7 | 1.3 | 76 |
| 2,3,4,6-Tetrachlorophenol | 1.1 | 0.82 | 75 |
| 4,5-Dichloroguaiacol | 16 | 15 | 94 |
| 4,5,6-Trichloroguaiacol | 3.2 | 2.7 | 84 |
| 3,4,5,6-Tetrachloroguaiacol | 1.1 | 0.75 | 68 |
| 3,4,5-Trichlorocatechol | 2.3 | 1.7 | 74 |
| 3,4,5,6-Tetrachlorocatechol | 2.9 | 1.2 | 41 |

[a]Amount added to 50 mL of effluent.
[b]Three determinations; effluent from softwood pulp.

The recovery efficiency of the resin adsorption method for low levels of chlorinated phenolics was evaluated by spiking a relatively large (10-L) volume of water with a sample of combined mill effluent from a bleached kraft pulp mill. Excellent recovery is indicated when comparing the amounts of three major chlorinated phenolic components found in the diluted and undiluted combined mill effluent samples (Table VIII). Even better recovery was obtained for those samples adjusted to pH 2 prior to passage through the resin.

**Table VIII. Recovery of Chlorinated Phenolics from "Riverwater" Samples by XAD-2 Resin**

| Chlorinated Phenolic | Amount Found ($\mu$g) | | |
|---|---|---|---|
| | BKME[a] (50 mL) | 0.5% BKME[b] (10 L) | |
| | | pH 2[c] | pH 7[c] |
| 2,4,6-Trichlorophenol | 0.28-0.32 | 0.28-0.36 (100%)[d] | 0.16-0.22 (56%) |
| Trichloroguaiacols | 2.8-3.6 | 3.1-3.4 (100%) | 2.5-3.0 (84%) |
| Tetrachloroguaiacol | 1.5-1.9 | 1.7-1.8 (106%) | 1.2-1.4 (81%) |

[a]Five determinations.
[b]Three determinations.
[c]pH of water samples prior to resin adsorption.
[d]Average % of chlorinated phenolics found as compared to those present in whole BKME.

*Concentration of Hexane Extract*

The application of the in situ acetylation technique for analyses of riverwater samples requires a concentration of the hexane extract to very low volumes, as low as 10 $\mu$L. For best results, severe losses of the chlorinated phenylacetates through evaporation must be avoided. The effect of evaporation losses on the quantitative results was assessed by performing a 100-fold dilution on a standard solution of chlorinated phenolic acetates (50 $\mu$L added to 50 mL of hexane) and then using the two-step concentration procedure (i.e., rotary evaporator and evaporation under nitrogen) to reduce the volume of the hexane extract to approximately 12 $\mu$L. It was evident from (1) visual inspection of the chromatograms for concentrated and unconcentrated hexane extracts and (2) a comparison of the quantitative results (Figure 13), that the concentration procedure would not give rise to serious losses of the acetate derivatives. Even if there were losses, there were no significant variations of the various chlorinated phenolics relative to the internal standard.

## APPLICATIONS

### Bleach Plant Effluents

Examples of the types of GC profiles obtained on analysis of spent softwood kraft bleaching liquors by the direct acetylation method are given in

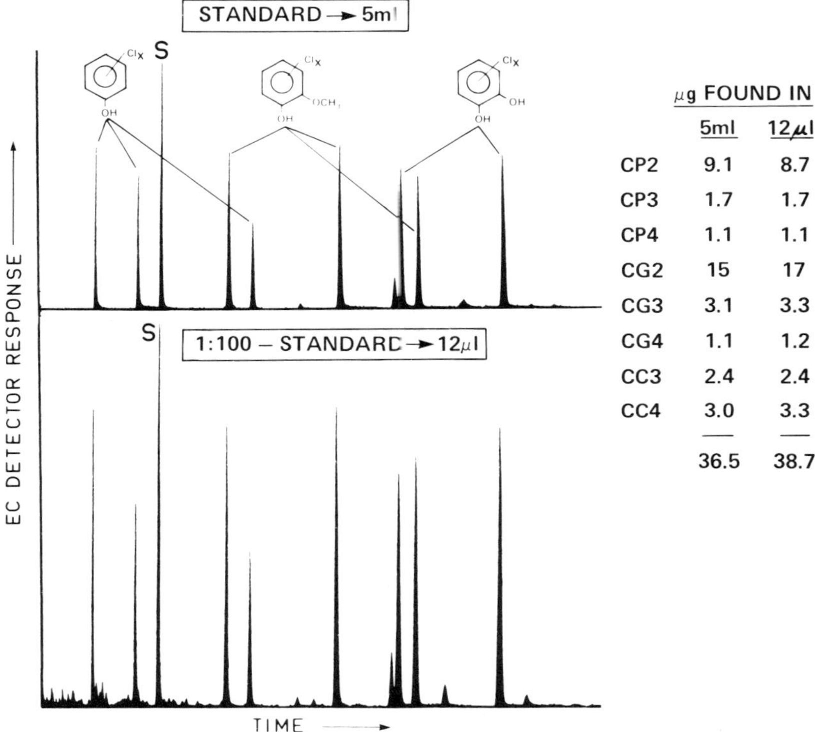


**Figure 13.** Effect of concentration of hexane extract to low volumes on the quantitative determination of chlorinated phenolic compounds as their acetate derivatives. CP2, CP3, CP4 = 2,4,-di-, 2,4,6-tri- and 2,3,4,6-tetrachlorophenol, resp.; CG2, CG3, CG4 = 4,5-di-, 4,5,6-tri- and 3,4,5,6-tetrachloroguaiacol, resp.; CC3 and CC4 = 3,4,5-tri- and 3,4,5,6-tetrachlorocatechol, resp. Column 2; same attenuation used for each chromatogram.

Figure 14. A total concentration of 3–4 mg chlorinated phenolics per liter means that the concentrations of the individual chlorinated phenolics are some hundreds of µg/L. The unusual distribution of the various classes of chlorinated phenolics in the two bleaching (C and E) liquors, i.e., predominnatly chlorinated catechols in the chlorination effluent, and mainly chlorinated guaiacols, phenols and vanillins in the caustic extraction effluent, has been similarly observed by Lindstrom and Nordin [12]. Pentachloroacetone, as indicated in Figure 14, is commonly observed in the analysis of C-stage effluents. This chlorinated neutral compound coextracts with the acetate derivatives. Mortimer and Wong [15] have identified chlorinated acetones in spent chlorination liquor as TMS derivatives of the enol forms. Expressing

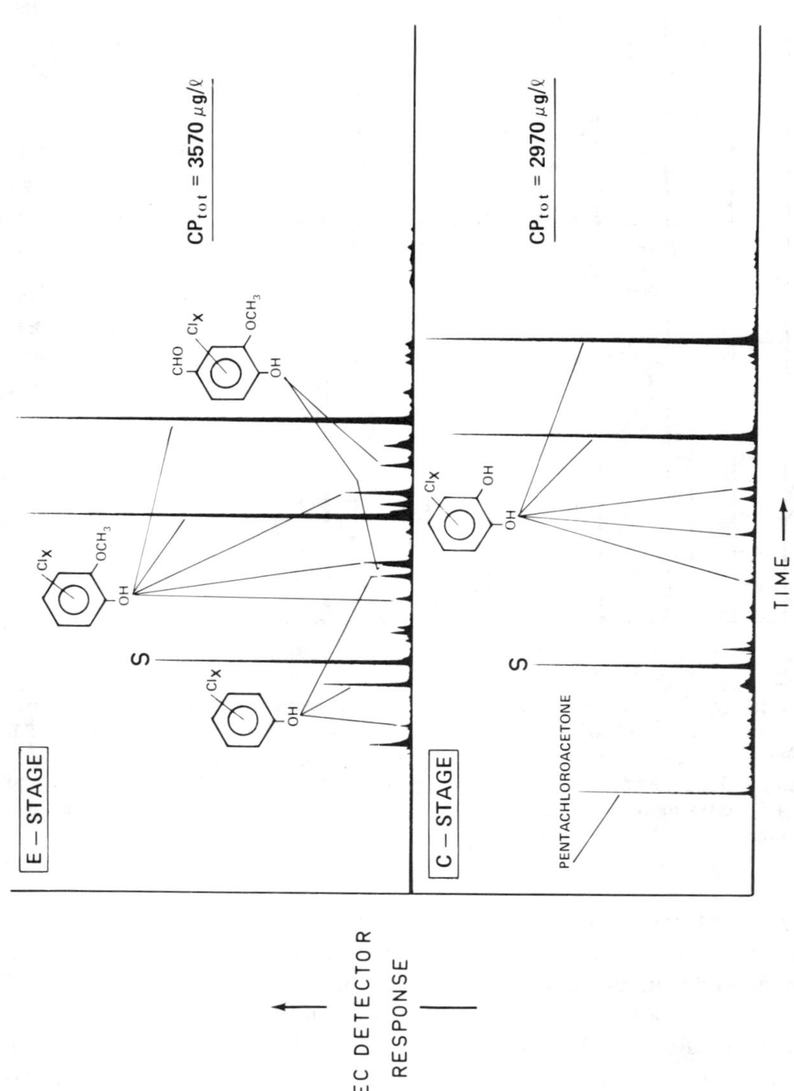


**Figure 14**. GC profiles for acetate derivatives of chlorinated phenolics isolated from (softwood) chlorination- and alkali-extraction-stage bleaching liquors. Column 2 used for GC analysis.

the chlorinated phenolics loadings according to a pulp mill's production of bleached kraft pulp is more appropriate for intermill comparisons. The results given in Table IX for three softwood bleached kraft pulp mills located in western Canada indicate loadings of 26–42 g of chlorinated phenolics per metric ton of pulp for the C-stage effluents and between 39 and 50 g/metric ton for the E-stage effluents.

## Bleaching Process Study

In a recent study supported through the CPAR program of Environment Canada we investigated the effects of pulp chlorination conditions on the formation of chlorinated phenolics in both chlorination and extraction-stage spent bleaching liquors [26]. The major bleaching variables investigated were $Cl_2$ dosage, $ClO_2$ substitution, chlorination temperature and chlorination final pH. Although the results of our study indicated that changes in all four bleaching variables could measurably affect the types and quantities of chlorinated phenolics present in the bleaching effluents, only substitution of chlorine dioxide for chlorine showed the most promise as an industrial measure for reducing the production of chlorinated phenolics. As shown in Figure 15 for extraction-stage effluent, a decrease in the quantities of chlorinated phenolics formed was obtained with increasing amounts of $Cl_2$ being replaced by $ClO_2$. (The chlorine demand referred to in Figure 15 may be regarded as the optimum dosage of $Cl_2$ to be applied as based on the lignin content of the unbleached pulp.) At 100% $ClO_2$ substitution, there were virtually no chlorinated phenolics present. The small quantity of 6-chlorovanillin re-

**Table IX. Typical Loadings of Chlorinated Phenolics in Kraft Bleaching Effluents**

| Mill[a] | Bleaching Sequence[b] | C-Stage (g/o.d. metric ton pulp) | $E_1$-Stage (g/o.d. metric ton pulp) |
|---|---|---|---|
| A | CEHDED | 42 | 50 |
| B | CDEHDED | 32 | 39 |
| C | CDEDED | 26 | 43 |

[a]Softwood kraft in all cases.
[b]Letters indicate the chemicals applied at each bleaching stage: C = chlorine, E = alkali extraction, H = hypochlorite, D = chlorine dioxide and $C_D$ = chlorine + chlorine dioxide mixture (typically 5–10% of total chlorine charge normally applied is added as chlorine dioxide to protect the viscosity of the pulp).

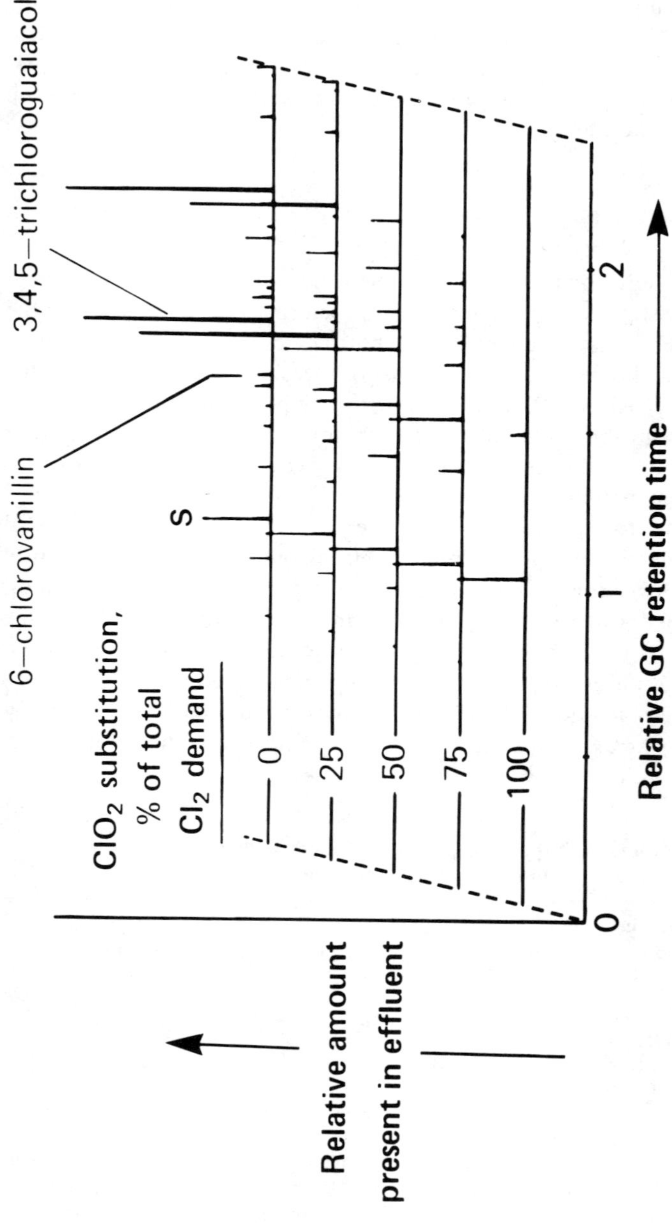


**Figure 15.** Computer-reconstructed GC traces illustrating the effect of $ClO_2$ substitution on the formation of chlorinated phenolics in extraction-stage effluent during the bleaching of softwood kraft pulp ($Cl_2$ applied = 99.6 kg/o.d. metric ton pulp).

maining may have arisen from the small amount of chlorine remaining in the $ClO_2$ during its preparation. A consideration of the formation of trichloroguaiacols in extraction-stage effluent as a function of $Cl_2$ dosage (with and without $ClO_2$ present) provides information about the mode of action of the $ClO_2$. The results shown in Figure 16 reveal that the $ClO_2$ plays a relatively passive role in the formation of chlorinated phenolics. The absolute amount of remaining $Cl_2$ used is the important controlling factor.

### Combined Mill Effluents

Biological treatment in an aerated (typically five-day) lagoon system is an external effluent treatment measure commonly used by pulp mills in the United States and Canada. By monitoring the chlorinated phenolics through the treatment system, we can identify which compounds are being effectively removed and which are resistant to biological treatment and thus likely to persist in the environment. The results of such a study in a softwood bleached kraft pulp mill located in western Canada are presented in Figure 17. The overall removal efficiency for chlorinated phenolics after 5-day aerated lagoon treatment was found to be about 30%. Several of the compounds, particularly the tri- and tetrachlorophenols and the tri- and tetrachloroguaiacols, appear to be unaffected by the biological treatment.

### Receiving Waters

Examples of chromatograms obtained for riverwater samples collected downstream of a softwood bleached kraft pulp mill and analyzed by the XAD-2 (preisolation)/acetylation procedure are shown in Figure 18. The riverwater volumes used for sample workup were 250 mL and 1 L, respectively, for the sampling stations located 30 and 100 km downstream from the pulp mill. A larger sample volume was taken at the 100-km station because by this point the first river had joined a second larger river which caused a further dilution of the effluent by about one-third. The GC profiles seem to indicate some reduction of the trichloroguaiacols during the transport of the effluent between the two river water stations. Although the data for Figure 18 were obtained using the XAD-2/acetylation procedure, the results in Figure 19 illustrate that the in situ acetylation procedure works equally well for the analysis of riverwater samples.

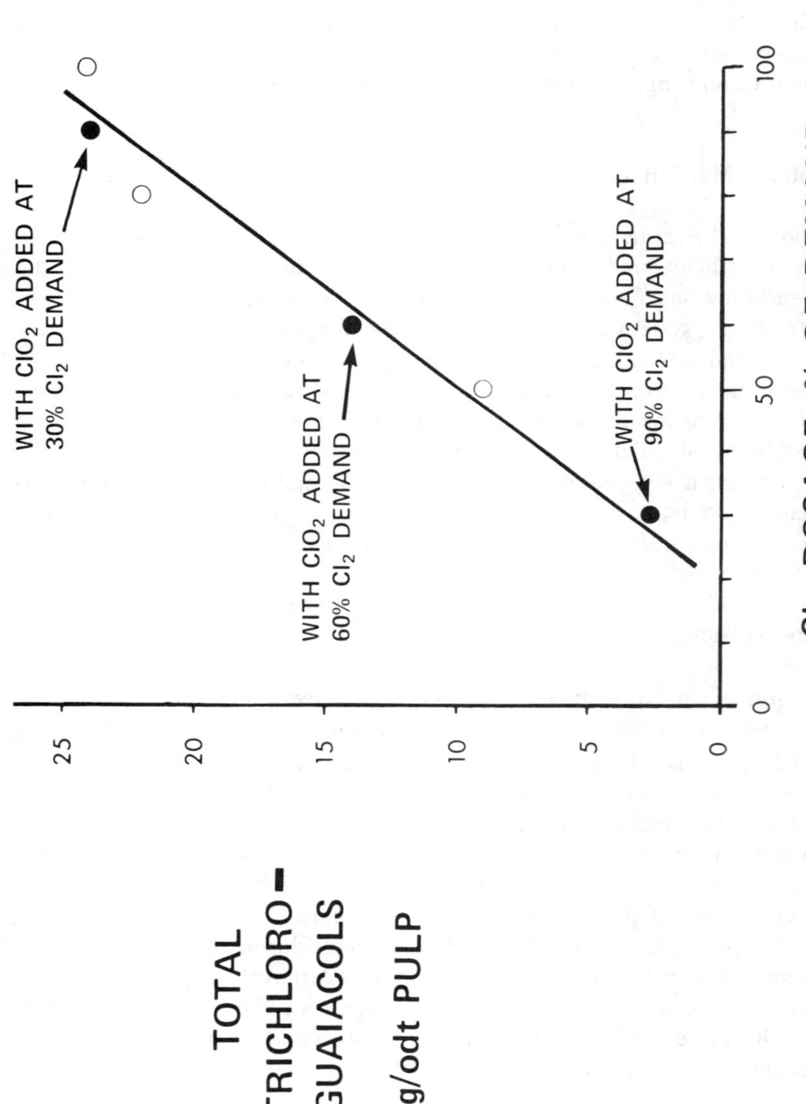


**Figure 16.** Effect of $Cl_2$ dosage, with and without $ClO_2$ substitution, on the formation of trichloroguaiacols in extraction-stage effluent during the bleaching of softwood pulp. (Total amount of bleaching agent applied = 120% of $Cl_2$ demand.)

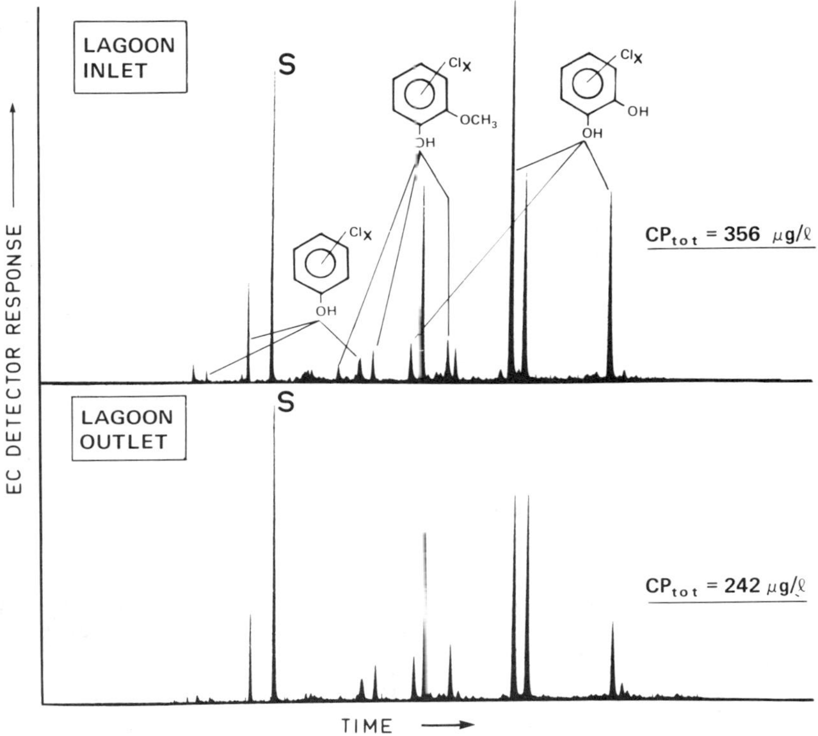


**Figure 17.** GC profiles of chlorinated phenolics (as acetate derivatives) through the biological treatment system of a bleached kraft pulp mill (column 3 used for GC analysis; S = internal standard, 2,6-dibromophenol).

## CONCLUSION

The in situ acetylation technique which we have developed provides a fast, sensitive and selective method for the determination of chlorinated phenolics in pulp mill effluents and in surface waters receiving such effluents. Although the method was developed specifically for pulp mill effluents, it probably can be used for the determination of chlorinated phenolics in other types of wastewaters and for the monitoring of riverwaters for chlorinated phenolics not necessarily of pulp mill origin.

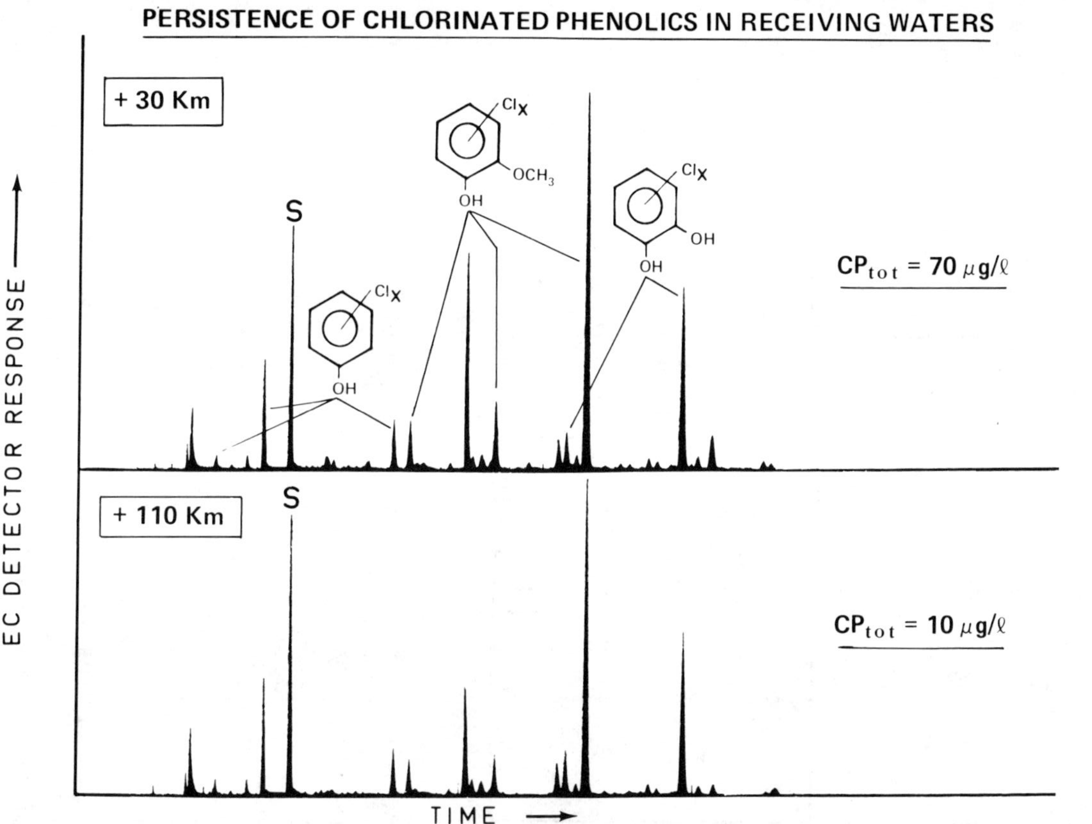


**Figure 18.** Capillary ECD gas chromatograms for riverwater samples analyzed by XAD-2 isolation/acetylation procedure (column 3 used for GC analysis; distances shown are distances along the river downstream from a bleached kraft pulp mill; S = internal standard).

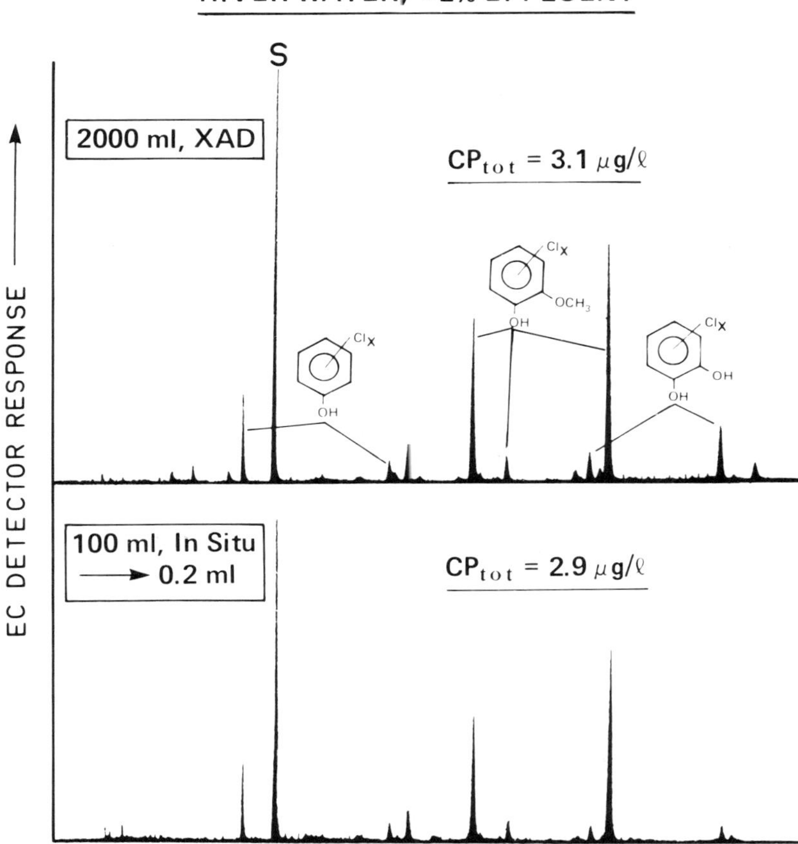


Figure 19. Comparison of capillary ECD gas chromatograms for a riverwater sample analyzed by XAD-2 isolation/acetylation procedure and in situ acetylation procedure (column 3 used for GC analysis; S = internal standard).

## ACKNOWLEDGMENT

This work was supported in part by Environment Canada through its Cooperative Pollution Abatement Research program. The authors are grateful to K. Parsons for her skillful laboratory assistance.

## REFERENCES

1. Rapson, W. H., and C. B. Anderson. "Mixtures of Chlorine Dioxide and Chlorine in the Chlorination Stage of Pulp Bleaching," *Pulp Paper Mag.* (Canada) (January 1966), p. T-47.
2. Kempf, A. W., and C. W. Dence. "Structure and Reactivity of Chlorolignin. II. Alkaline Hydrolysis of Chlorinated Kraft Pulp," *Tappi* 53(5):864 (1970).
3. Hardell, E.-L., and F. de Sousa. "Characterization of Spent Bleaching Liquors. Part 1. Spent Liquors from the Chlorine and Alkali Extraction Stages in the Prebleaching of Pine Kraft Pulp," *Sv. Papperstidn.* 80:110 (1977).
4. Pfister, K., and E. Sjöström. "Characterization of Spent Bleaching Liquors. Part 2. Composition of Material Dissolved During Chlorination (CEH Sequence)," *Paperi Puu* (1979), p. 220.
5. Leach, J. M., and A. N. Thakore. "Isolation and Identification of Constituents Toxic to Juvenile Rainbow Trout (*Salmo gairdneri*) in Caustic Extraction Effluents from Kraft Pulp Mill Bleach Plants," *J. Fish. Res. Board Can.* 32:1249 (1975).
6. Servizi, J. A. et al. "Toxicity of Two Chlorinated Catechols, Possible Components of Kraft Pulp Mill Bleach Waste," Int. Pacific Salmon Fish. Comm. Prog. Rep. No. 17 (1968).
7. Landner, L. et al. "Bioaccumulation in Fish of Chlorinated Phenols from Kraft Pulp Mill Bleachery Effluents," *Bull. Environ. Contam. Toxicol.* 18:663 (1977).
8. Leach, J. M. et al. "Removal of Toxic Constituents in Aerated Lagoons at B. C. Kraft Mills," paper presented at the 1978 Spring Conference, Technical Section, Pacific Coast and Western Branches, CPPA, Jasper, Alberta, Canada, May 17-21, 1978.
9. Leach, J. M. "Loadings and Effects of Chlorinated Organics from Bleached Pulp Mills," paper presented at the Third Conference on Water Chlorination, Colorado Springs, CO, October 1979.
10. Rogers, I. H., and L. H. Keith. "Organic Chlorine Compounds in Kraft Bleaching Wastes–Identification of Two Chlorinated Guaiacols," Fish. Mar. Serv. Res. Devel. Tech. Rep. No. 465, Environment Canada, Vancouver, British Columbia (1974).
11. Rogers, I. H., and L. H. Keith. "Identification of Two Chlorinated Guaiacols in Kraft Bleaching Wastewaters," in *Identification and Analysis of Organic Pollutants in Water*, L. H. Keith, Ed. (Ann Arbor, MI: Ann Arbor Science Publishers, Inc., 1976), pp. 625-640.
12. Lindström, K., and J. Nordin. "Gas Chromatography-Mass Spectrometry of Chlorophenols in Spent Bleach Liquors," *J. Chromatog.* 128:13 (1976).
13. Kachi, S. et al. "Low Molecular Weight Compounds Related to the Toxicity of Spent Bleaching Liquors," paper presented at the 1979 CPPA/TS Environmental Improvement Conference, Victoria, British Columbia, Canada, October 1979.

14. McKague, B., and C. C. Walden. "The Significance of Phenols in Pulp Mill Effluents," CPAR Project Report 881-1, Environment Canada (1979).
15. Mortimer, R. D., and A. Wong. "Isolation and Identification of Toxicants Present in Effluents Derived from the Pulping and Bleaching of Western Red Cedar," CPAR Project Reports No. 711-1 and 711-2, Environment Canada (1978 and 1979).
16. Rudling, L. "Determination of Pentachlorophenol in Organic Tissues and Water," *Water Res.* 4:533 (1970).
17. Krijgsman, W., and C. G. Van de Kamp. "Determination of Chlorophenols by Capillary Gas Chromatography," *J. Chromatog.* 131:412 (1977).
18. Chau, A. S., and J. A. Coburn. "Determination of Pentachlorophenol in Natural and Wastewater," *J. Assoc. Off. Anal. Chem.* 57:389 (1974).
19. Coutts, R. T. et al. "Gas Chromatographic Analysis of Trace Phenols by Direct Acetylation in Aqueous Solution," *J. Chromatog.* 179:291 (1979).
20. Fisher, A. et al. "Ionic Dissociation of 4-Substituted, 2,6-Dichlorophenols," *J. Chem. Soc.* (B):686 (1976).
21. Matell, M. "Halogenated Guaiacoxylalkylcarboxylic Acids of Plant Physiological Interest," *Acta Chem. Scand.* 9: 1007 (1955).
22. Raiford, L. C., and J. G. Lichty. "The Chlorine Derivatives of Vanillin and Some of Their Reactions," *J. Am. Chem. Soc.* 52:4576 (1930).
23. Fort, R. et al. "Sur la Chlorination de Gaiacol: Les Tri- et Tétrachlorogaiacols," *Bull. Soc. Chim. France* (1955), p. 810.
24. Lubbecke, H., and P. Boldt. "Halogenation and Oxidative Halogenation of Phenols with Hydrogen Halides/Hydrogen Peroxide. Synthesis of p-Chloranil and p-Bromanil," *Tetrahedron Lett.* 34:1577 (1978).
25. Junk, G. A., J. J. Richard, M. D. Grieser, P. Witiak, J. Witiak, M. D. Arguello, R. Vick, H. J. Svec, J. S. Fritz and G. V. Calder. "Use of Macroreticular Resins in the Analysis of Water for Trace Organic Contaminants," *J. Chromatog.* 99:745 (1974).
26. Voss, R. H. et al. "Effect of Pulp Chlorination Conditions on the Formation of Toxic Chlorinated Compounds," CPAR Project Report No. 828, Environment Canada (1979).
27. Dence, C., and K. Sarkanen. "A Proposed Mechanism for the Acidic Chlorination of Softwood Lignin," *Tappi* 43(1):87 (1960).
28. Rogers, I. H. "Chlorinated Guaiacols in Bleaching Wastes: Comments on a Paper by Leach and Thakore," *J. Fish Res. Board Can.* 33:1858 (1976).
29. Leach, J. M., and A. N. Thakore. "Response to the Comments of I. H. Rogers," *J. Fish. Res. Board Can.* 33:1860 (1976).
30. Leach, J. M., and A. N. Thakore. "Identification of Toxic Constituents in Kraft Mill Bleach Plant Effluents," CPAR Project Report No. 245-3 (1976).
31. Thakore, A. N., and A. C. Oehschlager. "Structure of Toxic Constituents in Kraft Mill Caustic Extraction Effluents from $^{13}C$ and $^{1}H$ Nuclear Magnetic Resonance," *Can. J. Chem.* 55:3298 (1977).

# CHAPTER 54

# OXIDIZED RESIN ACIDS IN DOUGLAS FIR WOOD EXTRACTIVES

**Ian H. Rogers and Harold W. Mahood**

Department of Fisheries and Oceans
West Vancouver Laboratory
West Vancouver, British Columbia
Canada

The toxicity of resin acids to fish and other aquatic organisms has been known for many years [1]. Studies conducted in several countries over the past decade have shown that resin acids contribute a major portion of the acute toxicity of wastewaters associated with the pulping of coniferous wood species. Fortunately these toxic compounds are biodegradable and are removed, usually by treatment of wastewater in five-day aerated stabilization ponds, before discharge to rivers which may contain valuable fish stocks such as Pacific salmon or trout.

Fox [2] reported on the composition and dispersal of a kraft pulping wastewater into Nipigon Bay in Lake Superior. As part of that study, Brownlee and Strachan [3] concluded that dehydroabietic acid (DHA) was moderately persistent in the water column and could be detected in water and sediment samples collected several kilometers away from the discharge point. Moreover, an oxidized derivative was found in the lakewater and was tentatively identified as 7-keto-DHA. This was subsequently confirmed by synthesis of the compound [4] by permanganate oxidation of DHA according to the earlier work of Pratt [5]. 7-Keto-DHA has also been identified [6] in the waters of Lower Fox River, which flows into Lake Michigan at Green Bay, WI. This river collects the wastewaters of 15 pulp and/or paper mills besides 11 municipal sewage treatment plants in a length of 64 km.

Oxidized resin acids have occasionally been identified as natural products. Norin and Winell [7] identified 15-hydroxy-DHA in the extractives from the bark of common spruce *Picea abies*. Carman and Marty [8] reported its co-occurrence with 15-hydroxyabietic acid in *Agathis* species in Australia. Ekman and Sjoholm [9] identified 1$\beta$,15-dihydroxy-DHA and 1$\beta$,16-dihydroxy-DHA formed by the action of the fungus *Fomes annosus* on DHA. Ekman has reported the mass spectra of a number of modified resin acids in the reaction zone of *Fomes annosus*-infected sapwood of Norway spruce [10]. All compounds identified in the latter report are derivatives of DHA.

Rogers et al. [11] examined the chemical composition of a hardboard plant wastewater that discharges indirectly to the Fraser River at New Westminster, British Columbia (BC). This effluent is acutely toxic to sockeye salmon (*Oncorhynchus nerka*), and much of the toxicity could be attributed to the resin acids present. However, a complex mixture of oxidized resin acids was noted in toxic resin acid fractions, and no information was available in the literature as to the toxicity of such compounds to aquatic life. The wood supply at the hardboard plant is about 90% Douglas fir (*Pseudotsuga menziesii*) and it was therefore not surprising to note the presence in Douglas fir oleoresin of oxidized resin acids with mass spectra similar to those observed in the extracts of hardboard plant wastewater.

The present study was conducted in an effort to separate and identify as many as possible of the oxidized resin acids in Douglas fir oleoresin and to measure their toxicity to aquatic life.

## EXTRACTION OF DOUGLAS FIR AND RECOVERY OF RESIN ACID FRACTION

A sample of Douglas fir woodchips, prepared from three trees grown near Crofton, BC, was obtained from Dr. K. Hunt of the Resource Evaluation Section of Pulp and Paper Research Institute of Canada. The air-dried chips were ground to a coarse wood-meal in a Wiley mill and extracted with methanol for 24 hours in an all-glass sohxlet apparatus. The extract was methylated with diazomethane and analyzed by gas chromatography/mass spectrometry (GC/MS) for the presence of oxidized resin acid methyl esters. The reconstructed gas chromatogram (RGC) is shown in Figure 1.

A large sample of Douglas fir resin acids was available in our laboratory and had been extracted in 1971 from a bark beetle-infested tree grown at

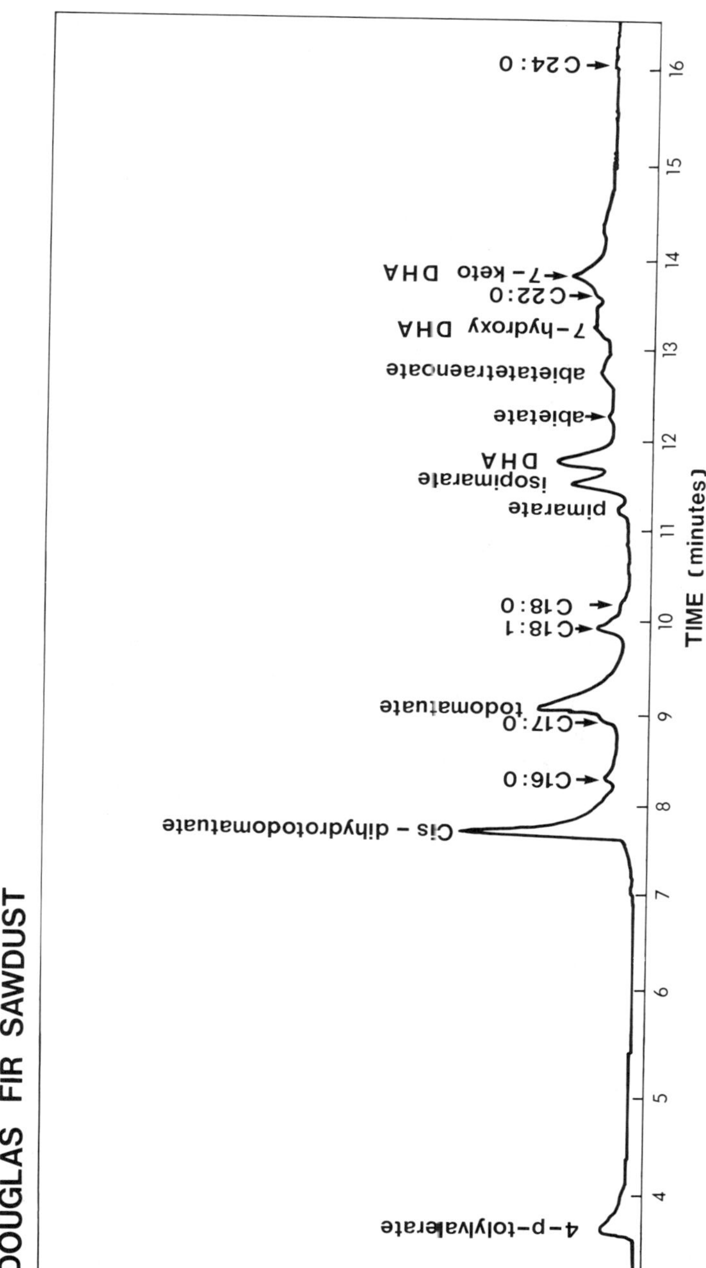


**Figure 1.** Reconstructed gas chromatogram of methylated wood extract from Douglas fir woodchips.

Savonna, BC. A total of 14,000 g of air-dried wood had been extracted in batches with petroleum ether (bp 65–110°C) in an all-glass sohxlet apparatus for 24 hr. A total of 1373 g of soluble resin (yield 9.8%) was recovered. This was dissolved in acetone, and cyclohexylamine was added slowly until no further precipitate was formed. The resin acid salts were removed by centrifugation, and the petroleum ether solution was checked for completeness of precipitation by further addition of cyclohexylamine. The resin acid salts were then suspended in petroleum ether and acidified with excess glacial acetic acid. After standing overnight, the petroleum ether phase was washed with water, dried and evaporated to yield the resin acid fraction (867 g, yield 6.2% on air-dried wood).

A sample of this resin acid fraction was methylated with diazomethane and analyzed by GC/MS to detect the presence of oxidized resin acid esters. Further work was concentrated on this extract because of the presence of the compounds of interest.

## PRELIMINARY SEPARATIONS OF THE RESIN ACID FRACTION

A 13-g sample of mixed resin acids and oxidized resin acids was methylated with diazomethane and applied as a solution in petroleum ether (bp 65–110°C) to a column of activity III neutral alumina (ICN Incorp., 152 g, bed depth = 30.5 cm). The column was eluted with 500-mL volumes of the following solvents: petroleum ether, petroleum ether/benzene (4:1), petroleum ether/benzene (1:1), benzene, benzene/diethyl ether (19:1), benzene/ether (4:1), benzene/ether (1:1), ether, methanol. The recovered subfractions were examined by thin-layer (TLC), gas (GC) and liquid (LC) chromatography. Various subfractions were bulked with the following results:

- Fraction A (4.388 g): 100% petroleum ether
- Fraction B (0.151 g): petroleum ether/benzene (4:1) and (1:1)
- Fraction C (0.176 g): petroleum ether/benzene (1:1)
- Fraction D (0.337 g): benzene
- Fraction E (0.214 g): benzene/ether (19:1)
- Fraction F (0.147 g): benzene/ether (4:1)
- Fraction G (0.287 g): benzene/ether (4:1)

Fraction A was a colorless syrup comprising the resin acid methyl esters. The other fractions contained mixtures of oxidized resin acid methyl esters of varying degrees of complexity, some of which did not allow the possibility of further resolution.

In another experiment a 50-g portion of Douglas fir resin acid fraction was methylated and treated with Girard T reagent in the usual manner. The nonketone fraction was treated a second time with the reagent and the ketone extracts bulked (yield 2.3 g).

## ISOLATION OF COMPONENTS BY PREPARATIVE LIQUID CHROMATOGRAPHY

A Hewlett-Packard Model 1084A liquid chromatograph equipped with automatic sample injector and ultraviolet (UV)-visible variable wavelength detector was fitted with a 25-cm ODS-2 Partisil Magnum 9 column purchased from Whatman, Inc. The detector was set at 220 $\mu$ and a flowrate of 8 mL/min established using gradient elution conditions with the solvent system $CH_3CN$/water, commencing with 10% $CH_3CN$ and increasing to 90% $CH_3CN$ over a 30-min run. Fraction G (0.287 g) was dissolved in 2-mL methanol and injected in 200-$\mu$L aliquots. A fraction was collected (compound 1, 55 mg), which showed only one component (GC and analytical LC).

Fraction B (0.151 g) was separated similarly into three subfractions, each of which contained one component. These were as follows: compound 2 (45 mg), compound 1 (25 mg) and methyl dehydroabietate (M+ 314, base peak 239, discarded).

Preliminary LC separation of the Girard T extract of the Douglas fir resin acid concentrate (isocratic conditions 90% $CH_3CN$, 10% $H_2O$) yielded a crude fraction of compound 1 (115 mg) and a pure sample of compound 2 (35 mg). The former was further purified using gradient elution conditions and yielded in addition to compound 1 an impure fraction of compound 3 (10 mg).

Preparative LC separation of the nonketone portion of the Douglas fir resin acids yielded a pure fraction of compound 4 (40 mg).

## IDENTIFICATION OF ISOLATED COMPONENTS

GC separations were performed on a Hewlett-Packard Model 5700 A gas chromatograph equipped with a flame ionization detector (FID) and temperature programmer. The 6-ft x 0.25-in. o.d. glass column was packed with 3% OV-101 on Chromosorb W (HP, 80–100 mesh). The temperature was programmed from 150 to 250°C at 8°C/min, and $N_2$ or He carrier gas flowrate was 40 mL/min.

GC/MS data were collected using a Hewlett-Packard Model 5992 A desk top unit with a membrane separator. Operating conditions were regulated via the "Autotune" program. Carrier gas was He at a flowrate of 30 mL/min. The glass-walled packed column was identical with that used in the GC separations and temperature conditions were reproduced as closely as possible. Infrared (IR) spectra were recorded with a Perkin Elmer Model 467 spectrophotometer. Samples were prepared as smears on KBr discs and absorptions were quoted in wave numbers ($cm^{-1}$). Proton magnetic resonance (PMR) spectra were recorded at 80 MHz on a Bruker spectrometer in the Chemistry Department, University of British Columbia, and signals are quoted in $\delta$ relative to tetramethylsilane as internal standard. High-resolution mass spectra were recorded in the Chemistry Department, University of British Columbia using a Kratos-A.E.I. MS 50 instrument.

## 7-Keto-DHA Methyl Ester (Compound 1)

Parent ion 328·2035 (60.9%) corresponds to $C_{21}H_{28}O_3$; satellite ions at:

1. 313·1798 (loss of $CH_3$, 6.7%);
2. 296·1781 (loss of $CH_3OH$, 15.1%);
3. 269·1898 (loss of $-COOCH_3$, 17.7%);
4. 268:1830 (loss of $HCOOCH_3$, 14.0%); and
5. 253·1569 (loss of $C_3H_7O_2$, 100%).

Low-resolution MS identical with published MS of 7-keto-DHA methyl ester [4,6,10]. IR:

1. 1725 (ester);
2. 1680 ($\alpha$,$\beta$ unsaturated carbonyl);
3. 1605, 1490, 1460 (aromatic);
4. 1250 (ester); and
5. 832 (aromatic).

PMR: 1.17, 1.22, 1.27, 1.32 (four methyl groups), 2.2-3.1 (complex multiplet), 3.62 (singlet, methyl ester), 7.25-8.0 (complex, 3H, aromatic).

## 7-Keto-15-hydroxy-DHA Methyl Ester (Compound 2)

Parent ion 344·1986 (5.2%) corresponds to $C_{21}H_{28}O_4$. Satellite ions at:

1. 329·1756 (loss of $CH_3$, 87.5%);
2. 326·1879 (loss of $H_2O$, 34.1%);

3. 294·1610 (loss of $H_2O + CH_3OH$, 4.34%);
4. 285·1861 (loss of $COOCH_3$, 1.7%);
5. 269·1552 (loss of $C_3H_7O_2$, 10.6%);
6. 267·1738 (loss of $H_2O + COOCH_3$, 6.0%); and
7. 266·1677 (loss of $H_2O + HCOOCH_3$, 4.7%).

Low-resolution MS (Figure 2) showed additional major fragments at m/e 203 (11.0%), 197 (10.5%), 187 (17.7%), 145 (10.3%) and 43 (25.5%). IR:

1. 3450 (–OH);
2. 1725 (saturated ester);
3. 1680 (aryl carbonyl);
4. 1605, 1490, 1460 (aromatic);
5. 1120 (doublet, tertiary OH); and
6. 830 (aromatic).

PMR (Figure 3):

1. 1.27, 1.35, 1.60 (4 $\underline{C}H_3$);
2. 2.2–2.9 (complex, multiplets);
3. 3.67 (singlet, $COO\underline{CH}_3$);
4. 7.37 (doublet, J 10 cps, $C_{11}$–H);
5. 7.77 (quartet, $J_1$ 10 cps, $J_2$ 2.5 cps, $C_{10}$–H); and
6. 8.08 (doublet, J 2.5 cps, $C_8$–H).

A sample of the ester (50 mg) was dissolved in 20 mL benzene, and to it was added 30 mg *p*-toluenesulfonic acid and 60 mg calcium chloride. The mixture was refluxed for 1 hr to give 40 mg dehydration product [12], which was purified by TLC. Low-resolution MS showed parent ion at m/e 326 (75%), with base peak at m/e 251 (loss of $C_3H_7O_2$ from parent ion). IR ($CHCl_3$ solution):

1. no hydroxyl absorption band;
2. 1725 (ester);
3. 1680 (aryl ketone); and
4. 1622 (double bond conjugated with aromatic ring).

PMR (Figure 4):

1. 1.27 and 1.35 singlets ($\underline{C}H_3$);
2. 1.77 ($\underline{CH_2}$);
3. 2.15 (see structure **I**, below);
4. 2.2–2.85 complex;
5. 3.63 singlet (–$COO\underline{CH}_3$);
6. 5.21 doublet (see structure **II**, below);
7. 7.23 doublet, J 9 cps ($C_{11}$–H);
8. 7.57 quartet, $J_1$ 9 cps, $J_2$ 2 cps ($C_{12}$–H); and
9. 7.98 doublet, J 2 cps ($C_{14}$–H).

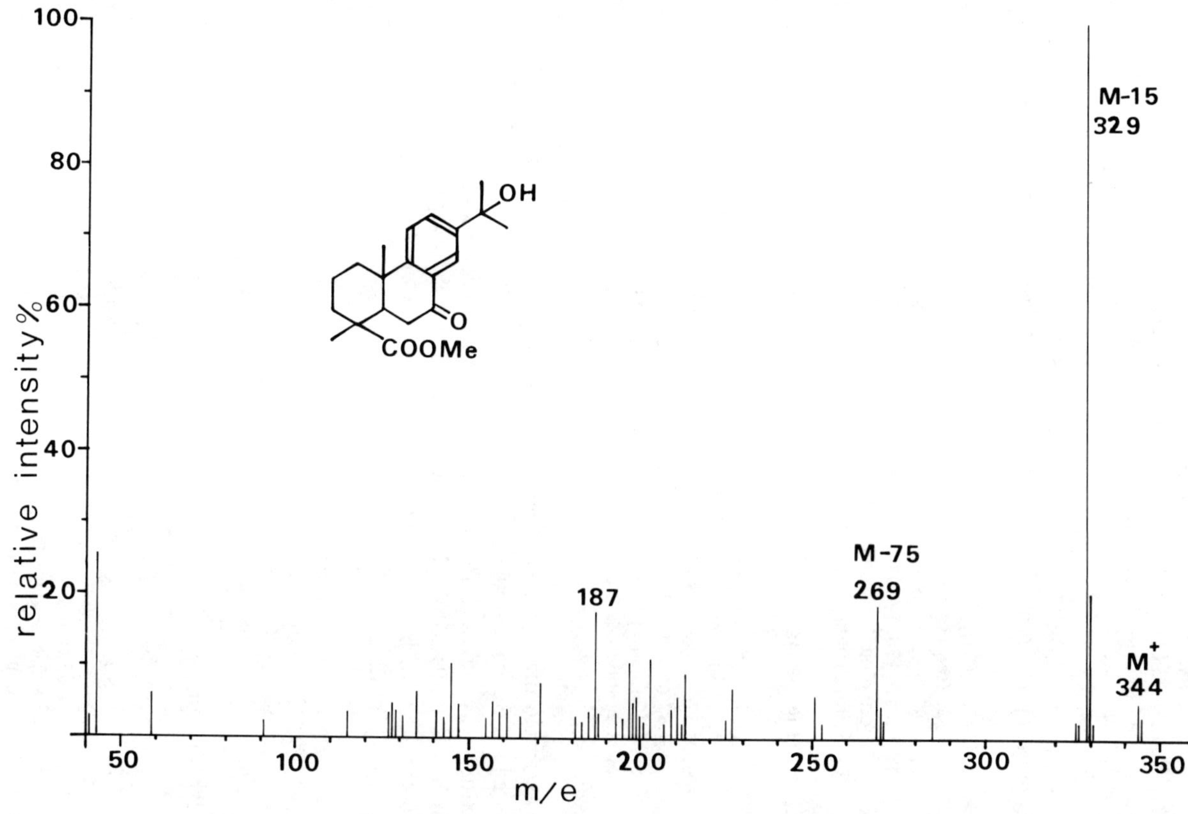


**Figure 2.** Low-resolution mass spectrum of methyl 7-keto-15-hydroxydehydroabietate.

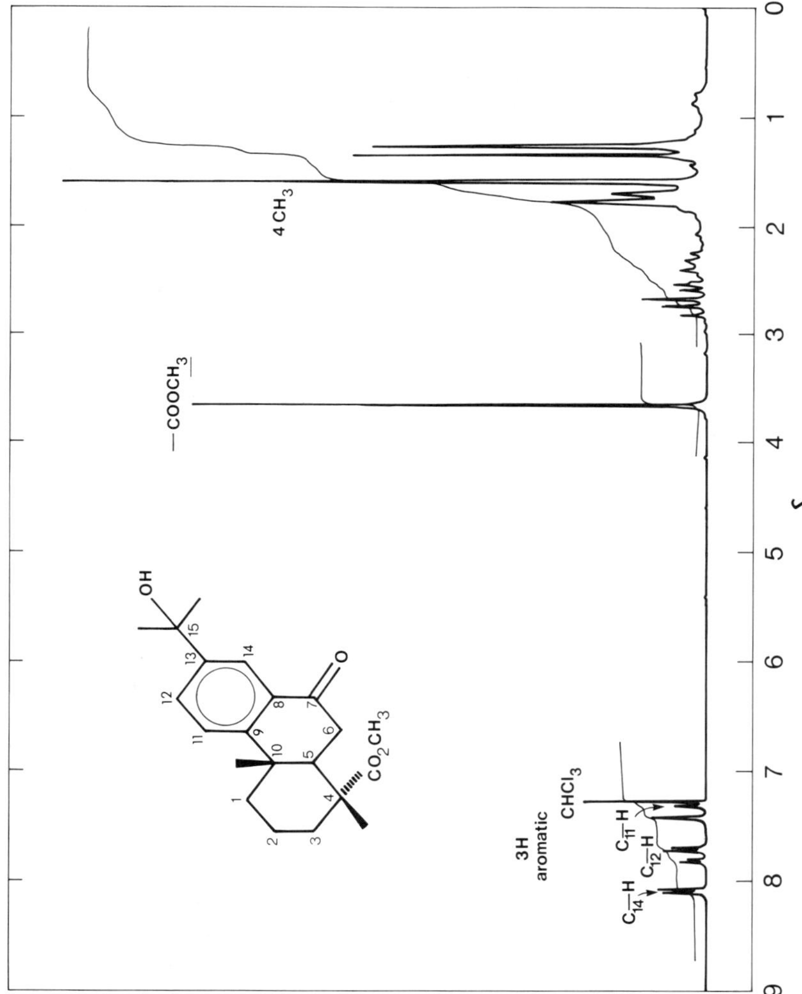


**Figure 3.** PMR spectrum of dehydration product recovered from methyl 7-keto-15-hydroxydehydroabietate.

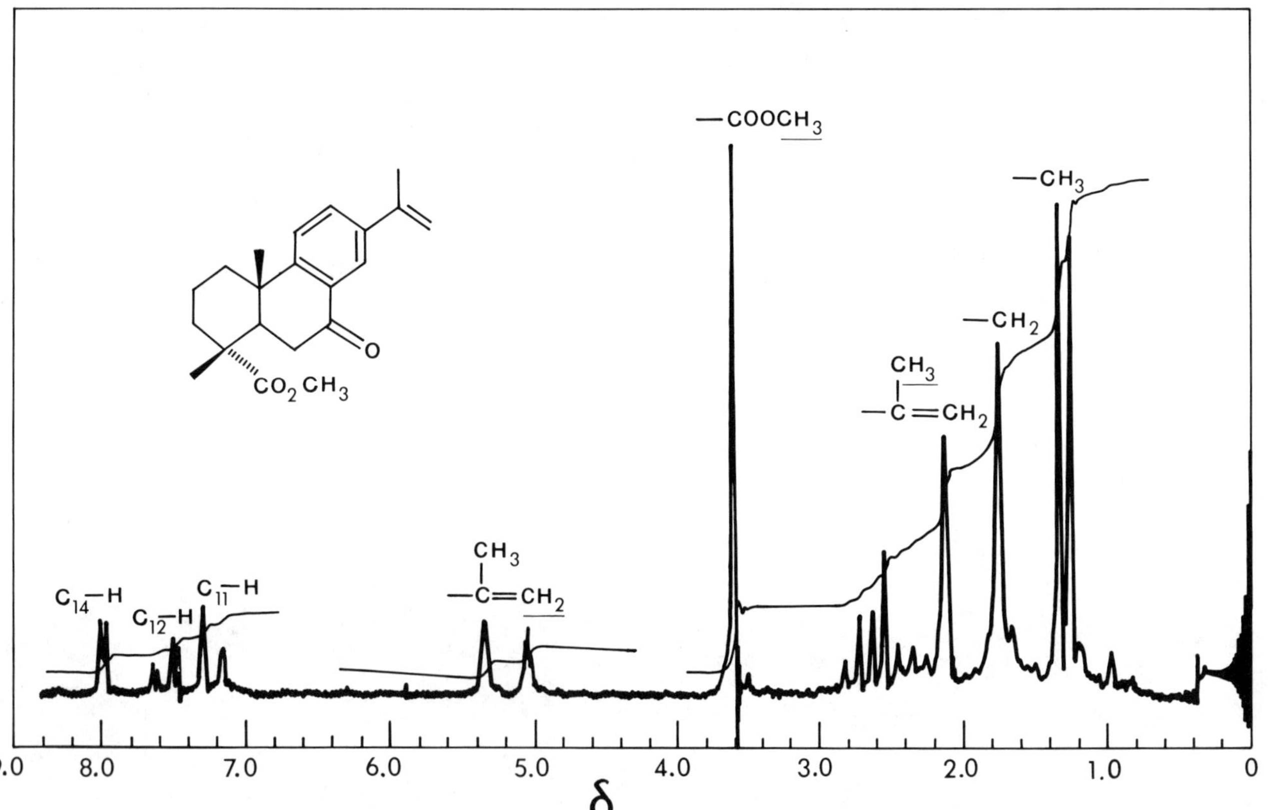


**Figure 4.** PMR spectrum of methyl 7-keto-15-hydroxydehydroabietate.

$$\begin{array}{c} CH_2 \\ \| \\ C\text{–}\underline{CH}_3 \end{array} \qquad\qquad \begin{array}{c} CH_3 \\ | \\ \text{-}C\text{=}\underline{CH}_2 \end{array}$$

I II

### Unknown (Compound 3)

Parent ion 302·151 (36.5%) corresponds to $C_{18}H_{22}O_4$. Satellite ions at:

1. 270·1258 (7.5%, loss of $CH_3OH$);
2. 243·1385 (4.5%, loss of $COOCH_3$);
3. 242·1321 (6.3%, loss of $HCOOCH_3$); and
4. 227·1069 (100%, loss of $C_3H_7O_2$).

Low-resolution MS showed additional major fragments at m/e 187, 177, 173, 161, 149, 137, 121, 118, 109, 95, 87, 69, 57 and 55.

### 7-Hydroxy-DHA (Compound 4)

Low-resolution MS parent ion m/e 330 (20.7%). Satellite ions at m/e:

1. 312 (M-18, 7.4%);
2. 287 (M-43, 5.4%);
3. 255 (M-75, 20%); and
4. 253 (M-77, 7.6%).

Other major fragments at m/e 238 (18%), 237 (100%), 195 (22%) and 162 (56%). Identical with published mass spectrum [10]. IR:

1. 3220 (hydroxyl);
2. 1725 (saturated ester); and
3. 825 (aromatic).

PMR:

1. 1.17, 1.27 (4 $\underline{CH}_3$);
2. 2.6–3.1 (septet, $C_{15}$–H);
3. 3.67 (singlet $COO\underline{CH}_3$);
4. 4.85 (triplet, $\underline{H}$–C–OH); and
5. 7.1–7.5 (complex, aromatic).

## Synthesis of 7-Keto-DHA and 7-Keto-15-hydroxy-DHA

Dehydroabietic acid was oxidized with alkaline permanganate at room temperature as described by Pratt [5]. The crude product was a mixture of two components, which were separated by column chromatography on Sephadex LH-20 with chloroform as solvent. This procedure yielded 7-keto-DHA as a colorless syrup. Methylation of a sample gave spectroscopic data identical with that recorded on compound 1 and with an authentic sample. From the column chromatography a mixture of 7-keto-DHA with the second compound (2:3) was recovered. This was separated by preparative thick-layer chromatography on silica plates using chloroform/methanol (9:1) as developing solvent. The second component was likewise recovered as a syrup and its methyl ester gave data identical with compound 2 from the oleoresin.

Samples of both these acids were bioassayed.

## Synthesis of 7-Hydroxy-DHA (Mixed $\alpha$ and $\beta$ Forms)

A sample of 7-keto-DHA was reduced with excess $NaBH_4$ in methanol at room temperature. After standing overnight, the reduction product was recovered by extraction with ether. The product was examined by TLC on silica with $CHCl_3$/MeOH (9:1) as developing solvent and the reaction was found to be incomplete. On repeating the reduction and allowing the product to remain in solution for another 24 hr, a colorless oil was obtained which showed only one component (TLC and GC). The methyl ester gave spectral data identical with that obtained from compound 4 for the oleoresin and the free acid was bioassayed.

## Static Bioassays with *Daphnia pulex*

With the exception of 7-hydroxy-DHA, compounds for testing were dissolved in absolute ethanol. Isopropyl alcohol was used to dissolve 7-hydroxy-DHA. Sodium carbonate solution (5 *N*) was added to the primary stock solutions, resulting in a pH of 11.7. The pH of bioassay solutions, made by diluting primary stock solutions, was adjusted to 7.3-7.6 by the addition of hydrochloric acid. Ethanol and isopropanol were bioassayed to ensure the concentrations present in tests were well below lethal levels.

Bioassays, except where noted, were single replicate exposures in 30-mL glass beakers owing to limited available amounts of test materials. Each beaker contained five individuals less than 24 hr old, and mortalities were checked once per day in each four-day test. Bioassay temperature was ambient (20–22°C). Dilution and control water was obtained from the laboratory Hatchery Creek water supply with hardness and alkalinity values of 85.2 and 57.4 mg/L, respectively, as $CaCO_3$.

## RESULTS AND DISCUSSION

### Chemical Identification

The reconstructed gas chromatogram illustrated in Figure 1 shows that similar oxidized resin acids occur in fresh extracts of Douglas fir wood as were known to be present in a large sample of resin acids recovered from this wood species several years previously. Moreover, we have observed these oxidized resin acids in hardboard plant wastewater [11] when Douglas fir woodchips are used as a raw material. It appears likely therefore that these modified resin acids are at least partially synthesized by the enzyme systems in the tree rather than formed solely by autooxidation or the attack of fungi on the corresponding resin acids. We have recently discovered the presence of similar compounds in a refiner groundwood wastewater sample from a mill which pulps spruce, pine and true fir, but is located well to the north of the growing range of Douglas fir. We believe therefore that these oxidized resin acids may prove to be of rather widespread occurrence in pulping wastewaters.

Isolation of the individual compounds from the wood extracts has been difficult. Removal of the resin acids from the extracts could be readily achieved either by column chromatography of the mixed methyl esters on deactivated alumina or of the free resin acids in ion exchange chromatography on DEAE-Sephadex A-25. Neither method was successful in removing the oxidized resin acids from a matrix of apparently polymeric acidic material present in the extract (see Figure 1). We were able to isolate some pure oxidized resin acids as their methyl esters from various subfractions by means of semipreparative reverse-phase LC. This technique did not however yield sufficient material to permit bioassays to be conducted, and it was necessary to synthesize some of the identified components from DHA.

The identification of 7-keto-DHA (compound 1) in Douglas fir oleoresin was not surprising, as this compound has been detected by other workers in the receiving environment near pulp mills. From the spectral data it was clear that the compound $C_{21}H_{28}O_3$ was the methyl ester of a carboxylic acid, contained an aryl ketone function, and three aromatic protons. Oxidation of

DHA with alkaline permanganate yielded a similar compound and the low-resolution mass spectra of both compounds were identical with the published fragmentation pattern for the methyl ester of 7-keto-DHA (4,6,10]. Since the C-7 position in the DHA ring system is allylically activated, it was anticipated that this position would be oxidized.

Inspection of the structural formula of DHA reveals that the C-15 position also should be activated because it is allylic to the aromatic ring. Compound 2 ($C_{21}H_{28}O_4$) contained an additional oxygen atom. Satellite ions observed in the high-resolution mass spectrum corresponded to loss of $H_2O$ (m/e 326), $H_2O + CH_3OH$ (m/e 294) and $H_2O + COOCH_3$ (m/e 267) from the parent ion. The compound also contained a carbomethoxy group and a labile methyl group as seen in the low-resolution MS (Figure 2), dominated by the M-15 peaked at m/e 329. The presence of an ester, a keto and a hydroxy function were revealed in the IR and PMR spectra. The aromatic protons in the PMR spectrum clearly showed the presence of a 1,2,5 trisubstitution pattern in the aromatic ring as in DHA itself (Figure 3). Definition of the hydroxyl group as tertiary in character followed from the facile dehydration of the methyl ester to the corresponding derivative of abieta-8,11,13,15-tetraenoic acid. The PMR spectrum of the dehydration product (Figure 4) clearly showed the presence of an isopropylidene group with two-proton doublet centered at 5.21 $\delta$ and a wide three-proton singlet at 2.15 $\delta$. The parent compound was formed by the permanganate oxidation of DHA and isolated by preparative TLC. Accordingly we assigned to the compound we isolated the structure 7-keto-15-hydroxy-DHA.

The third major component isolated from the oleoresin (compound 4) showed a parent ion at m/e 330 in the low-resolution MS and satellite ions at m/e 312, 287, 255 and 253 consistent with the elimination of $H_2O$, isopropyl, $HCOOCH_3 + CH_3$ and $H_2O + COOCH_3$. The mass spectrum was identical with the published fragmentation pattern of 7-hydroxy-DHA [10]. The presence of a secondary hydroxyl group was supported by an IR band at 3220 $cm^{-1}$ and a one-proton triplet at 4.85 $\delta$ in the PMR spectrum. The compound was synthesized by the borohydride reduction of 7-keto-DHA. This product was apparently a mixture of the $\alpha$ and $\beta$ forms at the C-7 position and was bioassayed as a mixture.

An unknown component (compound 3) gave a parent ion corresponding to $C_{18}H_{22}O_4$ when subjected to high-resolution MS. The satellite ions observed were consistent with the presence of a $COOCH_3$ group and a labile $CH_3$. Unfortunately the compound was isolated in low yield only and was not pure, so no further spectral information was gathered.

Various other components in the oleoresin could not be isolated but parent ions were observed at m/e 330, 332, 344, 358, 360 and 386.

Failure to isolate or observe the presence of 15-hydroxy-DHA deserves mention. Because of allylic activation of this position, this compound might be expected to occur. It has in fact been isolated from *Fomes annosus*-infected sapwood of Norway spruce and the MS fragmentation pattern has been published [10]. We have received two authentic samples of the methyl ester of this compound from Carman and Ayer. The mass spectrum recorded on the GC/MS instrument was that of methyl abieta-8,11,13,15-tetraenoate [10], suggesting that dehydration was taking place in the GC column before entry to the mass spectrometer. This was confirmed by recording the mass spectrum on a magnetic sector instrument with a solid sample insert. Under these conditions the spectrum recorded corresponded to the published spectrum of methyl 15-hydroxy-DHA [10]. Accordingly, we are unable to say whether or not this compound occurs in Douglas fir oleoresin. However, it is noteworthy that the mass spectrum of methyl abieta-8,11,13,15-tetraenoate was observed from nearly every oleoresin subfraction analyzed and was considered an artifact.

### Toxicity Tests

Results of the bioassays with *Daphnia* are summarized in Table I. DHA was bioassayed because of its ready availability and because it has been used as a standard toxicant for some years in our laboratories for work with sockeye salmon. It is our intention to define more accurately the 96-hr LC50 value for the mixed isomers of 7-hydroxy-DHA as a result of further tests. Isopropanol and ethanol, used as solvents in preparing the test substances for bioassay, were not acutely toxic at levels below 5000 and 7100 mg/L, respectively. The maximum concentrations of these alcohols present

**Table I.** ***Daphnia*** **96-hr LC50 Values**

| Test Material | 96-hr LC50 (mg/L) |
|---|---|
| DHA | |
| Test 1 | 6.5 |
| Test 2 | 5.4 |
| Test 3 | 5.3[a] |
| 7-Oxo-15-hydroxy-DHA | >50 |
| 7-Oxo-DHA | 42 |
| 7-Hydroxy-DHA (Mixed Isomers) | 10–20 |

[a]Triplicate concentrations.

in the bioassays were 575 and 1495 mg/L, respectively, and consequently were well below lethal concentrations. The ethanol concentration was approximately 150 mg/L at the LC50 concentration for DHA. The oxidized resin acids, especially 7-hydroxy-DHA, had marked surface active properties in aqueous solution. Although only a few compounds were available for testing, the results indicate a lower toxicity associated with the oxidized forms compared with DHA itself. We hope to be able to extend these tests as more compounds become available from this and related projects. We note that static bioassay tests with DHA against sockeye salmon using the same laboratory water supply give a 96-hr LC50 value of 2.15 mg/L [13]. Thus, *Daphnia* are more tolerant of resin acids than are sockeye salmon.

## CONCLUSION

A beginning has been made to the isolation, structural identification and aquatic organism toxicity testing of a series of oxidized resin acids that seem to be of widespread occurrence in softwood resins and various pulping wastes. The compounds isolated to date have all been derived from dehydroabietic acid, which is more stable chemically than other common resin acids because of the presence of the aromatic ring. However, a number of unknown oxidized resin acids were noted in Douglas fir oleoresin, and access to full-scale preparative LC may allow the isolation of some of these substances. In view of the significance of resin acid toxicity to valuable fishery resources, more information on the properties of the oxidized derivatives is desirable.

## ACKNOWLEDGMENTS

We thank Professor J. P. Kutney and Dr. Mahatam Singh of the Chemistry Department, University of British Columbia, for assistance with various aspects of this work and particularly for performing the dehydration experiment. The bioassays were conducted at the Sweltzer Creek Salmon Laboratory of the International Pacific Salmon Fishery Commission by Mr. D. Martens and Dr. J. Servizi. We thank Professor W. A. Ayer and Professor R. M. Carman for the gift of authentic samples of methyl 15-hydroxydehydroabietate and Dr. W. M. J. Strachan for a sample of 7-ketodehydroabietic acid. Dr. K. Hunt of Pulp & Paper Research Institute of Canada provided a sample of Douglas fir woodchips.

## REFERENCES


1. Ebeling, F. *Vom Wasser* 5:192-200 (1931).
2. Fox, M. E. "Fate of Selected Organic Compounds in the Discharge of Kraft Paper Mills into Lake Superior," in *Identification and Analysis of Organic Pollutants in Water*, L. H. Keith, Ed. (Ann Arbor, MI: Ann Arbor Science Publishers, Inc., 1976), pp. 641-660.
3. Brownlee, B., and W. M. J. Strachan. "Persistent Organic Compounds from a Pulp Mill in a Near-Shore Freshwater Environment," in *Identification and Analysis of Organic Pollutants in Water*, L. H. Keith, Ed. (Ann Arbor, MI: Ann Arbor Science Publishers, Inc., 1976), pp. 661-670.
4. Brownlee, B., and W. M. J. Strachan. *J. Fish. Res. Board Can.* 34:830-837 (1977).
5. Pratt, Y. T. *J. Am. Chem. Soc.* 73:3803-3807 (1951).
6. Wisconsin Department of Natural Resources, Water Quality Evaluation Section. "Investigation of Chlorinated and Non-chlorinated Compounds in the Lower Fox River Watershed," U.S. EPA Report EPA-905/3-78-004 (1978).
7. Norin, T., and B. Winell. *Acta Chem. Scand.* 26:2289-2296 (1972).
8. Carman, R. M., and R. A. Marty. *Aust. J. Chem.* 23:1457-1464 (1970).
9. Ekman, R., and R. Sjoholm. *Acta Chem. Scand.* B33:76-78 (1979).
10. Ekman, R. *Acta Acad. Abo. Ser. B* 39(6):7 (1979).
11. Rogers, I. H., H. W. Mahood, J A. Servizi and R. W. Gordon. *Pulp Paper Can.* 80(9):T286-290 (1979).
12. Wenkert, E., and T. E. Stevens. *J. Am. Chem. Soc.* 78:2318 (1956).
13. Davis, J. C., and R. A. W. Hoos. *J. Fish. Res. Board Can.* 32:411-416 (1975).

# CHAPTER 55

# HALOGENATED ORGANIC COMPOUNDS IN SPENT BLEACH LIQUORS: DETERMINATION, MUTAGENICITY TESTING AND BIOACCUMULATION

**A. Bjørseth, G. E. Carlberg, N. Gjøs, M. Møller and G. Tveten**

Central Institute for Industrial Research
Oslo, Norway

It is well known that effluents from the pulp and paper industry may cause an acute toxic effect in aquatic organisms. Some organic acids and phenols responsible for this effect have been identified [1-5]. Of particular environmental interest are the effluents from chlorine bleacheries. During the bleaching sequences, large amounts of chlorinated compounds are formed [6]. The environmental impact of chlorination was demonstrated by Pfister and Sjøstrøm [7], who showed that effluents from the chlorination stage are 10 times more toxic to fish than the effluents from other stages in the bleaching sequence.

Most studies so far have been focused on the acute toxic effects of the effluents. The development of short-term bioassays, such as the Ames *Salmonella* test, has provided an easy and convenient way for testing effluents as well as pure chemical compounds for their mutagenic and potential carcinogenic effects. In combination with chemical analysis, the short-term bioassays represent a good method for testing industrial and municipal effluents for the presence of such compounds. In recent years, therefore, increased attention has been given to persistent, lipophilic chlorinated organics and their mutagenic and potential carcinogenic effects [6,8-11].

Several chlorinated and brominated hydrocarbons have been detected in the effluents [1-18], some of which exhibit mutagenic effects. Some of the chlorinated phenols in the effluents have been shown to accumulate in fish [19].

This work summarizes some recent work in our laboratory and presents some data on mutagenic compounds in effluents from chlorine bleacheries and their identification. Furthermore, preliminary results of bioaccumulation studies of the lipophilic compounds are discussed.

## EXPERIMENTAL

### Effluent Studies

#### *Sampling*

Samples were collected from the different bleaching stages and from the total effluent from chlorine bleacheries in sulfite and sulfate plants. The samples were taken as integrated samples over a period of 30 min under normal process conditions. They were stored in glass containers at 4°C until analysis.

#### *Extraction*

A simple scheme of extraction divided the samples into nonpolar and polar extracts. By addition of KOH pellets, the pH was adjusted to 11. Nonpolar organic compounds were extracted twice with cyclohexane using a magnetic stirrer. The samples were then acidified to pH 2 with concentrated HCl and extracted twice with butylacetate. When necessary, the extracts were concentrated about 20-fold using a modified Vigreux distillation column. A total extract was obtained by adjusting pH to 2 with concentrated HCl and extracting twice with diethyl ether using magnetic stirring.

#### *Neutron Activation Analysis (NAA) of Organically Bound Chlorine*

NAA was performed on the unconcentrated extracts. The analysis was carried out at the Institute of Atomic Energy, Kjeller, Norway. Neutron activation was effected at a flux of about $1.5 \times 10^{13}$ n/cm$^2$-sec. The induced radioactivity was registered immediately after irradiation with a multichannel spectrometer equipped with a Ge(Li) detector. $^{38}Cl$ was used for the determination of chlorine [20].

*Gas Chromatography (GC) and Gas Chromatography/Mass Spectrometry (GC/MS) Analysis*

The nonpolar compounds were analyzed by glass capillary GC on a Carlo Erba Fractovap 2101 equipped with a flame ionization detector (FID) and splitless injector. The column was 50 m x 0.35 mm i.d. coated with SE-54. Injector and detector temperatures were 275°C, and the temperature programming was from 40 to 250°C at 3°C/min.

The GC/MS system consists of a Finnigan 4000 quadrupole mass spectrometer connected to a Finnigan 9610 gas chromatograph and an Incos 2000 data system. The column and GC conditions described above were used.

The polar constituents were silylated by injecting 3 $\mu$L of the extract and then immediately 3 $\mu$L of the silylating reagent, N,O-*bis*-trimethylsilyl-trifluoroacetamide (BSTFA) [21,22].

The derivatized compounds were analyzed by a GC/MS system consisting of a Hitachi Perkin-Elmer 990 gas chromatograph and an Incos 2000 data system. A 2-m x 3-mm stainless steel column packed with 3% Dexsil 300 on Gaschrom Q was used, and the temperature ranged from 70 to 300°C with a programming rate of 8°C/min, as described elsewhere [16].

*Mutagenicity Testing*

The histidine-requiring strains of *Salmonella typhimurium* TA98 and TA100 were kindly supplied by B. N. Ames. The strains are sensitive to mutagenic compounds, causing point mutations.

For compounds needing metabolic activation, liver homogenate was added. The 9000-g supernatant fraction (S-9) was prepared from male Wistar rats injected with Aroclor 1254, 500 mg/kg i.p. 5 days prior to preparation.

The mutagenesis assay was carried out as described by Ames et al. [23]. To each test tube containing 2 mL of molten top agar were added. 0.1 - 0.5 mL test solution, 0.1 mL of an overnight broth culture of the bacterial tester strain, and 0.5 mL S-9 mix containing per mL 0.1 mL S-9, 8 $\mu$mol $MgCl_2$, 33 mmol KCl, 5 $\mu$mol glucose-6-phosphate, 4 $\mu$mol NADP and 100 $\mu$mol sodium phosphate buffer, pH 7.4. The ingredients were mixed and poured onto minimal medium plates. After 2 days of incubation at 37°C, histidine revertant colonies were scored. Results are expressed as the average of five replicate plates from which the number of spontaneous revertants has been subtracted. Cyclohexane and diethyl ether extracts and single compounds have been tested.

## Fish Samples

### *Sampling*

Fish were caught downstream of four different pulp and paper mills (two saltwater and two freshwater recipients). Each sample consists of at least four fish with a combined weight of about 2 kg. For comparison, fish were also caught in areas free of pulp and paper mill effluents. The fish were kept frozen until they were analyzed. So far the following species have been or are under investigation: cod, flounder, perch, roach, burbot, freshwater herring and bream.

### *Extraction and Cleanup*

The extraction and cleanup procedures have been published previously [24,25] and they will only briefly be described. The fish oil was produced from a cyclohexane/isopropanol extraction of the homogenized fish. The oil was then divided in two for analysis of nonpolar and polar compounds.

*Nonpolar Compounds.* The nonpolar compounds were partitioned between acetonitrile and the fish oil by shaking the oil (100 g) with three successive portions of acetonitrile. After distilling off the acetonitrile a second partitioning was performed on the oil rest. The partitioned oil was transferred to an alumina oxide column and the column was eluted with two pentane portions (150 mL each), followed by pentane + 5% diethylether (200 mL), pentane + 30% diethylether (120 mL) and diethylether (120 mL). The compounds in the first pentane eluate were further fractionated on a silica gel column. This column was eluted with four pentane portions (30, 20, 65 and 125 mL).

*Polar Compounds.* The fish oil was dissolved in benzene and mixed with dilute NaOH and Sephadex QAE-A25 ion exchange resin. After discarding the liquid the resin was washed two times with benzene/dilute NaOH and distilled water. A mixture of dilute HCl and benzene was then added to the ion exchanger to transfer the phenols from the resin to the benzene. The phenols were acetylated using acetic anhydride [25].

### *GC and GC/MS Analysis*

The nonpolar compounds were analyzed by a Hewlett-Packard 5730 A gas chromatograph with simultaneous FID and electron capture detection (ECD) [26]. The column was 15 m x 0.35 mm i.d. coated with SE-54. The temperature programming used was 110–230°C at 5°C/min. For the GC/MS analysis the Finnigan system previously described was used.

The acetylated polar compounds were analyzed with the Hewlett-Packard 5730 A gas chromatograph described above and were identified by comparing retention times with those of standards.

## RESULTS AND DISCUSSION

### Effluent Samples

#### *Neutron Activation Analysis*

Typical values of the total amount of organohalogenated compounds in the effluents from sulfite and sulfate plants, as determined by NAA, are given in Table I. These values might vary depending on the raw material, the handling of chips/wood and differences in the cooking and bleaching processes.

In one sulfate plant, the discharge of polar and nonpolar organochlorine was determined to be 0.12 and 0.05 kg/ton of bleached pulp, respectively. This amounts to the total annual discharge of about 30 tons of organochlorine from this plant.

As shown in Table I we have found that sulfite plants discharge about equal amounts of nonpolar and polar chlorinated organic compounds, while sulfate plant effluents contain relatively less nonpolar and more polar constituents.

Brominated compounds have also been found in the effluents from a sulfite plant using seawater in the bleaching process.

Two combined effluent samples from a sulfate plant were used to study the distribution of organochlorine vs molecular size [27]. One sample consists of effluents from the 1 and 2 chlorination stages, and the other consists of effluents from the first chlorination and alkaline stages. The samples were filtered to remove particles (Millipore 0.45 m) and the filtrates were subjected to membrane filtration (cut size at mol wt 500 amu). The results

**Table I. Typical Total Organochlorine Concentrations (ppm) in Bleachery Effluent from Sulfite and Sulfate Plants**

| | Nonpolar (Cyclohexane) | Polar (Butylacetate) |
|---|---|---|
| Sulfite | 1.1 | 1.5 |
| Sulfate | 0.7 | 3.7 |

are shown in Table II and reveal that about 20% of the total organic chlorine are associated with compounds with molecular weight less than 500 amu. The table also shows that mixing the effluent streams from the chlorination and alkaline stages increases threefold the percentage of organic chlorine retained on the filter, compared to the effluent from the chlorination stages.

### *GC/MS Analysis of Nonpolar Compounds*

The individual compounds in different effluents have been identified by GC/MS. We have found that the compounds present and their relative abundance can vary considerably. This is probably due to differences in the raw materials and plant processes. We have, however, also found indications that there are differences as well as similarities in the types of compounds discharged from sulfate and sulfite bleacheries. In Table III we have summarized the main chlorinated compounds having been identified in the effluents from the two types of bleacheries.

In sulfite bleachery effluents, chlorinated cymene and other $C_{10}$, $C_{15}$ and $C_{20}$ derivatives like calamenene and naphthalene are among the main constituents. These compounds are probably derived from terpenes during cooking.

In sulfate bleachery effluents, chlorinated thiophene derivatives are among the main constituents. The thiophenes are probably synthesized during the cooking process and chlorinated during the bleaching. While small amounts of cymenes have been found in sulfate effluents, the thiophenes have so far not been found in sulfite effluents.

**Table II. Distribution of Organochlorine in Combined Effluents from Different Bleaching Stages in a Sulfate Plant**

| Sample | Total Organochlorine ($\mu$g/L) | On Particles >0.45 $\mu$m (%) | In Solution (<0.45 $\mu$m) (%) | Membrane Filtered (<500 amu) (%) | Total Accounted for (%) |
|---|---|---|---|---|---|
| 1 and 2 Chlorination Stages | 15.2 | 14 | 82 | 24 | 96 |
| 1 Chlorination and Alkaline Stages | 23.2 | 47 | 35 | 17 | 82 |

**Table III. Summary of Nonpolar Chlorinated Compounds Identified in the Bleachery Effluents from Sulfite and Sulfate Plants [6,12,16–18,31]**

| Chloro-compounds | Sulfite | Sulfate |
|---|---|---|
| Alkanes | Yes | Yes |
| Alkenes | Yes | Yes |
| Cymenes ($C_{10}$) | Yes | Yes |
| Calamenenes ($C_{15}$) | Yes | No |
| Naphthalenes ($C_{15}$) | Yes | No |
| Other Chloroaromatics | Yes | Yes |
| Thiophene Derivatives | No | Yes |
| Benzaldehyde | No | Yes |
| Hydroxycymenes | Yes | Yes |
| Pyrones | No | Yes |
| Trimethoxybenzenes | No | Yes |

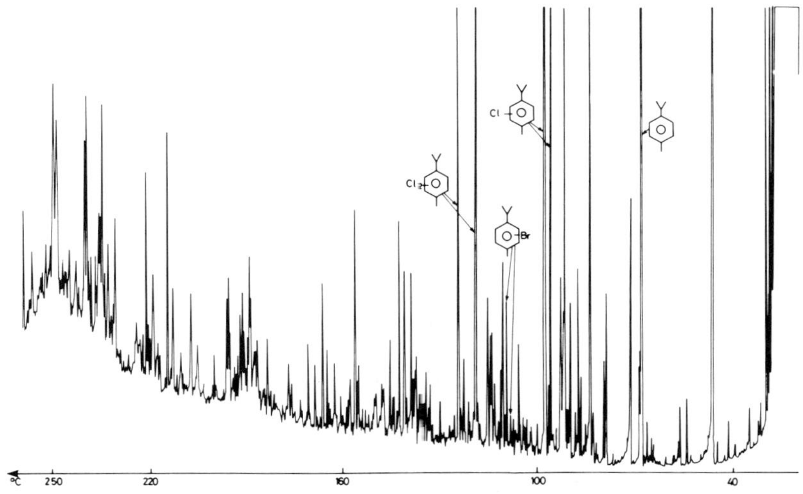


**Figure 1.** Glass capillary gas chromatogram of the nonpolar extract from the first chlorination stage in a sulfite plant.

Chlorinated aliphatics such as chloroform and hexachloroethane have been identified in both kinds of effluents.

A typical example of a gas chromatogram from the chlorination stage in a sulfite plant is given in Figure 1. As can be seen from the figure, chlorinated cymenes are among the main constituents of this effluent. By cochromatographing the sample with prepared standards of halogenated *p*-cymenes the concentration of mono- and dichlorocymene was determined to 150 and 100 μg/L, respectively.

From the quantitative data of the identified chlorinated hydrocarbons it is possible to calculate the contribution from these compounds to the total chlorine content determined by NAA. Hence it is possible to estimate the amount of chlorinated compounds not identified by the GC/MS. For the sample shown in Figure 1 it was found that the chlorinated *p*-cymenes are responsible only for about 10% of the total chlorine present. By including all other compounds identified by GC/MS still about 70-80% of the total amount of organochlorine compounds remains unidentified. This is in accordance with the molecular filtration data showing that about 80% of the sample had molecular weight larger than 500 amu.

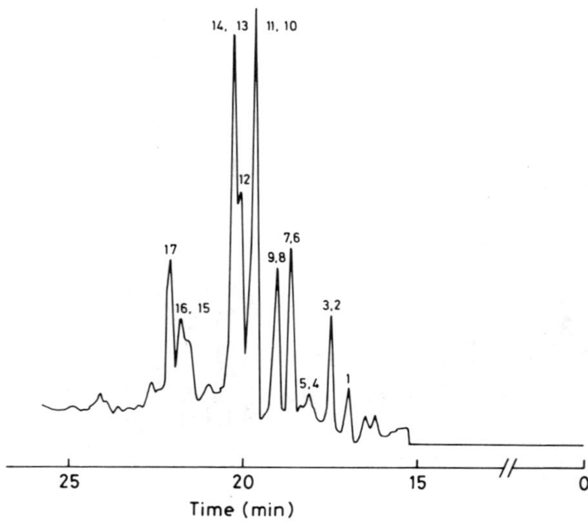


**Figure 2.** Total ion chromatogram of polar constituents of sulfite plant effluent. The peaks are dichlorocatechol (1,3,7); monochloro-1,2,3-trihydroxybenzene (2,5,6); monochloro-1,2,4-trihydroxybenzene (5,6); dichloro-1,2,3-trihydroxybenzene (8, 10); dichloro-1,2,4-trihydroxybenzene (10,13); dichloro-dihydroxy-metoxybenzene (9); trichlorocatechol (12); trichloro-1,2,3-trihydroxybenzene (16); trichloro-1,2,4-trihydroxybenzene (17). Peaks 4, 11, 14 and 15 are suggested to be tetrahydroxybenzenes; 11 and 14 are their mono- and 15 their dichloro derivates.

### *GC/MS Analysis of Polar Compounds*

The individual compounds in different effluents have been identified by GC/MS after on-column TMS derivatization. The main polar constituents in sulfite and sulfate effluents have been shown to be chlorinated derivatives of catechol, guaiacol and lesser amounts of phenol [1,4,14]. In addition we have found varying amounts of chlorinated trihydroxybenzenes in both types of effluents. Figure 2 shows the total ion current chromatogram of the polar constituents in a sulfite plant effluent. By comparing with synthesized standards, the main constituents in this effluent was shown to be chlorinated derivatives of 1,2,3- and 1,2,4-trihydroxybenzenes. Chlorinated tetrahydroxybenzenes have also tentatively been identified in this sample.

### *Mutagenicity Testing*

The effluents from the different stages in the bleaching process as well as pure compounds have been tested in the Ames *Salmonella* mutagenicity test using the strains TA1535, TA98 and TA100. The highest mutagenic response was obtained in the absence of liver microsomes with strain TA100. Very weak or no mutagenicity was obtained using TA1535 or TA98. The results show that effluents from all chlorination stages contain compounds with strong mutagenic effects. Furthermore, the effluents from the hypochlorite stage in a sulfate plant also exhibit strong mutagenic effects. The results from testing of ether extracts from effluents are shown in Table IV. Nonpolar extracts also contain mutagenic compounds in effluents from the chlorination stages and the hypochlorite stage. Addition of liver microsomes usually reduced the mutagenic effect, indicating that liver enzymes to some extent detoxify the mutagenic compounds. These results are in agreement with the results obtained in Sweden [8,9], where they found mutagenic activity in the chlorination stages and the hypochlorite stage.

Extracts from a sulfite plant were tested at different pH values to investigate the stability of the mutagens. The untreated extract showed highest activity in the absence of rat liver microsomes (see Table V). Addition of microsomes decreased the activity by approximately 40%. After treatment of the extracts at pH 9, the mutagenic activity without liver microsomes was reduced by 20%. Addition of liver microsomes reduced the activity further. No mutagenic activity was observed in the sample treated with 2 *N* NaOH with or without liver microsomes.

As the GC/MS analysis showed that the extracts contained the same components before and after alkaline treatment, at least two explanations can be

Table IV. Mutagenic Effects of Ether Extracts from Spent Bleach Liquor Effluents in the *Salmonella* Strains TA98 and 100 (Effluents Concentrated 400X)[a]

| Samples | Liver Microsomes Added | No. of Mutants per Plate (Mean of Five Plates)[b] | |
|---|---|---|---|
| | | TA98 | TA100 |
| Sulfite Plant | | | |
| First Chlorination Stage | + | 0 | 517[c] |
| | – | 83[c] | 713[c] |
| Second Chlorination Stage | + | 47[c] | 1360[c] |
| | – | 69[c] | 1344[c] |
| Alkaline Stage | + | 0 | 0 |
| | – | 0 | 4 |
| First Sewage Stage (second chlorination, | + | 56[c] | 1247[c] |
| alkaline hypochlorite) | – | 145[c] | 1746[c] |
| Second Sewage Stage (first chlorination, | + | 19 | 85[d] |
| $SO_2$ stage) | – | 89[c] | 175[c] |
| Sulfate Plant | | | |
| First Chlorination Stage | + | 143[c] | 450[d] |
| | – | 100[c] | 0 |
| Second Chlorination Stage | + | 60[c] | 154[c] |
| | – | 130[c] | 474[c] |
| Alkaline Stage | + | 0 | 42[c] |
| | – | 0 | 126[c] |
| Hypochlorite Stage | + | 0 | 183[c] |
| | – | 0 | 734[c] |

[a]Spontaneous revertants have been subtracted (35 for TA98, 120 for TA100). Statistical analysis with students t-test.

[b]The chlorination stages gave a 60-95% killing effect of the bacterial growth, making it difficult to compare mutually the mutagenic potential. Some of the extracts were diluted five times to reduce the influence of toxic compounds. A strong increase in the number of mutants per milliliter of concentrated extract was observed. The first chlorination stage in the sulfate process contained especially high amounts of toxic components (95% killing effect), and there was no detectable mutagenic activity in the original extract without liver microsomes. By fivefold dilutions, however, a strong mutagenic effect was observable with and without activation.

[c]Significance level, $p < 0.001$.

[d]Significance level, $0.001 < p < 0.01$.

given for the loss of the mutagenic activity: (1) the mutagenic compound(s) was not identified by GC/MS, or (2) synergistic effects of comutagens play an important role in the observed mutagenicity of the untreated extract. These factors may be lost during the alkali treatment.

**Table V. Mutagenic Effects of Untreated Extract from a Sulfite Plant, and Extracts Treated at pH 9 with 2 *N* NaOH on *Salmonella* Strain TA100 (100 μL sample (effluents conc. 5x) was added per plate)[a]**

| Extract Treatment | Number of Mutants per Plate (Mean of Five Replicates) | |
|---|---|---|
| | + Liver Homogenate | - Liver Homogenate |
| Untreated | 79 | 126 |
| pH 9, 40 min | 28 | 98 |
| 2 *N* NaOH, 40 min | 0 | 0 |

[a]The number of spontaneous mutants (about 120) have been subtracted.

The instability of the mutagens in the effluents at high pH is in accordance with previous results [6,8,9].

Some of the halogenated cymenes and chlorinated trihydroxybenzene derivatives have been synthesized and tested for mutagenic activity. The nonpolar compounds bromo-*p*-cymene and dichloro-*p*-cymene showed a weak mutagenic effect in the absence of liver microsomes. No effect was observed with chloro-*p*-cymene. Trichloro-1,2,3-trihydroxybenzene, identified in effluents from a sulfate plant, also showed a weak mutagenic effect.

These compounds can, however, only in part explain the observed mutagenicity of the extracts.

Very little work has been done in testing compounds identified in pulp mill effluents. Nestmann et al. [28] have tested ten resin acids, one of which was found to be mutagenic.

## Fish Samples

To study the potential environmental impact of the chlorinated compounds discharged from the pulp and paper industry, fish caught downstream of several plants have been analyzed for the total amount of chlorinated compounds and for individual nonpolar and polar compounds.

### *Neutron Activation Analysis*

The content of organochlorine compounds has been found to vary considerably between different fish species caught at the same location. Four different fish species have been caught in a freshwater lake at a location where a river containing effluents from a sulfite plant enters the lake. As a control, two of the species were also sampled from an uncontaminated area of the lake. Table VI shows that the total organochlorine contents in fish oil from the contaminated area is from about 3 (perch) to 11 (roach) times higher than similar values for fish from the uncontaminated area.

This clearly shows that fish from the contaminated area has accumulated organochlorine compounds. The large variation in the organochlorine content, from 220 ppm for perch to 1990 ppm for freshwater herring, might be due to a number of reasons. Differences in fat content, rate of metabolism, feeding habits and whether the fish is stationary or not might be important in this respect.

*Polar Compounds*

The identity and concentration of the polar compounds in fish oil have been found to have large local variations. Previous work in our laboratory has identified tri- and tetrachloroguaiacols, and in a current investigation chlorinated phenols have been identified [27]. This is in accordance with other studies where tri-, tetra- and pentachlorophenol have been identified in fish oil in addition to the chlorinated guaiacols [19,29]. These fish oil results indicate that of all the chlorinated phenolic compounds present in bleachery effluents only chlorinated guaiacol and phenol are sufficiently stable and lipophilic to accumulate in fish. The rest of the phenolics in the effluents are probably degraded in the river or cleared fairly rapidly by the fish.

In fish oil from bream caught in a river receiving discharges from both a sulfite and a sulfate plant, 450 ppb of trichloro- and 600 ppb of tetrachloroguaiacol were found.

**Table VI. Total Organochlorine Content in Fish Samples from an Area Contaminated with Pulp and Paper Plant Effluents**

| Fish Species | Number of Fish in Each Sample | Total Weight (kg) | Total Organochlorine (ppm in oil) |
|---|---|---|---|
| Contaminated Area | | | |
| Perch | 8 | 2.0 | 220 |
| Roach | 35 | 2.2 | 470 |
| Burbot | 5 | 5.5 | 520 |
| Freshwater Herring | 8 | 2.1 | 1990 |
| Uncontaminated Area | | | |
| Perch | 28 | 1.5 | 80 |
| Roach | 24 | 1.5 | 43 |

From laboratory bioaccumulation experiments with bleaks, concentration factors for tri- and tetrachloroguaiacol were found to be on the order of 400 after two weeks of exposure [30]. No steady state was, however, reached during the experiment. In another bioaccumulation experiment, where rainbow trout were exposed to bleachery effluents diluted 40 times, a steady state was reached after two weeks [19]. The trout were found to bioaccumulate trichlorophenol, and tri- and tetrachloroguaiacol.

The two experiments indicate a relatively rapid clearance from the fish tissue after discontinuation of exposure. The two guaiacols were completely eliminated after two weeks [30], and trichlorophenol showed similar behavior. As a comparison, pentachlorophenol in similar experiments was still present in the fish after 16 days of depuration in uncontaminated water [31].

### *Nonpolar Compounds*

From the extensive cleanup procedure each fish sample gives a number of different fractions, both from the alumina oxide and silica gel columns. From a number of experiments we have, however, found that most of the nonpolar compounds are present in the silica gel fractions, in particular in the first fraction. Figure 3 shows the gas chromatogram of the first silica gel fraction of a cod sample caught between a sulfite and a sulfate plant. The plants are situated about 1 km apart at the seaside. The main compounds in this fraction were found to be terpenes. In addition dichlorocymene and polychlorinated biphenyls (PCB) were identified.

The chromatogram in Figure 3 has been compared to chromatograms of the cyclohexane extracts of the bleachery effluents. Apart from the PCB, the chromatograms looked quite similar. In addition to the terpenes, many of the unidentified compounds in this fish oil fraction have previously been shown to be present in the effluents. The PCB found in the fish is probably a result of the global distribution of these compounds.

In other fish samples, both mono- and dichlorocymenes have been identified. In the bream sample mentioned in the previous section the concentration of mono- and dichlorocymene in the fish oil was determined to be 5 and 3.6 ppm, respectively.

The gas chromatographic profile from different fish samples varies according to fish species and sampling locations. All chromatograms seem, however, to include a pattern of peaks corresponding to peaks one to six in the chromatogram in Figure 3.

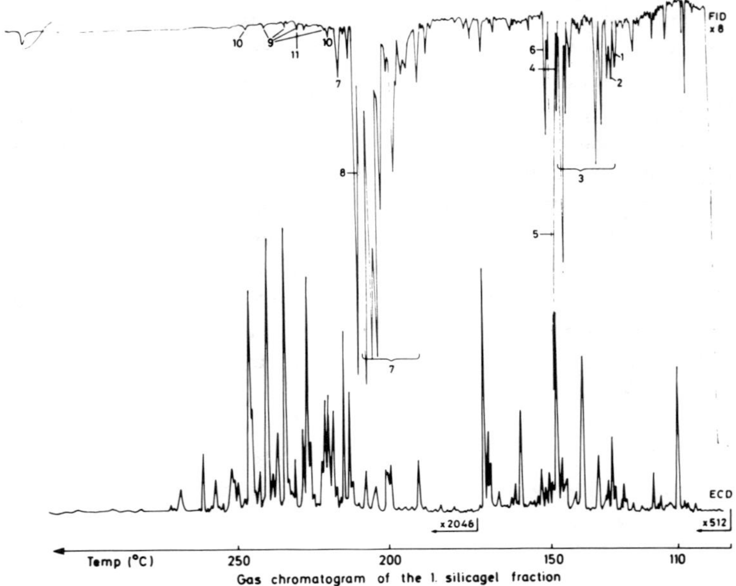


**Figure 3.** Gas chromatogram of nonpolar compounds in fish downstream of a pulp and paper plant. The peaks are (1) M = 172 (probably alkylated dihydronaphthalene); (2) dichlorocymene; (3) sesquiterpene ($C_{15}H_{24}$); (4) probably terpene ($C_{15}H_{24}$); (5) clalmenene (6) M = 200, (7) diterpene ($C_{20}H_{32}$), (8) M = 256, (9) pentachlorobiphenyl; (10) hexachlorobiphenyl; (11) DDE.

## CONCLUSIONS

It has been shown that besides the polar compounds causing acute toxic effects, effluents from chlorinated bleacheries also contain compounds exhibiting mutagenic effects. The compounds identified and tested so far can only in part explain the observed mutagenicity of the extracts. Metabolic activation and alkaline treatment reduces the mutagenicity. Only about 20% of the organochlorine compounds have molecular weight below 500 amu. Some of the compounds do bioaccumulate in fish. While the accumulated polar compounds have been shown to clear fairly rapid, little is known about the metabolic kinetics and the persistence of the accumulated nonpolar compounds.

## REFERENCES


1. Leach, J. M., and A. N. Thakore. *J. Fish. Res. Board Can.* 30:479 (1973).
2. Leach, J. M., and A. N. Thakore. *J. Fish. Res. Board Can.* 32:1249 (1975).
3. McKague, A. B., and C. C. Wolden. CPAR Project Report No. 245-5 (1978).
4. Voss, R. H., I. T. Wearing, R. D. Mortimer, T. Kovacs and A. Wong. Paper presented at the Third International Congress on Industrial Waste Water and Waste, Stockholm, Sweden, February 1980.
5. Kachi, S., et al. Paper presented at the Environmental Improvement Conference, Victoria, British Columbia, Canada, October 1979.
6. Bjørseth, A., G. E. Carlberg and M. Møller. *Sci. Total Environ.* 11:197 (1979).
7. Pfister, K., and E. Sjøstrøm. *Sv. Papperstidn.* 81:195 (1977).
8. Ander, P., K. E. Erikson, M. C. Kolar, K. Kringstad, V. Rannug and C. Ramel. *Sv. Papperstidn.* 80:454 (1977).
9. Erikson, K. E., M. C. Kolar and K. Kringstad. *Sv. Papperstidn.* 82:95 (1979).
10. Høglund, C., A.-S. Allard, A. H. Neilson and L. Landner. *Sv. Papperstidn.* 82:447 (1979).
11. Carlberg, G. E., N. Gjøs, M. Møller, K. O. Gustavsen, G. Tveten and L. Renverg. *Sci. Total Environ.* 15:3 (1980).
12. Harris, E. E., E. C. Sherrard and R. L. Mitchell. *J. Am. Chem. Soc.* 56: 889 (1934).
13. Ota, M., W. B. Durst and C. W. Dence. *Tappi* 56:139 (1973).
14. Lindstrøm, K., and J. Nordin. *J. Chromatog.* 128:13 (1976).
15. Keith, L. H. *Environ. Sci. Technol.* 10:555 (1976).
16. Bjørseth, A., G. Lunde and N. Gjøs. *Acta Chem. Scand.* B31:979 (1977).
17. Eklund, G., B. Josefsson and A. Bjørseth. *J. Chromatog.* 150:161 (1978).
18. Lindstrøm, K., and J. Nordin. *Sv. Papperstidn.* 81:55 (1978).
19. Landner, K., K. Lindstrøm, M. Karlsson, J. Nordin and L. Sørensen. *Bull. Environ. Contam. Toxicol.* 18:663 (1977).
20. Lunde, G., J. Gether and E. Steinnes. *Environ. Sci. Technol.* 9:155 (1975).
21. Street, H. V. *J. Chromatog.* 41:358 (1969).
22. Rasmussen, K. E. *J. Chromatog.* 120:491 (1976).
23. Ames, B. N., J. McCann and E. Yamasaki. *Mutat. Res.* 31:347 (1975).
24. Jensen, S. *Ambio* 5:257 (1976).
25. Renberg, L. *Anal. Chem.* 46:459 (1974).
26. Bjøseth, A., and G. Eklund. *J. High Resolution Chromatog. Chromatog. Commun.* 2:22 (1979).
27. Carlberg, G. E., N. Gjøs and G. Tveten (in preparation).

28. Nestmann, E. R., E. G. H. Lee, J. C. Mueller and G. R. Douglas. *Environ. Mutagen.* 14:361 (1979).
29. Lindstrøm, K., and L. Renberg. Report to the National Swedish Environment Protection Board (in Swedish) (1979).
30. Renberg, L., O. Svanberg, B. E. Bengtsson and G. Sundstrøm. *Chemosphere* 9:143 (1980).
31. Pruitt, G. W., and B. J. Grantham. *Trans. Am. Fish. Soc.* 106:462 (1977).

## CHAPTER 56

# ANALYSIS OF TOXICITY AND BIODEGRADABILITY OF ORGANOCHLORINE COMPOUNDS RELEASED INTO THE ENVIRONMENT IN BLEACHING EFFLUENTS OF KRAFT PULPING

**Mirja Salkinoja-Salonen, Maija-Liisa Saxelin and Jaakko Pere**

Department of General Microbiology

**Timo Jaakkola**

Department of Radiochemistry

**Juhani Saarikoski**

Department of Zoology
University of Helsinki
Helsinki, Finland

**Risto Hakulinen and Onni Koistinen**

Enso-Gutzeit Osakeyhtiö
Imatra, Finland

In Finland, as in North America, pulp is bleached by chlorine, using $Cl_2$, $ClO_2$ or a mixture of both. In 1979 about 2.8 million tons of pulp, involving some 140 million kg of chlorine, were bleached in Finland. As already stated by other authors [1,2] about 10% of this will appear organically bound.

Many organic chlorine compounds are toxic and recalcitrant to biodegradation. It is these properties of organic chlorides that are utilized in pesticides, the most widely used of which is pentachlorophenol (world production in 1978 over 25 million kg [3]). Spent bleach liquors are usually released into lakes, rivers or the sea, where toxic organic chlorides can harm aquatic life and accumulate in the food chain [4-8].

In Finland, 51 out of the 62 pulp mills discharge into shallow lakes or rivers which are less than 30 m deep. The total volume of bleaching effluents is about 170 million $m^3$ and it contains some 200-300 tons of chlorinated phenols, catechols, guaiacols and dimethoxyphenols [9]. Some chlorinated resin acids and extractives also occur in the bleach liquor. However, the bulk of the organically bound chlorine is of high-molecular-weight chlorolignin, the annual discharge of which in Finland is about 20,000 tons [10].

The aim of the studies reported in this chapter was to analyze and follow-up in the environment the organochloric compounds that are discharged from bleaching of pulp. We assayed their toxicity to different levels of life (bacteria, fungi, daphnids, fish) and studied biodegradability in situ (lakewater samples and bottom sediments) and in the laboratory.

The toxicity assays showed that acute toxicity of spent bleach liquor resides in the low-molecular-weight fraction. The macromolecular chlorolignin was found to be metabolically stable both in the lake sediments and in the laboratory tests. The smaller, toxic molecules, however, are biodegradable. A biological method was developed to remove biochemical oxygen demand (BOD), chlorophenolics, toxicity and mutagenicity from the bleaching effluents.

## MATERIALS AND METHODS

### Analysis of Chlorophenols

Chlorophenols, guaiacols and catechols were extracted from spent bleach liquor and lakewater samples as shown in Figure 1, ethylated with diazoethane [11] and analyzed by gas-liquid chromatography (GLC) at 175°C with an 80-m long capillary column coated with OV-101 as the liquid phase, and monitored simultaneously with flame ionization (FID) and electron capture ($^{63}Ni$) detectors (ECD) [12]. Recoveries of the different chlorophenolics in ether phase III are summarized in Table I. The reference compounds unavailable commercially (chlorinated catechols, guaiacols and dimethoxyphenols) were synthesized by J. Knuutinen, Department of Chemistry, University of Jyväskylä [13,14]. From lake sediments the chlorophenols were extracted by the QAE Sephadex absorption method of Rehnberg [15], ethylated with diazoethane [11] and analyzed by GLC in the same way as for the water samples. Recovery of chlorophenols, chlorocatechols and chlorodimethoxyphenols varied in the range of 6–90%. To monitor extraction losses, 2,3,6-trichlorophenol (50 $\mu$g/L) was added to all samples prior to extraction.

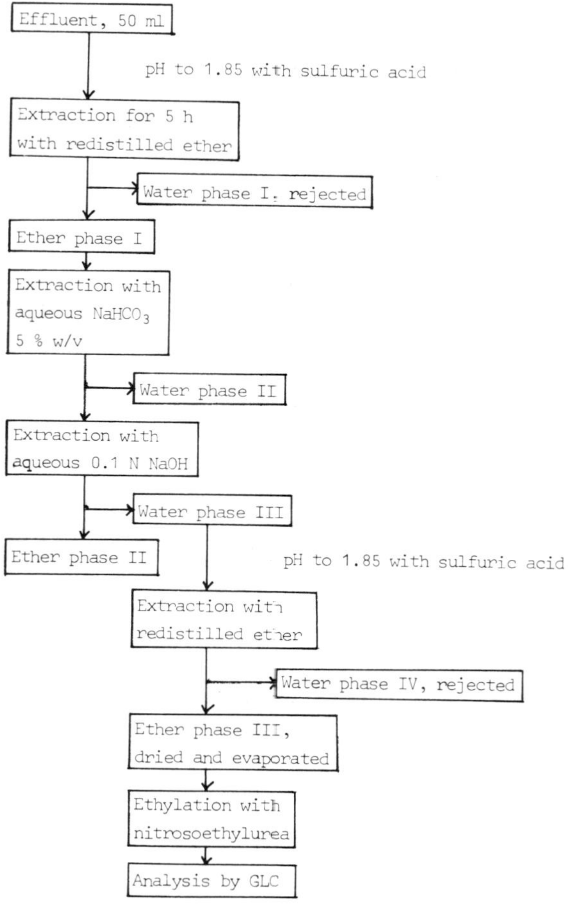


**Figure 1.** Steps in the analysis of chlorophenols of spent bleach liquor.

## Chlorine Measurements

One gram of accurately weighed, air-dried, homogenized sample was suspended in 10 mL of distilled water, sonicated six time for 10 sec (MSE ultrasonic desintegrator fitted with the linear rod) and diluted to 50 mL. $NaNO_3$ was then added to a final concentration of 100 m*M*, and chloride ions were measured with the ion-specific electrode of Orion Research (Model 94-17A) attached to a digital reading mV-meter (Orion 701A). The low-level procedure of the manufacturer's instruction manual was used since the levels were between 5 and 50 m*M*. $Ni(NO_3)_2$ was added to duplicate samples to 10 m*M* to monitor for disturbance by sulfide ions; in all cases this was found to be negligible.

For the measurement of organically bound chlorine, 1 g of accurately weighed air-dried, homogenized sample was suspended in 10 mL of distilled water; 1 mL of 1 *N* NaOH was added to increase pH to over 9; and this mixture was incubated in an unstoppered round-bottom flask in a water bath at 55-58°C overnight. The residue was then taken up in 5 mL of 10% $HNO_3$, 15 mL of distilled water was added (final pH $< 1.0$) and this mixture was then boiled under reflux for 5 hr with occasional shaking to avoid foaming. The mixture was allowed to cool under reflux and then neutralized to pH 6-8 with 1 *N* NaOH and diluted to 50 mL with distilled water. Concentration of chloride was then measurred as described above for inorganic chloride. BOD, total organic carbon (TOC), chemical oxygen demand (COD), and color were determined according to standard procedures [16].

**Table I. Recovery by Extraction of Some Chlorophenolic Compounds from Spent Bleach Liquor[a]**

| Compound | Retention Time (min) | Recovery (% ± S.D.) |
|---|---|---|
| 2,6-Dichlorophenol | 1.4 | 6.3 ± 2.1 |
| 2,4-Dichlorophenol | 1.8 | 9.7 ± 9.1 |
| 2,4,6-Trichlorophenol | 2.4 | 67 ± 24 |
| 2,4,5-Trichlorophenol | 3.4 | 91 ± 8 |
| 4,5-Dichloroguaiacol | 4.2 | 98 ± 3 |
| 2,3,4,6-Tetrachlorophenol | 5.2 | 90 ± 3 |
| 3,4,5-Trichloroguaiacol | 7.3 | 91 ± 4 |
| Tetrachloroveratrol | 8.2 | 14 ± 8 |
| Tetrachloromuconic Acid | 9.7 | 13 ± 4 |
| Tetrachloroguaiacol | 10.4 | 98 ± 3 |
| Pentachlorophenol | 10.9 | 75 ± 12 |
| Tetrachlorocatechol | 12.8 | 22 ± 6 |

[a]Extraction was performed as described in the section, Assays of Chlorophenols (ether phase III). Analysis was performed on capillary GLC after ethylation. Recoveries were assayed with synthetic reference compounds, of which 100-200 ng was added to 50-mL samples of lake-, river- and rainwater, while bleaching effluents were spiked with 1000 ng/50 mL.

### Assays of Toxicity

Toxicity tests with water fleas (*Daphnia magna*) less than 24 hr old were performed as follows:

1. effluent was centrifuged to remove suspended solids;
2. a series of dilutions was made with activated-carbon-filtered, aerated tapwater as diluent;
3. 50 mL of each dilution was preaerated, and 5 daphnids were suspended in each; dilutions were analyzed in duplicate;
4. tests were read after 24 hours; fleas that did not jump within 5 seconds were considered dead.

Tests with bacteria (*Pseudomonas putida*) were performed according to the German standard procedure [17]. The fish tests were performed with guppies (*Poecilia reticulata*) in 40-L aquaria for 96 hr. The fish were reared and the tests conducted in an aquarium room with temperature 25-27°C and photoperiod 16 hr light and 8 hr dark. Dechlorinated local tapwater contained (according to the Water Examination Laboratory of Helsinki City Waterworks) 25-30 mg Ca/L and 2.5-5.5 mg Mg/L (corresponding to a hardness of 80-100 mg $CaCO_3$/L). The pH value was adjusted before tests from the initial value of 7.9-8.1 to 7.0 with HCl, taking special care that the $CO_2$ balance was restored after acid additions. Groups of ten fishes (average wieght 40-60 mg) were placed in cylindrical glass vessels containing 5 L water. After a 24-hr acclimation period, phenols to be tested were added in varying concentrations and the mortality observed for 96 hr. Sufficient oxygen content was assured by constant aeration. After each 12 hr, 4 L of the solution was replaced. When the concentrations of phenols were checked during tests, the the content of the most volatile (2-chlorophenol) appeared to remain at a level which was at least 90% of the initial concentration. In toxicity tests with pulp mill effluents, the test solution was prepared by mixing equal amounts of chlorination step effluent and alkaline extraction step effluent and adjusting the pH of the mixture to 7 with NaOH or HCl approximately 16 hr prior to testing. The 96-hr $LC_{50}$ values and the 95% confidence intervals were calculated according to the maximum likelihood method as described by Finney [18].

### Sampling of Sediments and Water from the Recipient

Water samples were taken with a Ruttner sampler or with a 2-m acrylic tube of 2 in. diameter. The sediment samples were collected during winter, through an opening made in the ice cover with a device specially developed for this by Huttunen and Meriläinen [19] and made in a local workshop. It is a flat metal vessel, the back wall of which is insulated. The measures of the vessel were designed to fit the measures of the blocks of solid $CO_2$ that are produced by the local supplier. The pocket facing the front wall is filled with blocks of solid $CO_2$ just before the vessel is allowed to sink through the opening in the ice-covered lake. The leading edge of the vessel contains lead, to give enough weight for the vessel to sink as deep as possible into the soft sediment. Details of this sampling device have been described elsewhere [19] and will be sent on request.

After about 0.5 hr, the vessel is towed up, the remains of the blocks of $CO_2$ are poured out, and the transverse section of the sediment that is frozen onto the frontwall of the vessel is removed by warming the wall with a flame from inside (a portable blow-burner). The sediment is allowed to glide onto a plywood support and is stored in an insulated box with solid $CO_2$ until analysis. We used an ordinary timberman's saw to cut the frozen sediment into layers of 2.5 cm.

## Mutagenicity

Mutagenicity was assayed by the Ames test [20,21]. The *Salmonella typhimurium* mutant TA 100 reproducibily indicated mutagenicity in spent bleach liquor, whereas the other Ames strains gave variable results. In most tests only TA 100 was used.

## Followup of Wastewater Streams in the Recipient by Using Bacteriophage as Tracer

We preferred phages of *Salmonella typhimurium* instead of coliphages that were described in the literature [22,23] because coliphages sometimes occur naturally in polluted waters and possibly even could multiply in a polluted recipient. We routinely used phage P22 mutant C2, which we found to persist in undiluted spent bleach liquor (pH 4–8.5) for two weeks with less than 20% loss of viability.

The phages were propagated by adding a 20 mL inoculum of P22C2 containing over $10^{10}$ plaque forming units (pfu) per mL into 50 mL of a culture of *S. typhimurium* in the logarithmic phase, Klett value filter 54) 70 to 90 in a medium containing 3 g meat extract, 5 g pepton, 0.25 g $MgSO_4 \cdot 7H_2O$ and 5 g NaCl per liter. The cultivation was performed aseptically at 37°C in a shaker. After 40 min, 1 mL of chloroform was added to lyse the culture.

After shaking, the lysate was centrifuged for 10 min at 4000 rpm. The supernate was titered for phage using strain SL696 as indicator, and should have contained 5-10 x $10^{10}$ pfu/mL. The titered supernate was used to inoculate a larger culture of *S. typhimurium* SL696 as follows: a 5- to 10-L culture was prepared by adding 200 mL of a culture grown overnight into a warm (37°C) sterile broth. At this stage there was no need for aseptic working, and instead of a sophisticated fermenter an ordinary pail placed in a 37°C room was used. The culture was mixed and aerated by pressurized air. When it reached a turbidity of Klett-54 reading of 90, the phage supernate was added, aeration was continued for 40 min, and then the culture was lysed with chloroform and centrifuged as above. In order to get a high titer of phage it was important to use a sufficiently high multiplicity of infection (moi or the number of phages per one bacterium, 1.5–4). Because the phage suspension was saturated with chloroform, it disturbed the growth of the *Salmonella* host if more than 1/10 volume of phage suspension was poured into the culture. It was therefore essential to use a phage solution of a high titer ($10^{10}$/ mL or more). If the starting inoculum of phage was less dense, the first lot of phage was collected from the soft-agar plates (obtained from titering of the phage) as follows: soft agar was collected from the petri dishes with the highest number (over 100) of plaques, frozen at -20°C, and centrifuged for 10 min at 4000 rpm to remove agar and host bacteria. The supernate was ready for use as inoculum.

The amount of phage needed for the followup of the movements of pulp mill effluent in the recipient water depended on the volume of the effluent, factor of dilution by the recipient and the distance from the mill where the followup was to be continued. We used 5-10 x $10^{14}$ phages (5-15 L of lysate) for the followup of effluents from a mill producing 25,000 $m^3$/day of bleaching effluents, over a distance of 20 km.

## Radiometric Determination of Age in Sediment Profile

A 2- to 7-g, air-dried, homogenized sample (from the respective layers of the sediment) was weighed accurately, placed in a plastic tube of 25 mL, and measured for $^{137}Cs$ in a NaJ (Tl) well–crystal of 3 x 3 in. (the well 1 x 2 in.) of type Harshaw 125W12/E-W4 in a multichannel pulse-height analyzer of type LP484 (Nokia, Finland). The measured spectra were analyzed by the method [24] of least-squares fitting, using the Burrough's 6700 computer (Computing Center of University of Helsinki).

### Determination of Molecular Weight Distribution of Spent Bleach Liquor

This was studied by gel filtration and by ultrafiltration. Gel filtration was performed over a Biogel P-10 column, 0.9 x 60 cm, equilibrated with 0.5 *M* NaCl. One ml of alkaline extraction and chlorination stage liquor, or factory effluent, was made to 0.5 *M* in respect to NaCl (20 µg of glucose added) and layered on the column. Elution liquid was 0.5 *M* NaCl, and 1.2-mL fractions were collected. One mg/mL of dextran blue was added as a marked of void volume. The end of the elution profile was detected by the appearance of glucose in the fractions (glucose-oxidase test, Boehringer-Mannheim). The columns were calibrated at regular intervals with lysozyme (grade I, Sigma, mol wt 14,600), cytochrome c (type III, Sigma, mol wt 12,400), aprotinin (from bovine lung, Boehringer-Mannheim, mol wt 6500), insulin chain B (Boehringer-M, mol wt 3400) and basitracin (mol wt 1411). Its separation properties remained constant for 6 months or longer at room temperature.

Ultrafiltration was performed with Amicon Diaflo 402 apparatus with filters PM-10 (>10,000 daltons), and the permeate from this with UM-2 (retains >1000 daltons). The retentates were resuspended in their original volume in 0.5 *M* NaCl (for gel filtration) or distilled water (toxicity tests).

## RESULTS AND DISCUSSION

### Toxic Constituents of Spent Bleach Liquor

Table II and Figure 2 show the results of analysis of 14 chlorinated phenols, methoxyphenols and catechols of spent bleach liquor.

It is seen from the results shown here and elsewhere [9,10,12,25,26] that this liquor contains 2-8 g of these organochloric compounds per $m^3$ of wastewater, summing at the total of 100–250 g of these chlorinated phenolics released into the environment per ton of bleached pulp. Since we were able to identify only about half of the peaks in the gas liquid chromatogram of the phenol fraction (ether phase III), the total release of chlorophenolics in spent bleach liquor can be estimated at 200–500 g/ton of pulp. In Finland, where $2.8 \times 10^6$ tons are bleached annually, this comes to 200–1000 ton/yr.

The results referred to above were obtained with the diethylether-NaOH extraction procedure followed by ethylation and GLC (see Methods). The acetylation method [1] gave lower, but not essentially different yields for the various chlorophenols. Pentachlorophenol was not detected by this method.

Table II. Chlorinated Phenols, Methoxyphenols and Catechols (g/ton of pulp bleached), in Spent Bleach Liquor of Three Kraft Pulp Mills in Finland[a]

| Sample | Chlorophenols | | | | | | Chlorocatechols | | | Chloroguaiacols | | | | 2,6-Dimethoxy-3,4,5-tri-chlorophenol | Total |
|---|---|---|---|---|---|---|---|---|---|---|---|---|---|---|---|
| | 2,4-Di- | 2,6-Di- | 2,4,6-Tri- | 2,4,5-Tri- | 2,3,4,6-Tetra- | Penta- | 3,4-Di- | 3,4,5-Tri- | Tetra- | 4,5-Di- | 3,4,5-Tri- | Tri-[b] | Tetra- | | |
| Mill A, birch pulp | | | | | | | | | | | | | | | |
| C/D | 22 | 1 | 1 | 0.05 | 0.4 | 0.07 | 25 | ND[c] | 0.7 | 0.5 | 0.02 | 0.02 | 0.03 | 0.08 | 51 |
| E-1 | 13 | 5 | 3 | 0.01 | 0.07 | 0.1 | 170 | 1 | 0.2 | 6 | 0.8 | 0.7 | 0.8 | 0.6 | 200 |
| Total | | | | | | | | | | | | | | | 251 |
| Mill B, birch pulp | | | | | | | | | | | | | | | |
| C-1 | 14 | 12 | 1 | ND | 0.03 | 0.02 | 6 | ND | 0.5 | 2 | 0.09 | 6 | 0.08 | 0.2 | 42 |
| E-1 | 1 | 0.2 | 3 | 0.2 | 0.3 | 0.1 | 37 | 0.2 | 0.2 | 6 | 3 | 4 | 0.3 | 2 | 58 |
| Total | | | | | | | | | | | | | | | 100 |
| Mill C, pine pulp | | | | | | | | | | | | | | | |
| C/D | 52 | 48 | 1 | 0.5 | 0.7 | 0.03 | 50 | ND | 0.3 | 9 | ND | ND | 8 | ND | 170 |
| E-1 | ND | ND | 3 | ND | 0.03 | 0.1 | 80 | 1 | 0.3 | 5 | ND | ND | 0.1 | ND | 88 |
| Total | | | | | | | | | | | | | | | 258 |

[a]Bleaching sequences in the mills were C/D-E-H-E-D-$SO_2$ (mill A), C-E-H-H-D-E-D-$SO_2$ (mill B) and C/D-E-D-E-D (mill C). Amount of chlorine used per ton of pulp was 35 kg in mill A, 45–55 kg in mills B and C. Volume of effluent per ton of pulp bleached was 34 $m^3$ in mill A, 50–60 $m^3$ in mill B, 50 $m^3$ in mill C. Corrections for recoveries were made as shown in Table I. Recovery of catechols was assumed to equal that of tetrachlorocatechol, i.e., 22%. No correction was made for the recovery of 2,6-dimethoxy-3,4,5-trichlorophenol. Figures shown are means for duplicate analyses of ⩾10 samples from mills A and B, three from mill C. Percentage standard deviations ranged from 20 to 80. Figures for mill A show total output of chlorophenols, since effluent from stage H and later from stages E and C/D was recirculated. Later stages were not analyzed for mills B and C.

[b]Isomer unknown.

[c]ND = not detected.

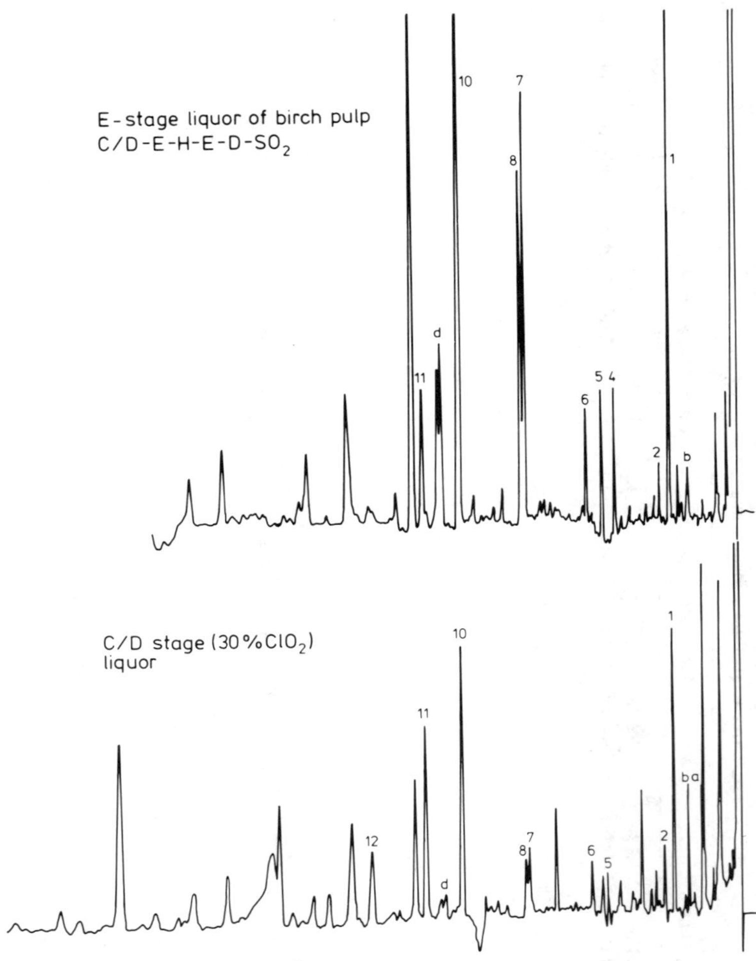


**Figure 2.** Capillary gas chromatograms of the phenol fraction (ether phase III of Figure 1) of bleaching effluents of kraft pulp of birch (left) and pine (right). The tracings are from an ECD which is sensitive to chlorine rather than to carbon, and so essentially detects the chlorinated constituents. The tracings represent equal volumes (0.05 mL) of the respective E and C(C/D) stage effluents. The compounds identified were the following: (1) 2,4,6-trichlorophenol; (2) 2,3,6-trichlorophenol (added as an internal reference); (3) 2,4,5-trichlorophenol; (4) 3,5-dichloro2,6-dimethoxyphenol and 4,5-dichloroguaiacol (these peaks coincide in the OV-101 column used); (5) 3,4-dichlorocatechol; (6) 2,3,4,6-tetrachlorophenol; (7) trichloroguaiacol, isomer unknown;

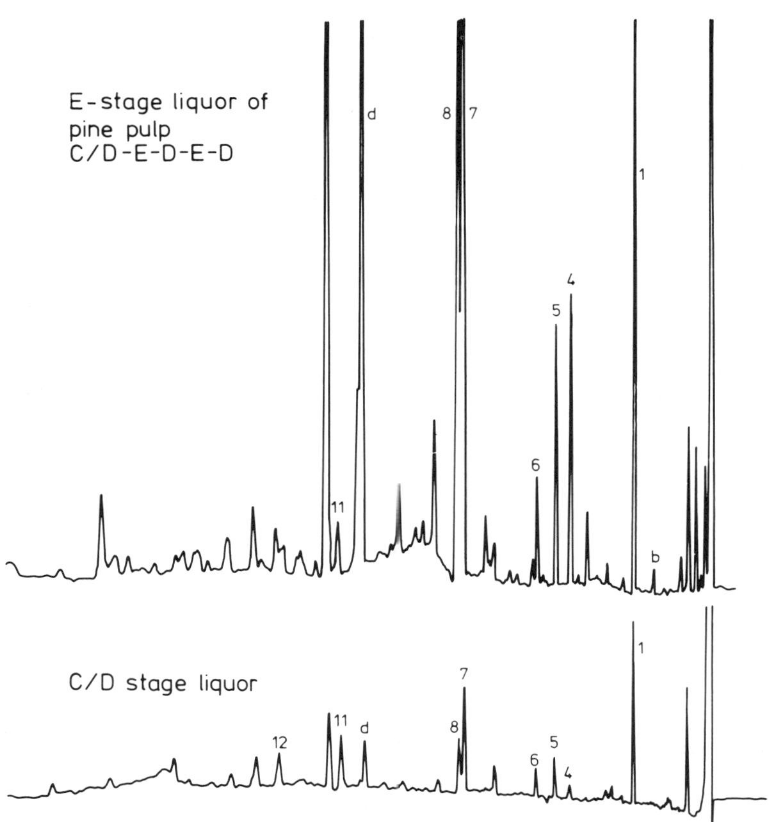


(8) 3,4,5-trichloroguaiacol; (9) trichlorocatechol, isomer unknown; (10) 3,4,5-trichloro-2,6-dimethoxyphenol; (11) pentachlorophenol; (12) tetrachlorocatechol; (a) 2,4-dichlorophenol; (b) 2,6-dichlorophenol; (c) 3,4,5-trichlorocatechol; (d) tetrachloroguaiacol.

Table III shows the results of toxicity tests for nine of the chlorinated phenolics found in spent bleach liquor, with phenol as reference. Tests were performed with fish, water fleas and bacteria. The correlation between toxicity and the degree of chlorination of the phenol is shown in Figure 3. It is seen that every new chlorine atom which is substituted into the benzene ring makes the compound more toxic to fish (*Poecilia reticulata*) and water fleas (*Daphnia magna*) but not to bacteria (*Pseudomonas putida*). Figure 4 shows

**Table III. Amounts (mg/L) of Chlorinated Phenols and Methoxyphenols Toxic at pH 7.0 to *P. putida, Daphnia magna* and *Poecilia reticulata***

| Compound | *P. putida* Inhibition | *Daphnia* $LC_{50}$ (24 hr) | *Poecilia* $LC_{50}$ (96 hr) |
|---|---|---|---|
| Phenol | | 25-50 | 43.3 |
| 4-Chlorophenol | 12.5-25 | 6.3-12.5 | 9.5 |
| 2,4-Dichlorophenol | 12.5-25 | 3.1-6.2 | 5.5 |
| 2,6-Dichlorophenol | 25-50 | 6.2-12.5 | 7.7 |
| 2,4,5-Trichlorophenol | 6.2-12.5 | 0.8-1.6 | 1.2 |
| 2,4,6-Trichlorophenol | 25-50 | 0.8-1.6 | 2.2 |
| 2,3,4,6-Tetrachlorophenol | 25-50 | 1.6-3.1 | 0.6 |
| Pentachlorophenol | 0.3-0.6 | 0.3-0.6 | 0.4 |
| 4,5-Dichloroguaiacol | 3.1-6.2 | 3.1-6.2 | 4.8 |
| 2,6-Dimethoxy-3,4,5-trichlorophenol | | 3.1-6.2 | 3.4 |

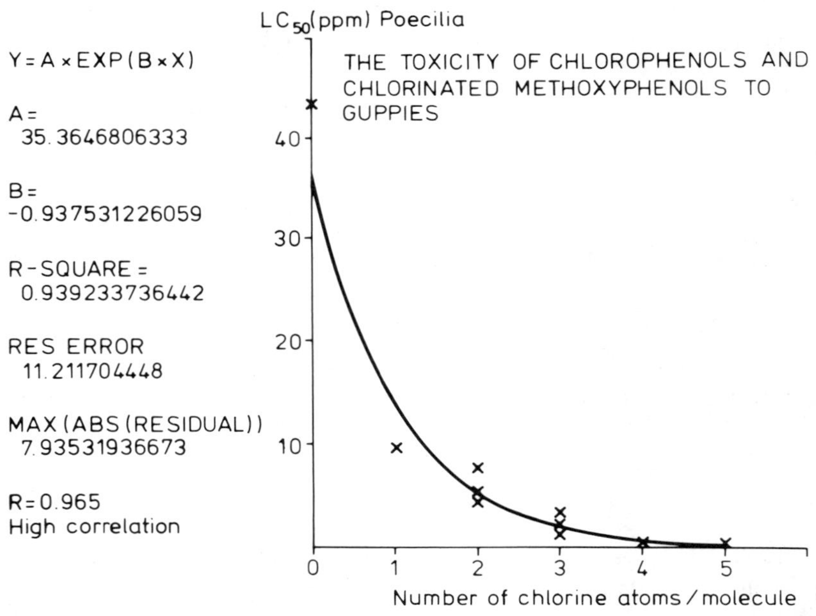


**Figure 3.** Dependence of the toxicity of individual chlorinated phenols, methoxyphenols and catechols to fish (A), water fleas (B) and bacteria (C) on the degree of chlorination of the phenolic compound. The compounds tested were those tabulated in Table III and phenol (number of chlorine atoms = 0).

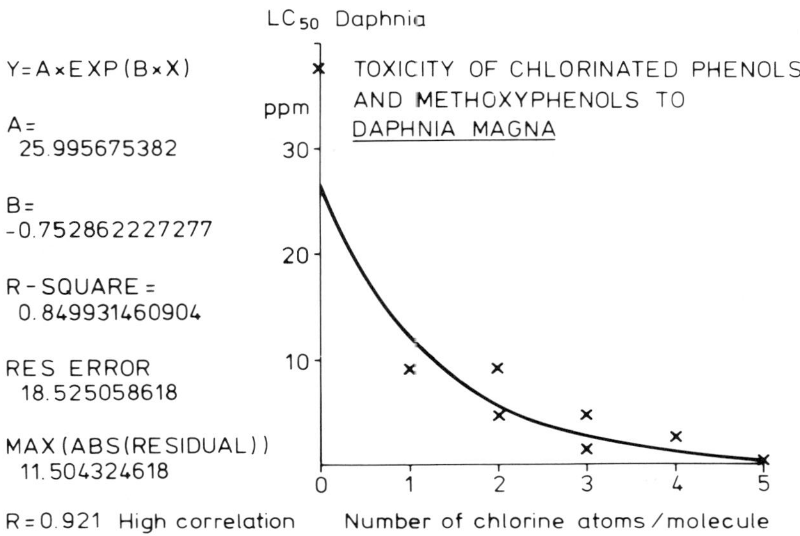


**Figure 3 (B), continued**

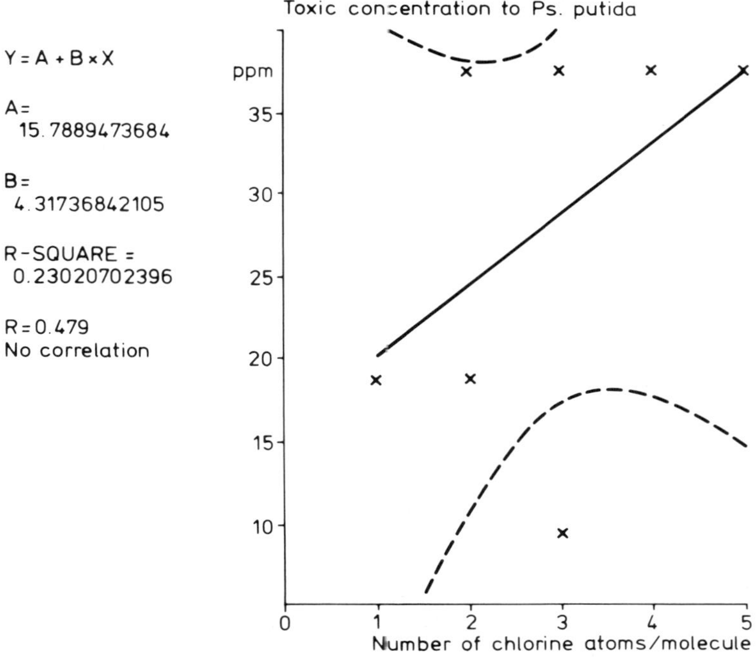


**Figure 3 (C), continued**

that the toxicity of chlorinated phenolics to *Daphnia magna* and *Poecilia reticulata* is highly correlated; the toxicity to fish thus seems to be well predicted by the toxicity to daphnids. This is of value since toxicity tests are cheaper and thus less time-consuming to perform with daphnids than with fish. The *P. putida* test does not predict toxicity to water animals (Figure 4B).

Figure 5 shows the correlation between daphnid toxicity of spent bleach liquor to its content of chlorophenolics, BOD and COD. It is seen that, whereas there is no correlation of toxicity to the chlorophenolics ($R = 0.52$), there is a significant correlation to BOD and COD ($R = 0.95$ and 0.96, respectively), i.e., the concentration of organic materials in the effluent. Therefore, there are also other toxic compounds than chlorophenols in these effluents.

Table IV shows the results of toxicity assays of spent bleach liquor after fractionation on the basis of molecular weight. It is seen that virtually all compounds toxic to daphnids as well as to the various fungi tested here, are of low molecular weight since it is found in the permeate of filter UM-2,

**Table IV. Toxicity to Several Fungi[a] and Daphnids of the Spent Bleach Liquor[b] of Birch Pulp in Various Molecular Weight Fractions**

| | Growth Inhibition | | | |
|---|---|---|---|---|
| Organism | Unfractionated | Below 1000 d | 1000 to 10,000 d | Over 10,000 d |
| *Alternaria* sp. | +++[c] | +++ | + | + |
| *Chaetomium globosum* | +++ | +++ | ++ | ++ |
| *Aureobasidium pullulans* | – | + | + | + |
| *Schlerophoma entoxylina* | + | ++ | + | + |
| *Ceratocystis* sp. | ++ | ++ | – | – |
| *Ceratocystis pilifera* | + | + | – | – |
| Daphnids[d] | <50 | 50 | 90 | 70 |

[a]Toxicity to fungi was measured as inhibition of growth on malt extract agar after substitution of spent bleach liquor, UM-2 permeate thereof (<1000 daltons) or resuspended retentates on filters UM-2 and PM-10 for the water in the medium. The plates were grown for 7 days at 28°C.

[b]Alkaline extraction-stage effluent from birch pulp bleaching was adjusted to pH 7.0 with HCl and fractionated with an Amicon Diaflo 402 ultrafiltration apparatus into fractions by molecular weight (see Methods). The ultrafilter retentates were resuspended into their original volume in activated-carbon–filtered tapwater and assayed for toxicity for daphnids. Molecular weight distributions of the fractions were verified by gel filtration on Biogel P-10 column, 0.9 x 60 cm (Figure 6).

[c]Symbols: +++ = strong; ++ = medium; + = weak; – = none.

[d]Nontoxic dilution (%).

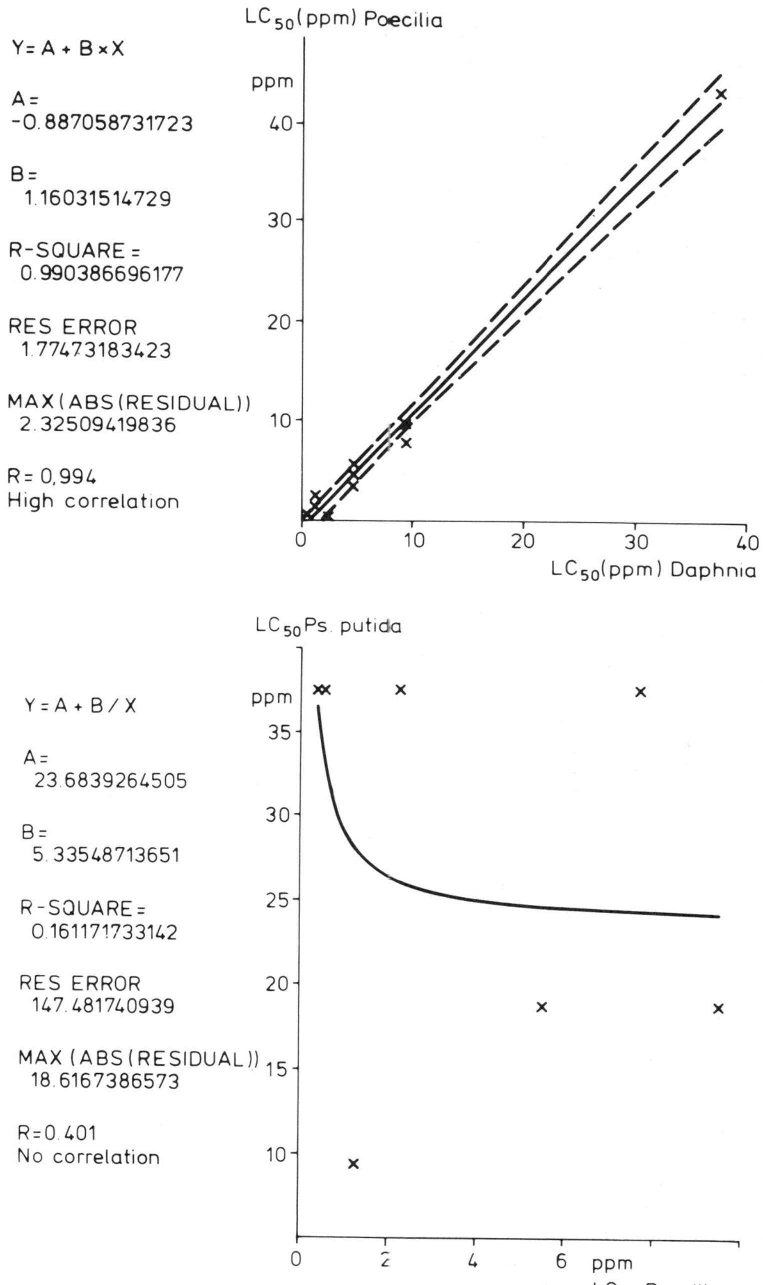


**Figure 4.** Correlation between toxicities of chlorophenolic compounds to experimental organisms: (top) between fish and water fleas; (bottom) between fish and bacteria.

which retains molecules of >1000 daltons in size. The permeate was clear and colorless. Color of spent bleach liquor is caused by molecules of higher molecular weight, named "chlorolignin." We verified the molecular weight distribution of ultrafiltration permeate and retentates by gel filtration. The results are shown in Figure 6.

The results of gel filtration thus confirm the fractionation by ultrafilters, and we are therefore apt to believe that the molecular weight distribution of spent bleach liquor as shown in Figure 6 is real. By comparing the data shown in Figure 6 to those shown in Figure 13 it can be concluded that the major part (over 80%) of the ultraviolet (UV)-absorbing material in spent bleach liquor is of high molecular weight, ranging from about 1500 to 10,000 daltons.

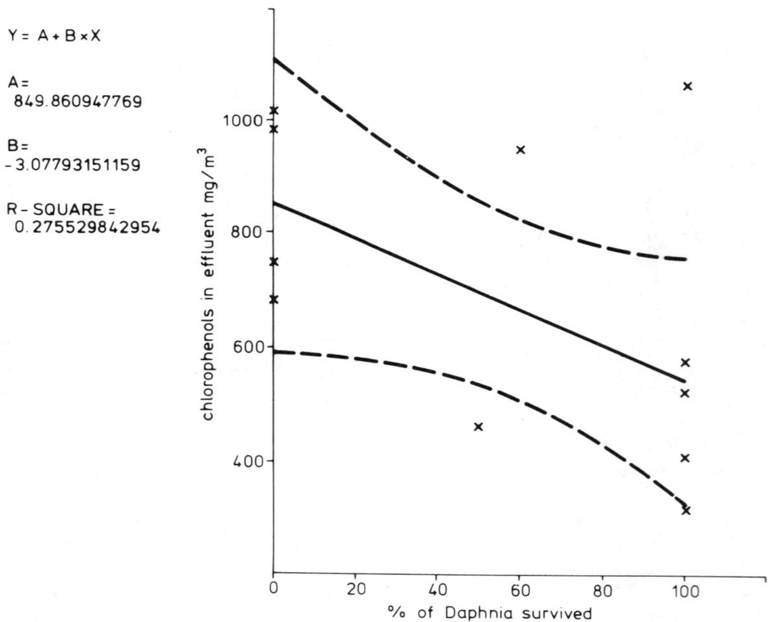


**Figure 5.** Correlations between toxicity of bleaching effluent to water fleas and (left) contents of chlorophenolic compounds; (upper right) biological oxygen demand; (lower right) chemical oxygen demand. The chlorophenolic compounds were calculated as the sum of those shown in Figure 2 and Table II.

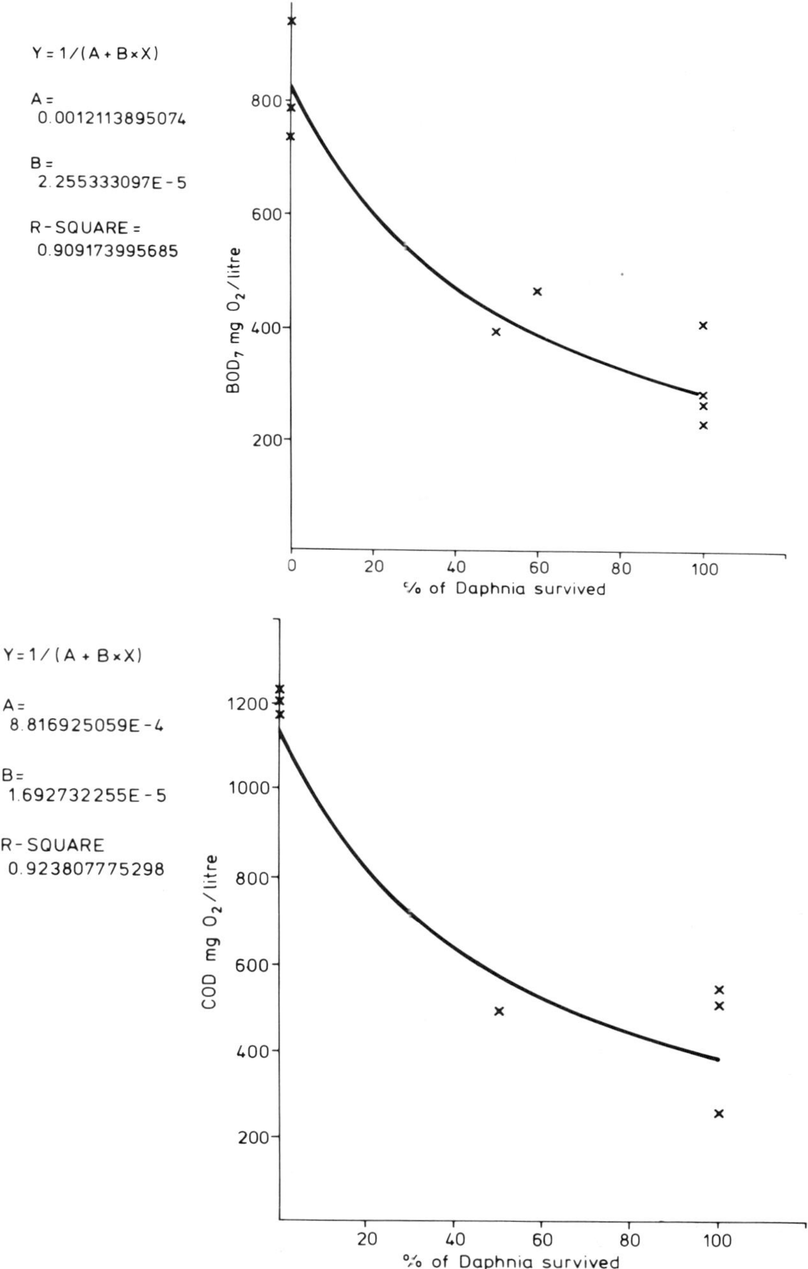


**Figure 5 (B), (C), continued**

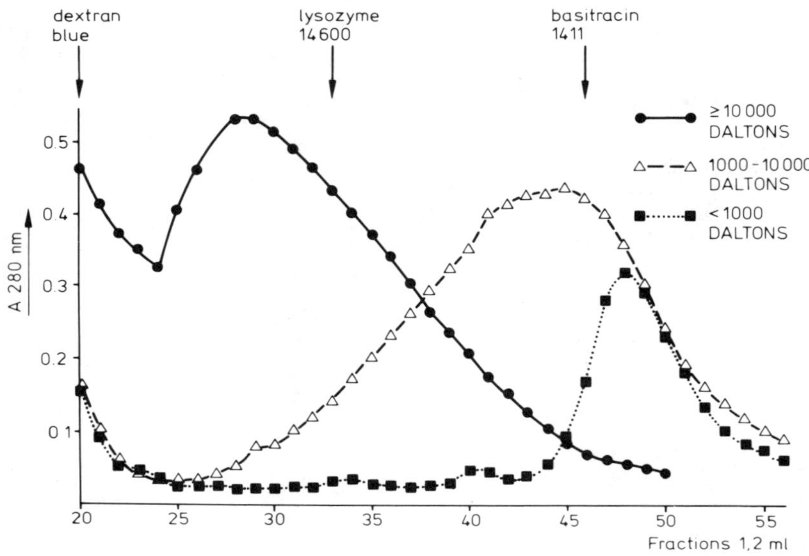


**Figure 6.** Results from gel filtration of ultrafiltered fractions of birch bleaching effluent from mill A (Table II). A biogel P-10 column of 0.9 by 60 cm was used in analysis of fractions obtained from PM-10 (retentate, >10,000 daltons) and permeate thereof over UM-2 (permeate, <1000 daltons, and retentate, between 1000 and 10,000 daltons). Positions obtained in a separate run for lysozyme and basitracin are indicated.

## Chlorophenols in Recipient Waters

Chlorophenol contents of recipient waters at a number of sites downstream and upstream from effluent pipes of two kraft pulp bleacheries are shown in Table V. A map over the area is reproduced in Figure 7. Control waters were taken at site 7, 4 km north of site 4 (not shown on the map).

With bacteriophage as tracer, it was discovered that effluent from the mill A bleachery flows via sites 1,2,3 and then down the river (site 6). We found bacteriophages a convenient and cheap method for the followup of wastewater streams. Several phages with different hosts may be used in a single experiment to visualize how different streams of water or waste water interact (Table VI).

All sampled waters contained chlorophenols (Table V), and comparison of the concentrations in Tables II and V shows three features. First, concentrations of 2,4,6-trichlorophenol and trichloroguaiacols in recipient waters

Table V. Concentrations (μg/L) of Some Chlorophenols in Surface (S) or Bottom (B) Waters of a Recipient Lake of a Kraft Pulp Mill in 1977–1980

| Site No. | Date of Sampling | 2,4-Dichlorophenol | | 2,4,6-Trichlorophenol | | Chlorinated Catechols[a] | | 2,3,4,6-Tetrachlorophenol | | Trichloroguaiacols | | 2,4,5-Trichloro-2,6-dimethoxyphenol | | Pentachlorophenol | |
|---|---|---|---|---|---|---|---|---|---|---|---|---|---|---|---|
| | | S[b] | B[b] | S | B | S | B | S | B | S | B | S | B | S | B |
| 1 | 20/05/77 | 1700 | | 2.5 | | | | 0.2 | | 0.5 | | 0.5 | | 0.2 | |
| | 24/08/77 | 140 | 280 | 3.6 | 2.0 | 197 | 0 | 0.7 | 0.8 | 0.2 | 0.2 | 0.3 | 0.5 | 0.6 | 0.4 |
| | 30/08/77 | 880 | 430 | 2.1 | 0.5 | | | 0.1 | 0.9 | 0.2 | 0.1 | 0.2 | 0.03 | 0.2 | 0.3 |
| | 22/04/80 | | 89 | 0.7 | 13.1 | 29 | 172 | 0.2 | 1.2 | 0.4 | 3.1 | 0.5 | 12.8 | 0.4 | 1.3 |
| 2 | 23/08/77 | 10 | 2 | 1.2 | 0.2 | 0.15 | 0 | 0.1 | 0.2 | 0.4 | 0.15 | | | 0.03 | 0.2 |
| | 28/03/78 | | | | 1.6 | | | | | | 0.8 | | | | 2.2 |
| | 11/10/78 | | | | | 9 | 7.7 | | | | | | | | |
| 4 | 08/02/78 | 250 | 1030 | 0.2 | 3.0 | 3.6 | 46 | 0.2 | 0.7 | 0.1 | 1.0 | 0.02 | 0.6 | 0.2 | 0.5 |
| | 14/05/80 | | | 0.3 | 0.3 | 0 | 0 | 0.2 | 0.2 | 0.2 | 0.1 | 0.2 | 0.1 | 0.7 | 2.2 |
| 5 | 24/08/77 | 10 | 10 | 0.1 | 0.4 | 0 | 0 | 0.3 | 0.3 | 0.1 | 0.1 | 0.02 | 0.0 | 0.2 | 0.6 |
| | 06/10/77 | | | | 0.3 | 0 | 0 | 1.3 | 1.3 | 1.2 | 1.3 | | | 0.8 | 0.6 |
| | 06/05/80 | | | 0.1 | 0.2 | 0 | 0 | 0.06 | 0.09 | | | | | 0.75 | 0.88 |
| 7 | 07/02/78 | | | 0.2 | | 0.1 | | 0.2 | | | | | | 0.5 | |
| | 11/10/78 | | | 0.01 | 0.01 | | | 1.0 | 0.01 | 1.3 | 0.4 | | | 0.9 | 1.7 |
| | 14/05/80 | | | 0.04 | 0.01 | | | 0.12 | 0.01 | | | | | 1.05 | 0.05 |

[a] 3,4-Dichloro-, 3,4,5-trichloro- and tetrachlorocatechol.

[b] S = surface, B = bottom water samples. Results from mixed samples (tube) are placed in between.

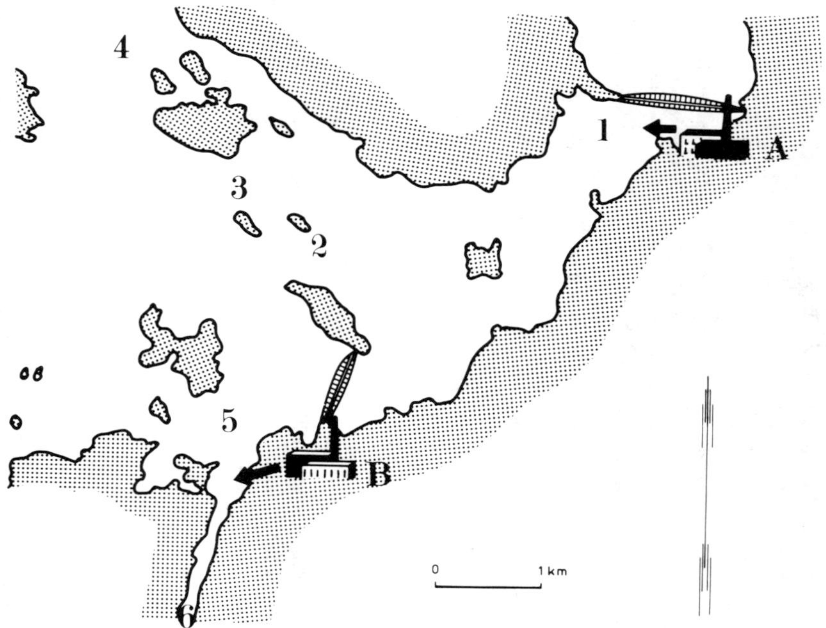


Figure 7. Map of the area relevant to this study. The mills and the sampling sites (numbers) are shown.

**Table VI. Flow of Bleaching Effluent from a Kraft Pulp Mill into Recipient Waters as Traced with Bacteriophage P22C2[a]**

| Days Since Release of Phage | Total Number of Phages ($10^6$ pfu) Recovered Location | | | |
|---|---|---|---|---|
| | 1 | 2 | 3 | 4 |
| 1 | 24,000 | 0 | 0 | 0 |
| 2 | 98,000 | 3,500 | 0 | 0 |
| 6 | 23,000 | 53,000 | 36 | 0.5 |

[a]A total of $3 \times 10^{14}$ plaque forming units (pfu) of phage was released into a bleachery effluent pipe of a pulp mill (A on the map in Figure 7). Samples of water from each meter of depth were collected 1, 2 and 6 days later at locations 1 and 2. At 1, samples were taken every 140 m, at 2, every 220 m across the water. The cross section of the water was 4320 $m^2$ at 1, 17,920 $m^2$ at 2. At locations 3,4 and 6, samples were taken from a single spot. Numbers shown for locations 1 and 2 were calculated for a 1-m thick slice across the water; those for 3 and 4 were calculated for a 1-$m^2$ column of water.

At location 6 a total of $9 \times 10^{13}$ pfu of phage was recovered 10 days since release of phage.

were 1/125 to 1/400 of those in undiluted bleaching effluent. Second, in all samples, concentrations of 2,3,4,6-tetrachlorophenol and pentachlorophenol were similar to those in undiluted bleaching effluent. Third, concentrations of chlorinated catechols were high in recipient waters. These discoveries indicate that in the natural aquatic environment chlorophenolics (2,4,6-trichlorophenol, chlorinated guaiacols) were biodegraded poorly or not at all. The dilution factor alone should lead to concentrations lower than those found: effluent from the mill A bleachery was 25,000 $m^3$/day, and it was diluted to less than 1/15 by effluents from mills A and B. During our measurements the daily flow in the stream (sites 5 and 6) varied between 30 and 60 x $10^6$ $m^3$.

It was rather surprising to find that in the recipient lake some chlorophenolics occurred at concentrations equal to or higher than those in undiluted bleaching effluent. Similar concentrations of 2,3,4,6-tetrachlorophenol and pentachlorophenol occurred in several of the 15 lakes that served as control (details not shown). This fact together with the finding that concentrations of 2,3,4,6-tetrachlorophenol and pentachlorophenol do not correlate with distances from the pulp mill, downstream or upstream, must mean that these compounds are part of widely spread pollution unrelated to the pulping industry. Other workers have discovered tetra- and pentachlorophenols in water reservoirs and tapwater [28,29].

Unlike the second group of compounds (tetra- and pentachlorophenol) the third group, chlorocatechols, seemed to be of pulp mill origin since their concentration in the recipient fell with an increasing distance from the mill. The concentration of chlorocatechols in the recipient was relatively high: for tetrachlorocatechol at sampling sites 1, 2 and 3 it was as high as that in undiluted effluent (about 4 mg/$m^3$, not shown in Table V) and for 3,4-dichlorocatechol it was in the range of 10 to 50 mg/$m^3$, equivalent to 1/10 to 1/5 of that in undiluted effluent. No tetrachloroguaiacol and little, if any, of 4,5-dichloroguaiacol was found in the samples. So it looks as if the chloroguaiacols were degraded into corresponding chlorocatechols in the recipient. In the sampled water, 90% or more of chlorocatechol was 3,4-dichlorocatechol, which would be the expected product from the degradation of 4,5-dichloroguaiacol, the most abundant chloroguaiacol in the effluent (Table II).

## Sediment Analyses

Figure 8 shows the results of chlorine measurements of a benthic sediment taken at Vatavalkama (location 3 on the map in Figure 7). It is seen that the top 20 cm contains a large amount of organically bound chlorine, 3-15 g Cl/kg of dry matter. The content of dry matter was 3 (first 5 cm) to 5% (in

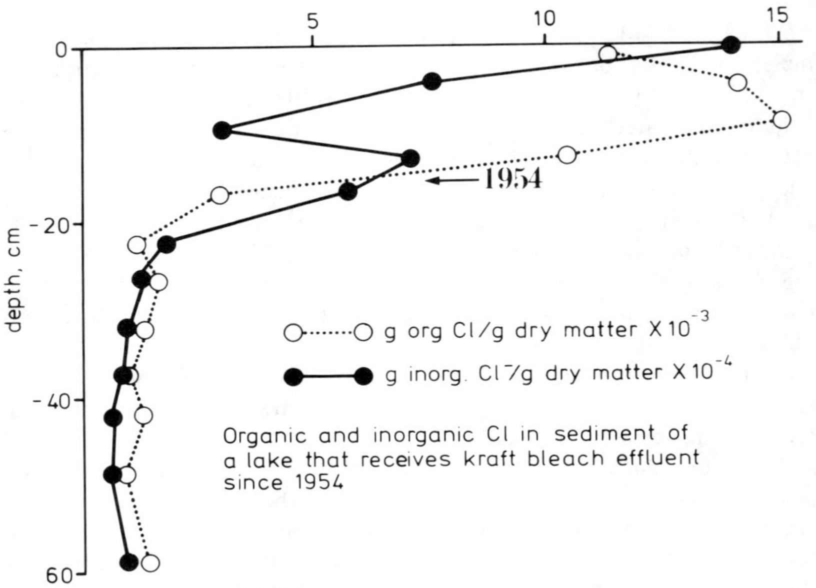


**Figure 8.** Concentrations of organically bound chlorine and inorganic chloride in a sample of benthic sediment taken at Site 3 (Vatavalkama) in Figure 7 in March 1979. For sampling procedure and analysis, see Methods.

the depth of 15 cm; results not shown). Therefore, the total content of organically bound chlorine was 7.8 $g/m^2$ for the top 15 cm of the 1979 sediment. New samples were taken in March 1980 and they contained 7.4 $g/m^2$ organically bound chlorine in the top 15 cm.

To see whether the steep increase or organically bound chlorine in the benthic sediment layers above –20 cm correlates with the start of bleaching with chlorine in mill A (Figure 7) in 1954, we determined the amount of $^{137}Cs$ in these layers. Atmospheric nuclear weapons tests have produced global fall-out of several radioisotopes, $^{137}Cs$ being one of the most important. $^{137}Cs$ has a high fission yield and relatively long physical half-life (31 yr). The deposition rate of $^{137}Cs$ onto the surface of the earth is well known, due to the intensive monitoring carried out all over the world. Significant amounts of $^{137}Cs$ were found for the first time in the environment in the mid-1950s. The annual deposition rate of $^{137}Cs$ was maximal during 1962 to 1964. Almost half of the cumulative $^{137}Cs$ was deposited during those three years. The rate of migration of $^{137}Cs$ in sediments is low, and this makes it possibe to use it for the dating of recently formed sediment layers. The results in Figure 9 show a

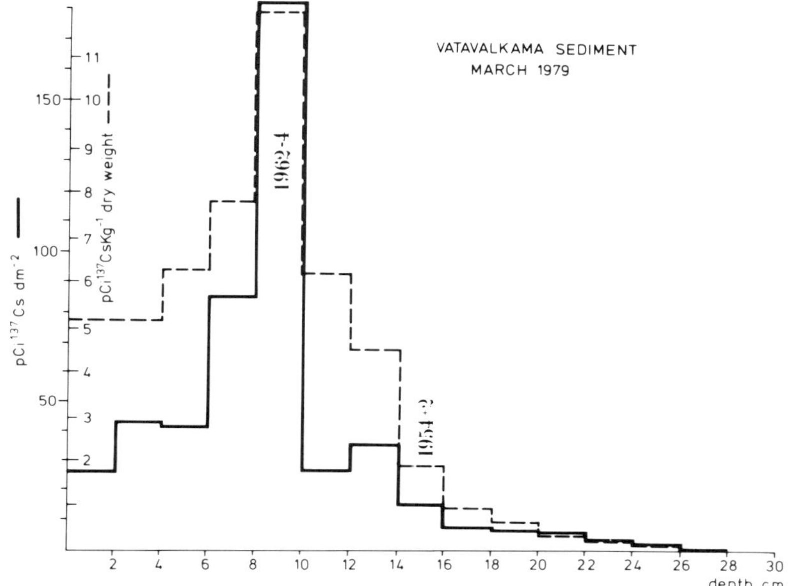


**Figure 9.** Vertical distribution of $^{137}$Cs in the sediment pictured in Figure 8.

clear-cut maximum of $^{137}$Cs concentration at the depth of 10 cm, which corresponds to the years 1962 to 1964. The decrease of $^{137}$Cs concentration found at the depth of 14 to 16 cm probably indicates the years 1954 to 1955. The accumulation of organically bound chlorine in the benthic sediment coincides with the start of bleaching in mill A in 1954. The other parts of the mill were in operation since 1948.

Because of the ecotoxic nature of chlorinated phenols, we considered it important to measure chlorophenols in the benthic sediment. Figure 10 shows the GLC of chlorophenols from the top 2.5-cm layers of benthic sediments taken at locations 3 and 4 in Figure 7 and at location 7 of Table V, 3 km north of and upstream from location 4. These chromatograms are very different from those of spent bleach liquor (Figure 2), in that for location 3, the most prominent peak stands for tetrachlorocatechol, in the two others for pentachlorophenol. Compared with the more heavily chlorinated compounds, the response factors of dichlorocompounds are very low (Table I), but one can see in Figure 10 that also dichlorophenols, especially 2,6-dichlorophenol, are abundant in the sediments from locations 3 and 4.

Sediment from each of three locations was analysed for chlorophenols for every 2.5 cm of depth and the results are recorded in Figure 11. Chlorophenols were found in the sediment to a depth of at least 15 cm.

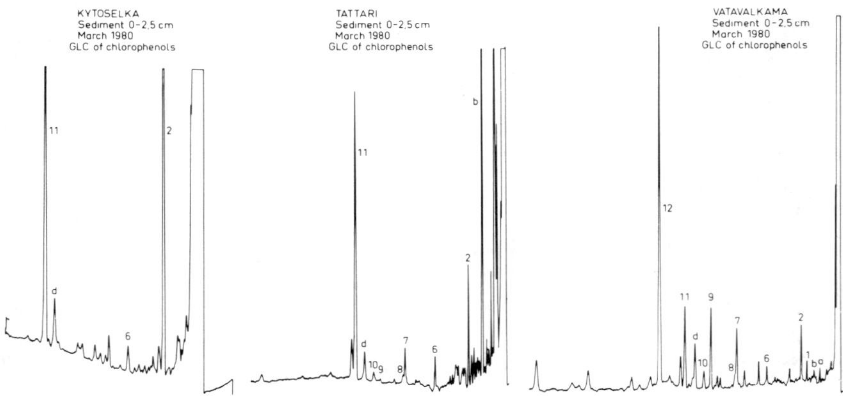


**Figure 10.** GLC tracings by an ECD from a phenol fraction (see Methods) obtained from the top 2.5 cm of benthic sediments collected in March 1980 at Sites 3 (Vatavalkama), 4 (Tattari) and at a control site (Kytöselkä, not shown on the map in Figure 7). For identification of the chlorophenols, see Figure 2.

The concentration of dichlorophenols at Vatavalkama (location 3) and Tattari (location 4) was 10 to 25 mg/kg of dry matter, corresponding to 0.3 mg dichlorophenols /L of wet sediment. The dry matter content was 3-5% in the Vatavalkama sediment, 5.5-12.5% in the Tattari sediment; it increased towards the deeper layers of the sediment. Individual sediment layers were measured separately, but the results are not recorded here. The concentration of dichlorophenols in the sediments was roughly equal to that in undiluted effluent (approx. 1 mg/L in mill A; Table II) or to that in the water samples from the same location (0.25 to 1 mg/L at Tattari; Table V). However, the 2,6 isomer that was only sporadically found in effluent from mill A (Table II) prevailed in the sediments. Therefore, it seems that 2,4-dichlorophenol may have disappeared, but 2,6-dichlorophenol has been formed from the other chloro compounds, possibly by metabolic action of microorganisms in the recipient waters and in the sediment.

Chlorocatechols were characteristic in the GLC of the Vatavalkama sediment, their concentration varied from 2 to 6 mg/kg dry matter or 0.1 to 0.2 mg/L in the top 10 cm of the sediment (Figure 11C). Most of it was tetrachlorocatechol (Figure 10) whereas 3,4-dichlorocatechol was the dominant chlorocatechol of the mill effluent. There was only about 0.02 mg of tetrachlorocatechol per liter of effluent, and in the lakewater samples more than 90% of the chlorocatechols were identified as dichlorocatechol (not specified in Table V). It seems that tetrachlorocatechol was formed from other chlor-

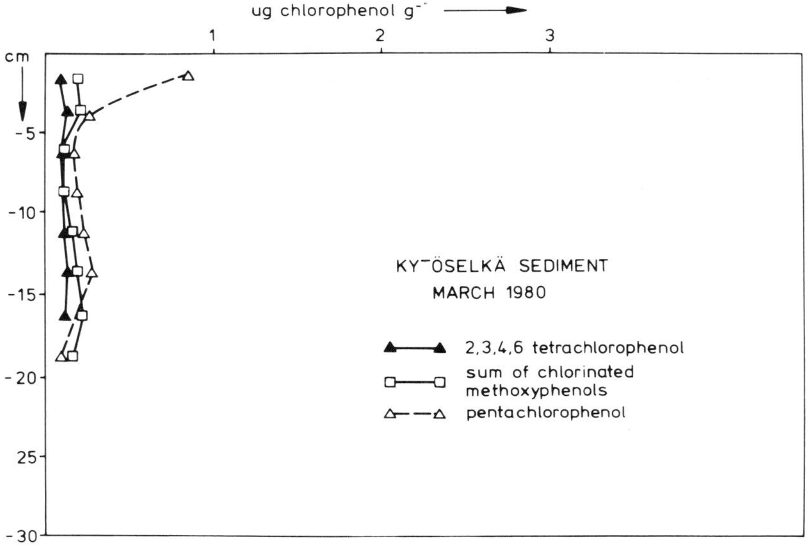


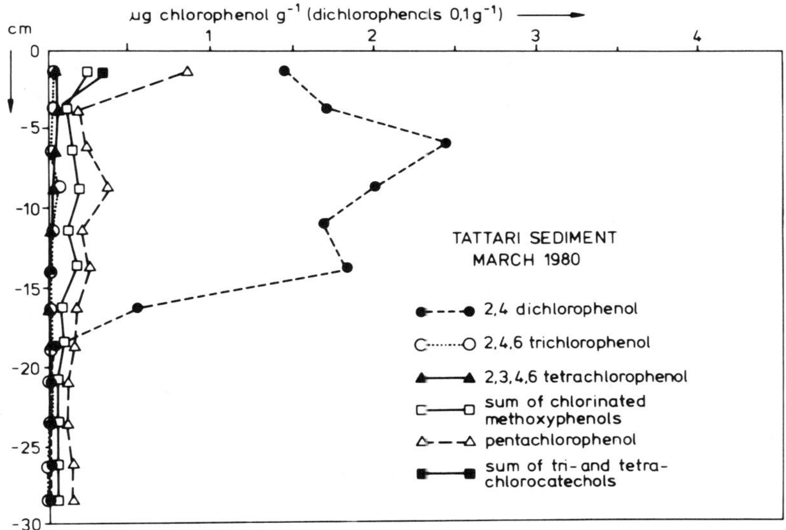


**Figure 11.** Contents of chlorinated phenols, methoxyphenols and catechols in the layers of benthic sediments collected in March 1980 at Sites 3 (Vatavalkama), 4 (Tattari) and at a control site (Kytöselkä). Concentrations are given as $\mu$g of chlorophenol per g (per 0.1 g for dichlorophenols) of dry weight.

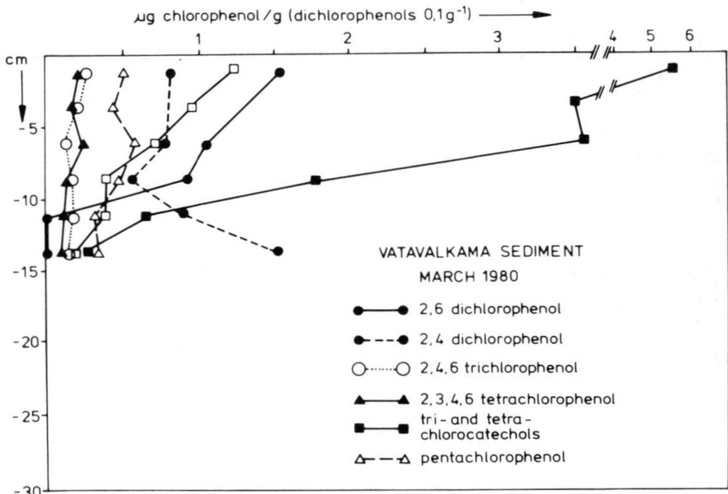


**Figure 11, continued**

inated compounds in the sediment, possibly by microbial action. Tetrachloroguaiacol is an obvious source for tetrachlorocatechol; however, there was only about 0.02 mg tetrachloroguaiacol/liter in the effluent (Table II), so other compounds must also be involved.

Chlorophenolic compounds represent only a very minor fraction of the organically bound chlorine present in the sediments: Vatavalkama sediment contained 10-15 mg of organically bound chlorine/g d.w. (Figure 8), in the top 10 cm, Tattari about 1 mg and Kytöselkä 0.5 mg (not shown). Of this chlorine less than 1% is contributed by the chlorophenolic compounds (Figure 11). In undiluted bleaching effluent also 1% of the chlorine is bound to chlorophenols [9]. The bulk of the organically bound chlorine in bleaching effluents is in chlorolignin. It thus seems that chlorolignin has not measurably decomposed into chlorophenols in the benthic sediments during the period of 25 years that bleaching effluents were led into the lake we studied.

The peak eluting at the retention time of tetrachlorocatechol was positively identified by mass spectrometry (two samples) and by measurement of the ratio of EC to the FID signal (all samples; see Methods) so we feel very sure of the significance of our findings. Paasivirta et al. [8] also reported tetrachlorocatechol (0.3 mg/kg) in sediment taken from an area polluted by pulping effluents in the lake Päijänne in central Finland.

In the top 2.5 cm of all the sediments studied the concentration of pentachlorophenol was 0.5-1 mg/kg. This holds also for samples taken in March

1979 (results not included in Figure 11). This means 0.015 mg at Vatavalkama, 0.044 mg at Tattari and 0.27 mg/L of sediment slurry at Kytöselkä (dry weight content 3% in the top 2.5 cm, 6-8% below that). In the effluent of mill A it was only 0.005 mg/L (Table II). The peak eluting at the retention time of pentachlorophenol was also positiviely identified by the ratio of EC to the FID signal (over 500 for pentachlorophenol, 100 to 150 for tetrachlorocompounds, much lower for less chlorinated compounds) and by mass spectrometry (all samples and two samples, respectively). Paasivirta et al. [8] found that the concentration of pentachlorophenol in benthic sediments in "nonpolluted" areas was higher than in areas polluted by pulp industry. In view of these findings it seems probable that the source of pentachlorophenol pollution is something else than the pulping industry or kraft bleaching.

Tetrachlorophenol behaved differently from pentachlorophenol. Part of the 2,3,4,6-tetrachlorophenol found in the sediments probably emanated from mill A, since the concentration of this compound was clearly higher in the top 8- to 10-cm of the sediment (the layer that contains bleaching effluent, see above) than below it (Figure 11C). At Tattari or Kytöselkä its concentration was very low or nil.

## Biological Purification of Bleaching Effluents

The picture emerging from the results discussed in the previous section is that, although the total amount of chlorophenols emanating from bleaching of kraft pulp is not very big (approx. 250 g/ton of bleached pulp in the case of the 14 chlorophenolics identified in the present work), these effluents are toxic, mutagenic and potentially harmful to water ecosystems since many of the chlorophenolic compounds proved biologically recalcitrant in the lake. The bleaching effluent contains little nitrogen and phosphorus (5-6 mg of N, 1-2 mg of P/L of effluent of mill A; Figure 7) and the ratio of C:N:P was 100:0.7:0.2, which is unfavorable for treatment by activated sludge or aerated lagoon. In fact, it was reported by Voss et al. [1] that only some 30% of the chlorophenols identified by them were removed during a five-day treatment in aerated lagoons.

We have experimented with biofilm reactors and obtained favorable results at laboratory scale [9,10] and pilot size reactor [26]. Figure 12 shows the setup of the combined reactor by which we obtained the best results. It consists of an anerobic fluidized bed reactor follwed by an aerobic trickling filter. We call the system the Fenox reactor [27]. Table VII shows that toxicity, chlorophenols and BOD were removed from the effluent; COD was only slightly removed, and color not at all. Chlorophenols and toxicity were re-

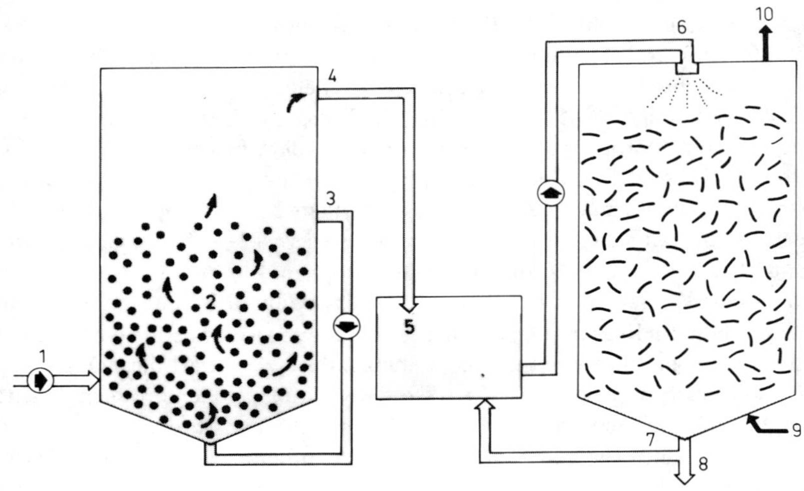


Figure 12. Fenox reactor operation scheme: (1) raw effluent feed; (2) fluidized bed; (3) recirculation pump; (4,5,6) flow via reservoir to aerobic trickling filter; (7) recirculation pump; (8) treated effluent; (9,10) inflow and outflow of air (7.5 $m^3$/hr per $m^3$ of filter volume). The bed in reactor 2 consists of resin particles of 0.1-0.5 mm in diameter. The trickling filter is filled with debarking waste of pine and plastic support.

**Table VII. Removal from Kraft Bleaching Effluents of BOD, COD, Color and Toxicity**

| Parameter | Input | Output | Percentage Removed |
|---|---|---|---|
| $BOD_5$ (mg/L) | 549 ± 121 | 241 ± 100 | 50 |
| $COD_{Mn}$ (mg/L) | 1102 ± 192 | 877 ± 228 | 21 |
| Color (mg/Pt) | 7850 ± 2355 | 6200 ± 2000 | |
| Nontoxic dilution $LC_{50}$ | <10-50% | 100% | 100 |

[a]The Fenox reactor (Figure 12) was fed with daily hydraulic loads of 1.5-8 reactor volumes at 30-35°C. Prior to feeding, effluent pH was adjusted to between 6 and 8. A dilution was considered nontoxic if 100% of the fleas survived for at least 24 hours. Means for seven experiments are shown.

moved by the anaerobic reactor [9], whereas BOD removal took place aerobically in the filter. In fact, BOD often increased during the anaerobic phase, probably reflecting decrease of toxicity.

Gel filtration analysis of the bleaching effluent before and after Fenox filtration showed that only lower-molecular-weight molecules were removed, up to 2000–3000 daltons (Figure 13). There was some increase of UV absorbance in the range of 10,000 daltons, possibly caused by polymerization of lower-molecular-weight phenolics during treatment. This is probably not undesirable since the high-molecular-weight molecules were found nontoxic or only weakly toxic (Table IV).

The mutagenic compounds of the bleaching effuent seem biodegradable, as mutagenicity of the effluent disappeared during biological treatment (Figure 14). Converse to other workers [30,31,32], we found that the mutagenicity of bleach effluent comes from the chlorination stage effluent (C in Figure 14), the alkaline extract being only weakly mutagenic. The contents of inorganic chloride decreased during the biological treatment of the effluent [9] by about 100 ppm. This means that chloroorganic compounds were mineralized by the reactor, not only chlorophenolics, of which there was only about 7 ppm, but other organic chlorine compounds as well. We were able to

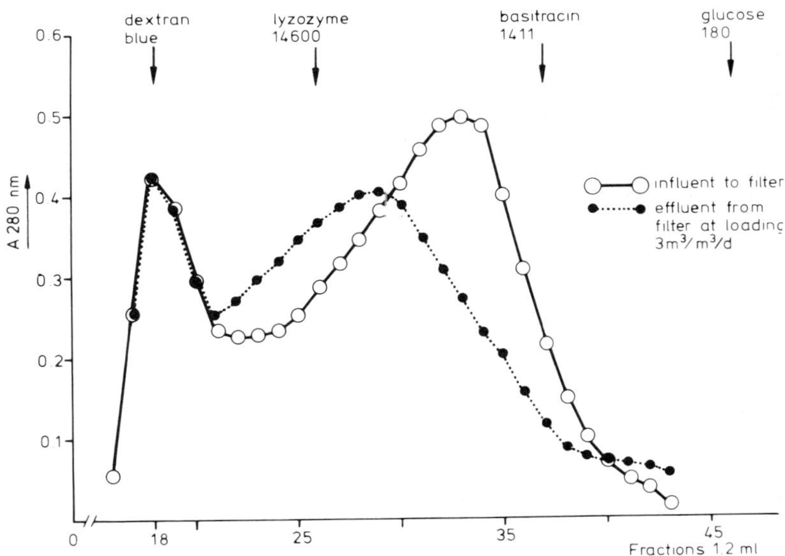


**Figure 13.** Changes in the molecular weight distribution of bleaching effluent during biological purification. The picture shows the gel filtration patterns of input and output effluents (see Table VI), of which 1.0 mL was loaded on a column of 0.9 by 60 cm of Biogel P-10. Positions obtained in a separate run for lysozyme, basitracin and glucose are indicated.

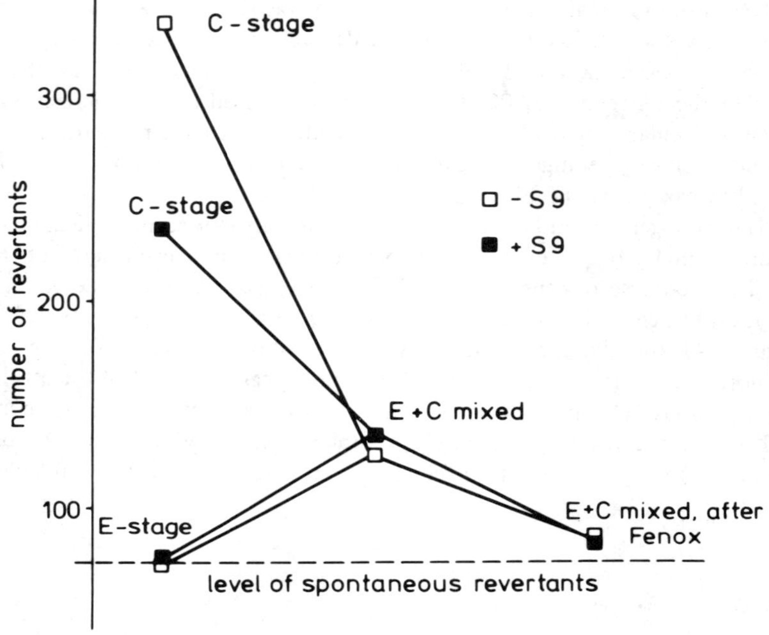


**Figure 14.** Mutagenicity to *Salmonella typhimurium* TA 100 [20,21] of birch bleaching effluent from stages C and E in mill A (Table II) separately, mixed in ratio 1:1, and after Fenox purification. + S9 = liver homogenate of barbiturate-treated mice added to the assay. The numbers of revertants recorded in the figure are for 1 mL of undiluted bleaching effluent.

identify only 14 of the about 40 chlorinated compounds (as judged by their ratio of EC to the FID signal) exhibited in Figure 2.

However, comparison of the GLC made of the phenol fraction of bleaching effluent before and after Fenox filtration showed that all different chlorophenols, both unidentified and identified, were efficiently removed (over 85%) when the hydraulic loading did not exceed 3 $m^3/m^3$ of reactor volume per day. When the hydraulic load was increased to six volumes, the efficiency of the removal of pentachlorophenol dropped to below 50%, whereas the others were well removed till the load was nine reactor volumes per day [9, 26]. We therefore conclude that even though some chlorophenolics are quite recalcitrant to biodegradation in the natural ecosystem, they can be removed by an adapted microflora in favorable conditions.

## SUMMARY

Bleaching effluents from three kraft pulp mills that bleach with different sequences pine and birch pulp with chlorine or chlorine supplemented with 30% chlorine dioxide were analyzed. Of the approximately 40 GLC peaks of the EC tracing of the GLC of the phenol fraction, 14 peaks were positively identified as chlorophenols, chlorocatechols and methoxylated chlorophenols. These summed up at 100–250 g of chlorophenolic compounds per ton of pulp bleached. Chlorophenols represent only about 1% of the organically bound chlorine found in the bleaching effluents. Most of the chlorine is bound to molecules of high molecular weight (chlorolignin), which are very stable in the environment: 25-year old layers of benthic sediments collected at a distance of 5 km downstream from kraft pulp mill, contained more than 10 mg of chlorine/g of dry matter.

Bleaching effluents were found acutely toxic to water fleas (*Daphnia magna),* fish *(Poecilia reticulata)* and bacteria *(Pseudomonas putida).* Toxicity of the mill effluent was somewhat variable, but no correlation was found in the individual samples between toxicity and contents of chlorophenols; instead, positive correlation was observed between toxicity and $BOD_5$ or COD.

Chlorophenolic compounds in spent bleach liquor were tested for toxicity to fish, water fleas and bacteria. Their acute toxicity to fish and water fleas, but not to bacteria, strongly correlated with the degree of chlorination of the molecule.

Toxicity of bleaching effluents resides in the molecules of low molecular weight (<1000 daltons).

Chlorophenolic compounds in a Finnish lake ecosystem were found recalcitrant to degradation. Chlorophenolic compounds of bleaching origin were found in lakewater and benthic sediment samples from up to several kilometers from the pulp mills.

In the bleaching effluent 2,6-dichlorophenol and tetrachlorocatechol were only barely, or not at all, detectable. In lake water and benthic sediment samples they often were the dominant chlorophenolic compounds. Their concentration decreased with distance from the mill, and they therefore probably represent transformation products of other chloroorganic compounds present in bleaching effluents.

Pentachlorophenol was found in all of the samples. Since its concentration did not decrease with increasing distance from the mill, it probably originates from a source other than the pulp mills.

Kraft bleaching effluents can be treated biologically to remove BOD, toxicity, chlorophenolic compounds and mutagenicity in a biological reactor consisting of an anerobic fluidized bed reactor and an aerobic trickling filter.

## ACKNOWLEDGMENTS

We thank Dr. J. Knuutinen and Dr. J. Paasivirta (Department of Chemistry, University of Jyväskylä) for the model compounds of chlorinated phenols, methoxyphenols and catechols, and for their help in building up a GLC system for the identification of the compounds. We also thank Dr. E. Toivanen (Water Research Laboratory, Helsinki City Water Works) for the mass fragmentographic analysis of some of the samples; Mrs. Riitta Boeck and Miss J. Backström for sediment analyses, and M. Viluksela for performing the fish toxicity studies. This study was supported by a grant from the Maj and Tor Nessling Foundation, and the Department of Trade and Industry, Finland.

## REFERENCES

1. Voss, R. H., J. T. Wearing and A. Wong. "A Novel Gas Chromatographic Method for the Analysis of Chlorinated Phenolics in Pulp Mill Effluents," Chapter 53, this volume.
2. Bjørseth, A., G. E. Carlberg, N. Gjøs, M. Møller and G. Tveten. "Halogenated Organic Compounds in Spent Bleach Liquors: Determination, Mutagenicity Testing and Bioaccumulation," Chapter 55, this volume.
3. Dickson, D. "PCP Dioxins Found to Pose Health Risks," *Nature* 238: 418 (1980).
4. Seppovaara, O., and T. Hattula. "The Accumulation of Chlorinated Constituents from Prebleaching Effluents in a Food Chain in Water," *Paperi Puu* 8:489-494 (1977).
5. Paasivirta, J., M.-L. Hattula, J. Särkkä, J. Janatuinen, M. Pitkänen and T. Kurkinen. "On the Analysis and Appearance of Organic Chlorine Compounds in the Lake Päijänne Ecosystem," *Org. Miljögifter Vatten* 2:439-462 (1976).
6. Landner, L., K. Lindström, M. Karlsson, M. Nordin and L. Sorensen. "Bioaccumulation in Fish of Chlorinated Phenols from Kraft Pulp Mill Bleachery Effluents," STFI Meddelande Serie B, No. 437 (B22), Scanforsk No. 126 (1976).
7. Holmbom, B., and K.-J. Lehtinen. "Acute Toxicity to Fish of Kraft Pulp Mill Waste Waters," *Paperi Puu* 11:673-684 (1980).
8. Paasivirta, J., J. Särkkä, T. Leskijärvi and A. Roos. "Transportation and Enrichment of Chlorinated Phenolic Compounds in Different Aquatic Food Chains," *Chemosphere* 9:441-456 (1980).
9. Hakulinen, R., and M. Salkinoja-Salonen. "The Use of Anaerobic Fluidized Bed Reactor for the Treatment of Effluents Containing Chlorophenols," in *Biological Fluidised Bed Treatment of Water and Waste Water,* P. F. Cooper and B. Atkinson, Eds. (Chichester, England: Ellis Horwood, 1981), pp. 374-382.

10. Salkinoja-Salonen, M., R. Paasivuo, R. Hakulinen and O. Koistinen. "A New Method for Biological Treatment of Chlorophenol-Containing Industrial Effluent," in *Aktuelle Probleme der Luftreinhaltung, Abgas, Abfall, Abwasser Recycling*, D. Behrens, Ed. (Weinheim, Germany: Verlag, Chemie, 1980), pp. 349-359.
11. Schlenk, H., and I. H. Gellerman. "Esterification of Fatty Acids with Diazomethane on a Small Scale," *Anal. Chem.* 32:1412-1414.
12. Salkinoja-Salonen, M., and V. Sundman. "Regulation and Genetics of the Biodegradation of Lignin Derivatives in Pulp Mill Effluents," in *Lignin Biodegradation: Microbiology, Chemistry, and Potential Applications, Vol. 2,* T. K. Kirk, T. Higuchi and H.-M. Chang, Eds. (Boca Raton, FL: CRC Press, 1980).
13. Knuutinen, J., R. Laatikainen and J. Paasivirta. "A Statistical Study of the Additive Substituent Effects in the $^{13}C$ NMR Chemical Shifts of Hydroxy- and Chloro-substituted Benzene," *Org. Magnetic Resonance* 4(5):360-365 (1980).
14. Knuutinen, J., and J. Paasivirta. "Thin Layer Chromatography of Chlorinated Guaiacols," *J. Chromatog.* 194:55-61 (1980).
15. Rehnberg, L. "Ion Exchange Technique for the Determination of Chlorinated Phenols and Phenoxy Acids in Organic Tissue, Soil and Water," *Anal. Chem.* 46:459-461 (1974).
16. *Standard Methods for the Examination of Water and Wastewater*, 14th ed. (New York: American Public Health Association, 1976).
17. Brinkman, G., and R. Kühn. "Comparison of the Toxicity Thresholds of Water Pollutants to Bacteria, Algae and Protozoan in the Cell Multiplication Inhibition Test," *Water Res.* 14:231-241 (1980).
18. Finney, D. J. *Probit Analysis* (Cambridge: Cambridge University Press, 1971).
19. Huttunen, P., and J. Meriläinen. "New Freezing Device Providing Large Unmixed Sediment Samples from Lakes," *Ann. Bot. Fennici* 15:128-130 (1978).
20. Ames, B. N., J. McCann and E. Yamasaki. "Methods for Detection of Carcinogens and Mutagens with the *Salmonella*/Mammalian Microsome Mutagenicity Test," *Mutat. Res.* 31:347-363 (1975).
21. McCann, J., N. E. Spingarn, J. Kobori and B. N. Ames. "Detection of Carcinogens as Mutagens: Bacterial Tester Strains with R Factor Plasmids," *Proc. Nat. Acad. Sci., U.S.* 72:979-983 (1975).
22. Kawata, K., M. Asec and V. P. Olivieri. "f2 Coliphage as Tracer in Waste Water Basin," *J. San. Eng. Div., ASCE* 100:1307-1310 (1974).
23. Kinnunen, K. "Tracing Water Movement by Means of *Escherichia coli* Bacteriophages," National Water Research Institute, National Board of Waters, Finland, Publ. 25 (1978).
24. Paatero, P. "Tietokoneohjelma Vipunen," Report A 46, Department of Physics, University of Helsinki, Helsinki, Finland (1978).
25. Hakulinen, R., O. Koistinen, M. Salkinoja-Salonen and M.-L. Saxelin. "Use of Anaerobic Filter for the Purification of Industrial Effluents," in *Proceedings of the Symposium on Anaerobic Digestion* (Cardiff, United Kingdom: Scientific Press, 1980), pp. 31-34.
26. Hakulinen, R., M.-L. Saxelin and M. S. Salkinoja-Salonen. "Purification of Kraft Bleach Effluent by an Anaerobic Fluidized Bed Reactor and

Aerobic Trickling Filter at Semitechnical Scale," Tappi Environmental Conference, New Orleans, LA, April 24-27, 1981.

27. Salkinoja-Salonen, M. S. "Waste Water Purifying Procedure," U.S. Patent No. 4,169,096 (1979).
28. Dietz, F., and J. Traud. "Zur Spurenanalyse von Phenolen, Insbesondere Chlorphenolen in Wässern mittels Gaschromatographie. Methoden and Ergebnisse," *Vom Wasser* 52:92 (1978).
29. Von Ninette, Z. "Desinfektionsmittel in Gewässern und Ihre Bedeutung für die Trinkwasseraufbereitigung," *Z. Wasser- Abwasserforsch.* 11(6): 187-193 (1978).
30. Ander, P., K.-E. Eriksson, M.-C. Kolar, K. Kringstad, U. Rannug and C. Ramel. "Studies on the Mutagenic Properties of Bleaching Effluents," *Sv. Papperstidn.* 80(14):454-459 (1977).
31. Eriksson, K.-E., M.-C. Kolar and K. Kringstad. "Studies of the Mutagenic Properties of Bleaching Effluents. Part 2," *Sv. Papperstidn.* 82(4):95-104 (1979).
32. Fevolden, S. E. "Ames Mutagenitetstest Anvendt på Avløpsvann," *Toxicitetstester* 2:85-94 (1978).

# CHAPTER 57

# CONSENSUS VOLUNTARY REFERENCE COMPOUNDS FOR ORGANIC WATER POLLUTANTS

**L. H. Keith**
Division of Environmental Chemistry
American Chemical Society
Austin, Texas

Only recently has it become possible to identify and analyze a large variety of specific organic (nonpesticide) environmental pollutants which exist at trace levels in the presence of thousands of other organic compounds. Advances both in techniques of separation (high-resolution gas and liquid chromatography) and in methods of identification (computerized mass spectrometry and, to a lesser extent, selective detectors and Fourier transform infrared spectrophotometry) have been key factors in this achievement.

The result has been a dramatic increase in the number of papers dealing with the identification and analysis of organic compounds in the environment in general. Most of the attention, however, has been focused on organics in water. An important and complex facet of this subject is the preparation of samples before instrumental analysis. Research involving sample preparation is often fragmentary by nature. Although many papers deal with the same aspect of sample preparation, they cannot be directly correlated because of differences in the samples, sample handling and/or model compounds used.

This became apparent when the first major symposium involving the identification and analysis of organic pollutants in water was held during the 1975

Chemical Congress of the North American Continent. At that meeting it was suggested by the late Aaron Rosen that a council be organized to establish a consensus voluntary system that could be used easily as a basis for establishing references against which variations or improvements in analytical methodology could be compared.

The American Chemical Society Division of Environmental Chemistry supported this concept. Subsequently, a division steering committee was established to direct a Council on Environmental Pollutants (CEP). The first objective was to formulate a list of voluntary reference compounds that could be used widely to compare improvements in new or modified techniques with existing methodology. This objective was met by convening an international panel of environmental chemists who chose by consensus a list of model reference compounds.

The 1979 Pacific Conference in Honolulu, sponsored by the American Chemical Society and the Chemical Society of Japan, was selected as the forum for that panel discussion. (The representatives who comprised that panel are noted in Table I.) Panel members, as well as interested members of the audience, nominated chemicals as consensus voluntary reference compounds (CVRC). This list is presented in Table II.

The guidelines for choosing these reference compounds were

1. to provide a wide selection of functional groups;
2. to represent, where possible, a range of pertinent physical properties such as volatility and polarity;
3. to choose compounds readily available in pure form at modest expense;
4. to avoid, where possible, multiple functional groups;
5. to choose, where possible, known water pollutants; and
6. to select, when possible, halogenated derivatives of several parent compounds.

In addition, extra weight was given to the selection of those compounds that have deuterated or carbon isotope isomers available, and to those compounds on the U.S. Environmental Protection Agency list of priority pollutants.

Known water pollutants were the prime candidates for reference compounds; the choice, however, was not restricted to these, for it was most important to obtain a wide variety of representative compounds. For example, relatively few water pollutants containing nitrogen and sulfur have been identified. Perhaps this is because they are not as abundant as organic pollutants that do not contain these hetero atoms. But a major contributing factor *could* also be that present technology is inadequate to concentrate, fractionate, separate and/or detect them even if they are present.

Some of the compounds chosen are toxic, and some are known or suspected mutagens or carcinogens. Therefore, appropriate safety and handling precautions should be used when handling these compounds or solutions of them.

Table I. CEP Panel Members

| Name | Representing |
| --- | --- |
| Lawrence Keith (Chairman) | American Chemical Society |
| Ryoshi Ishiwatari | Chemical Society of Japan |
| Krister Lindstrom | Swedish Chemical Society |
| Harry Hertz | U.S. National Bureau of Standards |
| James Lichtenberg | American Society for Testing and Materials and U.S. Environmental Protection Agency |
| Steven Heller | U.S. Environmental Protection Agency "WaterDROP" (Distribution Register of Organic Pollutants in Water) |
| Edo Pellizzari | U.S. Environmental Protection Agency "Master Analytical Scheme" |

Table II. CVRC by Chemical Class

| | |
| --- | --- |
| Alcohols | |
| Ethanol | Benzyl alcohol |
| 2-Chloroethanol | 3-Chlorobenzyl alcohol |
| Cyclohexanol | 2-Ethylhexanol |
| 2-Chlorocyclohexanol | Glucose |
| α-Terpineol | Glycerol |
| Cholesterol | |
| Aldehydes | |
| Acetaldehyde | Vanillin |
| Benzaldehyde | Chloral |
| 3-Chlorobenzaldehyde | Furfural |
| Crotonaldehyde | |
| Aliphatics | |
| Pentane | Chloroform[a] |
| *n*-Hexadecane | Bromoform[a] |
| Cyclopentadiene | Tetrachloroethylene[a] |
| Hexachlorocyclopentadiene[a] | 1,2-Dibromoethane |
| Cyclohexane | Heptachlor (1,4,5,6,7,8-Heptachloro-3a,4,7,7a-tetrahydro-4,7-methanoindene) |
| Chlorocyclohexane | |
| α-Pinene | |
| Amines | |
| Diethylamine | 3,3′-Dichlorobenzidene[a] |
| Aniline | Indole |

**Table II, continued**

| | |
|---|---|
| 2-Chloroaniline | 5-Chloroindole |
| Piperidene | Atrazine (2-Chloro-4-(ethylamino)-6-(isopropylamino)-*s*-triazine) |
| Quinoline | 3-Chloropyridine |
| 4-Chloroquinoline | |
| Benzidine[a] | |
| **Benzenoids** | |
| Benzene[a] | Decabromobiphenyl |
| Chlorobenzene[a] | Indan |
| Hexachlorobenzene[a] | 4-Isopropyltoluene |
| Bromopentafluorobenzene | 2,2-Diphenylpropane |
| Biphenyl | p,p′-DDE (1,1-dichloro-2,2-bis(*p*-chlorophenyl)ethylene |
| 4,4′-Dichlorobiphenyl | 1,2,4-Trichlorobenzene[a] |
| Decachlorobiphenyl | |
| **Carboxylic Acids** | |
| Acetic | Stearic |
| Dichloroacetic | Trimesic |
| Benzoic | Succinic |
| 4-Chlorobenzoic | Phenylalanine |
| Phthalic | 4-Chlorophenylalanine |
| 4-Chlorophthalic | Dehydroabietic |
| **Esters** | |
| Ethyl acetate | Methyl stearate |
| 2-Chloroethyl acetate | Methyl methacrylate |
| bis(2-Ethylhexyl) phthalate[a] | Methyl abietate |
| Phenyl benzoate | Glyceryl tripalmitate |
| 4-Chlorophenyl benzoate | |
| **Ethers** | |
| Diethyl ether | Pentachloroanisole |
| bis(2-Chloroethyl) ether[a] | Dibenzofuran |
| Diphenyl Ether | 1,4-Dioxane |
| Anisole | Tetrahydrofuran |
| **Ketones** | |
| Acetone | Anthraquinone |
| 1,1,3,3-Tetrachloroacetone | 2-Chloroanthraquinone |
| Acetophenone | Methyl isobutyl ketone |
| 4-Chloroacetophenone | 9-Fluorenone |
| Cyclohexanone | Fenchone |
| 2-Chlorocyclohexanone | Isophorone[a] |
| **Nitrogen Compounds** | |
| Urea | Acrylonitrile[a] |
| 1,2-Diphenylhydrazine[a] | Azobenzene |
| N-Nitrosodi-n-propylamine[a] | Caffeine |
| N-Nitrosodiphenylamine[a] | Carbazole |

**Table II, continued**

| | |
|---|---|
| Caprolactam | Uracil |
| Nitrobenzene | 5-Chlorouracil |
| 1,4-Nitrochlorobenzene | Uridine |
| Acrylamide | 5-Chlorouridine |
| Nitromethane | N-Methylcarbamate |
| **Organometallics** | |
| Tetraethyl lead | Tetrabutyl tin |
| Dimethyl mercury | Ferrocene |
| **Phenols** | |
| Phenol[a] | Bisphenol A |
| Pentachlorophenol[a] | Tetrabromobisphenol A |
| 4-Phenylphenol | 1-Naphthol |
| 4-Nitrophenol[a] | 4-Chloro-1-napthol |
| Guaiacol | 2,6-Di-tert-butyl-4-methylphenol |
| Catechol | 2,4-Dichlorophenol[a] |
| Tetrachlorocatechol | |
| **Phosphorus Compounds** | |
| Tri-n-butyl phosphate | Malathion (Diethyl mercaptosuccinate, *s*-ester of O,O-dimethylphosphorodithioate) |
| Triphenyl phosphate | tris(2-Chloroethyl) phosphate |
| Triethyl phosphine | Leptophos [(O-4-Bromo-2,5-dichlorophenyl) O-methyl phenylphosphonothionate] |
| Triphenyl phosphite | |
| Diethyl phosphoric acid | |
| **Polynuclear Aromatics** | |
| Naphthalene[a] | Benzo[a] anthracene[a] |
| 2-Chloronaphthalene | Benzo [a] pyrene[a] |
| Acenaphthylene[a] | Perylene |
| Anthracene[a] | Coronene |
| Pyrene[a] | Fluoranthene[a] |
| **Sulfonic Acids** | |
| Benzenesulfonic acid | 1-Naphthalenesulfonic acid |
| Methanesulfonic acid | 2,6-Naphthalenedisulfonic acid |
| **Sulfur Compounds (other than sulfonic acids)** | |
| Carbon disulfide | Thiophenol |
| Dimethyl disulfide | Ethylisothiocyanate |
| Diphenyl disulfide | Dimethyl sulfoxide |
| tert-Butyl mercaptan | Benzothiophene |
| Dimethyl sulfone | 1,4-Propanesulfone |
| 2-Acetylthiophene | α-Endosulfan[a] (6,7,8,9,10,10-Hexachloro-1,5,5a,6,9,9a-hexahydro-6,9-methano-2,4,3-benzodioxathiepin-3-oxide) |

[a]Compounds on the U.S. Environmental Protection Agency priority pollutant list.

Of course, no representative list of model compounds will be completely adequate. On the basis of our present knowledge and technology, however, the CEP list of CVRC should be adequate to handle most of the projected current needs.

It is hoped that this list will prove useful to environmental chemists and that it will provide a little cohesiveness to what is presently a very complex and uncoordinated technical area.

# INDEX